THE ENCYCLOPEDIA OF WORLD REGIONAL GEOLOGY, PART I

ENCYCLOPEDIA OF EARTH SCIENCES SERIES

Series Editor: RHODES W. FAIRBRIDGE
Columbia University

Published Volumes

THE ENCYCLOPEDIA OF OCEANOGRAPHY (Vol. I)
THE ENCYCLOPEDIA OF ATMOSPHERIC SCIENCES AND ASTROGEOLOGY (Vol. II)
THE ENCYCLOPEDIA OF GEOMORPHOLOGY (Vol. III)
THE ENCYCLOPEDIA OF GEOCHEMISTRY AND ENVIRONMENTAL SCIENCES (Vol. IVA)
THE ENCYCLOPEDIA OF WORLD REGIONAL GEOLOGY, PART 1: Western Hemisphere (Including Antarctica and Australia) (Vol. VIII)
THE ENCYCLOPEDIA OF WORLD REGIONAL GEOLOGY, PART 2: Eastern Hemisphere (Vol. VIII)

LIBRARIES
UNIVERSITY OF MAINE
AT ORONO

RAYMOND H. FOGLER LIBRARY

ORONO

ENCYCLOPEDIA OF EARTH SCIENCES, VOLUME VIII

The ENCYCLOPEDIA of WORLD REGIONAL GEOLOGY, PART 1:

Western Hemisphere (Including Antarctica and Australia)

EDITED BY

Rhodes W. Fairbridge

Professor of Geology
Columbia University
1975

Dowden, Hutchinson & Ross, Inc.
STROUDSBURG, PENNSYLVANIA

Distributed by
HALSTED PRESS   A Division of John Wiley & Sons, Inc.

Copyright © 1975 by **Dowden, Hutchinson & Ross, Inc.**
Library of Congress Catalog Card Number: 75-1406
ISBN: 0-470-25145-X

All rights reserved. No part of this book may be reproduced or transmitted in any form or by any means—graphic, electronic, or mechanical, including photocopying, recording, taping, or information storage and retrieval systems—without written permission of the publisher.

77 76 75   1 2 3 4 5
Manufactured in the United States of America.

LIBRARY OF CONGRESS CATALOGING IN PUBLICATION DATA

Library of Congress Cataloging in Publication Data

Fairbridge, Rhodes Whitmore, 1914–
  The encyclopedia of world regional geology.

  (Encyclopedia of earth sciences series ; v. 8)
  Includes bibliographical references.
  CONTENTS: A. Western Hemisphere including Australia and Antarctica.
  1. Geology–Dictionaries. I. Title. II. Series.
QE5.F33   550'.9   75-1406
ISBN 0-470-25145-X

Exclusive distributor: **Halsted Press**
A Division of John Wiley & Sons, Inc.

PREFACE

For a decade now your editor has been working on a series of single volume, autonomous encyclopedias dealing with the earth sciences. The published volumes include *Oceanography* (1966), *Atmospheric Sciences and Astrogeology* (1967), *Geomorphology* (1968), and *Geochemistry and Environmental Sciences* (1972). In our original plan there were to be eight volumes, but as the project has expanded, some have been subdivided into A and B, or Parts 1 and 2. With the ever-increasing interest of geologists, geophysicists, geochemists, physical geographers, and economists on global problems, it seemed appropriate that the next volumes should be devoted to world regional geology. Orginally planned to be no. VIII, we have had to separate them into Western Hemisphere (VIII, Part 1) and Eastern Hemisphere (VIII, Part 2). For reasons of space we include the Pacific, Australasia, and Antarctica with the Western Hemisphere.

We have selected the regions on a sliding scale of magnitude, starting with continental reviews (very generalized), regional analyses, country entries, and then the smaller spots, overseas territories and isolated islands. Certain special tectonic units (e.g., the Himalayas, the Caucasus, the Appalachian-Ouatchita belt) receive individual treatment, since they cross over most of the usual geographic boundaries.

The smaller the area, the more detailed, relatively speaking, will be its treatment. This might seem to be illogical, but it is generated by a desire to help the reader to uncover the key references to the most obscure or remote places, because they are often the most difficult to pinpoint. Major countries, for their part, are rather well furnished with regional volumes, geological, geotectonic, and mineral maps, and with richly endowed bibliographies. Some of the little places, however, are not so easily illuminated. Off-hand, where, would you say, should one start looking for the latest data on Andorra? Albania? Chad? Christmas Island? Nepal?

Then there are many parts of the world, rich in literature and surveys, but all in foreign languages. We have had the summary articles from those countries translated into English. The USSR, for example, or the People's Republics of China and Mongolia, are not the easiest places to find illuminated in every library. Many of our readers do not speak French or German; yet for many parts of the world 99 percent of the literature is in those languages. Now you can read about New Caledonia or Madagascar or Mali, and in English.

The critical reader may also notice a lack of uniformity in the various entries They have been edited, to be sure, but often only lightly. They are the contributions of different people, as indicated by the signatures, and show individual emphasis. Some of the entries represent the editor's summaries of available material; your editor has carried out professional field work in six continents and at least fifty countries, so that at times a few personal notes are introduced, but for the most part we have tried to be judicious and impartial.

The authorship is literally worldwide. In very many cases we have been fortunate in obtaining the generous cooperation of the leading specialists. Some of the articles represent condensations of longer works; others are specially generated reviews. But, whatever they are, they represent an introduction to a part of the world that the reader wants to investigate. No article can answer all the questions, but we hope to answer the basic ones and, if you have more, we hope to lead you to the literature, which can then be further pursued.

The Decade of Plate Tectonics

The present decade will go down in geological history as the decade of plate tectonics, representing the greatest revolution in geological methodology and philosophy since Darwin gave us *The Origin of Species.* For a hundred years we have been patiently mapping and collecting and analyzing. Within the last two decades marine geology has given us an insight into two-thirds of the earth that was formerly a closed book; paleomagnetics has given us an indication of the former positions of the continents; and marine magnetics has disclosed the "magnetic striping" of oceanic crust that, combined with paleontology and geochronology, provides us with the timing, the loci, and the rates of sea-floor spreading. Seismology completes the story by helping to locate the plate boundaries and the contemporary subduction zones. Today geologists are actively seeking the boundaries of ancient plates and the sites of long inactive subduction zones, together with the evidence of former collisions and their matching sides. They are using all the tried-and-true techniques of petrology and structure, stratigraphy, and paleontology, but with a new and quantitative working hypothesis

The idea of plates is not new. Edward Suess spoke of them in the 1880s. In his *Researches in China* of 1907, Bailey Willis described the underthrusting of the eastern Asiatic borders of the Pacific crust. As early as the 1890s, Hans Schardt was describing in the Alps the gravitational sliding engendered by the N-S collisions of Africa and Europe. Alfred Wegener and Emil Argand were insisting on the drift of continents in the 1910-1930 period. At that same time, the former director of the Geological Survey of Austria, Otto Ampferer, described the subduction zone of the Alps, but with an unpronounceable Germanic name—Verschluckungszone—from which we were rescued by the Swiss geologist Amstutz twenty years later with the perfectly pronounceable term "subduction zone." But it was all relatively hard to believe until three Columbia University seismologists, Oliver, Sykes, and Isaaks, analyzed the Tonga region of the southwestern Pacific to show how the Pacific plate seemed to be driving down along the "Benioff Zone" to melt away at depths around 700 km: a modern, dynamic subduction zone.

In short, our picture of world geology today is immensely exciting. To the "tried-and-true" traditional methods of our discipline come the new techniques of geophysics and geochemical chronology; together they are helping to build a coordinated picture of global geology. Never before has it been so necessary to delve into the problems of every part of the globe. Our volume tries, in its way, to help meet that need. Numbers of our contributors do not yet accept the challenge of plate tectonics, or, at any rate, they have not yet adjusted their traditional "fixist" thinking to the new "mobilistic" scene. In a way, there is no harm in that; it will make the new correlations and reconstructions all the more striking.

The Economic Challenge

The present decade also sees the beginning of a general recognition that the material resources of the earth are in finite quantities. Only twelve years ago, a distinguished oil geologist attacked your editor in most scathing terms for predicting, on the basis of careful calculations of the total volume of sedimentary rock on the earth, that the shrinking global supplies of petroleum would be introducing severe restrictions by the end of the 20th century. Yet today this change of heart, from limitless supplies to finite sobriety, is exercising the best minds in the industry.

The oil men have long regarded the world as their oyster, but today it is being increasingly recognized as *one* world, a world in which most of the anticlines have now been drilled, a world in which many of the shelves have no sediment. The advent of plate tectonics now provides a working hypothesis to explain the natural fractionation of hydrocarbons. The drifting of a plate over an ascending plume or "hot spot" is analogous to the bunsen burner in the chemist's world. Thus we have another clue to oil search.

The subduction of the oceanic plates also permits the recycling of superconcentrations of metals. All of a sudden, the economic geologists are beginning to see that the concentrations of metal ores show curious parallels to the ophiolite zones that identify ancient trenches, and to the granite aureoles that seem to derive from the melted "roots." Thus the theoretician, the oil men, the metal men, not to mention the foreign aid specialist, the engineering geologist, and the hydrologist, are all looking at global problems.

How To Use This Encyclopedia

The search for information in this volume should be simpler than with other volumes in this series, where many theoretical and terminological concepts were involved. This volume is frankly regional, geographical in its approach. If you want something on "Nepal," for example, you just flip the pages to "N" and there it is. But that is not all. At the end of each entry there are a number of *cross-references*. These are articles that deal with related subject matter. Now, Nepal is in the Himalayas, so we may also find some interesting material in *India* or in the special attached article on the *Himalayan Orogenic System*, or yet again in the general *China* article, or perhaps in the special item on *Tibet*. (The abbreviation q.v. is from the Latin *quod vide*, "which see.") Again, if one looks up *West Indies*, for example, it is helpful to learn the names of individual islands that also provide more detailed information.

Larger countries, such as the *United States* and *Canada*, are also provided with detailed regional analyses under their major physiographic subdivisions: e.g., *Appalachian Region* or *Canadian Shield*. In the case of *Australia*, in contrast, it is convenient to supply that breakdown under state headings since the states are few and the geological data are largely concentrated in similar administrative divisions. Cross-referencing to articles in other volumes of this earth science encyclopedia series is only made in exceptional instances.

Literature referencing is provided in the usual way, but with such a vast source of material, it has had to be curtailed severely. If a convenient recent summary is available, we shall be content to leave that as the main avenue available for future research. If there is no general synthesis, we have to gather more references so as not to skimp on certain areas of the field.

Indexing has been thorough, as in other volumes of this series, but again it has been an easier task because of the simplicity of a geographic framework. Not every town and village is indexed, because for such names the ordinary gazeteers are available (e.g., the *Columbia-Lippincott Gazeteer* or the *Times Atlas Gazeteer*). Some specific localities of exceptional interest are included, however, and the same is true for certain formation names or celebrated structures (e.g., the Dwyka Tillite, the Deccan Trap, the Parana Basin, the Cincinnatti Arch, the Old Red Sandstone).

An *author index* is a special innovation in this volume, because it is felt that the researcher often starts with the name of a scientist when tracking down a problem but when there is no specific country or state with which to identify.

Abbreviations and measures are handled on the basis of reasonable economy; for example, an orientation such as NNE-SSW is tedious to spell out and is perfectly clear without additional punctuation. The same is true for countries such as USSR, US, and UK. The metric system is used preferentially in most articles, although most authors have included equivalents, where a precise value (as distinct from a generalized figure) is given. Thus, if the author writes that the formation is 10,000 feet thick, the metric equivalent, in a similar "rounding," is given as 3000 m and *not* to decimals. Abbreviations used include mm, cm, m, km, km^2, etc., but the English equivalents—inches, feet, and miles—are usually spelled out, except for "sq mi," where the meaning is obvious. A simplified conversion table is supplied here, but for lengthy and very detailed reference, see our *Encyclopedia of Atmospheric Sciences and Astrogeology* (article: *Units, numbers,* etc.).

Abbreviations of geological *lithologic terms* are usually avoided in the text but are employed in tables, the favored system being: *ss* for sandstone, *sh* for shale, *ls* for limestone, *cgl* for conglomerate, *mrl* for marl, *dol* for dolomite. The use of Greek letters, formerly used in the UK and still widely used in France, for igneous rocks is avoided. Where Greek letters have been used, as in the Jurassic of Germany for stratigraphic classification, the best solution

Conversion Tables

Metric to English Units—Equivalents of Length

1 micron (μ) = 0.001 millimeter (mm) = 0.00004 inch (in.)
1 mm = 0.1 centimeter (cm) = 0.03937 in.
1000 mm = 100 cm = 1 meter (m) = 39.37 in. = 3.2808 foot (ft)
1 m = 0.001 kilometer (km) = 1.0936 yard (yd)
1000 m = 1 km = 0.62137 mile (mi)
1 in. = 2.54 cm
12 in. = 1 ft = 0.3048 m
1 cm = 0.39370 in. = 0.032808 ft
1 km = 10^5 cm = 0.62137 mile
1 fathom = 6 ft = 1.8288 m
1 nautical mile = 1.85325 km
1 in. = 2.54001 cm
1 ft = 30.480 cm
1 statute mile = 1.60935 km = 5280 ft
1 astronomical unit = 1.496 × 10^8 km = 92,957,000 miles
1 light year = 9.460 × 10^{12} km = 5.878 × 10^{12} miles
1 parsec = 3.085 × 10^{13} km = 1.917 × 10^{13} miles

Square Measures

1 square foot = 0.00002295684 acre
1 acre = 43560 ft² = 0.0015625 mi²
1 square yard = 0.836127 m²
1 hectare = 2.471054 acre
1 square mile (statute) = 640 acres
1 square cm = 0.1550 square in. = 0.0010764 square ft
1 square km = 10^{10} square cm = 0.3861 square mile
1 square in. = 6.452 square cm
1 square ft = 929.0 square cm
1 square mile = 2.5900 square km
1 mm² = 0.00155 in.²   1 in.² = 6.452 cm²
1 m² = 10.764 ft²   1 ft² = 0.09290 m²
1 km² = 0.3861 mi²   1 mi² = 2.5900 km²

Cubic Measures

1 gal (UK) = 4.5461 liters = 1.200956 gal (US)
1 liter = 0.219969 gal (UK) = 0.264173 gal (US)
1 gal (US) = 3.7854 liters = 0.832670 gal (UK)
1 cc = 0.0610 cu in. = 0.000035314 cu ft
1 cu in. = 16.387 cc
1 cu ft = 28317 cc
1 mm³ = 0.000061 in.³   1 in.³ = 16.387 cm³ (cc)
1 cm³ (cc) = 0.0610 in.³   1 ft³ = 0.028317 m³
1 m³ = 35.315 ft³   1 mi³ = 4.1681 km³
1 km³ = 0.239911 mi³

Statute Miles to Nautical Miles to Kilometers

Statute	Nautical	Kilometers	Statute	Nautical	Kilometers
¼	0.22	0.40	9	7.82	14.48
½	0.43	0.80	10	8.68	16.10
¾	0.65	1.21	20	17.36	32.20
1	0.87	1.61	30	26.05	48.30
2	1.74	3.22	40	34.74	64.35
3	2.61	4.84	50	43.42	80.45
4	3.48	6.45	60	52.10	96.55
5	4.35	8.05	70	61.00	113.00
6	5.22	9.65	80	69.60	129.00
7	6.08	11.27	90	78.16	145.00
8	6.96	12.90	100	87.00	161.00

Fathoms to Feet to Meters

Fathoms	Feet	Meters	Fathoms	Feet	Meters
¼	1.5	0.5	6½	39.0	11.9
½	3.0	0.9	6¾	40.5	12.3
¾	4.5	1.4	7	42.0	12.8
1	6.0	1.8	8	48.0	14.6
1¼	7.5	2.3	9	54.0	16.5
1½	9.0	2.7	10	60.0	18.3
1¾	10.5	3.2	11	66.0	20.1
2	12.0	3.7	12	72.0	21.9
2¼	13.5	4.1	13	78.0	23.8
2½	15.0	4.6	14	84.0	25.6
2¾	16.5	5.0	15	90.0	27.4
3	18.0	5.5	16	96.0	29.3
3¼	19.5	5.9	17	102.0	31.1
3½	21.0	6.4	18	108.0	32.9
3¾	22.5	6.9	19	114.0	34.7
4	24.0	7.3	20	120.0	36.6
4¼	25.5	7.8	30	180.0	54.9
4½	27.0	8.2	40	240.0	73.2
4¾	28.5	8.7	50	300.0	91.4
5	30.0	9.1	60	360.0	109.7
5¼	31.5	9.6	70	420.0	128.0
5½	33.0	10.1	80	480.0	146.3
5¾	34.5	10.5	90	540.0	164.6
6	36.0	11.0	100	600.0	182.9
6¼	37.5	11.4			

is to write them out "alpha," "beta," etc. So much of our work nowadays goes through the standard typewriter that any notation not normally provided is not a labor-saving device but a time-waster.

On tables and maps the internationally accepted geological *system abbreviations:* P€ (capital C with a bar), Є (with bar), O, S, D, C, P, Tr (or TR, TR written together), J, K, Te, and Q are favored. Lowercase subscripts may be added: e.g., Kl = lower Cretaceous, and a third letter designates the formation initial. In the US, the subsystems Mississippian and Pennsylvanian are indicated as Cm and Cp. The series within the Tertiary are indicated Tp, Te, To, Tm, Tpl. Quaternary stages are not usually identifiable and a simple lithographic notation is given: e.g., Qal (alluvium).

Although a long-time member of the International Subcommission that serves the International Union of Geological Sciences by recommending on Stratigraphic Classification, your editor has made only a desultory attempt

Inches and Millimeters

Inches	Millimeters	Inches	Millimeters	Inches	Millimeters	Inches	Millimeters	Inches	Millimeters
0.05	1.3	3.3	83.8	6.6	167.6	9.9	251.5	30.3	769.6
0.1	2.5	3.4	86.4	6.7	170.2	10.0	254.0	30.4	772.2
0.2	5.1	3.5	88.9	6.8	172.7	11.0	279.4	30.5	774.7
0.3	7.6	3.6	91.4	6.9	175.3	12.0	304.8	31.0	787.4
0.4	10.2	3.7	94.0	7.0	177.8	13.0	330.2	32.0	812.8
0.5	12.7	3.8	96.5	7.1	180.3	14.0	355.6	33.0	838.2
0.6	15.2	3.9	99.1	7.2	182.9	15.0	381.0	34.0	863.6
0.7	17.8	4.0	101.6	7.3	185.4	16.0	406.4	35.0	889.0
0.8	20.3	4.1	104.1	7.4	188.0	17.0	431.8	36.0	914.4
0.9	22.9	4.2	106.7	7.5	190.5	18.0	457.2	37.0	939.8
1.0	25.4	4.3	109.2	7.6	193.0	19.0	482.6	38.0	965.2
1.1	27.9	4.4	111.8	7.7	195.6	20.0	508.0	39.0	990.6
1.2	30.5	4.5	114.3	7.8	198.1	21.0	533.4	40.0	1016.0
1.3	33.0	4.6	116.8	7.9	200.7	22.0	558.8	41.0	1041.4
1.4	35.6	4.7	119.4	8.0	203.2	23.0	584.2	42.0	1066.8
1.5	38.1	4.8	121.9	8.1	205.7	24.0	609.6	43.0	1092.2
1.6	40.6	4.9	124.5	8.2	208.3	25.0	635.0	44.0	1117.6
1.7	43.2	5.0	127.0	8.3	210.8	26.0	660.4	45.0	1143.0
1.8	45.7	5.1	129.5	8.4	213.4	27.0	685.8	46.0	1168.4
1.9	48.3	5.2	132.1	8.5	215.9	28.0	711.2	47.0	1193.8
2.0	50.8	5.3	134.6	8.6	218.4	29.0	736.6	48.0	1219.2
2.1	53.3	5.4	137.2	8.7	221.0	29.1	739.1	49.0	1244.6
2.2	55.9	5.5	139.7	8.8	223.5	29.2	741.7	50.0	1270.0
2.3	58.4	5.6	142.2	8.9	226.1	29.3	744.2	51.0	1295.4
2.4	61.0	5.7	144.8	9.0	228.6	29.4	746.8	52.0	1320.8
2.5	63.5	5.8	147.3	9.1	231.1	29.5	749.3	53.0	1346.2
2.6	66.0	5.9	149.9	9.2	233.7	29.6	751.8	54.0	1371.6
2.7	68.6	6.0	152.4	9.3	236.2	29.7	754.4	55.0	1397.0
2.8	71.1	6.1	154.9	9.4	238.8	29.8	756.9	56.0	1422.4
2.9	73.7	6.2	157.5	9.5	241.3	29.9	759.5	57.0	1447.8
3.0	76.2	6.3	160.0	9.6	243.8	30.0	762.0	58.0	1473.2
3.1	78.7	6.4	162.6	9.7	246.4	30.1	764.5	59.0	1498.6
3.2	81.3	6.5	165.1	9.8	248.9	30.2	767.1	60.0	1524.0

(here and there) to adhere to this approved system of *stratigraphic terms*. Many unconventional uses of the terms "system," "series," and "formation" are found in different parts of the world and in the time available to edit this volume it was not possible to clean up the massive inconsistencies. The formal presentation of the Subcommission's work was only made to the International Geological Congress in Montreal in 1972, and the world will certainly take a little time to catch up. A summary of the recommendations may be found prepared by H. D. Hedberg in the journal *Lethaia* (1972, vol. 5, pp. 283-295).

We have tried to aim for consistency in writing formation and other stratigraphic names in uppercase letters, e.g., Manhattan Schist, Cretaceous System, Paleozoic Era. Similarly, divisions such as Upper Devonian Series or Epoch are capitalized. In distinction to the North American Code, (see *Bull. Am. Assoc. Petrol. Geologists,* 1961), however, we feel that "early" and "late" are informal, relative terms and should not get uppercase treatment.

Time units in the million- and billion ($\times 10^9$)-year range present problems. First, the billion ($\times 10^9$) of North America is not the billion ($\times 10^{12}$) of Europe, where 1×10^9 is a "milliard." In this volume we either express 1×10^9 as a "billion" or as "1000 million." Years are customarily given as y or yr. The world literature shows no consistency. Million years in Rankama's international series on the Precambrian are given as Myr, the billion as 1000 Myr. Some call them "eons" ("aeons"). Others again express these as my or by, both forms being objectionable because they are identical to common English words, m yr being preferred here.

In expressing years "before present" or "ago" the usual form is B.P., sometimes rendered YBP, but more clearly indicated as yr

PREFACE

B.P., but with multiple use B.P. is sufficient to imply years. (In radiocarbon chronology, the zero datum is A.D. 1950.)

Literature citations and journal abbreviations always present a problem. There are a series of International Standards Association recommendations concerning them, but few journal editors seem to have heard of them, so that in a broad cross section of world geologists such as are brought together in this encyclopedia, a dazzling variety of choices of both style and abbreviation have been offered. Our copyeditor is most familiar with the form employed by the American *Chemical Abstracts,* but there is nothing sacrosanct about that system. We do, however, try to insist on the internationally agreed system of ordering of citations—alphabetic (by author's surname), followed by the date, title, journal, volume, and pages.

A fellow encyclopedist, G. G. Hawley, who has gracefully instructed your editor in many of the "facts of life" for this type of publication, has drawn our attention to a sage remark by Samuel Johnson, the English language's first encyclopedist. He pointed out that people needed the encyclopedia less to be informed than to be reminded. In this way many of our entries will be used more for their references than their data. In any case the data tend to go stale with time, whereas the references will always constitute a valuable first approach to an area. The literary explorer, after an initial perusal of any article, could do worse than check over the list of references and then go to the modern bibliographies and, first, check on the country entry, and second, see what new things the recently referenced writers have published. For this purpose an ideal source is: *Bibliography and Index of Geology* (Geological Society of America, 12 numbers each year, plus cumulative volumes). Additional sources are available in French, German, and Russian literature indexes and abstracts. Much regional material, more of a geographic nature, is also to be found in: *Geo-Abstracts* (published at the University of East Anglia, England, and available in several series, including Series A, Landforms and the Quaternary (which is clearly "geology" for many people).

For general geographic updating one may refer to the *World Almanac* or the *Stateman's Yearbook.* A vast quantity of world statistics are contained in a recent volume: Victor Showers, *The World in Figures* (New York: Wiley-Interscience, 1973). For those concerned with conversions to and from the metric system, the following book is indispensable: D. H. K. Amiran and A. P. Schick, *Geographical Conversion Tables* (Zurich: Aschmann and Scheller A.G., for International Geographical Union, 1961).

Other useful references include: *Quantities, Units and Symbols* (Royal Society, London, 48p, 1971); *General Notes on the Preparation of Scientific Papers* (Royal Society, London, 2nd ed., 1965); and *Geowriting: A Guide to Writing, Editing and Printing in Earth Science* (by Wendell Cochran, Peter Fenner, and Mary Hill, 80p, American Geological Institute, Washington, D.C.).

Acknowledgments. It should be pointed out that the production of highly specialized encyclopedias of this sort is only marginally profitable for a publisher, and can only be achieved as a gesture of goodwill for our profession as a whole, by the unpaid and generous help of a large number of specialists whose names appear, alphabetically, in our list of contributors. In some cases, national or state geological surveys or universities have met the costs of drafting maps and diagrams; in others they have been met by the editor, who has also prepared many illustrations himself. The publisher and the editor have shared many of the costs of retyping. Most of the French and German translations have been made by the editor (most have been double-checked by the authors).

Many persons have aided our endeavors in various ways: by critical reading, by recommending authors, by sending reprints and pictures. To each and all, we express our appreciation.

Some of our readers and advisors may be specially mentioned: W. D. Brueckner (Newfoundland), Creighton A. Burk (Princeton), John H. Chronic (Boulder, Colo.), Robert W. Decker (Hanover, N.H.), Thomas W. Donnelly (Binghamton), Donald Green (Fiji), Erik Kjellesvig-Waering (Chicago), Roger L. Larson (Palisades, N.Y.), Arnold R. Lillie (New Zealand),

Allan Lowrie (Chesapeake Beach, Md.), B. T. Malfait (Corvallis, Oregon), Robert F. Schmalz (University Park, Pa.), Gernot C. Schmidt (Princeton), J. C. Schofield (New Zealand), Frank Shaffer (Brazil), and Lawrence L. Sloss (Evanston, Ill.).

We record with regret the deaths of two of our most distinguished contributors (and also good friends), Armin J. Eardley and Horacio J. Harrington.

Close observers will note that, over the years, the publishers' imprimatures of the various encyclopedias in "Fairbridge's series" have changed. This has been due to changes in corporate structures. The original publishing assistant editor, Charles Hutchinson, Jr., is now a partner in the newly developed firm of Dowden, Hutchinson & Ross. The other partners have also been closely associated with this highly specialized type of work in the past and we wish them well under the new banner.

RHODES W. FAIRBRIDGE

Publishers note:

The original concept of eight volumes to include all the subjects in the earth sciences has been modified because the volumes would be too voluminous and costly. Additional volumes now planned for future publication are as follows:

The Encyclopedia of Mineralogy
The Encyclopedia of Igneous and Metamorphic Petrology
The Encyclopedia of Sedimentology
The Encyclopedia of Soil Science
The Encyclopedia of Stratigraphy
The Encyclopedia of Paleontology
The Encyclopedia of the History of Geology
The Encyclopedia of Structural Geology
The Encyclopedia of Geophysics
The Encyclopedia of Ore Genesis and Metallogeny
The Encyclopedia of World Ore Deposits
The Encyclopedia of Applied Geology
The Encyclopedia of Petroleum Geology
The Encyclopedia of Beaches and Coastal Environments
The Encyclopedia of Volcanoes and Volcanology
The Encyclopedia of Snow, Ice, and Glaciology
The Encyclopedia of Geohydrology and Hydrology

CONTRIBUTORS

MICHAEL M. ANDERSON, Dept. of Geology, Memorial University of Newfoundland, St. Johns, Newfoundland, Canada. *Saint-Pierre and Miquelon.*

JACQUES AVIAS, Faculté des Sciences, Avenue Abbé P. Parguel, Montpellier, Hérault, France. *New Caledonia.*

MAHLON MARSH BALL, School of Marine Science, 10 Rickenbacker Causeway, Miami, Florida 33149. *Bahamas.*

MAXWELL R. BANKS, Dept. of Geology, University of Tasmania, Hobart, Tasmania, Australia. *Australia–Tasmania.*

PETER J. BARRETT, c/o British Phosphate Comm., Christmas Island, Indian Ocean, Australia. *Christmas Island.*

JOÃO JOSÉ BIGARELLA, Universidade do Parana, Instituto de Geologia, Caixa Postal, 756, Curitiba-Paraña, Brazil. *Brazil.*

JAN BONDAM, Gronlands Geolog. Undersogelse, Ostervoldgade 10, 1350 Kobenhavn K, Denmark. *Greenland.*

SAM BONIS, Instituto Geográfico Nacional, Avenida Las Americas, 5-76, Zona 13, Guatemala City, Guatemala. *Guatemala.*

ROBERT E. BOYER, Dept. of Geology, University of Texas, Austin, Texas 78712. *United States–Gulf Coastal Province.*

GERRIT C. BROUWER, Rijks Geologische Dienst, Postbus 157, Haarlem, The Netherlands. *French Guiana; Netherlands Antilles.*

DAVID A. BROWN, Dept. of Geology, Australian National University, P.O. Box 4, Canberra, A.C.T., Australia 2600. *Australia.*

JACQUES BUTTERLIN, ENS St. Cloud, 2, Avenue du Palais, 92 St. Cloud, France. *Dominican Republic; Haiti; Martinique.*

COLIN J. CAMPBELL, Shenandoah Oil Corporation, 29 Garrick House, Carrington Street, London, England. *Ecuador.*

JEAN-PIERRE CHEVALIER, Museum National d'Histoire Naturelle, Institut de Paléontologie, 8, Rue du Buffon, Paris 5, France. *Loyalty Islands; Society Islands; Tahiti; Tuamotu Islands; Tubuai (Austral) Islands.*

BORIS CHOUBERT, C.N.R.S., B. Palissy-10, 77-Avon, France. *Guiana Shield–Regional Review.*

ROBERT L. CHRISTIE, Institute of Sediments and Petroleum, 3303 33rd Street N.W., Calgary T2L-2A7, Alberta, Canada. *Canada–Arctic Archipelago.*

T. H. CLARK, Dept. of Geological Science, McGill University, Montreal 110, Quebec, Canada. *Canada–St. Lawrence Lowlands of Quebec.*

PATRICK J. COLEMAN, Dept. of Geology, University of Western Australia, Nedlands, W. A., Australia. *New Hebrides; Solomon Islands.*

G. W. D'ADDARIO, B.M.R., P. O. Box 378, Canberra, A.C.T., Australia. *Australia–Northern Territory.*

IAN W. D. DALZIEL, Dept. of Geological Sciences, Columbia University, New York, N.Y. 10027. *Sub-Antarctic Islands.*

ZOLTAN DE CSERNA, Apartado Postal 69-736, Mexico 21, D.F. *Mexico.*

GABRIEL DENGO, ICAITI, Avenida la Reforma 4-47, Zona 10, A.P. 1552, Guatemala. *Belize (British Honduras).*

R. H. DOTT, JR., Dept. of Geology and Geophysics, The University of Wisconsin, Madison, Wisconsin 53706. *United States–Pacific Cordilleran Region.*

J. WYATT DURHAM, University of California, Dept. of Paleontology, Berkeley, California 94720. *Galápagos Islands.*

A. J. EARDLEY (deceased), formerly of 2618 Skyline Drive, Salt Lake City, Utah 84108. *North America.*

CONTRIBUTORS

RHODES W. FAIRBRIDGE, Dept. of Geological Sciences, Columbia University, New York, N.Y. 10027. *Antarctica; Bahamas; Barbados; Bismarck Archipelago; Caroline Islands; Clipperton Island; Cocos (Keeling) Islands; Cook Islands; Costa Rica; Cuba; Dominica; Dominican Republic; El Salvador; Falkland Islands (Islas Malvinas); Fernando de Noronha, Rocas, Trindade, Martin Vaz, and Saint Paul Rocks; French Polynesia; Gambier (Mangareva) Islands; Gilbert and Ellice Islands; Grenada; Honduras; Juan Fernández; Leeward Islands; Line Islands; Marquesas Islands (Isles Marquises); Marshall Islands; Nauru; Nicaragua; Oceania; Pacific Islands Trust Territory; Phoenix Islands; Pitcairn Islands; Revillagigedo Islands; St. Lucia; St. Vincent; San Ambrosio and San Félix Islands; Tokelau Islands (Union Islands); Trinidad and Tobago; Tuamotu Islands; Tubuai (Austral) Islands; United States–Rocky Mountain Province; Wake Island; Windward Islands.*

L. KENNETH FINK, JR., I. C. Darling Center, University of Maine, Walpole, Maine 04573. *Antigua; Guadeloupe; Leeward Islands; St. Christopher–Nevis–Anguilla.*

JON B. FIRMAN, Dept. of Mines, Box 38, Rundle Street, P.O., Adelaide, S.A., Australia. *Australia–South Australia.*

J. PAUL FITZSIMMONS, Dept. of Geology, University of New Mexico, Albuquerque, New Mexico. *United States–Colorado Plateau Province.*

HUBERT GABRIELSE, Geological Survey of Canada, 100 West Pender Street, Vancouver 3, British Columbia, Canada. *Canada–Cordilleran Region, Interior and Western Belts.*

EDMUND D. GILL, National Museum, 285-321 Russel Street, Melbourne, Victoria 3000, Australia. *Australia–Victoria.*

HENRI GONORD, C.N.R.S., B.P. 5004, 34- Montpelier, France. *New Caledonia.*

F. GONZÁLEZ BONORINO, Castilla 138, Fundación Bariloche, San Carlos de Bariloche, Rio Negro, Argentina. *Argentina.*

MARCUS GORINI, Lamont-Doherty Geological Observatory, Columbia University, New York, N.Y. 10027. *Fernando de Noronha, Rocas, Trindade, Martin Vaz, and Saint Paul Rocks.*

HORACIO J. HARRINGTON (deceased), formerly Professor of Geology, University of Buenos Aires, J.M. Guitierrez 2585, Buenos Aires, Argentina. *South America.*

CHRISTOPHER G. A. HARRISON, School of Marine Science, 10 Rickenbacker Causeway, Miami, Florida 33149. *Bahamas.*

DOROTHY HILL, University of Queensland, St. Lucia, Brisbane, Queensland, Austrailia. *Australia–Queensland.*

WILLIAM F. JENKS, Dept. of Geology, 5 Old Tech Building, University of Cincinnati, Cincinnati, Ohio 45221. *Peru.*

DANIEL E. KARIG, Dept. of Geological Sciences, University of California, Santa Barbara, California 93106. *Mariana Islands.*

PHILIP B. KING, U.S. Geological Survey, Pacific Coast Branch, 345 Middlefield Road, Menlo Park, California 94025. *United States–Appalachian Region, New England Region.*

ROLF KÖSTER, Geologisches Institut, Kiel, Olsenhamsenstr. 40/60, West Germany. *Chile.*

LEENDERT KROOK, P.O. Box 2403, Paramaribo, Surinam. S.A. *Surinam.*

ERNST LÖFFLER, Division of Land Research, Box 1666, Canberra 2601, Australia. *Bismarck Archipelago; New Guinea.*

HANS H. LOHMANN, Stader Strasse 34, Buxtehude, West Germany. *Bolivia.*

ALVIN L. LUGN, 510 7th Avenue NE, Hickory, N.C. 28601. *United States–Great Plains Province.*

ALEXANDER R. McBIRNEY, Dept. of Geology, University of Oregon, Eugene, Oregon 97403. *Galápagos Islands.*

RICHARD B. McCONNELL, Streatwick, Streat near Hassocks, Sussex, England. *Guiana Shield–Regional Review; Guyana.*

FRED T. MacKENZIE, Dept. of Geology, Northwestern University, Evanston, Illinois 60201. *Bermuda.*

JOHN D. MATHER, Hydrogeological Dept., Institute of Geological Science, Exhibition Road, London SW7, England. *Antigua; Turks and Caicos Islands.*

JOHN D. MOODY, Socony Mobil Oil Co., Inc., 150 42nd Street, New York, N.Y. 10017. *Central America–Regional Review.*

CONTRIBUTORS

ERIC W. MOUNTJOY, Dept. of Geological Science, McGill University, Montreal 2, P.Q., Canada. *Canada–Rocky Mountains and Eastern Cordilleran Region.*

GORDON PACKHAM, Dept. of Geology, University of Sydney, Sydney, New South Wales, Australia. *Australia–New South Wales.*

PHILLIP E. PLAYFORD, Geological Survey of W.A., Mineral House, 66 Adelaide Terrace, Perth, W.A. 6000, Australia. *Australia–Western Australia.*

WILLIAM H. POOLE, Geological Survey of Canada, 601 Booth Street, Ottawa, Ontario, Canada. *Canada–Atlantic Provinces.*

HANNFRIT PUTZER, Bundesanstalt für Bodenforschung, Hannover-Buchholz, P.O.B. 54, West Germany. *Paraguay.*

WILLIAM J. REA, Dept. of Geology, Llandinam Building, University College of Wales, Aberystwyth, Wales. *Grenada; Montserrat.*

WILLIAM M. REID, Dept. of Geology, University of Texas, Austin, Texas 78712. *United States–Gulf Coastal Province.*

HORACE G. RICHARDS, Academy of Natural Sciences, 19th and Parkway, Philadelphia, Pa. 19103. *Cayman Islands; Easter Island and Sala y Gomez; United States–Atlantic Coastal Province.*

M. J. RICKARD, Australian National University, Dept. of Geology, Box 4, Post Office, Canberra, A.C.T., Australia. *Australasia–Regional Review.*

EDWARD ROBINSON, Dept. of Geology, University of the West Indies, Mona, Kingston 7, Jamaica. *Jamaica.*

PETER RODDA, Office of the Director, Geological Survey Dept., Private Mail Bag, G.P.O., Suva, Fiji. *Fiji.*

AMOS SALVADOR, Humble Oil and Refining Co., Exploration Dept.–Room 3977, Houston, Texas 77001. *Venezuela.*

BRUCE V. SANFORD, Geological Survey of Canada, E. Petrol. Geol., Bedford Institute, Dartmouth, Nova Scotia, Canada. *Canada–Ontario Basin.*

J. C. SCHOFIELD, New Zealand Geological Survey, Box 61-012, Otara, New Zealand. *Cook Islands; Niue Island.*

E. J. SEARLE, Dept. of Geology, University of Auckland, Auckland, New Zealand. *Kermadec Islands.*

COLIN W. STEARN, Dept. of Geology, McGill University, Montreal 110, P.Q., Canada. *Canada.*

HAROLD T. STEARNS, Apt. 445, 4999 Kahala Avenue, Honolulu, Hawaii 96816. *American Samoa; Guam; Saipan; Tonga; United States–Hawaii; Wallis and Futuna Islands; Western Samoa.*

CHARLES R. STELCK, Dept. of Geology, University of Alberta, Edmonton, Alberta, Canada. *Canada–Northern Great Plains Province.*

GRAEME R. STEVENS, Dept. of Scientific and Industrial Resources, Geological Survey, P.O. Box 368, Lower Hutt, New Zealand. *New Zealand.*

R. H. STEWART, Panama Canal Company, Administration Building, Canal Zone. *Panama.*

FRITZ R. STIBANE, Geologisch-Paläontologisches Institut, Landraf-Philipp-Platz 4-6, Giessen, West Germany. *Colombia.*

C. H. STOCKWELL, Geological Survey of Canada, Dept. of Mines, Ottawa, Ontario, Canada. *Canada–Canadian Shield.*

EDMUND STUMP, Institute of Polar Studies, Ohio State University, 125 South Oval Drive, Columbus, Ohio 43210. *Antarctica.*

ARATA SUGIMURA, Department of Earth Sciences, Faculty of Science, Kobe University, Nada-Ku, Kobe 657, Japan. *Izu-Ogasawara–Iwo Islands (Nampo-Shoto).*

JAAN TERASMAE, Dept. of Geology, Brock University, St. Catherine, Ontario, Canada. *Canada–Ontario Basin.*

JUAN CARLOS M. TURNER, Dept. de Ciencias Geológicas, Facultad de Ciencias Exactas, Ciudad Universitaria, Pabellon 2, Buenos Aires, Argentina. *Uruguay.*

H. LEN VACHER, Bermuda Biological Station, St. George's West, Bermuda. *Bermuda.*

JOHN J. VEEVERS, School of Earth Sciences, Macquarie University, North Ryde, New South Wales, Australia. *Norfolk Island.*

W. A. VISSER, Veeweg 93, Nootdorp, The Netherlands. *New Guinea.*

ROBERT E. WALLACE, Geological Survey, 345 Middlefield Road, Menlo Park, California 94025. *United States–Basin and Range Province.*

ARTHUR J. WARDEN, 25 Tanson Street, Attadale, W. A. 6156, Australia. *New Hebrides.*

JOHN D. WEAVER, College Station, P.O. Box 5232, Mayaguez, Puerto Rico 00708. *Puerto Rico; Virgin Islands.*

J. MARVIN WELLER, 65 Eliseo Drive, Greenbrae, California 94904. *United States–Midwestern Region.*

RICHARD WEYL, Geologisch-Paläontologisches Institut der Justus Liebig Hochschule, Bismarkstrasse, 30, Gissen, West Germany. *West Indies.*

LEE WILSON, Environmental Consultants, Inc., 13626 Neutron Road, Dallas, Texas 75240. *United States of America–General.*

C. GORDON WINDER, Dept. of Geology, University of Western Ontario, London, Ontario, Canada. *Canada–Ontario Basin.*

FREDERICK F. WRIGHT, University of Alaska, 142 East Third, Anchorage, Alaska 99501. *United States–Alaska.*

R. NIGEL YOUNG, Land Resources Division, O.D.A., Tolworth, Surbiton, Surrey, England. *Bahamas.*

A

AMERICAN SAMOA

American Samoa is comprised of five islands and five islets situated at 14°35′–14°12′S and 168°–171°W. This is in the South Pacific, about 840 km NE of Fiji. Pago Pago is the capital and principal city, located on Tutuila Island. The five main islands from W to E are as shown in Table 1.

TABLE I. American Samoa

	Area (km²)	Height (m)
Tutuila (with Aunuu Islet)	135	642
Ofu (with Nuu Islet)	5	477
Olosega	4	646
Ta'u	40	916
Rose	0.8	3

American and Western Samoa together form an archipelago that stretches 480 km trending WNW–ESE, partly en echelon. The basic study is by Daly (1924), a volume that also carries important contributions on the petrology by Chamberlin, and on the reefs by Mayor (1924b).

Tutuila was built by five volcanoes over two, or possibly over three, parallel rifts trending S70°E. A caldera 9.5 km across was formed by collapse of the top of Pago volcano. Pago Pago Bay is a drowned river valley cut by a stream along the curved base of the caldera wall. A thick series of tuffs and differentiated lavas partly filled the caldera. Several bulbous trachyte domes in a highly viscous condition pushed through these tuffs to the surface. One of them dragged a thick section of basalt upward with it for hundreds of feet. About 600 dikes and 40 faults are recorded. Several of the dikes are unusual; two contain oriented dunite xenoliths and one contains bedded ash. Drilling revealed that the fringing reefs of Tuituila are chiefly bedded calcareous sandstone and siltstone and are not coral colonies.

Tutuila is the top of a volcanic pile about 5 km high if measured from the adjacent ocean floor. The latter is of Cretaceous age to judge from the sea-floor spreading history. The oldest exposed rocks are believed to be Pliocene in age and show reversed magnetism. The youngest rocks are nepheline basanite and comprise the Leone volcanics of Holocene age. All the rocks belong to the alkali suite and range from picrite basalt to trachyte. Numerous plug domes of trachyte occur. Tholeiitic basalts are not exposed.

The Manu'a Group includes Ofu, Olosega, and Ta'u islands as well as Nu'utele and Nu'usilaelae islets off Ofu Island. The rocks of this group range in age from Pliocene to Holocene. A submarine eruption occurred off Olosega in 1866. Ta'u is a collapsed basaltic shield volcano. Ponded lavas and pyroclastics accumulated in its summit caldera to a thickness of more than 300 m. Two rift zones radiate from the summit, one to the NE and the other to the NW. Following a period of extensive erosion, the northeast corner was extended by dunite-bearing lava flows. A tuff complex containing large dunite xenoliths and coral blocks buries a former sea cliff near the village of Faleasao.

Ofu and Olosega islands are a complex of six volcanic cones aligned along the main Samoan rift. Two of these cones developed shields and, later, collapsed calderas. The caldera of northern Ofu became partly filled with later basalts. Following quiescence and erosion, a tuff cone was formed, remnants of which are left as Nu'utele and Nu'usilaelae islets. High cliffs, the result of marine erosion, surround much of the islands. Benches at 1.5 and at 7 m indicate eustatic shorelines, common to all the Samoan Islands. All rocks belong to the alkali basalt suite.

Rose atoll has sand beaches 3 m above sea level and has emerged reef remnants which indicate a 1.5-m drop in sea level.

HAROLD T. STEARNS

References

Daly, R. A., 1924. "The geology of American Samoa," *Carnegie Inst. Wash. Publ. 340*, 95–145.

Dana, J. D., 1849. "Geology," in: *U.S. Exploring Expedition during the Years 1838–42 under Command of Charles Wilkes, U.S.N.*, vol. 10. New York: Putnam, 307–336.

Macdonald, G. A., 1944. "Petrography of the Samoan Islands," *Bull. Geol. Soc. Am.*, **55**, 1333–1362.

———, 1968. "A contribution to the petrology of Tutuila, American Samoa," *Geol. Rundschau*, **57**, 821–837.

Mayor, A. G., 1924a. "Rose atoll, American Samoa," *Carnegie Inst. Wash. Publ. 340,* 73–91.

———, 1924b. "Structure and ecology of Samoan reefs," *Carnegie Inst. Wash. Publ. 340,* 1–25.

Stearns, H. T., 1944. "Geology of the Samoan Islands," *Bull. Geol. Soc. Am.,* **55,** 1279–1332.

Stice, G. D., and McCoy, F. W., Jr., 1968. "Geology of the Manu'a Islands, Samoa," *Pacific Sci.,* **22**(4), 427–457.

Cross-references: *Gilbert and Ellice Islands; Oceania; Tonga; Western Samoa.*

ANTARCTICA

Antarctica is the southernmost continent, centered roughly about the South Pole, with most of its boundaries falling within the 65°S parallel. Its area, not including ice shelves and connected islands, is 12,393,000 km^2. However, only 2% of this area is actually exposed bedrock. The rest is covered by the largest ice sheet in the world, having a total volume of around 24,000,000 km^3, representing 89% of the earth's glacial ice. If melted, this ice would raise the global sea level by 55 m.

The I.G.Y. program of 1957–1958 led to the initiation of a comprehensive mapping scheme of Antarctica by the American Geographical Society ("Antarctic Map Folio Series"). Folio 12 of the Geologic Maps of Antarctica was initiated in 1964 and published in 1970 in 18 sheets of 1:1 million or better. This collection made possible the compilation of a single map on a 1:5,000,000 scale in 1972. Folio 16, entitled "Morphology of the Earth in the Antarctic and Subantarctic" (B. C. Heezen, M. Tharp, and C. R. Bentley), includes three maps at 1:15,000,000, depicting the sub-ice topography and that of the surrounding sea floor, the physiographic provinces, and the survey tracks.

Historical Notes

In the early 19th century the world knew nothing of an Antarctic continent. There had been talk for centuries of a "Great South Land," some scientists arguing that there had to

FIGURE 1. Morphologic map of Antarctica (Harrington, 1965).

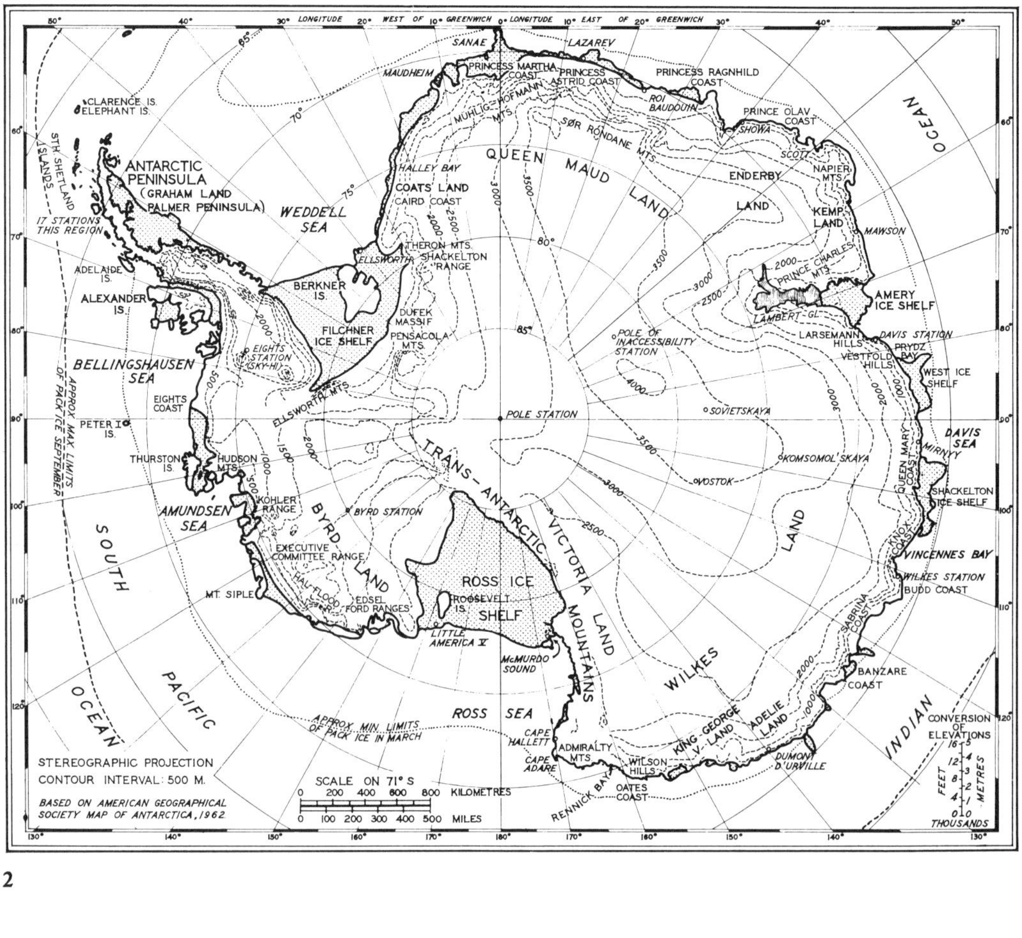

be one in order to maintain global balance. Captain Cook had once sailed to the margin of the shelf ice. The first substantive discoveries were by the French expedition under Dumont d'Urville (reported by Grange, 1848) and by U.S. Exploring Expedition under Wilkes (1845). They reported finding iceberg-carried erratic blocks of continental-type rocks such as granites and sandstones. (Early references are contained in reviews by Fairbridge, 1952; Harrington, 1965; Splettstoesser, 1966, and others.)

The first discovery of actual land and of those rocks *in situ*, in (South) Victoria Land, was by Sir James Ross after the two ships HMS *Erebus* and *Terror* anchored in McMurdo Sound in 1841. Earlier they had found granite erratic boulders on (volcanic) Possession Island offshore. The first fossil discoveries were on Seymour Island (off West Antarctica) by C. A. Larsen in 1893, and Borchgrevink collected granites *in situ* on Cape Adare (Murray, 1895). Nevertheless, at the end of the 19th century, skeptics still insisted on referring to "Murray's Hypothetical Antarctic Continent" (at a meeting of the Royal Society of London in 1898), for, clearly, very little had been actually delineated. The possibility of a giant "Antarctic Archipelago" was entertained, when setting up the program for Shackleton's National Antarctic Expedition (Gregory, 1901). This possibility would have been compatible with theories proposed earlier by Reiter and by Suess that there was a "Circum-Pacific fold belt," which continued from the Andes through Graham Land and what is now known as the Antarctic Peninsula to connect (with gaps) to New Zealand and Australia. Further it was reasoned that East Antarctica was a Precambrian Shield matching that of Western Australia.

These far-sighted speculations were to be remarkably confirmed, but not for almost a century. It should not be forgotten that Darwin had already remarked on the notable similarity of the Beacon Sandstone type facies of the southern continents he visited, and the Permo-Carboniferous tillites of those countries were well established by the 1870s, so Suess was able to coin the term "Gondwanaland," a hypothetical land mass that would have united all of them. The stage was then set for the continental drift reconstructions of Alfred Wegener in the decade before World War I.

The Swedish expedition then provided Nordenskjöld (1905, 1911) with material for actually distinguishing and naming *West Antarctica* (the younger fold-belt sector) and *East Antarctica* (the Precambrian shield sector). The Australians joined the British expeditions of these decades, at which time Sir Douglas Mawson (and Madigan) established the South Magnetic Pole, and, more importantly, mapped extensive sectors of the East Antarctic shore, bringing home very large suites of Precambrian morainic boulders, which today still provide the basis of what is known of that vast ice-covered area. Sir Edgworth David first used the term *Great Antarctic Shield* (1914).

A more hypothetical suggestion was that of a transantarctic rift, the *Ross-Weddell Graben* (Gregory, 1912), which was supposed to separate the two provinces. Clearly there are horst and graben features in this belt, but it is far from a simple rift. David and Priestley (1914) first used the term *Antarctic Horst*, which they explored in Victoria Land, but later when its continuity was proved, this became expanded to *Transantarctic Mountains* or *Transantarctic Escarpment* (e.g., in Harrington, 1965).

Explorations in Antarctica during the post-World War I phases were desultory and scattered, mainly limited to the work of Mawson. It was not until the development of new ice-sounding geophysical equipment and new transport and support logistics in the post-World War II period that systematic exploration began. This work reached a peak with the International Geophysical Year (1957–1959) and with the deployment of expeditions from a dozen countries or more. Most continuous has been the work of the U.S., Russian, and British teams. Apart from the vast improvement in systematic mapping, perhaps the two most important discoveries that bear on the question of continental drift and the breakup of Gondwanaland were (1) the discovery in the Horlick Mountains of Permo-Carboniferous tillites in 1957 (Long, 1962) and numerous subsequent confirmations (see, e.g., Frakes *et al.*, 1971); and (2) the tetrapod reptile (*Lytosaurus*) fossils in the Shackleton Glacier area, with its close South African affinities (Kitching *et al.*, 1972). These provided two critical keys in the linking together of the ancient Gondwana land mass in the early Mesozoic (see discussions, e.g., in Adie, 1964; and in the Second Gondwana Symposium: South Africa, 1970).

In the last decade it has been clearly shown that not all of West Antarctica (the *"Antarctandes"* of Arctowski, 1895; *"Andean Province"* of Adie; . . .) consists of very youthful alpine-type fold-belt material, for the Ellsworth and Whitmore mountains, for example, were found to lie athwart the Ross-Weddell depression. They contained Paleozoic metasediments for the most part, but in their principal early Mesozoic orogeny they are clearly comparable with the rest of the circum-Pacific, where this phase is remarkably widespread (e.g., Akiyoshi Orogeny of Japan).

ANTARCTICA

In East Antarctica, the Russians (e.g., Ravich and Grikurov, 1970) have been able to identify two divisions: (1) a region of pre-Baykalian consolidation that extends from about 105°E to 25°W, evidently the primeval nucleus, facing the African sector; and (2) a region with a late Baykalian folded basement that extends from that border to the Transantarctic Escarpment and even to include numbers of structural blocks within the West Antarctic region. The Russians also infer a broad platform-type area, assumed to overlie a "pre-Baikalian" (latest Precambrian) basement, in a broad wedge extending from 105–150°E inward to the area below the south geographic pole; this platform they believe to be a plains-type sector extensively underlain by little-disturbed Paleozoic-Mesozoic-Cenozoic sediments (in part, in broad downwarps or "syneclises").

Concerning the drift evolution of the Antarctic continent, the marine geological studies in the Southern Ocean play a key role. Sea-floor spreading, e.g., between Antarctica and Australia, has been confirmed as mid to late Cenozoic age (see, e.g., the work of Hayes, Talwani, Conolly, and others in Anon., 1972).

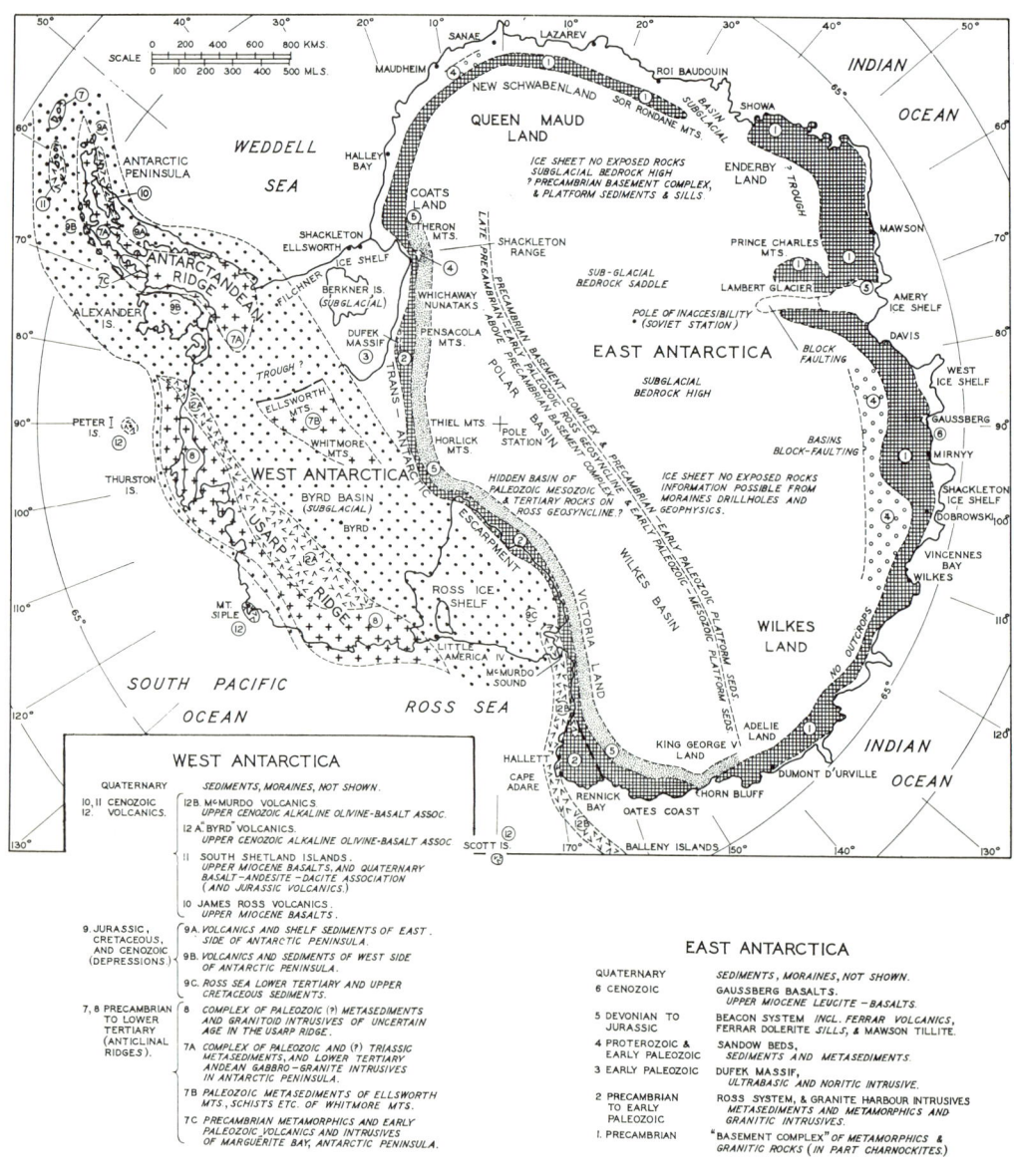

FIGURE 2. Geological sketch map of Antarctica.

The deep-sea drilling ship *Glomar Challenger* was able to bring invaluable confirmatory data to this problem.

Morphotectonic Subdivision

The continent is divided into East and West Antarctica on physiographic and structural grounds. *East Antarctica* is a Precambrian cratonic shield bounded on one side by younger orogenic belts of the Transantarctic Mountains. Its crustal thickness exceeds 40 km, and were the ice removed and glacial rebound to occur, the land surface on the cratonic side of the Transantarctic Mountains would have the form of a dissected, block-faulted plateau.

In contrast, *West Antarctica* is a group of younger continental crustal blocks, 30+ km thick, bounded by deep trenches. Were the ice removed, this area would remain beneath sea level except for several large islands. The rocks of West Antarctica are in general younger than those of the Transantarctic Mountains. They represent successive stages of compressive tectonic activity that characterized this margin of the continent from late Precambrian through early Tertiary. The continent is seismically quiet at present, but a few volcanos are active in the areas of McMurdo Sound and the Antarctic Peninsula.

Precambrian Shield

Continental Border. The bulk of East Antarctica is presumed to be underlain by meta-

FIGURE 3. Subglacial or bedrock morphology, after an inset in the American Geographical Society's map of Antarctica 1962. Areas below sea level are shown by a dotted ornament. The 500-m bathymetric contour indicates the approximate position of the outer edge of the continental shelf (Harrington, 1965).

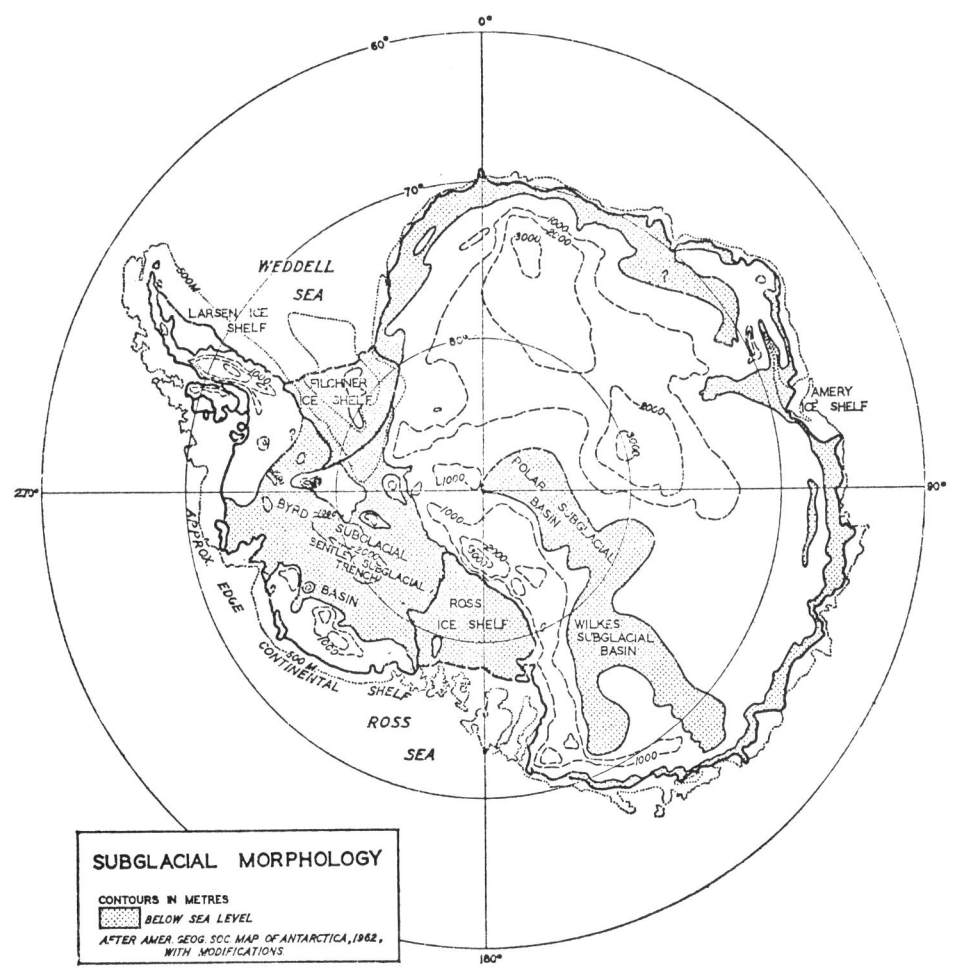

TABLE 1. Composite Section of Central Transantarctic Mountains

Age	Formation	Description	Thickness (m)
Late Tertiary	Sirius Fm. McMurdo Volcanics	Semilithified till, occasional intermixed stratified lenses. Olivine basalt flows, with associated explosive portions. (Much more extensive exposures in Victoria Land). —UNCONFORMITY—	
Jurassic	Kirkpatrick Basalt	Tholeiitic flows, rare sedimentary interbeds with conchostracans, holostean fish. (Correlative Ferrar Dolerite sills with a cumulative thickness of about 1000 m intrude the Beacon sequence.) DISCONFORMITY	600+
Triassic	Prebble Fm. [BEACON SUPERGROUP]	Mudflow debris, pyroclastic breccia, tuff, and tuffaceous sandstone.	3–460+
	Falla Fm.	Volcanic sandstone, shale; tuff dominates upper part. *Dicroidium*.	160–530
	Fremouw Fm.	Subarkose, volcanic sandstone, greenish-gray mudstone; *Lystrosaurus* near base; logs, coal, *Dicroidium* near top. DISCONFORMITY	About 650
Permian	Buckley Fm.	Arkonic and volcanic sandstone, dark-gray shale, coal, *Glossopteris*.	About 750
	Fairchild Fm.	Massive subarkose and arkose.	130–220
	Mackellar Fm.	Dark shale and fine-grained sandstone.	60–140
	Pagoda Fm.	Tillite, sandstone and shale. DISCONFORMITY	125–395
Devonian?	Alexandra Fm.	Quartz arenite and sandstone. —UNCONFORMITY—	0–330
Ordovician -Precambrian		Basement metasedimentary complex intruded by granitic rocks. —Ross Orogeny—UNCONFORMITY—	
Ordovician	Granite Harbour Intrusions	Granite to diorite, equigranular to porphyritic, syn- and post-deformation. Minor gabbro.	
Cambrian	Shackleton Limestone (also Leverett Fm., Henson Marble, Fairweather Fm., Taylor Fm.)	Limestone, archeocyathids (others include marble, quartzite, conglomerate, silicic volcanics). —Beardmore Orogeny—UNCONFORMITY—	9000+
Late Precambrian	Wisconsin Range Batholith Wyatt Fm.	Granite to quartz diorite, equigranular to porphyritic. Silicic, porphyritic volcanic rock. —UNCONFORMITY—	Several 100+
	Goldie Formation (incl. LaGorce Fm.)	Graywacke, argillite, quartzite. —Nimrod Orogeny—UNCONFORMITY—	6500+
Precambrian	Nimrod Group	Mica schist, banded gneiss, augen gneiss, marble, amphibolite, subordinate calc-silicate schist.	2500+

morphosed Precambrian rocks, for such is the case in exposures rimming the continent between 15°W and 145°E longitude, and judging from the morainic boulders. The predominant metamorphic grade of these rocks is granulite facies, indicating high temperatures and relatively anhydrous conditions. Of note also are the numerous charnockite bodies found scattered throughout the periphery. The oldest reported date is 3000 m yr for a granite in western Queen Maud Land. Age dates ranging from 1500 to 500 m yr characterize Wilkes Land. Polymetamorphism is indicated petrographically at numerous localities throughout the exposed shield, with many rocks being isotopically homogenized at about 500 m yr, corresponding to the Pan-African Event. This was particularly felt in the Queen Maud Land area.

Amphibolite and greenschist facies rocks occur at the head of the Lambert Glacier, at the most interior exposures of bedrock on that side of the continent. In addition, granite, both foliated and unfoliated, intrudes these rocks.

Western Queen Maud Land. In western Queen Maud Land Precambrian clastic rocks, including graywacke, arkose, and conglomerate, are found. A shallow-water environment of deposition is indicated. They are in part interbedded with mafic to intermediate lava flows, some of which exhibit pillow structure, indicating subaqueous extrusion. Regional metamorphism of these rocks is slight, shown only by sericitized feldspar. Large diabase bodies of Precambrian age (1600 m yr) are found intruding these, as well as the higher-grade rocks in western Queen Maud Land; plutons ranging from diorite to granite, and thought to be Precambrian in age, also occur in this region.

Transantarctic Mountains. One area of Precambrian rocks is exposed in the central Transantarctic Mountains in the Miller and Geologists ranges. The assemblage is a varied sequence of schists, gneisses, and marbles metamorphosed to the amphibolite facies. Radiometric dating indicates that these rocks are at least 1990 m yr in age with isotopic reequilibration occurring at about 600 and 450 m yr.

Mobile Belt of the Transantarctic Mountains and West Antarctica

Late Precambrian. Deposition began in the late Precambrian on the margin of this stable craton with graywacke-shale sequences. These are found all along the Transantarctic Mountains and possibly in western Marie Byrd Land. Silicic, pyroclastic volcanic rocks overlie the sediments in the central part of the range. Some of these units are intruded by calc-alkaline porphyries dated at 600± m yr. They are *syn*- and *post*-orogenic with respect to the deformation and metamorphism of the "Beardmore Orogeny."

Cambro-Ordovician. Cambrian limestones occur unconformably above the older rocks throughout the range, with particular areas being characterized by archeocyathid and trilobite assemblages. Pyroclastic volcanics overlie limestones in the Pensacola Mountains and are interbedded with them in the central Transantarctic Mountains. As before, calc-alkaline intrusions followed the depositional sequence. Both crosscutting and gradational relationships exist between the intrusions and country rock, with widespread anatexis indicated by migmatite and gneisses that occupy the axis of deformation of the "Ross Orogeny." Metamorphic and intrusive age dates group around 500 m yr. As mentioned above, in addition to the mobile margin, the older Precambrian rocks of the craton record a thermal event that reset K/Ar clocks at 500± m yr. The full significance of this feature is still unknown.

The Ellsworth Mountains form an exception to this pattern. There, conformable sediments from the late Precambrian through the Permian remained undeformed until they were folded in the Triassic. A late Cambrian fauna there dates a group of varied clastic rocks that overlie a carbonate sequence inferred from stratigraphic position to be late Precambrian.

Devonian. The Devonian marked the beginning of deposition of the Beacon Supergroup, a distinctive, nearly flat-lying sequence of clastic sediments including much feldspathic sandstone, a tillite, and overlain by basaltic lavas, which has often been correlated with similar rocks in Australia, Africa, and South America.

In the Devonian, following post-Ross Orogeny uplift and erosion throughout the Transantarctic Mountains, sedimentation was more provincial than in previous periods. Marine sandstones and shales are found in the Pensacola Mountains and the Ohio Range, as well as in the Ellsworth Mountains, and clastic deposits identify local paleobasins in the central Transantarctic ranges. In northern Victoria Land rhyolites accumulated while calc-alkaline plutons were intruded there and in western Marie Byrd Land. This activity is called the "Borchgrevink Orogeny." Some metamorphic rocks of Marie Byrd Land are postulated to have developed in that period, but as yet this awaits confirmation by radiometric age determination.

Permo-Carboniferous. A major ice sheet accumulated in the area of the Transantarctic and Ellsworth mountains during the late Carboniferous and early Permian. Thick glaciomarine deposits are found in the Ellsworth

Mountains, with continental tillites occurring in the central Transantarctic Mountains (Fig. 4). The Pensacola Mountains contain rock that is transitional between these two facies. Turbidite sequences in the Antarctic Peninsula area and Alexander Island record the first confirmed deposition in those areas.

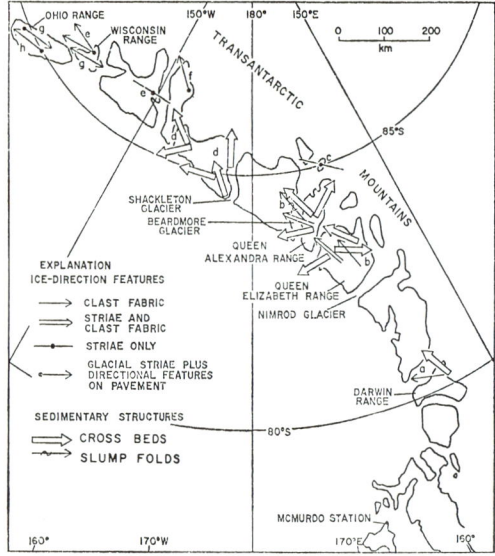

FIGURE 4. Permo-Carboniferous striae and related features in Transantarctic Mountains (Coates, 1972).

Later Permian. Permian fluvial and lacustrine deposits overlie the tillites in both the Transantarctic and Ellsworth mountains. The basin is known to have extended onto the craton in the area of the Shackleton Range and western Queen Maud Land, where similar clastic rocks exist. There is also an isolated occurrence of Permian clastics in the Lambert Glacier region. The *Glossopteris* flora characterizes all these deposits, with coal seams being abundant. Northern Victoria Land remained a source area for most of the period, although a few isolated outcrops have been reported. Stream-current markers indicate flow at this time away from southern Victoria Land along the present axis of the range.

Triassic. The regional paleoslope was reversed at the beginning of the Triassic by uplift which culminated in the folding of the "Gondwanide Orogeny" in the Ellsworth and Pensacola mountains. Fluvial deposits are found in the central Transantarctic Mountains and northern Victoria Land. These rocks are dated by the *Dicroidium* flora and the *Lystrosaurus* reptile fauna which are very similar to assemblages found in the other southern "Gondwana" continents.

An increasing fraction of the Permian and Triassic sediments consists of epiclastic volcanic material. Late Triassic tuffs and associated pyroclastic debris overlie the sediments in the central Transantarctic Mountains. In addition, tectonic activity at this time is indicated at the base of the Antarctic Peninsula and on Thurston Island by several radiometric dates.

Jurassic. Early to mid-Jurassic plutons occur in the area between the Ellsworth and Pensacola mountains and to a limited extent in western Marie Byrd Land and the Antarctic Peninsula. Mid-Jurassic shales and volcaniclastic

TABLE 2. Composite Section of Northern Antarctic Peninsula

Age	Formation	Description	Thickness (m)
Pleistocene	Pecten Conglomerate	Conglomerate. Pectens.	
		—UNCONFORMITY—	
Upper Miocene-Pliocene	James Ross Island Volcanic Group	Marine tuffs, pillow basalts conglomerate.	300+
		—UNCONFORMITY—	
Lower Miocene	Seymour Island Series	Soft sandstone, shale, conglomerate. Fossil penguins	160
		—UNCONFORMITY—	
Lower Tertiary	"Andean" Intrusions	Calc-alkaline plutons.	
Upper Cretaceous	Snowhill Island Series	Sands, gravels, some clays. Ammonoids, lamellibranchs.	3580
		—UNCONFORMITY—	
Middle Cretaceous	Intrusions	Acid-intermediate-basic plutons.	
Upper Jurassic	Volcanic Group	Andesites, rhyolites, agglomerates.	230+
Middle Jurassic	Intrusions	Acid plutons.	
	Mt. Flora Beds	Conglomerate, sandstone, abundant plant fossils.	300
		—UNCONFORMITY—	
Carboniferous-Permian(?)	Trinity Peninsula Series	Graywacke, siltstone, sandstone, shale.	10,000+(?)

TABLE 3. Stratigraphic Table of the Fossil Record in Antarctica

Quaternary	Sub-fossil penguins, many rookeries; subfossil shells, Falkland Islands, McMurdo Sound, etc. Shell beds, Deception Island (late Pleistocene?); shell beds, McMurdo Sound and Cockburn Island (early Pleistocene or late Pliocene).
Tertiary	Miocene plants in James Ross Volcanics of the James Ross Archipelago, and at the South Shetlands. Lower Miocene or Upper Oligocene plants and rich marine faunas at Seymour and Cockburn islands, east side Antarctic Peninsula. Spores and pollen (*Nothofagus,* etc.) and marine microfossils in erratics, McMurdo Sound. Lower Tertiary and Upper Cretaceous forminifera, Burdwood Bank, Scotia Arc; derived pollen in Upper Oligocene Seymour Island bed.
Cretaceous	Rich biota of Lower and Middle Campanian marine invertebrates, and plants, at James Ross Archipelago, Antarctic Peninsula, east side. Aptian plants and marine fauna, Alexander Island, Aptian ammonites, lamelli-branchs, etc., South Georgia; Cretaceous plants, brachiopods, etc. South Orkney Islands.
Jurassic	Plants and marine fauna, Alexander Island; wood, South Shetlands. Rich plant beds, Hope Bay, Antarctic Peninsula; plants in Victoria Land with bivalve crustacea (conchostracans).
Triassic	Plants, Beacon Group (*Dicroidium*)
Permian Carboniferous Devonian Silurian Ordovician Cambrian	Wood, leaves, microfossils, *Glossopteris* flora, and invertebrate trails many localities, Beacon System (Glaciation in Upper Carboniferous) Mid or Upper Devonian freshwater fish remains, and Lower Devonian plants, with marine problematica, Victoria Land. Lover Devonian spore assemblage and rich marine fauna. Horlick Mountains. Supposed Ordovician graptolite, South Orkneys, now considered obscure plant fragment. Cambrian Archaeocyathid sponges, trilobites, algae, etc., in limestones of Ross System; Cambrian microfossils in Sandow Beds.
"Proterozoic" "Archaean"	

Source: Harrington (1965).

sediments are found locally in parts of the peninsula, and widespread calc-alkaline volcanic rocks of the Late Jurassic cover much of the peninsula and a small area in western Marie Byrd Land.

Diabase sills and dikes of this period intrude the Beacon sediments of the Transantarctic Mountains, with tholeiite basalts capping the sequence. Related to these is the Dufek Massif—a large, stratiform, basic igneous intrusion that crops out in the Pensacola Mountains.

Cretaceous. Shallow and deeper-water marine rocks from the Early Cretaceous are found on Alexander Island, and shallow marine sediments of the Upper Cretaceous occur at the northern end of the Antarctic Peninsula. The major feature of this period, however, is an enormous volume of calc-alkaline intrusions from the Antarctic Peninsula to western Marie Byrd Land. These were emplaced in the Upper Cretaceous and Lower Tertiary during the "Andean Orogeny."

Basaltic Provinces

Victoria Land and Marie Byrd Land. Tertiary and Quaternary alkaline basalts exist over a considerable area in the McMurdo Sound region, northern Victoria Land, and Marie Byrd Land. Eruptions probably began in the Early Tertiary in Marie Byrd Land with fumarolic activity continuing at present on Mt. Erebus and Mt. Melbourne in the McMurdo Sound and northern Victoria Land sector. The extrusions formed stratiform shield volcanos, usually having associated parasitic cones, with an implied structural control of their eruptive patterns in specific areas. At the head of the Scott Glacier, Mt. Early is composed of basalts related to this province. This isolated occurrence is the southernmost volcano in the world.

Antarctic Peninsula. Mildly undersaturated basalt is found in the James Ross Island area of the northern Antarctic Peninsula. This is overlain by a *Pecten*-bearing conglomerate dated as Pliocene. The volcanics in places also overlie clastic rocks correlative with the Lower Miocene sediments of Seymour Island, containing a rich floral and faunal assemblage (of warm paleoclimatic character).

Tertiary and Quaternary Glaciation

Onset. The time of onset of glaciation in Antarctica was certainly pre-Quaternary. Earliest indications are from quartz grains recovered from deep-sea cores from the Southern Ocean, which have been interpreted as glacially derived. Dated as Eocene and Oligocene, this material was probably produced from calving glaciers, but not necessarily of continental-ice-sheet proportions. Tillite and associated volcanic rock from the Jones Mountains, dated radiometrically at 7 m yr, is a record of probable continental glaciation.

Fluctuations. Data for later fluctuations in the ice level have come from the ice-free Taylor, Wright, and Victoria valleys in the McMurdo Sound area. Once filled with glacial ice flowing from the polar plateau, these unusual valleys now contain only isolated alpine glaciers on their sides. Morainal deposits there indicate that the East Antarctic ice sheet spilled into the valleys at least five different times, with the original carving of them occurring prior to 4 m yr ago as indicated by radiometric dating. In addition, the Ross ice shelf became grounded several times in the past, causing ice to back up into the valleys, damming lakes whose strandlines now mark the valley walls. Glacial deposits in the central Transantarctic Mountains also record periods when ice stood higher, owing to grounding of the Ross ice shelf.

Much of the ice sheet in West Antarctica rests on bedrock considerably beneath sea level. It has been speculated that this body could surge into the surrounding ocean, causing marked sea-level changes in a relatively short period of time, but more study is needed to confirm or reject this hypothesis.

FIGURE 5. Position of Antarctic structural belts in relation to those of adjacent Gondwana land masses.

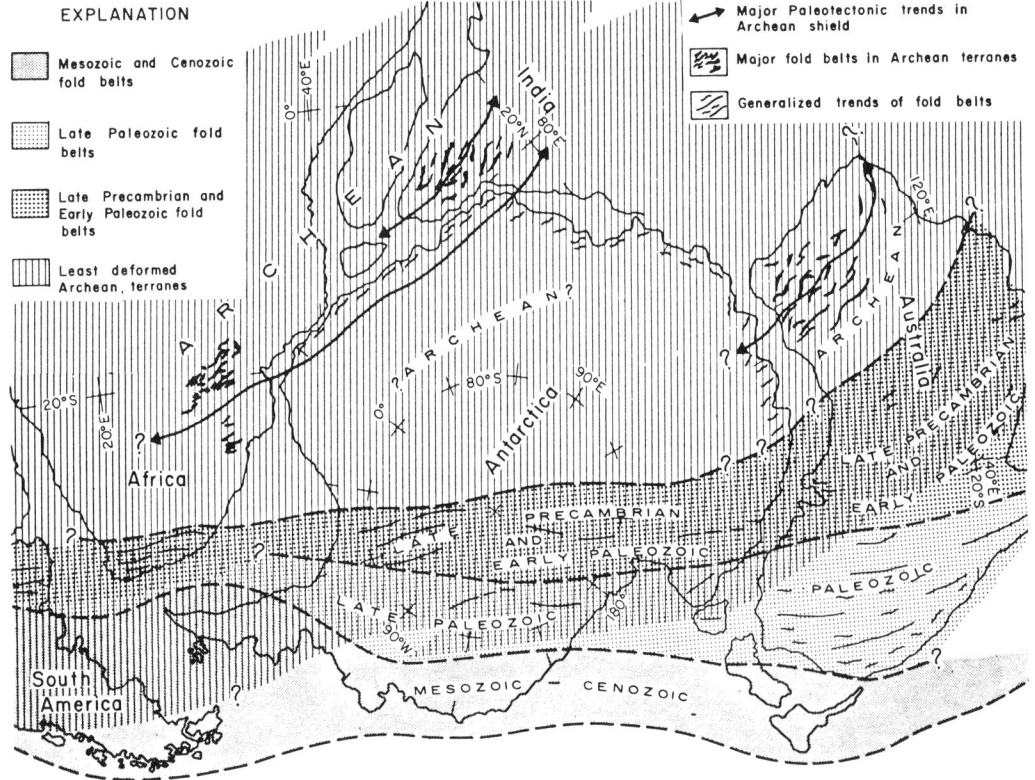

Antarctica's Position in Gondwanaland

Looking back through the record of Antarctica as it is today, one imagines a continental structure much the same as now throughout the Tertiary, with basaltic volcanos accumulating along the Pacific coastline. Cretaceous batholiths were intruded at the Pacific margin of West Antarctica and were preceded there by Jurassic calc-alkaline volcanics. During those periods, marine basins around the Antarctic Peninsula received sediments. Also at that time the blocks of West Antarctica were perhaps rifted from the craton into their present configuration. Triassic folding in the Ellsworth-Pensacola mountain area followed a marine basin there in the Devonian and Carboniferous. By contrast, the northern Victoria Land area in the Triassic received detritus from the Ellsworth-Pensacola uplift. This was preceded in the Carboniferous and Permian by a positive source area which had been tectonically effected in the Devonian. The entire range was adjusting isostatically following the deformations and intrusions of the Ordovician. These were preceded by a Cambrian marine environment of volcanic and carbonate accumulation in the Transantarctic Mountains area. Prior to this, orogenic activity was felt in the deeper-marine sediments of the late Precambrian continental rise in the area of the Transantarctic Mountains. Much earlier in the Precambrian the crust of the craton was consolidated.

Gondwanaland. This Antarctic cratonic nucleus was encompassed within the "supercontinent" of Gondwanaland and located centrally in it (Fig. 5). Late Precambrian and Phanerozoic orogenic belts from Australia, New Zealand, Africa, and South America merge with ones in Antarctica. The Adelaide Geosyncline of Australia and the Damara Geosyncline of southwestern Africa are linked by the late Precambrian–early Paleozoic Ross Geosyncline. The Devonian marine basin of the eastern Transantarctic Mountains also included parts of southern Africa, Patagonia, and the Falkland Islands. The major Devonian consolidation of the Tasman Geosyncline of Australia accompanied in time the igneous activity in northern Victoria Land and western Marie Byrd Land. Permo-Carboniferous glacial deposits are found in all the Gondwanaland fragments (Fig. 6), as are the nearly identical plant fossils from the Permian and Triassic and amphibian and reptile remains from the Triassic. In the same way, the Gondwanide Orogeny was felt along the Pacific margin during the Triassic, commencing locally in the Permian.

The breakup of Gondwanaland (Fig. 7) began in the Mesozoic, and the South America-Africa

TABLE 4. Stratigraphic Comparison Among East Antarctica, South Africa, and the Falkland Islands

Source: Harrington (1965).

FIGURE 6. Antarctica during the late Carboniferous and early Permian, illustrating inferred polar shift, with ice flow directions (Crowell and Frakes, 1972). Letters are initials of principal sedimentary basins (dotted).

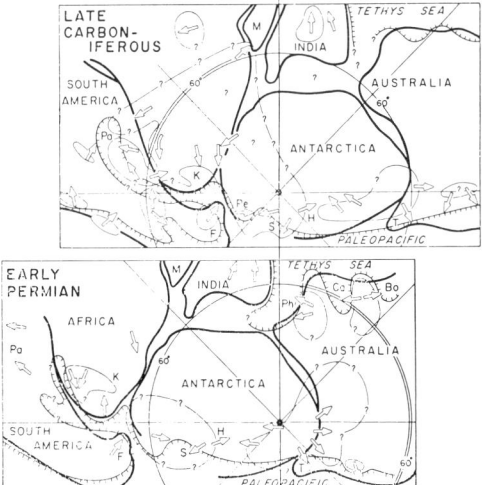

FIGURE 7. Comparison of selected Gondwanaland models (Ford, 1972).

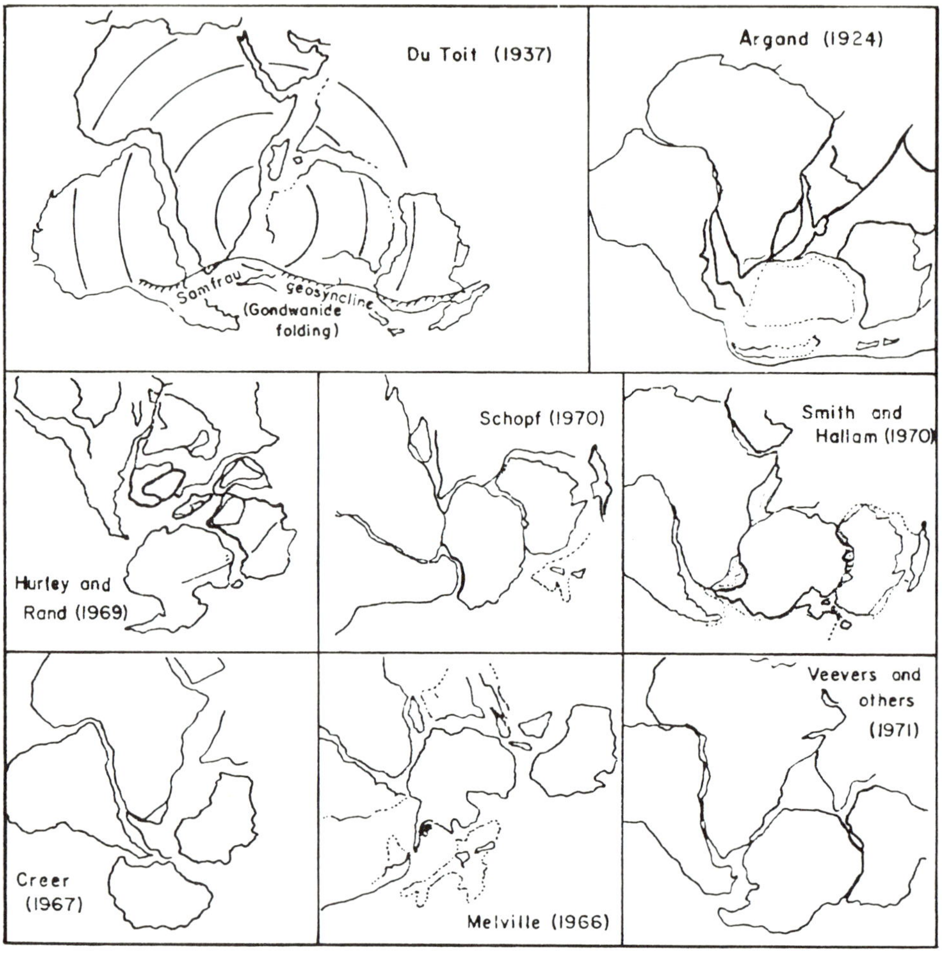

portion separated from the Antarctic-Australia part in the mid-Cretaceous. Compression, as evidenced by calc-alkaline igneous activity, continued along the New Zealand, West Antarctic, and South American front. This configuration required right-lateral displacement between East and West Antarctica and along the Alpine Fault in New Zealand. At this time West Antarctica became fragmented. New Zealand broke free from Antarctica in the late Cretaceous, and Australia separated in the Eocene. Oroclinal bending of the Antarctic Peninsula–Scotia arc area accompanied the separation of South America and Africa. Since the early Tertiary, Antarctica has remained tectonically calm as the other Gondwanaland fragments continue to move away from it.

EDMUND STUMP
RHODES W. FAIRBRIDGE

References

Adie, R. J., ed., 1964. *Antarctic Geology*. Amsterdam: North-Holland Publ.; New York: Wiley-Interscience, 758p.

———, ed., 1972. *Antarctic Geology and Geophysics*. Oslo: Universitetsforlaget (Symp. Antarctic Geol. Solid Earth Geophys., 1970), Int. Union Geol. Sci., Ser. B, 1, 876p.

Barrett, P. J., 1972. "Stratigraphy and paleogeography of the Beacon Supergroup in the Transantarctic Mountains, Antarctica," 2nd Gondwana Symp., Proc. Pap. 249–256.

———, Baillie, R. J., and Colbert, E. H., 1968. "Triassic amphibia from Antarctica," *Science,* 161(3840), 460–462.

Bentley, C. R., 1965. "The land beneath the ice," *in* T. Hatherton, ed., *Antarctica*. New York: Praeger, 259–277.

———, and Clough, J. W., 1972. "Antarctic subglacial

structure from seismic refraction measurements," in R. J. Adie, ed., *Antarctic Geology and Geophysics*. Oslo: Universitetsforlaget, 683–691.

Cailleux, A., 1963. *Géologie de l'Antarctique*. Paris: Expéditions Polaires Françaises, Publ. 242, 208p.

Calkin, P. E., and Nichols, R. L., 1972. "Quaternary studies in Antarctica," *in* R. J. Adie, ed., *Antarctic Geology and Geophysics*. Oslo: Universitetsforlaget, 625–643.

Craddock, Campbell, ed., 1970. *Geologic Maps of Antarctica*. Am. Geogr. Soc. Map Folio Series, **12**, 6p.

Crawford, A. R., and Campbell, K. S. W., 1973. "Large-scale horizontal displacement within Australo-Antartica in the Ordovician," *Nature, Phys. Sci.,* **241**(105), 11–14.

Crowell, J. C., and Frakes, L. A., 1972. "Ancient Gondwana glaciations," *Second Gondwana Symp., Proc. Pap.,* 469–476.

Dalziel, I. W. D., and Elliot, D. H., 1971. "Evolution of the Scotia Arc," *Nature,* **233**(5317), 246–252.

____, and Elliot, D. H., 1973. "The Scotia Arc and Antarctic margin," *in* F. G. Stehli and A. E. M. Nairn, eds. *The Ocean Basins and Continental Margins: 1, The South Atlantic*. New York: Plenum, 171–246.

Denton, G. H., Armstrong, R. L., and Stuiver, M., 1970. "The late Cenozoic glacial history of Antarctica", *in* K. K. Turekian, ed., *The Late Cenozoic Glacial Ages*. New Haven: Yale Univ. Press, 267–306.

Elliot, D. H., 1972. "Aspects of Antarctic geology and drift reconstructions," *in* R. J. Adie, ed., *Antarctic Geology and Geophysics*. Oslo: Universetetsforlaget, 849–858.

Fairbridge, R. W., 1952. "The geology of the Antarctic," *in* F. A. Simpson, ed., *The Antarctic Today*. Wellington: New Zealand Antarctic Society, 56–101.

Ford, A. B., 1972. "Fit of Gondwana continents; drift reconstruction from the Antarctic continental viewpoint," *24th Internat. Geol. Cong.,* Sec. 3, 113–120.

Frakes, L. A., Matthews, J. L., and Crowell, J. C., 1971. "Late Paleozoic glaciation: Part III, Antarctica," *Geol. Soc. Am. Bul.,* **82**, 1581–1604.

Grindley, G. W., and McDougall, I., 1969. "Age correlation of the Nimrod group and other Precambrian rock units in the central Transantarctic Mountains, Antarctica," *N.Z. J. Geol. Geophys.,* **12**, 391–411.

Grushinsky, N. P., and Frolov, A. L., 1967. "Some conclusions on the structure of Antarctica," *J. Geol. Soc. Australia,* **14**(2), 215–223.

Gunn, B. M., 1963. "Geological structure and stratigraphic correlation in Antarctica," *N.Z. J. Geol. Geophys.,* **6**(3), 423–443.

Hadley, J. B., ed., 1965. "Geology and paleontology of the Antarctic," *Am. Geophys. Union, Antarct. Res. Ser.,* **6**, 274 p.

Hamilton, W., 1967. "Tectonics of Antarctica," *Tectonophysics,* **4**, 555–568.

Harrington, H. J., 1965. "Geology and morphology of Antarctica," *in* J. Van Mieghem and P. Van Oye, eds., *Biogeography and Ecology in Antarctica. Monographiae Biologicae* **15**. The Hague: W. Junk, 1–71.

____, 1972. "Recent advances in the study of the Beacon Supergroup in the Transantarctic Mountains," *2nd Gondwana Symp., Proc. Pap.,* 257–264.

Hayes, D. E., ed., 1972. *Antarctic Oceanology II: The Australian–New Zealand Sector*. Washington: Am. Geophys. Union, Antarct. Res. Ser. **19**, 364p.

Kapitsa, A. P., 1967. "Antarctic glacial and subglacial topography, *in* T. Nagata, ed., *Proc. Symp. Pacific-Antarctic Sciences*. 11th Pac. Sci. Cong. Rep., Spec. Issue **1**, 82–91.

Katz, M. B., 1972. "Paired metamorphic belts of the Gondwanaland Precambrian and plate tectonics," *Nature,* **239**(5370), 271–273.

Kennett, J. P., and Brunner, C. A. 1973. "Antarctic late Cenozoic glaciation: evidence for initiation of ice rafting and inferred increased bottom-water activity," *Geol. Soc. Am. Bull.,* **84**, 2043–2052.

Kitching, J. W., Collinson, J. W., Elliot, D. H., and Colbert, E. H., 1972. "Lystrosaurus zone (Triassic) fauna from Antarctica," *Science,* **175**(4021), 524–527.

Knox, G. A., 1963. "Antarctic relationships in Pacific biogeography," *in* J. L. Gressitt, ed., *Pacific Basin Biogeography*. 10th Pacific Sci. Congr. Symp. Honolulu: Bishop Museum Press, 465–476.

Palmer, A. R., and Gatehouse, C. G., 1972. "Early and Middle Cambrian trilobites from Antarctica," *U.S. Geol. Surv. Prof. Pap.,* **456D**, 37p.

Plumstead, E., 1962. "Fossil floras of Antarctica," *Trans Antarc. Exped., 1955–58*. Sci. Rep. 9, 153p.

Priestley, R., Adie, R. J., and Robin, G. de Q., eds., 1964. *Antarctic Research*. London: Butterworth, 360p.

Ravich, M. G., and Grikurov, G. E., 1970. "Main features of the tectonics of Antarctica," *Internat. Geol. Rev.,* **12**(11), 1297–1309. Translated from *Soviet Geol.,* **1970**(1), 12–27.

Smith, A., and Hallam, A., 1970. "The fit of the southern continents," *Nature,* **225**, 139–144.

Splettstoesser, J. F., 1966. "Antarctic geological literature," *Geotimes,* **11**(4), 33–35.

Stevens, G. R., 1967. "Upper Jurassic fossils from Ellsworth Land, West Antarctica, and notes on Upper Jurassic biogeography of the South Pacific region," *N.Z. J. Geol. Geophys.,* **10**(2), 345–393.

Voronov, P. S., and Ushakov, S. A., 1965. "Some problems in studying isostatic processes in Antarctica," *Soviet Antarctic Exp.* **3**, 351–354.

Warren, G., 1965. "Geology of Antarctica," *in* T. Hatherton, ed., *Antarctica*. New York: Praeger, 279–320.

Woollard, G. P., 1962. "Crustal structure in Antarctica," *Am. Geophys. Union Geophys. Mono.* **7**, 53–73.

Zhivago, A. V., 1967. "Bottom morphology and tectonics of the Southern Ocean, *in* T. Nagata, ed., *Proceedings of the Symposium on Pacific-Antarctic Sciences*. 11th Pac. Sci. Congr., Rep., Spec. Issue **1**, 124–135.

Cross-references: *Australasia–Regional Review; South America; Sub-Antarctic Islands*.

ANTIGUA

The British Associated State of Antigua is formed of the islands of Antigua, Barbuda, and the uninhabited Redonda, collectively covering 445 km^2 (171 sq mi). These islands were formerly part of the Leeward Islands Federation, which broke up in 1956. They lie within the Lesser Antillean Island Arc in the eastern part of the Caribbean Sea.

The island of Antigua is 280 km^2 (108 sq mi) in area, roughly triangular in shape, and reaches a maximum elevation of 402 m above sea level in its more mountainous region. It is the emergent portion of an extensive submarine bank connecting Barbuda to Antigua. The climate is tropical and the average annual rainfall is about 1100 mm.

Antigua belongs to the eastern row of islands (Limestone Caribbees) that mark the crest of the main volcanic ridge of the Lesser Antilles Island Arc created by pre-Miocene volcanic activity.

Three main geological units are recognized on Antigua. Each unit corresponds to a main physiographic division of Antigua: the Southwestern Volcanic Hills, the Central Plain, and the Northeastern Limestone Region. In order of decreasing age, these main units are:

1. *Basal Volcanic Series:* These are the oldest rocks on Antigua and constitute some of the best exposures of volcanogenic products of the pre-Miocene volcanism. The sequence, approximately 2000 m thick, consists mainly of agglomerates and tuffs interbedded with or cut by basalt-andesite-dacite flows and minor intrusions. The age of the exposed volcanics is problematic; the only dates are based on fossil assemblages in the Seaforth Limestone, which may not be an intercalated unit but a lower Oligocene fringe reef perched on the flank of much older volcanics.
2. *Central Plain Succession:* Overlying the Basal Volcanics with no recognizable unconformity are 100–1500 m of stratified tuffs and agglomerates with cherts and limestones of both marine and freshwater origin. This unit is probably of Oligocene age. Two minor intrusions occur in this central low-lying belt, which cuts diagonally across the island.
3. *Antigua Formation:* The entire northeastern third of the island consists of the Antigua Formation, a series of limestones and marls, with some interbedded sandstones and clays. Almost 450 m thick, this formation rests with a slight unconformity on the Central Plains Succession. At one time this formation was considered the type section for Caribbean middle Oligocene, but recent debates over the Oligocene-Miocene boundary have pointed up the problems inherent in assigning a precise age to one site that would have worldwide applicability. The single foraminiferal zone of *Globigerina ciperoensis ciperoensis* is the zonal assemblage found in clay bands within the main Antigua Formation limestones and in the Central Plain tuffaceous series near Seaforth. The Antigua Formation is abundantly fossiliferous and is notable for its rich coral, echinoid, and larger foraminiferal assemblages.

The Oligocene rocks are overlain by patchy clays and marls which are encountered up to about 65 m above sea level and are of Pliocene or Pleistocene age. Structurally the island is tilted and has been cut by normal NW-SE-trending strike faults with downthrow toward the southwest and some subsidiary faulting at right angles.

Specific eruptive centers have yet to be designated on Antigua; more detailed studies of the southwestern district should reveal the remnant central volcanic domes. In general the igneous rocks are representative of a basalt-andesite-dacite series of calc-alkaline affinity. The general petrographic characteristics can be summarized as follows:

1. *Basalts:* porphyritic pyroxene basalt; plagioclase phenocrysts with bytownite-labradorite composition and augite.
2. *Andesite:* porphyritic with andesine phenocrysts dominant, but also phenocrysts of augite, diopside, hornblende, hypersthene, and enstatite. Ferromagnesian minerals are pseudomorphed by antigorite and chlorite. The groundmass is often chloritized and has minor amounts of finely crystalline quartz. The prevalent textures are pilotaxitic, hyalopilitic, and trachytic.
3. *Dacites:* porphyritic with phenocrysts of oligoclase-andesine and quartz (with resorption features). Groundmass consists of quartz, plagioclase microlites, and glass.

The specific geologic history of Antigua can be summarized as follows:

1. Accumulation of lower part of Basal Volcanic Series during main pre-Miocene volcanism concomitant with development of main arc ridge. Products of coeval volcanic centers coalesced to form topographic high presently marked, approximately, by edge of Barbuda Bank.
2. Periods of quiescence allowed accumulation of minor epiclastic deposits and development of minor fringing reefs followed by renewed volcanic and intrusive activity to complete the Basal Volcanic Series.
3. General decline of volcanism during Oligocene and accumulation of stratified tuffaceous rocks of Central Plain; depositional environment is mainly shallow-water marine but also includes minor freshwater areas.
4. Complete cessation of activity during late Oligocene-early Miocene and widespread erosion produced beach-type conglomerate of Cassada Gardens Gravels and deposition of shallow-water reefal facies of the Antigua Formation.
5. Subsequent uplift, tilting of island and minor faulting; general dissection of island by erosion and deposition of late Tertiary and Quaternary terrestrial detritus deposits.

The island of Barbuda lies 40 km N of Antigua and has an area of 155 km^2 (60 sq mi). It is composed of two distinct limestones, the older of which forms a table-like plateau in the east of the island. Both limestones are considered to be of Quaternary age (Martin-Kaye, 1959).

JOHN D. MATHER
L. KENNETH FINK, JR.

References

Christman, R. A., 1972. "Volcanic geology of southwestern Antigua, B.W.I.," *Geol. Soc. Am. Mem.* **132,** 439–448.

Cushman, J. A., 1931. "Cretaceous foraminifera from Antigua, B.W.I.," *Contrib. Cushm. Lab. Foram. Res.,* 7(2), 33–46.

Earle, K. W., 1923. *Report on the Geology of Antigua.* Leeward Islands: Govt. Printing Office, 28p.

Harris, D. R., 1965. "Plants, animals and man in the Outer Leeward Islands, West Indies: an ecological study of Antigua, Barbuda and Anguilla," *Univ. Calif. Publ. Geogr.,* **18,** 1–164.

Martin-Kaye, P. H. A., 1959. *Reports on the Geology of the Leeward and British Virgin Islands.*" St. Lucia, B.W.I.: Voice Publ. Co., 117p.

_____, 1969. "A summary of the geology of the Lesser Antilles," *Overseas Geol. Mineral Resources,* **10,** 172–206.

Russell, R. J., and McIntire, W. G., 1966. *Barbuda Reconnaissance* (Coastal Studies Ser. 16). Baton Rouge, La.: Louisiana State Univ., 53p.

Thomas, H. D., 1942. "On fossils from Antigua and the age of the Seaforth Limestone," *Geol. Mag.,* **79,** 49–61.

Trechmann, C. T., 1941. "Some observations on the geology of Antigua," *Geol. Mag.,* **78,** 113–124.

Wigley, P., 1973. "The distribution of strontium in limestones on Barbuda, West Indies," *Sedimentology,* **20,** 295–304.

Cross-references: *Barbados; Guadeloupe; Leeward Islands; St. Christopher-Nevis-Anguilla; Virgin Islands; West Indies.*

ARGENTINA

Argentina covers an area of 2,796,000 km^2 (1,079,520 sq mi). It is bounded on the W by *Chile* (q.v.), on the N by *Bolivia* (q.v.) and *Paraguay* (q.v.), on the NE by *Brazil* (q.v.) and *Uruguay* (q.v.), and on the E by the Atlantic Ocean. The most distinctive physiographic features are the Andean Cordillera, which divides Argentina from Chile; the flat plains of the Chaco-Pampas region to the E of the Andes; and the plateau region of Patagonia to the S. The country extends from 22 to 55°S latitude, which accounts for a wide variety of climatic zones between subtropical and subarctic.

History of Geological and Mining Activity

The history of geological investigations in Argentina begins with Alcides D'Orbigny and Charles Darwin, who visited the country in the 1820s and 1830s. Systematic studies were initiated only after several European geologists were invited to cooperate during the 1870s and 1880s and the Academy of Sciences of Córdoba was established as a center for their activity. Foremost among these scientists were A. Stelzner, L. Brackebusch, F. Kurtz, C. Burmeister, and G. Bodenbender. The first geological map of a large part of the territory, prepared by Brackebusch, was published in 1891. Most of the geologic activity was then concentrated in the mountainous northwest, but investigations into the stratigraphy and paleontology of the Pampas and Patagonia were also begun at the early period, mostly by Burmeister, A. Bravard, and A. Doering but also by some Argentinian scientists. Outstanding among them was Florentino Ameghino, who, with his brother Carlos, produced a monumental work on the Tertiary and Quaternary paleontology of Patagonia and the Pampas.

In 1904, a geological and mining survey, manned mostly by German geologists and French mining engineers, was founded under the Ministry of Agriculture. Almost simultaneously, the first schools for "naturalists," giving broad training in geology, zoology, and botany, were established in the Universities of Buenos Aires and La Plata, and the first graduates appeared during the 1910s. Among the foreign geologists were J. Keidel, R. Stappenbeck, R. Beder, P. Groeber, and G. Bonarelli, and among the local workers were F. Pastore and J. J. Nagera, who produced important work.

The systematic mapping of the country, on a scale of 1:200,000, was initiated in the 1910s and progressed significantly only after a new generation of Argentine geologists was incorporated into the geological survey (then *Dirección de Minas y Geología*) during the 1940s. This organization (now the *Instituto Nacional de Geología y Minería*) completed the mapping of 100 quadrangle sheets (each comprising about 4000 km^2) in addition to numbers of geologic and mining reports. In 1950, it also published geologic maps of Argentina on a scale of 1:2,500,000 (revised in 1964).

During the 1930s and 1940s, several government agencies set up and developed their own geological and mining services: *Yacimientos Petrolíferos Fiscales* (the government oil agency), *Yacimientos Carboníferos Fiscales* (the solid fuels agency), *Comisión Nacional de Energía Atómica,* and *Dirección General de Fabricaciones Militares.*

At first the training of geologists in Argentina was in liberal arts schools, unlike most other Latin American countries, where geologists generally were trained in mining schools. In addition to Buenos Aires, four more universities now have geology schools (La Plata, Córdoba, Sur, and Tucumán). The only mining engineering school is in San Juan. The number of geologists now active in the country, practically all locally trained, is approximately 700.

Geomorphic and Geotectonic Divisions

Combining geomorphic and geotectonic features, the following principal morphotectonic regions can be distinguished (see Fig. 1): (1) the Andean Cordillera, (2) the Puna, (3) the Precordillera, (4) the Sierras Subandinas, (5) the Sierras Pampeanas, (6) the Chaco-Pampean plains, (7) the Mesopotamian Plateau, (8) the Patagonian Plateau, and (9) the Sierras Bonaerenses. Some of these regions are complex and can in turn be divided into several subunits.

Andean Cordillera. The Andes form the western border of Argentina, although south of about 44°S, the Cordillera lies mostly on the Chilean side. The mean peak height is well above 6000 m N of about 34°, reaching the highest altitude at Mt. Aconcagua (6959 m, the highest in the Western Hemisphere). Southward, the altitude decreases gradually and in the Patagonian Andes the peaks are generally well below 3000 m. The Argentine Andes were formed essentially during the Tertiary orogeny, but several earlier geologic events gave rise to a number of complex morphotectonic units. Four sections can be distinguished:

1. Between 22° and 28°S, a chain of volcanic peaks consisting of andesitic strato-volcanoes (mostly inactive) of late Tertiary to Pleistocene age rise above the Puna Plateau along the Chilean and Bolivian border and extend far into Bolivia.
2. Between 28° and 39°S, the Andes consist essentially of a western belt of marine Mesozoic (Cordillera Principal) and an eastern belt of Lower Mesozoic-Permian volcanics unconformably overlying Carboniferous graywackes (Cordillera Frontal). A simplified structural picture of the

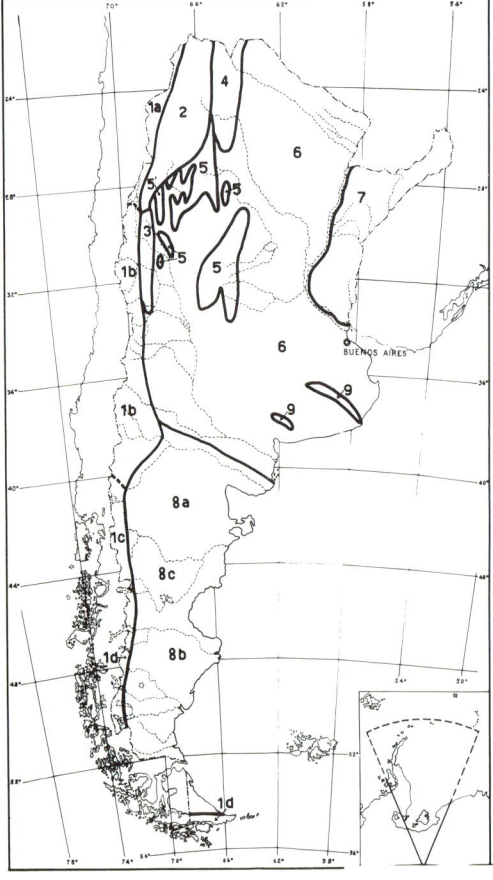

FIGURE 1. Morphostructural regions of Argentina: (1) Andean Cordillera, (2) Puna, (3) Precordillera, (4) Sierras Subandinas, (5) Sierras Pampeanas, (5') Sierras Transpampeanas, (6) Chaco-Pampean Plain, (7) Mesopotamian Plateau, (8) Patagonia, (9) Sierras of Buenos Aires. *Inset:* Argentina's claim in West Antarctica.

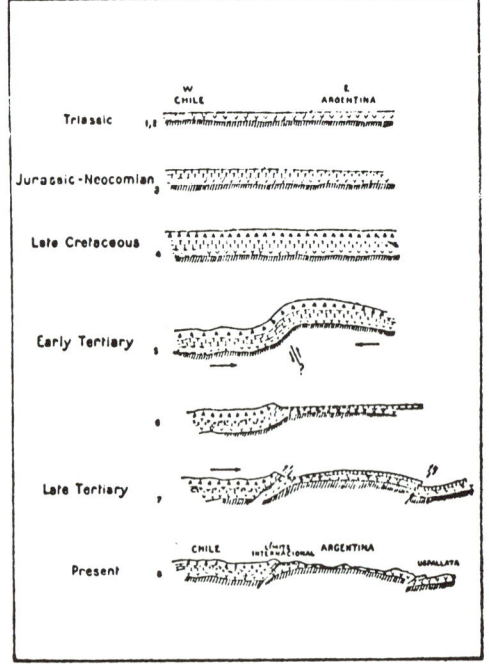

FIGURE 2. Stages in the development of the Andean Cordillera, latitude 33°S (from González Bonorino, 1950b).

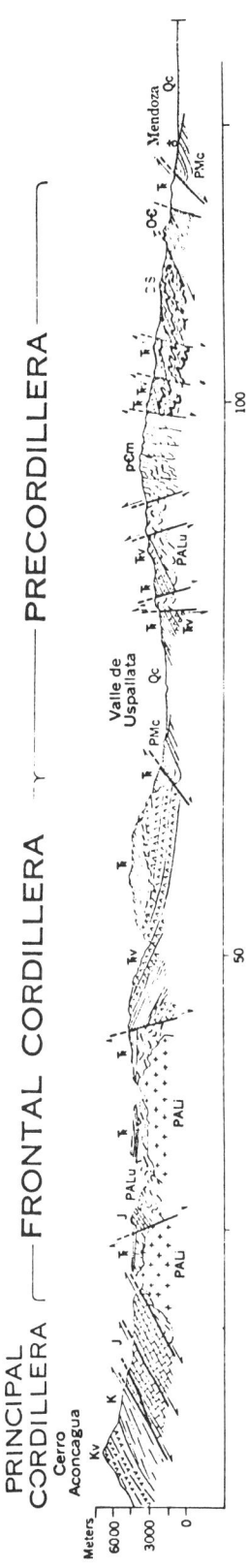

FIGURE 3. Geologic section across the Andean Cordillera (from Harrington, 1956).

Cordillera near Mt. Aconcagua shows a huge westward-dipping monocline of Mesozoic beds. Located on this section and sometimes mistaken for a volcano, Mt. Aconcagua consists of late Mesozoic to early Tertiary layered volcanics forming part of the monocline and partly thrust eastward and upward over the marine sediments below. In the southern part of this section, the Cordillera Frontal disappears and the marine formations extend eastward to form the Neuquén embayment, folded into oil-bearing anticlines and penetrated by shallow-seated Tertiary intrusives (Fig. 2).

3. Between 39° and 45°S, the Andes are characterized both by fault blocks of predominantly late Paleozoic metamorphic and igneous rocks and Tertiary layered volcanics, and by a string of late Tertiary to Pleistocene volcanoes (Mt. Lanin, 3775 m; Mt. Tronador, 3554 m).

4. Between 45°S and the eastern tip of Tierra del Fuego, the Andes (the axis running mostly on the Chilean side) are composed of Cretaceous granitic rocks intruded into Lower and Middle Mesozoic geosynclinal flysch sediments, covered unconformably by Upper Cretaceous and Tertiary molasse sediments. This section of the Cordillera swings sharply to the E along the southern border of Tierra del Fuego and continues in the Scotia Arc and the continent of *Antarctica* (q.v.) (Fig. 3).

Puna. The Puna is the high plateau that is an extension of the Bolivian Plateau. As a morphotectonic region, the Puna Plateau is characterized by wide, flat sedimentary basins (mean altitude, 4000 m) filled with Tertiary and Quaternary pyroclastics and divided by relatively low, N-S-trending ranges that are formed mainly by folded Lower Paleozoic geosynclinal sediments and low-grade Precambrian metasediments. A string of volcanoes, several exceeding 6000 m (Cerro Bonete: 6872 m), marks the Chilean border. Huge salt deposits occupy the bottoms of many of the closed basins.

Precordillera. This mountain range parallels the main Cordillera from 29 to 33°S and is separated from the Cordillera Frontal by a straight N-S tectonic valley. The Precordillera is characterized by the predominance of early Paleozoic marine (flysch) sediments overlain unconformably by late Paleozoic marine and terrestrial (molasse) sediments. Lower to Middle Paleozoic graywackes and limestones are typical of the Precordillera.

Sierras Subandinas. This unit, which extends N from 26°30'S far into Bolivia, is a system of N- to NNE-trending ranges paralleling the eastern border of the Puna Plateau and consisting of folded Tertiary and Cretaceous terrestrial sediments. The anticlines contain cores of low-grade Precambrian metamorphics.

To the N, the stratigraphic column, which includes Upper ("Gondwana") and Middle Paleozoic formations, is very thick but thins to the south as the Precambrian basement rises slowly from its sedimentary cover. The intensity of folding decreases to the E (imbricate structure predominates against the Puna Block); some of the anticlines toward the Chaco plain are oil-bearing.

Sierras Pampeanas. These are block mountains of late Precambrian-early Paleozoic crystalline rocks occupying a wide area in north-central and north western Argentina between the Precordillera, the Puna, and the Sierras Subandinas. They are bounded by N-S vertical or reverse faults and are tilted mostly to the E on the eastern Sierras and to the W on the western Sierras. In places they show well-preserved peneplain surfaces partly covered by late Paleozoic and Tertiary red beds. The Sierras Pampeanas merge with the Sierras Subandinas as the mountain blocks plunge northward to form the cores of anticlines; the blocks also come together to form the substratum of the Puna Plateau. A mountain chain (Sierra de Famatina) lying between the Sierras and the Precordillera is sometimes separated from the former as an independent, transitional unit (Sierras Traspampeanas). Wide intermontane basins, or bolsones, separate the mountain units. The Sierras reach their highest altitudes at Sierra de Famatina (6400 m) and Aconquija (5550 m).

Chaco-Pampean Plain. This low plain occupies the eastern part of the country from the Paraguayan border to the Colorado River in the S. The plain is a sedimentary basin with a thin cover of Quaternary fluvial and eolian sediments, underlain by a moderately thick (in parts, over 4000 m) sequence of flat-lying Tertiary to late Paleozoic ("Gondwana") sediments. The greatest part of the Chaco-Pampean Plain lies below 200 m. Structurally, this basin is a complex graben between the Sierras Pampeanas, the Sierras Subandinas and the Cordillera to the W, and the Brazilian craton to the E.

Mesopotamian Plateau. A region of rolling plains bounded by the Paraná and Uruguay rivers, the Mesopotamian Plateau is a slightly raised part of the Chaco-Pampean region. The surface formations are Lower Pleistocene fluvial sediments in the S, Tertiary sandstone red beds in the middle, and diabase sills in the NE panhandle, reflecting a gradual structural rise toward the Brazilian craton.

Patagonia. This is a very complex region characterized morphologically by extensive mesas (altitude 900 m in the W, 700 m near the Atlantic coast) separated by cañadones, valleys that have few water courses. Structurally, it is a low-lying shield with two positive areas separated by a negative area that corresponds to the San Jorge Gulf. In the positive areas the peneplain, formed on crystalline basement and Paleozoic metasediments, is largely covered by flat-lying Mesozoic volcanics and red beds, Tertiary sandstones, and Tertiary basaltic lava flows. The negative area is a basin that contains Mesozoic terrestrial and marine clastic beds underlain by Mesozoic volcanics and overlain by Tertiary terrestrial and marine tuffaceous sandstones and tuffs. This basin contains the largest petroleum and gas reserves of the country. Another oil-bearing sedimentary basin is located at the southern end of Patagonia (Tierra del Fuego). A thin pediment gravel deposit covers the mesas throughout Patagonia.

Sierras of Buenos Aires. This unit consists of two parallel, relatively inconspicuous mountain ranges emerging from the Pampas plains, some 300 and 500 km, respectively, to the SW of Buenos Aires. Their NW-SE trend appears anomalous in the predominantly N-S structural pattern of Argentina. The southernmost range (Ventana system) is formed of intensely folded Paleozoic quartzites, metagraywackes, and other clastic sediments (including a Permian tillite deposit). The northern range (Tandilia system) consists of crystalline rocks (the oldest known in the country, 1800 m yr) covered by marine beds (mostly quartzites, limestones, and dolomites) generally correlated stratigraphically with the lower levels of the Ventana system, but showing little deformation.

The Malvinas or *Falkland Islands* (q.v.), closely related geologically to Patagonia, are underlain partly by slightly folded Devonian marine beds, partly by "Gondwana" continental beds, which are underlain by a glacial Permian formation.

Structural Framework

The Argentina territory consists of a main body of negative to slightly positive cratonic areas fringed on the west by a succession of folded geosynclinal basins of decreasing age from late Precambrian to late Mesozoic (Fig. 4). The Brazilian Shield is represented by a negative unit comprising the Chaco-Pampean plains (including the restricted positive units of Sierras Pampeanas and Sierras Bonaerenses). A younger, semicratonic unit is Patagonia, which is positive except for the basins of San Jorge Gulf and southern Santa Cruz-northern Tierra del Fuego. The trend of primary structures (schistosity, fold, axis, etc.) in the basement rocks is essentially N-S, but in the east-central part the trend becomes increasingly eastward (E-W in the northern Sierras of Buenos Aires).

FIGURE 4. Positive areas and main sedimentary basins of Argentina: (1) Salta Basin, (2) Chaco-Parana Basin, (3) Salado Basin, (4) Cuyo Basin, (5) Neuquén Basin, (6) San Jorge Basin, (7) Magellanian Basin, (8) Colorado Basin.

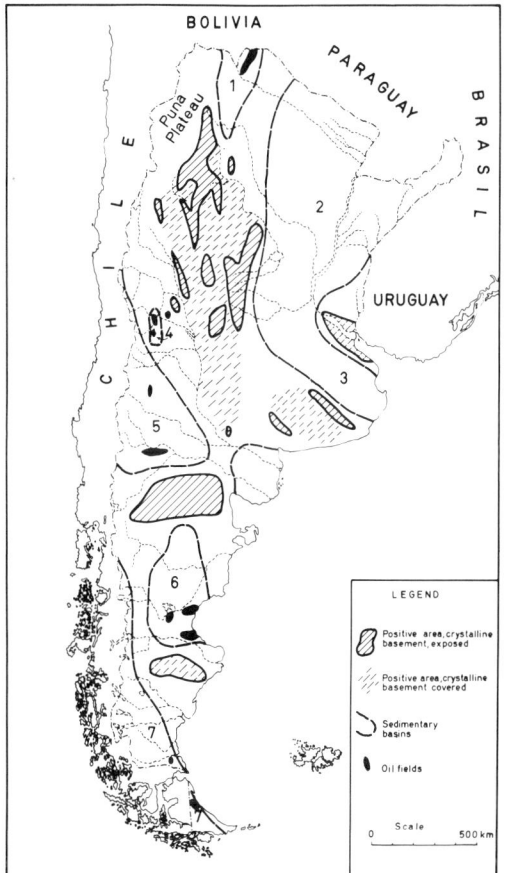

The exposed geosynclinal belts, oriented essentially N-S, are younger and longer to the W; their ages are successively late Precambrian(?) (W Sierras Pampeanas), early Paleozoic (Precordillera), late Paleozoic (Cordillera Frontal), and Mesozoic (Cordillera Principal). Reaching the Atlantic Ocean and becoming younger to the S, geosynclines wrap around the southern border of the cratonic region in divergent patterns.

Historical Geology

The oldest rocks in Argentina (1800 m yr) are in the crystalline basement underlying the Pampas and exposed in the Tandilia range. They represent an extension of the Brazilian craton and consist mainly of migmatitic and granitized schists. The next oldest rocks are those of the Sierras Pampeanas, which border the depressed craton to the W composed of medium- to high-grade metamorphics and migmatites of about 600 m yr intruded by granitic plutons, part of which are as young as early Paleozoic. Magmatism and, consequently, metamorphic grade decrease rapidly north of parallel 28° along the E border of the Puna; near the Bolivian territory, the Precambrian consists of slates and graywackes.

The first Paleozoic marine invasion, Middle Cambrian to Middle Devonian, covered the W border of the central cratonic area and at times invaded the foreland of the Chaco region. The geosyncline reached the Atlantic Coast in at least two recorded places (near Bahía Blanca and San Matías Gulf).

The Paleozoic formations were folded and eroded in the Mississippian (Caledonian revolution), and a new geosyncline became established W of the earlier one. The Upper Paleozoic basin was in turn deformed (Hercynian orogeny), and granitic stocks and small batholiths were emplaced. In the Precordillera, a moderate relief of Devonian and older sediments was covered by partly marine, partly continental Mississippian deposits, including a formation of glacial and glaciomarine conglomerates (Fig. 5). Farther to the east, on the foreland, extensive continental closed basins formed on the peneplaned crystalline basement, and thick red beds deposits (Paganzo Formation) were formed until the Triassic, when the whole area was elevated and subjected to a new erosion cycle.

"Gondwana" deposits are widespread in the subsurface of the Chaco-Pampean Plain and crop out again in the Paraná-Uruguay Basin. A *Glossopteris* flora was found in the Ventana Ranges of Buenos Aires province.

The early Mesozoic was a time of extensive volcanic and continental red beds deposition in western Argentina and Patagonia. It was followed by the development of a geosynclinal basin along most of the western Argentine border; between parallels 36° and 40°, the sea extended to the E in a wide embayment. A shallow sea (San Jorge Gulf embayment) covered the area between the two positive areas of Patagonia. In the late Cretaceous, forces from the Pacific folded and thrust the Jurassic and Lower Cretaceous marine sediments (first phase of Andes development). After a period of erosion, further thrusting from the W during the Middle Tertiary uplifted the main Cordillera to its present height and shaped the morphotectonic configuration of the present Argentine territory (Cordillera Frontal, Precordillera, Puna, Sierras Pampeanas, and Sierras Subandinas).

The Tertiary was characterized by extensive volcanism throughout the Cordillera and Pata-

FIGURE 5. Land and sea distribution in Argentina and adjacent areas in late Carboniferous time (from Herrero Ducloux, 1963).

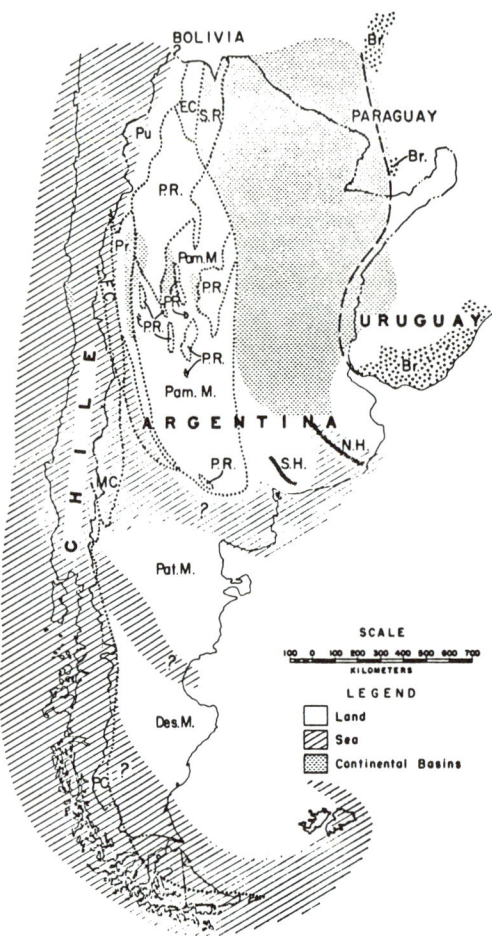

gonia as well as in scattered places of the Sierras Pampeanas. Shallow seas, connecting the Atlantic and Pacific oceans, invaded part of Patagonia and covered the eastern Pampas region.

The Pleistocene glaciation sculptured the mountain ranges and deposited thick moraine and fluvioglacial deposits. The regional snowline varied from about 5000 m on the Bolivian border to sea level in Patagonia. In the Patagonian Andes, valley glaciers formed, partly coalescing into piedmont glaciers along the E foothills. No continental ice mantle covered any area of the Argentine territory during the Pleistocene; the only Andean glacier reaching the Atlantic was on the Magellan Strait. Outside the Andes proper, cirque formation occurred in only a few of the highest Sierras Pampeanas.

Another interesting Pleistocene event is the deposition of a widespread loess mantle, only a few tens of meters in thickness, now underlying the major part of the Pampas region. This is a "warm" loess, as it was formed at the expense of the alluvial deposits occupying the semiarid area between the mountain slopes to the E, SW, and NW, and the Pampas (grassland) area to the E. Abundant volcanic shards are generally found mixed in the loess. The Pampas region was also the site of thick loess deposition during the Tertiary.

Despite extensive Pleistocene volcanic activity along the Andes, there are today few active volcanoes in the Argentine territory.

Fossil Localities. Rich Upper Cambrian to Devonian trilobite fauna is found in the Precordillera and along the eastern foothills of the Puna region. Mississippian and Pennsylvanian marine fauna also occurs in the Precordillera. Important Permian and Triassic vertebrate (reptile) localities are in red beds (also containing fossil plants) that fringe the Precordillera and Sierras Pampeanas; the famous locality at Ischihualasto has been studied by Argentine and American specialists. The Mesozoic marine beds of Neuquén and Mendoza provinces are well known for ammonite and molluscan fauna. The Lower Tertiary tuffaceous beds of Patagonia are noted for mammalian fauna, made famous by the classic studies by Florentino Ameghino and the Princeton expeditions. The Pleistocene loess deposits of the Pampas contain a characteristic mammalian fauna.

Mining Activity. Mining in Argentina, a relatively minor activity, represents only $400 million yearly (including petroleum), or about 1.5% of the gross national product. Argentina exports zinc, lead, silver and tungsten concentrates, borates, and salt; but the country is deficient in many of its required ores.

Low-grade iron is mined from Lower Paleozoic sedimentary deposits in the NW, but Argentina imports 88% of the ore for its mills. Plans for the development of a similar deposit in Patagonia are under way. No bauxite deposits are known, and all aluminum is imported; possible future sources are low-grade laterites in Misiones and alunite in northern Patagonia. There is very little copper mining. Lack of copper deposits, in contrast to the huge Chilean deposits, is attributed to the structural asymmetry of the Andes; the Upper Mesozoic copper-bearing formations of Chile are not generally represented in the eastern Andes. There is only one medium-sized base-metal mine operating (Aguilar mine, Jujuy province; Zn, Pb). Among the most interesting mineral localities are the Se and Te complex mineral veins in Sierras de Umango, La Rioja province, and the Li-bearing pegmatites in San Luis province.

Argentina is practically self-sufficient in oil

and gas. There are five petroleum-producing areas, situated in Patagonia, Salta, Mendoza, Neuquén, and Tierra del Fuego. Coal reserves are, however, poor. Only one coal mine, in southern Patagonia, is in operation.

F. GONZÁLEZ BONORINO

References

Angelelli, V. F., Lima, J. C., Herrera, A., et al., 1970. "Descripción del mapa metalogenetico de la Republica Argentina; minerales metaliferos," *Argent. Rep. Direc. Nacl. Geol. Mineria*, Anales, **15**, 183p.

Aubouin, J., and Borello, A. V., 1970. "Regard sur la géologie de la Cordillère des Andes; relais paléogéographiques et cycles orogéniques superposés; le Nord argentin," *Bull. Soc. Géol. France*, ser. 7(12), 246–260.

Auer, V., 1959. "The Pleistocene of Fuego-Patagonia. 3. Shoreline displacements," *Ann. Acad. Sci. Fennicae*, ser. A (60).

____, 1970. "The Pleistocene of Fuego-Patagonia. 5. Quaternary problems of southern South America," *Ann. Acad. Sci. Fennicae*, ser. A, pt. 3, 194p.

Borrello, A. V., 1972. "The Precordillera as a type of geosyncline in Argentina," *24th Intern. Geol. Congr. Sec. 3* (Canada), 293–299.

Cingolani, C. A., and Deutsch, S., 1973. "Âges rubidium-strontium des formations magmatiques de la chaîne de la Ventana (Sierras Australes, Province de Buenos Aires, Argentine)," *Ann. Soc. Géol. Belgique*, **96**, 263–274.

Creer, K. M., Mitchell, J. G., and Abou Deeb, J., 1971. "Paleomagnetism and radiometric age of the Jurassic Chon Aike formation from Santa Cruz Province, Argentina: implications for the opening of the south Atlantic," *Earth Planet. Sci. Lett.*, **14**, 131–138.

Cuerda, A. J., and Baldis, B. A., 1971. "Silurico-Devonico de la Argentina," *Ameghiniana*, 8(2), 128–162.

Feruglio, E., 1950. "Descripción geologica de la Patagonia," *Yac. Petrol. Fisc.* (Buenos Aires), **1**, 323; **2**, 344; **3**, 441.

Fleck, R. J., et al., 1972. "Chronology of late Pliocene and early Pleistocene glacial and magnetic events in southern Argentina," *Earth Planet. Sci. Lett.*, **16**, 15–22.

Gerth, H., 1955. *Der geologische Bau der Sudamerischen Kordillera*. Berlin: Gebr. Borntraeger, 264p.

González Bonorino, F., 1950a. "Algunos problemas geologicos de las Sierras Pampeanas," *Rev. Asoc. Geol. Argent.*, 5(3), 81–110.

____, 1950b. "Geologic cross-section of the Cordillera de los Andes at about 33° L.S.," *Bull. Geol. Soc. Am.*, **61**, 17–25.

Groot, J. J., Groot, C. R., Ewing, M., Burckle, L., and Conolly, J. R., 1967. "Spores, pollen, diatoms, and provenance of the Argentine Basin sediments," *Progr. Oceanog.*, **4**, 179–217.

Harrington, H. J., 1956. "Argentina," *Handbook of South American Geology* (Mem. 65). New York: Geol. Soc. Am., 129–165.

____, 1962. "Paleogeographic development of South America," *Bull. Am. Assoc. Petrol. Geologists*, **46**, 129–165.

Herrero Ducloux, A., 1963. "The Andes of western Argentina," in O. E. Childs and B. W. Beebe, eds., *Backbone of the Americas, a Symposium*. Tulsa, Okla.: Am. Assoc. Petrol. Geologists, 16–28.

Hormann, P. K., et al., 1973. "New data on the young volcanism in the Puna of NW-Argentina," *Geol. Rundschau*, 62(2), 397–417.

Leanza, A. F., ed. 1972. "Geologia regional Argentina," *Acad Nacl. Cs. Cordoba*, 869p.

Schwab, K., 1970. "Ein Beitrag zur jungen Bruchtektonik der argentinischen Puna und ihr Verhaltnis zu der angrenzenden Andenabschnitten," *Geol. Rundschau*, 59 (3), 1064–1087.

Thompson, R., 1972. "Paleomagnetic results from the Paganzo Basin of northwest Argentina," *Earth Planet. Sci. Lett.*, **15**, 145–156.

Turner, J. C. M., 1970. "The Andes of northwestern Argentina," *Geol. Rundschau*, 59(3), 1028–1063.

Urien, C. M., 1967. "Los sedimentos modernos del Rio de la Plata exterior," *Bol. Serv. Hidrografia Naval* (Buenos Aires), 4(2), 113–213.

Valencio, D. A., and Mitchell, J., 1972. "Paleomagnetism and K-Ar ages of Permo-Triassic igneous rocks from Argentina and the international correlation of Upper Palaeozoic–Lower Mesozoic formations," *24th Intern. Geol. Congr. Sec. 3* (Canada), 189–195.

Volkheimer, W., 1970. "Neuere ergebnisse der Andenstratigraphie von Sud-Mendoza (Argentinien) und benachbarter Gebiete und bemerkungen zur Klimageschichte des sudlichen Andenraums," *Geol. Rundschau*, 59 (3), 1088–1124.

Zambrano, J. J., and Urien, C. M., 1970. "Geological outline of the basins in southern Argentina and their continuation off the Atlantic shore," *J. Geophys. Res.*, 75(8), 1363–1396.

Cross-references: *Bolivia; Brazil; Paraguay; South America; Sub-Antarctic Islands; Uruguay.*

AUSTRALASIA–REGIONAL REVIEW

Australasia is customarily defined as the continental mass of Australia, together with New Guinea and the islands of Melanesia and New Zealand, which collectively have had a common geological evolution. The tectonic development of Australasia commenced with the development of an Archean nucleus in Western Australia. Continental fragmentation and consolidation characterized the Proterozoic, after which newly developing orogenic zones migrated northeastward until in the Mesozoic and Cenozoic the island complexes of New Guinea, the Solomons, New Hebrides, Fiji, New Caledonia, and New Zealand were formed. Seven cycles of orogenic activity leading to cratonization and platform development may be recognized (Figs. 1 and 2).

The development of the island complex (Fig. 3) is thought to result from or to accompany the northeastward movement of the Australian cratonic plate away from the Australian–Antarctic mid-ocean "spreading zone."

Details of the tectonic history are mainly found in general texts, which are listed in the entries for individual states and countries (q.v.). Only recent papers specifically dealing with tectonics are given here. (Permission to use material and ideas gleaned from the Tectonic Map Committee of the Geological Society of Australia is gratefully acknowledged.)

Development of the Australian Craton

A simplified tectonic map (Fig. 1) shows the distribution of the major orogenic provinces and the cratonic sedimentary basins and depicts their internal structures. The important tectonic relationships are illustrated in Fig. 2.

Archean metamorphic complexes constitute an ancient continental nucleus that occurs in the Yilgarn (Ia) and Pilbara (Ib) provinces in Western Australia; elsewhere Archean rocks can only be recognized where they are involved as basement to younger orogenic belts, e.g., Rum Jungle (Ic). Within the granitic metamorphic complex, elongate sedimentary geosynclinal troughs containing important greenstones can be recognized. Radiometric dating indicates two periods of metamorphism and granite intrusion, between 3050 and 2900 m yr and 2750 and 2600 m yr, followed by basic dyke intrusion at 2400 m yr. The break between the Yilgarn and Pilbara blocks indicates the early development of a major E–W mobile zone.

FIGURE 1. Sketch tectonic map of Australia, showing orogenic and platform domains. Abbreviations: PB, Perth Basin; CvB, Carnarvon Basin; CB, Canning Basin; AB, Amadeus Basin; OB, Officer Basin; GAB, Great Artesian Basin; MB, Murray Basin; BB, Bowen Basin; SB, Sydney Basin; BsB, Bass Strait Basin.

FIGURE 2. Diagrammatic representation of tectonic evolution of Australasia (modified from Geological Society of Australia, 1971).

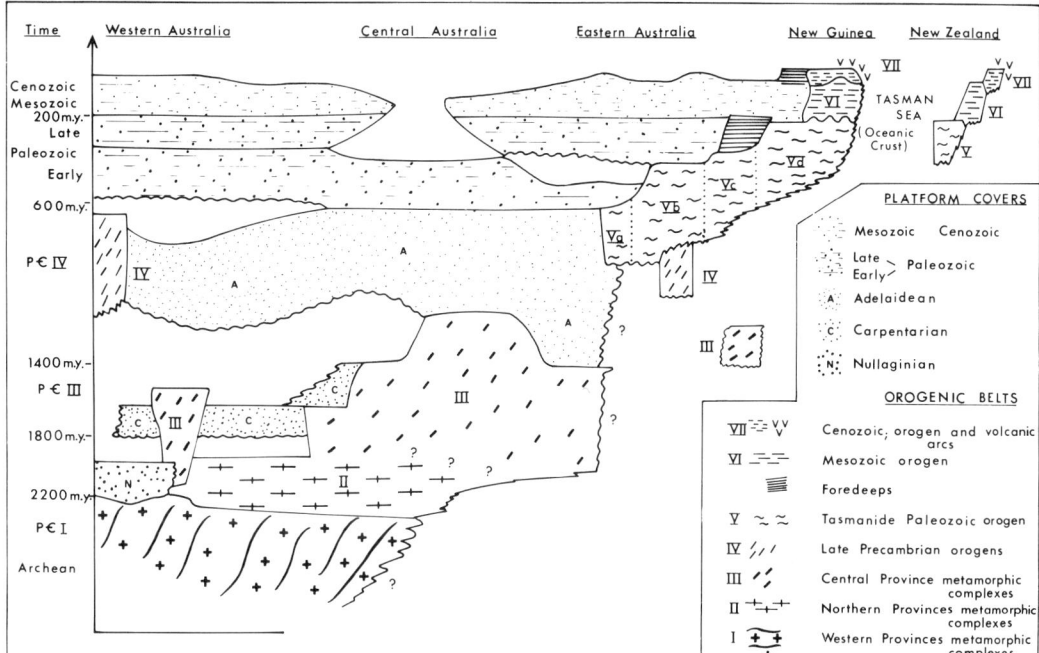

From 2200 m yr the Archean is overlain in the Hamersley-Nullagine district by sediments and volcanics, which constitute the oldest recognizable platform cover. To the northeast more mobile sedimentary troughs sank between basement blocks and underwent orogeny during 1900–1700 m yr. The metamorphic structures in these northern orogenic provinces (II) trend latitudinally, but relations between them are unknown.

A Carpentarian (1800 m yr) platform cover spreads over most of northern Australia and orogenic activity shifts southward, affecting new sedimentary troughs and reworking the basement. Structures in these central provinces (III) are complex, and trends vary from northeast to north; the several deformation episodes recognized cannot be correlated. Radiometric dates vary from 1900 to 1000 m yr and concentrate in groups: 1900–1700 m yr, 1550–1500 m yr, 1400–1350 m yr, 1100–950 m yr.

Older metamorphic rocks (2250 m yr and 1800 m yr) in the Arunta complex (IIIa) in the center of Australia were reactivated to high metamorphic grades at this time. The Georgetown province (IIIb) in northern Queensland possibly also belongs with the central provinces.

Stabilization of the Australian shield was then virtually complete and from 1200 m yr until 650 m yr ago stable platform conditions spread across the craton. Orogenic activity was restricted to small areas in Western Australia (IVa) and Tasmania (IVb) (Fig. 2), and geosynclinal downwarping commenced in the east.

During the Paleozoic the N-S Tasman Geosyncline developed as a series of more regular elongate troughs against the eastern side of the craton. Platform sedimentation continued on the craton itself, but the renewal of activity is manifest by the widespread Antrim Plateau (AP) basalts and by rifting and warping at the western edge of the craton to form the Canning Basin (CB) and later the Carnarvon (CvB) and Perth (PB) basins (Figs. 1 and 2). The Amadeus basin (AB) and northern Officer basin (OB) developed as deeply subsiding troughs, following ancient zones of weakness, that linked the Tasman Orthogeosyncline with the western basins. The craton also responded to important movements in the adjacent orogenic belt so that sedimentation was interrupted after the Devonian to be followed by spreads of terrestrial and glacigene sediments across the southern half of the craton in the Permian and new broad downwarp basins in the Mesozoic.

Within the Tasman Geosyncline, complex successions of troughs and ridges consolidated by orogenesis migrated northeastward: Kanmantoo (Va) deformed in the Cambrian to early Ordovician Delamerian Orogeny; Lachlan (Vb) deformed in the Ordovician Tyennan and Benambran orogenies and the Silurian to Devonian Bowning and Tabberabberan orogenies; Hodgkinson (Vc) deformed in the early Carboniferous Kanimblan Orogeny; and the New England-Brisbane districts (Vd) deformed in the Permian Hunter-Bowen Orogeny. Throughout these belts, deformation is only moderately intense, forming slate belts; higher-grade regional metamorphic zones are rare. Serpentine intrusions occur in the Lachlan and New England belts, and granite batholiths are abundant. Termination of each orogenic episode is commonly marked by molasse sediments and post-orogenic granites and volcanics. Foredeep downwarps developed adjacent to the Hunter-Bowen Orogen forming the Bowen (BB) and Sydney (SB) basins. The western edges of the geosynclinal belt are hidden because the orogenic belts formed earlier became cratonized and covered first by Upper Paleozoic, then Mesozoic, platform cover as the Great Artesian Basin (GAB) complex developed.

The margins of the craton also sank markedly during the Mesozoic with the breakup of Gondwanaland. Orogenic activity shifted to the island arcs, and the craton was restricted to minor deformations, with granite intrusions limited to a small marginal basin in northeastern Queensland.

Cenozoic activity was similar; the southern margin of the craton became flexured as the Murray (MB), Bass (BsB), and Officer (OB) basins developed. In the late Tertiary, strong uplift of the eastern margin of the Tasman Orogen was accompanied by abundant basalt extrusions; Tertiary peneplains became tilted to the west, and the seaward margin abruptly downfaulted. To the north, part of the orogen sank to form the now drowned Queensland Plateau in the Coral Sea and the basement of the Great Barrier Reef (Fig. 3). These epeirogenic movements reactivated older weaknesses in the craton to produce a system of grid-like

FIGURE 3. Tectonic elements of the Melanesian island complex, Southwest Pacific (modified after Cullen, 1970).

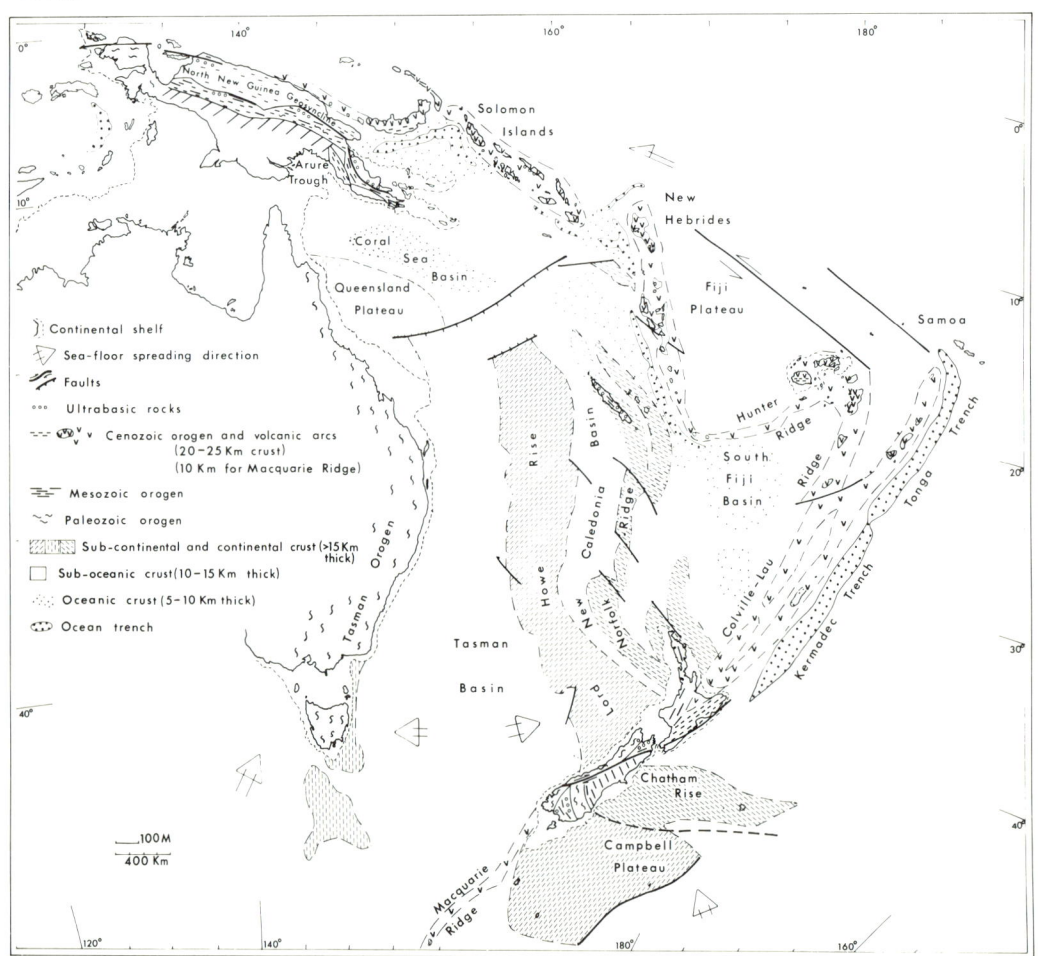

topographic lineaments outlining major fault blocks.

Mesozoic and Cenozoic Tectonics of the Melanesian Islands

The Paleozoic beginnings of the island histories are enigmatic. Remnants of an early Paleozoic orogen are found in West Irian and New Zealand, and a late Paleozoic-Mesozoic orogen (Maorian Geosyncline) can be traced from New Zealand through New Caledonia into New Guinea. These orogens are represented, in part, in the metamorphosed basement of a Cenozoic geosynclinal girdle that together with the presently active volcanic arcs mark the outer border of the Australasian region (Fig. 3).

New Zealand. After an early Devonian transgression over the Paleozoic basement, there was a break in sedimentation until the late Carboniferous, when thick geosynclinal sediments began to accumulate. The downwarp shifted eastward to the present island axis as thick Permian and Lower Jurassic sediments were deposited. Folding and basic volcanism in the early Permian were accompanied by ultrabasic intrusion. The rise of this Permian volcanic arc shed sediments eastward to form the New Zealand Geosyncline. Marginal and axial facies can be recognized throughout New Zealand and also in New Caledonia.

The Rangitata Orogeny began in Middle Jurassic time with basaltic eruptions and emplacement of ultrabasics, and in the southern part of the belt metamorphism produced the Haast Schist complex. The Alpine fault started to disrupt the older basement of South Island. Considerable volcanism associated with folding occurred again in the Cretaceous, following which terrestrial sediments accumulated in fault troughs and new downwarps developed on the eastern side. During the late Cretaceous a marine transgression extended westward, submerging the land by the Middle Oligocene. Interfingering troughs on unstable fault blocks then developed and increasing mobility culminated in the Plio-Pleistocene Kaikoura Orogeny.

North Island has been volcanically active since the early Miocene, with widespread andesites followed by dacite-rhyolite ignimbrites, and then by basalt outpourings, which are still continuing.

New Caledonia. Triassic to Lower Jurassic graywackes derived from the west overlie a basement of Devonian to Permian acid tuffs and argillites. Orogenic movements and basic volcanism interrupted sedimentation until the late Cretaceous transgression. Sediments of marginal facies passed eastward into oceanic sediments; basic volcanism and emplacement of large ultrabasic masses accompanied westward thrusting during terminal orogeny in the Oligocene. Miocene sediments then transgressed the eroded orogen and uplift continued to the present with no further volcanism.

New Guinea. Paleozoic metamorphics and granodiorites occur in West Irian and as cores in the Central Highlands. Since the Permian, this zone formed an emergent ridge at the outer margin of the Australian craton; during the Cretaceous it was probably a volcanic arc extending southeastward into the Owen Stanley Ranges. In the middle to late Miocene intrusion of mainly acid-to-intermediate plutons accompanied uplift along the central axis. Cover rocks, including parts of the Mesozoic platform sediments, were thrust southward. This uplift shed sediments southward into a foredeep trough at the margin of the platform, and to the southeast flysch accumulated from the Oligocene to Pliocene in the Aure Trough.

The northern margin of the central metamorphic complex is an intense fault system along which large ultramafic bodies were intruded during the Eocene or Oligocene. From the beginning of the Cretaceous, oceanic sediments were deposited to the north and were covered by clastics as the central axis emerged in the late Miocene. Basement faulting produced mild folding in this north New Guinea "Geosyncline."

Recent strong uplift and tholeiitic (basalt-andesite) volcanism has formed a volcanic arc along the northern part of eastern New Guinea extending eastward through the northern parts of New Britain.

Solomon Islands. A Mesozoic metamorphic basement was covered by pre-Oligocene andesites and basalts, followed by Miocene sediments and reef limestones and thick Pliocene sediments in fault troughs. To the west is an active calc-alkaline andesite chain becoming picritic in the New Georgia Group. To the east of a shear zone along the Upper Oligocene ultrabasics on Santa Isabel Island is a zone in which basalts were covered since late Cretaceous times by thick sediments. To the north the New Britain arc is truncated by a continuation of the Solomons chain in New Ireland.

New Hebrides. The active volcanic belt continues southeastward; here basalt volcanoes stand on Pliocene fault-trough fills, which in turn overlie pre-Miocene andesitic volcanics and sediments. Ultrabasic intrusions in the south of Pentecost Island are possibly related to those in the Solomons.

Fiji Islands. Eastward from the New Hebrides a submarine ridge leads to the Fiji Group.

Here a gabbro- and granodiorite-studded basement of Eocene to Upper Miocene metagraywackes and tholeiitic volcanics formed a northeast-trending axial ridge on which a series of andesitic volcanoes developed followed in Plio-Pleistocene times by a central zone of basalts. As in all the islands, block faulting is the major manifestation of tectonism, and in Fiji the intervolcano basins and fault troughs are filled with volcaniclastic debris.

The Lau Ridge and the Tonga chain continue the outer volcanic arc to link up with New Zealand (Fig. 3). The Lau Islands consist of latest Miocene andesites and younger olivine basalts, whereas Tonga has an Eocene sedimentary foundation on which tholeiitic basalt and andesite-dacite volcanoes have developed since Pleistocene times.

Oceanic Structure and the Development of the Melanesian Island Arc Complex

Ocean geophysics has shown that the old theory of a foundered Australasian ("Melanesian") continent is untenable; an explanation involving extensional rifting, sea-floor spreading, and drifting of the craton margins is now preferred. The main tectonic features are summarized in Fig. 3.

The submarine Campbell Plateau, Lord Howe Rise, and Queensland Plateau have subcontinental to continental crustal thickness (17–30 km or more) and are thought to be composed of Paleozoic geosynclinal rocks. The Norfolk Ridge has a slightly thinner crust (15–20 km) and is thought to continue the Mesozoic orogen from New Caledonia through New Zealand out onto the Chatham Rise. These ridges are separated from New Guinea by the Coral Sea and a series of rift faults. The outer volcanic arc from New Britain to New Zealand has a crustal thickness of 20 to 25 km; it is possibly continued by the thinner (10 km) Macquarie Ridge to join the Australian-Antarctic mid-ocean ridge. Deep ocean trenches and seismic zones characterize this volcanic arc. Under West Irian a broad zone of earthquake foci dips gently southward down to 200 km; the zone steepens through vertical in eastern New Guinea to dip steeply northward under the New Britain arc. An abundance of deeper foci down to 500 km marks the junction with the Solomon chain. A line of shallow foci connects New Ireland with New Guinea across the Bismarck Sea. The foci under the Solomons and New Hebrides lie in a vertical zone. This region is also unusual in that ocean trench deeps occur on the inner side of the arc.

Although the Hunter Ridge effects a connection, the relationship of the Solomons-New Hebrides arcs to the Fiji-Tonga-Kermadec arc is not clear. Probably a major transform fault following the Vitiaz trench, in part, carries the Tonga-Kermadec arc system westward to the Marianas Islands. Deep (660 km) earthquake foci occur beneath the Lau-Colville Ridge, intermediate (200 km) foci beneath the Tonga-Kermadec Ridge, and shallow foci beneath the ocean trench. These define a narrow seismic zone that steepens with depth. Dip-slip movements on this structure are related to the underthrusting and subduction of the Pacific crust; some strike-slip motions have also been proposed.

Volcanic magma types in this belt have been related to depth above the seismic zone, and a time and space relationship of tholeiitic, calc-alkaline, and shoshonitic magmas has been inferred for parts of New Guinea, New Britain, and Fiji. The potash-rich shoshonite suites in particular seem to develop with cessation of orogenic activity.

Current theories attribute the marginal seas behind island arcs to extensional rifting, and the Coral Sea, Tasman, New Caledonian, and South Fiji basins are thought to have developed in the Paleogene as segments of the eastern margin of the Australian craton were rifted off and drifted eastward. The Lord Howe Rise, Norfolk Ridge, and even the outer volcanic arcs show a remarkable similarity in length and angular form to the polygonal outline of the Australian craton margin. These deep ocean basins are no longer active, and the crustal thicknesses (10 km) and heat flow there are normal; magnetic patterns do not indicate simple spreading. The ocean floor between the Lau-Colville and Tonga-Kermadec ridges is a new sea-floor spreading rift with high heat flow.

In explaining the origin of these features it is necessary to refer to the migrations of the Australian plate (Fig. 4) and to the breakup of *Gondwanaland*. Australia may have been close to India prior to the latest Jurassic and then separated from Antarctica in the late Eocene; New Zealand separated from Antarctica somewhat earlier. The northward and eastward movements about the major spreading zones caused the Australian craton to impinge against the southeast Asian (Indonesian) island arcs in about Middle Miocene times. The Tasman, New Caledonian, and South Fiji basins represent latitudinal extensional zones overriding the Pacific plate at the Tonga-Kermadec Trench. The oceans S of Australia and New Zealand, the Coral Sea Basin, and possibly the Bismarck Sea represent zones of meridional spreading. Convergences resulting from interaction between

FIGURE 4. Paleolatitude of Australia through Phanerozoic time as indicated by paleomagnetic observations. The position of Australia with respect to other continental members of Gondwanaland is indicated by the dashed line; the region occupied at present by West Antarctica is shown dotted. In the diagram headed "Mesozoic," the outline of the remainder of Gondwanaland is not well established (from Embleton, 1973).

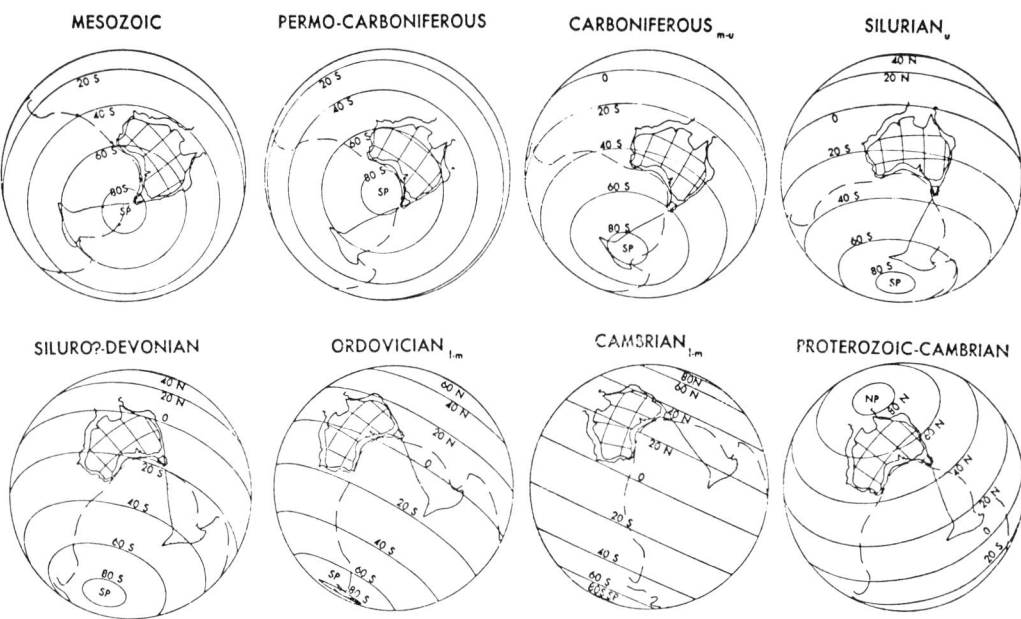

these two movement directions probably cause the contortions of the outer arc.

M. J. RICKARD

References

Burns, R. E., Andrews, J. E., and Scientific Staff, 1972. "Glomar Challenger down under," *Geotimes*, 17 (5), 14–16.

Carey, S. W., 1970. "Australia, New Guinea and Melanesia in the current revolution in concepts of the evolution of the earth," *Search*, 1, 178–189.

Coleman, P. J., 1970. "Geology of the Solomon and New Hebrides islands, as part of the Melanesian re-entrant in the southwest Pacific," *Pacific Sci.*, 24, 289–314.

Compston, W., and Ariens, P. A., 1968. "The Precambrian geochronology of Australia," *Can. J. Earth Sci.*, 5, 561–583.

Cullen, D. J., 1970. "A tectonic analysis of the southwest Pacific," *N. Z. J. Geol. Geophys.*, 13, 7–20.

Davies, H. L., and Smith, I. E., 1971. "Geology of eastern Papua," *Bull. Geol. Soc. Am.*, 2, 3299–3312.

Denham, D., 1969. "Distribution of earthquakes in the New Guinea–Solomon islands region," *J. Geophys. Res.*, 74, 4290–4299.

Doyle, H. A., 1971. "Seismicity and structure in Australia," in: *Recent Crustal Movements*, Wellington: Roy. Soc. N.Z. (Bull. 9), 149–152.

Embleton, B. J. J., 1972. "On the unity of Gondwanaland during the Lower Paleozoic," *Search*, 3(9), 338–339.

———, 1973. "The paleolatitude of Australia through Phanerozoic time," *J. Geol. Soc. Austral.*, 19(4), 475–482.

Geological Society of Australia (Tectonic Map Committee), 1971. *Tectonic Map of Australia*. Sydney, 1:5,000,000.

Gill, J. B., 1970. "Geochemistry of Viti Levu, Fiji, and its evolution as an island arc," *Contrib. Mineral. Petrology*, 27, 179–203.

Hayes, D. E., and Conolly, J. R., 1972. "Morphology of the southeast Indian Ocean," in D. E. Hayes, ed., *Antarctic Oceanology II: The Australian–New Zealand Sector*, Antarct. Res. Ser., No. 19, 125–145.

Heirtzler, J. R., et al., 1968. "Marine magnetic anomalies, geomagnetic field reversals and motions of the ocean floor and continents," *J. Geophys. Res.*, 73, 2119–2135.

———, et al., 1973. "Age of the floor of the eastern Indian Ocean," *Science*, 180, 952–954.

Isacks, B., Sykes, L. R., and Oliver, J., 1969. "Focal mechanisms of deep and shallow earthquakes in the Tonga-Kermadec region and the tectonics of island arcs," *Bull. Geol. Soc. Am.*, 80, 1443–1470.

Jakes, P., and White, A. J. R., 1969. "Structure of the Melanesian Arcs and correlation with distribution of magma types," *Tectonophysics*, 8, 223–236.

Johnson, R. W., Mackenzie, D. E., and Smith, I. E., 1971. 'Seismicity and late Cainozoic volcanism in parts of Papua-New Guinea," *Tectonophysics*, 12, 15–22.

Jones, J. G., 1971. "Australia's Caenozoic drift," *Nature*, 230(5291), 237–239.

Karig, D. E., 1971. "Origin and development of marginal basins in the western Pacific," *J. Geophys. Res.*, 76, 2342–2561.

Lillie, A. R., and Brothers, R. N., 1970. "The geology of New Caledonia," *N.Z. J. Geol. Geophys.,* **13,** 145–183.

Scheibner, E., 1973. "A plate tectonic model of the Paleozoic tectonic history of New South Wales," *J. Geol. Soc. Austral.,* **20**(4), 405–426.

Slater, R. A., and Goodwin, R. H., 1973. "Tasman Sea guyots," *Marine Geol.,* **14**(2), 81–99.

Solomon, M., and Griffiths, J. R., 1972. "Tectonic evolution of the Tasman Orogenic Zone, Eastern Australia," *Nature, Phys. Sci.,* **237**(70), 3–6.

Veevers, J. J., Heirtzler, J. R., et al., 1973. "Deep Sea Drilling Project, Leg 27, in the eastern Indian Ocean," *Geotimes,* April, 16–17.

Vogt, P. R., and Conolly, J. R., 1971. "Tasmantid guyots, the age of the Tasman Basin, and motion between the Australia plate and the mantle," *Bull. Geol. Soc. Am.,* **82,** 2577–2584.

Cross-references: *Antarctica; Australia; New Caledonia; New Guinea; New Zealand; Norfolk Island.*

AUSTRALIA

Australia is an "island continent" covering 7,667,080 km^2 (2,967,741 sq mi), situated between latitudes 10° and 44°S (including Tasmania) and longitudes 113° and 153°E. It is administered as a commonwealth of states (and territories) within the British Commonwealth. Its capital, Canberra, is also the headquarters of the Bureau of Mineral Resources, Geology and Geophysics, which issues national and regional maps and reports. Separate state geological surveys also exist (see under respective headings). The Commonwealth also administers *Christmas Island* (q.v.), *Cocos (Keeling) Islands* (q.v.), *Heard Island,* and *Macquarie Island* (see *Sub-Antarctic Islands*), and claims a sector of East Antarctica. *Papua New Guinea* was formerly administered by Australia and since 1975 has been independent (see *New Guinea*). The principal professional association is the Geological Society of Australia, which publishes a major journal that includes special numbers dedicated to each state, and a geotectonic map on 1:5 million scale (1971).

This account of the stratigraphy and paleogeographic development of the Australian continent should be read in conjunction with the article on *Australasia–Regional Review.*

Precambrian

Precambrian rocks, exceeding 600 m yr in age, are developed mainly in Western Australia, South Australia, and Northern Territory, but are known to occur in all states except Victoria. In Queensland and central Australia (Fig. 1) there are some large areas of Precambrian metamorphic rocks that cannot yet be subdivided (Einasleigh Metamorphics, Musgrave-Mann Complex, etc.).

Archeozoic (Archean)

In surface outcrops, structurally complex plutonic and high-grade metamorphic Archean rocks (older than 2200 m yr) occupy much of the southwestern part of Western Australia. The Yilgarn "System," which consists mainly of granitic gneiss and large intrusive granite bodies with local charnockitic granulites and belts of greenstones (metamorphosed basic igneous rocks), forms a Precambrian Shield.

Farther north, in the Pilbara district, similar lithologies have been grouped in the Pilbara "System," in which three similar rock units have generally been recognized. At several points in the Northern Territory and in the north of Western Australia, they also appear from beneath Proterozoic strata (e.g., the Arunta and Rum Jungle complexes).

The Archean greenstone bodies are the principal loci for gold mineralization in Western Australia.

Proterozoic

Although there are extensive areas in Western Australia and the Northern Territory where Proterozoic (ca. 600–2200 m yr) sediments are not yet amenable to stratigraphic subdivision, a threefold division of some of these rocks now seems possible.

The oldest, the Nullaginian System (ca. 1800–2200 m yr), is best developed in the Pilbara region of Western Australia and in the Katherine-Darwin region of the Northern Territory, where it comprises a thick succession of sedimentary and volcanic rocks. Metamorphosed equivalents are also known elsewhere in the Northern Territory and Queensland. It includes the banded ironstones of the Hamersley Ranges in Western Australia, as well as the Rum Jungle uranium ores.

The Carpentarian System (ca. 1400–1800 m yr) appears to be the most extensive of the Proterozoic systems, occurring as thick, virtually unmetamorphosed sequences throughout much of the northern half of the continent. In the Mt. Isa geosyncline, rocks of this age include the well-known copper-lead-zinc ores at Mt. Isa in Queensland. Metamorphic complexes of similar age are well developed in South Australia and western New South Wales. The iron ores of Yampi Sound in northwestern Australia also belong here.

The Adelaidean System (ca. 600–1400 m yr) consists of a sequence of thick shallow-water

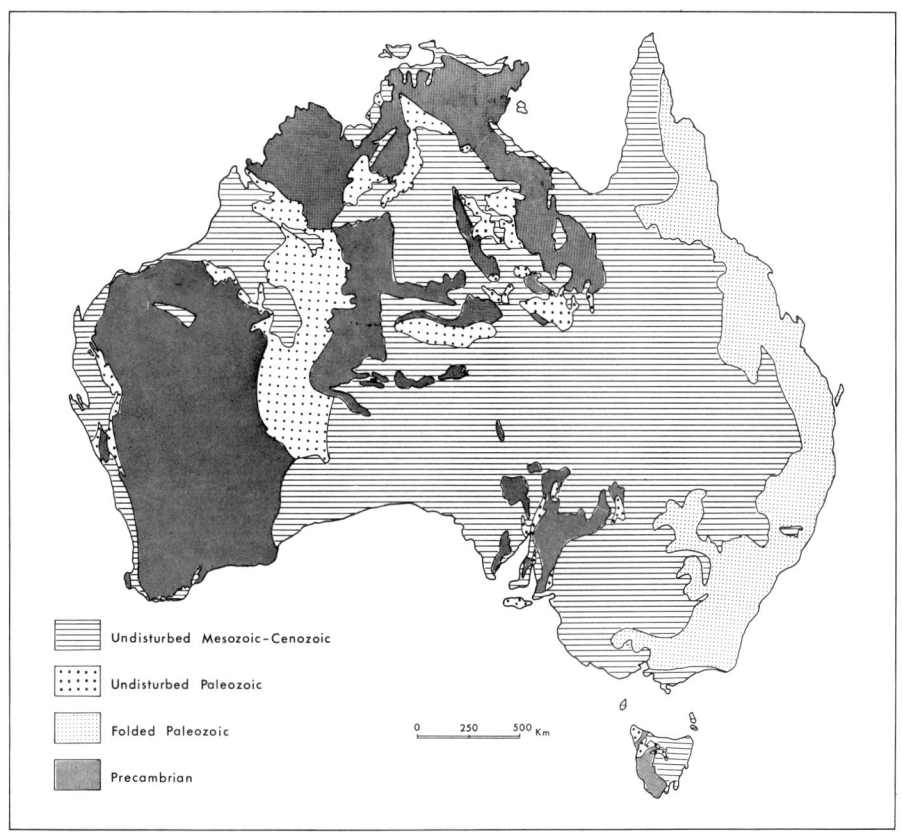

FIGURE 1. Simplified geologic map of Australia (data from Geological Map of Australia, Bureau of Mineral Resources, May 1971).

sediments of basinal type, including sandstones, shales, dolomites, limestones, red beds, and extensive glacial deposits. It lacks the turbidites and other features of a typical geosynclinal sequence to which the South Australian sections have often been referred in the past. Most other Adelaidean rocks consist of unaltered and unfolded sediments deposited in basins within older Precambrian rocks and are known both in outcrop and subsurface in South Australia (Officer Basin), Central Australia (Amadeus, Georgina, and Ngalia basins), and northern Australia (Arafura, McArthur, and South Nicholson basins).

On the whole the Adelaidean and undisturbed Carpentarian rocks appear to be less mineralized than those deposited earlier. Detailed mapping has, however, revealed the presence of some valuable deposits of iron ore, bauxite, and segregations of lead-zinc and uranium.

Paleozoic

Cambrian. Cambrian rocks occur mainly in northwestern and central Australia and South Australia. In the Adelaide Geosyncline there is conformity between the Cambrian and the Adelaidean rocks. Elsewhere there were few Lower Cambrian sediments, but a widespread marine inundation occurred in Middle Cambrian time, producing a shallow central sea in which limestones, shales, sandstones, and cherts were deposited. Upper Cambrian sediments occur in the Bonaparte Gulf, Georgina, and Amadeus basins. On the Antrim Plateau region of Western Australia, tholeiitic basalts were extruded over a wide area in the early Cambrian.

In the eastern part of the continent, the Tasman Orthogeosyncline was initiated, with graywackes and volcanics prominent among the deposits.

Mineralization of Cambrian rocks is apparent only in South Australia (e.g., the Nairne pyrite), but in this state and in Western Australia and the Northern Territory, rocks of this age are important sources of underground water.

Ordovician. At the very beginning of the Ordovician period sedimentation ceased in the Adelaide "Geosyncline," and the accumulated basinal sediments underwent strong deformation, with overfolding and overthrusting in the east (Delamerian Orogeny).

In most other parts of the continent this was

AUSTRALIA

a period of relative tectonic stability, but in the Tasman Orthogeosyncline the Benambran Orogeny caused reorganization of the trough and arch pattern and was accompanied by a small amount of igneous activity toward the close of the period.

Deposition continued without interruption from the Cambrian over large parts of the northwest of the continent, including the Canning Basin, and in the Amadeus Basin. The Tasman Orthogeosyncline was also the site of extensive sedimentation, with both shelly and graptolitic facies represented, the latter mainly in Victoria and New South Wales.

Ordovician rocks are important hosts for mineral deposits in the Tasman Orthogeosyncline (e.g., the gold reefs in Victoria and New South Wales), although the actual mineralization is associated with granitic rocks of younger age.

Silurian. Silurian deposits are confined primarily to the Tasman Orthogeosyncline. In one of its subdivisions, termed the Lachlan Geosyncline, acid and intermediate volcanism were followed by extensive shelf-limestone deposition, with rich invertebrate faunas. In the middle of the period another tectonic episode, accompanied by widespread acid volcanism, affected this area and continued into the early Devonian. Some large-scale granodiorite intrusion in New South Wales and central Victoria, possibly with some serpentinization, may also be of late Silurian age.

Some of the Silurian limestones (marbles) have produced excellent building stones, and others are used widely for cement and flux manufacture. Gold, silver-lead, and copper mineralization are associated with granitoids of possible late Silurian age.

Devonian. The Devonian system is notable for its wide range of sedimentary environments, ranging from reef-limestone platforms and deep-water troughs to arid terrestrial basins. The period commenced with tectonism and widespread acid volcanism (Snowy River Volcanics) in southeastern Australia. During Middle Silurian time, a major tectonic event (the Tabberabberan Orogeny) affected the eastern regions; following this, much of the eastern region of the continent, except Tasmania, became the site of terrestrial deposition along rather narrow, meridionally trending belts,

FIGURE 2. Map of elements of geology of Australia (original, slightly modified, provided by Bureau of Mineral Resources, Canberra).

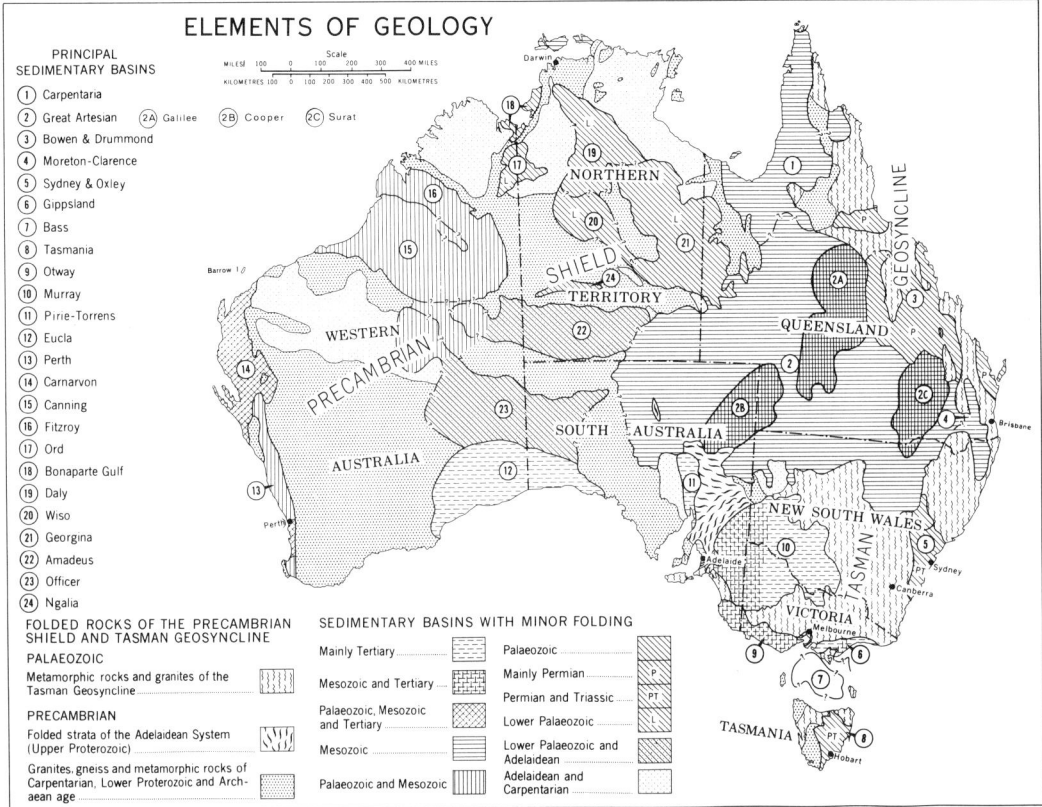

known as the Lambian Basins, which occupy the site of the former Lachlan Geosyncline.

In Western Australia, there are marked differences from the eastern regions. There had been a general recession of the early Paleozoic seas, and deposition did not resume until an inundation from the ancestral Indian Ocean took place in middle Devonian time, accompanied by the development of remarkable coral reefs in the Kimberley area. Marine sediments accumulated during the remainder of the period in the Bonaparte Gulf, Fitzroy, and Canning basins. No igneous activity has been recorded there (Fig. 2).

Mineralization accompanied the intrusion of middle Devonian granites in eastern Australia, although it is difficult to distinguish between deposits generated by these intrusions and those of early Carboniferous age. The gold and copper deposits at Cobar, New South Wales, and some Victorian occurrences of gold may belong here. Some Devonian limestone horizons in Western Australia (e.g., in the Fitzroy and Bonaparte Gulf basins) may be good prospects for oil and gas.

Carboniferous. In most areas deposition was continuous from the Devonian, and the boundaries between the systems are difficult to define. Diastrophic movements, collectively referred to as the Kanimblan Orogeny, affected most of the Tasman Orthogeosyncline at intervals throughout the Carboniferous period.

In the early part of the period, terrestrial sedimentation continued in the Lambian Basins in eastern Australia, but marine deposits are evident in the more easterly geosynclinal troughs (New England Geosyncline, Hodgkinson Basin, etc.).

In Western Australia, early Carboniferous marine sedimentation occurred in the Bonaparte Gulf, Fitzroy, and Canning basins, but except for a small area in the Fitzroy Basin where Upper Carboniferous strata have been identified in a bore, these basins appear to have been emergent during the later portion of the period.

During the late Carboniferous, a remarkable and abrupt change in latitude has been noted during studies of paleomagnetism in the eastern region, coincident with a marked deterioration in climatic conditions and the appearance of "cold-water" sediments and faunas, the precursors of the extensive Permian glaciation.

As expected in association with the Kanimblan Orogeny, igneous activity in the form of acid and basic plutons occurred in the Tasman Orthogeosyncline and, especially in the late Carboniferous epoch, there was extensive volcanism in New South Wales, Queensland, and the Bonaparte Gulf Basin. These igneous bodies are frequently associated with metallogenesis in the eastern region.

Permian. Rocks of Permian age are notable for a number of reasons: they occur over a vast area of the continent; they are the most important source of the country's coal and have prospects for petroleum; they show clear evidence of a widespread glaciation; and they contain numerous representatives of the typical Gondwana faunas and floras.

Lower Permian glacial features (tillites, varves, striated pavements, etc.) are represented widely on the surface and in the subsurface of the southern half of the continent west of the eastern seaboard ranges. In the north there is evidence of ice-rafted erratics. These deposits appear to be appropriate to a polar icesheet radiating northward from a paleopole situated 15°W of the present position of Tasmania. In Western Australia as the climatic conditions ameliorated in the late Permian, many of these glaciated areas were converted to basins of shallow marine and terrestrial sedimentation.

In the eastern regions the Permian period began with cold-water marine sedimentation with extensive volcanism in places (e.g., the Bowen Basin) but by the middle of the period, terrestrial conditions had set in, with coals forming in many places. Toward the end of the period, a major but localized tectonic movement (the Hunter-Bowen Orogeny) folded and uplifted the sequences in the Newcastle geosyncline and the basins to the west. This orogeny was accompanied by the intrusion of granitoids (New England Batholith) and probably a major portion of the Great Serpentine Belt.

Economically the Permian System in Australia is marked by important metalliferous mineralization (tin, molybdenum, tungsten, bismuth, antimony) in northern New South Wales and by the Mt. Morgan copper-gold deposits in Queensland. The period is most noted, however, for its vast resources of bituminous coal in the Bowen and Sydney basins in eastern Australia. There is minor Permian coal in Western Australia. It has also been an important source of oil and natural gas (Gidgealpa-Moomba, South Australia).

Mesozoic

Triassic. The widespread *Glossopteris* flora of the Permian was abruptly replaced by that dominated by *Dicroidium,* and some deposits have yielded a varied amphibian and reptilian fauna. Sedimentation during the Triassic was predominantly terrestrial, with only small occurrences of marine deposits. The main areas of terrestrial deposition were the intramontane basins associated with the Hunter-Bowen mountain chain. In Western Australia each of the major basins initiated during the Paleozoic contains marine to brackish-water sequences, with fluvio-lacustrine material in the Canning and Perth basins.

As elsewhere in Gondwanaland, coal is important in this system also (Ipswich coals of Queensland, the Leigh Creek Coalfield of South Australia), along with some excellent building stones (Hawkesbury Sandstone) and brick-making materials.

Jurassic. Sediments of the Jurassic are confined to the Canning, Carnarvon, and Perth basins in Western Australia and to the basins of the eastern central and northeastern parts of the continent. In Tasmania, the Jurassic System is notable for the intrusion of vast dolerite sequences, mainly sills.

In eastern Australia, as with the Triassic, the Jurassic deposits are chiefly nonmarine. Their

FIGURE 3. Paleogeographic evolution of Australia.

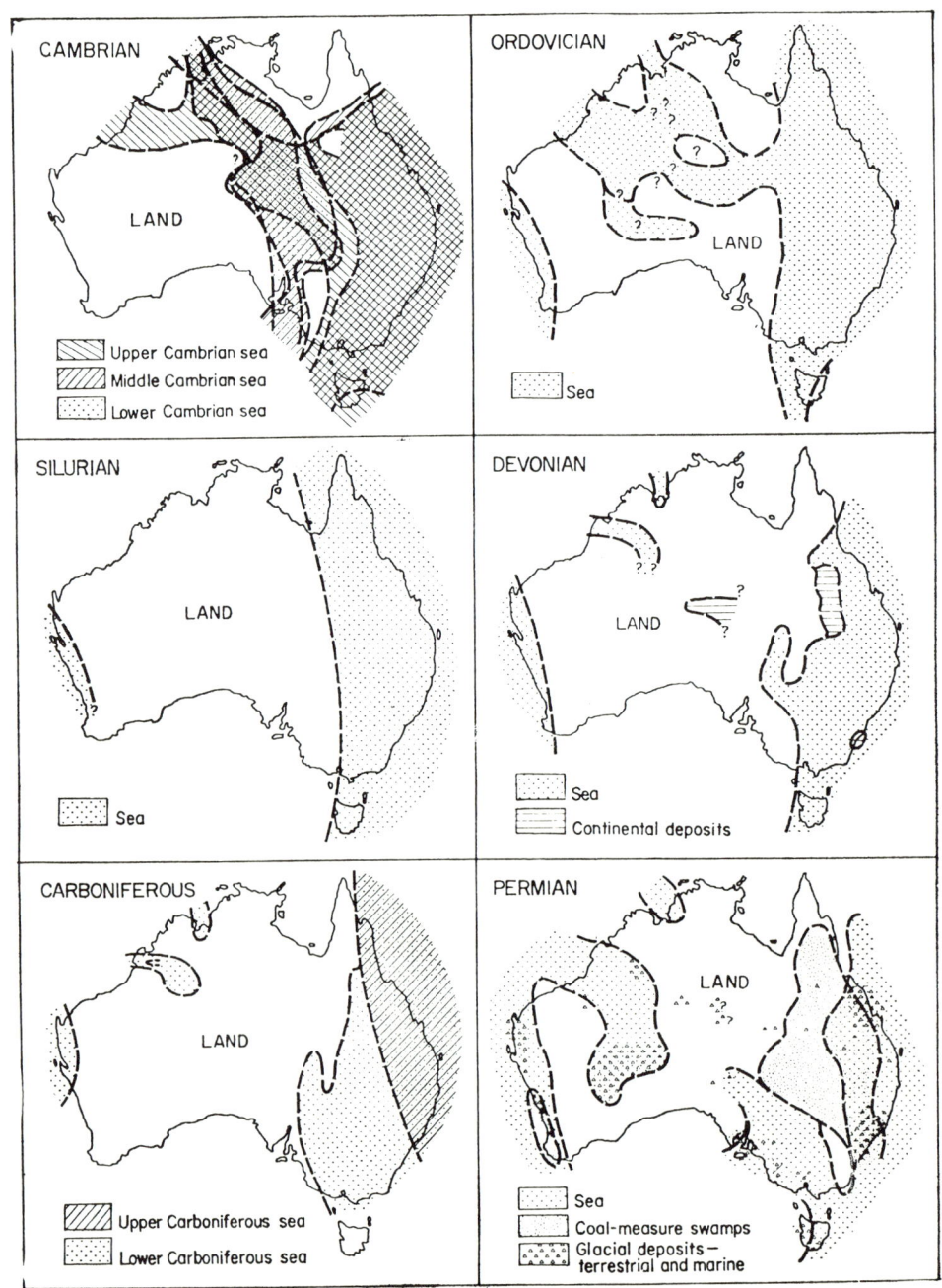

most extensive development, mainly in the subsurface, is in the Great Artesian Basin, where widespread crustal downsagging took place and allowed the deposition of up to 2000 m of strata in places. In Western Australia much of the sequence is marine, but only in the Carnarvon Basin is it of deep-water origin.

Bituminous coal, oil, and gas are known from rocks of this age, including the first commercial oil field in Australia (Moonie, Queensland). Jurassic sandstones also provide some of the most important aquifers in the Great Artesian Basin.

Cretaceous. In the early Cretaceous, shallow epeiric seas flooded widely across the continent, separating the present surface area into three distinct land masses. Three new basins (Eucla, Otway, and Gippsland) with thick sedimentation were initiated. Strata of this age are present in outcrop or subsurface over nearly one-third of the continent. The major transgression occurred in Aptian times, followed by a general retreat in the late Cretaceous, which time is represented only by marginal marine deposits in Western Australia and in northern and southeastern Australia. Extensive nonmarine sediments occur mainly in Queensland, where deposition continued in the Great Artesian Basin. By the end of the period virtually the entire continent, with the exception of the Otway Basin in Victoria, seems to have been emergent. A rather local tectonic movement (the Maryburian Orogeny) strongly deformed part of the eastern coastal region of Queensland at this time, the final orogenic spasm in the Tasman Orthogeosyncline. Although igneous activity was very limited during the period, radiometric dating has shown that both the Mount Dromedary shoshonitic complex of coastal New South Wales and the Point Cygnet syenite porphyry of Tasmania are of this age.

Some minor metalliferous mineralization is also associated with the Maryburian Orogeny in Queensland. Basal Cretaceous strata provide some important aquifers in the Great Artesian Basin, and oil accumulations in the Carnarvon Basin (Rough Range) and Barrow Island, Western Australia, are also found in sandstones of this period.

Cenozoic

Tertiary. Discontinuous and rather fragmentary marine deposits of Tertiary age are marginal to the continent, occurring mainly in the western Carnarvon and Eucla basins of Western Australia and in the St. Vincent, Murray, Otway, Bass, and Gippsland basins of southern Australia. The strata consist of limestones, calcarenites, and sandstones.

The terrestrial deposits comprise the small lacustrine and fluvial formations of the eastern highlands and coastal plains and those of the great internal drainage systems (Lake Eyre and Darling-Warrego basins) inherited largely from older basinal areas (Fig. 3).

The tectonic situation was generally epeirogenic in type, although large-scale volcanism in both early and late Tertiary times did occur in the east and southeast of the continent. Toward the end of early Tertiary time, continued uplift had drained the sea from most of the marginal basins and the continent was probably connected by "land bridges" to Tasmania and Papua. These connections were probably severed by downwarping in middle Tertiary time, but were restored temporarily during the Quaternary eustatic oscillations.

The climate appears to have been temperate to tropical throughout the Tertiary, with widespread development of lateritization and silcrete formation due to the deep leaching of the soils and underlying rocks.

An epeirogenic movement (the Kosciusko Uplift) affected the eastern highlands in the late Tertiary and prepared the relief for the establishment of glacier conditions in the Snowy Mountains region in the Quaternary.

Of primary economic importance in the Tertiary are the thick and extensive brown coals of Victoria, the bauxite deposits of Queensland, northern and Western Australia, and, recently, petroleum production on a substantial scale from the Tertiary sediments in the Bass Strait area.

Quaternary. The last 2 m yr have seen extensive fluviatile and eolian deposition in the eastern internal basins and marginal marine incursions. Stable tectonic conditions have prevailed with glaciation affecting only a small area of the southeastern highlands and Tasmania. Volcanism on a local scale occurred in southern Victoria and northern Queensland.

Summary of Continental Evolution

The earliest discernible features of continental development in Australia are seen in the west in the form of three cratonic nuclei (Yilgarn, Pilbara, and Arunta) with intervening geosynclinal troughs. These were consolidated to a single craton by a series of ep-Archean orogenies. There was a lengthy period of platform deposition during Proterozoic times, along with extensive volcanism from time to time.

Early in Paleozoic time the Tasman Orthogeosyncline developed along the eastern seaboard and after a complex history was consolidated to form part of the craton by the beginning of the Mesozoic. This was followed by a gradual spread of basinal areas eastward and the predominant deposition of continental sediments covering the platform area.

DAVID A. BROWN

References

Anon, 1973. "Australian code of stratigraphic nomenclature," *J. Geol. Soc. Austral.*, **20**(1), 105–112.
Bank of New South Wales, and Maxwell, W. G. H., 1971. *Offshore Australia. The Continental Shelf, the Slope, and Beyond.* Sydney: Bank of N.S.W.
Brown, D. A., Campbell, K. S. W., and Crook, K. A. W., 1968. *The Geological Evolution of Australia and New Zealand.* Oxford: Pergamon, 409p.
Conolly, J. R., 1970. "Sedimentary history of the continental margin of Australia," *Trans. N. Y. Acad. Sci.*, **32**(3), 364–380.
Crowell, J. C., and Frakes, L. A., 1971. "Late Paleozoic glaciation of Australia," *J. Geol. Soc. Austral.*, **17**(2), 115–155.
David, T. W. E. and Browne, W. R., eds., 1950. *The Geology of the Commonwealth of Australia,* 3 vols. London: Edward Arnold.
Dunn, P. R., Thomson, B. P., and Rankama, K., 1971. "Late Pre-Cambrian glaciation in Australia as stratigraphic boundary," *Nature*, **231**(5304), 498–502.
Embleton, B. J. J., 1973. "The palaeolatitude of Australia through Phanerozoic time," *J. Geol. Soc. Austral.*, **19**(4), 475–482.
Geological Society of Australia (State Volumes), 1958–1968. "Western Australia," *J. Geol. Soc. Austral.*, **4**(2), 1958; "South Australia," *Ibid.*, **5**(2), 1958; "Queensland," *Ibid.*, **7**, 1960; "Tasmania," *Ibid.*, **9**(2), 1962; "New South Wales," *Ibid.*, **16**(1), 1968.
Geological Society of Australia (Tectonic Map Committee), 1971. *Tectonic Map of Australia and New Guinea.* Sydney, 1:5,000,000.
Gill, E. D., and Hopley, D., 1972. "Holocene sea levels in eastern Australia–discussion," *Marine Geol.*, **12**, 223–242.
Griffiths, J. R., 1971. "Continental margin tectonics and the evolution of southeast Australia," *J. Austral. Petrol. Explor. Assoc.*, **11**(1), 75–79.
Howard, P. F., 1972. "Exploration for phosphorite in Australia–a case history," *Econ. Geol.*, **67**, 1180–1192.
Laseron, C. F., and Brunnschweiler, R. O., 1969. *Ancient Australia: the Story of Its Past Geography and Life.* Sydney: Angus & Robertson, 253p.
McAndrew, J., and Marsden, M. A. H., eds., 1968. *Regional Guide to Victorian Geology.* Univ. of Melbourne, Geol. Dept.
McLeod, I. R., ed., 1965. "Australian mineral industry: the mineral deposits," *Austral. Bur. Mineral Resources Geol. Geophys. Bull.*, **72**.
Mulvaney, D. J., and Golson, J., 1971. *Aboriginal Man and Environment in Australia.* Canberra: Australian Nat. Univ. Press, 389p.
Noakes, L. C., 1966. "Commentary to accompany map-sheet *Geology Second Edition,*" *Atlas of Australian Resources,* ser. 2, Canberra: Dept. Nat. Develop., 1: 6,000,000.
———, 1974. "Mineral resources of Australia," *Search*, **5**(1–2), 11–16.
Parkin, L. W., ed., 1969. *Handbook of South Australian Geology.* Adelaide: Geol. Surv. S. Austral., 268p.
Peterson, J. A., 1971. "The equivocal extent of glaciation in the southeastern uplands of Australia," *Proc. Roy. Soc. Victoria*, **84**(2), 207–212.
Pickett, J. W., Jell, J. S., Seddon, G., et al., 1972. "Correlation of the middle Devonian formations of Australia," *J. Geol. Soc. Austral.*, **18**(4), 333–347.
Roberts, J., Jones, P. J., Jell, J. S., et al., 1972. "Correlation of the Upper Devonian rocks of Australia," *J. Geol. Soc. Austral.*, **18**(4), 467–490.
Scheibnerova, V., 1971. "Implications of deep sea drilling in the Atlantic for studies in Australia and New Zealand; some new views on Cretaceous and Cainozoic palaeogeography and biostratigraphy," *Search*, **2**(7), 251–254.
Sinden, J. A., ed., 1972. *The Natural Resources of Australia.* Sydney: Angus & Robertson, 374p.
Slatyer, R. O., and Perry, R. A., eds., 1969. *Arid Lands of Australia.* Canberra: ANU Press, 321p.
Talent, J. A., Campbell, K. S. W., Davoren, P. J., et al., 1972. "Provincialism and Australian early Devonian faunas," *J. Geol. Soc. Austral.*, **19**(1), 81–97.
Warren, R. G., 1972. "A commentary on the metallogenic map of Australia and Papua New Guinea," *Austral. Bur. Mineral Resources Geol. Geophys. Bull.*, **145**.

Cross-references: *Antarctica; Australasia–Regional Review; Australia–New South Wales, Northern Territory, Queensland, South Australia, Tasmania, Victoria, Western Australia; New Caledonia; New Guinea; New Zealand; Norfolk Island; Sub-Antarctic Islands.*

AUSTRALIA–NEW SOUTH WALES

North South Wales extends from latitudes $28°$ to $37°$S and longitudes $141°$ to $153°$E. Its area is 497,997 km^2 (309,433 sq mi). It was only nine years after the first settlement was established on the continent at Sydney that coal was discovered at Newcastle, but it was not until 1851 that the first metalliferous deposits were worked. This was when payable quantities of gold were found at Ophir in the Orange district. A period of mineral prospecting on a grand scale was initiated and it culminated in the discovery of the enormous Broken Hill silver-lead-zinc deposit in 1882. This deposit and the coalfields are now the principal source of the state's mineral wealth. The value of minerals produced in 1969–1970 was (Aust) $385 million.

The Geological Survey of New South Wales was founded in 1875 and geological training at the undergraduate and postgraduate levels at all the universities in New South Wales (and the Australian Capital Territory). They are the Universities of Sydney, New South Wales, and Macquarie University (all situated in Sydney); University of New England (Armidale); Universities of Newcastle and Wollongong; and the Australian National University (Canberra, A.C.T.).

Tectonic Subdivisions and Geomorphology

Seven geological provinces can be recognized in New South Wales (Fig. 1). The North Western, Lachlan, and New England Fold Belts contain moderately to highly deformed rocks, while the remainder of the provinces are almost undeformed cross-cutting basins. The relationships among the three deformed provinces are not yet clear. The North Western Fold Belt containing early and late Proterozoic, early and middle Paleozoic deposits is separated from the early and middle Paleozoic rocks of the Lachlan Fold Belt by a zone of alluvium and poor

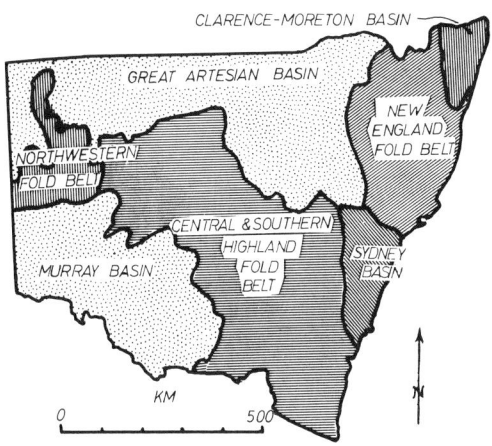

FIGURE 1. Geological provinces of New South Wales.

FIGURE 2. Geological map of New South Wales.

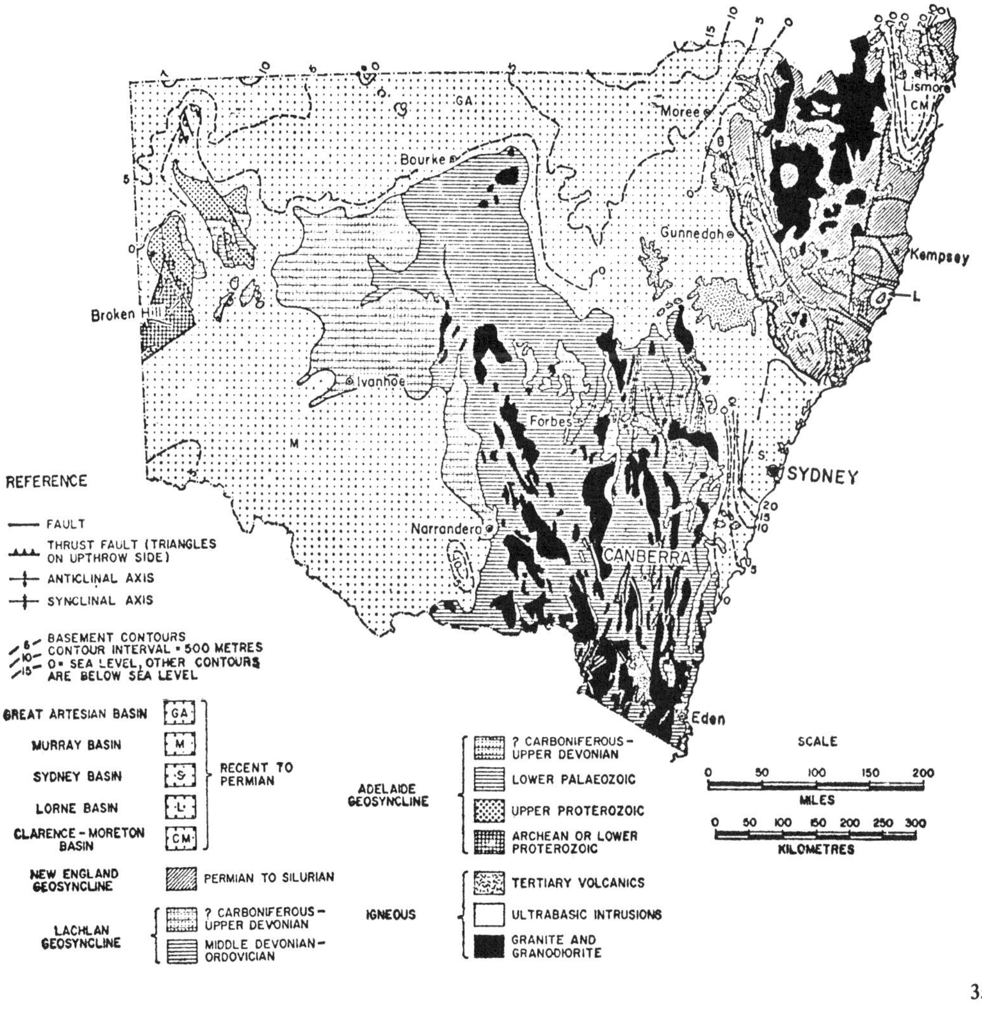

outcrops along the Darling River. Similarly, the New England Fold Belt, in which Devonian to Permian sediments are dominant, is isolated from the Lachlan Fold Belt by the Sydney Basin. Figure 2 shows the principal geological structural elements in the fold belts. A number of plate tectonic models have been proposed to explain the evolution of the region. These have been reviewed by Packham and Leitch (1973), none of these is fully satisfactory.

Three of the basins extend outside the state and, except for the Sydney Basin and the Great Artesian Basin, none of them comes into contact. Permian sediments in the lower part of the Sydney Basin thicken into the Hunter Valley and pass into the New England Fold Belt, while both the Permian and Triassic sediments thin to the N under the edge of the Great Artesian Basin, where they are overlapped by Jurassic sediments, which are in turn overlain by early Cretaceous. The Clarence-Moreton Basin occupies the northeast corner of New England, and the Triassic and Jurassic sediments it contains are continuous with those of the Moreton Basin in Queensland. In the opposite corner of the State, the Murray Basin, which extends into Victoria and South Australia, contains Tertiary and Quaternary sediments.

The divide between the easterly and westerly flowing streams is on the average about 80 miles inland, and it has probably been in something like its present position since soon after the Hunter-Bowen Orogeny (mid to late Permian). The sediments of the basins have been derived for the most part from the fold belts, which have undergone uplift and denudation. On the margin of the basins, for example the western margin of the Sydney Basin and the southern margin of the Great Artesian Basin, post-Cretaceous uplifts led to the stripping off of part of the sedimentary cover, revealing paleogeomorphic land forms with relief in excess of 100 m. Farther away from the basin margins the fold belts are more dissected. The most subdued landscape ("peneplanation"?) seems to have been in the mid-Tertiary. In the late Tertiary-early Pleistocene there were strong vertical movements (the Kosciusko Uplift) which resulted in considerable dissection, carving deep valleys into the fold belts and plateau-forming basins in the eastern third of the state. The amount of this uplift diminished rapidly westward. Flood basalts filled many of the valleys and today cap many plateaus. Pleistocene glaciation was limited to a small area in the Australian, and interesting cirques are found around

FIGURE 3. Cirque lake in the Australian Alps (36°S latitude). Lake Cootapatamba (Australian aboriginal word for "the frozen waters where the eagle swoops to drink") with the slopes of the summit of Mt. Kosciusko (7316ft elevation) rising on the right of the lake. In winter the Cootapatamba Valley or "saddle" is entirely blanketed in snow. The foreground is strewn with erratic blocks. (Photo: New South Wales government.)

Mt. Koseiusko (Fig. 3). Lower elevations were affected by periglacial action.

Stratigraphic History

North-western Fold Belt (Fig. 4) The oldest of the sequences, the Willyama Complex is thought to be early Proterozoic in age. It is composed of regionally metamorphosed sandy and shaly rocks outcropping in the Broken Hill Block. The regional metamorphism has produced schists and gneisses that contain cordierite, andalusite, chiastolite, garnet, staurolite, and sillimanite. A great variety of igneous-like rocks also occurs, including granite gneisses, aplites, amphibolites, and pegmatites. The original pattern of metamorphism is partly obscured by later deformation, which produced zones of retrogression.

The Torrowangee Beds (Upper Proterozoic) contain tillites and glacial boulder beds, laminated shales, limestones lenses, and quartzites. The Torrowangee Beds outcrop north of Broken Hill and overlie the Willyama Complex with a sharp unconformity.

The Wonominta Beds in the Tibooburra-Wonominta Block is a succession of low-grade metamorphic rocks that includes slates, phyllites, quartz-rich graywackes, and basic volcanics. The last-mentioned may prove to be early Cambrian.

Overlying the Wonominta Beds, 130 km NE of Broken Hill around Mt. Wright is a small area of basic to intermediate volcanics, limestones, and shales containing early and middle Cambrian faunas (mainly trilobites). Middle Cambrian shales, limestones, and tuffaceous sediments occur 125 km to the N at Mt. Arrowsmith. These are overlain unconformably by a thick sequence of cross-bedded quartzites, conglomerates, shales, and limestones (latest Cambrian to early Ordovician). These sediments form a fluvial deltaic-shallow marine complex in which the sequences become more marine eastward and northward. The Paleozoic successions and the Torrowangee Beds are overlain unconformably by Upper Devonian quartzites.

FIGURE 4. Structural geological map of New South Wales (after Scheibner, 1972). Abbreviations: C, Capertee Belt; M, Monaro anticlinorial zone; SC, South Coast anticlinorial zone.

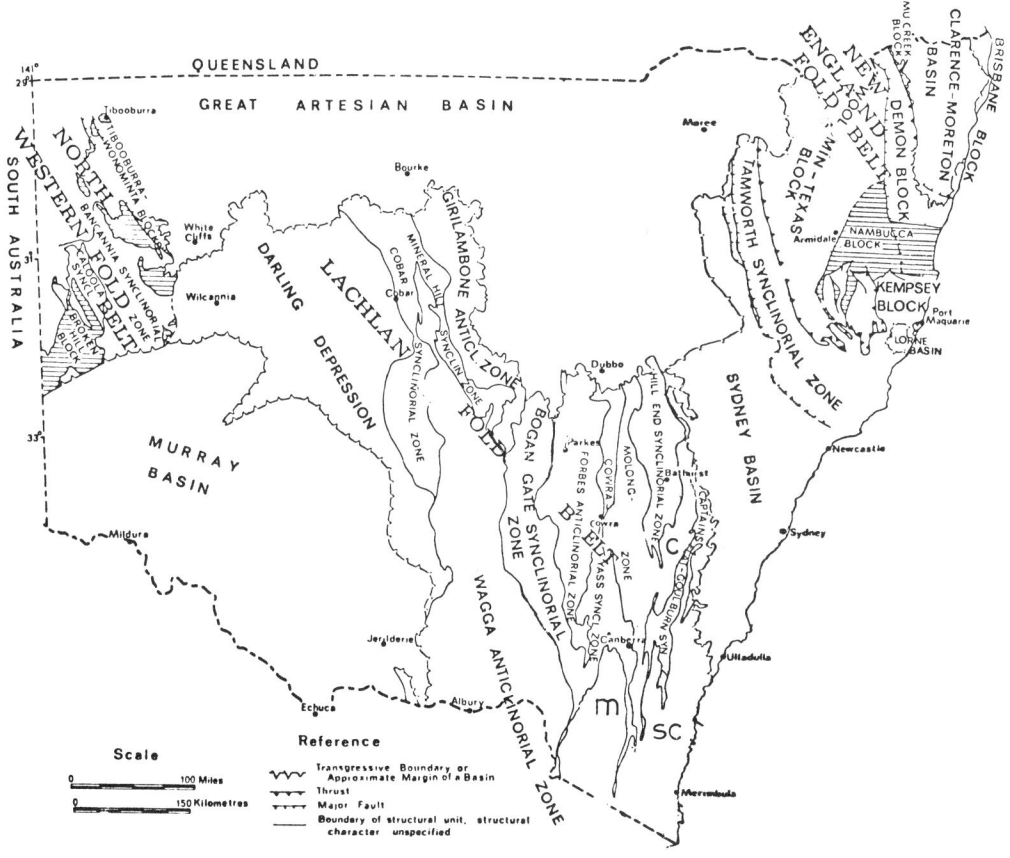

All the sequences above the Willyama Complex and the Wonominta Beds are gently to moderately folded. The movements between the Cambrian and Ordovician correspond generally to the Tyennan and Delamerian orogenics of Tasmania and South Australia, respectively. The Broken Hill Block is part of the Adelaide Fold Belt, which it resembles closely both stratigraphically and structurally. The Tibooburra-Wonominta Block has no direct analog in South Australia. There are many differences between it and the Kanmantoo, which margins the Adelaide Fold Belt.

Lachlan Fold Belt. This province is the older of the two folded tracts of country that comprise the Tasman Geosyncline in New South Wales. This older tract, the Lachlan Geosyncline, extends northward from Victoria through New South Wales, disappearing under the Great Artesian Basin to the north. Its history has been extremely complex and the study of the stratigraphy shows that four main orogenic movements have contributed to the progressive stabilization of the area, commencing in the early Paleozoic and continuing until the Carboniferous. In the New England area, the younger fold province (the New England Geosyncline), sedimentation, and tectonic movements ended for the most part at the end of the Permian.

No *Cambrian* fossils have been found in the Lachlan Fold Belt in New South Wales. The only rocks of suspected Cambrian age are the Girilambone Beds, which occupy most of the Girilambone anticlinorial zone. They are slates, phyllites, schists, quartz graywackes, and quartzites with minor basic volcanics. Their age relationships to Ordovician rocks are unknown.

Ordovician rocks (mostly late Ordovician) occupying large areas of the Southern Highlands and occupying much of the Wagga, Monoro, and South Coast anticlinorial zones, are composed of quartz-rich graywackes, gray slates, and black graptolite-bearing slates. These rocks may represent the fill of a marginal sea behind an island arc. At Albury, Wagga Wagga, Cooma, and other localities, an extensive regional metamorphic terrain that contains high-grade rocks and anatectic granites is developed.

In the Capertee, Molong, and Forbes anticlinorial zones, which extend a little S of the latitude of Cowra, is an area characterized by an abundance of early to late Ordovician andesitic volcanics. In the strip of country between Cowra and Wellington and W of Parkes, early to middle late Ordovician limestones containing shelly fossils occur in the sequence. These volcanics may represent part of the island arc E of the marginal sea. South of Cowra the arc appears to have been much farther E, off the present coastline.

Nowhere in New South Wales has a conformable sequence from Ordovician to *Silurian* strata been found. Even where no angular discordance exists, depositional breaks occur. Three breaks in the sequence have been detected. The oldest is between the late Ordovician and early Llandoverian (the Cobbler's Creek orogenic phase), the second is between the early and late Llandoverian (the Panuara orogenic phase), and the third is between the late Llandoverian and the mid Wenlock (the Quarry Creek orogenic phase). These movements jointly constitute the Benambran Orogeny. The intensity of the orogenic movements was greatest in southern New South Wales, where regional metamorphism and granite emplacement took place. Granites associated with these metamorphics are rich in potash, poor in lime, and relatively poor in soda. All these chemical features are common to the associated regionally metamorphosed sediments, and field relationships at some localities suggest an anatectic origin for the granites.

Metamorphism occurred earlier in the Wagga anticlinorial zone than to the east, probably as a result of the earlier two Benambran movements. Late Llandoverian quartz-rich sediments (in part turbidites), in all probability derived from the Ordovician sequence, were deposited in the Yass and Captains Flat-Goulburn synclinal zones and on part of the Monaro anticlinal zone. These early Silurian sequences were deformed in the Quarry Creek Orogenic Phase. Metamorphism and granite emplacement of the same style as the earlier movements took place in the Monaro and to a lesser degree in the South Coast anticlinorial zones. It appears that crustal shortening associated with the Benembran Orogeny converted the region from the suggested arc-marginal sea to an unstable platform.

The Silurian rocks of the state are typified by dacitic or rhyolitic volcanics and limestone. Both are present in most sections. Other common lithologies are slate, graywacke, and sandstone. The Silurian is thin in the Molong anticlinorial zone (the Molong High). To the E and W the section thickens, and to the E at least passes from shallow-water deposition into graywacke facies (Hill End Trough). This trough probably formed in the middle Silurian. There is some evidence for thinning again in the late Silurian to the E of Sofala (Capertee High). The thicker zone W of the Molong High (the Cowra Trough) is thrust against a more stable zone W of Forbes (the Parkes Platform) on which the Silurian section is again thin. The

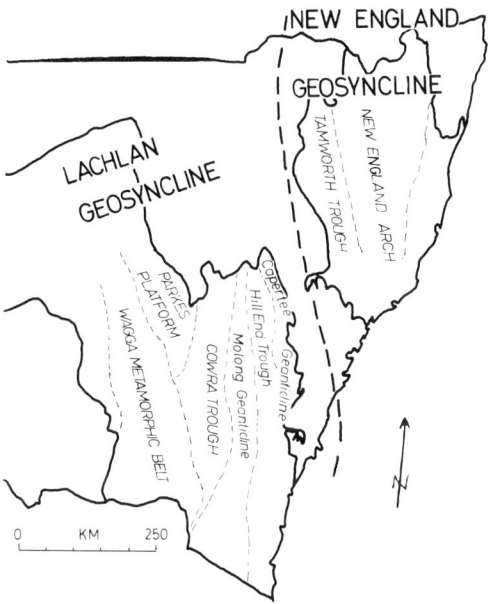

FIGURE 5. Geotectonic units of the eastern fold belts of New South Wales.

Cowra Trough could be as old as middle Ordovician. The Hill End Trough could pass E of Canberra; the highly fossiliferous sequence at Yass and that at Canberra were deposited on the southern extension of the Molong High. This structure broadens southward, passing into Victoria. The southern extension of the Cowra Trough is problematical. Perhaps it terminated in the southern highlands S of Tumut, bounded on the west by another geanticlinal structure, largely a hypothetical one, extending from the Girilambone anticlinorial zone through the Ordovician metamorphic area of the Wagga anticlinorial zone into the Omeo district of Victoria. The Parkes Platform lies to the NW of this structure. West of the last-mentioned high structure there was possibly an additional trough, corresponding approximately to the Cobar synclinorial zone, which may have extended S into central Victoria and connected with the Melbourne Trough. The Cowra Trough and the Hill End Trough (N of the Bathurst Granite; Fig. 5) could have formed as interarc basins with oceanic crust flows, but there is no direct evidence for this hypothesis.

Another series of orogenic movements occurred from late Silurian well into the early *Devonian;* these constitute the Bowning Orogeny. Large granite and granodiorite masses were associated with this orogeny, including the Kosciusko Granite, the Murrumbidgee Batholith, and the Gunning-Wyangala Batholith. In some cases at least it is certain that the foliation that is common in Bowning Granites is tectonically induced and is related to the structural pattern in the country rocks.

The Hill End Trough, and possibly the Cowra Trough, were unaffected by the Bowning Orogeny, sedimentation continuing conformably. In the early part of the Devonian, there were outpourings of acid and intermediate lavas, rhyolites being the most characteristic. These volcanics are found at Yarrangobilly, Yass, Canberra, S of Cowra near Tumut, NW of Forbes, and W of Molong. Tuffs and graywackes of volcanic origin were deposited in the Hill End Trough. On the Molong High from Molong northward, limestones and shales of the Garra Formation overlie the volcanics. Limestones, sandstones (in part turbidites), and shales were deposited in the Cowra Trough and quartzites, shales, limestones, and volcanics on the Parkes Platform. The calcareous sediments on the western edge of the Hill End Trough pass eastward into slates and graywackes overlying the volcanics, but E of the trough on the Capertee High shallow-water deposits (mainly limestones) occur from Mudgee southward for 60 miles. In the Yass district, a sequence of limestones conformably overlies the acid volcanics mentioned above and to the W is overlain by conglomerate and sandstones. Similar limestones occur at Yarrangobilly to the S. In the Cobar area, in the Darling Depression the Amphitheatre Group was deposited in the early Devonian. The shales, quartzites, and siltstones form a large fluvial-delta shallow-water marine complex that prograded from the SW. Fluvial sedimentation probably persisted through the middle Devonian.

The early Devonian depositional phase was followed by yet another series of orogenic movements, the Tabberabberan Orogeny. This appears to have occurred late in the Middle Devonian. The Cowra and Hill End troughs were folded at this time, and the large granite masses extending along the axis of the Cowra Trough from Dubbo to Tumut were emplaced; so, too, were some of the granitic masses in the eastern part of the southern highlands

The Tabberabberan Orogeny produced a marked change in the sedimentation style of the Lachlan Fold Belt and marked the end of all flysch-like sedimentation and the extension of paralic sedimentation from the W right across the fold belt. The sediments are mainly quartz sandstones, often cross-bedded, red, green, and gray shales, and conglomerates. The fossils found are plant remains, fish plates, and brachiopods. There are three main areas of these late Devonian sediments. East of the Hill End Trough and extending down the present coastline, representing a more tectonically active regime than that to the W, is a succession of considerable thickness. The second area is W of the Molong High, extending W to the eastern margin of the main Ordovician metamorphic belt. The western belt, in which sedimentation continued into the early *Carboniferous,* lies W of the metamorphic belt that occupies the Darling Depression. In the second and third areas the folds are broad and nowhere are the beds known to be intruded by granite. Cross-bedding directions indicate that these two areas were linked, with much of the sediment coming from the southwest from outside the fold belt. In the eastern area the folding is a little more intense, and granitic intrusions are known (e.g., the Bathurst Granite). This sedimentation ended deposition in the Lachlan Geosyncline in New South Wales. The folding of the beds took place during the Kanimblan Orogeny, which occurred about the middle of the Carboniferous.

The Lachlan Fold Belt can be explained generally by a marginal sea-island arc-interarc basin model. Many points of detail of the regional geology are not sufficiently well known to allow a definitive interpretation to be arrived at.

Although there were marked increases in the tectonic stability of the region associated with the Benambran and Tabberabberan orogenies, it appears that no significant mountain building took place. The supply of quartz-rich detritus from the W and SW continued throughout, the degree of its eastward penetration being limited by the distribution of troughs and highs. After the Tabberabberan Orogeny, when acid and intermediate volcanism ceased, the quartz-rich sand spread right across the fold belt. This is the time when calcalkali volcanism was established in the Tamworth synclinorial zone and probably marks the location of a trench farther E than the one existing before the orogeny.

New England Fold Belt. The Tamworth synclinal zone in the western part of New England contains gently to moderately deformed sediments, arcuate in outcrop, extending down the Hunter Valley to the coast. It is bounded on the E by the Peel Thrust and the Great Serpentine Belt. The sequence ranges, apart from minor occurrences of Ordovician and Silurian sediments along the Peel Fault, from early Devonian to Permian with minor unconformities.

The oldest Devonian beds (the Tamworth Group) consist of limestones, shales, radiolarian rocks, graywackes, keratophyres, and spilites. The Baldwin Formation (late Devonian), which overlies this, consists of graywacke, conglomerate, and breccia. The detritus is dominantly andesitic. Then follows a sequence of shales and graywackes, overlain by cross-bedded sandstones, limestones, and shales (all containing shelly fossils). The succession, which extends well into the Carboniferous, thins toward the W, where unconformities have been recognized. Nonmarine sedimentation commences first in the lower part of New England and the Hunter Valley with lithic and feldspathic cross-bedded sandstones, conglomerates, acid and intermediate lavas, and pyroclastics containing some marine intercalations. Some of the detritus is of granitic origin. Nonmarine sedimentation became general at the time of commencement of the glacial phase of the Upper Carboniferous. The glacial sequence contains tillites, varve shales, fluvioglacial sediments, tuffs, basalts, and rhyolites. In the lower part of the Hunter Valley, in the vicinity of Lochinvar and Pokolbin, the glacial sequence rests on granites of unknown age. Early Permian sediments are found in fault-bounded blocks on the Peel Fault and in the southern part of the Tamworth synclinorial zone. These sediments are marine and probably derived from the Woolomin-Texas Block. Permian sediments are extensive to the S and W in the Sydney Basin.

The Kempsey Block is made up of Carboniferous to Permian beds similar to those of the Tamworth synclinorial zone except that marine conditions persisted throughout. The folding in the Kempsey Block is not as strong as in the central part of New England but nevertheless produced a right-angle unconformity between Permian beds and horizontal Triassic sandstones of the Lorne Basin at Camden Haven.

The oldest fossils recorded from central New England are Silurian or early Devonian corals from the northern part of the Demon Block. Their stratigraphic relationships are unfortunately unknown. In the western part of the Woolomin-Texas Block the oldest beds are the unfossiliferous Woolomin Formation, composed of basic volcanics and cherts with some graywackes. The graywacke succession which appears to overlie these beds is very lithic and thought to span the Devonian and part of the Carboniferous. A few Carboniferous fossil localities are known, most of them are in the N. Possible older Paleozoic graywackes and slates occur on the coast in the Brisbane Block.

In central New England most of the rocks are strongly deformed. At least one orogenic movement occurred in this region prior to the main folding in the Permian (Hunter-Bowen Orogeny). There is an unconformity between Permian sediments and the Woolomin Beds near Armidale and the Permian Ashford Coal Measures and early Carboniferous sediments near the Queensland border. This uplift formed the New England Arch, lying to the E of the Tamworth Trough, in which the sediments of the western part of the New England Arch were deposited. Much of the arch was apparently submerged during the early Permian since deposits of this age are widespread. Very thick deposits of graywackes, slates, and pebbly mudstones were deposited in the Nambucca Block in the early Permian. The trough in which they were deposited may have been developed by extensional rifting.

The early Permian sediments were involved in the main folding, and indeed E of Armidale have become involved in high-grade regional metamorphism. Thrusting was accompanied by the injection of serpentinite to the W and E of this central zone. This was followed by the intrusion of the enormous mass of the New England Batholith.

As in the Lachlan Fold Belt, the regional geology can be interpreted generally by an island arc model, but the amount of crustal thickening associated with the Mid-Permian Orogeny that destroyed the arc system appears to have been substantially greater than in the Lachlan Fold Belt. The many points of uncertainty in the regional geology make it impossible to arrive at a detailed geodynamic interpretation of the fold belt.

Sydney Basin. The Sydney Basin sequence overlies Ordovician, Silurian, and Devonian rocks unconformably on the southern and western margins. The Permian sediments pass northward into the Hunter Valley, which is transitional between the Sydney Basin and the New England Fold Belt. Much of the sediment in the basin has been derived from the fold belt. To the NW the Sydney Basin passes into the Oxley Basin, a southeastern attenuation of the Great Artesian Basin.

In the Hunter Valley, the Permian System commences with the Dalwood Group, a succession of marine polymictic conglomerates, lithic sandstones, and some shales, which is followed by the Greta Coal Measures. Above this is a second marine phase, represented by sediments of the Maitland Group, and finally there are two coal measure sequences, the Tomago and Newcastle. The Permian sequence can be traced along the western margin of the New England Fold Belt to beyond Gunnedah. In the northwestern part of the basin the Early Permian contains acid, intermediate, and basic volcanics.

Except for a local thin coal measure succession NW of Ulladulla, the basal outcropping beds of the basin are Permian marine beds. In the S there are equivalents of both the Dalwood and Maitland groups. Along the western margin, sandstones, frequently with glacial erratics, dominate and correlate with the Dalwood Group. On the south coast, the sediments are thicker and more fossiliferous and finer grained. Large faunas have been collected from localities such as Gerringong and Ulladulla. Overlying are the extensively mined fluvial and deltaic Permian Illawarra Coal Measures, which are equivalent to the Newcastle Coal Measures.

The Triassic comprises a fluvial, deltaic, and marginal marine complex. The Narrabeen Group at the base contains lithic sandstones, conglomerates, and gray and red shales. Along the coast there is a high proportion of lithic detritus, but inland the sandstones are quartzose and form the cliffs of the western part of the Blue Mountains. The overlying Hawkesbury Sandstone, outcropping around Sydney, is a glistening cross-bedded quartz sandstone. A shale lens in the sandstone at Brookvale has yielded a large fish fauna. The Wianamatta Group at the top of the sequence is shaly at the base, but lithic sandstones are common in the upper part.

Clarence-Moreton Basin. The fluvial and deltaic succession in the Clarence-Moreton Basin is unconformable on Permian and earlier rocks of the New England Fold Belt. The beds range from Triassic to (?)Cretaceous, attaining a maximum thickness in New South Wales of the order of 3000 m. Tectonic disturbances that took place during the history of the basin divided the sequence into three parts. The lowest division included the Nymboida Coal Measures, which, in addition to quartz-lithic sandstones, polymictic conglomerates, shales, and a little coal, contains some rhyolitic tuffs. The Bundamba Group (quartz sandstone, gray shale, and siltstone), the Marburg Formation (more shaly), and the Walloon Coal Measures (gray shale, thin coal seams, and lithic sandstones), a sequence of Upper Triassic to Jurassic nonmarine formations extending south from Queensland, overlie the Nymboida Coal Measures unconformably. Unconformities have been observed between the Walloon Coal Measures and the overlying Kangaroo Creek Sandstone, a glistening cross-bedded quartz sandstone. The highest unit, the Grafton Formation, consists of interbedded sandstones, siltstones, and shales.

Great Artesian Basin. This basin comprises fluvial to shallow marine sediments of Triassic to early Cretaceous age. The Triassic sediments are unconformable on the Permian in the subsurface connection between the Sydney Basin and the Bowen Basin (in Queensland) but pass into the Sydney Basin to the S. The Triassic sediments comprise sandstones, shales, and conglomerates. Early Triassic sediments are confined to the SE of the basin in New South Wales. Middle Triassic deposition was more widespread in the E, but the middle Triassic beds are disconformably overlain by early or middle Jurassic sediments. They and the Triassic beds are overlapped by the Jurassic sequence, which rests directly on the Paleozoic basement. The Garrawilla Lavas (basaltic) in the SE of the basin lie between the Triassic and Jurassic sequences. The Jurassic contains sandstones and shales with some conglomerate horizons. Sandy rocks are more prominent in the western half of the basin, where they show marked thickening between basement "highs." This Jurassic sequence, which has been correlated with the early to middle Jurassic Precipice Sandstone, Evergreen Shale, Hutton Sandstone, and Injune Creek Group of southern Queensland, is overlain by sandstones, shales, and sandy shales of the Blythesdale Group (late Jurassic to early Cretaceous). These sediments are dominantly deltaic to marginal marine.

Blue and gray marine shales of the Rolling Downs Group follow. The outcrop of these beds at White Cliffs and Lightening Ridge contains the precious opal for which these localities are famous. In the deeper parts of the basin along the Queensland border the marine sediments are overlain by more sandy deltaic beds of the Rolling Downs Group.

Murray Basin. Although the sequence in the basin consists of only about 300 m of Tertiary and Quaternary sediments, it is underlain in places by Permian and Cretaceous beds. On the western margin near Jerilderie, a sequence of about 300 m of Permian glacial sediments, marine sandy shales and coal measures, rests on Ordovician schists. A section in the western part of the basin consists of Eocene carbonaceous sands and sandy clays; marine Oligocene and Miocene limestones, clays, silts, and conglomerates; Pliocene carbonaceous sandy clay; and Pleistocene and Holocene sands. In the eastern part of the basin lacustrine sediments underlie Quaternary fluviatile sands and clays that are known to reach 100 m in thickness.

Post-Orogenic Igneous Activity. Throughout the eastern half of the state are widespread occurrences of flows and minor intrusives which range in age from Mesozoic to Tertiary. Based on scant data there appear to be maxima in the Eocene and Miocene. Most of the flows are alkali olivine basalts and tholeiites, but in the W, leucite basalts have been found at several localities. The largest known post-orogenic intrusive is the Cretaceous monzonite complex at Mt. Dromedary on the south coast. This predated the rifting off of the Lord Howe Rise (a microcontinental block) and the formation of the Tasman Sea (80 to 60 m yr). Minor intrusions of teschenite and syenite are common in the Sydney Basin. Some of these are Jurassic to Cretaceous, according to potassium-argon dates. Three Tertiary volcanic centers contain trachytic lavas. They are Mt. Canobolas (near Orange), the Warrumbungle Mts. (250 km N), and the Nandewar Ranges (160 km farther N). In the NE corner of the state, overlying the sediments of the Clarence-Moreton Basin, are Tertiary volcanics of rather different composition. The extrusive rocks include basalts and rhyolites. A ring complex at Mount Warning (Tweed Complex; Fig. 6), which contains granophyric rocks, represents the stump of an enormous central volcano whose lavas are found in the nearby

FIGURE 6. Mount Warning (1156 m), on the Tweed River, northern New South Wales, the trachyandesite plug that intrudes the eroded stump of a giant shield volcano of Miocene age. The core is an exposed magma chamber of syenite within a belt of gabbro, and parthy surrounded by a ring-dike of syenite (Ewart *et al.,* 1971). Its lavas and a basalt-rhyolite-basalt alternation are seen in the Lamington Plateau and McPherson Range in the background. The low-lying country surrounding this center, drained by the Tweed River, is a giant "erosion caldera," 30 km in diameter; the entire cone was over 100 km in diameter. The exposed floor of the erosion caldera is composed of Jurassic and Silurian rocks unrelated to the volcanism. (Oblique air photos: R. W. Fairbridge.)

McPherson Range and the Lamington Plateau, extending into Queensland over 100 km across.

Mineral Resources

The total value of mineral production for the fiscal year 1969—1970 was (Aust) $385 million. Although tremendous quantities of gold were won in the early days of mining in New South Wales, it is now relatively unimportant, largely a by-product of other mining activities. The gold found was very largely alluvial, concentrated, no doubt, during the long post-Paleozoic denudation of the fold belts, especially that of the southern and central highlands. The Paleozoic mineral deposits are very varied, but with two important exceptions the deposits are small. These are the Captain's Flat lead-zinc-copper deposit and the Cobar copper and gold deposits.

The Captain's Flat deposits occur in a sharp syncline of Silurian beds near a major shear. The mineralized horizons are shale beds within acid to intermediate volcanics and calcareous and tuffaceous shales below them. Over 4 million tons of ore (lead, zinc, copper, silver, gold) were produced prior to the closing of the mine in 1962.

At Cobar, almost vertically following stratigraphic horizons within the C.S.A., occur siltstone, the Great Cobar Slate, and, at the junction of the Great Cobar Slate and the Chesney Graywacke, all units of the Silurian (?) Cobar Group. Some 8 million tons of ore have been produced.

The Broken Hill ore deposit lies within the Precambrian Willyama Complex. Over 100 million tons of ore has been mined. The average composition is 13.2% lead, 7.3 oz/ton silver, and 9.1% zinc. The ore body plunges to the N and S and is known to extend for a distance of over 6 km. The main minerals are sphalerite, galena, pyrrhotite, marcasite, chalcopyrite, arsenopyrite, and loellingite in a gangue that contains garnet, rhodonite, calcite, and quartz. The lodes are separated by mica-sillimanite schist, gneiss, and metaquartzite, all complexly folded.

Beach sand deposits have yielded considerable quantities of rutile, zircon, and ilmenite. These deposits are common on the north coast, where they are often concentrated by wave action at the north ends of individual beaches. Alluvial tin deposits occur in New England in the Woolomin Texas Block and in central New South Wales in the Wagga anticlinorial zone.

Approximately half the total value of mineral production for the state is derived from coal (Aust $179 million). All of this is from the Permian coal measures, notably the Newcastle and Illawarra coal measures. The coal is bituminous, occurring in nearly horizontal seams that average 2—3 m. The coal is used both for coking and the generation of power.

GORDON PACKHAM

Appendix: A Model of the Paleozoic Tectonic History of New South Wales

(*Editorial Note:* Although plate tectonic analysis of the Australasian complex is still in its infancy and no universally agreed pattern is yet established, it is thought helpful at this stage to provide one model in summary. Many of its postulates remain to be tested in the field. R.W.F.)

The theoretical principles and basis (modified plate tectonics) on which the Paleozoic tectonic history of New South Wales was analyzed are dealt with elsewhere (Scheibner, 1972), but it is necessary to mention here a new model based on Elsasser's model of plate motion and retrograde motion of the sinking lithospheric slab at the Benioff Zone. Tensile stresses in the lithosphere at leading plate margin are created by retrograde motion of the Benioff Zone. Transmitted tensile stresses at the leading plate margin will trigger off the mechanism of crustal separation and the coupled process of crustal generation, i.e., either axial or nonaxial sea-floor spreading will start in fields of tensile stresses. This is an alternative hypothesis for the origin of marginal seas, differing from the diapiric pull-apart hypothesis of Karig.

On the leading edge of the advancing plate in the trench (upper and lower slope), oceanic sediments are scraped off the sinking plate, and continental rise or frontal-arc-basin sediments accumulate to form a flysch wedge. The flysch wedge is characterized by strong repeated deformation and the presence of basic volcanics and ultrabasics. The oceanward rotation of a continental plate accounts for most structural deformations in the marginal mobile zone.

The model of orogenic granite derivation as proposed by Hamilton accounts for the orogenic granites; however, one subduction zone cannot account for all the Paleozoic granites in New South Wales. Some granites and acid volcanics have been derived during subduction of oceanic crust of marginal seas, quite far from the major trench situated at the major plate margin.

Serpentinite belts in New South Wales mostly represent obduction zones rather than subduction zones. The obduction zones were formed during quick advancement of leading edges of a continental plate, the neighboring oceanic crust and upper mantle being encountered and upthrust.

In early Paleozoic time the studied area was postulated to be a continental margin of the Proterozoic Australian-Antarctic Continent which had been in contact with the hypothetical Paleo-Pacific Ocean. In the early Cambrian, decoupling of the Paleo-Pacific oceanic crust from the neighboring continental crust occurred. A system of marginal seas, island arcs, and trenches originated on the present eastern margin of the Australian plate.

In the early Cambrian in New South Wales, at the margin of the Australian plate, the following tectonic units can be recognized: Kanmantoo Marginal Sea and its continuation, the Bancannia Trough; the Mount Wright Volcanic Arc; the Gnalta Shelf, separated by the Koonenberry Fracture Zone from the White Cliffs Deeper Terrace and both underlain by the Proterozoic Wonominta Complex and the Girilambone Flysch Wedge (trench). These tectonic units were parts of the Kanmantoo-Tyennan Pre-Cratonic Province, which was sufficiently cratonized during the Delamerian Orogeny for the Gnalta Transitional Province to be formed above it.

The Kanmantoo-Tyennan Pre-Cratonic Province is represented in Victoria by the Heathcote Zone and in Tasmania by the Mount Read Volcanic Arc and Dundas Trough, and a similar situation can be observed on New Zealand (Cambrian volcanic arc), helping to position the old core of New Zealand south of Tasmania for the Cambrian. A similar situation possibly also existed in Queensland (see work of Heidecker).

During the Lower-Middle Ordovician a new strong extension occurred at the margin of the Australian plate. A substantial part of the Girilambone Flysch Wedge was separated from the main plate and the Wagga (Ballast Chert) Marginal Sea was formed. On the leading edge of the separated Girilambone Complex and on the oceanic basement, the Molong Volcanic Rise (volcanic arc) developed. Eastward the new flysch wedge accumulated (Monaro Basin). Oceanward rotation of the Australian plate caused the Benambran Orogeny. After this orogeny, a new location was established in the lowermost Silurian. This again caused a strong extension during which three marginal basins were created, from west to east the Cobar Trough, the Cowra Trough, and the Hill End Trough and its southern continuation, the Captains Flat Trough. The central and eastern troughs originated by splitting of the Molong Volcanic Rise into three segments. The western segment was left attached to the Parkes Shelf, the central segment formed the Molong-Canberra Shelf, and the eastern segment contributed to the formation of the frontal arc area.

During a later part of the Lower Silurian, the Quidong Orogeny occurred. During this orogeny high tectonic granites were emplaced, and the Capertee Rise formed. Also, acid volcanism started, and orogenic plutonism lasted throughout the Silurian and Lower-Middle Devonian. Sedimentation in the three marginal basins continued.

During the Middle and Upper Silurian a new location of the trench was established. A new inter-arc basin developed, the Murruin Basin east of the Capertee Rise. A frontal volcanic arc, which developed at the end of the Silurian, is concealed at present, but it supplied volcanic detritus into the Tamworth Frontal Arc area. Within this area a deeper frontal arc trough developed, the Tamworth Trough, characterized by deep-water sedimentation and spilite volcanism. On the upper and lower slope the Woolomin Beds accumulated to form the new flysch wedge.

During the Upper Silurian and Lower Devonian, the Australian plate rotated oceanward. First the Cobar Trough was closed and then the Cowra Trough. The oceanic crust was subducted, but in the terminal phases also obduction zones developed, the Coolac and Tumut Pond serpentinite belts. Shelf sedimentation spread over the area of the deformed troughs, while the Hill End Trough was further filled by flysch of terrestrial and volcanic provenance. The volcanic island arc east of the Murruin Basin shed quantities of volcanic debris into the Tamworth Unstable Shelf (frontal arc basin) area. At the leading plate margin the Woolomin Flysch Wedge was further accumulated. The main subduction zone accounted for the volcanic frontal arc, but the acid and intermediate volcanics and extensive plutonics were derived by subduction of the oceanic crust of marginal seas and by partial fusion of accreted crust.

During the Middle Devonian Tabberabberan Orogeny extensive areas of former marginal seas and rises, forming the Lachlan Pre-Cratonic Province, were structurally deformed, metamorphosed, intruded by plutons, and so cratonized. The Tamworth Frontal arc area and the Woolomin Flysch Wedge were also cratonized, to a lesser extent, however. Above the Lachlan Pre-Cratonic Province the Lambie Transitional Province was established. The area that continued in its pre-cratonic development might be called New England Pre-Cratonic Province, and here two tectonic units can be recognized: the Mandowa Shelf and the Texas Unstable Shelf (frontal arc area). The Lambie Transitional

Province was composed of the Ravendale and Hervey Terrestrial Basins on the west and the Lambie Intramontane Depression on the east, with the short-lived Lambie Shelf during Frasnian time.

During the Kanimblan Orogeny (Upper Visean) the Lachlan Fold Belt (craton) was formed. Quick movement of the Australian plate caused the oceanic crust and mantle of the Woolomin-Texas complex to be encountered and an obduction zone (Peel Thrust) formed. In the marginal part of the Lachlan Fold Belt, the discordant Bathurst granites were emplaced during the Namurian. Subsequently, subduction terminated at the Woolomin-Texas trench. The cratonic Woolomin-Texas complex emerged to form a land mass (New England Arch), while to the west and south the Kullatine Shelf was formed, and farther west the terrestrial-paralic Ayr Basin.

Retrograde motion of the subduction zone started to cause extension in the future Sydney-Bowen Basin in the Upper Carboniferous. While an island arc probably developed in Queensland, a volcanic rift existed in New South Wales.

In Lower Permian time new extension was expressed in formation of the Sydney-Bowen Basin, and inland epicratonic basins. Crustal separation had cut across the older structures and the Nambucca Basin was in contact with two zones: the Drake Shelf, with the Woolomin-Texas complex, and the Yessabah Shelf, with the Tamworth Group as a basement. It is indicated that the Nambucca Beds with some ophiolites were laid down on oceanic crust, possibly in a marginal sea and repeatedly deformed.

During the Hunter Orogeny (mainly Upper Artinskian) the Carboniferous Peel Thrust (obduction zone) was a strike-slip displaced in a clockwise sense. The strike-slip movement amounted to about 150 km. The northern section of the Woolomin-Texas Block was displaced along a series of meridionally trending strike-slip faults. Along one of these faults, the Demon Fault, the Demon Block moved in a clockwise sense southward, squashing the Nambucca Beds against another rigid block, the Kempsey Block. South of the Demon Block a minor obduction zone was formed. An enormous amount of granite was derived from the subduction zone and by fusion of crustal material. A few granites were emplaced during the structural deformation, while most of New England granites are post-kinematic and post-strike-slip movement. The point of "collision" migrated northward, the orogenic granites being derived in Permian time in New South Wales but in Triassic time and later in Queensland. Similarly, structural deformation migrated northward with time.

A transitional province was established in the Upper Permian in New South Wales and definitive cratonization occurred during the Bowen Orogeny.

ERWIN SCHEIBNER

References

Conolly, J. R., and Ferm, J. C., 1971. "Permo-Triassic sedimentation patterns, Sydney Basin, Australia," *Bull. Am. Assoc. Petrol. Geol.*, 55(11), 2018–2032.

Crook, K. A. W., Bein, J., Hughes, R. J., and Scott, P. A., 1973. "Ordovician and Silurian history of the southeastern part of the Lachlin Geosyncline," *J. Geol. Soc. Austral.*, 20(2), 113–144.

David, T. W. E., and Browne, W. R., 1950. *Geology of the Commonwealth of Australia*, 3 vols., London: Edward Arnold.

Ewart, A., Paterson, H. L., Smart, P. G., and Stevens, N. C., 1971. "Binna Burra, Mount Warning," in: *Geol. Excursions Handbook*, Victoria: 43rd Congr. Geol. Soc. Austral., 63–78.

Faniran, A., 1970. "The Sydney duricrusts; their terminology and nomenclature," *Earth Sci. J.*, (Waikato Geol. Soc.), 4(2), 117–128.

Flinter, B. H., Hesp, W. R., and Rigby, D., 1972. "Selected geochemical, mineralogical and petrological features of granitoids of the New England Complex, Australia, and their relation to Sn, W, Mo, and Cu mineralization," *Econ. Geol.*, 67, 1241–1262.

Goldbery, R., and Holland, W. N., 1973. "Stratigraphy and sedimentation of redbed facies in Narrabeen Group of Sydney Basin, Australia," *Bull. Am. Assoc. Petrol. Geol.*, 57(7), 1314–1334.

Hind, M. C., and Helby, R. J., 1969. "The Great Artesian Basin in New South Wales," *J. Geol. Soc. Austral.*, 16(1), 481–497.

Hockley, J. J., 1973. "Differentiation trends in the Warrumbungle Volcano, New South Wales, Australia," *Geol. Rundschau*, 62(1), 179–187.

Jennings, J. N., 1972. "The age of Canberra landforms," *J. Geol. Soc. Austral.*, 19(3), 371–378.

McAndrew, J., ed., 1965. "Geology of Australian ore deposits," *Publ. 8th Commonwealth Mining Met. Congr.*, 1.

Packham, G. H., ed., 1969. "The geology of New South Wales," *J. Geol. Soc. Austral.*, 16(1), 656p.

____, and Leitch, E. C., 1974. "The role of plate tectonics in the interpretation of the Tasman Mobile Zone," *Geol. Soc. Austral. Spec. Publ. 5.*

Pels, A., 1969. "The Murray Basin," *J. Geol. Soc. Austral.*, 16(1), 499–511.

Scheibner, E., 1972. "Structural map of New South Wales," *Records Geol. Surv. N.S.W.*, 42.

____, 1973. "A plate tectonic model of the Paleozoic tectonic history of New South Wales," *J. Geol. Soc. Austral.*, 20(4), 405–426.

Standard, J. C., 1969. "The Sydney Basin, Triassic system Hawkesbury Sandstone," *J. Geol. Soc. Austral.*, 16(1), 407–417.

Stanton, R. L., 1972. "A preliminary account of chemical relationships between sulfide lode and banded iron formation at Broken Hill, New South Wales," *Econ. Geol.*, **67**, 1128–1145.

Strusz, D. L., 1971. *Canberra, Australian Capital Territory and New South Wales* (1:250,000 Geol. Ser. Sheet, SI/55-16). Canberra: Austral. Bur. Mineral Resources Geol. Geophys., 47p.

Voisey, A. H., 1969. "The New England region, stratigraphy, lower Paleozoic systems," *J. Geol. Soc. Austral.*, **16**(1), 227–231.

Ward, C. R., 1972. "Sedimentation in the Narrabeen Group, Southern Sydney Basin, New South Wales," *J. Geol. Soc. Austral.*, **19**(3), 393–409.

Warner, R. F., 1970. "The early Tertiary landscape in southern New England, New South Wales, a reappraisal," *Austral. Geogr.*, **11**(3), 242–258.

Webby, B. D., 1972. "Devonian geological history of the Lachlan Geosyncline," *J. Geol. Soc. Austral.*, **19**(1), 99–123.

Young, R. W., 1970. "A probable post-uplift age for the duricrust on the south coast of New South Wales," *Search*, **1**(4), 163–164.

Maps: Geological Map of N.S.W., 1962 N.S.W. Geol. Surv. Tectonic Map of Australia, Geol. Soc. Austral. 1:250,000 Geological Map Series, N.S.W. Geol. Surv.

Cross-references: *Australasia–Regional Review; Australia–Queensland, South Australia, Victoria.*

AUSTRALIA–NORTHERN TERRITORY

The Northern Territory occupies that part of Australia N of 26°S and between 129 and 138° E. It is bounded by the Timor and Arafura Seas and the Gulf of Carpentaria to the N, and by the adjacent states of Queensland, South Australia, and Western Australia. The area is 1,347,520 km^2, measuring about 1700 km N-S and about 1000 km E-W. The low flat coastline, mainly with sandy beaches and mud flats thickly fringed with mangroves, is about 1700 km long. It is broken here and there by headlands of sandstone, marl, laterite, and granite that form cliffs up to 30 m and is indented by inlets and rivers, few of which are navigable. The largest of the many islands are the Bathurst and Melville islands in the NW and Groote Eylandt in the Gulf of Carpentaria. Port Darwin and Melville Bay, Gove Peninsula, are the only sheltered deep-water bays developed for shipping.

From the coast the land surface gradually rises to a height of about 300 m by 18°S. Here the higher lands form the divide with the streams that flow—ephemerally—to the endorheic interior. Over a wide area toward the S the land elevation is about 600 m; there are several mountain ranges, which trend generally E-W. Mount Zeil (1510 m) in the MacDonnell Ranges is the highest peak. In the SW there are a number of usually dry salt pans, the largest of which is Lake Amadeus. Large areas of the interior are semidesert, particularly in the west. Part of the dune-covered Simpson Desert overlaps the southeastern portion of the Territory. Extensive regions of good cattle pasture land are in the N and on the Barkly Tableland in the NE.

There are two main seasons, a wet one from November to April and a dry one from May to October. During the wet season, when the northwest monsoon affects the northern part of the Territory, there are occasionally tropical thunderstorms in the interior. The vegetation in the higher-rainfall regions is similar to that of Indonesia, but there is no rain forest. Cypress, ironwood, bloodwood, and paperback trees are cut and milled for local use on the coastal areas. The Barkly Tableland and the Victoria River district possesses grasslands and shrub savanna. The most common grass is annual sorghum, which may grow to 3 m high during the wet season. Most of the sandy plain of the interior and the ranges in the Alice Springs area are covered by spinifex grass.

Darwin, the capital of the Northern Territory, is the administrative center and the principal sea port and international airport for traffic from Europe and Asia. It was settled in 1869. Other principal centers are Alice Springs, Tennant Creek, and Katherine. A 1530-km-long bitumen-sealed road, the Stuart Highway, links Darwin to Alice Springs. The Barkly Highway, 648 km long, connects Mt. Isa in western Queensland to Tennant Creek. Railway lines connect Darwin to Birdum (510 km) and Alice Springs to Marree in South Australia (800 km).

Malay trepang-fishers used to visit the coast in the past, but the first European contact was in 1623 by the Dutch ship "Arnhem." Tasman surveyed the coast in 1644, and the first inland explorers were Leichhardt (1845), Gregory (1855–1856), and Stuart (1860–1962). In 1872 the overland telegraph line was completed from Adelaide to Darwin. The western country was explored by Giles (1872–1874) and by Forrest (1879). The Territory was first administered by New South Wales (1860–1863), then by South Australia (1863–1910), and subsequently by the Commonwealth Government.

History of Geological Investigations

The first geologist to visit central Australia was Chewings in 1886, followed in 1889 by East, and Brown in 1889. Brown and Thornton found early Paleozoic fossils in the Amadeus Basin in 1890. Other geological pioneers included Tate (1882), Tenison Woods (1886), and Carnegie (1897).

In 1901 the South Australian Government sent prospecting expeditions into the southwestern corner of the Territory, followed by Basedow in 1903 and in 1905 and 1908 by George. In 1912, after the Commonwealth had taken control, Woolnough reconnoitered the Katherine-Darwin region and a small geological survey was set up. Many investigations have been carried out since by Commonwealth and company geologists and geophysicists. In 1935 the Commonwealth, Queensland, and Western Australian governments jointly initiated the Aerial, Geological and Geophysical Survey of Northern Australia, a work interrupted by World War II.

Noakes of the Bureau of Mineral Resources of Australia (B.M.R.) compiled the first geological map of the Katherine-Darwin region in 1949, and since then BMR field parties have carried out reconnaissance surveys, regional mappings, airborne and gound magnetometer surveys, and regional gravity surveys of the whole Territory.

The first recorded find of gold was by Litchfield on the Blackmore River a few kilometers south of Darwin in 1865. Uranium was first discovered at Rum Jungle in 1949, and additional deposits were discovered by B.M.R. geologists near Katherine and at the South Alligator River.

Geomorphology

The Northern Territory can be divided into the following seven physiographic zones from north to south:

The Coastal Plain. The coastal plain is almost 150 km wide SE of Darwin, and in the Joseph Bonaparte Gulf and the Gulf of Carpentaria it reaches 80 km; it narrows in the NE, and in places the coast is marked by low cliffs. Inland elevations range up to 80 m. Relief is low, with scattered low hills. Streams meander across the plain and generally merge into mangrove swamps.

Central Plateau. Extensive plateaus and plains cover more than two-thirds of the Terri-

FIGURE 1. Satellite photograph of part of the northern margin of the Amadeus Basin, showing steeply dipping Proterozoic and Paleozoic formations in the MacDonnell Ranges (trending E-W). The ranges in the SW are cut by the Finke River. The canoe-shaped range in the bottom right is the Waterhouse Range Anticline. The width covered in the photograph is about 220 km. (By courtesy of the U.S. National Aeronautical and Space Administration.)

FIGURE 2. Gently dipping beds of Mount Currie Conglomerate at Mt. Olga. Ayers Rock in the background. (By courtesy of Australian News and Information Bureau.)

tory, forming gently undulating surfaces ranging from 180 to 700 m in the center. North of Alice Springs drainage is mainly endorheic.

Dissected Margin. The central plateau is bounded to the N by a dissected margin 150 to 300 km wide. This belt includes the Victoria River and the Daly River begins. On the plateau margins local relief is up to 180 m, with narrow, steep-sided structurally controlled valleys.

Central Ranges and Uplands. The subdued "ranges" in the center and in the west rise above the central plateau with heights of only 15 to 150 m, though in places bounded by steep scarps. Beveled plateaus extend E-W north of Alice Springs, with relief not exceeding 150 m.

Southern Ranges. Located in the southwestern corner of the Territory, and near Alice Springs these ranges form an almost unbroken E-W belt of parallel-strike ridges 400 km long and 100 km wide (Fig. 1). Relief is commonly up to 250 m, and the highest mountain in central Australia is Mt. Zeil (1510 m).

Southern Deserts. The landscape for the remaining southern portion of the Territory is more than three-fourths covered by dunes and includes part of the Simpson Desert, with its SSE-NNW sand dunes. In the W there are isolated plateaus and ranges (Fig. 2), with salt pans and limestone plains farther W, sand plains in the N, and a fringe of granite hills and plains of the Mann and Musgrove ranges in the S. The Finke River is the only stream (ephemeral) that maintains its drainage beyond the area (reaching the closed basin of Lake Eyre at -16m below mean sea level); all the others disperse their water within the area.

Land Surfaces

Hays (1967) has distinguished four mature erosion surfaces (Figs. 3 and 4).

Early Cretaceous Erosion Surface. This is the oldest and highest surface, occurring within the central plateau, and ranging in height from 250 to 550 m. Residuals are irregularly distributed. The surface is best developed in the ranges around Tennant Creek. It is absent from large areas of the plateau but is represented in the summits of scattered ranges in the SW and in the extreme W. The general original relief of the surface was less than 60 m.

Late Cretaceous to Middle Tertiary Surface. This surface is most extensively developed on the central plateau. Exposures are

FIGURE 3. Land surfaces of the Northern Territory.

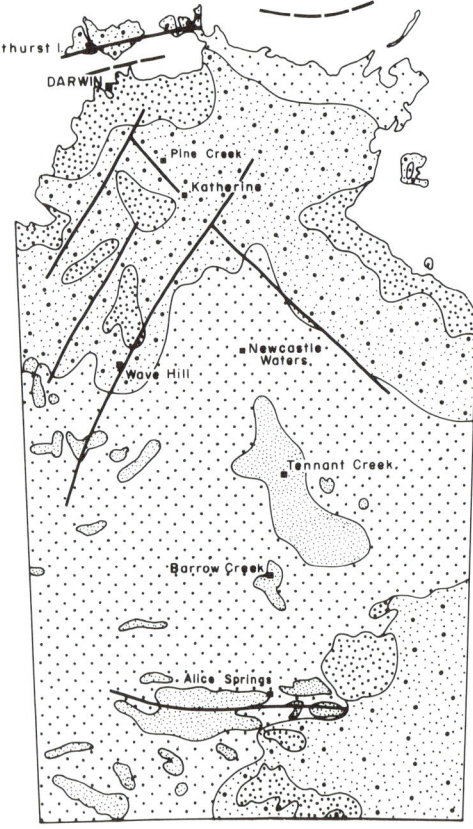

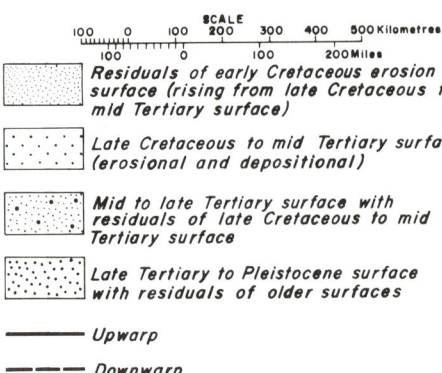

abundant in areas of active erosion, such as the central ranges and on the northern rim. The original relief was less than 15 m over long distances, today ranging from 350 m down to 180 m and below.

Miocene-Pliocene Surface. This is developed in the dissected margin, where it encroaches upon the late Cretaceous to mid-Tertiary surface. Local relief is restricted to resistant structures. Around Wave Hill it stands at 180 m. To the N the surface has been dissected by headwaters of the Victoria and Daly rivers, in the NW sloping gently toward the Joseph Bonaparte Gulf. To the NE the surface reaches 300 m.

Pliocene-Pleistocene Surface. This multicyclic surface with lateritization still active is developing at present in the coastal areas, in a few inland basins on rocks of all ages from Lower Proterozoic upward.

Warping. Mild deformation and broad warping have affected the Northern Territory since Cretaceous sedimentation began. In Fig. 3, two major upwarp axes, which parallel the boundary of the late Cretaceous to middle Tertiary surface, mark tectonic features that have been active since the Lower Proterozoic and have controlled the north Australian coastline. Axes of warping, to the W of the two main converging axes, coincide with axes of gentle folds. The Van Diemen Gulf is the result of ENE-WSW downwarping, and a closed depression is found in Joseph Bonaparte Gulf. The dissected margin NE of Katherine is due to uplift which postdated the late Cretaceous to Middle Tertiary surface. In the south the Miocene-Pliocene surface is tilted due to late Tertiary subsidence of the Lake Eyre basin in South Australia.

Stratigraphy

The major tectonic domains in the Northern Territory are shown in Fig. 5.

Archean. Archean rocks older than 2300 m yr outcrop here and there in the NW and N as inliers in Proterozoic rocks. Rock types include migmatite, gneiss, and altered basic volcanics. Rum Jungle, an eroded dome of low-grade metasediments, is one of a few places in Australia where the contact of the Archean with Lower Proterozoic is exposed. The metasediments were domed around the granite basement (granite: 2450 m yr).

The Arnhem Complex (Archean-Lower Proterozoic) is a basement "high" of granite,

FIGURE 5. Upwarped Mesozoic erosion surface, seen in a low oblique air photo, extensively dissected during the Cenozoic, in the MacDonnell Ranges (Precambrian to Ordovician sediments). The summit plane is here about 800 m.

FIGURE 4. Schematic section showing the Northern Territory land surfaces and their residuals.

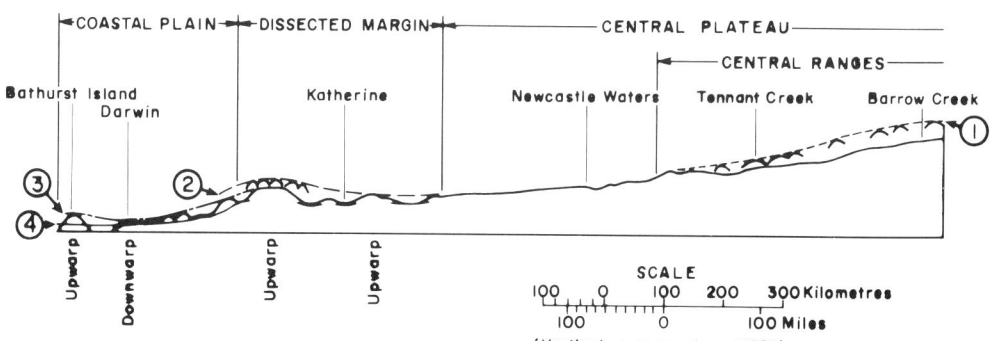

① Residuals of early Cretaceous erosion surface rising from late Cretaceous to mid Tertiary surface

② Late Cretaceous to mid Tertiary erosional and depositional surface

③ Mid to late Tertiary surface with residuals of late Cretaceous to mid Tertiary surface

④ Late Tertiary to Pleistocene surface with residuals of older surfaces

gneisses, and granulite exposed NE of the McArthur Basin.

The Litchfield Complex (Archean or Lower Proterozoic) is an elongated belt W of the Pine Creek Geosyncline, formed of schist, granulite and amphibolite intruded by granite, migmatite and dolerite.

Proterozoic. *Lower Proterozoic* (2400–1800 m yr). Lower Proterozoic rocks crop out in three main areas: between Katherine and Darwin, in the Granites-Tanami Complex, and in the Tennant Creek area. In the area between Katherine and Darwin Lower Proterozoic sediments laid down in the Pine Creek Geosyncline, a composite, fairly shallow intracratonic trough that developed in stages over the Rum Jungle Complex (Walpole et al., 1968), include arkose, quartz-graywacke, siltstone, chert, and dolomite 3000 m thick. Later, secondary troughs developed and graywacke 2800 m thick accumulated in the western trough, together with about 6000 m of dolomite, chert, and carbonaceous rocks overlain by siltstone and graywacke siltstone in the eastern trough. Finally, 1300 m of sandstone was deposited on a newly developed platform extending SW into the Victoria River area. The rocks of the Granites-Tanami Complex (Archean-Lower Proterozoic) comprise schists, phyllite, chert, dolomite, shale, and quartzite with some acid and basic volcanics and graywacke, locally intruded by granites. In the Tennant Creek area the Warramanga "Geosyncline" deposits consist of 1500 m of a graywacke-shale assemblage intruded by massive, foliated granite and adamellite complexes, granite porphyry, and overlain by ignimbrites.

In the Davenport Range, S of Tennant Creek, 7500 m of quartz sandstone, argillaceous toward the base of the sequence with porphyritic rhyolite, andesite, and basalt flows, has been laid down. In the Ashburton Range, N of Tennant Creek, a sequence more than 3300 m thick is very similar and coeval to the sequence of the Davenport Range. Those rocks were probably formed in secondary basins (Davenport "geosyncline") marginal to the Warramanga "Geosyncline."

The Arunta Complex consists of metamorphic rocks deformed before and after the deposition of the sediments in the Amadeus Basin. Rock types present are granulite, schist, gneiss, amphibolite, calcsilicate rocks, and quartzite, with acid and basic igneous intrusives (granulites: 2250 m yr metamorphism; granites: 1800–1300 m yr).

Carpentarian and Adelaidean. The Carpentarian System has a base dated at about 1800 m yr, and the base of the Adelaidean is taken at about 1400 m yr (Dunn et al., 1966). Rocks of the Carpentarian System crop out over wide areas, are little folded, and rest on steeply dipping Lower Proterozoic beds. They consist mainly of sandstone, shale, dolomite, and limestone with volcanics prominent toward the base and are well developed near the Gulf of Carpentaria. More than 8500 m of arenite and carbonate sedimentary rocks occurs in the McArthur Basin, the type sequence.

In the South Nicholson Basin over 6000 m of mudstone, silt, and glauconitic sand accumulated, equivalent to the upper part of the McArthur Basin sequence.

In the Musgrave Complex the oldest rocks are granulites and granites overlain unconformably by a sequence of acid and basic volcanics and quartzite. All these rocks are intruded by mafic and ultramafic bodies, extending easterly for 320 km. The rock types include dunite, norite, olivine norite, pyroxenite, and occasional picrite, troctolite, and gabbro (granulites: 1600 and 1300 m yr old).

Adelaidean sediments were laid down in the Arafura and Victoria river basins and possibly over the Granites-Tanami Complex and consti-

FIGURE 6. High oblique air photograph of superimposed drainage on the heavily folded and peneplaned MacDonnell Ranges (Mesozoic planation surface).

tute the basal formations of the Ngalia, Georgina, and Amadeus basins. In the Arafura Basin more than 1350 m of Adelaidean terrigenous stable shelf sediments lies unconformably on older units of the McArthur Basin. They consist of quartz sandstone and shale.

An unnamed basin W of the Arunta Complex contains over 5000 m of sedimentary rocks, which include sandstones, conglomerate, shale, siltstone, and silicified stromatolitic beds with chert. It is unconformable on a lower Carpentarian granite and Archean-Lower Proterozoic basement (Granites-Tanami Complex) and affected by irregular deformation and overlain unconformably by the flat-lying Cambrian of the Wiso Basin.

In the Ngalia Basin, an E-W intracratonic depression within the Arunta Complex, much of the Proterozoic is concealed by Cenozoic deposits. Some 6500 m of thickness has been indicated by aeromagnetic and seismic surveys (Cooper et al., 1971). The basal formations consist of quartzite, siltstone, and fine sandstone in its lower part, overlain by sandstone, siltstone, tillite, and dolomite. The Georgina Basin, NE of the Arunta Complex, also contains Adelaidean sediments. A glacigene unit at the base consists of siltstone with some striated boulders up to 1 m in diameter, quartz sandstone, and dolomite. A sequence of shale, quartz sandstone, siltstone, and dolomite 1500 m thick overlies the glacial unit (Smith, 1972). The sequence contains Cambrian fossils near the top. The Amadeus Basin, bounded on the N by the Arunta Complex and on the S by the Musgrave Complex, contains about 2500 m of quartzite and carbonate overlain by sandstone, siltstone, shale, limestone, dolomite, and glacigene sediments equivalent to those in the Georgina Basin (Wells et al., 1970).

Paleozoic. Basic volcanics, up to 1000 m thick, flooded across much of the northwestern part of the Northern Territory soon after the beginning of the Cambrian. These volcanics are overlain by Middle Cambrian marine limestone, shale, and sandstone of the Daly River Basin and Middle Cambrian interbedded shales, thin-bedded limestone, and Devonian sandstone of the Ord River Basin. Breaks in sedimentation from Silurian to Middle Devonian were caused by emergence and erosion without deformation, and during that time the area was mostly land with limited relief.

In the Bonaparte Gulf Basin during the Devonian and Carboniferous a carbonate and arenite sequence was laid down on the basic volcanics, followed by a paralic sequence in the Permo-Triassic. In the Georgina Basin Middle Cambrian carbonate rocks are widespread but Upper Cambrian, Ordovician, and Devonian freshwater sequences are restricted to the southern part of the basin. The northwestern margin is obscured by Mesozoic sediments. Cambrian and Middle Ordovician are mainly carbonate rocks, the Middle Ordovician sequence sandstone and siltstone, and Devonian formation sandstone. In the Wiso Basin the oldest rocks are dolomite, limestone, and dolo-

FIGURE 7. Giant inselberg, Ayers Rock, near Alice Springs, consisting of silicified arkosic Ordovician sandstone, an erosional relic of a Mesozoic land surface. (Photo: C. R. Twidale.)

mitic siltstone of lower middle Cambrian age. In a trough about 3000 m thick in the southern half of the basin, Lower to Middle Ordovician dolomite and sandstone are overlain unconformably by Devonian sandstone.

In the Amadeus Basin during the Cambrian, thick conglomerate and arkose were deposited in the SW and sand, shale, and some marine sediments to the NE. The sandstone was succeeded by 2000 m of carbonate rocks, lutite, and evaporite deposits, which were followed by penesaline and marine stromatolitic carbonate rocks. Ordovician sandstone, shale, and minor carbonate rocks up to 2500 m overlie the basal formations conformably in the N and unconformably in the S. Broad vertical movements started in the Silurian. Devonian sandstone up to 900 m thick, partly shallow marine and partly fluvial and eolian, crops out extensively. A major upheaval was followed by Late Devonian molasse facies over 3000 m thick with fluvial and lacustrine sedimentation which persisted into the early Carboniferous. These deposits, as a result of a major orogeny in the Carboniferous, are unconformably overlain by Permian glacials (Wells et al., 1970). In the Ngalia Basin Cambrian dolomite, siltstone, and minor sandstone are over 300 m thick, Ordovician sandstone about 1000 m thick, and Carboniferous sandstone 2000 m thick.

Mesozoic. Mesozoic rocks are widespread over most of the northern part of the Territory and in the SE on the fringe of the Eromanga Basin, the sediments of which overlap the late Proterozoic-Paleozoic succession in the Amadeus Basin. In the N they crop out mainly as isolated residual mesas or as broad tablelands. They are Lower Cretaceous in age, with a maximum thickness of about 60 m, and have not been folded, although in a few places they are cut by minor faults. Both nonmarine and marine facies are present and consist of conglomerate, sandstone, siltstone, and claystone but are strongly lateritized. In the SE the Eromanga Basin sequence includes up to 1500 m of flat-lying to gently inclined Lower Jurassic to Upper Cretaceous fluvial and shallow marine sedimentary rocks, in places represented only by highly weathered residuals separated by large tracts of fluvial and eolian sand cover.

Cenozoic. The emergence of the Cretaceous sedimentary rocks in late Mesozoic or early Tertiary time was accompanied by little differential warping. All earlier rocks became deeply weathered, laterite and soil profiles developed, and alluvial detritus accumulated along the stream courses, and in coastal areas. Numerous planation surfaces and inselbergs remained (Figs. 5, 6, 7, 8). In addition, sand, colluvial rubble, and in limited areas marine sediments have been deposited. In some places there are large salt lake basins in which a few hundred meters of Mesozoic and Tertiary continental sediments occur.

Structure

The Tectonic Map of Australia and New Guinea (Geological Society of Australia, 1971) depicts relationships among orogenic, transitional, and platform units which suggest that

FIGURE 8. Mt. Connor, a large mesa capped by subhorizontal Precambrian quartzite, an erosional relic left during Cenozoic erosion. (Photo: C. R. Twidale.)

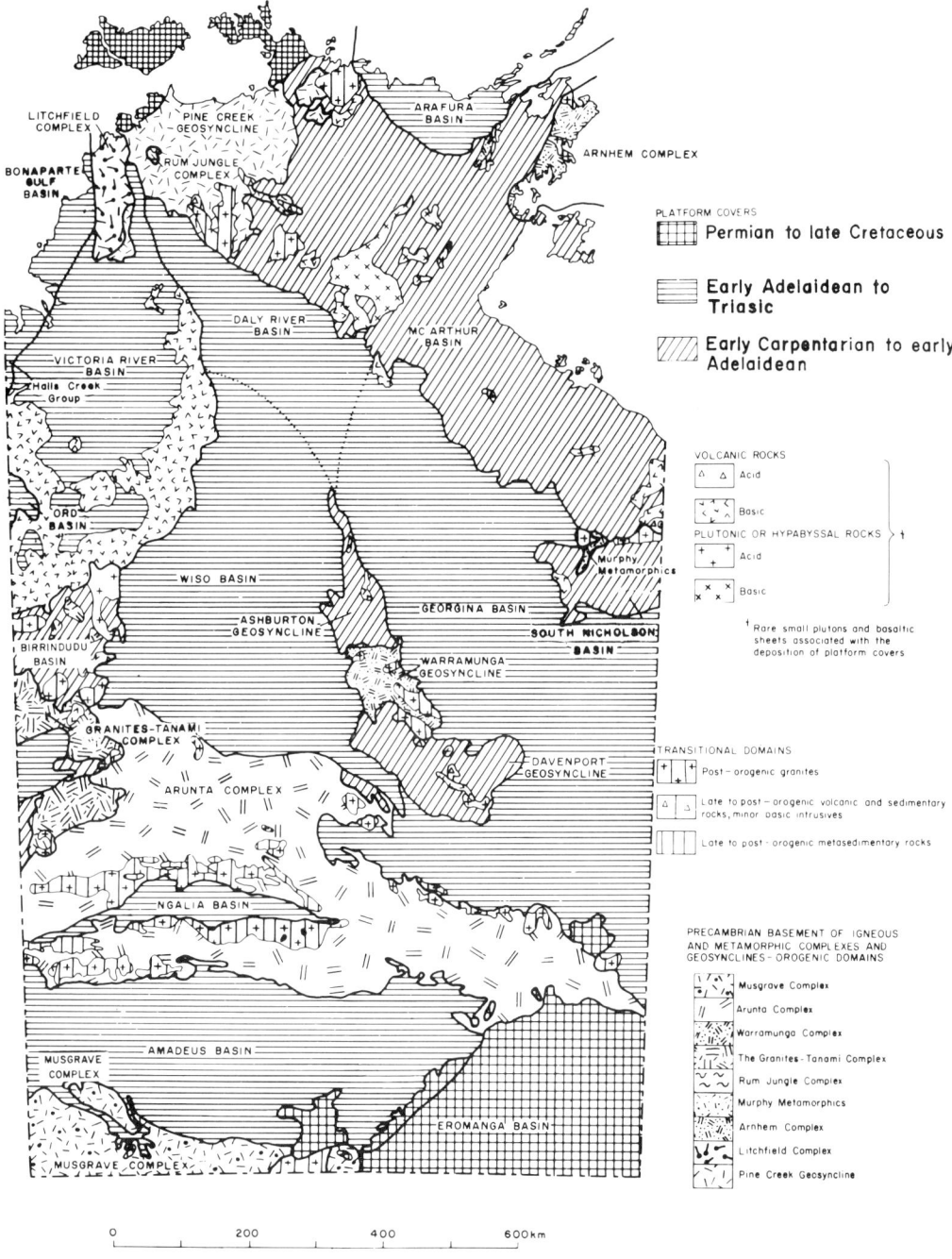

FIGURE 9. Structural map of the Northern Territory which shows outcrops of Precambrian basement complexes, transitional (Precambrian) domains, volcanics, and platform covers (sediments).

the Australian continent developed through a series of cycles, each of which culminated in cratonization.

The orogenic domains, with geosynclines forming belts adjoining their separate cratons, occur in the Northern Territory in the Rum Jungle, Granites-Tanami, Arnhem, Litchfield, Arunta, and Musgrave "complexes and the Pine Creek and Warramunga geosynclines."

The transitional domains are marked by late to post-orogenic deformation, abundant plutonic and volcanic rocks, and rare metamorphism. Such rocks unconformably overlie the immediately preceding orogenic domain and are associated with the Granites-Tanami Complex, the Pine Creek and Warramunga "geosynclines" and in the NE are represented by some post-orogenic granites.

Following cratonization, mild deformation led to the development of three platform cover domains, in separate groups of sedimentary basins with rare plutons and associated basaltic sheets: the Davenport "Geosyncline," the unnamed basin which surrounds the Granites-Tanami Complex, and the Victoria River, Arafura, Amadeus, Ngalia, Daly River, Georgina, Wiso, Ord, and Bonaparte Gulf basins. Development in each group is similar and contemporaneous, and their deposits overlie unconformably the transitional domains and older orogenic provinces.

In northern Australia the geosynclines and basins were deformed at the end of the Lower Proterozoic. Rocks of the Pine Creek Geosyncline were folded along axes generally trending between NW-SE to N-S, usually following the trend of the geosyncline margins but was locally modified by basement inliers. High-angle reverse and normal faulting accompanied the folding and was mainly parallel to the fold axes. Regional metamorphism of the sediments was generally very low grade. Dolerite sills intruded Carpentarian and Adelaidean sedimentary rocks both before and after folding. Rocks of the Warramunga "Geosyncline" are separated by a major NW-SE shear zone.

Acid volcanism and granite intrusions followed by basic volcanism and platform sedimentation occurred in Carpentarian time, with rocks being very little folded in some places and strongly in others.

The Bonaparte Gulf Basin is a simple and fairly symmetrical structure marked by block faulting, and folds are rare. The crustal sagging that initiated the basin in the early Cambrian may have been caused by the withdrawal of basaltic magma from the mantle (Veevers, 1967). In the early Late Devonian there was a major upheaval along the southeastern edge of the Bonaparte Gulf Basin, and upthrown fault blocks shed thick coarse clastics (conglomerate) into the basin.

In the Georgina Basin faulting also predominates over folding. Faults that affected the Middle Cambrian along the northern margin of the basin trend WSW, parallel to the basement fault systems. Faulting likewise predominates in the post-Devonian tectonism. These major faults have produced two major asymmetrical synclines in the basin with steep dips on the southwestern flanks. Conditions in the Wiso Basin were rather similar.

In the Ngalia and Amadeus basins warping and deposition began at about the same time. Faulting processes have been related to the epeirogenic movements that produced the observed unconformities. A major disturbance occurred during the Carboniferous, when the sedimentation ceased. In the northern Ngalia Basin the rocks became folded and overthrust; in the S they were merely tilted and block-faulted. Regional anticlinora separate the Amadeus Basin from the Ngalia Basin. On the southwestern, northern, and northeastern margins of the Amadeus Basin there are nappe structures, involving both basement rocks and sedimentary cover rocks. All the nappes front toward the basin, and the rocks are strongly folded. There are two major unconformities: a folded unconformity between Late Proterozoic and Cambrian sedimentary rocks and another between the folded Devono-Carboniferous (?) rocks and flat-lying Permian and Mesozoic sediments. Major thrusts occur within the anticlinoria of basement rocks N and S of the Amadeus Basin. Throughout the Northern Territory Mesozoic sedimentary rocks are mostly flat-lying or have been gently warped during the Cenozoic.

Paleontology

Substantial numbers of faunal assemblages are known in the Northern Territory, but many localities have yet to be studied in detail, especially for the possible occurrence of microfossils.

Few of the primitive forms of life have left recognizable traces in the Proterozoic rocks. Stromatolites (alga structures) are common in the Lower Proterozoic of the Pine Creek Geosyncline and the Carpentarian of the McArthur Basin, where large oncolites (also presumed to be algae) have been found. Some forms were also indentified in the Adelaidean sequence of the Amadeus Basin.

Lower Cambrian archaeocyathids, algae, and sponges are common in the S. Middle Cambrian trilobites are widespread, in the Bonaparte

Gulf, Ord, Daly River, and Georgina basins. Ordovician graptolite-bearing rocks occur in the Georgina, Amadeus, and Bonaparte Gulf basins. Lower Ordovician nautiloids occur in the Georgina and Amadeus basins as well as trilobites, brachiopods, and conodonts. Upper Devonian brachiopods and plant fossils occur in the Bonaparte Gulf Basin. Permian brachiopods, marine molluscs, and land plants have been described from the Bonaparte Gulf Basin. Lower Cretaceous ammonites were collected near Darwin. Tertiary marine deposits contain bryozoa, foraminifera, pelecypods, gastropods, and echinoderms. Vertebrates are well represented in deposits by fish and reptiles, by sharks' teeth, and bird and marsupial remains.

Mineral Deposits

Remoteness from potential markets has hampered mineral development in the Northern Territory. The main mineral province is located between Katherine and Darwin. Most of the economic deposits of gold, copper, silver, lead, zinc, tin, tungsten, uranium, and iron ore are in Proterozoic rocks and, except for uranium, are related to pegmatite and granite intrusions. In the mineral province between the Barkley tablelands and the Gulf of Carpentaria a large zinc-lead deposit has been found in pyritic shales at McArthur River. It contains 200 million tons of ore averaging 9% zinc and 4% lead, but metallurgical problems delay commercial exploration.

Uranium reserves put Australia among the world's leading uranium sources. Copper (since 1952) is produced in mines near Tennant Creek that were originally worked for gold. Additional copper-gold discoveries show reserves of 3,500,000 tons, which average 2.6% copper and 1.2 dwt gold per ton.

Bauxite deposits occur at Gove and Marchinbar islands on the Gulf of Carpentaria. Proved reserves at Gove are about 250 million tons; and at Marchinbar, 9 million tons. Lateritic nickel deposits are found near the southwestern corner of the Territory. Several manganese deposits are present on Groote Eylandt in the Gulf of Carpentaria, of considerable extent in gently inclined seams of pisolitic ore, with reserves in excess of 50 million tons. Large-scale production began in 1966.

Extensive deposits of phosphate occur as beds up to 20 m thick, in lower Middle Cambrian rocks around the northwestern margin of the Georgina Basin and in the Amadeus Basin; however, they are of low grade and distance from a market makes them uneconomical at present.

Coal seams have been intercepted in bore holes and oil exploration wells in Lower Permian rocks between the mouths of the Fitzmaurice and Daly rivers. Lignite has been found at Santa Teresa area 70 km SE of Alice Springs. At Mereenie and Palm Valley, W of Alice Springs, gas and oil have been encountered in Cambro-Ordovician rocks of the Amadeus Basin; proven gas reserves are estimated at 1.57 trillion ft^3; proven oil reserves at Mereenie are estimated at 60 million barrels. Exploration for oil and gas is continuing, particularly in the offshore Bonaparte Gulf Basin.

G. W. D'ADDARIO*

*Published with the permission of the Director, Bureau of Mineral Resources, Geology and Geophysics, Canberra.

References

Australian Bureau of Mineral Resources, 1970. *Australian Mineral Industry Review.* Canberra: Austral. Bur. Mineral Resources Geol. Geophys., 357p.

Cook, P. J., 1971. "Illamurta diapiric complex and its position on an important Central Australian structural zone," *Bull. Am. Assoc. Petrol. Geologists,* **55,** 64–79.

Cooper, J. A., Wells, A. T., and Nicholas, T., 1971. "Dating of glauconite from the Ngalia Basin, Northern Territory, Australia," *J. Geol. Soc. Austral.,* **18**(2), 97–106.

Dunn, P. R., Plumb, K. A., and Roberts, H. G., 1966. "A proposal for stratigraphic subdivision of the Australian Precambrian," *J. Geol. Soc. Austral.,* **13**(2), 593–608.

Folk, R. L., 1971. "Longitudinal dunes of the northwestern edge of the Simpson Desert, Northern Territory, Australia. 1. Geomorphology and grain size relationships," *Sedimentology,* **16,** 5–54.

Forman, D. J., 1971. "The Arltunga Nappe Complex, MacDonnell Ranges, Northern Territory, Australia," *J. Geol. Soc. Austral.,* **18,** 173–182.

Geological Society of Australia, 1971. *Tectonic Map of Australia and New Guinea.* Sydney.

Hays, J., 1967. "Land surfaces and laterites in the north of the Northern Territory" in J. N. Jennings and J. A. Mabbutt, eds., *Landform Studies from Australia and New Guinea.* Canberra: ANU Press, 182–210.

Hodgson, E. A., 1965. "Devonian spores from the Pertnjara Formation, Amadeus Basin, Northern Territory," *Austral. Bur. Mineral Resources Geol. Geophys. Bull.,* **80,** 65–83.

Jones, B. G., 1971. "Upper Devonian to lower Carboniferous stratigraphy of the Pertnjara Group, Amadeus Basin, Central Australia," *J. Geol. Soc. Austral.,* **19,** 229–249.

———, 1973. "Sedimentology of the Upper Devonian to Lower Carboniferous Finke Group, Amadeus and Warburton basins, Central Australia," *J. Geol. Soc. Austral.,* **20**(3), 273–294.

Leitch, E. C., Fisher, D., and Mason, D. R., 1970. "A

suggested orogenic break within the metamorphic rocks of the Arunta Complex, Central Australia," *Search,* **1**(4), 159.
Smith, K. G., 1972. "Stratigraphy of the Georgina Basin," *Austral. Bur. Miner. Resources Geol. Geophys. Bull.,* **111**, 156p.
Tonkin, P. C., 1973. "Discovery of shatter cones at Kelly West near Tennant Creek, Northern Territory, Australia," *J. Geol. Soc. Austral.,* **20**(1), 99–102.
Veevers, J. J., 1967. "Phanerozoic of Northern Australia," *J. Geol. Soc. Austral.,* **14**(2), 253–271.
____, 1971. "Shallow stratigraphy and structure of the Australian continental margin beneath the Timor Sea," *Marine Geol.,* **11**, 209–240.
Walpole, B. P., Crohn, P. W., Dunn, P. R., and Randal, M. A., 1968. "Geology of the Katherine-Darwin region, Northern Territory," *Austral. Bur. Mineral Resources Geol. Geophys. Bull.,* **82**, 304p.
Wells, A. T., Forman, D. J., Ranford, L. C., and Cook, P. J., 1970. "Geology of the Amadeus Basin, Central Australia," *Austral. Bur. Mineral Resources Geol. Geophys. Bull.,* **100**, 222p.
Wopner, H., and Milton, B. E., 1972. "Illamurta Diapiric Complex and its position on an important central Australian structural zone, discussion, *Bull. Am. Assoc. Petrol. Geologists,* **56**(5), 964–972.

Cross-references: *Australasia–Regional Review; Australia–Queensland, South Australia, Western Australia.*

AUSTRALIA–QUEENSLAND

The State of Queensland occupies the northeastern 22% of the Australian continent; it has an area of 1,738,000 km², more than half of which lies in the tropics. The State Geological Survey, with a staff of 70 geologists and geophysicists, is based at Brisbane, the capital city. Nearly all of the 116 sheet areas on the 1:250,000 scale have been mapped since 1950, by cooperation of the survey with the Federal Bureau of Mineral Resources, Geology, and Geophysics.

Gold was the first metal to be won in important quantities in Queensland, but today ranks only eighth in value; famous mines of the gold period were Gympie, Charters Towers, and Mt. Morgan; the total gold extracted exceeds 20 million fine ounces. In 1931 Mt. Isa came into production as a silver-lead-zinc mine, but copper is now its main metal and it has the largest output of the underground mines in Australia. Mary Kathleen produced 4010 tons of uranium oxide between 1958 and 1963. Exploitation of the Weipa bauxite deposits began in 1961.

Coal mining began in the 1850s, and reserves in steam, coking, and metallurgical coal are immense. Petroleum was proved in the Moonie field in 1961 and there is a pipeline to Brisbane.

Artesian water, proved in the Great Artesian Basin in 1887, is fundamental to the rich grazing industries of the state.

The Charters Towers School of Mines, founded in 1900 and closed in 1920, provided the first geological teaching in the state; but after the University of Queensland was founded in 1911, teaching shifted to Brisbane. Here the Department of Geology and Mineralogy of the University at St. Lucia has a professorial and lecturing staff of 16; there is also the James Cook University of North Queensland at Townsville, founded in 1960, which has a small department of geology.

Geomorphic and Structural Framework

Queensland is an area of mature topographic forms and subdued relief as a result of long erosion and peneplanation; it has been entirely above sea level since the end of the Lower Cretaceous. Except in parts to the SE and NE, there are no high mountains or escarpments and the highest point is only 1602 m (5287 ft). The Great Divide is of inconspicuous relief over most of its length.

The main geomorphic divisions correspond also to the main geotectonic divisions, which are three: the cratonic Precambrian Shield exposed to the NW and N; the Tasman Geosyncline to the E, which became stabilized and uplifted at the end of the Paleozoic; and, overlapping them both, the Mesozoic and Cenozoic Great Artesian Basin.

Geomorphically the dominant feature is the Great Artesian Basin, which occupies about two-thirds of the state. Its surface is a great plain that drains in the SE into the River Murray System, in the SW it drains by braided streams into Lake Eyre, which is below sea level, and in the N flows into the Gulf of Carpentaria. From the plain rise low mesas of dissected lateritic duricrust.
plain rise low mesas of dissected lateritic duricrust.

The site of the Tasman Geosyncline is now a belt of relative highland, mainly with a NNW grain; it includes more plain than high country, and in several regions Cenozoic basaltic plateaus have been built up over its dissected terrain. Faulted and cuesta landforms occur particularly along the coastal belt, where the alluvium-filled plains are backed by steep and extensive escarpments. The continental shelf, facing the Coral Sea as far south as 24°S, carries the Great Barrier Reef in a magnificent contemporary carbonate province.

Where the older Precambrian rocks of the shield are exposed to the NW they form low,

FIGURE 1. Structural elements of Queensland (Hill and Denmead, 1960).

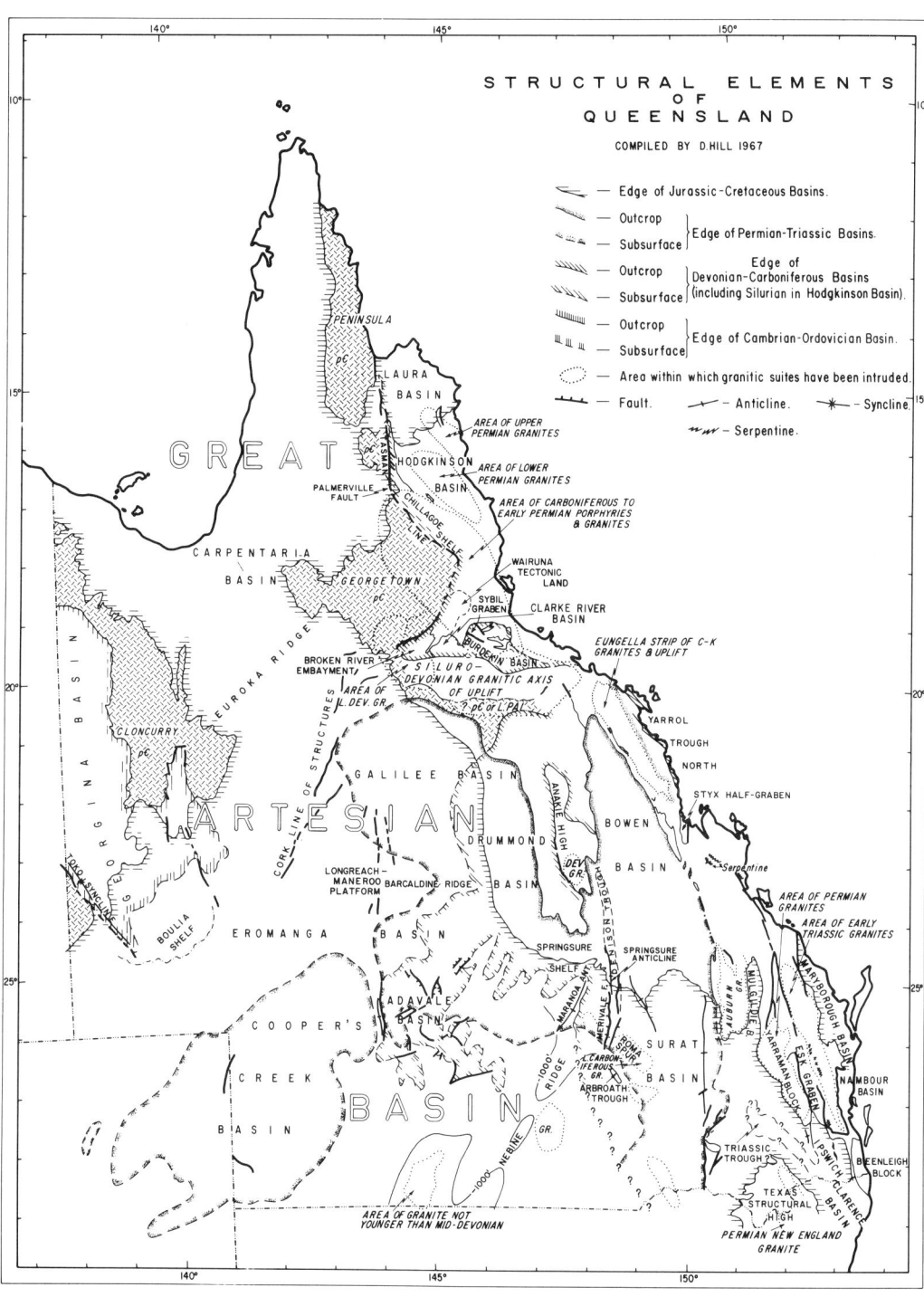

roughly N-S-trending ranges with alluvial plains between.

The Precambrian rocks of the shield are exposed in three regions—Cloncurry, Georgetown, and the Peninsula—separated by the cover rocks of the Great Artesian Basin. In the Cloncurry and Peninsula regions, fold axes and faults are dominantly N-S, but in the Georgetown region the trends of the Proterozoic conform with the arcuate foliations in the Archean and also to the boundary with the Tasman Geosyncline.

The folded older Proterozoic belt of the Cloncurry region, which carries much of the metal wealth of the state, passes to the W under the very gently folded late Proterozoic, Cambrian, and Ordovician strata of the area marked in Fig. 1 as the northwestern shelf.

In the Queensland portion of the Tasman Geosyncline the oldest fossils known are early Ordovician, but its pre-Silurian history is as yet obscure. The youngest marine fossils are early Cretaceous. Today the geosynclinal belt is occupied by NNW- or N-S-trending basins of sediments and volcanics separated by narrow structural highs of low-grade metamorphic rocks, which are either the basement rocks of the basins or the older sediments of the basins slightly metamorphosed and uplifted in horsts or anticlinal axes. The major orogenic movements affecting paleogeography in the geosyncline appear to have been mid-Devonian, when several basins developed; further orogeny occurred in late Permian, when the main part of the geosyncline ended its history as a region of marine deposition; and in mid-Mesozoic there were vertical adjustments involving large-scale faulting.

The main subdivisions of the geosyncline, shown in Fig. 1, with the ages of their strata, are as follows: Chillagoe Shelf and Broken River Embayment—Silurian and Devonian(?) marine; Hodgkinson Basin—Silurian, Devonian, and mid-Carboniferous, marine and fresh water; North Coastal Structural High—lower or middle Paleozoic; Clarke River, Star Basin—Devonian to mid-Carboniferous, marine and freshwater; Charters Towers—granitic axis of uplift; Drummond Basin—Devonian to mid-Carboniferous, marine and freshwater, the latter predominant; Anakie Structural High—lower or middle Paleozoic; Springsure Shelf—Permian connection between Great Artesian and Bowen basins; Bowen Basin—Permian and some Triassic; Auburn and Eungella Granitic complexes—pre-Permian and Permian; Gogango Structural High—Paleozoic; Yarrol Basin—Devonian to Permian marine, some Triassic and Jurassic freshwater; Texas Structural High—Silurian to Permian marine; South Coastal Structural High—Precambrian(?) Silurian(?) and early Devonian marine; Maryborough Basin—Permian to Cretaceous, marine and freshwater; Ipswich Basin—freshwater Triassic and Jurassic; Esk Graben—Permian marine and freshwater, Triassic freshwater; Styx Basin—Cretaceous coal measures; Laura Basin—freshwater Jurassic and marine Cretaceous.

The Great Artesian Basin is divided into the Carpentaria, Eromanga, and Surat basins by ridges in the Precambrian or Paleozoic basement. Its strata include Permian and Triassic terrestrial beds along the eastern edge of the Eromanga Basin, and the Permian sequence of the Bowen Basin underlies the Triassic freshwater strata at the bottom of the Surat Basin sequence. Beneath parts of the Eromanga Basin lie the Devonian(?) Adavale Basins and the Permian-Triassic Cooper and Galilee Basins. Everywhere above the Triassic, the Great Artesian Basin has terrestrial Jurassic sandstones and coal measures and Lower Cretaceous marine sandstones and shales followed by freshwater Upper Cretaceous and, centrally in the sub-basins, by freshwater Cenozoic.

Stratigraphic History

Precambrian strata range from high-grade schists and gneisses and quartzites to limestones and shales; both basic and acid volcanics are interbedded. About 50 formations, as well as seven granites and unconformities, have been mapped in the Cloncurry Complex. Eleven formations, five granites, and two unconformities have been mapped in the Georgetown massif. In the Brisbane area the Rocksberg Greenstones and the Bunya Phyllite may be Precambrian. One Upper Proterozoic formation in the Cloncurry region contains oolitic iron ores like the Wabana deposits.

Cambrian and Ordovician strata outcrop only to the NW; they are marine (dolomites, limestones, shales, and fine-grained sandstones), nearly horizontal and almost unaltered and rich in trilobites, as at Beetle Creek 20 miles W of Mt. Isa.

Silurian fossiliferous marine strata, including coral limestones with reefal developments, are known mineralized in the Chillagoe Shelf and steeply folded in the Broken River embayment. Graywacke, slates, and quartzites in the North and South Coastal and Texas highs may be Silurian.

Lower and early(?) Middle Devonian andesitic volcanics, radiolarian cherts, shales, and coralline limestones in the Texas, South Coastal, and Anakie structural highs are thought to predate a mid-Devonian orogeny that radically altered the paleogeography of the geosyncline. Serpentinites in the South Coastal High may be mid-Devonian. Small lenses of late Middle Devonian coralline limestone in the Yarrol Basin are associated with andesitic volcanics. In the Star-Clarke River Basin thick Middle Devonian limestones transgress a granitic and metamorphic terrain. In the Broken River Embayment Lower to early Upper Devonian limestones including very rich stromatoporoid and coral reefs are interbedded with siltstones. The Hodgkinson Basin sequence of conglomerates, graywackes, shales, and lenticular limestones is largely Devonian. In the Drummond Basin much variegated terrestrial shale and sandstone of late Middle and Upper Devonian age was deposited. Some marine Devonian is known on the eastern side of Cape York Peninsula.

The Carboniferous is fully developed only in the Yarrol Basin, where it is marine throughout except in marginal areas, and conformable with both Devonian and Permian. Oolitic Lithostrotion limestones are characteristic in the Visean. Lithostrotion limestones are also known in the western part of the Texas Structural High. In the Star and Clarke river basins Tournaisian marine shales and sandstones are conformable with Upper Devonian marine below and *Rhacopteris*-bearing beds above, following which sedimentation ceased. In the Drummond Basin early and Middle(?) Carboniferous sediments are all terrestrial and, like those of the Star and Clarke River basins, were folded before the Permian, the structures reflecting differential vertical movement rather than strong compression. The Hodgkinson Basin was subjected to four phases of Carboniferous folding, mainly under east-west compression. Some of the north Queensland granites may be Carboniferous. There are no Carboniferous coal measures.

Permian strata are characterized by the *Eurydesma-Ingelarella* fauna and the *Glossopteris* flora. Bowen Basin sedimentation began with intermediate vulcanicity, continued with marine flooding from the S or E, and ended with a thick sequence of coal measures. Across the Springsure Shelf into the Great Artesian Basin, the earliest Permian deposits are considered to be of fluvioglacial origin. The Esk Graben, Texas Structural High, and Maryborough Basin contain Permian marine and freshwater sediments with volcanics; the Gympie gold reefs are in Permian strata. The Yarrol Basin sedimentation ended with intermediate volcanics in mid-Permian time.

Permian and perhaps early Triassic compression from the E and NE folded and thrust the Paleozoic sequences of the geosyncline, with some serpentinite intrusion, and granites invaded the highs and the margins of the basins. One result was a narrow belt of close folding and faulting that passes NW through the generally NNW trending Bowen Basin and has provided variation in the rank of the Permian coals. Differential uplift occurred, marine sedimentation ceased, and the geosynclinal site became part of the shield craton. Many of the eastern Queensland granites are Permian in age. During the late Carboniferous, Permian, and possibly early Triassic, a remarkable series of central igneous complexes, including ring complexes, were developed on the Georgetown massif and the neighboring parts of the relatively stabilized Tasman Geosyncline.

Triassic deposits are terrestrial except for Scythian(?) marine siltstones discovered in the Maryborough Basin. In the Esk Graben, the Ipswich, Maryborough, and Bowen basins, and parts of the Great Artesian Basin, particularly the Surat Basin, volcanic material formed a large part of the initial sediments, and the youthful topography of the uplifted geosynclinal belts provided abundant immature terrigenous sediment; later sedimentation rates slowed, and coal measures were formed in the Ipswich Basin, where rich *Dicroidium* flora and insect faunas have been recovered. Some differential vertical movement continued.

Jurassic sediments are entirely terrestrial. All sections of the Great Artesian Basin, the Laura, Ispwich, and Maryborough basins, and part of the Yarrol Basin received large volumes of immature sand at first and coal measures with an *Otozamites-Ptilophyllum* flora later. The Precipice Sandstone of the Surat Basin is the Moonie reservoir. Intrusive igneous activity occurred along the western margin of the Maryborough Basin, and extensive activity in southeast Queensland. Earth movements occurred along old lines, emphasizing older structures.

Cretaceous sedimentation followed that of the Jurassic without break, except in the Ipswich Basin, which received none. Marine seas flooded the Laura and part of the Yarrol basins (Stanwell area) at the beginning of the period, and during the Aptian flooded all parts of the Great Artesian, Laura, and Maryborough basins but withdrew from the Stanwell area. Albian marine sediments, fine grained and with the small carbonate content mainly concentrated in concretions like those of the Albian, formed only in the Great Artesian Basin and withdrew everywhere at the end of the stage. Coal measures formed in the Maryborough Basin after the withdrawal, and terrestrial sandstones and shales formed in the Eromanga Basin. Cretaceous sands, like those of the Jurassic, are aquifers, the interbedded marine clays being aquicludes. The Styx Coal Measures are possibly Albian.

Folding along NNW axes and possibly of early upper Cretaceous age has affected the Maryborough Basin, but flank dips are 20° or less except where faulting has occurred. The Ispwich Basin also suffered some crustal movement that was possibly of this date.

Cenozoic sediments of Queensland are difficult to date, owing to both the absence of marine strata and the evidence of Pleistocene glaciation. Some basis of division is provided by distinctive statewide surfaces of lateric duricrust that developed mainly during the Miocene (Fig. 2).

Lower Tertiary freshwater sandstones and shales are extensive in the Eromanga Basin, and near the South Australian border lie in very low, elongated domes and anticlines, probably due to Cretaceo-Tertiary movement in the basement. East of the Great Divide fluviatile and lacustrine sediments are more local in development, and in many areas are being dissected by the present rivers; they include oil shales, brown coal, and diatomites. Igneous rocks occur only on or to the E of the Great Divide, and mainly in south Queensland, where several basaltic plateaus with interbedded rhyolites greatly modify the topography and where three great central igneous complexes and ring complexes are known. A series of alkaline trachyte plugs are famous as the "Glass Houses" first reported by Captain Cook.

Upper Cenozoic (post-laterite) volcanics also are predominantly plateau-forming basalts, on or to the E of the Great Divide; at least four provinces are known in north Queensland.

Upper Cenozoic (mainly Pleistocene) sediments include quartz gravel, sandstones, and clays, outwash from the dissection of the laterite. They also include the brownish sands and the alluvial "black soils" of the Darling and other downs; these contain the celebrated fossil marsupial fauna, including the giant *Diprotodon*; early Holocene alluvium has yielded the Talgai human skull.

Inland sand dunes in the SE of the State, belonging to the glacial phases, migrated here from the Simpson Desert system (of South Australia); clay pans spread out between the dunes. In many other inland areas there are very extensive plain lands, today usually densely forested with brigalow or gidgea and consisting of up to 3 or 6 m of uniform dense clays that are quite distinct from riverine clays.

Quaternary coastal features include raised shell beds and beaches of the last (mid-Holocene) emergence variably estimated as 3–6 m. There are incoherent sands extensively developed in the coastal lowlands. The Great Sandhills of the coastal islands and parts of the mainland are almost entirely siliceous and of two superimposed systems, the upper being entirely wind piled from the present beaches, the lower being Pleistocene and possibly derived in part from the land. Finally, the Great Barrier Reefs, rising from the continental shelf, have a proved thickness of reef-rock of 112–183 m (378–506 ft) that is in part post-Pleistocene but lies above older carbonate rocks, including Miocene. No dolomitization is known.

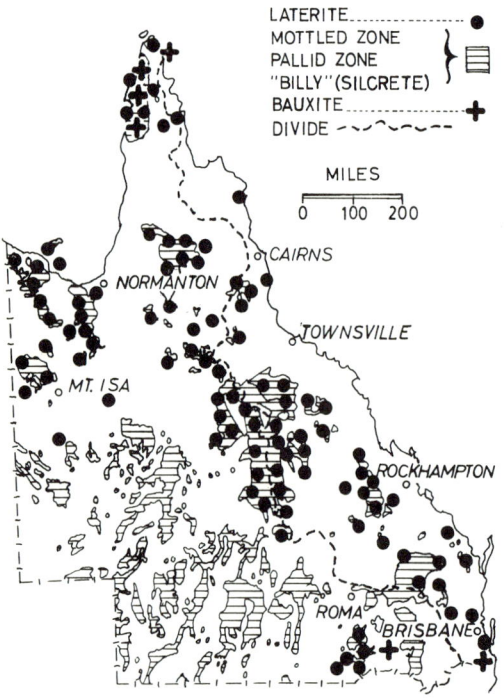

FIGURE 2. Distribution of laterites and companion materials in Queensland (Hill and Denmead, 1960).

Mineral Production

Queensland's mineral production for 1973 exceeded $500,000,000. Gold, produced mainly at Mt. Morgan and Cracow, exceeded 1800 kg; copper, 146,000 tons; lead, 126,000 tons; zinc, solely from Mt. Isa, 108,000 tons; tin, mainly from the Cooktown district, 2500 tons; silver, 305,000 kg; from the beach-sand industries of the south coast, rutile, 112,000 tons and zircon, 90,000 tons; bauxite, mainly from Weipa, 8,980,000 tons; coal production was nearly 20,000,000 tons. The first oil field, Moonie in the Surat Basin, came into production in the early 1960s. Extensive gas fields have been developed in the Roma district.

Other mineral production includes brick and pipe clay, building stones, fireclay, ironstone, pyrite, manganese, salt, and silica. In addition, gemstones—opals and sapphires—have been recovered to an estimated value of $10,000,000 in 1973.

DOROTHY HILL

References

Black, L. P., and Richards, J. R., 1972. "Isotopic composition and possible genesis of ore leads in northeastern Queensland, Australia," *Econ. Geol.,* 67, 1168–1179.

____, Morgan, W. R., and White, M. E., 1972. "Age of a mixed Cardiopteris–Glossopteris flora from Rb-Sr measurements on the Nychum Volcanics, North Queensland," *J. Geol. Soc. Austral.,* 19(2), 189–196.

Clarke, D. W., and Paine, A. G. L., 1970. *Chargers Towers, Queensland* (Geol. Ser. Sheet SF/55-2). Canberra: Austral. Bur. Mineral Resources, Geol. Geophys., 1:250,000, 34p.

Cook, P. J., 1972. "Petrology and geochemistry of the phosphate deposits of northwest Queensland, Australia," *Econ. Geol.,* 67, 1193–1213.

David, T. W. E., and Browne, W. R., 1950. *The Geology of the Commonwealth of Australia,* 3 vols. London: Edward Arnold.

Davis, A., ed., 1971. "Proceedings of the 2nd Bowen Basin Symposium held at Brisbane, 7th–9th October 1970," *Queensland Geol. Surv. Rept. 62,* 610p.

Evans, J. H., 1965. "Bauxite deposits of Weipa," *Eighth Commonw. Mining Metall. Congr.,* 1, 396–401.

Gardner, J. V., 1970. "Submarine geology of the western Coral Sea," *Bull. Geol. Soc. Am.,* 81, 2599–2614.

Glikson, A. Y., 1972. "Structural setting and origin of Proterozoic calc–silicate megabreccias, Cloncurry region, northwestern Queensland," *J. Geol. Soc. Austral.,* 19(1), 53–63.

Green, D. C., 1969. "Transitional basalts from the eastern Australian Tertiary province," *Bull. Volcanol.,* 33(3), 930–941.

Greenwood, R. H., 1955. "Die geographische Gliederung von Queensland," *Die Erde,* pts. 3–4, 261–272.

Grubb, P. L. C., 1971. "Genesis of the Weipa bauxite deposits, N. W. Australia," *Mineral Deposita* (Berlin), 6(4), 265–274.

Hill, D., ed., 1953. *Geological Map of Queensland."* Brisbane: Dept. of Mines, 1:2,534,400.

____, and Denmead, A. K., eds., 1960. "The geology of Queensland," *J. Geol. Soc. Austral.,* 7, 1–474.

Hopley, D., 1971. "The origin and significance of north Queensland island spits," *Z. Geomorphol.,* 15(4), 371–389.

Marsden, M. A. H., 1972. "The Devonian history of northeastern Australia," *J. Geol. Soc. Austral.,* 19(1), 125–162.

Maxwell, W. G. H., 1968. *Atlas of the Great Barrier Reef.* Amsterdam: Elsevier, 268p.

Meyers, N. A., 1969. "Carpentaria Basin," *Queensland Geol. Surv., Rept. 34,* 36p.

Mollan, R. G., Dickins, J. M., Exon, N. F., *et al.,* 1969. "Geology of the Springsure 1:250,000, sheet area, Queensland," *Austral. Bur. Mineral Resources, Geol. Geophys. Rept. 123,* 114p.

Olgers, R., 1970. *Buchanan, Queensland* (1:250,000 Geol. Ser. Sheet SF/55-6). Canberra: Austral. Bur. Mineral Resources, Geol. Geophys., 12p.

Power, P. E., 1967. "Geology and hydrocarbons, Denison Trough, Australia," *Bull. Am. Assoc. Petrol. Geologists,* 51(7), 1320–1345.

Scheibnerova, V., 1971. "The Great Artesian Basin, Australia, a type area of the Austral Biogeoprovince of the Southern Hemisphere, equivalent to the Boreal Biogeoprovince of the Northern Hemisphere," *Proc. Planktonic Conf.,* 2(2), 1129–1138.

Smart, J., Grimes, K. G., and Doutch, H. F., 1972. "New and revised stratigraphic names; Carpentaria Basin," *Queensland Govt. Mining J.,* 73(847), 190–201.

Walpole, B. P., ed., 1960. *Tectonic Map of Australia.* Canberra: Austral. Bur. Mineral Resources Geol. Geophys., 1: 2,534,400.

____, ed., 1962. *Geological Notes in Explanation of the Tectonic Map of Australia.* Canberra: Austral. Bur. Mineral Resources Geol. Geophys., 72p.

Cross-references: *Australasia–Regional Review; Australia–New South Wales, Northern Territory, South Australia; New Guinea.*

AUSTRALIA–SOUTH AUSTRALIA

South Australia occupies an area of 984,000 km^2 in the south-central portion of the Australian continent. Except for its Indian (southern) Ocean coastline, it has no natural boundaries. Elsewhere its limits coincide with 26°S, 129°E, and 141°E.

The major structural units were established early in geological time. Among the geological features well known in the world literature are the Proterozoic rocks, with their fine glacial sequences, the Cambrian faunas, the Permian sediments and their associated glacial land forms, the Mesozoic and Tertiary basin sediments, and surficial deposits of Quaternary age.

The first work to deal exclusively with the geology of South Australia was that of Howchin (1918). This was followed by a comprehensive treatment by Glaessner and Parkin in the *Journal of the Geological Society of Australia,* Vol. 5, Pt. 2, published in 1958. The rate of accumulation of geological information has increased rapidly in recent years, as indicated by the *Handbook of South Australian Geology* prepared by the Geological Survey of South Australia (Parkin, 1969), on which the present contribution is largely based (mainly on sections prepared by N. H. Ludbrook, B. P.

Thomson, H. Wopfner, and the writer). The *Handbook* contains important references not quoted here.

Major Structural and Morphological Units

Much of the state belongs to the Australian Precambrian Shield, which is comprised of the Musgrave Block in the NW and the Gawler Block, which extends 300 km NW from Spencer Gulf. The rocks of both blocks are post-Archean in age. The Musgrave Block forms part of the topographic backbone of the continent, whereas the Gawler Platform is subdued except for the hilly terrain of the Gawler Ranges (Fig. 1).

East of the Gawler Platform is a less stable

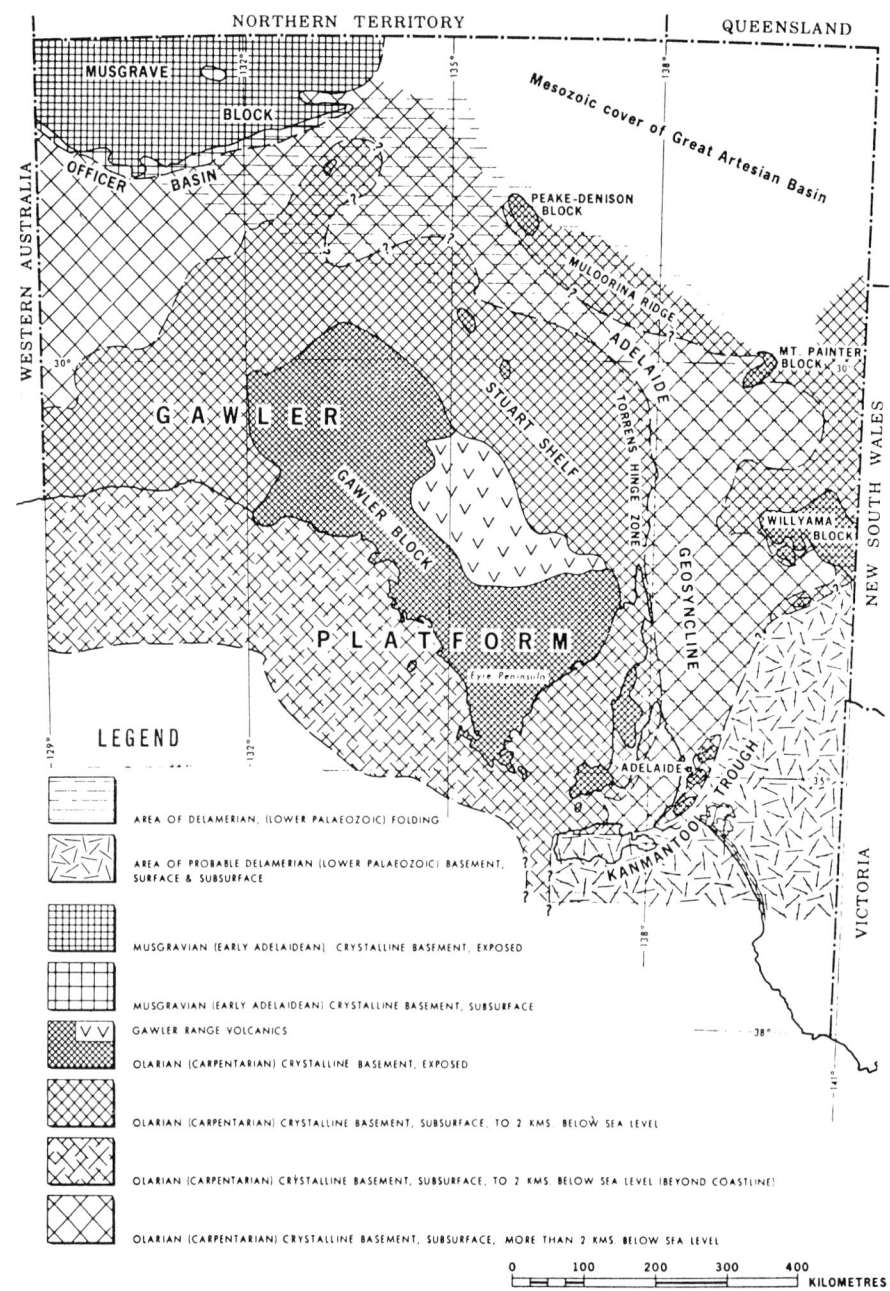

FIGURE 1. Precambrian basement structural units of South Australia (S. Australian Dept. Mines; in Parkin, 1969).

region that developed into a great geosynclinal trough, the Adelaide Geosyncline, containing more than 15,000 m of Proterozoic sediments. Conditions of sedimentation were generally shallow and there are widespread glacial accumulations. Sedimentation continued into the early Paleozoic. In the SE in the Kangaroo Island-Kanmantoo area, a deep trough developed and fine-grained deep-water sediments were deposited. The youngest deposits here are Cambrian.

During the Lower Paleozoic Delamerian Orogeny there was major folding followed by the intrusion of granite. This early mountain belt extended N and NW from Kangaroo Island to the Musgrave Ranges, and N and NE to the Barrier Ranges in New South Wales.

During the closing phases of the orogeny, Ordovician sediments were deposited in the Great Artesian Basin area, followed by possible Silurian. Sedimentation continued into the Devonian in the deeper parts of the basin. Following the orogeny there was a long period of relative stability. In the southern portion of the state, the general absence of post-Cambrian and pre-Permian sediments suggests that the area was uplifted and formed a source area for Middle Paleozoic sediments deposited to the N and E.

During Permian time, the pre-Permian basement was scoured and downwarped portions were infilled with glacial deposits. In the SW of the Great Artesian Basin, Permian sediments occupy infrabasins now overlain by younger rocks. Block faulting appears to have been widespread in the southern part of the state at this time. Evidence from the Lake Phillipson Trough, the Arkaringa Basin, the Gulf St. Vincent region, and the Murray Basin indicates the existence of glacial valleys and troughs into which the sea gained access for a short time. The Cooper Basin appears to be entirely non-marine.

During the Mesozoic, widespread downwarping of the crust occurred, reaching its maximum development before the late Cretaceous. The oldest Mesozoic strata are Triassic nonmarine, partly evaporitic, which have been found at depth in the Cooper Basin. Terrestrial sediments of Jurassic age occur throughout most of the South Australian portion of the Great Artesian Basin.

There was a major marine transgression in early Cretaceous time. A regression is indicated by coal interbeds in the Winton, the last formation with wide distribution in the Great Artesian Basin. By the end of Cenomanian, all of South Australia was land and scattered fluviatile deposits form the only sedimentary record.

Sediments of Lower Tertiary age are widespread in the interior of South Australia and commonly occur as thin sands and gravels capped with silcrete. In many places in the Great Artesian Basin the sediments show a minor erosional disconformity with the underlying Mesozoic rocks.

On the southern margins of the state shallow basins developed at the beginning of the Tertiary. This was followed by uplift of the basin margins and complementary subsidence of the present basin areas, leading to marine transgressions. Relatively stable conditions prevailed in the southern basins during the mid-Tertiary.

FIGURE 2. Recently active fault systems in South Australia, in part based upon seismic analysis by Stewart and Mount (1972). Note the intersection of two conjugate trends, the NW-SE Fitzroy-Spencer Fracture Zone (becoming WNW-SSE in the Lake Torrens Lineament), and the NE-SW Mabaina-Darling Lineament, and the "zone of diapirism" that occurs around the ancient triple point near the intersection of the Flinders Ranges, Mt Lofty Ranges, and Olary Upland. Right-lateral strike slip and clockwise crustal rotation are involved. The active opening and collapse of the Spencer Gulf-St. Vincent Gulf grabens suggests an incipient aulacogen that has developed subsequent to the separation of Australia and Antarctica (mid-Tertiary).

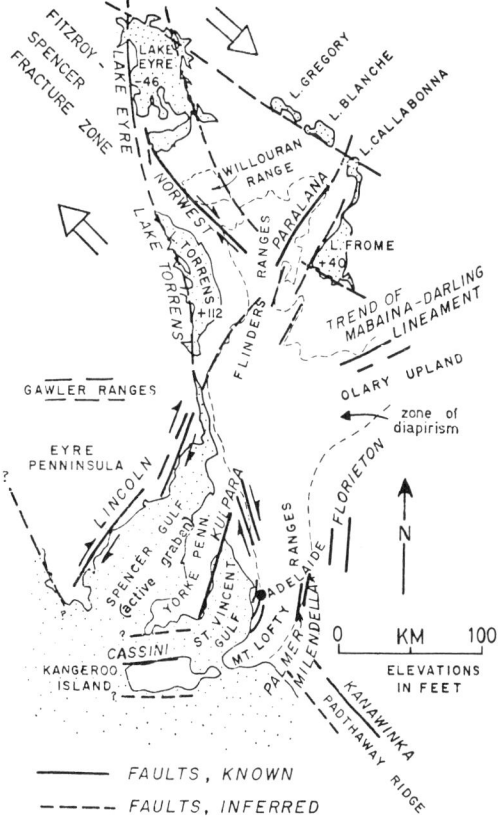

The breakup of the basins and a withdrawal of the sea along the southern continental margin commenced in the late Miocene. Renewal of tectonic activity, mainly taphrogenic, in the Pliocene during the Kosciuskan Uplift produced a well-defined pattern of lineaments and reduced the extent of the sedimentary basins. The movements continued into the Pleistocene (Fig. 2).

Block faulting during the Kosciuskan Uplift produced the present highland chain stretching from Kangaroo Island through the Mt. Lofty Ranges to the latitude of Port Pirie and NE to the Barrier Ranges, and from near Port Pirie through the Peak and Denison ranges to the Musgrave Ranges in the far NW. The Musgrave and Tomkinson ranges together with the Everard Ranges and the Mann Ranges form an E-W-

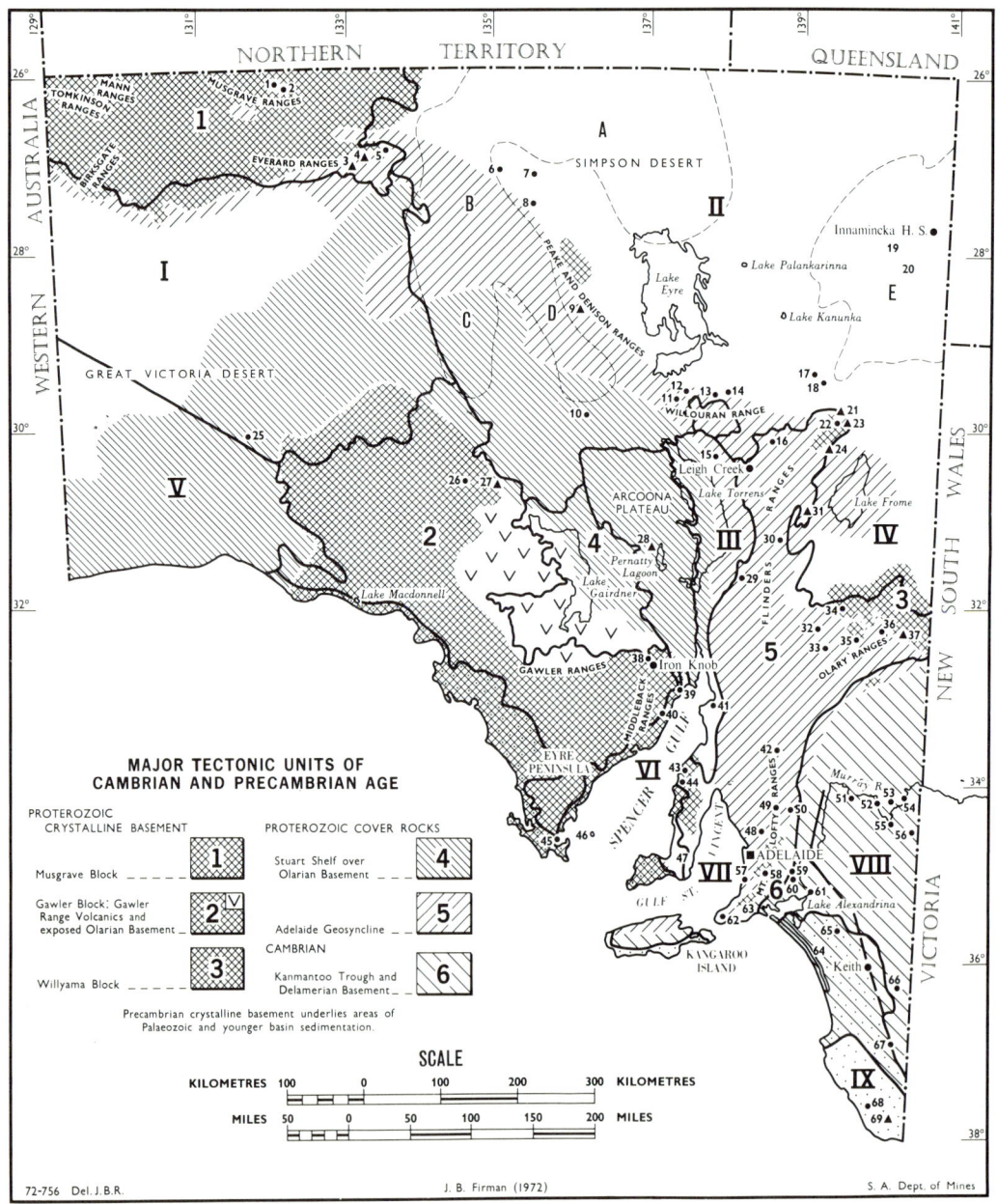

FIGURE 3. Principal morphological subdivisions of South Australia. Eolianite distribution on the southern coastal margin is indicated by dots. See also the table on the next page.

Western Basins	Western Shield	Central Basins	Highland Chain	Eastern Basins
I. Officer Basin Province			1. Musgrave Ranges Province Peak and Denison Ranges	II. Great Artesian Basin Province
	2. Gairdner Province		5. Central Ranges Province	
	Lake Gairdner area	III. Torrens Basin Province	Flinders Ranges	IV. Frome Embayment Province
V. Eucla Basin Province	Eyre Peninsula	VI. Spencer Basin Province	Olary "Ranges"	VIII. Murray Basin Province
		VII. St. Vincents Basin Province	Mt. Lofty Ranges	
				6-IX. Padthaway Ridge
			Kangaroo Island	IX. Otway Basin Province

Basins
I. Officer Basin
II. Great Artesian Basin
III. Torrens Basin
IV. Frome Embayment
V. Eucla Basin
VI. Spencer Basin
VII. St. Vincent Basin
VIII. Murray Basin
IX. Otway Basin

Infrabasins
A Pedirka Basin
B Arkaringa Basin
C Lake Phillipson Trough
D Boorthanna Trough
E Cooper Basin

Towns and Other Localities
1. Ernabella Mission
2. Kenmore Park H.S.
3. Chambers Bluff
4. Mt. Chandler
5. Granite Downs H.S.
6. Todmorden H.S.
7. Alberga
8. Oodnadatta
9. Mt. anna
10. Billa Kalina
11. Coward Cliff
12. Alberrie Creek
13. Callanna H.S.
14. Marree
15. Ediacara H.S.
16. Mt. Lyndhurst
17. Reedy Spring
18. Blanchewater
19. Gidgealpa
20. Moomba
21. Prospect Hill
22. Mt. Fitton H.S.
23. Mt. Babbage
24. Mt. Painter
25. Maralinga
26. Tarcoola
27. Mt. Eba
28. Mt. Gunson
29. Hookina
30. Oraparinna H.S.
31. Mt. John
32. Waukaringa
33. Teetulpa H.S.
34. Crocker Well
35. Manna Hill
36. Olary
37. Radium Hill
38. Corunna H.S.
39. Whyalla
40. Moonabie H.S.
41. Pt. Pirie
42. Burra
43. Wallaroo
44. Moonta
45. Pt. Lincoln
46. Spilsby Island
47. Yorke Peninsula
48. Gawler
49. Kapunda
50. Dutton
51. Waikerie
52. Kingston
53. Monash
54. Renmark
55. Loxton
56. Nadda
57. Noarlunga
58. Echunga
59. Woodside
60. Kanmantoo
61. Tailem Bend
62. Delamere
63. Fleurieu Peninsula
64. The Coorong
65. Coonalpyn
66. Bordertown
67. Naracoorte
68. Tantanoola
69. Mt. Gambier

trending feature about 500 km long, the ranges rising abruptly to about 1500 m with typical inselberg landscapes.

Complementary graben structures were formed marginal to the ranges in the Lake Torrens, Lake Frome, Spencer Gulf, and Gulf St. Vincent areas. Elsewhere, within the major sedimentary basins, minor shallow Cenozoic basins developed. The largest of these is the Lake Eyre Basin, which is a vast depression reaching 15 m below sea level.

In the Lower Pleistocene, uplift and dissection of the Tertiary cover mass and weathered basement rocks in the ranges led to deposition of clays, sands, and gravels in adjoining basins and grabens. As a result of repeated uplift, older deposits are higher in the landscape and closer to the range axes than younger deposits.

The shift from the Pliocene to the Pleistocene was accompanied by ferruginization and silicification of older surfaces. The change in environment is reflected in various ways, including a change in grain size and extent of deposits from widespread Pliocene sands to Pleistocene sandy clays of the Lower Pleistocene. Middle Pleistocene deposits record strong shifts in sea level and the repeated appearance in southern Australia of periglacial climates. In the SE successive raised beach deposits alternate with extensive littoral dune belts. Faulting at the end of the Middle Pleistocene and a milder climate are recorded in the extensive deposits of a younger alluvial sequence deposited in the Upper Pleistocene. Toward the end of the Pleistocene, inland eolian deposits were laid down.

Stratigraphic Succession

Precambrian. The oldest rocks in Australia (ca. 2300 m yr), established from isotopic dating in Western Australia and the Northern Territory, have not yet been identified in South Australia. Sequences found are of Lower Proterozoic and Carpenterian age.

The Musgrave Block. This is an immense crystalline basement terrain that extends into the adjacent states. The Hinkley, Mann, Davenport, and Ferdinand faults in this area appear to be part of great global shear systems.

The Musgrave-Mann Metamorphics in the southern Musgrave and Tomkinson ranges belong to the granulite facies, including glassy quartzite with sillimanite, garnet-bearing quartzite, a variety of gneissic granulite, and, in some places, lenses of marble. The original sediments belonged to a thick shale sequence that contains minor dolomite and sandstone units. Jaspilitic iron formations occur in the Mt. Caroline-Kenmore Park area, and may correlate with similar units in the Gawler Block.

Pre-Adelaidean tectonism is suggested by erratic pre-Adelaidean isotopic ages from granitoid rocks in the Birksgate Ranges area; northeast trends parallel to the Kimban folding of the Gawler Block are also thought to be of this age.

High temperatures associated with granulite metamorphism during Adelaidean time, about 1400-1100 m yr may have eliminated most isotopic evidence of earlier events. This later metamorphism is named the Musgravian orogenic cycle. A number of phases and associated events have been recognized, from oldest to youngest: the Ernabellan Phase, marked by folding and metamorphism of granulite facies rocks in the Ernabella area; the Everardian plutonic Phase, indicated by isotopic ages of about 1300 m yr from granites in the Everard and Birksgate ranges; the Kulgeran Phase of folding of the Adelaidean cover with intrusion of granite (1130 m yr) at Kulgera; and the Winanyan Phase of fracturing, mafic intrusion, uplift, and thrusting. The mafic intrusives, named the Giles Intrusive Complex, which contain dunites, norites, olivine norites, pyroxenites, and some picrites, troctolites, and gabbro, form a chain extending E-W for over 300 km.

The Gawler Block. The Cleve Metamorphics include the Flinders "Gneiss," comprised of hypersthene granulites and gneisses near Port Lincoln on Eyre Peninsula, and the Hutchison Group of amphibolite and amphibolitic schist, which includes an iron formation with dolomite and marble named the Greenpatch Metajaspilite. The latter is thought to be equivalent to the Middleback Group, which includes jaspilite, schist, and dolomite of the Middleback Ranges.

The Cleve Metamorphics were folded into a system of mountain chains during a period of folding and regional metamorphism called the Kimban Phase. Isotopic ages for granulite metamorphism, granite gneiss, and granite range from 1600 to 2000 m yr. The Kimban Phase was active in early Carpentarian time.

Overlying the older basement deformed by Kimban folding in the Whyalla district is the Moonabie Formation of metaquartzite and conglomerate. The formation is intruded in the Moonabie Range area by the Moonabie Porphyry, a porphyritic rhyolite with minor spilite.

The Moonabie Formation was folded during the Charlestonian Phase of folding and plutonism. Massive granite intrusions associated with this phase extend across northern Eyre Peninsula. The Charleston Granite in the Moonabie

Range intrudes dolomite, shale, and thick quartzite and conglomerate beds. The shale is dated at about 1550 m yr. The Tarcoola Beds, which are comprised of conglomerate, sandstone, slate, and dolomite, and the quartzites of the Wallabying Range are possibly equivalent formations of the same age as the Corunna Conglomerate.

The Corunna Conglomerate is moderately folded. Extrusion of the Gawler Range Volcanics occurred at about 1535 m yr, continuing until about 1470 m yr, when the Moonta Porphyry of Yorke Peninsula and young dikes near Corunna were emplaced. This is the Wartakan Phase of folding and acid volcanism that ended the Carpenterian orogenic cycle in the Gawler Block.

Other Basement Rocks. Precambrian crystalline basement rocks are known from the Mt. Lofty Ranges, the Flinders Ranges, the Peake and Denison ranges, and the Willyama Block, which extend E from the Olary Ranges to the Barrier Ranges in New South Wales. These rocks, comprised of schists, pegmatitic and granitoid rocks, and augen gneisses, are exposed in isolated anticlinal inliers. In the Flinders Ranges small outcrops of diapiric breccias contain blocks of granite gneiss and jaspilite.

Anticlinal cores of a metasedimentary basement rock in the NE of the Flinders Ranges between Mt. Painter and Prospect Hill are called the Radium Creek Metamorphics. In the N the metasediments have been intruded by the pre-Adelaidean Yerila Granite and the Watleowie Granite. In the S a great laccolith is composed of the Mt. Neill Granite-Porphyry and the Terrapinnan Granite. The deformation, granite intrusion, and deformation are referred to as the Terrapinnan Phase. Isotopic ages range from 1410 to 1900 m yr and include a total rock age of 1650 m yr, suggesting correlation with the Charlestonian Phase of the Gawler Platform.

A structurally complex area of basement inliers occurs in the Peake and Denison ranges between the Mt. Painter area and the Musgrave Block and contains slate, phyllite, mica schist, quartzite, amphibolitic rocks that grade into migmatites, and granite gneiss.

The basement rocks of the Willyama Block, the Willyama Complex, appear to begin with high-grade metamorphic sillimanite garnet gneiss that extends E from the Radium Hill area into New South Wales, where it forms the major rock unit containing the Broken Hill lode.

The central part of the province is dominated by adamellite, pegmatite, and granitoid rocks, and also contains some intrusions of basic plugs and sills. Some sediments also occur, the sericitic Weekeroo Schist and the Outalpa Quartzite, with thin iron formations such as the Koolka Metajaspilite.

Early folding, metamorphism, and igneous events are referred to as the Olarian Orogeny after the Olary district. The Broken Hill metamorphism has been dated at 1650–1700 m yr, the age of lead mineralization at Broken Hill and of the uranium deposits at Radium Hill and Crocker Well. A date of 1580 m yr for the Binberrie Adamellite approximates the terminal Charlestonian phase of the Gawler Platform, and a date of 1520 m yr for the Mundi Mundi Granite corresponds to the Wartatakan Phase of the Gawler Platform.

Proterozoic Cover Rocks

Mawson and Sprigg (1950) (see Parkin, 1969) and Sprigg (1952) defined the terms

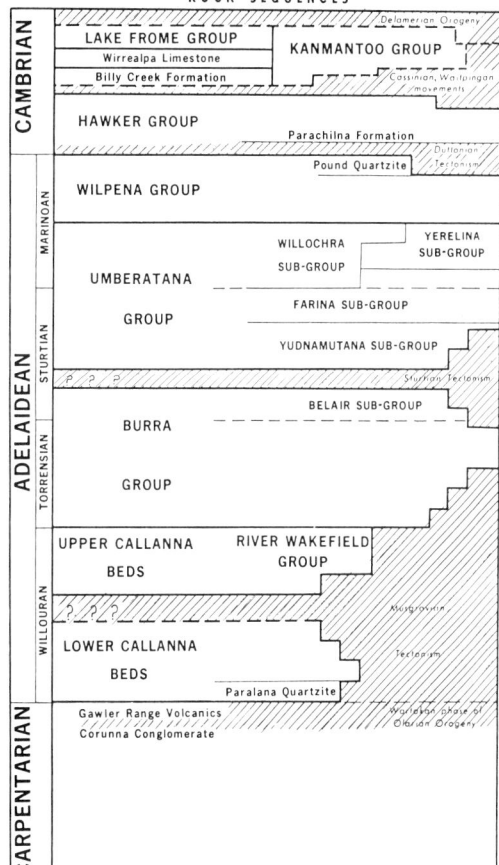

FIGURE 4. Cambrian and Precambrian time and rock terms used in the Adelaide Geosyncline (S. Australian Dept. Mines; in Parkin, 1969).

Wilouran, Torrensian, Sturtian, and Marinoan for stratigraphic units in the Adelaidean sequence (Fig. 4). These names have since been adapted as time terms, and four major rock units, the Callanna Beds, Burra Group, Umberatana Group, and Wilena Group have been defined by Thomson et al. (1964). The sequence from oldest to youngest is summarized below.

Willouran Callanna Beds. The Lower Callanna Beds are quartzites and metamorphosed dolomitic rocks (including marble and amphibolite) at the base, basaltic and andesitic lavas associated with basic intrusives, and an upper sequence of sandstone and quartzite interbedded with red and green shales up to 2000 m thick. The sequence has been described in the Mt. Painter area of the Flinders Ranges. The Lower Callanna Beds intrude the younger cover rocks as discordant diapir structures. Diapirs are widespread, but a salt lubricant is largely missing, although halite casts are common (Fig. 5).

The type sequence of the Upper Callanna Beds is in the Willouran Ranges, about 2000 m thick, and consists of alternating flaggy sandstone and green siltstone with thin basal dolomites.

Torrensian-Early Sturtian Burra Group. The rocks of the Burra Group extend from the Mt. Lofty Ranges to the Peake and Denison ranges in the N. The group is about 6000 m thick and occupies the western part of the Adelaide Geosyncline. It was deposited in a lagoonal to shallow marine environment and is the result of a major sedimentary cycle beginning at least 1000 m yr ago. The sequence commences with the Rhynie Sandstone and its equivalents, includes a dominantly carbonate siltstone succession, and ends with shallow-water clastics of the Belair Subgroup.

Sturtian-Marinoan Umberatana Group. Sedimentation in this interval was preceded by mild folding and diapirism followed by subaerial erosion. It was marked by the Sturtian Glaciation, and tectonic movement continued.

The Umberatana Group extends throughout the Adelaide Geosyncline and onto the marginal basement. The group is generally 3500–4500 m thick but increases to over 6000 m in the SE. The oldest sequence is the tillitic Yudnamutana Subgroup of Sturtian age, then follows a thick nonglacial shallow marine and continental succession called the Interglacial Sequence, and this is followed by the tillitic Yerelina Subgroup of Marinoan age. The Sturtian glaciation has isotopic dates that indicate a minimum age of about 750 m yr.

Marinoan Wilpena Group. This group, which includes the youngest Precambrian sequences in South Australia, was deposited after the Marinoan glaciation. It began with a carbonate-sandstone-siltstone sequence, followed by red and green siltstones, and by limestone and sandstone. The group formed an extensive blanket of shallow-water marine and continental sediments extending from the

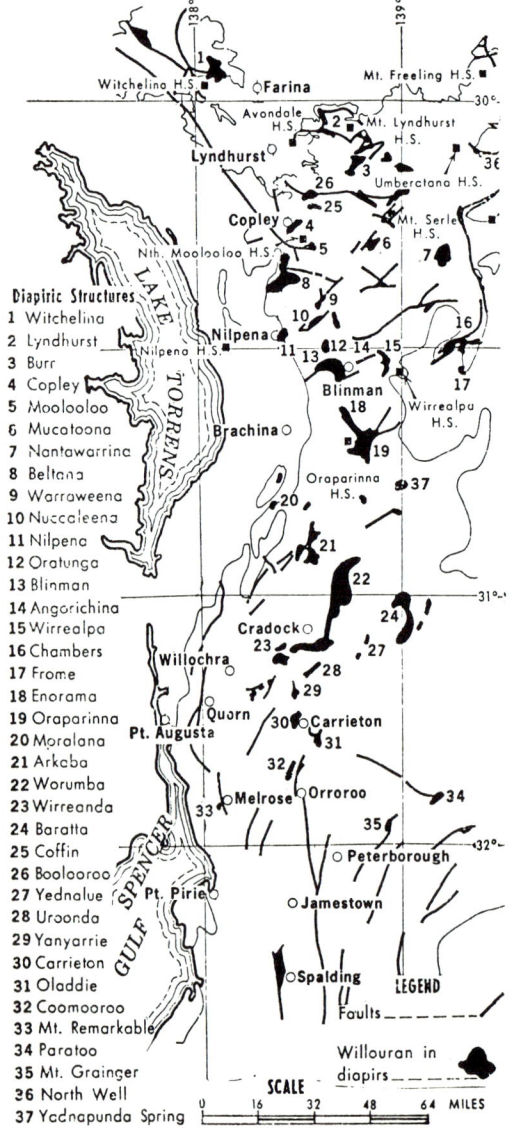

FIGURE 5. Diapirs in the Adelaide Geosyncline of the Flinders Ranges and northern Adelaide Ranges. A salt lubricant (with carbonate and shaley interbeds) may have existed in the Lower Callanna Beds (Willowran). Many of the Precambrian lithic members occur as "rafts" in the diapirs. An extensive copper mineralization is associated with the diapirs of the northern Flinders Range. (S. Australian Dept. Mines; in Parkin, 1969).

Diapiric Structures
1. Witchelina
2. Lyndhurst
3. Burr
4. Copley
5. Moolooloo
6. Mucatoona
7. Nantawarrina
8. Beltana
9. Warraweena
10. Nuccaleena
11. Nilpena
12. Oratunga
13. Blinman
14. Angorichina
15. Wirrealpa
16. Chambers
17. Frome
18. Enorama
19. Oraparinna
20. Moralana
21. Arkaba
22. Worumba
23. Wirreanda
24. Baratta
25. Coffin
26. Boolooroo
27. Yednalue
28. Uroonda
29. Yanyarrie
30. Carrieton
31. Oladdie
32. Coomooroo
33. Mt. Remarkable
34. Paratoo
35. Mt. Grainger
36. North Well
37. Yednapunda Spring

Stuart Shelf of the Gawler Block E across the geosyncline and into New South Wales.

The youngest unit is the Pound Quartzite, which was once thought to mark the base of the Cambrian but now is regarded as the youngest Precambrian (Adelaidean) unit. At Ediacara on the western margin of the geosyncline, the unit contains a remarkable fauna of fossil jellyfish, sea pens, the segmented worm *Spriggina* (named after the discoverer, R. C. Sprigg), and several other organisms previously unknown.

Paleozoic

In the northeastern Mt. Lofty Ranges near Dutton, about 600 m of the basic and intermediate Truro Volcanics unconformably overlie Umberatana Group sediments and underlie Lower Cambrian sediments, marking the Duttonian Folding, which took place from later Marinoan to the early Cambrian. This folding influenced a zone extending into the area of the present Murray Basin and along the eastern and southern flanks of the present Lofty Ranges.

Within the Adelaide Geosyncline, Cambrian deposition was preceded by a regressive period late in the Adelaidean. A stable shelf, the Stuart Shelf, was exposed along Yorke Peninsula and to the W of Lake Torrens at this time.

The Cambrian Succession. At the onset of the Cambrian the sea transgressed the shelf and the floor of the geosyncline subsided. Because of the mobile conditions during basin development, the Lower Cambrian sequence is characterized by marked changes in thickness and facies. The most complete sequences occur in the Flinders Ranges and are called the Hawker Group. Important stratigraphic units in it are, from oldest to youngest, the Parachilna Formation, Wilkawillina Limestone (a reef-type limestone with abundant *Archeocyatha*), Parara Limestone, Bunkers Sandstone, Oraparinna Shale, and the Narina Graywacke.

Toward the end of the Lower Cambrian, positive movements occurred throughout the Adelaide Geosyncline, and this led to local disconformities.

In the Cooper Basin, sediments of Middle to Upper Cambrian age consist of tuffaceous shale, dolomite, and a limestone that contains volcanic bombs and small ejactamenta.

In the Adelaide Geosyncline, shallow marine red beds of the Billy Creek Formation, containing red and green shale and siltstone, were deposited. Pink tuffaceous siltstones in the lower parts of the formation may indicate volcanicity in the Cooper Basin. At the beginning of the Middle Cambrian, shallow-water carbonate deposits were laid down, including the Wirrealpa and Aroona Creek Limestone in the Flinders Ranges and the Ramsay Limestone on Yorke Peninsula. The period of basin-wide carbonate deposition was followed by deposition of the Lake Frome Group, containing red and green clastics about 4000 m thick. The lower formations include red beds with some dolomites—the Moodlatana, Balcoracana, and Pantapinna formations. The youngest formation of the Lake Frome Group is the Grindstone Range Sandstone.

At the close of Hawker Group sedimentation, the great Kanmantoo Trough developed in the eastern Mt. Lofty Ranges Kangaroo Island area. This tectonic event is named the Waitpingan Subsidence. Compensating upward movement in adjacent basement areas to the N and W are termed the Cassinian Uplift.

The rapid sedimentation in the Kanmantoo Trough was marked by complex intertonguing of units and facies variation. The sequence, which has an "apparent thickness" of about 18,000 m, is named the Kanmantoo Group and contains siltstone with graywacke, arkose, phyllite, and carbonate rocks.

The Delamerian Orogeny. Structural and stratigraphic relationships in the Delamere area of the Fleurieu Peninsula provide evidence of an orogeny that probably began in the Upper Cambrian. This event is called the Delamerian Orogeny.

Metamorphism at this time of the crystalline basement and of the sediments in the Adelaidean Geosyncline and the Kanmantoo Trough is recorded in the Peake and Denison ranges, the Flinders Ranges near Mt. Painter, Kangaroo Island, and also in the great metamorphic arc that sweeps NE from Kangaroo Island to the Willyama Block. The orogeny reached its climax with the intrusion of the Anabama Granite and the Palmer Granite in Lower Ordovician time. The Mudnawatana Granite at Mt. Painter, the Victor Harbour Granite on Fleurieu Peninsula, and other granites on Kangaroo Island were probably intruded during the closing phases of the orogeny.

Ordovician, Silurian, and Devonian Rocks. During the closing phases of the Delamerian Orogeny, Ordovician sediments were deposited in the Great Artesian Basin. Outcrops occur in a small area in northern South Australia near Mt. Chandler and have been encountered in bores, usually at depths of 1000 m or more. Graptolites of early Ordovician age were deposited in a pelagic environment.

A prolonged period of relative stability marked by faulting followed the Delamerian Orogeny. The general absence of post-Cambrian and pre-Permian sediments in the southern portion of state suggests that the area was uplifted and formed a source area for middle Paleozoic sediments deposited in basins to the N and E.

Sediments deposited in the deeper parts of the Great Artesian Basin include possible Silurian and certainly Devonian. Upper Devonian dolomite, anhydrite, and magnesitic shale indicate an evaporite environment, probably developed in a marginal area of an embayment. Upper Devonian sediments generally underlie the Pedirka Basin in the west-central Simpson Desert and include the Innamincka Red Beds, also thought to be Devonian.

Permian Sediments. Early in the Permian a considerable area of the state was covered by continental ice (Fig. 6). Permian sediments in the Lake Phillipson Trough and the Arckaringa Basin, together with evidence from Gulf St. Vincent and the Murray Basin, disclose the presence of glacial valleys and troughs to which the sea gained access for a short time. Tillite is found at the base of the sequence. The brief marine incursion was followed by lacustrine conditions, during which carbonaceous siltstones and mudstones with a rich microflora were deposited. Coal interbeds occur in this basin in the Mt. Toodina Beds. The Cooper Basin was apparently nonmarine.

The Lake Phillipson Trough contains a sequence of 900 m of Lower Permian (Sakmarian to Artinskian) sediments, identified by their microflora and by foraminifera in the lower part. The lower beds of the sequence are the "Lake Phillipson Beds"—first recognized by Jack (1930) as Permo-Carboniferous till—which are about 75 m thick and contain pebbles, pyrite, biotite, and garnet. Overlying the "Lake Phillipson Beds" are the Stuart Range Beds, which are about 500 m in thickness. The lowest mudstones contain foraminifera and rare molluscs, ostracodes, and vertebrate remains.

In the Arckaringa Basin, an inlier of Permian age 90 m thick has coal interbeds containing *Glossopteris, Gangamopteris,* and other leaf remains.

The Boorthanna Trough contains the lower part of the Lake Phillipson sequence. The "Lake Phillipson Beds" are probably synchronous with pockets of Permian till outcropping on the western flanks of the Peake and Denison ranges (Reyner, 1955).

The Permian sequence in the Cooper Basin is composed of two units, the Merrimelia Formation and the overlying Gidgealpa Group. The Merrimelia Formation has a thickness of 360 m in the subsurface type section at Merrimelia No. 1 Well, where it is dated on microflora as Lower Permian (Sakmarian). The Gidgealpa Formation, over 300 m thick, contains coal interbeds and unconformably overlies the Merrimelia Formation, or older.

The marine, fluvioglacial, and other deposits of this time include the Permian till, sands, and clays of the Fleurieu Peninsula, named the Cape Jervis Beds, and equivalents that have been revealed in oil exploration wells near Renmark and elsewhere.

A deep basement trough, the Renmark Trough, extends between Monash and Renmark and contains Permian sediments with foraminifera up to 450 m thick that are believed to be of a marine glacial origin. O'Driscoll (1960, p. 24) states that the trough appears to have a steep-sided U-shaped section, and probably marks the course of a glacier, the infill being usually a greenish- or bluish-gray clay.

Mesozoic

During the Triassic and most of the Jurassic, the present land area of the state was above sea level and sedimentation was restricted to thin terrestrial and lacustrine deposits. In the Upper Jurassic, subsidence occurred in the Gambier Embayment. Drilling has not penetrated below the basal Cretaceous, so Jurassic sediments may occur below the basal Cretaceous.

Triassic and Jurassic Continental Sediments. The Triassic sediments of the Cooper Basin, which are termed the Nappamerri Formation, have been encountered in most wells drilled there. In the South Australian portion of the Cooper Basin, they are up to 200 m thick and appear to overlie conformably an Upper Permian Gidgealpa Group, which suggests a Lower Triassic age. The absence of marine fossils, together with the dolomitic nature and

FIGURE 6. Deeply grooved early Permian glacial pavement (Selwyn's Rock, Inman Valley, South Australia). Ice movement was from SE to NW (direction of hammer handle). The Precambrian rock surface is secondarily scoured by fluvial abrasion (right to left fluting). (Photo: R. W. Fairbridge.)

maroon color, suggest deposition in a landlocked basin under mildly evaporitic conditions. A thin sequence of Upper Triassic sediments dated by fossil spores was encountered below 2060 m in Pandieburra No. 1.

Triassic sediments also occur in intermontane basins within the Flinders Ranges. The best known are the Leigh Creek Coal Measures, which occur in four small basins near Leigh Creek. Other examples include the Springfield and Boolcunda basins about 250 km S of Leigh Creek.

Sediments of Jurassic age underlie Cretaceous sediments at depth throughout most of the South Australian portion of the Great Artesian Basin. The Jurassic sediments represent an extensive terrestrial freshwater depositional cycle which, in places, may have continued into earliest Cretaceous time. They comprise a lower and an upper sandstone unit separated by shale and coal. The oldest of those deposits is a lithological equivalent of the Hutton Sandstone of the Mulgildie and Surat basins in Queensland. On palynological evidence, the Hutton Sandstone is of Lower to Middle Jurassic age.

In South Australia, equivalents of the Birkhead Formation and the Walloon Coal Measures of Queensland follow conformably upon the Hutton Sandstone but overlap the latter in places. The Birkhead Formation is 30–90 m thick in some wells. It thins to the S and SE. Palynological evidence suggests a middle to late Jurassic age. The sediments are indicative of a freshwater swamp environment with intercalations of lacustrine deposits.

The uppermost Jurassic is subdivided in Queensland into Gubberamunda Sandstone, Fossil Wood Beds, and Mooga Sandstone. This subdivision may be recognized in the southeastern part of the Cooper Basin. However, toward the center of the Cooper Basin, the whole interval becomes increasingly sandy and is referred to as the "Mooga Formation." The upper part of the sequence may extend into the early Cretaceous. The Mooga Formation is the most widespread of the Jurassic sediments and can be traced to surface outcrops along the southern and western margins of the Great Artesian Basin. Its thickness varies from 210 m in the N to 350 m in the central parts of the basin.

In the northern Flinders Range the Mooga Formation is well-sorted sandstone, 20–30 m thick, and contains abundant plant remains of possibly early Cretaceous age; W of the Birdsville Track Ridge, this sandstone, the Village Well Formation, thickens to about 300 m and then thins considerably W of Lake Eyre, to N and S of Oodnadatta. Here it has been described as Algebuckina Sandstone, resting on highly weathered and kaolinized basement rocks. This deep weathering of the underlying rocks is indicative of the existence of a stable land surface prior to the deposition of the Algebuckina Sandstone. Stripped surfaces in the Peake and Denison ranges have been correlated with the base of the Jurassic Algebuckina Sandstone in the adjoining basin.

The Algebuckina Sandstone includes a lower kaolinitic sandstone unit and an upper clean, well-sorted sandstone. The lower unit consists of white, fine to coarse-grained quartz conglomerate, part of which is auriferous; it is current bedded throughout, with much cut and fill and clay pebble breccias. The upper unit is a well-sorted quartz sandstone, with large-scale concave current bedding. Silicified plant impressions at the top of the unit near Mt. Anna are very similar to the Mt. Babbage flora and indicate a later Jurassic to early Cretaceous age.

Dark lignitic clays of upper Jurassic age are known from the Polda Basin, which extends latitudinally from western Eyre Peninsula into the Great Australian Bight, possibly connecting with the Eucla Basin at the edge of the continental shelf.

In the deeper parts of the Gambier Embayment of the Otway Basin, Jurassic sediments may exist in the considerable thickness of sediments known to exist below the bottom of drill holes in basal Cretaceous sediments.

Cretaceous. Cretaceous strata are the most widespread of the Mesozoic rocks in South Australia. Downwarping began early in the Cretaceous, and large areas were inundated by the sea. These movements reached their maximum development before the late Cretaceous. By late Cretaceous time most of the basins had become filled with sediments and the environment gradually changed from marine to fluviolacustrine. The Gambier Embayment of the Otway Basin is an exception; marine conditions became important there during the late Cretaceous.

In the Great Artesian Basin there was a major transgression of the early Cretaceous time; the basal sediments reflect a transition from the terrestrial freshwater environment that prevailed in the Jurassic.

In the Marree area and around the north and northeastern margin of the Flinders Ranges, the Lower Cretaceous transgression is represented by the Pelican Well Formation, the Trinity Well Sandstone Member of the Marree Formation, the Parabarana Sandstone—which is a marine facies equivalent of the Mt. Anna Sandstone Member of the Cadnaowie Formation—and the Wilpoorinna Breccia Member of the Marree Formation.

In the western part of the Great Artesian Basin, including the Oodnadatta area, sedimentation commenced with the deposition of the Cadna-owie Formation. The various lithologies can be attributed to specific marginal marine environments such as backwater lagoons, sand bars, deltaic conditions, and shallow, agitated waters.

The Mt. Anna Sandstone Member, which is developed along the western and southern margins of the Great Artesian Basin, consists of well-sorted current-bedded sandstones with lenses and layers containing clasts of porphyry conglomerate and is thought to be of deltaic and fluviatile origin, the porphyry apparently derived from the Gawler Ranges. In places there are large amounts of fossil wood. This sandstone interfingers with the other rock types of the Cadna-owie Formation. The age of the Cadna-owie Formation is generally regarded as early Cretaceous. The formation comprises one of the most persistent seismic reflection horizons in the basin. The thickness is rather uniform, ranging from 45 to 74 m.

Conformably overlying the Cadna-owie Formation is the Bulldog Shale; some 60–150 m above its base is a horizon of ellipsoidal limestone concentrations which contains a rich marine fauna of bivalves, gastropods, and crinoid stems. In the basin center the Bulldog Shale is up to 460 m thick but thins toward the margins and near Oodnadatta is only 150 m thick. Equivalent claystones and shales which overly basal sandstones in the Marree area, together with the overlying Oodnadatta Formation, are named the Maree Formation. The lower part contains a Roma fauna, rich in molluscs; arenaceous foraminifera are common. In the middle of the sequence is a thin member, a green glauconitic sandstone. The upper part of the Marree Formation contains an Albian (Tambo) fauna with ammonities and both arenaceous and calcareous foraminifera.

The Coorikiana Member of the Oodnadatta Formation consists largely of silty shale and some lenticular limestone. The Wooldridge Limestone Member of the Oodnadatta Formation, near Fossil Creek, contains abundant ammonities and celestite. The uppermost part of the formation consists of 50 m of glauconitic sandstone, the Mt. Alexander Sandstone Member. The thickness of the Oodnadatta Formation varies from 140 m near the margin to 600 m in the deeper parts of the basin. Its fauna is Albian. In the next younger formation, the Winton Formation, a regression of the Cretaceous sea is indicated by increased sand and an abundance of coal. In places large calcified trunks of fossil trees are common. The thickness in the deeper parts of the Cooper Basin is over 600 m. It is regarded as a freshwater deposit and paleontological evidence suggests a Cenomanian age. It was the last of the Cretaceous formations, with widespread distribution throughout the Great Artesian Basin, and the top is an erosional surface.

By the end of Cenomanian time, all of the South Australian portion of the basin was land, and the sedimentary record of this time interval is restricted to occasional fluviatile deposits such as the channel fills of large, low-gradient streams, examples of which are the Mt. Howie Sandstone in the far NE of the state, and the post-Winton Upper Cretaceous sands encountered in Kalladeina No. 1 Well.

In the Eucla Basin, Lower Cretaceous sediments have been intersected in drill holes put down near the Trans-Australia Railway and in the southern part of the basin. Coarse basal arenites, resting on crystalline basement or Permian sediments, are lithological equivalents of sandstones on the western margin of the Great Artesian Basin.

In the Murray Basin the basal nonmarine quartz sandstones are generally regarded as of early Cretaceous age. The overlying marine beds consist of siltstones, followed by a dark gray marine shale of Aptian age. Marine Cretaceous sediments younger than Albian are not known here, and were followed by a fluviolacustrine environment. A marine environment continued in the Gambier Embayment, where the sediments constitute the Otway Group, which probably exceeds 3000 m in thickness, and consists of green to gray lithic or feldspathic siltstone and fine to medium-grained sandstones. Coarser sediments are not uncommon. Volcanic and other igneous rock fragments occur as detrital components. Carbonaceous material occurs throughout, and some of the mudstones grade into coal. In situ carbonized plant roots indicate a nonmarine environment for part of the sequence. The sediments are usually finely interbedded, but thicker beds, which show cross-bedding, slump structures, and scouring, are also present. The Otway Group is informally subdivided into two units on the basis of an unconformity detected from seismic data. The lower unit is Upper Jurassic to lowermost Cretaceous, whereas the upper ranges from lowermost Cretaceous, possibly up to the Cenomanian.

Cenozoic

Lower Tertiary. The terms "Lower," "Middle," and "Upper" Tertiary are only used in a relative sense. Sediments of Lower Tertiary age are widespread in the inland parts of the state. Within the Great Artesian Basin they commonly occur as thin sands and gravels in erosional disconformity with the underlying Cretaceous beds and are capped with silcrete.

In the NE of the state near Innamincka, unfossiliferous cross-bedded sandstone overlies Cretaceous sediments. The sandstone, later silcreted, is 3 m up to 25 m thick. In the Oodnadatta area the Macumba Sandstone is probably an equivalent.

Near Marree the Lower Tertiary sediments consist of sands with a distinctive basal conglomerate composed of various forms of quartz, named the Murnpeowie Formation. Its usual thickness is 3–12 m, exceptionally 80 m. Sediments probably of equivalent age are the lacustrine beds found at 45–80 m depth in bores in the bed of Lake Eyre, silicified shales with leaf remains in the Callanna area, and silicified sandstones with fossil leaves at Mt. Eba.

In the south coast basins, at the beginning of the Tertiary, uplift of the ranges and complementary subsidence was followed by marine transgression. Conditions of deposition were paralic, and carbonaceous sands and silts, lignitic beds, and low-rank coals were laid down. In late Eocene time normal marine limestones and marls were deposited.

In the Eucla Basin, Tertiary sedimentation began in the Middle Eocene with carbonaceous and pyritic clays, sands, and silts of the Pidinga Formation. In some bores a coarse limonitic and glauconitic sand, the Hampton "Conglomerate," is present which may correlate with the Kongorong Sand (see below) of the Gambier Embayment.

The St. Vincent Basin is part of a complex graben structure that contains a number of smaller basins. The largest is the Adelaide Plains Basin, which links with the deposits of the Noarlunga and Willunga Basins offshore in St. Vincent Gulf. The youngest strata beneath the Tertiary beds are of Permian age.

The oldest Tertiary beds of St. Vincent Basin are the Middle to Upper Eocene North Maslin Sands, which are partly marine but also contain lignite beds and low-rank coal. They are followed disconformably by glauconitic South Maslin Sands of Upper Eocene age.

FIGURE 7. Stratigraphic correlation table for the South Australian Tertiary. (S. Australian Dept. Mines.)

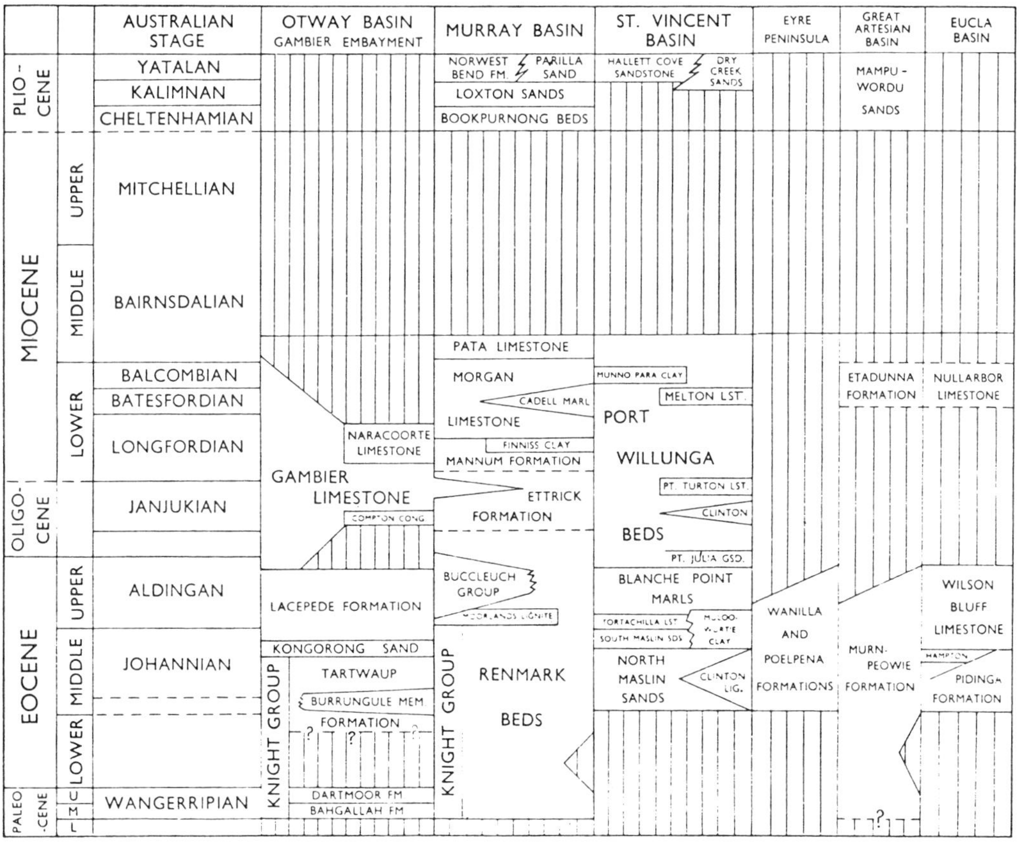

In the Murray Basin the name "Knight Group" has been used for undifferentiated sands, clays, and siltstones of Paleocene to Eocene age (Fig. 7). The break at the top of these beds is marked by a green or gray glauconitic clay-stone with pisolitic ironstone. Other nonmarine and paralic beds of Paleocene to Eocene age occurring in the subsurface have been named the Renmark Beds by Harris (1966).

The "Knight Group" beds are overlain by the transgressive Buccleuch Beds, which contain important Upper Eocene marine faunas and are about 50 m thick. The bryozoal limestone at the base formed in clear and rather deep water apparently localized in the corridor between the Marmon Jubuk Fault and the island chain SW of Coonalpyn.

Middle Tertiary. Silicified rocks are widespread in the Great Artesian Basin. Various kinds of silicification, ranging from case hardening, through formation of discontinuous layers of opaline silica, to massive more-or-less continuous sheets of siliceous duricrust (silcrete), are known. These siliceous materials were widely developed in the time interval from the Lower Tertiary to the Quaternary. The most prominent silcretes are those stratigraphically associated with the Murnpeowie Formation. This duricrust is commonly a hard, very fine-grained siliceous rock, gray red-brown or yellowish and up to 8 m thick. Generally below the duricrust there is a zone of white, bleached rocks near the base of which some ferruginous mottling is common. This is underlain in some places by strongly ferruginized rocks or a zone of ironstone concretions.

Silcreted rocks older than the Murnpeowie Formation are uncommon, but in some places the Cretaceous Marree Formation and the Precambrian Umberatana Group are silicified. The stratigraphic association of silcrete with Tertiary sediments suggests that the silcrete may have been formed in lowlands where drainage conditions strongly influenced accumulations of silica.

A variety of reasons points to polygenetic features in the classic silcrete profile: two separate layers of silcrete in the Murnpeowie Formation merge and vary in thickness from place to place; both silcretes include conglomerate; the upper part of the lower bed shows a columnar structure normal to the bedding; two separate silcrete layers can also be recognized in logs of bores put down in the Frome Embayment; in some places, ferruginous and silicified materials occur in separate horizons in the one profile, in others, varying amounts of silica and iron may be present in the one horizon. Not all of the weathering zones underlying the silcrete cap were necessarily formed at the same time as the cap itself and some, such as the underlying bleached and ferricreted zones, may be considerably older. Bands and lenses of limonitic or hematitic shale and sandstones interstratified with kaolinitic material may represent separate events preceding the deposition and modification of younger beds in the silcrete sequences.

Evidently the "duricrust profile" in the Great Artesian Basin consists of various layers formed in different ways at different times, the basal layers being much older than the upper layers.

South of the Warburton River and E of Lake Eyre, along the courses of the Warburton and Cooper Creek, and in cliffs marginal to lakes and elsewhere, "limonitic grits" mark a surface of disconformity on the early Tertiary, and these are succeeded by a sequence of mudstones and dolomites called the Etadunna Formation. Intraformational breccias are common, indicating repeated exposure and drying of the shallow-water lakes in which the sediments were deposited. Vertebrate remains in lacustrine gypsiferous marls and clays of this formation E of Lake Eyre near Lake Kanunka were first recorded in 1882. An Oligocene or Miocene age for the Etadunna Formation has been suggested, but it may be younger.

In the Eucla Basin the major part of the sequence consists of the Wilson Bluff Limestone, which has a maximum thickness of about 120 m and a glauconitic bed at the base which contains a Middle Eocene fauna. It is followed by the Nullarbor Limestone, about 30 m thick, which was deposited late in the Lower Miocene.

The Tortachilla Limestone and equivalents in St. Vincent Basin mark an important marine transgression during the Middle and Upper Eocene. The Tortachilla Limestone grades into the Upper Eocene Blanche Point Marls. Equivalent of the marls overlie the Eocene Muloowurtie Clays of Yorke Peninsula.

The Port Willunga Beds of Upper Eocene to Miocene age are an important stratigraphic unit equivalent in age to the Gambier Limestone. In the Murray Basin the Oligocene marine transgression is marked by equivalents of the Compton Conglomerate and by the Ettrick Formation. The Ettrick Formation grades up in places into white bryozoal limestone, indicating a marine connection with the Gambier Embayment.

The Mannum Formation, which is overlain in places by Finniss Clay, is transgressive onto bedrock on the western margin of the basin. This unit and the Morgan Limestone—which with the Cadell Marl Lens attains a thickness of 90 m—indicate warm shallow seas. The units are neritic to littoral. In the deeper parts of the basin, where deposition was continuous, the limestone is named the "Morgan-Mannum limestone."

In the Gambier Embayment the Compton Conglomerate marks the Oligocene marine transgression, which extends far into the Murray Basin. The overlying Gambier Limestone is diachronous. In the western part of the embayment, its lowest part is of Upper Eocene age. The limestone extends well into the Lower Miocene.

Upper Tertiary. Tectonic activity, which led to the breakup and elevation of the Tertiary basins, commenced in the late Miocene following the relatively stable conditions that were marked by deposition of marine limestone during the Middle Tertiary.

Deformation of the Port Willunga Beds and the older sediments was followed by deposition of the Hallett Cove Sandstone and the Dry Creek Sands, which are equivalent in part.

Fluviatile sands in the Barossa Valley and possibly those near Gawler are of Miocene to Pliocene age.

An early Pliocene transgression in the Murray Basin is represented by the Bookpurnong Beds. The overlying Loxton Sands were deposited under estuarine and fluviolacustrine conditions.

A minor downwarp in the late Pliocene led to marine transgression and deposition of estuarine Norwest Bend Formation in a tract between Tailem Bend and Waikerie now occupied by the Murray River. The Parilla Sand is a clayey and micaceous quartz sand equivalent in part to the Norwest Bend Formation, into which it grades. The unit was probably deposited in a fluviolacustrine environment.

In Victoria (Quambatook) sediments equated with Parilla Sand contain marine molluscs and bryozoa and may include the initial littoral phase of the Murravian transgression.

Late Cenozoic Tectonics and Volcanism. A sequence of tectonic events that began with the breakup of the mid-Tertiary surface led to extensive clastic deposition throughout the remainder of the Tertiary. By the close of the late Pliocene it had produced a well-defined pattern of structural lineaments, including faults and major joint systems. Some lineaments parallel older faults and have throws up to at least 60 m. The lineaments trend roughly NW and NE, occurring in sets, with intersections forming rectangular and rhomboid blocks. The pattern of lineaments probably reflects deep-seated structures and tectonic elements in the crystalline basement.

A renewal of faulting and warping and an increase in volcanic activity mark the Kosciuskan Uplift, accompanied by a different style of sedimentation in the basins marginal to the uplifted block mountains of this time. The development of the earlier pattern of lineaments throughout the late Cenozoic is associated with upwarping of the inland margins of the basins and downwarping of the seaward margins, with the uplift of the ranges and the stranding of Tertiary and older Pleistocene littoral sediments high on their margins. In the SE, within the Gambier Embayment of the Otway Basin, there was volcanic activity that may be also associated with the lineaments.

Quaternary Sediments. Deposits overlying late Pliocene marine sediments and underlying a thin Holocene cover can be subdivided into three (unequal) major sequences of presumed Pleistocene age. Sedimentary evidence suggests a very long time span for the lower sequence.

Lower Pleistocene. Lower Pleistocene deposits are widespread inland. They include saprolite developed on (?)Tertiary carbonaceous and older rocks on the southern margin of the Musgrave Ranges; an oxidized and silicified layer of detritus possibly marking the Karoonda Surface, and occurring near the base of an olive clay sequence on the western margin of the Great Artesian Basin; and the Avondale Clay on the margin of the Flinders Ranges, etc.

At Lake Eyre North, a possible equivalent of the Avondale Clay is underlain by a sequence of dolomite and mudstone (King, 1956). Similar dolomite and mudstone sequences contain the Ngapakaldi fauna of pelicans, flamingoes, ducks, crocodiles, lungfish, and an ancestral koala at Lake Palankarinna and Lake Kanunka, named the Etadunna Formation. An Oligocene age has been suggested but it could possibly be younger.

Grayish purple or reddish clays overlie the duricrust at various scattered exposures. Limestones with alga-like structures at the base of the clays may be of lacustrine origin.

Grabens within Lake Torrens, Spencer Gulf, and Gulf St. Vincent all contain thick sequences of clay, sand, and gravel assigned to Avondale Clay or Hindmarsh Clay. Hindmarsh Clay is typically developed within Gulf St. Vincent, where it overlies Carisbrooke Sand, which contains a fauna thought to be of Plio-Pleistocene age.

In the Murray Basin the thin fluviolacustrine Blanchetown Clay is at the base of the Pleistocene sequence. Thin lenses of dolomitic micrite occur within it in some places; near the present coast they include fossiliferous limestone. The Karoonda Surface near the base of the clay sequence in South Australia is marked by structured, oxidized, and silicified detritus. This is possibly stratigraphically equivalent to similar deposits near Bordentown, Adelaide, Boston

Bay, and to similar inland deposits. A lensing channel deposit of yellow sand, called the Chowilla Sand, is derived from the oxidized detritus on the Karoonda Surface and is intercalated in the base of the Blanchetown Clay within the late Cenozoic Loxton Basin.

The Quaternary glaciers that affected the highlands of eastern Australia and Tasmania did not extend into South Australia. However, the fluctuating sea level caused by glacioeustasy had a profound effect on climate, coastal configuration, and landscape development.

On the Padthaway Archipelago and in the Otway Basin, early Pleistocene seas left a fossiliferous sandy limestone and calcareous sand, the Coomandook Formation. Spatial relationships suggest that Coomandook Formation is at least as old as shelly and gravelly sands at the base of the Bridgewater Formation. Other marine deposits of uncertain age are found at the stranded marine cliff at Up and Down Rock, near Tantanoola. Evidence of marine submergence is provided by pockets of reef shells scattered over the dolomitized Mount Gambier Limestone up to 75 m above present sea level.

Middle Pleistocene. Thin discontinuous deposits of lacustrine limestone overlie Lower Pleistocene clastics and older rocks in many parts of the state. They contain ostracodes, the gastropod *Coxiella*, the freshwater plant Chara, and reed casts. In some places they are silicified. Most are pale gray, olive, or brown micrites with varying amounts of dolomite. Tentatively included in this stratigraphic position are limestones of the Eateringinna area; the Mangatitja Limestone of the Officer Basin; the Mt. Willoughby Limestone on the western margin of the Great Artesian Basin; other limestones near Billakallina, Callana, and Maralinga; the lower part of Telford Gravel in the Lyndhurst area; the Nilpena Limestone on the western flanks of the Northern Flinders Ranges; the Bungunnia Limestone in the Murray Basin; and fossiliferous shallow marine and aeolian deposits in the Bridgewater Formation on the southern coastal margin.

Overlying the lacustrine limestones and older rocks is a distinctive duricrust called Ripon Calcrete, which is very widespread. The calcrete marks an extensive paleosurface, the Ripon Surface, at the base of an ancient brown soil profile. On the coast it separates the Lower from the Upper Member of the Bridgewater Formation.

The sequence up to Ripon Calcrete is eroded in many places on the western margin on the Lake Eyre Basin. The saprolitic and sedimentary green clays formed on the younger surface within the basin are referred to the Bopeechee Clay. They contain fossil skeletons of giant kangaroo, emu, wombat, *Diprotodon*, and abundant plant debris; they may be of Middle Pleistocene age.

Moderately hard layers of calcrete younger than Ripon Calcrete are widespread throughout the state (Figs. 8 and 9). These duricrusts characterize an ancient brown soil called the Bakara Soil and mark the paleosurface called the Bakara Surface. The calcrete is stratigraphically associated with the top of the Telford Gravel and other fluvial deposits inland, lacustrine limestones of the SE, eolianates of the Upper Member of the Bridgewater Formation, and the marine Anadara-bearing Glanville Formation of the coastal margin.

Upper Pleistocene. Following uplift and deep dissection related to the last low stand of the sea prior to the Flandrian transgression, the Pooraka Formation was laid down as alluvial fans and as valley fill. Older calcrete pans were reworked in erosional areas and buried in depositional areas. The formation is found on the western and northern margins of the Murray Basin as colluvial and alluvial material. The alluvial facies consist of reddish-brown clayey sands, sands and gravels occupying stream chan-

FIGURE 8. Table of Cenozoic soils and related continental stratigraphic units in South Australia. Calcretes and other paleosol crusts constitute essentially stratigraphic horizons which can be widely correlated.

AGE					STRATIGRAPHIC UNITS
QUATERNARY	RECENT			Se	Semaphore Sand and equivalents : Leached sandy slope deposits (not shown on diagram)
				M	Molineaux Sand, Bunyip Sand (Murray Basin)
				Br	Fulham Sand (St. Vincent Basin)
				ST	St. Kilda Formation
	PLEISTOCENE	UPPER		C	Thin red, grey or yellow clay (not shown on diagram)
				LO	Loveday Soil carbonate
				W	Woorinen Formation (Murray Basin)
				Pc	Pooraka Formation (St. Vincent and Spencer Basin) and equivalents
		MIDDLE		K	Calcrete in Bakara Soil
				L	Loess and carbonate silt - quartz sand mixtures
					Older carbonate horizons
				Bu	Bridgewater Formation — upper member
				R	Ripon Calcrete
				Bl	Bridgewater Formation — lower member
		LOWER			Olive clay (Keswick Clay and equivalents) over bedrock or at top of Hindmarsh Clay (not shown on diagram)
				B	Blanchetown Clay (Murray Basin)
				H	Hindmarsh Clay (St. Vincent and Spencer Basin)
				O	Ferricreted Parilla Sand and equivalents
TERTIARY	PLIOCENE			P	Parilla Sand over Loxton Sands
	OLDER			La	Laterite profile with ferricrete
				T	Tertiary sediments (undifferentiated)
				X	Basement rocks

FIGURE 9. Soil/stratigraphic profiles across the Mt. Lofty Ranges (above) and Wesern Eyre Peninsula (below), showing sequence of Tertiary and Quaternary continental sediments (accumulative soils) and crusts. Letter code refers to the table in Fig. 8. Number code: 1, lateritic podzolic soils; 2, yellow podzol; 3, black earth; 4, solonized brown soil (over calcrete); 5, red-brown earth; 6, calcareous coastal sand; 7, podzol (sandy ironstone); 8, desert sandhills; 9, solonized brown soils (or dunes); 10, siliceous sands, etc.

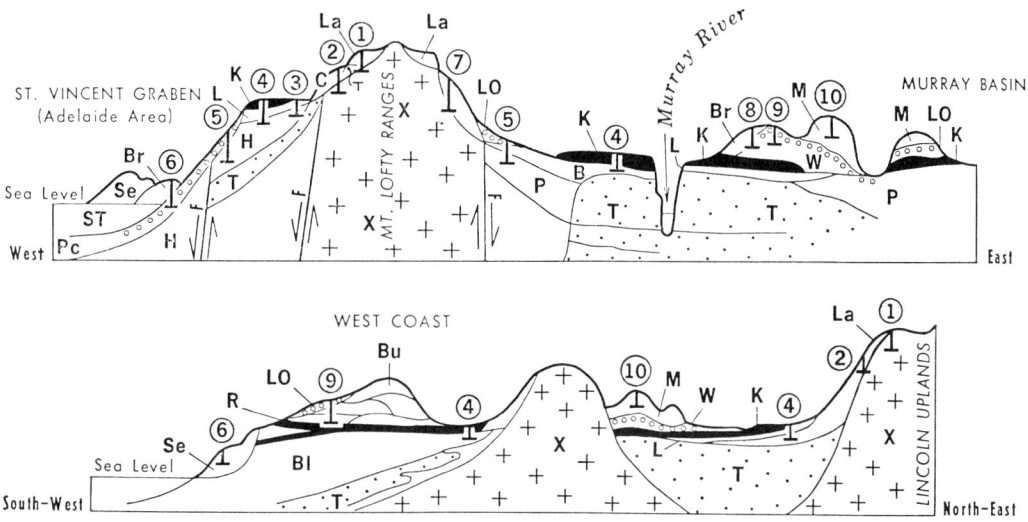

nels, incised through older materials on the flanks of the ranges, or overlying calcrete or older alluvials in the basin.

Horizons of nodular carbonate are found near the top of the Pooraka sequence. The carbonates and other relic features in soils mark the red-brown earths and younger brown soils that were a feature of the late Pleistocene landscape. Extensive fans are developed near the range margins, but these do not persist far into the inland basins. Here the Pooraka Formation occurs as a valley fill in stream courses incised through middle Pleistocene and older deposits.

During the Upper Pleistocene (Figs. 10 and 11), widespread eolian deposition began in the southern part of the state with deposition of Woorinen Formation, a pale red-brown quartz sand mixed with carbonate silt. A thin layer, the Callabonna Clay, is superimposed on the Pooraka Formation and its equivalents to form red-brown earths and stony red duplex soils (Callabonna Soil). This layer is taken to mark the Pleistocene-Holocene boundary in southern Australia.

Holocene. Among the Holocene deposits of the inland are talus of the range margins and

FIGURE 10. Late Pleistocene sand-ridge plain in northwestern part of Eyre Peninsula. Inselberg in distance is Pildappa Hill, a Precambrian (Carpentarian) granite. (Photo: C. R. Twidale.)

FIGURE 11. Detail of granite weathering on Ucontichie Hill (compare with Fig. 10), Eyre Peninsula. Soil from the granite hills was stripped off during Pleistocene arid phases, and during more humid phases the hill slope plain contact zone suffered alternate chemical weathering and evaporation with crust development. Similar "wave-hill" features are also found in semiarid parts of Western Australia. (Photo: C. R. Twidale.)

consisting of gypsiferous silts with halite crusts. Several generations of mound springs occur on the margins of the Great Artesian Basin. Only the youngest are of Holocene age. The deposits consist mainly of reworked Cretaceous sediments, black soft mud, and minor amounts of travertine.

The Simpson Sand of the great dune fields (Fig. 12) is one of the most prominent inland units. Materials immediately adjacent, or in alluvial spreads reworked into lee-side dunes, control the color and composition of the sands. Some dunes are layered and three layers commonly occur: the upper, yellow; the middle, darker reddish yellow; and the lower, a brown somewhat clayey sand with nodules of soil carbonate. The structure and basal position in sequence of the carbonate suggest a correlation with Upper Pleistocene carbonates in older dunes of the Woorinen Formation.

Gypsum dunes near the margins of the inland lakes are probably of late Pleistocene to Holocene age because they lie close to the present lake surface. A subfossil horizon of gypsite commonly occurs near the dune surface and is present elsewhere in the landscape as an eolian layer or as an impregnation in older units.

Stratigraphic relationships of the various deposits of the Holocene are better displayed on the coastal margins and in the southern basins.

FIGURE 12. Linear dunes of the Simpson Desert.

alluvium within the basins. Polymict gravels containing well-rounded materials are derived from older deposits. Monomict gravels, usually subangular rather than rounded, are common adjacent to older basement rocks and weathering silcrete pans and in the modern stream courses.

Within the Lake Eyre drainage basin, present streams have cut through older Holocene deposits. The Tingana Clay of the Strzelecki, Cooper, Warburton, and other streams draining into Lake Eyre is one of the older and more widespread of the layers exposed by this dissection.

Modern sediments of Lakes Eyre, Torrens, Frome, Gairdner, and other smaller lakes of northern Eyre Peninsula are thin evaporites

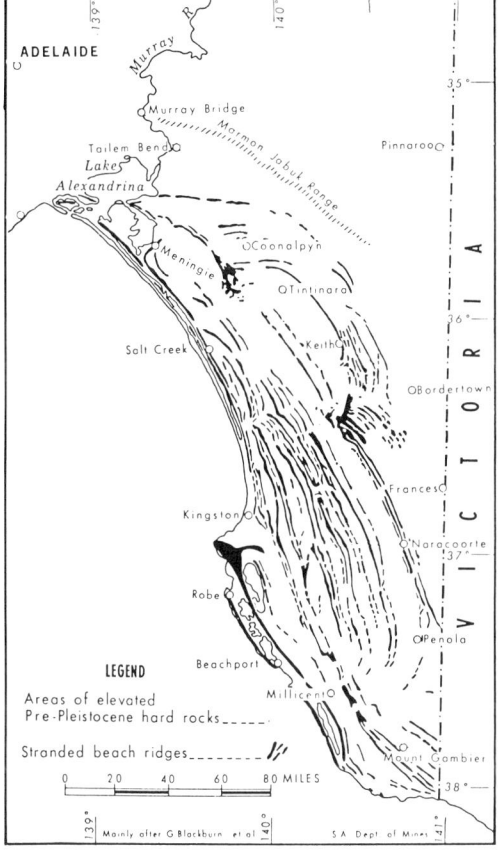

FIGURE 13. Sketch map of the southeastern part of South Australia, showing the stranded beach ridges of Pleistocene age (black). They are in part warped, as near Mt. Gambier. Most of the white parts of the map is a veneer of Holocene sands, continental or marine. In an offshoot of Lake Alexandrina running SE, the Coorong Lagoon, there are contemporary carbonate muds (with dolomitization in progress).

The deposits of the Flandrian transgression are distinctive. The most widespread is the St. Kilda Formation, which contains shallow marine sands, estuarine muds, and shelly deposits of the beach ridges. These deposits intertongue with other sequences, including the gypsiferous Yamba Formation and its equivalents on the west coast, dune sands marginal to St. Vincent Basin, Monoman and Coonambidgal Formations of the Murray River, and carbonate deposits of the Coorong lagoon and shallow lakes parallel to the coastline in the SE of South Australia.

The southern dune fields overlie Upper Pleistocene Woorinen Formation and older units. They include an inland belt of reddish Fulham Sand of the western and central basins, and Bunyip Sand of the Murray Basin, a southern belt of yellow Molineaux Sand, and a coastal belt of Semaphore Sand.

The Fulham Sand occurs on the margin of St. Vincent Gulf. Here shelly deposits marking the highest stand of the Flandrian transgression overlie lower-phase Fulham Sand with carbonate pipes and are overlain by younger layers of dune sand. A similar sequence occurs in the Murray River valley, where old aeolian sands with carbonate pipes are apparently overlain by the Coonambidgal Formation.

Eolian gypsum dunes overlying or marginal to lacustrine beds of the Yamba Formation are interbedded with the upper phase of Bunyip Sand and contain a subfossil horizon of gypsite tentatively correlated with a similar layer in the inland dunes.

Semaphore Sand marks the modern coast. It appears to be derived from and deposited upon the Bridgewater Formation, beach ridges of the St. Kilda Formation, and Recent offshore bars. In the Coorong area, basal layers intertongue with younger layers of carbonate (dolomitic) sediments in the Coorong lagoon.

Geological Exploration and Mining

Even before the founding of the first colony in 1836, there was a trade in salt from salt lakes on Kangaroo Island, collected by crews of whaling and sealing vessels. Silver-lead at Glen Osmond and copper ore at Montacute were discovered and mined in the early 1840s. Copper was discovered at Kapunda in 1842, followed by Burra and Wallaroo-Moonta. For many years South Australia was one of the world's foremost copper producers, and for 80 years copper was its principal mineral resource. Discoveries of gold, silver, lead, and zinc also contributed. Gold discoveries at Echunga in 1852, followed by those at Barossa, Waukaringa, Woodside, Mannahill, Teetulpa, and Tarcoola (1868–1900), resulted in comparatively small production.

Production of iron ore at Iron Knob to serve the newly established steel industry at Newcastle began in 1915, and at that time the output of industrial minerals (gypsum, salt, and limestone) was also greatly increased.

Since World War II, mineral developments have included the Leigh Creek Coalfield, which provided a local source of fuel for electric power generation; uranium at Radium Hill; talc at Gumeracha and Mount Fitton; barite at Oraparinna; and gypsum at Lake MacDonnell. At Nairne, pyrite deposits have been exploited for sulfuric acid manufacture, and calcareous sand deposits at Coffin Bay have been developed. The most significant hydrocarbon production to date has been the natural gas field in the Moomba-Gidgealpa area.

JON B. FIRMAN*

*Published by permission of B. P. Webb, Director of Mines, South Australian Department of Mines (and acknowledgments to R. K. Johns, B. G. Forbes, and B. Daily).

References

Campana, B., and Wilson, R. B., 1955. "Tillites and related glacial topography of South Australia," *Eclog. Geol. Helv.*, 48, 1–30.

Daily, B., 1956. "The Cambrian in South Australia," *Rep. 20th Internat. Geol. Cong., Mexico, 1956: El sistema Cambrico, su Paleogeografia y el problema de su base*, 2, 91–147.

____, and Milnes, A. R., 1971a. "Discovery of late Precambrian tillites (Sturt Group) and younger metasediments (Marine Group) on Dudley Peninsula-Kangaroo Island, South Australia," *Search*, 2 (11–12), 431–432.

____, and Milnes, A. R., 1971b. "Stratigraphic notes on lower Cambrian fossilferous metasediments between Campbell Creek and Lunkalilla Beach in the type section of the Kanmantoo Group, Fleurieu Peninsula, South Australia," *Trans. Roy. Soc. S. Austral.*, 95(4), 199–214.

____, Gostin, V. A., and Nelson, C. A., 1973. "Tectonic origin for an assumed glacial pavement of late Proterozoic age, South Australia," *J. Geol. Soc. Austral.*, 20(1), 75–78.

Dasch, E. J., Milnes, A. R., and Nesbitt, R. W., 1971. "Rubidium–strontium geochronology of the Encounter Bay granite and adjacent metasedimentary rocks, South Australia," *J. Geol. Soc. Austral.*, 18 (3), 256–266.

Dunn, P. R., Thomson, B. P., and Rankama, K., 1971. "Late Precambrian glaciation in Australia as a stratigraphic boundary," *Nature*, 231(5304), 498–502.

Firman, J. B., 1967. "Stratigraphy of late Cainozoic deposits in South Australia," *Trans. Roy. Soc. S. Austral.*, 91, 165–180.

____, 1971. "Paleosols–a stratigraphic definition," in: *Etudes sur le Quarternaire dans le Monde*. Paris: Union Internat. pour l'etude du Quat. 8th INQUA Cong., 1969.

Forbes, B. G., 1966. "The geology of the Marree 1:250,000 map area," *Rep. Invest. Geol. Surv. S. Austral.*, 28, 47p.

Glaessner, M. F., and Parkin, L. W., eds., 1958. "The geology of South Australia," *J. Geol. Soc. Austral.*, 5(2), 148p.

____, and Wade, M., 1966. "The Late Precambrian fossils from Ediacara, South Australia," *Paleontology*, 9, 599–628.

Harris, W. K., 1966. "New and redefined names in the South Australian Lower Tertiary stratigraphy," *Quart. Geol. Notes, Geol. Surv. S. Austral.*, 20, 2.

Howchin, W., 1907. "Report on Cambrian and ? Permo-Carboniferous glaciations in South Australia," *Rep. Austral. Assoc. Adv. Sci.*, 11, 264–272.

____, 1918. *The Geology of South Australia*. Adelaide: Govt. Printer, 515p. (2nd ed., 1929).

Jack, R. L., 1930. "Geological structure and other factors in relation to underground water supply in portions of South Australia," *Bull. Geol. Surv. S. Austral.*, 14, 48p.

Jessup, R. W., and Norris, R. M., 1971. "Cainozoic stratigraphy of the Lake Eyre Basin and part of the arid region lying to the south," *J. Geol. Soc. Austral.*, 18(3), 303–331.

Johnstone, M. H., Lowry, D. C., and Quilty, P. G., 1973. "The geology of southwestern Australia–a review," *J. Roy. Soc. W. Austral*, 56(1), 5–15.

King, D., 1956. "The Quaternary record at Lake Eyre North and the evolution of existing topographic forms," *Trans. Roy. Soc. S. Austral.*, 79, 93–103.

McColl, D. H., and Williams, G. E., 1970. "Australite distribution pattern in southern central Australia," *Nature*, 226(5241), 154–155.

Mawson, D., 1942. "The structural character of the Flinders Ranges," *Trans. Roy. Soc. S. Austral.*, 66, 262–272.

____, 1947. "The Adelaide Series as developed along the western margin of the Flinders Ranges," *Trans. Roy. Soc. S. Austral.*, 71, 259–280.

____, 1949. "The Elatina glaciation. A third recurrence of glaciation evidenced in the Adelaide System." *Trans. Roy. Soc. S. Austral.*, 73, 117–121.

Mulcahy, M. J., 1973. "Landforms and soils of southwestern Australia," *J. Roy. Soc. W. Austral.*, 56(1), 16–22.

O'Driscoll, E. P. D., 1960. "The hydrology of the Murray Basin Province in South Australia," *Bull. Geol. Surv. S. Austral.*, 35, 148p.

Parkin, C. W., ed., 1969. *Handbook of South Australian Geology*. Adelaide: Govt. Printer, 268p.

Reyner, M. L., 1955. "The geology of the Peake and Denison region," *Rept. Invest. Geol. Surv. S. Austral.*, 6, 23p.

Smith, R., and Kamerlings, P., 1969. "Geological framework of the Great Australian Bight," *J. Austral. Petrol. Explor. Assoc.*, 9(2), 60–66.

Sprigg, R. C., 1949. "Early Cambrian 'jellyfishes' of Ediacara, South Australia, and Mount John, Kimberly district, Western Australia," *Trans. Roy. Soc. S. Austral.*, 73, 72–99.

____, 1952. "The geology of the Southeast Province, South Australia, with special references to Quaternary coast-line migrations and modern beach developments," *Bull. Geol. Surv. S. Austral.*, 29, 120p.

Stewart, I. C. F., and Mount, T. J., 1972. "Earthquake mechanisms in South Australia in relation to plate tectonics," *J. Geol. Soc. Austral.*, 19(1), 41–52.

Stuart, W. J., Jr., 1970. "The Cainozoic stratigraphy of the eastern coastal area of Yorke Peninsula, South Australia," *Trans. Roy. Soc. S. Austral.*, 94, 151–178.

Thomson, B. P., et al. 1964. "Precambrian rock groups in the Adelaide Geosyncline. A new subdivision," *Quart. Geol. Notes, Geol. Surv. S. Austral.*, 9, 19p.

____, 1970. "A review of the Precambrian and lower Palaeozoic tectonics of South Australia," *Trans. Roy. Soc. S. Austral.*, 94, 193–221.

Wade, M., 1970. "The stratigraphic distribution of the Ediacara fauna in Australia," *Trans. Roy. Soc. S. Austral.*, 94, 87–104.

Wopfner, H., and Douglas, J. G., eds., 1971. *The Otway Basin of Southeastern Australia.* Adelaide: Geol. Surv. S. Austral. and Victoria, 464p.

Cross-references: *Australasia—Regional Review; Australia—New South Wales, Northern Territory, Queensland, Tasmania, Victoria, Western Australia.*

AUSTRALIA—TASMANIA

Tasmania is a rugged, triangular island of 68,000 km^2 (26,215 sq mi), with a relief up to 1600 m. It is a southerly projection of the eastern Australian mountain belt and is separated from the mainland by Bass Strait. The strait is a basin 90 m deep which is underlain by continental material. The King Island Rise and the Tail Bank are on the W and the Bassian Rise is on the E. Downwarping of the Bass Strait area began possibly in the Triassic and continued at an increasing rate during the Upper Cretaceous and Tertiary. Bass Strait was closed at times (eustatically) during the Pleistocene. The east and west coasts of Tasmania are fault-controlled.

History of Geological Work

Coal was discovered in Tasmania in 1793 by French scientists during the D'Entrecasteaux Expedition. A. W. Humphrey was official mineralogist to the state from 1804 to 1812. Later, the island was visited by such prominent geologists as Darwin, Jukes, and Strzelecki between 1830 and 1845. Local geological work began with J. Milligan's (1849) study of the coal resources and with the work by A. R. C. Selwyn on Tasmanian oil shales and coals in the 1850s. Gold was discovered in 1852, followed by the visit of W. B. Clarke in 1855, and this led to the appointment of Charles Gould as geological surveyor between 1859 and 1869. Tin was discovered at Mt. Bischoff in 1871, and this was followed by the discovery of zinc, lead, and copper at Rosebery and Mt. Lyell. R. M. Johnston worked as an amateur geologist in Tasmania between 1870 and 1919. His *Geology of Tasmania* (1888) is classic and is the first reasonably comprehensive statement on the geology. The Geological Survey was initiated with W. H. Twelvetrees in charge in 1899. After 1918, Loftus Hills began underground water, coal, and oil shale surveys. Later the survey was less active, until 1954, when J. E. Symons was appointed as Director of Mines and regional surveys were begun at a scale of 1 inch to the mile, with an active research program. From 1922 to 1946, A. N. Lewis, an amateur geologist, made significant contributions to the knowledge of Pleistocene glaciation and the Permian and Ordovician stratigraphy of Tasmania.

At the turn of the century, the Zeehan School of Mines taught geology as part of their courses in metallurgy. Geology was taught at the University of Tasmania from 1899 to 1915 and sporadically from 1926 until 1936. The department of geology was instituted in 1946 under S. W. Carey. Research initially concentrated on regional mapping and geotectonics.

Geological History

Precambrian. Silts, sands, conglomerates, and dolerite intrusions of the olivine basalt suite total 6000 m and represent the oldest sequence (Fig. 1). Quartzites, phyllites, and slates, deformed conglomerates, and amphibolites with eclogitic nodules have been metamorphosed to the greenschist facies. They were affected late in the Precambrian by the "Frenchman Orogeny," which occurred in two phases. Basic dikes with concentrations of magnetite were intruded between the phases. The folds and a NNE-trending geanticline are probably related to the second orogenic phase.

On both sides of this geanticline, quartzite, slate, and phyllites, some conglomerate, and dolomite with basalt and tuff are 4500 m thick. These rocks were gently folded during the "Penguin Movement." Intrusion of dolerite sills and dikes, one of which has been dated at 700 m yr, occurred before and after the folding.

Following the Penguin Movement, a long interval of erosion was succeeded by gentle downwarping along a NNW-trend, with deposition in shallow water of oolitic limestones and dolomites and a small proportion of clastic sediments such as silt and sand. The dolomitic sequence reached a thickness of 1200 m, and a low source area is indicated during its formation. Deposition of the dolomites was followed by uplift of a meridional ridge, the Tyennan Geanticline.

Cambrian. Cambrian rocks rest unconformably on the metamorphosed Precambrian but conformably on the dolomitic sequence west of the Tyennan Geanticline, where there was a change in the type of sedimentation and in the rate of sinking. The first deposits were mainly siltstones, with thin graywackes, dolomitic siltstones, polymict conglomerates, chert, and minor spilitic and keratophyric lavas and tuffs, totaling 3000 m.

This succession is overlain by fossiliferous Middle and Upper Cambrian siltstones, gray-

FIGURE 1. Outline geological map of Tasmania.

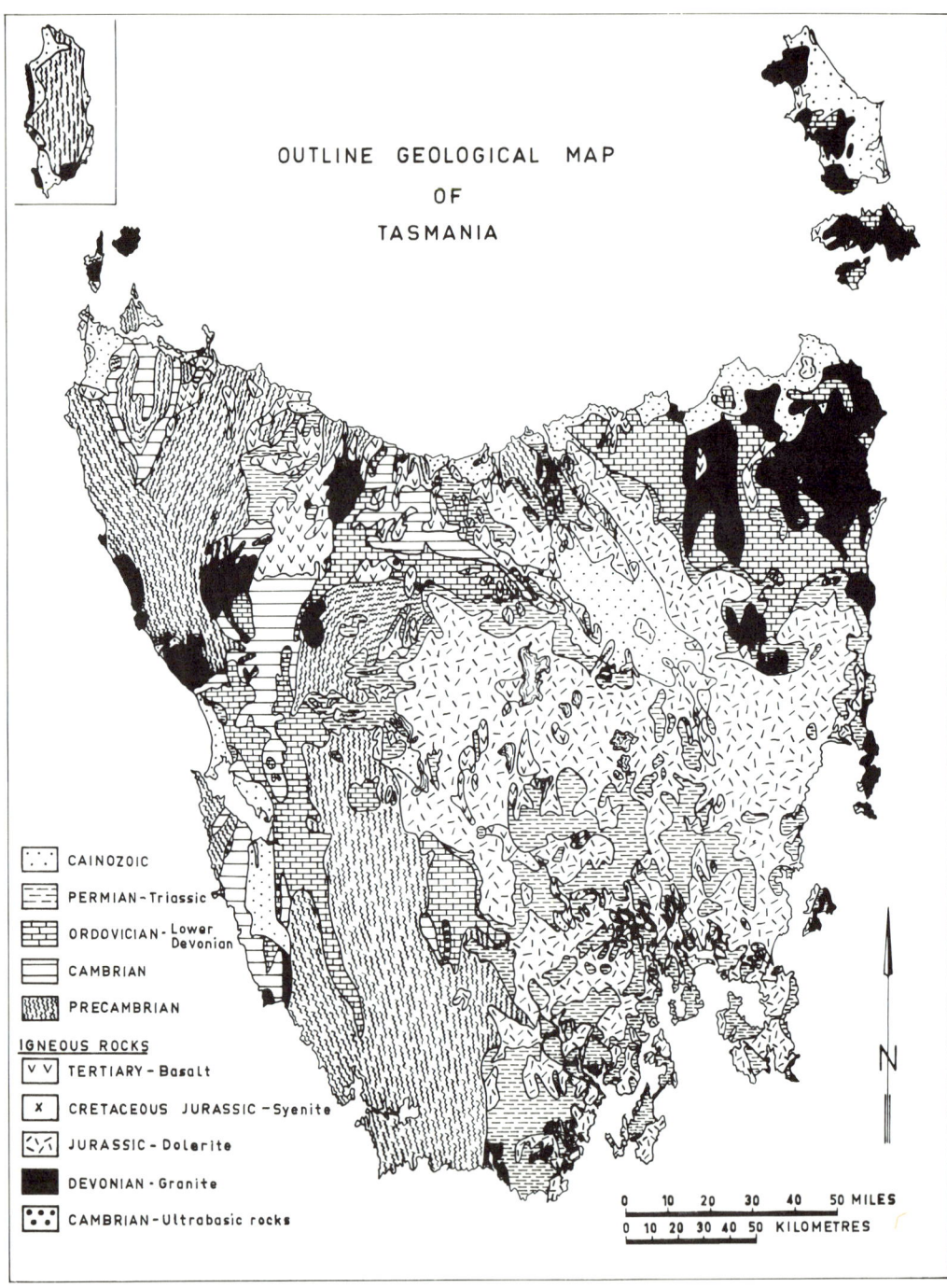

wackes, paraconglomerates, some limestone and dolomite beds, acid and basic lavas, and tuffs. These sediments show cyclic deposition, indicating intermittent uplift of the source areas, but in general they become coarser upward. Mostly acid lavas and pyroclastic rocks, totaling 3000 m and belonging to the spilite suite, occur in an arc bordering the geanticline on the west, north, and east.

Movements during deposition are shown by unconformities in the Middle Cambrian and at the base of the Upper Cambrian. A marine trough with some islands surrounded the Tyennan Geanticline during the Middle and Upper Cambrian. Trilobites, brachiopods, gastropods, sponges, and graptolites comprise the fauna.

Serpentinite derived from pyroxenite and other ultrabasic and basic rocks, intruded as concordant sills, as slightly discordant sheets, or as dikes into the Cambrian sequence. These intrusions generally were close to the Precambrian-Cambrian boundary on the eastern side of uplifted areas and form two roughly meridional belts 50–80 km apart. They contain segregations of osmiridium and other heavy metals and copper and nickel sulfides. Osmiridium occurs also in a Lower Franconian beach placer deposit. Several small concordant granitic intrusions into the core of the Mount Read arc volcanic pile were emplaced before the Ordovician.

Late in the Cambrian, rejuvenation of the Tyennan Geanticline during the "Jukesian Movement" produced a zone of tight folding near the geanticline and gentle folding elsewhere.

Ordovician. Local fanglomerates up to 420 m thick, developed from volcanic and other rocks, flank the Tyennan Geanticline. Later, alluvial fans of siliceous gravel and sand up to 720 m thick and derived from the Precambrian rocks of the Tyennan, Rocky Cape, and Asbestos Range geanticlines spread over the lowlands. Early in the Ordovician, some faulting occurred close to the margins of the Tyennan Geanticline. As the geanticlines eroded, the sea encroached and up to 300 m of shallow-water sand was deposited, passing up and toward south-central Tasmania into fossiliferous siltstone 300 m thick. The sandstones and siltstones contain Middle and Upper Arenigian trilobites, graptolites, and other fossils. With further erosion of the source area and further transgression of the shallow sea, limestone up to 1500 m in thickness covered much of Tasmania. The base of this limestone is Upper Canadian, the top Cincinnatian; and it is overlain by an Upper Ordovician siltstone about 150 m thick. Deposition of the sandstone, siltstone, and limestone occurred on a slowly sinking shelf in a warm sea. These shelf sediments pass eastward into siltstone with thin sandstone bands and contain rare Lower Ordovician graptolites.

Silurian and Devonian. In the late Ordovician or early Silurian uplift in northwestern Tasmania and faulting in northern Tasmania initiated a period of instability ending in the Middle Devonian. In western Tasmania, 800 m of pebbly sand and sand was deposited, followed by alternating units of sandstone and siltstone with some coralline limestone. The fauna consists predominantly of brachiopods and trilobites, but Lower Llandoverian, Wenlockian, and Lower Ludlovian graptolites occur. These shelf sediments pass NE and E into geosynclinal turbidites. They contain a few fossils (Lower Devonian graptolites) and are at least 1800 m thick. In southwestern Tasmania, a basin formed in the Lower Devonian in which accumulated 640 m of shallow-water conglomerate, sandstone and siltstone, and coralline limestone.

Devonian Orogeny (Table 1). There were two main phases of Devonian orogenic activity: the first formed long folds parallel to the margins of the geanticlines; the second created NNW-trending minor folds which were later cut by thrust and wrench faults. The Ordovician-Lower Devonian geosynclinal beds of northeastern Tasmania were folded into anticlinoria and synclinoria, while the Cambrian to Lower Devonian deposits to the W were faulted, the compression being from the east-northeast.

Following the orogeny, Upper Devonian to Lower Carboniferous alkaline to calcalkaline plutonic rocks were intruded along zones of inflection between major anticlinoria and synclinoria. The intrusions were often multiple, composed mainly of adamellite but with some basic differentiates.

Underground (karst) drainage developed in northern Tasmania after the folding, and Upper Devonian cave deposits were formed; these remain unfolded.

Permian. A hiatus occurs following the Devonian orogeny. During the Upper Carboniferous a large ice sheet extended from the W to cover all of Tasmania. The retreating ice formed a promontory in Cradle Mountain and a peninsula in the east coast separated by an island-dotted gulf. In the gulf, carbonaceous pyritic siltstone containing algal oil shale, glendonites, and iceberg-rafted erratics were deposited during the early Sakmarian. These beds coarsen upward and to the E. They are overlain by richly fossiliferous siltstone and limestone that contains bryozoans and large numbers of *Eurydesma*. The limestone is overlain by less fossiliferous siltstone deposited in shallow and possibly brackish water. Early in the Artinskian in northwestern Tasmania, well-sorted sand, silt, and coal were deposited on the coastal plains behind advancing deltas. *Glossopteris, Gangamopteris,* and *Noeggerathiopsis* were abundant. The sea advanced briefly over this coastal plain and formed a NNW-trending gulf in central and southern Tasmania. Later, in the Artinskian, the sea transgressed the plains and the highland areas and deposited richly fossiliferous siltstone and limestone together with ice-rafted erratics. Productids and bryozoans were the main fossils,

TABLE 1. Stratigraphic Table for Tasmania

Period (and Epoch)	Rock Types	Thickness (meters)	Fossils	Igneous Activity	Tectonism	Economic Deposits	Other Features
Quaternary Holocene Pleistocene	Alluvium; dunes; beaches etc. High-level beaches; swamp, lake deposits; alluvium in terraces; fossil dunes; till; rhythmites		*Euryzygoma, Thylacoleo, Acacia*		Normal faulting Normal faulting; some uplift	Alluvial tin, gold, osmiridium	Glaciation (at least two eposides)
Tertiary Pliocene				Basaltic vulcanism		Very minor bauxite	Warm, humid climate Disconformity
Miocene and U. Oligocene	Bryozoal calcarenites and fluviatile and paludal deposits	80 ≃100	*Trybliolepidina, Wynyardia* Coniferous and broad-leafed dicotyledonous flora	Basaltic vulcanism	Some uplift	Some alluvial tin	
Paleogene	Fluviatile and paludal deposits	≃300	Coniferous and broad-leafed dicotyledonous flora	Basaltic vulcanism			Warm, humid climate
	angular unconformity				Tensional faulting		
Cretaceous Upper Cretaceous and Paleogene	Laterite and bauxite	≃10				Minor bauxite	Warm, humid climate
Middle Cretaceous ?Cretaceous	Syenitic rocks			Syenitic stocks and dikes Appinitic intrusions and flows		Minor gold	95 m yr (K/Ar age)
Jurassic Middle Jurassic	Dolerite			Sheets, sills, dikes	Tensional faulting		165 m yr (K/Ar age)

Triassic	Coal; lithic arenites; claystones; quartz sandstone, siltstone	1000	*Cladophlebis, Dicroidium, Phoenicopsis, Pachypteris* (cycads); *Blinasaurus* (reptile)	?Andesitic tuffs	Gentle downwarping	Coal *brick "clays"*	Humid climate *(?cold)*; monsoonal climate
					?Gentle warping		
	conformity to low-angle angular unconformity						
Permian	Pebbly siltstone; sandstone; conglomerate; tillite; limestone; coal; oil shale	800	*Taeniothaerus, Martiniopsis, Wyndhamia, Stenopora, Eurydesma, Glossopteris, Tasmanites*	?Ash bed Metabentonite	Gentle down warping with some broad, low-amplitude upwarps	Subeconomic coal, oil shale, limestone	Two major cycles of shallow marine and nonmarine sediments; cyclothems; glacial influence from late Carboniferous to late Permian
Upper Carboniferous	Tillite, rhythmites	>600	*"Rhacopteris," Tasmanadia*				
Lower Carboniferous to Upper Devonian	nonconformity or angular unconformity						
	Granite rocks			Batholiths, stocks, sheets, dikes		Gold, tin, tungsten, lead-zinc, copper deposits	Granite rocks younging westward 365 to 340 m yr (K/Ar ages)
Middle Devonian	Cave deposits		*Radiospora*				Post-orogenic
	angular unconformity				Some folding in Lower Devonian		
Lower Devonian to Lower Silurian	Siltstone; sandstone; rare limestone	≃1500	*Martinophyllum, Squameofavosites, Australocoelia, Pleurodictyum, Notoconchidium, Monograptus, aequabile* and spp., *Cyrtograptus, Rostricellula*				Major sandstone-siltstone alternation; shelf deposits passing east to slope deposits
	?disconformity						

(table continues on next page)

Period (and Epoch)	Thickness (meters)	Rock Types	Fossils	Igneous Activity	Tectonism	Economic Deposits	Other Features
Ordovician to Upper Cambrian	≃2000	Limestone, siltstone, sandstone, conglomerate	*Ningkianolithus, Palaeophyllum, Tetradium, Foerstephyllum, Lichenaria, Machurites, Manchuroceras, Asaphopsis, Carolinites, Didymograptus, Clonograptus, Proceratopyge*		Local faulting in Lower Ordovician and Upper Cambrian	Limestone Fossil placer deposits with osmiridium	?Disconformity; Cambrian marine siltstones pass up into sandstone, overlain by fanglomerate; this followed by marine L. Ordovician sandstone, siltstone and by M-U Ordovician shelly limestone; in Ordovician shelf deposits pass east to slope deposits "Eugeosynclinal" association
Upper and Middle Cambrian		Siltstone; lithic wacke and conglomerate; chert; rare limestone; acid, intermediate and basic volcanic rocks, including ignimbrites; ultrabasic rocks; serpentinites; rare granites	*Glyptagnostus, Ptychagnostus, Lejopyge, Centropleura, Nepea*, sponges, dendroids	Granite stocks; ultrabasic complexes, sheets, sills, dikes, flows, including pillow lavas; ash beds; keratophyric and spilitic	Some folding and faulting	Copper deposits; zinc-lead	
Pre-Middle Cambrian	3000	Diamictite quartzite, slate, dolomite		Basic dikes and sills	Folding		?Glaciation 700 m yr (K/Ar age)
	≫6000	Quartzites; phyllites; schists, including garnetiferous schists; eclogites; amphibolites; dolomite		Amphibolite sheets, eclogite masses	Folding and metamorphism	Magnetite deposits	

but the limestone is also rich in crinoids and has a few corals such as *Euryphyllum*. The fauna is distinctly Western Australian. Early in the Kungurian, the highlands emerged from the sea. Deposition of sand and pebbles was followed by silt, but further uplift late in the Kungurian or early in the Kazanian caused the succeeding siltstones to be poorly fossiliferous. Late in the Permian, probably in the Tatarian, areas in western and northeastern Tasmania were uplifted and freshwater sands, silts, and coal with *Glossopteris* and *Vertebraria* were deposited on a flood plain.

The Permian rocks of Tasmania are only 750 m thick and were deposited on an unstable shelf in a frigid to cool temperate zone. The Permian now has gentle dome and basin structure superimposed on a SSE-plunging major syncline.

Triassic. The Triassic rests conformably, locally unconformably, on the Permian rocks and consists of nonmarine quartz sandstone, siltstone, and granule conglomerate up to 400 m thick. These rocks are cross-bedded and contain vertebrates, particularly amphibians, some fish and reptiles, and many plants, including equisetales, *Dicroidium,* and *Cladophlebis.* These rocks were overlain by 200 m of lithic arenite with chlorite, feldspar, and volcanic fragments as well as pieces of Permian mudstone. The lithic arenite is found with carbonaceous claystone and coal and is tuffaceous in places. Liverworts, ferns, seed ferns, cycads, and other plants flourished, showing that these Rhaetic beds formed in a cool humid climate. The Triassic rocks now form a major SSE-plunging syncline.

Igneous Activity (Fig. 2). Early in the Middle Jurassic, about 8000 km^3 of tholeiitic quartz dolerite intruded the pre-Permian rocks as pipes and dikes and cut the Permian and Triassic rocks as conesheets, sheets, sills, and dikes. These intrusions attained thicknesses of 400 m. Associated with the dolerites are granophyric and pegmatitic differentiates.

An alkaline syenite body with radial dike swarms intruded Jurassic dolerite and Permian sediments in southeastern Tasmania. The stock is associated with small gold deposits and gives a radiometric age of 100 m yr.

Tertiary. Late in the Mesozoic or early in the Tertiary NNW-trending normal faults developed horsts and grabens.

In the grabens, lacustrine, fluvial, and palustrine silts, sands, clays, gravels, and lignite were deposited during the Paleogene. These nonmarine sediments are up to 275 m thick and show asymmetrical cyclic sedimentation, probably representing repeated uplift of the source areas along faults. The sediments contain abundant plant fossils, remains of native pines, trees with broad thin leaves, cycads, fungi, ferns, casuarinas, banksias, and other trees. *Eucalyptus* and *Acacia* have been reported from Neogene nonmarine beds. Alluvial tin occurs in some of these nonmarine sediments in subbasaltic deep leads in northeastern Tasmania. Bauxite is present below and on top of the Tertiary nonmarine succession.

Around the margin of Bass Strait and along the western coast of Tasmania, almost as far S as Macquarie Harbour, marine calcarenites were deposited during an incursion of the sea beginning in the Upper Oligocene, reaching a maximum in the Middle Miocene, and then retreating. Readvance of the sea onto Flinders Island took place in the late Pliocene. The marine beds, only a few tens of meters thick, contain many invertebrate fossils as well as whales, sharks, and the skeleton of an opossum-like creature, the earliest known marsupial in this area.

Saturated and unsaturated olivine basalt, flowing from vents situated close to fault junctions, filled preexisting valleys to depths of 400 m. In places, the lava crossed valley divides and flowed into adjacent valleys, producing extensive lava fields. The valleys were cut in early Tertiary nonmarine sediments and in Oligocene-Miocene marine sediments. Some pre-Upper Oligocene basalts have been recognized, but most basalts are later than Middle Miocene.

Quaternary. Faulting occurred in the late Tertiary and Quaternary, and a few earth tremors still shake Tasmania. At times, during the Quaternary, alluvial deposits accumulated on valley floors, eolian deposits formed around the coastline, and glacial sediments were deposited on the highlands and in some valleys. Gold, tin, and osmiridium have been concentrated in some of the alluvial deposits. There is evidence of only one widespread Upper Pleistocene glacial phase, but there is scattered evidence of an earlier phase. During the principal phase, a highland icecap covered central Tasmania and spread N, W, and S as valley glaciers. Mountain glaciers occurred in southwest Tasmania, and valley glaciers were present in parts of south-central Tasmania. The glaciation is probably equivalent to the Wisconsin of North America. Pleistocene fluctuations in sea level are recorded as sea caves, inland cliffs, raised beaches, shore platforms, and as submerged river valleys. The fauna included some large extinct marsupials, such as *Nototherium* and *Diprotodon,* which probably reached Tasmania during a period of low sea level when the Bassian Rise formed a land bridge.

FIGURE 2. Outline tectonic map of Tasmania.

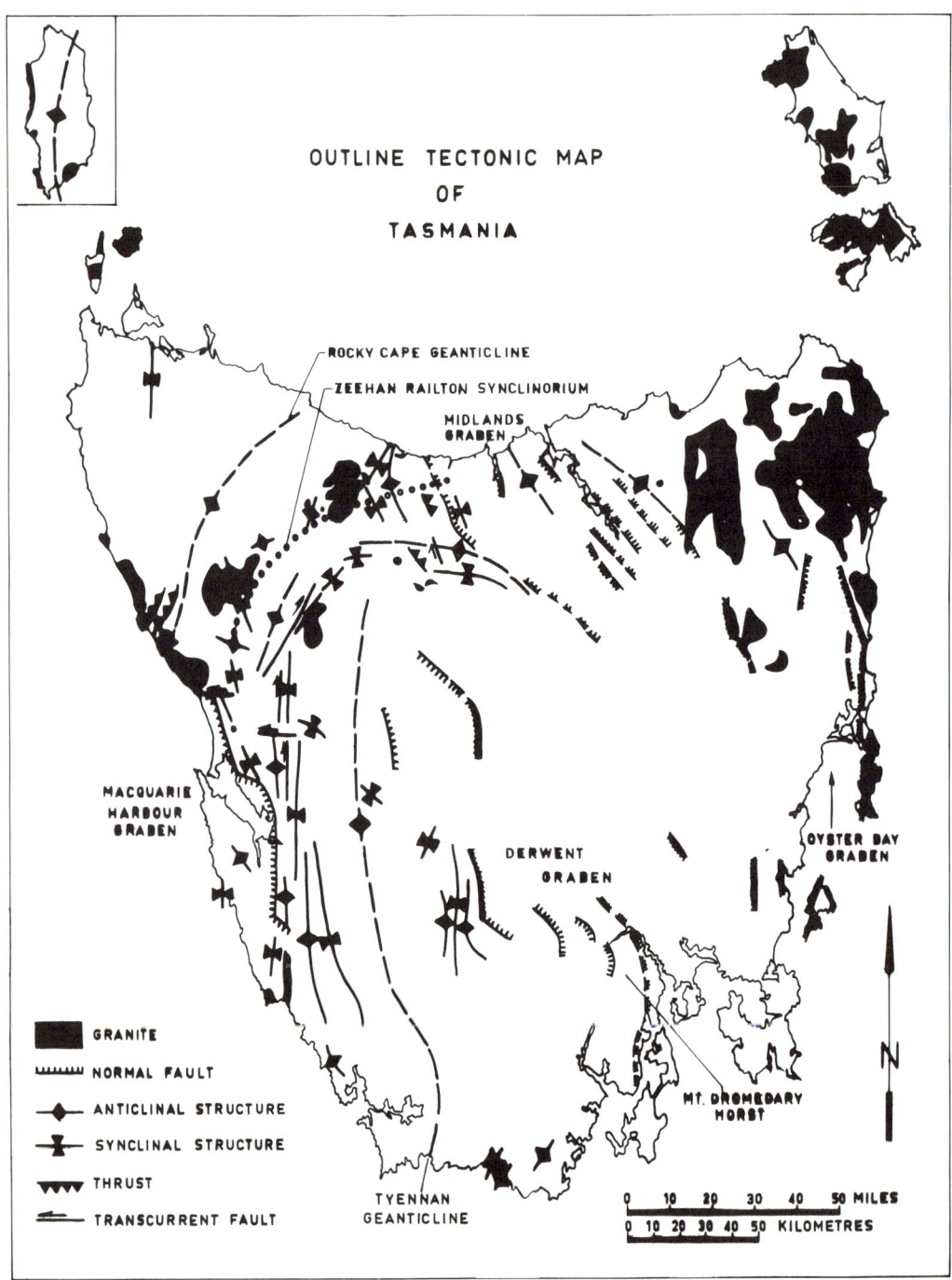

Tectonic Divisions

The folded pre-Permian rocks may be divided into both geanticlinal areas, such as those of the Tyennan, Rocky Cape, and Asbestos Range in which Precambrian rocks are exposed, and intervening synclinorial areas. These pre-Permian rocks are overlain by Permian and Triassic sediments gently folded into a major syncline with a SSE plunge and flanked by domes or plunging anticlines in northeastern and northwestern Tasmania. The Heemskirk anticlinorium, trending NW just off the west coast of Tasmania, was raised during the Devonian orogeny. Anticlinoria and synclinoria underlie northeastern Tasmania. The synclinal Permian and Triassic areas have been broken by faults, forming large grabens.

Morphological Divisions

Two main types of land form can be recognized (Fig. 3). The folded pre-Permian rocks crop out as resistant quartzite and conglomerate strike ridges and streams following the

FIGURE 3. Physiographic divisions of Tasmania.

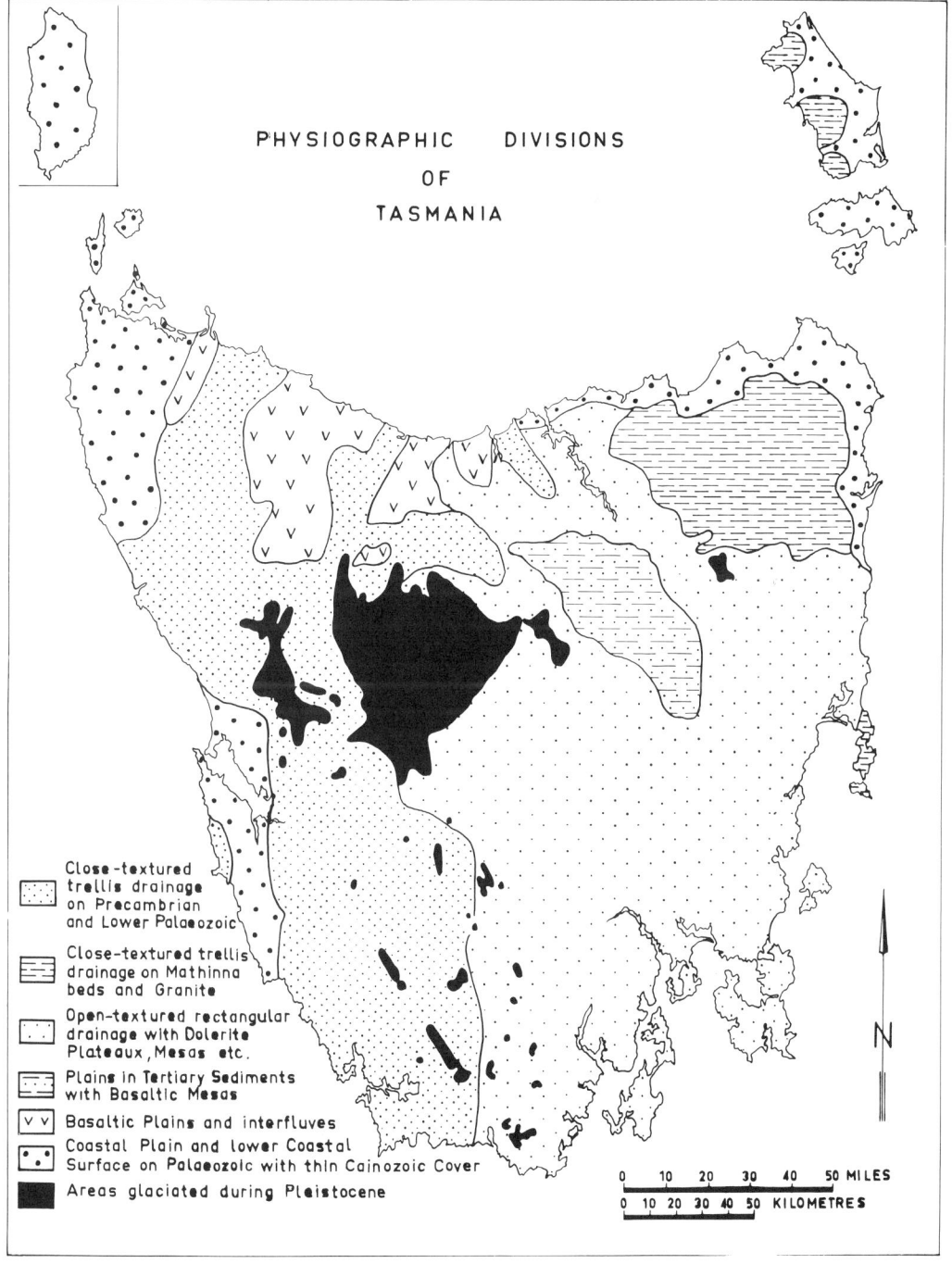

strike of softer shales, schists, and limestones have a trellised drainage. In northeastern Tasmania, the rocks are more homogeneous and the streams have a dendritic pattern with some minor joint control. The block-faulted, dolerite-intruded, subhorizontal Permian and Triassic sediments of central and southeastern Tasmania occur as tabular or plateau-like mountains, almost invariably capped by columnar dolerite and flanked by dolerite scree (Fig. 4). In this area the main drainage patterns are fault-controlled, and the minor streams follow a rectangular pattern.

The island is in a youthful stage of physiographic development. A series of accordant levels cut across both the folded pre-Permian rocks and the Permian and Triassic rocks at 1190–1340 m, 900–1070 m, 730–820 m, 360–460 m, and 90–270 m. This suggests uplift of 610–670 m, probably since the Miocene. The coastline is deeply indented due to post-glacial submergence, but higher sea levels at 20 m, 15 m, 6 m (last interglacial), and approximately 0.8–1.8 m (Holocene) have been recognized.

Metallogenesis. Most of the economic metalliferous deposits of Tasmania are Devonian and occur in and around granite. Veins, dikes, pipes, and nodules of greisen with cassiterite or quartz, cassiterite, and tourmaline occur in granite in northeastern and western Tasmania. On King Island scheelite occurs in bands in Upper Precambrian contact metamorphic marble. Cassiterite-wolframite veins occur in a zone over a granitic cupola which intrudes Silurian-Devonian beds at Aberfoyle in northeastern Tasmania, and shallow-dipping veins occur also at nearby Storeys Creek.

Cassiterite and wolframite occur in the fissure veins of faults, along joints and bedding planes in Ordovician sandstone, and in impure limestone close to a granite at Moina in north-central Tasmania. Cassiterite, pyrite, and pyrrhotite occur as fissure lodes, concordant sheets, and replacements in folded Upper Pre-

FIGURE 4. Monadnocks of subhorizontal glacigene Permian sediments intruded by Triassic dolerite sills and resting unconformably upon peneplaned Precambrian basement, Barn Bluff and Cradle Mt. Most of Tasmania was further glaciated during the Quaternary, the glaciers dissecting the former mature erosion surfaces and excavating glacial valleys and lake depressions. [Photo (oblique air): "The Mercury", Hobart; by permission.]

cambrian dolomites, sandstones, and shales intruded by quartz porphyry dikes at Mount Bischoff and Renison Bell in western Tasmania. Cassiterite occurs in pyrite-pyrrhotite lenses replacing beds of Cambrian slate and tuff at Mount Cleveland in northwestern Tasmania.

Lead, zinc, and silver occur as galena-sphalerite bodies in fissure veins along faults in Precambrian to Devonian rocks at Zeehan, in veins in a shear zone and in a porphyrite dike at Magnet near Mount Bischoff, and as saddle reefs in Ordovician sandstone at Round Mountain in northern Tasmania.

Granites at Renison Bell, Mount Bischoff, and Dolcoath near Moina have a halo of tin deposits close to the granite, with zinc, lead, and silver deposits farther from the granite.

Gold-bearing quartz reefs occur in Ordovician sandstone at Beaconsfield in northern Tasmania and in the Silurian and Devonian geosynclinal beds at Lefroy and Mathinna in northeastern Tasmania.

Galena and sphalerite occur as fissure lodes in shales and Cambrian tuffs at Mount Farrell and as a folded lens in a sericite schist overlain by dark gray shale beneath the Cambrian volcanics in western Tasmania. Chalcopyrite, bornite, and pyrite are disseminated in altered Cambrian volcanic rocks close to their contact with Ordovician conglomerates in western Tasmania. All these ore bodies are syngenetic or early post-depositional, although there has also been some later remobilization.

MAXWELL R. BANKS

References

Cleary, J. R., and Simpson, D. W., 1971. "Seismotectonics of the Australian continent," *Nature,* **230** (5291), 239–241.

Conolly, J. R., 1970. "Sedimentary history of the continental margin of Australia," *Trans. N.Y. Acad. Sci.,* Ser. 2, **32**(3), 364–380.

Crowell, J. C., and Frakes, L. A., 1971. "Late Palaeozoic glaciation of Australia," *J. Geol. Soc. Austral.,* **17**(2), 115–155.

Derbyshire, E., and Peterson, J. A., 1971. "On the status and Correlation of Pleistocene glacial episodes in southeastern Australia," *Search,* **2**(8), 285–288.

Gee, R. D., Marshall, B., and Burns, K. L., 1970. "The metamorphic and structural sequence in the Precambrian of the Cradle Mountain area, Tasmania," *Geol. Surv. Tasmania Rept. 11,* 26p.

Gill, E. D., Manser, W., Hopley, D., *et al.,* 1971. "Latest research on the Quaternary shorelines of Australasia," *Search,* **2**(2), 58–63.

Griffiths, J. R., 1971a. "Continental margin tectonics and the evolution of southeast Australia," *J. Austral. Petrol. Explor. Assoc.,* **2**(1), 75–79.

____, 1971b. "Reconstruction of the southwest Pacific margin of Gondwanaland," *Nature,* **234** (5326), 203–207.

Jago, J. B., Reid, K. O., Quilty, P. G., Green, G. R., and Daily, B., 1972. "Fossiliferous Cambrian limestone from within the Mt. Read Volcanics, Mt. Lyell mine area, Tasmania," *J. Geol. Soc. Austral.,* **19**(3), 379–382.

____, 1973. "Paraconformable contacts between Cambrian and Junee Group sediments in Tasmania," *J. Geol. Soc. Austral.,* **20**(3), 373–398.

McAndrew, J., ed., 1965. "Geology of Australian ore deposits," *Publ. 8th Commonwealth Mining Met. Congr.,* 547p.

Quilty, P. G., 1966. "The age of Tasmanian marine Tertiary rocks," *Austral. J. Sci.,* **29**(5), 143–144.

Solomon, M., and Griffiths, J. R., 1972. "Tectonic evolution of the Tasman orogenic zone, eastern Australia," *Nature Phys. Sci.,* **237**(70), 3–6.

Spry, A. H., and Banks, M. R., eds., 1962. "The geology of Tasmania," *J. Geol. Soc. Austral.,* **9**(2), 107–362.

Sutherland, F. L., 1971. "The question of late Cainozoic uplifts in Tasmania," *Search,* **2**(11–12), 430–431.

____, 1973. "The shoshonitic association in the Upper Mesozoic of Tasmania." *J., Geol. Soc. Austral.,* **19**(4), 487–496.

____, Green, D. C., and Wyatt, B. W., 1973. "Age of the Great Lake Basalts, Tasmania, in relation to Australian Cainozoic volcanism," *J. Geol. Soc. Austral.,* **20**(1), 85–94.

Additional References

Publications of the Tasmanian Department of Mines and the Papers of the Royal Society of Tasmania.

Cross-references: *Antarctica; Australasia–Regional Review; Australia–Victoria; New Zealand.*

AUSTRALIA–VICTORIA

Victoria is the second smallest state in Australia, with an area of scarcely 130,000 km². European adventurers seeking new pastures for their sheep in 1834 were the first settlers in Victoria. When gold was found in 1851, the economics of the colony changed, and one result was the establishment of the Geological Survey of Victoria. The rapid growth of the population was a result of gold rushes and led to the severance of the new colony from New South Wales.

The thickest deposits of brown coal in the world are found in the Latrobe Valley in eastern Victoria; they provide briquettes and electricity, while hydroelectric power is generated in the mountains. Natural gas, first found offshore commercially in 1965, is piped to Melbourne, and a giant offshore oil province has been developed in Bass Strait which separates Tasmania from the mainland. It provides 60% of Australia's oil needs.

In 1854 two scientific societies were established and became the Royal Society of Victoria, which has exercised an important influence on the development of science in the state. The Royal Society made a survey of the natural resources of the colony, studied the acclimatization of animals and plants to the area, and sent out the Burke and Wills Expedition to cross the continent from south to north. In 1854 the University of Melbourne and the National Museum were founded.

With increasing population and educational needs after World War II, Monash University was opened in 1961 and the Latrobe University in 1964. There is a fourth University at Geelong.

The study of the earth sciences is also advanced by the Bureau of Mineral Resources, the C.S.I.R.O. Division of Soils and Division of Applied Geomechanics, the University of Melbourne, government departments (Department of Agriculture, Mines Department, National Museum, Soil Conservation Authority), and technical schools and colleges.

Geomorphology

Victoria's geology and geomorphology (Fig. 1) are intimately linked to a complex geosynclinal belt (the Tasman Geosyncline) east of the Australian Shield. The geosyncline is hundreds of kilometers wide and over 4000 km long and contains Cambrian to Mesozoic sediments. The folded and uplifted sediments of the Tasman Geosyncline form the Great Dividing Range of Eastern Australia. Upper Mesozoic and Cenozoic tectonic movements reached two maxima, in the early Tertiary (Bass Strait Epoch) and in the Plio-Pleistocene (Kosciusko Epoch), associated with the "Older Basalts" and the "Newer Basalts," respectively. The Great Dividing Range is a long, broad arch; the highest point, Mt. Kosciusko, is only about 2200 m. The range follows the east coast of Australia and turns W in Victoria, forming highlands more or less parallel with Bass Strait.

Geomorphologically, Victoria consists of four areas (Fig. 2):

FIGURE 1. Geological sketch map of Victoria (Hills, 1940). Note that most of the Cenozoic ("Cainozoic") basalts of western Victoria are of Quaternary and younger Tertiary age, whereas those east of the Melbourne meridian are older Tertiary. The structural grain of the Paleozoic rocks is mainly N-S.

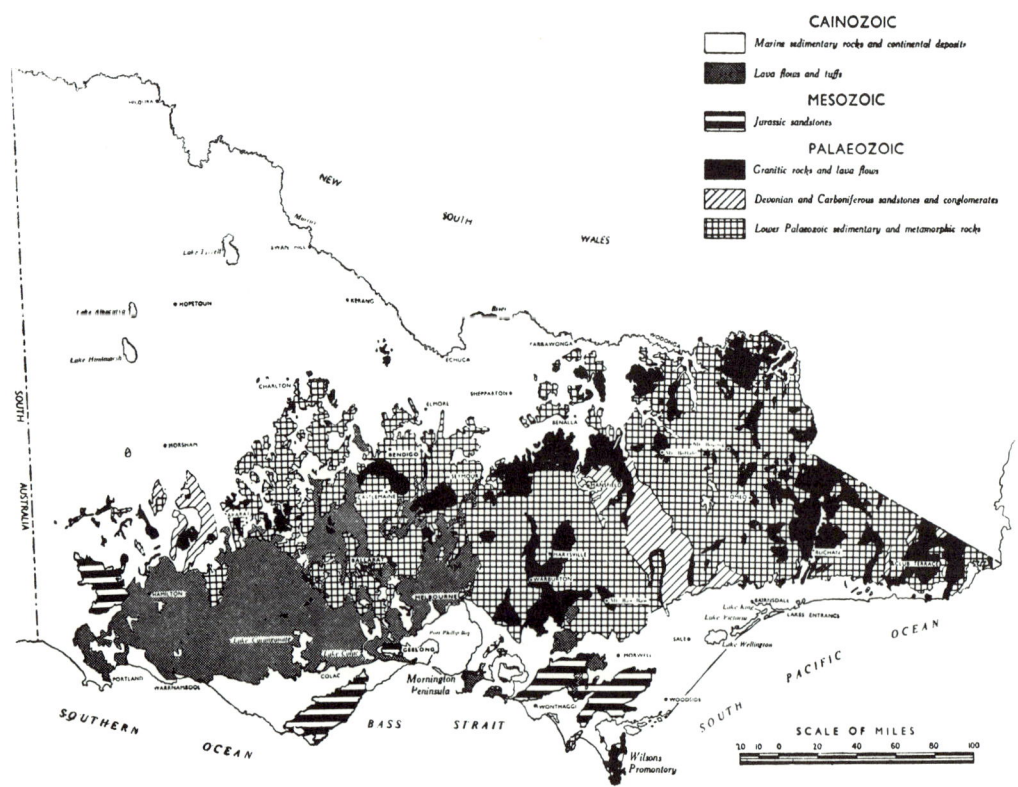

FIGURE 2. Geomorphologic divisions of Victoria (Hills, 1940). Compare with Fig. 1.

1. The Northern or Murray Basin Plains, which consist of flood plains of the Murray River (largest river in Australia) and its tributaries.
2. The Great Dividing Range, which in Victoria may be divided into the Western Highlands and the Eastern Highlands.
3. The Southern Plains, which consists of the basalt plains of western Victoria (with associated coastal plains) and the Gippsland Plains.
4. The Southern Highlands, which consists of the Otway Ranges and the South Gippsland Highlands, which are horsts of Lower Cretaceous nonmarine rocks.

Northern Plains. This is the driest of the four geomorphic areas, the isohyets rising from 250 mm per annum in the northwest to 500 mm near the Dividing Range. As a result there has been much accumulation of soil carbonates, "copi" (gypsum), and salt. These substances, combined with the low rainfall, make it impossible to obtain enough freshwater. Therefore, water from the Murray River is pumped over a considerable area.

The terrain is generally flat with late Pleistocene eolian ridges trending mainly E-W. Climatic cycles in the Quaternary were both wetter and drier than the present (see Fig. 3). The dune ridges are of either sand or parna (eolian clay). Some sand dunes are entirely Holocene, while others have cores of older red compacted sand (Pleistocene). Small quantities of parna (eolian clay) are blown about each summer, but much larger quantities were transported in the past. Parna dunes in western Vic-

FIGURE 3. Landscape of the semiarid northern plains near Mooralka, in northwest Victoria (rainfall, 250 mm). Erosion has reverted a freshwater ostracod limestone of Plio-Pleistocene age associated with lake clays, evidence of a much wetter paleoclimatic stage. (Photo: E. D. Gill).

toria have been dated as late Pleistocene. Similar dunes of earlier cycles appear to have been destroyed by the growth of lakes in the wetter periods and by rill erosion.

In the southeastern part there are many granite hills, but in most of the region there is a thick Cenozoic cover. The western part is underlain by Mesozoic sediments. The Murray Basin contains Tertiary marine rocks; these pass eastward into nonmarine beds. Tectonic movements were very slight in the Quaternary, but a slow sinking has continued in the NW.

Great Dividing Range. This belt consists of Paleozoic sediments and intrusive rocks. The sediments are largely marine, but there are some nonmarine Devonian and Carboniferous and important Permian tillites. Lower Tertiary basalts with thin fluviatile and lacustrine sediments are preserved on some of the plateaus.

Elevations reach 1800 m, and plateau areas above 1400 m are known as the "High Plains." Examples of these areas (remnants of an early Tertiary land surface) include the Monaro, Nunyong, and Baw Baw plateaus; the Bogong and Dargo high plains; the Wellington Tablelands; and the Bennison Plain. Contributing to the structure of the high country are the Devonian acid lavas, which are resistant to erosion.

Rainfall of 750 to 1500 mm is common in the Dividing Range, which means that the carbonates are leached and soils tend to be acid. The high areas are snow-covered in the winter.

Southern Plains. These form the southern edge of the Dividing Range. The relief is low and the rainfall is mostly between 600 and 750 mm. The western plain, which has an area of 23,400 km^2, is the third largest flood basalt region in the world (Fig. 4). Most of these extensive late Cenozoic basalt flows, ash spreads, and scoria cones possess young fertile soils. Some of the mid-Holocene tuffs are exceptionally rich (e.g., Tower Hill district). Beyond the volcanic plain are extensive sandy plains, and also country belonging to a lateritized Lower Pliocene landscape called the Timboon Terrain (Fig. 5).

Beneath these plains, Tertiary marine rocks are extensive and are underlain by Cretaceous rocks. The far west of Victoria is part of the Mount Gambier Sunkland (which also involves the southeast of South Australia), which has up to 550 m of Cenozoic rocks and a considerable thickness of Mesozoic. This basin is important for both groundwater and oil.

In eastern Victoria (Gippsland) there is a similar Cenozoic-Mesozoic basin also with important oil and gas production. In addition, there is the smaller Port Phillip–Western Port basin near the center of the southern coast of the state.

FIGURE 4. Columnar basalt (Tertiary), at the "Organ Pipes," Jackson's Creek, Sydenham, Victoria, Columns reach 20 m in height. (Photo: Victoria Mines Dept.)

Southern Highlands. Set between the three south coast basins are two horsts—the Otways W of Melbourne and the South Gippsland Highlands E of Melbourne. Both consist of Lower Cretaceous nonmarine arkoses and siltstones in which occur beds of bituminous black coal, contrasting with the Lower Tertiary brown coal of the Latrobe Valley, among other places. The black coal had to be mined using shafts, whereas the brown coal is obtained by open-cut methods.

The horsts rise only about 600 m, but they are high enough to catch 750–1500 cm of rain from the prevailing southwesterly winds.

FIGURE 5. Volcanic lake in the Tower Hill caldera, western Victoria. Holocene age: radiocarbon date approximately 7300 yr B.P. (Photo: E. D. Gill.)

Stratigraphy

No Precambrian rocks are known in Victoria. Extensive schists are known to have been metamorphosed Cambrian and Ordovician sediments. The bedrock of the state consists of Cambrian, Ordovician, Silurian, and Devonian marine strata with a meridional strike. They vary from consolidated sediments in open folds to chiastolite slates.

Cambrian. Some thousands of feet of faulted lavas, tuffs, and agglomerates with interbedded cherts comprise the oldest sequence in central Victoria, where they are altered to greenstones. This is the "diabase complex" of early writers. In terms of plate tectonics, it is probably related to an island arc of Cambrian time (Oversby, 1971). This complex is overlain by a sequence of siltstones and sandstones with intercalated beds of volcanic ash, which is highly fossiliferous in places and presents fairly well-preserved fauna, mostly trilobites and hydroids.

At Phosphate Hill near Mansfield, there is an unusual facies consisting of dark cherts, siltstones, and fine sandstones along with brown phosphatic rock. A number of Upper Cambrian brachiopods and other fossils have been found.

Ordovician. In central Victoria, there is a succession of some 4900 m of slates and sandstones with a remarkable graptolite fauna spanning the whole of the Ordovician (Fig. 6). There are few places in the world where there is such a good sequence of graptolite forms. Although antipodal to the classic series described in Europe, they demonstrate more fully the same succession of forms. Fine-grained sediments prevail, and these are now altered mostly to slates. The sandstones provide numerous examples of graded bedding, indicating their history in the Tasman Geosyncline. The Ordovician strata are for the most part tightly folded into anticlinoria and synclinoria (Fig. 7).

In contrast, the Upper Ordovician Riddell Grits present a shell-bearing facies instead of the usual graptolitic facies. Meridional-striking faults are common; then rapidly evolving the graptolites make it possible to pick out these complex structures with precision. The faults are reverse faults of steep hade, and slickensides are common.

Ordovician rocks in the Melbourne Trough are either conformably or only slightly unconformably overlain by the Silurian. Epi-Ordovician movements (Benambran Orogeny) were thus absent in some areas but appreciable in the eastern part of the state (former island arc site).

Silurian. As in the Ordovician, all the Silurian deposits are marine. They are unknown in

FIGURE 6. Correlation table for the Ordovician rocks of Victoria (from Brown et al., 1968).

EUROPEAN STAGES & ZONES		CHARACTERISTIC GRAPTOLITES (AUSTRALIAN)	VICTORIAN STAGES	CENTRAL TASMANIA
ASHGILLIAN	15	Dicellograptus complanatus	BOLINDIAN	
	14			
	13	Pleurograptus sp.		
CARADOCIAN	12	Dicranograptus hians	EASTONIAN	
	11	Climacograptus wilsoni		
LLANDEILIAN	10	Climacograptus peltifer	GISBORNIAN	GORDON LIMESTONE
	9	Nemagraptus gracilis		
	8	Glyptograptus teretiusculus	D4	
LLANVIRNIAN	7	Diplograptus decoratus	D3	
	6	Diplograptus intersitus	DARRIWILIAN D2	
	5	Diplograptus austrodentatus	D1	
ARENIGIAN	4	Cardiograptus Oncograptus	Ya2 YAPEENIAN Ya1	
		Isograptus caduceus var. maximus	Ca3	
		Isograptus caduceus var. victoriae	CASTLEMAINIAN Ca2	
		Isograptus caduceus var. lunatus	Ca1	
		Didymograptus balticus	Ch3	
		Didymograptus protobifidus	Ch2 CHEWTONIAN	
		Didymograptus protobifidus and Tetragraptus fruticosus	Ch1	
		Tetragraptus fruticosus 3-br	Be4	—?—?— FLORENTINE RIVER MUDSTONE
		T. fruticosus, 3-br + 4-br	Be3 BENDIGONIAN	
		T. fruticosus, 4-br	Be2	
		T. fruticosus and T. approximatus	Be1	—?—?— SANDSTONE
	3	T. approximatus	La3	
TREMADOCIAN		Adelograptus and Dictyonema	LANCEFIELDIAN La2	—?—?— TIM SHEA CONGLOMERATE
	2	Staurograptus and Dictyonema	La1	

the western part of the state and are best developed in the central part, but they occur also in eastern Victoria (Fig. 8). Siltstones and sandstones are common, and there are occasional conglomerates. As in the Ordovician, limestones are practically absent, contrasting with the De-

FIGURE 7. Diagrammatic profile (E-W) across the Campbelltown district. Note the very striking type of chevron folding in the Lancefieldian, Bendigonian, and Chewtonian (Lower Ordovician). Near the Campbelltown fault there is a thin veneer of Permain glacial material (tillite, etc.) and at Yandoit Hill there is a basaltic plug (or possibly a diatreme) preserving a Tertiary continental lens with lignite (Harris and Thomas, 1948).

0 1 2 3 4 km.
W.
E.
CAMPBELLTOWN FAULT
Perm. Glacial
Basalt
YANDOIT HILL
Tertiary Lignite
MUCKLEFORD FAULT
Lancefieldian
Yapeenian

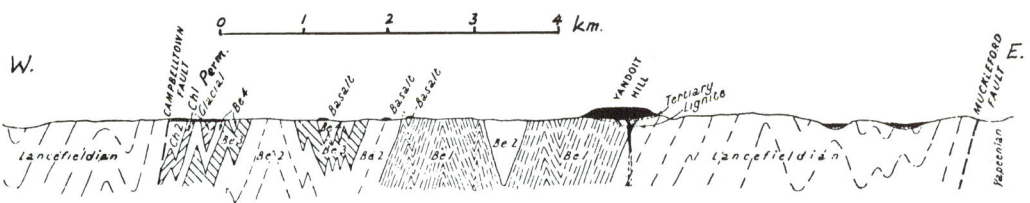

vonian (see below). In central Victoria the Lower Silurian is found at Keilor, W of Melbourne, where the first graptolites in Australia were discovered. The beds are mostly siltstones, and the graptolites are plentiful only in certain horizons. In Melbourne, the Upper Silurian strata are alternating siltstones and sandstones with graptolites and brachiopods. Most of the series is folded with moderate dips, but in certain areas there are crush zones with steep folds that pitch. These zones are generally invaded with seams of milky quartz which have occasionally yielded gold.

FIGURE 8. Victorian stratigraphic stages for the Silurian with principal graptolite zones (from Brown *et al.*, 1968).

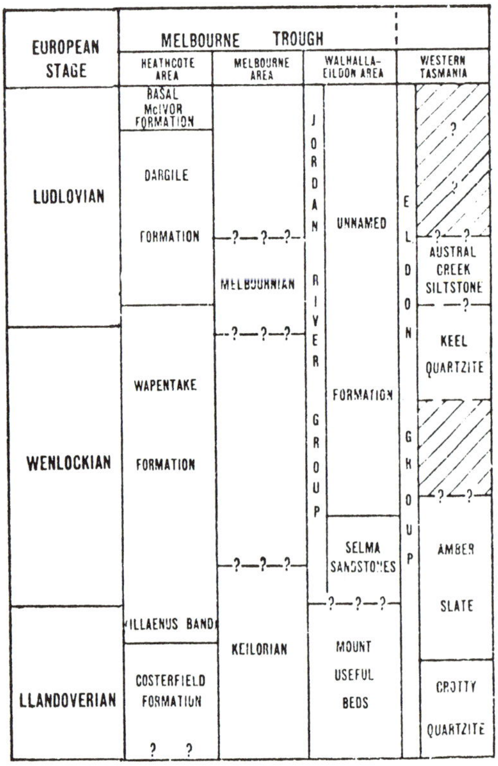

EUROPEAN STAGE
MELBOURNE TROUGH
HEATHCOTE AREA
MELBOURNE AREA
WALHALLA-EILDON AREA
WESTERN TASMANIA
LUDLOVIAN
WENLOCKIAN
LLANDOVERIAN
BASAL McIVOR FORMATION
DARGILE FORMATION
WAPENTAKE FORMATION
ILLAENUS BAND
COSTERFIELD FORMATION
MELBOURNIAN
KEILORIAN
JORDAN RIVER GROUP
UNNAMED FORMATION
SELMA SANDSTONES
MOUNT USEFUL BEDS
EILDON GROUP
AUSTRAL CREEK SILTSTONE
KEEL QUARTZITE
AMBER SLATE
CROTTY QUARTZITE

In the Upper Yarra district east of Warburton, it is not yet clear what part of the sequence is Silurian and what part is Devonian, owing to a lack of fossils and a change in facies. There are siltstones that appear to be pelagic in facies, and among these there are graywackes that appear to be due to submarine sliding. The two types of lithology are quite out of character with one another. The graywackes contain broken shallow-water brachiopods which are randomly oriented in such a way as to suggest turbidites. The paleogeographic picture in terms of plate tectonics is illustrated in Fig. 9.

Certain horizons of the Silurian are rich in fossils, including brachiopods, pelecypods, gastropods, cephalopods, trilobites, corals, starfish, carpoids, merostomes, graptolites, and other forms. Of interest is the occurrence of land plants, although the classic locality is now considered to be Devonian. In the Seymour and Yea districts, there are superposed horizons with land plants extending from the Silurian up into the Devonian.

Devonian. A considerable thickness of rocks originally classified as Silurian is now regarded as Devonian, i.e., if the base of the Ludlow Bone Bed in England is accepted as the base of the Lower Devonian. Thick siltstones and sandstones make up most of the sediments, but there are also many limestones significant for paleoclimatic reasons. In central Victoria there was a miogeosyncline; in east Gippsland, there was a eugeosyncline with widespread and thick volcanics–the Snowy River porphyries (Fig. 10). This belt was affected in earliest Devonian or latest Silurian by the Bowning Orogeny, but most extensively in the middle Devonian by the Tabberabberan Orogeny.

Lower Devonian strata occur in the Lilydale district east of Melbourne and in the Walhalla Synclinorium in eastern Victoria along the Walhalla–Woods Point–Eildon line. At Buchan and Bindi in eastern Victoria, middle Devonian strata occur associated with the Snowy River Volcanics. The limestones vary in lithology and facies and are rich in fossils. For example, that

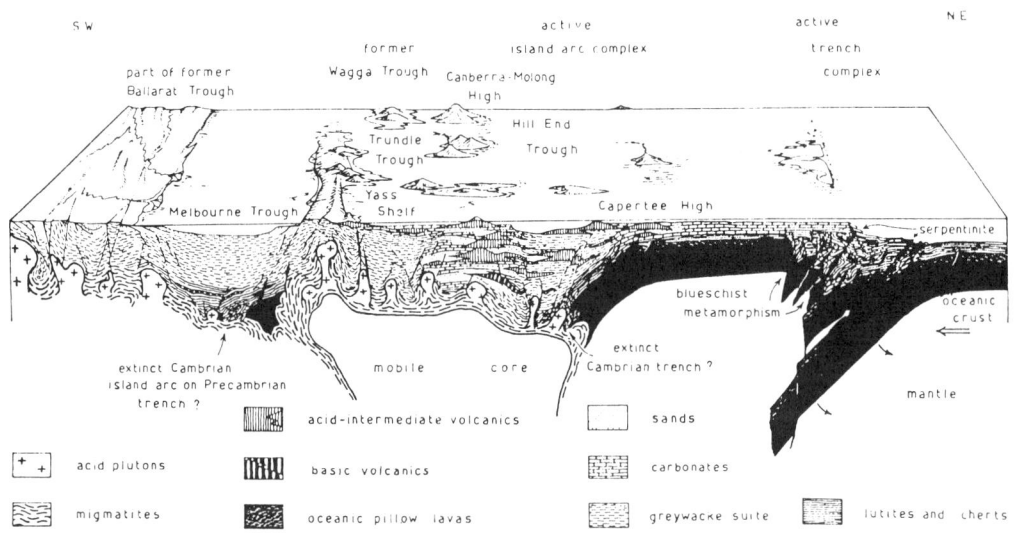

FIGURE 9. Paleogeographic sketches of southeastern Australia during the Ordovician, Silurian, and Devonian (adapted, after Packham, 1960).

at Lilydale is a coral-stromatoporid biostrome; some at Buchan are lime-muds with brachiopods.

The Devonian rocks contain some of the richest paleontological sites in the state. At Lilydale there are layers abundant in brachiopods and other forms, while at Kinglake there are layers with massed crinoids and starfish plus occasional carpoids, cystoids, and other rarer categories. At other sites, brachiopod-trilobite associations occur. The Nineteen Mile Quarry in the Upper Yarra district on the Woods Point Road is the classic locality in Victoria for ancient land plants, including *Baragwanathia* (named after a former chief of the Geological Survey), *Yarravia,* and others. The graptolites formerly found with these plants are now believed to be *Monograptus praehercynicus* and to be early Devonian.

Devonian marine sedimentation concluded with the Tabberabberan Orogeny, an important event throughout the Tasman Geosyncline. Sedimentation was interrupted at varying times in the Lower and Middle Devonian. Folding occurred toward the end of the Middle Devonian and in the Upper Devonian. It is thus contemporaneous with the Acadian phase of eastern North America. Granite and other intrusions were associated with this orogeny.

Nonmarine Upper Devonian and Carboniferous beds were laid down (without interruption) over considerable areas, and these strata were associated with extensive and often thick acid volcanic rocks. Classic ring-dikes and cauldron subsidences are a feature of this volcanism. Upper Devonian fish date the lower part of the volcanic succession.

Carboniferous. The Upper Devonian and Carboniferous nonmarine sediments already mentioned occur in fairly extensive areas of eastern Victoria and in the Grampian Mountains of western Victoria. In the Grampians there are deltaic beds and a marine transgression with the brachiopod *Lingula borungensis,* as well as ostracods, worm trails, and fish remains. The Grampians Group is about 6000 m thick and middle Carboniferous granites and porphyries intrude these rocks.

Permian. Over 600 m of sediments, with tillites at a number of horizons, are known in the Bacchus Marsh district of central Victoria. The siltstones and sandstones are considered to be glacigene. *Gangamopteris, Calamites,* and spores are known. Throughout the state there are small patches of continental Permian sediments remaining to indicate that these formations were widespread.

Mesozoic. Near Bacchus Marsh, continental Lower Triassic leaf beds disconformably overlie the Permian rocks. Some Jurassic nonmarine beds are probably present, but most of the beds in the Southern Highlands formerly referred to as Jurassic are now placed in the Lower Cretaceous. Rich plant beds are known, and at Koonwarra there are strata rich in leptolepid and other fish, insects, conchostraca, and plants. A king crab (*Victalimulus*) is also present. Upper Cretaceous marine beds are known from deep bores in western Victoria, the first reappearance

FIGURE 10. Hypothetical reconstruction of the Tasman geosyncline during the Middle Paleozoic (Oversby, 1971).

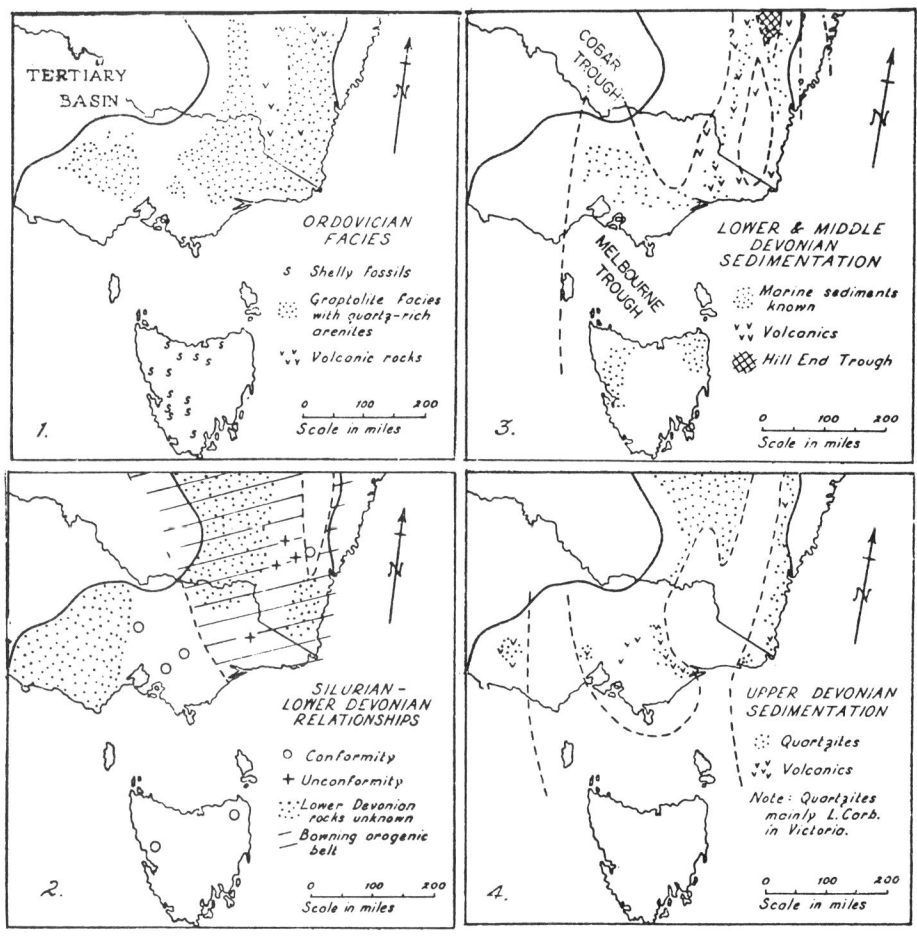

of marine sediments in the area since early Carboniferous.

Tertiary. Marine beds of all Tertiary epochs are known in Victoria. They occur in the southern plains and under the western half of the Murray (Murravian) Basin. The central part sagged down by more than 500 m (Fig. 11) along distinctive NE-SW and NW-SE lineaments (Hills, 1956). The Bass Strait Phase is the principal episode of these earth movements that began in the Upper Cretaceous and Lower Tertiary. Associated with these warping and taphrogenic movements were the Older Basalts, and Bass Strait was also formed as a result of these movements. The sea transgressed the Bass Strait area and the Murray Basin, reaching a climax in the Miocene. The climate warmed up to a maximum in the Miocene, and then temperatures fell once more until the Quaternary, as is shown by both the fossils and oxygen-isotope paleotemperature measurements. Crocodilians reached as far south as Victoria when the Miocene climate was warm. The marine strata are up to 1500 m thick and occupy the Mt. Gambier Sunkland, the Port Phillip Sunkland, and the Gippsland Basin. The nonmarine rocks include extensive deposits of brown coal, which reach their maximum development in the Latrobe River valley in eastern Victoria, where three seams total 300 m. Both marine and nonmarine strata have rich fossil horizons.

Quaternary. In the Upper Pliocene and Lower Pleistocene, there was a series of epeirogenic movements, the Kosciusko Phase, which led to the uplift of the Great Dividing Range and the Southern Highlands. The associated volcanism was responsible for the Newer Basalts (Pliocene to Holocene). Between the two phases there was a quiescent period with rather tropical conditions when deep leaching and kaolinization resulted in the Nunawading Terrain.

FIGURE 11. The Murray (Murravian) Basin, which formed during the late Cretaceous and early Tertiary, covering adjacent areas of Victoria, New South Wales, and South Australia. Edge of surrounding bedrock, solid black; bores, crosses. Contours show generalized pre-Tertiary basement in feet below sea level.

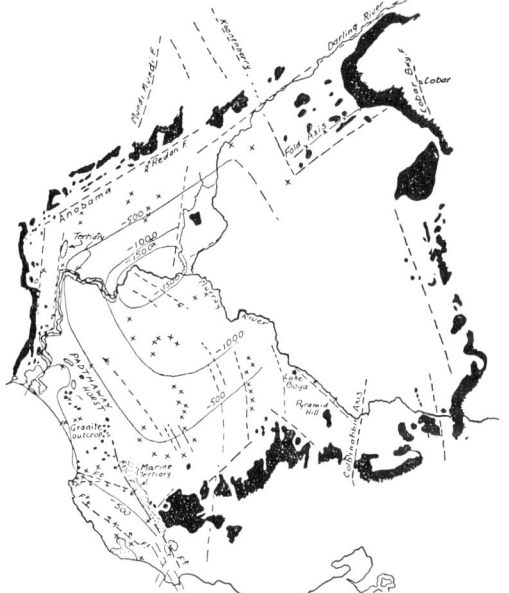

The Quaternary was characterized at times by a Mediterranean-type climate and formation of solods and podsols S of the Dividing Range, with red earths north of it. Marine Quaternary beds have been left by higher sea levels or exposed by tectonic uplift. Along the coast are formations of eolianite, chiefly referable to the regressive interglacial seas (Fig. 13). In the Warrnanbool district of western Victoria, three eoli-

FIGURE 13. Miocene marine limestone forms coastal cliffs at Stanhope Bay in western Victoria. It is irregularly dissected and channels filled by continental clayey sands of Pliocene age (left-hand side). The cliff is capped by landward-dipping eolianites of Pleistocene age, evidently emplaced prior to the erosion of the present sea cliff. (Photo: E. D. Gill.)

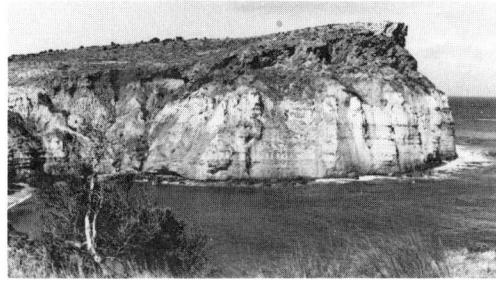

FIGURE 12. Correlation table for the Victorian Tertiary.

	MURRAY BASIN	OTWAY BASIN: GAMBIER SUNKLANDS	OTWAY BASIN: PORTLAND SUNKLANDS	OTWAY BASIN: CAPE OTWAY	AUSTRALIAN STAGES	BASS BASIN: ANGLESEA	BASS BASIN: GEELONG	GIPPSLAND BASIN
PLIOCENE	NORWEST BEND FORMATION; LOXTON SANDS; BOOKPURNONG BEDS		NORMANBY GROUP: WHALER'S BLUFF FORMATION; GRANGE BURN FORMATION		WERRIKOOIAN; ? ; KALIMNAN; ?		MOORABOOL VIADUCT SANDS	HAUNTED HILL GRAVELS (LOWER PART); JEMMY'S POINT FORMATION
MIOCENE	PATA LIMESTONE; MORGAN LIMESTONE; MANNUM FORMATION	GAMBIER LIMESTONE	GLENELG GROUP: MUDDY CREEK FORMATION; GAMBIER LIMESTONE	FISHING POINT MARLS; UPPER GLEN AIRE CLAYS	CHELTENHAMIAN; ? ; MITCHELLIAN; BAIRNSDALIAN; BALCOMBIAN; BATESFORDIAN; LONGFORDIAN	PUEBLA FORMATION	? ? ? ? FYANSFORD CLAY; UPPER MAUDE FM. / MAUDE BASALT / LOWER MAUDE FM. — BATESFORD LIMESTONE	TAMBO RIVER FORMATION; GIPPSLAND LIMESTONE: BAIRNSDALE LIMESTONE; WUK WUK MARL — GLENCOE LIMESTONE; LONGFORD LIMESTONE
OLIGOCENE	ETTRICK FORMATION	GAMBIER LIMESTONE	GLENELG GROUP: GAMBIER LIMESTONE	CALDER RIVER LIMESTONE; LOWER GLEN AIRE CLAYS	JANJUKIAN	POINT ADDIS LIMESTONE — JAN JUC FORMATION		LAKES ENTRANCE FORMATION
EOCENE	BUCCLEUCH GROUP; KNIGHT GROUP: MOORLANDS COAL MEASURES	GREENSANDS; KNIGHT GROUP: DARTMOOR FORMATION	GLENELG GROUP: NELSON FORMATION; WANGERRIP GROUP	CASTLE COVE LIMESTONE; BROWNS CREEK CLAYS; JOHANNA RIVER SANDS	JOHANNIAN	DEMONS BLUFF FORMATION; ? ; BOONAH SANDSTONE; ?		LATROBE VALLEY COAL MEASURES
PALAEOCENE	KNIGHT GROUP; ? ; ?	KNIGHT GROUP: BAHGALLAN FORMATION; PAARAATE FORMATION	WANGERRIP GROUP	ROTTEN POINT SANDS		EASTERN VIEW COAL MEASURES		LATROBE VALLEY COAL MEASURES; ? ? ?

an coastal formations have been described (Gill, 1967):

1. Present mobile dunes of calcareous sand.
2. Dunes of similar sand with a hard crust of calcrete that rest on marine beds of the Last Interglacial.
3. Well-cemented eolianite that rests on Penultimate Interglacial marine beds.

River terraces can be seen along the main streams. Above the present unconsolidated and unoxidized terrace of the Maribyrnong River at Melbourne, there is a consolidated and oxidized terrace that contains wood deposited between about 20,000 and 6000 years ago. The Keilor Cranium (an early aboriginal man) came from this terrace. Above that is an older fairly continuous terrace. On the walls of the valley there are traces of still older terraces.

EDMUND D. GILL

References

Aziz-ur-Rahman, 1971. "Paleomagnetic secular variation for recent normal and reversed epochs, from the Newer Volcanics of Victoria, Australia," *Roy. Astron. Soc. Geophys. J.,* 24(3), 255–269.

Brown, D. A., Campbell, K. S. W., and Crook, K. A. W., 1968. *The Geological Evolution of Australia and New Zealand.* London: Pergamon, 409p.

Douglas, J. G., 1971. "Progress of geological mapping in Victoria," *Mining Geol. J.* (Victoria Mines Dept.), 7(1), 33–35.

Franklin, E. H., and Clifton, B. B., 1971. "Halibut Field, southeastern Australia," *Bull. Am. Assoc. Petrol. Geologists,* 55(8), 1262–1279.

Geological Society of Australia, Victorian Branch, in prep. *Geology of Victoria.*

Gill, E. D., 1965. "Paleontology of Victoria," *Victorian Yearbook,* 24p.

____, 1967. "Evolution of the Warrnambool–Port Fairy Coast and the Tower Hill eruption, Western Victoria," in J. N. Jennings and J. A. Mabbutt, eds., *Landform Studies from Australia and New Guinea.* Austral. Nat. Univ., 341–364.

Griffiths, J. R., 1971. "Continental margin tectonics and the evolution of southeast Australia," *J. Austral. Petrol. Explor. Assoc.,* 11(1), 75–79.

Harris, W. J., and Thomas, D. E., 1948. "The geology of Campbelltown." *Victoria Mining Geol. J.,* 3(3), 46–54.

Hills, E. S., 1940. *Physiography of Victoria.* Melbourne: Whitcombe and Tombs, 292p.

____, 1956. "A contribution to the morphotectonics of Australia," *J. Geol. Soc. Austral.,* 3, 1–15.

Leeper, G. W., 1948. *Introduction to Soil Science.* Melbourne: Melbourne University Press, 253p.

Lovering, J. F., Mason, B., Williams, G. E., and McColl, D. H., 1972. "Stratigraphical evidence for the terrestrial age of australites," *J. Geol. Soc. Austral.,* 18(4), 409–418.

McAndrew, J., ed., 1965. "Geology of Australian ore deposits," *Publ. 8th Commonwealth Mining Met. Congr.* (Melbourne), 1, 547p.

____, and Marsden, M. A. H., 1968. *Regional Guide to Victorian Geology.* Melbourne: University of Melbourne.

Moore, B. R., 1971. "Paleogeographic and tectonic significance of diachronism in Siluro-Devonian age flysch sediments, Melbourne Trough, southeastern Australia," *Bull. Geol. Soc. Am.,* 82, 1087–1094.

Ollier, C. D., and Joyce, E. B., 1964. "Volcanic physiography of the western plains of Victoria," *Proc. Roy. Soc. Victoria,* 77(2), 357–376.

Oversby, B., 1971. "Paleozoic plate tectonics in the southern Tasman Geosyncline," *Nature Phys. Sci.,* 234(46), 45–47.

Packham, G. H., 1960. "Sedimentary history of part of the Tasman Geosyncline in southeastern Australia," *Intern. Geol. Congr. 21st Session,* pt. 12, 74–83.

Ward, W. T., 1971. "Postglacial changes in level of land and sea," *Geol. Mijnbouw,* 50(5), 703–718.

Weeks, C. G., and Hopkins, B. M., 1967. "Geology and exploration of three Bass Strait basins, Australia," *Bull. Am. Assoc. Petrol. Geologists,* 51, 742–760.

Cross-references: *Australasia–Regional Review; Australia–New South Wales, South Australia, Tasmania.*

AUSTRALIA—WESTERN AUSTRALIA

Western Australia is the largest state in the Commonwealth of Australia, occupying nearly one-third of the land area of the continent, or some 2,525,000 km^2 (975,000 sq mi). Approximately 55% of this area is occupied by Archean and Proterozoic crystalline and sedimentary rocks; the rest consists of Phanerozoic sedimentary basins that contain representatives of every system above the Proterozoic. The continental shelf adjoining Western Australia is up to 300 km wide (in the north) and covers an area of about 640,000 km^2.

The first European settlement in Western Australia was at Albany in 1827, but the colony was formally established at Perth, the capital, in 1829. The first official geologist was F. von Sommer (1847–1851), and a number of other geologists were employed in subsequent years. In 1896 the Geological Survey of Western Australia was organized under A. Gibb Maitland, and it now employs 55 professional scientists.

Systematic geological mapping of the colony began after a series of sensational gold discoveries. The first was in the Kimberley district in 1885, followed by the much larger discoveries in the Murchison district and at Coolgardie and Kalgoorlie in 1892–1893. Since then much of the economy of the state has been based on mining, and until 1966 the value of gold produced each year was greater than that of any other mineral. However, during the

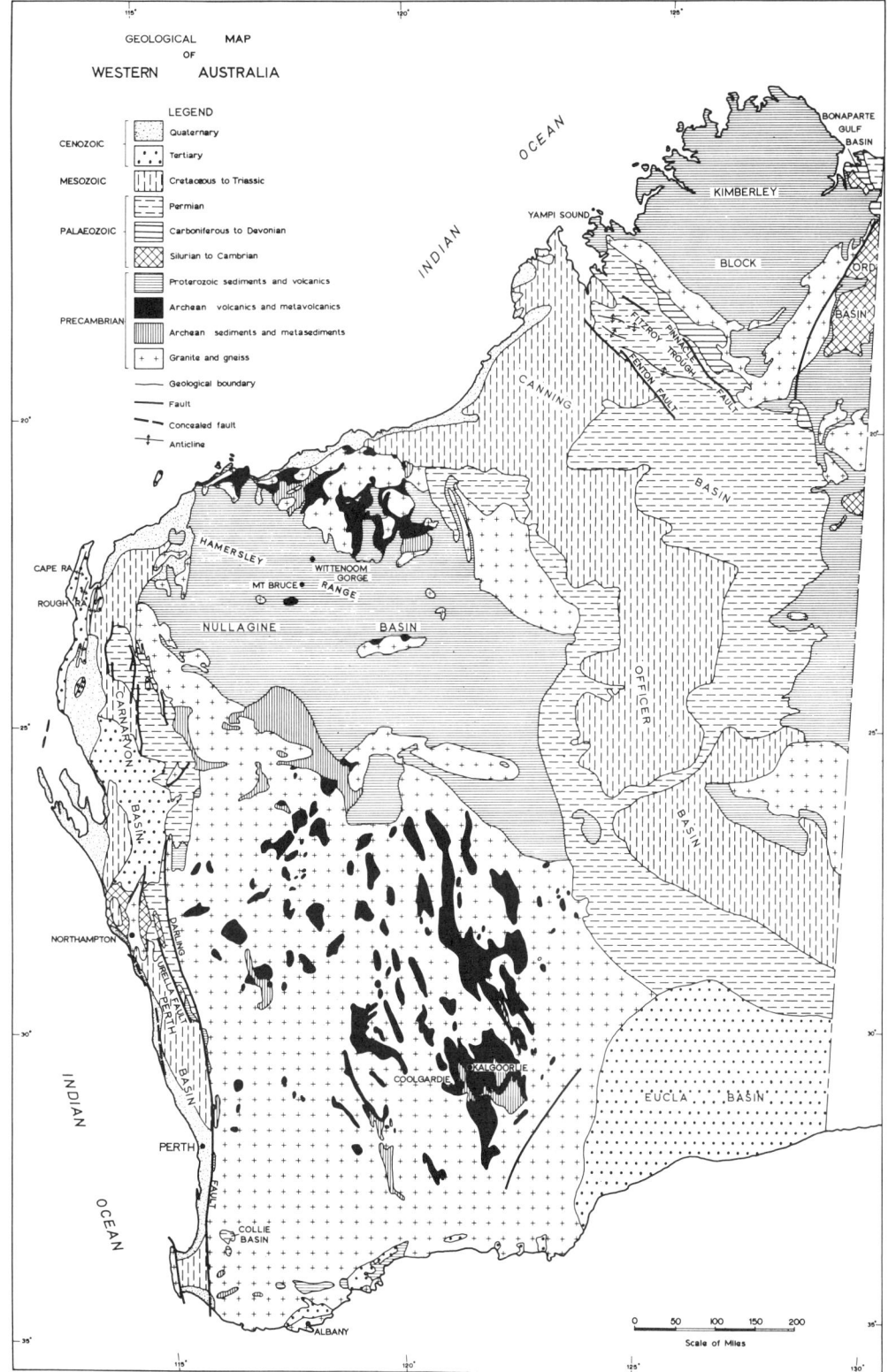

FIGURE 1. Geological map of Western Australia. (NB. This drawing was prepared several years ago: for Kimberley Block, read "Basin"; the mainly Archean area north of Wittenoom Gorge is the "Pilbara Block"; for Hamersley Range, read "Basin"; for Nullagine Basin, read "Bangamall Basin"; the mainly Archean area of the center and southwest is the "Yilgarn Block".)

1960s there was a series of major mineral discoveries in the state: iron ore, nickel, bauxite, and petroleum, and each of these has now outstripped gold in annual value of production.

The three teaching centers for geology in Western Australia are the University of Western Australia and the W.A. Institute of Technology, situated in Perth, and the Kalgoorlie School of Mines.

Geomorphology

Western Australia is a land largely of plains and plateaus. There are few mountain ranges, and the highest point, Mount Meharry, is only 1251 m. Most of the interior of the state is occupied by a broad, featureless plain that averages 300–450 m above sea level. This has been described as a near-perfect peneplain, which has undergone slight epeirogenic uplift in late Tertiary times (Jutson, 1934).

The climate of the greater part of Western Australia is arid and semiarid. Nearly half of the state has a rainfall of less than 250 mm per annum. The climate is suitable for intensive agriculture only in the SW, where the annual precipitation (mainly in the winter) is from 400 to 1250 mm, and in the tropical northern part, with rainfall of 650–750 mm (mainly in the summer).

Geotectonics

The tectonic framework of Western Australia is relatively simple. The main Precambrian nucleus is known as the West Australian Shield. It consists of two Archean blocks (the Yilgarn and Pilbara blocks) cut by NW-trending troughs of Proterozoic sediments and volcanics known as the Hamersley and Bangemall basins. In the N the Kimberley Basin is another large area of Proterozoic sedimentary rocks. The Precambrian basement is overlapped marginally by a number of Paleozoic and younger sedimentary basins.

The Archean rocks were intensely deformed some 2400 to 2800 million years ago by widespread orogenic movements, associated with granitic intrusions and extensive metamorphism. The regional trends of the major fold axes are N-NW. The Archean metamorphic, sedimentary, and volcanic rocks occur in synclinorial belts (known as "greenstone belts") flanking granitic domes.

Deformation of the Proterozoic rocks is very variable. In certain areas the Proterozoic rocks are strongly folded and metamorphosed and are in places intruded by granite. However, much of the Proterozoic sequence is little deformed or metamorphosed, for example the Lower Proterozoic sequences of the Hamersley and Kimberley basins.

No major orogenic movements have taken place in the state since the late Precambrian. Some mild folding has occurred in places and there has been widespread faulting, but no intense deformation or plutonic intrusion. Volcanics are found in the Lower Cambrian or uppermost Proterozoic of the Ord and Bonaparte Gulf basins; the Ordovician of the Officer Basin; the Upper Permian of the Carnarvon Basin; the Jurassic of the Canning, Bonaparte Gulf, and Browse basins; and the Lower Cretaceous of the Perth Basin.

Faulting, most of which apparently has normal movement, dominates the structure of most of the Phanerozoic sedimentary basins. However, some important folding occurred in the N part of the Carnarvon Basin in the late Tertiary, and the largest structure (the Cape Range) is some 100 km long and has 450 m of closure. In the Canning Basin the folding occurred during the Middle Triassic to Lower Jurassic, associated with right-lateral strike-slip movements. Both basins are extensively faulted.

The Darling Fault is one of the dominant structural features of the Australian continent. It extends N-S for nearly 1000 km and defines the eastern margin of the Perth Basin for most of its length. It is believed to be a normal fault having a throw of up to 12,000 m.

The Urella Fault is another very large fault in the Perth Basin. It is parallel to the Darling Fault and is believed to have a normal displacement of at least 6000 m. Faulting is widespread throughout the Perth Basin, mainly trending N-S. The basin is an asymmetric graben, although the western side is not marked by a single large fault comparable with the Darling or Urella faults.

The Paleozoic rocks of the Carnarvon Basin are extensively faulted, and most of the faults are believed to be normal. Faulting is also widespread in the northern part of the Canning Basin, the most important examples being the Pinnacle and Fenton faults, which mark the northern and southern boundaries of the Fitzroy Trough. This deep trough of sediments is filled largely by Lower Permian and Carboniferous deposits.

Diapiric salt intrusions are known in the northern Canning Basin (Silurian salt) and in the Officer Basin (Proterozoic salt).

The small Collie and Wilga basins are believed to have been scoured out by ice during the Early Permian glaciation. They are filled with continental Permian sediments, covered by thin lacustrine deposits of probable Tertiary age.

FIGURE 2. Typical Precambrian granite-gneiss topography on the south coast (Pascoe Island, Recherche Archipelago). Domal weathering and exfoliation of the granites is not to be confused with major domal structures. Subject to salt air, and exsudation, granite surfaces are often pock-marred by cavernous depressions locally known as "gnamma holes." (Photo: Australian Geographical Society.)

Stratigraphy

Precambrian. During the past decade stratigraphic studies of the Precambrian rocks have been following the systematic approach recommended by the Australian Code of Stratigraphic Nomenclature (Geological Society of Australia, 1964). Coupled with isotopic age determinations, there has been a radical revision of earlier regional correlations (Geological Survey of Western Australia, in press). Best known are the three main Proterozoic basins, the Hamersley, Kimberley, and Bangemall basins, which are very little deformed or metamorphosed.

The Archean rocks closely follow the typical Archean pattern found on other continents. Synclinorial belts ("greenstone belts") of sedimentary and volcanic rocks, usually metamorphosed to a low or medium grade, separate irregular ovoid granitic domes (Fig. 2). There is characteristically a high proportion of volcanic rocks, both extrusive and intrusive, ranging from acidic to ultramafic. They are interbedded with sedimentary rocks that are often of immediate volcanic derivation; cherts and cherty iron formations also occur (Glover, 1971).

The Hamersley Basin developed in the NW of the State between about 2300 and 1800 m yr ago. Some 30 named rock units are included within its three constituent groups (MacLeod, 1966; Trendall and Blockley, 1970). The basal Fortescue Group attains a maximum thickness of 4600 m and consists largely of basaltic lava and tuff, with intercalated shallow-water sediments. It is followed conformably by the Hamersley Group, about 2500 m thick, characterized by thick banded iron formations, with important acidic lavas and tuffs, and subordinate shale, dolomite, and dolerite sills. The uppermost Wyloo Group has a wide local range of basalt, dolomite, and quartzite in the lower

FIGURE 3. Steeply dipping late Precambrian metasediments (including important iron ores) exposed in the red cliffs of Cockatoo Island in the North-West Kimberley. (Photo: W. Pedersen, Australian Official Photograph.)

part, which gives way upward to a uniform and monotonous graywacke and shale succession.

In the extreme N the Kimberley Basin began to develop as deposition ended in the Hamersley Basin (Fig. 3). After the fairly rapid deposition of some 5000 m of sedimentary and volcanic rocks in the main basin, comprising some 23 formally named rock units, deposition continued locally in minor marginal fault-controlled basins throughout the remainder of Precambrian time. The late Precambrian succession comprises an additional 20 rock units. Compared to the Hamersley Basin, the Kimberley sedimentation contains more clastics, mainly sandstone and shale, although the sedimentation was initiated by the acidic Whitewater Volcanics, and the Carson Volcanics and Hart Dolerite followed shortly afterward. Glacigene rocks form the uppermost part of the succession (Dow and Gemuts, 1969).

The Bangemall Basin, of approximate age 1000 m yr, extended in a broad belt from south of the Hamersley Basin westward through the Warburton-Blackstone area, probably to join with the Adelaide Geosyncline of South Australia. The southern edge of the trough forms the northern boundary of the Archean Yilgarn Block. The dominant rocks are sandstone, shale, graywacke, and chert, but volcanics occur in the eastern area. In the Warburton-Blackstone area, according to Daniels, the sequence is dominantly rhyolitic, at first followed by a mixed sequence passing to an uppermost glacial succession that is correlated with that of the Kimberley Basin.

Phanerozoic. The most recent reference on the Phanerozoic stratigraphy of Western Australia is the volume by Geological Survey of Western Australia (1975).

Cambrian. Sediments of this age are known from the Bonaparte Gulf and Ord basins. They consist of limestone, shale, sandstone, and conglomerate of middle to late Cambrian age and are up to 900 m thick. They overlie basaltic flows of probable early Cambrian age.

Ordovician. These rocks are present in the Bonaparte Gulf and Canning basins and in a small outlier near the border adjacent to the Adameus Basin of the Northern Territory.

The thickest Ordovician sequence is in the Canning Basin, where some 1500 m of Lower and Middle Ordovician shale, dolomite, and limestone are known. In the Bonaparte Gulf Basin the Lower Ordovician is represented by glauconitic sediments more than 150 m thick.

Silurian. Silurian rocks are known from the Carnarvon Basin and possibly from the Canning Basin. An Upper Silurian unit of limestone, dolomite, sandstone, and evaporites more than 1300 m thick was intersected in a number of wells drilled in the Carnarvon Basin, but it is not exposed at the surface. It overlies 3000 m of an unfossiliferous sandstone, thought to be at least partly of Silurian age. In the Canning Basin the Silurian may be represented by an unfossiliferous sequence of shale, sandstone, limestone, dolomite, and evaporites some 1700 m thick.

Devonian. Devonian sedimentary rocks are known from the Bonaparte Gulf, Canning, and Carnarvon basins.

The sequence in the Bonaparte Gulf Basin consists of sandstone and limestone about 1800 m thick. Middle and Upper Devonian rocks exposed along the northern margin of the Canning Basin are mainly limestones. They form a series of well-exposed reef complexes more than 2000 m thick, richly fossiliferous and precisely zoned by means of goniatites, conodonts, and brachiopods. Elsewhere in the basin the Devonian is represented by limestone, shale, dolomite, conglomerate, and red beds more than 2000 m thick.

In the Carnarvon Basin the Devonian sequence includes richly fossiliferous limestone together with sandstone and conglomerate about 1600 m thick. Reefal deposits also occur offshore beneath the continental shelf.

Carboniferous. Carboniferous rocks have been recognized in the Bonaparte Gulf, Canning, and Carnarvon basins. The sequence in the Bonaparte Gulf Basin consists of more than 2000 m of limestone and sandstone. The Carboniferous of the Canning Basin is represented by about 300 m of Lower Carboniferous limestone, sandstone, and shale, and by more than 1500 m of Lower and Upper Carboniferous sandstone, shale, and siltstone. In the Canning Basin there are about 900 m of Lower Carboniferous limestone, dolomite, and siltstones.

FIGURE 4. Ice-rafted Permian tillite including faceted erratics and armored mudball (Irwin River District). (Photo: R. W. Fairbridge.)

FIGURE 5. Glacially striated rock surface, Lower Permian, Irwin River District. (Photo: R. W. Fairbridge.)

Permian. The Permian System is thicker and more widespread here than in any other Australian state and is represented in the Bonaparte Gulf, Browse, Canning, Carnarvon, Perth, Collie, Wilga, and Officer basins. In each of these the basal Permian unit (Sakmarian) is of glacial origin.

The sequence above the glacial unit is almost wholly marine in the Canning and Carnarvon basins, is mixed marine and continental in the northern Perth Basin (Figs. 4,5), and is wholly continental in the southern Perth, Collie, and Wilga basins. The thickest sections are in the Canning and Carnarvon basins, where each totals about 4500 m. After a richly fossiliferous interval which includes corals, there was a return of cooler conditions with ice-rafted erratics.

The only commercial coal deposits in the State occur in the Permian of the Collie Basin, where they range from Early to Late Permian in age. The continental facies are characteristically marked by a rich *Glossopteris* flora.

Triassic. Triassic sediments are present in the Perth, Carnarvon, Canning, Browse, and Bonaparte Gulf basins. The sequence normally consists of shale at the base in the Lower and (in some basins) Middle Triassic, overlain by a predominantly arenaceous Middle and Upper Triassic sequence. The thickest known succession is in the Perth Basin and is more than 2500 m thick.

Jurassic. The Jurassic System is represented in the Perth, Carnarvon, Canning, Browse, and Bonaparte Gulf basins. Most of the 3600-m sequence in the Perth Basin consists of terrigenous clastics with minor coal in the lower part and is continental, except for a thin marine Bajocian unit. A similar thickness is developed in the Carnarvon Basin, but there it is mainly marine shale, siltstone, and sandstone. In the other basins the section is thinner but still mainly marine. Basaltic lavas occur in the Jurassic of the Browse Basin and in the offshore Bonaparte Gulf Basin. Leucite lamproite plugs (Fig. 6) and flows in the northern Canning Basin are also believed to be of this age.

Cretaceous. Cretaceous sediments occur in the Bonaparte Gulf, Browse, Canning, Carnarvon, Perth, Eucla, and Officer basins. The sequence typically consists of marine, continental, and paralic sandstone, shale, and siltstone in the Lower Cretaceous, with marine carbonates, green sand, chalky marls, and some siltstones in the Upper Cretaceous. The thickest section known is in the offshore Carnarvon Basin, amounting to about 2000 m.

Tertiary. Marine Tertiary sediments occur in all coastal and offshore sedimentary basins of the state. Most of the section consists of limestone, with some shale and minor sandstone. The thickest known sequence is in the offshore Browse Basin, amounting to more than 3000 m. The most widespread occurrences onshore are in the Eucla Basin, where Tertiary limestones about 350 m thick underlie the flat, treeless Nullarbor Plain.

Quaternary. Thin Quaternary deposits blanket much of Western Australia. Of special interest is the "Coastal Limestone," a Pleistocene eolianite that is up to 300 m thick (Fig. 7). It contains some thin intercalated marine horizons. This formation occurs in many coastal areas of Western Australia, especially south of the tropics. It is believed to be a sensitive indicator of eustatic changes of sea level during the Quaternary. Parallel ridges of eolianite and marine bands indicate successive strand lines, and raised wave-cut platforms in the limestone are evidence of Pleistocene and Holocene changes of sea level.

Mineral Deposits

Western Australia possesses great mineral wealth, at first dominated by gold mining; but by 1970 gold had declined to only 2% of the

FIGURE 6. Machell's Pyramid, a leucitite volcanic neck, forming a residual hill above a characteristic tropical "discordant plain" in the Kimberley region, where folded Permian sediments directly underly a surface marked by laterite soil (note termite "hills," red in color). One of many leucitite plugs in the Fitzroy Valley, its intrusion may be mid-Mesozoic to early Tertiary. (Photo: R. W. Fairbridge.)

annual value of mineral production, being surpassed by iron ore, bauxite, nickel, and oil.

Iron Ore. Western Australia has the largest reserves of iron ore in the continent. The deposits discovered during the 1960s in the Pilbara district rank among the largest known in the world. Hematitic ores occur in both the Archean and Proterozoic and are mainly near-surface concentrations due to desilication of banded iron formations. Reserves of better than 60% iron approach 15,000 million tons. In addition there are some 20,000 million tons containing 50% or more iron (including limonite-goethite ores).

Bauxite. Bauxite deposits of lateritic type are widely developed over Archean granitic rocks in the SW and over Proterozoic volcanics in the northern Kimberleys. Reserves amount to 2,000 million tons.

Nickel. A number of important nickel discoveries have been made in the Eastern Goldfields district. Reserves at Kambalda, near Kalgoorlie, amount to some 20 million tons of ore averaging 3.4% nickel. A number of other

FIGURE 7. "Coastal Limestone" often marks Western Australia's littoral. Subaerial and intertidal induration of this eolian calcarenite is developed, in the upper level as a typical calcrete (caliche or soil travertine) that conforms to the topography of the original calcareous sand dunes (Pleistocene). It is punctuated by calcrete-lined (karst) solution pipes, now exposed by differential weathering. A wide intertidal platform is commonly exposed at low tide, and the soft eolianite indurated here by daily exposure and covering by the ocean. Traces of a 3-meter mid-Holocene marine terrace, believed to have been cut and indurated about 5000 years ago are also widely exposed. Example: Garden Island, The "Organ Pipes," Pt. Atwick. (Photo: R. W. Fairbridge.)

ore bodies are also being developed, while there are also some large low-grade sulfide and lateritic deposits that are not economic at present. The nickel mineralization in these areas is associated with ultrabasic intrusions in the Archean rocks.

Petroleum. Several oil and gas fields have been discovered in Western Australia since 1964. The Barrow Island oilfield has reserves in Cretaceous sandstone of some 200–250 million barrels, and began production in 1967. The Dongara gas field was linked to Perth by a pipeline in 1971. Reserves in this field amount to some 500 billion cubic feet, in Triassic and Permian sandstones. Several smaller gas fields have been found in the Perth Basin, and these will also be linked to the pipeline.

Major discoveries of natural gas were made on the continental shelf during 1971, at North Rankin, Rankin, and Goodwyn (Triassic) and Angel (Jurassic) in the Carnarvon Basin and at Scott Reef (Triassic and Jurassic) in the Browse Basin. Reserves of these fields will probably amount to many trillions of cubic feet.

Gold. Gold mineralization is generally associated with Archean volcanic and metavolcanic "greenstones" close to their contacts with granitic masses. Gold production reached a peak of 2,064,800 ounces in 1903, and since then has declined steadily as the various fields were worked out, so that by 1971 it was down to 348,000 oz.

Mineral Sands. Heavy-mineral sands are being worked at a number of localities on the coast S of Perth. They are associated with modern beaches and with Pleistocene strandlines up to several kilometers inland and at various elevations. Minerals present in economic concentrations are ilmenite, rutile, leucoxene, zircon, and monazite.

Large reserves of rutile-bearing sands were discovered in 1971 near Eneabba, 250 km N of Perth. The deposit is associated with an early Pleistocene or late Tertiary strandline.

Coal. The only commercial coalfield in Western Australia is in the Collie Basin. The coal is of Permian age and is of the noncoking subbituminous type. Total reserves amount to some 270 million tons, and annual production is now over 1 million tons.

Salt. Solar salt is being produced in large quantities from a number of localities in Western Australia, principally at Shark Bay, Port Hedland, Dampier, and near Carnarvon. Climatic conditions for solar-salt production are ideal in the northwestern part of the state, the only limiting factor at present being the availability of markets.

Other Minerals. Western Australia also produces amounts of manganese, tin, copper, lead, zinc, tantalite, columbite, beryl, gypsum, glass sand, limestone, feldspar, talc, and barite.

PHILLIP E. PLAYFORD

References

Ayres, D. E., 1972. "Genesis of iron-bearing minerals in banded iron formation mesobands in the Dales Gorge Member, Hamersley Group, Western Australia," *Econ. Geol.,* **67,** 1214–1233.

Condon, M. A., 1965. "The geology of the Carnarvon Basin, Western Australia," *Bur. Mineral Resources Geol. Geophys. Bull.,* **77,** pt. 1, 82p.; pt. 2, 191p.; pt. 3, 68p.

Daniels, J. L., 1974. "The geology of the Blackstone area, Western Australia," *W. Austral. Geol. Surv. Bull.,* **123.**

Dow, D. B., and Gemuts, I., 1969. "Geology of the Kimberley region, Western Australia—the East Kimberley," *W. Austral. Geol. Surv. Bull.,* **120,** 135p.

Fairbridge, R. W., 1953. *Australian Stratigraphy.* Univ. W. Australia Text Books Board.

Geological Society of Australia, 1964. "Australian code of stratigraphic nomenclature, 4th ed.," *J. Geol. Soc. Austral.,* **11,** 165–171.

Geological Survey of Western Australia, 1975. "The geology of Western Australia," *W. Austral. Geol. Surv. Mem.* **2.**

Glover, J. E., ed., 1971. "Symposium on Archaean rocks held at Perth, 1970," *Geol. Soc. Austral., Spec. Pub.* **3,** 469p.

Jutson, J. T., 1934. "The physiography (geomorphology) of Western Australia," *W. Austral. Geol. Surv. Bull.,* **95,** 366p.

Logan, B. W., Davies, G. R., Read, J. F., and Cebulski, D. E., 1970. "Carbonate sedimentation and environments, Shark Bay, Western Australia," *Am. Assoc. Petrol. Geologists Mem.* **13,** 223p.

Lowry, D. C., 1970. "Geology of the Western Australian part of the Eucla Basin," *W. Austral. Geol. Surv. Bull.,* **122,** 201p.

MacLeod, W. N., 1966. "The geology and iron deposits of the Hamersley Range area, Western Australia," *W. Austral. Geol. Surv. Bull.,* **117,** 170p.

McWhae, J. R. H., Playford, P. E., Lindner, A. W., Glenister, B. F., and Balme, B. E., 1958. "The stratigraphy of Western Australia," *J. Geol. Soc. Austral.,* 4(2), 161p.

Martison, N. W., McDonald, D. R., and Kaye, P., 1973. "Exploration on continental shelf off Northwest Australia," *Bull. Am. Assoc. Petrol. Geologists,* **57** (6), 972–989.

Perry, W. J., and Roberts, H. G., 1968. "Late Precambrian glaciated pavements in the Kimberley region, Western Australia," *J. Geol. Soc. Austral.,* 15(1), 51–56.

Playford, P. E., and Lowry, D. C., 1966. "Devonian reef complexes of the Canning Basin, Western Australia," *W. Austral. Geo. Surv. Bull.,* **118.**

Trendall, A. F., and Blockley, J. G., 1970. "The iron formations of the Precambrian Hamersley Group, Western Australia," *W. Austral. Geol. Surv. Bull.,* **119,** 366p.

Veevers, J. J., 1967. "The phanerozoic geological history of northwest Australia," *J. Geol. Soc. Austral.,* **14**(2), 253–271.

——, and Wells, A. T., 1961. "The geology of the Canning Basin, Western Australia," *Bur. Mineral Resources Geol. Geophys. Bull.,* **60,** 322p.

Cross-references: *Australasia–Regional Review: Australia–Northern Territory, South Australia.*

B

BAHAMAS

The Bahamas (also once known as the *Lucayos*) are an extensive archipelago in the western Atlantic covering about 11,406 km² (4404 sq mi), SE of Florida, often classified with the Caribbean islands. The former British colony has been autonomous, since July 1973, entitled "The Commonwealth of the Bahamas." The physiographic province of the Bahamas (11,836 km², 4570 sq mi) is larger and includes a small SE group under a separate (colonial) administration (*Turks and Caicos Islands,* q.v.).

There are 18 main islands, around 700 islets and cays, and over 2300 miscellaneous rocks and reefs. The islands were first discovered by Columbus, who is reputed to have made his landfall in the Americas on Watling Island in October 1492.

The structural basis of the islands, the Bahama Platform, lies E and SE of Florida and is closely related to the submarine Blake Plateau, which lies off the shore of Georgia. It covers over 100,000 km² (40,000 sq mi), and extends from $20°50'$ to $27°25'$N and $72°37'$ to $80°32'$W. On this platform, embayed by deep oceanic troughs (e.g., Tongue of the Ocean, Exuma Sound), are several distinct "banks": (a) the *Little Bahama Bank,* which lies in the NW (principal islands: *Grand Bahama, Great* and *Little Abaco*); (b) the *Great Bahama Bank* in the middle (with *Andros, Exuma, New Providence, Long, Eleuthera, Cat*); and (c) the *Navidad, Caicos,* and *Silver Banks* in the SE, as well as several smaller ones along the northeastern margin facing the open Atlantic (notable islands: *San Salvador* or *Watling, Rum Cay, Mayaguana, Crooked, Acklins, Great* and *Little Inagua*). The group is separated by the 80-km-wide Florida Strait or New Bahama Channel from Florida and by the Old Bahama Channel from Cuba to the south. The most important island is New Providence with the capital, Nassau.

All the islands are low-lying, of shallow marine, coral and eolianite limestones (often with calcrete crusts, paleosols, etc.), carbonate sands, and many living reefs. Dunes reach a maximum elevation (on Cat Island) of about 67 m. These limestones were deeply affected by karst erosion during low sea-level episodes of the Quaternary, and there are numerous caves and sinks. The smaller pipes are known as "banana holes" (because they hold soil and water), whereas the larger sinks (which go down vertically to 100 m and more) are called "blue holes" from the color of the water in contrast to the pale green over the banks and reefs.

The climate is mild (January–March) to tropical, and much influenced by the NE Trade Wind, and also somewhat by the Gulf Stream on the western side. Rain usually falls in May to October and hurricanes occur from July to October. Precipitation at Nassau averages 1175 mm. North Abaco and other northerly islands receive up to 1500 mm, while those in the SE may get very much less (Ragged Island, about 500 mm).

History of Research

Geological research in the area has mostly been done by North American geologists, many of whom have had connections with the oil industry. The research gained impetus in the late 1940s, when contemporary carbonate sedimentation attracted most attention, but little was directed at the land areas. The Bahama Banks have been researched as modern sedimentary analogs of the extensive, frequently oil bearing, shelf carbonate deposits found in the geological column. With the exception of the landform studies of Doran and of Lind, there are only scattered references to onshore geology in the literature. In 1968 the Bahamas commissioned an integrated land resources survey from a British government agency. By 1974 the thirteen major islands had been investigated in this survey, and the resulting reports include landform and groundwater studies accompanied by large- and medium-scale maps. Good-quality topographic maps at large scale and air photos are available for most of the islands.

Economic Geology

Geologic materials contribute very little to the economy of the Bahamas, except for the celebrated beach sands, the value of which as a tourist "substrate" is considerable. Oil investigations have not to date proved any economic reserves, although a number of deep wells have been drilled. An oil refinery at Freeport, Grand Bahama, handles 140,000 barrels per day, but the oil is of foreign origin. Salt is produced commercially in solar salt pans on the island of Inagua and by a smaller industry on Long Island. Cement production in Grand Bahama uses local limestone and supplies the home market. There is no good source of mechanically strong crushed aggregate, with the consequence that hard rock aggregates are sometimes imported; the limited "hard" limestone occurrences are mostly remote from major areas of construction. Sawn dune limestone (eolianite), once used extensively for building, is not used today. Oolitic sands, commercially referred to as aragonite, are dredged off Cat Cay (W of Andros) and shipped to the United States.

Potable water supplies are pumped from wells sunk into the freshwater lenses that occur in the limestone aquifer. The freshwater lenses float with limited mixing above the salt water, which permeates the rock at depth below each island. The freshwater lenses are best developed in the broader flat islands (Andros, Abaco) and may attain a maximum thickness of 40 m but are commonly much less (Mather, 1971; Buckley, 1973: see references after "Young, 1972," p. 115).

General Geology and Geomorphology

The exposed coral and shallow-water marine limestones of the islands are of Pleistocene age, certainly pre-classic Wisconsin, as shown by "greater than" radiocarbon dates and the deep karst features. These limestones were formed when sea level was appreciably higher than the present. These are extensively capped by Pleistocene eolianites, some of which represent the windblown carbonate sands that were blown up onto the older platform when the sea level began to drop during glacial emergence (Fig. 1). It appears that there are multiple cycles represented by interdune paleosols containing indigenous snail populations, reminiscent of those in *Bermuda* (q.v.). Such soils can form within a few centuries during interstadial episodes.

There are no volcanic rocks and no rocks older than Pleistocene occur at the surface. Deep wells (one of 4446 m on Andros and another of 5700 m on Cay Sal Bank) have penetrated thick sequences of Tertiary and Upper Cretaceous carbonates with some dolomites and anhydrite: both ended in the Lower Cretaceous. The general character of these older deposits is similar to today's surface deposits on the platform. Thus there has been an overall history of shallow-water carbonate sedimentation keeping pace with regional subsidence. This has generated one of the world's largest carbonate provinces with an approximate volume of 1.5×10^6 km^3 (Dietz *et al.*, 1970).

A wide variety of Holocene accretionary bedforms can be seen in clear shallow water throughout the Banks; some are mobile and some partly fixed. Their form and distribution is influenced by the submerged pre-existing Pleistocene topography and by contemporary tidal fluctuations on the Banks.

The Pleistocene limestones exposed on land represent a wide range of ancient sedimentary environments the modern counterparts of which can be observed in and around the islands today. A prime distinction can be made between flat relatively lowland areas of shallow marine origin and the chains of low hills of eolian origin which form the "spines" of the long, thin islands such as Cat Island, Eleuthera, Long Island, and the eastern margins of Mayaguana and Inagua. A further notable distinction is between the suite of dune rocks and the extensive beach-ridge deposits which form large tracts of low ridgeland in Mayaguana, Cat Island, North Abaco, and elsewhere. Within these environments (marine, eolian, and littoral) there is a spectrum of subenvironments which reflect local (Pleistocene) energy conditions and sediment supply (Young, 1971). The rocks of marine origin range from poorly sorted, burrowed, pelletic, muddy sediment to clean, well-sorted, graded oolite. Wide expanses of coral-rich lagoon deposits, now strongly cemented and containing in situ corals, occur behind beach-ridge deposits in Mayaguana and Inagua, but such deposits are rarer in the northern Bahamas. Most of the rocks of littoral origin are of uniformly well graded oolite, although hurricane boulder ramparts are not uncommon where the shore lies adjacent to deep waters. The rocks of eolian origin have characteristics varying between poorly rounded fine-grained skeletal sediments, very fine well-sorted oolite, and shelly ooid—peloid sediment of intermediate grain size. The dunes are mostly transverse and have a wide variety of shapes, slopes, and altitudes. In addition, limestones of semilacustrine, lacustrine, and marsh origin occur, notably in Andros but also elsewhere.

Most of the land forms in the Bahamas are constructional forms or their still recognizable relics, and photogeologic methods show a clear picture of the complex accretionary history of each island, in which the formation of one deposit has frequently modified or controlled succeeding deposition. At least six different styles of carbonate dune growth separated in time have been distinguished by Young (1972) in Cat Island and similar episodic accretion can

FIGURE 1. Map showing the principal Bahama Banks and islands. Indicated in black are the major eolianite ridges; within them the dips are always away from the seaward beaches (simplified from Ball, 1967).

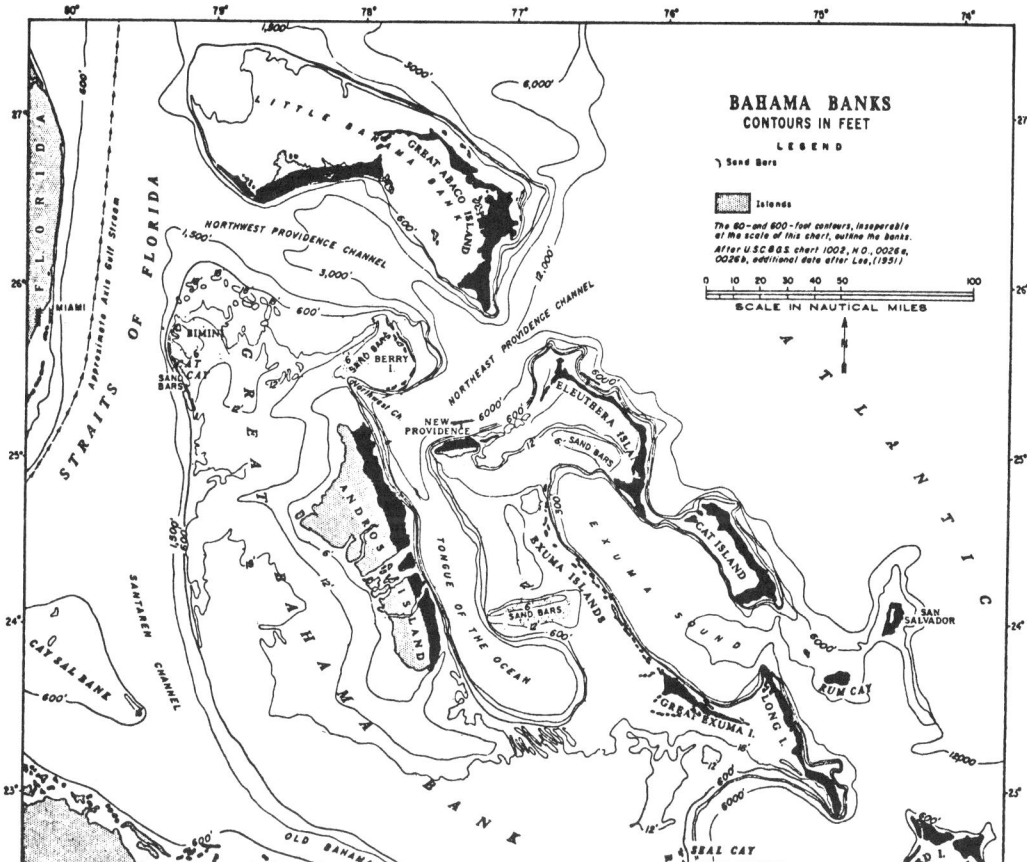

be discerned within the littoral and marine facies. (Similar patterns of dune accretion have been described from Bermuda by Land et al., 1967—see: *Bermuda*).

Karst weathering has modified the constructional land forms to varying degrees; the effect on the more recently deposited limestones is relatively minor. Mainly the karstification is dependent on lithofacies and in some cases results in soil-covered rocky plains of high agricultural capability (Young, 1974). Some of the older limestones show the effects of deep penetration by percolating water during times of low Pleistocene sea level when effective relief was large. There are at least 119 major sinkholes (blue holes or in Yucatan, cenotes) dotted over the land area of Andros while others occur in other islands. Of 78 holes depth-sounded on Andros by the Bahamas Land Resource Survey, four exceeded 100 m, the deepest being 110 m. All are water-filled today. The outer platform margins are in places notched and terraced by intertidal marine erosion at various depths from −4 to −30 m.

Lind (1969) has investigated the Holocene sandy landforms around Cat Island and comparable geomorphic features also occur as appendages around the coasts of the other islands. The question of a mid-Holocene high sea level is controversial, but evidence of high stands at 2, 1, and 0.5 m has been described by Lind from the low-energy Exuma Sound coast of Cat Island. A mangrove peat from Bimini, today at −3 m, is taken to be evidence of negative oscillation of Holocene sea level and was dated at 4370 B.P.; it was used to designate a "Bahama Emergence." An emerged beach rock at +2 m on Bimini dates to 2300 B.P., and the emerged carbonate muds on Andros date to 2330–2660 B.P. (the Dunkirk or "Abrolhos Submergence" of Fairbridge, 1961).

The Bahama Banks

The major part of the Bahama Banks is covered today with a mantle of calcium carbonate sand. At the extreme edges of the banks, the sediment is composed mainly of skeletal material, including the tests of mollusca, foraminifera, material from calcareous algae, and madreporarian corals, with minor contributions from other organisms (Fig. 2). Typical analyses of edge deposits reveal that as much as 92% is composed of skeletal material. In contrast to this, in the interior of the banks, the percentage of recognizable skeletal material averages only 12% (Illing, 1954). The muddy material found in the interior of the banks is thought to be caused by breakdown of calcium carbonate precipitating algae such as *Penicillus* (Stockman *et al.*, 1967; Neumann, 1969). Variations in water circulation play a large part in the distribution patterns of all the surface sediments (Ginsburg *et al.*, 1963).

One of the main problems concerning the geology of the Bahamas have been to explain the relationship of the carbonate platforms to the intraplatform basins, such as Exuma Sound, Tongue of the Ocean, and the two Providence channels (Fig. 1), and to discover the nature of the crystalline basement. One clue is the observation by Talwani (1960) that there is a striking correlation between gravity anomaly and platform topography (Fig. 3). By and large, the topographically low intraplatform basins are areas of negative gravity anomaly. The adjusted gravity anomaly must result from density contrasts below the depth of the floors of the basins. This interpretation was based in part on seismic refraction. These data show that correlatable seismic layers are depressed beneath the basins. For instance, the 6–6.2 km/sec layer is at a depth of 4 km beneath the Florida Straits and in the NW Providence channel, but at a depth of only 2.5 km beneath Great Bahama Bank (Sheridan *et al.*, 1966), and has been identified as Lower Cretaceous on the basis of interval velocities measured in well surveys in south Florida. Thus it seems probable that the original formation of the basins in the Bahamas was tectonic, causing the basins to be faulted below the shallow-water depth zone of vigorous carbonate formation, i.e. less than 20 m. The amplitude of faulting necessary to do this might be quite small.

The present relief contrast between platforms and basins is as much as 2000 m. It follows that the deposition of allochthonous shallow-water sediments and pelagic material in the basins was unable to match the regional subsidence rate so that the relief contrast between platform and basin has increased with time.

Subsequent to the original formation of the basins, secondary faulting or infilling has allowed shallow-water carbonate formation to accumulate in two places which were once structural lows. These are at the south of the southern end of Tongue of the Ocean and the center of Little Bahama Bank. Both places are characterized by negative gravity anomalies. At least nine diapiric structures have been located in Exuma Sound. Sedimentation rates seem to have greatly accelerated in the late Quaternary (Lidz, 1973).

The key to understanding the basement is the realization that the Bahamas are situated in a region of overlap if the continents around the Atlantic Ocean are cartographically fitted together to their early Mesozoic position (Bullard *et al.*, 1965). Since the area of West Africa onto which the Bahamas overlap would have to have been present before the Atlantic started to open, it would seem that some, at least, of the basement of the Bahamas originated after opening started. Leg 11 of the Deep Sea Drilling Project included four holes in and near the

FIGURE 2. Map of the main Bahama Banks, indicating the principal sediment types (modified after Chenoweth, 1970).

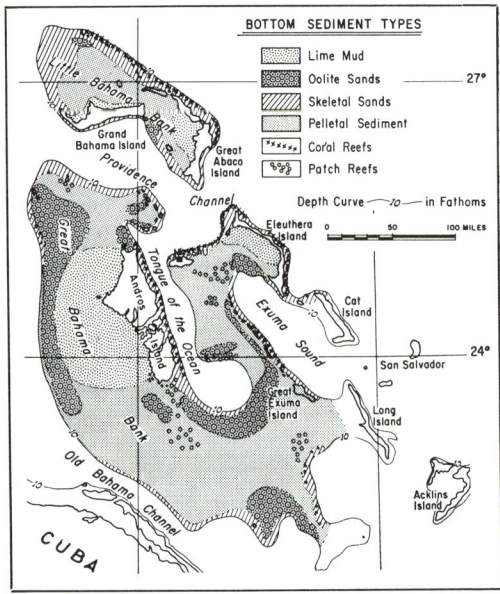

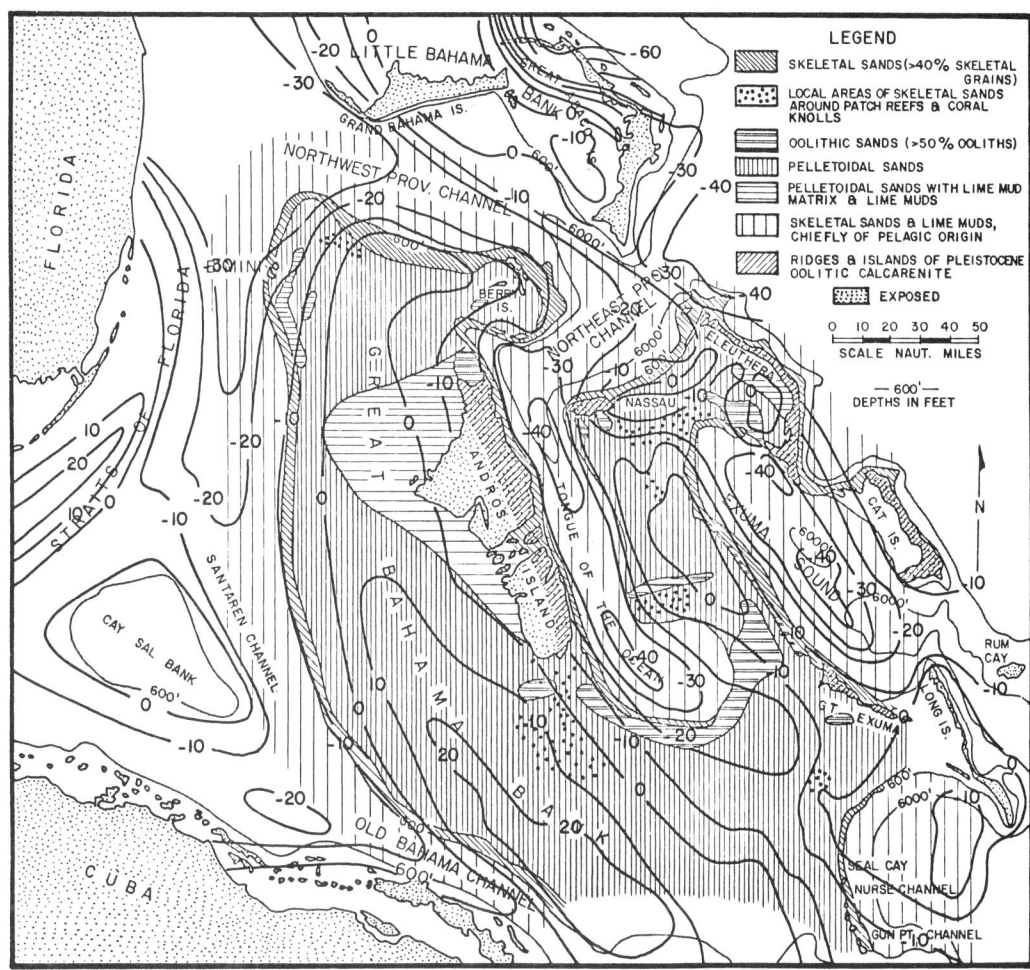

FIGURE 3. Map of the main Bahama Islands showing lithologic facies. Superimposed on this is the residual gravity anomaly map of Talwani (1960), which took into account density contrasts between water in the deep intraplatform straits and basins and the laterally adjacent carbonates of the shallow-water platforms.

Bahamas (98–101) which disclosed Mesozoic pelagic carbonate facies, strikingly comparable with those of the Alpine Tethys (Bernoulli, 1972). The alternation between shallow banks and deep channels is particularly well duplicated in the Apennines or in the Dinarides (d'Argenio, 1970).

One possible origin for the Bahamas calls for the infilling of the narrow proto-Atlantic Ocean with large volumes of clastic sediments from the adjacent continents. These sediments might build up to sea level, allowing the commencement of rapid carbonate production. Alternatively, a slow stretching apart accompanied by evaporite or carbonate sedimentation could have continued from the start. Further spreading of the Atlantic with faulting could cause the structural complexity of the platform which results in the present deep basins (Dietz et al., 1970; see also the discussion by Walper, 1971).

A second possibility is that at the beginning of spreading, the oceanic crust was at sea level, possibly because of anomalous mantle properties below the recently formed midoceanic ridge. Oceanic type crust could exist at sea level at the beginning of spreading; for example, the Afar Depression of *Ethiopia* (q.v.) is thought to be such a region. Further spreading to the east would open the whole area to an oceanic environment and allow the start of rapid carbonate buildup.

A third possibility, discussed by Uchupi *et al.* (1971), is that the basement structures of the NW and SE Bahamas are different. The NW Bahamas were formed on a subsiding oceanic crust, whereas the SE Bahamas were formed on a ridge associated with an old fracture zone which offset the early mid-Atlantic ridge crest.

FIGURE 4. Schematic east-west section across Florida Coastal Plain and Blake Plateau just north of the Bahamas showing hypothetical deep structures (from W. S. Olsen, 1974).

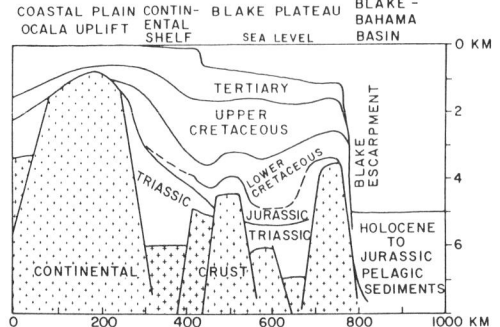

Further geophysical observations will probably help clarify the situation.

Some oceanic crust and/or "stretched" and thinned continental crust can be therefore considered. The case for a continental basement is argued by Meyerhoff and Hatten (1974) from a "fixistic" viewpoint. A "mobilistic" model is offered by Olsen (1974), whose profile of the Blake Plateau may be closely comparable to one through the Bahamas (Fig. 4). The argument for an oceanic volcanic foundation, a "neocraton" over which taphrogenic Jurassic clastics and salt accumulated during the early stretching phase, most closely accords with modern plate tectonic thought (Glockhoff, 1973; discussion by Dietz and Holden).

RHODES W. FAIRBRIDGE
R. NIGEL YOUNG
CHRISTOPHER G. A. HARRISON*
MAHLON MARSH BALL*

*The section on the Bahama Banks represents a contribution from the University of Miami, Rosenstiel School of Marine and Atmospheric Science. This research was supported by the Office of Naval Research Contract N00014-67-A-0201-0013 and National Science Foundation Grants GA-27465 and GB-27252. The sections on geomorphic, general, and economic geology were based on studies for the Bahama Land Resource Survey, and published by permission of the Bahamas Government.

References

Andrews, J. E., 1970. "Structure and sedimentary development of the outer channel of the Great Bahama Canyon," *Bull. Geol. Soc. Am.,* 81, 217–226.

_____, Shepard, F. P., and Hurley, R. J., 1970. "Great Bahama Canyon," *Bull. Geol. Soc. Am.,* 81, 1061–1078.

Ball, M. M., 1967. "Carbonate sand bodies of Florida and Bahamas," *J. Sed. Petrology,* 37(2), 556–591.

Bathurst, R. G. C., 1971. "Recent carbonate environments: 1. General introduction and the Great Bahama Bank," in *Carbonate Sediments and Their Diagenesis* (Dev. Sedimentol., 12, 620p). Amsterdam: Elsevier.

Bernoulli, D., 1972. "North Atlantic and Mediterranean Mesozoic facies; a comparison," *in* C. D. Hollister, J. I. Ewing, *et al.,* eds. *Initial Reports of the Deep Sea Drilling Project,* vol. 11. Washington, D.C.: Govt. Printing Off., 801–871.

Broecker, W. S., and Thurber, D. L., 1965. "Uranium series dating of corals and oolites from Bahaman and Florida Key limestones," *Science,* 149, 58–60.

Bullard, E. C., Everett, J., and Smith, A. G., 1965. "The fit of the continents around the Atlantic," *Phil. Trans. Roy. Soc., Lond.,* 258A, 41–51.

Chenoweth, P. A., 1970. "Bahama waters provide new carbonate rock data," *World Oil,* 170, 93–100.

D'Argenio, B., 1970. "Evoluzione geotettonica comparata tra alcune piattaforme carbonatiche dei Mediterranei Europeo ed Americano," *Atti Accad. Pontaniana* (Naples), 20(243), 1–34.

Dietz, R. S., Hodden, J. C., and Sproll, W. P., 1970. "Geotectonic evolution and subsidence of Bahama platform," *Bull. Geol. Soc. Am.,* 81, 1915–1928.

Fairbridge, R. W., 1961. "Eustatic changes in sea level," *Physics and Chemistry of the Earth,* vol. 4. London: Pergamon, 99–185.

Ginsburg, R. N., Lloyd, R. M., Stockman, K. W., and McCallum, J. S., 1963. "Shallow-water carbonate sediments," *in* M. N. Hill, ed., *The Sea,* vol. 3. New York: Wiley-Interscience, 554–582.

Glockhoff, C., 1973. "Geotectonic evolution and subsidence of the Bahama platform: discussion," *Bull. Geol. Soc. Am.,* 84, 3473–3476. (see discussion: Dietz and Holden, *ibid.,* 3477–3482.)

Illing, L. V., 1954. "Bahaman calcareous sands," *Bull. Am. Assoc. Petrol. Geologists,* 38, 1–95.

Lind, A. P., 1969. "Coastal landforms of Cat Island, Bahamas," *Res. Pap.* 122, Univ. Chicago, Dept. Geogr., 156p.

Lidz, B., 1973. "Biostratigraphy of Neogene cores from Exuma Sound Diapirs, Bahama Islands," *Bull. Am. Assoc. Petrol. Geologists,* 57(5), 841–857.

Lynts, G. W., Judd, J. B., and Stehman, C. F., 1973. "Late Pleistocene history of Tongue of the Ocean, Bahamas," *Bull. Geol. Soc. Am.,* 84, 2665–2684.

Meyerhoff, A. A., and Hatten, C. W., 1974. "Bahamas salient of North America: tectonic framework, stratigraphy and petroleum potential," *Bull. Am. Assoc. Petrol. Geologists,* 58(6), 1201–1239.

Muller, G., 1970. "Petrology of the cliff limestone (Holocene), North Bimini, Bahamas," *Neues Jahrb. Min., Monatsh.,* 11, 507–523.

Neumann, A. C., 1969. "Algal production and lime mud deposition in the Bight of Abaco: a budget," *Geol. Soc. Am. Spec. Pap.* 121, 219.

_____, and Moore, W. S., in press. "Sea level events and Pleistocene coral ages in the northern Bahamas," *Quat. Res.*

Newell, N. D., 1955. "Bahamian platforms," *Geol. Soc. Am. Spec. Pap.* 62, 303–316.

_____, Rigby, J. K., Whiteman, A. J., and Bradley, J. S., 1951. "Shoalwater geology and environments, Eastern Andros Island, Bahamas," *Bull. Am. Mus. Nat. Hist.,* 97, 1–29.

Purdy, E. G., 1963. "Recent calcium carbonate facies of the Great Bahama Bank: 2. Sedimentary facies," *J. Geol.,* 71(4), 472–497.
Richards, M. G., 1954. "Pleistocene mollusks from Andros Island, Bahamas," *Nautilus,* 67(4), 120–121.
Sheridan, R. E., 1971. "Geotectonic evolution and subsidence of Bahama platform, discussion," *Bull. Geol. Soc. Am.,* 82(3), 807–809.
_____, Drake, C. L., et al., 1966. "Seismic refraction study of continental margin east of Florida," *Bull. Am. Assoc. Petrol. Geologists,* 50, 1972–1991.
Shinn, E. A., Ginsburg, R. N., and Lloyd, R. M., 1965. "Recent supratidal dolomite from Andros Island, Bahamas," *Soc. Econ. Paleontol. Mineral., Spec. Publ.* 13, 112–124.
Stockman, K. W., Ginsburg, R. N., and Shinn, E. A., 1967. "The production of lime mud by algae in South Florida," *J. Sed. Petrology*, 37, 633–648.
Talwani, M., 1960. "Gravity–anomalies in the Bahamas and their interpretation," Ph.D. thesis, Columbia Univ.
Uchupi, E., Milliman, J. D., Luyendyk, B. P., Bowin, C. O., and Emery, K. O., 1971. "Structure and origin of southeastern Bahamas," *Bull. Am. Assoc. Petrol. Geologists,* 55, 687–704.
Walper, J. L., 1971. "Geotectonic evolution and subsidence of Bahama platform: discussion," *Bull. Geol. Soc. Am.,* 82(4), 1129–1130.
Young, R. N., 1972. "The application of carbonate facies analysis to landform studies for development in Cat Island and Abaco Island, Bahamas," *Mem. 6th Conf. Geol. del Caribe,* Margerita, Venezuela, 163–165.

In addition, island-by-island reports of the *Bahama Land Resource Survey* by J. D. Mather, D. K. Buckley, and R. N. Young are currently being published by the Ministry of Overseas Developement, Surbiton, England.

Cross-references: *Bermuda; Cuba; Dominican Republic; Haiti; Puerto Rico; United States–Atlantic Coastal Province.*

BARBADOS

One of the most interesting geologically of the world's smaller islands, Barbados has become celebrated both for its uplifted deep-sea facies and for its uplifted Pleistocene coral reefs. The structure discloses an asymmetric uplift on the E and a flattening on the N and S.

The core of the island consists of the New Scotland Beds, at least 200 m thick, outcropping on the east coast; these are possibly Cretaceous to Eocene in age. The upper part contains *Nummulites.* The facies is bathyal to abyssal with mainly terrigenous material in turbidite relationships, showing much slumping and penecontemporaneous disturbance, in short, a flysch facies. It has been interpreted as a major olistostrome (gravity slide) by Daviess (1971). In these sands and clays there are some asphaltic beds.

Overlying these beds unconformably follows the Oceanic series, particularly in the southwestern part of the island and on Mt. Hillaby, with 190 m of deep marine facies, in five formations: (1) *Globigerina* chalk, apparently a deep-sea ooze; (2) *Radiolarite,* almost exclusively tests of radiolaria and diatoms and sponge spicules also containing a deep-sea echinoid; (3) a second *chalk horizon;* (4) a mottled *red and white clay,* comparable to deep-sea red clay, containing a few radiolaria; (5) a bed of *gray volcanic ash and mudstone.* There are now known to be rather similar deep-water sequences in Haiti, Cuba, and Trinidad. There has been much dispute about their age, interpretations ranging from late Eocene through Miocene.

At Bissex Hill in the N they are overlain unconformably by Globigerina marls and limestone, 30 m thick, containing echinoids and shark's teeth, possibly of Oligocene or Miocene age. Again following unconformably are the *Coral Limestones,* a series of uplifted Pleistocene reefs that cover most of the island. These Pleistocene reefs rise in 10–30 m steps and are tilted up toward the east-center part of the island, the result of a combination of progressive uplift and glacio-eustasy (cf. *Netherlands Antilles*). Absolute dating by Th^{230} (Mesolella et al., 1969) of the three lowest reefs gives 82,000 ("Bermuda I"), 105,000 ("B II"), and 125,000 ("B III") years, the higher ones possibly going back to over 300,000 years. This sequence has played a very important role in helping to substantiate the Milankovitch theory of climatic modulation, for each of the last three peaks is exactly as predicted by him (Broecker *et al.,* 1968). Submerged barrier reefs offshore at 70 and 20 m suggest pre-Holocene transgression still stands at 16,000 and 12,000 B.P. (Macintyre, 1967).

RHODES W. FAIRBRIDGE

References

Beckmann, J. P., 1953. "Die Foraminiferen der Oceanic Formation (Eocaen–Oligocaen) von Barbados, Kleinen Antillen," *Eclog. Geol. Helv.,* 46(2), 301–412.
Broecker, W. S., Thurber, D. L., Ku, T. L., Matthews, R. K., and Mesollela, K. J., 1968. "Milankovitch hypothesis supported by precise dating of coral reefs and deep sea sediments," *Science,* 159, 297–300.
Bronnimann, P., 1949. "Notes on the ecologic interpretation of fossil Globigerina from the West Indies," *Micropaleontologist,* 3(2), 23–27.
Cizancourt, M. de, 1948. "Nummulites de l'ile de la

Barbade," *Mém. Soc. Géol. France,* Ser. 57, **27,** 36p.
Daviess, S. N., 1971. "Barbados: a major submarine gravity slide," *Bull. Geol. Soc. Am.,* **82,** 2593–2602.
James, N. P., Mountjoy, E. W., and Omura, A., 1971. "An early Wisconsin reef terrace at Barbados, West Indies, and its climatic implications," *Bull. Geol. Soc. Am.,* **82,** 2011–2018.
Macintyre, I. G., 1967. "Submerged coral reefs, west coast of Barbados, West Indies," *Can. J. Earth Sci.,* **4**(3), 461–474.
____, 1970. "Sediments off the west coast of Barbados: diversity of origins," *Marine Geol.,* **9**(1), 5–23.
Mesolella, K. J., 1967. "Zonation of uplifted Pleistocene coral reefs on Barbados, West Indies," *Science,* **156**(3775), 638–640.
____, et al., 1969. "The astronomical theory of climatic change: Barbados data," *J. Geol.,* **77,** 250–274.
____, et al., 1970. "Facies geometries within Pleistocene reefs of Barbados, West Indies," *Bull. Am. Assoc. Petrol. Geologists,* **54,** 1899–1917.
Senn, A., 1940. "Paleogene of Barbados and its bearing on history and structure of Antillean–Caribbean region," *Bull. Am. Assoc. Petrol. Geologists,* **24**(9), 1548–1610.
____, 1947. "Die Geologie der Insel Barbados (Kleine Antillen) und die Morphogenese der umliegende marine Gross formen," *Eclog. Geol. Helv.,* **40**(2), 199–222.
Steinen, R. P., Harrison, R. S., and Matthews, R. K., 1973. "Eustatic low stand of sea level between 125,000 and 105,000 B.P.: evidence from the subsurface of Barbados, West Indies," *Bull. Geol. Soc. Am.,* **84,** 63–70.
Trechmann, C. T., 1925. "The Scotland Beds of Barbados," *Geol. Mag.,* **62,** 481–504.
____, 1933. "The uplift of Barbados," *Geol. Mag.,* **69,** 19–47.

Cross-references: *Antigua; Cuba; Haiti; Netherlands Antilles; Trinidad and Tobago; West Indies; Windward Islands.*

BELIZE (BRITISH HONDURAS)

This former British colony, previously known as British Honduras, now has self-governing status. It has been politically claimed by Guatemala as the "Departamento de Belice." Situated in Central America at the western end of the Caribbean Sea, it covers 22,965 km^2, including 370 km^2 of cays and islands.

Hardwood timbers, agriculture, and fishing are the major exploited resources; minerals are few and have not been developed.

Structurally, its northern portion forms part of the Yucatan Peninsula (see *Mexico*) and the southern part, which includes the Cockscomb or Maya Mountains, is an extension of the NE trends of northern Guatemala (see *Central America; Guatemala*).

The oldest rocks, which crop out in the Maya (or Cockscomb) Mountains, consist of a Pennsylvanian-Permian metasedimentary and sedimentary sequence, correlative with the Santa Rosa Group of Guatemala. This sequence is intruded by granitic bodies and, in its middle part, has a volcanic unit of rhyolitic composition.

This belt is followed to the S by a strongly folded series of Cretaceous dolomites and limestones which exceed 1500 m in thickness. Bituminous manifestations as well as anhydrite intercalations found at the subsurface are characteristic of the Cretaceous sequence. Paleocene-Eocene calcarenites, sandstones, and shales overlie the Cretaceous carbonates and occur usually along synclinal throughs. North of the Maya Mountains also occur Cretaceous carbonate rocks, extensively covered by nearly flat-lying Paleocene-Eocene limestones and Pliocene white marls and clays.

There is block-faulting offshore toward the Gulf of Honduras and the margin of the continental shelf is marked by a barrier reef with many intermediate patch reefs and shelf atolls. The "blue holes," sink-holes with drowned stalactites, explored by Cousteau on one of the *Calypso* cruises, provide evidence of Pleistocene sea levels down to –120 m.

GABRIEL DENGO

References

Bateson, J. H., 1972. "New interpretation of geology of Maya Mountains, British Honduras," *Bull. Am. Assoc. Petrol. Geologists,* **56**(5), 956–963.
____, and Hall, I. H. S., 1971. "Revised geologic nomenclature of pre-Cretaceous rocks of British Honduras," *Bull. Am. Assoc. Petrol. Geologists,* **55**(3), 529–530.
Dickerson, R. E., and Weisbord, N. E., 1931. "Cretaceous limestone in British Honduras," *J. Geol.,* **39,** 483–486.
Dill, R. F., 1971. "The blue hole–a structurally significant sink hole in the atoll off British Honduras," *Geol. Soc. Am. Abstr.,* **3**(7), 544.
Dillon, W. P., and Vedder, J. G., 1973. "Structure and development of the continental margin of British Honduras," *Bull. Geol. Soc. Am.,* **84,** 2713–2732.
Dixon, C. G., 1956. *Geology of Southern British Honduras, with Notes on Adjacent Areas.* Belize: Govt. Printer, 85p.
Flores, G., 1952. "Geology of northern British Honduras," *Bull. Am. Assoc. Petrol. Geologists,* **36**(2), 404–413.
Hall, I. H. S., and Bateson, J. H., 1972. "Late Paleozoic lavas in Maya Mountains, British Honduras, and their possible regional significance," *Bull. Am. Assoc. Petrol. Geologists,* **56**(5), 950–956.

Kesler, S. E., Bateson, J. H., Josey, W. L., Cramer, G. H., and Simmons, W. A., 1971. "Mesoscopic structural homogenity of Maya series and Macal series, Mountain Pine Ridge, British Honduras," *Bull. Am. Assoc. Petrol. Geologists,* **55**(1), 97–123.

___, Kienle, C. F., and Bateson, J. H., 1974. "Tectonic significance of intrusive rocks in the Maya Mountains, British Honduras," *Bull. Geol. Soc. Am.,* **85**, 549–552.

Ower, L. H., 1928. "Geology of British Honduras," *J. Geol.,* **36**, 494–509.

Ross, C. A., 1962. "Permian Foraminifera from British Honduras," *Paleontology,* **5**, 297–306.

Scholle, P. A., and Kling, S. A., 1972. "Southern British Honduras: lagoonal coccolith ooze," *J. Sed. Petrology,* **42**(1), 195–204.

Stoddart, D. R., 1962. "Three Caribbean atolls: Turneffe Islands, Lighthouse Reef, and Glover's Reef, British Honduras," *Atoll Res. Bull.,* **87**, 151p.

___, 1963. "Physiographic studies on the British Honduras reefs and cays," *Geogr. J.,* **128**(2), 161–173.

Cross-references *Central America–Regional Review; Guatemala; Jamaica; Mexico.*

BERMUDA

Bermuda lies 1250 km (675 nautical miles) southeast of New York City, near latitude 32°N and longitude 65°W, and consists of over 150 islands and islets covering an area of approximately 50 km^2. The topography is rolling hills up to 50 m in elevation. Most of the land is less than 30 m above sea level. The Bermuda islands are situated on the southeastern edge of a 650 km^2 platform, the top of the *"Bermuda Pedestal,"* which is the northernmost and largest of three volcanic seamounts on the *Bermuda Rise* (Fig. 1).

The climate is semitropical, the average monthly temperatures ranging from 16.5°C in February to 26.7°C in August. Surface sea temperatures range from 15.6°C in winter to 29.4°C in summer. Average annual rainfall is 1470 mm. Bermuda is the northernmost limit of reef-building corals in the Atlantic.

A site of limestone deposition since the Tertiary, Bermuda marks an isolated carbonate province within the North Atlantic basin. Around the Bermuda Rise, pedestal-derived carbonate sediments at the base of the seamount pass into clays. The clays, in turn, are surrounded by continent-derived sediments covering the Hatteras, Sohm, and Nares abyssal plains.

The physiography of the Bermuda platform is suggestive of an atoll (see Fig. 2) in that carbonate islands, shoals, and reefs surround a central, shallow-water lagoon. However, many of the Bermudian shoals and reefs are not constructional but consist of eroded and submerged eolian limestones thinly veneered with encrusting organisms (Stanley and Swift, 1967). Bermuda, therefore, is not a true atoll.

The Quaternary limestones cap the volcanic pile that constitutes the bulk of the Bermuda Pedestal. Little is known of the volcanics or of the pre-Pleistocene history of Bermuda. Sub-

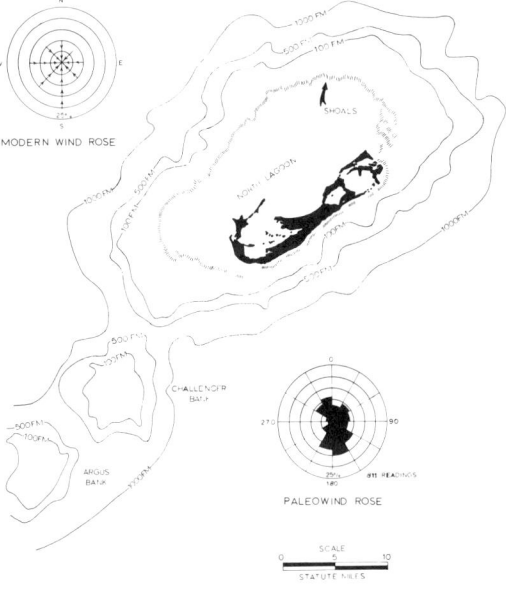

FIGURE 1. The Bermuda islands and adjacent banks. The modern wind rose illustrates the percentage frequency distribution of hourly wind direction for the eight prime points of the compass recorded at St. George's, Bermuda, during the period August 1932, to April 1954. The paleowind rose pictures the distribution of cross-bedding dip azimuths in the Bermuda eolianites. This distribution is an approximation of the direction of Pleistocene paleowinds.

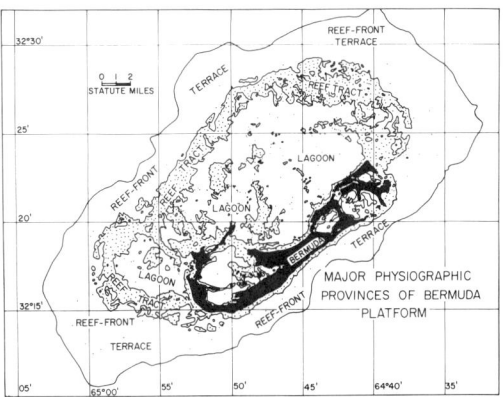

FIGURE 2. Major physiographic provinces of Bermuda platform.

TABLE 1. Geological Features of the Bermuda Pedestal

Top of volcanics	Subhorizontal, erosional (?) surface with average depth –76 m.	Officer *et al.*, 1952
	Probable buried topographic highs beneath Great Sound and Castle Harbour (–36 m).	Vacher, 1971
Pre-Pleistocene sediments	Data from bore hole in Southampton Parish on flank of volcanic pile. Calcarenites to –75 m, underlain successively by 60 m of weathered volcanic material containing Miocene foraminifera; 30 m of rounded, water-worn volcanic sediments; and then basalts.	Pirsson, 1914
Volcanic pile		
Lithology	Amygdaloidal, microcrystalline alkali basalts	Pirsson, 1914
Structure	Interpretation of two circular Bouguer anomalies, one centered over Great Sound (amplitude 55 mgal, and other over lagoon N of Castle Harbour (amplitude 25 mgal.	Stuart, 1970
Age	Uppermost volcanics dated (K-Ar technique) at 30–55 m yr.	Gees, 1969
Crustal features	Undulating basement decreases in amplitude out from the base of the Pedestal. Possibly due to loading of the basement by the Pedestal.	Officer *et al.*, 1952

surface relationships are summarized in Table 1 and shown in Fig. 3. Sea-floor spreading history suggests a Lower Cretaceous age for the underlying crust, and bores to the top of the volcanics have furnished samples dated 30–55 m yr. The limestone-basalt contact, according to seismic-refraction studies (Officer *et al.*, 1952), is subhorizontal and at an average depth of about 75 m below sea level, but borings show it as shallow as –37 m (Newman, 1959).

Bermudian geology is important because (1) it represents a case study of limestone sedimentation and diagenesis (see Table 2), and (2) Bermuda may be tectonically rather stable, and thus serve as a possible "standard" for Pleistocene and Holocene sea-level history (Vacher, 1971). Oceanic islands like Bermuda are largely free of the isostatic effects of water loading. Hence, former positions of "relative sea level" in Bermuda may approximate "absolute sea level."

Pleistocene Rocks

The Pleistocene section of Bermuda is a cyclic alternation of limestones and paleosols. Eolianites constitute over 90% of the limestone volume and represent lithified transverse dune ridges. They consist principally of large-amplitude (>10 m), high-angle (>35°), convex-upward, leeward-dipping foresets. Foresets are overlain on the windward flank of the lithified ridge by complex trough cross-bedding formed by eolian scour-and-fill processes (Mackenzie,

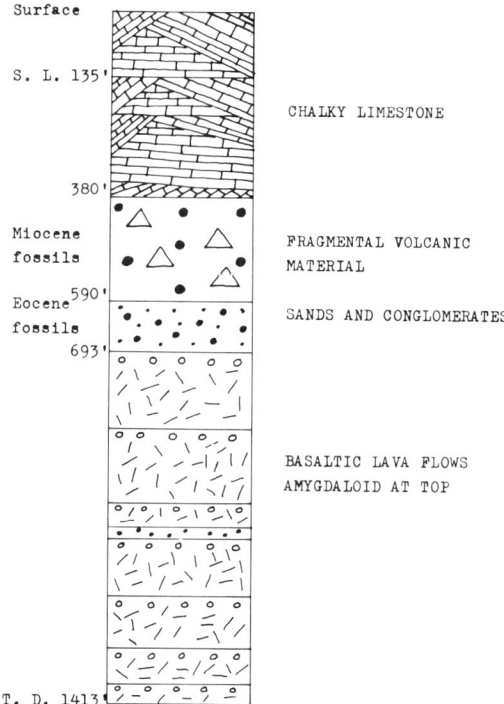

FIGURE 3. Diagrammatic columnar section of a water-well boring drilled in 1912 in Southampton Parish, Bermuda.

TABLE 2. Recent Sediments of Bermuda

Environment	Physiography	Deposits	Reference
Reef-front terrace	Irregular surface, ranging from 15–30 m depth; 5–8 m depth at platform edge.	*Calcarenite* Abundant coral heads; calcarenite consisting of reef suite.	Upchurch (1971)
Reef tract	On outer margin, coalesced reefs in anastamosing ridges separated by channels up to 20 m deep. Lagoonward, reefs are more isolated and form pinnacles.	*Reefs* (1) Built from a coral framework and coralline algae cement, and (2) built of an intergrowth of encrusting coralline algae and vermetid gastropods. Internal sediments partially filling the cavities and cement are important elements of the fabric of both reef types.	Ginsburg and Schroeder (1969a)
		Synsedimentary cement of internal sediments is 15 mole% Mg-calcite. Cementation most extensive on periphery of reef tract.	Ginsburg and Schroeder (1969b)
		Calcarenite Mixture of reef (coral, foraminifera, eolianite) and lagoon suites (*Halimeda* and bivalves).	Upchurch (1971)
Lagoon	Comprised of several topographic basins with scattered pinnacle reefs and shallow flats. Depth of lagoon floor about 15 m, maximum 25 m.	*Calcarenite* Composed of lagoon suite.	Upchurch (1971)
Shoreline		*Calcarenite*	
	North shore Erosional coastline, pocket embayments and beaches alternating with cliffs.	Sediment of beaches derived from contributors in bays and from headland erosion, bay suite identical to lagoon suite.	
	South shore Depositional coastline, extensive beaches, shoreline dunes up to 10 m high (inactive).	Sediment of beaches derived principally from algal-vermetid reefs offshore.	
Harrington Sound	3-km-diameter circular basin with average depth about 20 m, closed except for narrow inlet.	*Calcilutite* Coral (*Oculina*) and pelecypod fragments in fine-grained matrix to depth of about 15 m. Fine-grained fraction Silt-sized aragonite derived from algae (e.g., *Penicillus*); fine silt- and clay-sized low-Mg calcite derived from bioerosion (e.g., by the sponge *Cliona lampa*).	Neumann (1965)

1964b). Interfingering of eolianites and beach deposits and superposition of cross-bed units of separate eolianite ridges show that (1) each ridge represents a distinct sedimentation episode (the parallel ridges are not waves of migratory dunes); (2) the ridges are formed parallel to the strandline; and (3) the sand was supplied from beaches by onshore winds (Bretz, 1960; Mackenzie, 1964a; Land et al., 1967). Beach-dune transitions include transgressive, stillstand, and regressive relationships, indicating a source of sand regardless of the movement of sea level. The present-day wind pattern (Fig. 1) produces onshore winds capable of forming dunes around the perimeter of the Bermuda islands. Cross-bedding in the eolianites suggests patterns not very different at times during the Pleistocene.

Superposed calcarenite bodies occur in a consistent pattern of seaward accretion; oldest units lie farthest inland, and younger units lie progressively closer to the present shoreline. Superposed calcarenite units show a systematic change from an unconsolidated sediment of metastable magnesian calcite and aragonite to a tightly cemented rock composed solely of calcite. Stable-isotope ratios (Gross, 1964) as well as laboratory simulation show that the diagenesis occurred in the subaerial environment by freshwater percolation.

Terra-rossa paleosols of Bermuda occur as island-wide layers dividing the limestone cap into a succession of limestone formations. The soils developed on the limestones during freshwater diagenesis (Land *et al.*, 1967); however, they are not solely residuum of weathered limestone, because (1) there is insufficient reduction of the carbonate dunes to account for the thickness of overlying soils; and (2) the soils contain abundant quartz that could not be derived from either the limestones or the buried volcanics. Clearly the soils formed principally from accumulation of atmospheric dust (Bricker and Mackenzie, 1970), and the mineralogy of dust collected today in Bermuda is identical to that of the silt- and clay-sized fraction of the soils; furthermore, the present influx of atmospheric dust to Bermuda is sufficient to account for the soil thickness, assuming comparable dust concentrations of the Pleistocene and present-day atmospheres. Bermudian terra rossa, therefore, is analogous to the "red-clay facies" formed in areas of extremely slow sedimentation on the deep-sea floor.

Sea-Level History

The Bermudian sea level versus time curve is shown in Fig. 4. The Holocene curve (Neu-

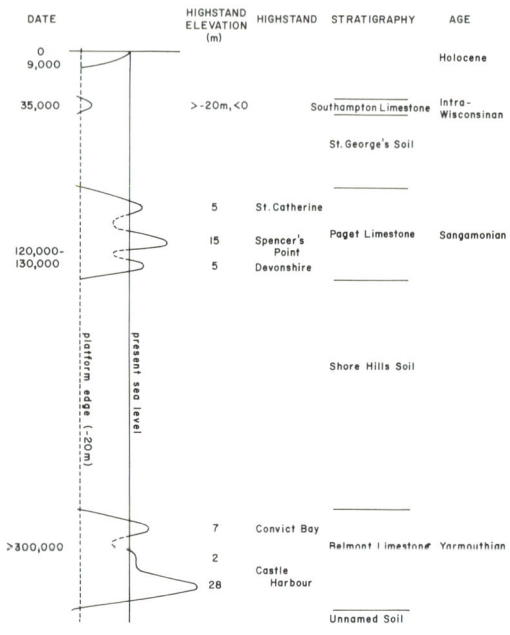

FIGURE 4. Sea level versus time curve.

mann, 1971) is based on radiocarbon dating of peats resting on bedrock in Harrington Sound and in various coastal ponds. The late Pleistocene curve (Vacher, 1971) is derived from the stratigraphy of eolian and marine limestones and terra-rossa paleosols. The 35,000-year date is based on radiocarbon ages of skeletal limestone (Neumann, 1969; Land *et al.*, 1967) and may be spurious, owing to contamination. The older dates are uranium-series dates on aragonitic corals (Land *et al.*, 1967; Richards *et al.*, 1969). Elevation of Pleistocene highstands is taken from the maximum elevation of marine deposits.

The Holocene curve suggests that sea level rose rapidly (3.4 m/1000 yr) from 9200 to 4000 years B.P., and then more gradually (1 m/1000 yr) from −4 m to present position during the past 4000 years. Sea level was below its present position throughout the rise. There is evidence of a reversal between 9000 and 8500 years B.P.

A sea-level rise may have peaked below the present position of sea level 35,000 years ago, as suggested by a *Chione* layer at 25 m depth in Harrington Sound (Neumann, 1969) and by the youngest eolianite in the Bermudian stratigraphic column. Although a mid-Wisconsin age for the Southampton formation is consistent with the available data, other ages for this unit cannot be discounted.

FRED T. MACKENZIE
H. LEN VACHER

[Contribution 600, Bermuda Biological Station]

References

Bretz, J. H., 1960. "Bermuda: a partially drowned, late mature Pleistocene karst," *Bull. Geol. Soc. Am.,* **71,** 1729–1754.

Bricker, O. P., and Mackenzie, F. T. 1970. "Limestones and red soils of Bermuda, discussion," *Bull. Geol. Soc. Am.,* **81,** 2523–2524.

Gebelein, C. D., 1969. "Distribution, morphology, and accretion rate of Recent subtidal algal stromatolites, Bermuda," *J. Sed. Petrology,* **39,** 49–69.

Gees, R. A., 1969. "The age of the Bermuda sea mount," *Maritime Sediments,* **5,** 2–4.

Ginsburg, R. N., and Schroeder, J. H., 1969a. "Introduction to the growth and diagenesis of Bermuda reefs," manuscript prep. for the Bermuda Conf. on Carbonate Cements, Bermuda Biol. Station, 13p.

____, and Schroeder, J. H., 1969b. "Recent synsedimentary cementation in subtidal Bermuda reefs," in O. P. Bricker, R. N. Ginsburg, L. S. Land, and F. T. Mackenzie, eds., *Carbonate Cements* (Spec. Publ. 3). Bermuda: Biological Station for Research, 31–34.

Gross, M. G., Jr., 1964. "Variations in the O^{18}/O^{16} and C^{13}/C^{12} ratios of diagenetically altered limestones in the Bermuda Islands," *J. Geol.,* **72,** 170–194.

Land, L. S., Mackenzie, F. T., and Gould, S. J., 1967. "The Pleistocene history of Bermuda," *Bull. Geol. Soc. Am.,* **78,** 993–1006.

Mackenzie, F. T., 1964a. "Bermuda Pleistocene eolianites and paleowinds," *Sedimentology,* **3,** 52–64.

____, 1964b. "Geometry of Bermuda calcareous dune cross-bedding," *Science,* **144,** 1449–1450.

Neumann, A. C., 1965. "Processes of Recent carbonate sedimentation in Harrington Sound, Bermuda," *Bull. Marine Sci.,* **15,** 987–1035.

____, 1966. "Observations on coastal erosion in Bermuda and measurements of the boring rate of the sponge, *Cliona lampa*," *Limnol. Oceanog.,* **11,** 92–108.

____, 1971. "Quaternary sea-level data from Bermuda," *Quaternaria,* **14,** 41–43.

Newman, W. S., 1959. "Geological significance of recent borings in the vicinity of Castle Harbour, Bermuda," *Am. Assoc. Adv. Sci. Preprints Intern. Oceanog. Cong.,* 46–47.

Officer, C. B., Ewing, M., and Wuenschel, P. C., 1952. "Seismic refraction measurements in the Atlantic Ocean. 4. Bermuda, Bermuda Rise, and Nares Basin," *Bull. Geol. Soc. Am.,* **63,** 777–808.

Pirsson, L. V., 1914. "Geology of Bermuda Island: the igneous platform," *Am. J. Sci.,* **38,** 189–206, 331–334.

____, and Vaughan, T. V., 1913. "A deep boring in Bermuda Island," *Am. J. Sci.,* **36,** 70–71.

Richards, H. G., 1971. "Sur le Pléstocène des Bermudes," *Quaternaria,* **15,** 67–69.

____, Abbott, R. T., and Skymer, T., 1969. "The Marine Pleistocene mollusks of Bermuda," *Acad. Nat. Sci. Philadelphia. Notulae Nuturae,* **425,** 1–10.

Sayles, R. W., 1931. "Bermuda during the Ice Age," *Proc. Am. Acad. Arts Sci.,* **66,** 381–467.

Stanley, D. J., and Swift, D. J. P., 1967. "Bermuda's southern aeolianite reef tract," *Science,* **157,** 677–681.

____, and Swift, D. J. P., 1968. "Bermuda's reef-front platform: bathymetry and significance," *Marine Geology,* **6,** 479–500.

Stuart, W. D., 1970. "Bermuda gravity anomalies and pedestal evolution," *Trans. Am. Geophys. Union,* **51,** 316.

Upchurch, S. B., 1971. "Sedimentation on the Bermuda Platform," *Diss. Abstr. Intern.,* **31**(7), 4140B.

Vacher, H. L., 1971. "Late Pleistocene sea-level history: Bermuda evidence," Ph.D. thesis, Northwestern Univ.

____, 1973. "Coastal dunes of younger Bermuda," in Coates, D. R., ed., *Coastal Geomorphology.* Binghamton, N.Y.: SUNY, Publ. in Geomorph., 355–391.

Verrill, A. E., 1907. "The Bermuda Islands: 4. Geology and paleontology, and 5. An account of the coral reefs," *Trans. Conn. Acad.,* **12,** 45–348.

Cross-references: *Bahamas; Barbados; United States–Hawaii.*

BISMARCK ARCHIPELAGO

A collective term for all the islands lying north of the island of *New Guinea* (q.v.), the Bismarck Archipelago extends from 1–7°S, and 146–153°E. Together with northeastern New Guinea (then Kaiser-Wilhelms-Land), they came under the German New Guinea protectorate from 1884 to 1914, and after World War I were placed under Australian mandate and administered along with the Territory of Papua and New Guinea.

The principal islands include *New Britain,* with Rabaul its chief town; *New Ireland, New Hanover* (also called Lavongai), *St. Matthias Group* (notably Musau), the *Admiralty Islands* (notably Manus), and others in the far west (Hermit Island, the Ninigo Group, and Vuvulu). There is also a minor chain of active volcanic islands close to the north coast of northeastern New Guinea (going from E to W: Umboi, Long, Karkar, Bam, Manam, Blupblup, and several others). Long Island has an interesting crater lake, which boils occasionally.

Structurally these islands form an oval ring about the Bismarck Sea (see *Encyclopedia of Oceanography*). They are all on volcanic foundations and represent two major structural units, the southern Bismarck island arc, which is part of the northern New Guinea arc (Thompson and Fisher, 1965), and the northern Bismarck arc or New Ireland arc, which is the western extension of the Solomon chain. The southern Bismarck island arc includes New Britain, except for the Gazelle peninsula, which is structurally part of the New Ireland arc, and the chain of volcanic islands off the north coast of New Guinea. The New Britain arc has the distinctive features of an island arc: it is arcuate in shape; it is fronted at its convex side by an oceanic trench, the New Britain Trench, which descends to 7880 m; it has a well-defined seismic "Benioff zone" dipping steeply northward to a depth of over 500 km (Denham, 1969); and it has a belt of active volcanoes mainly along the north coast. The New Ireland island arc is not as well developed. There is no associated trench and no active volcanism; however, there is a chain of Quaternary volcanics running parallel to the north coast and also a zone of seismic epicenters dipping to the north.

Analysis of earthquake slip vectors discloses the existence of three secondary plates in this region, between the Pacific and Australian plates (Johnson and Molnar, 1972). These are referred to as the North Bismarck Plate, South Bismarck Plate, and Solomon Sea Plate (see Fig. 1).

New Britain, formerly known as *Neu-Pommern,* or *Biarara* in the native language, is an elongate crescent-shaped island, nearly 500 km in length and 45–80 km across. The climate is wet tropical with maximum rainfall between December and March along the north coast when the area is under the regime of the perturbation belt of the I.T.C. and between May and August along the south coast when the NE trades dominate the wind regime.

New Britain consists of a core of Eocene volcanics, mainly basic to intermediate lavas and pyroclastics, and associated intrusives, overlain by younger volcanics and limestone (Miocene). These rocks form the central and southern parts of the island. The present topography is a typical "selva" or tropical ridge and ravine landscape (see *Encyclopedia of Geomorphology*) with sharp crested ridges separated by V-shaped valleys and is the result of very young block faulting and uplift accompanied by rapid erosion. In areas of limestone, spectacular cone karst land forms have developed. Uplift was greatest along the southern coast, where series of raised coal reefs are present in many places. Coral reefs are now forming along most of the coastline.

Quaternary and recent volcanism is restricted to a relatively narrow belt along the central northern coast and the eastern (Rabaul) and western (Cape Gloucester) ends of the island. The volcanic arc is, however, continued to the W by the chain of islands off the northern coast of New Guinea. The volcanoes are mostly central-type strato-volcanoes (Fisher, 1957) and are built up of lava flows and pyroclastics. Calderas are present on several volcanoes, and Vulcan, Tavurvur, Ulawun, Lalobau, Pago (all on New Britain) and Manam and Long Island volcanoes have all erupted this century. The eruption of the two post-caldera volcanoes of Tavuvur (The Mother) and Vulcan near Rabaul in 1937 caused temporary evacuation of the town and killed over 500 people. The curiously shaped N-S Talasea ridge, tipped by the Willaumez Peninsula, which contains a large crater lake (Lake Dakataua), is a line of extinct volcanoes and probably delineates a major fracture at right angles to the long axis of New Britain. About 50 km offshore from Talasea Ridge there is a cluster of coral islands probably on volcanic stumps, the *Vitu Islands,* which include one perfect atoll, *Unea.*

New Ireland, formerly known as *Neu-Mecklenburg* (*Tombara* in the native language), is situated N of New Britain and forms a long narrow NW-trending island with several narrow necks. Like New Britain the island is underlain by older volcanics (Oligocene), predominantly andesitic pyroclastics and related intrusives. The overlying rocks include limestone, which forms extensive areas of tropical cone karst in the western half of the island, and volcanogenic sediments. The southeastern part of the island is a mountainous ridge and ravine landscape, whereas most of the northwestern part is formed by tropical cone karst.

The present asymmetric form of the island, with the watershed and the area of maximum relief running close to the south coast, is due to NE tilting with the maximum rate of uplift along the south side. This is well expressed on the karst plateaus, which dip at 5–10°NE. There are extensive raised coral reef terraces along the northern coast, and most of the coastline is fringed by recent coral reefs.

New Hanover is also of volcanic origin, but severe erosion has destroyed most of the vol-

FIGURE 1. Inferred plate motions in the New Guinea Bismarck Archipelago area. Arrows indicate the direction of relative motion determined by earthquake slip vectors. Double arrows show strike-slip faulting, single arrows show underthrusting direction, and the dashed arrow shows extension (Johnson and Molnar, 1972).

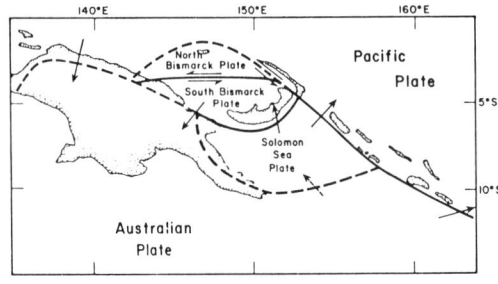

canic landforms. It is formed by low mountains and hills that rise steeply from the south but slope gently to the north, where large areas of dissected fans and swampy alluvial plains occur. Uplifted coral reefs veneer most of the coast and a barrier reef is also developed. Structurally it represents an extension of the New Ireland trend.

Running parallel to the New Ireland-New Hanover trend, 20–40 km to the NE, is a row of volcanic islands, including Feni, Tanga, Lihir, and Tagar islands and coral islets and reefs. Lihir and Ambitle (Feni Island) have active solfatara fields. To the SE this trend continues in the *Solomon Chain,* two islands of which, Buka and Bougainville, are embraced in Papua New Guinea. To the NW the same trend is represented by the *St. Matthias Group,* the principal island being Mussau, another mature volcanic complex.

The northwestern part of the Bismarck Archipelago is occupied by the *Admiralty Islands,* which comprise one large island, *Manus,* and some very much smaller islands grouped south of Manus in a rough semicircle. Manus Island has a moderate relief but is deeply dissected in its central areas. It has a base of Oligocene volcanics flanked by Miocene limestone and other sediments. Basaltic and dacitic flows and tuffs form the western third of the island. Raised coral terraces form the east and west coast and much of the island is surrounded by a complex barrier-reef system.

The islands grouped around Manus are mostly volcanic and the smaller ones are coral islets and sand cays. *Lou* and *Baluan islands* have well-preserved volcanic landforms and south of *Lou Island* there is an active submarine volcano, *Tuluman.*

Manus Island was visited by the *H.M.S. Challenger* in 1875 but its principal claim to fame comes from the anthropological studies by Margaret Meade.

Some 90 km WNW are the *Hermit Islands,* which rim islets of coral sands and limestones on an "almost atoll." Farther west again is the *Ninigo Group,* which includes seven atolls, ranging up to 30 km in diameter.

ERNST LÖFFLER
RHODES W. FAIRBRIDGE

References

Bureau of Mineral Resources (Australia), 1972. *Geology of Papua New Guinea.* Canberra: Bur. Mineral. Res. 1:1,000,000.

Denham, D., 1969. "Distribution of earthquakes in the New Guinea–Solomon Island region," *J. Geophys. Res.,* 74, 4290–4299.

Fisher, N. H., 1957. "Melanesia," *Catalogue of Active Volcanoes of the World including Solfataric Fields.* Naples: Intern. Volcanol. Assoc., pt. 5, 105p.

Johnson, T., and Molnar, P., 1972. "Focal mechanisms and plate tectonics of the Southwest Pacific," *J. Geophys. Res.,* 77(26), 5000–5032.

Ollier, C. D., and Bain, J. H. C., 1972. "Geology of Papua New Guinea," *Encyclopedia of Papua New Guinea.* Melbourne: Melbourne Univ. Press.

Thompson, J. E., and Fisher, N. H., 1965. "Mineral deposits of New Guinea and Papua and their tectonic setting," *Proc. 8th Commonwealth Mining Met. Congr. Austral. New Zealand,* 6, 115–148.

Cross-references: *Australasia–Regional Review; New Guinea; Solomon Islands.*

BOLIVIA

The Republic of Bolivia, which covers an area of 1 million km^2 (424,163 sq mi), occupies a landlocked position within the South American continent. Its wealth consists of many minerals and a growing output of hydrocarbons. The silver ore concentration of the Cerro Rico of Potosí is considered to have been one of the richest in the world. Its wealth may have led to a silver inflation in medieval times.

Bolivia's first geological explorer was Alcide d'Orbigny, who traveled the Andes in the years 1826–1832 and gave a good description of Ordovician, Devonian, and Permian strata and fossils. For example, *Cruziana* sp. was named by him in honor of the first president of Bolivia, José Antonio de Santa Cruz. Scientific mining geology started about 1920 and petroleum geology about 1930.

Geology is taught at the *Universidad Mayor de San Andres,* La Paz. Geological research is centered at the *Servicio Geológico Nacional* and with the state oil company, *Yacimientos Petrolíferos Fiscales Bolivianos,* La Paz.

Bolivia earns about 75% of its foreign exchange through mineral exports, especially from tin ore concentrates (30,000 tons in 1970). The importance of tin ore to the Bolivian economy is shown by the following figures for a recent year: value of tin ore exports, $93 million; all other minerals, $35 million (silver, lead, copper, gold, tungsten, zinc, sulfur). Production of oil in 1973 was 17 million barrels, and is still rising.

Geotectonic and Geomorphic Subdivisions

Bolivia occupies a central position within the South American Andes and is on the southwestern edge of the Brazilian Shield. A generalized E-W section consists of (1) the Brazilian Shield;

FIGURE 1. Tectonic map of Bolivia.

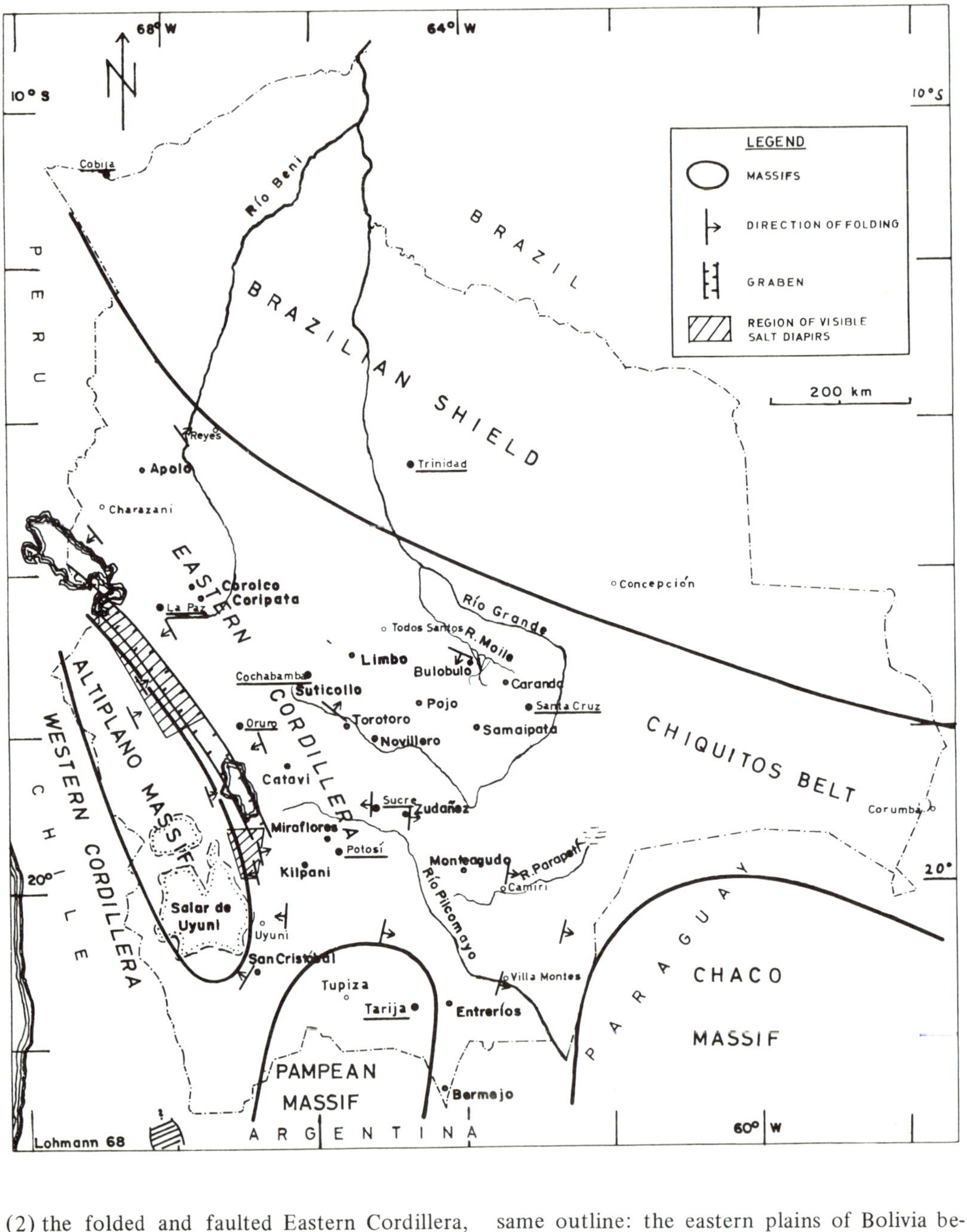

(2) the folded and faulted Eastern Cordillera, which is partly thrust over the southwestern edge of the Brazilian Shield; (3) the pre- or early-Andean granitic Altiplano Massif, which served as a buttress for Andean folding; and (4) the Western Cordillera (boundary line with Chile along its high peaks), mostly covered by young volcanics (Fig. 1).

The geomorphic subdivisions follow the same outline: the eastern plains of Bolivia belong to the Amazon and Paraná basins. The rivers are only now cutting through the Eastern Cordillera to the central high plains (3800 m) between the Eastern and Western Cordilleras. These central high plains (Spanish: Altiplano) still possess a local river system, with the Desaguadero River running from Lake Titicaca to the evaporation basin of Lake Poopó. The West-

FIGURE 2. Stratigraphic outline of Bolivian Andes (all thicknesses in meters).

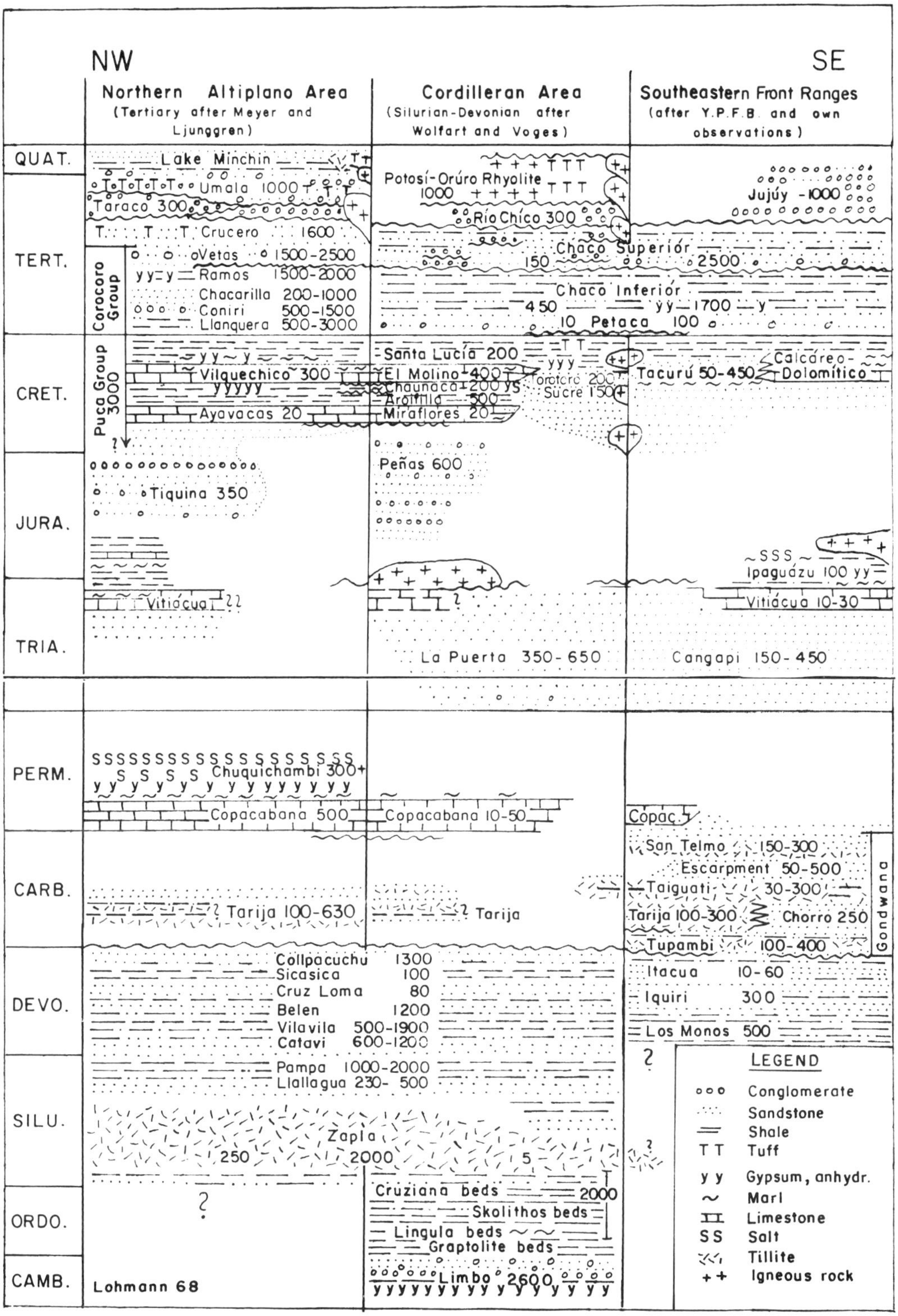

	NW Northern Altiplano Area (Tertiary after Meyer and Ljunggren)	Cordilleran Area (Silurian-Devonian after Wolfart and Voges)	SE Southeastern Front Ranges (after Y.P.F.B. and own observations)
QUAT.	Lake Minchin	Potosí-Orúro Rhyolite 1000	Jujúy 1000
TERT.	Umala 1000; Taraco 300; Crucero 1600; Coracoro Group: Vetas 1500-2500, Ramos 1500-2000, Chacarilla 200-1000, Coniri 500-1500, Llanquera 500-3000	Río Chico 300; 150; Chaco Superior; Chaco Inferior 450; Petaca 10-100	2500; yy 1700 y
CRET.	Puca Group 3000: Vilquechico 300, Ayavacas 20	Santa Lucía 200; El Molino 400; Chaunaca 200; Aroifilla 300; Miraflores 20; Ororo 200; Sucre 150	Tacurú 50-450; Calcáreo-Dolomítico
JURA.	Tiquina 350	Peñas 600	SSS Ipagúazu 100 yy
TRIA.	Vitiácua 22	La Puerta 350-650	Vitiácua 10-30; Cangapi 150-450
PERM.	SSSS Chuquichambi 300+; yyyyyy; Copacabana 500	Copacabana 10-50	Copac.
CARB.	? Tarija 100-630	? Tarija	San Telmo 150-300; Escarpment 50-500; Taiguati 30-300; Tarija 100-300; Chorro 250; Tupambi 100-400 (Gondwana)
DEVO.	Collpacuchu 1300; Sicasica 100; Cruz Loma 80; Belen 1200; Vilavila 500-1900; Catavi 600-1200		Itacua 10-60; Iquiri 300; Los Monos 500
SILU.	Pampa 1000-2000; Llallagua 230-500; Zapla 250	Zapla 2000; 5	?
ORDO.	?	Cruziana beds 2000; Skolithos beds; Lingula beds; Graptolite beds	
CAMB.	Lohmann 68	Limbo 2600	

LEGEND
- ooo Conglomerate
- .... Sandstone
- = Shale
- T T Tuff
- y y Gypsum, anhydr.
- ~ Marl
- ⊥⊥ Limestone
- S S Salt
- ∿ Tillite
- + + Igneous rock

ern Cordillera is more like the erosional edge of the Altiplano, with a number of very young volcanoes aligned along it.

Stratigraphic and Structural History

The geology of the Bolivian part of the Brazilian Shield is virtually unknown. The geology of the Chiquitos belt, which leads to the early Paleozoic iron ore deposit of Mutún near Corumbá, has been described by Chamot (1963).

As shown in Fig. 2, one can distinguish in the Bolivian Andes a thick early Paleozoic sequence, a late Paleozoic through Jurassic sequence, and a thick Cretaceous and Tertiary unit consisting mostly of red beds. Bolivia has evidence of two Paleozoic glaciations: (1) the Zapla tillite near the Ordovician-Silurian boundary, originating from the granitic Altiplano Massif; and (2) the Carboniferous Gondwana tillites, at least partially originating from the La Rioja region of northwestern Argentina. The Bolivian Gondwana glaciation came to an end before the transgression of the Wolfcampian Copacabana limestone.

The major orogenic activity in the Bolivian Andes took place during the late Devonian, late Triassic, and throughout the late Tertiary and Quaternary. The Coniri-Copacabana fault with about 10 km throw to the west separates the Eastern Cordillera from the Altiplano block. It seems to have been active throughout the Mesozoic and Tertiary. On the downthrown western side there are Neogene or Quaternary salt diapirs; the salt may be Permian. The main uplift of the Andes appears to be Pliocene and Pleistocene. Near La Paz there is evidence of a 200-m post-glacial (Holocene) differential uplift. The gas-producing structure of Bulobulo near Santa Cruz is undergoing uplift and is changing the course of the Ichilo River (Lohmann, 1970).

Important igneous and metallogenetic activity took place during the emplacement of (1) the pre-Silurian Altiplano granite; (2) granites and granodiorites of the Eastern Cordillera, especially the Cordillera Real east of La Paz, with (pneumatolytic) tin mineralization in the late Triassic; (3) Miocene to Pliocene volcanic activity. The overlapping of (3) over (2) has led to the hybrid character of the "tin-silver formation," which is therefore two-phased, Triassic and Tertiary, in its genesis.

Mineral Deposits

Tin, Silver, Antimony. Cassiterite mining is concentrated between Oruro and Potosí, where N-S-striking anticlines formed of early Paleozoic rocks are mineralized at the contact with Tertiary latites. A late Tertiary rhyolite cover has been slightly folded about the anticlinal axes, so future prospecting must therefore follow these axes. In Colavi-Canutillos near Potosí the mineralogically distinctive "wood tin" appears in layers within a Mesozoic red sandstone; this may indicate synsedimentary mineralization. Redeposited cassiterite ore in Quaternary placers is only of minor importance. Silver and antimony minerals are usually a by-product of tin mining. In the Cordillera Real region E of La Paz tin is usually associated with wolframite.

Lead, Zinc. The richest galena and sphalerite deposits are in the Matilda mine NE of Lake Titicaca. They seem associated with a more calcareous facies of early Paleozoic siltstones.

Copper. The well-known native copper, chalcocite, and malachite deposits of Corocoro and Chacarílla occurring in Tertiary red beds of the Altiplano may be products of ionic interaction between salt brines (see *salt diapirs* mentioned above) that carry copper ions and emanate hydrocarbon gas. Near Desaguadero on the south shore of Lake Titicaca, chalcocite and malachite are precipitated around Tertiary plant remains and asphaltite.

Gold. The main gold production derives from dredging of Pliocene and Pleistocene (Cangalli) river terraces on the east slope of the Andes, east of La Paz (Yani, Teoponte).

Iron. Near Mutún-Corumbá on the Bolivian-Brazilian border there is an unexploited major sedimentary iron ore complex, possibly of Cambrian age.

Fuel, Hydrocarbons

Bolivia has no coal deposits except for a few local coal seams in the Gondwana sequence near Lake Titicaca.

Boliva's crude oil production is becoming a major factor in the economy. The reserves are estimated at 3,600,000 tons of crude oil and 142,000,000 m^3 of gas. Presently only small quantities of minor oil and gas are exported to neighboring countries.

Bolivian oil and gas production comes from the Subandean eastern front range between Santa Cruz and Bermejo on the border with Argentina. The main source rocks are Lower Paleozoic shales and sometimes Mesozoic limestone beds. The reservoir rocks are Devonian sandstone, Carboniferous intraglacial sand bodies, and Upper Mesozoic sandstones and sandy limestones. Production from the well-known Camiri oil field is from a N-S anticline of Palezoic rocks occuring beneath an overthrust sequence of the same age. Production is derived only from the underthrust side.

Further prospective petroleum areas are the front ranges from Santa Cruz to the Peruvian border and parts of the Altiplano. On the Peruvian side of Lake Titicaca there was some oil production at Pirin, near Puno, but this is now exhausted. In the Bolivian Altiplano, Paleozoic, Jurassic, and Cretaceous units contain oil, and there is a wide range of Paleozoic to Tertiary reservoir rocks. Salt, presumably Permian, may serve as cap rock and oil trap.

HANS H. LOHMANN

References

Ahlfeld, F., 1967. "Metallogenetic epochs and provinces of Bolivia," *Mineral. Deposita* (Berlin), 2, 291–311.
___, 1970. "Zur Tektonik des andinen Bolvien," *Geol. Rundschau,* 59(3), 1124–1140.
___, and Branisa, L., 1960. *Geologia de Bolivia.* La Paz: Inst. Boliviano del Petroleo, 245p.
___, and Schneider-Scherbina, A., 1964. "Los yacimientos minerales y de hidrocarburos de Bolivia," *La Paz, Bol. Dep. Nacl. Geol.,* 5, 388p.
Barth, W., 1972. "Das Permokarbon bei Zudañez (Bolivien) und eine Übersicht des Jungpaläozoikums in zentralen Teil der Anden," *Geol. Rundschau,* 61, 249–270.
Chamot, G. A., 1963. "Esquisse géologique de la plateforme du bouclier Bresilien dans l'Oriente de Chiquitos, Bolivie," *Eclog. Geol. Helv.,* 56, 817–851.
Fricke, W., and Voges, A., 1968. "Beitrag zur Kenntnis der andinen Sedimentationsräume in Bolivien, Sud-Peru und Nord-Chile," *Geol. Jahrb.,* 85, 941–972.
Helwig, J., 1972. "Stratigraphy, sedimentation, paleogeography, and paleoclimates of Carboniferous ("Gondwana") and Permian of Bolivia," *Bull. Am. Assoc. Petrol. Geologists,* 56(6), 1008–1033.
Lohmann, H. H., 1970. "Outline of tectonic history of Bolivian Andes," *Bull. Am. Assoc. Petrol. Geologists,* 54, 735–757.
Radelli, L., 1966. "New data on tectonics of Bolivian Andes from a photograph by Gemini 5 and field knowledges," *Travaux Lab. Géol. Fac. Sci. Univ. Grenoble,* 42, 237–261.
Schlatter, L. E., and Nederlof, M. H., 1966. "Bosquejo de a geologia y paleogeografía de Bolivia," *La Paz, Bol. Serv. Geol. Bolivia,* 8, 49p.
Turneaure, F. S., 1960. "A comparative study of major ore deposits in Central Bolivia," *Econ. Geol.,* 55(2), 217–254; 55(3), 574–606.
Zeil, W., 1970. "Zur geologie der Anden," *Geol. Runschau,* 59(3), 827–834.

Cross-references: *Argentina; Brazil; Chile; Paraguay; Peru; South America.*

BRAZIL

Brazil, located in central and eastern South America, occupies nearly half the continent. The country has an area of 8,513,844 km^2 (3,286,488 sq mi), of which 36% is Precambrian shield rock, comprising crystalline plateaus with some mountain ranges. Sedimentary plateaus are located between the shield areas with 23% of the territory, and the lowlands and plains (elevations below 200 m) make up 41% of the country.

Owing to the vast territory and the small number of geologists, knowledge of the geology of Brazil is still in a stage of reconnaissance, with the exception of small areas that have been well studied for economic purposes or have been mapped in detail.

During the 18th century, the interior of Brazil was widely searched for gold and gems. At the end of that century, J. V. Couto published a paper on mining and related subjects. At the beginning of the 19th century, J. Bonifacio de Andrada e Silva and his brother, M. F. Ribeiro de Andrada, pioneered geological studies in Brazil.

The first stage in the geologic reconnaisance of the country ended in about 1870. At this time the land was visited by many scientists, mining engineers, naturalists, and geographers, including W. L. Eschwege, A. d'Orbigny, C. Darwin, P. W. Lund, F. Castelnau, J. L. R. Agassiz, C. F. Hartt, G. S. de Capanema, and J. M. da Silva Coutinho. Hartt summarized the knowledge of that period in the *Geology and Physical Geography of Brazil,* published in Boston in 1870.

In 1875 the *Comissão Geológica do Império* was established under the direction of C. F. Hartt. In this commission O. A. Derby and C. Branner worked intensively in surveying large areas. Unfortunately, the commission was dissolved in 1877, at which time Derby joined the *Museum Nacional* as head of the Department of Geology and Mineralogy.

In 1876 the well-known *Escola de Minas* was organized at Ouro Preto under the direction of H. Gorceix. Most of Brazil's famous geologists, including L. F. Gonzaga de Campos, F. de Paula Oliveira, E. P. de Oliveira, M. A. Lisboa, and J. C. da Costa Sena were graduated from this school.

In 1886 the *Comissão Geográfica e Geológica da Provincia de São Paulo* was created and was directed by Derby. Today it is the *Institute Geografico e Geologico de São Paulo.* In 1892 the *Comissão de Estudos das Minas de Carvão-de-Pedra do Brasil* was organized under the direction of I. C. White and the cooperation of F. de Paula Oliveira and C. de Campos. Its final report, 1906, offers fundamental knowledge of southern Brazilian geology.

In 1907 the *Servico Geológico e Mineralógico do Brasil* was established. O. Derby was the first director of this survey and the country's most productive geologist at the time. In 1933 this survey was included in the *Departamento Nacional da Producāo Mineral.*

From the beginning of this century to World War II, important geological works included those of L. F. Gonzaga de Campos, E. P. de Oliveira, A. B. Lamego, A. I. de Oliveira, C. Washburne, J. Pacheco, D. Guimaraes, L. Jaques de Moraes, L. F. de Moraes Rego, O. Barbosa, O. H. Leonardos, P. F. de Carvalho, V. Leinz, and R. Maack.

During World War II and after, geologic researches were carried out at several universities (São Paulo, Paraná, Rio Grande do Sul, Recife). Major surveys were done by *Conselho Nacional do Petróleo* and *Petrobrás,* and the U.S. Geological Survey collaborated in surveying several areas of economic value.

Most Brazilian universities have departments of geology (and also geography, where some geomorphology is done).

A marine geological and geophysical effort has developed in recent years. This marine research has been spearheaded by Petrobrás, the Departmento de Hidrografia e Navegacāo of the Navy, and by the universities, including the Instituto Oceanografico da Universidade de São Paulo, the geology departments of the Universidade Federal of Rio de Janeiro and of Rio Grande do Sul, and the Laboratario de Sciencias do Mar of the Universidade Federal de Pernambuco.

Mineral Resources and Mining Activities

For more than three centuries, Brazil was searched for gold and gems. The first mineralogic survey was started by the Andrada brothers. At this time several new minerals were described. Some of the minerals named after Brazilian localities are brazilianite, goyazite, and tripuhyite.

Toward the end of the 19th century, more technical ore prospecting began following the work of several geological commissions. By World War II intensive ore prospecting had started; for instance, iron (reserve 20 billion metric tons), manganese (reserve 17 million metric tons), coal, petroleum, atomic minerals. Nevertheless, a great part of certain mining activities is still done by gold washers (*garimpeiros*), who search for diamonds, gold, beryl, scheelite, rock crystal, tantalite, rutile, and gems. Besides these activities, modern mines produce coal, hematite, manganese ore, and others.

Several minerals (including andradite, arrojadite, ascoryalite, tavorite, and florencite) were named after mineralogists, some Brazilian, who studied these minerals. Although the mineral resources of Brazil are not well known, the country is among the richest in the world in high-grade ore reserves and is by now able to provide the world demand with 10% of manganese ore, 10% of wolframite ore, 70–80% of beryl, 70–90% of tantalum ore, 40–50% of monazite high in thorium, 99% of rock crystal, 40–50% of pure mica, and 100% of pure zircon.

The distribution of the mineral wealth in Brazil is not uniform. Some minerals are very abundant, such as those already mentioned as well as arsenopyrite ("mispickel"), cobaltian wad ("asbolan"), columbite, garnierite, spodumene, scheelite, magnesite, calcite (limestone), dolomite, and barite gypsum. Others are only sufficient, such as monazite, chromite, manganese ore, bauxite, rutile, and "amianthus" (chrysotile asbestos).

Pennsylvanian coal, mined in Santa Catarina, Rio Grande do Sul, and Paraná, is high in ash and sulfur content and not very good for making coke. However, coal production in Brazil is increasing. The reserve is estimated at 500 million metric tons. After 1954, petroleum production was greatly increased by the state-controlled *Petrobrás.* The most productive area of Brazil is located in the Reconcavo Basin in Bahia. Several offshore basins have also been discovered. Production in 1973 was 63 million barrels. The second largest oil shale reserves of the world (after the United States) are found in Brazil. The greatest oil reserve, however, is in the Irati oil shales of the southern Brazilian Permian sequence and there are further huge reserves in the Plio-Pleistocene Tremembé Graben between São Paulo and Rio de Janeiro.

Geomorphic Divisions

In spite of the large area, the general relief of Brazil is relatively low. About 41% of the country is below 200 m, and only 3% is above 900 m. Low erosional surfaces and pediplains between 200 and 300 m comprise 17% of the territory. The land above 600 m (15%) is located in two structural provinces: (1) shield areas uplifted by deep-seated warping; (2) intercratonic sedimentary basins and basaltic plateaus uplifted at the same time as the shield areas by post-Cretaceous epeirogenic movements.

Ab'Saber (1968) identifies six major geomorphic units in Brazil: (1) Guiana Plateau; (2) Brazilian Plateau; (3) Uruguay-South Rio Grande Palteau; (4) coastal plains and lowlands;

FIGURE 1. Geomorphic divisions of Brazil (based on Ab'Sáber, 1968).

(5) Amazonian plains and lowlands; and (6) Pantanal Plain (Fig. 1).

The large Brazilian Plateau can be divided into: (1) Central Plateau, comprising large areas of crystalline plateau and sedimentary plateau (*chapadas* and *chapadões*), such as Roncador and Parecis; (2) Southern Plateau with large areas of sedimentary and basaltic plateau; (3) Maranhão-Piauí Plateau corresponding to the uplifted sedimentary Parnaiba Basin; (4) Northeastern Plateau of crystalline shield rocks, and isolated sedimentary *chapadas*; (5) Oriental and South-Oriental Plateau, which is the more complex and mountainous part of the Brazilian Plateau and which comprises crystalline ranges as well as great escarpments and rift valleys among other features.

The highest elevations in Brazil are as follows:

Pico de Neblina:	3014 m
Pico da Bandeira:	2890 m, between Minas Gerais and Espírito Santo
Monte Roraima:	2875 m, at the border of Rio Branco territory with Venezuela and British Guiana
Pico dos Cruzeiros:	2861 m, between Minas Gerais and Espírito Santo
Pico do Cristal:	2798 m, in Minas Gerais
Agulhas Negras:	2787 m, between Minas Gerais and Rio de Janeiro

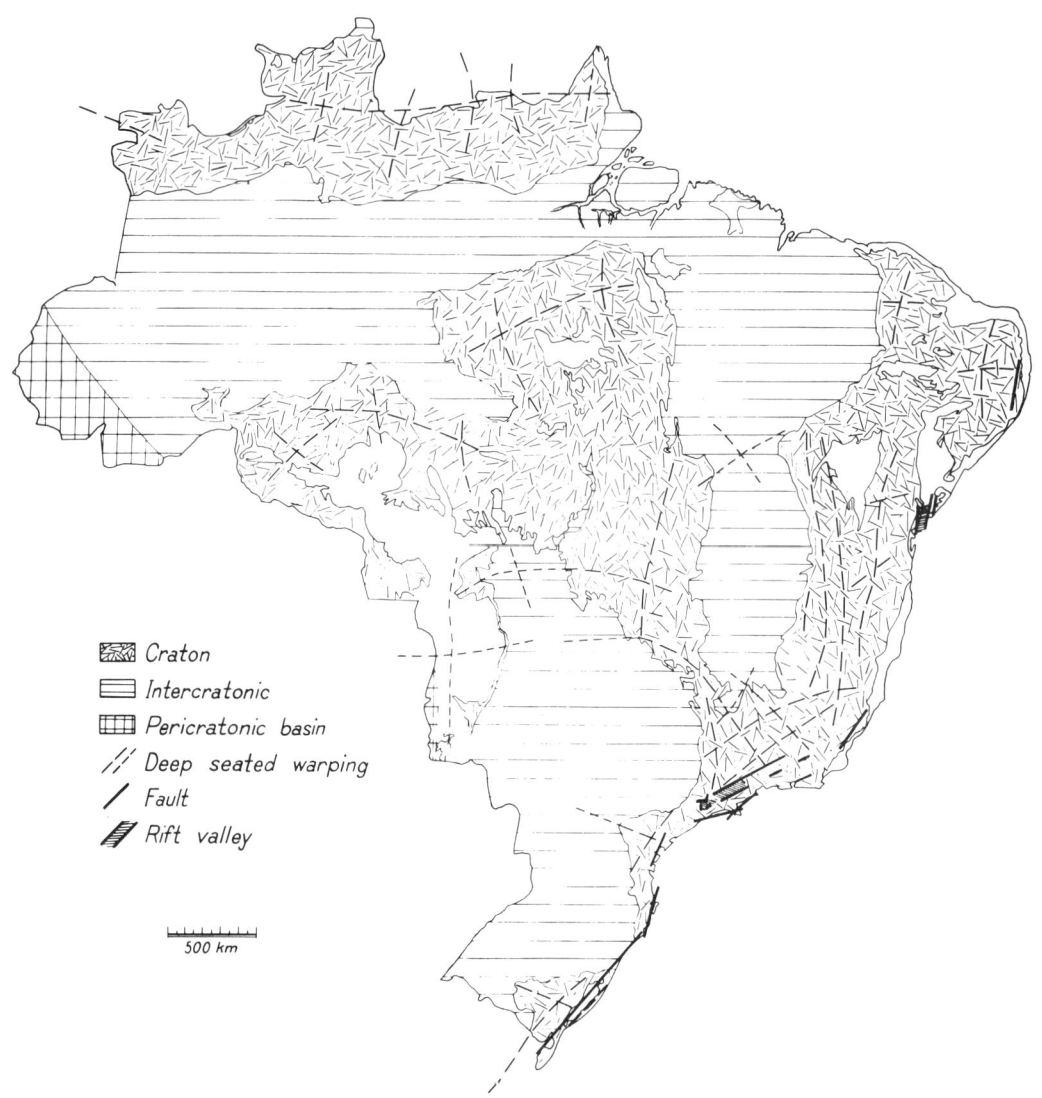

FIGURE 2. Paleotectonic map of Brazil (Bigarella and Ab'Sáber).

In Brazilian hydrography, the Amazon River basin and São Francisco River are the largest. The Amazon River basin drains 56% of the land, having an hydraulic power of about 4.4 million hp. At the town of Obidos, the discharge of the river is 0.1 million m^3/sec. In 1 m^3 there is 54 g of dissolved material and 135 g of sediment load. The famous Paulo Afonso Falls (with important hydroelectric development) are part of the São Francisco River. The Paraná River basin drains about one-tenth of the country and provides much of the hydraulic power of Brazil. The Guaira or Sete Quedas Falls and the 70-m-high Iguaçu Falls are the most important in this basin. The Brazilian hydraulic power potential is fourth largest in the world.

Structural Framework and Geotectonic Divisions

Since the early Cambrian, the Brazilian sedimentary and tectonic history has been controlled by the following major geotectonic units: craton, intercratonic basin, and pericratonic basin (Fig. 2).

The Guiana, Brazilian, and Uruguay shields are stable, positive, and little-deformed cratonic units. The Brazilian Shield is large and discontinuous, comprising several subunits with local names. The cratonic areas have not been disturbed since the Silurian. However, they show deep-seated warping (*plis de fond*) and faulting. Fig. 2 shows the main axes of the deep-seated warpings, the main faults and rift valleys, and the intercratonic basins. The Amazonas, Parnaíba, São Francisco, and Paraná basins are semistable, slightly negative, units between the cratonic areas. They have been superimposed on former geosynclines.

Only a small part of Brazil (Acre and Pantanal) is in the semimobile, slightly negative, and semiresistant South American pericraton unit. These pericratonic areas, located between the craton and the orogenic Andean belt, are characterized by a large fault system and by extensive Upper Cenozoic sedimentation. The Andean folding reaches Brazil only in the western part of Acre.

For the Precambrian, the main geotectonic feature was the Espinhaço geosyncline, which has been filled with several thousand meters of late Precambrian detrital and chemical sediments including some well-known iron deposits. Recent studies by Rosier and Ebert in the state of Rio de Janeiro seem to indicate that the "Serra do Mar" area is comparable to the central part of a large orthogeosyncline of Alpine type, the Paraíba geosyncline.

According to Ebert (1970) the lower part of the Middle Precambrian Rio das Velhas Series was affected by the important Ontarian orogeny. The orogeny was marked by intrusions of granite, pegmatite (1100 m yr), diorite, basic, and ultrabasic rocks and predated the deposition of the upper formation of the Rio das

FIGURE 3. Geologic map of Brazil (based on Lamego, 1960).

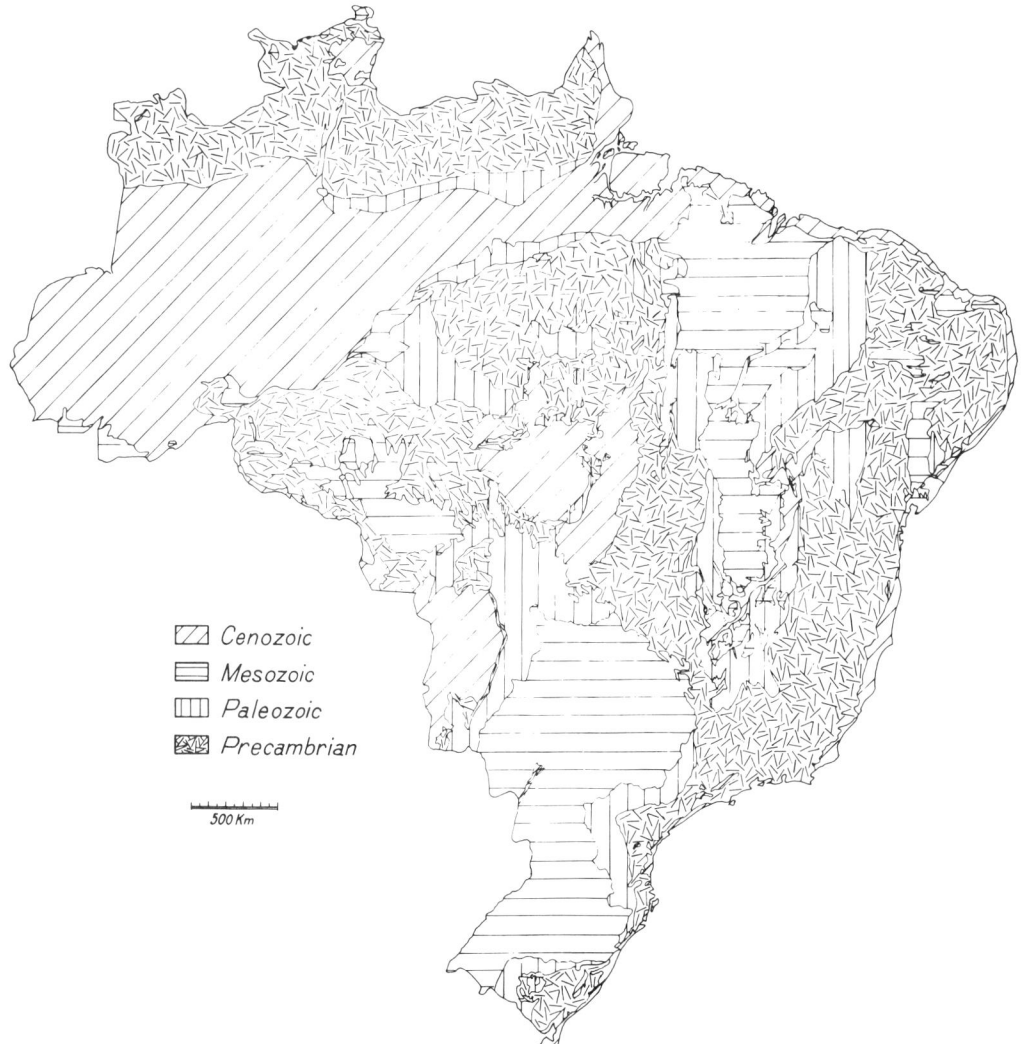

Velhas Series. The sediments of the Espinhaco geosyncline (late Precambrian) were affected by at least two orogenic phases. An intrusive cycle occurred 400–600 m yr ago. During the Precambrian there was a gradual decrease in orogenic intensity from the coastal region toward the interior. According to Ruellan (1953) the predominant structural directions in the Brazilian Precambrian areas are as follows:

1. Brazilian trend (NE-SW to NNE-SSW)
2. Caribbean trend (NW-SE)
3. San Franciscan trend (N-S)
4. Amazonian trend (E-W)

The deposition of the Lavras Series began after a diastrophic phase, which correlates with the Penokean revolution.

The beginning of the Paleozoic, or even the end of the Precambrian, was characterized by the extrusion of quartz porphyritic rhyolitic lavas in Amazonia and at the eastern border of the Paraná Basin. In Rio Grande do Sul there was an intense andesitic volcanism associated with the Camaquã Series (early Paleozoic). The Taconic disturbance has possibly affected several areas in Brazil. However, both the nonfossiliferous character of the sequences affected and the lack of age determinations make correlation very difficult and uncertain.

Important volcanic activity occurred in the Rhaetic, as in the Mesozoic. The basalt lavas extruded through numerous fissures (dikes) striking NW-SE. Local magmatic differentiation led to alkaline and ultrabasic volcanics (Poços de Caldas and Araxá in Minas Gerais, Lajes in Santa Catarina, etc.). Post-Cretaceous basaltic lava flows occur in several places in northeastern Brazil, as well as in the oceanic islands of Fernando de Noronha and Trindade.

Precambrian. Precambrian rocks make up the cratonic areas and were emergent during most of subsequent geologic time. However, in restricted areas they have undergone negative movements due to faulting at various times (Fig. 3).

The Precambrian is known in Brazil as "the crystalline complex," which is usually divided into Archean and Algonkian according to the degree of metamorphism. For a long time it was almost impossible to separate the several Brazilian Precambrian sequences. Criteria based on the degree of metamorphism are very imprecise. The main difficulty was the deep weathering, which developed a regolith that hid the uncomformities and contacts. The granitic gneiss rocks were often considered Archean, whereas the epimetamorphic rocks were placed in the Algonkian. However, in the 1950s this concept was completely abandoned and, after detailed studies in Minas Gerais and Rio de Janeiro, these complex sequences were divided into early, middle, and late Precambrian.

Despite their wide occurrence, the early Precambrian rocks have been only poorly studied in Brazil. Many of the areas formerly supposed to be early Precambrian are actually late Precambrian, in spite of the high degree of metamorphism. This is especially true of the eastern part of the Brazilian Shield. The early Precambrian gneisses are represented by the Mantiqueira Series.

In the middle Precambrian, the area now occupied by the Paraná, São Francisco, and Parnaíba basins was a NE-SW trending eugeosyncline. These sediments were intensely folded, metamorphosed, in part granitized, and intruded by granitic bodies. The middle Precambrian rocks are in the Rio das Velhas Series (Fig. 4).

In the late Precambrian this area again showed geosynclinal characteristics. These rocks, unconformably overlying the Rio das Velhas Series, are known as the "Minas Series" with its well-known iron formations. Also late Precambrian and separated from the Minas Series by an angular unconformity is the Itacolumi Series, which contains the famous flexible quartzite (itacolumite).

In the late Precambrian the Amazon Basin area seems to have been a miogeosyncline of E-W trend. The rocks were strongly folded and the degree of metamorphism is slight.

Paleozoic. The Parnaíba, São Francisco, Paraná and Amazon intracratonic basins subsided during the Lower Paleozoic in the area of the former Precambrian geosyncline. During the Paleozoic and Mesozoic, these basins were subjected to repeated periods of subsidence and sedimentation alternated with periods of emergence and erosion. The areas of these basins were about the same during both eras. The uplift was gentle without strong folding and only with slight epeirogenic warping. The slightly negative tendency of the basins has become reduced in more recent times. The basins were mostly marine in the Lower Paleozoic, becoming continental in the Upper Paleozoic.

The Uatumã Series in the Amazon Valley and the Lavras Series in Bahia and Minas Gerais are thought to be Cambrian. Some deposits from the Lower Paleozoic (Bebedouro formation) are thought to be of glacial origin, although this is still uncertain. The tillites in the Iapó formation are also problematical. The Ribeira Series in São Paulo and the Corumbá Series in Mato Grosso are also early Paleozoic. The Ordovician record is very uncertain. Perhaps the sediments of the Jacadigo Series in Mato Grosso are of this age.

In the Amazon Basin, the first well-

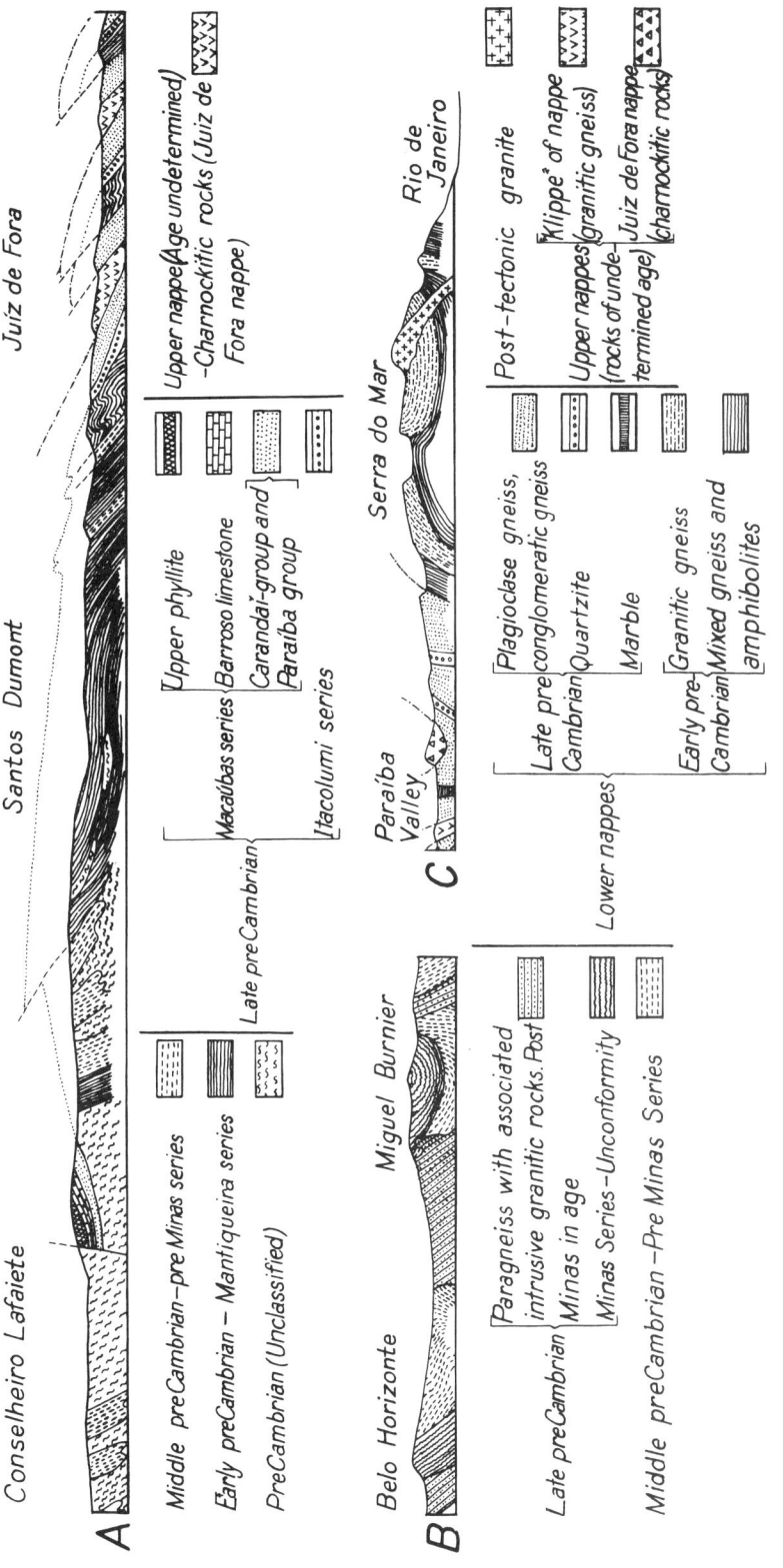

FIGURE 4. Geologic profiles of Precambrian areas, between Rio de Janeiro and Belo Horizonte, based on Ebert, Dorr II and Rosier (Oliveira, 1956).

established Paleozoic sequences are marine Silurian with fossil worm tubes, graptolites, brachiopods, pelecypods, cephalopods, ostracods, and porifera. The Bambui Series, cropping out mainly in Mina Gerais and Bahia, once considered to be Upper Silurian is now known to be Precambrian.

In southern Brazil, and in the states of Mato Grosso and Goiás the youngest Paleozoic rocks with fossils are Devonian. Before this period, in Rio Grande do Sul and Santa Catarina, there were restricted basins originated by faulting and filled with continental sediments containing volcanics (Camaquã and Itajaí series).

Most of the Precambrian areas at the beginning of the Devonian were a peneplain, probably formed under humid climatic conditions. The Devonian sea transgressed this erosion surface leaving thick deposits of cross-bedded sandstone and fossiliferous shale. In the Amazon Basin, *Protosalvinia brasiliensis* is a typical fossil plant. Among other fossils are brachiopods, pelecypods, gastropods, trilobites, etc. The Devonian facies in the Paraná Basin have been attributed by some writers to semiglacial or glacial conditions. This could not be proved, however, apart from the probable cold-water environment suggested by the fossil assemblage (Fig. 5).

The sea then retreated, and the land was eroded for a long period of time. This was the "Gondwana" cycle of erosion, widely recognized in the southern hemisphere. In the Amazon Basin the Itaituba Series is Carboniferous, *Productus* being one of the index fossils. In the Paraná Basin the Gondwana facies may be called the Santa Catharina "Supergroup." It started with the Tubarão Series, which is considered to be Upper Carboniferous. The lower part of this series is composed of glacial sediments (tillite, varvite, fluvioglacial sandstone, etc.). These are part of the famous glacial sequences that characterize the late Paleozoic in the Southern Hemisphere (Gondwanaland). The upper part of the Tubarão Series is generally not glacial, although some coal horizons may interfinger with tillites. The *Glossopteris* flora is well represented.

The Permian sequence in the Paraná Basin is represented by the Passa Dois Series, which includes the Iratí oil shales (the most important Brazilian fuel reserve), with *Mesosaurus brasiliensis* as the guide fossil. The environment is continental with some brackish-water sediments. Increasing aridity toward the top of the Permian is evident. Eurydesma fauna and *Glossopteris* flora are present. Some Permian formations, such as the Terezina and the Serrinha, contain a rather important pelecypod fauna composed of autochthonous genera and species.

Mesozoic. Most of Brazil was above sea level during the Mesozoic. Only in the Cretaceous did the sea invade several areas in the north and northeast.

The Triassic sediments are continental and were deposited in an arid environment. The sedimentation was very extensive in the Paraná and Parnaíba basins, and seas transgressed far over the shield areas. In the Paraná Basin, the Upper Triassic Santa Maria Formation has a reptilian fauna composed of *Rhynchocephalia, Dicynodontia, Cynodontia,* and *Pseudosuchia,* and it also contains a *Thinnfeldia-Dicroidium* flora. In spite of being geographically restricted (Santa Maria, RGS), this area is stratigraphically very important. It is overlain by the Botucatú Formation, formed under arid conditions and representing one of the largest fossil desert environments in the world. Much volcanic activity occurred in the Rhaetic and covered the Paraná Basin with several thick basaltic lava flows (1560 m in Pôrto Epitácio, SP). The total area of lava flows is about 1,200,000 km^2, the largest in the world. Volcanism occurred also in northern Brazil (Roraima) and in the Parnaiba Basin. During and after the lava flows, arid conditions prevailed in the Paraná Basin (Fig. 6).

Continental Cretaceous deposits were laid down under semiarid and humid climates and cover large areas of the country. Denudation in the source areas produced a well-developed peneplain, one of the oldest preserved erosion surfaces in Brazil (Summitar surface). (*Editorial note:* The question of terminology, peneplain or "pediplain," is resolved if peneplain is simply defined as an extensive or continent-wide erosion surface, regardless of genesis. R.W.F.) The Baurú Series in the Paraná Basin is Upper Cretaceous with a reptilian fauna containing *Chelonia, Arcosauria,* and *Titanosaurus,* remnants of fish, molluscs, and ostracods. In northwestern Mato Grosso (Parecis sandstone) silicified wood was found (*Araucarioxylon* and *Cupressionoxylon*).

The Albanian-Aptian marine transgression, covering probably the eastern half of Brazil, was only briefly important. Deposition was thin and mostly destroyed during post-Cretaceous epeirogeny and erosion, but enough widely scattered outliers remain to witness the extensive transgression. The thick sections of gypsum of the Chapada do Araripe in the western part of the State of Pernambuco date from this marine incursion.

During the Turonian and Maestrichtian, marine transgressions occurred over large areas of Ceará and Rio Grande do Norte. In Paraiba and Pernambuco, Senonian marine sediments are present.

FIGURE 5. Geologic profiles through sedimentary basins (Oliveira, 1956). A, Amazon Basin; B, Parnaíba Basin (Plummer, modified by Kegel); C, Section between Serra Grande-Terezina in Parnaíba Basin (Kegel); D, Section in Paraná Basin between Campo Grande and Serra do Jacadigo at Bolivian border (Almeida, 1963). E, Section in Paraná Basin between São Sebastiao island and Botucatu (Almeida, 1963); F, Section between Paraná and Paraguay rivers (Almeida, 1963). 1, Basement complex; 2, Cambrian; 3, Ordovician; 4, Silurian; 5, Devonian; 6, Carboniferous; 7, Permian; 8, Mesozoic; 9, Cenozoic.

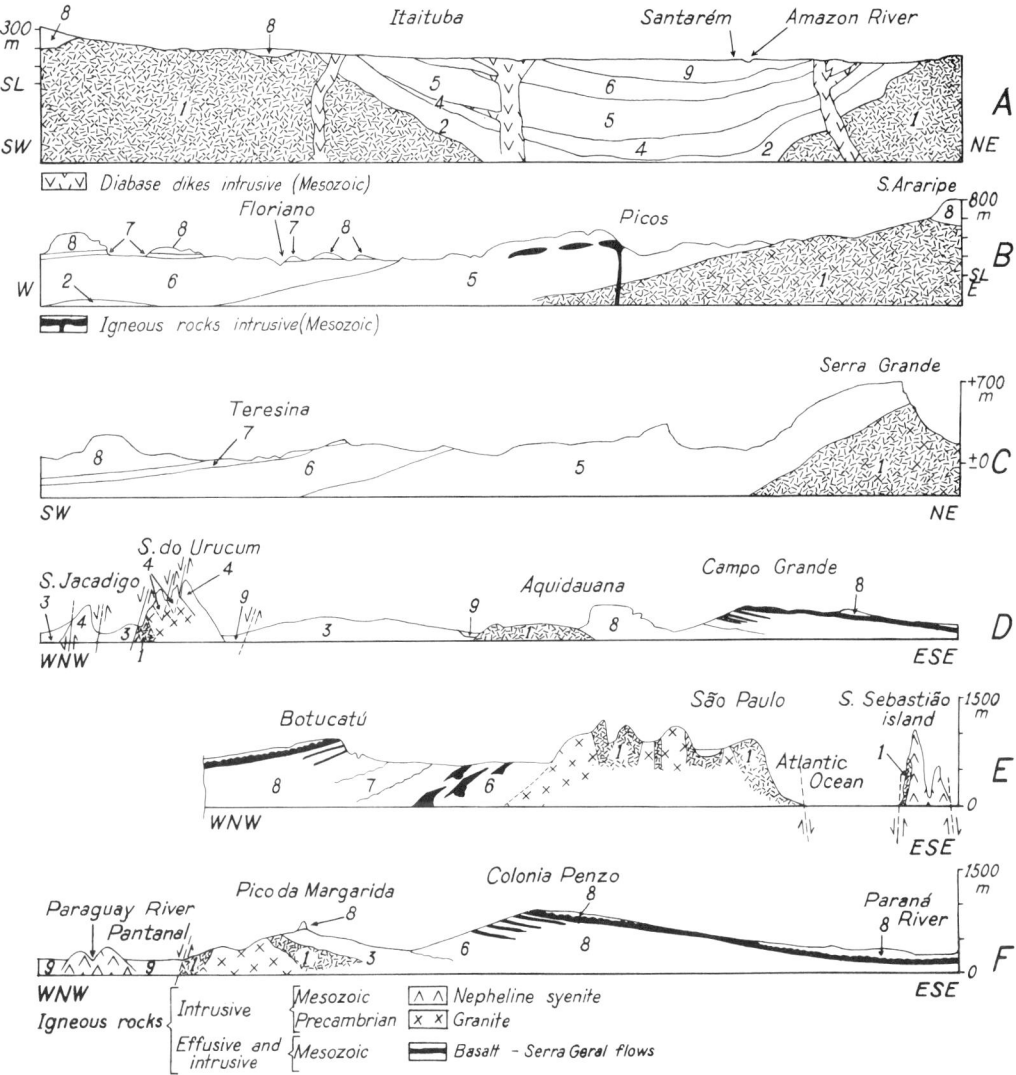

From an economic and petroleum exploration point of view, the Cretaceous coastal, marginal, taphrogeosynclinal basins are very important. They date from the separation of the South American continent, which began in the Lower Cretaceous (130 m yr) and initiated the opening of the South Atlantic. The correlation of the sediments of these basins with their counterparts on the African margin has now been well established.

All the contemporary petroleum production is derived from these marginal basins. The major supply at present comes from Bahia's Reconcavo Basin, a long N-S-trending graben (over 4000 m deep) joined by the nonproducing Tucano and Jatobá basins farther north. Other oil fields have been found in the narrow coastal Sergipe-Alagôas basin. All sedimentary basins are being intensively explored, both on and offshore by Petrobras, the government-operated oil monopoly, and new strikes may be expected in the presently nonproducing basins.

FIGURE 6. Schematic column illustrating Gondwana formations in southern Brazil. (From Caster, 1952.)

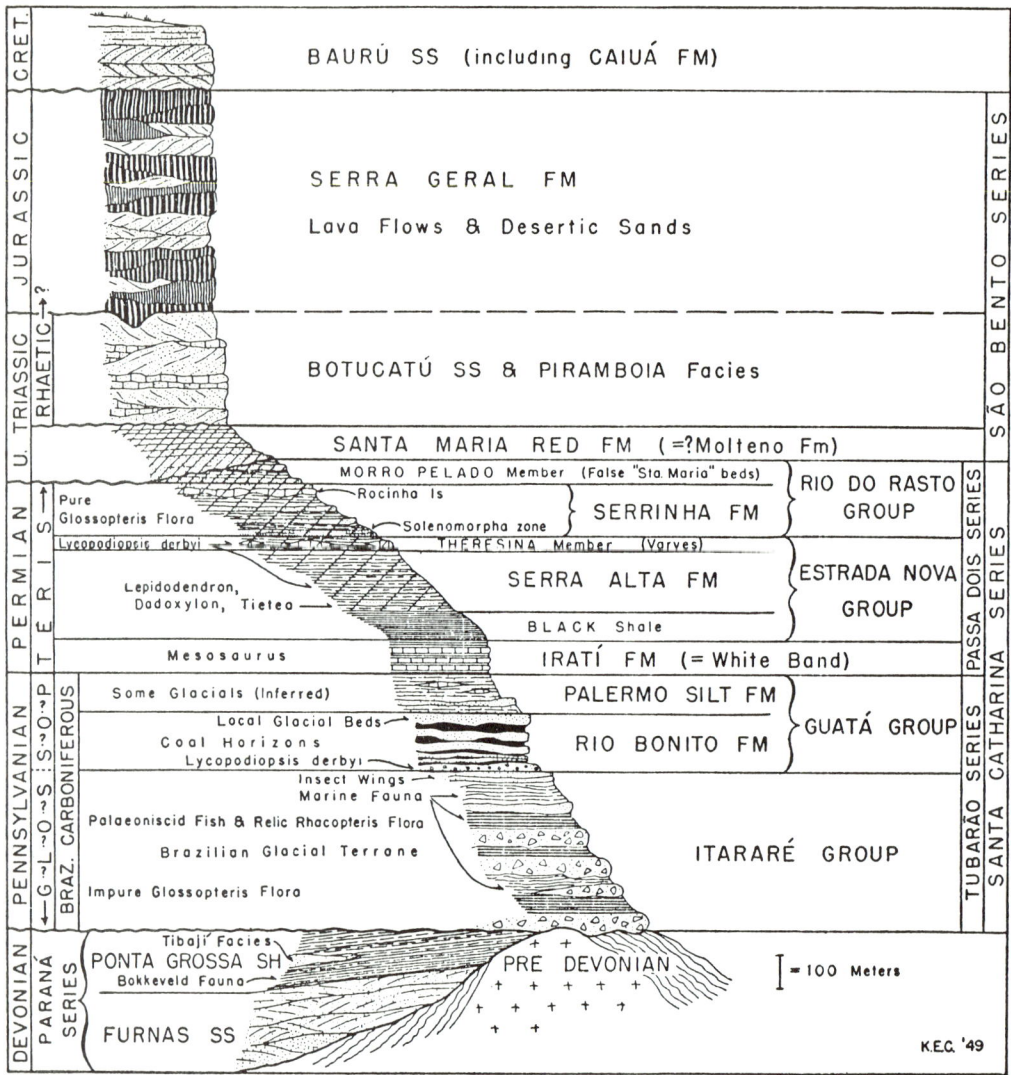

Salt has been drilled or detected geophysically in most of the marginal basins north of Parana. These Cretaceous salt occurrences, besides offering excellent structural traps for oil, are the raw material for a developing chlorine and caustic soda industry in the State of Alagoas. Deposits in Sergipe of carnallite and sylvite, associated with the salt, are awaiting development.

Cenozoic. Erosion was the main feature of the Cenozoic in Brazil. However, Cenozoic sediments cover large areas in Amazonia, Pantanal, the northern part of the Parnaíba Basin and parts of the coast, as well as small basins such as the Curitiba, São Paulo, Gandarela, and Itaboraí. Many of the Cenozoic sediments were deposited under semiarid conditions during the development of peneplain surfaces. Faulting or deep-seated folding were responsible for small steep tectonic basins and for larger shallow and flat basins.

The Barreiras Series, considered Pliocene, is the most important Cenozoic sequence in Brazil. It is present in the Amazon Valley and contains *Dinosuchus terror,* an enormous fossil crocodile about 10 m long. The Barreiras sediments follow the coast from the north of Brazil to as far south as Rio de Janeiro state. The small tectonic Paleocene basin of Itaboraí, at the foot of the Serra do Mar near Niteroi (RJ), has a very rich fossil fauna (mammals, reptiles, and molluscs) and flora. The deposits of the small basins of São Paulo and Curitiba have been considered Pliocene, but they may be Plio-Pleistocene or even Pleistocene. Marine Paleocene deposits are known in the Pernambuco coastal area and from the Miocene in Pará and Piauí.

The present landscape was developed mostly in the Cenozoic. The erosive phase began after the Cretaceous-Eocene peneplain was formed. One of the correlative deposits for this peneplain is the Baurú Series. During the early Tertiary this surface was uplifted, warped, and faulted leading to the development of the Serra do Mar, which in many areas is composed of a complex system of block-faulted mountains. After a long period of erosion in the Middle Tertiary, a new erosion surface developed, probably by peneplanation processes, but without correlative deposits. In Paraná this is known as the "Alto Iguaço surface" and in São Paulo as the "Paleogene surface." Another peneplain developed toward the end of the Pliocene or even in the Lower Pleistocene. It is known in Paraná as the "Curitiba surface," and the correlative deposits of semiarid environment are in the Curitiba Basin.

Remnants of Pleistocene pediments have been found in the coastal area, as well as inland from Rio Grande do Sul to the northeast of Brazil. Certainly pediments formed in more than one period. These stages of semiaridity are correlated with Pleistocene glaciers, because they can be traced down the valleys to match the low sea-level stands.

Sedimentary History

In summary, the marine sedimentary history of Brazil can be subdivided into six sequences of deposition with corresponding interruptions marked by regional unconformities:

1. Late Precambrian erosion surface:
   Latest Precambrian or Cambrian deposition
2. Erosion interval of regional extent but inadequately known:
   Silurian to early Carboniferous deposition
3. Middle Carboniferous (Namurian), unconformity of regional extent:
   Late Carboniferous to Permian (Gondwana) depositional phase
4. Triassic and a large part of Jurassic, interval of erosion:
   Lower Cretaceous (Wealdian) depositional phase—a wide swath of sediments deposited northward of Bahia
5. Post-Wealdian—Pre-Aptian unconformity—very extensive:
   Aptian-Albian deposition—widespread over the eastern half of Brazil
6. Upper Cretaceous-Tertiary unconformity:
   Tertiary Deposition

In the intracratonic depressions there was extensive continental deposition during marine regressions. A schematic section illustrating the above for southern Brazil was presented by Caster in 1952 (Fig. 6).

During the Cenozoic, a cyclic alternation between semiarid and humid conditions is evident, as well as an active epeirogenic uplift represented by several erosion levels. The tropical, humid phases were probably very warm; the semiarid conditions most likely were cooler.

JOÃO JOSÉ BIGARELLA

References

Ab'Sabér, A. N., 1968. "O relevo brasileiro e seus problemas," in A. de Azevedo, ed., *Brasil—a Terra e o Homem, Coleção Brasiliana,* vol. 1, 2nd ed. São Paulo: Cia Ed. Nacl., 135–218.

Almeida, F. F. M. de, 1963. "Os fundamentos geologicos," in A. de Azevedo, ed., *Brasil—a Terra e o Homem, Coleção Brasiliana,"* vol. 1, São Paulo: Cia Ed. Nacl., 55–120.

____, Ruy Derze, G., and Vinha, C. A. G. da, 1972. *Mapa Geologico do Brasil, 1:5,000,000.* Rio de Janeiro: Dept. Nacl. Prod. Mineral., Minist. Minas e Energ.

American Geological Institute, 1966. *Guidebook.* Brazil: Intern. Field Inst.; Washington, D.C.: A.G.I.

Asmus, H. E., and Porto, R., 1972. "Classificação das bacias sedimentares Brasileiras segundo a tectonica de placas," *26th Congr. Brasileiro Geol.,* Belem Para.

Azevedo, A. de, ed., 1968. *Brasil: a Terra e o Homem, Coleção Brasiliana,* vol. 1, 2nd ed. São Paulo; Ca. Ed. Nacl., 607p.

Bauer, E. J., 1967. "Genesis of lower Cretaceous a sandstone, Reconcavo Basin, Brazil," *Bull. Am. Assoc. Petrol. Geologists,* 51(1), 28–54.

Beurlen, K., 1970. *Geologie von Brasilien.* Berlin: Gebr. Borntraeger, 444p.

Bigarella, J. J., and Ab'Saber, A. N., 1964. "Palaeogeographische und palaeoklimatische Aspekte des Kaenozoikums in Sudbrazilien," *Z. Geomorphol.,* N.F. 8, 286–312.

____, and Andrade, G. O. de, 1965. "Contribution to the study of the Brazilian Quaternary," *Geol. Soc. Am. Spec. Pap.* 84, 433–451.

Branner, J. C., 1903. *A Bibliography of the Geology and Mineralogy and Paleontology of Brasil,* vol. 12. Rio de Janeiro: Arch. Mus. Nacl., 116p.

Caster, K. E., 1952. "Stratigraphic and paleontologic data relevant to the problems of Afro-America migration during the Paleozoic and Mesozoic," *Bull. Am. Museum Nat. Hist.* 99(3), 105–152.

Cloud, P., and Dardenne, M., 1973. "Proterozoic age of the Bambui Group in Brazil," *Bull. Geol. Soc. Am.,* 84, 1673–1676.

Ebert, H., 1970. "The Precambrian geology of the Borborema Belt (states of Paraiba and Rio Grande do Norte; northeastern Brazil) and the origin of its mineral provinces," *Geol. Rundschau,* 59(3), 1292–1326.

Fairbridge, R. W., and Richards, H. G., 1970. "Eastern coast and shelf of South America," *Quaternaria*, **12**, 47–55.

Ferreira, E. O., 1972. *Carta tectonica do Brasil, 1:5,000,000*, vol. 1. Rio de Janeiro: Dept. Nacl. Prod. Mineral. Minist. Minas e Energ., 19p.

Guimarães, D., 1964. "Geologia do Brasil," *Div. Fomento Prod. Mineral., Mem. 1* (Rio de Janeiro).

Harrington, H. J., 1962. "Paleogeographic development of South America," *Bull. Am. Assoc. Petrol. Geologists*, **46**(10), 1773–1814.

Jones, F. O., 1973. "Landslides of Rio de Janeiro and the Serra das Araras escarpment, Brazil," *U.S. Geol. Surv. Prof. Pap.* **697**, 42p.

Klein, G. de V., Melo, U. de, and Favera, J. C. della, 1972. "Subaqueous gravity processes on the front of Cretaceous deltas, Reconcavo Basin, Brazil," *Bull. Geol. Soc. Am.*, **83**, 1469–1492.

Lamego, A. R., 1960. *Mapa Geológico do Brasil.* Rio de Janeiro: Dept. Nacl. Prod. Mineral.

Maack, R., 1960. *Kontinentaldrift und Geologie des südatlantischen Ozeans.* Berlin; de Gruyter, 164p.

Meister, E. M., and Aurich, N., 1972. "Geologic outline and oil fields of Sergipe Basin, Brazil," *Bull. Am. Assoc. Petrol. Geologists*, **56**(6), 1034–1047.

Oliveira, A. I., 1956. "Brazil," *Geol. Soc. Am. Mem.* **65**, 1–62.

Padula, V. T., 1969. "Oil shale of Permian Iratiformation, Brazil," *Bull. Am. Assoc. Petrol. Geologists*, **53**(3), 591–602.

Pommerene, J. B., 1964. "Geology and ore deposits of the Belo Horizonte, Habirite and Macacos quadrangles, Minas Gerais, Brazil," *U.S. Geol. Surv. Prof. Pap.* **341-E**, 58S.

Putzer, H., 1956. *Mineralmacht Brasilien.* São Paulo: Deutsch Brasil. Handels.

___, 1971. "Das Itabirit-revier, Serra dos Carajas, Para, Brasilien," *Erzmetall.*, **24** (1), 7–11.

Renger, F., 1970. "Fazies und magmatismus der Minas-Serie in der südlichen Serra de Espinhaço, Minas Gerais, Brasilien," *Geol. Rundschau*, **59**(3), 1253–1292.

Rolff, P. A. M. A., 1956. "Minerios do Brasil," *Rev. da Escola de Minas*, **20**(3–5), 1–22 (sintese).

Ruellan, F., 1953. *O escudo Brasileiro e os dobramentos de fundo*, Rio de Janeiro: Univ. Brazil, Fac. Nacl. Fil. Curso de Espec. em Geomorfologia.

Scheibe, E. A., 1932. "Über die Entstehung brasilianischer Itabirite," *Z. Deutsch. Geol. Ges.*, **84**, 36–47.

Simmons, G. C., 1968. "Geology and iron deposits of the western Serra do Curral, Minas Gerais, Brazil," *U.S. Geol. Surv. Prof. Pap.* **341-6**, 57S.

Suszczynski, E. F., 1970. "La géologie et la tectonique de la Plateforme Amazonienne," *Geol. Rundschau*, **59**(3), 1232–1253.

Trendall, A. F., 1968. "Three great basins of Precambrian banded iron-formation: a systematic comparison," *Bull. Geol. Soc. Am.*, **79**, 1527–1544.

Tugarinov, A. I., and Bibikova, E. V., 1971. "Geochronology of Brazil," *Geochem. Intern.*, **8**(4), 495–503.

Tyler, S. A., 1948. "Itabirite of Minas Gerais, Brazil," *J. Sed. Petrology*, **18**(2), 86–87.

Wallace, R. M., 1965. "Geology and mineral resources of the Pico de Itabirito district, Minas Gerais, Brazil," *U.S. Geol. Surv. Prof. Pap.* **341-F**, 68S.

Cross-references: *Argentina; Bolivia; Fernando de Noronha, etc.; French Guiana; Guiana Shield–Regional Review; Guyana; Paraguay; Peru; South America; Surinam; Uruguay.*

C

CANADA

Canada, the second largest country in the world, includes 9.8 million km² (3.8 million sq mi) and stretches from about latitude 41°N to 82°N, from the temperate climatic zone to the polar. A country so large and diverse includes a great variety of geological structures, climates, landscapes, and a full range of sedimentary rocks from early Precambrian (more than 3 b yr ago) to Holocene. Within its borders the geology is as diverse as the climate and topography and includes the most extensive exposure of Precambrian rocks in the world, three folded mountain chains, little-deformed platform sediments, and continental shelves.

The four major tectonic units in Canada are a central shield of exposed Precambrian rocks, platforms that embrace the shield on three sides, folded mountain chains bordering the platforms, and coastal plains between the mountains and the sea (Fig. 1).

Canadian Shield

The area of 5 million km² of Precambrian rocks that occupies much of eastern and central Canada has a peneplaned surface shaped roughly like a very flat saucer. The center of the shield is occupied by Hudson Bay and a basin containing about 3000 m (10,000 ft) of Phanerozoic sediments. The eastern edge of the shield reaches a relief of 1500 m in the Torngat Mountains of Labrador and in Baffin Island and is deeply indented by fiords. The glaciation of

FIGURE 1. Major structural units of Canada.

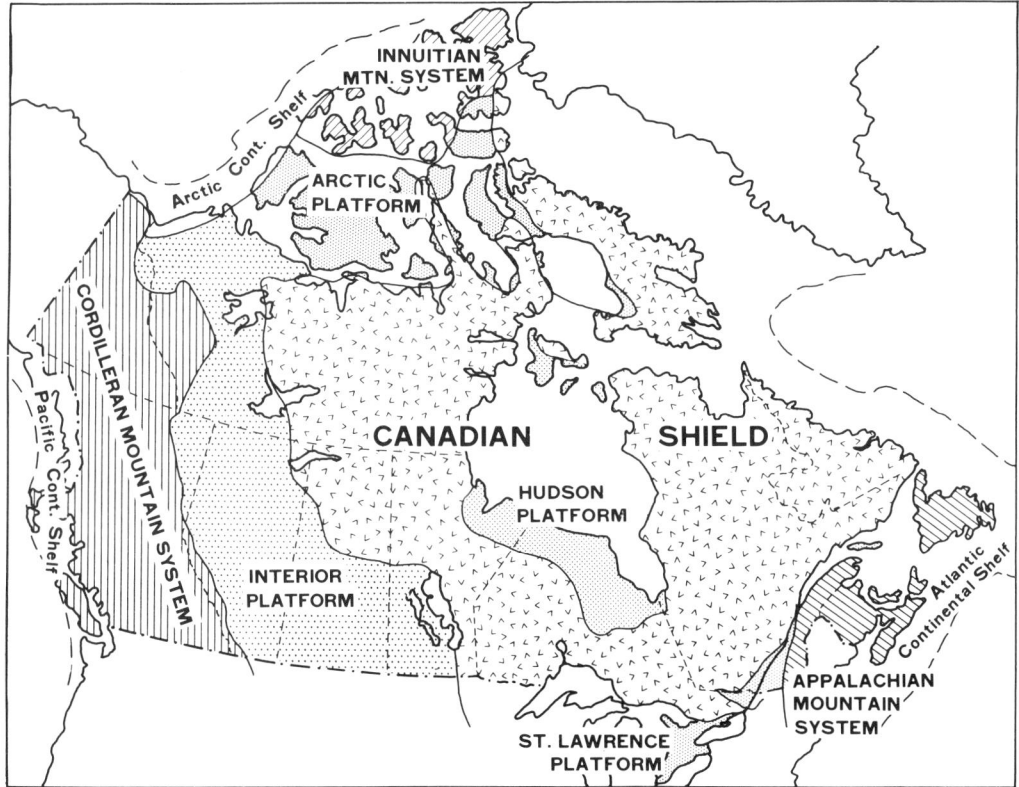

the shield modified the preglacial peneplaned surface by erosion and deposition, deranged the drainage, and impounded many streams to form lakes. The largest lakes, which are among the greatest masses of freshwater in the world, have been excavated along the boundary between the shield and the overlapping Phanerozoic rocks (Lakes Great Bear, Great Slave, Athabaska, Winnipeg, Huron).

The Canadian Shield has been divided into geological provinces (shown on Fig. 1 of the *Canadian Shield* article) on the basis of structural trends and the age of deformation. In each province the basement rocks consist of metamorphosed graywacke-greenstone assemblages intruded by granite gneiss and are interpreted as the roots of geosynclinal mountains. Locally these assemblages are unconformably overlain by little-deformed sedimentary rocks of the limestone-quartzite suite, which were deposited under stable platform conditions. The oldest geosynclinal rocks, called *Archean,* occur in the Superior and Slave provinces and were deformed in the Kenoran orogeny of about 2500 m yr ago (Fig. 2). Platform rocks of the next younger stratigraphic division of the Precambrian, the Aphebian (Fig. 3), cover the Archean rocks of the Slave and Superior provinces, and their equivalents in the geosynclinal facies underlie narrow deformed belts in the Churchill, Bear, and Southern provinces. These Aphebian geosynclinal rocks were deformed in the Hudsonian orogeny about 1700 m yr ago. Helikian rocks of the platform facies overlie the Aphebian rocks of the Churchill province unconformably, but their equivalents in the Grenville province have been highly deformed by the Grenville orogeny of about 950 m yr ago. Little-deformed platform rocks lying on the rocks folded in the Grenville orogeny are classed as Hadrynian.

Although during Precambrian time the shield was the site of repeated orogenies, in Phanerozoic time it acted as a stable block subsiding to receive the flood and rising to drain off the ebb of continental seas but resistant to the compressive forces that acted along the margins of the continent. Such stable behavior is said to be cratonic and the shield is the core of the North American craton.

Platforms

Around the shield, Paleozoic and Mesozoic sediments generally overlie Precambrian rocks unconformably. Through borings and geophysical evidence, the surface of the Precambrian crystalline rocks can be traced downward beneath the covering sedimentary rocks to the marginal orogenic belts. The areas in which the Phanerozoic rocks form a veneer from a few feet to a few thousand feet thick on the Precambrian basement are called the *platforms* and constitute the outer part of the North American craton. In Canada platform areas include the St. Lawrence lowland, the Great Plains, and parts of the Arctic islands bordering the shield (Fig. 4). The eastern part of the St. Lawrence lowland is part of a rift system that can be traced from the Gulf of the St. Lawrence southwestward through the north-central United States. That section of it in Ontario is a gently dipping homocline of Cambrian to Devonian rocks divided by a low NE-trending arch.

The platform of the Great Plains in Canada is divided into basins, in which the Paleozoic and Mesozoic sedimentary rocks reach thicknesses of 3000 m, and arches over which they thin. The largest of these basins, the Western Canada sedimentary basin, underlies much of Alberta, Saskatchewan, and a small part of southwestern Manitoba and is bounded by the Peace River arch on the northwest and by the Transcontinental arch on the Southeast. A division into basins and arches is also characteristic of the Arctic segment of the platform, but the Paleozoic sedimentary rocks that underlie it are much thinner than those in the plains.

Orogenic Belts

The *Appalachian Mountain system* occupies the whole of Nova Scotia, New Brunswick, and Newfoundland and the southeastern part of Quebec. The Appalachian geosyncline received sediments from late Precambrian to Devonian time. It was deformed and intruded in several episodes—the complexity of which is only now

FIGURE 2. Precambrian divisions of Canada.

Approx. Age million years	Stratigraphic Division	Orogeny
	HADRYNIAN	
950		Grenville
1400	HELIKIAN	Elsonian
1700		Hudsonian
	APHEBIAN	
2500		Kenoran
	ARCHEAN	

FIGURE 3. Canadian Shield: vertical air photograph showing diabase dikes cutting gently folded Aphebian strata, Bear Province, N.W.T.

being realized—during the early Paleozoic and was extensively deformed, intruded by granite batholiths, and metamorphosed in the middle and late Devonian Acadian orogeny. During Carboniferous time part of the system in Newfoundland, Nova Scotia, and New Brunswick was broken into fault-bounded basins that received marine deposits and later terrestrial sediments with extensive coal beds. Block faulting and terrestrial deposition in basins resumed in Triassic time.

The section of the *North American Cordillera* in Canada is about 700 km wide and can be divided from E to W into Foothills, Rocky Mountains, Plateau, and Coastal Ranges zones. The Foothills belt on the eastern side of the system consists of intricately thrust-faulted Mesozoic sandstones and shales overlying Paleozoic limestones. The ranges of the Rocky Mountains are thrust-bounded fault blocks of Paleozoic limestones and quartzites that have moved eastward against the platform along a deep décollement over the crystalline Precambrian basement. The central plateau region of the Cordillera is separated from the Rocky Mountains to the east by a straight valley called the *Rocky Mountain Trench* in the south and the *Tintinna Trench* in Yukon Territory. The origin of these trenches is unknown, but most hypotheses involve a combination of faulting and glaciation. The Plateau is underlain by metamorphic and igneous rocks of complex structure ranging in age from Precambrian to Cenozoic. The Coast Ranges of British Columbia are sculptured from middle and late Mesozoic batholiths but also contain patches of metamorphosed sediments that are remnants of the original geosynclinal filling and Mesozoic and Cenozoic continental sedimentary rocks deposited in postorogenic basins.

The Cordilleran geosyncline was a complex of basins and swells that received sediments from late in the Precambrian to the Cenozoic. Early periods of deformation appear to have taken place in middle and late Paleozoic.

In the Mesozoic the geosyncline was affected by a series of orogenies that progressively reduced the basins in the trough and extended the uplifts. Important periods of deformation, not all of which affected the whole geosyncline, include the *Tahltanian orogeny* (mid-Triassic), the *Inklinian orogeny* (late Triassic and early

FIGURE 4. Devon Island. Nearly flat-lying Paleozoic strata unconformably overlying Precambrian rocks. (Photo: No. A-14233, National Air Photo Library, Surveys and Mapping Branch, Dept. Energy, Mines and Resources. Photo flown at altitude of 30,000 ft, with foreground scale of 1 inch = 5000 ft.)

Jurassic), the *Nassian orogeny* (mid-Jurassic), and the *Columbian orogeny,* which represents the climax of deformation in the coastal and central sector. The Columbian orogeny is divided into late Jurassic, early Cretaceous, and earliest late Cretaceous phases. The Rocky Mountains and the foothills were deformed in latest Cretaceous and early Cenozoic in the Laramide orogeny.

The *Innuitian orogenic belt* crosses the Parry Islands and swings northward across Ellesmere Island and the northern coast of Greenland. The axial part of the orogenic belt, characterized by metamorphism, intrusion, graywacke-suite sedimentation, and volcanism, is covered by Mesozoic and Cenozoic sediments in the western islands but emerges in northern Ellesmere Island. In the early Paleozoic the site of the Innuitian belt was occupied by the *Franklin* (or *Franklinian*) *geosyncline.* The folding of this belt took place in the Ellesmeran orogeny at the end of the Devonian period, although earlier movements are recorded.

Coastal Plains and Continental Shelves

Much of the sediment that accumulated around the margins of Canada is covered by the sea and constitutes the continental shelves. The Atlantic continental shelf is a northward extension of the Atlantic coastal plain and shelf of the United States. Off Nova Scotia, where the shelf is about 200 km wide, it is underlain by a wedge of seaward-dipping Cenozoic and Mesozoic beds about 10,000 m thick. A group of shallow banks culminating in Sable Island is located along the seaward edge of the shelf. Piercement structures indicated by seismic profiling suggest that salt intrusions, possibly from an early Mesozoic evaporite series, are present beneath the shelf.

The topography off the west coast of British Columbia is too irregular to be described as a shelf. However, late Mesozoic and Cenozoic sediments accumulated in deep basins offshore. Much of the early filling of these basins is volcanic ejecta and flows; later, marine and nonmarine sandstones predominated.

The *Sverdrup Basin,* which covers the western end of the Innuitian orogenic belt, contains sediments of late Paleozoic to Cenozoic intruded by masses of evaporites. The evaporites at the surface are gypsum and anhydrite, but gravity measurements suggest that salt may be present at depth. The basin beds were deformed in the *Eurekan orogeny* of the mid-Cenozoic by forces that produced folds and faults and intruded the evaporites into the anticlines and injected them along fault planes.

The poorly consolidated clastic sediments of the *Arctic coastal plain* are late Cenozoic. They are exposed in a narrow belt along the northwestern edge of the Arctic archipelago; similar sediments are believed to underlie the Arctic continental shelf, which extends for 160 km into the Arctic Ocean.

Glaciation

The surface of Canada has been modified by successive glaciations during the Pleistocene epoch. Over most of Canada multiple glaciation is difficult to demonstrate, and the deposits record only the retreat of the last glaciers. The continental glacier that centered in the Hudson Bay region is called the *Laurentide ice sheet.* The Cordillera was covered by a complex of glaciers that formed an ice sheet and buried all but the highest mountains. Only a few square miles of Yukon Territory escaped glaciation.

Resources

Among the earliest discoveries of petroleum on this continent were those made in southwestern Ontario in the late 1850s. The Ontario fields, producing from Devonian dolomites, Silurian reefs, dolomitized Ordovician limestone, and Cambrian sandstone, have long since been surpassed in production by the oil and gas found in the western Canada sedimentary basin. Most of the western fields are located in Alberta, but some oil and gas are produced from Saskatchewan and southwestern Manitoba and the Northwest Territories. In 1970 the reserves were estimated at 8½ billion barrels of oil and 53×10^{12} ft^3 of natural gas. Sixty percent of the oil reserves of western Canada are located in reef bodies of middle and late Devonian age beneath the western plains. Other important reservoirs are found in Mississippian limestones (11%) and Cretaceous sandstones (20%). In the foothills gas and some oil are trapped in Paleozoic limestones in complex anticlines. The Athabaska Tar Sands of Cretaceous age in northeastern Alberta are estimated to contain 626 million barrels of heavy oil and to represent one of the largest reservoirs in the world. The sands are now being exploited by strip mining, but extraction has been difficult. Future petroleum provinces in Canada are likely to be the Innuitian region, the Sverdrup Basin, the platforms of the Yukon and Northwest Territories, and the Atlantic continental shelf.

The most extensive bituminous coal deposits in the country are in the Cretaceous sediments of the plains, foothills, and Rocky Mountains of Alberta. Coal is also mined from small fields along the Pacific Coast. Bituminous coal is mined in the Carboniferous basins of the Atlantic coastal provinces, largely in Nova Scotia. Lignite is mined by stripping methods from Cenozoic rocks in Saskatchewan.

From the Canadian Shield, the Cordillera, and the Appalachian region have been mined metallic and nonmetallic resources that have made Canada a leading world producer of nickel, asbestos, copper and other base metals, iron, gold, and uranium. In the Appalachian system the principal ores are the base metal deposits around Bathurst in New Brunswick, but mines in the southeastern corner of Quebec in serpentinized ultrabasic rocks have made the country the leading producer of asbestos in the noncommunist world.

The most valuable commodities mined from the Canadian Shield are iron ore, nickel, copper, zinc, gold, and uranium. The largest deposits of iron ore are mined in the banded iron formations of the Labrador trough of Quebec and Labrador. Much of the nickel comes from mines around the Sudbury Basin in central Ontario but more-recently discovered deposits at Lynn Lake and Thompson in Manitoba are now making important contributions to nickel production. Copper is recovered from the copper-nickel ores of Sudbury and from many copper-zinc deposits in Quebec, Ontario, and Manitoba. Gold deposits are largely in the greenstone belts of the Superior and Slave areas of the shield, but gold is also obtained as a by-product of other metal mines. Most of the uranium production is from conglomerates of the area north of Lake Huron in Ontario.

Metals mined in quantity in the Cordilleran region are, in order of value, lead, gold, copper, silver, and zinc; some tin and mercury have also been produced. The largest lead-zinc mine in this area, the Sullivan Mine, has produced nearly half of the mineral wealth of British Columbia. Placer gold of the Cariboo district started the first Canadian gold rush in the 1870s. At the end of the century gold was discovered in the Klondyke area of the Yukon. From these deposits, in an unglaciated part of the country, 250 million dollars worth of gold has been extracted to date.

Exploration

Although explorers of the 17th and 18th centuries made some cursory geological observations, the systematic surveying of the country did not begin until William Logan was appointed first director of the Geological Survey of Canada in 1842. Logan gathered around him a few assistants, organized a field program, and at the end of 20 years, four years before the confederation of Canada, published in the *Geology of Canada* the first outline of the bedrock of the central part of the country. After Confederation (1867) the Survey grew and reconnaissance parties were sent westward and northward under such men as J. B. Tyrrell, D. B. Dowling, A. P. Low, and G. M. Dawson to establish the broad outlines of the geology along waterways and roads. Until about 1950 the mapping of Canada proceeded slowly and much of the efforts of the Survey were concentrated in detailed work in areas of economic interest. During the 1950s and 1960s extensive use of helicopters to transport large parties of geologists working in a single area enabled the Survey to complete mapping practically the entire country on a scale of 4 inches to the mile.

The progress of the search for mineral wealth in Canada can be measured by the contrast of the lonely prospector of yesterday, tramping the bush and panning the creek, and the geophysicist of today, interpreting aerial surveys made with complex remote sensors. The use of many geophysical instruments for mineral exploration was first made in Canada, and radiometric, electromagnetic, magnetic, and gravimetric surveys are now the routine tools of the modern prospector. The mapping of the distribution of trace amounts of metals in soils, vegetation, and natural waters has also been extensively used in Canada by geochemists to locate mineral deposits.

Organization of Geology in Canada

The largest group of earth scientists in Canada is employed by the Geological Survey, a division of the federal Department of Energy, Mines, and Resources. The Survey's main office is in Ottawa, but branches are located in several centers across the country, the largest of which is the Institute of Sedimentary and Petroleum Geology in Calgary, Alberta. The Mines Branch of the same department in Ottawa also employs geologists and mineralogists in the study of mineral deposits. All provinces have scientific organizations concerned with field and laboratory studies in earth science under various departments and different names.

Some 27 universities in Canada offered undergraduate training in geological sciences by 1970 and 18 offered graduate degrees. University research in earth sciences is largely sponsored by the National Research Council, but grants are also made from a variety of other federal and provincial departments.

Three general and national associations represent Canadian geologists: the Geological Association of Canada, the Canadian Institute of Mining and Metallurgy, and the Canadian Society of Petroleum Geologists. Each of these publishes a periodical. In addition, the National Research Council publishes the bimonthly *Canadian Journal of Earth Sciences*.

Although local museums exist in many centers, only the National Museum in Ottawa and the Royal Ontario Museum in Toronto have large geological, mineralogical, and paleontological collections.

COLIN W. STEARN

References

Anon., 1970a. *Index of Publications, 1959–1969*. Ottawa: Dept. of Energy, Mines and Resources.

———, 1970b. *Principal Mineral Areas of Canada* (Map 900A). Ottawa: Dept. Energy, Mines and Resources.

Baird, D. M., 1964. "Geology and landforms as illustrated by selected Canadian topographic maps," *Pap 64–21*. Ottawa: Dept. Energy, Mines and Resources, 59p.

Clark, T. H., and Stearn, C. W., 1968. *The Geological Evolution of North America,* 2nd ed. New York: Ronald Press, 434p.

Douglas, R. J. W., ed., 1970. *Geology and Economic Minerals of Canada* (Econ. Geol. Rept. 1, 5th ed.). Ottawa: Geol. Surv. Can., 838p.

Locke, D. H., n.d. *Selected Guides for Geologic Field Study in Canada and the United States* (ESCP Ref. Ser. 9). Englewood Cliffs, N.J.: Prentice-Hall, 56p.

McCrossan, R. G., and Glaister, R. P., eds., 1964. *Geological History of Western Canada.* Calgary: Alberta Soc. Petrol. Geologists.

McGee, B. A., 1971. *Thesaurus of the Canadian Index to Geoscience Data, Edition 71/1*. Ottawa: Dept. Energy, Mines and Resources, 201p.

Matthews, W. H., III, 1965. *Selected Maps and Earth Science Publications for the States and Provinces of North America* (ESCP Ref. Ser. 4). Englewood Cliffs, N.J.: Prentice-Hall, 42p.

Nelson, S. J., 1970. *The Face of Time.* Calgary: Alberta Soc. Petrol. Geologists, 133p.

Paterson, W. S. B., 1972. "Laurentide ice sheet: estimated volumes during Late Wisconsin," *Rev. Geophys. Space Phys.,* 10(4), 885–917.

Price, R. A., and Douglas, R. J. W., eds., 1972. "Variations in tectonic styles in Canada," *Geol. Assoc. Can. Spec. Pap.* 11, 688p.

Smith, C. Y., ed., 1970. *Background Papers on the Earth Sciences in Canada.* Ottawa: Dept. Energy, Mines and Resources, 318p.

Walcott, R. I., 1972. "Late Quaternary vertical movements in eastern North America: quantitative evidence of glacio-isostatic rebound," *Rev. Geophys. Space Phys.,* 10(4), 849–884.

Wanless, R. K., 1969. *Isotopic Age Map of Canada.* Ottawa: Dept. Energy, Mines and Resources, 1:5,000,000.

Cross-references: *Canada–Arctic Archipelago, Atlantic Provinces, Canadian Shield, Cordilleran Region, Northern Great Plains Province, Ontario Basin, Rocky Mountains and Eastern Cordilleran Region, St. Lawrence Lowlands of Quebec.*

CANADA–ARCTIC ARCHIPELAGO

The islands of the Canadian Arctic archipelago lie north of the Canadian mainland, and for convenience in this discussion will include the large Boothia and Melville peninsulas. This sparsely vegetated and nearly unpeopled region is one of the great archipelagos of the world; it extends about 2500 km E to W and 2000 km from the mainland to the northern tip. Sea access to most of the islands is restricted, because of ice, to the summer season, and the "polar pack" ice of the Arctic Ocean presses the year round against the northwestern margin of the archipelago.

In the past the Arctic islands were inhabited only by Eskimos and very few others. Fur trapping and handicrafts were the principal economic activities. Since about 1960, however, the region has become increasingly active as the search for petroleum has accelerated, and this has resulted in a marked increase in services and accessibility.

Physiography and Structural Provinces

The generalized surface of the archipelago may be likened to a giant warped, triangular platter, tilted so that the northwestern edge is submerged and the eastern edge raised, the northern apex crumpled, and upthrust to form the highest part, and the central part ribbed by subradiating raised welts (archs or uplifts) that separate shallow depressions. Plains, lowlands, uplands and plateaus, highlands, and mountains are all present. Most elevated surfaces are rolling and dissected by valleys and deep channels.

The boundaries between some of the physiographic areas are irregular and many are gradational. Nevertheless, there is a strong coincidence among physiographic, structural, and stratigraphic provinces (see Fig. 1). Thus the plains, lowlands, and sedimentary plateaus mainly coincide with relatively undeformed sedimentary basins, whereas uplands, plateaus, and mountains are expressions of uplifted arches of basement metamorphic rock of the Canadian Shield and belts of relatively severely folded younger sedimentary rocks.

The structural provinces of the Arctic islands can be grouped into two broad regions: the central stable region to the south, and the Innuitian orogenic region to the north (Table 1). The central stable region includes the basins and arches and the Canadian Shield in the southern half of the archipelago and is part of the central stable region of North America. The sinuous belts of folded rocks of the northern islands form the Innuitian region, in which major orogenies occurred in late Precambrian, mid-Paleozoic, and mid-Tertiary times. Included in this region are the Franklinian Geosyncline and the Sverdrup Basin; the geosyncline is a major tectonic depression that was the site of nearly continuous sedimentation from late Precambrian to late Devonian times, and the Sverdrup Basin is a regional depression superposed on the folded geosyncline and containing an essentially concordant succession of strata ranging from late Mississippian (Viséan) to early Tertiary.

The Arctic coastal plain is a narrow strip of land of very low relief underlain by late Tertiary and early Quaternary rocks. It overlies Mesozoic strata on the northwestern side of the Sverdrup Basin and is presumably continuous with the continental shelf, which extends about 160 km into the Arctic Ocean.

The existing form or pattern of the Arctic archipelago probably derives from late Cenozoic and Pleistocene times. A peneplain postdates a major, early Tertiary orogeny, and this surface now stands at various, sometimes considerable, elevations. However, the system of channels that now separates the islands is evidently a drowned valley system modified by Pleistocene glaciation. The larger channels, many straight- or arcuate-walled, are controlled by graben or rift-valley structures; such a structure also evidently separates the archipelago from Greenland and formed Baffin Bay. The

TABLE 1. Generalized Stratigraphic Table for the

		CENTRAL STABLE REGION	
		WESTERN Banks, Victoria Islands	EASTERN Prince of Wales Island, Boothia Peninsula, Somerset Island, Baffin Island
Cenozoic			Gravel, sands, silt, peat
		~~ unconformity ~~~~~~~~~~~~~~~~	
		Sandstone, shale, coal 1,500'	
Mesozoic	Cretaceous	disconformity	
		Shale, sandstone 1,500'	Sandstone, coal
	Jurassic		
	Triassic		
Paleozoic	Permian		
	Pennsyl- vanian		
		~~	
	Mississ- ippian	angular unconformity	
		~~	
	Devonian	Sandstone, shale	Conglomerate, sandstone 2,000'
			local angular unconformity
	Silurian	Dolomite, limestone, shale 3,600'	Dolomite, limestone, shale, anhydrite, intraformational conglomerate, sandstone 4,000'-5,000'
	Ordovician		
		disconformity	
	Cambrian	Sandstone 400'	Sandstone, dolomite, intraformational conglomerate 300' - 2,000'
		~~ unconformity	
Pre Cambrian	Hadrynian	diabase Basalt 1,000' Limestone, sandstone, siltstone, shale, gypsum 11,000'	diabase Quartzite, sandstone, dolomite, shale 10,000' - 15,000'
	Helikian		diabase Quartzite, volcanic rocks 5,500'
	Aphebian	CANADIAN SHIELD: Granite, granodiorite, pegmatite, migmatite, gneiss, quartzite, marble	
Archean			

Various Regions of the Canadian Arctic Archipelago

INNUITIAN REGION		
WESTERN AND CENTRAL Western Queen Elizabeth Islands, Axel Heiberg Island, & central Ellesmere Island	**NORTHERN** Northern Ellesmere Island, northern Axel Heiberg Island	
up to 250'		
Sandstone, shale, coal 2,000 - 10,000'		SVERDRUP BASIN
disconformity		
Sandstone, siltstone, shale, basalt, tuff, gabbro up to 10,000'		
Sandstone, shale, siltstone, conglomerate up to 5,000'		
disconformity		
Sandstone, shale, siltstone, conglomerate up to 17,000'		
disconformity		
Limestone, sandstone, conglomerate, shale, anhydrite, basalt up to 6,000'		
	Sandstone, shaly sandstone 1100'	
Sandstone, siltstone, shale, limestone 15,000'	Sandstone, siltstone, conglomerate, greywacke, limestone, slate, volcanic rocks 20,000'	FRANKLINIAN GEOSYNCLINE
Shale, siltstone, dolomite, limestone, anhydrite 6,000 - 16,000'	Sandstone, limestone, impure sandstone, volcanic rocks, shale 2,000-4000'	
	Shale, shaly limestone, dolomite, sandstone, conglomerate, volcanic rocks 7,500'	
Sandstone, shale, limestone, dolomite 3,000'	unconformity	
Dolomite, conglomerate, sandstone, shale 5,600'	Gneiss, schist, quartzite, phyllite, greenschist, limestone, marble, slate, sandstone, chert, volcanic rocks	

FIGURE 1. Structural-stratigraphic provinces, Canadian Arctic archipelago.

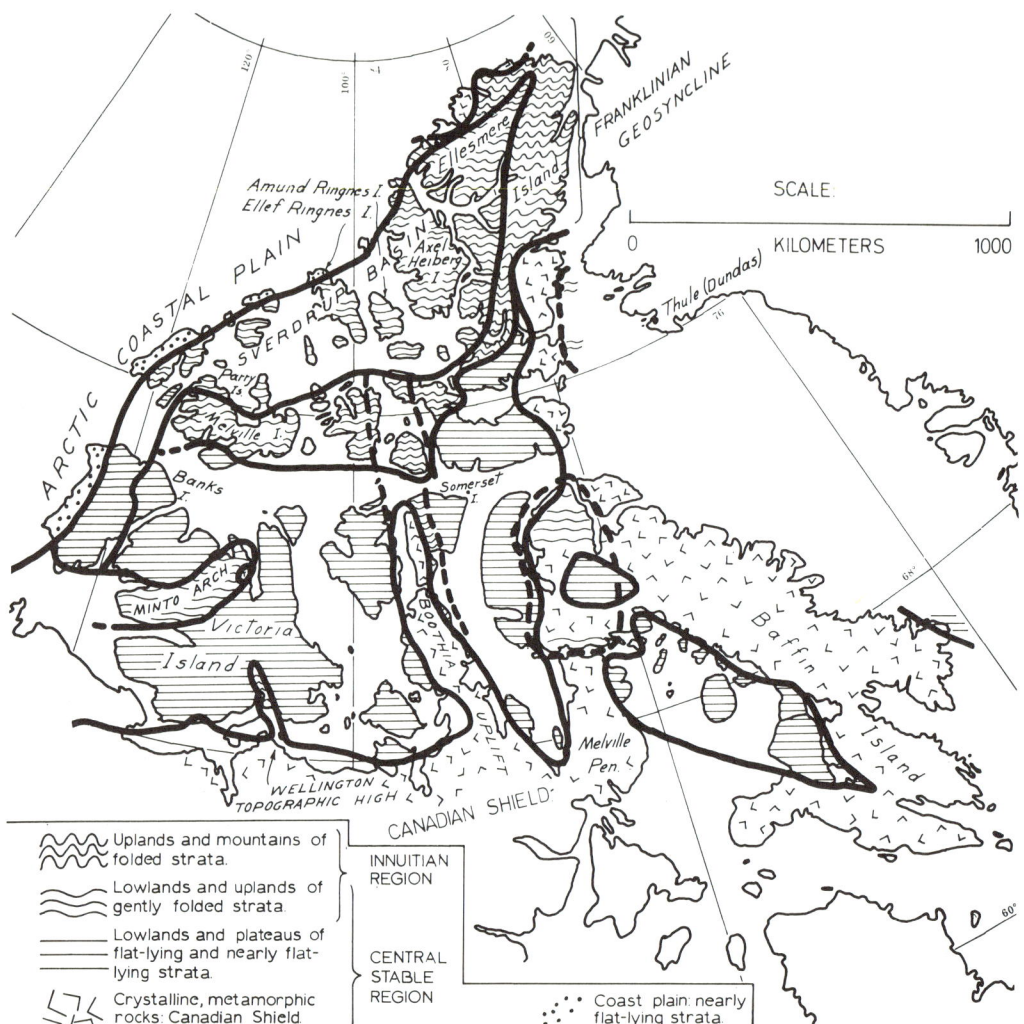

variation in land levels from island to island suggests that the major, now drowned, drainage system also is fault controlled.

Widespread raised beaches indicate appreciable post-Pleistocene isostatic uplift throughout the islands.

Geology

In a general way the oldest rocks of the islands lie to the SE, and the age of exposed rocks decreases toward the NW. The Canadian Shield of Precambrian metamorphic rocks forms a basement complex along the southern and eastern edges of the archipelago. Overlying the shield are unmetamorphosed late Precambrian and, more continuous, early Paleozoic beds that now form scattered sedimentary basins (see Fig. 1). The early Paleozoic formations pass northward into a vastly thicker geosynclinal sequence, the Franklinian, which is folded, truncated, and in turn overlain by a thick basinal sequence of late Paleozoic, Mesozoic, and Tertiary sediments, the Sverdrup Basin.

The Sverdrup Basin and older units were folded and faulted by the last orogeny to affect the archipelago, the Tertiary Eurekan orogeny.

Precambrian Metamorphic Rocks of the Canadian Shield. The Precambrian basement rocks of the southern part of the Arctic archipelago are part of the Churchill province of the Canadian Shield. The Churchill province consists of Archean and early Proterozoic sedimentary and plutonic rocks, mainly metamorphosed and converted to granitic gneisses during the Hudsonian orogeny. Radiometric ages (K-Ar) range between 1650 and 1850 m yr. The shield rocks are prevailingly heterogeneous, banded, and complexly folded gneisses, with limited areas of massive rocks, mainly granitic. Granitic gneisses, often biotitic and interbanded with biotite-garnet-quartz-feldspar gneiss, are predominant. Irregularly distributed throughout the gneisses are areas, belts, and bands of felsic to mafic gneisses and schists containing variously amphibole, hypersthene, epidote, and chlorite. The massive rocks vary from alaskite to ultrabasic rock and on southern Baffin Island include bodies of granulite and charnokite.

Bands and zones of paragneiss or metasedimentary rocks occur throughout the basement complex; gneissoid biotitic quartzite, both with and without garnet, is most abundant. Interbedded with these at various localities are bands of marble, iron formation, amphibole schist, relicts of pillowed lava flows, biotite-garnet-sillimanite schist, and other rocks derived from sediments of various compositions. In some areas lenses and bands of amphibolite are thought to be derived from dykes and sills.

A prominent, NW-trending swarm of diabase dykes intrudes the Archean rocks on Baffin Island. Radiometric age determinations suggest an age of 1100–1200 m yr.

An outcrop of granite only 5 km long on northern Victoria Island is inferred to intrude a quartzite formation and has yielded an isotopic age of 2405 m yr. These rocks are thus a distant inlier of Archean Canadian Shield rocks. A ridge of basement rock connecting with the shield is suggested by the narrow, N-trending Wellington topographic high of Precambrian rocks of southern Victoria Island.

Metamorphic Rocks of Northern Ellesmere and Axel Heiberg Islands. A wide variety of metamorphic rocks of uncertain age is exposed in northern Ellesmere and Axel Heiberg islands. Formations of more than one age may be included, and evidently some of the rocks are exposed basement terrane of the Franklinian Geosyncline.

On Ellesmere Island, biotitic, garnetiferous, and amphibolitic feldspar gneisses, mica schists, quartzite, marble, and a variety of granitic gneisses form a metamorphic complex. A pre-Ordovician age suggested by an unconformity is confirmed by an isotopic age of 550 ± 35 m yr. On Axel Heiberg Island a complex includes both sedimentary and volcanic rocks, with metamorphism advanced to the greenschist facies. A Cambrian or earlier age is suggested from stratigraphic correlations, and a radiometric age of 535 ± 49 m yr appears to confirm this. The two complexes appear to be Cambrian or older rocks that underwent a metamorphic event in Cambrian or Hadrynian time.

Massive, fresh-appearing granite occurs in the metamorphic terrane, as does also some norite, peridotite, and dunite; these may relate to the period of orogeny in which the Franklinian geosyncline was deformed.

Late Precambrian Sedimentary Rocks of the Canadian Shield. Remnants of bedded, relatively little disturbed Precambrian rocks are exposed as both large and small areas in the central stable region, both isolated in the Canadian Shield terrane and as inliers where the thin cover of Paleozoic beds has been stripped away. The Precambrian beds overlie the eroded basement gneisses with profound angular unconformity, and are in turn overlain by early Paleozoic beds, often with only slight angular discordance. Hadrynian and Neohelikian ages are assigned from these relationships and from radiometric age determinations on associated volcanic flows and intrusive dykes and sills.

Hadrynian and conformably underlying, probably Neohelikian, beds in northwestern Baffin Island total at least 6000 m in thickness and include, at the base, an andesitic and basaltic volcanic formation. Overlying are thin- to thick-bedded, mainly shallow-water units of sandstone, shale, siltstone, dolomite, and conglomerate. Northwest-trending diabase dykes transect the late Precambrian beds.

A concordant sequence of limestone, sandstone, siltstone, shale, and gypsiferous rocks with a total thickness of about 3500 m forms the Minto arch on northwestern Victoria Island and southern Banks Island (see Fig. 1). Disconformably overlying this Proterozoic group is a basaltic volcanic formation about 300 m thick. The sedimentary and volcanic sequence is similar in many respects, including the presence of numerous and voluminous diabase-gabbro sills, to the upper part of the Coppermine Series of the mainland south of Victoria Island, and indeed the two sedimentary sequences may be connected beneath the Paleozoic cover.

Beds about 2000 m thick of cross-bedded quartzite, red fissile shales, and dark sills or volcanic layers occur in a small area on the southeastern edge of Ellesmere Island. These rocks probably relate to a late Precambrian basin that is represented by similar rocks widely exposed at Thule (Dundas), Greenland.

About 3000 m of quartzite, red sandstone, dolomite, and shale on northwestern Somerset Island may be of Hadrynian age. These beds rest unconformably on gneisses of the Churchill province, are overlain conformably and transitionally by Paleozoic beds, and contain

FIGURE 2. Structural trends, Arctic archipelago.

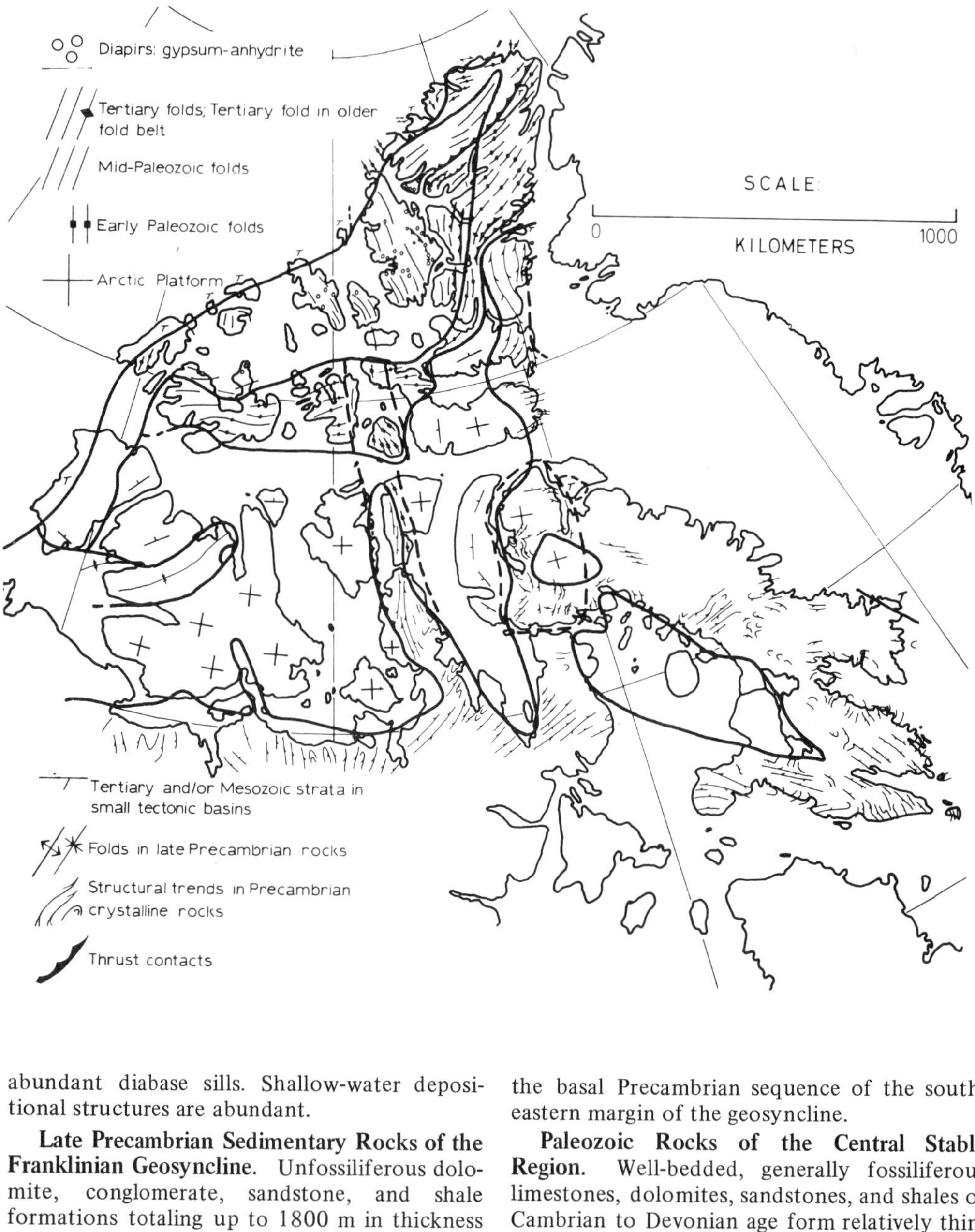

abundant diabase sills. Shallow-water depositional structures are abundant.

Late Precambrian Sedimentary Rocks of the Franklinian Geosyncline. Unfossiliferous dolomite, conglomerate, sandstone, and shale formations totaling up to 1 800 m in thickness are disconformably overlain by Lower Cambrian to Siluro-Devonian beds in east-central Ellesmere Island, the Precambrian and Paleozoic accumulations together forming the Franklinian Geosyncline. The Precambrian units thin and are overstepped by Cambrian beds toward the Canadian Shield, and the base of the thicker, basinal part of the column is not known.

Some unfossiliferous beds of northernmost Axel Heiberg Island are possibly equivalent to the basal Precambrian sequence of the southeastern margin of the geosyncline.

Paleozoic Rocks of the Central Stable Region. Well-bedded, generally fossiliferous limestones, dolomites, sandstones, and shales of Cambrian to Devonian age form relatively thin, widespread accumulations that rest with angular unconformity upon the eroded Canadian Shield and remnants of late Precambrian sedimentary basins. The Paleozoic beds occur as structural basins separated by arches, uplifts, and topographic highs of Precambrian rocks (see Figs. 1 and 2). Paleozoic outliers are preserved on the major arches, which would indicate that evidently these beds were once continuous and the arches were subsequent features. It is possible that the basal or younger

Paleozoic beds thin toward the arches, but there is as yet no evidence for this, and indeed, on northern Baffin Island a region that has been shown to be a depocenter in Cambrian time is now part of an arch.

The Paleozoic rocks of the central stable region are nearly everywhere flat lying or only slightly inclined but are displaced by numerous faults along many basin edges.

The Paleozoic sequence of the central stable region generally comprises Lower and Middle Cambrian sandstone, dolomite, and intraformational conglomerate; and Lower Ordovician to Lower Devonian limestone, dolomite, intraformational conglomerate, and anhydrite. Middle and Upper Devonian limestones and sandstones occur in basins flanking the Boothia uplift, but elsewhere they are absent except where adjacent to the Franklinian Geosyncline. Major stratigraphic breaks occur between Middle Cambrian and early Ordovician strata and between Lower and Middle Devonian. Other, perhaps local unconformities, are recognized; there is a mid-Ordovician uplift and disconformity documented in northern Baffin Island, and Middle Devonian beds locally rest with angular unconformity on older Paleozoic beds near the Boothia uplift.

The Lower and Middle Cambrian clastic and carbonate sediments are of variable thickness, from a few hundred meters in the central and western regions to over 650 m on northern Baffin Island. In parts of these regions, however, the Cambrian is absent, perhaps removed by erosion during a long hiatus in Upper Cambrian time. A probable Cambrian sedimentary basin has been defined in northern Baffin Island from thickness (isopach) data; northwestern and southern sources are indicated by paleocurrent data.

The Ordovician, Silurian, and Lower Devonian rocks are mainly carbonates with some anhydrite. The latter mark a long period of stable conditions during which steady slow subsidence allowed more or less continuous deposition. Lower Ordovician beds, including limestone, dolomite, and shale, are evidently widespread in the central stable region and occur as units about 300 m in total thickness. These units are characterized by a general paucity of fossils. Overlying mid-Ordovician and Silurian beds, in contrast, are extremely fossiliferous, and especially notable is the mid-Upper Ordovician *Arctic Ordovician fauna*. The Ordovician-Silurian beds are mainly limestone and dolomite, with shale and sandstone interbeds locally developed. Thicknesses of up to 1200 m are common.

The Ordovician-Silurian succession is represented, in the western basin, by a uniform sequence of dolomite, up to 1000 m thick, that rests disconformably on Lower Cambrian beds and in places rests directly on Precambrian rocks.

Paleozoic Rocks of the Franklinian Geosyncline. A thick geosynclinal sequence of Proterozoic to mid-Paleozoic beds, the Franklinian Geosyncline, occupies the northern part of the Arctic archipelago. It is well exposed in the northeast and along its southeastern and southern margins, but much of the geosyncline lies buried beneath younger Sverdrup Basin rocks (see Fig. 1). Formations of the geosyncline are continuous with those of the central stable region but attain a total thickness of about 12,000 m compared to maxima of 1800–3000 m in the platform region. The geosynclinal region was folded and faulted during orogenic episodes extending from early Devonian to Mississippian times.

Most rocks of the geosyncline can be grouped into six major facies, some of these complexly related. The major facies, with their tectonic relationships, are (1) a carbonate facies, often with thick evaporites, mainly representing a shelf marginal to the more rapidly sinking parts of the geosyncline; (2) a graptolitic shale facies deposited in the deeply subsiding, axial part of the geosyncline; (3) a flysch facies, representing the deeply subsiding axial part of the basin adjacent to tectonic lands; (4) a variable sandstone-siltstone-shale facies with evaporites and carbonates, these deposited as thick, basin-axial units under varying conditions of uplift and subsidence; (5) a syntectonic clastic facies derived from local tectonic lands; and (6) a sandstone facies consisting of thick units of nonmarine and marine sandstone with minor coal and conglomerate deposited during the closing, neo-orogenic era of the geosynclinal cycle.

Cambrian, Ordovician, and Silurian Carbonate Rocks. Cambrian to Devonian carbonate rocks with abundant evaporites and minor sandstones occur as basin-marginal formations along the southeastern and southern edge of the geosyncline, the units thickening rapidly northwestward toward the axis of the geosyncline. The lowest Paleozoic beds of the geosynclinal basin may also be of carbonate facies, but this is uncertain because of lack of exposure.

Cambrian beds of the geosyncline are known only on eastern Ellesmere Island, where about 1000 m of sandstone and conglomerate overlie late Precambrian beds, separated from them by a regional unconformity. Up to 2000 m of Cambrian, Ordovician, and Silurian limestone, gypsum-anhydrite, dolomite, and shaly limestone overlies the Cambrian unit. Gypsum-anhydrite occurs at many horizons in the sequence and forms two thick formations, one of them up to 800 m thick. These conspicuous, thick evaporite units are characteristic of the shelf or miogeosynclinal regions.

Ordovician, Silurian, and Devonian limestones, dolomites, and shaly limestones are most typical of the southern part of the geosynclinal margin. There the carbonate units intefinger with graptolitic rocks of the basinal facies.

Reefal carbonate banks occur at many localities in the basin-margin zone, both within the carbonate facies and, basinward, in equivalent graptolitic rocks.

Ordovician to Devonian Graptolitic Rocks. Shales and shaly siltstones with graptolites outcrop over much of the exposed length of the Franklinian Geosyncline. The boundary between the graptolitic and the carbonate facies is distinct in places, but obscure or complex elsewhere due to thick reefal masses or, as near the Boothia uplift, as a result of local tectonic activity. The graptolitic units thicken northward, and in northern Ellesmere Island pass laterally into the flysch facies of the geosyncline.

Cambrian to Devonian Clastic Rocks of the Geosynclinal Region. Thick basinal units exposed in the northernmost part of the archipelago comprise impure sandstone, subgraywacke, graywacke, micaceous gritty shale, and impure limestone of a flysch facies. These rocks, equivalents of the graptolitic facies to the south, evidently were derived from tectonic lands in the vicinity of the present northern continental shelf. The flysch rocks pass northward into a complex assortment of volcanic rocks and sandstones, limestones, and conglomerate that apparently represent a shelf environment.

Silurian and Devonian Clastic Rocks of Tectonic Regions. A belt of uplifted Precambrian shield and folds related to it, respectively, the Boothia uplift and the Cornwallis fold belt, extend northward across the southern margin of the geosyncline and disappear beneath the Sverdrup Basin. Red-weathering Devonian conglomerates and sandstones derived from exposed shield and lower Paleozoic rocks flank the tectonic belt and are themselves folded and faulted with the older units. The earliest conglomerate appears in a Silurian formation. This clastic syntectonic facies passes laterally away from the uplift into normal marine sandstones and limestones of the geosyncline.

Devonian red conglomerate, sandstone, shale, and breccia with tuffaceous arenite, tuff, and keratophyric and spilitic volcanic flows occur on northern Axel Heiberg Island. The N-trending folds there may represent a northern expression of the tectonism that affected the Boothia region.

Devonian Clastic Rocks of the Geosyncline. The youngest rocks of the geosyncline are Middle and Upper Devonian sandstones, carbonates, and shales that form a dominantly clastic facies, marine in the lower part and terrigenous in the upper. In most places strata within this facies succeed the basinal graptolitic shale sequence conformably. Locally, such as near the Boothia uplift, an angular unconformity underlies the Middle Devonian beds. A second, younger unconformity in the Boothia region indicates late Middle Devonian tectonism.

The lower part of the late Devonian sequence is composed of drab-colored shales, siltstones, and sandstones, with carbonates and evaporites; some formations grade laterally into a carbonate shelf facies. The marine Devonian rocks pass upward into calcareous sandstone and sandy limestone, shale, and siltstone with some marine fossils and some carbonaceous remains, then, upward again, into terrigenous, cross-bedded sandstone with coal seams. The upper clastic part of the sequence is a clastic wedge that heralded a profound change in the tectonic conditions of the geosyncline. These youngest sediments of the geosyncline derive from tectonic lands to the north, now perhaps submerged beneath the continental shelf.

Deformed beds of the Franklinian Geosyncline as young as late Devonian age are unconformably overlain by late Mississippian nonmarine basal beds of the Sverdrup Basin. The principal deformation of the geosyncline, named the *Ellesmerian orogeny,* evidently took place in the interval between these ages, and resulted in E- and NE-trending folds of the Parry Islands and Ellesmere Island (see Figs. 1 and 2).

The Sverdrup Basin. The Sverdrup Basin overlies the deformed Franklinian Geosyncline with profound angular unconformity. The basin contains a sequence of marine and nonmarine beds ranging in age from Lower Carboniferous (Viséan) to early Tertiary (Eocene), the composite thickness totaling about 15,000 m in axial regions in the eastern part of the basin, where deepest subsidence evidently occurred. The locus of maximum subsidence in the eastern part of the basin evidently moved westward from western Ellesmere Island during late Paleozoic time to western Axel Heiberg Island during late Mesozoic time. Thus nowhere does the section attain the total composite thickness.

The oldest beds of the Sverdrup Basin, perhaps only locally deposited, are late Mississippian (middle Viséan) sandstones and shaly sandstones with coal seams; these are disconformably overlain by Upper Carboniferous clastic rocks.

Upper Carboniferous (Namurian) and Permian beds of the Sverdrup Basin include limestone, conglomerate, sandstone, shale, chert, and gypsum-anhydrite, the total thickness ranging from a few hundred meters to over 3300 m. Volcanic rocks also occur locally.

Between late Carboniferous and early Permian times an essentially conformable succession of marine beds accumulated. Nearly everywhere the basal beds of the succession are red conglomerate and sandstone with minor limestone, but higher beds are developed as four facies belts conforming to the axial trend of the Sverdrup Basin. The belts are a marginal sandstone facies, the lowest beds of which are continuous with the widespread basal clastic unit just noted; a central shale, siltstone, and chert facies; and, flanking the axial facies, two belts of limestone, sandy limestone, and argillaceous limestone. Two anhydrite-gypsum units occur in the Sverdrup Basin: one, of Pennsylvanian age, overlies the basal conglomerates in the axial region and is the source of evaporites in the numerous piercement structures or domes characteristic of that part of the basin; the other, of Permian age, occurs in the southeastern carbonate facies belt and evidently does not form intrusions.

The marginal regions of the Sverdrup Basin are characterized by rapidly thinning formations, by minor unconformities, and by pinching out and overstepping of units.

Late Permian, mainly clastic formations are separated from earlier ones by a regional unconformity, and the younger units are markedly transgressive, some extending over two or more of the underlying facies belts.

Basaltic lavas occur within the Carboniferous and Permian sequence on northern Axel Heiberg Island and on nearby Ellesmere Island; the presence of these volcanic rocks in the center of the basin is indicated by blocks of basic pillow lava and breccia in gypsum piercement domes on Ellef and Amund Ringnes islands.

Mesozoic beds form a thick, structurally conformable succession in the Sverdrup Basin, resting upon Permian beds but separated from them by a regional unconformity. Nondeposition or erosion in Permian time is indicated by the absence of youngest Permian stages beneath a complete Triassic sequence. Disconformities also occur at several levels in the Jurassic.

Mesozoic formations are of marine shale and siltstone facies in the axial regions of the basin, with marine sandstones intertonguing from the margins, particularly on the south and east. Sandstone formations with cross-bedding and ripple marks and including dark micaceous shales, siltstones, and coal or carbonaceous beds are widely distributed, and this nonmarine lithofacies recurs, intercalated with the marine facies in Mesozoic and Tertiary times. The composite thickness of Triassic to Cretaceous formations in the deeper parts of the basin is about 11,000 m; the units thin rapidly marginward, however, and many pinch out and are overstepped by younger beds. Cretaceous sandstones and shales transgress the edge of the basin to the SW and evidently were continuous with beds of this age in the vicinity of the Mackenzie delta and even in Alaska.

Basalt flows and thin bentonitic and tuffaceous beds are locally intercalated in the Cretaceous marine sandstones and shales of the Sverdrup Basin. Diabase and gabbro sills are thick and abundant in parts of the basin, and ring dikes and massive intrusions are associated with some gypsum domes. The intrusive rocks are found only in formations older than an Upper Cretaceous shale unit that contains, in only two areas, some few hundreds of feet of basalt flows.

The youngest beds of the Sverdrup Basin are sandy coal measures of early Tertiary age. These nonmarine beds, up to 3000 m thick and including sandstone, shale, and lignite, overlap the Sverdrup Basin and occur as downfaulted outliers at widely scattered localities throughout the archipelago. Tertiary beds overlie the extensive Cretaceous rocks of the western edge of the Arctic central stable region.

Some of the early Tertiary beds of the Sverdrup Basin are conglomeratic and synorogenic, thus marking the end of the period of uniform subsidence of the basin. Flat-lying basalts and associated fragmental beds on the eastern extremity of Baffin Island may be related to or be part of a Greenland Tertiary volcanic province.

The Sverdrup Basin was deformed in early Tertiary time by essentially compressive forces of the Eurekan orogeny. Resulting structural trends closely follow older trends of the mid-Paleozoic Ellesmerian orogeny (see Fig. 2). Folds in the western part of the basin are broad and open, and follow, but may not be related to, the N-trending Boothia-Cornwallis structures of Lower Devonian age. Folding is more intense to the NE, and there it parallels the early Carboniferous Ellesmerian structures of the Franklinian Geosyncline, so that Franklinian beds have undergone both orogenies. The effects of the Eurekan orogeny, however, extend farther east than do those of the Ellesmerian in the central Ellesmere Island fold belt. Faults are prominent in the Eurekan belt, and much or all movement on giant N-dipping thrusts may relate to this orogeny.

The diapiric gypsum-anhydrite bodies char-

FIGURE 3. Mesozoic sedimentary rocks of the Sverdrup Basin are folded and faulted on western Axel Heiberg Island. Diapiric intrusions of white-weathering gypsum-anhydrite are emplaced along anticlinal axes and faults. (T-489L-38, oblique air photo: Canada Dept. Energy, Mines and Resources.)

acteristic of the axial part of the Sverdrup Basin usually are associated with the planes of thrust faults and the crests of anticlines (Fig. 3). The connection between the intrusive domes and thick evaporite beds of Pennsylvanian age near the base of the Sverdrup Basin sequence is confirmed by Carboniferous fossils in limestone blocks embedded in the brecciated and contorted gypsum rock.

The Arctic Coastal Plain. Gravel, sand, and silt deposits of late Tertiary or earliest Pleistocene age underlie the Arctic coastal plain (see Fig. 1) and rest unconformably upon Mesozoic rocks in the northwestern limb of the Sverdrup Basin synclinorium. These beds, up to 80 m thick, are characterized by abundant fresh uncompressed logs and sticks. The source of the detritus lay to the south and east; evidently the sands and gravels, which are dominantly coarse grained and cross-bedded, were a continuous fluvial plain and delta deposit, derived probably during and after the period of Tertiary orogeny.

The gently sloping surface of the coastal plain lies nearly, but not quite parallel to the bedding of the late Tertiary rocks, which in places dip seaward more steeply than the plain; landward dips also occur. The sediments may be slightly deformed, although the dips may be partly or wholly depositional features. In any case, the emergent plain is evidently a physiographic surface that truncates the youngest beds, which are also dissected by the existing system of valleys, channels, and straits. Fault scarps in the young beds indicate Tertiary or Holocene movement, perhaps due to reactivation of older faults in the underlying, deformed Sverdrup Basin.

Both deposition and erosion probably are

taking place on the continental shelf, the seaward extension of the coastal plain.

Probable equivalents of the coastal plain alluvial formation are unconsolidated gravel, sand, and silt deposits, locally with abundant wood, that cap high-level terraces on Axel Heiberg and Ellesmere islands. The terraces now lie 100–600 m above existing valleys and evidently are remnants of mature preglacial valleys. The woody formation contains buried soil profiles, peat layers, and peaty beaver-pond deposits with gnawed wood.

Glaciation. Glacial land forms, till, and erratics are scattered throughout the archipelago, although such features are relatively scarce in the low-relief islands in the NW. In the higher eastern islands, on the other hand, there are abundant relatively fresh and striking glacial features, such as grooved and rounded rock surfaces, end moraines, esker complexes, and, in the mountains, glacial valleys. Two main ice sheets existed during the last glaciation: the *Laurentide* ice sheet, flowing northward from the mainland; and a coalescing series of glacier complexes that were especially well developed in the highlands and mountains of the eastern part of the archipelago. The Laurentide ice affected especially the lowlands of the southern part of the archipelago, where the landscape is characterized by markedly linear land forms, such as drumlins and eskers.

Stratified terrestrial sediments with plant and woody layers underlie or occur between tills or marine sediments at several localities in the Arctic islands; these probable interglacial deposits are dated by radiocarbon at greater than 30,000 years. Certain organic beds contain pollen of grasses, sedges, and other herbaceous plants; some peaty beds contain wood from small trees and some beaver-gnawed sticks. Evi-

FIGURE 4. Ice fields and glaciers characterize the mountain belt of northern Ellesmere Island. (T-409L-196, oblique air photo: Canada Dept. Energy, Mines and Resources.)

dently the interglacial climate was somewhat more favorable than at present.

Shell-bearing till and stratified deposits at some localities and scattered shell debris at others at elevations well above the postglacial marine limit are evidence of marine events prior to the last major glaciation. Finite isotopic dates on this material range from 20,000 to 40,000 years and are considered minimal. Glacial ice is believed to have transported and dispersed older marine material so that shells of great age may now be found at elevations, such as on upland surfaces, higher than any of the marine levels.

The northern edge of the last continental (Laurentide) ice sheet was at its northernmost position on the south coast of Melville Island about 11,000 to 10,000 years ago. By about 9000 years ago deglaciation was well advanced, as dated by shells from the highest beaches at many localities. The larger existing ice caps, which include the largest in Canada, may be remnants from the last glaciation (Fig. 4).

Deglaciation resulted in exposure of a widespread pattern of flow features and moraines and the formation of numerous beaches by large and small proglacial lakes and by marine incursion. The highest raised marine features range from a few meters to about 200 m or more above present sea level, and a more or less regular and general increase in height of these features can be discerned from the periphery to the central islands. Tectonic uplift has affected some islands, but the elevated marine features are, at least in part, due to isostatic depression of the archipelago by the glacial ice load; uplift during and following wastage of the ice is recorded by "flights" of beaches above the present shore. Absolute rise of sea level during deglaciation and areal variation in time of ice retreat (and hence of marine invasion) have resulted, however, in a complex pattern of emerged marine features.

ROBERT L. CHRISTIE

References

Andrews, J. T., McGhee, R., and McKenzie-Pollock, L., 1971. "Comparison of elevations of archaeological sites and calculated sea levels in Arctic Canada," *J. Arctic Inst. N. Am.,* 24(3), 210–228.

Dineley, D. L., 1971. "Arches and basins of the southern Arctic Islands of Canada," *Proc. Geol. Assoc. Lond.,* 82, 411–443.

Douglas, R. J. W., Norris, D. K., Thorsteinsson, R., and Tozer, E. T., 1963. "Geology and petroleum potentialities of northern Canada," *Geol. Surv. Can. Pap.* 63-31, 28p.

Fortier, Y. O., 1963. "Geology of the north-central part of the Arctic Archipelago, Northwest Territories (Operation Franklin)," *Geol. Surv. Can. Mem.* 320, 671p.

____, McNair, A. H., and Thorsteinsson, R., 1954. "Geology and petroleum possibilities in Canadian Arctic Islands," *Bull. Am. Assoc. Petrol. Geologists,* 38(10), 2075–2109.

Jeletzky, J. A., 1970. *Marine Cretaceous Biotic Provinces and Paleogeography of Western and Arctic Canada.* Ottawa: Dept. Energy, Mines and Resources, 92p.

Pitcher, M. G., ed., 1973. *Arctic Geology.* Tulsa: Am. Assoc. Petrol. Geologists, Mem. 19, 747p.

Prest, V. K., 1970. "Quaternary geology of Canada," *Geology and Economic Minerals of Canada* (Econ. Geol. Rept. 1, 5th ed.). Ottawa: Geol. Surv. Can., 677–764.

Raasch, G. O., ed., 1961. *Geology of the Arctic: Proceedings of the First International Symposium on Arctic Geology, Calgary, Canada, 1960,* vol. 1. Toronto: Univ. Toronto Press, 339–595.

Sproule, J. C., and Associates Ltd., 1972. *Generalized Geological Map of Northern Canada and Adjacent Areas.* Calgary: Intern. Geol. Eng. Consultants.

Stockwell, C. H., 1961. "Structural provinces, orogenies, and time-classification of rocks of the Canadian Precambrian Shield," *Geol. Surv. Can. Pap.* 61-17, 108–118.

____, 1964. "Fourth report on structural provinces, orogenies, and time-classification of rocks of the Canadian Precambrian Shield," *Geol. Surv. Can. Pap.* 64-17, 1–21.

Taylor, A., 1955. "Geographical discovery and exploration in the Queen Elizabeth Island," *Geogr. Branch, Dept. Mines and Tech. Surv. Can. Mem.* 3, 172p.

____, 1956. *Physical Geography of the Queen Elizabeth Islands, Canada,* 12 vols. New York: Am. Geogr. Soc.

Thorsteinsson, R., 1974. "Carboniferous and Permian stratigraphy of Axel Heiberg Island and western Ellesmere Island, Canadian Arctic Archipelago," *Geol. Surv. Can., Bull.* 224, 115p.

____, and Tozer, E. T., 1970. "Geology of the Arctic Archipelago," *Geology and Economic Minerals of Canada* (Econ. Geol. Rept. 1, 5th ed.). Ottawa: Geol. Surv. Can., 548–590.

Cross-references: *Canada; Canada–Canadian Shield, Cordilleran Region, Northern Great Plains Province, Rocky Mountains and Eastern Cordilleran Region; Greenland; North America; United States–Alaska.*

CANADA—ATLANTIC PROVINCES

The Atlantic Provinces of Canada consist of four political units: Newfoundland, Nova Scotia, Prince Edward Island, and New Brunswick. The last three are traditionally known as the "Maritime Provinces." The four comprise a more or less distinct morphotectonic division— the Canadian sector of the Appalachian orogenic belt, although a narrow part of the latter also extends along the southeastern borders of the province of Quebec.

Sources of geological information, apart from the Geological Survey of Canada (Ottawa), include the provincial agencies: Department of Mines and Energy, located at St. John's, Newfoundland; Department of Mines, Halifax, Nova Scotia; Department of Industry and Commerce, Charlottetown, Prince Edward Island; and Department of Natural Resources, Fredericton, New Brunswick. The principal centers of geological education in the region are at Memorial University of Newfoundland (St. John's), Dalhousie University (Halifax, Nova Scotia), and the University of New Brunswick (Fredericton). There are also smaller universities with geology departments (St. Francis Xavier, Antigonish, Nova Scotia; St. Mary's University, Halifax, Nova Scotia; Acadia University, Wolfville, Nova Scotia; and Mt. Allison University, Sackville, New Brunswick). Further, there is the important Bedford Institute of Oceanography, in Dartmouth, Nova Scotia. Dalhousie University also has an Institute of Oceanography. Beiological oceanographic centers are found in St. John's, Newfoundland, and St. Andrews, New Brunswick.

Physiography

The Canadian Appalachian region (Fig. 1) is subdivided into a rather large number of distinctive units, in part depressed below the waters of the Gulf of St. Lawrence, Cabot Strait, and the seaward continental shelves of Nova Scotia and Newfoundland (including the Grand Banks). The region is embellished by a complex geological ancestry but unified by a common history since late Paleozoic time. The landscape is dominated by a well-developed peneplain, essentially of Cretaceous age; it is partly tilted down to the SE (and overlapped offshore by Cretaceous and Tertiary sediments).

Newfoundland consists of the *Newfoundland Highlands*, peneplaned remnants of Precambrian and Paleozoic rocks ranging from 200 to 800 m (in the NW). To the SE the same type of rock complex is tilted down to form the *Atlantic Uplands* (of Newfoundland), 200=300 m. The *Central Lowland* is rolling country mainly less than 200 m, and also modeled in Paleozoic rocks, but largely cloaked in Pleistocene till and related deposits.

Nova Scotia also includes comparable features, the *Nova Scotia Highlands* (including Cobequid Mountains, Antigonish Highlands, Cape Breton Highlands), reaching 300–500 m, and deeply dissected near the margins. The *Atlantic Uplands* (of Nova Scotia) occur in the SE, where the peneplain dips seaward from a maximum elevation of 200 m. There is also a distinctive upland adjacent to the Bay of Fundy, where *North Mountain* is a flat-topped ridge of Triassic basalt (at 170 m). Important areas of minor relief include the *Cumberland Lowland,* the *Minas Lowland,* and the *Annap-*

FIGURE 1. Geological elements, Appalachian region of Atlantic Provinces and adjacent Quebec (adapted from Poole *et al.,* 1970).

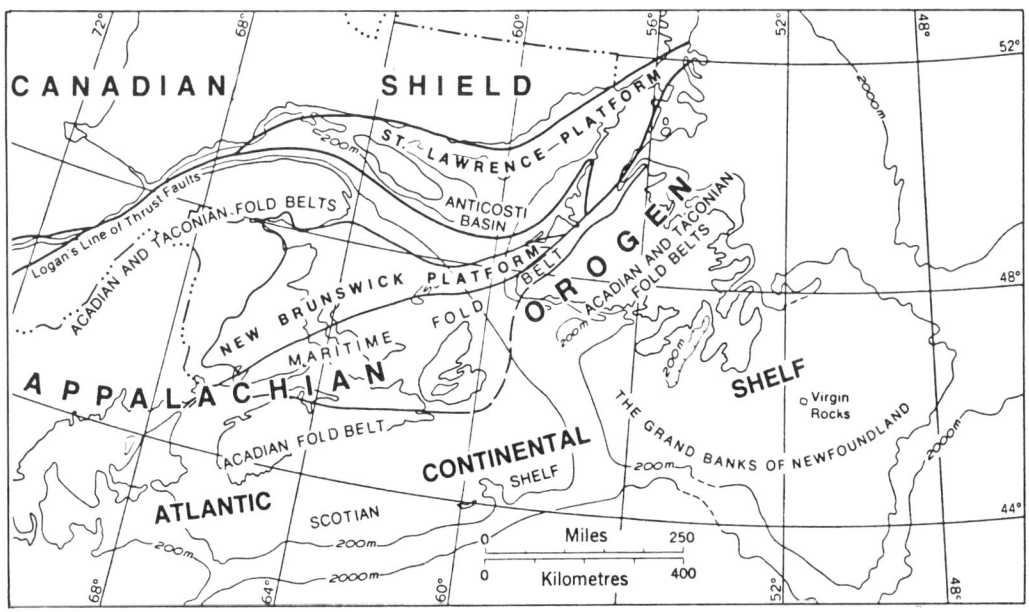

olis Valley, all carved in soft late Paleozoic or Triassic sediments.

New Brunswick also slopes SE with a high region, the *New Brunswick Highlands,* U-shaped, open to the NE, and ranging from 700 m down to sea level at the Bay of Fundy. The *Chaleur Uplands* straddle the border with Quebec, rise to 350 m, and are developed in folded Paleozoic rocks with deeply entrenched valleys. The *Maritime Plain* extends around the coast from Chaleur Bay into Nova Scotia and includes the whole of Prince Edward Island and the Magdalen Islands. It is controlled by mainly soft, flat-lying rocks and does not exceed 200 m in elevation (Picton Group).

Many of the valleys in this region were cut in preglacial times as were numbers of coastal terraces. The region was entirely covered by continental ice during Quaternary glaciations and extensively mantled by till and outwash deposits of the last retreat. Particularly in New Brunswick there are extensive areas of soft preglacial soils and regolith which show extremely little evidence of glacial scour or stripping. Interglacial deposits (?Sangamon) have been identified in Nova Scotia, notably on Cape Breton Island.

The classic Wisconsin was marked by the growth of a distinctive Appalachian glacier complex (Prest, 1970; McDonald, 1971). Its retreat was much influenced by flooding of the Bay of Fundy and the Gulf of St. Lawrence in late Wisconsinan time, isolating individual ice centers from about 13,000 B.P. onward. Radiocarbon dating of the retreat stages shows an important interplay with rising sea level. The marine limit, now isostatically uplifted, reaches over 80 m in places. A remarkable feature of Appalachian glaciation was the establishment of at least one local maritime ice center, on what is now continental shelf and which may then have represented an isostatically responsive "marginal bulge." Newfoundland developed its own ice sheet during Wisconsinan time; only the northernmost tip of the island was overridden by Laurentide ice. Newfoundland also developed several local ice centers during deglaciation.

Geosynclinal Units. The Atlantic Provinces occupy the NE end of the Appalachian Geosyncline and the bordering parts of the St. Lawrence Platform. The Appalachian Geosyncline in Canada extends about 1600 km and up to 600 km wide, not counting the continental shelves on its SE and NE sides, which would add another 40% to the area (in all about

FIGURE 2. Geological map of the Canadian Appalachian region.

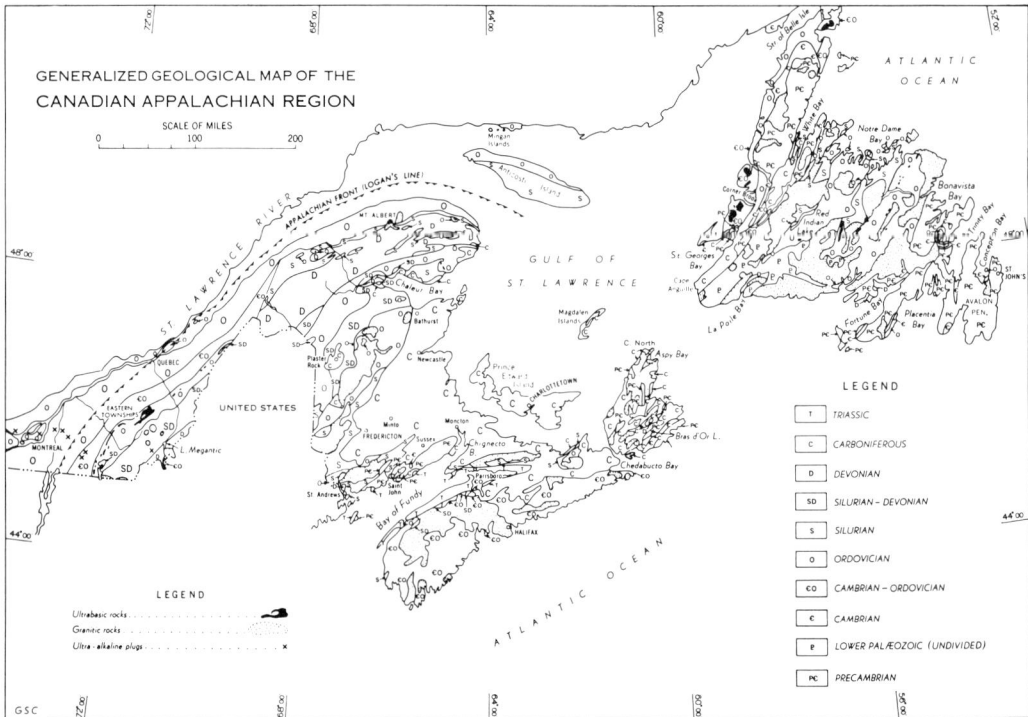

285,000 km², or 110,000 sq mi). The term "Appalachian Geosyncline" is retained as a general term for the belt of folded and unfolded Paleozoic and some Precambrian rocks lying southeast of the St. Lawrence Platform. Several second-order geosynclines are identified and their history may be traced from birth to conversion into geanticlines and/or platforms.

Precambrian rocks within the southeastern part of the Appalachian Geosyncline constituted the *Trinity Trough* (formerly "*Avalon Geosyncline*"), which was ancestral to the Appalachian Geosyncline and which prevailed until latest Hadrynian (latest Proterozoic) and then behaved as a platform throughout Phanerozoic time. The flanking, mainly early Paleozoic geosynclines, named the *Notre Dame Trough* (formerly "*Acadian Geosyncline*") and the *Meguma Trough,* developed up to the Devonian. The Notre Dame Trough (but not the Meguma) underwent three phases of the Ordovician *Taconian Orogeny,* during which ultramafic bodies were emplaced, some rocks were folded and some metamorphosed, a few granites were emplaced, geanticlines were developed, and two klippen moved from the geosyncline onto the St. Lawrence Platform. In mid-Devonian times both the Notre Dame and Meguma sequences were tectonized and converted to an immense platform during the paroxysmal *Acadian Orogeny.*

The *Fundy Geosyncline* (epieugeosyncline) of Devonian to Permian strata developed upon a zone of disrupted Acadian orogen and became filled almost exclusively with molasse-type sediments during the *Maritime Disturbance*. The "hard" thrust faults making *Logan's Line* probably developed in late Devonain; and perhaps the northern mainland part of the Notre Dame Trough is an immense allochthon. Upper Triassic continental sediments were deposited in a fault basin that formed along the south side of part of the earlier Fundy Geosyncline. Mesozoic and Cenozoic were mainly times of erosion; the resulting detritus lie beneath the waters of the continental shelves.

This report is largely a condensation of a synthesis based upon the data and interpretations of many geologists (Poole, 1967; Poole *et al.,* 1970; Williams, 1969; Rodgers, 1970; Rose *et al.,* 1970). In the last few decades, Alcock (1947) and Weeks (1957) described the geology of the Appalachian Geosyncline, mainly from a stratigraphic point of view. Later, Neale *et al.* (1961) published generalized geological and tectonic maps with an accompanying short account of the tectonic history (Fig. 2). New interpretations in the light of plate tectonics have recently been proposed by Bird and Dewey (1970) and by Schenk (1971).

Precambrian Stratigraphy

Helikian and Older. Rocks as old as Archean form the Grenville structural province, which unconformably underlies the flat-lying Paleozoic cover of the St. Lawrence Platform. The schist, gneiss, marble, quartzite, granite, and anorthosite, typical of the Grenville, also appear as inliers in the platform in West Newfoundland (Clifford, 1969). One large anorthosite body in southwestern Newfoundland, presumed to be Precambrian, lies within a Paleozoic metamorphic terrane along the northwestern margin of the geosyncline. Faults and folds affect the Grenville rocks and their Lower Paleozoic cover in West Newfoundland, and two large klippen of Ordovician and older rocks derived from the geosyncline rest upon the cover rocks of approximately the same age.

Hadrynian. Known and probable Hadrynian rocks occur in two different parts of the Appalachian Geosyncline: in the Trinity Trough, and along the northwestern margin of the Notre Dame Trough. No Hadrynian rocks are known on the St. Lawrence Platform in southern Quebec. Some red beds of the Bradore Formation underlying Lower Cambrian carbonate and shale on the platform in West Newfoundland may be Hadrynian.

Trinity Trough (formerly identified as Avalon Geosyncline). Rocks of this suite outcrop on the Avalon Peninsula and adjacent parts of southeastern Newfoundland, on Cape Breton Island, and in southern New Brunswick. The name is derived from the Trinity Bay in the Avalon Peninsula, where the rocks are best exposed. By late Hadrynian, the Trinity Trough had evolved to a platform, the Avalon Platform, upon which were deposited platform-type sediments of latest Hadrynian(?) and Paleozoic age. The early Hadrynian rocks of Cape Breton Island and southern New Brunswick are very similar to each other but differ markedly from the southeastern Newfoundland rocks.

In SE Newfoundland, the lower assemblage consists of weakly metamorphosed, thick volcanics and shallow- and deep-water clastic sediments (Harbour Main, Love Cove) of presumed early Hadrynian age. Some of the volcanics are subaerial (McCartney, 1967). The Harbour Main in central Avalon Peninsula became folded and intruded by the Holyrood granite batholith during the Avalonian Orogeny (Lilly, 1966; McCartney *et al.,* 1966). The upper assemblage grades from fine marine clastic sediments (Conception, Connecting Point) upward into nonmarine coarse arkosic sediments, red beds, rhyolites, and basalts (Musgravetown, Hodgwater, Cabot). These are typical terrestrial deposits showing evidence of rapid lateral changes in facies and thickness, and they probably represent deposition from fault-bounded blocks. The coarse clastics are capped by a blanket of latest Hadrynian(?) quartz sand (Random and equivalents), which represents the base of a platformal succession of Cambrian shale, siltstone, and minor limestone passing upward into Lower Ordovician shale and sandstone with Clinton-type hematite beds.

In southeastern Newfoundland there is what appears to be a poorly developed geosynclinal cycle: from volcanics and clastic sediments in a shallow- and deep-water and subaerial environment, to local deformation and granite intrusion of the Avalonian Orogeny, to marine deposition grading upward into continental deposition and volcanism, and finally to stable platform-type deposition during latest Hadrynian(?) and Early Paleozoic. On the other hand, the rocks may represent the growth of a volcanic island complex followed by erosion and quiescence (Hughes and Brückner, 1971).

In contrast, the Trinity Trough rocks of Cape Breton Island and southern New Brunswick are different from those of southeastern Newfoundland, although the Cambro-Ordovician platform-type formations and fauna are strikingly similar (Hutchinson, 1952, 1962). In Cape Breton Island and southern New Brunswick, the lower assemblage consists of quartzite, argillite, and limestone, now largely metamorphosed to schist, quartzite, and marble (George River, Green Head). The age of these rocks is much in doubt, possibly Hadrynian but probably older than the lower assemblage in SE Newfoundland. Some of the Green Head limestones are stromatolitic.

The upper assemblage on Cape Breton Island and in southern New Brunswick comprises thick, dominantly acid volcanic breccias, flows, and sediments (Fourchu, Coldbrook). The volcanics are overlain by red beds which grade upward into late Hadrynian(?) quartzite at the base of the Cambro-Ordovician platform sequence.

The upper assemblage of the three areas has been correlated (Weeks, 1957), and a recent Rb-Sr study supported the correlation (Fairbairn *et al.*, 1966), although the dates obtained are younger than geologically anticipated. Correlation of the groups in the lower assemblage is tenuous: they all may be Hadrynian.

No Precambrian rocks are exposed between the Grenville craton on the NW and the Trinity Trough on the SE except along the northwestern margin of the Notre Dame Trough, but both troughs probably rest on a basement of Grenville or older.

FIGURE 3. Latest Hadrynian, Cambrian and Lower Ordovician Deposits, Atlantic Provinces and adjacent Quebec (Poole, 1967).

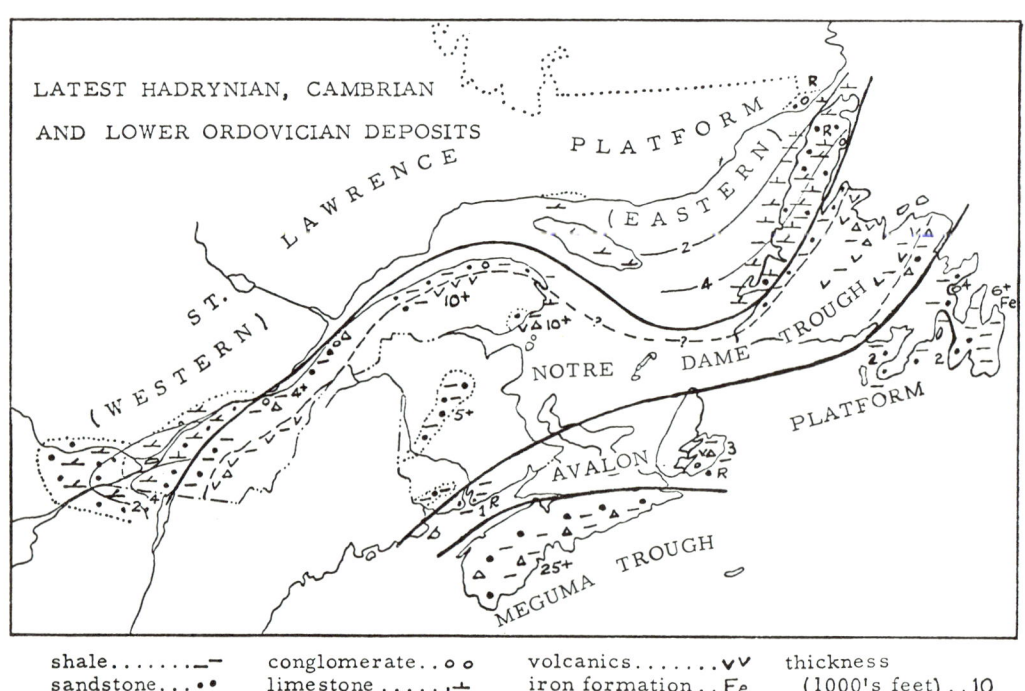

Notre Dame Trough (formerly referred to as Acadian Geosyncline). Late Hadrynian ("Eocambrian") clastic sediments and minor volcanics occur in the Notre Dame Trough only aong its northwest margin in Eastern Townships of southern Quebec (lower parts of Oak Hill Group and Sutton Schists) and possibly in western Newfoundland (lower parts of Fleur de Lys Group and southwestern equivalents, Mount Musgrave Formation and/or Grand Lake Brook Group). To the SE the sediments probably graded into volcanic-bearing geosynclinal assemblages which are covered by younger rocks; northwestward, they probably wedged out completely. In West Newfoundland, they were overlapped by the Lower Cambrian(?) coarse red beds of the Bradore Formation. These Hadrynian strata were regionally metamorphosed to schist, quartzite, greenstone, gneiss, and amphibolite probably during Ordovician and possibly in places in earlier times. In Newfoundland, the metamorphic rocks were deformed and intruded by granites in the Devonian as well.

Pre-Hadrynian(?) basement blocks in the southeastern parts of the Maritime region were probably parts of an African shelf (Schenk, 1971). Owing to repeated opening and closing of the proto-Atlantic the borders of NW Africa progressively lost increments to Nova Scotia and Newfoundland.

Cambrian and Early Ordovician

Cambrian and Lower Ordovician rocks occur in widely separated belts of the Notre Dame and Meguma troughs and on the St. Lawrence and Avalon platforms (Fig. 3). The Cambrian strata occur generally conformably above Hadrynian strata and grade upward into Lower Ordovician rocks. No orogenies interrupted the sequences, but the early phase of the Taconian Orogeny brought deposition to a close on the Avalon and St. Lawrence platforms by uplift.

Notre Dame Trough. Along the northwest edge of the Notre Dame Trough in southern Quebec, a two-zone belt developed in which the outer zone, nearest the craton, received feldspar-bearing sandstone, shale, and limestone-conglomerate of the Quebec Group while the inner zone, toward the southeast, developed as a volcanic and graywackeshale zone ("Caldwell," Shickshock). Rocks of the Quebec Group have many similarities to Newfoundland strata preserved in the lower two-thirds of the klippen now resting on the St. Lawrence Platform (Stevens, 1970; Kindle and Whittington, 1958). In the axial region of the Notre Dame Trough in New Brunswick, orthoquartzite and shale of possible Cambrian to early Middle Ordovician age, the oldest rocks exposed in the Miramichi Geanticline, represent a para-platformal environment. The late Middle Ordovician Tetagouche volcanic and graywacke belt was developed upon these quartzites and shales. Lower Ordovician (and older?) black shale and orthoquartzite (Cookson Formation) were deposited along the southeast margin of the Notre Dame Trough, adjacent to the Avalon Platform. In western Newfoundland, Lower Ordovician basaltic pillow lavas, graywacke, graphitic slate, and chert of the inner zone lie unconformably(?) on the Fleur de Lys paraschist and gneiss of the outer zone which bound them on the west. Metamorphism of the older rocks probably took place in latest Cambrian or earliest Ordovician. On the east side of the Notre Dame Trough, the graywacke-slate assemblage of the lower Gander Lake Group of presumed Early Ordovician and (?)Cambrian age lies next to the Hadrynian Love Cove Group apparently without upper Hadrynian strata intervening.

Meguma Trough. An estimated 9000 m of quartzose graywacke and dark slate of the Cambrian and Lower Ordovician Meguma Group were deposited in the Meguma Trough of Nova Scotia, southeast of the Avalon Platform. The lower unit of the group, the Goldenville Formation (probably late Cambrian), consists of 6000 m or more of graywacke and slate. These sediments contain sole markings indicating NE- to E-flowing turbidity currents aligned generally parallel to the Devonian fold trends. The upper unit, the Halifax Formation, consists of dark slate and siltstone 500–4000 m thick which contain lowermost Ordovician (Tremadocian) graptolites. The Meguma Group, formerly called the "Gold-Bearing Series," contains gold-quartz veins mined many years ago. It seems probable that the sediment source lay to the SE, either beneath the Scotian Shelf or farther SE (Schenk, 1971).

Avalon Platform. Dark gray shale and minor limestone, 1000 m or more thick, were deposited on the Avalon Platform during Cambrian and Early Ordovician. The sediments contain abundant brachiopods, trilobites, and early Ordovician graptolites. The lithostratigraphy and paleontology of the three outcrop areas of the platform—southeastern Newfoundland, Cape Breton Island, and southern New Brunswick—are remarkably similar. The Cambrian shale is underlain by a thin quartzite of probable late Hadrynian age which was deposited from a sea transgressing a tectonically quiet area. The quartzite is an essentially transitional sequence. A Middle Cambrian graywacke-volcanic sequence was deposited locally on Cape Breton Island (Weeks, 1954). In Newfoundland, Lower Ordovician beds at the top of the sequence contain sandy and oolitic hematite beds which were mined at Wabana for many decades. The distribution of the formations indicates nearby stable land areas. In Newfoundland, from the basinal axis near Trinity Bay, succeeding younger formations transgressed eastward and westward in the early part of the Cambrian, whereas upper Cambrian and Ordovician sediments probably covered all the platform and were continuous with bordering deposits of the Notre Dame Trough to the west (Hutchinson, 1962). The stratigraphic record on Cape Breton Island indicates transgression northwestward onto a highland.

St. Lawrence Platform. Cambrian and Lower Ordovician quartz sandstone, carbonate, and shale are the oldest strata covering the Grenville crystalline craton. In West Newfoundland (Whittington and Kindle, 1969; Swett and Smit, 1972; Strong and Williams, 1972), the Bradore Formation, comprising Lower Cambrian(?) red arkosic conglomerates and sandstone, rests directly upon basement; these are succeeded by an incomplete sequence of Cambrian shales and limestones and thence by the distinctive and widespread Lower Ordovician St. George dolomite. Most of these strata thicken to the SE and near the edge of the platform pass into sandy rocks resembling the Fleur de Lys Group and Mount Musgrave Formation along the adjoining western margin of the Notre Dame Trough.

On the mainland north of Anticosti Island (see the article *St. Lawrence Lowlands*), the oldest exposed platform strata are Lower Ordovician dolomite resting with an intervening thin basal clastic facies upon the Grenville. Cambrian strata presumably underlie the northern Gulf of St. Lawrence, and have probably been overlapped by the dolomite. Early Ordovician on the platform represented a period of shallow dolomite-producing seas transgressing a very stable craton.

Taconian Orogeny, Early Phase. The term "Taconian Orogeny" describes the orogenic events that affected the Appalachian Geosyncline and bordering St. Lawrence Platform during Ordovician time. "Taconian" is preferred to "Taconic" in order to extend the term to orogenies beyond the time limits and style of deformation of the type area in the Taconic Mountains of eastern New York and adjacent states. Tectonic events known mainly from the St. Lawrence Platform and designated as post-Canadian (post-Early Ordovician) are considered here the early phase of the Taconian Orogeny; the middle and late phases occurred during Middle and Late Ordovician, respectively. Ultramafic and associated mafic bodies (peridotite and dunite, variously serpentinized, and gabbro) and minor granitic bodies were emplaced in volcanic terranes of the Notre Dame Trough mainly along the northwest margin and to lesser extent along the southeast margin during Early to Middle Ordovician. They contain asbestos deposits in Eastern Townships and northern Newfoundland. In the Meguma Trough pre-Middle Ordovician volcanics are not known. The early phase of the Taconian Orogeny produced epeirogenic movements on the St. Lawrence and Avalon platforms. On the St. Lawrence Platform, epeirogenic uplifts and southward withdrawal of the seas, led to erosion, which stripped back the cover sediments extensively on the western platform, whereas the eastern platform recorded only mild effects with little erosion. Deposition on the Avalon Platform came to an end with black shales in southern New Brunswick and on Cape Breton Island, and hematitic sediments in southeastern Newfoundland. Middle and Upper Ordovician strata are not known to occur on the platform (with the exception of the probable Ordovician Browns Mountain Group in northern Nova Scotia).

Middle and Late Ordovician

Middle and Late Ordovician times were tectonically active within the Appalachian Geosyncline with the buildup of deep-water volcanic and graywacke-rich troughs, emplacement of ultramafic-mafic-granitic bodies in some parts and granites in others, rise of geanticlinal belts, emplacement of two klippen upon the St. Lawrence Platform of West Newfoundland, and extensive folding along the geosyncline's northwestern margin and probably elsewhere.

Early Middle Ordovician Deposits. Within the Notre Dame Trough, older geosynclinal graywacke-volcanic deposition probably continued from Early Ordovician through to early Middle Ordovician. The volcanic rocks contain the copper sulfide deposits of Eastern Townships of Quebec and northern Newfoundland. Similarly, para-platformal deposition of quartz sand and shale continued in central New Brunswick.

In the Meguma Trough, deposition of gray shale in the early Ordovician changed sometime during the Ordovician or Silurian to volcanics, graywacke quartz sand, and shale of the White Rock Formation of uncertain age. On the St. Lawrence Platform, early Middle Ordovician limestone and dark shales were deposited disconformably upon Lower Ordovician dolomites as the seas reoccupied the platform. The Avalon Platform is not known thus far to have strata of early Middle Ordovician age.

FIGURE 4. Geological sketch map of Newfoundland (Kay, 1972).

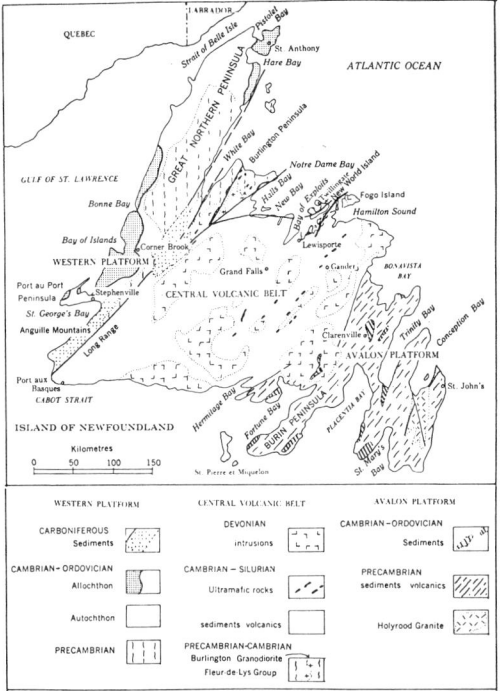

Taconian Orogeny, Middle Phase. Orogenic events of mid-Middle Ordovician age, the middle phase of the Taconian, include uplift of the ultramafic-mafic-granitic intrusions, the emplacement of geosynclinal sequences upon the St. Lawrence Platform, and probably also the folding of the mainland part of the Avalon Platform and intrusion of mixed granitic bodies.

The northwest margin of the Notre Dame Trough, during or immediately after the emplacement of ultramafic-mafic bodies, became uplifted. Some belts (Sutton, Bennett, Shickshock, Mount Musgrave, Fleur de Lys) within the marginal zone were probably regionally metamorphosed during early or mid-Ordovician. Middle Ordovician strata in the vicinity of the belt contain some detritus probably originating from the fold belt.

One of the more important tectonic events of the middle Taconian orogenic phase was the westward sliding or thrusting of what are now two klippen in West Newfoundland, one on the northern end of the Great Northern Peninsula and the other in southwestern Newfoundland. The klippen hypothesis, given prominence by Rodgers and Neale (1963), was challenged by Lilly (1964), but is now generally accepted (Figs. 4 and 5). The klippen consist of Cambro-Ordovician, and (?)Eocambrian, clastic sediments as well as some thin carbonate beds and breccias, plus grayeacke, volcanics, and ultramafic-mafic-granitic bodies in the upper parts (Brückner, 1966). The clastic rocks were deposited along the margin of the Notre Dame Trough, where platform-type sediments thicken and grade into a clastic assemblage off the edge of the St. Lawrence Platform. The volcanics and plutonic rocks resemble those now occurring not far southeast on the inner side of the marginal zone. The two klippen were probably at one time connected as one large, perhaps irregular, sheet which became separated by erosion resulting from Late Ordovician uplift of the Grenville basement to form the present-day inlier.

The age of the westward sliding of the southwestern klippe is well established. The klippe moved over platformal lowermost Middle Ordovician (Whiterock-Marmor) Table Head limestone and black shale, and the west end became unconformably overlain by autochthonous mid-Middle Ordovician (Porterfield and Wilderness) Long Point limestone, sandstone and shale. The classic Taconic klippe in New York was emplaced slightly later in late Middle Ordovician (Barneveld) time. Thus, in mid-Middle Ordovician the volcanic belts containing intruded ultramafic-mafic-granitic rocks were folded and uplifted. The allochthon in West Newfoundland moved westward from the bordering Notre Dame Trough onto the St. Lawrence Platform and came to rest before the early stage of late Middle Ordovician time.

Elsewhere in the Notre Dame Trough, tectonic events of the middle Taconian phase are less spectacular. In central New Brunswick, deposition of para-platformal quartz sand and shale gave way abruptly to eugeosynclinal deposition of the Tetagouche Group. In central Newfoundland, deposition of eugeosynclinal strata continued on from Early Ordovician time. The southeast margin of the Notre Dame Trough

FIGURE 5. Faults and fault blocks in north-central Newfoundland (Kay, 1972).

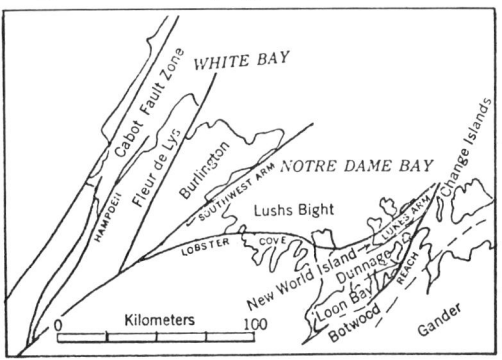

behaved differently from that of the northwest margin. In southern New Brunswick, Lower Ordovician black shale and quartzite (Cookson Formation) became folded, uplifted, and eroded. Ultramafic-mafic bodies in east-central Newfoundland (Gander Lake belt) were probably emplaced at this time. It is postulated that during mid-Middle Ordovician the Avalon Platform in New Brunswick and on Cape Breton Island was folded, intruded by mixed granitic rocks, then uplifted and exposed to erosion. The St. Lawrence Platform near the Notre Dame Trough was slightly uplifted, the seas withdrew, and then soon thereafter readvanced.

Late Middle Ordovician Deposits. During late Middle Ordovician (Wilderness and Barneveld) time, deposition of geosynclinal sediments and volcanic rocks continued in the Notre Dame and possibly in the Meguma Trough. The Avalon Platform lacks deposits of this age and presumably remained a positive area. The seas readvanced onto the St. Lawrence Platform for the most widespread transgression in the history of the platform. Late Middle Ordovician deposition was followed by the late phase of the Taconian Orogeny.

Late Middle Ordovician deposits along the northwest margin of the Notre Dame Trough are dominantly thick graywacke and dark slate; volcanics are rare. The rocks form two belts, one (Quebec Group) on the northern side of the volcanic belt (Cloridorme Formation) and one on the south (Beauceville, Magog). Detritus from ultramafic rocks and volcanic rocks have been recognized in places in both belts. In central New Brunswick, the Tetagouche volcanic belt was initiated upon an earlier para-platform. There, a very thick, rapidly varying assemblage of volcanics, graywacke, black carbonaceous shale and chert and minor iron formation, in part manganiferous, was formed almost completely devoid of carbonate sediments. The rocks contain the rich base metal sulfide deposits of the Bathurst camp. At the same time, a belt of clastic deposition (shale, graywacke, quartz sandstone), lying between the Tetagouche belt on the S and the Quebec belt on the N, changed over to deposition of ribboned argillaceous limestone and shale. This, the Matapedia belt, was the site of dominantly carbonate deposition from late Middle Ordovician through to earliest Silurian. Late Middle Ordovician deposits, so widespread in the northern half of the mainland Notre Dame Trough, have not been identified in the S half; middle Taconian deformation and intrusion apparently left this area one of highlands undergoing erosion. Deposits similar to the Tetagouche Group were formed in central Newfoundland between the older rocks on the margins of the geosyncline.

In the Meguma Trough, volcanics and sediments of the post-Lower Ordovician, pre-Upper Silurian White Rock Formation may have been deposited at this time. On the St. Lawrence Platform, limestone deposition was widespread. Dark colored Trenton and Utica shales deposited in the Ottawa-Quebec basin graded westward into limestones, some of which have produced petroleum. In southwestern Newfoundland, limestone, sandstone, and shale of the Long Point Formation of Middle Ordovician (Porterfield and Wilderness) age were deposited unconformably upon the klippe and represent a resumption of platform-type deposition following interruption during emplacement of the klippe.

Upper Ordovician Deposits and Taconian Orogeny, Late Phase. Upper Ordovician deposits are rare in the Appalachian Geosyncline and abundant on the St. Lawrence Platform, a pattern resulting from the late phase of the Taconian Orogeny which affected much of the geosyncline.

Limestone-shale deposition continued in the Matapedia belt of the Notre Dame Trough whereas the northwest margin was folded, thrust toward the St. Lawrence Platform, and uplifted as the Quebec Geanticline. The Tetagouche belt was intruded by granite, folded, and uplifted as the Miramichi Geanticline. Deposition in the axial region of the geosyncline in Newfoundland continued beyond Middle Ordovician but coarser clastics were deposited and the seas became more shallow than previously. Deposition probably continued in the Meguma Trough, while the Avalon Platform probably remained a positive area. Detritus shed from the geanticlinal belt of the northwest margin of the Notre Dame Trough spread westward across the St. Lawrence Platform, displacing limestone deposition as far west as Lake Superior.

In the Notre Dame Trough of Newfoundland, the northwest margin was probably uplifted and eroded before being covered by Silurian volcanics. Farther to the SE in the axial region of the geosyncline, deposition of graywacke and conglomerate appears to have continued through late Ordovician. The southeast margin of the trough may have been uplifted.

According to plate-tectonic analysis (Bird and Dewey, 1970), Africa and North America collided in the Devonian and the proto-Atantic remained closed until the present-day Atlantic began to open in Triassic and Jurassic. This collision brought representatives of North Africa's Upper Ordovician to Nova Scotia (Schenk, 1971).

Silurian and Early Devonian

Deposition of Silurian and Lower Devonian sediments and volcanics was widespread in the Appalachian Geosyncline and on the St. Lawrence Platform. The deposits were the last in the Appalachian Geosyncline before the mid-Devonian climactic Acadian Orogeny in which almost the entire area was folded, intruded by granitic batholiths, uplifted, and eroded, to be followed by deposition during Late Devonian to Permian times of mainly continental deposits.

Most Silurian and Lower Devonian rocks are easily distinguished from Cambro-Ordovician rocks. They consist of volcanics, commonly acidic in composition with plentiful tuffs and breccias, sandstone and conglomerate, shale and limestone. Some of the coarse sediments are red and apparently non-marine. The Siluro-Devonian strata were deposited in troughs and intermontane basins in shallow seas lying between late Taconian geanticlinal uplifts (Fig. 6). For short times during the late Silurian and early Devonian on the mainland, the seas withdrew from parts of the troughs, and continental deposition prevailed. Episodes of significant folding and of plutonism did not occur during this period on the mainland, but may have done so during the late Silurian and early Devonian in central Newfoundland. Fine clastic sediments were deposited on parts of the Avalon Platform. Limestone with shale was deposited on the St. Lawrence Platform, and farther west, in southern Ontario, evaporitic basins were developed.

Notre Dame Trough. Silurian and Lower Devonian strata of the mainland were deposited in the Gaspé and Fredericton troughs lying between the Quebec and Miramichi Geanticlines and the Avalon Platform. In Newfoundland, Silurian deposits and in one locality Lower Devonian strata are contained in a trough between the Burlington Geanticline and the Avalon Platform. Rocks of the Gaspé Trough in northern New Brunswick and southern Quebec consist of two belts of contrasting lithologic assemblages in each of which lithological facies vary from place to place. The northern belt consists mainly of siltstone and limestone with some quartzite and volcanics deposited in a more stable environment. Base metal sulfide deposits of the Gaspé Copper type occur in skarns developed in Lower Devonian calcareous rocks near Devonian granites. In the southern belt, Lower to Upper Silurian thin-bedded sandstone, siltstone, and shale were deposited gradationally above lowermost Silurian ribboned limestone of the Matapedia belt in western New Brunswick. Rocks of the southern belt are apparently thicker and contain abundant acidic volcanics with lesser andesite and basalt, accom-

FIGURE 6. Silurian and Lower Devonian tectonic elements, Atlantic Provinces and adjacent Quebec (Poole, 1967).

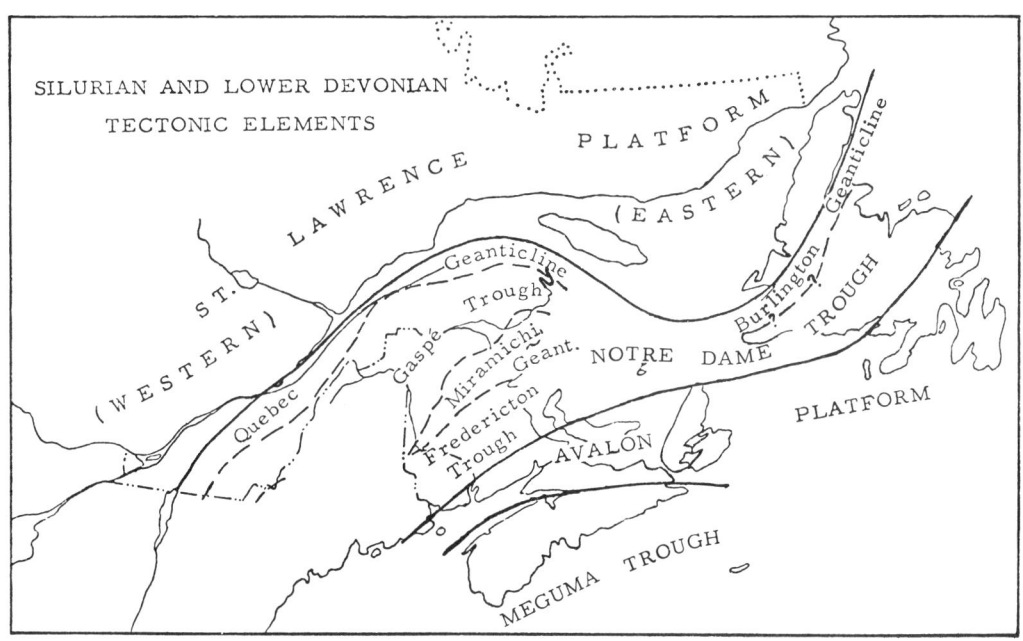

panied by generally less well differentiated formations of siltstone, quartzite, and minor limestone. Volcanism occurred here and there at times. Lower Devonian sandstone and shale commonly have a sparse marine fauna and specks of carbonaceous plant matter. Strata of the Fredericton Trough are poorly fossiliferous. A turbidite sequence along the north side of the trough consists of graywacke and slate containing middle and late Silurian graptolites; the sequence passes SE into shallow-water sediments and volcanics.

The Silurian rocks of the Notre Dame Trough in Newfoundland are not everywhere well dated, but contain a shelly fauna which indicate an early and middle Silurian age (Williams, 1967). They form four lithostratigraphic units, which in order of decreasing age are: (1) mainly graywacke, (2) mainly conglomerate, (3) volcanic rocks, and (4) red to grey micaceous sandstones. They reflect depositional environments beginning in earliest Silurian with deep water and changing to shallow water and thence to terrestrial conditions in early Wenlock. Terrestrial plant-bearing clastic rocks of early Devonian age occur in southwestern Newfoundland isolated by schists and gneisses into which they grade.

Meguma Trough. Deposition of volcanic rocks, shale, and quartz sand of the poorly dated White Rock Formation probably continued throughout the early and middle Silurian. During late Silurian, similar siltstone, shale, and quartzite of the Kentville Formation was deposited and, locally, a volcanic-pebble conglomerate. During Early Devonian, the seas became shallow and some shelly argillaceous limestone and hematitic iron formation were formed. These were the last to be deposited in the Meguma Trough.

Avalon Platform. Silurian and lower Devonian sediments were deposited on the Avalon Platform in northern mainland Nova Scotia and on Cape Breton Island. Strata of the Arisaig Group in northern mainland Nova Scotia consist of 1000 m or more of siltstone, shale, and sandstone with a shelly fauna ranging from earliest Silurian to earliest Devonian. Younger, Lower Devonian green and red sandy beds (Knoydart) conformably overlie the Arisaig strata and together with red beds in the latest Silurian of the Arisaig Group represent terrestrial deposition heralding the coming Acadian Orogeny. On eastern Cape Breton Island, a presumably Silurian or Lower Devonian quartzite with feldspathic sandstone and volcanic-pebble conglomerate (Middle River) rests unconformably upon Cambrian rocks (Weeks, 1954).

St. Lawrence Platform. Silurian sediments are lacking in Quebec Basin, but probably at one time covered this area. To the W in Ontario, they are dominantly dolomites, in which reef structures are common. Evaporite deposits were formed in late Silurian basins. Silurian limestone on Anticosti Island conformably overlies Ordovician. Younger Silurian strata and Devonian strata may underlie the northern Gulf of St. Lawrence south of Anticosti Island.

Upper Silurian to Lower Devonian red and green sandstone, and limestone in southwestern Newfoundland lie conformably upon Middle Ordovician sediments, above the west edge of the Ordovician klippe (Rodgers, 1965). On the east side of the platform, along the west shore of White Bay, acidic volcanics, sandstone, and granite-bearing conglomerate of Silurian age (Williams, 1967) appear to be an encroachment of Notre Dame Trough deposition upon the St. Lawrence Platform.

Middle and Late Devonian and Acadian Orogeny

Middle and late Devonian was the time of the climactic or paroxysmal orogeny in the Appalachian Geosyncline. The change from marine to nonmarine sedimentation was completed by about the end of early Devonian. Geosynclines, troughs, platforms, and geanticlines of pre-Devonian times ceased to exert controls on post-Devonian deposition; a new set of tectonic elements was born and a new style of deposition related to nearby uplifts began. Strata of middle and late Devonian age occur in only a few places in the Appalachian Geosyncline, and all are continental coarse sandstones and conglomerates with locally intercalated basic volcanic flows. On the St. Lawrence Platform, limestone and shale were deposited in southern Ontario.

The main phase of the Acadian Orogeny was characterized by folding, faulting, metamorphism, granite intrusion, uplift, and erosion. Almost the entire Appalachian Geosyncline, as far east as the edge of the continental shelf, as well as some adjacent parts of the St. Lawrence Platform, was converted to a stable craton, which thereafter was deformed mainly by faults, warps, and gentle basinal subsidence in decreasing degree of intensity throughout the late Paleozoic. The depositional record is sparse for this period.

Late Early and Middle Devonian Deposits. In the Notre Dame Trough on the mainland, marine deposition continued from Silurian to about early Devonian (Siegenian) when continental deposition of plant-bearing gray sandstones began. In Gaspé, the Lower Devonian marine calcareous deposits of the Gaspé Limestone became overlain by Lower Devonian (Emsian) feldspathic sandstones (York River) grading upward through sandstones and conglomerates (Battery Point) to probable early Middle Devonian (Eifelian) red conglomerates (Malbaie) (Mc Gerrigle, 1950). The lower sandstones contain both plants and marine shells and record the transition from marine to nonmarine environments. These clastic sediments filled a deep trough with well over 3000 m of strata in which conglomerates become more and more abundant in the younger beds. The trough rocks can be considered a synorogenic flysch deposited during the earliest stage of Acadian Orogeny, with sediments derived from the south, perhaps from the rejuvenating Miramichi Geanticline in which Devonian granites were being emplaced. Maybe related to the trough rocks are the Frontenac quartzites and volcanics in Eastern Townships of Quebec. Along Chaleur Bay on the south shore of Gaspé Peninsula, coarser clastic sediments, red sandstone and conglomerate of the Pirate Cove Formation, were deposited closer to the rising orogen.

In the Meguma Trough, deposition came to an end in early or middle early Devonian, during which fine marine clastics became mixed with hematitic formations indicative of a shallow-water, near-shore environment. On the Avalon Platform of the mainland, terrestrial red and gray sandstone and siltstone of the middle early Devonian age (Knoydart) was deposited conformably upon the marine Arisaig Group in northern mainland Nova Scotia. Probably during latest early Devonian and middle Devonian, red sandstone, conglomerate, and acidic and basic volcanics (River John Group) were deposited apparently during the early phase of the Acadian Orogeny. On Cape Breton Island, the continental, plant-bearing conglomerate, arkose and shale with minor tuff of the McAdam Lake Formation were deposited unconformably upon Cambrian strata.

Strata in the Notre Dame Trough of Newfoundland of early and/or early middle Devonian age outcrop only in the southwestern part of the Island. There, the Bay du Nord Group consists of continental plant-bearing slate with minor quartzite and conglomerate over 3000 m thick. These rocks grade laterally into schist and gneiss and are cut by Devonian granites. Their limited distribution and nonmarine character reflect the general subaerial conditions of the geosyncline during at least late Silurian and Devonian. The Avalon Platform of southeastern Newfoundland lacks strata of this age, also suggesting a period of nondeposition or erosion. Lower and middle Devonian strata on the St. Lawrence Platform occur in southern Ontario, but were eroded elsewhere.

Acadian Orogeny. The term "Acadian Orogeny" is used here to describe the orogenic event in which the Appalachian Geosyncline and parts of the bordering St. Lawrence Platform were intensely folded, faulted, metamorphosed, intruded by granite stocks and batholiths, uplifted, and subjected to erosion during the middle and early late Devonian (Fig. 7). Essentially, the term refers to the paroxysmal orogenic phase in the geosyncline, and in a narrow sense. The lack of Upper Silurian strata in the Notre Dame Trough inhibits a satisfactory analysis of the tectonic history in Newfoundland during this period, although one may assume that the island was uplifted and undergoing erosion. Isotopic dating of granites and metamorphic rocks clearly points to a paroxysmal orogenic stage in mid-Devonian at the same time as on the mainland, and similarly, the Carboniferous clastic strata can be interpreted as products of the waning stage of the Acadian Orogeny.

According to one model of plate-tectonic theory, an E-dipping subduction zone was affecting the region from Ordovician till late Devonian times, during which interval the eastern Maritimes became secured to Morocco. The Acadian Orogeny includes an early phase in

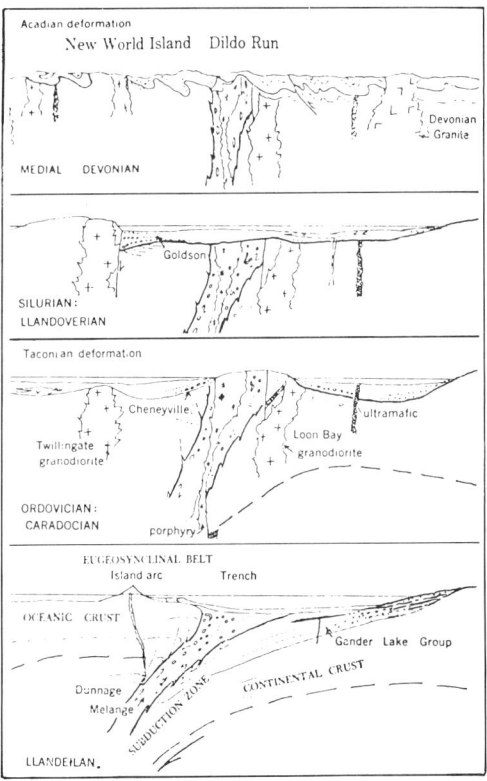

FIGURE 7. Hypothetical restored sections in central Newfoundland (Kay, 1972).

early and early middle Devonian, and possibly locally in late Silurian, and a main phase in late middle and early late Devonian. The term *"Maritime Disturbance"* refers to the waning phase of the Acadian Orogeny. It is marked by coarse and fine terrestrial deposits derived from block uplifts of the Acadian orogen and deposited unconformably upon this orogen in fault basins and downwarped basins from late Devonian to Permian. Deposition of Windsor carbonates and evaporites in middle Mississippian marked a temporary cessation of tectonic activity. The local low-grade metamorphism (slate grade) marks the only known deep-seated process associated with the disturbance.

Such terms as "Appalachian Revolution," "Appalachian Disturbance," and "Allegheny Disturbance" have been used in the Appalachian region of United States to depict tectonic events mainly involving folding and thrust faulting of the Carboniferous and older strata. However, the different character of the tectonism and the epieugeosynclinal position of the post-Acadian strata in the Appalachian region of Canada require a different term.

Appalachian Geosyncline. During the main phase of the Acadian Orogeny in late middle and early late Devonian, the Notre Dame and Meguma troughs and Avalon Platform were folded and faulted; parts were metamorphosed to schist and gneiss; much was intruded by granite batholiths; and the entire area was uplifted and eroded. All the Notre Dame Trough was folded. Some of the faults in Gaspé and Eastern Townships are steep thrusts directed northerly toward the St. Lawrence Platform. Almost everywhere in the Appalachian Geosyncline, folds have steep axial planes and axes plunge at varying degrees from horizontal to steeply NE and SW. The folds form regional arcuate patterns which curve from northeast to east to northeast in the shape of immense open right-lateral drag-folds. Such arcuate patterns appear in the Meguma Trough rocks; in the rocks extending from southern New Brunswick to northern Nova Scotia and Capte Breton Island; in the rocks of Gaspé Peninsula, of southern Newfoundland and of northern Newfoundland (Williams et al., 1969); and indeed in the trace of the boundary between the St. Lawrence Platform and the Notre Dame Trough. Major transcurrent faults associated with these arcs follow a pattern curved in a similar sense (Webb, 1969). These fold and fault patterns appear to be related to the original borders of the geosynclines and troughs. The folds and thrust faults involving Devonian and older strata of eastern Gaspé are cut at an acute angle by NW-trending faults. These faults appear to be younger than the folds and thrusts, and probably are mainly transcurrent.

The N-directed thrusts of *Logan's Line,* located along the boundary between the Appalachian Geosyncline and the St. Lawrence Platform, probably were formed, or at least were rejuvenated, in late Devonian time. Thrusting toward the platform in a late orogenic stage is characteristic of most orogenic belts. Most if not all of the movement along the zone of Logan's Line is probably late Devonian. Logan's Line curves from E to SE between Anticosti Island and Gaspé, and may pass into a transcurrent fault extending SE through Cabot Strait between Cape Breton Island and Newfoundland and beyond to the continental margin. Accordingly, movement on this fault is right-lateral, similar to that of the proposed late Devonian transcurrent fault that intersects the New Jersey coast (Drake and Woodward, 1963).

Wilson (1962) identified the *Cabot Fault* (see Fig. 5) and speculated that it was a continuation of the left-lateral Great Glen Fault of Scotland. Black (1964) provided additional paleomagnetic data to support the hypothesis that Newfoundland had rotated about $30°$ anticockwise relative to the mainland during middle or late Devonian; geological evidence of the supposed rotation is lacking.

To explain the existence of "Pacific" and "Atlantic" Cambrian faunal realms in the northwestern and southeastern parts of the Appalachian Geosyncline, respectively, Wilson (1966) proposed that these two land areas during the Precambrian and Cambrian were separated by an ocean, the *Proto-Atlantic Ocean.* Following the supposed closure of the ocean by continental drifting from latest Ordovician to Devonian times, the North American and European-African continents remained as one unit until about Triassic or Jurassic when the continents pulled apart on a new line of fracture corresponding to the present-day continental shelf edges to produce the Atlantic Ocean. During closure of the proto-Atlantic, uplifts developed and various Ordovician to Carboniferous clastic wedges spread westward onto the continental interior platform (i.e., the St. Lawrence Platform). These uplifts supposedly caused the klippen masses to slide onto the platform during Ordovician time. Closure produced continental environments of deposition typified by Carboniferous and Permian strata. Wilson's postulated line of closure in Newfoundland falls along the boundary between the Notre Dame Trough and the Avalon Platform. He suggested that the Hermitage Bay fault on the south coast may be that line. Of corroborative interest are the NW overturned folds and subhorizontal to SE-dipping cleavage in the Ordovician(?) clastic sediments in one locality adjacent to the northwest edge of the Avalon Platform.

Wilson postulated that before closure, the Notre Dame Trough was an island arc system, open to the proto-Atlantic Ocean on the SE and bordering a continent (the St. Lawrence Platform). Implicit in this model is that the Grenville crustal rocks do not underlie the entire trough (Williams, 1964), a fact yet to be established. The Meguma Trough must have its counterpart and source area in Europe or NW Africa.

Most of the strata of the Appalachian Geosyncline, which were unmetamorphosed at the onset of the Acadian Orogeny, were only weakly metamorphosed to slate-grade during the event. However, some areas of schists and gneisses were developed. Plutonic rocks were intruded in most parts of it. Most of the Devonian plutonic rocks on the mainland are granitic; biotite granodiorite and quartz monzonite are the most common types. Those of New Brunswick and Nova Scotia reach batholithic proportions, whereas those few in southern Quebec are mainly stocks. Almost all are regionally discordant bodies with sharp contacts and steep walls, and have hornfelsic thermal aureoles. In a few places, younger granites of the sequence are muscovitic and are associated with such elements as fluorine, boron, beryllium, tin, tungsten, and molybdenum. Mineralization in some skarns yields base metals, as in northern Gaspé. The Devonian granitic bodies that intruded parts of the mainland Avalon Platform are similar petrologically to those in the troughs, but of smaller dimensions. A few Devonian plutons are composed of diorite and gabbro, and one small stock in southern New Brunswick is peridotite and gabbro.

The plutonic rocks of Newfoundland are somewhat different from those of the mainland. The usual biotite granodiorite and quartz monzonite are common in western, southern and eastern parts of the Notre Dame Trough and western half of the Avalon Platform. Composite intrusions with cores of hornblende granodiorite and quartz monzonite and border zones of mafic phases, gabbro and diorite, were emplaced as stocks and batholiths in the axial parts of the trough. The composite granitic bodies are clearly younger than Silurian rocks but some K-Ar dates on the mafic marginal rocks are as old as middle Ordovician. Granites emplaced in the western half of the Avalon Platform are granodiorite and quartz monzonite extending from the adjacent geosyncline. Some contain fluorite and one in the St. Lawrence district on the southern tip of Burin Peninsula is mined for fluorite.

Folding, regional metamorphism, and granite intrusion of the main phase of the Acadian Orogeny converted the Appalachian Geosyncline into a stable orogen which was uplifted and extensively eroded in late Devonian and later times. Coarse red clastic sediments of late Devonian or more commonly of Carboniferous age overlie rocks of the orogen with spectacular angular unconformity.

St. Lawrence Platform. The deformational effects on the St. Lawrence Platform were slight in southern Quebec and moderate to strong in West Newfoundland. Strata adjacent to Logan's thrusts of presumed Devonian age were folded. The strata overlying the Grenville inlier in West Newfoundland are faulted, folded, and in southwestern Newfoundland metamorphosed. The geosynclinal klippen strata of southwestern and northern Newfoundland are folded, faulted, and metamorphosed. How much of this deformation and metamorphism can be attributed to processes before, during and after emplacement is difficult to determine.

Upper Devonian Deposits. A zone of steep NE-trending faults developed on the stable Acadian Orogen from Bay of Fundy to White Bay in northern Newfoundland. The older tectonic elements no longer influenced sedimentation; a new set of tectonic elements was initiated which prevailed from the late Devonian to Permian during which the intensity of tectonism gradually diminished. The zone of disrupted Acadian orogen, within which was deposited thick assemblages of continental clastics interrupted by one short stage of marine deposition in middle Mississippian, is also the zone of trough downwarping and intermittent movement of faults bounding internal linear uplifts. This zone is referred to as the *"Fundy Epieugeosyncline"* and is bordered on the NW by the New Brunswick Platform and on the SE by the Nova Scotia and Newfoundland Platforms. The epieugeosyncline and the platforms were called the Fundy Basin by Kelley (1967). The zone of thickest deposits and of mild Carboniferous deformation is called a "geosyncline" and not a trough or basin because it actually consisted of a composite of individual troughs, basins, and intervening uplifted source areas which had a complex non-uniform history of positive and negative movements. The boundaries of the geosyncline are placed in the zone across which relatively thin, essentially undeformed sediments of the platforms grade into thicker, deformed sediments of the geosyncline. Compared to the platformal areas, strata in the geosyncline have the greatest variations in facies, thicknesses, and environments and were deposited throughout a longer period of time.

Upper Devonian strata occur only in a few localities within the Appalachian Geosyncline and only in southern Ontario on the St. Lawrence Platform. Those within the geosyncline are continental coarse clastics and shales with some volcanic rocks. On the New Brunswick Platform, along Chaleur Bay, the Fleurent polymictic conglomerate, 14 m thick, is overlain by 110 m of the Escuminac gray shale and sandstone which have yielded excellent fish fossils and plant remains (Alcock, 1936). The conglomerate contains pebbles and boulders of granite and a variety of other rocks. Along the southern edge of the New Brunswick Platform, in southwestern New Brunswick, are about 800 m of plant-bearing Upper Devonian red conglomerates and few basaltic flows, the Perry Formation. No exposures of Upper Devonian strata are known on the Nova Scotia Platform. Upper Devonian strata occur in a few places in the Fundy Epieugeosyncline, commonly as red sandstone and conglomerate bearing spores.

On the Newfoundland Platform, Upper Devonian granite-bearing conglomerates and arkosic sandstones unconformably overlie Cambrian quartzites, and are cut by the Belleoram granite stock of presumed late Devonian age. Farther east, Upper Devonian conglomerate is overthrust from the SE by Hadrynian rocks.

Carboniferous and Permian and the Maritime Disturbance

Detritus eroded from the Acadian orogen appears in Carboniferous and Permian formations in all provinces of the Atlantic region. The stratigraphy on the three platforms is relatively simple, but within the Fundy Epieugeosyncline the facies are the result of a highly complex interplay of positive uplifts and negative troughs. For details, refer to Kelley (1967), Kelly (in Poole *et al.*, 1970), and Belt (1965, 1969) as well as to Gussow (1953) and Howie and Cumming (1963) for regional syntheses. The strata have been dated by plant fossils, by spores, by invertebrates, and in the Permian sediments by vertebrates. Carboniferous fossils are correlated with European forms, and consequently European time terms are often used in preference to North American terms.

Carboniferous and Permian Deposits. Sediments of the Carboniferous and Permian fall into three broad rock-stratigraphic groups. Some of the fauna and possibly some flora, may have been controlled by environment and thus may not be true time-stratigraphic indicators (Schenk, 1967). The oldest group is called the Horton in Nova Scotia (Bell, 1929) and other names in New Brunswick and Newfoundland. These rocks are mainly Tournaisian (early Lower Mississippian). In general the Horton and its correlatives consist of nonmarine red and gray conglomerate, sandstone, and shale which covered the uneven erosion surface on the Acadian orogen. Thicknesses vary up to about 3000 m. Some beds in southern New Brunswick are bituminous, contain veins of the hydrocarbon, albertite, and are the source and reservoir rocks of the small Stony Creek oil and gas field.

The next youngest group is the distinctive Windsor, which consists of up to nearly 1000 m of limestone, gypsum, salt, red sandstone, shale, and conglomerate. They are of mainly Viséan age, i.e., middle Mississippian. These rocks contain the only marine strata recognized in the post-Acadian Paleozoic sequences, and indicate a temporary cessation of tectonic activity.

Rocks of the Horton and Windsor groups are mainly restricted to the Fundy Epieugeosyncline although they have encroached upon the platforms in southern New Brunswick, southwestern Newfoundland, and southern Nova Scotia. Two large outliers of Windsor are known: the red beds, limestone and gypsum beds in the Plaster Rock area of western New Brunswick and the limestone beds along the coast S of Halifax. Thick Horton and Windsor rocks may be expected in the deep basins of the Gulf of St. Lawrence and on the shelves east of the epieugeosyncline, where seismic refraction has detected great depths to basement. Such areas may exist, for example, in the central part of the Gulf of St. Lawrence, between Prince Edward Island and Cape Breton Island and east of Cape Breton Island.

Overlying the Windsor Group is a thick sequence of continental clastic rocks of late Viséan (middle Mississippian) to early Permian They are referred to in order of decreasing age as the Canso, Riversdale, Cumberland, and Pictou groups. Together, they comprise over 6000 m of strata.

Tectonic activity during deposition of these Upper Paleozoic strata gradually decreased so that Upper Pennsylvanian and Lower Permian sediments, although immensely thick in some areas as a result of basinal downwarping, are only gently warped and cut by only a few faults. Extensive coal swamps developed in middle and late Pennsylvanian, particularly in central New Brunswick (Minto), northern Nova Scotia (Pictou and Cumberland), and Cape Breton Island (Sydney). Volcanic rocks of both silicic and mafic varieties occur here and there throughout the assemblage. Most of them are interlayered with the sediments of the epieugeosyncline probably because of deep faults. One notable center of Mississippian silicic volcanicity is in the Mount Pleasant area of southern New Brunswick in which various base metal sulfides and tin minerals are associated with intense sericitization, silicification, and kaolinization.

Maritime Disturbance. The older formations in the Carboniferous and Permian assemblage of the Fundy Epieugeosyncline were openly folded to varying degrees of intensity from time to time during the *Maritime Disturbance*. Major faults apparently controlled the development of troughs and uplifts. Folded strata commonly border the faults.

Metamorphism of Carboniferous and Permian sediments is negligible except for a zone extending roughly from the south coast of New Brunswick E along the fault zone on the north shore of Minas Basin to Chedabucto Bay and out to sea. Along this zone, the folded Carboniferous strata have developed a fracture cleavage found nowhere else in the Carboniferous of the Atlantic Provinces. It is interesting to note that the zone of metamorphism, almost certainly of middle or late Pennsylvanian age, is also a zone of late Triassic faulting, deposition, and volcanism.

Mineralization. Carboniferous mineralization reflects mineral assemblages and structural controls characteristic of postorogenic stages. Barite, galena, sphalerite, and silver minerals occur mainly in basal beds of the Windsor limestone. Fluorite, barite, and strontium minerals are present in a number of places. There is sulfide metallization in the Mississippian silicic volcanic rocks. Manganese oxide deposits and copper and lead sulfides are associated with Carboniferous sediments apparently near uplifts of older rocks from which these sediments were derived. Uraniferous minerals and hydrocarbons occur in a few places.

Mesozoic

No sediments of late Permian to middle Triassic age are known from the Atlantic Provinces. The region was probably undergoing erosion at that time with the products accumulating on the continental shelves.

Triassic Deposits and Palisades Disturbance. In late Triassic, deep faults extending from Bay of Fundy E through to Chedabucto Bay and on across the Scotian Shelf caused a downfaulted and downwarped trough to develop (*Palisades Disturbance*) in which were deposited 1000 m or more of continental red and gray conglomerate, sandstone, and shale (Wolfville, Blomidon) overlain by several hundred feet of tholeiitic basalt (North Mountain Basalt) (Klein, 1962) locally with a spectacular variety of zeolites (Aumento, 1966). The sediments represent fluviatile and lacustrine facies. Fossils are all nonmarine and include shells, reptilian bones, vertebrate teeth, and fish scales.

General taphrogeny along the Appalachian belt reflects the crustal stretching that heralded the present Atlantic opening in Jurassic times. The Triassic grabens represent a "basin-range province" both in tectonics and sedimentology. The Triassic strata have been flexed into a broad syncline with limbs dipping a few degrees. The zone of Carboniferous metamorphism and the Triassic faults pass from Chedabucto Bay eastward across the Scotian Shelf where a negative gravity anomaly, coupled with seismic refraction profiles and aeromagnetic surveys, indicates a deep trough filled with sediments.

Jurassic and Cretaceous Dikes. Jurassic deposits have not been recognized in the Canadian Appalachians but exist in subsurface on the shelves. Undeformed lamprophyre dikes which cut folded Ordovician and Silurian rocks in Notre Dame Bay, Newfoundland, have yielded K-Ar dates suggestive of a late Jurassic to early Cretaceous age.

Cretaceous Deposits, Intrusion and Erosion. Cretaceous sediments occur in one place on shore within the Canadian Appalachian region, and alkaline Monteregian plugs in southern Quebec were emplaced during the Cretaceous. Clay, silt, sand, and lignitic beds of probable early Cretaceous age have been identified near Truro in northern Nova Scotia. Cretaceous sediments underlie large parts of the Scotian Shelf and the Grand Banks and are in turn underlain by Jurassic (McIver, 1972; Bartlett and Smith, 1971) evidently dating back to the opening of the Mesozoic Atlantic (Fig. 8). Oscillations of sea level caused considerable channeling and refilling of the shelf deposits.

Cretaceous would seem to be a time of epeirogenic uplift and faulting in eastern Canada. Kumarapeli and Saull (1966) speculate that rifting along the St. Lawrence River valley and northwestward across the Grenville craton took place at this time. Many of the faults they postulate are based only on physiography. Nevertheless, it is interesting to speculate that the lignitic sediments of Nova Scotia, the Cretaceous lignitized wood and fossil leaves found far away near Schefferville, Labrador, and the

FIGURE 8. Profile showing Mesozoic-Cenozoic sediments and geomorphic surfaces in Nova Scotia and the Scotian Shelf (King, 1972).

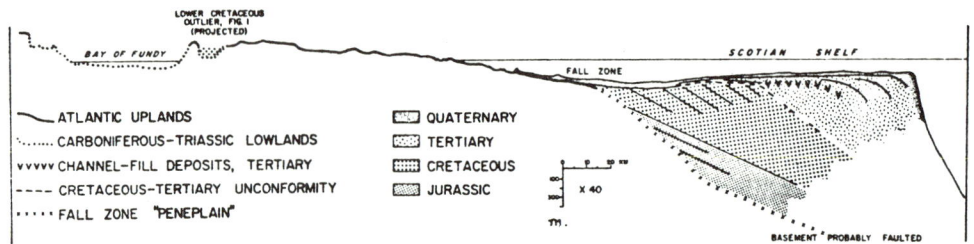

emplacement of Monteregian plugs all indicate epeirogenic uplift coupled perhaps with normal faulting. At this time, too, the Canadian Shield may have been largely stripped of its Paleozoic cover.

Cenozoic

Tertiary deposits are common on the Scotian Shelf and Grand Banks but are unknown on shore. Tertiary mudstone, silt, and muddy sand are more than 1000 m thick beneath Sable Island near the edge of the Scotian Shelf off Nova Scotia (McIver, 1972).

WILLIAM H. POOLE

References

Alcock, F. J. 1936. "Geology of the Chaleur Bay region," *Geol. Surv. Can., Mem. 183.*
_____, 1938. "Geology of Saint John region," *Geol. Surv. Can., Mem. 216.*
_____, 1947. "The Appalachian region," *in* R. J. W. Douglas, ed., *Geology and Economic Minerals of Canada* (Econ. Geol. Rept. 1, 3rd ed.). Ottawa: Geol. Surv. Can., 98–155.
Aumento, F., 1966. "Zeolite minerals, Nova Scotia," *in* W. H. Poole, ed., *Guidebook, Geology of Parts of Atlantic Provinces.* Toronto: Geol. Assoc. Can. and Mineral. Assoc. Can., 71–78.
Bartlett, G. A., and Smith, L., 1971. "Mesozoic and Cenozoic history of the Grand Banks of Newfoundland," *Can. J. Earth Sci.,* 8(1), 65–84.
Bell, W. A., 1929. "Horton-Windsor District, Nova Scotia," *Geol. Surv. Can. Mem. 155.*
Belt, E. S., 1965. "Stratigraphy and paleogeography of Mabou Group and related Middle Carboniferous facies, Nova Scotia, Canada," *Bull. Geol. Soc. Am.,* **76,** 777–802.
_____, 1969. "Newfoundland Carboniferous stratigraphy and its relation to the Maritimes and Ireland," *Am. Assoc. Petrol. Geologists, Mem.* **12,** 734–753.
Bird, J. M., and Dewey, J. F., 1970. "Lithosphere plate–continental margin tectonics and the evolution of the Appalachian orogen," *Bull. Geol. Soc. Am.,* **81,** 1031–1060.

Black, R. F., 1964. "Palaeomagnetic support of the theory of rotation of the western part of the island of Newfoundland," *Nature,* 202 (4936), 945–948.
Brückner, W. D., 1966. "Stratigraphy and structure of west-central Newfoundland," *in* W. H. Poole, ed., *Guidebook, Geology of Parts of Atlantic Provinces,* Toronto: Geol. Assoc. Can. and Mineral. Assoc. Can., 137–151.
_____, 1969. "Geology of eastern part of Avalon peninsula, Newfoundland; a summary," *Am. Assoc. Petrol. Geologists, Mem.* **12,** 130–138.
Church, W. R., 1969. "Metamorphic rocks of Burlington peninsula and adjoining areas of Newfoundland, and their bearing on continental drift in north Atlantic," *Am. Assoc. Petrol. Geologists, Mem.* **12,** 212–233.
_____, and Stevens, R. K., 1971. "Early Paleozoic ophiolite complexes of the Newfoundland Appalachians as mantle–oceanic crust sequences," *J. Geophys. Res.,* **76,** 1460–1466.
Clifford, P. M., 1969. "Evolution of Precambrian massif of western Newfoundland," *Am. Assoc. Petrol. Geologists, Mem.* **12,** 647–654.
Douglas, R. J. W., ed., 1970. *Geology and Economic Minerals of Canada* (Econ. Geol. Rept. 1, 5th ed.). Ottawa: Geol. Surv. Can., 838p.
Drake, C. L., and Woodward, H. P., 1963. "Appalachian curvature, wrench faulting, and offshore structures," *Trans. N.Y. Acad. Sci.,* **26,** 48–63.
Eastler, T. W., 1969. "Silurian of Change Islands and eastern Notre Dame Bay, Newfoundland," *Am. Assoc. Petrol. Geologists, Mem.* **12,** 425–432.
Fairbairn, H. W., et al., 1966. "Whole-rock age and initial $^{87}Sr/^{86}Sr$ of volcanics underlying fossiliferous Lower Cambrian in the Atlantic provinces of Canada," *Can. J. Earth Sci.,* **3,** 509–521.
Gussow, W. C., 1953. "Carboniferous stratigraphy and structural geology of New Brunswick, Canada," *Bull. Am. Assoc. Petrol. Geologists,* **37,** 1713–1816.
Helwig, J., and Sarpi, E., 1969. "Plutonic–pebble conglomerates, New World Island, Newfoundland, and history of eugeosynclines," *Am. Assoc. Petrol. Geologists, Mem.* **12,** 443–466.
Horne, G. S., 1970. "Complex volcanic–sedimentary patterns in the Magog Belt of northeastern Newfoundland," *Bull. Geol. Soc. Am.,* **81,** 1767–1788.
Howie, R. D., and Cumming, L. M., 1963. "Basement features of the Canadian Appalachians," *Geol. Surv. Can. Bull.,* **89,** 18p.
Hughes, C. J., and Brückner, W. D., 1971. "Late

Precambrian rocks of eastern Avalon peninsula, Newfoundland—a volcanic island complex," *Can. J. Earth Sci.,* **8,** 899–915.

Hutchinson, R. D., 1952. "The stratigraphy and trilobite faunas of the Cambrian sedimentary rocks of Cape Breton Island, Nova Scotia," *Geol. Surv. Can. Mem. 263.*

———, 1962. "Cambrian stratigraphy and trilobite faunas of southeastern Newfoundland," *Geol. Surv. Can. Bull.,* **88.**

Kay, M., 1967. "Stratigraphy and structure of northeastern Newfoundland bearing on drift in North Atlantic," *Bull. Am. Assoc. Petrol. Geologists,* **51,** 579–600.

———, 1972. "Dunnage mélange and lower Paleozoic deformation in northeastern Newfoundland," *24th Intern. Geol. Congr.,* **3,** 122–133.

———, 1973. "Tectonic evolution of Newfoundland," in K. A. DeJong and R. Scholten, eds., *Gravity and Tectonics.* New York: Wiley-Interscience, 313–326.

Kelley, D. G., 1967. "Some aspects of Carboniferous stratigraphy and depositional history in the Atlantic Provinces of Canada," *Geol. Assoc. Can., Spec. Pap.* **4,** 213–228.

Kennedy, M. J., Neale, E. R. W., and Phillips, W. E. A., 1972. "Similarities in the early structural development of the northwestern margin of the Newfoundland Appalachians and Irish Caledonides," *24th Intern. Geol. Congr.,* **3,** 516–531.

Kindle, C. H., and Whittington, H. B., 1958. "Stratigraphy of the Cow Head region, western Newfoundland," *Bull. Geol. Soc. Am.,* **69,** 315–342.

King, L. H., 1972. "Relation of plate tectonics to the geomorphic evolution of the Canadian Atlantic Provinces," *Bull. Geol. Soc. Am.,* **83,** 3083–3090.

———, MacLean, B., and Drapeau, G., 1972. "The Scotian Shelf submarine end-moraine complex," *24th Intern. Geol. Congr.,* **8,** 237–249.

Klein, G. de V., 1962. "Triassic sedimentation, Maritime Provinces, Canada," *Bull. Geol. Soc. Am.,* **73,** 1127–1145.

Kumarapeli, P. S., and Saull, V. A., 1966. "The St. Lawrence Valley System: a North American equivalent of the east African rift valley system," *Can. J. Earth Sci.,* **3,** 639–658.

Lilly, H. D., 1964. "Possible 'Taconic' klippen in western Newfoundland," *Am. J. Sci.,* **262,** 1130–1135.

———, 1966. "Late Precambrian and Appalachian tectonics in the light of submarine exploration on the Great Bank of Newfoundland and in the Gulf of St. Lawrence, preliminary views," *Am. J. Sci.,* **264,** 569–574.

McCartney, W. D., 1967. "Whitbourne map-area, Newfoundland," *Geol. Surv. Can. Mem. 341.*

———, et al., 1966. "Rb/Sr age and geological setting of the Holyrood granite, southeast Newfoundland," *Can. J. Earth Sci.,* **3,** 947–957.

McDonald, B. C., 1971. "Late Quaternary stratigraphy and deglaciation in eastern Canada," in *The Late Cenozoic Glacial Ages.* New Haven: Yale Univ. Press, 331–353.

McGerrigle, H. W., 1950. "The geology of eastern Gaspé," *Quebec Dept. Mines, Geol. Rept. 35.*

McIver, N. L., 1972. "Cenozoic and Mesozoic stratigraphy of the Nova Scotia Shelf," *Can. J. Earth Sci.,* **9,** 54–70.

Neale, E. R. W., Béland, J., Potter, R. R., and Poole, W. H., 1961. "A preliminary tectonic map of the Canadian Appalachian Region based on age of folding," *Bull. Can. Inst. Mining Met.,* **54,** 687–694.

Poole, W. H., 1967. "Tectonic evolution of Appalachian region of Canada," *Geol. Assoc. Can. Spec. Pap.* **4,** 9–51.

———, et al., 1970. "Geology of southeastern Canada," in R. J. W. Douglas, ed., *Geology and Economic Minerals of Canada* (Econ. Geol. Rept. 1). Ottawa: Geol. Surv. Can., 228–304.

Prest, V. K., 1970. "Quaternary geology of Canada," in R. J. W. Douglas, ed., *Geology and Economic Minerals of Canada* (Econ. Geol. Rept. 1). Ottawa: Geol. Surv. Can., 677–764.

Rodgers, J., 1965. "Long Point and Clam Bank formations," *Proc. Geol. Assoc. Can.,* **16,** 83–94.

———, 1970. *The Tectonics of the Appalachians.* New York: Wiley-Interscience, 271p.

———, and Neale, E. R. W., 1963. "Possible 'Taconic' Klippen in western Newfoundland," *Am. J. Sci.,* **261,** 713–730.

Rose, E. R., et al., 1970. "Economic minerals of southeastern Canada," in R. J. W. Douglas, ed., *Geology and Economic Minerals of Canada* (Econ. Geol. Rept. 1). Ottawa: Geol. Surv. Can., 306–364.

Schenk, P. E., 1967. "The significance of algal stromatolites to paleoenvironmental and chronostratigraphic interpretations of the Windsorian Stage (Mississippian), Maritimes Provinces," *Geol. Assoc. Can. Spec. Pap.* **4,** 229–243.

———, 1971. "Southeastern Atlantic Canada, northwestern Africa and continental drift," *Can. J. Earth Sci.,* 8(10), 1218–1251.

Stevens, R. K., 1970. "Cambro-Ordovician flysch sedimentation and tectonics in west Newfoundland and their possible bearing on a proto-Atlantic ocean," *Geol. Assoc. Can., Spec. Pap.* **7,** 165–177.

Strong, D. F., and Williams, H., 1972. "Early Paleozoic flood basalts of northwestern Newfoundland; their petrology and tectonic significance," *Proc. Geol. Assoc. Can.,* 24(2), 43–54.

Swett, K., and Smit, D. E., 1972. "Paleogeography and depositional environments of the Cambro-Ordovician shallow marine facies of the North Atlantic," *Bull. Geol. Soc. Am.,* **83,** 3223–3248.

Webb, G. W., 1969. "Paleozoic wrench faults in Canadian Appalachians," *Assoc. Petrol. Geologists, Mem.* **12,** 754–786.

Weeks, L. J., 1954. "Southeast Cape Breton Island, Nova Scotia," *Geol. Surv. Can., Mem. 277.*

———, 1957. "The Appalachian region," in C. H. Stockwell, ed., *Geology and Economic Minerals of Canada* (Econ. Geol. Ser. 1, 4th ed.). Ottawa: Geol. Surv. Can., 123–205.

Whittington, H. B., and Kindle, C. H., 1969. "Cambrian and Ordovician stratigraphy of western Newfoundland," *Am. Assoc. Petrol. Geologists, Mem.* **12,** 655–664.

Williams, H., 1964. "The Appalachians in northeastern Newfoundland—a two-sided symmetrical system," *Am. J. Sci.,* **262,** 1137–1158.

, 1967. "Silurian rocks of Newfoundland," *Geol. Assoc. Can. Spec. Pap.* **4,** 93–137.

, 1969. "Pre-Carboniferous development of Newfoundland Appalachians," *Am. Assoc. Petrol. Geologists, Mem.* **12,** 32–58.

, Kennedy, M. J., and Neale, E. R. W., 1970. "The Hermitage Flexure, the Cabot Fault, and the disappearance of the Newfoundland central mobile belt," *Bull. Geol. Soc. Am.,* **81,** 1563–1568.

Wilson, J. T., 1962. "Cabot Fault, an Appalachian equivalent of the San Andreas and Great Glen faults and some implications for continental displacement," *Nature,* **195**(4837), 135–138.

, 1966. "Did the Atlantic close and then reopen?" *Nature,* **211**(5050), 676–681.

Cross-references: *Canada–St. Lawrence Lowlands of Quebec; United States–Appalachian Region, New England Region.*

CANADA–CANADIAN SHIELD

The Precambrian rocks of the Canadian Shield cover 4,828,000 km² (1,864,000 sq mi) of Canada and the Lake Superior and Adirondack regions of the United States of America. Except on the northeastern side, facing the Greenland Shield, the Canadian Shield is surrounded by relatively undisturbed Phanerozoic rocks which were protected by the stable Precambrian basement. The Phanerozoic rocks may have once covered the whole shield, but today they are only outliers, as in the Hudson Bay platform and in many widely scattered small downfault basins. For the best general account and more detailed accounts and maps, refer to Douglas (1970).

Topography and Glaciation

The Canadian Shield is an ancient peneplain that is now saucer shaped; from sea level on the western shore of Hudson Bay the shield rises to an elevation of about 300 m along the western boundary of the shield and to 1500 m and more along the Atlantic coast. The warped peneplain has a local relief of only a few hundred feet or less, except along deep fiords on the Atlantic Coast, where the country is mountainous.

The whole shield was glaciated during the Pleistocene, and ice caps still remain on some of the Arctic islands. The ice removed most of the weathered material and in many places left hummocky outcrops interspersed with low areas of drift and swamp. Extensive areas are covered with glacial-lake clay and thus rock outcrop is scarce. The glaciation disrupted the preglacial drainage system and left innumerable lakes, rapids, falls, and diverted rivers.

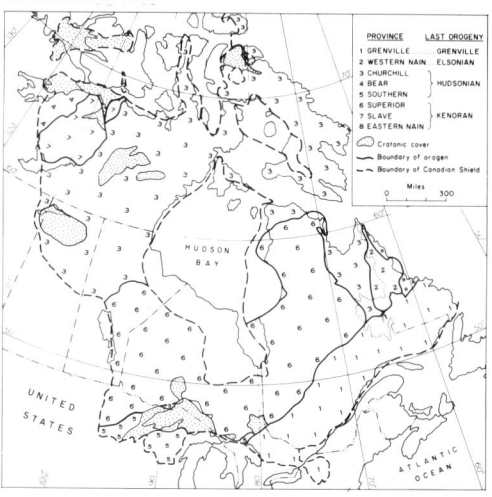

FIGURE 1. Structural provinces and orogenies, Canadian Shield.

Structural Provinces

The Canadian Shield has been divided into a number of structural provinces. Boundaries between provinces are drawn where one trend is truncated by another either along major unconformities or, in their absence, along orogenic fronts. The several structural provinces and orogenies are outlined and named in Fig. 1.

Time Classification

Geologic maps have been prepared for most of the shield although much of the field work was of a reconnaissance nature. These geological studies have solved many of the local problems dealing with the orogenic activity and stratigraphic succession. Until the application of isotopic methods for determining age, however, long-range correlation studies were fraught with difficulties. By about 1970 some 2400 isotopic dates had been obtained, mostly by the K-Ar method and few by the U-Pb and Rb-Sr methods.

Four main periods of folding, metamorphism, and granitic intrusion have been found, each of which was followed by epeirogenic uplift; deep erosion was followed by the deposition of unconformably overlying sedimentary and volcanic rocks. Thus it is now possible to construct a natural time classification system that shows the succession of events (Stockwell, 1964, 1973). The orogenies are the most important and most widespread events in the geological history of the shield and the major unconformities are the most important map boundaries, marking fundamental changes in geological conditions. Accordingly, the closing dates of orogenies are most useful for marking time boundaries between the main divisions of eon, era, and subera (Table 1). Names for the orogenies and time units are given in the table, which also shows the estimated age of their boundaries in millions of years. The time of closing of each orogeny is closely fixed within the short interval between the youngest dateable orogenic phase and the oldest dateable epeirogeny. For example, pegmatite is chosen as commonly representing the youngest orogenic phase, and its age was determined by the U-Pb method on uraninite and monazite. The oldest dateable metamorphic and intrusive cooling horizon of the epeirogeny is the K-Ar age of orogenic amphibole, which retains argon at a higher temperature than do the micas. Because of the uncertainty in the Rb^{87} half-life, results of Rb-Sr dating are not included in the table. Following the recommendations of the American Commission on Stratigraphic Nomenclature (1961), the orogenies are defined by rock in type regions rather than by isotopic dates. The type regions are (1) the Superior province for the Kenoran orogeny, (2) the Churchill for the Hudsonian, (3) the Western Nain for the Elsonian, and (4) the Grenville province for the Grenvillian. The ages given in Table 1 are based on averages for each mineral and method used in each of the type provinces and are subject to change as more data become available and as age methods and interpretations improve.

Stratigraphic and Orogenic History

The earliest history of the shield is imperfectly known, but events over the succeeding 2000 m yr have left a record indicating the following cycle repeated four times: deposition, orogeny, epeirogeny, and deep erosion. Each orogeny was followed by relatively minor post-orogenic intrusions.

The generalized geological map (Fig. 2) illustrates the distribution of rocks of the several time units, both in the type regions and in other parts of the shield. Not all the rocks can be fitted precisely into the framework of the time classification.

Archean. The Archean was a period of much volcanic activity and of the deposition of immense thicknesses of sediments. These rocks are best preserved in the Superior and Slave provinces where they were deformed by the Kenoran orogeny. They are also found in parts of other provinces, where they have been partially reworked by younger orogenies.

In the Superior and Slave provinces most of the volcanic material is of basic composition, but more acid compositions are also found. In places the basic lavas have been metamorphosed to amphibolites, especially near contacts with granitic intrusions; but more commonly these lavas are only slightly metamorphosed and primary structures, such as pillows, flow lines, amygdules, and the original texture, are still well preserved. Ash beds, tuff, and agglomerate are interlayered with the flows and iron formations are found with them. Basic sills are common and probably related to the flows. The sedimentary rocks are usually poorly sorted and consist mostly of altered graywacke and slate, conglomerate, minor arkose, quartzite, and calcareous beds. Primary structures, such as bedding, graded bedding, and cross-bedding, are often well preserved. Almost everywhere the volcanic rocks and the sedimentary rocks are conformable with one another and are often interbedded. Major unconformities, however, have been recognized at several widely separated localities. Conglomerates above the unconformities contain pebbles and boulders of volcanic rocks, iron formation, and granites. Beneath the unconformities, but only rarely recognized, are older granites that intruded volcanic rocks. However, it is not possible to define earlier events or to make detailed time subdivisions of the Archean.

The Kenoran orogeny (late Archean) was characterized by intense folding accompanied by the emplacement of granitic and more basic magma. The granites are now exposed over a larger area than the sediments and volcanics that they intruded. The sediments and volcanics remain in isolated, irregularly shaped areas and in belts.

The supposition that Archean rocks exist in certain parts of the Churchill province is based

TABLE 1. Precambrian Time Classification for Type Regions of the Canadian Shield*

EON	ERA	SUB-ERA	OROGENY	AGE OF BOUNDARY m.y.
PROTEROZOIC	HADRYNIAN			570
				1000
	HELIKIAN	NEOHELIKIAN	GRENVILLIAN	
				1400
		PALEOHELIKIAN	ELSONIAN	
				1800
	APHEBIAN		HUDSONIAN	
				2560
ARCHEAN			KENORAN	

GSC

*Estimates of age of boundaries are based on determinations by the U-Pb and K-Ar methods.

mainly on geological reasoning but finds support in a few Kenoran-age dates. Within this province the effects of the Hudsonian orogeny are widespread, and it is possible that some of the Archean rocks were involved in both orogenies. Where the Hudsonian orogeny had slightly affected the rocks, they resemble the low-grade metamorphic rocks of the Superior province; but where the orogenic forces were strong, the rocks have been converted into gneisses and are difficult to distinguish from Aphebian gneisses, which are also present in the province.

Archean rocks have been traced from the Superior province across the Grenvillian front into the Grenville province, where they are more highly metamorphosed and deformed due to reworking during the Grenvillian orogeny. At present their areal extent within the Grenville province can only be conjectured.

Aphebian. The sedimentary rocks of the Aphebian period show a marked change in depositional environment, for they are better sorted than the Archean sediments, as evidenced by the greater abundance of quartzite and limestone and by less graywacke (Pettijohn,

FIGURE 2. Generalized geological map, Canadian Shield.

1943). Many of the limestones contain stromatolites. Also present are slate, conglomerate, and iron formation. Volcanic rocks are common in parts of the succession, and basic sills are found in both the volcanics and the sediments.

Low-grade metamorphic rocks are usually found in geosynclinal troughs along the borders of the Superior and Slave provinces. The rocks are widespread, lie unconformably on deeply eroded rocks of the Kenoran orogen, and form gently dipping homoclines at the contact with these provinces. Away from the Kenoran mobile belt, the rocks were folded during the Hudsonian orogeny. Aphebian sedimentary and volcanic rocks are also found in scattered belts and in irregular areas within the main body of the Churchill province. In places they were only slightly metamorphosed during the Hudsonian, but more commonly they have been converted into schist, gneiss, and migmatite. In such high-grade metamorphic rocks the basal unconformity has not generally been recognized, and the Aphebian gneisses are difficult to distinguish from the Archean gneisses. However, the presence of more quartzite and limestone in the Aphebian may serve to distinguish the two.

Virtually unfolded Aphebian sandstone and conglomerate locally cover the Kenoran orogen of the Superior province.

Low-grade metamorphic Aphebian rocks in the Churchill and Southern provinces have been traced into the Grenville province, where they have been converted into gneisses during the Grenvillian orogeny. They are probably strongly folded with the other Grenville province rocks, and they are difficult to identify in the field. Therefore, Aphebian rocks are shown only locally on the geological map (Fig. 2).

The Hudsonian orogeny (late Aphebian) was accompanied by extensive granitic intrusions, but in the Churchill province Hudsonian granites are not readily distinguished from those of the reactivated Kenoran granites.

Helikian and Paleohelikian. Rocks of the Helikian and Paleohelikian ages consist predominantly of continental sandstone, conglomerate, and lava flows. They lie unconformably on the Hudsonian orogens, where they form areas of virtually unfolded cover rocks. The Sudbury norite-micropegmatite is Paleohelikian.

In the western Nain province the Elsonian orogeny (late Paleohelikian) is accompanied by discordant intrusions of anorthosite, gabbro, and granitic rocks. In the Grenville province many bodies of anorthosite and gabbro are probably late Paleohelikian, but they were remobilized and metamorphosed during the Grenvillian orogeny.

Neohelikian. In one small area at the north boundary of the Grenville province, Neohelikian sedimentary and volcanic rocks lie unconformably on the Elsonian orogen. To the S these rocks were folded during the Grenvillian orogeny. Well within the Grenville province a small area of similar rock, which was folded during the Grenvillian, is probably Neohelikian.

The Grenvillian orogeny (late Neohelikian) was a period of folding, metamorphism, and granitic intrusion. It is probable that much of the material in the Grenville province was deformed by older orogenic activity and then reworked during the Grenvillian so that the resulting Grenville structures are very complex.

Neohelikian and Hadrynian. In the western part of the shield, which was unaffected by the Grenvillian orogeny, the Keweenawan lava flows and sediments and the Duluth gabbro are Neohelikian-Hadrynian, as are the Coppermine River flows and sediments and the Muskox intrusion. The sediments and volcanics are overlain without marked unconformity by Phanerozoic rocks.

Small areas of nearly horizontal sandstone and conglomerate (Hadrynian or possibly younger) lie unconformably on the Grenvillian orogen. Hadrynian rocks are well preserved outside the boundaries of the Canadian shield in the Appalachian and Cordilleran provinces.

Postorogenic Intrusions. Postorogenic intrusions include extensive swarms of diabase and gabbro dikes of various Proterozoic ages (Fahrig and Wanless, 1963). Other postorogenic intrusions include bodies of granite and syenite, some associated with carbonatite. The alkaline syenites were emplaced throughout the Proterozoic and intruded the orogens after they were stable. Circular volcanic and cryptovolcanic structures (some Phanerozoic) also occur in the shield.

Mineral Deposits

The major deposits include gold, nickel, copper, zinc, silver, and iron. A brief description of the most important deposits is given below.

Gold. Gold deposits occur mainly in the Kenoran orogens of the Superior and Slave provinces and are found chiefly in Archean sedimentary and volcanic rocks, although some occur in granitic rocks or other intrusives. Most deposits occur as veins or stockworks that follow shear zones, fractures, or faults. Some gold is obtained from mineralized schist or breccia. Native gold and tellurides are associated with pyrite and other sulfides in a quartz gangue. Silver is a by-product.

Most production has come from an unusually large belt of slightly metamorphosed sedimentary and volcanic rocks in the south-central part of the Superior province; these include the mines of the Porcupine and Kirkland Lake camps and of the Larder Lake-Malartic belt, where most deposits occur in subsidiary fracture and shear zones adjacent to a major fault. Numerous deposits are also found in the western part of the Superior province and in the Slave province. A few mines in the Churchill and Grenville provinces have produced only small amounts of gold. Large amounts of gold have been obtained as a by-product of copper and nickel refining.

Nickel and Copper. The Sudbury area of Ontario has long been the main source of the world's nickel. The deposits are closely related to a body of Paleohelikian norite-micropegmatite near the border between the Superior and Southern provinces. The norite crops out as an elliptical ring, with micropegmatite forming an inner zone. In places the norite grades outward into mineralized quartz diorite. The ore may also occur in quartz diorite pods which are separate from the main quartz diorite body. Much of the ore is closely associated with the quartz diorite and occurs along the outer contact of the norite. It is also commonly found as rock fragments cemented by sulfides (mainly pyrrhotite, pentlandite, chalcopyrite, and cubanite) in breccia zones. In addition to nickel and copper, the ores yield important amounts of gold, silver, and metals of the platinum family.

In Manitoba the Thompson nickel mine (Zurbrigg, 1963) and several others lie within a NE-trending belt of the Churchill province several miles from the border of the Superior province. The country rock consists of intensely folded gneisses intruded by granite, pegmatite, and numerous lenticular bodies of peridotite. At the Thompson mine sulfides are found in peridotite, but the main ore body follows a zone of schist that is conformable with the folded metasedimentary formations. The ore contains numerous rock inclusions and consists mainly of pyrrhotite, pentlandite, and pyrite with minor chalcopyrite and other minerals.

Also in Manitoba, but well within the Churchill province, are the Lynn Lake nickel deposits. This ore is found in breccia zones surrounding diorite to peridotite stocks and consists mainly of pyrrhotitie, pentlandite, and chalcopyrite.

Copper-Zinc and Zinc-Lead. Copper-zinc and zinc-lead deposits are found mainly in the Superior province but also occur in the Churchill province and, less importantly, in the Grenville province. The ore bodies are mainly lenticular to irregular or pipe-like in shape, and they contain disseminated to massive pyrite, pyrrhotite, chalcopyrite, sphalerite, and galena in various proportions. Important amounts of gold, silver, and, locally, cadmium are also recovered. Most of the more important deposits are found in Archean volcanic rocks and tend to follow their structure. In the Noranda and nearby districts of the Superior province the ore is found mainly in acid volcanic flows and tuff; however, some of the ore follows contacts with the more basic flows. In the Flin Flon district of the Churchill province, most of the ore is in basic volcanics (probably Archean), and in the largest deposit it follows a sheared contact zone between flows and agglomerate. In both provinces deposits of a different type occur in gneisses and form long bodies that follow the regional trend. In these deposits pyrrhotite is more prominent and precious metals are less important. A few are also found in metamorphosed limestones and gneisses of the Grenville province. In the Chibougamau district of the Superior province the ore follows shear zones and faults in gabbro and anorthosite. Native copper has been mined from Neohelikian-Hadrynian basaltic flows in the Southern province.

Silver. A large amount of silver has been produced from the Cobalt and surrounding districts, where the ore is closely associated with sills of diabase that cut the nearly flat-lying sediments and the underlying Archean rocks. The sediments, the diabase, and presumably the silver mineralization are all Aphebian. The ore occurs as veins filling fractures, faults, and shatter zones in the above three rock types and consists of native silver, smaltite, niccolite, and other valuable minerals, all in a gangue of calcite.

Uranium. Uranium deposits were formed under a variety of geological conditions and are found in the Southern, Churchill, Bear, and Grenville provinces. In the Blind River district of the Southern province, large deposits of uranium ore occur in gently dipping Aphebian conglomerate at or near the contact with unconformably underlying Archean rocks. Pyrite, brannerite, pitchblende, uraninite, and monazite occur in the matrix of the conglomerate. At the Eldorado mine in the Bear province veins containing pitchblende, cobalt, nickel, and silver follow shear and shatter zones cutting Aphebian sedimentary and volcanic rocks. In the Beaverlodge area of the Churchill province, pitchblende ore follows fault zones that cut gneisses, and it also occurs in a body of brecciated syenite. In the Bancroft region of the

Grenville province, uraninite and uranothorite occur in pegmatites and gneisses.

Iron Ores. Iron ore is produced mainly from Archean and Aphebian sedimentary iron formation. Large amounts of ore, most probably enriched by a leaching process, have been mined in the Lake Superior region and in the Labrador trough. Taconite ore has been mined in recent years in the Lake Superior region. At Steep Rock Lake the ore is primarily goethite and appears to be the gossan of an ancient surface of weathering. Siderite ore is produced from the Helen mine. The Aphebian iron formation of the Labrador trough has been traced into the Grenville province, where it was highly folded and metamorphosed. The iron formation was transformed into large, low-grade hematite and magnetite ore bodies that are easily beneficiated. Some ores in the Grenville province are of contact metasomatic origin (e.g., magnetite in skarn), whereas other ores are mined for both iron and titanium. In addition, iron is obtained as a by-product in the treatment of nickel and copper sulfide ores.

Other ores. Some other materials produced from the shield include asbestos from ultrabasic rocks; molybdenum from pegmatitic quartz veins; lithium, mica, and feldspar from pegmatite; magnesium from meta-dolomite; magnesia from brucitic limestone; talc from dolomite limestone; and sulfuric acid from the treatment of sulfide ores.

C. H. STOCKWELL

References

American Commission on Stratigraphic Nomenclature, 1961. "Code on stratigraphic nomenclature," *Bull. Am. Assoc. Petrol. Geologists,* 5(5), 645–665.

Darnley, A. G., Grasty, R. L., and Charbonneau, B. W., 1972. "A radiometric profile across part of the Canadian Shield," *Geol. Surv. Can. Pap.* 70-46, 42p.

Dimroth, E., 1970. "Evolution of the Labrador geosyncline," *Bull. Geol. Soc. Am.,* 81, 2717–2742.

Douglas, R. J. W., ed., 1970. *Geology and Economic minerals of Canada* (Econ. Geol. Rept. 1, 5th ed.). Ottawa: Geol. Surv. Can., 838p.

Fahrig, W. H., and Wanless, R. K., 1963. "Age and significance of diabase dyke swarms of the Canadian Shield," *Nature,* 200(4910), 934–937.

Pettijohn, F. J., 1943. "Archean sedimentation," *Bull. Geol. Soc. Am.,* 54, 925–972.

Stockwell, C. H., 1964. "Fourth report on structural provinces, orogenies, and time classification of rocks of the Canadian Precambrian Shield," *Geol. Surv. Can. Pap.* 64-17, 1–21.

———, 1973. "Revised Precambrian time scale for the Canadian Shield," *Geol. Surv. Can. Pap.* 72-52, 4p.

———, in prep. "Time classification of Precambrian rocks and events of Canada and of adjacent shield areas of the United States," *Geol. Surv. Can.*

Zurbrigg, H. F., 1963. "Thompson mine geology," *Trans. Can. Inst. Mining Met.,* 66, 227–236.

Cross-references: *Canada–St. Lawrence Lowlands of Quebec; North America.*

CANADA–CORDILLERAN REGION, INTERIOR AND WESTERN BELTS

The interior and western belts of the Canadian Cordillera include all of western Canada west of the Richardson and Mackenzie mountains, the Northern Rocky Mountain Trench, and the Columbia Mountains. The regions occupy a large part of British Columbia, most of the Yukon Territory, and a small part of the southwestern Mackenzie district; and they cover an area of about 960,000 km^2 (375,000 sq mi.)

History of Geological Investigations

The earliest geological studies of the region were dominated by G. M. Dawson, R. G. McConnell, and R. A. Daly. These geologists, in particular Dawson and McConnell, established a broad outline of the western Cordilleran geology through extensive field investigations during the latter part of the 19th century and the early part of the 20th century. Not until the late 1950s, however, were sufficient data available for the compilation of a comprehensive synthesis of the geology of the entire Cordillera. The Geological Survey of Canada obtained much of this information by helicopter surveys. In British Columbia mapping by the Geological Survey has been supplemented by the British Columbia Department of Mines and Petroleum Resources, particularly in areas of current mineral exploration and near mining camps.

Physiography

The interior belt of the Canadian Cordillera contains diverse morphogenetic regions, including areas of high rugged relief (Selwyn, Wernecke, Ogilvie, Pelly, Cassiar, Omineca, and Skeena mountains), as well as areas of low relief, although they may be deeply incised locally [Porcupine, Yukon, and the interior plateaus (*see* Fig. 1)]. Old Crow, Eagle, and Liard plains are large intermontane areas of low relief, largely covered by overburden. Several great, remarkably linear, NW-trending valleys

FIGURE 1. Physiographic elements of the Canadian Cordillera, after Geological Survey of Canada, 1969.

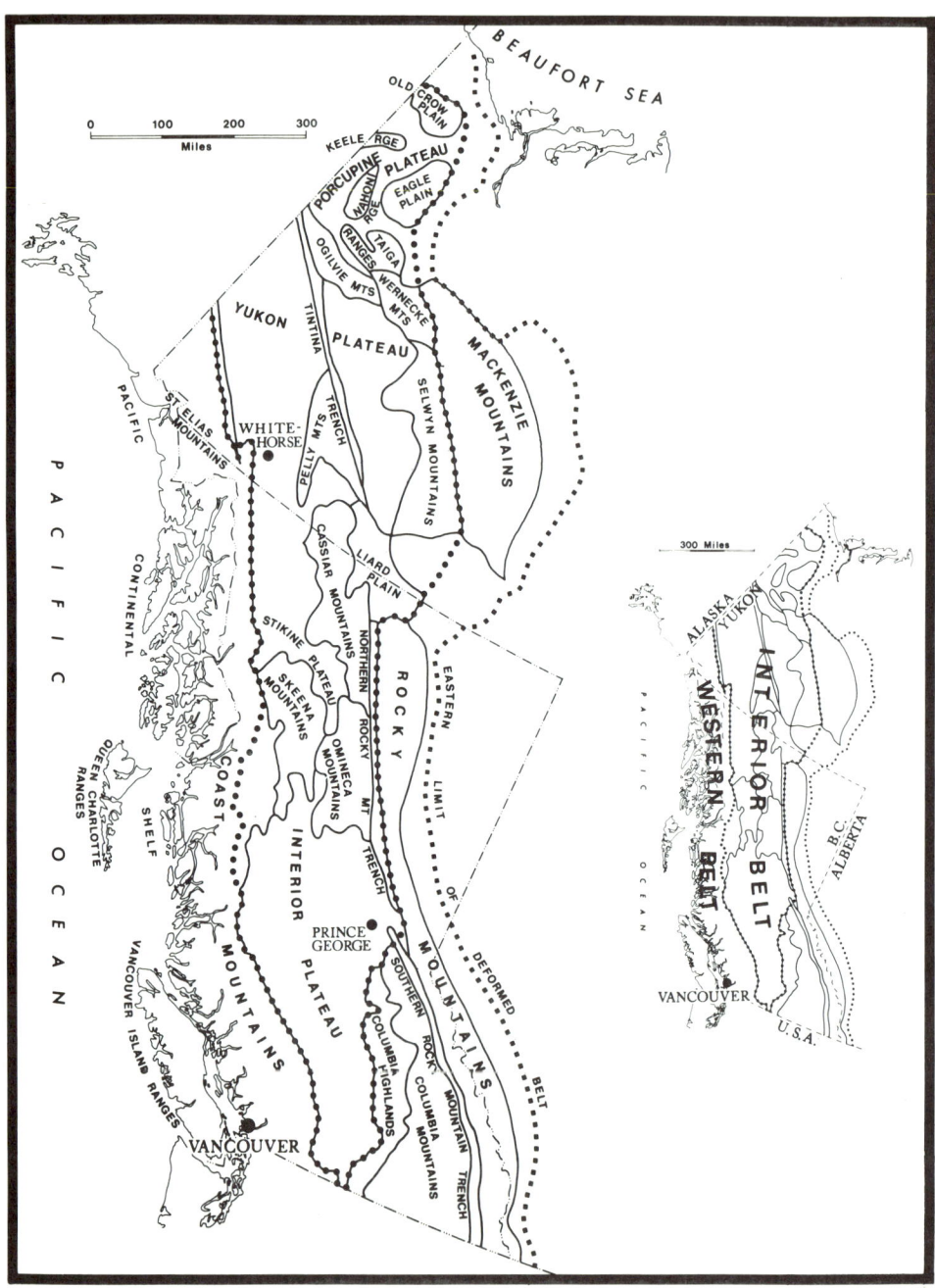

traverse the southwestern Yukon Territory and northern British Columbia. The Shakwak Trench in southwestern Yukon forms the boundary between the interior and western belts. The Northern Rocky Mountain Trench in north-central British Columbia separates the interior and eastern belts. The Tintina Trench lies in part within the Yukon plateau and in part separates the plateau from the Ogilvie and Pelly mountains.

The western belt of the Canadian Cordillera includes the very rugged St. Elias and Coast mountains with their numerous ice fields and glaciers, and it also includes the less rugged and lower ranges of Queen Charlotte and Vancouver islands. Mount Logan in the St. Elias Mountains, which reaches 6108 m (19,850 ft), is the highest peak in Canada. The ruggedness and scenic beauty of the Coast Mountains is enhanced by deeply incised fiords that extend well into the central parts of the mountains.

Drainage of the interior and western belts is effected by numerous large rivers such as the Yukon, Liard, Peace, Stikine, Nass, Skeena, and Fraser, most of which drain into the Pacific Ocean. Exceptions are rivers, including the Liard and Peace, which are tributaries of the Mackenzie River and drain into the Arctic Ocean.

Geotectonics

The region is divided into a number of tectonic areas that were important at various times during the evolution of the Cordilleran orogen (see Fig. 2). Most are predominantly the products of Mesozoic and Cenozoic geologic history, but some, particularly in northern Yukon, reveal tectonic events dating back to the Proterozoic. Thus the Ogilvie arch and the Selwyn Basin in northern and eastern Yukon contain sediments of contrasting thickness and facies ranging from the Proterozoic to the Middle Devonian.

Since at least the Middle Proterozoic, it appears that the Yukon Territory north of Selwyn Basin was underlain by a sialic crystalline basement which was a westward extension of the North American craton. The geology of the remaining region northeast of the Tintina Trench suggests the deposition of a sedimentary prism on a westwardly thinning basement. The Pelly-Cassiar platform either was connected to the North American craton or it represents an early Paleozoic island south and west of a Proterozoic and early Paleozoic trough (eastern Selwyn Basin and Kechika trough). The western extent of crystalline basement under the interior and western belts during the Proterozoic and Paleozoic is debatable. It probably did not extend west of the eastern margin of a belt of late Paleozoic oceanic rocks that crop out within the interior belt. Two crystalline belts, the Coast and Omineca geanticlines, influenced the distribution and character of Mesozoic strata. At the same time the Queen Charlotte and Vancouver islands may have evolved on an oceanic basement.

According to plate-tectonic theory, the geology of the interior and western belts of the Canadian Cordillera indicates a Proterozoic and early Paleozoic Atlantic-type coast (trailing edge of a crustal plate) changing to a tectonically active area subjected to collisions with a Pacific crustal plate. The compressional phase

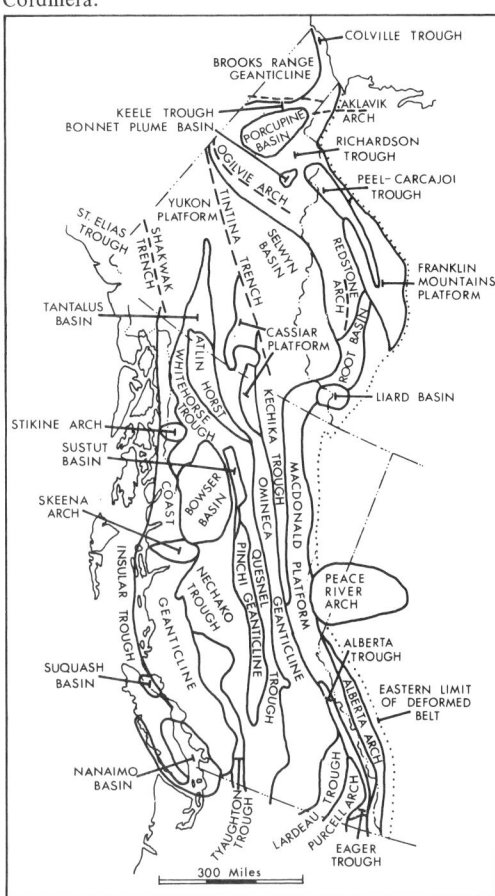

FIGURE 2. Tectonic elements of the Canadian Cordillera.

was accompanied by subduction zones, volcanic arcs, and plutonism during the late Paleozoic, Mesozoic, and Cenozoic.

Structure

Interior Belt. Folds and faults in the interior belt of northern Yukon have diverse trends and structural styles. Some N-S-trending high-angle faults in Richardson Mountain, having right-lateral displacements, swing southeasterly into the Wernecke Mountains, where the faults have northeasterly directed thrust displacements. The Wernecke Mountains are essentially fault-block mountains with NW-trending structures and transect the W-trending structures of the Mackenzie Mountains. The Ogilvie Mountains and Taiga Ranges are characterized by W- and NW-trending upright open folds that are cut by moderate-to-steep dipping faults. The folds of the western Taiga Ranges plunge westerly into a trough that separates the Taiga and

Nahoni ranges. Nahoni Range and Eagle Plain are underlain by strata exposed in N-trending open folds. West of the Nahoni Range is a belt of N-NE-trending, W-dipping thrust faults. All these structures are truncated by the Aklavik arch, a tectonic feature intermittently active since at least the Middle Paleozoic.

In northwestern Selwyn Basin a major southerly dipping imbricate thrust belt contains isoclinal folds and strongly cleaved strata. Elsewhere in Selwyn Basin tight, upright folds are common; although in the southeastern part northeasterly directed asymmetric folds are prevalent.

Southwest of Tintina Trench NW-trending structures are dominant. A strong foliation is common and the bedding has often been obliterated.

The well-stratified rocks of the Pelly, Cassiar, and Omineca mountains are strongly deformed. Structures in the Pelly Mountains have a NE-directed asymmetry; those in the Cassiar Mountains have both a NE- and a SW-directed asymmetry, and those in the Omineca Mountains have a SW-directed asymmetry. All are truncated at acute angles by the Tintina and Rocky Mountain trenches.

Tightly folded rocks of the Atlin horst are bounded on the SW by a major NE-dipping thrust fault which, with a similar fault immediately to the SW, separates the structural province of eastern and northeastern British Columbia from the rest of the interior belt. Broad, open NW-trending folds are characteristic of Whitehorse trough and parts of the northern Omineca geanticline. The well-bedded rocks of Bowser basin are strongly folded, mainly along NW trends, but locally folds conform in trend to the margins of the basin. In the rest of the southern interior belt, faults play a dominant role. Nechako and Quesnel troughs are extensively block faulted. The Tyaughton trough has a system of NW-trending fault slices called the *Yalakom fault zone*, which, analogous to the Pinchi fault zone, has effected significant transcurrent displacement.

Western Belt. In the St. Elias Mountains SW-dipping thrust faults occur near the Shakwak Trench. Several low-angle NE-dipping thrust faults are present where one of the major faults and the trench converge. On the southwest side of the St. Elias Mountains, several thrust faults and folds have a SW-directed asymmetry. Thus in the western belt the strata have been thrust outward from a central core.

The coast plutonic complex of the Coast geanticline has structures that are related to the emplacement of the steep-walled plutons that contain pendants of foliated metasedimentary and metavolcanic rocks. The largest belt of

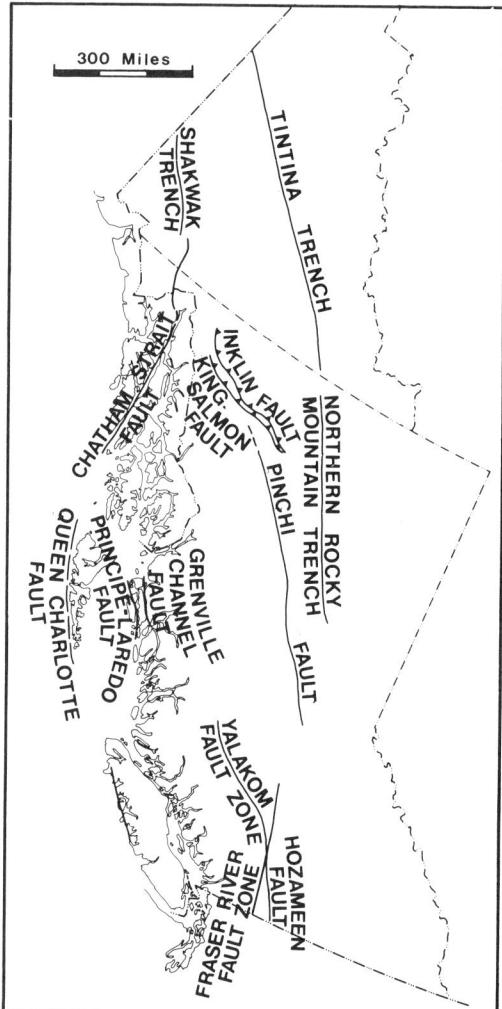

FIGURE 3. Major lineaments and faults of the interior and western belts.

metasediments, occurring as steep-walled pendants, trends obliquely across the complex from southeastern Alaska. Over large areas on the east side, gneisses and migmatites, which appear to underlie the metasediments, have gently dipping foliations. Recumbent fold axes trend easterly and northerly. In the southern part of the complex northwesterly elongated roof pendants are partly bounded by faults. The western part of the Cascade Mountains are similar in structure to the St. Elias Mountains in containing westerly directed recumbent folds and shallow-dipping thrusts, whereas in the eastern part the thrust faults are easterly directed.

Strata on the Queen Charlotte Islands in the northern insular trough were deformed by NW-trending right-lateral faults. This motion has been combined with a downward movement of the eastern blocks. Vancouver Island has a complex mosaic of fault blocks bounded by N-, NW-, and NE-trending faults.

Major Lineaments and Faults. Structurally controlled topographic lineaments and faults of regional importance affect the distribution of the geological units in the interior and western belts (see Fig. 3).

Except for the SW-directed King Salmon and Inklin thrust faults in northern British Columbia, most of the faults are probably right-lateral transcurrent. In a few cases it has been possible to measure the magnitudes of the displacement.

Times of Regional Deformation. Middle Proterozoic rocks in the Wernecke and Ogilvie mountains were folded and faulted before the deposition of the Upper Proterozoic strata. Most of the folds and thrust faults in the northernmost interior belt were formed during a late Cretaceous and early Cenozoic deformation. Structures throughout the Selwyn Basin and parts of the Yukon plateau are probably middle Cretaceous. Widespread folding and thrust faulting are tentatively dated as late middle or early late Jurassic in the Pelly, Cassiar, and Omineca mountains and in parts of the Stikine plateau. Middle Jurassic deformation was important in the insular belt and in the southern interior belt, whereas in the Cascade Mountains middle Cretaceous and early Cenozoic folding and thrusting occurred. Late Tertiary thrust faulting was important in the St. Elias Mountains. Block faulting was common during the late Cretaceous and early Cenozoic.

Stratigraphy

The oldest known rocks are exposed in the northern part of the interior belt. These and succeeding strata, ranging from the Proterozoic to the Middle Devonian, are well-bedded miogeosynclinal rocks with few volcanics. Younger sequences, exposed throughout the western and interior belts, include abundant volcanic rocks and are thus eugeosynclinal. During the Mesozoic, these rocks were intruded by granitic plutons and large areas were regionally metamorphosed. Also, during the Mesozoic and early Cenozoic, thick nonvolcanic marine and nonmarine assemblages were deposited. Subaerial volcanism was widespread in the late Mesozoic and Cenozoic. Marine Upper Cretaceous and Cenozoic rocks are confined to the Pacific continental shelf or to the surrounding areas. Most of the region was glaciated during the Pleistocene.

Proterozoic. Strata generally considered to be correlative with the middle Proterozoic Purcell-Belt rocks of the southeastern Cordillera are exposed in the Wernecke and Ogilvie mountains and in the Keele Range (Table 1). The oldest rocks are a monotonous sequence of dark-weathered slate, argillite, and phyllite, as much as 1500 m (5000 ft) thick. They are overlain by more than 1500 m (5000 ft) of distinctive, orange-weathered dolomite and locally by massive cherty dolomite. Diabase dikes and sills are present in the Keele Range. The Racklan orog-

TABLE 1. Canada—Cordilleran Region: Stratigraphic Table, Simplified (thicknesses in thousands of feet)

eny subjected the Middle Proterozoic rocks to low-grade regional metamorphism, folding, tilting, block faulting, and uplift. The regional unconformity developed at this time is one of the most conspicuous in the northern Cordillera.

Upper Proterozoic rocks, widespread in the eastern part of the northern interior belt, underlie large areas in most of the mountain ranges and probably in parts of the western Yukon plateau. Throughout the region sequences as great as 4500 m (15,000 ft) thick include a clastic lower unit consisting of argillite, phyllite, slate, and siltstone and are characterized by beds of feldspathic pebble conglomerate; and an upper unit, commonly a few thousand feet thick, consist of limestones and maroon, red, and green slates. A distinctive group of rocks 1500 m (5000 ft) thick contains banded jasper-hematite iron formation and is found in the Wernecke Mountains and farther west. The lower and, particularly, the middle parts of this unit contain poorly sorted and very coarse conglomeratic mudstone or diamictite.

The source of the large volume of detrital Upper Proterozoic rocks remains in doubt, although a large portion of these sediments were probably derived from the craton to the east. Certain aspects of the conglomeratic mudstone and iron formation suggest deposition in a basin that was partly bounded by faults.

Paleozoic. Cambrian rocks crop out along the Alaska-Yukon border near the Yukon River, in the Wernecke Mountains, and in the Pelly, Cassiar, Omineca, and eastern Selwyn mountains. The Lower Cambrian is predominantly quartz sandstones and is the thickest such sequence in the Cordillera. *Olenellus* trilobites are common in the uppermost shaly sandstones. In the Selwyn Mountains the sandstones grade southwesterly and westerly into siltstones and shales. The overlying beds are carbonates and contain abundant *Archaeocyathus* fossils. Thicknesses of the Lower Cambrian range from 300 m (1000 ft) to more than 2000 m (6500 ft).

Lower Ordovician to Middle Devonian rocks reflect the Cambrian structural pattern. Graptolitic shale and siltstone are characteristic of the troughs and basins, whereas well-bedded carbonates form the platforms and arches. There is a well-defined facies change that roughly coincides with the northern and eastern margins of the Selwyn Basin. In many places the argillaceous units are thinner than those of correlative carbonate units. Several classic localities show the interfingering of graptolite shales with shell-bearing carbonate units. Middle Devonian carbonates are known in the St. Elias Mountains. The southernmost part of the western belt contains a small outcrop of Devonian limestone overlying a diorite gneiss basement complex of unknown age.

Late Devonian detrital sediments indicate uplift in the northernmost part of the interior belt (Ellesmerian orogeny) and west of the Cassiar and Omineca mountains (Caribooan orogeny). Black shale, siltstone, and sandstone are abundant, and thicknesses of up to 1000 m (3300 ft) are not uncommon.

Mississippian rocks are continuous with Upper Devonian sandstones. In the northernmost part of the interior belt, the Mississippian contains carbonate and shale. Elsewhere the carbonate is interbedded with a wide variety of rocks, including thick chert-pebble conglomerates in the western Selwyn Mountains, and mafic volcanics, tuff, volcanic arenite, chert-pebble conglomerate, chert, graywacke, and slate in the remainder of the interior belt and in the St. Elias and Cascade mountains. In the Cassiar Mountains ultramafics, gabbro, basic volcanics, and bedded chert form an ophiolite suite. Total thicknesses are generally more than 1000 m (3300 ft).

The Pennsylvanian is represented by limestone and sandstone in the Keele Range. A regional unconformity separates these rocks from the overlying Lower Permian shale, siltstone, and chert conglomerate which are exposed in the Keele Range and Ogilvie Mountains. Succeeding Permian strata include limestone with interbedded black shale and chert.

Farther south in the interior belt, Pennsylvanian and Permian rocks form thick sequences of bedded chert, argillite, limestone, basic volcanics, and ultramafics. Similar sequences, but lacking ultramafics, are present on Vancouver Island in the western belt. In the northern Coast Mountains, the St. Elias Mountains, and the Cascade Mountains more variable lithologies include volcanic arenite, andesite, tuff, conglomerate, and limestone. In most places, and particularly where bedded chert forms part of the sequence, Pennsylvanian and Permian strata are as much as several thousands of meters thick.

Mesozoic. Lower Triassic rocks are unknown in the interior and western belts. Middle Triassic rocks are found in the northern Coast Mountains, the St. Elias Mountains, and the southern part of the interior belt; their lithologies are similar to those of late Paleozoic strata. Pre-Middle Triassic rocks in the Stikine plateau region were deformed and regionally metamorphosed. Basalts and andesites, ranging in thickness from 1000 to more than 3000 m (3300–10,000 ft), characterize the Upper Triassic except in the northernmost part of the interior belt, where there are thin sequences (partly Middle Triassic) of limestone and shale. Limestone commonly forms the youngest Upper Triassic beds, thus indicating a general quiescence.

Throughout the length of the interior belt and in the eastern Coast Mountains and the Stikine plateau region, Lower Jurassic granodiorite plutons, the oldest recognized granitic rocks in the region, intrude Upper Triassic rocks and are unconformably overlain by Lower Jurassic strata. The northern Coast geanticline and Omineca geanticline were source areas for Lower Jurassic graywacke, shale, and conglomerate, of which several thousand meters were deposited in Whitehorse trough. Turbidites occur locally and some sequences contain abundant volcanic rocks. Middle Jurassic volcanics are important units on Queen Charlotte Islands and in the central and southern interior belt.

Granite intrusion and uplift segmented the interior belt during the Middle Jurassic. Several thousand meters of Upper Jurassic graywacke and shale were deposited in the Tantalus basin, Bowser basin, and the Tyaughton trough. Shales with minor amounts of sandstone are exposed in the northernmost part of the interior belt. Middle Jurassic granites intruded parts of Vancouver Island and possibly the Queen Charlotte Islands. Late Jurassic deformation and regional metamorphism were widespread in the northern Omineca geanticline.

Marine Lower Cretaceous rocks occur in the northern interior belt, the insular belt, and in St. Elias trough. Nonmarine sediments containing coal accumulated in the Tantalus and Bowser basins and in the basal section of the St. Elias sequence. Large volumes of post-tectonic granite were emplaced in the Omineca and Coast geanticlines during the early and middle Cretaceous. Uplift during the Aptian produced regional unconformities throughout much of the western Cordillera. Thick coarse sediments fill the Tyaughton trough, indicating uplift of the surrounding area. In this trough and in southern Bowser and Selwyn basins, Lower to Middle Cretaceous volcanic rocks of variable composition are well exposed. The Upper Cretaceous is represented by nonmarine, coal-bearing clastic sediments, in part very coarse and locally intercalated with andesite and basalt volcanics. These sediments were deposited in the northwestern Selwyn and the Sustut basins, and along major lineaments such as the Tintina and Rocky Mountain trenches. The coal-bearing rocks are interbedded with marine strata in Nanaimo basin.

Cenozoic. Local nonmarine sedimentation took place throughout the Cenozoic and volcanism was widespread. Intermediate to acid volcanics are particularly abundant in Eocene sequences, whereas plateau basalts characterize the Miocene to Holocene. Pleistocene glacial deposits and erosional features (Fig. 4) are typical of much of the western Cordillera except in the unglaciated part of the northern interior belt.

Metamorphism and Granitic Intrusion. Regional metamorphism and plutonism were concentrated along the Cassiar and Coast geanticlines during the Mesozoic and Cenozoic (see Fig. 5). In the Omineca geanticline and the Whitehorse trough a succession of granitic rocks ranging from early Mesozoic quartz diorites or granodiorites to early Cenozoic potassic granites have been emplaced at intervals of about 30 m yr. The highest-temperature phase of the regional metamorphism, which is not directly related to the granite intrusions, took place during the middle Triassic in parts of the Stikine plateau and during the late middle or early late Jurassic in the Pelly, Cassiar, and Omineca mountains.

Evolution of the Coast geanticline throughout the Mesozoic and early Cenozoic involved a complex series of metamorphic and plutonic events. Quartz diorite and diorite are common in the western parts of the geanticline, whereas more potassic rocks occur along the eastern margin. Roof pendants of metasedimentary rocks are widespread.

In the Cascade Mountains granitic intrusion continued until the late Tertiary.

Mineral Deposits

The interior and western belts of the Canadian Cordillera are rich in mineral deposits, which have been mined since the mid-19th century. Many settlements and transportation

FIGURE 4. Hudson Bay Mountain west of Smithers, British Columbia. (Photo: British Columbia Telephone Company.)

FIGURE 5. Simplified metamorphic map of the Canadian Cordillera (Monger and Hutchinson, 1971).

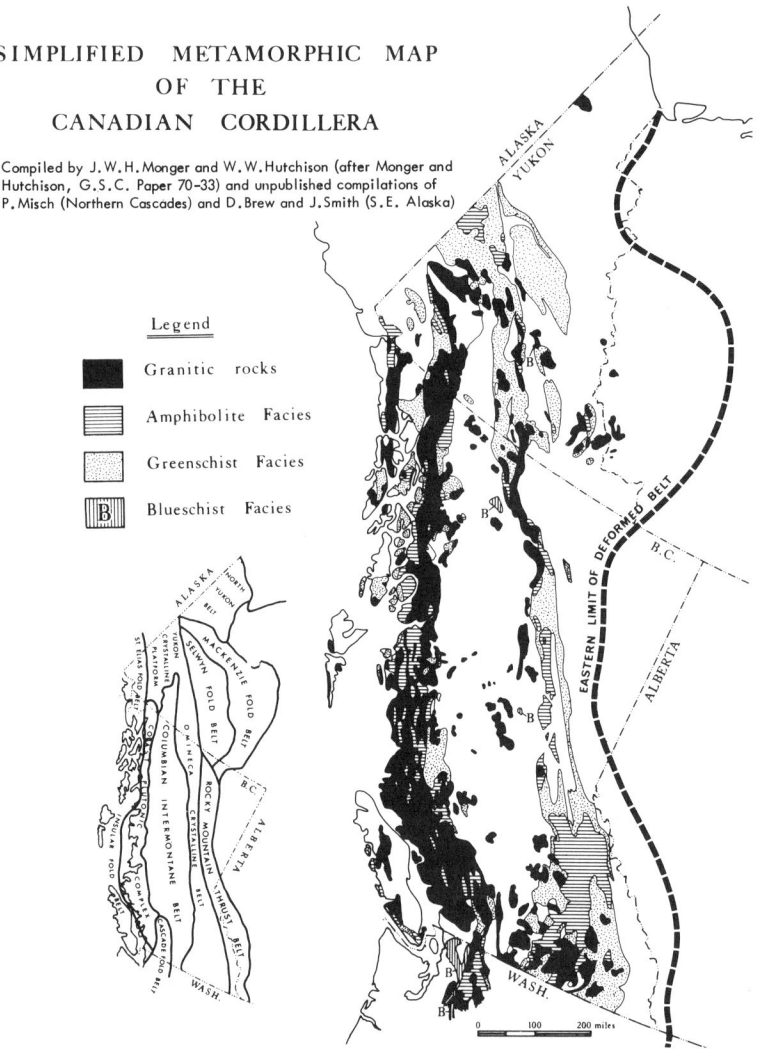

routes throughout the region owe their existence to the mineral industry.

Fuels. Since 1852 high-volatile bituminous coal has been mined from Upper Cretaceous strata in Nanaimo basin. This coal also occurs in rocks of similar age in Suquash basin and in the Queen Charlotte Islands. Lower Cretaceous coal has been mined near Telkwa in southern Bowser basin. Another coal field farther north in Bowser basin is possibly late Jurassic. In the Yukon Territory coal has been mined at Carmacks in Tantalus basin. Most Tertiary clastic rocks contain lignite, but it is produced only from the southern part of the interior belt.

Nonmetallics. Asbestos has been found in two ultramafic bodies at Cassiar and Clinton Creek in the northern part of the interior belt. The former contains exceptionally long fiber of high quality.

Barite, witherite, fluorite, perlite, talc, gypsum, and nephritic jade occur but have not been extensively mined.

Metallics. *Gold.* The early history of the mineral industry in the Cordillera was dominated by placer gold. Since the 1850s the interior belt in British Columbia has produced more than 5 million ounces, mainly from the Cariboo and Cassiar regions, whereas since the late 1890s the Yukon Territory with its famous Klondike region has produced more than 11 million ounces. Lode gold, generally in quartz veins, has been recovered from a number of mines, the most important of which are the Bralorne-Pioneer in the southern Coast Ranges, the Cariboo Gold Quartz in the Cariboo region, and the Hedley Camp in the southernmost part of the interior belt. The latter is unique in that it occurs within a skarn zone.

Silver. Many mines have been worked for silver and accompanying minerals in the Stikine arch region and in northern Selwyn Basin, where United Keno has been the largest silver-producing mine in Canada. Most deposits are in Mesozoic rocks.

Lead–Zinc. The largest known lead–zinc deposits are those near the Ross River in west-central Selwyn Basin. There, several roughly concordant ore bodies occur in Cambrian sedimentary strata. Many mines have yielded lead and zinc in combination with silver and/or copper.

Copper. In recent years extensive exploration has resulted in the discovery of a number of large copper deposits, particularly in the southernmost interior belt, in the Stikine area, and on Vancouver Island. Copper is commonly associated with Upper Triassic and Lower Jurassic volcanic rocks which were intruded by granitic plutons. Several, such as Copper Mountain and the recently opened Bethlehem mine in the southernmost interior belt, are porphyry copper deposits. The mine at Britannia in the southern Coast Ranges is in a replacement deposit along shear zones in a sedimentary roof pendant within granitic rocks.

Molybdenum. Several large, low-grade molybdenum deposits have recently been developed in the Skeena arch region and farther south in the interior belt. They commonly occur in leucocratic phases of granite plutons. Endako mine in central British Columbia is the largest molybdenum mine in Canada.

Iron. Pyrometasomatic magnetite deposits, some containing large amounts of copper, have been mined on Queen Charlotte, Vancouver, and Texada islands in the insular belt. All occur in skarn zones formed in Upper Triassic limestone which has been intruded by granite plutons. It is thought that the underlying Upper Triassic volcanic rocks were the source for the ore minerals.

Tungsten. Minor tungsten deposits are known but only two deposits have been mined. The Red Rose mine in the Skeena arch yields tungsten from wolframite, ferberite, and scheelite occurring in quartz veins that cut granitic rocks. At Canada Tungsten mine in the eastern Selwyn Basin, scheelite is mined from a skarn zone within Lower Cambrian limestone.

Mercury. The main producer of mercury in the Cordillera has been the Pinchi Lake mine, where cinnabar of either primary or secondary origin is found in carbonate rocks along the Pinchi fault.

Nickel and Chromium. Minor amounts of nickel and chromium are found in ultramafic and ultrabasic bodies throughout the Cordillera. The only commercial production, however, has come from an ultrabasic body mined for nickel in the southernmost part of the Coast geanticline.

HUBERT GABRIELSE

References

Brown, A. S., 1968. "Geology of the Queen Charlotte Islands, British Columbia," *B.C. Dept. Mines Petrol. Res. Bull.,* **54,** 226p.

Douglas, R. J. W., 1970. *Geology and Economic Minerals of Canada.* (Econ. Geol. Ser. 1, 5th ed.). Ottawa: Geol. Surv. Can., 838p.

Gabrielse, H., 1967. "Tectonic evolution of the northern Canadian Cordillera," *Can. J. Earth Sci.,* 4(2), 271-297.

Geological Survey of Canada, 1968. *Geological Map of Canada* (Map 1250A). Ottawa: Geol. Surv. Can., 1:5,000,000.

___, 1969. *Physiographic Regions of Canada* (Map 1254A). Ottawa: Geol. Surv. Can., 1:5,000,000.

Gunning, H. C., ed., 1966. "Tectonic history and mineral deposits of the western Cordillera, a symposium," *Can. Inst. Mining Met.,* Spec. Vol. 8.

Monger, J. W. H., and Hutchinson, W. W., 1971. "Metamorphic map of the Canadian Cordillera," *Geol. Surv. Can. Pap. 70-33,* 61p.

Wheeler, J. O., ed., 1970. "Structure of the southern Canadian Cordillera," *Geol. Assoc. Can. Spec. Pap.* 6, 166p.

Cross-references: *Canada; ,Canada–Rocky Mountains and Eastern Cordilleran Region; United States–Alaska, Pacific Cordilleran Region.*

CANADA–NORTHERN GREAT PLAINS PROVINCE

The Northern Great Plains Province of Canada continues to the Arctic Ocean and is an extension of the Great Plains Province of the United States of America. It is bounded on the east by the Canadian Shield, on the west by the Rocky Mountains and Mackenzie Mountains, and on the north by the Arctic Ocean and covers an area of about 2,200,000 km^2 (850,000 sq mi). Geographically it is contained within southern Manitoba, southern Saskatchewan, Alberta, northeastern British Columbia, and part of the Northwest Territories.

Historical Notes

Much of the original geological mapping centered around the discovery of coal (1857) in southern Saskatchewan; petroleum at Norman Wells (1920) in the Northwest Territories; natural gas (1883) at Langevin, Saskatchewan; and the discovery by Peter Pond in 1788 of bituminous sands at McMurray, Alberta. Development of the transcontinental railroads created a demand for coal, and subbituminous and lignite coals were mined in the areas around Estevan in Saskatchewan and in the Lethbridge, Drumheller, and Edmonton districts in Alberta.

In 1825–1827 Sir John Richardson conducted the first geological studies of the eastern margins of the interior plains while with the Sir John Franklin expedition. Additional data came later while working for the Franklin expedition in 1848. Subsequent history includes 1855, A. K. Isbister (Devonian at McMurray, Alberta); 1857, James Hector (Palliser expedition, southern plains); 1857–1858, H. Y. Hind (expeditions into Manitoba and Saskatchewan, with paleontology description by E. Billings); 1873, A. R. C. Selwyn (crossed the plains). The International (49th parallel) Boundary Survey brought G. M. Dawson west in 1874, and he later mapped the Bow and Belly rivers with R. G. McConnell. Beginning in 1881 McConnell traversed the Mackenzie, Peace, Athabasca, and Liard rivers. Around the turn of that century J. B. Tyrrell and D. B. Dowling worked out the stratigraphy of the Manitoba escarpment, and Dowling's subsequent work defined the general stratigraphy of the southern plains.

In the early part of the 20th century the plains were mapped by geologists J. A. Allan, A. E. Cameron, D. B. Dowling, C. O. Hage, G. S. Hume, E. M. Kindle, B. Kirk, F. H. McLearn, L. S. Russell, R. L. Rutherford, J. O. G. Sanderson, P. S. Warren, E. S. Whittaker, R. T. D. Wickenden, M. Y. Williams, and T. A. Link and his associates on the wartime Canol project.

Early mapping was dominated by the Geological Survey of Canada. From 1945 on oil exploration drill cores have revealed the deeper strata of the plains. The Alberta Society of Petroleum Geologists and the Saskatchewan and Edmonton Societies have published extensively on the subsurface geology. There is a new interest by provincial and federal surveys and conservation boards.

Important geology schools in the plains area are located at the provincial University of Manitoba (Winnipeg and Brandon), the University of Saskatchewan (Saskatoon and Regina), the University of Alberta (Edmonton), and the University of Calgary.

Geomorphology

The northern interior plains area is drained mainly by the Mackenzie River system, which

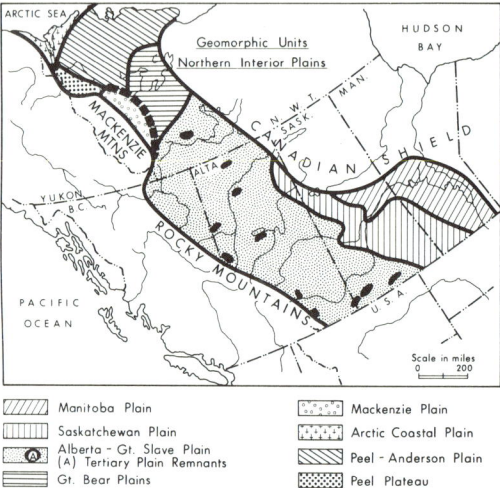

FIGURE 1. Geomorphic units of the northern interior plains, Canada.

flows into the Arctic Ocean, and by the Nelson River, which flows into Hudson Bay. Only the southernmost part of Alberta is in the Missouri River system. On the Arctic slope, the Anderson, Horton, and Hornaday rivers drain the area north of Great Bear Lake.

There are several natural divisions of the Great Plains (Fig. 1). The flat plains of Manitoba (first prairie level) represent the old bottom of glacial Lake Agassiz. Lakes Winnipegosis, Winnipeg, and Manitoba are the present remnants of that lake. The Saskatchewan Plain between the Manitoba escarpment and the Missouri coteau (second prairie level) has moraines, postglacial lake prairies, and long *coulées* dating back to when glacial waters flowed out through the Mississippi River system.

Southwestern Saskatchewan, Alberta, northeastern British Columbia, and the Northwest Territories south of the Liard River are parts of another natural physiographic unit. This unit has terminal moraines left by both Cordilleran and Keewatin glaciations. It is crossed by the S-trending late Pleistocene *coulées*, redissected by present rivers, and studded with remnants of the former Tertiary plain levels. Some of these former high plain levels are the Wood Mountain, Cypress Hills, Hand Hills, Porcupine Hills, Swan Hills, Clear Hills, Buffalo Head Hills, and Horn Mountains, as well as numerous uplands that jut out in front of the foothills, such as the Cochrane plateau and Nose Mountain. Proglacial lake bottoms form the flat Fort Nelson lowlands and the Peace River lowlands within the dissected uplands of northern Alberta and northeastern British Columbia.

North of the Liard River the central Mackenzie-Great Bear lowlands are bounded by

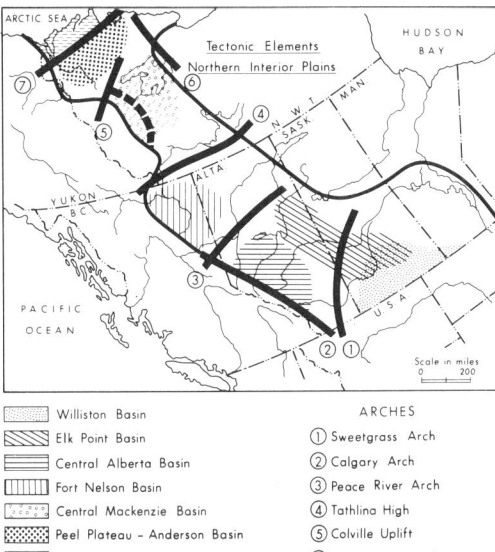

FIGURE 2. Tectonic elements of the northern interior plains, Canada.

the Mackenzie Mountains and are interrupted by the N-S-trending Franklin Mountains and by a series of E-W-trending crossfolds north of Norman Wells.

North of the Mackenzie and Franklin mountains and south of the Mackenzie River, the Peel plateau grades down to the Peel-Anderson-Horton plain into the Arctic coastal plain, with the Mackenzie delta at its widest point.

Geotectonics. Geotectonic divisions are not apparent from the surface features because they are based primarily on subsurface information (Fig. 2). Beneath the plains the W-dipping Precambrian basement reaches a depth of 6000 m along the western foothills. There are arches and basins in the basement which include the Williston basin in southern Saskatchewan; the Sweetgrass arch; the central Alberta basin; the Peace River arch; the Fort Nelson basin; the Tathlina high; the central Mackenzie low, with its disruptive welt of the Franklin Mountains; the Colville uplift; and Peel plateau—Anderson Basin; Inuvik-Minto arch; Beaufort Basin margin (Arctic coastal plain, Mackenzie delta); and the Coppermine arch. In the Devonian another basement arch (Calgary arch) existed along western Alberta and bounded the Elk Point basin of eastern Alberta and central Saskatchewan.

Structure

The interior plains have little surface structure. Salt collapse structures are known in the subsurface of southern Saskatchewan and are suspected in the central Mackenzie region. The Alberta syncline, which extends from southwestern Alberta with *en échelon* extensions up into northeastern British Columbia, controlled the distribution of the sedimentary beds in Alberta.

The plains in northeastern British Columbia are crossed by the E-W-trending Peace River anticlinal nose that plunges E.

From the southern Mackenzie Mountains a few structures plunge S under the plains to be reflected at the surface by the Petitot and Bovais Lake anticlines. Subsurface faults, downthrown to the E, roughly parallel the mountain front and continue S throughout northeastern British Columbia as far south as the Peace River.

In the Northwest Territories the Tertiary and late Cretaceous beds are usually preserved within discontinuous basins which formed between fault blocks and folds that extend to the Mackenzie and Franklin mountains. These mountains were deformed before the late Cretaceous. The Peel plateau is a gentle E-W-trending syncline, and the Inuvik uplift on the eastern margin of the Mackenzie delta seems to mark a pre-Mesozoic structural trend.

In the south the stratigraphic history indicates eustatic and gentle diastrophic warpings from the late Proterozoic to about the Eocene. Eocene uplift and folding accentuated middle Cretaceous structures in the central Mackenzie region. Igneous intrusions are not known within the plains area except for a few minette dikes that extend to the Sweetgrass buttes in Montana and for basaltic sills and dikes intruding the Ordovician-Silurian in the northern plains near the Arctic coast.

Historical Geology

Pre-Cretaceous sedimentary rocks under the plains of western Canada were deposited during multiple transgressions of the Pacific Ocean onto the margins of the Precambrian shield. Successive and more extensive transgressions penetrated eastward toward the present Rockies in Alberta, with the Precambrian Belt and Windermere clastic sediments defining a former shoreline with the Precambrian shield.

Paleozoic. Middle Cambrian seas deposited carbonates as far E as southern Saskatchewan, and supersaline basins fringed the shield in the Norman Wells area (Fig. 3). Late Ordovician to middle Silurian seas reached Manitoba and joined with a sea coming through Hudson Bay. Arctic flooding deposited Ordovician-Silurian carbonates over most of the plains area north of Great Slave Lake. A late Silurian regression to

FIGURE 3. Paleozoic stratigraphic terminology for the northern Great Plains, Canada.

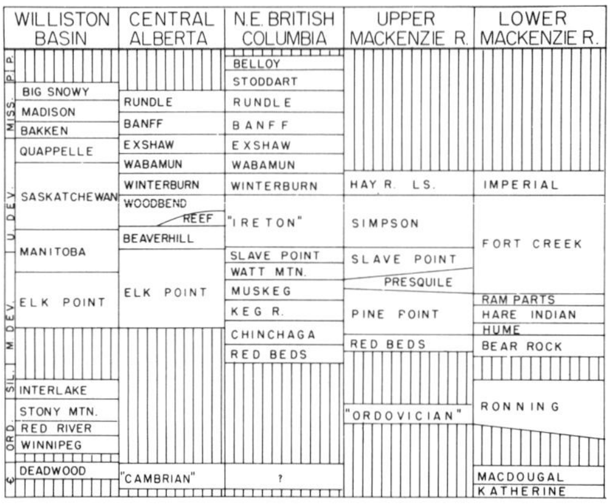

the Cordilleran area exposed most of western Canada to erosion. The middle Devonian sea, with marginal coral reefs, returned through an embayment in the Liard River region, transgressing all the plains area north of Great Slave Lake to the Arctic Ocean. The middle Devonian sea penetrated around the northern end of a peninsula (Calgary arch) and as far E as Manitoba. The Calgary arch extended from Montana to the Peace River area along the present Rocky Mountains and the foothills of Alberta. The embayment E of the Calgary arch became a salt basin (Elk Point basin) late in the Middle Devonian; however, by the late Devonian, shallow seaway connections were established and the Alberta basin became the site of a reef complex which accumulated with petroleum. At the end of the Devonian, uplift of the Arctic margin deposited shallow-water sands on the northern plains region. The late Devonian and Mississippian Pacific seas inundated the Calgary arch, and shallow-water Mississippian carbonates bear oil in southeastern Saskatchewan and southwestern Manitoba. Pennsylvanian and Permian regression caused the shoreline to return to western Alberta and northern Yukon.

Mesozoic. In the Triassic the Pacific seas transgressed the southern Peace River area. The shallow-water Triassic carbonates and sandstones in northeastern British Columbia produce petroleum (Fig. 4).

The Jurassic brought a renewed advance of the Pacific seas into the southern Canadian plains, and some of the oil fields in southwestern Saskatchewan produce from shoreline sands of this transgression. Sandy beds mark the withdrawal of the Jurassic seas, and the shoreline once again moved to western Alberta by the end of the Jurassic. In the early Cretaceous, uplift in central British Columbia restricted the seas to an embayment in the Peace River area. This embayment connected with the Pacific Ocean between the Omineca and Selkirk mountains.

Uplift in early to Middle Cretaceous, along with emplacement of batholiths in British Columbia, caused the Pacific seas to withdraw from the Cordilleran region. To the east the Arctic Ocean transgressed from the north, penetrating as far south as south-central Alberta. Advancing and retreating sandy shorelines of this Arctic sea formed many of Alberta's smaller oil fields and the McMurray tar sands. The rising Mackenzie Mountains restricted the Arctic embayment and the interior seaways then connected with the Gulf of Mexico. In the late Cretaceous broad seas extended from western Alberta to Manitoba and from the Mackenzie Mountains to beyond Great Bear Lake. Clastic debris eroded from rising mountains in British Columbia and formed a series of deltas along western Alberta in the late Cretaceous and oil and gas accumulated in the sands. Expulsion of waters from the Gulf of Mexico took place in the late Cretaceous when the Belly River and Edmonton formation deltas covered Alberta and the seas finally withdrew through southern Saskatchewan.

Cenozoic. Western Canada was emergent at the start of the Cenozoic, except for the Mackenzie delta, and whether or not the Paleocene sea entered central Canada is open to question. The continental Paleocene sands of the Paskapoo formation are the last extensive deposits in western Canada. The land rose and only terrace and river gravels were deposited, to be later redissected by further uplift and erosion.

Pleistocene advancing glaciers blocked established stream patterns and the preglacial deposits are closely related to the overlying glacial deposits. Except during the maximum advance

FIGURE 4. Mesozoic stratigraphic terminology for the northern Great Plains, Canada.

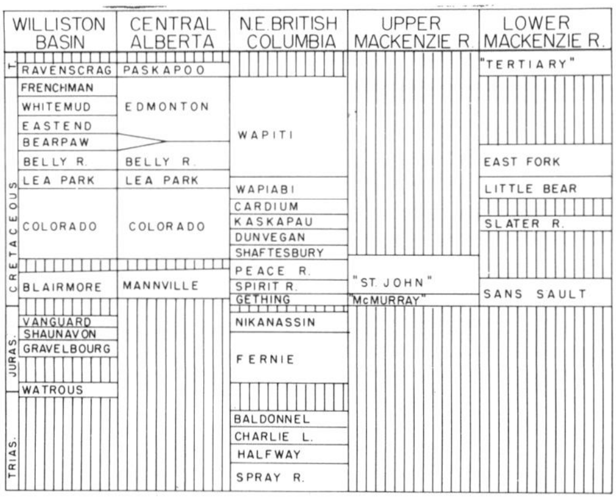

of the Wisconsin ice, a "corridor" was left through Alberta and northeastern British Columbia between the Cordilleran and the Keewatin ice sheets, permitting faunal migration. Postglacial lakes created much of the present flat farmlands of western Canada, and reestablishment of the drainage created many of the picturesque *coulées* in the country. Postglacial uplift caused reentrenchment of the river systems in the plains area, thus a majority of the valleys are quite young.

Stratigraphy

In many places the surface of the Precambrian basement has weathered to an arkose or to a similar clastic deposit ("granite wash") which ranges in age from Cambrian to Devonian.

There is about 650 m of Cambrian rock under the western plains. The Cambrian isopachs thin to zero in the subsurface along the eastern side of the Williston basin and the southern side of the Peace River arch. The Lower Cambrian of the Mackenzie River area consists of clastic rocks, and the Middle Cambrian is represented by red beds and salt deposits. The Upper and Middle Cambrian strata of the prairie provinces are mainly reddish sandstones with carbonates appearing on the western side of the plains.

The Ordovician strata are confined to the Williston basin and to the Mackenzie River valley north of Great Slave Lake. There are mostly Upper Ordovician carbonates in the south, with both Middle and Upper Ordovician found in the lower Mackenzie area and on the Arctic slope.

The Silurian is confined to the Williston basin, to the Manitoba escarpment in the S, and to the Mackenzie portion N of the Horn Mountains. Silurian shallow-water carbonates conformably overlie the Upper Ordovician and are difficult to distinguish from the Ordovician.

Devonian strata were originally deposited over all the plains of western Canada, but now they crop out only along the margins of the interior plains. An erosional hiatus separates the Silurian from the Middle Devonian. The Middle Devonian is represented by fine sandstones, carbonates, and salt deposits in the Williston basin and in eastern Alberta; dolomites, carbonates, and shales in northern Alberta and in the Mackenzie valley; and black shales in the western part of northeastern British Columbia. Upper Devonian shallow-water carbonates, supersalines, and fine sandstones are found in the Williston basin. Reef carbonates and shales occur in Alberta; and with an increase in sands to the NW the sequence becomes dominantly sandstones in the Mackenzie valley, with coarse clastics appearing in the lower Mackenzie area.

The Mississippian of the Williston basin is composed of shallow-water carbonates, fine clastics, and occasionally anhydrite beds. In southern Alberta the lower part of the sequence is shale, although the upper Mississippian is carbonate. In northeastern British Columbia the lower part appears as a black shale facies with sandstones upward. The Mississippian does not crop out in the Mackenzie valley proper.

Pennsylvanian and Lower Permian are generally not represented except in the Peace River area, where coarse sandstones and occasional carbonates represent interrupted succession.

The Triassic extends under the plains in the westernmost portion of central Alberta and in the Peace River area of Alberta and British Columbia. The Lower Triassic is a marine calcareous siltstone, but the Middle Triassic is composed of beach sands, shallow-water carbonates, red beds, and anhydrite. The uppermost Triassic is missing from the plains.

The Jurassic is restricted to the Williston basin and to western Alberta and northeastern British Columbia. In southern Saskatchewan, it consists of sands, shales, and thin limestones; and in Alberta and British Columbia, the Jurassic is represented by marine shale and fine sand.

The latest Jurassic and earliest Cretaceous are absent except on the Arctic slope. The Lower Cretaceous consists of fluvial sands and shales in southern Saskatchewan and southern Alberta but has marine tongues in central Alberta. Lower Cretaceous marine shales predominate the Mackenzie valley with sand tongues occurring in northern Alberta and northeastern British Columbia. In Manitoba and Saskatchewan Upper Cretaceous shales at the Saskatchewan-Alberta boundary are in facies with sand tongues that were deposited in a brackish environment, the upper portion becoming delta sands in western Alberta. In northeastern British Columbia-Liard River area, a hiatus occurred during the Upper Cretaceous. Late Cretaceous beds in the Northwest Territories are present as erosional remnants as far north as the Arctic Ocean.

Tertiary deposits are entirely continental except in the Mackenzie delta, and gravels of fluvial origin are remnant highlands in Alberta and southern Saskatchewan. In the lower Mackenzie valley the Tertiary continental beds remain in the lowlands of the basin and range complex south of the Norman Wells area.

Quaternary beds are mainly glacial tills, sands, and gravels, along with preglacial and postglacial lake deposits. On the Arctic coast the Quaternary delta deposits of the Mackenzie

River occupy the northwestern extension of the interior plains.

Celebrated Upper Cretaceous dinosaur fossils, made famous by the Sternbergs and Barnum Brown, are found in the Belly River and Edmonton formations in the badlands of the Red Deer River in southern Alberta. The Swift Current plateau Oligocene beds have yielded excellent *Titanothere* and *Brontothere* faunas.

Orogenies

Diastrophic uplifts in the mid-Ordovician, early Devonian, late Paleozoic, and early Cretaceous only slightly deformed the plains area. Middle-Tertiary uplift of the Rocky Mountains, however, did fold beds in western Alberta, forming the Alberta syncline.

Late Eocene plutonism (Sweetgrass intrusions) along the southernmost Alberta boundary formed small minette dikes. A Caledonian orogenic phase affected the Arctic coastal region, but most of this mobile belt is covered by later sediments. In northeastern British Columbia and the Peace River area of Alberta, late Paleozoic basement faulting (reactivated sporadically in the Mesozoic) folded the Paleozoic at depth.

In the central Mackenzie valley a basin and range terrain exists, with the Paleozoic rocks cropping out through the Tertiary and the Upper Cretaceous which cover the evidence of a Middle Cretaceous orogeny.

Economic Geology

Mineral Deposits. There are important mineral deposits in the plains area.

Iron Ore. A bedded medium-grade oolitic iron ore occurs in the Upper Cretaceous in the Clear Hills-Peace River district and contains an estimated 60 billion tons of an exploited hematite-goethite ore.

Lead-Zinc. An epithermal replacement deposit in Middle Devonian reef carbonates occurs on the south shore of Great Slave Lake. The 12% galena-sphalerite ore content is estimated at 120 million tons (in production since 1965).

Potash and Salt. The Middle Devonian salines of the Elk Point formation contain bitter salts, and shafts have been sunk to exploit the K-salts at Esterhazy and in basins near Saskatoon, near Regina, and near Yorkton, Saskatchewan. Reserves to 1150 m depth may contain 6.4 billion tons of recoverable potash ore. Salt has been recovered through drill holes at Elk Point and McMurray in Alberta.

Nonmetallics. Some pottery clays have been mined in the Medicine Hat and Eastend area from the Upper Cretaceous (Whitemud) beds. The only building stone being quarried is the Tyndall stone, a mottled Ordovician limestone from Manitoba. Gypsum deposits are known, but not exploited, from the lower Peace River and west of Great Bear Lake along the Mackenzie. Sulfur reserves from natural gas are about 80 million tons.

Coal. Coal reserves of the plains are estimated at about 25 billion tons. Alberta and Saskatchewan have late Cretaceous and Paleocene deposits of lignite to subbituminous-rank coal. Cretaceous coal is mined from the Lethbridge area in southern Alberta, from the Drumheller badlands of the Red Deer River, and from the Edmonton district of central Alberta. Paleocene coal is found just east of the foothills in central Alberta and Tertiary lignites are mined at Estevan in Saskatchewan.

Tar Sands (Oil Sands). Heavy oil-bearing sands of the Lower Cretaceous McMurray formation crop out along the lower Athabasca River, and other Lower Cretaceous "tar" sands underlie much of northern central Alberta. Estimated reserves are on the order of 100–300 billion barrels. Similar sands are also reported north of Great Bear Lake.

Oil and Gas. The proved oil reserves of the plains are estimated at 4 billion barrels (1963), and gas reserves about 40 trillion ft^3. In southeastern Saskatchewan and southwestern Manitoba there are numerous Mississippian oil fields on the northeastern pinchouts of the Williston basin. In southwestern Saskatchewan shoreline Jurassic beds have pinchout traps on the northwestern margin of the Williston basin. The truncation of the westward-dipping Mississippian in western Alberta formed a series of oil and gas reservoirs that are joined by Permian and Triassic pinchout traps in the Peace River area of Alberta and British Columbia.

The present principal oil fields are in the Devonian reefs that surround the Edmonton (western Alberta) basin. Successively more extensive Devonian transgressions across the Peace River "high" to the south left a series of younger reefs.

Small oil fields in the Lower Cretaceous have been developed in central Alberta. A marine wedge of Upper Cretaceous (Cardium) sand formed the largest (in areal extent) oil field in Canada, covering about 1300 km^2 (Pembina field).

CHARLES R. STELCK

References

Alberta Society of Petroleum Geologists, 1960. *Oil Fields of Alberta.* Calgary: Alberta Soc. Petrol. Geologists, 272p.

___, 1969. *Gas Fields of Alberta.* Calgary: Alberta Soc. Petrol. Geologists, 407p.
Caldwell, W. G. E., 1970. "The Cretaceous system in Saskatchewan," *Saskatchewan Geol. Soc. Mesozoic Core Seminar Pap.* **1,** 6p.
Clark, T. H., and Stearn, C. W., 1968. *Geological Evolution of North America.* New York: Ronald Press, 570p.
Douglas, R. J. W., ed., 1970. *Geology and economic minerals of Canada,* (Econ. Geol. Ser. 1, 5th ed.). Ottawa: Geol. Surv. Can., 838p.
___, et al., 1963. "Geology and petroleum potentialities of northern Canada," *Geol. Surv. Can. Pap.* **63-31,** 26p.
Geological Survey Canada, 1969. *Geological Map of Canada* (Map 1250A). Ottawa: Geol. Surv. Can., 1:5,000,000.
McCrossan, R. C., ed., 1958. *Annotated Bibliography of Geology of the Sedimentary Basin of Alberta and of Adjacent Parts of British Columbia and Northwest Territories, 1845–1955.* Calgary: Alberta Soc. Petrol. Geologists, 499p.
___, and Glaister, R. P., eds., 1964. *Geological History of Western Canada.* Calgary: Alberta Soc. Petrol. Geologists, 232p.
Nelson, S. J., 1970. *The face of time: the Geological History of Western Canada.* Calgary: Alberta Soc. Petrol. Geologists, 133p.
Oswald, D. H., ed., 1967. *International symposium on the Devonian system, Calgary, 1967,* vol. 1. Calgary: Alberta Soc. Petrol. Geologists, 675–878.
Prather, R. W., and McCourt, G. B., 1968. "Geology of gas accumulations in Paleozoic rocks of Alberta Plains," *in* B. W. Beebe, ed., *Natural Gases of North America.* (Mem. 9, vol. 2). Tulsa: Am. Assoc. Petrol. Geologists, 1238–1284.
Stockwell, C. H., 1957. *Geology and economic minerals of Canada* (Econ. Geol. Ser. 1, 4th ed.). Ottawa: Geol. Surv. Can., 517p.
Webb, J. B., 1964. "Historical summary," *Geological History of Western Canada.* Calgary: Alberta Soc. Petrol. Geologists, 218–232.

Cross-references: *Canada; Canada–Arctic Archipelago, Canadian Shield, Rocky Mountains and Eastern Cordilleran Region; North America; United States–Great Plains Province.*

CANADA–ONTARIO BASIN

The Ontario Basin of Canada occupies approximately 67,000 km^2 (26,000 sq mi) of the southern part of the province of Ontario, adjacent to the lower Great Lakes (Huron, Erie, and Ontario). The Ontario Basin includes the most densely populated part of Canada, occupying a key position in agriculture, industry, and transportation, notably as the base of operations for much of northern Ontario's valuable mineral resources.

The following summary is based partly on the material in Guidebooks A42 and A45 of the International Geological Congress, Canada, 1972 (see references: Terasmae *et al.*, 1972; Winder and Sanford, 1972.).

The principal centers of geological studies and research in the Ontario Basin area are at the universities: Toronto, Queens (Kingston), Waterloo, McMaster (Hamilton), Brock (St. Catherines), Western Ontario (London), and Windsor. Systematic investigation and mapping is carried out by the Geological Survey of Canada (Ottawa). The provincial authorities (Ontario Ministry of Natural Resources) deal with economic deposits. The Canada Center for Inland Waters, Burlington, has a geological section and is concerned with lake problems.

Physiography and Bedrock Geology

Southern Ontario comprises three major physiographic regions: The *Central St. Lawrence Lowland* (an area bordered by the Ottawa and St. Lawrence rivers, extending eastward from Ottawa and Brockville), the *West St. Lawrence Lowland* (southwestern Ontario bordered by lakes Ontario, Erie, and Huron), including Manitoulin Island, and the *Frontenac Arch,* which trends NW-SE and separates the two lowlands.

Both the Central and West St. Lawrence lowlands are underlain by sedimentary rocks of Paleozoic age. The Frontenac Axis is part of the Canadian Shield (Grenville Province).

The Niagara Escarpment, extending from the western end of Lake Ontario to Georgian Bay and Manitoulin Island, transects the West St. Lawrence Lowland, as a northeasterly-facing cuesta, formed by resistant carbonates of Silurian age that dip gently southwest into the Michigan Basin. In the south, and on Manitoulin Island, the escarpment attains an elevation of about 195 m and a relief of 90 m, whereas near Georgian Bay the relief is up to 300 m. The lowland below the escarpment is underlain by Cambrian and Ordovician strata up to 600 m thick near the western end of Lake Ontario. Farther west Cambrian, Ordovician, and younger Devonian strata reach a thickness of 1500 m at the south end of Lake Huron and beneath Lake Erie at the U.S./Canadian boundary (see Fig. 1).

The Algonquin Arch, trending in a southwest direction beneath southern Ontario, forms a broad positive basement lineament (Precambrian) from which overlying Paleozoic strata either dip westward into the Michigan Basin, where they reach a maximum thickness of about 4300 m in central Michigan, or southward into the Allegheny Trough, where they attain an ultimate thickness of more than 6000 m in Pennsylvania and West Virginia.

Farther south, the north-trending Findlay Arch extends through western Ohio and beneath the extreme western part of southern

FIGURE 1. Geological map of the Paleozoic sequence of the Ontario Basin (Winder and Sanford, 1972).

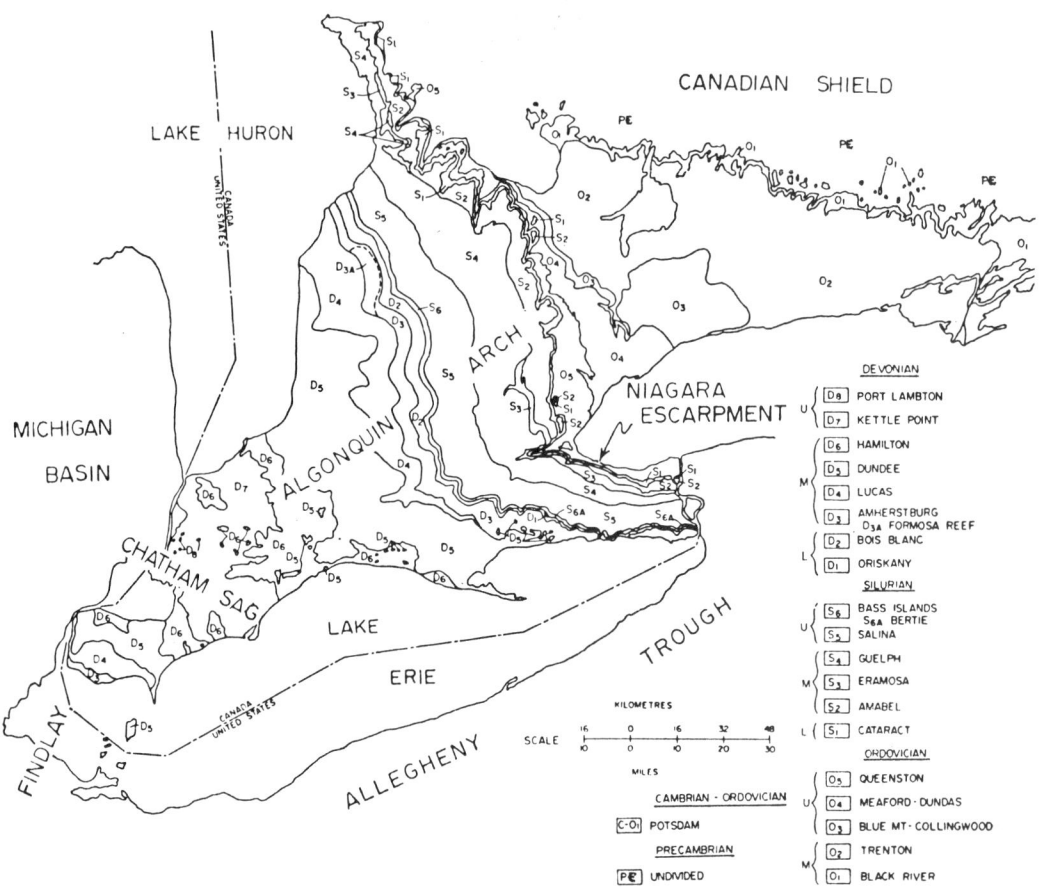

Ontario, and adjacent region of southeastern Michigan. The topographical prominence of the Findlay Arch decreases in Ontario, and its present surface expression is apparent only in the vicinity of Essex County and the adjacent Bass Islands in western Lake Erie.

Although Paleozoic strata generally appear to be flat lying, regional dips are generally consistent at 5.5–8.5 m/km into the Michigan Basin and Allegheny Trough, respectively. Numerous local reversals of regional dip and normal faults have been delineated in subsurface by drilling in exploration for petroleum and natural gas. Minor faulting can also be seen locally at the surface, and well-developed joint systems are conspicuous in most outcrop sections (Brigham, 1972).

Historical Notes

The first systematic geological investigations of southern Ontario were begun soon after the inauguration of the Geological Survey of Canada in 1842. Sir William Logan, the Survey's first director, assisted by such eminent scientists as Alexander Murray, T. Sterry Hunt, Robert Bell, and Elkanah Billings, made the initial classification of the Paleozoic rocks here. In 1843 Alexander Murray investigated the region between Georgian Bay and Lake Erie, where he established a stratigraphic framework similar to the New York section, comprising some 10 lithostratigraphic divisions.

In 1861 T. Sterry Hunt proposed the "Anticlinal Theory of Accumulation" as a result of observations of oil seepages in Gaspé, Quebec. This basic concept was almost immediately put to practical use in the early development of the oil industry in Ontario, as well as in the United States. Credit for much of the early biostratigraphy and correlation of Paleozoic rocks must go to Elkanah Billings, Survey Paleontologist, to whom fell the laborious task of describing the fossil collections that were submitted by the field geologists. Accounts of these investigations can be found in Logan's 1863 report, "The Geology of Canada."

In the present century further detailed investigations have been carried out. In the Ordovician system, significant contributions were made by Johnston (1912, 1914), Parks (1923, 1928), Okulitch (1939), Winder (1960), and Liberty (1969). Classic studies of the Silurian Niagara Escarpment were made by Williams (1919) and Bolton (1957). Devonian rocks were studied in detail by Stauffer (1915) and Best (unpubl. thesis).

The first systematic geological mapping of bedrock formations in southern Ontario was carried out by Caley during the years 1936 to 1947. The Geological Survey of Canada memoirs are a valuable guide to the geology of the region. The investigations of Caley were continued by Sanford during the years 1949–1966.

A number of recent contributions to the economic geology of southern Ontario were made by Hewitt (e.g., 1960). A standard subsurface grid was designed by Beards (1967). The nomenclature followed here is mainly that of the Geological Survey of Canada (*in* Douglas et al., 1970).

Stratigraphy

Precambrian. Bordering the Paleozoic terrain of southern Ontario are the metamorphic and igneous rocks of the Canadian Shield (Grenville Province). Peneplanation of the Precambrian surface was thorough but left residual hills. In places deep weathering profiles are preserved, for example, at Kingston in a roadcut on Highway 401 that exposes Paleozoic sediments overlapping Precambrian rocks that had weathered to some depth prior to the deposition of the basal conglomerate.

During the initial phase of Paleozoic sedimentation (Late Cambrian to Early Ordovician), the peneplaned surface of the broad Precambrian Algonquin Arch extended southwestward across southern Ontario, thence due S into west-central Ohio. The depositional and erosional truncation of Cambrian and Early Ordovician strata against the flanks of the Algonquin Arch and subsequent onlap and overlap of initial Middle Ordovician strata across its axis indicates that it was a topographically positive feature during the early Paleozoic.

In contrast, the Findlay Arch seems to have evolved during or after Middle Devonian time. A broad depression, the "Chatham Sag," developed after Late Silurian time between it and the Algonquin Arch.

Upper Cambrian and/or Lower Ordovician. Except for scattered outliers preserved along

FIGURE 2. Classification and correlation of the Paleozoic rocks of the Ontario Basin (simplified, from Winder and Sanford, 1972). Note that the Windsor-Sarnia . . . column is in part separated from the Hamilton-Niagara sequence by the Algonquin Arch.

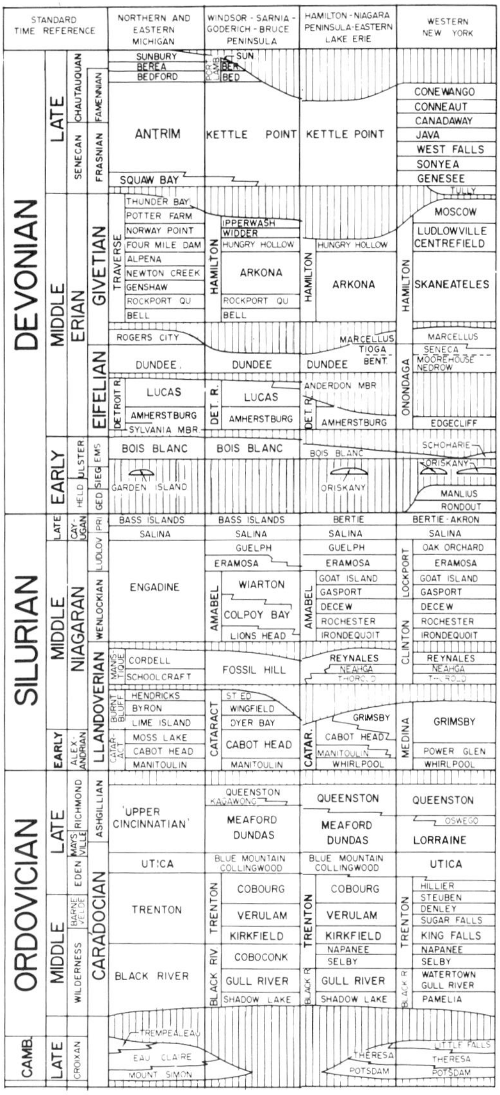

the southern flank of the Frontenac Arch in a region N of Kingston, Upper Cambrian or Lower Ordovician strata are confined to the subsurface, where their beveled edges are overlapped unconformably by Middle Ordovician rocks (see Figures 2 and 3). Flanking the Algonquin Arch, Cambrian or younger rocks gradually increase in thickness from their erosional zero edge to 155 m beneath Lake Erie and ultimately to 2000 m in the Allegheny Trough. In this part of Ontario they are referred to as the Potsdam, Theresa, and Little Falls formations in ascending order of sequence, a terminology

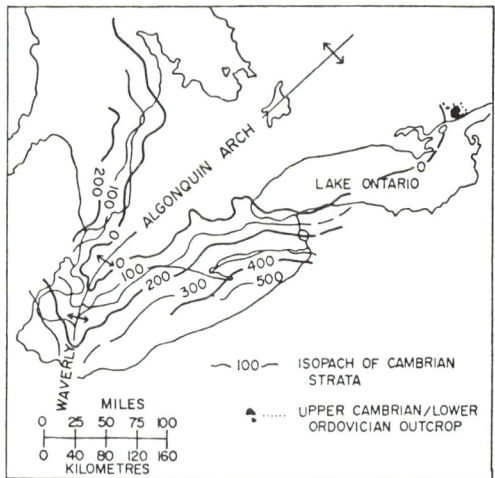

FIGURE 3. Isopach map of Upper Cambrian rocks.

adopted from the lithologically similar Upper Cambrian succession bordering the Adirondack Mountains in New York State. The basal Cambrian strata which form the Potsdam Formation are comprised of arkosic conglomerate, and red, white, and gray ortho-quartzitic sandstone up to 46 m thick. These grade upward to alternating orthoquartzitic sandstone and gray or gray-tan finely crystalline oolitic dolomites and gray shaly and glauconitic dolomite, up to 100 m thick, of the Theresa Formation. The youngest Cambrian strata encountered in drill holes beneath central Lake Erie are dolomites up to 31 m thick of the Little Falls Formation. The formations are highly diachronous as they are traced into the Allegheny Trough.

On the western side of the Ontario peninsula, bordering the Michigan Basin, Cambrian strata thicken rapidly to 620 m in southern Michigan. The Mount Simon and Eau Claire formations have been traced into Ontario from Michigan and Wisconsin, and for the most part are lithologically similar to their counterparts eastward, the Potsdam and Theresa formations.

Fossils have not yet been recovered from these rocks in Ontario, but their age is considered to be Late Cambrian on the basis of subsurface stratigraphic continuity with rocks at the St. Croixan type section in Wisconsin and in New York State. The Potsdam sandstones at the Hughes Farm "Park of Pillars" 16 km N of Kingston, are unfossiliferous and are assumed to be of Late Cambrian or Early Ordovician age. Nearby they rest directly on Precambrian rocks with angular unconformity, but are overlapped by strata of Middle Ordovician age. On the northeast side of the Frontenac Arch (Ottawa Valley), however, Wilson (1946) favored an Ordovician age for the unfossiliferous Nepean (Potsdam) because of its gradational relationship with succeeding Lower Ordovician carbonate rocks. It is not unreasonable to assume that the Potsdam may also be of Early Ordovician age, at least in part, at many localities bordering highland areas of the Canadian Shield and Adirondack Mountains. Terrigenous sedimentation began in Dresbachian time and continued upward through the Franconian and Trempealeauan, and into the early Canadian epoch as the shoreline gradually transgressed the Precambrian highland surface (Fisher, 1956).

Middle and Upper Ordovician. Rocks that form the Paleozoic-Precambrian boundary between eastern Lake Ontario and Georgian Bay are comprised largely of carbonates of Middle Ordovician age. Over this distance of some 330 km they form a low cuesta. The escarpment on the north is 3–15 m high, rising above the Precambrian peneplaned surface. South of the escarpment Middle and Upper Ordovician rocks form the bedrock surface over a wide region of south-central Ontario, and eventually disappear beneath the Silurian rocks of the Niagara Escarpment.

The Ordovician system is readily divisible into two contrasting gross lithological divisions; a lower limestone succession (Black River and Trenton groups) up to 300 m thick, and an upper terrigenous succession comprised mainly of shales (Collingwood, Blue Mountain, Meaford-Dundas and Queenston formations) up to 550 m thick.

The Black River Group, 30–155 m thick, comprises the oldest Middle Ordovician rocks in southern Ontario and is divisible into the Shadow Lake, Gull River, and Coboconk formations in ascending order of succession. The relatively unfossiliferous Shadow Lake overlaps Upper Cambrian strata to rest with angular unconformity on the Precambrian. The formation is comprised of maroon, red, and green silty and calcareous shales and siltstones alternating with argillaceous, dolomitic limestone.

In the subsurface, the basal Shadow Lake commonly contains arkose and coarse rounded and frosted quartz grains, the latter informally referred to as "golf ball sand" from their appearance under the microscope. Because of the irregular configuration of the Precambrian basement surface the thickness is highly variable over short distances. The Gull River Formation locally overlaps the Shadow Lake to lap onto the Precambrian. It is a microcrystalline limestone (micrite) containing interdigitating beds of dolomite; black chert and metabentonite seams are common. Except for ostracods, fossils are relatively rare in lower Gull River strata; in the middle and upper beds distinctive forms include *Cryptophragmus, Tetradium, Bathyurus,* endocerid and actinocerid cephalopods, a few coiled gastropods, brachiopods, and occasional stromatolites. The Coboconk Formation is thick-bedded crystalline limestone (calcisiltite and calcarenite). It locally contains chert nodules. Characteristic fossils are such distinctive forms as *Stromatocerium, Tetradium, Doleroides, Sowerbyella, Triplesia, Eodinobolus, Maclurites, Gonioceras,* and *Receptaculites.*

Primary sedimentary features, such as flat pebble conglomerate, cross-bedding, desiccation polygons, and worm burrowing, are well preserved in the Black River Group. Reconstructed environments appear to vary from subtidal to intertidal and possibly supratidal conditions (Winder, 1960; Mukherji, 1969).

In contrast to the dominantly microscopic textures of the Black River Group are the 90–170 m of argillaceous bioclastic limestones of the Trenton Group. These strata begin with shaly limestones alternating with bioclastic limestones and calcarenites that form the Kirkfield Formation. The formation is characterized by the brachiopod *Resserella* and crinoids. Overlying it are the bioclastic limestones, calcarenites, and interbedded shaly limestones and thin shales of the Verulam Formation. The base contains a profusion of the bryozoa *Prasopora.* Other forms include *Cryptolithus,* and ostracods such as *Basslerata, Thomasatia,* etc. The upper Trenton strata are represented by the argillaceous microcrystalline limestones of the Cobourg Formation, which contains such common Trenton genera as *Isotelus, Hormotoma, Fusispira, Rafinesquina,* and some distinctive new genera, including *Probillingsites, Plectatrypa, Catazyga, Oxoplecia, Pseudogygites,* and *Triarthrus.*

The Trenton limestones are generally more shaly than the Black River, and more fossiliferous. They probably represent shallow shelf deposits which were forming marginal to the Allegheny Trough (Winder, 1960).

Succeeding the Trenton Group is a variable succession of shales, sandstones, and carbonate rocks that thicken from 155 m in Bruce Peninsula to 560 m beneath Lake Erie. At the base are dark shales of the Collingwood and Blue Mountain formations, they contain a similar fauna, including *Pseudogygites* and *Triarthrus.* The succeeding Meaford-Dundas Formation forms a succession of gray shales, siltstones, and some limestones. The Queenston Formation, maroon shales and siltstones, represents the youngest Ordovician strata in southern Ontario. Northwestward it gradually changes to the gray shales and dolomites of the Meaford and Kagawong formations, respectively. The Meaford and Dundas fauna is similar to that of the Trenton but contains distinctive pelecypods, such as *Byssonychia,* and bryozoa. The Queenston is only sparsely fossiliferous bordering the Allegheny Trough, but enclosed biostromes in the Georgian Bay area contain a faunal assemblage diagnostic of the Richmond stage.

The stratigraphic nomenclature, regional correlation, and biochronology of the Ordovician system in eastern North America has long been encumbered by controversy and lack of agreement based on subjective considerations. In more recent years, however, the North American time-stratigraphic nomenclature (stages) has become relatively standardized, as shown in Fig. 2, on the basis of benthonic faunas and conodonts.

Silurian. The Silurian system in southern Ontario is bounded on the north and east by the prominent Niagara Escarpment that extends from the vicinity of Rochester, New York, to Manitoulin Island. The Middle Silurian rocks which form the crest of the escarpment are hard, resistant carbonates, and these are underlain by less resistant shales (Lower Silurian and Upper Ordovician). As the shales erode, the carbonate rocks break off along the well-developed joint systems, thus preserving the steep face of the escarpment. The Silurian system is lithologically heterogenous to a high degree and has been described by Bolton (1957), and in the subsurface by Sanford (1969).

The Silurian rocks of this region originated for the most part as shallow platform sequences, deposited marginal to two sedimentary basins: (1) the elongated Allegheny Trough on the south, extending into the Niagara Peninsula and beneath eastern Lake Erie, and (2) the Michigan Basin on the west, its eastern rim passing through western and central Lake Erie, thence N to parallel the present axis of the Algonquin Arch. The Allegheny Trough was the site of clastic sedimentation during the Early Silurian, whereas during the Late Silurian both the Allegheny Trough and the Michigan Basin became dominated by carbonates and evaporites.

Lower and Early Middle Silurian. The oldest Silurian rocks comprise the Cataract Group (see Fig. 2). In the eastern Niagara Peninsula this term embraces the white and gray orthoquartzitic Whirlpool sandstone, the gray, green, and red Cabot Head shales, and the red protoquartzitic sandstones of the Grimsby, in ascending order. The Cataract Group would appear to be bounded at base and at top by regional uncomformities. The upper surface has been considerably leveled by erosion. The Group is, in large part, of marine subtidal and intertidal origin. It grades rapidly eastward in New York State to coarse clastic red beds of deltaic origin, referred to the Medina Group. Paleocurrent structures are well displayed in most of the formations, particularly the Grimsby, which contains cross-bedding, cut and fill and roll structures, shale chip conglomerate, etc. (Martini, 1971).

Characteristic fossils of the Cataract Group include *Paleofavosites, Brockocystis, Virgiana, Resserella, Coelospira, Helopora, Phaenopora, Liocalymene, Leperditia, Zygobolba,* and the trail *Arthrophycus alleghaniensis.*

Middle Silurian. Overlying the Cataract Group is a complex succession of crinoid bank and complex reef deposits with associated shales and minor sandstone that fall into three genetically related lithostratigraphic divisions: (1) the Fossil Hill and equivalent Thorold, Neahga, and Reynales formations, (2) the Amabel Group, and (3) the Guelph and partly equivalent Eramosa formations.

High-energy shelf-edge deposits occur along the margin of the Michigan Basin in Ontario from Bruce Peninsula to the Algonquin Arch; thence extending beneath central and western Lake Erie. Throughout this belt the rocks are 130–170 m thick and rapidly thin to 30 m in the lower part of the Michigan Basin and to 100 m in Niagara Peninsula and eastern Lake Erie.

Halysites and *Pentamerus* are common fossils to both the Reynales and Fossil Hill formations. In addition, the Fossil Hill Formation locally contains an abundant fauna, including such genera as *Pentameroides, Costistricklandia, Camarotoechia, Favosites, Alveolites,* and many others.

The Amabel Group unconformably succeeds the Fossil Hill-Reynales formations and comprises a variety of carbonate and shale facies, deposited in a number of contrasting environments. Fringing the Michigan Basin is a thick succession of dolomites composed largely of echinoderm ossicles and plates of the Colpoy Bay and Wiarton formations. These intertongue with cherty dolomites (calcisiltite and micrite) of the Lions Head Formation in the lower part of the Michigan Basin. The crinoid bank deposits of the Colpoy Bay and Wiarton formations have wide distribution in Bruce Peninsula and on the Algonquin Arch, but decrease to a narrow band in the southwest.

The Amabel Group is largely a product of a shallow subtidal environment. The Wiarton-Colpoy Bay crinoid bank deposits are of high-energy origin and formed the shelf platform between two contrasting depositional environments: deeper waters of the Michigan Basin to the northwest; and the shelf lagoon (Allegheny Trough) to the south.

Fossils are common to abundant in the Amabel Group. They include the genera, *Dicoelosia, Salopina, Eocoelia, Conchidium, Costistricklandia, Stephanocrinus, Caryocrinites, Paraechmina, Fletcheria, Scutellum,* and *Cheirurus.*

The Guelph Formation overlies the Amabel and consists in large part of clean platformal carbonate rocks that form a thick reef complex around the margins of the Michigan Basin. These consist of tan dolomite that grades to dark bituminous and cherty dolomite in the Michigan Basin. Immediately adjacent to the platform are large bioherm patch reefs with a vertical relief of as much as 50 m. Toward the center of the Michigan Basin, where subsidence was more rapid, pinnacle reefs developed with a relief of 100–150 m. Dolomitization has generally obliterated fossils, but where preserved they usually occur as external molds, such as the pelecypod *Megalomus,* the brachiopod *Trimerella,* and the coral *Pycnostylus.*

Upper Silurian. The youngest Silurian rocks of southern Ontario are comprised of dolomite, limestone, salt, anhydrite, gypsum, and shale of the Salina and Bass Islands (Bertie) formations. They have a maximum thickness of 500 m bordering Lake Huron (Fig. 4), and thin to 100 m over the Algonquin Arch. The reconstructed depositional edge of Salina salts bordering the Michigan Basin roughly coincides with the edge of the thick carbonate bank deposits of the subjacent Guelph Formation. The gradual migration of the salt edge to its present position much lower in the basin is the result of leaching (see Figs. 2 and 3) (Sanford, 1965).

The extensive megafauna of the Silurian is illustrated by Williams (1919), Bolton (1957, 1966), and Norford *et al.* in Douglas *et al.* (1970).

In European terminology the Llandoverian stage embraces the Cataract Group and the Thorold, Reynales-Fossil Hill formations; the Wenlock stage, the Amabel Group, which includes the Irondequoit to Goat Island formations; and the Ludlow-Pridoli, the Guelph, Salina and Bass Islands of southern Ontario Correlation of the relatively unfossiliferous Salina and Bass Islands of southern Ontario with the late Ludlow and Pridoli stages of Berdan *et al.* (1969) is largely based on the stratigraphic position of these formations.

Studies of conodonts by Rexroad and Rickard (1965) and Pollock *et al.* (1970) have resulted in a biostratigraphic succession which can be correlated with the standard English series.

Devonian. If the Silurian-Devonian boundary of southern Ontario was not largely obscured by thick Quaternary deposits, an abrupt

FIGURE 4. Profile of Silurian strata, N-S, along the eastern flank of the Michigan Basin in SW Ontario (B. V. Sanford).

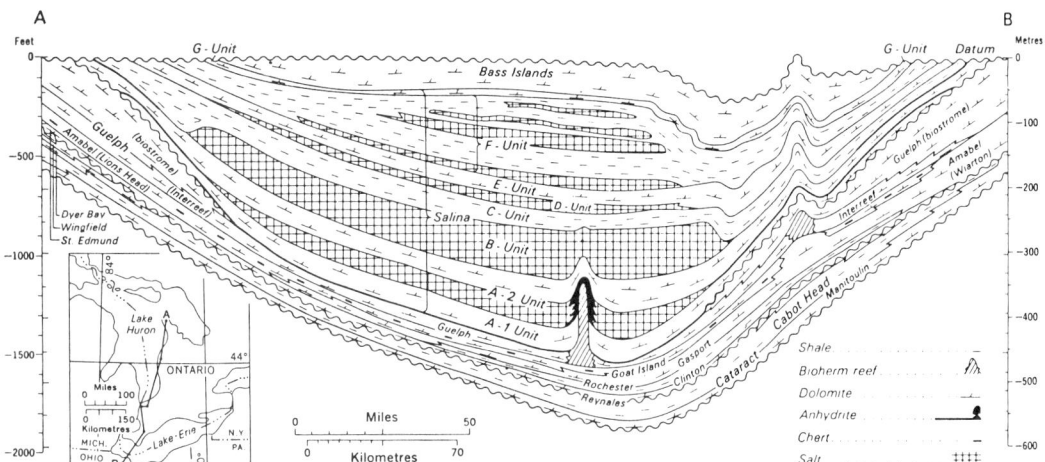

topographical feature comparable to the Niagara Escarpment could be observed to extend from the Niagara River to Bruce Peninsula, thence beneath Lake Huron to Mackinac Island in northern Michigan. This cuesta has resulted from the erosion of soft Upper Silurian shales that are overlain by hard resistant carbonates (Bois Blanc Formation) of Lower Devonian age (see Fig. 5).

Lower Devonian. The oldest Devonian strata of southern Ontario are orthoquartzitic sandstones of the Oriskany Formation that are preserved as erosional remnants in depressions on the irregular Upper Silurian surface in Niagra Peninsula. Because the Oriskany was in turn subjected to erosion, its distribution is unpredictable and nowhere is it known to exceed 6 m in thickness. The formation contains a typical Oriskany fauna, including *Rensellaeria,* *Acrospirifer, Hypparionyx,* and *Plethorhyncha.*

The late Lower Devonian Bois Blanc Formation unconformably succeeds the Oriskany. It consists of dolomitic limestone, characterized by an abundance of nodular chert. The Bois Blanc is dominated by brachiopods, including *Amphigenia* and many other genera. It also includes such characteristic rugose corals as *Aemulophyllum exiquum, Edaphophyllum sulcatum,* and *Acrophyllum oneidense.*

Middle Devonian. Succeeding the Bois Blanc Formation is a variable succession of limestones and dolomites of Middle Devonian age. These strata are 180 m thick at the foot of Lake Huron and thin southwestward to 120 m beneath Lake Erie. The Amherstburg Formation initiates this succession; it is comprised of brown sucrosic dolomites. Intertonguing with it along the southern part of the Michigan Basin

FIGURE 5. Profile of Devonian strata, NW-SW, across the Algonquin Arch (B. V. Sanford).

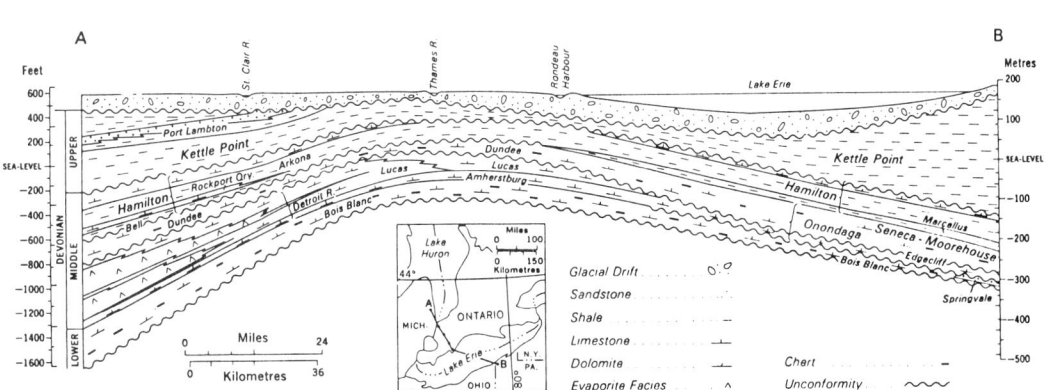

(Ontario and Michigan) are strandline and eolian deposits comprised of well-sorted, rounded, and faceted orthoquartzitic sandstone comprising the Sylvania Member. The Amherstburg grades southeastward to coral and crinoidal limestones in Niagara Peninsula. In adjacent areas of New York State, equivalent beds form the Edgecliff Member of the Onondaga Formation. Bioherm reefs occur in the Amherstburg in Ontario, bordering the Michigan Basin. The Amherstburg contains a prolific rugose coral fauna of a variety of ptenophyllid, disphyllid, and cystiphyllid genera.

Succeeding the Amherstburg are microcrystalline dolomites of the Lucas Formation; these grade to anhydrite and halite in the Michigan Basin. To the SE they give place to aphanitic limestone, locally referred to the Anderdon Member. Carbonate rocks of the Lucas Formation became emergent immediately following their deposition and were truncated by erosion over a wide area bordering the Michigan Basin. They were thus either removed by erosion or were never deposited in the Allegheny Trough.

The Lucas Formation contains a relatively meager fauna characterized by *Prosserella*, gastropods, stromatoporoids, and stromatolites. An upper sandy unit of the Anderdon contains *Paraspirifer, Brevispirifer, Hexagonaria,* and *Eridophyllum.*

The succeeding Dundee Formation rests unconformably on the Lucas. The formation is comprised of brown crystalline limestones that are commonly crinoidal and cherty at the base, becoming fine grained at the top. The Tioga bentonite, which forms the boundary between the Moorehouse and Seneca members of the Onondaga Formation in New York State, has been recognized 9–12 m above the base of the Dundee Formation in southern Ontario.

The Dundee limestones of Ontario contain a relatively sparse fauna; some of the most common species include *Atrypa elegans, A. costata,* and *Brevispirifer lucasensis.*

The erosional edge of the overlying Hamilton Group is marked by a buried escarpment that extends beneath Lake Erie, crossing Ontario to Lake Huron, and thence to the Thunder Bay, Michigan. The oldest beds of the Hamilton Group are the black bituminous shales of the Marcellus Formation. The formation is nowhere exposed at the surface in Ontario, but loose slabs found along the shoreline of Lake Erie have yielded *Orbiculoides* and *Leiorhynchus.*

On the eastern margin of the Michigan Basin, shales and limestones of the Traverse Group overlap the beveled edges of the Rogers City and Dundee formations and merge with the Hamilton Group in southern Ontario.

Exposures of the Hamilton Group in southern Ontario are confined to the south end of Lake Huron.

The Hamilton strata are highly fossiliferous. The list given in Stumm and Wright (1958) includes 987 species found in the outcrop.

Upper Devonian. The Upper Devonian of southern Ontario is represented by the Squaw Bay and Kettle Point formations, and Port Lambton Group in ascending order of succession. The type section of the Kettle Point is on the southeast shore of Lake Huron, where it consists of thin, evenly bedded black shales, with numerous spherical calcareous concretions up to 125 cm in diameter that are commonly referred to as "kettles." Pyrite nodules, up to 7 cm are also common. The Kettle Point contains conodonts and pyritized radiolaria (Winder, 1966), the brachiopod *Lingula,* and a profusion of amber colored *Tasmanites.*

The Kettle Point is succeeded by the Bedford (soft, gray shales), Berea (gray argillaceous sandstones), and Sunbury (black shales) formations, in ascending order, that reach a combined thickness of 62 m along the shore of the St. Clair River.

Regional correlation and biostratigraphy of the Devonian system in Ontario and Michigan has been reviewed by Sanford (1967). The sequence contains several unconformities of both local and regional significance.

The isolated outcrop of Oriskany sandstone, near Nelles Corners in Niagara Peninsula, contains an Early Devonian (Siegenian) fauna. The Bois Blanc Formation is also Late Devonian (Emsian) on the basis of corals and is equivalent to the Schoharie Formation of New York State.

The early Middle Devonian (Eifelian) is represented by the Amherstburg, Lucas, and Dundee formations which pass laterally into the Edgecliff, Nedrow, Moorehouse, and Seneca members of the Onondaga Formation in New York State (Sanford, 1967). The Dundee limestone at St. Marys contains conodonts of Eifelian age (Ferrigno, 1971). The Hamilton Group is considered Givetian, although there is some disagreement about the lowermost unit, the black Marcellus shale, which might be in part Eifelian (see Oliver et al.; 1967). The Hungry Hollow Formation contains a diagnostic fauna which permits direct correlations with the Centerfield of New York State, the Four Mile Dam of northern Michigan, and the Ten Mile Creek of northwestern Ohio.

The basal strata of the Late Devonian (Frasnian) are nodular black chert (possibly a remnant of the Squaw Bay Formation) in which Winder (1967) found a "conodont hash" spanning a considerable range of time. The type section of the Kettle Point Formation is basal

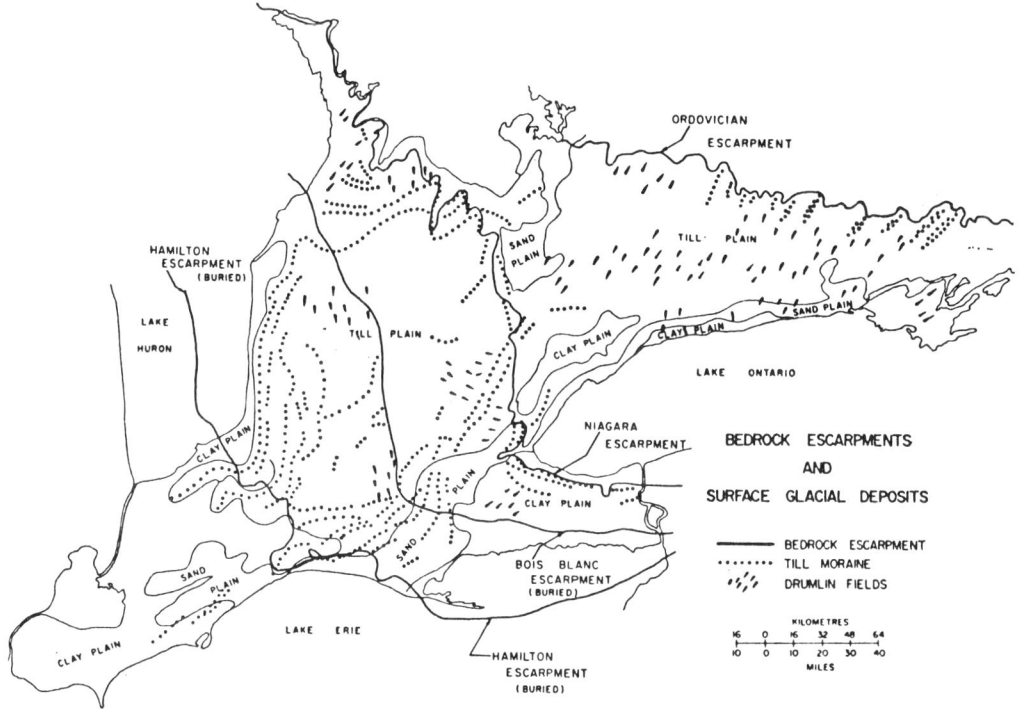

FIGURE 6. Bedrock escarpments and surface glacial features (Terasmae et al., 1972).

Famennian (Winder, 1962). However, the condonts recovered from drill core suggests that the entire Upper Devonian succession is represented in southern Ontario (Winder, 1966). On the basis of spores, D. C. McGregor has established that the youngest Paleozoic strata in southern Ontario (Bedford, Berea, and Sunbury formations) are of Late Devonian age (Sanford, 1967, p. 994).

Late Paleozoic-Mesozoic-Cenozoic. During post-Devonian time, the entire region has been remarkably stable, subject only to gentle upwarping and tilting along preexisting trends. During the Mesozoic and Cenozoic the region must have been involved in the climatic and peneplanation cycles of the rest of the continent, but little is known of this history from local evidence.

Late Paleozoic-Mesozoic. Uplift and erosion was accompanied by extensive leaching of Silurian salts, resulting in local deformation, minor faulting, and development of joint fracture systems.

Quaternary. Southwestern Ontario has a relatively continuous cover of glacial deposits relating to the last glaciation (Wisconsin). Numerous end-moraines were formed by ice lobes occupying the different Great Lakes basins and extensive glacio-lacustrine sediments cover areas adjacent to the Great Lakes.

In the Central St. Lawrence Lowland, marine sediments of late Wisconsin age (the Champlain Sea) cover the glacial deposits and extend up the Ottawa River valley to Pembroke. The cover of Quaternary deposits on the Canadian Shield is generally discontinuous because of greater relief of the bedrock surface.

Quaternary deposits older than the last glaciation are rare on the surface in southern Ontario and are only found in or adjacent to deep preglacial valleys that dissect the sedimentary rocks of the West St. Lawrence Lowland.

Glaciers probably advanced into Ontario from the N and E several times during the Pleistocene. The present surface expression of the region is a reflection of the last glacial cycle which began about 100,000 yr B.P. and terminated during the ice recession about 10,000 yr B.P. (Wisconsin)

The Paleozoic rocks are for the most part obscured by Quaternary glacial deposits. Four major cuestas have been carved into the bedrock surface; two of these are prominently displayed and the third and fourth are buried beneath glacial sediments (see Fig. 6). These are: (1) *Black River Escarpment,* which extends from Kingston to Georgian Bay, with elevations rising to 15 m or more above the Precambrian surface. A succession of small lakes occur along this lineament. This escarpment has been modified to a large degree by glacial erosion. (2) *Niagara Escarpment,* 70–300 m high, which parallels the south shore of Lake Ontario as far W as Hamilton, thence swinging northwestward to Owen Sound and Bruce Peninsula. This escarpment has also been modified by glacial scouring. (3) *Bois Blanc Escarpment,* for the most part obscured beneath glacial deposits, extends from the Niagara River to the vicinity of Douglas Point on Lake Huron. (4) *Hamilton Escarpment,* which extends from the east end of Lake Erie, crossing the Ontario peninsula, thence NW beneath Lake Huron to Thunder Bay in northeastern Michigan.

Glacial deposits have rather limited thickness and distribution in Precambrian terrain, but from the Paleozoic contact they increase in thickness generally toward the SW, reaching about 125 m along the north shore of Lake Erie. In some buried valleys and interlobate moraines they may exceed 250 m.

The glacial deposits, as illustrated in Fig. 6, largely consist of ground moraine with many end-moraines and outwash deposits, with drumlin fields in certain areas. In the vicinity of Peterborough, about 400 drumlins have been recognized. The numerous sand plains and clay plains are usually related to the succession of glacial lakes formed during the recession. Successions of beach deposits have been established adjacent to and surrounding all the Great Lake basins (Sly and Lewis, 1972).

Traveling, for example, from Montreal to Toronto (see International Geological Congress Guidebook A42: Terasmae *et al.,* 1972), three very distinctive physiographic divisions can be seen in relationship to their Quaternary origin:

1. The flat marine clay plains and terraces of the Champlain Sea (11,800–10,200 yr B.P.: Mott, 1968) in the East St. Lawrence Lowland of Quebec that extend up the St. Lawrence Valley as far as Brockville (Henderson, 1970) and up the Ottawa Valley as far as Petawawa (above Pembroke). Since the sea followed the retreating ice, the "marine limit" is not synchronous, rising from around 120–170 m above present MSL near Montreal to around 200 m in the Ottawa Valley. Numerous richly fossiliferous localities with youngest Pleistocene and earliest Holocene marine faunas are known, mostly associated with beaches, cobble ridges, and littoral dunes that formed along the shores and islands of this great temporary arm of the sea. Typical fossils include *Hiatella arctica* (L.), *Mya arenaria* (L.), *Balanus crenatus* (Bruguiere), *Mytilus edulis* (L.), and *Macoma balthica* (L.). These marine and estuarine deposits may exceed 50 m thickness in places and were progressively uplifted during the postglacial glacio-isostatic rebound, and terraced, at first by wave action and later (early and middle Holocene) by the fluvial action of the St. Lawrence and Ottawa rivers, and their tributaries. Each terraced phase appears to represent an interruption caused by either a temporary halt in isostatic crustal uplift or a temporary eustatic drop of sea level. A characteristic feature of the salty marine silty clays ("*Leda* Clay") at the terrace margins is the extensive landsliding (the "quick-clay" phenomenon).

2. Between Brockville and Kingston the St. Lawrence Valley crosses the Frontenac Arch, which joins the Precambrian of the Canadian Shield to the Adirondack Uplift of New York State. The Quaternary geology of this section has been studied in detail by Terasmae (1965) in conjunction with the St. Lawrence Seaway Project during the years 1956–1959. The highway from Cornwall to Morrisburg traverses the clay plain of the Champlain Sea, interrupted by low ridges and knobs of glacial till. West of Prescott the Paleozoic sedimentary rocks rise above the clay and till plain and can be observed in shallow roadcuts. This change is due to structural geology of the bedrock; here the Paleozoic sediments rise on the eastern flank of the Frontenac Arch. The cover of surficial deposits thins and becomes discontinuous, and bedrock outcrops are the predominant feature. Glacial erosional phenomena are well displayed on the Precambrian outcrops.

3. The surface relief of the Precambrian rocks is much greater than that of the nearly flat-lying Paleozoic rocks. Between Gananoque and Kingston the cover of surficial deposits, and also the Paleozoic sediments on the Precambrian rocks, thickens and becomes continuous on the west flank of the Frontenac Arch. Between Kingston and Belleville the Paleozoic rocks are relatively flat-lying and show little deformation. The cover of surficial deposits is generally thin, as indicated by frequent roadcuts showing bedrock outcrop. However, there are many valleys cut into this bedrock plain. At least some of these valleys are preglacial in origin, and the largest of them are occupied by the present Napanee, Salmon, and Trent rivers. The bedrock surface is commonly grooved and striated, and a good example of this is shown just east of Napanee.

Here, in the West St. Lawrence Lowland there are no marine deposits but only widespread accumulations of the proglacial lake Iroquois, dating from the last deglaciation (11,500–12,500 yr B.P.), an ancestor of the present Lake Ontario, but much larger.

Pre-Wisconsin Quaternary sediments were mainly covered over or scraped off by the last ice advance except on and near the preglacial valleys, e.g., in the Toronto area.

The Pleistocene sequence of the Toronto area is one of the most well known in North America. It gained its fame largely through numerous papers by A. P. Coleman between 1894 and 1941, although an earlier, less-publicized account by G. J. Hinde (1877) described many of the major features. Main features are the extensive exposures of interglacial sediments overlying, and overlain by, layers of glacial till. These interglacial beds were found to contain abundant remains of fossil plants and animals, many of which were believed to be extinct.

FIGURE 7. Time and rock-stratigraphic names of the Wisconsin stage in southern Ontario. (*Note:* Nissouri corresponds to Tazewell in the Midwest sequence, and Point Bruce to Cary.)

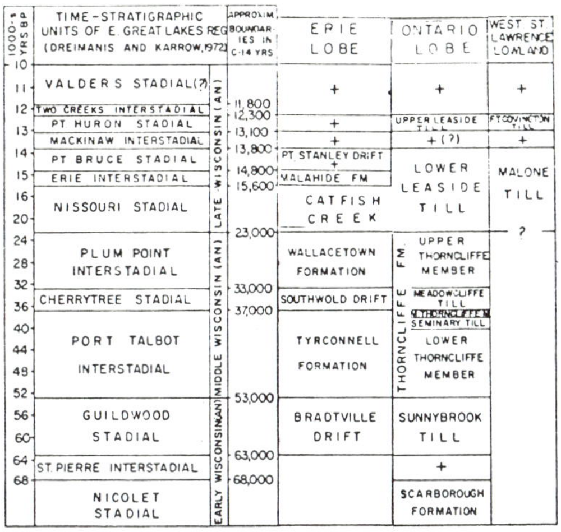

1000 YRS BP	TIME-STRATIGRAPHIC UNITS OF E. GREAT LAKES REG (DREIMANIS AND KARROW,1972)	APPROX. BOUNDARIES IN C-14 YRS	ERIE LOBE	ONTARIO LOBE	WEST ST LAWRENCE LOWLAND
10					
11	VALDERS STADIAL(?)		+	+	+
12	TWO CREEKS INTERSTADIAL	11,800			
13	PT HURON STADIAL	12,300	+	UPPER LEASIDE TILL	FT COVINGTON TILL
14	MACKINAW INTERSTADIAL	13,100	+	+ (?)	+
15	PT BRUCE STADIAL	13,800	PT. STANLEY DRIFT	LOWER LEASIDE TILL	MALONE TILL
16	ERIE INTERSTADIAL	14,800	MALAHIDE FM		
20	NISSOURI STADIAL	15,600	CATFISH CREEK		
24		23,000		UPPER THORNCLIFFE MEMBER	?
28	PLUM POINT INTERSTADIAL		WALLACETOWN FORMATION		
32		33,000		MEADOWCLIFFE TILL	
36	CHERRYTREE STADIAL	37,000	SOUTHWOLD DRIFT	THORNCLIFFE SEMINARY TILL	
40					
44	PORT TALBOT		TYRCONNELL FORMATION	LOWER THORNCLIFFE MEMBER	
48	INTERSTADIAL				
52		53,000			
56	GUILDWOOD STADIAL		BRADTVILLE DRIFT	SUNNYBROOK TILL	
60		63,000			
64	ST PIERRE INTERSTADIAL	68,000		+	
68	NICOLET STADIAL			SCARBOROUGH FORMATION	

The age of the interglacial beds has been a major problem inasmuch as they are beyond the reach of radiocarbon dating. Recent paleomagnetic and other studies suggest that they are in the range 120,000-80,000 yr B.P. (Mörner, 1972), that is, approximately Sangamon (in the U.S. scale) or Eemian (European). Beneath the interglacial beds is an earlier till (York Till) usually classified as Illinoian. Striations on the underlying Ordovician limestone (exposed in the Don Valley Brickyard in Toronto) indicate ice movement from the E and SE.

Over the York Till in the classic section in the Don Valley Brickyard, and also in the Scarborough Bluffs (modern cliffs of Lake Ontario), is the Don Formation, comprising stratified sand, silt and clay, and containing fossil shells (freshwater), wood, leaves, pollen, and diatoms (Coleman, 1933; Terasmae, 1960), suggesting an interglacial lakeside climate with average temperatures several degrees above that of today. The lake was more than 18 m higher than today. The upper surface of the formation is deeply weathered. The Don Formation is followed by the Scarborough Formation, a clay followed by a sand, a deltaic sequence that contains abundant floral and faunal remains; fossil diatoms, wood, pollen, ostracods, insects, and molluscs are found and a cool climate is inferred (Karrow, 1967). These beds extend several miles and have a maximum elevation of about 120 m above sea level. They are believed to represent a large delta, formed in a lake which stood more than 45 m above present Lake Ontario. This lake of Scarborough time had almost exactly the same level as glacial Lake Iroquois, which had its outlet via Rome, N.Y., to the Hudson Valley and the Atlantic. It was held up at this level by a dam of glacial ice in the middle St. Lawrence Valley.

Following the high-water stage of the Scarborough Formation, there was a low-water stage when valleys were cut in the delta; this may have occurred during a period of ice retreat which allowed drainage down the St. Lawrence Valley. Immediately on top of the Scarborough Formation is an extensive till sheet named the Sunnybrook Till. Fabric studies indicate ice movement to the west out of the Ontario Basin. This till is assigned to the Early Wisconsinan (about 50,000-65,000 years old). As this ice retreated, water was ponded in front of it and varved clays were deposited. The outlet for these glacial waters was probably again at Rome, because they reached an elevation similar to that of the later Lake Iroquois.

During a major nonglacial interval, the mid-Wisconsinan (30,000-50,000 years ago), a complex of sediments named the Thorncliffe Formation was deposited on top of the Sunnybrook Till; it consists of stratified clay, silt, and sand. In a few places plant remains have been found in the beds, and a cool climate is indicated. The nature of the sediments suggests that lacustrine and fluvial environments prevailed. Later in this interval, ice readvanced from the E to deposit till layers alternating with waterlaid deposits; two major till units have been traced along the bluffs but have not been found to extend far inland. Farther north, beyond the limit of these minor advances, sediments of the Thorncliffe Formation continued to accumulate. Thus, two tills, the lower named the Seminary Till (a thin layer of clayey sand till), and the upper named the Meadowcliffe Till (a silty clay till), interfinger with sediments of the Thorncliffe Formation.

As colder conditions returned, the major ice advance of the "classical" or Late Wisconsinan covered the area, depositing a thick layer of silty sand till known as the Leaside Till. In many locations N of the Bluffs, this till layer can be subdivided into two layers of similar lithology, commonly separated by sand and gravel of kame origin. For about 17,000 years there was continuous ice cover over the Toronto area. Then, during the time of the Mankato (Port Huron) substage, the ice retreated from this area, exposing a drumlin-covered and fluted till plain. Meltwaters ponded in front of the ice formed lakes at various temporary levels, the most prominent being that of Lake Iroquois, with its outlet at Rome. This lake stage has an average radiocarbon age of about 12,000 years (Karrow et al., 1961).

The effect of Lake Iroquois was to erode a prominent terrace and shorecliff in the glacial and interstadial deposits. At the foot of the cliff, blocks of till and varved clay fell into the lake and were covered by lacustrine silt. Farther offshore varved clay was formed. Gravel bars were built across embayments in the shoreline, and headlands were eroded. Continued ice retreat opened the St. Lawrence Valley and the lake was drained (in several stages) to a very low level. The outlet had been much depressed by the weight of the ice and the spillway debouched into the rising Champlain Sea, beginning about 12,000 yr B.P. (Prest, 1970).

Corresponding close to the beginning of the Holocene epoch, about 10,000 years ago, the lowest water level was reached and Lake Ontario began; since then isostatic rebound has gradually raised the outlet to its present level of 75 m above sea level. Along the old Iroquois bluff, gulleys were eroded, from the mouths of which small alluvial fans were built on the level surface of the Iroquois terrace. Molluscs, vertebrate bones, and charcoal have been found in these deposits; radiocarbon dates of about 5000 yr B.P. have been obtained. Recent wave erosion by Lake Ontario is causing cliff retreat at about 30 cm/yr.

The next Quaternary province to the W and S is that formerly occupied by the Huron and Erie lobes of the continental ice sheet. Interlobate moraines are preserved in the center and there are numerous extensive recessional endmoraines looped around the southwest part of the province.

Beneath the late Wisconsin tills on the southern side of the southwestern Ontario peninsula there is a region of interest comparable to that of the Scarborough-Don interglacial deposits. This is the Plum Point-Port Talbot area along the north shore of Lake Erie. In this case a Middle Wisconsin interstadial sequence is displayed (Dreimanis and Karrow, 1965; Dreimanis et al., 1966; Terasmae et al., 1972). Early Wisconsin tills (Bradtville Drift, around 53,000–63,000 yr B.P.) are found in bores below lake level, and are overlain by the Tyrconnell Formation (Port Talbot Interstadial), consisting of varved clays, lacustrine silt, gyttja, and beach deposits (dated around 42,000–47,000 yr B.P.). These are separated by a minor readvance indicator, the Southwold Drift (Cherrytree Stadial, approximately 33,000–37,000 yr B.P.), from a second relatively warm phase, the Wallacetown Formation (Plum Point Interstadial, approximately 23,000–33,000 yr B.P.). All is covered by the Catfish Creek Drift, representing the main advance of the late Wisconsin continental ice (Nissouri Stadial).

The Port Talbot and Plum Point interstadials are also represented in all probability by some 22 m of pollen-bearing sand, silt, and clay with organic matter revealed by drilling in a buried channel of the Niagara Gorge that leads from St. Davids to the Whirlpool (Hobson and Terasmae, 1969). Much of the present gorge has been eroded during the last 12,000 years, although at times (when the principal Great Lakes drainage was deflected north) it may have been very greatly reduced. Thus, an *average rate* of erosion for that period is a meaningless symbol. The present gorge was partially eroded at a very rapid rate for about 1000 years and later reactivated on a weaker scale since the closing of the North Bay spillway (around 5500 yr B.P.; see below).

During deglaciation of the Erie and Huron basins in late Wisconsin time a series of pro-

FIGURE 8. Map showing limits of glacial lakes Algonquin and Iroquois about 12,500 yr B.P. (with elevations in feet). (Map: Geol. Surv., Canada.)

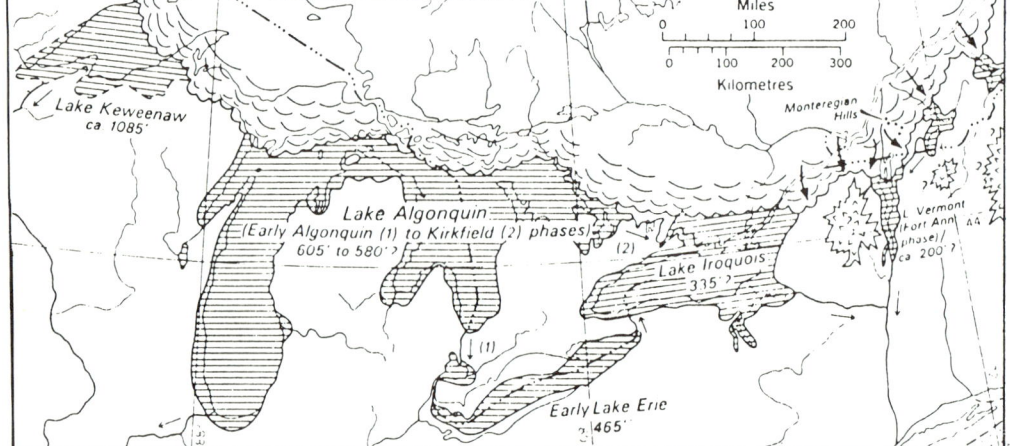

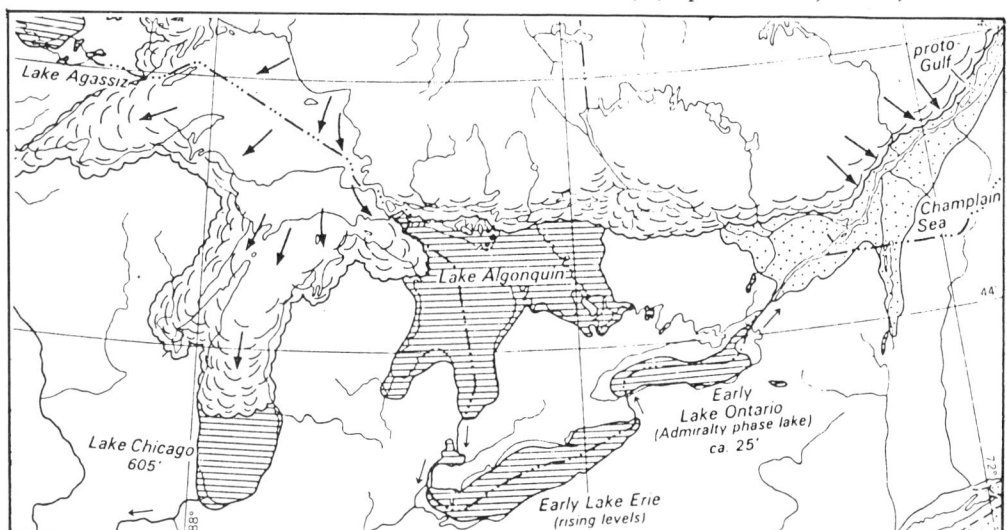

FIGURE 9. Map showing limits of glacial Lake Algonquin, with others in the Erie and Ontario basins, and part of the Champlain Sea about 11,800 yr B.P. (with elevation in feet). (Map: Geol. Surv., Canada.)

glacial lakes occupied the Huron and Erie depressions (Hough, 1958; Dreimanis, 1969; Prest, 1970). Around 14,500 yr B.P. glacial Lake Maumee began to form around the southwest end of the present Lake Erie; its outflow was to the SW (Gulf of Mexico). In the interlobate area glacial Lake London developed in the wide valley of the Thames River, leaving lacustrine and deltaic deposits, in successively lowered terraces.

At approximately 13,500 yr B.P. (the Mackinaw interstadial, or Bölling in Europe), the warming trend accelerated sharply, producing glacial Lake Arkona, which spread from the Erie basin into lower Lake Huron and overflowed into the southern part of Lake Michigan (glacial Lake Chicago) and thence to the Gulf of Mexico.

There was a brief readvance stage around 13,100–12,300 yr B.P. (Port Huron Stadial), when the lakes became somewhat more restricted, the Lake Whittlesey stage in the Erie Basin. The outlet was still to the W at first, but later the melting in the Finger Lakes-Mohawk-Hudson area of New York permitted an overflow to the E, into the Atlantic (Fig. 8).

General warming followed in the post-12,000-yr-B.P. phase (beginning of Alleröd of Europe) and the shape of the present lakes began to emerge; the ice front, however, still lay across northern Lake Huron, the Frontenac Arch, and the Adirondacks and there ensued the glacial Lake Algonquin stage in the Huron basin and the Lake Iroquois stage in the Ontario sector, with the overflow directed into the Hudson. Meantime melting back from the northern Lake Huron region and the Frontenac Arch led to a brief phase when the upper Great Lakes overflowed into Lake Ontario via Georgian Bay and Lake Simcoe in the vicinity of Belleville. At the same time the Upper St. Lawrence became ice-free and Lake Iroquois spilled over into the Champlain Sea, to assume a new low level, the early Lake Ontario (Fig. 9).

An important, but brief glacial readvance followed, the Valders Stadial (climax about 10,800 yr B.P., equivalent to the Salpausselkä or Late Tundra stage of northern Europe). Ice readvanced from the north into the Great Lakes region. Meanwhile the spillway from the youthful Lake Ontario into the Upper St. Lawrence and Champlain Sea apparently remained open. The retreating Valders ice left an end-moraine and an extensive till plain north of Lake Ontario, where some of the earlier glacial features are severely deformed.

Progressive ice retreat followed with a brief pause, possibly of early Valders age, at the Dummer Moraine that extends from Lake Simcoe to Kingston. The retreat now exposed all of southern Ontario, which has remained ice-free since the beginning of the Holocene (10,300 yr B.P.). Postglacial isostatic rebound initially led to uparching and changes in the extent of glacial lakes in the S, so Lake Erie became separated from Lake Huron and overflowed NE into Lake Ontario. As the ice retreated, so the axis of the marginal upwarp moved northward, cutting off the Huron-Belleville outlet. Before 10,000 yr B.P. the Huron outlet had shifted to a new channel through Georgian Bay and North Bay to the Ottawa Valley, either via Lake Nipissing and Mattawa or via Fossmill and Petawawa, gaining waters in great volume from the northern glacial lakes (Lake Barlow-Ojibway). The Ottawa River outlet formed a large delta at Petawawa and downstream a little, at Pembroke, radiocarbon-dated marine shells indicate its entrance into Champlain Sea at 10,800 yr B.P.

Approximately in mid-Holocene time (around 5500 yr B.P.) upwarp along the Ottawa Valley closed that outlet and the disappearance of the former crustal rebound bulge across the axis of southernmost Ontario permitted Lake Huron to overflow via Lake St. Claire into Lake Erie (glacial Lakes Nipissing-Algoma, dating 5500–2800 yr B.P.). Further isostatic adjustment continues today (Lewis, 1969), the crust continuing to rise differentially in the north. The zero isobase now runs through southern Lake Erie–Troy (N.Y.)–St. John (N.B.).

Economic Geology

Southern Ontario has long been one of the higher-ranking industrial regions of Canada. The Ontario oil fields have the distinction of being the oldest in North America. Production of natural gas has also been very important. Using depleted Silurian reef reservoirs for storage during the low consumption periods of the summer months, it has proved practicable and economical to transport western Canada gas to eastern Canada markets (southern Ontario and western Quebec).

Salt deposits occur over a wide region bordering the Michigan Basin, and its recovery from artificial brine fields has been carried out in southern Ontario for more than a century. In recent years two salt mines have come into operation (for highway salt) and increased the annual production to about 80% of the total Canadian output.

Gypsum deposits occur locally in the Niagara Peninsula and some of these have been under active exploitation for 150 years. Brick and tile are the chief products manufactured from the Ordovician and Devonian shale and clay deposits of southern Ontario. Because of the wide variety of uses in the construction industry, the limestones and dolostones constitute the most valuable economic products for the manufacture of lime and cement, crushed stone and dimension stone. Small quantities of Cambrian and Silurian sandstones are used for building stones. The Paleozoic rocks have also provided, through glacial action, the primary source of rich agricultural soils in southern Ontario. Besides their importance as agricultural soils, the Quaternary deposits provide very substantial economic resources in the form of sand and gravel, and groundwater reservoirs. The importance of geology in the context of establishing recreational areas and parks has become well recognized in recent years. Provincial legislation exists which protects and controls the operation of all pits and quarries along the Niagara Escarpment.

C. GORDON WINDER
BRUCE V. SANFORD
JAAN TERASMAE

References

Baker, M. B., 1916. "The geology of Kingston and vicinity," *Ont. Dept. Mines 25th Rep.,* pt. 3, 71p.
Beards, R. J., 1967. "Guide to the subsurface Paleozoic stratigraphy of southern Ontario," *Ont. Dept. Energy Res. Manag. Pap. 67-2.*
Berdan, J. M., *et al.,* 1969. "Siluro-Devonian boundary in North America," *Bull. Geol. Soc. Am.,* **80,** 2165–2174.
Bolton, T. E., 1957. "Silurian stratigraphy and palaeontology of the Niagara escarpment in Ontario," *Geol. Survey Canada Mem.* 289, 145p.
____, 1966. "Silurian faunas of Ontario," *Geol. Survey Canada Pap. 66-5,* 45p.
Brigham, R. J., 1972. "Structural geology of southwestern Ontario," *Ont. Dept. Mines and North Affairs, Pap. 71-2,* 110p.
Caley, J. F., 1940. "Palaeozoic geology of the Toronto-Hamilton area, Ontario," *Geol. Survey Canada Mem. 224,* 284p.
____, 1941. "Palaeozoic geology of the Brantford area, Ontario," *Geol. Surv. Canada Mem. 226,* 176p.
____, 1943. "Palaeozoic geology of the London area, Ontario," *Geol. Surv. Canada Mem. 237,* 171p.
____, 1945. "Paleozoic geology of the Windsor-Sarnia area, Ontario," *Geol. Survey Canada, Mem. 240,* 227p.
Chapman, L. J., and Putman, D. F., 1966. *The Physiography of Southern Ontario.* Toronto: University of Toronto Press, 386p.
Coleman, A. P., 1933. "The Pleistocene of the Toronto region," *Ontario Dept. Mines,* **41,** pt. 7, 1–55.
Douglas, R. J. W., *et al.,* 1970. *Geology and Economic Minerals of Canada.* Geol. Surv. Canada Econ. Geol. Rep. 1, 838p.
Dreimanis, A., 1960. "Pre-classical Wisconsin in the eastern portion of the Great Lakes region, North America," *21st Internat. Geol. Cong.,* Copenhagen, pt. 4, 108–119.
____, 1961. "Tills of southern Ontario," *in* R. F. Legget, ed., Soils in Canada. *Roy. Soc. Canada, Spec. Pub. 3,* 80–96.
____, 1969. "Late Pleistocene lakes in the Ontario and Erie Basins," *12th Conf. Great Lakes Res. Proc.,* 170–180.
____, 1970. "Last ice age deposits in the Port Stanley map-area, Ontario (40 I/II)," *Geol. Surv. Canada Pap. 70-71,* pt. A, 167–169.
Dreimanis, A., and Karrow, P. F., 1965. "Southern

Ontario," in *Internat. Assoc. Quaternary Res. 7th Congr. Guide Book.* Nebraska Acad. Sci., 90–110.

———, 1972. "Glacial history in the Great Lakes-St. Lawrence Region and the classification of the Wisconsin Stage," *24th Internat. Geol. Congr.*, pt. 12, 5–15.

Dreimanis, A., Terasmae, J., and McKenzie, G. D., 1966. "The Port Talbot interstade of the Wisconsin glaciation," *Canada J. Earth Sci.*, 3, 305–325.

Ferrigno, K. F., 1971. "Environmental influences on the distribution and abundance of conodonts from the Dundee limestone (Devonian), St. Marys, Ontario," *Can. J. Earth Sci.*, 8, 378–386.

Henderson, E. P., 1970. "Surficial geology: Brockville and Mallorytown," *Geol. Surv. Canada*, Map 6-1470.

Hewitt, D. F., 1960. "The limestone industries of Ontario," *Ont. Dept. Mines Indust. Mineral Circ. 5*, 177p.

Hill, J. V., 1966. "Silurian reef carbonates," *Ont. Petrol. Inst., 5th Ann. Conf.*

Hinde, G. J., 1877. "Glacial and interglacial strata of Scarborough Heights and other localities near Toronto, Ontario," *Can. J.*, 15(5), 388–413.

Hobson, G. D., and Terasmae, J., 1969. "Pleistocene geology of the buried St. Davids Gorge, Niagara Falls, Ontario: geophysical and palynological studies," *Geol. Surv. Canada Pap. 68-67*, 16p.

Hough, J. L. 1958. *Geology of the Great Lakes.* Urbana, Ill.: University Illinois Press, 313p.

Johnston, W. A., 1912. "Geology of the Lake Simcoe area, Ontario; Brechin and Kirkfield sheets," *Geol. Surv. Canada Rept. 1911*, 253–261.

———, 1914. "Geology of the Lake Simcoe area, Ontario; Beaverton, Sutton, and Barrie Sheets," *Geol. Surv. Canada Rept. 1912*, 294–300.

Karrow, P. F., 1963. "Pleistocene geology of the Hamilton-Galt area," *Ont. Dept. Mines Geol. Rept. 16*, 68p.

———, 1967. "Pleistocene geology of the Scarborough area," *Ont. Dept. Mines Geol. Rept. 46*, 108p.

———, Clark, J. R., and Terasmae, J., 1961. "The age of Lake Iroquois and Lake Ontario," *J. Geol.*, 69, 659–667.

———, and Terasmae, J., 1970. "Pollen bearing sediments of the St. Davids buried valley fill at the Whirlpool, Niagara River gorge, Ontario," *Canada J. Earth Sci.*, 7, 539–542.

LaSalle, P., 1966. "Late Quaternary vegetation and glacial history in the St. Lawrence lowlands, Canada," *Leidse Geol. Mededel.*, 38, 91–128.

Lewis, C. F. M., 1969. "Late Quaternary history of lake levels in the Huron and Erie Basins," *12th Conf. Great Lakes Res. Proc.*, 250–270.

Liberty, B. A., 1969. "Paleozoic geology of the Lake Simcoe Area, Ontario," *Geol. Surv. Canada Mem. 355*, 201p.

Liberty, B. A., and Bolton, T. E., 1971. "Paleozoic geology of the Bruce Peninsula Area, Ontario," *Geol. Surv. Canada Mem. 360*, 163p.

Logan, W. E., 1854. "On the physical structure of the western district of Upper Canada," *Can. J.*, 3, 1–2.

———, 1863. "Geology of Canada: report of progress from its commencement to 1863," *Geol. Surv. Canada Rept.*, 983p.

Martini, I. P., 1971. "Grain orientation and paleocurrent systems in the Thorold and Grimsby sandstones (Silurian)," *J. Sed. Petr.*, 41, 425–435.

McDonald, B. C., 1971. "Late Quaternary stratigraphy and deglaciation in eastern Canada," in K. K. Turekian, ed., *The Late Cenozoic Glacial Ages.* New Haven: Yale University Press, 331–353.

Mörner, N. A., 1972. "When will the present interglacial end?," *Quat. Res.*, 2(3), 341–349.

———, and Dreimanis, A., 1970. "The Erie interstadial; type section, lake level, ice recession and correlations," *Geol. Soc. Am; Abs.*, 2(7), 631.

Mott, R. J., 1968. "A radiocarbon-dated marine algal bed of the Champlain Sea episode near Ottawa, Ontario," *Canada J. Earth Sci.*, 5(2), 319–324.

Mukherji, K. K., 1969. "Supratidal carbonate rocks in the Black River (Middle Ordovician) Group of southwestern Ontario, Canada," *J. Sed. Pet.*, 39, 1530–1545.

———, and Winder, C. G., 1970. "Regional microfabric analysis and insoluble residues of the Black River Group (Middle Ordovician) in Southern Ontario," *Canada J. Earth Sci.*, 7, 1437–1448.

Okulitch, V. J., 1939. "The Ordovician section at Coboconk, Ontario," *Royal Canadian Inst.*, 22, 319–339.

Oliver, W. A., et al., 1967. "Devonian of the Appalachian Basin," in *Internat. Symp. Devonian System*, vol. 1. Calgary: Alberta Soc. Petr. Geologists, 1001–1040.

Pollock, C. A., Rexroad, C. B., and Nicoll, R. S., 1970. "Lower Silurian conodonts from northern Michigan and Ontario," *J. Paleont.*, 44, 743–764.

Prest, V. K., 1970. "Quaternary geology of Canada," in R. J. W. Douglas, ed., *Geology and Economic Minerals of Canada.* Geol. Survey Canada Econ. Geol. Rept. 1, 677–764.

Rexroad, C. B., and Rickard, L. V., 1965. "Zonal conodonts from Silurian strata of the Niagara gorge," *J. Paleont.*, 39, 1217–1220.

Sanford, B. V., 1961. "Subsurface stratigraphy of Ordovician Rocks in southwestern Ontario," *Geol. Surv. Canada Pap. 65-9*, 54p.

———, 1965. "Salina salt beds, southwestern Ontario," *Geol. Surv. Canada Pap. 65-9*, 7p.

———, 1967. "Devonian of Ontario and Michigan," in *Internat. Symp. Devonian System*, vol. 1. Calgary: Alberta Soc. Pet. Geol., 973–999.

———, 1969. "Silurian of southwestern Ontario," *Ont. Petrol. Instit. Proc., 8th Ann. Conf.*

———, 1969. "Geology, Toronto-Windsor area," *Geol. Surv. Canada*, Map 1263 A.

———, and Quillian, R. G., 1959. "Subsurface stratigraphy of Upper Cambrian rocks in southwestern Ontario," *Geol. Surv. Canada Pap. 58-12*, 33p.

Sly, P. G., and Lewis, C. F. M., 1972. "The Great Lakes of Canada–Quaternary geology and limnology," *24th Internat. Geol. Congr.*, Field Excursion A43, 92p.

Stauffer, C. R., 1915. "Devonian of Southwestern Ontario," *Geol. Surv. Canada Mem. 34.*

Stumm, E. C., and Wright, J. D., 1958. "Check list of fossil invertebrates described from the Middle Devonian rocks of the Thedford-Arkona region of southwestern Ontario," *Mich. Univ. Mus. Paleont. Contr.,* **14**(7), 81–132.

Sweet, W. C., and Bergstrom, S. M., 1971. "The American Upper Ordovician standard: XIII. A revised time-stratigraphic classification of North American Upper Middle and Upper Ordovician rocks," *Bull. Geol. Soc. Am.,* **82**, 613–628.

Terasmae, J., 1960. "A palynological study of Pleistocene interglacial beds at Toronto, Ontario," *Geol. Surv. Canada Bull.,* **56**, 24–60.

———, 1965. "Surficial geology of the Cornwall and St. Lawrence Seaway project areas, Ontario," *Geol. Surv. Canada Bull.,* **121**, 54p.

———, Karrow, P. F., and Dreimanis, A., 1972. "Quaternary stratigraphy and geomorphology of the eastern Great Lakes region of southern Ontario," *24th Internat. Geol. Congr.,* Field Excursion A42, 75p.

Williams, M. Y., 1919. "The Silurian geology and faunas of the Ontario Peninsula, Manitoulin Island and adjacent islands," *Geol. Surv. Canada, Mem. 111*, 195p.

Wilson, A. E., 1946. "Geology of the Ottawa-St. Lawrence lowland, Ontario and Quebec," *Geol. Surv. Canada, Mem. 241*, 65p.

Winder, C. G., 1954. "Burleigh Falls and Peterborough map-areas, Ontario," *Geol. Surv. Canada, Pap. 53-27*, 10p.

———, 1960. "Paleoecological interpretation of Middle Ordovician stratigraphy in southern Ontario, Canada," *21st Internat. Geol. Congr.,* Copenhagen, pt. 7, 18–27.

———, 1961. *Lexicon of Paleozoic Names in Southwestern Ontario.* Toronto: University of Toronto Press, 121p.

———, 1962. "Upper Devonian age of the Kettle Point shale," *Roy. Soc. Canada Trans.,* **56**, 85–95.

———, 1966. "Conodont zones and stratigraphic variability in Upper Devonian rocks, Ontario," *J. Paleont.,* **40**, 1275–1293.

———, 1967. "Micropaleontology of the Devonian in Ontario," in *Internat. Symp. Devonian System,* vol. II. Calgary: Alberta Soc. Petr. Geologists., 711–719.

———, and Sanford, B. V., 1972. "Stratigraphy and paleontology of the Paleozoic rocks of southern Ontario," *24th Internat. Geol. Congr.,* Field Excursion A45-C45, 74p.

Cross-references: *Canada–Canadian Shield, St. Lawrence Lowlands of Quebec; North America; United States–Appalachian Region, Midwestern Region, New England Region.*

CANADA–ROCKY MOUNTAINS AND EASTERN CORDILLERAN REGION

This account summarizes the geology of the eastern Canadian Cordillera and includes the Rocky Mountains proper, southern Rocky Mountain Trench, Columbia Mountains, eastern Mackenzie Mountains, Franklin Mountains, and Richardson Mountains (*Physiographic Regions of Canada, G.S.C. Map 1254A*). These mountains occupy portions of western Alberta, eastern British Columbia, eastern Yukon Territory, and western Northwest Territories.

History of Geological Investigations

The earliest geological investigations were carried out by pioneers of the Geological Survey of Canada—G. M. Dawson, R. G. McConnell, J. McEvoy, and C. Camsell, who mapped large portions of the Cordillera during the latter part of the 19th and early part of the 20th centuries. This was followed by a period of more detailed mapping with emphasis on local areas of economic importance. Only recently has it been possible to prepare relatively complete syntheses of the region, based on data obtained by helicopter-supported parties of the Geological Survey of Canada and the investigations of oil and coal companies, of the Alberta Research Council, and of the British Columbia Department of Mines and Petroleum Resources. Large parts of this region, nevertheless, have been mapped only at 1:250,000 scale, although by 1972 this coverage was nearly complete.

Physiography

The Rocky Mountains province generally consists of rugged topography attaining relief of more than 2500 m near the Continental Divide. The southern Rocky Mountains are generally divisible from E to W into foothills, front ranges, main ranges, and locally western ranges on the basis of morphotectonics. In the Peace River-Pine Pass area only the foothills and mountains can be differentiated, plunging northward to the intermontane Liard Plateau. To the N the Mackenzie Mountains form a broad arcuate fold belt which continues northwestward to the Wernecke and Ogilvie mountains. The Richardson Mountains are N-trending and are separated from the Mackenzie Mountains by the Peel plateau. The Columbia Mountains form a distinct series of rugged glacier-covered mountain ranges west of the Rocky Mountain Trench.

The southern Rocky Mountains are renowned for their scenic, rugged, snow- and glacier-covered mountains in which Kootenay, Banff, Yoho, and Jasper National Parks are located (Fig. 1). The highest peak of this region is Mt. Robson 3956 m (12,972 ft), about 100 km west of Jasper.

Geotectonics

Tectonically most of the region can be classified as a foreland thrust or fold belt, separated

FIGURE 1. Mount Kitchener and shoulder of Suttfield Peak on left underlain by Middle Cambrian carbonates in eastern Main Ranges, capped by the northern portion of the Columbia Icefields, as viewed from the Banff-Jasper highway.

from an alpine-type orogen or core zone (eastern crystalline belt) to the west. West of the core zone is the structurally complex interior belt and intermontane plateau region. These structures largely resulted from Cenozoic and Mesozoic orogenic activity, representing the western segment of a westward-thickening wedge of sediments deposited on the cratonic margin (Figs. 2 and 3). Strata as old as Proterozoic overlie a sialic crystalline basement, an extension of the Churchill and Slave provinces of the Canadian Shield. This wedge is an extension of the cratonic shelf sequence, which presumably grades westward into eugeosynclinal facies, but the transition is obscured by structural complexities. The wedge is often classified as a miogeosyncline, but the term "miogeocline" is preferred because the depositional basin appears to have been one sided and not synclinal.

The deep structure beneath the southern Rocky Mountains has been revealed in series of seismic sections (Bally et al., 1966); and contours on the basement established by deep drilling (King, 1969) suggest that the westward dip steepens beneath the foothills and mountains. Geophysical evidence indicates that the crust near the 49th parallel is 45-50 km thick and suggests that the Precambrian continental margin extends to the Rocky Mountain Trench, and possibly as far W as the Kootenay arc (Western Columbia Mountains, Berry et al., 1971) as does the sedimentary record (Price and Mountjoy, 1970). The core zone gneisses of the Shuswap complex may represent remobilized Precambrian Hudsonian cratonic rocks, or may have been derived from Windermere (Hadrynian) pelitic sediments.

The sedimentary history and structure of the Rocky Mountains and foothills are interrelated with the core zone and interior belts to the west. Most of the Rocky Mountains are allochthonous and have been detached from the southwesterly sloping basement surface and displaced eastward up onto the flank of the craton along interleaved and overlapping thrust faults and associated folds (Fig. 3). A transition occurs across the eastern part of the Cordillera from essentially brittle deformation in the foreland margin of the unmetamorphosed sedimentary wedge of the Rocky Mountains to large-scale penetrative flow in the hot ductile rocks of the mobile infrastructure in the core zone.

Stratigraphy

The sedimentary sequence consists of (1) a prograded continental terrace (miogeocline to platform) of middle Proterozoic to late Jurassic

FIGURE 2. Distribution of main areas of sedimentary, volcanic, granitic, and metamorphic rocks in the Cordillera of North America. The sedimentary rocks are divided into miogeosynclinal (miogeocline of this report) and eugeosynclinal, respectively, without and with volcanic rocks. The insert map illustrates physiographic and geologic subdivisions of the Canadian Cordillera: (1) Rocky Mountain belt; the Rocky Mountain Trench occurs along the boundary between 1 and 2; (2) Omineca crystalline belt or core zone, Columbia Mountains in the south; (3) Intermontane belt; (4) Coast Plutonic complex (western fold belt); (5) Insular belt. (Monger et al., 1972.)

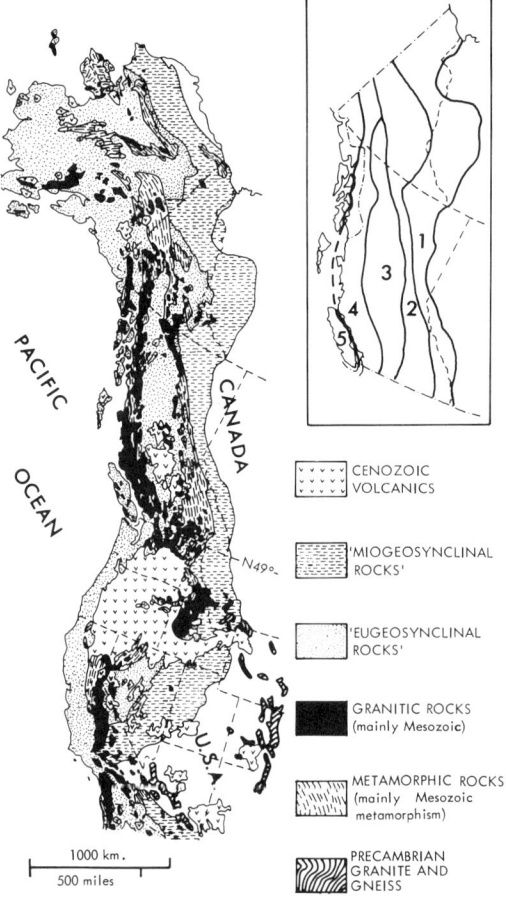

age, derived from the craton to the east, and (2) a clastic wedge from Late Jurassic to Paleocene ("molasse" facies) restricted to the eastern part of the belt, derived from uplifted Columbia and in part Rocky Mountains to the west. The Rocky Mountain succession forms the thicker, western portion of the sedimentary wedge that overlaps northeastward onto the Hudsonian crystalline basement of early Proterozoic (Aphebian) crystalline rocks that extends under the interior plains from the Churchill and Slave provinces.

The miogeocline-platform sediments consist mainly of carbonates, shales, and mature sandstones and are more than 14,000 m thick. The platform is divisible into six distinct assemblages:

1. Middle Proterozoic (Helikian, Purcell, and equivalent sequences)
2. Late Proterozoic (Hadrynian, Windermere, and equivalent sequences)
3. Late Proterozoic and early Cambrian (clastic shelf or terrace wedge)
4. Early Paleozoic carbonate shelf (Middle Cambrian to Silurian)
5. Middle to late Paleozoic carbonate shelf (Devonian to Permian)
6. Early Mesozoic clastics (Triassic to Middle Jurassic)

Proterozoic. Middle Proterozoic (Helikian) strata, the oldest known in the Cordillera (base not observed), comprise a thick sequence of fine-grained clastics and carbonates mainly of shallow-water marine origin. They crop out in four areas: (1) the Purcell sequence in the southernmost Rocky Mountains and Purcell Mountains (southeastern Columbia Mountains); (2) the northern Rocky Mountains; (3) near the headwaters of the South Redstone River in the Mackenzie Mountains; and (4) in the Franklin Mountains.

These middle Proterozoic strata consist of quartzites, shales, and carbonates with stromatolites. Often they contain shallow-water features such as mud cracks and coarse cross-bedding. Gypsum and gypsiferous siltstones occur in the Mackenzie Mountains. They range from about 3000 m in the Rockies to a maximum composite thickness of 14,000 m in the Purcell Mountains. In the northern Rocky Mountains thicknesses range from 2800 m to more than 4500 m in the western Mackenzie Mountains.

Uplift, tilting, and blocking faulting, referred to the *Racklan orogeny,* followed in the Mackenzie Mountains and northern Rocky Mountains. Uplift, gentle folding, tilting, and faulting (termed the *East Kootenay orogeny*) also occurred in the Purcell Mountains, with the addition of granitic intrusion and regional metamorphism (dated at 675 to 790 m yr). Elsewhere only a regional unconformity separates these rocks from overlying Upper Proterozoic Windermere strata.

Upper Proterozoic (Hadrynian) sediments are more widespread and extend almost as a continuous belt in the Rocky Mountains. These strata are mostly of marine origin, comprising unsorted, feldspathic, fine-pebble conglomerates, sands, and silty pelites; local carbonates; coarse, poorly sorted conglomerates; and diamictites at or near the base (e.g., Toby Con-

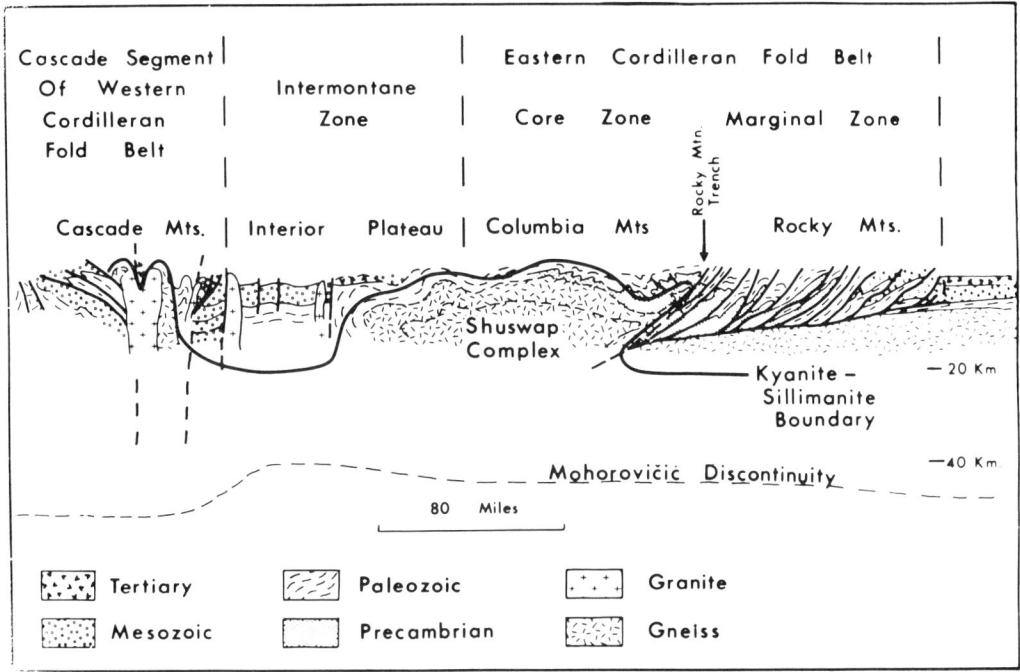

FIGURE 3. Highly schematic cross section across the southern Canadian Cordillera showing inferred relationship of Rocky Mountains and Columbia Mountains core zone to the intermontane and western fold belt of the coast ranges (from Wheeler, 1970, p. 156).

glomerate). Bedded, chert-hematite iron formation associated with conglomerates and mudstones occur in the Rapitan group in the northern Mackenzie Mountains. Thicknesses range from 1800 to 5500 m in the southern Rocky Mountains, up to about 2500 m in the northern Rocky Mountains and northern Mackenzie Mountains, and up to 300 m in the Franklin Mountains. They therefore comprise a large volume of impure clastic sediments.

Upper Proterozoic strata were deposited mostly to the W and SW of the exposed Purcell sediments in local uplift areas there and in the Mackenzie Mountains. Deposition probably occurred in marine environments of moderate depth that permitted little opportunity for sorting and reworking.

Paleozoic. A new sedimentary cycle began with quartz sandstones in latest Precambrian time along most of the eastern Cordillera and continued to latest early Cambrian or early Middle Cambrian time. Olenellid trilobites first appear in shaly clastic or carbonate rocks in the upper part of this sequence. These sands are locally feldspathic and conglomeratic and derived from the craton to the east. These sands thicken westward to as much as 3000 m in the western main ranges near Jasper and in the Columbia Mountains (Fig. 4).

In the southern Rocky Mountains the early Cambrian clastic sediments grade upward into the Middle Cambrian through Silurian (early Paleozoic) carbonate shelf sediments up to 3000 m thick. They reflect an abrupt change in sedimentary regime, presumably the result of peneplanation of the Canadian Shield and sedimentary onlap of its western margin. The Cambro-Ordovician carbonates consist of a cyclical sequence of shallow-water platform or bank facies that thicken westward. Many of the peaks of the southern Rocky Mountains main ranges have been carved from these carbonates. To the SW they pass into calcareous shales and slates in the western main ranges, and again to carbonates that lie unconformably on Proterozoic strata along the eastern edge of the Purcell arch in the Columbia Mountains. Eastward they grade into a red bed facies bordering the craton.

The Upper Ordovician is unconformable over all older systems and is, in turn, conformably overlain by the Silurian. Remnants of the Silurian and Ordovician carbonates and minor sandstones are truncated beneath the Devonian and Upper Ordovician and more complete sections occur near the Rocky Mountain Trench. In the Richardson Mountains the Ordovician and Silurian are represented by a thick sequence of thinly bedded graptolitic shale, argillaceous limestone, limestone turbidite, and chert.

FIGURE 4. Mt. Geikie, seen from the N, in the main ranges of the Rockies (elevation 3200 m), showing 1500 m of early Cambrian quartz sandstones.

The middle to late Paleozoic carbonate shelf (Devonian to Permian) sequence included carbonate reef complexes bordering local uplifts and lying unconformably on older Paleozoic sediments. Lower Devonian carbonates only occur in the western Mackenzie Mountains. A deeper-water shale basin occurred to the N and W. With deepening waters in Upper Devonian times a shale basin also formed over the previous carbonate shelves in southwestern Alberta. A thick Upper Devonian clastic sequence in the Richardson and the northern Mackenzie mountains suggests uplift N and NE as a result of the *Ellesmerian orogeny* in the Arctic archipelago. Clastic sediments apparently derived from the W were deposited in the front ranges of the southern Rocky Mountains. Latest Devonian strata only occur in the southern Mackenzie Mountains, with nondeposition or erosion elsewhere. These deposits are overlain by a thin transgressive black shale that may straddle the Devono-Mississippian boundary.

During Mississippian time the platform persisted with diverse carbonate facies in cyclic alternations. Crinoidal carbonates grade westward into cherty limestones and shales. Carboniferous and Permian strata do not occur in the Mackenzie Mountains whereas nearly 1800 m of these strata occur in southern Franklin Mountains and Liard plateau. Sands, siltstones, and shales are present in the southern Richardson Mountains. Pennsylvanian rocks are largely absent except in the southern Rocky Mountains, where sandstones and dolomites were deposited. Permian strata are restricted mostly to the southern and northern Rocky Mountains and are represented by sandstones, chert, and phosphatic beds. The *Cariboo orogeny* took place in early Mississippian time and affected the northern (Cariboo) and southern (Selkirk) Columbia Mountains. Thrusting and uplift appear to have taken place in both areas.

Triassic to Middle Jurassic clastics with minor carbonates mark a new cycle (up to 1500 m of shallow-water platform deposits derived from the craton. They conformably overlie Permian and older rocks but are mostly restricted to the southern and northern Rocky Mountains. What is now the core zone in the Columbia Mountains was uplifted (*Tahltanian orogeny*).

From mid-Middle Jurassic onward there was extensive deformation and uplift in much of the Cordillera. The core zone formed a large land mass W of the present Rocky Mountain Trench and shed large volumes of clastic sediments eastward. The main ranges were involved in this uplift by mid-Cretaceous time. The resultant sediments form a clastic wedge sequence up to 6000 m thick in the southern Rocky Mountains. Several distinct unconformity bounded clastic wedges suggest a NE-migrating foredeep (Bally et al., 1966; Price and Mountjoy, 1970) as deformation progressed intermittently eastward.

The clastic wedge consists of four main sequences: (1) late Upper Jurassic and (?) early Lower Cretaceous, nonmarine, coal-bearing sediments and near-shore marine clastics; (2) dominantly nonmarine clastics of Lower Cretaceous age which extends NE far onto the craton; (3) marine shales of early Upper Cretaceous age that interfinger westward with tongues of coarser clastics that are partly nonmarine; and (4) nonmarine clastics of late Upper Cretaceous and Paleocene age that extend from the foothills over the western plains.

No Upper Jurassic and Cretaceous sediments occur in the Mackenzie Mountains other than some Lower Cretaceous strata near the northern boundary and in the Franklin Mountains which extend N into Peel plateau and plain. A thick succession of mostly marine Upper Jurassic and Lower Cretaceous clastic sediments occurs in the northern Richardson Mountains.

Granitic intrusions were emplaced in the Columbia Mountains during the Jurassic and earliest Upper Cretaceous and generally postdate regional deformation.

The westward increase in rates of thickening and the rate of increase of maximum size of clasts in the foothills and front ranges can be related to a palinspastic reconstruction (unfolding and unfaulting the structure, Fig. 3). This indicates that the source of the largest clasts must have been derived from nearby thrust sheets just to the W. Thus the sediments accumulated in progressively eastward migrating clastic wedges eroded from periodically advancing thrust sheets in what are now the Rocky Mountains main ranges and western front ranges (Price and Mountjoy, 1970). Deformation terminated with compression and uplift of the foothills belt in the Paleocene. In summary, the deformation in the southern and northern Rocky Mountains occurred as a series of intense pulses spread over an interval of 100 m yr from late Jurassic to Paleocene, making it difficult to separate individual orogenies such as the Columbian and Laramide.

Postorogenic Clastics. The final phases of deformation in much of the Canadian Rocky Mountains appear to have ended during the Paleocene, but some uplift and moderate folding continued. Several small structural basins were superimposed on earlier structures. The sediments are generally nonmarine and local. They include (1) Flathead valley, southern Rocky Mountains—2000 m of nonmarine coarse clastics (Kishenehn Formation) in a structural basin, latest Eocene or earliest Oligocene, that lie with angular unconformity on and postdate the Lewis thrust; (2) Fort Norman, Franklin, and Mackenzie mountains—about 1800 m of nonmarine clastics, some lignite, that cover the middle part of the Franklin Mountains and border the eastern flank of the Mackenzie Mountains, Paleocene or Eocene, and rest unconformably on lower Cretaceous to middle Devonian strata, moderately folded; (3) Bonnet Plume basin, southern Richardson Mountains—more than 1500 m of nonmarine clastics and minor peat, in a structural depression, latest Upper Cretaceous (Maestrichtian) to early Tertiary (Paleocene), that are gently folded and postdate Richardson Mountain structures; (4) Moose Channel formation, western margin of Mackenzie delta and northern end of Richardson Mountains—more than 500 m of nonmarine clastics together with underlying Upper Cretaceous marine shales that unconformably overlap older structures.

A few small intrusions were emplaced during the early Tertiary in the Columbia Mountains. During the mid-Tertiary, erosion was dominant. Some of the late Tertiary basalt flows of the interior plateau reached into the Columbia Mountains. Miocene silts and gravels along the Fraser and Thompson valleys and the southern Rocky Mountain Trench indicate that the major drainage routes were established by the Miocene. Little of the debris produced during the late Tertiary erosion is preserved, owing to extensive erosion and Pleistocene glaciation.

Structure

The Rocky Mountains are divided from south to north into five segments: Columbia Mountains, southern Rocky Mountains (south of Peace River), northern Rocky Mountains, Mackenzie and Franklin mountains, and Richardson Mountains (*Physiographic Regions of Canada, G.S.C. Map 1254A*).

Columbia Mountains (Core Zone). The Columbia Mountains, a gentle, convex eastward arc, form the southern part of the eastern core zone (Wheeler, 1970) and consist of four structural elements. The Shuswap metamorphic complex forms the heart of the core zone and is flanked on the N by the Cariboo Mountains and on the E by the Kootenay arc, which is in turn flanked on the E by the Purcell anticlinorium (Figs. 3 and 5).

Underlain mainly by Purcell Middle Proterozoic sedimentary rocks, the Purcell anticlinorium consists of several N-plunging segments separated by faults as old as Cambrian. In cross section the anticlinorium is box shaped, the central part dominated by a broad, gently undulating anticline transected by thrust and normal faults. Westward the structures merge with the Kootenay arc, a complex belt of metamorphic rocks ranging from Proterozoic to Middle Jurassic. The structure resulted from

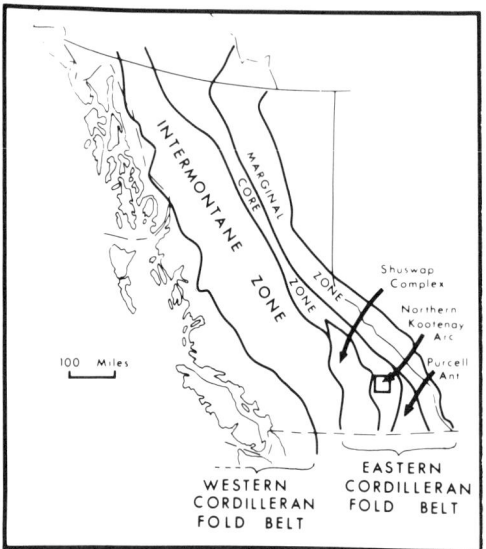

FIGURE 5. Locality sketch map showing situation of Shuswap complex, northern Kootenay arc, and Purcell anticlinorium.

polyphase deformation that began in mid-Paleozoic and was completed by mid-Jurassic prior to the emplacement of discordant granites. The earliest folds are isoclinal and were accompanied by large-scale *décollement* thrusting.

The Shuswap metamorphic complex contains high-grade metamorphic rocks (upper amphibolite facies) and consists in its eastern part of a series of gneiss domes spaced at intervals of 50–100 km. The domes are characterized by cores of migmatitic granitoid gneiss enveloped by mantles of metasedimentary gneiss and, locally, nepheline syenite gneiss, in turn fringed with metasedimentary gneiss riddled with pegmatite. The metasedimentary gneiss contains a lower sequence that is generally similar in lithology, although not in detail, to the Lower Cambrian–Hamill Quartzite–Badshot Limestone succession in the Kootenay arc and an upper sequence broadly resembling the Carboniferous–Permian(?) Milford group. Strata as young as Upper Triassic and Lower Jurassic can be traced into the Shuswap complex.

The principal deformation and metamorphism of the Shushwap complex were post-late Triassic or early Jurassic. These began in the eastern part with intense metamorphism and migmatization accompanying large scale E-W-trending interfolding the gneiss dome cores and their mantle.

To the N the Cariboo Mountains form a series of complexly folded anticlinoria and synclinoria in relatively incompetent Upper Proterozoic strata, which in the deepest levels are deformed into isoclinal folds and intensely metamorphosed. These fold structures plunge or grade northward into faulted structures in more competent late Proterozoic and Cambrian rocks at a higher structural level.

Southern Rocky Mountains and Foothills. The southern Rocky Mountains and foothills are among the world's best exposed and best known arcuate foreland thrust and fold belts (Bally et al., 1966; Dahlstrom, 1970; and Price and Mountjoy, 1970). Thrust faults and concentric folding are the dominant tectonic elements. The thrust faults are generally SW-dipping, concave upward, and locally folded. They form slip surfaces along which the well-layered sedimentary wedge was stripped from its basement and displaced to the NE, up the flank of the craton, to which it is now stacked up as imbricate thrust slices and associated folds. The northeastward thrust relative to the basement is as much as 200 km and the stacked thrust sheets have surficially thickened the cover up to 8000 m. Thrust faults are most numerous in the incompetent Cretaceous shales and sandstones of the foothills, more widely spaced in the front ranges, which are formed largely of Paleozoic carbonates, and are less frequent in the thick competent lower Paleozoic successions of the eastern main ranges (Fig. 6).

South of Valemount the Malton Gneiss (Shushwap-type) above the Purcell fault crosses the Rocky Mountain Trench and is locally thrust over Cambrian and the older rocks of the westernmost Rocky Mountains. The structures in the main ranges to the E outline a culmination and deflection of the structural grain apparently related to the northeastward emplacement of this gneiss (dated as 111 m yr). Evidently the main ranges had formed before mid-Cretaceous time and probably began to develop as early as late Jurassic, considering the record of Mesozoic sedimentation to the E (Price and Mountjoy, 1970). Structures in the front ranges and foothills formed later.

Southern Rocky Mountain Trench. The Rocky Mountain Trench is one of the most prominent pseudolinear (Figs. 2 and 3) topographic features of the Canadian Cordillera, the origin of which has led to much speculation. The exact relationships are often covered by Miocene to Quaternary fill. Faults bordering the western side of the southern Rocky Mountain Trench cause older Cambrian and Precambrian strata to be thrust over younger strata exposed on the eastern side or floor of the trench. Between latitudes 51 and 53° the W-dipping Purcell fault truncates structures on

FIGURE 6. Maligne Lake, looking NW, with Front Range structure in Upper Paleozoic strata on right, and Lower Cambrian and Precambrian rocks of main ranges in upper left displaced eastward along the Pyramid thrust. Mt. Robson is the high white-capped peak in the upper left.

both sides of the trench (Fig. 7). In addition, the Precambrian and Cambrian successions are readily correlated across the trench at several places. Thus the southern Rocky Mountain Trench is not a strike-slip feature, as has been claimed, but nearly everywhere is related to thrust-fault structures on one or both sides.

Northern Rocky Mountains and Foothills.
The southern Rocky Mountains around latitude 54°N to the Peace River are transitional in structural style to the northern Rocky Mountains. Here they gradually change character, becoming narrower, plunging northward, the fault displacements decreasing, and folding becoming more predominant.

Foothill folds are long, narrow, tightly compressed in Triassic sandstones, steeply inclined, and broken by forelimb thrusts. The folds become less persistent and more numerous in the overlying Cretaceous shales. These structures lie discordantly over Paleozoic structures, along a *décollement* zone in Devono-Carboniferous black shales. To the W there are Paleozoic and Precambrian thrust sheets similar to the front and main ranges of the southern Rocky Mountains. To the N Precambrian strata form very broad, long, and high-amplitude folds that have gently dipping western flanks and vertical or faulted eastern flanks. The main thrust zone occurs west of these folds.

Mackenzie Mountains. The Mackenzie Mountains form a broad fold arc, one of the most striking in the entire North American Cordillera, which swings the N-S trend to an almost E-W trend in the northwest (Fig. 2). They comprise a series of broad short folds, the crests and troughs nearly flat with sharply upturned and faulted flanks in an *en échelon* pattern. Most thrust faults are minor and commonly end at tear faults. The larger fold struc-

FIGURE 7. Physiographic subdivisions of the southern Canadian Rocky Mountains and Columbia Mountains and location of discovered hydrocarbon accumulations (from Dahlstrom, 1970, p. 336).

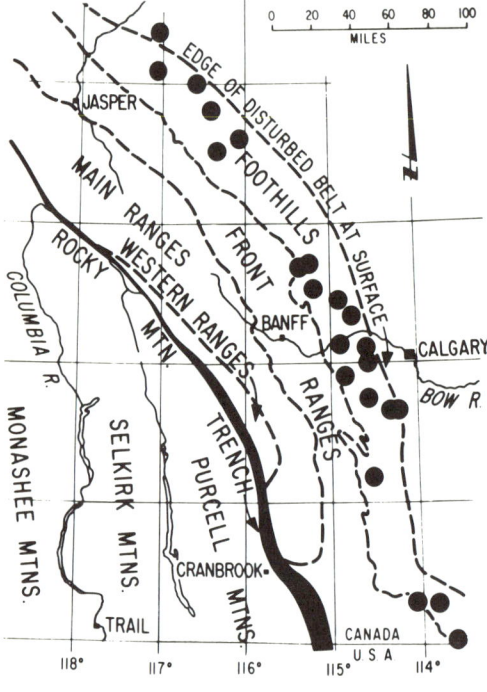

tures with Precambrian sedimentary cores probably formed with a *décollement* close to the crystalline basement.

Franklin Mountains. The Franklin Mountains form a broad arc of folds E of the Mackenzie Mountains, plunging at both ends to disappear beneath Lower Cretaceous strata. Often there are both E- and W-dipping thrust faults on the same structure. These mountains are similar structurally to the Mackenzies but are separated from them by a narrow Mesozoic depression. The eastern margin is flanked by prominent thrust faults. Although cratonic basement block faulting has been interpreted for these structures, they probably resulted from superficial detachment at depth along a Cambrian salt and anhydrite horizon.

Richardson Mountains. The northern Richardson Mountains are broad, faulted folds and domes with lower Mesozoic or Paleozoic cores. The southern Richardson Mountains consist of a single broad, north-plunging, faulted anticlinal structure about 80 km across with Cambrian rocks in the core. The faults trend N-S or NW-SE, often with a large dextral transcurrent displacement and have the eastern side downthrown. These faults probably began during the Paleozoic with their eastern sides up and were reactivated during the Laramide deformation. Southward they merge with the E-W-

FIGURE 8. Schematic cross section of foothills west of Calgary showing structures, field types, and hydrocarbon reserves in billions of cubic feet (millions of barrels of oil in Turner Valley field) (from Dahlstrom, 1970, p. 338).

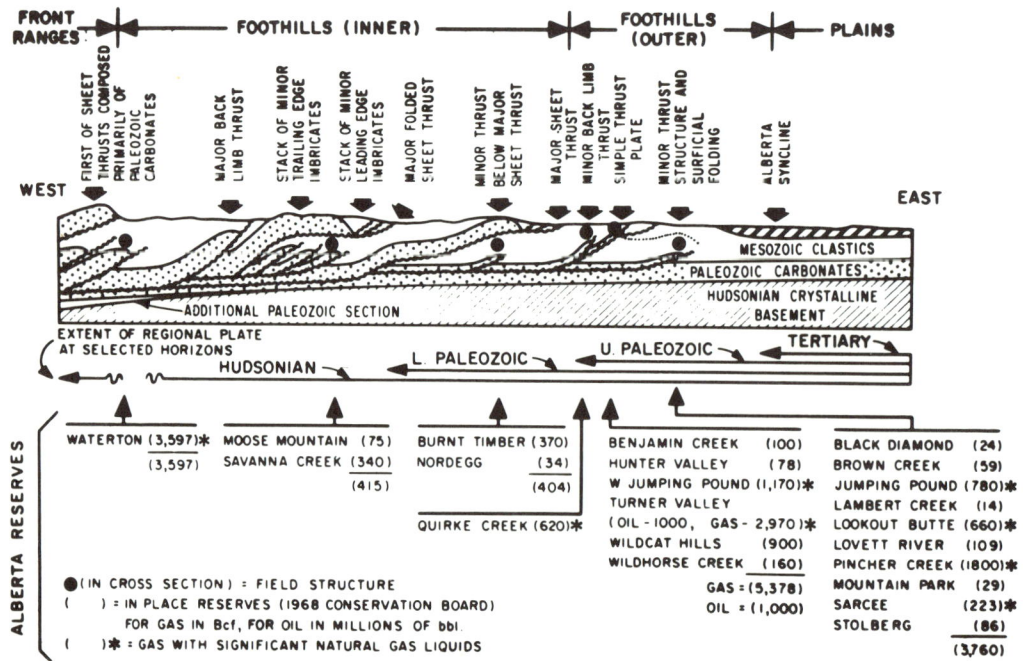

trending thrust faults of the northern Mackenzie Mountains.

Economic Minerals

Because the Rocky Mountain region is characterized by relatively simple folds and thrust structures, little or low-grade metamorphism, and sedimentary rocks of cratonic facies, the metallic mineral deposits are few and economic interest centers on petroleum, natural gas, coal, and nonmetallic deposits such as evaporites and structural materials.

Petroleum. Since the discovery of oil at Turner Valley in 1914, the most easterly thrust sheet of the foothills SW of Calgary, the foothills, and adjoining front ranges have been closely studied, but despite much drilling, no major oil field has been discovered. Instead, natrual gas, distillate, and H_2S from which sulfur is extracted have been found on a large scale (Fig. 8; Dahlstrom, 1970). The main fields are in southwestern Alberta mainly in Mississippian crinoidal limestones, where the reservoirs are cut off by underlying thrusts and occur in simple anticlines or fault slices. In one Devonian reservoir H_2S forms 87% of the gas. The Norman Wells oil field on the southwestern flank of the Franklin Mountains is a stratigraphic reef trap (Middle Devonian Kee scarp). Middle Devonian carbonates produce natural gas in two anticlines of the Liard plateau. Just west of the Richardson Mountains beneath Eagle Plain oil has been discovered in Pennsylvanian sandstones.

Coal. Extensive bituminous coal deposits occur and most of the coal is found in nonmarine Lower Cretaceous strata in the foothills between the 49th and 54th parallels, and in the Peace River area.

Deposits of gypsum, limestone, phosphate, and barite occur in some areas, but only limestones and barite are at present being quarried or mined.

Metallic Minerals. Sedimentary iron formation about 40 m thick occurs in the northern Mackenzie Mountains in the Precambrian Rapitan formation. It consists of jasper and hematite and averages about 46% iron.

Placer gold has been produced from several rivers draining the east side of the southern Rocky Mountains, mainly from the North Saskatchewan. It also occurs on the western side of the Cariboo Mountains and once precipitated a minor gold rush. Gold-bearing quartz veins and replacement bodies in limestone are also known.

West of the southern Rocky Mountains Trench a large number of lead-zinc deposits, some containing considerable silver, occur in the Purcell anticlinorium and are of Precambrian age. One of these, the Sullivan mine, is the largest lead-zinc producer in the Canadian Cordillera. The Sullivan deposit is unique and is a high-temperature replacement of thin-bedded Purcell argillites. Antimony, tin, gold, silver, copper, and cadmium are also produced. Other lead-zinc deposits are mined in the Kootenay arc in Cambrian rocks, and also occur in calcareous layers in the mantle of gneiss domes in the Shuswap complex.

ERIC W. MOUNTJOY

References

Bally, A. W., Gordy, P. L., and Stewart, G. A., 1966. "Structure, seismic data, and orogenic evolution of southern Canadian Rocky Mountains," *Bull. Can. Petrol. Geologists,* **14**(3), 337–381.

Berry, M. J., Jacoby, W. R., Niblett, E. R., and Stacey, R. A., 1971. "A review of geophysical studies in the Canadian Cordillera," *Can J. Earth Sci.* **8**, 1–14.

Dahlstrom, C. D. A., 1970. Structural geology in the eastern margin of the Canadian Rocky Mountains, *Bull. Can. Petrol. Geologists,* **19**(3), 332–406.

Douglas, R. J. W., ed., 1970. *Geology and Economic Minerals of Canada.* (Econ. Geol. Ser. 1, 5th ed.) Ottawa: Geol. Surv. Can., 838p.

Gabrielse, H., 1967. Tectonic evolution of the northern Canadian Cordillera," *Can. J. Earth Sci.* **4**, 271–297.

____, 1972. "Younger Precambrian of the Canadian Cordillera," *Am. J. Sci.,* **272**, 521–536.

King, P. B., compiler, 1968. *Tectonic Map of North America.* Washington D.C.: U.S. Geol. Surv., 1:5,000,000.

____, 1969. "The tectonics of North America – a discussion to accompany the *Tectonic Map of North America Scale 1:5,000,000*," U.S. Geol. Surv. Prof. Pap. **628**, 95p.

McCrossan, R. G., and Glaister, R. P., ed., 1964. *Geological History of Western Canada.* Calgary: Alberta Soc. Petrol. Geologists, 232p.

Monger, J. W. H., Souther, J. G., and Gabrielse, H., 1972. "Evolution of the Canadian Cordillera: a plate–tectonic model," *Am. J. Sci.,* **272**(7), 577–602.

Price, R. A., and Mountjoy, E. W., 1970. "Geologic structure of the Canadian Rocky Mountains between Bow and Athabasca Rivers, a progress report," *Geol. Assoc. Can., Spec. Pap.* **6**, 7–26.

Reesor, J. E., 1970. "Some aspects of structural evolution and regional setting in part of the Shuswap Metamorphic Complex," *Geol. Assoc. Can., Spec. Pap.* **6**, 73–86.

Shaw, E. W., 1963. "Canadian Rockies – orientation in time and space," *in* O. E. Childs, ed., *Backbone of the Americas* (Mem. 2). Tulsa: Am. Assoc. Petrol. Geologists, 231–242.

Wheeler, J. O., ed., 1970. "Structure of the southern Canadian Cordillera," *Geol. Assoc. Can., Spec. Pap.* **6**, 166p.

Cross-references: *Canada; Canada–Cordilleran Region, Northern Great Plains Province.*

CANADA—ST. LAWRENCE LOWLANDS OF QUEBEC

The lowland bordering the St. Lawrence River in Quebec is floored with Paleozoic sedimentary rocks which form an extension of the Interior Basin Province of North America (Fig. 1). It is hemmed in by the Adirondacks on the S, the Laurentians on the NW, and the Appalachians on the SE. It is divisible into two parts, here called the *Quebec Basin* and the *Anticosti Basin*. The Quebec Basin (name used by Geological Survey, rather than Montreal Basin) is at its widest (roughly at right angles to the river), just N of the international boundary, where, from Philipsburg NW to St. Jerome it is about 110 km across. Going downstream it narrows irregularly, and near Quebec it is scarcely more than 3—5 km across. Thence downstream, save for some basal sandstones exposed along the N shore of the St. Lawrence SW of Murray Bay, there are no exposures of Interior Basin rocks until the Mingan Islands and Anticosti Island are reached. There the Anticosti Basin is about 160 km wide, narrowing again northward to include the northwestern part of Newfoundland and the adjacent eastern margin of the Precambrian Shield along the Strait of Belle Isle. Originally doubtless a wide-shelf seaway, 300 km or more across, merging along its eastern margin with the western edge of the Appalachian geosyncline, it has reached its present restricted state through successive diastrophic constrictions along both eastern and western margins.

Because the borders cannot be precisely located, only an approximate figure of 27,000 km^2 (10,500 sq mi) can be given for the area of the Quebec Basin. The Anticosti Basin is largely offshore, but is roughly approximated at 40,000 km^2 (15,000 sq mi).

The early field work was carried out by the Geological Survey of Canada in the 1840s, but from the turn of the century on, geological mappings have been taken over by the provincial survey (the Quebec Department of Natural Resources). Private explorations had been pursued intermittently. Since about 1950, oil and gas possibilities have been assessed with the aid of geophysical and geochemical procedures. Quarrying for building stone, crushed stone, and concrete aggregate is general throughout the Quebec Basin. The high-silica Chateauguay sandstone is quarried for silicon alloys; suitable horizons of the Chazy and Trenton limestone formations support a flourishing cement industry. The Utica and Lorraine shales and the postglacial clays have been extensively used for brick-making; and postglacial sands, wherever suitable, are used in the production of concrete. Drilling for oil and gas has resulted in a hundred or so deep wells throughout the lowland, although only marginally economic production has resulted.

Because of the lack of population in the Anticosti Basin, there has been little exploitation of the resources there. However, considerable drilling activity both on Anticosti Island and in the surrounding waters of the Gulf of St. Lawrence has taken place during the past decade.

Montreal and Quebec have been the leading intellectual centers. In Montreal McGill University has the longest record of devotion to geology and mining, and the University of Montreal has, since 1960, established full departments of geology and geological engineering. The Université de Québec, with several divisions in the province, is now also established in Montreal. Laval University in Québec City, directly descended from the Collége des Jésuites founded in 1635, has throughout most of its history been devoted to the furtherance of the humanities, but in the last few decades has expanded to include geology and mining engineering. Sir George Williams University and Loyola College in Montreal also give instruction in geology and related subjects. There are also the "classical colleges," such as those at Nicolet and Trois Rivières (Three Rivers), where the general intellectual activities have been carried on over more than 300 years.

Geomorphologically speaking, both basins are marked by very little relief, the southern being wholly above sea level, whereas the northern basin is largely below sea level except for the Mingan and Anticosti islands. Structurally the Quebec Basin is an elongated canoe-shaped syncline in older Paleozoic rocks with its axis

FIGURE 1. Map showing the location of the Quebec Basin (QB) and the Anticosti Basin (AB) (stippled area).

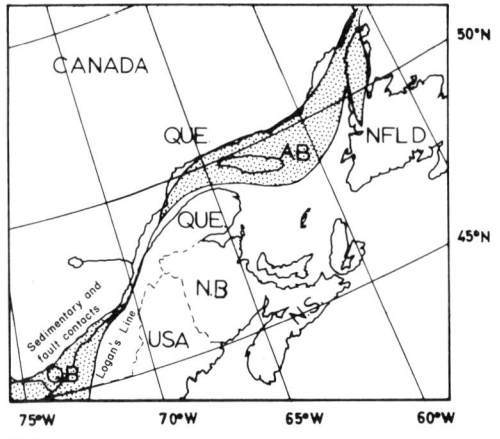

TABLE 1. Stratigraphic Correlation Table for St. Lawrence Lowlands

ERA	SYSTEM	SERIES	STAGE	GROUP	ST. LAWRENCE LOWLANDS FORMATIONS				EUROPEAN STAGES
					ONTARIO OTTAWA B.	MONTREAL BASIN* SOUTH	MONTREAL BASIN* NORTH	ANTICOSTI BASIN	
P A L E O Z O I C	DEV.	M	ERIAN			ST. HELEN'S ISLAND BRECCIA BLOCKS			GIVETIAN
		L	ULSTERIAN						SIEGENIAN
			HELDERBERGIAN						GEDINNIAN
	SILURIAN	U	CAYUGAN						LUDLOVIAN
		M	NIAGARAN		LOCKPORTIAN TONAWANDIAN ONTARIAN	CLINTON			CHICOTTE JUPITER GUN RIVER	WENLOCKIAN
									LLANDOVERIAN
		L	ALEXANDRIAN			MEDINA			BECSCIE
	ORDOVICIAN	U	CINCINNATIAN	GAMACHIAN					ELLIS BAY	ASHGILLIAN
				RICHMONDIAN	QUEENSTON	QUEENSTON	BECANCOUR R.	BECANCOUR R.	VAUREAL	
				MAYSVILLIAN	LORRAINE	RUSSELL CARLSBAD	PONTGRAVE R. NICOLET R.	PONTGRAVE R. NICOLET R.	--- ENGLISH HEAD	
				EDENIAN	UTICA	BILLINGS EASTVIEW	LACHINE	LOTBINIERE	MACASTY	
		M	CHAMPLAINIAN	SHERMANIAN	TRENTON	COBOURG SHERMAN FALL --- HULL ROCKLAND	TETREAUVILLE MONTREAL --- DESCHAMBAULT MILE END OUAREAU	NEUVILLE GRONDINES ST. CASIMIR DESCHAMBAULT --- PONT ROUGE		CARADOCIAN
				KIRKFIELDIAN ROCKLANDIAN						
				BLACKRIVERAN	BLACK RIVER	LERAY LOWVILLE PAMELIA	LERAY LOWVILLE PAMELIA	LERAY LOWVILLE PRECAMBRIAN		LLANDEILIAN
				CHAZYAN	CHAZY	ROCKCLIFFE	LAVAL		MINGAN	LLANVIRNIAN
		L	CANADIAN		BEEKMANTOWN	OXFORD	BEAUHARNOIS		ROMAINE PRECAMBRIAN	ARENIGIAN TREMADOCIAN
C A M B R I A N		U	CROIXAN	TREMPEAULEAU'N FRANCONIAN DRESBACHIAN	POTSDAM	MARCH NEPEAN PRECAMBRIAN	CHATEAUGUAY COVEY HILL PRECAMBRIAN		LABRADOR BASIN HAWKE BAY FORTEAU BRADORE PRECAMBRIAN	
		M	ALBERTAN							
		L	WAUCOBAN		LABRADOR					

*The Geological Survey of Canada now uses "Quebec Basin" for "Montreal Basin."

more or less parallel to the St. Lawrence River but crossing it a few miles above the city of Quebec and crossing the international boundary south of Hemmingford. The syncline is comparable in some ways to the easternmost valley and ridge subprovince of Pennsylvania, and is likewise bounded on the SE by thrust faults along which overriding slices of the Appalachian terrane were shoved or slid to the NW during the Paleozoic orogenies. The lower sedimentary units of the synclinal succession are unconformable on the Precambrian of the Laurentian and Adirondack massifs, and in part are faulted against the former along en échelon NE-SW-trending normal faults. In addition, in the Montreal vicinity and elsewhere there are numerous E-W faults, and others have been detected by geophysical investigations.

Conditions are different in many ways in the Anticosti Basin. There the exposed rocks dip more or less uniformly southward from their unconformable sedimentary contact with Precambrian rocks on the mainland adjacent to the Mingan Islands, and are presumably continuous beneath Anticosti Island and to the south. Although no representative of this older Paleozoic series appears on the north shore of the Gaspé Peninsula, it is likely that the same synclinal structure continues here as in the Quebec Basin. Comparable en échelon faults are unknown here, but drill hole data and geophysical surveys show Anticosti Island to be traversed by a network of normal faults.

In summary, the sedimentary series is all that remains of a once-more-extensive shelf development resting unconformably upon the

FIGURE 2. Geological map of the Quebec Basin (courtesy of Quebec Department of Natural Resources).

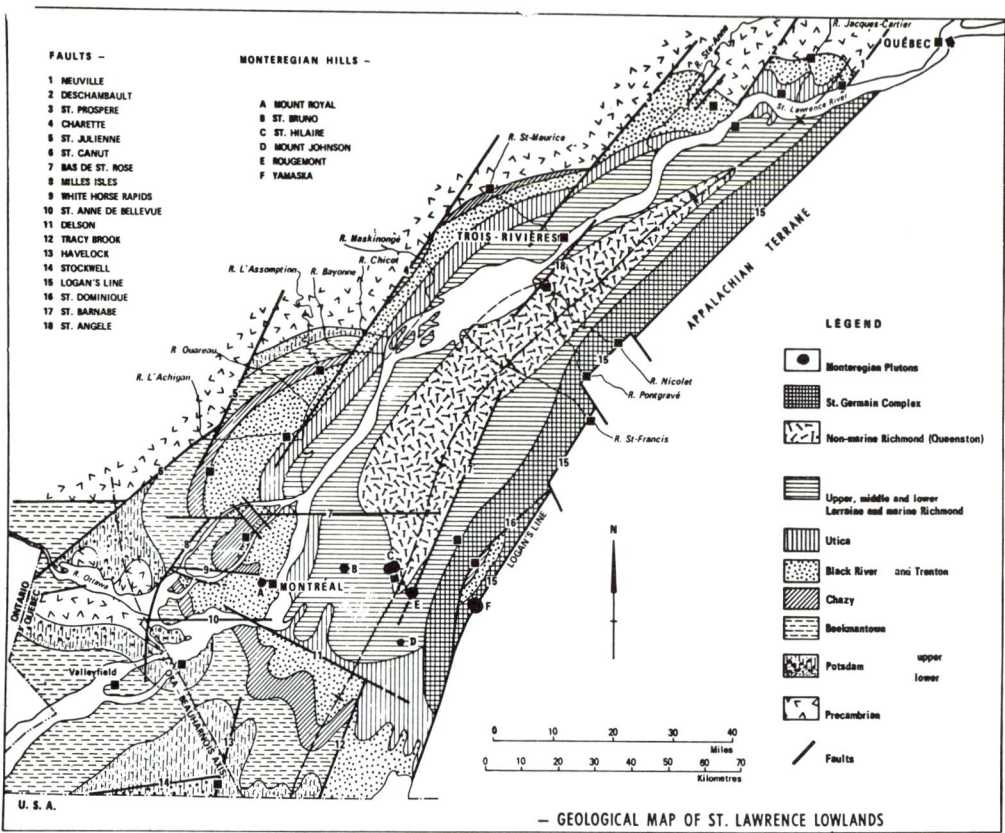

Quebec Basin

Stratigraphic Succession. *Potsdam Sandstone.* The stratigraphic succession differs considerably from that of the Anticosti Basin (see Table 1). In the Quebec Basin (also called Montreal Basin) the lowest formation is the thick Covey Hill Sandstone, well exposed at a few places in the southwestern corner of the area (Fig. 2). This facies is interpreted as derived from the mechanical erosion of material prepared by a long-continued period of weathering of the Canadian Shield since the Grenville orogeny. These products, largely reduced to quartz, K-feldspars, and heavy residue minerals, make up a sandstone in part conglomeratic and reddish, especially in its lower part, that is over 600 m thick. This is followed by up to 250 m of white quartz sandstone, the Chateauguay Formation, devoid of pebbles and red color, Precambrian Shield, and later deformed by orogenic disturbances along its southeastern borders.

and with very little feldspar. This is taken to be the result of further weathering of both the Covey Hill sands not at that time consolidated and an influx of more mature sands from the supplying area. There are no fossils in the Covey Hill beds, which have thus been interpreted as nonmarine, but the Chateauguay sandstones contain *Lingulepis acuminata, Climactichnites, Protichnites, Skolithos,* etc.— the first indicating a Late Cambrian age from correlative strata in New York State. Neither is known north of the St. Maurice or St. Francis rivers. At Melocheville Chateauguay sandstone is quarried as a source of silica (up to 99.5% SiO_2) and is used in the manufacture of ferrosilicon alloys.

Beekmantown Dolomite. Following the deposition of these Potsdam-type sandstones the basin continued to sink, allowing the waters from the Appalachian geosyncline to transgress farther over the basement. Carbonates accumulated to more than 600 m, to be followed by over 1300 m of marine and nonmarine shales and sandstones. At the base the Chateauguay formation is succeeded by the Beauharnois

Dolomite (Beekmantown Group), 250 m thick in the Montreal area, but unknown N of the Potsdam rocks. The dolomite shows a great diversity of textures and structures, from dense through fine to coarse-grained crystalline, in places massive, and in others with paper-thin stratification. Parts are vuggy, the cavities lined or filled with quartz, calcite, dolomite, fluorite, etc. Much of it is rich in quartz grains. Fossils are scarce in the dolomite except for the alga *Cryptozoon* and the gastropod *Lecanospira*. In the rare limestone beds, usually restricted to a few centimeters in thickness, a small fauna characterized by the trilobite *Hystricurus* has been found. This is a correlative of the Beekmantown dolomite of New York State. This rock is quarried for crushed stone and concrete aggregate.

Middle Ordovician Limestones. Unconformable above the Beauharnois lies the Chazy Group, consisting of rocks of mixed lithology, known locally as the *Laval Formation*. It begins with a few meters of a basal sandstone which passes upward to 100 m of shaly, sandy, and dolomitic limestone, with intercalations of pure calcarenite up to 15 m thick. Fossils are common throughout, particularly the brachiopods *Mimella* and *Rostricellula*. Bryozoa abound together with the earliest corals (*Billingsarea* and *Eofletcheria*), built reefs at least as old as those on Valcour Island, N.Y., and thus the world's oldest known coral reefs. The dolomitic facies of this formation is associated with the reefs or with an abundance of the reef-building fossils. The calcarenite is made up of sand-grain-sized fragments of cystid, and possibly crinoid, plates and columnals. The commonest recognizable echinoderms are *Malocystites* and *Bolboporites*. Up to half a century ago the calcarenite provided most of the local stone used in buildings in Montreal and the surrounding towns. Production of dimension stone is now practically nil. It is used for crushed stone and for concrete aggregate. Although the Chazy rocks overlap the Beekmantown beds westward, none are known NE of Trois Rivières.

The Black River Group follows but is nowhere more than 25 m thick and consists of a lower Pamelia dolomite followed by the Lowville and Leray limestones. The Pamelia is unfossiliferous. The Lowville contains an abundant fauna, especially the coral *Tetradium*, which is also sparingly present in the Leray. In the latter, brachiopods such as *Rafinesquina* first become important and there are large (up to 60 cm across) heads of the coral *Foerstephyllum*. *Stromatocerium* is abundant enough to make up 15-cm beds. Black River strata crop out along the north side of the St. Lawrence River as far as Quebec City.

The Trenton Group comes next, with basal lenses of limestone (Ouareau, Fontaine, St. Alban, etc.) only a few meters thick, apparently filling hollows left on the surface of the Black River beds. Then follows the Deschambault Limestone, a calcarenite composed of dissociated fragments of crinoids. This rock is almost indistinguishable in the hand from the Laval calcarenite. It extends from Montreal to Quebec City and is 25–30 m thick but thickens in deep wells toward the center of the basin. At St. Marc des Carrières (formerly Deschambault) a score of quarries formerly produced dimension stone which was widely used in Quebec and Montreal, but there is little activity today. This limestone emits a strong odor of hydrocarbons when freshly broken and oozes small droplets of petroleum. Fossils of Hull affinities are abundant with bryozoans, particularly *Rafinesquina* and the alga *Solenopora*. Chert nodules usually occur in this formation. Otherwise, it is a remarkably pure limestone, reaching 99.5% $CaCO_3$ in places. It has long been providing material for lime kilns, and today it supplies a large cement plant at Joliette.

Above the Deschambault in the southern part of the basin comes up to 120 m of irregularly interbedded dark gray limestone and shale known as the *Montreal Formation*. This formerly provided much dimension stone for Montreal. The fauna is abundant and varied. *Rafinesquina, Dalmanella, Sowerbyella,* and *Zygospira* are the commonest brachiopods, and the trilobites *Isotelus, Flexicalymene, Ceraurus,* and in the lower beds *Cryptolithus*, are ubiquitous. Bryozoa are particularly abundant in certain beds, as are crinoids, gastropods, bivalves, and ostracods.

Next comes the Tetreauville Limestone, a dark gray rock also 120 m thick, more or less regularly interbedded with black shale, the proportions of limestone to shale ranging from 10:1 to 2:1. It is admirably suited for the manufacture of cement, and in the largest cement plant in the basin it is used with only minor additives. Fossils are much the same as in the Montreal Formation but are much scarcer.

In the northern part of the basin both Montreal and Tetreauville formations are correlated with the Neuville Limestone, which combines their various characteristics. The Neuville Formation is up to 120 m thick and supports cement plants at St. Basile and Charlesbourg.

Upper Ordovician Clastics. Almost all the Late Ordovician deposits are clastics, beginning with the black Lachine shales (south) and Lotbinière shales (north) belonging to the Utica Group. These shales are normally 90–120 m thick and are widely distributed throughout the basin. In the extreme eastern part they may be as much as 330 m thick. The facies is a typical euxinic "Black Shale," for the most part unfossiliferous, but some bedding planes are crowded with graptolites and other plankton. It was formerly quarried at Delson for brick making.

Succeeding the Utica shale comes the Nicolet River Formation of the Lorraine Group, a thick (800 m) series of calcareous shales and sandstones with thin intercalated limestone beds. It begins everywhere with a shale sequence up to 300 m thick resembling the underlying Utica but gray instead of black and in places rich in marine fossils. There are also lenses of fine-grained sandstone (up to 10 m) and limestone (up to 1 m) commonly in the upper part. The lower shale part was exposed during the excavation for the St. Lawrence Seaway, and is quarried at Laprairie for brick making. It is most extensively exposed in the northern part of the basin, on both sides of the river. Fossils are abundant with the brachiopods *Strophomena* and *Catazyga* throughout. The trilobite *Cryptolithus* occurs in the basal shales, and *Proetus* is characteristic of the middle part. The gastropod *Lophospira* and the bivalve *Pholadomorpha* occur especially in the upper part. It is well exposed along the Nicolet River below St. Léonard, where the whole 800 m thickness is seen without a gap, one of the most spectacular exposures in this basin.

Lying conformably upon the Nicolet River Formation is the Pontgravé River Formation (50 m) belonging to the Richmond Group; it is lithologically similar to the underlying beds but is more calcareous. Fossils are very common and contain both Lorraine species and elements belonging to beds of Richmond age elsewhere in North America. Large bivalves such as *Whitella*, the solitary coral *Streptelasma*, and at the very top the brachiopod *Zygospira* are particularly abundant. Although rather soft, it is well exposed along many of the streams flowing westward into the St. Lawrence.

Succeeding this formation there comes a thick, red, nonmarine, subaerial deltaic accumulation of siltstone known as the Bécancour River Formation, originating as a rapidly eroded wash from the rising Appalachians. One deep well proved over 600 m thickness of it. Except for a few trails it is unfossiliferous, but correlatives in Ontario carry an Upper Ordovician (Richmond) fauna. It replaces the marine Richmond progressively westward. By the end of the Ordovician most of southern Quebec and Ontario were covered by a blanket of this widespread deltaic deposit.

Devonian Limestones. In the Quebec Basin there are no other lithified formations, although Lower and Middle Devonian blocks of limestone in a diatreme breccia on St. Helen's Island (Montreal) betray the one-time presence there of Devonian formations, now completely eroded away.

Postglacial Deposits. Postglacial deposits (very late Pleistocene and Holocene) blanket most of the basin. Unsorted morainal material are succeeded by nonmarine sands and clays deposited during times of impounded meltwater, and later by marine clays (*Leda* clay) up to 50 m thick, and sands (*Saxicava* sand) rarely more than 6 m thick deposited during the Champlain submergence of the basin. Isostatic readjustment of the crust locally has almost completely drained the sea from the Quebec Basin, allowing the consequent streams entering the St. Lawrence to carve deeply incised valleys, thus producing the countryside as it exists today.

Structure. The Montreal Basin is synclinal in structure (Fig. 3). The western limb, apart from minor folding near Montreal, dips about 2° toward the axis. On the eastern limb the dip increases from the axis eastward, becoming vertical and even overturned as a result of the overriding of Taconic slices coming from the east, with the glide plane inclined upward toward the west. However, for a few kilometers W of the westernmost thrust (i.e., Logan's Line) the beds have been so broken up as to lose most of their stratigraphic order. This strongly deformed zone is called the *St. Germain Complex.*

The *en échelon* western border faults, downdropping to the E, were probably in existence and active in Precambrian time, and may have been re-activated many times since, and even today seem to be responsible for some earthquakes affecting the lowland. Other normal

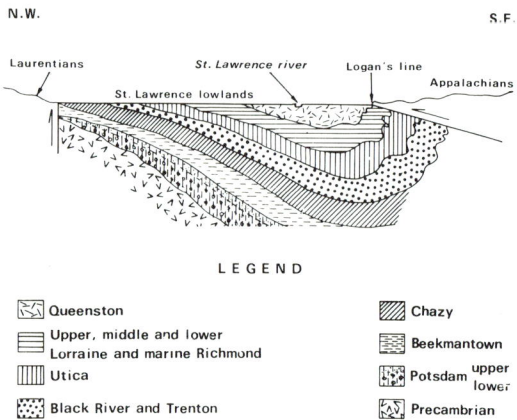

FIGURE 3. Geological cross section across the St. Lawrence Lowlands of Quebec (courtesy of Quebec Department of Natural Resources).

faults, particularly those in the vicinity of Montreal, like the small folds referred to above, may reflect a progressive graben development.

Igneous Activity. The only igneous activity in this belt has been the intrusion of alkaline magmas in early or mid-Cretaceous time (95–125 m yr), resulting in the Monteregian plutons and a network of satellitic bodies, e.g., Mounts Royal, St. Bruno, St. Hilaire, Johnson, Rougemont, and, straddling Logan's Line, Yamaska. These are distributed irregularly along an E-W band (see Fig. 2). Beyond the limits of this area, Brome, Shefford, and Megantic mountains lie within the Appalachian terrane. The form of the plutons is not known; they have been interpreted as volcanic necks, laccoliths, and stocks. There is no evidence that any of them broke through to the surface, except in the case of the St. Helen's Island diatreme. Their present physiographic form, subdued conical hills rising 150–300 m above the surrounding plain, is the result of differential erosion.

The plutons are composed of nepheline syenite and an older gabbro. Several localities have yielded rare or new minerals, e.g. in the syenite of Mount St. Hilaire. The old Corporation quarry, now within the grounds of the University of Montreal, was a famous collecting ground for many species, including native arsenic. Dawsonite, weloganite, and dresserite are associated with syenitic dikes and sills.

The Anticosti Basin

The separation of the Quebec and Anticosti basins is only an approximation. The lithologic differences in most of the formations indicate more than facies changes and point to some sort of discontinuous barrier. The outcrops along the northern shore of the St. Lawrence River for 100 km below Quebec City are essentially similar lithologically and faunally with those of the Quebec Basin probably as far as the Saguenay River. Beyond that point there is no evidence of lowland rocks until the Mingan and Anticosti islands are reached.

Southward the Anticosti Basin extends presumably as far as the Appalachian front (Logan's Line, see Fig. 4). Eastward it continues to the Strait of Belle Isle, where on the northern shore coastal outcrops of Cambrian sandstones and limestones are its last expression.

Memoirs by Schuchert and Dunbar, by Twenhofel, and more recent papers by Bolton are the main sources of information obtained from the surface exposures. In recent years deep wells drilled on Anticosti have yielded

FIGURE 4. Map of part of the Gulf of St. Lawrence showing the Anticosti Basin outlined by heavy lines. (a) Diagrammatic cross section along the line ABC to show, with exaggerated dip, the known structure of the basin and the supposed structure of the supra-Chicotte beds and their relationship to the Appalachian terrane of Gaspé: 1, Romaine to Macasty formations; 2, English Head to Ellis Bay formations; 3, Becscie to Chicotte formations; 4, Supra-Chicotte formations.

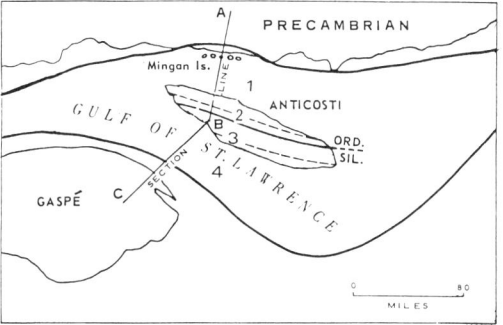

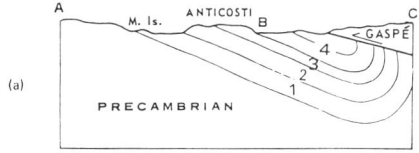

data discussed by Roliff. The lowest rocks exposed belong to the Bradore Group on the Labrador side of the Strait of Belle Isle, but the continuous section begins with the Romaine Formation of Canadian age, and progresses upward through the Ordovician and into the Middle Silurian with no significant break. In all about 2000 m of beds are known to be present, and nearly 1000 m of additional Silurian and Devonian beds lie beneath the waters of the Gulf of St. Lawrence between Anticosti and Logan's Line.

Strait of Belle Isle. The lowest beds exposed, of Lower Cambrian age, occur on both sides of the Strait of Belle Isle (see Table 1). The basal Bradore Formation consists of red, unfossiliferous arkose, conglomerate and sandstone with white quartz sandstone at the top, and the overlying Forteau Formation consists of marine shales and limestone with archaeocyathid reefs. The Bradore Formation resembles the Covey Hill and Chateauguay of the Lowland proper, and may be of Middle or even early Cambrian age.

No equivalent of the Forteau Formation is known in the Quebec Basin. On the Newfoundland coast a third formation occurs, the Hawke Bay, composed largely of sandstone but with lesser amounts of limestone and shale, but its inclusion within the lowland is debatable.

Mingan Islands and Anticosti. North of Anticosti, and fringing the mainland coast, lie the Mingan Islands, a score or more in number. They are composed wholly of Ordovician rocks, which also outcrop in a narrow band along the mainland coast. Two formations are involved—the Romaine (Beekmantown) dominantly dolomite, and Mingan (Chazy) dominantly limestone.

Romaine Formation. Above a basal bed of sandstone the rest of the formation is largely dolomitic, in places with chert and vugs. As with the Beauharnois Formation in the Quebec Basin fossils are both poorly preserved and rare. *Cryptozoön steeli, Syntrophia lateralis,* and *Bathyurus amplimarginatus* are the only forms in a fauna of 37 species that are known outside the Anticosti Basin, and indicate a correlation with the upper part of the Beauharnois Formation. On Mingan Islands 80 m is the probable thickness, but southward it increases markedly, reaching 380 m on Anticosti.

Mingan Formation. The Mingan formation lies unconformably upon the Romaine beds and begins with a basal conglomerate or a sandstone that grades quickly into limestone, and in the well logs there are minor amounts of sand toward the top of the formation. It is similar to the Laval Formation of the Quebec Basin. Some beds are replete with fossils. Of the 111 species recorded 25 are closely related to or identical with elements of the Laval fauna. Twenhofel gave a section (necessarily incomplete) of 43 m; the average thickness in the Anticosti wells is 106 m.

Middle Ordovician Limestones. Beds of Black River age do not crop out but in Anticosti wells up to 2 m of unfossiliferous pinkish fine-grained limestone occurring between recognizable Chazy and Trenton strata is assigned to this group.

No Trenton Group strata are known in outcrop but in the Anticosti wells 300 m of limestone and shale can be assigned to this group. *Sowerbyella sericea, Homotelus stegops,* and *Ceraurus pleurexanthemus* indicate a close faunal relationship with the Trenton fauna of the Quebec Basin.

Upper Ordovician Formations. Again not known in outcrop, the Utica Group (Macasty Formation) has been recognized in hundreds of loose blocks of black bituminous shale with characteristic Utica fossils littering the northwestern shore of Anticosti. In the cores black Utica shale ranges from 15 to 90 m in thickness. Graptolites abound in certain layers. *Leptobolus* sp., *Triarthrus becki,* and *Geisonoceras* sp. are common in both surface blocks and the cores.

Above the Macasty shale lie nearly 1000 m of shale, shaly limestone, and limestone. These beds make up the English Head Formation (below), mainly shale up to 300 m thick, and the Vauréal formation (above), over 600 m of mixed limestone and shale. Above the Vauréal limestone comes 73 m of limestone referred to the Ellis Bay Formation. The lower part of the English Head probably belongs to the Lorraine Group, the rest of the English Head, the Vauréal, and the Ellis Bay to the Richmond Group.

Bioherms abound throughout the Upper Ordovician section. Fossils in general are abundant. The bivalves *Cyrtodonta anticostiensis* and *Rhytimya emma;* the brachiopods *Strophomena hecuba, Dalmanella meeki,* and *Leptaena rhomboidalis;* and the coelenterate *Halysites catenularia* are among common forms.

The Silurian Formations. The Silurian section is not as thick as the Ordovician. It begins with the Becscie Formation, in which a basal conglomerate is composed of limestone pebbles and boulders. This is followed by dense dove-gray and fine-grained crystalline limestone, up to 80 m thick. Fossils are extremely abundant in places and include the brachiopods *Virgiana barrandei* and *Camarotoechia neglecta,* and the bryozoan *Phaenopora superba.*

The Gun River and Jupiter Formations both consist largely of thin-bedded ash-gray limestone. The Gun River (100 m thick) contains abundant coral reefal developments and intraformational conglomerate beds. The Jupiter Formation (180 m) lacks the conglomerate beds. Among the fossils brachiopods are the most abundant, *Brachyprion leda, Hyattidinia congesta junea, Triplesia anticostiensis,* and *Virgiana barrandei.* Of the trilobites *Phacops orestes* is common.

The highest exposed beds in the Anticosti Basin make up the Chicotte Formation exposed along the southern shore of the island. These are crinoidal and reefal limestones and only 22 m are exposed. Fossils, of which the coral *Chonophyllum canadense* and the brachiopod *Cyrtia myrtia* are characteristic, indicate a correlation with the Rochester and part of the Lockport Formations of New York.

The development of Upper Silurian and Devonian above the Chicotte is not known, but there is room for at least 1000 m of them between the southern coast of Anticosti and Logan's Line, assuming that the same structural pattern obtains here as in the Quebec Basin. Accordingly, the total sedimentary section in the Anticosti Basin would be about 3000 m.

Structure. On the Mingan Islands there is a regional dip of $1-2°S$, carrying these formations to positions about 1500 m beneath the surface of Anticosti Island (Fig. 4a), where they have been found in deep wells. On that island both Ordovician and Silurian beds have a strike of about 100° and dips up to 2°S, except around bioherms where quaquaversal dips up to 20° occur.

Minor faults are recognizable in surface exposures on both Mingan and Anticosti islands. Geophysical explorations and photogeology reveal on Anticosti Island a complex of faults many of which are parallel to the strike.

Igneous Activity. Igneous rocks are confined to two diabase dikes on the northern shore of Anticosti Island, more or less parallel with the strike. Potassium-argon determinations give an age of 178 ± 8 m yr, which is early Jurassic.

Economic Potential. Little use has been made of the limestone resources. Deep wells drilled on Anticosti Island did not yield encouraging prospects for natural gas, and other wells offshore in the Gulf of St. Lawrence have been disappointing.

T. H. CLARK

References

Bolton, T. E., 1971. "Geological map and notes on the Ordovician and Silurian litho- and biostratigraphy, Anticosti Island," *Geol. Surv. Can. Pap.* **71-19**, 44p.

Clark, T. H., 1956. "Oil and gas in the St. Lawrence Lowland of Quebec," *Trans. Can. Inst. Mining Met.*, **59**, 278-282.

———, 1964. *Yamaska–Aston Area* (Geol. Rept. 102). Quebec: Quebec Dept. Nat. Res., 192p.

———, 1972. *Montreal Area* (Geol. Rept. 152). Quebec: Quebec Dept. Nat. Res.

Douglas, R. J. W., ed., 1970. *Geology and Economic Minerals of Canada* (Econ. Geol. Rept. 1). Ottawa: Geol. Surv. Can., 838p.

Dreimanis, A., and Karrow, P. F., 1972. "Glacial history of the Great Lakes–St. Lawrence region, the classification of the Wisconsin(an) stage, and its correlatives," *24th Intern. Geol. Congr.*, **12**, 5-15.

Gadd, N. R., McDonald, B. C., and Shilts, W. W., 1971. "Deglaciation of southern Quebec," *Geol. Surv. Can. Pap.* **71-47**, 19p.

MacPherson, J. B., 1967. "Raised shorelines and drainage evolution in the Montreal Lowland," *Cahiers Geogr. Québec*, **23**, 343-361.

Roliff, W. A., 1968. "Oil and gas exploration–Anticosti Island, Quebec," *Proc. Geol. Assoc. Can.*, **19**, 31-36.

Twenhofel, W. H., 1929. "Geology of Anticosti Island," *Geol. Surv. Can. Mem.* **154**, 481p.

———, 1938. "Geology and paleontology of the Mingan Islands, Quebec," *Geol. Soc. Am., Spec. Pap.* **11**, 132p.

Wagner, F. J. E., 1970. "Faunas of the Pleistocene Champlain Sea," *Bull. Dept. Energy, Mines and Resources, Ottawa*, **180**, 104p.

Cross-references: *Canada; Canada–Atlantic Provinces, Canadian Shield; North America; United States.*

CAROLINE ISLANDS

Lying in the western Pacific Ocean, the Caroline Islands comprise an archipelago made up mainly of atolls; Kusaie and Ponape islands are the largest of the volcanic islands. The most important island groups are Palau, Truk, and Yap. There are 37 atolls and 963 islands embracing a total land area of 1200 km² (460 sq mi). (*Note:* There is also a single "Caroline Island," an atoll in the *Line Islands,* q.v., which should not be confused with those discussed here.)

The Carolines are situated between 0 and 10°N, and 133 and 165°E. The mean annual temperature is 27°C (80°F), and the average annual rainfall is 2670 mm (105 in.). The islands are subject to highly destructive typhoons. In winter they are within the belt of the NE trade winds, but in summer they come under the influence of the intertropical convergence.

Administratively the islands fall into the *Pacific Islands, U. S. Trust Territory* (q.v.) under United Nations mandate, along with the *Marianas* and the *Marshall Islands* (q.v.). The people are mainly Micronesians. Originally claimed by Spain, in 1885 the Carolines became a German colony (von Prowazak, 1913) later to become part of the Japanese Empire (1914-1945). Most of the trade is in copra, trochus, and fishing. Phosphate, bauxite, and iron are mined.

Geologically the Carolines are divided roughly into three belts or groups: the western, Yap-Palau orogenic belts (connected island arcs); the central or main atoll area (centered on Truk); and the eastern mainly volcanic belt (centered on Ponape). A general report on the coral reefs of the Carolines was given by Nugent (1946).

Yap (also given as *Jap, Wap, Yappu, Ruul*), famous as the island of the "stone money" (discs of aragonitic limestone), is a deeply weathered andesitic volcanic peak, mountainous and very fertile, measuring about 20 by 8 km. It is maturely dissected with a broad fringing reef, is flat, but has no barrier. The Yap people are of Malay origin. There are modest accumulations of bauxite (est. 50,000 tons).

The Yap island arc trends NE-SW, with submerged connections (displaced by transform faults) to the Mariana arc to the NE, and to the Palau arc to the SW. Fronting it on the east is the Yap Trench (8597 m, 4662 fm). Ngulu atoll rises about 100 km SW of Yap.

The Palau (Pelew) islands (6°50′ to 8°00′N, straddling 134°30′E)—notably Babelthuap (Babeltop), Koror, and Peleliu—form a second arc SW of Yap. The main islands are, like Yap, deep weathered, late Tertiary andesitic volcanics but capped by extensive uplifted coral-reef platforms. These are tilted and faulted and are weathered into a complex karst landscape. In pockets there are rich phosphate accumulations. These have been penetrated by deep borings for phosphate search.

To the SW the Yap arc swings to the W and splits into three N-S ridges, the westerly one being represented in the Sonsorol Islands, Pulo Anna, Merir, Itelau Reef, and Tobi. Very little is known geologically of this area.

In the Truk or central group most of the islands are coral atolls, although Fais (Feis), in the extreme W, is a slightly uplifted (20 m) coral platform, 2 km across, steeply cliffed, and containing phosphate. Truk itself is a semiatoll, with 10 small volcanic islands, Shinchiyo and Shiki (up to 452 m on Tol), within an immense lagoon 70 km across, with 60 coral sand or limestone islets on the rim (Stark and Hay, 1955; Stark et al., 1958). The volcanics are basaltic with some trachyte and are believed to be centers of a large shield volcano of approximately Miocene age.

Ifaluk atoll, 800 km W of Truk, has been described in detail by Arnow (1965), and by Tracey et al. (1961). Other interesting atolls in this scattered area from W to E include Ulithi, Sorol, Eaurupik, Woleai, Faraulep, Gaferut, Pigailoe, Lamotrek, Puluwat, Pulap, Ulul, and Magur.

The eastern islands (sometimes labeled "Senyavin Islands") are dominated by olivine basalt cones and a few atolls. Ponape or "Ascension Island" (865 m, 2595 ft, in Mt. Totolau) and Kusaie (634 m, 2079 ft) are mountainous and very fertile. Both are deeply dissected, with drowned valley embayments filled by alluvium and encircled by barrier reefs. There are many archeological traces of former (?ancestral Polynesian) cultures in the now scarcely populated interiors of these volcanic islands. Kusaie is situated 1250 km ESE of Truk. On a broad submerged ridge extending south of Ponape (Caroline-Solomon Ridge), there are several scattered atolls—Ngatik, Nukuoro, and, farthest south, Kapingamaringi. The last, just 1°N of the equator, has been studied in detail by McKee (1958) and McKee et al. (1959).

The Caroline Islands have been closely studied in connection with late Quaternary eustatic changes in sea level. The 1967 Scripps' expedition, *Camarsel*, reported by Shepard (1970) and Bloom (1970a, 1970b), demonstrated a lack of evidence for the middle Holocene high terrace cited from many other regions.

RHODES W. FAIRBRIDGE

References

Arnow, T., 1965. "The hydrology of Ifalik atoll, western Caroline Islands," *Atoll Res. Bull.*, **44**, 1015.

Bloom, A. L., 1970a. "Holocene submergence in Micronesia as the standard for eustatic sea-level changes," *Quaternaria*, **12**, 145–154.

———, 1970b. "Paludal stratigraphy of Truk, Ponape, and Kusaie, eastern Caroline islands," *Bull Geol. Soc. Am.*, **81**, 1895–1904.

Bridge, J., 1948. "A restudy of the reported occurrence of schist on Truk, eastern Caroline Islands," *Pacific Sci.*, **2**, 216–222.

Davis, W. M., 1928. "The coral reef problem," *Am. Geog. Soc. Spec. Publ. 9*, 283–307.

McKee, E. D., 1956. "Geology of Kapingamarangi Atoll, Caroline Islands," *Atoll Res. Bull.*, **50**, 1–38.

———, 1958. "Geology of Kapingamarangi Atoll, Caroline Islands," *Bull. Geol. Soc. Am.*, **69**, 241–277.

———, Chronic, J., and Leopold, E. B., 1959. "Sedimentary belts in lagoon in Kapingamaringi Atoll," *Bull. Am. Assoc. Petrol. Geologists*, **43**, 501–562.

Nugent, L. E., Jr., 1946. "Coral reefs in the Gilbert Marshall and Caroline Islands," *Bull. Geol. Soc. Am.*, **57**, 735–780.

Prowazak, S. von, 1913. *Die Deutschen Marianen, ihr Natur und Geschichte.* Leipzig.

Schlanger, S. O., and Brookhart, J. W., 1955. "Geology and water resources of Falalop Island, Ulithi Atoll, Western Caroline Islands," *Am. J. Sci.*, **253**, 553–573.

Shepard, F. P., 1970. "Lagoonal topography of Caroline and Marshall Islands," *Bull. Geol. Soc. Am.*, **81**(7), 1903–1914.

Stark, J. T., and Hay, R. L., 1955. "Xenoliths in pyroclastic breccia of Truk islands, Western Pacific," *Bull. Geol. Soc. Am.*, **66**, 1621.

———, and Hay, R. L., 1963. "Geology and petrology of Truk," *USGS Prof. Pap. 409*.

———, et al., 1958. "Military geology of Truk Islands, Caroline Islands," U.S. Army Chief Eng., Intell. Dir., H.Q.U.S. Army Pacific, Tokyo, 1–205.

Tayama, R., 1952. "Coral reefs in the South Seas," *Japan Hydrographic Office Bull.*, **11**, 292p.

Tracey, J. I., Jr., Abbott, D. P., and Arnow, T., 1961. *Bernice P. Bishop Mus. Bull.*, **222**, 1–75.

Cross-references: *Mariana Islands; Marshall Islands; Pacific Islands Trust Territory.*

CAYMAN ISLANDS

The Cayman Islands, consisting of Grand Cayman, Little Cayman, and Cayman Brac, lie in the Caribbean Sea S of Cuba and W of Jamaica. Cayman Brac is the farthest E and lies 216 km W of Cabo Cruz, Cuba. It is about 21 km long by 2.5 km wide. Little Cayman, the same size, lies 8 km W of Cayman Brac; and Grand Cayman, 32 km long by 6.5–13 km in width, lies some 96 km farther to the SW.

Grand Cayman is about 290 km SSE of the Isle of Pines, Cuba.

The islands are part of the British West Indies and were formerly a dependency of Jamaica. The population is approximately 5000, the greater percentage being on the island of Grand Cayman. The largest town is Georgetown, the capital, other towns being Boddentown, West Bay, and East End. Grand Cayman is low and flat and is protected by coral reefs. The others are similar, although Cayman Brac rises to 40 m. The climate is hot and humid.

The three islands lie on an ancient submarine ridge or mountain range which, before submergence, may have extended eastward to the Sierra Maestra of Oriente province, Cuba, and westward to the Misteriosa Bank and toward the Central American coast, possibly to the Coxcombe Mountains of Belize (British Honduras). It is possible that the islands, as well as Jamaica and Haiti, were once much closer to the Central American coast and that they reached their present position by left-lateral strike-slip faulting (Matley, 1926; Hess and Maxwell, 1953).

The *Cayman Trench* (sometimes known as the *Bartlett Trough*), which reaches a depth of 7119 m (23,356 ft) lies south of the islands, and extends ENE-WSW for 1600 km from the southeastern tip of Cuba to Guatemala, where it continues as a major lineament.

Marine geological studies along and across the Cayman Trench indicate that it marks the boundary of the Caribbean and North American plates. There are two distinct transcurrent fault trends, the master set being ENE-WSW (left-lateral) and a second-order set with NW-SE trends (and right-lateral slip). There is evidence that the trench developed as a megashear since early Tertiary times, and it is still marked by high seismicity (Pinet, 1971; Erickson et al., 1972). Structures suspected to be salt diapirs have been identified along the Honduras sector of the trough, and these are highly suggestive of incipient plate-opening. The most active movement was probably Miocene-Pliocene. The total transcurrent shift may exceed 400 km.

Each of the three Cayman Islands represents a fault block of white limestones of Middle Tertiary age, called the *Bluff Formation,* which may be of various ages (Oligocene to Miocene) on the different islands. Along the margins of each island is a fringe of younger limestone, of Pleistocene age, called the *Ironshore Formation,* in which marine fossils are numerous (Matley, 1926; Richards, 1955; Rehder, 1962; Brunt et al., 1973). This formation rises only to only 3 to 4 m above high-tide level.

There is evidence that Grand Cayman is rising at the present time, possibly at a rate of as much as 1 cm per year during the past 50 years. Areas can be seen in which there are now streets, but which were below sea level within the memory of residents; also the harbor at Boddentown has been abandoned because of the uplift. However, it is doubtful that the island has been rising continuously at this rapid rate. The weathering of the raised coral reefs, for example, at the region known as "Hell" near the western end of the island, certainly suggests greater age than a few hundred or thousand years (Richards, 1955; Folk et al., 1973). Recently some corals from the Ironshore Formation have been shown to be older than 35,000 years (Brunt et al., 1973).

HORACE G. RICHARDS

References

Bowin, C. O., 1968. "Geophysical study of the Cayman Trough," *J. Geophys. Res.,* 73(16), 5159–5173.

Brunt, M. A., Giglioli, M. E. C., Mather, J. D., Piper, D. J. W., and Richards, H. G., 1973. "The Pleistocene rocks of the Cayman Islands," *Geol. Mag.,* 110(3), 209–221.

Doran, E., 1955. "Land forms of Grand Cayman, British West Indies," *Texas J. Sci.,* 6, 360–377.

Eggler, D. H., et al., 1973. "Ultrabasic rocks from the Cayman Trough, Caribbean Sea," *Bull. Geol. Soc. Am.,* 84, 2133–2138.

Erickson, A. J., Jr., et al., 1972. "Heat flow and continuous seismic profiles in the Cayman Trough and Yucatan Basin," *Bull. Geol. Soc. Am.,* 83, 1241–1260.

Folk, R. L., Roberts, H. H., and Moore, C. H., 1973. "Black phytokarst from Hell, Cayman Islands (B.W.I.)," *Bull. Geol. Soc. Am.,* 84(7), 2351–2360.

Gough, D. I., and Heirtzler, J. R., 1969. "Magnetic anomalies and tectonics of the Cayman Trough," *Geophys. J. Roy. Astronom. Soc.,* 18, 33–49.

Hess, H. H., and Maxwell, J. C., 1953. "Caribbean Research Project," *Bull. Geol. Soc. Am.,* 64, 1–6.

Matley, C. A., 1926. "The geology of the Cayman Islands (British West Indies) and their relation to the Bartlett Trough," *Quart. J. Geol. Soc. Lond.* 82, 352–387.

Pinet, P. R., 1971. "Structural configuration of the northwestern Caribbean plate boundary," *Bull. Geol. Soc. Am.,* 82, 2027–2032.

Rehder, H., 1962. "The Pleistocene mollusks of Grand Cayman Island, with notes on the geology of the island," *J. Paleontol.,* 36(3), 583–585.

Richards, H. G., 1955. "The geological history of the Cayman Islands," *Acad. Nat. Sci. Phila, Notulae Naturae,* 284, 11p.

Vaughan, T. W., 1926. "Species of *Lepidocyclina* and *Carpenteria* from the Cayman Islands," *Quart J. Geol. Soc. Lond.,* 82, 388–400.

Cross-references: *Vol. I, Caribbean Sea; Cuba; Guatemala; Jamaica; Mexico; West Indies.*

CENTRAL AMERICA—REGIONAL REVIEW

Central America includes the countries of Guatemala, British Honduras, El Salvadór, Honduras, Nicaragua, Costa Rica, and Panamá (including the Panama Canal Zone), covering 550,000 km². (Some geographers extend the region to the isthmus of Tehuantepec in Mexico, but this is awkward for practical purposes.) Table 1 shows pertinent geographical data. (For place names and other geographic data, refer to the American Geographical Society's 1:1,000,000 map series of Latin America.)

The area first became known to European eyes in 1502 when Christopher Columbus landed at Cape Honduras, and its rocks were first studied by the early Spanish explorers, who soon located and began exploiting the rich mines of the Audencias of Castilla de Oro and Guatemala. In spite of this 16th century beginning, much of Central America is even now relatively little known geologically.

The pioneer work of Sapper, Gabb, Olsson, Lohman, Dixon, and Terry has led to synoptic works by Schuchert (1935), Sapper (1937), Roberts and Irving (1957), and Weyl (1961). Since 1920 exploration for oil has resulted in a substantial increase in geological knowledge of the area, and mining has contributed much information in other parts. Central American geology has enormously benefited from the aid of the U.N.D.P. and other international missions during the last decade (Weyl, 1969–1970; see especially summary of cartographic and mission activities by Bohnenberger).

Geotectonic Provinces

Central America is divisible into three major geotectonic or tectonogeomorphic provinces: (1) the *volcanic terrane,* comprised of the Pacific cordillera of volcanic ranges extending from the Mexican border to western Panamá, the Nicaraguan volcanic upland, the Pacific coastal plain, and the Isthmian link; (2) the *fault-block mountainous massifs* of Honduras, southern Guatemala, and southern British Honduras, which have been referred to as "nuclear Central America" and in which are exposed most of the pre-Cretaceous rocks; and (3) the *Petén lowland* of northern Guatemala and northern British Honduras, which is continuous with and a part of the Campeche-Yucatan basin of Mesozoic and Tertiary deposition (see the article on *Mexico*).

The Volcanic Terrane. *The Pacific Cordillera.* The most striking feature of Central American geology is the nearly continuous line of volcanoes which begins at Volcán de

TABLE 1. Geographical Data for Central America

Country	Area (sq mi)	Highest Mountain	Elevation	Major Rivers	Major Lakes
Guatemala	42,042	V. de Tajumulco	4211 m (13,814 ft)	Motagua, Usumacinta, Sarstun, Chixoy	Atitlán, Petén, Izabal
Belize	8867	Victoria Peak	1122 m (3670 ft)	Belize, Hondo	New R. Lagoon
El Salvadór	8260	V. de Sta. Ana	2386 m (7802 ft)	Lempa	Ilopango, Guija, Camalotal
Honduras	43,277	Cerros de Culmi	2590 m (8470 ft)	Patuca, Ulua, Choluteca	Yojoa, Laguna, Caratasca
Nicaragua	51,660	Cerro Chachagón	1804 m (6300 ft)	Coco	Nicaragua, Managua
Costa Rica	19,695	Chiripó Grande	3837 m (12,533 ft)	San Juan, Sixola, Reventazon	Arenal
Panamá	28,753	V. de Chiriquí	3374 m (11,410 ft)	Bayano, Chucunaque, Tuira	Gatún, Madden

FIGURE 1. (*above*) geology of Central America; FIGURE 2. (*below*) tectonics of Central America.

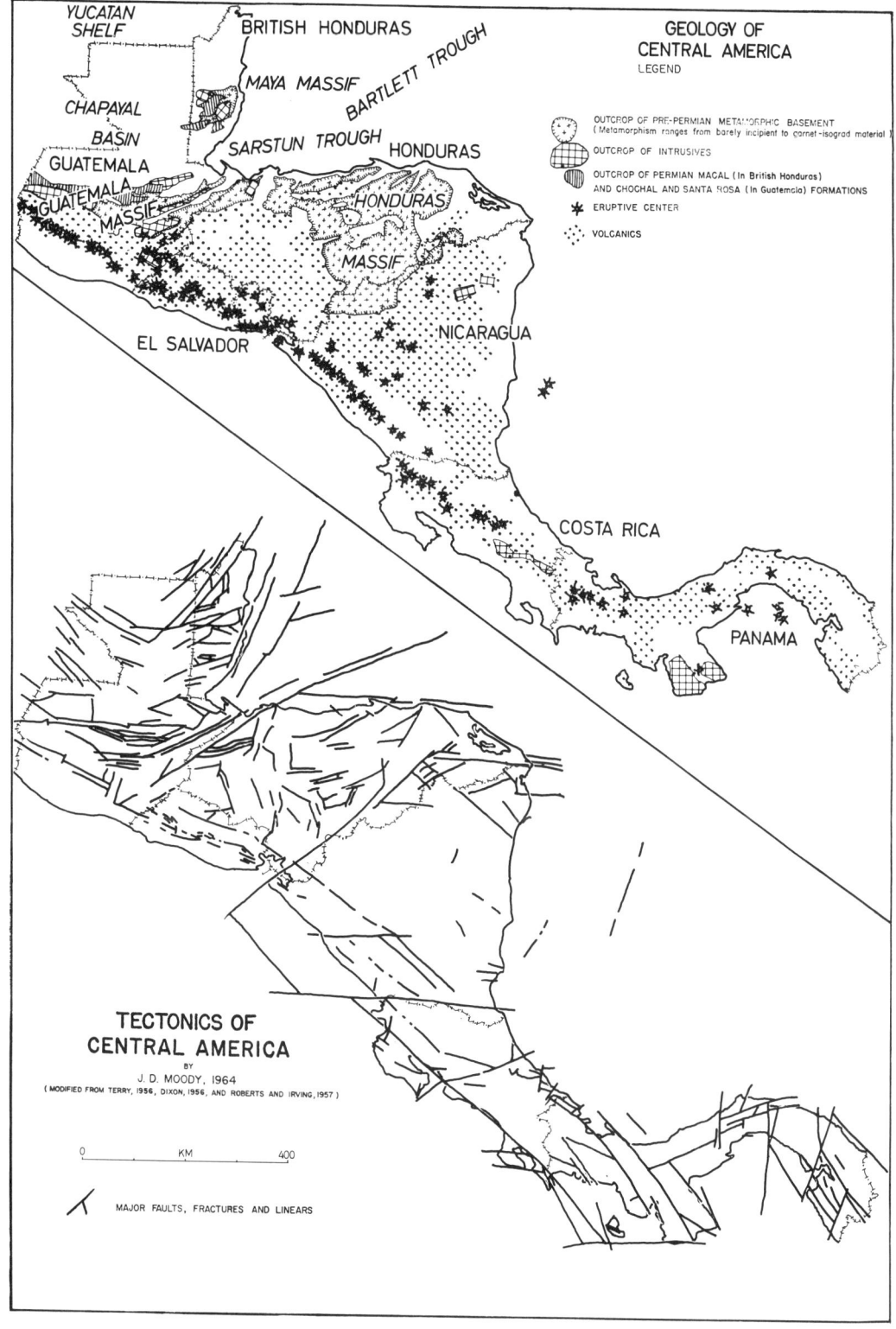

Boquerón in Mexico just N of the Guatemalan border and continues across Guatemala, El Salvadór, Nicaragua, and Costa Rica to Cerro de Santiago in western Panamá, a distance of over 1400 km. Although this chain is practically continuous, it is actually made up of a number of straight-line segments (Fig. 1).

The first segment extends from Boquerón to V. de Pacayá, SW of Ciudad de Guatemala, a distance of 200 km. The strike of this segment is S60°E, and the elevations of the peaks range from about 2550 to 4211 m and include the highest peak in Central America, Volcán de Tajumulco. This line of volcanoes lies on extension of the western border fault of the Chiapas Massif of Mexico, and is separated from it by the westward extension of the E-W fault system of Huehuetenango. Pacayá lies approximately on the western extension of the Motagua valley shear zone.

Southeast of Pacayá the volcanic chain continues S70°E to the Gulf of Fonseca with many active volcanoes, none of which exceeds 2400 m in height. This second volcanic segment, 320 km in length, terminates with the intersection of the N-S Honduras graben system, which transects Honduras at about longitude 87°45'W. Another major shear zone, which follows the valleys of the Choluteca and Patuca rivers, strikes out of the Gulf of Fonseca at about N45°-50°E and reaches the Caribbean coast in the vicinity of Punta Patuca.

The third link in the volcanic chain extends 280 km from V. de Coseguina on the Gulf of Fonseca to V. de Madera in Lago de Nicaragua. The strike is S50°E and the highest elevation is 1763 m at V. de Viejo.

There is a right-lateral offset between the Nicaraguan volcanoes and the Cordillera Volcánica of western Costa Rica, the next linear volcanic element. This offset coincides with a major crustal lineament, which includes the Clipperton fracture zone and the southern termination of the Guatemala trench. The Cordillera Volcánica strikes S45°-50°E, includes peaks rising to 1840 m, and extends 150 km from V. de Orosi to the Rio Grande. An offshoot of the Cordillera Volcánica is the Cordillera Centrál, 80 km in length, which strikes S65°E and includes the high cones north of San José, all over 2800 m.

The Cordillera de Talamanca extends from the Rio Grande south to the Panamanian border, a distance of 140 km, and contains a number of peaks over 3000 m. Not much is known of this range, but it appears to be comprised predominantly of intrusive masses and older, possibly pre-Cretaceous, crystalline rocks rather than young volcanics. However, from Cerro Picacho and Volcán de Chiriqúi to Cerro Santiago, the high volcanic cones reappear. From the northern end of the Talamancas to Santiago is 270 km, and the strike is S60°E.

Thus the entire volcanic chain is comprised of five straight segments, separated from each other by tectonic features of regional importance. The younger cones are built up on older series of volcanics, which in the northern part of the area can be seen to rest unconformably on pre-Tertiary rock units. Many of the volcanoes have been active in historic time. The effusives include rhyolites, basalts and latites, but andesites predominate.

The Nicaraguan Volcanic Upland. Paralleling the main volcanic ranges on the NE is a topographically low area which includes the Lempa valley in El Salvadór, the Fonseca lowland, Lakes Managua and Nicaragua, and the San Juan lowland. This regional low separates the main line of young volcanoes from a broad, hilly volcanic plateau, here named the *Nicaraguan volcanic upland,* which lies to the NE, bounded by a major scarp.

The volcanics of the upland are thought to be of the same general age as the foundation on which the younger cones to the W have been built. Mostly they are unfolded and subhorizontal except in the more strongly faulted areas. A few inliers of Cretaceous sediments and crystalline rocks have been mapped in this terrane. There has been variable dissection by erosion.

In Nicaragua, the plateau is highest along the western edge, where elevations of 1500 m or more are known, and slopes gradually to the E toward the Caribbean coastal plain. Along the Honduras-El Salvadór border, elevations reach over 2000 m. The upland stretches from Nicaragua across western Honduras into southern Guatemala.

To the N the volcanic upland overlaps onto the Honduras Massif; outliers of the volcanics are known almost to the Gulf of Honduras. To the E the upland merges into the Caribbean coastal plain; to the SW the upland terminates in a high scarp at the edge of the longitudinal lowland. The chain of active volcanoes, at least in El Salvadór and Nicaragua, appears to have been built up on a major downfaulted block of the Nicaraguan volcanic upland.

The Pacific Coastal Plain. From the Mexican border to the latitude of Managua, an alluvial coastal plain borders the volcanics of the Pacific Cordillera. In places this plain is up to 50 km wide, and is part of a piedmont accumulation which separates the volcanic chain from the Guatemala trench. The average width of this belt is 120 km. Vertical relief from the high peaks of the range to the depths of the trench approaches 7000 m.

Except for the Gulf of Fonseca, the northern part of the Pacific coast line is relatively straight. From the latitude of Managua south, however, nonvolcanic rocks appear, and five peninsulas jut out into the Pacific from Costa Rica. These peninsulas expose a suite of mixed sediments and basic intrusives and extrusives, which has been called the *Nicoya complex*. Serpentines and metamorphics outcrop on Sta. Helena peninsula. These rocks differ markedly from those exposed to the E and appear to be separated from them by a major zone of faulting.

The Isthmian Link. The *"Isthmian link"* is here considered to include all of Panamá east of Cerro de Santiago (long. 81°45'W). Although volcanic rocks are much in evidence, the link itself is by no means a young volcanic arc, as commonly thought. The volcanic rocks of much of central and eastern Panama are definitely older than those of the Pacific Cordillera or the Nicaraguan volcanic upland and are probably the same age as the Nicoya complex of the western Costa Rica peninsulas. The dominant form of tectonism is major block faulting (see Fig. 2). Although a few youthful eruptive centers are known, none has been active in post-Pleistocene times.

Other Volcanic Areas. Off the eastern coast of Nicaragua are four small islands–*Great Corn, Little Corn, San Andrés,* and *New Providence*–all of which appear to be volcanic necks or eruptive centers similar to Cerro Tortuguero in northeastern Costa Rica. On the Pacific side, practically all the islands in the Gulf of Panamá and the Gulf of Chiriqúi, as well as Caño off Osa peninsula, are volcanic in origin. The *Cocos Islands* (q.v.) at 87°W, 5°30'N, are also volcanic, as of course are the *Galapágos* (q.v.) and *Malpelo*. There is evidence of submarine volcanism along the southwestern side of the Guatemala trench and at its southern termination.

Nonvolcanic Areas in the Volcanic Terrane. A number of relatively small inliers of Cretaceous and older rocks are known in the northern part of the terrane. A typical example is at Metapán, near the common corner of Guatemala, El Salvadór, and Honduras. Here the Cretaceous sediments are strongly faulted, folded, and mineralized.

Separating Lake Nicaragua from the Pacific is the *Rivas arch,* which exposes a considerable thickness of dominantly clastic sedimentary rocks with a varying admixture of volcanics. These sediments range in age from Oligocene to Upper Cretaceous and are not heavily involved tectonically; similar outcrops are found as far S as the Gulf of Nicoya. The older Cretaceous sediments of the Costa Rica peninsulas were mentioned briefly above. From the Gulf of Nicoya south to the Sona peninsula of Panamá is an area of clastic Tertiary sediments with relatively little admixture of volcanics. The same situation exists on the Caribbean side from Puerto Limón to S of the Chiriqúi lagoon. These sediments are strongly faulted and locally folded. Oligocene and Eocene limestones outcrop W of Puerto Limón, and Upper Cretaceous limestones are known on the Rio Changuinola in northern Panamá. The Santa Maria valley and Panama Canal Zone of central Panamá contain thick Tertiary clastics, again with volcanic components varying from considerable to negligible. In Darién, the Bayano and Chucunaque valleys and the San Miguel-Garachiné areas contain notable accumulations of Tertiary clastic sediments, likewise strongly faulted and locally folded.

From about Bluefields north, eastern Nicaragua east of the volcanic upland consists of an alluvial coastal plain up to 100 km wide. Wells drilled near Punta Gorda have revealed the presence of 2000 m or so of Tertiary clastic sediments.

Massifs of Nuclear Central America. *Honduras Massif.* The *Honduras Massif* is a roughly diamond shaped crustal block (see Fig. 2) which occupies most of Honduras, reaching marginally into Nicaragua, El Salvadór, and Guatemala. Crystalline rocks of pre-Triassic age outcrop in the eastern part of the block. Regional dip is to the W, and in the western half Cretaceous sediments and Tertiary volcanics are exposed to the Lempa valley. A N-S fault system, the Honduras graben, breaks the massif into two blocks, and extends from the Gulf of Fonseca on the S to Puerto Cortéz on the north. Other major faults with strong E-W component strikes have broken the Honduras Massif into a series of horsts and grabens, with resultant rough topography. A major E-W fault system parallel to the coast bounds the block on the N; the Motagua and related shear zones bound it on the NW; the Pacific Cordillera and the Lempa valley bound it on the SW; and the southeastern boundary, less distinct, is a series of faults along the Choluteca and Patuca valleys.

The crystalline rocks of the massif are made up of metasediments of various types, and intrusives; the degree of metamorphism is variable and ranges from barely incipient to garnet isograd. In most contacts either Cretaceous sediments or Tertiary volcanics rest unconformably on the crystalline basement, but in the Rosário mining district E of Tegucigalpa the crystalline rocks are overlain by a thick section of dark shales and sandstones from which a late Triassic or early Jurassic flora has been described. Mineralization is not uncommon in the crystallines or the overlying sediments and volcanics.

On the larger islands in the Gulf of Honduras are exposed crystalline rocks similar to those outcropping in central and eastern Honduras; these islands appear to be part of a down-dropped block of the Honduras Massif.

There is a sizable coastal plain east of the crystalline outcrop which extends from Cape Camarón south to the Nicaraguan border. Test wells for oil have revealed the presence of a considerable section of Tertiary sediments; at Punta Gorda Eocene clastics were found to be resting unconformably on granite, suggesting that the massif extends under the coastal plain and has had a relatively long history as a positive element. To the E of the coastal plain and extending halfway to Jamaica is a broad shallow shelf dotted with coral reefs, cays and shoals, called the Moskitia shelf. A well drilled on this shelf encountered over 5100 m (14,980 ft) of sediments and bottomed in either late Cretaceous or early Tertiary beds.

Guatemala Massif. Separated from the Honduras Massif by the onshore continuation of the southern flank of the Bartlett trough is another area of pre-Cretaceous rocks called the *Guatemala Massif,* which extends from Amatique Bay across Guatemala to the Mexican border. As in the Honduras Massif, dominantly E-W faults have broken the massif into a series of horsts and grabens. Three sets of faults predominate, with E-W, ENE-WSW, and NW-SE strikes. Metamorphic rocks like those of Honduras form the cores of the ranges.

Additionally, a thick series of Permian or perhaps in part upper Pennsylvanian sediments—limestones, sandstones, and locally phyllitized dark shales—occur in some of the fault blocks. Large serpentine and granitic intrusives have been mapped. Mineralization is associated with the faulting and intrusion. The Alto de los Cuchumatanes is a large horst of Cretaceous and Permian marine sediments N of the main zone of tectonism of the massif. The nearest thing to a fold belt in all Central America occurs just N of the zone of major faulting of the Guatemala Massif and forms a belt of foothills between the massif and the Petén lowland to the N. Several large anticlinal folds, more or less complicated by faulting, have been mapped in this belt. The granitic rocks and Permian sediments of the western part of the massif continue into Mexico as the Chiapas Massif.

Maya Massif. The *Maya Massif* is an uplifted fault block exposing pre-Cretaceous rocks in central British Honduras. Here a thick series of graywackes, quartzites, slates, and schists that carries fragmentary Paleozoic invertebrate remains is widely exposed. These metasediments, called the *Maya Series,* are unconformably overlain by the Permian *Macal Series;* the base of the Maya is not seen. The Maya Series has been strongly folded, faulted, and intruded by granites, and metamorphism varies from incipient to garnet-isograd. Granitic rocks that intrude the Maya Series and appear to be unconformably overlain by the Macal Series have been dated radiometrically at 235 ± 35 million years. The Macal Series, well over 3000 m thick and consisting of shales, sandstones, conglomerates, and limestones, has itself been intruded by porphyry. The Maya Massif is bounded on the N by a major E-W fault system; on the E by NNE faults down to the E; on the S by ENE faults down to the S, which form the northern border of the Sarstun trough; and on the W the older rocks of the massif dip W under Cretaceous sediments. Although the massif is extensively faulted and intruded, mineralization is negligible.

Motagua Shear Zone. Major faults in southern British Honduras and Guatemala which strike E-NE are thought to be on-shore continuations of the suspected major faults of the Bartlett trough. The faults bounding the Maya Massif on the S are the elements of this fault group that are farthest north, and the shears of the Motagua valley appear to tie into the south scarp of the Bartlett. Thus the onshore portion of the Bartlett trough is about 110 km wide, and grabens such as the Sarstun trough, the Motagua valley, and the Izabal depression are considered to be extensions of the trough. Mapping in the Motagua valley indicates that here at least the faulting originated before Cretaceous time and continued intermittently into the Holocene.

Several lines of reasoning suggest that the Motagua and related shear zones, and probably the entire Bartlett trough, comprises one of the most important tectonic features of Central America. For example, in Central America no Tertiary or younger volcanics are known N of the Motagua but are widespread S of it. Permian sediments of great thickness are known N of the Motagua but not S of it. In a sense the Motagua forms the boundary between Gulf of Mexico-type geology, as in the Petén lowland, and Caribbean-type geology, as in the Tertiary volcanics of southern Central America. Serpentine intrusives (with the exception of the core of the Sta. Helena peninsula) and related exotic differentiates such as massive hypersthene, biotite, and nephrite are restricted to the central part of the shear zone; chromite mineralization is essentially restricted to the shear zone, lead-zinc-silver mineralization is most abundant

N of it, and gold mineralization occurs almost exclusively S of it.

Petén Lowland. North of the Guatemala and Maya massifs lies a broad, fairly flat terrane in which are exposed generally flat lying sediments of Cretaceous and Tertiary age. This flat terrane is called the *Petén lowland*. The sediments are mostly carbonates, marls, and evaporites, with a basal Cretaceous clastic zone of variable thickness. Elevations in the Petén rarely exceed 200 m.

The Rio Usumacinta drains the southern part of the Péten through the *Chiapas depression*, a broad gentle graben dropped down between the Chiapas highlands and the Yucatan shelf. This structural and topographic lowland has been called the Chapayal basin in Guatemala.

Along the northern edge of the Chapayal basin is the *La Líbertad arch,* which is a gentle regional high that strikes about S65°E and separates the Usumacinta and Naranjo drainage systems. East of La Líbertad and south of Lago de Petén, the west plunge of the Maya Massif interferes tectonically, and the strike of the arch appears to change to E and then NE.

Lago de Petén itself lies in a graben between the La Líbertad arch and the southern edge of the Yucatan shelf; the graben apparently results from the intersection of the La Líbertad arch and associated faults with the NNE faults of the Bacalair or Rio Hondo fault zone. North of Lago de Petén the terrane is thought to be relatively structureless into Yucatan.

East of the Rio Hondo fault zone and north of the Maya Massif is a similar tectonic element, here considered to be a part of the Petén lowland. The eastern part of this coastal element includes a series of large faults parallel to the present coastline, the easternmost of which marks the eastern edge of the Yucatan shelf. That these faults have been active over a long period of time is indicated by facies changes observed in the Cretaceous section in wells drilled in northern British Honduras.

Tectonic Summary. The dominant tectonic feature of Central America is the series of volcanic ranges that extend along the western coast of Central America from V. de Boquerón in southern Chiapas to Cerro Santiago in western Panamá, a distance of over 1400 km. These volcanic ranges parallel the Guatemala trench and appear to be related to it and to a series of onshore faults that were activated at least early in Tertiary and probably in pre-Tertiary time. Volcanic rocks associated with this major zone of tectonism are widespread in Central America.

The connection between the volcanic ranges and South America is effected by a zone of upfaulted blocks with only minor related volcanism, and is not a volcanic arc, as commonly thought.

The second most important tectonic feature of Central America is the onshore continuation of the Bartlett trough. The area of intersection of the tectonic elements of the Bartlett with the tectonic zone of the volcanic ranges coincides with the tectonically complex "nuclear Central America." The Honduras, Guatemala, and Maya massifs are in the area of intersection and in a marginal and/or terminal relation to the Bartlett trough. North of nuclear Central America is the tectonically undisturbed Petén lowland. South of the Honduras Massif is the extensive volcanic terrane.

The absence of fold belts in Central America is notable. A few anticlinal structures are known in the border zones between the massifs and in southeastern Costa Rica and Darién. Faulting, volcanism, and intrusion are the principal tectonic manifestations.

Stratigraphic Patterns

Pre-Cretaceous. The oldest dated rocks in Central America are the Maya Series of British Honduras. These metasediments contain Paleozoic fossils and are older than granitic rocks which intrude the Maya and have been dated at 235 ± 35 million years radiometrically. Similar metamorphics in the Guatemala Massif can be shown to be pre-Permian. In the Honduras Massif the metamorphics are pre-late Triassic or early Jurassic and may well include equivalents of the Permian and pre-Permian sequences of the Guatemala and Maya massifs. Pre-Cretaceous metamorphics also occur as a small inlier on the Sta. Helena peninsula. No definite Precambrian rocks are known from Central America, but may possibly exist in the Honduras Massif.

The Macal Series of British Honduras and Santa Rosa and Chochal formations of Guatemala make up the known Permian and Pennsylvanian rocks of Central America. Their outcrop is limited to the Maya and Guatemala massifs; similar rocks have been encountered in wells in northern British Honduras. Paleontologic studies indicate the presence of a fairly complete sequence from Upper Pennsylvanian into Guadalupian.

The existence of pre-Cretaceous Mesozoic sediments in Central America is indicated by the presence of late Triassic and/or early Jurassic plant remains in a thick dark shale sequence near Tegucigalpa, and in the Todos Santos red beds in northwestern Guatemala. Jurassic molluscs have been described from the

Tegucigalpa area. The Todos Santos and its equivalents form the basal transgressive unit of the Mesozoic and include pre-Cretaceous beds in the western part of the area. This basal unit is time-transgressive, and to the E in British Honduras and in eastern Honduras is Lower Cretaceous in age. Bedded salt may exist in the Todos Santos in western Guatemala. Jurassic volcanism is mentioned in some reports.

Cretaceous. Lower Cretaceous sediments are widespread in northern Central America, but are not known S of northern Nicaragua except possibly in the Nicoya complex. Massive limestones of Lower Cretaceous age commonly overlie a basal transgressive clastic unit, the upper part of which at least is also Lower Cretaceous in age. On the Yucatan shelf a major development of anhydrite is probably partly of Lower Cretaceous age. This sequence, which locally is over 2500 m thick, continues into the Upper Cretaceous. Equivalent reef facies are known in British Honduras.

Upper Cretaceous marls and limestones are widespread in Guatemala, British Honduras, and to a lesser extent in El Salvadór and Honduras. Sediments of Upper Cretaceous age are also known on the Rivas arch in southwestern Nicaragua and extending S to the Gulf of Nicoya, and in northwestern Panamá on the Rio Changuinola. Cretaceous sediments are absent in eastern Honduras.

Volcanic rocks of Cretaceous age have been described in the Tegucigalpa area, but have not been reported farther N. The Cretaceous of the Rivas arch includes volcanic components, as does the Nicoya complex of Costa Rica; the volcanic basement complex of Panamá is considered to be at least partly of Cretaceous age.

Cenozoic. Sediments of Paleocene and lowest Eocene age have been reported from Central America only in the Petén lowland. Middle and Upper Eocene beds, however, are fairly widespread and have their greatest development in the Sarstun trough and Izabal graben. Here the Eocene Toledo Formation is thick, clastic, and variable in lithology. It seems likely that the Lower Cretaceous limestone hills near Punta Gorda in southern British Honduras are olistoliths in the Toledo Formation. A thick Toledo section was drilled below the Miocene Rio Dulce limestone W of Livingstone, Guatemala. The lithology of the Toledo as well as isopachs support the idea that the onshore continuations of the Bartlett trough were strongly negative areas in Eocene times. Thick Eocene sediments with considerable volcanic admixture are known in southeastern Costa Rica. The Eocene of the Petén lowland is relatively thin and consists of marls, limestones, and evaporites.

Oligocene and Miocene beds occur in considerable thickness in eastern and western Costa Rica, western, central and eastern Panamá, and in the Rivas arch area of Nicaragua. These sediments are clastic with varying admixtures of volcanic material, and some thin limestone development. Locally, thick limestones are developed—e.g., the Rio Dulce and Dacli limestones. In the Petén lowland Oligocene and Miocene sediments are thin, calcareous, and evaporitic. Most of the volcanics of the Nicaraguan volcanic upland are probably of this age.

Marine Pliocene and Pleistocene formations are rare, but have been described from the Panama Canal Zone. Thick alluvial and sheetwash deposits filling flood plains, as in the San Juan lowland, are generally thought to be Pliocene. The youngest volcanics are Pleistocene and even Holocene.

Mineral Deposits

Fuels. Although there has been considerable prospecting for oil in Central America, no commercial deposits have yet been discovered. Some 66 wildcat wells have been drilled in the areas deemed to have some oil potential. The only significant seep in Central America is near the town of Garachiné in Darién, Panamá. A significant show of oil was encountered in a well in southern Costa Rica in a fractured andesite flow in a thick sedimentary section; minor shows have been reported from several wells in the Petén lowland. Lignite and coal have been reported in Oligocene and Miocene sediments in Panamá, but are of no commercial importance.

Metallic Minerals. Gold mineralization is widespread from the Maya Mountains to the Colombian border. The most important mine in all Central America, at least in colonial days, was the famous Espiritu Santo gold mine at Caña in Darién. El Rosário mine east of Tegucigalpa is a significant present-day gold producer. Most of the gold mineralization is in quartz veins in the volcanic rocks. There seems to be a relation of the gold occurrences to the volcanic trends (Roberts and Irving, 1957).

Lead-zinc and lead-zinc-silver mineralization is important at several localities in the Guatemala Massif, and in the Metapán area.

Chromite occurs in association with the serpentine intrusive masses of the Motagua shear zone. Stibnite mineralization is known at several places on the Honduras Massif. Manganese occurrences in Central Panamá are of some importance. Iron, copper, tungsten, tin, and mercury deposits are known, mostly on the Honduras Massif, but are of little importance.

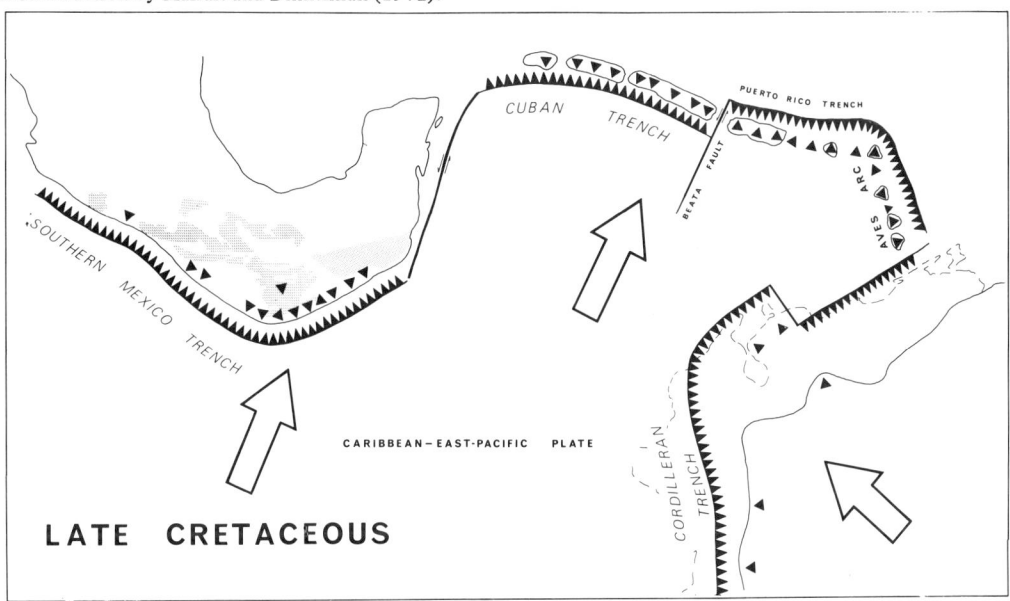

FIGURE 3. Central America and the Caribbean area in late Cretaceous times, according to a plate tectonic reconstruction by Malfait and Dinkelman (1972).

Nonmetallic Minerals. Quartz crystals and muscovite of commercial importance occur in Guatemala. Barite has been mined in British Honduras. Sand, gravel, clay, and limestone of potential commercial importance occur in a number of areas, but are exploited only locally.

JOHN D. MOODY

Central America: Appendix (Editorial Note)

Plate Tectonic Interpretations. During the preparation of this encyclopedia the plate tectonic theory has been applied, with several different interpretations, to the problems of Central America and the Caribbean. In 1958 Carey proposed a "Gulf of Mexico Diothesis," which involved a sundering of a former land mass ("Llanoria") in that region, clockwise rotation, and eventual separation of the Guatemala and Honduras blocks in a V shape about the present Cayman Trough ("Honduras Sphenochasm").

This purely geometric "jigsaw" treatment was superseded by subsequent models supported by more field geological information as well as submarine and geophysical data. Although conclusions are still in the category of working hypotheses, a useful model is that offered by Malfait and Dinkelman (1972). The North and South American plates are seen as converging during the late Mesozoic, coincident with the westward drift of both continents, but with a relative motion of the East Pacific-Caribbean crust to the east (Fig. 3). There is widespread evidence of Laramide orogenic phenomena.

Major transcurrent faulting has long been recognized in the northern and southern Caribbean, left lateral in the north, and right lateral in the south, along the borders of South America. This transcurrent faulting (e.g., in the Motagua Fault and Cayman Trough) appears to have been initiated during the late Eocene, and coincides with extensive volcanism in Central

FIGURE 4. Stress patterns during the Guatemala-Honduras tensional phase (post-late Miocene), following Malfait and Dinkelman (1972).

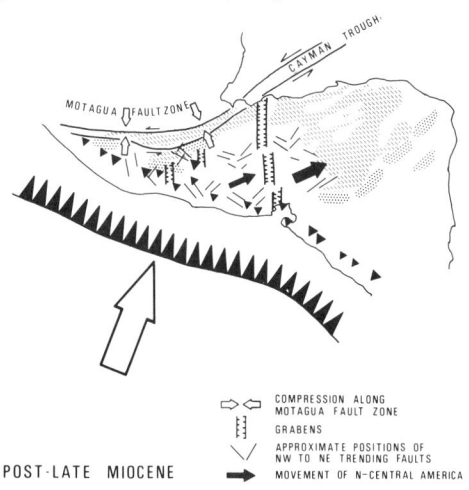

FIGURE 5. Present-day geotectonic elements of Central America and the Caribbean area, as interpreted in terms of plate tectonics (from Malfait and Dinkelman, 1972).

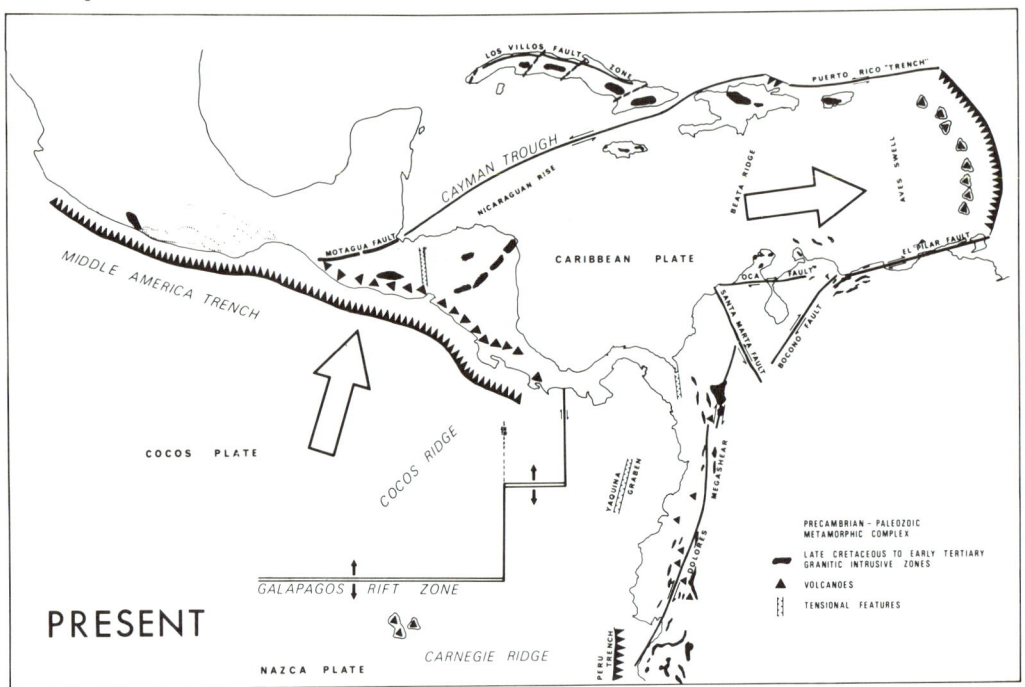

America, which may be interpreted as caused by a decoupling from the East Pacific Plate at the close of the Laramide phase. Subduction of the Pacific crust was taking place in the Southern Mexico Trench and elsewhere. In the early Oligocene that subduction extended all the way across the SW "front" of Middle America to intersect with the Bolivar Trench subduction of South America.

Volcanism associated with subduction along what became the Middle America Trench continued, off and on, until today. Tensional stress with N-S grabens developed in the Guatemala-Honduras sector in post-Late Miocene time (Fig. 4), and transform faults leading to left-lateral (N-S) displacements started to develop along the Cocos Ridge (Fig. 5), cutting off the Panama Isthmus from the Middle America Trench after the early Pliocene. Paleontologic evidence (from pelagic foraminifera) shows that the isthmus, as an Atlantic-Pacific seaway, was closed after about 5.7 m yr B.P. (Emiliani et al., 1972).

RHODES W. FAIRBRIDGE

References

Carey, S. W., 1958. *Continental drift: a Symposium.* Tasmania: Hobart, 375p.

Carpenter, R. H., 1954. "Geology and ore deposits of the Rosario mining district and the San Juancito Mountains, Honduras, Central America," *Bull. Geol. Soc. Am.,* **65,** 23–38.

Dengo, G., and Bohnenberger, O. H., 1969. "Structural development of northern Central America," *Am. Assoc. Petrol. Geologists Mem.* **11,** 203–220.

Dixon, C. G., 1956. *Geology of Southern British Honduras, with Notes on Adjacent Areas.* Belize: Govt. Printer, 85p.

Emiliani, C., Gartner, S., and Lidz, B., 1972. "Neogene sedimentation on the Blake Plateau and the emergence of the Central American isthmus," *Palaeogeogr., Palaeoclim., Palaeoecol.,* **11**(1), 1–10.

Flores, G., 1952. "Geology of northern British Honduras," *Bull. Am. Assoc. Petrol. Geologists,* **36**(2), 404–413.

Graham, A., ed., 1973. *Vegetation and Vegetational History of Northern Latin America.* New York: Elsevier, 416p.

Hoffstetter, R., 1960. "Amerique Central," *Lexique Stratig. Intern.* (Paris), **5,** fasc. 2.

Lloyd, J. J., 1963. "Tectonic history of the south Central American orogen," *in* O. E. Childs, ed., *Backbone of the Americas* (Mem. 2). Tulsa: Am. Assoc. Petrol. Geologists, 88–101.

Malfait, B. T., and Dinkelman, M. G., 1972. "Circum-Caribbean tectonic and igneous activity and the evolution of the Caribbean plate," *Bull. Geol. Soc. Am.,* **83,** 251–272.

Meyerhoff, H. A., and Meyerhoff, A. A., 1973. "Circum-Caribbean tectonic and igneous activity and the evolution of the Caribbean Plate, discussion," *Bull. Geol. Soc. Am.*, **84**, 1101–1104.

Mills, R. A., et al., 1967. "Mesozoic stratigraphy of Honduras," *Bull. Am. Assoc. Petrol. Geologists,* **51**, 1711–1786.

Pinet, P. R., 1972. "Diapirlike features offshore Honduras: implications regarding tectonic evolution of Cayman Trough and Central America," *Bull. Geol. Soc. Am.,* **83**, 1911–1922. (Discussion in Meyerhoff, A. A., 1973, *Ibid.,* **84**, 2147–2152.)

Roberts, R. J., and Irving, E. M., 1957. "Mineral deposits of Central America," *U.S. Geol. Surv. Bull.,* **1034**, 205p.

Sapper, K., 1937. "Mittelamerika," *Handbuch Reg. Geol.* (Heidelberg), 8(4), 160.

Schuchert, C., 1935. *Historical Geology of the Antillean–Caribbean Region.* New York: Wiley, 811p.

Terry, R. A., 1956. "A geological reconnaissance of Panama," *Calif. Acad. Sci., Occ. Pap. 23,* 91p.

Uchupi, E., 1973. "Eastern Yucatan continental margin and western Caribbean tectonics," *Bull. Am. Assoc. Petrol. Geologists,* **57**(6), 1075–1085.

Vinson, G. L., 1962. "Upper Cretaceous and Tertiary stratigraphy of Guatemala," *Bull. Am. Assoc. Petrol. Geologists,* **46**(4), 425–456.

———, and Brineman, J. H., 1963. "Nuclear Central America, hub of Antillean transverse belt," *in* O. E. Childs, ed., *Backbone of the Americas* (Mem. 2). Tulsa: Am. Assoc. Petrol. Geologists, 101–113.

Weyl, R., 1961. "Die Geologie mittelamerikas," *Regionalen Geologie der Erde.* Berlin: Gebr. Borntraeger, 226p.

———, 1969–1970. "Geologische Bilder aus Mittelamerika," *Natur und Museum,* **99**, 415–423, 559–570; **100**, 120–128, 269–278, 362–370.

———, 1970. "Mittelamerika," *Zentr. Geol. Paläontol.,* **1970**, 1003–1051.

Woodring, W. P., 1954. "Caribbean land and sea through the ages," *Bull. Geol. Soc. Am.,* **65**(8), 719–732.

———, 1957. "Geology and paleontology of Canal Zone and adjoining parts of Panama," *U.S. Geol. Surv. Prof. Pap. 306-A,* 145p.

Cross-references: *Belize; Colombia; Costa Rica; Cuba; El Salvador; Guatemala; Honduras; Jamaica; Mexico; Nicaragua; Panama; West Indies.*

CHILE

Chile is a narrow elongate country (741,767 km^2 or 292,256 sq mi) on the Pacific side of the South American continent. The capital is Santiago. The country extends from about 18° to 56°S, a distance of over 4225 km. The breadth varies between 90 and 390 km (Fig. 1). Several small islands in the Pacific also

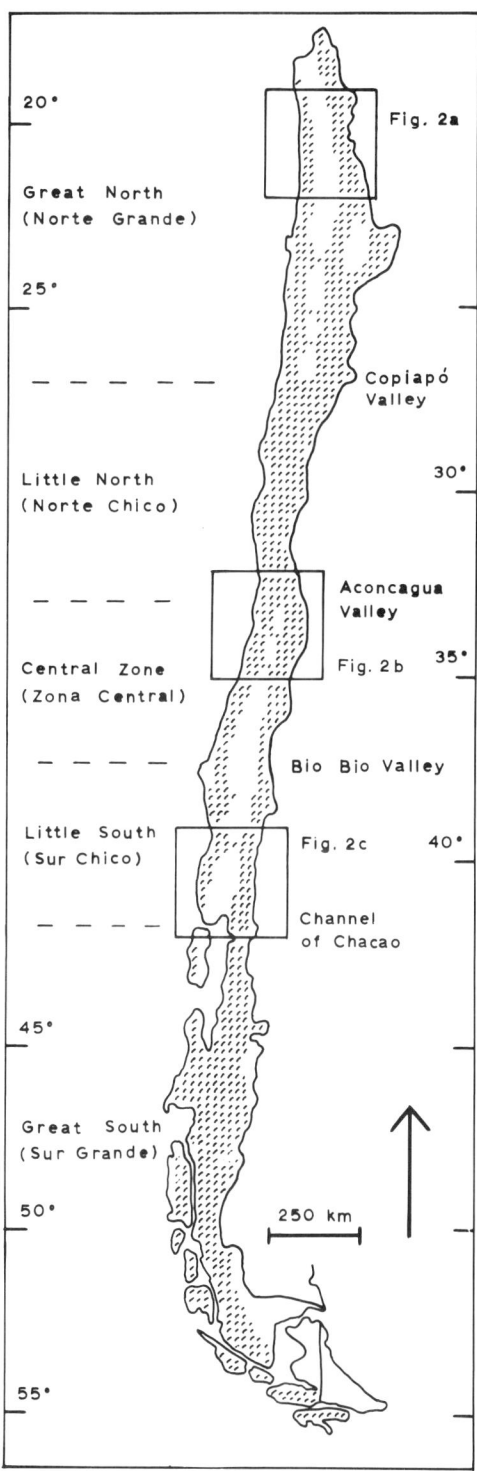

FIGURE 1. Geomorphological divisions of Chile. The Coastal Range and the High Cordillera are dotted. The Longitudinal Valley and the Chilean part of the Patagonia tableland in the Great South are white. The parts in insets are the sectors in Fig. 2.

FIGURE 2. Simplified sectors of the geologic map of Chile (Instituto de Investigaciones Geológicas, 1960): (a) Great North (with additions according to Zeil, 1964b); (b) transition zone between the Little North and the Central Zone; (c) Little South (with additions by the author).

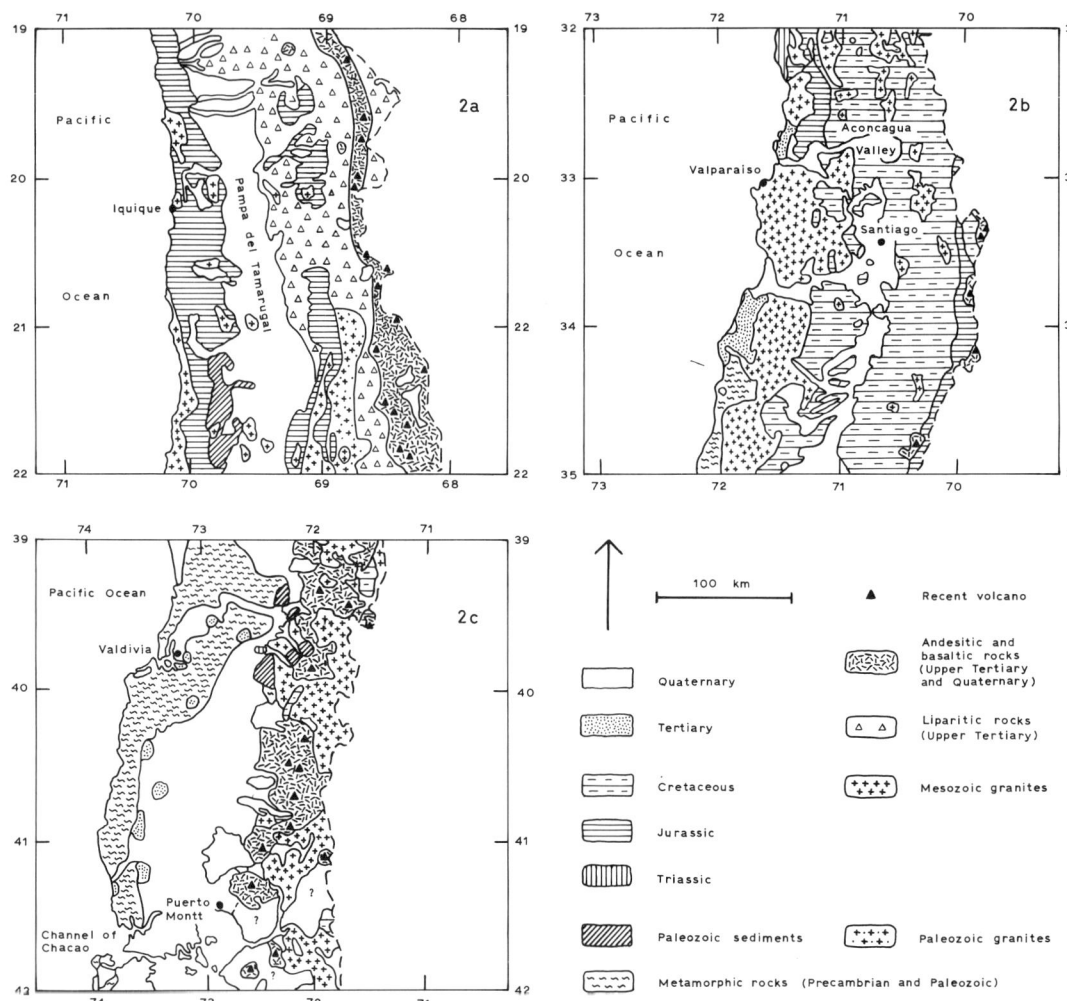

belong to Chile (see *Juan Fernandez Island, Easter Island*). Since 1940, Chile claims the sector of the Antarctic lying between 43 and 90°W, an area of 1,250,000 km^2.

With the exception of small areas in the south, Chile is dominated by the Cordillera. The mountains are rich in mineral deposits. The initial geological activities were therefore confined to mining and instruction in industrial mining.

The oldest geological training center is the *Instituto de Geologia* of the State University of Santiago. It chiefly trains mining engineers. At the same university a geology department was opened in 1958, the *Escuela de Geologia*. Geological survey work has been carried out by the *Instituto de Investigaciones Geologicas* since 1957.

The first summarizing description of the geology of Chile was written by J. Brüggen (1934). As a supplement to a second edition of this work (Brüggen, 1950), the first geologic map of Chile was included. Further summaries were written by J. Muñoz Cristi (1950, 1956), H. Fuenzalida (1950), and W. Zeil (1964a). In 1960 a considerably improved geological map on the scale 1:1,000,000 was compiled and published by the *Instituto de Investigaciones Geologicas*.

Geomorphologic Divisions

Chile has three geotectonic units from west to east: the Coastal Range, the Longitudinal Valley, and the High Cordillera; all strike N-S.

The Longitudinal Valley is discontinuous in some areas, and there the Coastal Range and the High Cordillera are joined by intervening mountains. These mountainous structures are crossed by different climatic belts, the result being that there is extreme aridity in the northern area and extreme wetness in the southern area. The above factors lead to the natural division of the country into five zones.

Far North (Norte Grande). The Far North desert (Fig. 1) reaches from the boundary with Peru (18°S) to the valley of the Rio Copiapo (about 27°S). The mountains are half buried in a huge accumulation of detritus. Basins are common and filled with halite and gypsum and partially with saltpeter. Here the threefold structural division is distinctly developed (Fig. 2a). The Coastal Range has a height of up to 1500 m, and the Longitudinal Valley forms the Pampa del Tamarugal in the north and a number of depressions in the south.

The High Cordillera gradually rises up to the plateau of the Puna de Atacama (4000 m), where there are volcanoes rising more than 6000 m. The snow line occurs at 5800–6000 m.

Near North (Norte Chico). The Near North semidesert (Fig. 1) extends between the valley of the Rio Copiapo (about 27°S) and the valley of the Rio Aconcagua (33°S). There is less talus here and there are no salt basins. The Coastal Range and the High Cordillera are connected and form a rugged mountainous country (Fig. 2b). Here the Cordillera reaches the greatest height of 6958 m on the Argentine side in the Aconcagua. The snow line lies at 5000 m, and there are no volcanoes.

Central Zone (Zona Central). The Central Zone (Fig. 1) reaches from the valley of the Rio Aconcagua (33°S) to the valley of the Rio Bio Bio (about 37–38°S). It forms the core of the country and is characterized by moderate winter rains and dry summers. The threefold division is again evident (Fig. 2b). The Coastal Range reaches a height of between 700 and 2000 m. In the N the Longitudinal Valley contains a number of small basins; but toward the S, the valley becomes more uniform in structure and broader and lower. The High Cordillera rises steeply to about 4000 m, where again there are volcanoes. The snow line lies at 4000 m.

Little South (Sur Chico). The Little South (Fig. 1) stretches from the valley of the Rio Bio Bio (between 37 and 38°S) to the Channel of Chacao (nearly 42°S). It rains here all year, but in winter the rainfall is especially heavy. Together with the comparatively warm climate, this results in very deep weathering.

The threefold division remains in spite of the decrease in height of the mountains (Fig. 2c). The Coastal Range continues southward and then disappears entirely in the Channel of Chacao. The Longitudinal Valley is broken off at Valdivia (39–40°S) by a range of mountains. The High Cordillera still reaches a height of 2000 m, and there are high volcanoes. The snow line falls below 2000 m.

Great South (Sur Grande). The Great South reaches from 42°S to the southern end of the continent (Fig. 1). The climate is cold and stormy with heavy rains all year. The snow line is at about several hundred meters. The Patagonian ice cap forms an extensive cover on the mountains. The threefold division is indistinct beyond 44°S. The mountains become more isolated, forming peninsulas and islands separated by deep fjords. In the far S, Chile extends toward the E into the Patagonian tableland.

Geotectonic Divisions

The threefold division of Chile into a Coastal Range, a Longitudinal Valley, and the High Cordillera is the result of vertical movements. These began in the Upper Tertiary and continue in the Holocene.

The structures of the Coastal Range and the High Cordillera were formed during at least two orogenic cycles with intrusions of magma. We can differentiate between a Preandean geosyncline, folded during the Upper Paleozoic, and an Andean orogeny in the Upper Mesozoic. These two orogenies are recognizable in nearly all parts of the country, but in varying intensity.

The Coastal Range forms the basement and was intensively folded in the Upper Paleozoic. These rocks were metamorphosed and intruded with granite. The minor structures trend NW-SE. The Mesozoic sedimentary cover was only slightly affected by the Upper Mesozoic orogeny, although granites were intruded into the Coastal Range at this same time.

In the High Cordillera the comparatively rare Paleozoic rocks and the Mesozoic sediments are intensively folded. The most important orogenic phases with extensive intrusions of granite appear to be in the Middle and Upper Cretaceous. Locally there was also folding in the Lower Tertiary; the preferred direction of the structures was N-S. Occasionally the older Preandean trends are also seen.

Stratigraphic History

A detailed summary of the structural and stratigraphic events is not possible today. They have been closely investigated only in scattered sections of the country. A comparison of these isolated areas is difficult and there are still many points that are not clear.

The Basement Complex. Large parts of the country consist of a metamorphic basement complex. These are principally epizonal to mesozonal schist, phyllite, quartzite, and gneiss, and at several places there are plutonic rocks. The age of the basement complex is unknown, but large areas probably belong to the Precambrian. This appears to be the case with the Coastal Range of the Central Zone and the Little South (Fig. 2c).

Paleozoic. Much of the Paleozoic has been metamorphosed. In some cases, especially in the Little North, a slightly metamorphic unit was found containing sparse remnants of plant and marine fossils. The most interesting sequence lies in the High Cordillera of the Great North, where Ordovician sandy shales are found with a rich graptolite fauna. In the other parts of the Great North there are graywackes, sandstones, and shales with a few fossil plants that probably belong to the Pennsylvanian or the Permian.

In the Great South other shales, sandstones, limestones, and marbles of considerable thickness with fossils were also deposited in the Pennsylvanian or Permian. This sequence lies in the coastal regions between 50 and 52°S. In the Upper Paleozoic, an orogeny took place that produced tight folds trending NW-SE. It occurred in at least two phases and metamorphosed some of the older rocks. Some of the Chilean granites were radiometrically dated as Upper Paleozoic.

Mesozoic. With the beginning of the Mesozoic, the Upper Paleozoic mountains were rapidly eroded. In Chile two separate troughs were formed: the *Andean Geosyncline* in the north and in the Central Zone, and the *Magallanes Geosyncline* in the southern Great South. These two troughs differ as to the time of their formation, as to their sedimentary filling, and as to the time of subsequent folding. During the Mesozoic the Little South and the northern part of the Great South remained a stable area between the geosynclines.

Triassic. The Andean Geosyncline contains few Triassic deposits. They consist of terrestrial sediments with plant fossils and alternate with extensive acid volcanics. During short periods of marine transgressions from the Pacific Ocean, a small amount of marine sediments was deposited. There has been no Triassic discovered in the Magallanes Geosyncline.

Jurassic. The thick Jurassic marine sediments in the Andean Geosyncline contain many fossils, and large amounts of intercalated intermediate and mafic lava, tuffite, and agglomerate (the Porphyry Formation). The Jurassic Porphyry Formation forms the greatest part of the Coastal Range of the Great North (Fig. 2a). Upper Jurassic sediments are found especially in the High Cordillera of the Little North and the Central Zone (Fig. 2b). In the Upper Jurassic in the Magallanes Geosyncline, eruption began of great amounts of volcanic rocks, especially quartz porphyry.

Cretaceous. In the Andean Geosyncline the same conditions prevailed in the Lower Cretaceous as in the Jurassic. Lower Cretaceous rocks are also found in the High Cordillera of the Little North and the Central Zone (Fig. 2b).

The Lower Cretaceous main phase of the Andean orogeny produced uplift and N-S-striking folds. Toward the E the intensity of the folds increases so that there are complicated structures near the Argentine border. The influence of the older Paleozoic structures is seen at times. The Andean orogeny formed the foundation of the present mountain structures and thereby completely altered the paleogeographic situation.

The intrusion of the Andean batholiths accompanied the folding with great volumes of granite and diorite. Earlier intrusions occurred between the Jurassic and the Cretaceous, and later intrusions perhaps in the Tertiary. It is probable that these are connected with weaker orogenic pulses. The Andean batholiths can be traced over the whole length of Chile. The largest batholiths are found in the Coastal Range of the Little North and the High Cordillera of the Little South (Fig. 2c), as well as in the northern part of the Great South and in parts of the mountains of the remote Great South.

In the Upper Cretaceous a new transgression took place in the Senonian. These sediments (conglomerates, sandstones, and marls of the Quiriquina beds) are restricted to a small strip near the coast along the Coastal Range of the Central Zone and the northern Little South. With an unusually rich fauna and flora, as described by W. Wetzel (1927), they are the most famous fossil-bearing series in the country.

In the Magallanes Geosyncline considerable subsidence began in the Cretaceous. The thickness of the sedimentation reached up to 10,000 m. In the Middle Cretaceous, folding with intrusions of granite took place probably at the same time as the major folding of the Andean Geosyncline. But the main folding of the Magallanes Geosyncline occurred in the Upper Cretaceous-Tertiary, with broad fold structures, the intensity of which decrease from west to east.

Cenozoic. *Tertiary.* Conditions during the Tertiary on the Pacific coast were similar to those in the Upper Cretaceous. The first transgression took place in the Eocene with a transgression similar to the Quiriquina marked by frequent oscillations from marine to continental facies. On the peninsula of Arauco (37–38°S), there are also deposits of coal.

The sediments of the Miocene transgression are found along the Coastal Range of the Central Zone, the Little South, and the northern part of the Great South. Its terraces can be followed along the Coastal Range for long distances. The Miocene sea was probably connected to the Atlantic Ocean in the Little South area, which had been stable in the Mesozoic.

The marine Pliocene can be followed along the Coastal Range from the Great North near 23°S to the Little South with few interruptions. The formation of the continental Tertiary is very different in different parts of the country. In the Great North there are Lower Tertiary conglomerates, red clay, and salt deposits which were probably folded in the Oligocene.

The Upper Tertiary of the Great North is closely connected with the formation of the Longitudinal Valley. Beginning with the Miocene and Pliocene, extrusions of acid liparites and dacites flowed from fissures in a fracture belt over 1000 km long. In the Pliocene the eruption began of andesitic and basaltic lavas from the earliest volcanoes in the High Cordillera.

In the Central Zone and in the Little South there were basins with terrestrial and lacustrine sedimentation with local deposition of coal. Subsidence of the southern Longitudinal Valley started in the Miocene, affecting the whole structure by the Pliocene. At the same time many volcanoes became active, at first with eruptions along fissures on the periphery of the High Cordillera and also from old volcanoes. In this southern volcanic zone there are only andesites and basalts. After the main folding had taken place, the Magallanes Geosyncline became a shallow depression. Further folding with granite intrusions took place probably in the Oligocene. The Upper Tertiary sediments are continental with some layers of coal and basaltic flows.

The volcanic islands in the eastern Pacific which belong to Chile (*Juan Fernandez,* q.v.; *Easter Island,* q.v.; etc.) were also probably formed during the Upper Tertiary.

Quaternary. In the Quaternary the elevation of the Coastal Range and the High Cordillera and the subsidence of the Longitudinal Valley continued, thereby developing the present morphology. The mafic volcanism produced andesitic and basaltic eruptions which culminated in the construction of numerous volcanic cones. The volcanic activity gradually diminished so that only a few are still active, especially in the High Cordillera of the Central Zone and the Little South. The Quaternary of the northern Longitudinal Valley consists of thick detritus which still continues to accumulate. Salt basins formed in numerous depressions in the north and include deposits of Chile saltpeter. In the N, Quaternary glaciation is restricted to the higher elevations of the High Cordillera. From the Central Zone southward, the glacial formations in the High Cordillera increase progressively to the fjords and channels of the Great South. In the Little South the numerous lakes lying on the slopes of the High Cordillera are surrounded by moraine amphitheaters. The southern Longitudinal Valley is filled with moraines and glaciofluvial sands. The Würm moraines in the High Cordillera begin about 39°30'S; from there to the S they expand into the Longitudinal Valley and reach the Pacific at the Channel of Chacao (near 42°S). On the coast there are many interglacial strand lines at various heights, located between the Great North at about 23°S and the Little South. Postglacial rise in sea level is also recognizable at different places.

Seismic Activity

Tectonic movements continue up to the present time, as is evident by very frequent and violent earthquakes. The epicenters always seem to be related to surface faults. Earthquakes occur along the borders of the Coastal Range, the Longitudinal Valley, and the High Cordillera, and also on the continental shelf.

Along the coast of Chile, the continental shelf is narrow and the continental slope is steep. The Atacama Trench, which has a depth of nearly 8000 m, lies in front of the Great

North and the Little North. Toward the S the trench becomes more shallow and gradually dies out.

The greatest frequency of earthquakes is in the N, in the area of the world's greatest relief contrast (14,000 m) between the Atacama Trench and the High Cordillera. Toward the south, earthquakes become less frequent; and in the deep south, they are quite exceptional.

One of the last earthquake sequences was in May 1960, in the Little South and in the north of the Great South. There was great damage to settlements and a considerable number of landslides in the mountains. Large areas of the coast were raised and other places subsided 1 to 2 m.

Mining Industry

Chile is rich in mineral resources. The majority of the large deposits are to be found in the north.

The emplacement of the ore deposits is always closely connected to the Andean batholiths. Today copper mining is the chief mineral industry. The largest mines are Chuquicamata, Potrerillos, and El Salvador in the north, and El Teniente in the Central Zone. Iron and manganese mines are also important. Gold and silver mines were important in the last century.

Saltpeter mines are located in the desert of the Great North. In earlier times Chile had a world monopoly in saltpeter. Since the discovery of artificial saltpeter, production has greatly decreased. A by-product in the mining of saltpeter is iodine. One other speciality of the Great North is guano, which is the dried excrement of birds and is collected in large quantities near the coast. There are sulfur mines on the high volcanoes in the Great North.

Petroleum and coal are found in Cretaceous and Tertiary sediments in the south.

ROLF KÖSTER

References

Brüggen, J., 1934. "Grundzüge der geologie und lagerstattenkunde Chiles," Heidelberg, 362p.

———, 1950. "Fundamentos de la Geología de Chile," Santiago de Chile: Inst. Geogr. Militar, 374p.

Carter, W. D., and Aguirre, L., 1965. "Structural geology of Aconcagua Province and its relationship to the Central Valley graben, Chile," *Bull. Geol. Soc. Am.,* 76(6), 651–664.

Cecioni, G., 1970. *Esquema de Paleogeografia Chilena.* Santiago de Chile: Ed. Univ., 144p.

Clark, A. H., Mayer, A. E. S., Mortimer, C., Sillitoe, R. H., Cooke, R. U., and Snelling, N. J., 1967. "Implications of the isotopic ages of ignimbrite flows, southern Atacama Desert, Chile," *Nature,* 215, 723–724.

Corvalan Diaz, J., 1965. "Geología general," *Geografia Economica de Chile,* 2nd ed. Santiago de Chile, 35–97.

Dalziel, I. W. D., and Cortes, R., 1972. "Tectonic style of the southernmost Andes and the Antarctandes," *24th Intern. Geol. Congr.,* 3, 316–327.

Farrar, E., Clark, A. H., Haynes, S. J., Quirt, G. S., Conn, H., and Zentilli, M., 1970. "K-Ar evidence for the post-Paleozoic migration of granitic foci in the Andes of northern Chile," *Earth Planet. Sci. Lett.,* 10, 60–66.

Fisher, R. L., and Raitt, R. W., 1962. "Topography and structure of the Peru-Chile trench," *Deep-Sea Res.,* 9, 423–443.

Fuenzalida H., 1950. "Orografía," *Geografia Económica de Chile.* Santiago de Chile, 1–54.

———, Cooke, R. U., Paskoff, R., Segerstrom, K., and Weischet, W., 1965. "High stands of Quaternary sea level along the Chilean Coast," *Geol. Soc. Am. Spec. Pap.* 84, 473–496.

Gonzalez-Bonorino, F., and Aguirre, L., 1970. "Metamorphic facies series of the crystalline basement of Chile," *Geol. Rundschau,* 59(3), 979–994.

Guest, J. E., 1969. "Upper Tertiary ignimbrites in the Andean Cordillera of part of the Antofagasta Province, northern Chile," *Bull. Geol. Soc. Am.,* 80, 337–362.

Halpern, M., 1973. "Regional geochronology of Chile south of 50° latitude," *Bull. Geol. Soc. Am.,* 84, 2407–2422.

Harrington, H. J., 1961. "Geology of parts of Antofagasta and Atacama Provinces, northern Chile," *Bull. Am. Assoc. Petrol. Geologists,* 45(2), 169–197.

Hayes, D. E., 1966. "A geophysical investigation of the Peru-Chile trench," *Marine Geol.,* 4, 309–351.

Herm, D., and Paskoff, R., 1967. "Vorschlag zur Gliederung des marinen Quartärs in Nord-und Mittel-Chile," *Neues Jahrb. Geol. Palaeontol., Monatsh.,* 10, 577–588.

Hollingworth, S. E., and Rutland, R. W. R., 1968. "Studies of Andean uplift. 1. Post-Cretaceous evolution of the San Bartolo area, north Chile," *Geol. J.,* 6(1), 49–62.

Instituto de Investigaciones Geológicas, 1960. *Mapa Geológico de Chile.* Santiago de Chile: Inst. Inv. Geol., 1:1,000,000.

Katz, H. R., 1970. "Randpazifische Bruchtektonik am Beispiel Chiles und Neuseelands," *Geol. Rundschau,* 59(3), 898–926.

———, 1971. "Continental margin in Chile–is tectonic style compressional or extensional?" *Bull. Am. Assoc. Petrol. Geologists,* 55(10), 1753–1758.

Kausel, E., and Lomnitz, C., 1969. "Tectonics of Chile," *Intern. Upper Mantle Symp. 22-B,* Inst. Geofis., UNAM, Mexico.

Levi, B., 1970. "Burial metamorphic episodes in the Andean geosyncline, central Chile," *Geol. Rundschau,* 59(3), 994–1013.

Lomnitz, C., 1969. "Sea-floor spreading as a factor of tectonic evolution in southern Chile," *Nature,* 222, 366–369.

———, 1970. "Major earthquakes and tsunamis in Chile during the period 1535 to 1955," *Geol. Rundschau,* 59(3), 938–960.

Miller, H., 1970. "Das problem des hypothetischen

"Pazifischen Kontinentes" gesehen von der chilenschen Pazifikküste," *Geol. Rundschau,* 59(3), 927–938.
Morgan, W. J., Vogt, P. R., and Falls, D. F., 1969. "Magnetic anomalies and sea-floor spreading on the Chile Rise," *Nature,* 222, 137–142.
Mortimer, C., 1973. "The Cenozoic history of the southern Atacama Desert, Chile," *J. Geol. Soc. Lond.,* 129, 505–526.
Mueller, G., 1968. "Genetic histories of nitrate deposits from Antarctica and Chile," *Nature,* 219, 1131–1134.
Muñoz Cristi, J., 1950. "Geologia," *Geografia Económica de Chile.* Santiago de Chile, 55–187.
____, 1956. "Chile," in Jenks, ed., *Handbook of South American Geology* (Mem. 65). New York: Geol. Soc. Am., 191–214.
Ocola, L., 1966. "Earthquake activity of Peru," in J. S. Steinhart and T. J. Smith, eds., *Geophysical Monograph 10.* Washington, D.C.: Am. Geophys. Union, 509–528.
Park, C. F., 1972. "The iron ore deposits of the Pacific Basin," *Econ. Geol.,* 67, 339–349.
Paskoff, R., 1967. "Recent state of investigations on Quaternary sea levels along the Chilean coast between Lat. 30° and 33° S.," *J. Geosci.* (Osaka City Univ.), 10, 107–113.
____, 1970. *Recherches Géomorphologiques dans le Chili Semi-aride.* Bordeaux: Biscayne Frères Imprimeurs, 420p.
Plafker, G., and Savage, J. C., 1970. "Mechanism of the Chilean earthquakes of May 21 and 22, 1960," *Bull. Geol. Soc. Am.,* 81(4), 1001–1040.
Ruiz, C., 1965. "Geologia yacimientos metaliferous de Chile," *Inst. Invest. Geol. Chile,* (Santiago), 305.
____, Aguirre, L., Corvalan, J., Rose, H. J., Jr., Segerstrom, K., and Stern, T. W., 1961. "Ages of batholithic intrusions of northern and central Chile," *Bull. Geol. Soc. Am.,* 72, 1551–1560.
Rutland, R. W. R., 1971. "Andean orogeny and sea-floor spreading," *Nature,* 233, 252–255.
____, Guest, J. E., and Grasty, R. L., 1965. "Isotopic ages and Andean uplift," *Nature,* 208, 677–678.
Scholl, D. W., Huene, R. von, and Ridlon, J. B., 1968. "Spreading of the ocean floor: undeformed sediments in the Peru–Chile trench," *Science,* 159, 869–871.
____, Christensen, M. N., Huene, R. von, and Marlow, M. S., 1970. "Peru–Chile trench sediments and sea-floor spreading," *Bull. Geol. Soc. Am.,* 81, 1339–1360.
Scott, K. M., 1966. "Sedimentology and dispersal pattern of Cretaceous flysch sequence, Patagonian Andes, southern Chile," *Bull. Am. Assoc. Petrol. Geologists,* 50(1), 72–107.
Segerstrom, K., 1964. "Quaternary history of Chile: brief outline," *Bull. Geol. Soc. Am.,* 75, 157–169.
____, 1967. "Geology and ore deposits of central Atacama Province, Chile," *Bull. Geol. Soc. Am.,* 78(3), 305–318.
Sillitoe, R. H., 1973. "Geology of the Los Pelambres porphyry copper deposit, Chile," *Econ. Geol.,* 68, 1–10.
____, Mortimer, C., and Clark, A. H., 1968. "A chronology of landform evolution and supergene mineral alteration, southern Atacama Desert, Chile," *Trans. Inst. Mining Met.,* 77(744), B166–169.
Stiefel, J., 1970. "Das Andenprofil im Bereich des 45 südlichen Breitengrades," *Geol. Rundschau,* 59(3), 961–979.
Thomas, N. A., 1970. "Tektonik Nordchiles," *Geol. Rundschau,* 59(3), 1013–1027.
Vegara, M., and Gonzalez-Ferran, D., 1972. "Structural and petrological characteristics of the late Cenozoic volcanism from Chilean Andean region and west Antarctica," *Krystalinikum* (Prague), 9, 157–182.
Watters, W. A., and Fleming, C. A., 1972. "Contributions to the geology and paleontology of Chiloe Island, southern Chile," *Phil. Trans. Roy. Soc. Lond.,* 263B(853), 370–407.
Wetzel, W., 1930. "Die Quiriquina-Schichten als Sediment und Paläontologisches Archiv." *Palaeontolographica,* 73, 49–106.
Zeil, W., 1964a. "Geologie von Chile," *Regionale Geologie der Erde,* vol. 3. Berlin: Gebr. Borntraeger, 233p.
____, 1964b. "Die Verbreitung des jungen Vulkanismus in der Hochkordillere Nordchiles," *Geol. Rundschau,* 53(2), 731–758.

Cross-references: *Argentina; Bolivia; Peru; South America; Sub-Antarctic Islands.*

CHRISTMAS ISLAND

An Australian territory in the Indian Ocean, situated at 10°25'S, 105°43'E, about 2620 km (1425 n mi) NW of Perth, and 410 km (22 n mi) S of Java Head, Christmas Island has an area of 143 km^2 (55 sq mi). The climate is monsoonal. Relief is hilly with a generally thick soil cover supporting a tropical rain forest typical of the Malay Archipelago. The interior is a plateau attaining a height of 357 m at Murray Hill, whereas the coast and border escarpments are precipitous. Cocos-Keeling, the only other islands in the northeastern Indian Ocean, lie 1000 km (540 n mi) WSW. (A second Christmas Island, in the Pacific Ocean, is the largest atoll in the Pacific and is one of the *Line Islands,* q.v.)

Deposits of phosphate on the island have been worked since 1900. A monograph of Christmas Island was published by Andrews (1900) with later geological investigations by Trueman (1965) and Barrie (1967).

The best sequence is exposed in Flying Fish Cove, with the oldest rocks (Andrews' "lower volcanic" series) comprising andesites and trachybasalts. These probably form the foundation of this oceanic island, and are capped by "Yellow Limestones" 20–45 m thick, of Eocene age (Nuttal, 1926). Some 58 m of "upper volcanics" follow, mainly palagonite tuffs and submarine basalitic lava flows, which include limburgites. On these rest Miocene orbitoidal limestones 52 m thick, in turn capped by Pliocene-Pleistocene reef limestones which were originally in the form of an atoll, and have subsequently been uplifted and marginally terraced. The final phase of vulcanism is marked by the intrusion of basaltic dykes into these cap limestones, and the extrusion of vesicular (now phosphatized) lavas in the vicinity of Murray Hill.

The island is today in the form of a three-pointed star, although offshore the shape of the foundation is circular. This pattern, as well as the structural details, is interpreted as indicating a history of radial landsliding of the volcanic foundation (Fairbridge, 1950), probably during the tuff-eruption stage. Local adjustment faulting has occurred around the margins of the younger reef limestone. Modern reefs are not important because of the steep offshore gradient.

The island rises from 4500-m depths and the cone is an isolated seamount not directly connected to known ridges or trends of seamounts. An active fracture zone generally trending N-S may have served as the guide for upwelling magmas. The oceanic crust in this area is of Lower Cretaceous age.

Phosphatic deposits of considerable economic importance are present on the interior plateau and on parts of the wave-cut terraces. Upper layers are of crandallite-millisite (Fe, Al-phosphates) underlain by apatite (Ca-phosphate) contained within a karrenfeld (karst) limestone profile. The origin of the phosphate is believed to be by deposition of guano in an atoll environment. Subsequent elevation and aerial weathering has led to mobilization of the phosphate and conversion of the carbonate floor to calcium phosphate.

PETER J. BARRETT

References

Andrews, C. W., 1900. *A Monograph of Christmas Island (Indian Ocean)*. London: British Mus. Nat. Hist., 337p.

Barrie, J., 1967. *The Geology of Christmas Island*. Canberra: Australian Bur. Min. Res. Geol. Geophys., Record, **1967**/37.

Fairbridge, R. W., 1950. "Landslide patterns on oceanic volcanoes and atolls," *Geogr. J.*, **115**, 84–88.

Nuttal, W. L. F., 1926. "A revision of the orbitoides of Christmas Island (Indian Ocean)," *Q. J. Geol. Soc. Lond.*, **82**, 22–43.

Trueman, N. A., 1965. "The phosphate, volcanic and carbonate rocks of Christmas Island (Indian Ocean)," *J. Geol. Soc. Austral.*, **12**(2), 261–283.

Cross-references: *Australia; Cocos (Keeling) Islands; Line Islands; Nauru; Vol. I, Indian Ocean; Vol. III, Atolls, Coral Reefs.*

CLIPPERTON ISLAND

Clipperton, administered by France, is an isolated islet 1100 km off the coast of Mexico in the eastern equatorial Pacific at $10°19'$N, $109°13'$W. It is virtually an atoll, having a ring reef with sand banks and a lagoon measuring 2 by 4 km, with an area of 1.6 km^2; but it includes a small volcanic pinnacle reaching an elevation of 29 m. The island has also been called *Isla de la Pasión,* and *I. de los Medanos* (sand banks).

The geology and natural history have been described by Wharton (1898), Teal (1898), Obermuller (1959), and Sachet (1960, 1962). The base of the atoll is a seamount rising from near the Clipperton Fracture Zone to the west of its intersection with the East Pacific Rise. It is the only emergent member of the Clipperton Seamount Chain. Sea-floor spreading evidence suggests that the underlying crust is of Miocene age.

The volcanic material is a deeply weathered alkaline trachyte, partly phosphatized, and associated with guano deposits. Presumably Tertiary in age, the volcanic stump is capped by an atoll with a remarkably deep (100 m) lagoon. There is an emerged reef limestone terrace.

The ring reef appears to have completed a closure of the lagoon quite recently and now there is a great variation in the salinity and pH (7–10) that appears to favor a weak dolomitization.

RHODES W. FAIRBRIDGE

References

Bourrouilh-Le Jan, F., 1971. "Existance et conditions d'une dolomitisation precoce à partir d'aragonite dans les eaux isolées de l'atoll de Clipperton," *8th Intern. Sed. Congr. Addl.*, Abstr., Heidelberg, 2.

Obermuller, A. G., 1959. "Contribution à l'étude géologique et minérale de l'île de Clipperton," in *Recherches géologiques et minérales en Polynésie française.* Paris: Inspection Générale des Mines et de la Géologie, 45–60.

Sachet, M. H., 1960. "Histoire de l'île Clipperton," *Cahiers Pacifique* (Paris), **2**, 1–22.

____, 1962. "Monographie physique et biologique de l'île Clipperton," *Ann. Inst. Océanogr.*, **40**(1), 89–97.

Sclater, J. G., Anderson, R. N., and Bell, M. L., 1971. "Elevation of ridges and evolution of the central eastern Pacific," *J. Geophys. Res.*, **76**(32), 7888–7915.

Teal, J. J. H., 1898. "Phosphatized trachyte from Clipperton Island," *Quart. J. Geol. Soc. Lond.*, **54**, 230–232.

Wharton, W. J., 1898. "Note on Clipperton Atoll

(northern Pacific)," *Quart. J. Geol. Soc. Lond.,* 54 228–229.

Cross-references: *Galápagos Islands; Easter Island and Sala y Gómez; Revillagigedo Islands.*

COCOS (KEELING) ISLANDS

These two atolls, comprising some 27 coral islands, are in the northeastern Indian Ocean about 1100 km (700 miles) SW of Sumatra and about 2775 km (1725 miles) NW of Perth, Australia. The southern atoll, the largest, is Cocos (or Keeling), and the northern one is North Keeling. The main islands are West, Home, Direction, South, and Horsburgh. Total land area is 13 km^2 (5 sq mi). Rainfall is moderate and the climate is pleasant. No volcanic rocks crop out, but the atoll foundations are surely volcanic seamounts, rising from oceanic crust of Cretaceous age. The coral reefs, in classic atoll form, have long attracted attention (Darwin, 1842; Guppy, 1889; Wood-Jones, 1910; Davis, 1928).

There is another *Cocos Island* (Isla del Coco) belonging to Costa Rica, in the Pacific, 960 km SW of Panama and 500 miles (805 km) N of Galapagos. It is situated on the submarine Cocos ridge that joins the Galapagos to Central America.

RHODES W. FAIRBRIDGE

References

Darwin, C., 1842. *The Structure and Distribution of Coral Reefs.* 3rd ed., New York: Appleton, 1889, 344p.
Davis, W. M., 1928. "The coral reef problem," *Am. Geogr. Soc., Spec. Publ.* 9, 596p.
Guppy, H. B., 1889. "The Cocos-Keeling Islands," *Scottish Geogr. Mag.,* 5, 281–279, 457–474, 569–588.
Wood-Jones, F., 1910. *Corals and Atolls.* London: Lovell Reeve, 392p.

Cross-references: *Australasia–Regional Review; Christmas Island; Vol. I, Indian Ocean; Oceania.*

COLOMBIA

Colombia comprises an area of 1,138,934 km^2 (439,737 sq mi) between the latitudes 12°N and 4°S, situated at the northwestern limit of the South American continent. Some 1500 km of coast face the Pacific Ocean, as against about 1760 km that face the Atlantic Ocean (Caribbean).

The earliest geologic descriptions originate from Alexander von Humboldt in 1801. He was followed in that century by only a few natural scientists: Boussingault, Brongniart, D'Orbigny, Karsten, Hettner. In the present century geologic data have been derived from the exploration for mineral deposits and hydrocarbons by Ospina, Scheibe, Stutzer, Grosse, Hubach, and others. Trümpy (1943) published material resulting from exploration by several oil companies. The most significant regional and general work was drawn up during the 1950s and 1960s by Bürgl, who also published the first *Historia Geologica de Colombia* (Bürgl, 1961). The most recent general map, scale 1:1,500,000, was printed in 1962.

Subsequently a systematic geological mapping program was initiated. By 1971 about 10% of the entire area was mapped and published at a scale of 1:100,000. The investigations are guided by the Instituto Nacional de Investigaciones Geologico Mineras (Ingeominas, Bogotá) and published in part in *Boletin Geologico*. Significant activity is also being generated by the Colombian Petroleum Company and by foreign oil companies, whose investigations, unfortunately, are not made public. Oil production in the early 1970s was 73 million barrels annually.

The principal center of geologic training in Colombia is the Departamento de Geologia of the Universidad Nacional de Colombia in Bogotá. In addition there are the Faculdad de Minas in Medellin associated with the Universidad Nacional and the Departamento de Geologia of the Universidad Industrial de Santander in Bucaramanga. All these institutions occasionally publish geological reports.

Geomorphic and Structural Divisions

Colombia consists essentially of two large geomorphological regions (Fig. 1): (1) the Llanos, a lowland whose elevations is generally under +200 m and rises to +500 m only at the margin of the Andes (the Llanos form the eastern and still mostly unaccessible and underdeveloped part of Colombia); and (2) the Andes, which dominate the western portion of the country with their 1500 km long N-S extension.

The Andean mountain chain, still a discrete entity in Ecuador to the south, fans out in Colombia into three ranges, the western Cordillera Occidental, the Cordillera Central, and the eastern Cordillera Oriental, divided by the wide longitudinal valleys of the Rio Cauca in the west and the Rio Magdalena in the east. Each is

FIGURE 1. Principal orographic units of Colombia.

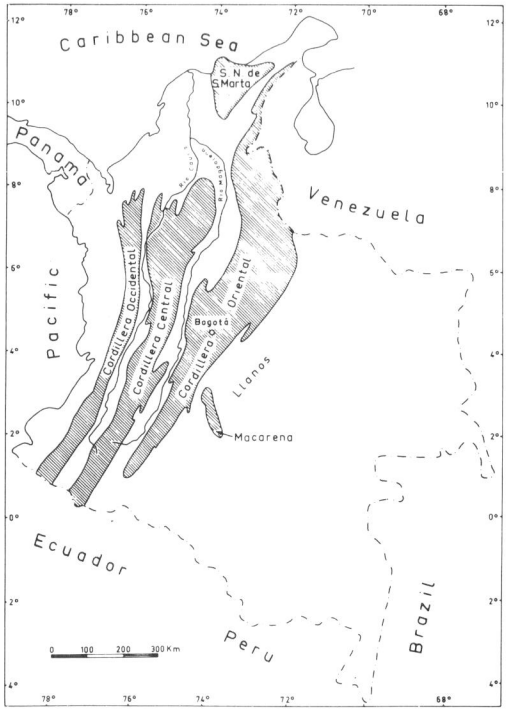

marked by distinct structural and geomorphic features.

Regional Geology

Llanos. The substructure of the Llanos is part of the Precambrian crystallines of the Guiana Shield. Almost no details of the basement are known. The sedimentary section in the western part begins with an incomplete sequence of early Paleozoic rocks which have undergone minor tectonic disturbance. This sequence outcrops in the Sierra de la Macarena, an isolated massif protruding from the Llanos in the vicinity of Uribe. These Paleozoics were reached by boreholes near the headwaters of the Rio Meta and Rio Casanare. Triassic and Jurassic appear to be missing and only the Cretaceous transgressed far to the east. The greatest thickness of Cretaceous sediments is in the sub-Andean trough from which they pinch out toward the east. The Tertiary is also restricted to this depression. Although marine conditions still prevailed during the Cretaceous, the area became positive later, as indicated by intercalated coal seams beginning in the Eocene.

Cordillera Oriental. The backbone of the Cordillera Oriental is made up of "macizos antiguos." These include some crystalline rocks of the Guiana Shield, but also slightly metamorphosed rocks of the Quentame Series, considered by Trümpy (1943) to be the metamorphic facies of the unaltered Cambro-Ordovician of the Llanos (Macarena area). They are overlain by an incomplete sequence of Devonian and Carboniferous, fully marine in character at the base but interspersed with brackish or terrestrial intercalations toward the top. During the Permian, a narrow bay seems to have extended from the Caribbean to the south of Bogotá into the Cordillera Oriental. Faunally documented Permian is rare and occurs only in the Cordillera Oriental.

In general, younger sediments are not preserved on the massifs, but around their edges. They are terrestrial, mostly red, sandstones and graywackes of great thickness. According to current interpretation, they are of early Jurassic age (Neogiron, according to Bürgl, 1964). Wide areas of the Cordillera Oriental are covered by fine clastic Cretaceous deposits. This Cretaceous series alone attains thicknesses of cover 16,000 m in the central portion of the Eastern Cordillera, in the Cuenca de Cundinamarca (Bürgl, 1961). The most vigorous and youngest uplift occurred in the eastern half of the range. Here, in the Nevado del Cocuy, an altitude of +5493 m is attained. The Eastern Cordillera is dissected by deep valleys which are tectonically controlled by N-S-trending fault zones. Headward erosion from these valleys incise into the +4000 m high Paramos. Valleys filled with talus and colluvial material form the Sabanas, the most important one being the Sabana of Bogotá.

Cordillera Central. This segment of the Andes is clearly distinguished from the preceding one by its geologic structure and morphology. In place of the Paleozoic and Mesozoic sediments of the Eastern Cordillera, the Central Cordillera is composed primarily of metamorphics and igneous rocks (Fig. 2). The degree of metamorphism increases from S to N. Although the metamorphics of the southern Cordillera Central consist predominantly of phyllites, in the northern portion they also include micaschists and gneisses. The age of these Cajamarca series has been disputed but is probably Cambro-Ordovician sediments and volcanics metamorphosed during the Caledonian Orogeny based on the investigations of Stibane (1970). Hence they correspond to the Quetame Series of the Eastern Cordillera. Other authors have attributed the metamorphism to a Mesozoic orogeny (Radelli, 1967; Butterlin, 1969).

The Cordillera Central is interspersed with numerous magmatic rocks of various ages (Barrero, et al., 1969). The Cretaceous Andean

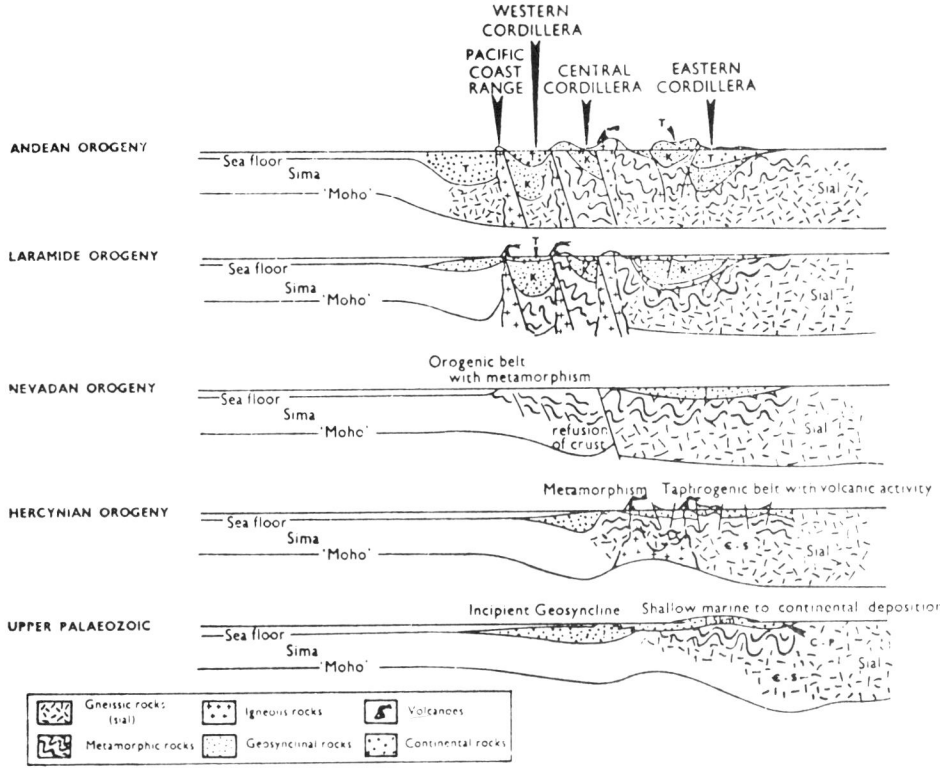

FIGURE 2. Diagrams suggesting the evolution of the Andes in Colombia (from the records of the British Petroleum Co. and drawn by C. J. Campbell; simplified, after Stoneley, 1969).

batholiths are especially prominent. Smaller intrusives and volcanics are concentrated along the tectonically affected margins of the Cordillera Central. The range is today topped by a number of young, usually snow-covered volcanoes. Some of them are still active, for instance the highest (+5439 m), the Nevado del Huila.

The morphology differs noticeably from that of the Eastern Cordillera. The range is not divided into various N-S-striking chains, but forms a single, less rugged feature. Because of the deep weathering of the crystalline rocks, it forms a more gentle single mountain ridge which dips at low angles to both the east and the west.

The moderately elevated Sierra Nevada de Santa Marta, located in the northern portion of the country, is entirely isolated and surrounded only by Tertiary plains and the Caribbean. It is considered to be a northerly extension of the Cordillera Central. Here, in the deeply dissected, rugged massif, the Colombian Andes attain their greatest elevation of +5780 m in the Pico Cristobal Colon. Metamorphic and magmatic rocks of various ages constitute most of the Sierra. Its base is made up of Precambrian rocks, interspersed with a number of Mesozoic intrusives and extrusives. Of special interest for the geological history of the northern Andes is a small occurrence of Devonian-Carboniferous rocks in the central part. Mesozoic sediments are limited to the southern foothills. The youngest orogenic modification, not observed elsewhere in the Colombian Andes, is documented in the northwestern part of the Sierra Nevada. Here sediments of questionable age were transformed to micaschists (Grupo Esquistos de Santa Marta) during the Eocene and subsequently penetrated by an enormous batholith.

Cordillera Occidental. The Western Cordillera is fundamentally different from the other parts of the Colombian Andes. Current knowledge of the geologic history begins only with the Mesozoic, since older rocks do not outcrop. The oldest rocks of the "Grupo de Dagua" are metasediments. Their age is assumed by Ospina (1911) to be Triassic-Jurassic, although no faunal or radiometric evidence is available. They consist of phyllites and micaschists, metamorphosed from fine clastic sediments and tonalites. Radiolarites and siliceous limestones of great thicknesses occur in the upper portion of the sequence lasting until Middle Jurassic (by paleontological data). At the same time the first volcanics appear, initiating the "Grupo Diabasico." The latter is composed of more than 10,000 m of diabases and basalts with pillow structures and other indications of submarine origin interbedded with lydites. Paleontological data from these marine intercalations show that volcanic activity continued throughout the Cretaceous.

Based on facies, origin, and the relationship with the Grupo Diabasico, the Grupo de Dagua has to be regarded as eugeosynclinal. However, in this case a complete orogenic cycle did not occur; i.e., no significant tectogenesis followed the geosynclinal stage. Rather, the observable metamorphism that gradually decreases in intensity toward the top of the section can be solely attributed to overburden pressure of the approximately 15,000-m-thick sequence (see Nelson, 1957, on load metamorphism).

Tertiary Basins. Throughout its geological history the three parts of the Colombian Andes became increasingly accentuated by a gradual uplift of the Cordilleras and concurrent subsidence of the intervening depressions. Four large Tertiary basins were formed, from W to E: (1) the *Sinclinorio del Pacifico,* which extends from Ecuador to Panama; (2) the *Sinclinorio del Cauca,* which widens and deepens considerably in the lower course of the N-flowing Rio Magdalena; (3) the *Sinclinorio del Magdalena,* whose southern portion closely follows the Rio Magdalena; at the point at which the river swings W (across the N-plunging Cordillera Central), the Sinclinorio del Magdalena turns NE and contains the valley of the Rio Cauca; and (4) the *Sinclinorio pre-Andino* along the eastern foot of the Cordillera Oriental.

The Tertiary sediments are predominantly marine in the Sinclinorio del Pacifico and in the north of the country, close to the Caribbean, whereas the thick sequence in the south and east of the country are essentially of terrestrial origin. Since the beginning of the Tertiary, the rivers followed the depressions in a S-N direction. They filled the upper valleys with conglomerates and coarse sands, and in the lower stretches they deposited finer-grained sediments which intertongued northward with marine deposits. During the Tertiary, the facies boundaries between marine and terrestrial sedimentation shifted from S to N and back several times.

Simultaneously with these continuously shifting Tertiary facies boundaries, two large-scale transgressions occurred and can be observed as far inland (S) as the middle Magdalena Valley. These occurred in the Middle Eocene and in the Upper Oligocene. This last transgresssion was reversed by a regression that persists until today.

Geologic History

Active tectonism in the Llanos, for all practical purposes, ended during the Precambrian. The last orogenic pulse affecting the Guiana Shield probably occurred during the late Algonkian. The Shield area was largely peneplaned by the Ordovician. During the Paleozoic and later in the Mesozoic, there were short-lived transgressions across the western margin of the passive shield.

The observable geologic history of the Colombian Andes begins with the Cambrian. At this time a geosyncline developed along the western edge of the Guiana Shield. During the late Ordovician or possibly early Silurian, the geosyncline underwent intense folding and metamorphism in the Caledonian orogeny. In the late phases of this cycle granitic magmas intruded the western and central Cordillera. The base of the Devonian where it outcrops in the Cordillera Oriental is made up of metasediments of the Quetame Group. Their time equivalent in the Cordillera Central is the Cajamarca Series. They also are metasediments and metavolcanics, which, however, underwent considerably stronger metamorphism. In Antioquia, in the northern part of the Cordillera Central, all facies transitions from phyllite to gneiss are present. The great thicknesses, original rock compositions, and numerous intercalations with initial volcanics indicate that the Cambro-Ordovician eugeosyncline proper was located in the area of the Cordillera Central.

Following metamorphism and uplift the early Andes were dissected into belts paralleling the previously developed N-S strike. This dissection produced a relatively small low area in the position of today's Cordillera Oriental into which the Devonian sea transgressed from the north. In all likelihood the Cordillera Central was a land mass at this time. This geotectonic relationship remained essentially unchanged until the Mesozoic. The distribution of Carboniferous and Permian marine areas was similar to that of the Devonian. However, the Devonian sea extended far to the south into the Cordillera Oriental, and the shoreline gradually moved northward during the Carboniferous and Permian. Although the subsidence still persisted, the facies became progressively more terrestrial. Magmatic activities occurred along fault zones that formed the block margins.

The same pattern continued into the early Mesozoic. Clastic, predominantly red, terrestrial sediments were deposited over the entire Eastern Cordillera. A slight change in structural conditions is indicated by a marine invasion (Payande) from the west over a small area in the Cordillera Oriental. Subsidence began again during the Jurassic in the area of the Eastern Cordillera, enabling almost 14,000 m of marine Cretaceous sediments to be deposited (Bürgl, 1961).

The late Mesozoic saw the broadest inundation in the northern Andes. During the Cretaceous not only was the Central Cordillera invaded but also a large portion of the Guiana Shield. Early in the Tertiary this broad transgression withdrew toward the present coastal area. The broad Paleozoic structural features became increasingly accentuated during the Mesozoic and Cenozoic.

The geologic history of the Western Cordillera can be traced from the beginning of the Mesozoic. An important depression was formed seaward of the Central Cordillera. It was subsequently filled with clastic and siliceous sediments, followed by an extremely thick sequence of volcanics. At a later date more acidic intrusions were emplaced. The Western Cordillera thus had a significantly different geologic evolution than that of the contemporaneous miogeosyncline of the Eastern Cordillera. The large Andean batholiths in the central portion of the range are of Cretaceous age.

In the Tertiary the entire Andes were uplifted, although there were no important orogenic events. The one exception is the northern flank of the Sierra Nevada de Santa Marta and the Peninsula Guajira located immediately to the north, where at the turn of the Mesozoic-Cenozoic, an intensive orogenic phase occurred. It is, however, restricted to the northern margin of the continent.

The last expression of geotectonic activity and one which has existed since the Tertiary is that of active volcanism, limited to a narrow strip of the Central Cordillera. In addition, the country frequently suffers from earthquakes, an indication of the mobility of the earth's crust along the margins of two plates moving toward each other.

FRITZ R. STIBANE

References

Anderson, T. A., 1972. "Paleogene nonmarine Gulanday Group, Neiva Basin, Colombia, and regional development of the Colombian Andes," *Bull. Geol. Soc. Am.*, **83**, 2423–2438.

Anon., 1971. *Guidebook to a Geologic Section between Bucaramanga and La Uribe, Middle Magdalena Valley, Field Trip No. 12.* Bogota: Colombian Soc. Petrol. Geol. Geophys., 23p.

———, 1972. *Guidebook to the geology of the Eastern Cordillera between Aguazul–Sogamoso–Villa de Leiva.* Bogota: Colombian Soc. Petrol. Geol. Geophys., 67p.

Barrero, D., Alvarez, J., and Kassem, T., 1969. "Actividad ignea y tectónica en la cordillera central durante el meso-cenozoic," *Colomb., Serv. Geol. Nac., Bol. Geol.*, **17**(1–3), 145–173.

Bürgl, H., 1961. "Historia geologica de Colombia," *Rev. Acad. Colombiana Cienc. Exact. Fis. Nat.*, **11**(43), 136–191.

———, 1964. "El 'Jura-Triasico' de Colombia," *Bol. Geol.*, **12**(1–3), 5–31.

———, 1973. "Precambrian to middle Cretaceous stratigraphy of Colombia," Bogota: priv. publ., 214p., transl. from Spanish by C. G. Allen (rev. *Bull. Am. Assoc. Petrol. Geologists*, 1973, 1583).

Butterlin, J., 1969. "A propos de la géologié des Andes de Colombie," *Rev. Géogr. Phys. Géol. Dyn.*, **11**(1), 65–76.

———, 1972. "La posición estructural de los Andes de Colombia," *6th Congr. Geol. Venezol.*, Caracas, 1969, **2**, 1185–1200.

Case, J. E., and MacDonald, W. D., 1973. "Regional gravity anomalies and crustal structure in northern Colombia," *Bull. Geol. Soc. Am.*, **84**, 2905–2916.

———, et al., 1971. "Tectonic investigations in western Colombia and eastern Panama," *Bull. Geol. Soc. Am.*, **82**, 2685–2712.

Hammen, T. van der, 1966. "The Pliocene and Quaternary of the Sabana de Bogota (the Tilata and Sabana formations)," *Geol. Mijnb.*, **45**, 102–109.

———, and Gonzales, E., 1960. "Upper Pleistocene and Holocene climate and vegetation of the Sabana de Bogota, Colombia, South America," *Leidse Geol. Mededel.*, **25**, 261–315.

Julivert, M., 1970. "Cover and basement tectonics in the Cordillera Oriental of Colombia, South America, and a comparison with some other folded chains," *Bull. Geol. Soc. Am.*, **81**, 3623–3646.

Krummerbacher, R., 1973. "Sur la formation des Andes colombiennos," *Eclogae Geol. Helv.*, **66**(2), 325–337.

MacDonald, W. D., and Opdyke, N. D., 1972. "Tectonic rotations suggested by paleomagnetic results from northern Colombia, South America," *J. Geophys. Res.*, **77**(29), 5720–5730.

McLaughlin, D., Jr., 1972. "Evaporite deposits of Bogata area, Cordillera Oriental, Colombia," *Bull. Am. Assoc. Petrol. Geologists,* **56**(11), 2240–2259.

Nelson, H. W., 1957. "Contribution to the geology of the central and western Cordillera of Colombia in the sector between Ibague and Cali," *Leidse Geol. Mededel.*, **22**, 1–76.

Ospina, T., 1911. *Reseña Sobre la Geología de Colombia y Especialmente del Antiguo Departamento de Antioquia.* Medellín.

Pichler, H., Stibane, F. R., and Weyl, R., 1974. "Basischer Magmatismus und Krustenbau im südlichen, Mittelamerika, Kolumbien und Ecuador," *N. Jb. Geol. Paläont. Mh.*, **1974**(2), 102–126.

Radelli, L., 1967. "Géologie des Andes colombiennes," *Trav. Lab. Geol. Fac. Sci. Univ. Grenoble Mem. 6,* 470p.

Stibane, F., 1970. "Beitrag zum Alter der Metamorphose der Zentral-Kordillera Kolumbiens," *Mitt. inst. Col.–Alem. Inv. Cient.,* **4,** 77–82.

Stoneley, R., 1969. "Sedimentary thicknesses in orogenic belts," *in* P. E. Kent, G. E. Satterthwaite, and A. M. Spencer, eds., *Time and Place in Orogeny,* London: Geol. Soc., 215–238.

Trümpy, D., 1943. "Pre-Cretaceous of Colombia," *Bull. Geol. Soc. Am.,* **54,** 1281–1304.

Cross-references: *Brazil; Central America–Regional Review; Ecuador; Panama; Peru; South America.*

COOK ISLANDS

The Cook Islands are situated in the South Pacific about midway between Tonga and Tahiti, and some 2700 km (1700 miles) NE of New Zealand, which formerly administered them. They have been independent since 1965. They extend from 8–23°S and 156–167°W. (*Niue,* q.v., formerly included in the administrative area, is now semi-independent.) There are 15 principal islands, embracing about 240 km^2 (90 sq mi). (There is also a Cook Island in the South Sandwich Group; see *Subantarctic Islands.*)

There are two main chains trending NW-SE: the *Northern Chain,* which are atolls, emerged or near-atolls, includes Aitutaki, the Hervey Islands, Takutea, Mitiaro, and Mauke, together with scattered ones farther north; and the *Southern Chain,* which includes volcanic islands, atolls, and emerged atolls. Most important is the volcanic Rarotonga (principal town, Avarua) with its rich soil and central peak, Te Manga, reaching 652 m. It has a mean temperature of 23.6°C, and receives 2150 mm precipitation yearly. Under the influence of the SE trade winds most of the year, during the southern summer there are occasional hurricanes.

The scattered northern atolls do not follow any distinct trend. Most easterly is Penrhyn Atoll (Tongareva), which approaches the Line Islands, NNW of Tahiti; it has a fine natural anchorage in the 20 X 12 km lagoon, and some 15 rim islands. Manihiki Atoll has a closed lagoon as has the nearby Rakahanga Atoll. The three Danger Islands are also on a closed lagoon atoll, and Nassau Island, sharing the same submarine plateau, is only 600 m across and has a fringing reef. Suwarrow Atoll (name corrupted from the Russian Suvorov) is about 12 km in diameter. Palmerston Atoll is on a line between Samoa and Aitutaki.

The geological record is most complete for *Rarotonga,* which contains many formations.

There are two main divisions: basaltic eruptives followed by younger phonolitic eruptives (Wood and Hay, 1970). The latter include rocks, 2.3 and 2.8 m yr old (Tarling, 1967). The sea-floor spreading history shows that the crust here is Cretaceous. Weathering is deep and dissection leaves knife-edged ridges and deep gorges, which are graded to a low sea level. There is a fringing reef, but no major emerged or submerged benches, apart from +2 and +3 m eustatic platforms. The latter has a minimum radiocarbon date of 28,200 ± 50 years B.P. (Schofield, 1970).

Mangaia is one of the most interesting islands in the whole Pacific (Marshall, 1927, 1929). Some 8 km across and rising to 160 m, the island was probably initiated in late Cretaceous or early Tertiary times by basalt eruptions. It has a central volcanic cone, deeply degraded, which is surrounded by a former barrier reef, now uplifted. The emerged lagoon—now swampy or agricultural land, the "taro flats"—is now 6–12 m above sea level and up to 500 m wide. The rim, known technically (from the Polynesian word) as a "makatea" forms a fortress-like wall all around, 50–60 m high and steeply cliffed both on the exterior and interior. There are the usual eustatic benches. The makatea is eroded into a wildly jagged karst (see *Aldabra* in the *British Indian Ocean Territory*). This limestone may be Miocene or Oligocene.

Aitu is somewhat like Mangaia, with a volcanic core surrounded by an emerged reef rim (makatea). Some 5 km across, the degraded volcano core reaches 80 m in elevation and is surrounded by a swampy former lagoon up to 400 m wide. The makatea rises 20 m, up to 1200 m across, and like Mangaia is marked by a deep karst erosion. The reef rock is largely dolomitized and is an important piece of evidence against the pressure theory that had been proposed for the Funafuti dolomitization (see *Gilbert and Ellice Islands*).

Another of the northern line is *Mauke,* 5 km across, with a very small volcanic center and emerged coral reef up to 30 m in elevation. Aitutaki is a near atoll with a wide lagoon (only 4–5 m in depth) and a small volcanic island 5 km across and about 150 m high. There may be eustatic erosion benches at 6–76 m on several of the islands (Schofield, 1966, 1967).

The reefs, lagoons, and sand cays ("motus") of the Cook Islands range in age from mid-Tertiary to the present. Absolute dating of the 2-m eustatic platform at Mangaia gave 90,000 and 110,000 years (Veeh, 1966).

RHODES W. FAIRBRIDGE
J. C. SCHOFIELD

References

Chubb, L. J., 1927. "Mangaia and Rurutu; a comparison between two Pacific Islands," *Geol. Mag.*, **64**, 518–522.

Crossland, C., 1928. "Coral reefs of Tahiti, Moorea and Rarotonga," *Linn. Soc. Lond. J. Zool.*, **36**, 577–620.

Gibbs, P. E., Stoddart, D. R., and Vevers, H. G., 1971. "Coral reefs and associated communities in the Cook Islands," *Roy. Soc. N.Z. Bull.*, **8**, 91–105.

Grange, L. E., and Fox, J. P., 1953. "Soils of the lower Cook Group," *N.Z. Soil Bur. Bull.*, **8**.

Gregory, H. E., 1925. "Tongareva (Penrhyn) Island," *Bernice P. Bishop Mus. Bull.*, **21**.

Hochstein, M. P., 1967. "Seismic measurements in the Cook Islands, southwest Pacific Ocean, *N.Z. J. Geol. Geophys.*, **10**(1), 499–521.

Marshall, P., 1908. "Geology of Rarotonga and Aitutaki," *N.Z. Inst. Trans.*, **41**, 98–100.

_____, 1911a. "Alkaline rocks of the Cook and Society Island," *Australasian Assoc. Adv. Sci. Rept. 13*. 196–201.

_____ 1911b. "Coral reefs of the Cook and Society Islands," *Australasian Assoc. Adv. Sci. Rept. 13*, 140–145.

_____, 1927. "Geology of Mangaia," *Bernice P. Bishop Mus. Bull.*, **36**.

_____, 1929. "Mangaia and Rurutu, a comparison between two Pacific Islands," *Geol. Mag.*, **66**, 385–389 (discussion of Chubb, 1927).

_____, 1930. "Geology of Rarotonga and Atiu," *Bernice P. Bishop Mus. Bull.*, **72**, 1–74.

Schofield, J. C., 1966. "Evidence for Quaternary sealevels from the Cook Islands and the effect of density changes on post-glacial sea-level rise," *Proc. 11th Pacific Sci. Congr.*, Tokyo, **4** (abstracts of papers, geological sciences), 15.

_____ 1967. "Pleistocene sea-level evidence from Cook Islands," *J. Geosci.* (Osaka City Univ.), **10**, 115–120.

_____, 1970. "Notes on Late Quaternary sea levels, Fiji and Rarotonga," *N.Z. J. Geol. Geophys.*, **13**(1), 199–206.

Stoddart, D. R., 1972. "Reef islands of Rarotonga," *Atoll Res. Bull.*, **160**, 7p.

Summerhayes, C. P., 1967. "Bathymetry and topographic lineation in the Cook Islands," *N.Z. J. Geol. Geophys.*, **10**(1), 382–399.

Tarling, D. H., 1967. "Some paleomagnetic results from Rarotonga, Cook Islands," *N.Z. J. Geol. Geophys.*, **10**(1), 400–406.

Veeh, H. H., 1966. "Th^{230}/U^{238} ages of Pleistocene high sea level stand," *J. Geophys. Res.*, **71**(3), 379–386.

Wood, B. L., 1967. "Geology of the Cook Islands," *N.Z. J. Geol. Geophys.*, **10**(6), 1429–1445.

_____, and Hay, R. F., 1970. "Geology of the Cook Islands," *N.Z. Geol. Surv. Bull.*, **82**, 103p. (geol. map, 1:25,000).

Cross-references: *American Samoa; Line Islands; Society Islands, Tahiti; Tokelau Islands; Tubuai Islands.*

COSTA RICA

Situated in Central America, Costa Rica has an area of 50,700 km^2 (19,575 sq mi). It is bordered by Nicaragua on the N, Panama on the SE, the Caribbean Sea on the E, and the Pacific Ocean on the W. The coasts are lined with swampy lowlands which extend into the central highlands, a plateau about 1000 m (3300 ft) above sea level. The Guanacaste, the Central, and the Talamana mountains cross the country from the NW to the SE. The highest mountain is Mt. Chirripó, SE of Cartago, with an elevation of 3820 m. The Poas Volcano has the largest volcanic crater in the world.

Gold, manganese and sulfur are mined. Aluminum laterite reserves are proved and will probably be exploited within the next decade.

Geologically the central highlands consist of an extension of the Nicaragua Volcanic Uplands, a more or less stablized region of Oligocene to Miocene volcanics overlying an ancient volcanic foundation of Mesozoic age (probably Cretaceous or older). Faulted down to the SW is a block-faulted zone marked by the San Juan lowlands, partly filled by late Cenozoic clastics or tuffs, and by the principal belt of Pleistocene-Holocene volcanoes, the NW-SE-trending Cordillera Volcánica. This sector is offset in the NW from the young volcanic belt of Nicaragua by a right-lateral strike-slip lineament that corresponds to the E-W-trending Clipperton Fracture Zone.

On the south coast, in five peninsulas, there is the Nicoya complex, a zone of volcanics, ultrabasic igneous rocks, and serpentines of probably early Cretaceous age.

SE of the Central Valley, the backbone of the country is defined by the Cordillera de Talamanca, where granodioritic and other intrusions of Miocene age cut Cretaceous and Tertiary sedimentary rocks.

The San Juan Lowlands are part of the main graben or central depression of Central America, which separates the older volcanic belt on the NE and the younger and contemporary volcanics on the SW. These structural belts parallel the offshore of the Middle America Trench, which is associated with shallow seismicity and intermediate on the landward side (Molnar and Sykes, 1969), proving the existence of an active belt of plate subduction ("Benioff Zone") in this sector. It dies out around longitude 84°W, that is, toward the Panama border.

The main graben is believed to have been initiated already in the Miocene, for in northwestern Costa Rica the Upper Miocene Gatun Formation indicates a shallow marine facies (Dengo *et al.*, 1970). The older volcanics associated with the faults on the northeastern side of the graben zone are small and include some extinct cones in the Torguguero area. Along the main fault line on the southwestern side there is the Guanacaste Volcanic Range. There are several large calderas here, partly concealed by younger cones (Healy, 1969). Coinciding with transverse faults along this line are the small cones N of Barba and Poas.

RHODES W. FAIRBRIDGE

References

Dengo, G., 1962. "Tectonic-igneous sequence in Costa Rica," in A.E.J. Engel et al., eds., *Petrologic Studies: A volume in Honor of A.F. Buddington.* New York: Geo. Soc. Am., 133–161.

―――, 1962. *Estudio Geológico de la Región de Guanacaste, Costa Rica.* San José: Instituto Geografico de Costa Rica, 112 p.

―――, G., Bohnenberger, O., and Bonis, S., 1970 "Tectonics and volcanism along the Pacific marginal zone of Central America," *Geol. Rundschau,* **59,** 1215–1232.

Healy, J., 1969. "Notas sobre los volcanes de la Sierra volcanica de Guanacaste, Costa Rica," *Inst. Geogr. Nacl. Costa Rica,* Inf. Semest. (E./J.), 37–47.

Henningsen, D. W., 1968. "Stratigraphy and paleogeography of upper Cretaceous and Tertiary sediments in southern Costa Rica," *Trans. 4th Caribbean Geol. Conf.,* 353–356.

Hoffstetter, R., et al., 1960. "Amerique Central," *Lexique Stratig. Intern.* (Paris), **5,** fasc. 2.

Kesel, R. H., 1973. "Notes on the lahar landforms of Costa Rica," *Z. Geomorph.,* **18,** 78–91.

Krushensky, R. D., and Escalante, G., 1968. "Activity of Irazu and Poas volcanoes, Costa Rica, November 1964–July 1965," *Bull. Volcanol.,* **31,** 75–84.

Melson, W. G., and Saenz, R., 1968. *The 1968 Eruption of Volcan Arenal, Costa Rica.* Washington, D.C.: Smiths. Inst. Cent. Sh. Phen., 35p.

Molnar, P., and Sykes, L. R., 1969. "Tectonics of the Caribbean and Middle America regions from focal mechanism and seismicity," *Bull. Geol. Soc. Am.,* **80,** 1639–1684.

Murata, K. J., Dondoli, C., and Saenz, R., 1966. "The 1963–65 eruption of Irazu volcano, Costa Rica," *Bull. Volcanol.,* **29,** 765–796.

Roberts, R. J., and Irving, E. M., 1957. "Mineral deposits of Central America," *U.S. Geol. Surv. Bull.,* **1034,** 205p.

Vinson, G. L., and Brineman, J. H., 1963. "Nuclear Central America, hub of Antillean transverse belt," in O. E. Childs, ed., *Backbone of the Americas* (Mem. 2). Tulsa: Am. Assoc. Petrol. Geologists, 101–113.

Weyl, R., 1961. *Die Geologie Mittelamerikas, Regionalen Geologie der Erde.* Berlin: Gebr. Borntraeger 226p.

―――, 1966. "Tektonik, Magmatismus und Krustenbau in Mittelamerika und Westindien," *Geotekton. Forsch.,* **23,** 67–109.

Williams, H., 1952. "Volcanic history of the Meseta Occidental, Costa Rica," *Univ. Calif. Publ. Geol. Sci.,* **29**(4), 145–180.

Cross-references: *Central America–Regional Review; El Salvador; Nicaragua; Panama.*

CUBA

The island of Cuba is the largest of the Antilles, or Caribbean islands, with an area of approximately 110,922 km^2 (44,218 sq mi). Its official name is "República de Cuba." The capital is Habana (Havana) and the chief cities include Santiago de Cuba, Camagüey, Guantánamo, and Santa Clara. The island is separated from U.S. territory on the NW by a 160-km (90-mile) channel, the Straits of Florida. A large shallow intermediate bank (Cay Sal) belongs to the Bahamas. The separating Nicholas Channel is 1500 m deep. The much narrower Old Bahama Channel (600 m deep) separates Cuba on the NE from the Great Bahama Bank. To the E it is separated by the 3000-m-deep Windward Channel from Hispaniola (Haiti). On the W the Yucatan Channel and the main course of the Gulf Stream separates it from Mexico.

The main island of Cuba is 1290 km long (E-W) and 70–200 km across. There are numerous smaller offshore islands; these are mainly coral or sandcays (e.g., on the northern coast the Archipelago de Sabana; on the southeastern coast the Jardines de la Reina; on the SW the Archipelago de los Canarreos and the large sandy island, the Isla de Pinos, blocking the Golfo de Barabano).

The coastline of Cuba alternates between rocky cliffed sectors and sandy beaches. There are numerous small coves and bays ("bolsas"), with some extensive mangrove swamps.

In the interior the country is largely undulating, but mostly at less than 100 m elevation. There are four rather mountainous areas: (a) in the NW, the Sierra de los Organos, a limestone terrain characterized by tropical cone karst ("mogotes" and poljes, etc.), 300–700 m; in the N the complex structures of the Havana and Matanzas Highlands; in the center the Santa Clara Hills (200 m) and Escambray Mts. (700 m); in the E there is the 250-km Sierra Maestra (over 1000-m elevations, max. 1972 m in the Pico Turquino). The Sierra Maestra is separated by the Valle Central from the Baracos Highlands in the NE.

The climate of Cuba is tropical. In August, the warmest month, the average temperature in Havana is 27.7°C, in January 21.3°C. Most of the year the northeast trade wind blows, bringing precipitation to the eastern and higher

FIGURE 1. Tectonic sketch maps of Cuba (Khudoley, 1967). I, Cretaceous magmatic area (Zaza tectonic unit); II, Tertiary magmatic area (Cauto tectonic unit); III, Upper Cretacerous basic and ultrabasic intrusives; IV, Cretaceous granitoids; V, Tertiary granitoids; VI, Tertiary depressions; VII, deep fault; VIII, boundary between facies-structural zones (tectonic units). Principal structures (north-south): *parageosyncline; miogeosyncline:* 1, Old Bahamas Channel depression; 2, Cayo Coco tectonic unit; 3, Remedios tectonic unit; 4, Sierra de Jatibonico deep fault; 5, Las Villas tectonic zone or marginal elevation; *eugeosyncline and intrageanticlines of Zaza tectonic unit:* 6, Las Villas deep fault; 7, Bahía Honda tectonic unit; 8, Consolación del Norte deep fault; 9-Pinar del Río tectonic unit; 10, Pinar del Río deep fault; 11, Palacios depression; 12, San Diego de los Baños tectonic unit; 13, Isla de Pinos tectonic unit; 14, Cochinos depression; 15, Trinidad tectonic unit; 16, Central basin depression; 17, La Trocha deep fault; 18, Ana Maria depression; 19, Cauto depression; 20, Nipe depression; 21, Guantánamo depression; 22, Oriente tectonic unit; 23, North Bartlett deep fault (Cayman Trench).

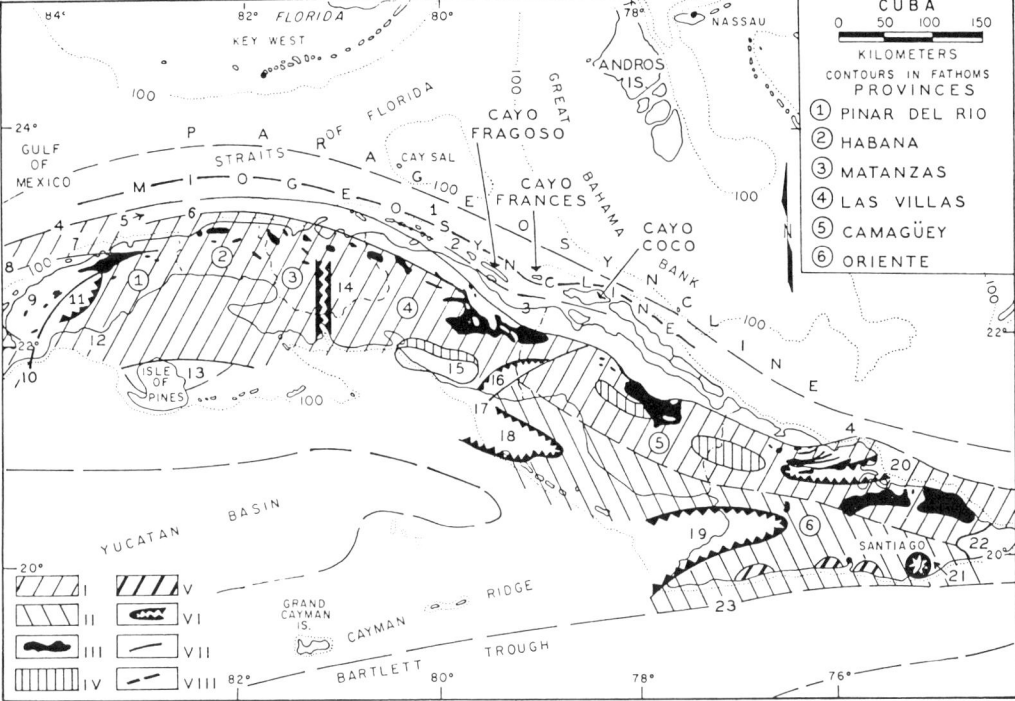

parts, with drier conditions on the lee sides. Average rainfall is 1000—2000 mm. Parts of the lee side of the Sierra Maestra are semiarid. The northwestern coast is subject to hurricanes, averaging one every 2 to 5 years.

Cuba's mineral resources include what may well be the world's largest deposit of nickel, still under development. There is considerable production of iron (mostly lateritic), manganese, copper, chrome, barite, gold, and silver. There are minor petroleum and asphalt reserves.

Structurally Cuba lies in the Greater Antillean sector of the Caribbean island arc that sweeps eastward from the North American continent near Guatemala to join the South American continent near Grenada. The main island, with offlying shelf and islets (on which there has been considerable drilling), permits a general classification into geotectonic units that largely parallel the physiographic extension of the island and correspond to former geosynclinal ridges and troughs. The sedimentation in the separate troughs differs in varying degrees (Fig. 1).

On the foreland side, to the N, lies the parageosynclinal platform of the Bahamas (q.v.), which is largely stable but has suffered progressive downwarping through time. The profile is well established in a well on Cay Sal disclosing limestone, dolomite, and anhydrite facies—1200 m of Tertiary, 4500 m of Cretaceous and Upper Jurassic, with another 4800 m of earlier sediments indicated by seismic survey.

Along the northern shore, shelf, and offlying islands (e.g., Cayo Fragoso, Cayo Frances, and Cayo Coco) there is a miogeosynclinal belt, containing the following units: the Old Bahamas Channel, Cayo Coco unit, Remedios unit, the Sierra de Jatibonico border fault, and Las Villas tectonic zone—the southern limit of the miogeosynclinal facies. The *Remedios unit* is equivalent to the Camajuani zone of Meyer-

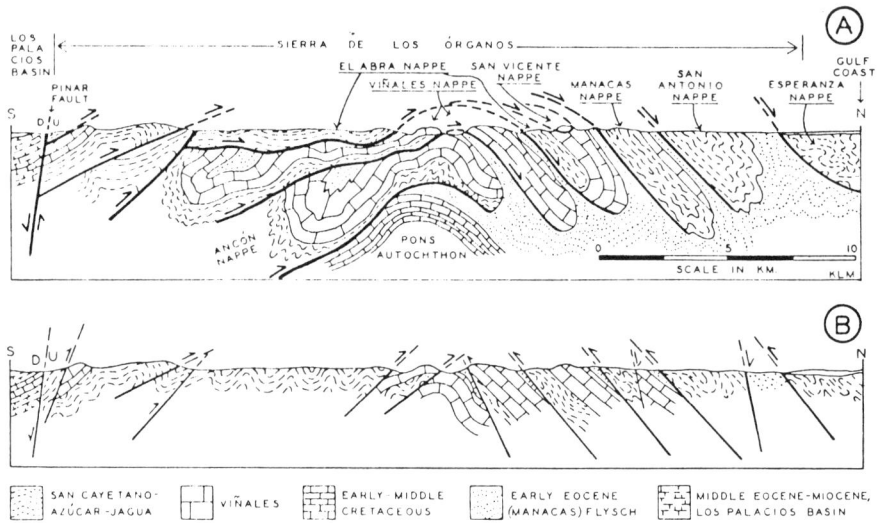

FIGURE 2. Two interpretations of the Sierra de los Organos, in northwestern Cuba (from Khudoley and Meyerhoff, 1971). Model "A," above, based largely on the work of Hatten and of Furrazola, indicates strongly developed nappe tectonics, with seven superimposed thrust sheets, directed from S to N. Metamorphism is progressively greater toward the uppwer nappes. Model "B," below, favored by Khudoley (representative of the Russian "verticalist" tectonic school), is today generally regarded as traditionalist.

hoff and Hatten (1968) and the Cayo Coco unit is their Remedios zone. Within this province there are up to 14 km of Jurassic, Cretaceous, and Tertiary sediments, mainly carbonates, anhydrite, and salt. It is gently folded for the most part, with some salt domes that are diapiric into the Miocene in places with considerable fault zones.

This foredeep province is bounded to the south by the Sierra de Jatibonico Fault, over 450 km in length, vertical or steeply dipping S. The *Las Villas* zone (Placetas zone of Meyerhoff and Hatten, 1968) is interpreted as an intrageosynclinal welt or median belt. It contains only 900 m of Upper Jurassic and Lower Cretaceous carbonates and siltstones and nothing younger, suggesting that it was a structural high for extended periods. It is intruded by ultramafic and granitic rocks. Its southern boundary is the Las Villas Fault, which can be traced 800 km along the northern side of the island, partly offshore. It is, like the Jatibonico line, vertical to steeply S-dipping, and represents an overthrust of the eugeosynclinal over the miogeosynclinal units (Fig. 2).

In the eugeosynclinal Province, the *Zaza Unit* (Santa Clara zone of Meyerhoff and Hatten, 1968) occupies much of the island, its 7–8 km of deposits ranging from Lower Cretaceous through Miocene in a variety of volcanic, terrigenous, and carbonate facies. Late Cretaceous ultramafic rocks are particularly important on the northern boundary. Tectonic breaks occurred in the Turonian ("Subhercynian phase"), pre-Paleocene, pre-Eocene, mid-Eocene ("Cuban" phase of the Laramide orogeny), pre-Oligocene, and pre-Miocene. Intrusives range from ultramafics to granitic types, mainly Cretaceous and Eocene. Some absolute dating of basement rocks and types of metamorphic rocks from Isla de Pinos (Isle of Pines) and Sierra de la Trinidad suggests late Paleozoic ages and there is evidence here of a Variscan orogenic cycle. Russian workers, however, consider all these rocks to be middle Mesozoic (Khudoley, 1967).

In westernmost Cuba the Zaza zone splits in two, with the *Pinar del Rio unit* representing a structural high with shales, limestones, and intrusive rocks from ultramafics to granitic types. In west-central Cuba the Cochinos zone is a transverse graben containing up to 1000 m of Tertiary sediments and trending N-S.

In the eastern half of the island there are several more depressions of this sort: the *central basin depression* (Camaguey-Las Villas provinces) with terrigenous Upper Cretaceous and Eocene; the *Nipe depression,* with mainly argillaceous sediments of Eocene and Oligocene age; and on the southern coast, the *Ana Maria depression, Cauto depression,* and *Guantanamo depression,* containing Cretaceous and Lower and Upper Tertiary sediments with tuffs, limestones, silts, and shales (Fig. 3). In this southern belt the Cauto tectonic unit and magmatic belt, the igneous activity was mainly Tertiary.

FIGURE 3. Paleogeographic sketch maps of Cuba for Upper Tertiary (Khudoley and Meyerhoff, 1971). Miocene (including Aquitanian) and Pliocene facies complexes of Cuba (from Iturralde-Vinent, 1969). 1, Carbonate rocks, neritic facies; 2, marly rocks, pelagic facies; 3, shale, shallow-water brackish facies; 4, shaly terrigenous rocks; 5, basal conglomerate; 6, land areas; 7, faults. (*Note:* Late Miocene was marked by a worldwide regression.)

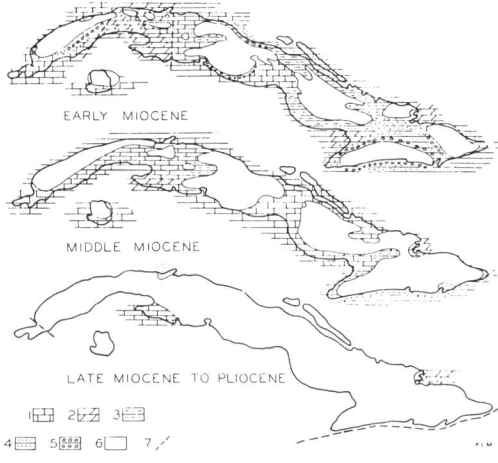

Most observers agree that Cuba originally lay northward of a former land mass that furnished sediments to this complex orthogeosynclinal belt. Although one hypothesis would explain its disappearance by "oceanization," and another by "back-arc basin" extension, the most popular view is that Cuba has migrated during the Tertiary in a NW direction from S of Mexico during the plate tectonic evolution of *Central America* (q.v.) and the Caribbean (see *West Indies*).

On the southeastern coast of Cuba the land is cut off abruptly by the line of the Cayman Trough. The region is seismic and Taber has described the adjacent Sierra Maestra as some of the finest examples of fault-block mountains. The raised coral reef terraces (known as "seboruco" the Jaimanitas Formation) here, especially at Punto Escalereta represent a classic "staircase" system, comparable with those of northern New Guinea (Taber, 1934). The cone karst of western Cuba is also one of the classic geomorphic sites of the world (Meyerhoff, 1938; Lehmann et al., 1956; Acevedo-Gonzales, 1967).

RHODES W. FAIRBRIDGE

References

Acevedo-Gonzales, M. J., 1967. "Classificación general y descripción del Carso Cubana," *Inst. Nacl. Hydr., Publ. Esp.* **4**, 33–64.

Bermudez, P. J., 1950. "Contribution al estudio del Cenozoico cubano," *Mem. Soc. Cubana Hist. Nat.*, **19**, 205–375.

Butterlin, J., 1956. *La Constitution Géologique et la Structure des Antilles*. Paris: Centre Nat. Rech. Sci., 453p.

Corral, J. I., 1939. "La union de Cuba con el continente Americano," *Soc. cubana ing. Rev.*, **33**(7), 581–681.

Daetwyler, C. C., and Kidwell, A. L., 1959. "The Gulf of Batobano, a modern carbonate basin," *Proc. 5th World Petroleum Congr.*, New York, **1**, 1–21.

Darton, N. H., 1926. "Geology of the Guantanamo Basin, Cuba," *Wash. Acad. Sci. J.*, **16**, 324–333.

Ducloz, C., and Vuagnat, M., 1962. "A propos de place des serpentinites de Cuba," *Genève Arch. Sci.*, **15**, 309–332.

Furrazola-Bermudez, G., et al., 1964. "La Habana," in *Geológia de Cuba*, vol. 1. Ed. Consejo Nac. Univer., 239p.

Hatten, C. W., and Meyerhoff, A. A., 1970. "The Caribbean area: a case of destruction and regeneration of a continent, discussion," *Bull. Geol. Soc. Am.*, **81**, 1855–1862.

Itturalde-Vinent, M. A., 1972. "Principal characteristics of Oligocene and lower Miocene stratigraphy of Cuba," *Bull. Am. Assoc. Petrol. Geologists*, **56**(12), 2369–2379.

Khudoley, K. M., 1967. "Principal features of Cuban geology," *Bull. Am. Assoc. Petrol. Geologists*, **51**(5), 668–677.

_____ and Meyerhoff, A. A., 1971. "Paleogeography and geological history of Greater Antilles," *Geol. Soc. Am. Mem.* **129**, 199p.

Kozary, M., 1954. "Conglomerates associated with the Cubitas Plateau, Cuba," Ph.D. thesis, Columbia Univ., 168p.

_____, 1956. "Ultramafics in the thrust zones in northwestern Oriente, Cuba," *20th Intern. Geol. Cong., Resumenes*, 138–139.

_____, 1968. "Ultramafic rocks in thrust zones of northwestern Oriente province Cuba," *Bull. Am. Assoc. Petrol. Geologists*, **52**(12), 2298–2317.

Kuhlmann, D. H. H., 1970. "Die Korallenriffe Kubas. 1. Genese und Evolution," *Intern. Rev. Hydrobiol.*, **55**(5), 729–756.

Lehmann, H., Krommelbein, K., and Lotschert, W., 1956. "Karstmorphologische, geologische und botanische Studien in der Sierra de los Organos auf Cuba," *Erdkunde*, **10**(3), 185–203.

Lewis, J. W., 1932. "Geology of Cuba," *Bull. Am. Assoc. Petrol. Geologists*, **16**(6), 533–555.

MacGillavry, H. J., 1970. "Geological history of the Caribbean," *Proc. Kon. Ned. Akad. Wetensch.* (Amsterdam), Ser. B, **73**(1), 64–96.

Meyerhoff, H. A., 1938. "Texture of karst topography in Cuba and Puerto Rico," *J. Geomorphol.*, **1**, 279–295.

_____ and Hatten, C. W., 1968. "Diapiric structures in central Cuba," in J. Braustein and G. D. Brien, eds., *Diapirism and Diapirs* (Mem. 8). Tulsa: Am. Assoc. Petrol. Geologists, 315–357.

_____, et al., 1969. "Geologic significance of radiometric dates from Cuba," *Bull. Am. Assoc. Petrol. Geologists*, **53**(12), 2494–2500.

Mitchell, R. C., 1955. "The ages of the serpentinized

peridotites of the West Indies," *Proc. Kon. Ned. Akad. Wetensch. (Amsterdam)*, Sr. B, **3,** 194–212. 194–212.

Núñez-Jimenez, A., *et al.,* eds., 1962. *Mapa Geológico de Cuba.* La Habana: Minist. Indust., Inst. Cub. Rec. Min., 1:1,000,000.

Palmer, R. H., 1945. "Outline of the geology of Cuba," *J. Geol.,* **53**(1), 1–34.

Rigassi-Studer, D., 1961. "Quelques vues nouvelles sur la géologie cubaine," *Chron. Mines et Rech. Minière,* **29,** 3–7.

Rutten, L., 1940. "On the age of the serpentines in Cuba," *Proc. Ned. Akad. Wetensch. (Amsterdam),* **43,** 542–547.

Škvor, V., 1969. "The Caribbean area: a case of destruction and regeneration of a continent," *Bull. Geol. Soc. Am.,* **80**(6), 961–968 (with discussion by Hatten, C. W., and Meyerhoff, A. A., 1970, *Ibid.,* **81,** 1855–1862).

Taber, S., 1931. "Structure of Sierra Maestra, Cuba," *J. Geol.,* **39,** 532–557.

_____, 1934. "Sierra Maestra of Cuba, part of the northern rim of the Bartlett Trough," *Bull. Geol. Soc. Am.,* **45,** 567–620.

Cross-references: *Bahamas; Guatemala; Haiti; Jamaica; Mexico; United States–Gulf Coastal Province; West Indies.*

D

DOMINICA

A British Associated State in the West Indies, Dominica is the largest and northernmost of the British Windward Islands. (It was administratively transferred from the Leeward Islands in 1940.) It was discovered by Columbus in 1493. Its area is 790 km² (290 sq mi), which makes it also the third largest of the Lesser Antilles. The island measures 47 by 24 km. Situated S of *Guadeloupe* (q.v.) and N of *Martinique* (q.v.), it is one of the inner volcanic arc of the eastern Caribbean belt. Its highest elevations are the inactive volcano of Morne Diablotin (1421 m) in the N, with Morne Trois Pitons (1427 m), and the active cone of Grand Soufrière (729 m) in the S.

Dominica lies in the belt of the northeast trade winds, and in August and September there is usually very heavy rainfall. Precipitation varies greatly in different parts of the island, the southern (lee side) being healthiest. Sharford (at 170 m) receives 4700 mm per year but the average is about 2000 mm. Temperatures range from maxima of 36°C down to minima of 18°C.

Most of the mountainous and forested slopes are deeply dissected into gorges, with torrential streams, waterfalls, and lakes. The coasts are steeply cliffed and offshore drop away abruptly. The 500-m isobath is only 1 km or so from the coast.

The rocks are predominantly andesitic but tend to be enriched in residual trace elements and radiogenic isotopes compared with those from other islands in the Lesser Antilles. A partly welded pyroclast flow (Sigurdsson, 1972) is unique in the Lesser Antilles. Volcanic activity has been very subdued in historic times. No truly magmatic eruptions have been recorded, although a steam explosion occurred in 1880 at Grand Soufrière (Robson and Tomblin, 1966). Present-day activity is confined to four main solfataric areas, including the famous Boiling Lake at Grand Soufrière.

RHODES W. FAIRBRIDGE

References

Davis, W. M., 1926. *The Lesser Antilles* (Spec. Publ. 2). New York: Am. Geogr. Soc., 207p.

Earle, K. W., 1928. "Geological notes on the island of Dominica," *Geol. Mag.,* **65,** 169–187.

Hodge, W., 1943. "The vegetation of Dominica," *Geogr. Rev.,* **33,** 349–375.

Hovey, E. O., 1904. "Boiling Lake of Dominica," *Bull. Geol. Soc. Am.,* **16,** 570.

Martin-Kaye, P. H. A., 1969. "A summary of the geology of the Lesser Antilles," *Overseas Geol. Mineral Res.,* **10**(2), 172–206.

Robson, G. R., and Tomblin, J. F., 1966. *Catalogue of the Active Volcanoes of the World Including Solfataric Fields.* Naples: Intern. Assoc. Volcanol., pt. 20, 56p.

Sapper, K., 1903. "Besuch der Insel Dominica," *Centralbl. Geol. Min.,* 305–314.

Sigurdsson, H., 1972. "Partly-welded pyroclast flow deposits in Dominica, Lesser Antilles," *Bull. Volcanol.,* **36,** 148–163.

Spencer, J. W., 1902. "Geological and physical development of Dominica," *Quart. J. Geol. Soc.,* **58,** 345–353.

Weyl, R. 1966. *Geologie der Antillen.* Berlin: Gebr. Borntraeger, 410p.

Cross-references: *Guadeloupe; Leeward Islands; Martinique; West Indies; Windward Islands.*

DOMINICAN REPUBLIC

Occupying the eastern two-thirds of the island of Hispaniola in the West Indies with an area of 48,735 km² (18,816 sq mi), the Dominican Republic is bordered on the N by the Atlantic Ocean, on the S by the Caribbean, and on the E by the Mona Passage, which separates the Dominican Republic from Puerto Rico. Haiti, on the W, occupies the other third of Hispaniola. This island is the second largest of the Greater Antilles and lies between Cuba on the W and Puerto Rico on the E (Mona Passage). It was discovered by Columbus in 1492.

Four wooded mountain ranges cross the Dominican Republic from NW to SE, with two principal ones. The Cordillera Central is the highest and longest and here Pico Duarte rises to a height of 3175 m (10,477 ft), the highest peak in the West Indies. It is separated from the Cordillera Septentrional by the valley of La Vega Real, occupied by the Cibao River, one of the most fertile belts. In the SW of the Dominican Republic there are two short mountain ranges separated by a graben which encloses

Lake Enriquillo, a saline body of water 50 m (163 ft) below sea level.

The climate is tempered by ocean breezes and the temperature rarely goes above 32°C (90°F). The trade winds from the NE bring moisture to the windward side of the mountains but to the W and S are dry with deserts. Rainy seasons are May–June and September–November. Coolest months are November–March.

Bauxite, iron ore, salt, gypsum, and nickel are mined. The iron occurs in contact deposits at Hatillo, at the northern foot of the Cordillera Central, with reserves of 45 million tons of 67% iron. Similar deposits occur on the south side of the Cordillera de Monte Cristi at Sabana Grande. Lateritic nickel is associated with serpentines on the southern side of the Cordillera Central (Falconbridge Dominicana mining complex near Bonao). Extensive bauxite contains 4 million tons of reserves. In the Serros de Sal of the Cordillera de Neiba there is one of the world's greatest salt deposits. It is in the Lower Miocene evaporites and contains 400 million tons of halite reserves. There is also gypsum associated with it. Hydropower (60,000 kw) is provided by the Valdesia dam on the Nizao River, and others are planned on this river. A second important source is the Tavera dam on the Bao River (80,000 kw).

Stratigraphy

The key geologic elements of the eastern part of Hispaniola were established by Gabb (1873), who observed (1) a highly tectonized core or basement of Cretaceous age, (2) flanking belts of Tertiary sedimentary formations, and (3) a post-Pliocene overlay of limestone and gravel which he called "antillite" or the "Coast Formation." Vaughan et al (1921) discovered that this last actually included fossiliferous Miocene and that the Quaternary formations were only thin veneers. The formation of surface travertine duricrusts (calcrete) on the calcareous rocks of all ages makes it difficult, in reconnaissance surveys, to distinguish them (Barrett, 1962). Butterlin (1956), with experience in *Haiti* (q.v.), showed a high degree of similarity with the western part of the island, and there are also clearly close affinities with *Puerto Rico* (q.v.).

The oldest fossiliferous rocks have been established as Aptian-Albian in age (Koschmann and Gordon, 1950; Bowin, 1966). Butterlin believes that at this stage the Cuban geosyncline extended through Hispaniola, and sedimentation is characterized by spilites and keratophyres, interbedded with silicified limestones and radiolarites, clearly a deep-water facies (*Duarte Formation, Maimon Formation:* Bowin, 1966). These rocks are metamorphosed (greenschist facies) and covered unconformably by submarine basaltic and andesitic rocks (*Siete Cabezas Formation, Peravillo Formation:* Bowin, 1966). Upward toward the top of the Cretaceous there is a gradual appearance of pelagic carbonates (*Macaya Formation* of Haiti). About the beginning of the Tertiary there was a Laramide orogenic phase, with E-W to WNW-ESE folds.

Early Tertiary was transgressive again with clastics to start with, passing up into carbonates, with submarine basalts and tuffs here and there. Calcarenites are important in the *Hidalgo Limestone.* Upper Eocene was regressive in connection with upper Middle Eocene orogenic activity, but sedimentation returned in the early Oligocene, bringing marls and limestones in the *Sombrerito Formation* in the lower part, *Cevicos* and *Florentino* formations in the middle.

Heavy folding began in late Oligocene and early Miocene, leading to flysch development. Further uplift, but with synchronous basin development, produced molasse facies (*Yague Group*). The Dominican Miocene is subdivided into three: *Cercado Formation* in the Lower Miocene with conglomerates, sandstones, and siltstones; the *Gurabo Formation* in the Middle or Upper Miocene (Bermudez and Seiglie, 1970) with sandy limestones and marl; and the Lower Pliocene *Mao Formation* at the top, with sandstones and limestones.

In the Pliocene there was general emergence, marked by peneplanation, locally passing to littoral formations with conglomerates and reefs. Block faulting caused irregular uplift in the late Tertiary and early Quaternary.

The Pleistocene is characterized by the widespread presence of sheet gravels, commonly 1–2 m thick, in places with mottled clays, covering the coastal plain belt below about 120 m and sloping gently seaward, and also thinning in that direction. These are interpreted as regression facies of glacial stages. They are superimposed by multiple marine terraces—with sea caves and raised beach and coral reef deposits—up to 82 m elevation, suggesting interaction between glacioeustatic oscillations and uplift (Barrett, 1962). Barrett mapped the terraces E and W of Ciudad Trujillo from field surveys with the aid of airphotos and a U.S. Army Map topographic series of the coastal area on 1:50,000 scale. The Instituto Geografico Militar also began a 1:25,000 series.

Coastal stream valleys display similar Pleistocene terracing cut into reddish-brown alluvial fill and older clay or limestones. The valleys are deepened far below sea level, and there are submerged terraces along the coastal cliffs as well as drowned offshore reefs. The Rio La Romana is a ria 100 m in depth at its mouth, dropping away steeply into a submarine canyon with 1400 m depth only 300 m offshore. Further evidence of Pleistocene aridity and regression is in the form of extensive calcareous eolianites, notably at Cabesota de Barlovento, which is presumably Wisconsin in age inasmuch as it shows none of the early Pleistocene higher sea-level marks, except for evidence of small Holocene oscillations. The cross-bedding shows a dominant northeast trade wind at the time.

Quaternary limburgites similar to related rocks in the Republic of Haiti (nepheline basalts) occur in the San Juan valley and the region of Constanza (Macdonald and Melson, 1969). Donnelly (in Anon., 1971) considers that they are related to a Benioff zone below the northeastern part of Hispaniola Island.

JACQUES BUTTERLIN
RHODES W. FAIRBRIDGE

References

Anon., 1971. *6th Conf. Geol. del Caribe, Resúmenes*, 41p. (mimeograph with suppl.).

Barrett, W., 1962. "Emerged and submerged shorelines of the Dominican Republic," *Revista Geogr.* (Inst. Pan-Am. Geogr. Hist., Brazil), 30 (56), 51–77.

Bermudez, P. J., 1949. "Tertiary smaller Foraminifera of the Dominican Republic," *Cushm. Lab. Foram. Res. Spec. Publ.* 25, 322p.

____, and Seiglie, G. A., 1970. "Age, paleoecology, correlation and Foraminifers of the uppermost Tertiary formation of northern Puerto Rico," *Caribbean J. Sci.*, 10(1,2), 17–24.

Bowin, C. O., 1966. "Geology of central Dominican Republic (a case history of part of an island arc)," *Geol. Soc. Am. Mem.* 98, 83p.

Butterlin, J., 1956: *La constitution géologique et la structure des Antilles.* Paris: Centre Nat. Rech. Sci., 453p.

Gabb, W. M., 1873. "On the topography and geology of Santo-Domingo," *Trans. Am. Phil. Soc. Phila.*, 15, 49–259.

Koschmann, D. G., and Gordon, M., Jr., 1950. "Geology and mineral resources of the Maimon Hatillo district, Dominican Republic," *U.S. Geol. Surv. Bull.*, 964-D, 307–359.

Macdonald, W. D., and Melson, W. G., 1969. "A late Cenozoic volcanic province in Hispaniola," *Caribbean J. Sci.*, 9(3–4), 81–91.

Nagle, F., 1966. *Geology of the Puerto Plata Area, Dominican Republic.* Nat. Sci. Found., Dept. Geol., Grant G 14217, 171p.

____, 1971. "Geology of the Puerto Plata Area, Dominican Republic, relative to the Puerto Rico trench," *Trans. Caribbean Geol. Conf.*, 5, 79–84.

____, 1972. "Chaotic sedimentation in north-central Dominican Republic," *Geol. Soc. Am. Mem.* 132, 415–428.

Palmer, H. C., 1963. *Geology of the Moncion–Jarabacoa Area, Dominican Republic.* Princeton, N.J.: Princeton Univ., Dept Geol., 256p.

Vaughan, T. W., et al., 1921. *A Geological Reconnaissance of the Dominican Republic.* Geol. Surv. Dom. Repub., 268p.

Cross-references: *Vol. I, Caribbean Sea; Cuba; Haiti; Puerto Rico; West Indies.*

E

EASTER ISLAND AND SALA Y GÓMEZ

Easter Island (Isla de Pascua), or to use its Polynesian names, Rapa Nui (Large Island) or Te Pita Te Henua (Navel of the World), is the most isolated island in the Pacific and one of the most isolated islands in the world. It lies at 27°8'5W, some 3700 km (2300 miles) W of the coast of Chile and 2500 km (1400 miles) E of the nearest inhabited island (Pitcairn). The island is administered by the Republic of Chile, although the inhabitants are basically Polynesian. Until very recently, it was visited only once a year by a supply ship from Chile; the trip took 9–10 days from Valparaiso. At the present time there is regular air service from Santiago, Chile.

The area is about 130 km² (50 sq mi). Its climate is subtropical. There is a scarcity of freshwater since the island has no streams and there is dependence on the moderate rainfall that collects in the extinct volcanic crater lake of Rano Raraku.

The island is of volcanic origin and owes its triangular shape to extinct volcanic cores that form its three corners. The highest, Terevaka, in the north, reaches 556 m (1767 ft). Poike or Katiki, on the east, was formerly a separate island surrounded by cliffs. The southerly of these volcanoes (Rano Raraku) was the main source of the material used in making the celebrated stone statues. The island is principally ash, but there are numerous lava flows, notably on the southern side of Terevaka, where there are also more than 20 parasitic tuff cones in radical lines, one of which contains obsidian. There is, however, no record of volcanic activity within recorded time. Detailed studies on the geology and petrology have been made by Bandy (1937) and Baker (1967). The lavas are comparable to those of the Galápagos, Pitcairn, and the Marquesas in being rather siliceous and nepheline free; they include oligoclase-andesite. Potassium argon determinations suggest an age of about 3 m yr for the exposed portion of Poike and 300,000 years for Terevaka (Baker, 1967). Sea-floor spreading data suggest a Pliocene age for the oceanic crust here.

The island has become famous because of the gigantic stone statues carved from the native volcanic rock by the early inhabitants. Who these inhabitants were and where they came from has been debated by archeologists and historians for many years. There are two main theories. One, as expressed by Thor Heyerdahl in the popular volume *Aku Aku,* is that the island was settled by paleo-Indians from South America; the other view, more widely held, is that the island was colonized by Polynesians, most likely coming from the Marquesas, some 1800 miles to the NW.

Much has been written on the archeology. A selected list of references would include Heyerdahl (1958, 1968), Heyerdahl and Ferdon (1961), Metraux (1940), and a recent excellent summary by Father Sebastian Englert (1970) edited by William Mulloy.

There have been different opinions regarding the dates of the culture of the island. Until recently, the earliest generally accepted date has been about 857 A.D., although it was realized that there were probably people on the island before then. Ayres (1971) has obtained some radiocarbon dates indicating that the time range for the use of the island as a ceremonial center was from before 690 ± 130 A.D. to the late 1800s. A date of about 400 A.D. for the initial settlement of the island is suggested.

The shores of Easter Island are cliffed, ranging from 20–300 m, but no evidence of terracing is reported. There are no coral reefs, the winter seawater temperature being too low.

Sala y Gómez is administered by Chile along with Easter Island. The name is that of its discoverer, Sala y Gómez (1793) and a poem was written about it by A. von Chamisso. The only recent note on it is by Fisher and Norris (1960). It is uninhabited, covering only 0.12 km², and reaching an elevation of 30 m. There are two rocky points joined by a low isthmus. It is situated about 26°28'S, 105°28'W. It represents the peak of an isolated volcanic seamount. There is a submarine terrace with a shelf break at 130 m which is 2–3 km offshore. It is situated on a ridge, marked by a line of submerged seamounts extending over 1500 km E of Easter Island and may reach *San Ambrosio and San Felix* (q.v.), which follows the southern border of the Easter Island Fracture Zone.

HORACE G. RICHARDS

References

Ayres, W. S., 1971. "Radiocarbon dates from Easter Island, East Polynesia," *J. Polynesian Soc.,* **5.**
Baker, P. E., 1967. "Preliminary account of recent geological investigations on Easter Island," *Geol. Mag.,* **104,** 116–122.
Bandy, M. C., 1937. "Geology and petrology of Easter Island," *Bull. Geol. Soc. Am.,* **48,** 1589–1610.
Chubb, L. J., 1933. "Geology of Galápagos, Cocos, and Easter Island," *Bernice P. Bishop Mus. Bull.,* **110,** 1066.
Englert, S., 1970. *Island at the Center of the World.* New York: Scribner's 191p., trans. and ed. by W. Mulloy.
Falke, H., 1941. "Die Insel Sala y Gómez," *Natur Volk,* **71**(3), 146–150.
Fisher, R. L., 1958. *Preliminary Report on Expedition Downwind, University of California, S.I.O.* (Gen. Rept. Ser. 2). Washington, D.C.: IGY.
____, and Norris, R. M., 1960. "Bathymetry and geology of Sala y Gómez, Southeast Pacific," *Bull. Geol. Soc. Am.,* **71,** 497–502.
Herron, E. M., 1972. "Two small crustal plates in the South Pacific near Easter Island," *Nature Phys. Sci.,* **240**(98), 35–37.
Heyerdahl, T., 1958, *Aku-Aku.* Skokie, Ill.: Rand McNally (also numerous reprints).
____, 1968. "The prehistoric culture of Easter Island," in I. Yawata and Y. H. Sinoto, eds., *Prehistoric Culture in Oceania.* Honolulu: Bishop Mus. Press, 133–140.
____, and Ferdon, E., eds., 1961. *Archaeology of Easter Island* (Monogr. School of Am. Res. and Mus. of N. Mex. 24). Stockholm: Forum Publ. House, vol. 1, 559p.
Lacroix, A., 1939. "Composition mineralogique des roches volcaniques d l'Ile de Paques," *C. R. Acad. Sci.,* **202,** 527.
Luke, H., 1954. "Easter Island," *Geogr. J.,* **120,** 422.
Metraux, A., 1940. "Ethnology of Easter Island," *Bernice P. Bishop Mus. Bull.,* **160.**

Cross-references: *Chile; Galápagos Islands; Juan Fernández; Marquesas Islands; Oceania; Pitcairn Islands; San Ambrosio and San Félix Islands; Society Islands; United States–Hawaii.*

ECUADOR

The Republic of Ecuador straddles the equator on the Pacific coast of South America and is bordered by Colombia to the north and Peru to the south and east. The country, which covers an area of 263,777 km² (109,483 sq mi), is made up of three physiographic provinces: the coastal lowlands, the Andean mountain chains, and the Oriente province in the Amazon headwaters.

Geologically, the country extends from the oceanic province of the Pacific, across the Andean mobile belt to the margins of the Precambrian Guayana Shield in the east. It forms the southern part of the Northern Andes, a fold belt that swings northward through Colombia and Venezuela to connect with the Antillean arc. An important saddle—the Marañón Portal, corresponding with the Huancabamba deflection in northern Peru—separates the Northern Andes from the remainder of the chain.

After the early work in the last century by Alexander von Humboldt and pioneers such as Bouguer and Condamine, geological investigations in Ecuador were generally neglected. The coastal and Oriente provinces have been investigated in moderate detail through exploration for oil, but the Andes are still known only in the most general terms; and large areas, especially in the southeastern part of these ranges, are still virtually unexplored.

Major Geological Provinces

Ecuador is divisible into two major provinces (Fig. 1). In the Eastern province a basement of Precambrian igneous and metamorphic rocks belonging to the Guayana Shield is overlain by a relatively complete sequence of Phanerozoic sediments. In the Western province, on the other hand, the basement appears to be represented by a Mesozoic sequence of eugeosynclinal and oceanic rocks, capped by a comparatively thin Tertiary succession. The boundary between the two provinces is drawn along a postulated megashear, comprising the Guayaquil and Dolores faults (Campbell, 1968, p. 260), which may ultimately connect with the Oca-Pilar faults to make up a major dextral shear zone bordering the South American continent. A possible crustal model, discussed further in the section on structure, is illustrated in Fig. 1.

Geological Evolution

A definitive stratigraphic classification for Ecuador has not yet been worked out, and many of the formation terms in common use are informal and lack proper definition or type section. Furthermore the age range of many of them is still subject to discussion. Volume V of the *Lexique Stratigraphique International* is the standard reference for stratigraphic nomenclature in Ecuador. A stratigraphic diagram is given in Fig. 2.

Precambrian. Precambrian igneous and metamorphic rocks forming the western margin of the Guayana Shield lie at a depth of about 1000 m below a cover of continental sediments on the eastern margin of the Oriente Basin.

FIGURE 1. (*Top*) major geological provinces and fault zones of Ecuador and (*bottom*) a postulated crustal model.

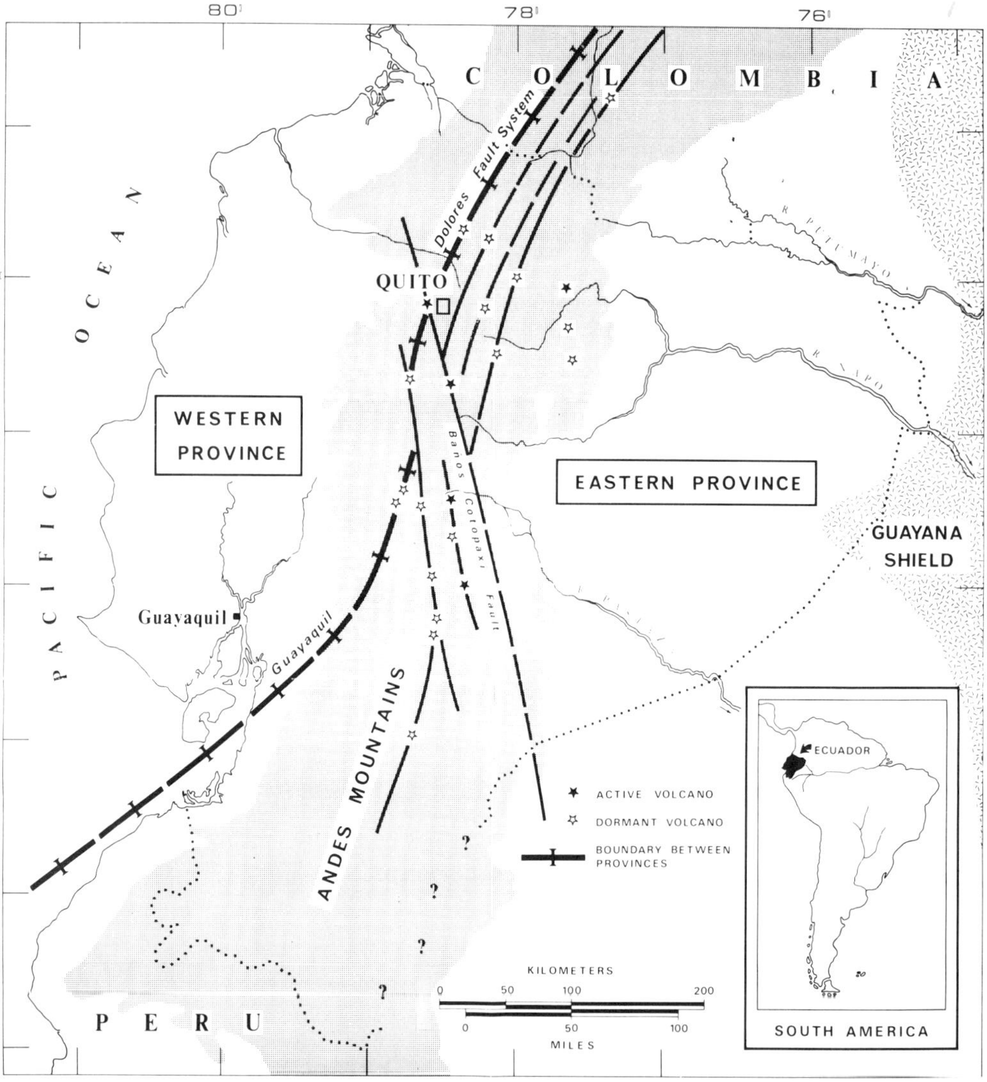

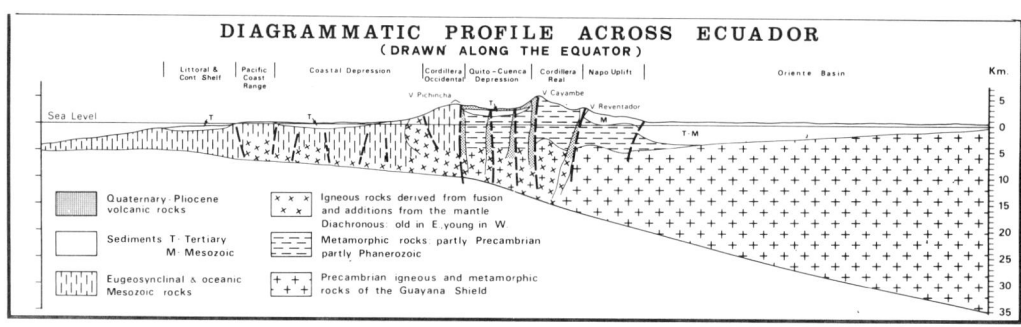

FIGURE 2. Stratigraphic diagram.

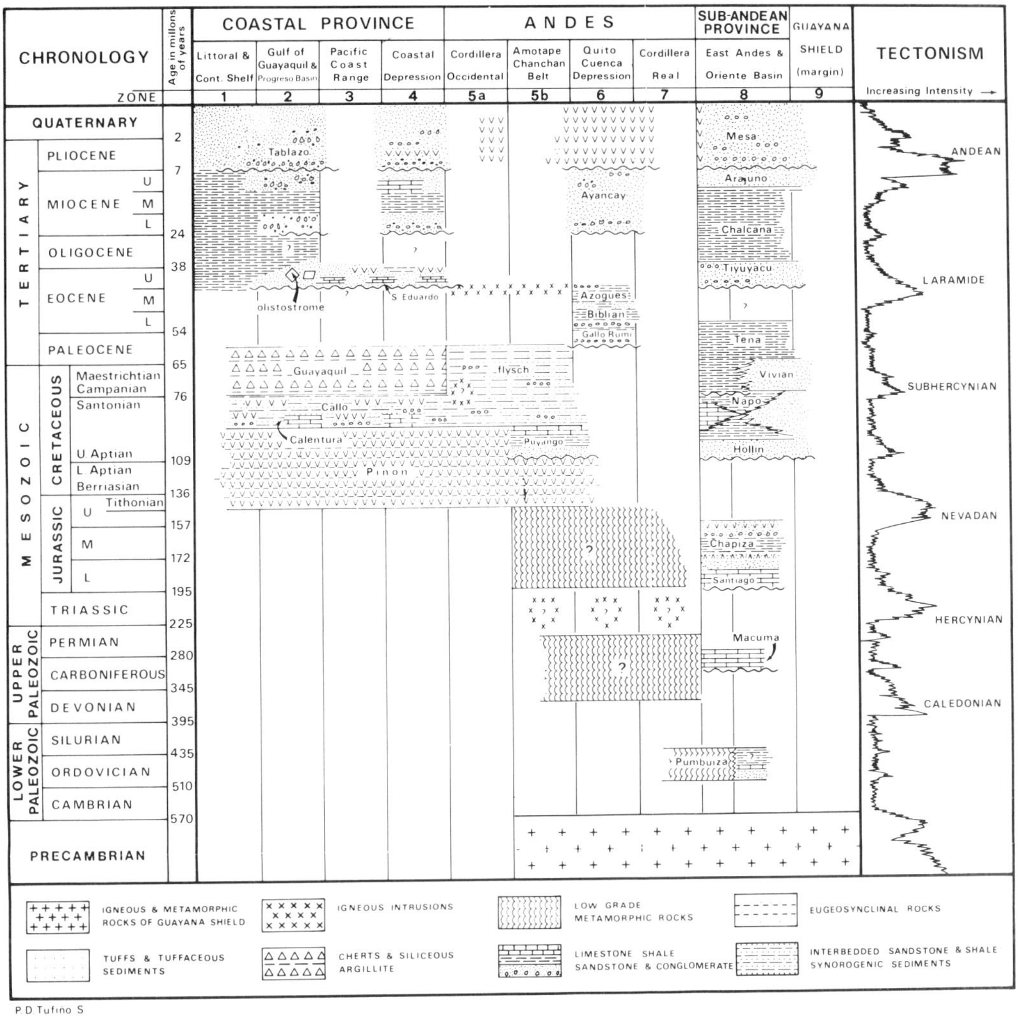

In the absence of radiometric age determinations it is difficult to determine the extent to which Precambrian rocks occur in the Andes themselves. The metamorphic rocks of the Cordillera Real may be partly of Precambrian age, although younger developments probably predominate. Gneissic rocks, however, are found in part of the Amotape-Chanchan belt, especially in the cordilleras east of Machalá. They are tentatively ascribed to the Precambrian, since they are evidently older than the low-grade metamorphic rocks in the vicinity that correlate with the Carboniferous *Amotape Formation* of Peru (q.v.).

Lower Paleozoic. Lower Paleozoic rocks occur in isolated localities along the eastern margin of the Andes from Colombia to Peru and have also been penetrated in boreholes to the east. The relationships suggest that a miogeosyncline followed the approximate course of the present Eastern Andes, whereas a shelf paralleled the margins of the Guayana Shield. The geosynclinal province was deformed and subjected to low-grade metamorphism during Middle Paleozoic ("Caledonian") earth movements, whereas the shelf province was not so affected. Fossils have been obtained from the shelf but the age of the metamorphic sequences is based on inference and stratigraphic position. Such paleontological evidence as is available suggests that the geosyncline reached its greatest extent during the Ordovician, for neither Cambrian nor Silurian rocks have been identified.

In Ecuador the Lower Paleozoic is represented by the *Pumbuiza Formation,* a sequence of dark, partly graphitic slates and quartzitic sandstones cropping out in the Serranía de

Cutucú (Tschopp, 1953, p. 2310). Other low-grade metamorphic rocks along the eastern margin of the Cordillera Real may also be of this age. It is possible that unmetamorphosed Lower Paleozoic rocks occur at depth in the Oriente, for such rocks have been found in the Marañon 110-1 borehole in eastern Peru.

Middle Paleozoic (Caledonian) Orogeny. The Lower Paleozoic cycle of sedimentation was brought to a close by important earth movements during the early Devonian, which may be classified, as far as their age is concerned, as Caledonian. The orogeny caused the deformation and metamorphism of the Lower Paleozoic geosynclinal sequence and was followed by a period of sustained regional uplift.

Upper Paleozoic. In Colombia and in other parts of the Andean region, a new cycle of sedimentation began in the Middle Devonian, and it is possible that rocks of this age may be present at depth in the Oriente Basin of Ecuador. In outcrops the Lower Paleozoic succession is overlain directly by a Pennsylvanian sequence of siliceous limestones, shales, and minor calcareous sandstones, forming the *Macuma Formation* (Tschopp, 1953, p. 2310). The type section is in the Serrania de Cutucú in the Andean foothills, but the formation almost certainly extends eastward at depth.

Whereas a shelf environment continued in the eastern part of the country, geosynclinal conditions may have prevailed in the Andean province, for a general westerly migration of structural events has been noted in the Colombian extensions of the chain. Consequently, undated low-grade metamorphic rocks in the Cordillera Real and the Amotape-Chanchan belt may represent Upper Paleozoic geosynclinal deposits. Support for this hypothesis comes from the fact that low-grade metamorphic rocks in the Cordillera Tahuin apparently connect with the *Amotape Formation* of Peru that has yielded Pennsylvanian fossils.

Late Paleozoic (Hercynian) Orogeny. Renewed orogenic activity returned to the Andean region at the end of the Paleozoic era and persisted into the early Triassic. As discussed above, these movements may have caused the metamorphism of the Upper Paleozoic geosynclinal rocks in the Andes. The movements were accompanied by widespread igneous activity in the Andean region characterized by the emplacement of alkali granites. Granites of this composition on the eastern margin of the Cordillera Real, as, for example, that near the town of Baños, may thus be tentatively ascribed to the Triassic.

Early Mesozoic. *Upper Triassic and Lower Liassic.* Renewed subsidence in the late Triassic led to a marine transgression represented by a sequence of limestones cropping out in northeastern Peru. These beds are in turn overlain by a Lower Liassic succession of dark siliceous limestones, calcareous sandstones, and black shales with some volcanic intercalations, comprising the *Santiago Formation,* which extends into eastern Ecuador (Tschopp, 1953, p. 2312).

Lower and Middle Jurassic. After that brief transgression the sea again retreated from the east Andean region and a succession of red beds were deposited there during the succeeding early and middle Jurassic periods. These sediments, which make up the *Chapiza Formation,* consist of reddish-colored shales, arkosic, and partly tuffaceous sandstones and conglomerates, capped in certain areas by lavas and breccias. The formation was evidently deposited under arid continental conditions, providing an environment suitable for the deposition of evaporites; in Ecuador such developments are limited to thin beds of anhydrite in the lower part of the succession, but in neighboring Peru strata assigned to the Chapiza include diapirs containing salt. As it is unfossiliferous, the age of the Chapiza Formation is still subject to discussion; some authors consider that it may range into the Cretaceous.

In the Western Andes of Colombia low-grade metamorphic rocks, comprising the *Dagua Group,* are thought to represent Lower Mesozoic geosynclinal rocks metamorphosed by Nevadan earth movements, and it is possible that a similar development may be present in the Andes of Ecuador, although it has not been distinguished from other metamorphic sequences.

Late Mesozoic. The late Mesozoic was characterized by the development of geosynclines in the Andean region generally. In Ecuador the Cordillera Real formed a welt that separated a eugeosynclinal province in the west from a miogeosynclinal province in the east.

Eugeosynclinal Province. Late Mesozoic rocks of the eugeosynclinal province are widely exposed in the Western Andes, the Amotape-Chanchan belt, and the coastal region. The succession is at least 5000 m thick. Volcanic rocks, including submarine lavas, predominate in the lower part, whereas siliceous sediments with minor volcanic intercalations characterize the upper part of the sequence.

In the coastal region the succession is divisible into two main groups: the *Piñon Group* below and the *Callo Group* above. The Piñon Group consists of over 1000 m of porphyritic basic lavas, spilites, tuffs, and agglomerates with which are interbedded minor intercalations of tuffaceous sandstone and siliceous argillite. The overlying Callo Group consists of up to 3000 m of sandstone, shale, conglomerate, chert, tuff, and volcanic agglomerate. A basal siliceous limestone, known as the *Calentura Limestone*, is developed in certain areas. Cherts with interbedded shales and minor tuffs are abundant at the top of the succession and are distinguished as the *Guayaquil Formation*.

Different conditions obtain in the Andean region, for the equivalent of the upper part of the Callo formation is represented by a succession of synorogenic sediments made up of sandstones, conglomerates, and shales. This development may be due to the effects of the Subhercynian orogenic movements (discussed below).

In general the eugeosynclinal sediments are poorly fossiliferous and the age ranges of the various formations, as indicated in Fig. 2, are uncertain.

Miogeosynclinal Province. The east Andean Cretaceous miogeosyncline borders the Guayana Shield from Venezuela to Peru. In parts of Colombia and Peru subsidence commenced in the Tithonian (latest Jurassic), whereas other regions, including eastern Ecuador, were not submerged until mid-Cretaceous times, when the geosyncline reached its greatest extent.

In Ecuador the first deposits of this new cycle of sedimentation consist of marine to epicontinental quartzose sandstones with minor interbedded carbonaceous shales and thin coals, forming the *Hollin Formation*. These beds, which attain a thickness of about 200 m, are exposed in the Andean foothills and occur at depth throughout the Oriente Basin. Albian fossils have been obtained from the top of the formations, and the base is ascribed to the Upper Aptian on palynological evidence.

As subsidence continued and the relief of the source areas was reduced by erosion, progressively finer grained deposits were laid down in the miogeosyncline during the succeeding Cenomanian, Turonian, Coniacian, and Santonian stages. This interval is represented by the *Napo Formation*. In the Andean foothills it consists of up to 400 m of euxinic limestones and shales with minor interbedded glauconitic sandstones. The proportion of sandstone increases eastward across the Oriente Basin at the expense of the limestones and shale.

Subhercynian Movements. Important orogenic movements in Santonian times brought the preceding geosynclinal phase of sedimentation to a close. The movements were strongest in the Andean region, where they led to the deposition of synorogenic sediments discussed previously. In the east Andean region they caused regional uplift, the withdrawal of the sea from the Oriente Basin, and locally the erosion of the upper part of the Napo Formation. The movements were probably also indirectly responsible for the change in sedimentary environment in the coastal region mentioned previously. Elsewhere in the Andes the movements were associated with the emplacement of batholiths, the Peruvian batholith and the Antioquia batholith of Colombia, and it is thus possible that some of the undated intrusions in the Ecuadorian Andes may be of this age.

Cretaceous-Tertiary Transition. In the E the Subhercynian movements uplifted provenance areas in the Guayana Shield, which in turn led to a return to the deposition of coarse clastic sediments in the Oriente Basin. In the Peruvian and Colombian portions of the basin these beds are represented, respectively, by the *Vivian* and *Guadalupe Formations*. Equivalent beds are lacking in the Ecuadorian foothills, but the results of drilling indicate that they occur in the subsurface in the eastern part of the basin.

These sandstones are overlain by a succession of mainly reddish partly siliceous and partly carbonaceous clays, thin limestones, and minor sandstones making up the *Tena Formation* (Tschopp, 1953, p. 2325). The sequence was deposited under paralic conditions that marked a transition from the marine conditions of the Cretaceous to the nonmarine conditions of the Tertiary. The age range of the formation is uncertain, but it is tentatively thought to represent Campanian to Paleocene or perhaps early Eocene.

In the coastal region the sediments deposited during this interval were largely removed by erosion during the succeeding Laramide orogeny.

Laramide Orogeny. The onset of Laramide orogenic movements during the Eocene caused important paleogeographic changes in the Andean region generally. In Ecuador the Andean ranges were uplifted and replaced the Guayana Shield as the principal source of sediment. East of these ranges lay a wide depression drained by rivers that united to flow through a gap in the Andes in northern Peru, known as the *Marañón Portal*, and to debouch into the Pacific from a delta that built up in the Gulf of Guayaquil and adjoining areas to the south. The coastal basin subsided and was filled with mixed clastic and tuffaceous sediment. It was flanked by an uplift forming the Pacific Coast Range, an unstable element affected by faulting which not only provided lines of weakness by which igneous rocks reached the surface but also led to the development of widespread submarine slumping in the adjacent basins.

Furthermore, the Laramide movements caused the development of major fault systems, including the Guayaquil-Dolores megashear discussed previously, and were associated with widespread igneous activity in the Andes themselves.

In summary, then, the Laramide orogeny provided the framework for the present structure of the country. In Peru and the southern part of the continent emphasis is given to these movements, which are sometimes designated as "Andean," a term restricted here to the late Tertiary movements.

Tertiary of the Eastern Province (Eocene–Miocene). The Laramide orogeny heralded a new phase of nonmarine sedimentation in the east Andean region. It opened with the deposition of conglomerates and coarse sandstones (the *Tiyuyacu Formation* and equivalents) derived from the Laramide chains; but as the source areas were reduced by erosion, these deposits became progressively finer grained. The sediments, which consist principally of red clays and silts, were mainly laid down under fluviatile conditions, although large lakes with some connection with the open sea through the Marañón Portal developed periodically, being represented by strata bearing brackish-water microfossils. Many local formations have been established to describe this sequence, but they do not warrant detailed consideration here.

Nonmarine Tertiary rocks also occur in the Quito-Cuenca depression, which essentially forms a continuation of the Cauca Basin of Colombia. They are exposed in the vicinity of Cuenca, but elsewhere are mainly obscured below a cover of Pliocene and Quaternary volcanic rocks.

Tertiary of the Western Province. In general terms the Tertiary rocks of western Ecuador make up a succession of marine sandstones and clays with which are interbedded numerous tuffs and tuffaceous layers. The sequence is characterized by a high content of cherts and siliceous material. Conglomerates, limestones, and olistostromes (gravity slides) are also developed locally.

Over most of the region, the Tertiary succession begins with a series of coarse-grained sandstones and conglomerates of Eocene age. These sediments are generally rich in siliceous material and contain intercalations of tuff and lava in many areas. Limestones, the *San Eduardo Formation* and equivalents, are locally developed at the base of the sequence and have yielded fossils of middle Eocene age.

Special conditions obtained on the Santa Elena Peninsula, where what was previously considered a highly disturbed zone of normal sediments has been found to be an olistostrome complex (Colman, 1970). The area is known in detail from drilling. Evidently huge rafts of sediments slumped into the area during upper Eocene times, and some have provided reservoirs for oil accumulations. The source of the slumped masses is thought possibly to lie in the vicinity of the Colonche-Chongón hills. Between the olistostromes lie autochthonous turbidites which display many fine examples of the sedimentary structures characteristic of such facies.

It appears that rocks attributable to the Oligocene (s. s.) are either thin or missing in western Ecuador.

The Miocene succession opens with coarse clastics made up of tuffaceous sandstones and conglomerates that pass upward into deepwater shales and siltstones. It closes with a return to shallow-water deposition characterized by conglomeratic sands, silts, and limestones, mostly tuffaceous, pointing to continued contemporaneous volcanic activity that is probably centered in the Cordillera Occidental.

Little is known concerning Tertiary stratigraphy offshore. The Gulf of Guayaquil contains a great thickness of Miocene marine and deltaic strata, and there is some evidence that the continental shelf off the west coast may be formed largely of plastic clays of the type that gives rise to diapiric structures.

Andean Orogeny. The appearance of coarse clastic sediments in late Miocene time points to a renewal of orogenic activity that reached a climax at the end of that period. The present structure of the country derives principally from movements associated with this Neogene "Andean" orogeny, of which the details are given in the section on structure.

Pliocene and Quaternary. The Andean orogeny was succeeded both by uplift of the Andean ranges and by the outbreak of volcanic activity. The volcanoes are located on lines of weakness associated with major transcurrent faulting, and most of the central part of the Andean chain is covered by andesitic tuffs and lavas. The positions of the volcanic centers, both active and dormant, are illustrated on Fig. 1. They give rise to the highest mountains in the country: Chimborazo reaches an altitude of 6267 m, and Cotopaxi, whose crater is still intact, stands at 5897 m above sea level.

The rapid uplift of the Andes and the associated outbreak of volcanic activity led to the growth of huge alluvial fans and terraces on both sides of the Andes. These deposits contain a high proportion of volcanic material, in places including blocks several meters in diameter.

In the coastal region the Pliocene-Quaternary is represented by flat-lying fine conglomerates and calcareous sandstones containing abundant megafossils. These beds are collectively referred to the *Tablazo Formation* although in detail they are found to include at least three different erosion surfaces. In the Esmeraldas region in the north the sequence is represented by the *Cachabi Formation,* made up of clays and coarse sands interbedded with tuffs and volcanic ash. The latter locally contain commercial deposits of placer gold.

Structure

The structural provinces of Ecuador are depicted in Fig. 3 and are described below.

Guayana Shield. The Guayana Shield is a stable block of Precambrian igneous and metamorphic rocks capped by a thin veneer of Phanerozoic continental sediments. The shield, which lies mainly outside Ecuadorian territory, is bordered by a series of arches (the Vaupes swell, Cononaco arch, and Marañón arch) that project westward into the Subandean province. The Vaupes swell impinges on the Andes, but the other arches are more subtle features that have not been clearly defined.

Oriente Basin. The Oriente Basin is a segment of the Subandean province that separates the Guayana Shield from the Andean mobile belt. The basin is asymmetrical with a wide eastern flank on which basement dips at less than $1°$. However, the section above basement reaches a thickness of as much as 5000 m in the axial zone in front of the Andes. The basin plunges regionally southward, the degree of inclination increasing south of the Cononaco arch. The western part of the basin is affected by faults and related flexures associated with the Andean front; the axial region is deformed into broad folds of Laramide age, and the eastern flank is a slightly disturbed platform. It is locally cut by basement faults that have in places warped the overlying sediments.

Andean Foothills. The Andean foothills form a zone of moderate folding and reverse faulting. Although from a stratigraphic standpoint the Andean foothills represent the uplifted western flank of the Oriente Basin, structurally they form part of the Andes, being in many ways analogous with the Eastern Andes of Colombia. In the N is the Napo uplift, a broad positive feature of Mesozoic rocks that is capped by three volcanic centers, including Reventador (3485 m) and Sumaco (3900 m), which are remarkable as being the only active volcanoes in the Eastern Andes. In the S is the Cutucú uplift, a more complex feature bringing Paleozoic strata to the surface. It marks the northern extremity of the virtually unexplored Cordillera de Condor, which continues into Peru. Between the Napo and Cutucú uplifts lies

FIGURE 3. Structural zones of Ecuador.

Zone 1: *Western Littoral and Continental Shelf*
1a. Esmeraldas-Muisne Basin
1b. Manta Basin
Zone 2: *Gulf of Guayaquil Province*
2a. Progreso Basin
Zone 3: *Pacific Coast Range*
3a. Rio Verde Arch
3b. Jama Hills
3c. Tosagua Arch
3d. Colonche-Chongon Hills
Zone 4: *Coastal Depression*
4a. Borbon Basin
4b. Quininde Basin
4c. Jipijapa Basin
4d. Daule Basin
Zone 5: *Western Andes*
5a. Cordillera Occidental
5b. Amotape-Chanchan Belt
Zone 6: *Quito-Cuenca Depression*
6a. Quito Basin
6b. Cuenca Basin
Zone 7: *Cordillera Real*
Zone 8: *Eastern Andes and Oriente Basin*
8. Oriente Basin
8a. Garzon Massif, Colombia
8b. Napo Uplift
8c. Puyo Reentrant
8d. Serrania de Cutucu
8e. Cordillera de Condor
8f. Santiago Basin, Peru
8g. Cerros de Campanguiz, Peru
Zone 9: *Guayana Shield* (margin)
9a. Vaupes Swell
9b. Cononaco Arch?
9c. Marañon Arch

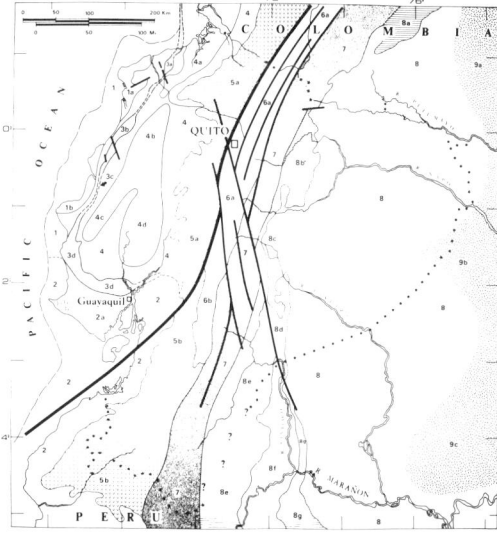

the Puyo reentrant, a complex depression formed in part by the postulated Baños-Cotopaxi wrench fault, and possibly reflecting thinning over the Cononaco arch.

The foothills zone is bordered to the east by a series of high-angle reverse faults that dip toward the Andes, which in places also bound asymmetrical anticlines (Tschopp, 1953, Fig. 7). Dextral wrench faulting truncates the northern end of the Napo uplift, and some of the other larger faults may also have a transcurrent component.

Cordillera Real. The Cordillera Real forms the core of the Andes, built of high-grade metamorphic rocks, cut by intrusions, and in places capped by Quaternary volcanoes. A major fault zone forms the eastern border of the range, which is also cut by large alkali-granite intrusions. A sinistral wrench fault, cutting the Cordillera Real and here named the *Baños-Cotopaxi fault*, is postulated to explain an orogenic deflection (the Quito-Riobamba deflection, Campbell, 1970), the alignment of volcanic centers, and other features. It is thought to extend from the Pichincha volcano (near Quito) through Cotopaxi to Yaupi on the Peruvian border.

Quito–Cuenca Depression. The Quito-Cuenca depression is an intermontane graben separating the Cordillera Real from the Western Andes. It is cut by numerous faults belonging to the Dolores-Guayaquil fault system discussed previously. It is an extension of the Cauca Basin of Colombia, and likewise is broken into sub-basins by a series of crosshighs, which are, however, largely obscured below Pliocene and Quaternary volcanic rocks. Faults along the margins have caused lines of weakness along which volcanic rocks have reached the surface. The depression is flanked on each side by a series of snow-capped volcanic peaks–the "Avenue of Volcanoes" of Alexander von Humboldt (Fig. 4).

Western Andes. The Western Andes are a branch of the Andes made up principally of Cretaceous eugeosynclinal rocks. The Dolores-Guayaquil fault cuts through the Western Andes at the latitude of Guayaquil and divides the mountain chain into two discrete elements.

In the north is the *Cordillera Occidental,* an extension of a range of the same name in Colombia. It consists of moderately to strongly deformed Mesozoic eugeosynclinal rocks cut by an arcuate belt of granodiorite intrusions that are probably of Eocene age. According to the proposed crustal model, basement in the range consists of oceanic crust of Mesozoic age.

In the S lies what may be termed the *Amotape-Chanchan Belt,* a somewhat ill-defined element forming an extension of the Amotape Mountains of Peru. High-grade metamorphic rocks, possibly of Precambrian age, form the western border of the zone, whereas to the E lie Cretaceous strata in both eugeo-

FIGURE 4. Sangay Volcano viewed from the Andean foothills near Puyo. (Photo: Gottfried Hirtz.)

synclinal and shelf facies, in places capped by Quaternary volcanic rocks. This part of the Western Andes is also intruded by granodiorites, but whereas in the Cordillera Occidental they form an arcuate, structurally controlled belt, in the Amotape-Chanchan belt the intrusions are more circular in outline and occur at random following no recognizable structural trend. The significance of these differences is still obscure, but they nevertheless emphasize the importance of the Guayaquil-Dolores line in separating major structural provinces.

Coastal Depression. The Coastal Depression is a shallow structural depression in front of the Andes. Cretaceous eugeosynclinal basement rocks either outcrop or lie at shallow depths over much of the Coastal Depression. However, two sedimentary basins filled with a maximum of 3000 m of marine Tertiary strata have developed–the Borbon Basin in the north and the trilobed Daule-Quininde-Jipijapa basin in the south. The structure of the region is dominated by faulting, with broad flat-lying or gently synclinal regions being separated from one another by highly deformed faulted uplifts. The faults are linear and, although causing intense local deformation, do not in general give rise to much stratigraphic separation, possibly an indication that they have a dominant transcurrent component. This structural pattern has been designated "shear folding" and appears to be characteristic of the circum-Pacific region.

Pacific Coast Range. The Pacific Coast Range is an arcuate belt of partly buried uplifts. It appears to be an extension of a chain of the same name that enters Colombia from Central America and follows the coast to Cabo Corrientes, where it heads offshore, being represented by Gorgona Island. It crosses the coast again in northern Ecuador, where it forms an arcuate belt of uplifts—the Rio Verde arch, the Jama Hills, the Tosagua arch, and the Colonche-Chongon Hills. This structural element is apparently related more to the Central American orocline than to the Andes, and it is striking that the southern extremity, comprising the last-named hills, impinges on the Andes almost at right angles.

Western Littoral and Continental Shelf. The western littoral and continental shelf is a little-known border zone. The Pacific Coast Range is bordered by another structural depression, most of which lies offshore and is virtually unknown. Two small coastal embayments of marine Tertiary rocks, the Manta Basin and the Esmeraldas-Muisne Basin, do, however, belong to this province.

Gulf of Guayaquil Province. The Gulf of Guayaquil province is a unique development on the west coast of South America. The Gulf of Guayaquil is the only major indentation on the west coast of South America and appears to be due to the interaction of two major fault systems. First, it lies on a great E-W transverse line that cuts not only the South American continent but also the adjacent oceans. It is represented principally by the Romanche fracture zone in the Mid-Atlantic, the Amazon Basin, which separates the Guayana Shield from the Brazilian Shield, and the Galapagos fracture zone in the Pacific. Possibly belonging to this major system are the important faults that parallel the coast of the Santa Elena peninsula. Second, the Gulf of Guayaquil marks the point where the Dolores-Guayaquil faults system runs offshore, apparently causing a dextral offset to the Peru Trench.

Ecuador lies in the circum-Pacific belt of seismic activity and has experienced numerous earthquakes of strong intensity. Earthquake centers lie in two principal belts: one follows the Dolores-Guayaquil megashear, and the other lies offshore, being possibly related to the northern end of the Peru Trench. As elsewhere in the Andean chain the epicenters lie at progressively greater depths toward the E; the offshore belt has shallow epicenters (h: $<$ 70 km), whereas those in the Andes are of moderate depth (h: 130–180 km). Deep-seated earthquakes (h: 550–650 km) have been recorded from the western Guayana Shield beyond the Ecuador frontier.

Galapagos Islands. Finally, no account of the structure of Ecuador would be complete without reference to the *Galapagos Islands* (q.v.) lying nearly 1000 km off the coast. These islands, with their unique fauna and flora, first studied by Charles Darwin, not only contributed data essential to modern theories of evolution but are also of considerable interest from a geological and oceanographic standpoint. The islands, which are of volcanic origin, lie on the western end of a submarine uplift, the Carnegie Ridge, and are bordered to the N by the Galapagos fracture zone, a major E-W fault system. Volcanic cones, crater lakes, and lava fields dominate the scenery of the archipelago.

Crustal Model

Geophysical studies capable of revealing the deep-seated structure of the country have not yet been undertaken, and it is consequently possible to consider only a very tentative crustal model. A hypothetical profile is, however, given in Fig. 1. It shows that continental crust of the Guayana Shield attains a thickness of about 35 km in the eastern part of the country but thins and dies out in the Andes. To the W lies a belt of oceanic crust thought to be in the process of being welded to the shield by accretion. Positive gravity anomalies in the Colombian extensions of the Central Andes suggest that the chain lacks a root of low-density sediments that characterize many other orogenic belts. The Ecuadorian Andes seem to be formed of a thin layer of Precambrian sialic rocks overlain by Phanerozoic igneous rocks and metamorphosed geosynclinal rocks. Again partly based on inferences from Colombia, it is probable that both igneous and metamorphic rocks become progressively younger westward across the Andes.

Nappes and gravity structures, as found, for example, in the Alps and possibly in parts of the Andes of Peru and Bolivia, are conspicuously absent in Ecuador. Instead the structure appears to be dominated by major linear fractures interpreted as transcurrent faults. The Dolores-Guayaquil system is the most important of these, and in terms of modern theories of global plate tectonics may have marked during the Mesozoic and early Tertiary a segment of the junction between the continental plate of northern South America and the oceanic Cocos and Caribbean plates that have moved northward and eastward relative to the continent. Posthumous displacements along this line have probably persisted to the present day, being responsible for continuing volcanic and seismic activity.

Economic Geology

Surface seepages of oil and gas drew early attention to the petroleum possibilities of the Santa Elena peninsula. The Ancon field was discovered in 1913 and by the end of 1969 had produced 94,389,145 barrels of oil. Several other small fields in the vicinity bring the total cumulative production of Ecuador to 102,729,781 barrels (December 31, 1969). These fields are, however, now approaching exhaustion and in the early 1970s production has only been about 1.5 million barrels annually.

In 1967 a successful wildcat stimulated an extensive exploration program in the Oriente province which has resulted in the discovery of several large oil fields in the northern part of the basin. Exploratory drilling is continuing southward, and there can be little doubt that the opening of a trans-Andean pipeline will place Ecuador firmly on the list of important petroleum-exporting countries.

Placer gold has been worked since before the Spanish Conquest, and some small-scale operations continue in the Esmeraldas region and in the Oriente. A number of mineralized zones are known, especially in the Amotape-Chanchan belt, and preliminary investigations to develop copper prospects are under way. Plate-tectonic studies suggest that the disseminated copper (and molybdenum) deposits are Mio-Pliocene in age.

COLIN J. CAMPBELL

References

Bristow, C. R., 1973. *Guide to the Geology of the Cuenca Basin, Southern Ecuador.* Quito: Ecuadorian Geol. Geophys. Soc., 54p.

Campbell, C. J., 1968. "The Santa Marta wrench fault of Colombia and its regional setting," *Trans. 4th Caribbean Geol. Conf.,* Trinidad, 247–261.

____, 1970. *Guide to the Puerto Napo Area, Eastern Ecuador, with Notes on the Regional Geology of the Oriente Basin.* Quito: Ecuadorian Geol. Geophys. Soc., 40p.

____, 1974. "Ecuadorian Andes," *in* A. M. Spencer, ed., *Mesozoic-Cenozoic Orogenic Belts.* London: Geol. Soc., Spec. Publ. **4,** 725–732.

Canfield, R. W., and Bucaram, P., 1966. *The Geology of the Coastal Zone of Ecuador.* Quito: Minist. Indus. y Comercio, 90p.

Colman, J. A. R., 1970. *Guidebook to the Geology of the Santa Elena Peninsula.* Quito: Ecuadorian Geol. Geophys. Soc., 34p.

Gubler, Y., and Ortynski, I., 1966. *Informe Geologico Preliminar sobre las Posibilidades Petroliferas de las Cuencas Sedimentarias del Ecuador.* Quito: Minist. Indus. y Comercio, 94p.

Ham, C. K., and Herrera, L. J., Jr., 1963. "Role of Subandean fault system in tectonics," *in* O. E. Childs and B. W. Beebe, eds., *Backbone of the Americas* (Mem. 2). Tulsa: Am. Assoc. Petrol. Geologists, 47–61.

Lewis, G. E., Tschopp, H. J., and Marks, J. G., 1956. "Ecuador," in *Handbook of South American Geology* (Mem. 65). New York: Geol. Soc. Am., 251–288.

Marchant, S., 1961. "A photogeological analysis of the structure of the western Guayas province, Ecuador," *Quart. J. Geol. Soc. Lond.,* 115(6), 317–338.

Nygren, W. E., 1950. "Tertiary geosyncline in western Ecuador and Colombia," *Bull. Geol. Soc. Am.,* 61(12), pt. 2, 1540.

Putzer, H. von, 1968. "Tertiäre Lignite im interandinen Graben von Ecuador als Beispiel für synorogene Kohlebildung in intramontanen Becken," *Geol. Jahrb.,* 85, 461–488.

Sauer, W., 1971. "Geologie von Ecuador," *Regionalen Geologie der Erde,* vol. 11. Berlin: Gebr. Borntraeger, 316p.

Tschopp, H. J., 1953. "Oil explorations in the Oriente of Ecuador 1938–1950," *Bull. Am. Assoc. Petrol. Geologists,* 37(10), 2303–2347.

Cross-references: *Brazil; Columbia; Galápagos Islands; Peru; South America.*

EL SALVADOR

Smallest of the Central American countries, and the only one facing the Pacific and without an Atlantic coastline, El Salvador has an area of 21,390 km² (8260 sq mi). It is bordered by Honduras on the N and E, Guatemala on the W, and by the Pacific Ocean on the SW. There is a narrow Pacific coastal belt, a central plateau between two mountain ranges, and northern lowlands formed by the Lempa River valley bounded by a high mountain range that rises to the border of Honduras. Major lakes include Ilopango, Guija, and Camalotal. El Salvador is a country of many volcanoes and has frequent eruptions and earthquakes. Mount Izalco (the "Lighthouse"), with an elevation of 2386 m (7828 ft) had a major eruption in 1967, as did Mt. San Miguel (elevation 2132 m, 6994 ft). Izalco's constant red glow makes it visible to sailors at sea. El Salvador lies entirely in the tropics, but the heat is modified in the interior by the elevation.

FIGURE 1. Geological sketch map of El Salvador [modified after Dürr and Klinge (1960), and Tricart (1961)]. 1, Littoral alluvium and fluvio-volcanic deposits; 2, late Quaternary and active volcanoes; 3, hydro-volcanic deposits of the central graben (Late Quaternary); 4, Quaternary basaltic volcanoes associated with the northern border of the central graben; 5, lacustrine deposits of the middle Lempa valley (block by volcanic dam); 7, pre-Pliocene volcanics; 8, Metapan sedimentary beds (? Cretaceous).

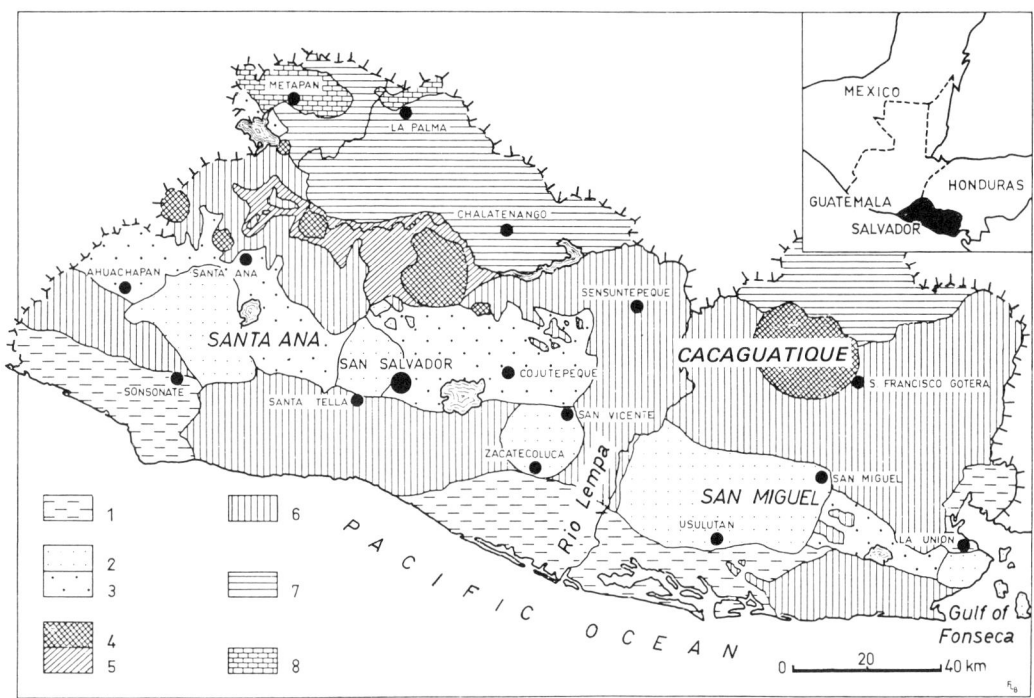

Small amounts of gold, silver, zinc, lead, and sulfur are mined. Limestone is quarried for cement and building stone.

Geologically, both of the mountainous belts are volcanic, trending WNW-ESW, separated by a complex graben zone, the Central Depression (of Williams and Meyer-Abich, 1955). All structures are cut off at the SE of the country, at the Gulf of Fonseca, by the N-S-trending Honduras Graben ("Comayagua Graben" of Sapper, 1937). Dengo (1968) recognized that it is a major lineament of complex fractures.

The northern belt represents the southerly limit of the Honduras Massif, which has a foundation (outside El Salvador) of crystalline rocks of pre-Triassic age, mainly metasediments, which may be Paleozoic or even possibly Precambrian. In the El Salvador part it is largely buried by late Tertiary or youthful volcanics (Fig. 1).

The oldest sedimentary rocks appear to be Upper Cretaceous red sandstones, marls, and limestones associated with a regionally block-faulted belt that lies S of the Honduras Massif and is overlain by Pliocene volcanics, a westerly extension of the Nicaragua Volcanic Upland (see *Central America*), and exposed in the Lempa Valley. Outcropping around Metapan and La Palma, in the northernmost part of El Salvador near the frontiers of Guatemala on Honduras, this Cretaceous sequence is highly disturbed and mineralized.

The Lempa valley was largely blocked by Pliocene volcanic activity, so that thick lake deposits and water-laid tuffs are now widespread there. The northern border of the central valley is punctuated by Pleistocene volcanoes; the southern border is marked by Holocene and contemporary centers.

The block-faulted segments reach the south coast in the Balsam and Jucuaran blocks (Pliocene volcanics), but are covered elsewhere by a coastal plain, by the broad Lempa delta, or by coastal lagoons with mangrove swamps. Their thickness and stratigraphy is unknown. Alternation of climates from humid to semiarid during the Quaternary is shown by the weathering of tuffs, paleosols, and the development of carbonate duricrust in places (Tricart, 1961).

According to Dengo *et al.* (1970) the volcanics are distributed in three belts, corresponding to the basement structures. The northern belt is marked by the earliest eruptions (pre-Pliocene), notably the giant extinct volcanoes Masahuat (in the NW) and Cacaguatique (in the NE). The Central Depression has smaller and medium-sized extinct cones, including Carrizal (S of Masahuat), Cerro Guazapa, Cerro Nejapa, and Yayantique. Also in this belt are Tigre and Zacate Grande, which are islands in the Gulf of Fonseca, probably also related to the N-S Honduras graben zone. Largest of all is the southern belt, which includes all the youngest and active centers, rising above Pliocene volcanic foundations. These are mostly located near the southern boundary faults in the central graben and include five large strato volcanoes: Santa Ana, San Vicente, San Salvador, Usulutan, San Miguel, and other smaller ones. Older and more deeply eroded centers occur in the Sierra de Jacuaran (in the SE). This southern belt also includes the major volcano-tectonic collapse features, such as the Coatepeque Caldera and the Ilopango depression.

Genetically it can be seen that the volcanoes of El Salvador follow the trend of the Middle America Trench offshore, which suggests a subducting plate boundary, here complicated by major transverse fractures zones. Minor arcuate fault systems related to the general stress field are accompanied by smaller groups of volcanic cones, as in the SW, where seven volcanoes occur in the Apaneca Range W of Santa Ana. In the central part, the Apastepeque area, there are several maars and cinder cones.

An important geothermal field is being developed at Ahuachapan. It has proved to be an important source of microseismic activity (Ward and Jacob, 1971). Studies of volcanic temperatures and the fumarole gases were begun in 1966 with the eruption of Izalco (Rose and Stoiber, 1969).

RHODES W. FAIRBRIDGE

References

Dengo, G., 1968. *Estructura Geológica, Historia Tectónica y Morfología de América Central.* Mexico: Centro Regional de Ayuda Tecnica, 1–50.

____, Bohnenberger, O., and Bonis, S., 1970. "Tectonics and volcanism along the Pacific marginal zone of Central America," *Geol. Rundschau,* 59(3), 1215–1232.

Dürr, F., and Klinge, H., 1960. "Beiträge zur Stratigraphie und zur Paläopedologie des mittleren Salvador," *Neues Jahrb. Geol., Palaeontol, Monatsh,* 3, 111–132.

____, and Klinge, H., 1969. "Die Verbreitung der jüngsten Bims-Aschen um San Salvador, El Salvador, Zentral-Amerika," *Neues Jahrb. Geol. Palaentol., Abh.,* 110, 393–396.

Gierloff-Emden, H. G., 1957. "Vier Karten zur physischen Geographie von El Salvador," *Erdkunde,* 11, 58–64.

____, 1959. "Die Küste von El Salvador. Eine morphologisch-ozeanographische Monographie," *Acta Humboldtiana,* Ser. Geogr.–Ethnogr., 22, 183p.

Meyer, J., 1964. "Stratigraphie der Bimskiese und Aschen des Coatepeque-Vulkans in westlichen El Salvador (Mittel-amerika)," *Neues Jahrb. Geol. Palaeontol. Abh.,* 119(3), 215–246.

Meyer-Abich, H., 1956. "Los volcanes activos de Guatemala y El Salvador," *An. Serv. Geol. Nacl. El. Salvador,* 3, 1–102.

____, 1960. "El Salvador," *Lexique Stratig. Intern., Amerique Centrale* (Paris), 5, fasc. 2a, 99–128.

Rose, W. I., and Stoiber, R. E., 1969. "The 1966 eruption of the Izalco volcano, El Salvador," *J. Geophys. Res.,* 74(12), 3119–3130.

Sapper, K., 1937. "Mittelamerika," *Handbuch Reg. Geol.,* 8(4), 160p.

Stirton, R. A., and Gealey, W. K., 1949. "Reconaissance geology and vertebrate paleontology of El Salvador, Central America," *Bull. Geol. Soc. Am.,* 60(11), 1731–1754.

Tricart, J., 1961. "Aperçu sur le Quaternaire du Salvador (Amerique centrale)," *Bull. Soc. Géol. France,* Ser. 7, 3, 59–68.

Ward, P. L., and Jacob, K. H., 1971. "Microearthquakes in the Ahuachapan Geothermal Field, El Salvador, Central America," *Science,* 173, 328–330.

Weyl, R., 1961. "Die Geologie Mittelamerikas," *Regionalen Geologie der Erde,* vol. 15. Berlin: Gebr Borntraeger, 226p.

Williams, H., and Meyer-Abich, H., 1955. "Volcanism in the southern part of El Salvador with particular reference to the collapse basins of Lakes Coatepeque and Ilopango," *Univ. Calif. Publ. Geol. Sci.,* 32(1), 1–64.

Cross-references: *Central America–Regional Review; Guatemala; Honduras; Nicaragua.*

F

FALKLAND ISLANDS (ISLAS MALVINAS)

The Falkland Islands (a British dependency, claimed by Argentina) have an area of 16,655 km^2 (6430 sq mi). Situated in the South Atlantic Ocean, around 52°S, 60°W, they lie approximately 800 km NE of Cape Horn. There are some 100 islands in the Falklands proper, but 200 islands in the administrative group, which includes not only the major islands of East Falkland and West Falkland, but also islands of the Scotia Arc, notably South Georgia and South Sandwich (for the last two, see *Subantarctic Islands*).

In the Falklands proper, the topography is low, up to 705 m (2315 ft, Mt. Adam); the islands are rocky, treeless, and climatically are harried by strong winds and cold rains. Mean temperature is 5.0°C, with a range from 0.5°C in August to 10.5°C in January.

There are many peat swamps and much peaty soil, which provides for a rich growth of tussock grass, supporting a limited economy of sheep farming. The population is largely of Scottish and Welsh origin.

Geomorphologically the islands are largely controlled by structure and differential weathering. The region was peneplaned throughout the late Mesozoic and most of the Tertiary, interrupted in the late Cenozoic by repeated eustatic lowering, so that the coasts are strongly indented and marked by numerous rias, with skerries offshore.

The geology was first studied in 1833 (the visit of HMS *Beagle*) and briefly reported on by Charles Darwin (1846). Later, after a visit by the Swedish South-Polar Expedition of 1901–1903, it was described in much more detail by Andersson (1906, 1916) and by Halle (1912). A comprehensive survey was made for the British colonial service by Baker (1923), and the plant fossils were studied by Seward and Walton (1923). The Dwyka-type glacial beds and "Gondwana" correlations have been discussed by Caster (1952), Frakes and Crowell (1967, 1969), and many others.

A Precambrian basement outcrops at Cape Meredith in the SW of West Falkland, with granite, gneiss, schist, and pegmatite. Baker (1923) suggested a correlation with the Nama System of South Africa. In the course of Leg 36 of the Deep Sea Drilling Project, high-grade metasediments and granite were cored on the Falkland Plateau approximately 800 km E of the islands (Barker *et al.*, 1975). The Falklands are clearly on an old structural platform of continental rocks belonging physically to the SE of South America. The Falkland Plateau was attached to South Africa until the opening of the South Atlantic in the Cretaceous when it subsided below sea level.

The Precambrian basement is overlain by a Paleozoic sequence which is mostly flat or gently tilted, but in places strongly folded, that is closely correlated with that succession in South Africa. The early part is Devonian and possibly older. It consists of 3000 m of quartzites, sandstones, and shales (locally altered to slates). It rests on the Precambrian with a local basal conglomerate. Devonian fossils, resembling those of the Bokkeveld beds of South Africa, include the trilobite *Homolonotus, Leptocoelia, Chonetes, Spirifer,* and other brachiopods (first collected by Darwin). Many of the species are also identical with beds of the same age in Bolivia and Brazil. There are biogeographic affinities in this "austral" province (with connections to New Zealand and Australia), called "Malvinokaffric" by Rudolf Richter, and there are well-marked distinctions from the "boreal" province. The lowest part of the sequence may reach into the Ordovician and may be, in part, glaciomarine. In the upper part of the sequence are the Port Philomel beds, with plant remains resembling the Witteberg of the Cape.

Unconformably overlying the early–middle Paleozoic succession follows a 3500-m-thick Permo-Triassic sequence, the Lafonian Series of Halle (1912). The base is marked by tillites, comparable to the Dwyka, followed by varved shales and sandstones with waterborne erratics. The upper part includes the Bay of Harbours beds, which contain a rich *Glossopteris* flora typical of the Lower Gondwanas (see *South Africa; India; Australia*).

The Permo-Triassic and older beds are cut by dikes and sills, diorites, and dolerites, comparable to those of the Paraná Basin and the Karroo, and presumably Late Triassic–early Jurassic.

There is a fine Quaternary interglacial

deposit at West Point Island of clays, containing a luxuriant warm-climate flora (*Podocarpus, Libocedrus, Fitzroya*), and overlain by a diamictite, possibly a till, but perhaps more likely a mudflow deposit. For much of the Pleistocene the island remained in a largely unglaciated, but extreme periglacial condition. A special feature is that of the extensive "stone-rivers," first noted by Andersson (1906) and Halle (1912). From 100 to 1000 m across and many kilometers in length, they are regarded as products of solifluction and mudflows under extreme, periglacial conditions. There are also extensive raised beaches (Adie, 1953).

Structurally the Paleozoic and Triassic formations of the Falklands are deformed into a series of folds trending NW-SE in the W, swinging to E-W on the east coast. The folds are largely asymmetric, increasing in intensity to the S.

An exceptional trend is observed on the western shore of the Falkland Channel (separating the two main islands) with a single NE-SW fold, presumably related to a transverse fault zone here.

The folding is apparently Cimmerian, possibly comparable in time to the Palisadian taphrogeny of eastern North America (180–220 m yr). There are numerous dolerite ("diabase") dikes and sills comparable to those of the Karroo and Parana. Several reconstructions indicate an extension of the Patagonide arc of Argentina, swinging through the Falklands, to continue in the Cape Ranges ("Capides") of South Africa (see discussion in Mitchell-Thomé, 1970, p. 331). Adie (1952a, 1952b), in contrast, sees no relationship with Patagonia, while Du Toit's (1927) predrift reconstruction would bring the Falklands closer to the Parana Basin.

RHODES W. FAIRBRIDGE

References

Adie, R. J., 1952a. "The position of the Falkland Islands in a reconstruction of Gondwanaland," *Geol. Mag.*, **89**, 401–410.
———, 1952b. "Representatives of the Gondwana System in the Falkland Islands," *Symposium sur les séries de Gondwana, 19 Intern. Geol. Congr.*, Algeria, 385–392.
———, 1953. "New evidence of sea level changes in the Falkland Islands," *Falkland Islands Dependencies, Surv. Sci. Rept. 9*, 8p.
———, 1958. "Falkland Islands," *Lexique Stratig. Intern., Amerique Latine* (Paris), **5**, fasc. 9c, 35–59.
Andersson, J. G., 1906. "Solifluction," *J. Geol.*, **14**, 97–104.
———, 1916. "Geology of the Falkland Islands," *Wiss. Ergebn. Schwed. Sudpol. Exped., 1901–3*, **3**, 1–38.

Baker, H. A., 1923. "Final report on geological investigations in the Falkland Islands 1920–22," *Quart. J.*, **79**(1), iv.
Barker, P. F., et al., 1975. "Initial report of the Deep Sea Drilling Project, Leg 36, Ushvaia, Argentina to Rio de Janeiro, Brazil," National Science Foundation, Washington, D.C.: Govt. Printing Office.
Borello, A. V., 1963. *Sobre la geologia de las Islas Malvinas*. Buenos Aires: Ed. Cultur. Argent., Minist. Educ. y Justicia, 70p.
Caster, K. E., 1952. "Stratigraphic and palaeontologic data relevant to the problem of Afro-American ligation during the Palaeozoic and Mesozoic," *in* E. Mayr, ed., *The Problem of Land Connections Across the South Atlantic, with Special Reference to the Mesozoic*. Bull. Am. Mus. Nat. Hist., **99**, Art. 3, 105–152.
Clarke, H. A., 1919. "Falklandia," *Proc. Nat. Acad. Sci.*, **5**, 102–103.
Darwin, C., 1846. "Geology of the Falkland Islands," *Quart. J. Geol. Soc.*, **2**, 267–274.
Du Toit, A. I., 1927. "A Geological comparison of South America with South Africa," *Carnegie Inst. Publ. 381*, 158p.
Frakes, L. A., and Crowell, J. C., 1967. "Facies and paleogeography of the late Paleozoic Lafonian diamictite, Falkland Islands," *Bull. Geol. Soc. Am.*, **78**, 37–58.
———, and Crowell, J. C., 1969. "Late Paleozoic glaciation. 1. South America," *Bull. Geol. Soc. Am.*, **80**(6), 1007–1042.
Halle, T. G., 1912. "Geological structure and history of the Falkland Islands," *Bull. Geol. Inst., Upsala*, **11**, 115–226.
Mitchell-Thomé, R. C., 1970. "Geology of the South Atlantic Islands," *Regionalen Geologie der Erde*, vol. 10. Berlin: Gebr. Borntraeger, 1–367.
Seward, A. C., and Walton, J., 1923. "On a collection of fossil plants from the Falkland Island," *Quart J. Geol. Soc. Lond.*, **79**, 313–333.

Cross-references: *Argentina; Bolivia; Brazil; Paraguay; South America; Sub-Antarctic Islands.*

FERNANDO DE NORONHA, ROCAS, TRINDADE, MARTIN VAZ, AND SAINT PAUL ROCKS

There are five oceanic islands to the NE and SE of the northeastern point of Brazil, Cape São Roque, and part of Brazilian territory. They have all been admirably described in detail by Mitchell-Thomé (1970).

Fernando de Noronha, at $3°50'S$, $32°25'W$, part of a small archipelago, covers 18.4 km^2. The oldest rocks, phonolites, trachytes, and ultrabasics, were once claimed to be Upper Cretaceous, but radiometric dates indicate Miocene (8–11 m yr). They are deeply weathered and eroded, and were followed by explosive pyroclastics with nephelinite intrusions, and in turn by ankaratrites (1.7–3.2 m yr), apparently in three cycles. The island has been studied by Branner (1889), Almeida (1955), and others. The Peak (Pico) of Fernando de Noronha is, to quote Branner (1889), "the most striking landmark in the South Atlantic Ocean; it is 1000 feet high, with the upper portion perpendicular...." It is remarkably similar in appearance to the spine of Mt. Pelée when it was freshly emerged, but, according to Almeida, this is coincidental and is due to erosion of the joint system.

The area seems to be rather stable now and Quaternary "raised beaches" up to 60–70 m have been claimed by Branner. Almeida, however, considers that some, at least, are artificial in view of the presence there in a number of places of Portuguese ceramics. The most prominent level merges into what appears to be an abrasion platform 2–5 km wide, the site of an airfield today. This surface, the Central Plain, is according to Almeida, an erosional (pediment) surface due to subaerial processes associated with the development of mountain slopes under a former semiarid climate or, at least, with lesser rainfall than today.

These erosion surfaces known as "mesettas," vary in level from place to place around the archipelago (Quixaba Plain, 20–40 m; "Dois Abraços" High, 171m; three others ranging from 60 to 100 m: Floresta High, 140 m; Santo Antonio Mountain, 105m; Curral Mountain, 126 m, at the eastern portion of the island). They are related in part to differential erosion in the subhorizontal layering of lava flows (nepheline basalts or ankaratrites).

Pleistocene eolianites were discussed by Branner (1890a) and others. Known as the Caracas Sandstone, they reach 20 m in thickness at Atalaia and Caracas points. The eolianites are secondarily notched by several low eustatic stands. There are also active carbonate dunes today and extensive *Lithothamnium* (algae) reefs and smaller vermetid reefs.

The Quaternary raised beaches, as far as Almeida's work indicated, are not higher than some early marine deposits at the 9–11 m level (early Quaternary). They consist of conglomerates, sandstones, and gravels. They can be found on the Santo Antonio and Pontinha peninsulas, Atalaia Cove, and between the Cachorro and Conceicão beaches. The first two overlie ankaratritic flows; the last ones are over tuffs and volcanic breccias.

There are phosphates on Rata Island overlying Caracas Sandstone. The phosphate is recent guano from the numerous sea birds of the island.

Fernando de Noronha Islands and Atol das Rocas belong to a very conspicuous E-W lineament of seamounts, guyots, and basement highs. This lineament extends into the continental shelf off the Brazilian Ceara State as a buried magmatic body. Also on the strike of the lineament is the Mecejana phonolite (30 m yr), at the city of Fortaleza. Recent oceanographical work associated this lineament to a prominent fracture zone to the south of Chain Fracture Zone. It was accordingly named Fernando de Noronha Fracture Zone (Gorini et al., 1974).

Atol das Rocas

Also called the "Baixo das Cabras" because of the great number of shipwrecks here, the "Atol das Rocas" at 3°52'S, 33°49'W, lies 160 km W of Fernando de Noronha and 200 km N of Cape São Roque. It has been claimed to be an atoll, the only one in the open Atlantic, but it is not a true atoll. It is a ring-shaped algal carbonate bank, 3.5 km across, with a shallow lagoon, crowned by calcareous eolianites (Ottmann, 1962), and, though much smaller, it is thus analogous to *Bermuda* (q.v.).

"Ilha do Farol" (106,000 m^2) and "Ilha do Cemiterio" (53,200 m^2) are the two islands of Rocas. They are mounds consisting of detrital carbonates that do not even reach 3 m in altitude. A beach rock in the Cemiterio Island and pinnacles of dead algal banks lying 3–4 m above the present-day reef level suggest a former sea-level stand at +2.5 m (Mabesoone and Coutinho, 1970). The foundation is a seamount, probably similar to that of Fernando de Noronha and to seamounts and guyots lying westward.

Trindade

Trindade (also spelled "Trinidad") is located at 20°30'S, 29°20'W, about 1140 km E of the Brazilian mainland and 48 km W of Martin Vaz. Its length is 4.8 km and it reaches about 600 m in elevation. Mitchell-Thomé (1970) has provided an excellent map, based on a 1:10,000 survey by Almeida (1960). There is a youthful but extinct volcanic cone at the southeastern end of the island, where there is a remarkable tunnel 130 m long crossing the peninsula at sea level. Spectacular domes and necks of phonolite punctuate the landscape, which is deeply ravined.

The island was first visited by Sir Edmund Halley in 1700 and later by Hooker in 1839, but little geological work was done till Almeida's survey (1961). The oldest formations, the Trindade complex, consist mainly of pyroclastics and breccias of phonolite and tannbuschite. There is evidence of ignimbrite flows. There are 16 phonolite domes and necks, the largest being 450 m in diameter. There are numerous ultramafic dikes. The older complex is followed by the Desejado Sequence, consisting of phonolite, grazinite, and nephelinite flows and pyroclastics up to 400 m thick. These are followed by the Morro Vermelho Formation, probably middle Pleistocene in age, products of an explosive volcanic phase of ankaratrite, mainly lapilli-tuffs, and breccias, up to 230 m thick. A smaller accumulation, also youthful, is the Valado Formation, which is related to a single vent providing tannbuschite flows and pyroclastics. Finally, there is the youngest lava, probably late Pleistocene, from the Paredão volcano, which produced ankaratrite lavas, tuffs, and breccias.

There are well-developed marine platforms and raised beaches corresponding to about +3 m, correlated by Almeida with a mid-Holocene high eustatic stand. There is a well-developed insular platform to -110 m with submarine terraces at -47, -59, and -77 m.

Almeida believes the volcanic activity began in the Tertiary, and radiometric dates indicate the oldest complex to range from about 2.5 to 3.5 m yr, thus to be Pliocene (see Cordani in Mitchell-Thomé, 1970).

The famous tunnel was cut by wave action in pyroclastics of the Paredão volcano following the strike of vertical fracturing. This joint system is responsible for the almost vertical cliff ("paredão") that has its base on a wave-cut platform. The sea is responsible for erosional features probably originated during a sea level stand 2–3 m high. This is suggested by the height of the tunnel at its northern side, a cutting of its probable base level, and a high wave-cut platform that is not continuously reached by the waves. However, the southern mouth of the tunnel is very often completely engulfed by the waves. Inside the tunnel, the in- and outflow of sea water carried by the swell is remarkable (and dangerous).

The wall of the tunnel shows a very well stratified sequence of pyroclasts with volcanic bombs of varied shapes, lapilli, ashes, and volcanic blocks. This deposit is so easily eroded that the beaches immediately to the north of "Paredão Volcano" are mostly constituted by pyroclastic sands.

Calcareous sandstones (calcarenites) that consist mostly of algal-derived fragments with very unstable igneous minerals such as sanidine are found in places, such as at Andradas beach near the Morro das Tartarugas. The detrital component of these sandstones is identical to the sands in the present-day beaches as well as to the recently formed beach rock calcarenites in the reef area. They constitute flat-lying platforms restricted in area and lying 2–3 m above the sea level.

There are barchan dunes of carbonate sands at Tartarugas beach. The sands, under the action of SE trade winds, climb partway up the Pico N.S. de Lourdes and the Morro das Tartarugas.

Numerous blocks derived from vertically jointed phonolitic domes and necks are spread over large areas, mostly in the vicinity of the domes. Some of them are found far away downslope, suggestive of formerly more active Pleistocene pediment slope processes, aided no doubt by earthquake activity.

Trindade Island rises from a 5500-m sea floor, and topographic profiles suggest a moat-like depression around it. It is perfectly in line with an E-W lineament of flat-topped seamounts known as "banks," the Columbia Seamounts. The degree of leveling and the surface area of the top of these guyots increase westward. Three recent earthquake epicenters have been located near Jaseur Bank. No apparent E-W trend has so far been detected in the continent that would match the Trindade Mountains lineament but oceanward it corresponds almost certainly to a fracture zone as indicated by a change in basement level.

Martin Vaz

Martin Vaz (sometimes given as "Martin Vas") is located around 20°30'S, 28°51'W. Actually an archipelago, there is one main island, about 600 m across and 175 m in elevation, with two smaller islands and several rocky islets. They rise from a truncated seamount several kilometers across. The islands have never been mapped, but samples collected on Ilha do Norte, studied by Scorza (1964), disclosed the presence of two alkaline volcanics, ankaratrite and hauynite; the latter is remarkably unusual in an oceanic setting. Radiometric ages of 60 m yr were found for a block of ankaratrite collected here. Despite the fact that the minerals present in the rock seem to have crystallized primarily, Cordani (1970) prefers to disregard this age until new data substantiate the dating.

Martin Vaz constitutes the eastward limit of the Trindade lineament so far detected.

São Pedro and São Paulo Rocks

This is the official Brazilian name of what is better known as *St. Paul Rocks* (also given as "St. Paul's Rocks" and "St. Peter and St. Paul"). They are really no more than a dozen isolated rocks, located on an E-W-trending ridge (of depths under 100 m) at 00°56'N, 29°22'W. The largest only reaches 23 m in elevation, and all rise from a −5-m platform about 0.5 km across. Possibly the most celebrated midoceanic islets, they have been visited by Darwin on HMS *Beagle* in 1832, by HMS *Challenger* and by the *Quest* in 1921, by *Meteor* in 1925, by a Brazilian Navy ship *Belmonte* in 1931, by HMS *Owen* in 1960 (Wiseman, 1965), and by others.

What makes St. Paul Rocks so famous is that the principal rock is a dunite, an olivine-rich peridotitic type believed to be characteristic of the mantle. It is cataclastically brecciated and mylonitized. For a century or more there has been controversy as to whether the rock was of volcanic, plutonic, or metamorphic origin. Wiseman (1965) collected new material and reexamined the older collections, proposing subspecies names for four types: challengerite, owenite, paulite, and questite.

The strong brecciation and mylonitization becomes understandable when the geotectonic situation of St. Paul Rocks is recognized, i.e., on a tranverse ridge connected with the St. Paul's Fracture Zone, that lies just N of the Romanche Fracture Zone, and the Romanche Trench, among the major E-W lineaments (Wilson's transform faults) that characterize the Mid-Atlantic Ridge. It is thus concluded that they represent a structural injection of a sliver of ultramafic mantle material along the line of a transform fault mobilized under sea-floor spreading. Rubidium−strontium dating suggests ages of 3.5 to 4.5 billion years (discussed by Mitchell-Thomé, 1970), and thus apparently St. Paul Rocks are representatives of the earth's earliest crust, or nonmobilized mantle, in either case a feature—if correctly interpreted—which makes it a unique spot on the globe.

In contrast to Wiseman's results, Djalma Guimarães had claimed a volcanic origin for St. Paul Rocks in 1932 when he received samples collected from there by a Brazilian Navy Lieutenant. The Navy's mission with the ship *Belmonte* was to install an automatic lighthouse (1931). When they were there, they felt an earthquake that was strong enough to roll blocks downhill. In fact, permanent damage of this lighthouse two years after is claimed to be due to earthquakes.

Among the petrographic types described in the Guimarães paper are:

Nepheline basalt. Similar to one from Fernando de Noronha. It is constituted by phenocrysts of olivine, augite, nepheline, and magnetite (magnetite-ilmenite). The olivine is sometimes altered to iddingsite.

Olivinite. Black-gray color, no visible crystals. Under the microscope big crystals of olivine occur in a microgranular matrix of olivine, etc. The texture is porphyroblastic and the matrix has a fluidal aspect. Serpentine occurs in some fractures in the samples.

Volcanic tuffs. Light-gray vesicular rock with white lithic fragments completely filled with veinlets, sometimes brecciated. The tuff consists of crystals and fragments of olivine, biotite, augite, basaltic glass, and rare gastropoda shells dispersed in a noncrystalline mass having very thin calcite veinlets. The olivine shows several stages of alteration, from iddingsite to serpentine, although unaltered crystals are abundant. It is possible that the tuff was produced by explosions of basaltic lava and, in particular, of nepheline basalt. Angular fragments of a translucent white rock with veinlets of vitreous-like mineral were identified as a phosphatic rock with veinlets of opal. The phosphatic rock shows the presence of ammonium, water, and iron and magnesium carbonates. The opal veinlets are restricted to the phosphatic rock.

In the olivinites Guimarães pointed out the probable effect of tangential stress deformation.

RHODES W. FAIRBRIDGE
MARCUS GORINI

References

Almeida, F. F. de, 1955. "Geologia e petrologia do arquipelago de Fernando de Noronha," *Brasil, Div. Geol. Mineral.*, **13**, 181p.

———, 1961. "Geologia e petrologia da Ilha da Trindade," *Brasil, Div. Geol. Mineral.*, **10**, 18.

Branner, J. C., 1889. "The geology of Fernando de Noronha, Pt. 1," *Am. J. Sci.*, **37**, 145–161.

———, 1890a. "The aeolian sandstone of Fernando de Noronha," *Am. J. Sci.*, **39**, 247–257.

———, 1890b. "Geologia de Fernando de Noronha," *Rev. Hist. Geogr. de Pernambuco*, 20–22.

———, 1903. "Is the peak of Fernando de Noronha a volcanic plug like that of Mont Pelée?" *Am. J. Sci.*, **16**, 442–444.

Cordani, U. G., 1970. "Potassium-argon ages of rocks from the Brazilian South Atlantic Islands," *Symposium on Continental Drift.* Montevideo: UNESCO.

Davies, T., 1890. "The natural history of the island of Fernando de Noronha," *J. Linn. Soc.* (Lond.), **26**, 86–94.

Flavio, F. (*see* Almeida, F. F. de)

Gorini, M. A., 1969. "Geologic observations in the 'Comissão Oceanográfica Leste I' aboard the Research Vessel 'Almirante Saldanha'," *Ann. Acad. Brasil. Cienc.*, **41**(4), 642R–643R.

———, Damuth, J. E., and Bryan, G. M., 1974. "The Fernando de Noronha Ridge and its relationship to equatorial Atlantic fracture zones," *Geol. Soc. Am., Abstr.* **6**(7), 762.

Gulmarães, D., 1932. "Notas petrographicas," *Ann. da Acad. Brasileira de Sci.,* **4**(1), 29–32; [see also (2)].

Lobo, B., 1919. "Ilha da Trindade," *Arq. Mus. Nacl.,* **22**, 105–158.

Mabesoone, J. M., and Coutinho, P. N., 1970. "Littoral and shallow marine geology of northern and northeastern Brazil," *Trab. Oceongr. Univ. Fed. Pe., Recife,* **12**, 1–212.

Melson, W. G., Hart, S. R., and Thompson, G., 1972. "St. Paul's Rocks, equitorial Atlantic: petrogenesis, radiometric ages, and implications on sea floor spreading," *Geol. Soc. Am. Mem.* **132**, 241–272.

Mitchell-Thomé, R. C., 1970. "Geology of the South Atlantic Islands," *Regionalen Geologie der Erde,* vol. 10. Berlin: Gebr. Borntraeger, 1–367.

Ottmann, F., 1962. "L'atol das Rocas dans l'Atlantique sub-tropical," *Rev. Géogr. Phys. Géol. Dyn.,* **5**(2), 101–106.

Renard, A., 1889. "Report on the rock specimens collected on Oceanic Islands during the voyage of H.M.S. Challenger during the years 1873–76," *Rept. on Sci. Results* (Lond.), **2**.

Scorza, E. P., 1964. "Duas rochas alcalinas das Ilhas Martin Vaz: notas preliminares e estudos," *Div. Geol. Mineral., Dept. Nacl. Prod. Mineral. Num.* **121**, 1–7.

Smith, A. C., and Burri, C., 1933. "The igneous rocks of Fernando de Noronha," *Schweiz. Mineral. Petrogr. Mitt.,* **13**, 405–434.

Tilley, C. E., and Long, J. V. P., 1967. "The porphyroclast-minerals of the peridotite-mylonites of St. Paul's Rocks (Atlantic)," *Geol. Mag.,* **104**, 46–48.

Veltheim, R. V., 1950. "Contribucao à geologia de ilha da Trindade," *Anais Acad. Brasil. Cienc.,* **22**(4), 463–469.

Washington, H. S., 1930. "The petrology of St. Paul's Rocks (Atlantic)," in *Report on the Geological Collections Made During the Voyage of the "Quest," Shackleton–Rowett Expedition in 1921–22.* London: Brit. Mus. Nat. Hist., 126–144.

Wiseman, J. D. H., 1965. "Petrography, mineralogy, chemistry and mode of origin of St. Paul Rocks," *Proc. Geol. Soc.,* **1626**, 146–147.

Wright, R., 1965. "Ramification of extreme age of St. Peter and St. Paul Rocks," *Bull. Am. Assoc. Petrol. Geologists,* **49**, 1709–1712.

Cross-references: *Bermuda; Brazil; South America.*

FIJI

Fiji is a group of more than 300 islands (Fig. 1). Suva, the capital, is about 3200 km NE of Sydney and 1230 km SW of Western Samoa. Ninety-five of the islands are greater than 1 km² in area, and 108 are permanently inhabited. Fiji was ceded to Britain in 1874; Rotuma, a Polynesian island about 640 km north of Suva, was likewise ceded in 1881 and was thereafter administered as part of the Crown Colony of Fiji. The colony received its independence in 1970 as a dominion within the British Commonwealth.

The largest island, Viti Levu, has an area of 10,384 km² and rises to a height of 1323 m. The Rewa River, the largest in Fiji, drains almost one-third of the island and enters the sea near Suva. One branch, the Wainimbuka, is Fiji's longest river (172 km). The second largest island, Vanua Levu, is also mountainous, with several large rivers. Many peaks on these islands and on Taveuni are higher than 900 m, and sixteen other islands rise above 300 m. The largest islands are mainly volcanic in origin, but some of the small ones consist wholly of limestone. Atoll reefs are common in the Lau Group (Fig. 2).

The climate is tropical but not extremely hot. There are two seasons, wet and dry, which correspond to the Southern Hemisphere's summer and winter. The two main islands have pronounced wet and dry zones, with marked vegetational contrast; the northwest areas are dry grass country, whereas the southeast is mostly forest-covered. The annual rainfall in the dry zones and on the islands of the Yasawa and Lau groups averages 170 cm, whereas in the wet zones the range is from 300 cm (e.g., Suva receives 320 cm) to more than 650 cm in the high country. The average temperature range is about 20 to 30°C. The extremes officially recorded in Fiji are 39 and 7°C, but temperatures of less than 4°C are known high in the interior of Viti Levu.

Maps at the scale of 1:50,000 are available for all the main islands and are based on aerial photographs taken in 1953; some have 50-ft contours but most have 100-ft form lines. Further photography at larger scales has since been carried out, and some large-scale instrument plots have been made.

Following earlier, localized work, systematic geological mapping at the scale of 1:50,000 was begun by the Fiji Geological Survey in 1957, and Kandavu is the only main area not yet mapped. The geological notes that follow are based largely on the work of Band, Bartholomew, Coulson, Hindle, Hirst, Houtz, Ibbotson, Rickard, Rodda, and Woodrow (references in Duberal and Rodda, 1968; and Rodda, 1967). The geology of the Lau Group is described by Ladd and Hoffmeister (1945), and that of Rotuma by Gardiner (1898) and Ibbotson (1960, unpublished report). Most of the paleontological work has been done by Cole, Coleman, Ladd, and Todd, and radiometric age determinations have been carried out mainly by Snelling and the Australian Mineral Development Laboratories. Houtz and Phillips (1963), although somewhat out of date, is the only publication dealing in detail with the economic geology of Fiji. The University of the South Pacific is situated at Suva, but at present offers

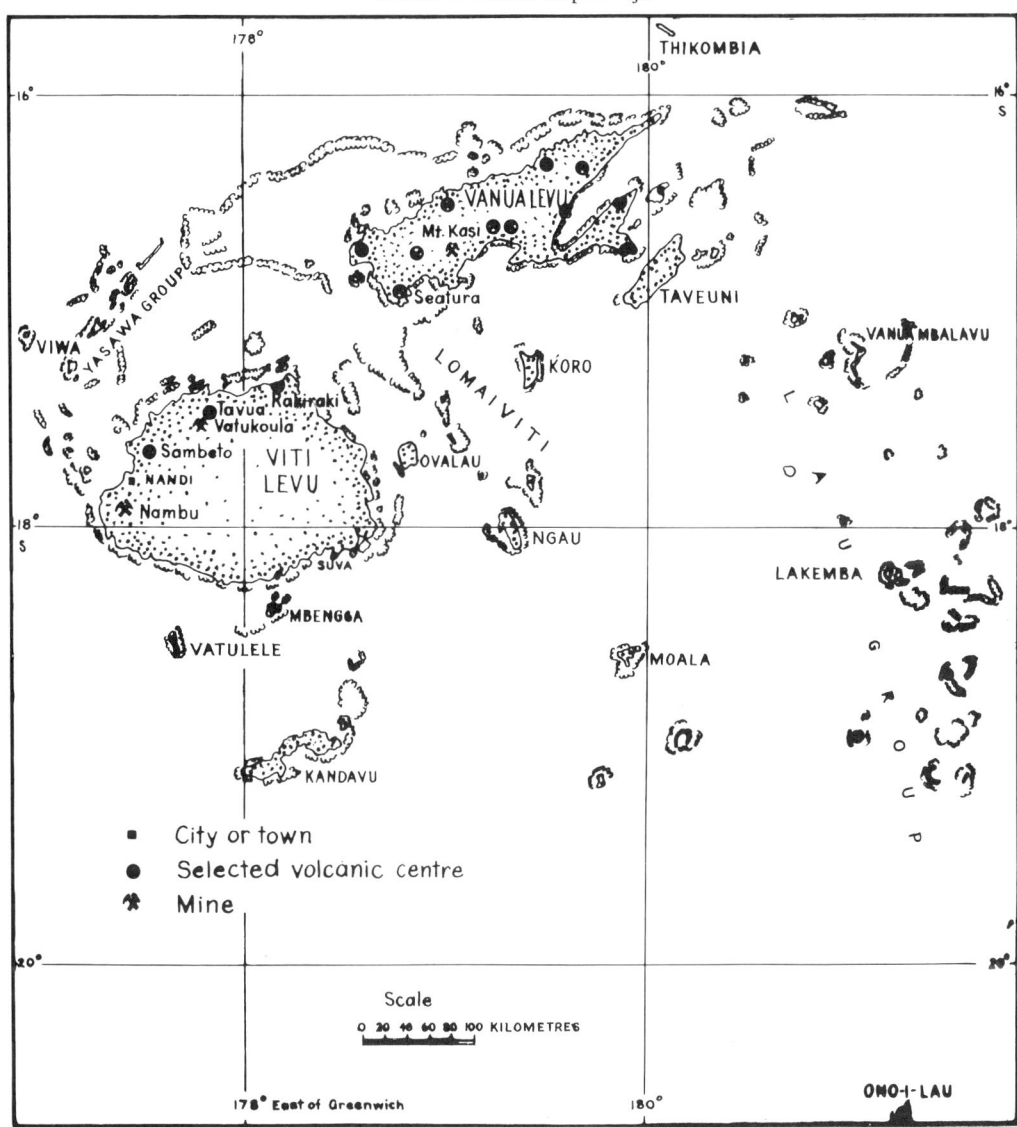

FIGURE 1. Sketch map of Fiji.

no program in geology, although some is included in high school training.

Regional Structural Setting

Crustal movements in the Fiji region have been quite complex. Chase (1971) has summarized the general picture. The Fiji Plateau is an area of oceanic crust that is relatively hot and elevated. It lies between the north end of the Tonga-Kermadec Trench and the south end of the New Hebrides Trench. The Hunter fracture zone curves eastward from the New Hebrides Trench and dies out near Fiji. This whole area lies between the Pacific and Australian crustal plates; the Pacific Plate is moving west with respect to the Australian Plate, with a relative velocity of about 10 cm per year. Fiji is close to a "triple point," with six smaller plates in the Melanesian region. The Tonga and New Hebrides island arcs are thought to have been continuous at the beginning of the Tertiary. A transform fault developed and separated the northern half (now the New Hebrides) from the southern half (now Tonga). The volcanism that built the Fiji Platform was most likely associated with this faulting. Other faults developed later, and sea-floor spreading has occurred in several places. The portion of the Fiji Plateau

FIGURE 2. Map of Lau Archipelago showing main islands (stippled) and reefs (black). (From U.S. Hydrographic Office Charts, nos. 2851 and 2852; Ladd and Hoffmeister, 1945.) This sketch map was based on the older hydrographic charts; modern spellings may be noted from north to south: Naitaumba, not Naitamba; Nggilanggila, not Ngilangilla; Namalata, not Malatta; Thakau Lasemarava, not Thakau Lasemarawa; Nayau, not Naiau; Wanggava, not Wangava; Namuka-i-lau, not Namuka.

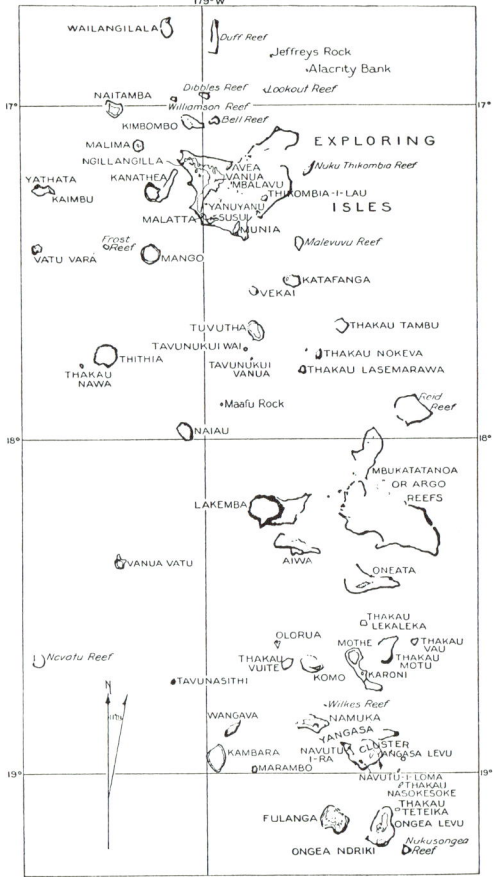

west of the Fiji Platform has probably been created within the last 10 m yr, and spreading during the last 5 m yr has separated the Lau Ridge from the Tonga Ridge. These are areas of generally high heat flow. Anticlockwise rotation of Viti Levu may have occurred, in addition to Australia and the Melanesian region (Van der Linden, 1969), and the Fiji Platform is now bounded on the north by an active transform fault. Much of Fiji is subject to earthquakes.

To the north of the Fiji Plateau is a zone of reefs and complex submarine topography known as the *Melanesian Border Plateau* (Fairbridge and Stewart, 1960). This plateau is marked by left-lateral shear (see also *New Guinea*) and block faulting, and the horsts or seamounts are covered by reefs. Some of these (e.g., Alexa Bank) are drowned atolls. The Border Plateau has been moved into its present position by the movement of the Pacific Plate, of which it is part. It is older than any of the sea floor adjacent to the south and was probably a Cretaceous Pacific Ocean archipelago. Younger volcanic islands occur on the plateau, among them Rotuma. An older idea of a now-dismembered "Melanesian Continent" or "Archeofijia" is no longer accepted.

General Geology and Stratigraphy

Viti Levu is the oldest island in the Fiji Group. The oldest known rocks are exposed in the west, SE of Nandi, and consist of andesitic volcanics and foraminiferal limestone (Tertiary b, Upper Eocene) intruded by a tonalite stock dated at 33 m yr (McDougall; see Duberal and Rodda, 1968). A thick sequence of flows and volcaniclastic rocks, ranging from tholeiitic basalt to dacite, extends across part of southern Viti Levu; two gabbro stocks intruding this sequence appear to be Eocene (K–Ar dates). From early Miocene to early middle Miocene time, epiclastic strata, including limestone lenses, were deposited with minor intercalated volcanic rocks. The total thickness of these Eocene and Miocene rocks is about 10 km. A major orogenic period occurred in the Middle and early Upper Miocene (Rodda, 1973). Folding and faulting took place, and tonalite stocks were intruded about 11 to 8 m yr ago. Uplift was followed by erosion, which exposed the tonalite in some areas.

Widespread sedimentation took place in the Viti Levu area during the later part of the Upper Miocene. Andesite was erupted from three main centers, providing part of the detritus in some sedimentary basins. Sedimentation continued into the Pliocene, and basaltic volcanoes (largely submarine) formed in the N. The Sambeto and Tavua volcanoes produced rocks of the shoshonite association, whereas the Rakiraki volcano produced basalt. The Tavua volcano is the largest of the three; a large caldera formed, and the last stages of activity took place between 5 and 4 m yr ago. The volcanic pile contains many sills, including a monzonite sill up to 300 m thick. Viti Levu is maturely dissected with many Quaternary fluvial and littoral formations. Former shorelines are emerged, up to 60 m above sea level.

The *Yasawa Group* and associated islands west of Viti Levu are formed largely of submarine tholeiitic basalt of older Miocene age, with scattered centers of dacitic to rhyolitic activity. Some andesitic volcaniclastic strata (Upper Miocene?) are also present. Viwa, a small island to the W, consists wholly of uplifted coral-algal reef of probable Pleistocene age. The island of Vatulele, south of Viti Levu, consists of uplifted and tilted limestone, with a small area of basaltic rocks which may be shoshonitic. Mbengga, to the E, is a single shield volcano, possibly Pleistocene in age, of subaerial basaltic rocks of the shoshonite association. Kandavu and its associated islands are almost wholly volcanic and seem to be made up of the products of about eight main volcanic centers, some basaltic but most intermediate.

Vanua Levu is formed largely of Miocene submarine flows and volcaniclastics. Tholeiitic basalt and basic andesite make up the major part of the island, and dacite is common in the NE. About 10 main centers of eruption have been recognized. Plugs of hypersthene and hornblende andesite are widespread, and breccia fans extend from some of them into sedimentary basins. Two main basins exist, the larger being the Ndreketi Basin (N of Mt Kasi). During the Pliocene a major change occurred; uplift took place and alkalic lavas ranging from hawaiitic basalt to trachyte were erupted subaerially to form the Seatura shield volcano. Beach deposits, coral reefs, and other shoreline features have emerged in places along the southern coast of the island.

Taveuni and *Koro* are the youngest volcanic islands in the Fiji Group. Taveuni has a core of differentiated rocks ranging from alkali basalt to trachyte; these are largely covered by subaerial flows of alkali basalt. Pahoehoe and aa lavas are common, with well-preserved lava tunnels and levees. The last stage of activity was the formation of cinder cones, of which more than 200 exist. Radiocarbon dating of an archeological site shows that activity continued up to less than 2000 years ago. On Koro, to the SW, three phases of activity have also been recognized; the rocks there are all alkali basalt, containing occasional peridotite nodules.

The islands of the *Moala Group* and Lomaiviti (apart from Koro) are mostly single volcanoes of alkali basalt, with intermediate intrusive rocks in some of the craters and calderas. Naingani, NW of Ovalau, is formed largely of massive dacite. Pillow lava makes up all of a small island just over 180 m high (Wakaya, east of Ovalau) and occurs for a short distance above sea level on several others, but most of the rocks are subaerial. Ngau has two main volcanic centers, and Ovalau has several minor centers in addition to the main one. All these islands are Upper Miocene and Pliocene, possibly ranging into the Pleistocene.

The islands and atoll reefs of the *Lau Group* have probably all had a similar history. In general, the islands are higher in the N and the W of the group; the youngest rocks are found in the S and the oldest farther N. The atoll reefs occur mostly in the E. The known history begins with a period of volcanic activity which produced olivine-free basalt, andesite, and some dacite. During the Lower Miocene (Tertiary *f*), extensive deposits of limestone were laid down. This limestone occurs on most islands. After some erosion, olivine basalt was erupted in several places, and limestone of Mio-Pliocene age overlies this on two islands. Vanua Mbalavu is the only island on which all four of these formations occur. Uplifted Pleistocene coral-algal reefs are found on several of the islands in the S of the group, reaching 27 m (90 ft) on Fulanga (Ladd and Hoffmeister, 1945). A higher reef limestone reported at 233 m (770 ft) on Tuvutha (Schofield, 1971) is actually Tertiary *f* zone. Basalt of probable Pleistocene age occurs on one island (Mango) in the N, and several islands show considerable subsidence that preceded the mid-Holocene emergence of about 2 m.

Rotuma, situated at $12°30'S$, $177°05'E$, is 12 km long and 47 km^2 in area. It is surrounded by a barrier reef 1 to 10 km offshore with depths of up to 50 m in the lagoon. Flows and minor fragmental rocks make up the major part of the island, and were apparently erupted from E-W fissures, forming a plateau. Cones of glassy tuff and agglomerate rise from this plateau to about 260 m above sea level. About 20 craters exist; lava tunnels extend from the bottoms of some. All the exposed rocks are basalt and olivine basalt and are probably Pleistocene.

Thermal springs occur at 25 localities on Vanua Levu; some of them boil and occasionally display geyser activity. Others, none of them close to 100°C, occur at 15 localities on Viti Levu and at places on five other islands. All consist of heated meteoric water.

Economic Geology

Metallic mineralization is widespread on Viti Levu and Vanua Levu and occurs on various other islands. On Viti Levu, much of the mineralization is related to the Miocene plutonic activity, but some is associated with younger volcanic centers. The Tavua volcano has important deposits of gold and silver tellurides; faults associated with the caldera, and dikes, are the main controls of deposition. Mineralization in Vanua Levu and elsewhere is largely related to volcanic centers, and a copper deposit in northeast Vanua Levu appears to have been formed by volcanic exhalations.

Many deposits of secondary manganese ore have been found and most are associated with limestone, which apparently caused the precipitation of manganese minerals from streams and groundwater. Bauxite has formed on basalt plateaus in southwestern Vanua Levu. Phosphate occurs on some islands in the Lau Group; it is oolitic or nodular, or occurs as phosphatic clay.

Gold has dominated the mining industry since 1932, when the Mt. Kasi mine on Vanua Levu produced 9.67 kg. Before it finally closed in 1946, the mine produced 2040.1 kg of gold and 147.7 kg of silver. The Tavua goldfield at Vatukoula on Viti Levu came into production in 1933, and had produced just over 95,510 kg of gold and 31,475 kg of silver (just over 3,000,000 oz and 1,000,000 oz, respectively) by the end of 1971. It is currently producing at the rate of about 3100 kg of gold and 900 kg of silver per year. Manganese has been important in the past, 161,906 metric tons of ore having been exported from 1949 to 1969, inclusive. Nambu mine, on Viti Levu, was worked for almost all this period and produced about two-thirds of the total. Surficial magnetite boulders totaling 57,850 metric tons were collected from a nearby area from 1957 to 1963. Minor amounts of copper ore have been produced from several small mines. There is current exploration for base metals and oil (offshore).

PETER RODDA*

*Published by permission of the Director of Mineral Development.

References

Chase, C. G., 1971. "Tectonic history of the Fiji plateau," *Bull. Geol. Soc. Am.*, **82**(11), 3087–3110.
Duberal, R. F., and Rodda, P., 1968. *Bibliography of the Geology of Fiji.* Suva: Oceania Printery, 81p.
Fairbridge, R. W., and Stewart, H. B., Jr., 1960. "Alexa Bank, a drowned atoll on the Melanesian border plateau," *Deep-Sea Res.*, **7**, 100–116.
Gardiner, J. S., 1898. "The geology of Rotuma," *Quart. J. Geol. Soc. Lond.*, **54**(213), 1–11.
Houtz, R. E., and Phillips, K. A., 1963. "Interim report on the economic geology of Fiji," *Econ. Rept. Geol. Surv. Fiji*, **1**.
Ladd, H. S., and Hoffmeister, J. E., 1945. "Geology of Lau, Fiji," *Bernice P. Bishop Mus. Bull.*, **181**.
Rodda, P., 1967. "Outline of the geology of Viti Levu," *N.Z. J. Geol. Geophys.*, **10**(5), 1260–1273 (with color-printed geological map).
____, 1973. "Fiji," *in* A. M. Spencer, ed., *Mesozoic-Cenozoic Orogenic Belts.* Geol. Soc. Lond. Spec. Publ. **4**, 425–432.
Schofield, J. C., 1970. "Notes on Late Quaternary sea levels, Fiji and Rarotonga," *N.Z. J. Geol. Geophys.*, **13**(1), 199–206.
____, 1971. "Note on high sea-level evidence from Lau Islands, South-West Pacific," *N.Z. J. Geol. Geophys.*, **14**(1), 240–241.
Shor, G. G., Jr., Kirk, H. K., and Menard, H. W., 1971. "Crustal structure of the Melanesian area," *J. Geophys. Res.*, **76**(11), 2562–2586.
Van der Linden, W. J. M., 1969. "Rotation of the Melanesian complex and of West Antarctica—a key to the configuration of Gondwana?" *Palaeogeogr., Palaeoclim., Palaeoecol.*, **6**, 37–44.

Cross-references: *Australasia—Regional Review; New Caledonia; New Hebrides; New Zealand; Tonga.*

FRENCH GUIANA

This former French colony, "La Guyane française," now an overseas department of France (since 1947), covers 91,000 km^2 (35,135 sq mi) on the northeastern coast of South America. Bounded to N by the Atlantic Ocean, on the E and S by Brazil and on the W by Surinam, this territory also includes three small offshore islands, the "Isles du Salut," one of which especially—Devil's Island—has had some notoriety.

The coastal area is the product of accumulation, partly of muddy Amazonian sediments, partly of sandy deposits from the interior; it is divided into wet and dry savannas, sand ridges, and mangrove swamps which extend more than 30 km inland. To the S the land rises gradually in a succession of residual rocky erosion surfaces at altitudes varying from 25 m to more than 600 m.

With the exception of the coastal area, the whole country (90%) is covered by rain forest. The climate is equatorial, moderated by trade winds from the E. Mean temperature is 26°C. The principal rainy season is from April to July; a second rainy period is between February and December. Mean annual precipitation is about 3000 mm.

Gold and bauxite are the most important mineral resources.

Geology

French Guiana is part of the *Guiana Shield* (q.v.) of Precambrian age. It is only covered in the coastal area by Tertiary and Quaternary deposits. Several periods of deposition, and magmatic and tectonic activity can be distinguished (Fig. 1).

The earliest rocks of kata- and mesometamorphic facies (amphibolites, quartzites, gneisses) of about 3000 m yr ago were affected by the first Guiana granitization (sodic metamorphism) at approximately 2550–2600 m yr

FIGURE 1. Geological sketch map of French Guiana (after Carte Géol. Dét. de la France, Cayenne, 1959; 1/2,000,000 scale). *Explanation:* 1, Quaternary; 2, dolerites; 3, granites ("caraibes") and migmatites; 4, Upper Precambrian (Orapu, Bonidoro); 5, granites ("guyanais") and migmatites; 6, Lower Precambrian (Paramaca, l. de Cayenne); 7, diorite, gabbro, amphibolite.

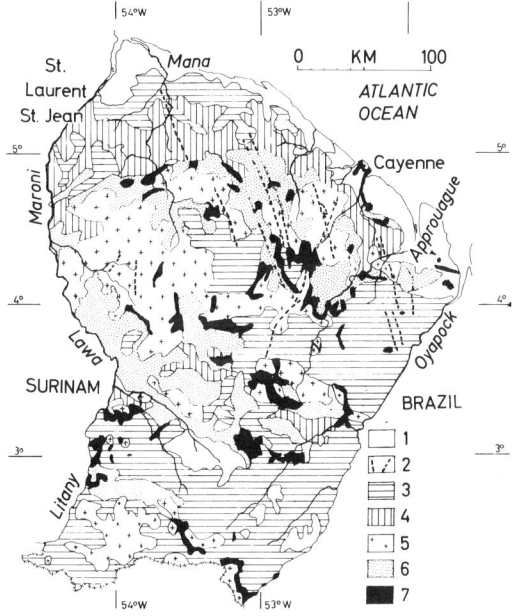

starting in the Lower Cretaceous and continued up into the Quaternary, determined the morphological evolution of the country.

The coastal plain of French Guiana is much more restricted than in Surinam and Guyana, and Precambrian bedrock even outcrops at the coastline and in the offshore islands (Iles du Salut). It has a maximum thickness near the mouths of the Maroni and Mana rivers in the NW. It is underlain by Paleocene marine and fluviomarine beds with a maximum thickness of less than 50 m, succeeded by Plio-Pleistocene sandy deposits, marine in the lower part and deltoid in the upper part. At the surface residual and fluviatile sands (Upper White Sands equivalents—Zanderij) are also of Plio-Pleistocene age. Fluviomarine and deltoid (lower part) and marine (upper part) deposits of the Série de Coswine (Old Coastal Plains equivalents—Coropina) are of Upper Pleistocene age and correspond with the Riss-Würm and Würm I-Würm II interglacial epochs. These are succeeded by the Série de Demarara (Young Coastal Plains equivalents—Demarara) marine kaolinitic clays with sandy intercalations and deposited in littoral lagoons of early Holocene (Flandrian) age. Recent and sub-Recent deposits are formed by marine and littoral muds and sandy littoral barriers. The coast is rapidly prograding today because of longshore drift ("Guiana Current") from the mouth of the Amazon.

GERRIT C. BROUWER

ago. This magmatic period, with the formation of the old "granites guyanais" (Guiana Granites), was succeeded by the Série du Paramaca (Paramaca equivalents—Barama) with a sedimentary lower part and a volcanic upper part. After a period of rejuvenation and remobilization of the "granites guyanais" at about 2000–2350 m yr ago, deposition started again with a flysch-type facies, the Série du Bonidoro (Rosebel equivalents—Cuyuni), succeeded after a short interruption by a molasse facies, the Série de l'Orapu (Armina equivalents—Haïmaraka). At approximately 1850–2050 and 2100–2200 m yr ago, this whole complex was affected by the Caribbean or trans-Amazonian granitization with the formation of metamorphic and paratectonic granites (Granite 3 equivalents—South Savanna and Younger Granites), followed by the intrusive "granites galibi." The Roraima deposits, having an age of 1500–1600 m yr, well known in Surinam and Guyana do not exist in French Guiana. Precambrian history ends with doleritic injections at approximately 1700 m yr. A second period of doleritic injections took place in Permo-Triassic times.

Several periods of peneplanation, probably

References

Aubert de la Rüe, E., 1953. *Reconnaissance Géologique de la Guyane Française Meridionale.* Paris: O.R.S.T.O.M., 127p.

Barruol, J., 1966. "Sur l'établissement de la carte géologique en Guyana française—méthodes—résultats," *Brazil, Div. Geol. Mineral., Avulso,* **41**, 153–156.

Boyé, M., 1960. "La géologie des plaines basses entre Organabo et le Maroni," *Mém. Carte Géol. Dét. de la France.* Paris: Dept. de la Guyane Fr., 148p.

Choubert, B., 1949. *Géologie et Petrographie de la Guyane Française.* Paris: O.R.S.T.O.M., 120p.

———, 1955. "Corrélation entre les phases de laterisation des Guyanes et les glaciations quaternaires," *C. R. Acad. Sci.,* **241**(1), 75–76.

———, 1957. "Essai sur la morphologie de la Guyane," *Mém. Carte Géol. Dét. de la France.* Paris: Dépt. de la Guyane Fr., 43p.

———, 1960. "Les granites precambriens des Guyanes et leur origine probable," *Mém. Carte Géol. Dét. de la France.* Paris: Dept. de la Guyane Fr., 176p.

———, 1964. "Ages absolus du Précambrian guyanais," *C. R. Acad. Sci.,* **258**, 631–634.

———, 1969. "Les Guyano-éburneïdes de l'Amérique du Sud et de l'Afrique occidental (essais de comparaison géologique)," *Bull. B.R.G.M.* (Paris), Ser. 4(4)2, 36–68.

———, 1974. "Le Précambrien des Guyanes," *Mém. B. R. G. M.* (Paris), **81**.

Lelong, F., 1969. "Nature et genèse des produits d'alteration de roches cristallines sous climat tropical humide (Guyane française)," *Sci. de la Terre Mem. 14* (Nancy), 188p.

Priem, H. N. A., et al., 1966. "Isotopic age determinations on Surinam rocks," *Geol. Mijnb.,* **45**(1), 16–19.

Spooner, C. M., et al., 1971, "Rb-Sr whole-rock age of the Kanuku complex, Guyana," *Bull. Geol. Soc. Am.,* **82**, 207–210.

Cross-references: *Brazil; Guiana Shield–Regional Review; Guyana; South America; Surinam.*

FRENCH POLYNESIA

The Territory of French Polynesia consists of about 130 islands and atolls, covering about 4000 km^2 (1500 sq mi), and widely scattered in the south-central Pacific Ocean, approximately between 7 and 27°S and 134 and 155°W. These island groups are the *Society Islands* (q.v.), which include *Tahiti* (q.v.) and Moorea, the *Marquesas Islands* (q.v.), the *Tuamotu Islands* (q.v.), the *Gambier Islands* (q.v.), and the *Tubuai or Austral Islands* (q.v.). *Clipperton Island* (q.v.), also part of the territory, lies far to the NE, at about 10°N and 109°W, southwest of Mexico. All the islands, except the Tuamotus and Clipperton, are of volcanic origin, usually surrounded by coral reefs and lagoons.

The climate is generally hot and humid but is moderated somewhat by the steady S.E. trade winds, except in the southern summer season (December–March).

RHODES W. FAIRBRIDGE

References

Aubert de la Rüe, E., 1958. "Observations aux le volcanisme tertaire et quaternaire de quelques îles de la Polynesia Française," *Bull. Volcanol.,* Ser. 2, **19**, 159–178.

———, 1959. "Etude géologique et prospection minière de la Polynésie française," *Recherche Géologique et Minérale en Polynesie Française.* Paris: L'Inspection Générale des Mines et de la Géologie, **7–42**.

Chubb, L. J., 1927. "The geology of the Austral or Tubuai Islands," *Quart. J. Geol. Soc. Lond.,* **83**, 291–316.

———, 1930. "Geology of the Marquesas Islands," *Bernice P. Bishop Mus. Bull.,* **68**, 1–71.

Crossland, C., 1928. "Coral reefs of Tahiti, Moorea, and Raratonga," *Linn. Soc. Lond. J. Zool.,* **36**, 577–620.

Davis, W. M., 1928. "The coral reef problem," *Am. Geogr. Soc. Spec. Publ.* **9**, 596p.

Guilcher, A., Berthois, L., Doumenge, F., 1969. "Les recifs et lagons coralliens de Mopelia et de Bora-Bora (Iles de la Société) et quelques autres recifs et lagons de Comparsison (Tahiti, Scilly, Tuamotu occidentales); morphologie, sedimentologie, fonctionnement hydrologique," *Fr. Off. Rech. Sci. Tech. Outre-Mer Mem.* **38**, 103p.

Lacroix, A., 1928. "La constitution lithologique des îles Polynesie australe," *C.R. Acad. Sci.,* **187**, 368.

Obelliane, J. M., 1955. "Contribution à l'étude géologique des îles des Establisments français de l'Océanie," *Sci. de la Terre* (Nancy), 3(3), 1–146.

Tercinier, G., 1955. "Rapport d'une mission aux etablissements français de l'Océanie," *Etude des Sols–Leurs Propriétés et Vocations.* Nouméa: Inst. Français d'Océanie, fasc. 1, 1–129.

Wiens, H. J., 1962. *Atoll Environment and Ecology.* New Haven: Yale Univ. Press, 532p.

Williams, H., 1933. "Geology of Tahiti, Moorea, and Maiao," *Bernice P. Bishop Mus. Bull.,* **105**, 1–90.

Cross-references: *Clipperton Island: Gambier Islands; Marquesas Islands; Oceania; Society Islands; Tahiti; Tuamotu Islands; Tubuai Islands.*

G

GALÁPAGOS ISLANDS

The Galápagos Islands, or Las Islas Encantadas ("The Enchanted Islands") as they were often called by the early Spaniards, are a possession of Ecuador, officially named the *Archipiélago de Colón,* situated in the eastern Pacific Ocean astride the equator about 925 km W of the mainland. The archipelago (Fig. 1), consisting of some 15 islands, as well as numerous rocks and islets, extends from about longitude 89°15'30" to 92°1'W and latitude 1°24'S to 1°40'N according to the 1959 revision of U.S. Hydrographic Office chart H.O. 1798. The official names of the islands are Spanish but all have English names as well. The name *Galápagos* is derived from the Spanish *galápago* (tortoise), used for the large land tortoises that formerly abounded in the islands.

The largest island, Isabela (or Albemarle), is about 145 km long and 80 km wide but has few inhabitants. Five large volcanoes are present on Isabela, with "Volcán Wolf" in the N rising to an elevation of about 1700 m, the highest point in the archipelago. The volcanic cone of Fernandina (Narborough) Island rises to nearly 1500 m and on San Salvador (James) and Santa Cruz (Indefatigable) the volcanic peaks reach nearly 900 m. On the other islands the maximum elevations are less.

Santa Cruz Island, more or less oval in outline, has a maximum diameter of about 42 km and includes the settlement of Academy Bay (with the Charles Darwin Research Station). San Cristóbal (Chatham), the easternmost island, measures 48 by 18 km. Wreck Bay, at the southwestern end of San Cristóbal, is the seat of government for the islands. Most of the inhabitants of the archipelago reside on the southern end of San Cristóbal and around Academy Bay on Santa Cruz, although a few are found on the southern end of Isabela, around the military airfield on Baltra, the salt mine on western San Salvador, and on Santa María.

Although the archipelago is astride the equator, the climate is not tropical but subtropical, with the lowland vegetation being mostly of thorn scrub and cactus types. Recorded data on precipitation and temperatures are few. Precipitation in the lowlands is very low in most years, although there commonly are low clouds or fog at night. Mean sea surface temperatures at Wreck Bay, San Cristóbal, varied from a high of about 25°C in February to a low of 18.5°C in August over a recent 2-year period. In some years the warm "El Niño" current (q.v., Vol. II) shifts southward and concurrently the tropical rainfall belt moves with it, so heavy rainfall may occur in the Galápagos in those years. Even in normal years large but unknown amounts of precipitation occur in the higher parts of Fernandina, Isabela, Santa Cruz, San Cristóbal, and Santa María, the islands that have the higher volcanoes. In consequence the vegetation may be very lush locally on these islands.

Because the volcanic rocks are mostly porous, much of the water sinks into the ground and surface runoff occurs only during the time of precipitation and shortly thereafter. Probably because so little water is retained at the surface, the usual tropical weathering pro-

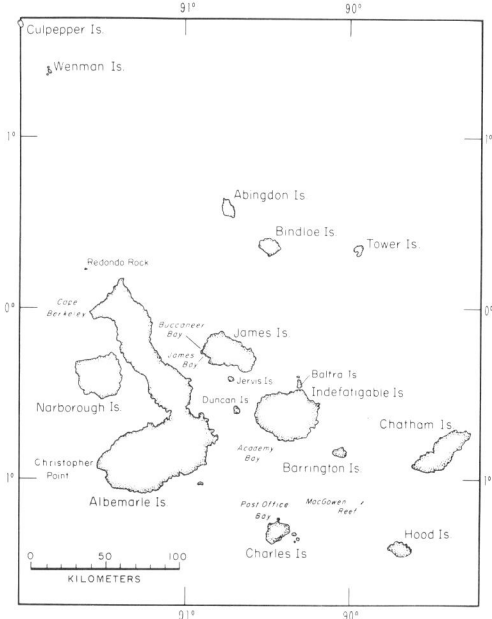

FIGURE 1. Galápagos Archipelago. Most of the islands have at least two names. Those given here are the early English names, which are often used in scientific literature, but equivalent Spanish names are given in the text.

cesses have been slower than normal and soils are usually thin. For the same reason there are few inland springs and only one presumably permanent small stream (at Freshwater Bay on southern San Cristóbal). However, there are brackish water springs at or near the intertidal zone around the larger islands. This virtual absence of potable water has been a major deterrent to settlement of the islands.

Structural Setting

The Galápagos Islands rise as peaks from the submerged Galápagos platform, which is at the western end of the elongated E-W-trending submarine Carnegie Ridge and at the eastern end of the Galápagos submarine fracture zone. The Carnegie Ridge, in turn, is separated from the mainland by the deep Peruvian Trench. The submarine Cocos Ridge extends southwest from Costa Rica to the Galápagos with average depths to the top being of the order of about 1800 m. The Galápagos platform is nearly divided by a deep reentrant submarine valley opening to the northwest. This valley seems to be related to the eastern termination of the Galápagos submarine fracture zone as well as to the major NW-trending fracture pattern of the region. South of this valley the depth to the top of the platform is generally less than 180 m, although it increases eastward. North of the valley the depths are greater and seem to be related to those of the southern end of the Cocos Ridge. Darwin, Wolf, Pinta, Marchena, and Genovesa islands are N of this valley. It seems probable that the platform is largely formed by the coalescence of aprons from the major volcanoes (and probably other submarine volcanoes as well). The southern and western edges of the platform are steep escarpments nearly 2900 m (10,000 ft) high that rise out of oceanic depths of about 3200 m (1800 fathoms). Individually the northern islands rise as prominently, about 1800-2100 m (600-7000 ft) above the sea floor, as the southern islands rise above the platform even though their subaerial heights are much less.

The major islands of the Galápagos have been built by the volcanoes rising from the Galápagos platform. Isabela Island is formed by the coalescence of six large volcanoes, but all the other larger islands, except San Cristóbal, have been built by a single large volcano with its accompanying satellites. San Cristóbal seems to have two principal volcanic centers. All but one of the seven large volcanoes on Isabela and Fernandina islands are of the shield type and have large summit calderas encircled by small spatter and cinder cones most of which occur along arcuate lines. Lower down on the flanks there are swarms of similar small cones along radial fissures. On the other islands most of the volcanoes have passed beyond the mature stage of development so well displayed on Isabela and Fernandina and lack the summit calderas and surrounding regularly aligned small cones. Instead, abundant large cinder cones are usually scattered around their flanks.

As in other oceanic islands the materials erupted by the volcanoes are almost all basaltic (mostly tholeiitic and alkalic basalts), although locally there may be considerable diversity. McBirney and Williams (1969) have noted that the Galápagos suite differs from the Hawaiian suite (often considered typical of oceanic rocks) "in that the principal differentiates are tholeiitic and are uniformly more siliceous..." and have hypothesized that the Galápagos magmas may have originated at shallower depths than those of the Hawaiian Islands. On Isabela and Fernandina prominent streams of dark lavas of recent date flow down the flanks of the volcanoes. On the other islands conspicuous but less abundant fresh flows are common. As a result of these flows the terrain is rough and jagged and travel by land is difficult. Along the coasts and on the smaller islands, beds of palagonitic tuff and pillow basalt are common.

Age of the Islands

Prior to the 1964 Galápagos International Scientific Project, few marine fossils had been found except on terraces or elevated beaches. The recorded occurrences had usually been interpreted to indicate purely local volcanogenic movements such as those at Urvina Bay on the western coast of Isabela Island a few years ago. Here, early in 1954 an uplift of about 4.5 m (15 ft) occurred, probably due to movements of the magma in nearby Volcán Alcedo, which erupted later in the year. As a consequence, molluscan shells and remains of other marine invertebrates abound in the uplifted area. A major discovery of the 1964 expedition was the finding by J. W. Durham and V. A. Zullo (Durham, 1965) of several fossiliferous late Miocene limestones interstratified with volcanic layers along the northeastern coast of Santa Cruz Island for a distance of about 12 km; the dips are generally low: nevertheless the thickness must be 1000 m or more. Here the oldest rocks are reddish-brown lapilli tuffs at least 600 m (several hundred feet) thick, containing blocks of lava of all sizes and blocks of fossiliferous marine limestone (see Fig. 2). A fossiliferous marine limestone, up to 3 m thick, overlies

FIGURE 2. Cerro Colorado, northeast Santa Cruz (Indefatigable) Island: oldest (late Miocene) known rocks in Galapagos reddish-brown lapilli tuffs with large clasts of fossiliferous limestone and lavas. Overlain by fossiliferous limestone and later volcanics on far right.

the lapilli tuff and in turn is overlain by pillow basalts and palagonitic tuffs of subaqueous origin. The latter are in turn overlain by thin flows of basalt and still other limestones. The fossils in the lowest limestone bed are numerous, of shallow-water facies. Their suggested late Miocene age is supported by the 600–700m (several thousand feet)-thick sequence of rocks that are stratigraphically located above the fossiliferous limestone and below a lava for which Cox and Dalrymple (1966) have reported a radiometric age of about 1.47 m yr (potassium–argon).

On Baltra Island much of the surface is a westward-tilted marine erosional surface. A similar surface is present N of Cerro Colorado on Santa Cruz truncating the section described above, and it may be continuous from there N to Baltra. Northeast Santa Cruz and Baltra islands have been uplifted and tilted slightly westward since the marine planation.

Paleomagnetic data from N and NE of the Galápagos have led Herron and Heirtzler (1967) to postulate a sea-floor spreading zone, trending E-W, north of the islands. Allan Cox (1971), using known ages of paleomagnetic reversals, has suggested that their data indicates a maximum age of 9 to 10 m yr for the islands. This is consistent with the late Miocene age suggested by the fossils. The paleomagnetic data reported by Cox and Dalrymple (1966) suggest that the oldest rocks of the archipelago are in the central area, including Santa Cruz, Santa Fe, and Baltra islands.

The observations on northeast Santa Cruz and the paleomagnetic data demonstrate that the islands have not been formed by a simple build-up of lava on the sea floor, but that there have been local uplifts, submergences, tilting, and erosion; nevertheless, there was no evidence indicating regional uplift or subsidence of the islands as a whole. The submarine topography of Cocos and Carnegie ridges and the apparent lack of guyots in this area support this conclusion.

Terrestrial Biota

The terrestrial biota of the islands is unbalanced; that is, it does not include representatives of all the major groups of plants and animals that would be expected in continental areas of similar size. Likewise the major groups that are present have in many instances developed endemic taxa on each of the major islands or they have evolved types to occupy ecologic niches that in continental areas would be occupied by nonrelated competitors. Notable instances of these types of evolution occur among the giant land tortoises (*Geochelone*), the land iguanas (*Conolophus*), the unique marine iguana (*Amblyrhyncus*), Darwin's finches (*Geospizidae*), and the tree-like composite *Scalesia* (a close relative of the garden sunflower).

Among the notable absentees in the native land biota are many plants [the vascular flora, recently revised by Wiggins and Porter (1971), totals approximately 700 taxa], including all gymnosperms; all carnivorous land mammals; all herbivorous mammals; all rodents except a woodrat; and most passerine birds (excepting mockingbirds, a warbler, a swallow, and a flycatcher as well as Darwin's finches). The snakes are represented by but a single genus (*Dromicus*), and lizards other than the iguanas by the genera *Tropidurus* and *Phyllodactylus*. Amphibians are absent. When one compares the Galápagos biota with that of adjacent South America or Central America, its lack of overall diversity is striking.

When the Galápagos terrestrial biota is examined in detail, the diversification that has occurred in certain stocks is amazing. Presumably these highly diversified groups represent either early immigrants to the islands or else types that had the genetic potential for such evolution. The diversification that has occurred in groups such as the tree-like *Scalesia*, the iguanas, the land tortoises, and the finches is marked and suggests a considerable antiquity for the original immigrants. The late Miocene age suggested by the oldest fossils is not discordant with the amount of evolution that has occurred, whereas the previously suggested Pliocene or Pleistocene ages necessitate a much faster rate of evolution.

Charles Darwin was the first naturalist to spend any substantial time in the Galápagos, staying there for 5 weeks in September and October of 1835. His acute observations there were to lead to a crystallization of ideas that changed the thinking of the scientific world. Two years later, in 1837, he wrote in his diary, "In July opened first notebook on Transmutation of species. Had been greatly struck from about the month of previous March on character of South American fossils, and species of Galápagos Archipelago. These facts (especially latter) origin of all my views." These ideas led ultimately (1859) to his *The Origin of Species*. Darwin believed that the Galápagos had never been connected to the mainland and had always been oceanic islands. In subsequent years a few investigators believed that it would have been impossible for the tortoises and some other elements of the biota to have reached the islands by sea and considered that the archipelago must have been part of a larger land mass extending westward from the American continents which had subsequently submerged to a major degree. Among the more notable believers in this theory were George Baur (1891), John van Denburg (1913), William Beebe (1924), and lately Léon Croizat (1958). Beebe considered that the biological data required a terrestrial connection to the Central American region and felt that the then known data on Cocos Ridge suggested that it had formerly been above sea level and served as the pathway to the Galápagos, which had been a single land mass instead of an archipelago before the foundering. However, most biogeographers, like Darwin, have considered the islands oceanic and accounted for the colonizing by chance dispersal over the sea. Some of these authors have favored dispersal along a route via Cocos Island following the pathway of the submarine Cocos Ridge, but others feel that the immigrants have largely been transported by chance from various sites. McBirney and Williams (1969) have recently suggested that masses of floating volcanic pumice, carried westward by oceanic currents, may well have served as a transporting agent for many of the migrants.

The data now available, both geological and biological, strongly support Darwin's view that the Galápagos Islands have always been separated from one another and from the American continents, although the data suggest a somewhat greater antiquity than Darwin postulated. The geological history of the Galápagos appears to be closely similar to that of the even more isolated Hawaiian Islands, save that the latter have a greater antiquity, probably dating from the late Mesozoic. The peculiarities of the Galápagos biota are like the Hawaiian, but developed to a slightly lesser degree, probably corresponding to the younger age of the Galápagos.

Volcanoes

The Galápagos Archipelago (Fig. 1) constitutes one of the world's largest and most active groups of oceanic volcanoes. The spacing and trends of the islands, volcanic vents, and fault scarps are controlled by two strong sets of fractures and faults. Many of the volcanoes are arranged in a more or less rectilinear manner, some along lines that trend NNW and others ENE.

The recent volcanoes fall into three general geologic and petrographic groups, one composed of the two large western islands, another the line of islands along the northeastern side of the archipelago and, third, the central and southern islands. The western islands of Fernandina (Narborough) and Isabela (Albemarle) are built of fluid tholeiitic lavas erupted from the summits and flanks of large steep-sided shield volcanoes, each of which has a summit caldera. Fernandina (Narborough) is a single large shield, but Isabela (Albemarle) (Fig. 3) consists of five major volcanoes: Cerro Azul, Sierra Negra, Alcedo, Darwin, Wolf, and the faulted remnant of a sixth at Cape Berkeley. The calderas and youngest vents are controlled by circumferential fractures around the summit; radial fractures are also common but do not form prominent rift zones as they do on Hawaiian volcanoes. The steep slopes cannot be attributed to the viscosity of the lava flows, which have been very fluid, but may result from unusual amounts of pyroclastic ejecta during earlier stages of growth or injection of sills.

Many eruptions have been recorded from the Fernandina (Narborough) volcano. In 1968 a small lava flow and strong explosions were followed by collapse of the caldera floor. On Isabela (Albemarle) Island, each of Wolf, Alcedo, Sierra Negra, and Cerro Azul have had

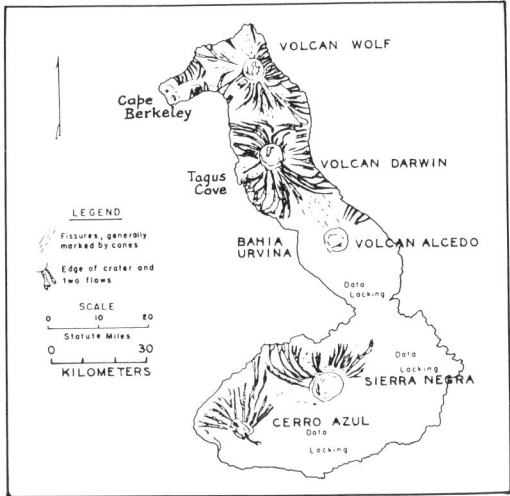

FIGURE 3. Diagrammatic map of Isabela (Albamarle) Island showing major volcanoes and lava flows (compiled by Banfield et al., 1956).

SE, consists of a cluster of pyroclastic cones and lava flows that have almost completely buried a summit caldera and the surrounding slopes of an earlier shield volcano. The lavas resemble the nonporphyritic alkali-olivine basalts of Pinta (Abingdon) Island. Genovesa (Tower) Island is a low cone with two craters, one near the summit and another on the southern coast. San Cristobal (Chatham) Island consists of two quite different halves. The older, western half is formed by a single large volcano that rises to a height of more than 700 m, with a few parasitic cones on its flanks; the eastern half lies at elevations of less than 150 m and is formed by clusters of minor cones and attendant lavas. The boundary between these two halves may be a concealed fault approximately in line with the profound submarine scarp west of Isabela and Bindloe islands.

The central islands are the most diverse in the archipelago. They have complex histories and contain the greatest variety of rock types, including strongly differentiated lavas and accidental plutonic blocks. Santiago (James) Island consists of a major volcano that occupies most of the northwestern part and a swarm of younger, relatively minor cones, mostly built along NW-trending fissures, from which floods of basaltic lava have been erupted during Holocene time. Many eruptions have been reported from the island during the last century. The small island of Rabida (Jervis) a few kilometers to the S, consists of interlocking domes and cones of strongly differentiated lavas and pyroclastic rocks. The lavas include ferrobasalts, icelandites, and siliceous trachytes, and among the accidental debris of a cinder cone on the north shore a wide range of plutonic blocks can be found that roughly parallel the compositional variations of the lavas. Pinzon (Duncan) Island, also in this central group, has similar lavas but has a longer history of activity. It has a small caldera and a broad shallow crater near the summit. Santa Cruz (Indefatigable) Island is a shield of basaltic lavas with many parasitic cones. It is uncertain whether or not it ever developed a summit caldera. The lavas are mainly alkaliolivine basalts, mostly of pahoehoe type. Floreana (Charles) Island is unlike any other in the archipelago. It is a broad shield with numerous scoria and lava cones scattered randomly over its slopes. Lava eruptions have been recorded in the last century, and all the rocks tested so far are normally polarized. The rocks are olivine-rich alkali basalts, some of which contain abundant ultramafic inclusions.

one or more recorded eruptions, and Darwin has probably had very recent eruptions that were not recorded. Alcedo has strong fumarolic activity. The lavas of all the western volcanoes are aluminous tholeiites and ferrobasalts, except for those of Cape Berkeley, which are olivine rich and slightly alkaline. Siliceous hedenbergite trachyte pumice was erupted from Alcedo volcano. Coarse-grained mafic inclusions, mainly eucrites, have been ejected from Beagle and Tagus cones, two small pyroclastic cones on the west shore of Isabela (Albemarle).

The smaller islands along the northeastern side of the archipelago have a more complex history; they show at least two periods of activity, and their rocks include both tholeiitic and magnesium-poor alkali basalts that commonly contain extremely abundant plagioclase phenocrysts. Darwin (Culpepper) and Wolf (Wenman) islands are the small eroded tops of large volcanoes that rise from depths of more than 1800 m. They lie in line with the base of a submarine fault scarp west of Marchena (Bindloe) and Pinta (Abingdon) islands. Two submarine volcanoes lie on this same line between Wolf (Wenman) Island and Pinta (Abingdon) Island. The latter consists of two distinct parts. The older part, which forms a narrow strip bordering the western coast, represents the remnant of a large volcano the western part of which has been downfaulted into the sea. The younger volcano rises to a height of 765 m and had already reached this height, principally by eruptions from the summit vent, when the latest eruptions began from fissures that trend NNW across its flanks. Bindloe, the next island to the

Chemically the rocks of the Galapagos volcanoes resemble those of other volcanoes close to the East Pacific Rise. The tholeiitic series is less magnesium rich than that of Hawaii; it displays extensive differentiation in both the effusive and intrusive rocks, with strong enrichment of iron, total alkalis, and excess silica. Alkali basalts are less common and show little differentiation.

J. WYATT DURHAM
ALEXANDER R. McBIRNEY

References

Banfield, A. F., Behre, C. H., Jr., and St. Clair, D., 1956. "Geology of Isabela (Albemarle) Island," *Bull. Geol. Soc. Am.,* **67,** 215–234.

Baur, G., 1891. "On the origin of the Galapagos Islands," *Am. Naturalist,* **25,** 217–229, 307–326.

Beebe, W., 1924. *Galapagos: World's End.* New York: Putnam, 443p.

Bowman, R. I., ed., 1966. *The Galapagos, Proceedings of the Symposia of the Galapagos International Scientific Project.* Berkeley: Calif. Univ. Press, 318p.

Chubb, L. J., 1933. "Geology of the Galapagos, Cocos, and Easter Islands," *Bernice P. Bishop Mus. Bull.,* **180,** 67p.

Colinvaux, P. A., 1969. "Paleolimnological investigations in the Galapagos Archipelago," *Mitt. Intern. Verein. Limnol.,* **17,** 126–130.

Cox, A., 1971. "Paleomagnetism of San Cristobal Island, Galapagos," *Earth Planet. Sci. Lett.,* **11,** 152–160.

_____, and Dalrymple, G. B., 1966. "Paleomagnetism and potassium-argon ages of the Galapagos Islands," *Nature,* **209** (5025), 776–777.

Croizat, L., 1958. *Panbiogeography.* Vol. 1. *The New World.* Caracas, by author, 1018p.

Darlington, P. J., Jr., 1957. *Zoogeography: The Geographical Distribution of Animals.* New York: Wiley, 675p.

Darwin, C. R., 1839. *Journal of Researches into the Geology and Natural History of the Various Countries Visited by HMS "Beagle," under the Command of Captain Fitzroy, R. N., from 1832 to 1836.* London: Henry Colburn, 615p.

_____, 1859. *On the Origin of Species by Means of Natural Selection, or the Preservation of Favored Races in the Struggle for Life.* London, 502p.

_____, 1876. *Geological Observations on Volcanic Islands Visited During the Voyage of HMS "Beagle",* 2nd ed. London, 647p.

Delaney, J. R., Colony, W. E., Gerlach, T. M., and Nordlie, B. E., 1973. "Geology of the Volcan Chico area on Sierra Negra Volcano, Galapagos Islands," *Bull. Geol. Soc. Am.,* **84,** 2455–2470.

Denburg, J. van, 1913. "The gigantic tortoises of the Galapagos Archipelago. Expedition of the California Academy of Sciences to the Galapagos Islands, 1905–1906," *Proc. Calif. Acad. Sci.,* Ser. 4, **2,** 203–374.

Dorst, J., 1970. "The Charles Darwin Foundation for the Galapagos Islands," *Nature Res.,* **6**(3), 11–14.

Durham, J. W., 1965. "Geology of the Galapagos," *Pacific Discovery,* **18,** 3–6.

Eibl-Eibesfeldt, I., 1961. *Galapagos, the Noah's Ark of the Pacific.* Garden City, N.Y.: Doubleday, 192p., trans. from the German by A. H. Brodrick.

Herron, E. M., and Heirtzler, J. R., 1967. "Sea-floor spreading near the Galapagos," *Science,* **6**(158), 775–780.

Larrea, C. M., 1960. *El Archipielago de Colon (Galapagos),* 2nd ed. Quito: Casa de la Cultura Ecuatoriana 423p. (with extensive bibliography).

McBirney, A. R., and Williams, H., 1969. "Geology and petrology of the Galapagos Islands," *Geol. Soc. Am. Mem. 118,* 197p.

Nordlie, B. E., 1973. "Morphology and structure of the western Galapagos Volcanoes and a model for their origin," *Bull. Geol. Soc. Am.,* **84**(7), 2931–2956.

Richards, A., 1962. "Active volcanoes of the Archipelago de Colon (Galapagos)," *Catalogue of the Active Volcanoes of the World.* Naples: Intern. Assoc. Volcanol., pt. 14, 1–33.

Slevin, J. R., 1931. "Log of the Schooner Academy, on a voyage of scientific research to the Galapagos Islands, 1905–1906," *Calif. Acad. Sci. Occ. Pap. 17,* 162p.

_____, 1959. "The Galapagos Islands, a history of their exploration," *Calif. Acad. Sci. Occ. Pap. 25,* 150p.

Stover, C. W., 1973. "Seismicity and tectonics of the east Pacific Ocean," *J. Geophys. Res.,* **78**(23), 5209–5220.

Swanson, F. J., et al., 1974. "Geology of Santiago, Rábida, and Pinzón Islands, Galápagos," *Bull. Geol. Soc. Am.,* **85,** 1803–1810.

Swarth, H. S., 1931. "The avifauna of the Galapagos Islands," *Calif. Acad. Sci. Occ. Pap. 18,* 299p.

Wiggins, I. L., and Porter, D. M., eds., 1971. *Flora of the Galapagos Islands.* Stanford, Calif.: Stanford Univ. Press, 998p.

Wolf, T., 1892. "El Archipelago de Galapagos," *Geografia y Geologia del Ecuador.* Leipzig: Publicado por orden del Supremo Gobierno de la Republica, pt. 5, 469–493.

Cross-references: *Ecuador; Oceania; Peru; South America; United States–Hawaii.*

GAMBIER (MANGAREVA) ISLANDS

The Mangareva group or Iles Gambier is the most southeasterly of the groups in French Polynesia, in the south-central Pacific. The *Pitcairn* group follows farther southeast in the same trend (q.v.).

The islands occur within a single barrier reef ring, 35 km across (N-S), which is submerged 10 m on the S side. Included inside the ring are eight small volcanic islands. The chief ones are Mangareva, Taravai, Aukena, and Akamaru, the maximum elevation being 441 m at Mt. Duff. The principal rocks are olivine basalts and picrites. The former lava flows are deeply dissected and the valleys, graded to low sea levels, are now partly infilled. There are small rim islets on the barrier reef along the northeastern side. The lagoon reaches a maximum depth of 80 m.

In addition to the principal barrier ring complex, there is a small atoll nearby, Timoe.

The sea-floor spreading history suggests that the oceanic crust here is of Eocene age and this therefore represents a maximum age for the volcanic foundations.

RHODES W. FAIRBRIDGE

References

Abrard, R., 1923. "Contribution à l'étude géologique des îles Marquises et Gambier," *C. R., Congrès Soc. Savantes, Dijon*, 4–158.

Agassiz, A., 1906. "The eastern tropical Pacific," *Mem. Mus. Comp. Zool.*, **33**, 1–75.

Aubert de la Rüe, E., 1959. "Etude géologique et prospection minière de la Polynésia Francaise," in *Recherche Géologique et Minérale en Polynésie Française*. Paris: L'Inspection Generale des Mines et de la Geologie, 17–42.

Davis, W. M., 1928. "The coral reef problem," *Am. Geog. Soc. Spec. Publ.* **9**, 596p.

Wiens, H. J., 1962. *Atoll Environment and Ecology*. New Haven: Yale Univ. Press.

Cross-references: *Oceania; Pitcairn Islands; Society Islands; Tahiti.*

GILBERT AND ELLICE ISLANDS

Lying in the NW-SE trend that runs through the *Marshall Islands* (q.v.), *Wallis* (q.v.), *Western Samoa* (q.v.), and *American Samoa* (q.v.) are two important chains—the *Gilbert Islands* (3°N–3°S, 173–177°E) and farther south the *Ellice Islands* (5–11°S, 176–179°E). The group is administered by the United Kingdom. Also included is an isolated phosphate island, *Ocean Island*, which lies 500 km to the west (see below). The best known of the Gilberts is the Tarawa Atoll, scene of a particularly violent battle in World War II. It is now the seat of government.

In the Gilberts there are eleven atolls and five coral islands without lagoons (Agassiz, 1903). In the atolls the lagoon depths decrease southward (Nugent, 1946). The Ellice Islands are all atolls from 2 to 20 km in diameter. Funafuti is the largest and has 30 islets on the rim reef.

Funafuti became famous in scientific circles when it was made the site of the first deep borings in a coral reef, organized by the Royal Society of London to test the Darwinian theory of atolls (Sollas, 1904; David and Sweet, 1904). The maximum depth penetrated was 345 m (1140 ft).

The Funafuti cores obtained have been widely studied and discussed, the lower part (below 193 m, 638 ft) being almost completely dolomitized, the whole section consisting of fragmented coral, foraminiferal sands, and lagoon carbonates. The uppermost 66 m (220 ft) includes aragonitic material, decreasing downward and considerably leached. The next part (66–193 m) discloses important mineral changes due to chemical reactions with sea water (Schmalz, 1956). Fairbridge (1950, 1957), following suggestions by Udluft and Reuling, believed that the dolomitization was in part triggered by burial pressure on biogenic high-magnesium calcites in the metastable lagoonal deposits, a process followed by general

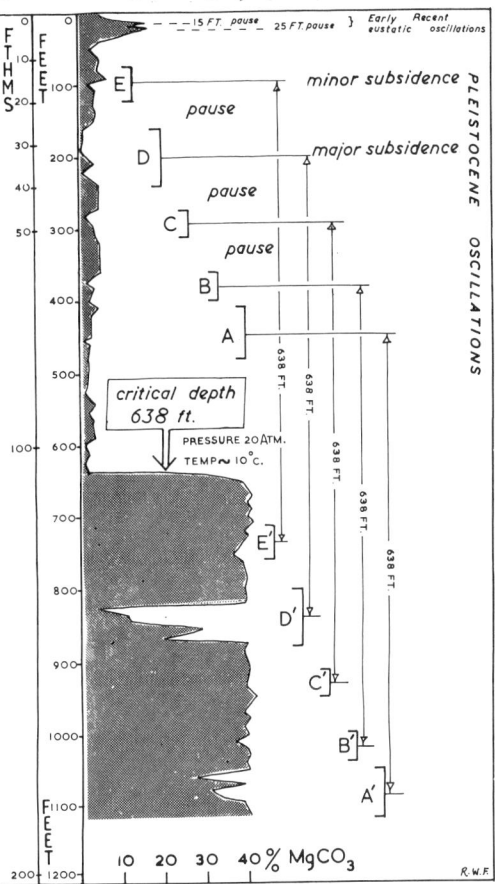

FIGURE 1. Magnesium carbonate analyses of the Funafuti atoll bore, as interpreted by Reuling. Note how the $MgCO_3$ curve is mostly under 5% in the upper part of the bore, but below 195 m (638 ft) is mostly about 40%. Coincidences of minimal peaks of $MgCO_3$ are always separated by 195 m (638 ft) ±6 m, i.e., suspected range of dolomitization). Minima may correspond with times of most rapid submergence and maxima with times of stability or slow submergence. The curious peaks at 4.5 and 8 ft correspond to known low stands of the sea in early Holocene time (from Fairbridge, 1957).

replacement. The porosity of the surrounding and underlying reef framework renders possible a diagenetic reflux of sea water that brings in additional magnesium (see Fig. 1). Different opinions exist concerning this reflux: possibly it is due to the evaporation of lagoon waters leading to the downward migration of high-density supersaline solutions, which carry out dolomitization on reaching a certain critical pressure with depth (nearly 200 m); the critical depth is quite abrupt but not necessarily related to the present sea level.

By seismic refraction, Gaskell and Swallow (1954) established that the truncated top of the volcanic cone of Funafuti was now at a depth of about 760 m (2500 ft). According to seafloor spreading theory and paleomagnetic dating, the oceanic crust here may be early Cretaceous, and the volcanic cone late Cretaceous. Presumably it is capped by Tertiary pelagic carbonates as at Eniwetok. The coral accumulation occupies the upper 520 m (1800 ft) and is believed to be largely Quaternary. These authors concluded that the submergence is due to isostatic subsidence because of the volcanic and sedimentary load (see comments under *Marshall Islands;* in view of the discontinuous *drowning,* the process is believed by Fairbridge to be dominated by the progressive but irregular shifts of pole position and the Pacific Plate during the Tertiary period; the relative pole shift was *away* from the central Pacific quadrant), causing geodetic rises of sea level in these latitudes. At the beginning of the Tertiary period the Gilbert and Ellice part of the Pacific Plate may have been around 25°S latitude.

Ocean Island (0°52'S, 169°36'E) is a phosphate island in an isolated position, a reef-capped volcanic seamount far removed from the main trend of the Gilbert-Ellice chain. It is now an uplifted atoll reaching 81 m, which was elevated in steps, producing slightly tilted terraces separated by steep cliffs. A shark's tooth in the limestone outcrop suggests a late Tertiary age (Owen, 1923). The phosphate was emplaced during one or more low sea-level stages and averages 15 m in thickness. It rests on a jagged karst of limestone pinnacles. There is also a contemporary fringing reef.

RHODES W. FAIRBRIDGE

References

Agassiz, A., 1903. "The coral reefs of the tropical Pacific," *Bull. Mus. Comp. Zool.,* **28,** 1–410.

Brigham, W. T., 1900. "An index to the islands of the Pacific," *Mem. Bernice P. Bishop Mus.,* **1.**

David, T. W. E., and Sweet, G., 1904. "The geology of Funafuti," *The Atoll of Funafuti.* London: Roy. Soc., Coral Reef Comm., Sec. 5, 61–124.

Ellis, A. F., 1936. *Ocean Island and Nauru.* Sydney.

Fairbridge, R. W., 1950. "Recent and Pleistocene coral reefs of Australia," *J. Geol.,* **58,** 330–401.

——, 1957. "The dolomite question," *Regional Aspects of Carbonate Deposition, Soc. Econ. Paleontologists Spec. Publ.* **5,** 125–178.

Gardiner, J. S., 1898. "The coral reefs of Funafuti, Rotuma and Fiji," *Proc. Camb. Phil. Soc.,* **9,** 417–503.

Gaskell, T. F., and Swallow, J. C., 1954. "Seismic experiments on two Pacific atolls," *Challenger Soc. Occ. Pap. 3,* 1–8.

Grimsdale, T. F., 1952. "Cycloclypeus (Foraminifera) in the Funafuti boring, and its geological significance," *Challenger Soc. Occ. Pap.* **2,** 11p.

Kaplan, P. A., Leontyev, O. K., Medvedev, V. S., and Nikiforov, L. G., 1972. "Some geomorphological features of atolls of the Ellice, Fenix, and Gilbert Archipelagos," Moscow Univ., *Herald,* **6,** 45–52 (in Russian).

Ladd, H. S., 1958. "Fossil land shells from western Pacific atolls," *J. Paleontol.,* **32**(1), 183–198.

——, 1968. "Fossil land snail from Funafuti, Ellice Islands," *J. Paleontol.,* **42**(3), 837.

Nugent, L. E., Jr., 1946. "Coral reefs in the Gilbert, Marshall and Caroline Islands," *Bull. Geol. Soc. Am.,* **57,** 735–780.

——, 1948. "Elevated phosphate islands in Micronesia," *Bull. Geol. Soc. Am.,* **59,** 977–994.

Owen, L., 1923. "Notes on the phosphate deposits of Ocean Island with remarks on the phosphates of the equatorial belt of the Pacific Ocean," *Quart. J. Geol. Soc. Lond.,* **79,** 1–15.

Power, F. D., 1925. "Phosphatic deposits of the Pacific," *Econ. Geol.,* **20,** 266–281.

Reed, P. R. C., 1903. "Notes on Ocean Island," *Geol. Mag.,* **10,** 298–300.

Schmalz, R. E., 1956. "The mineralogy of the Funafuti drill cores and its bearing on the physiochemistry of dolomite," *J. Sed. Petrology,* **26,** 185–186.

Skeats, E. W., 1918. "The coral reef problem and the evidence of the Funafuti boring," *Am. J. Sci.,* **45**(4), 81–90.

Sollas, W. J., 1904. "Narrative of the expedition in 1896," in *The Atoll of Funafuti.* London: Roy. Soc., Coral Reef Comm., 1–28.

Weber, J. N., and Woodhead, P. M. J., 1972. "Carbonate lagoon and beach sediments of Tarawa Atoll, Gilbert Islands," *Atoll Res. Bull.,* **157,** 21p.

Cross-references: *Caroline Islands; Fiji; Marshall Islands; Nauru; Oceania; Wallis and Futuna Islands; Western Samoa.*

GREENLAND

Greenland is the largest island in the Arctic Archipelago of North America. It is administered by Denmark, the principal settlement being Godthaab. The total area of Greenland,

GREENLAND

FIGURE 1. Sketch map of Greenland with locality names.

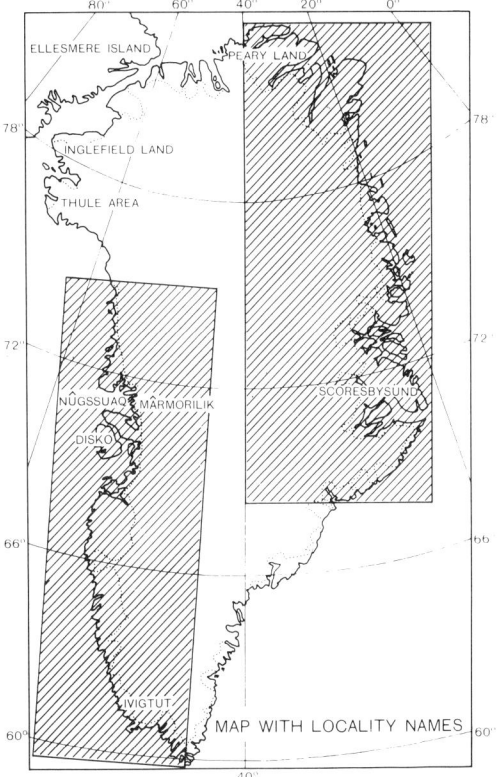

FIGURE 2. Sketch map of the Precambrian of western Greenland.

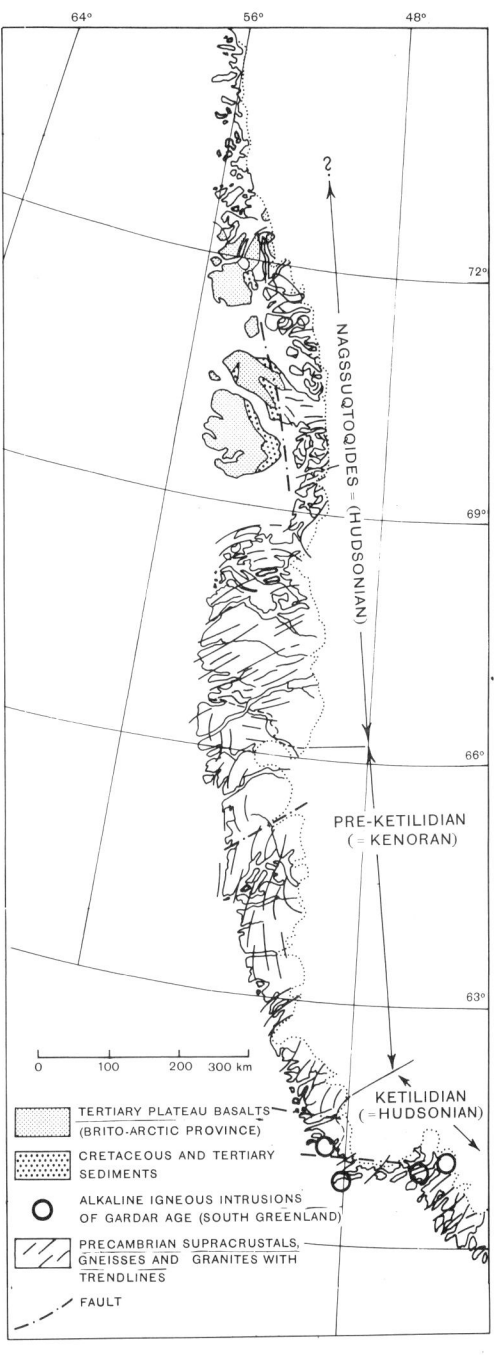

mobile belts of similar age have been recognized.

The central (Archean) block extends from 61°45' to 66°30'N on the west coast and from 62°15' to 64°10'N on the east coast. The dominating its islands, is 2,186,000 km² (844,000 sq mi), of which 1,802,400 km² is covered by the "Inland Ice" and peripheral glaciers, leaving 383,600 km² of coastal land surface. The country ranges from 60 to 84°N, and 10 to 70°W (Fig. 1). Major papers on the geology of Greenland are published primarily in *Meddelelser om Grønland* (Bulletins on Greenland), Copenhagen, or appear in the bulletins, reports, and miscellaneous papers of the Geological Survey of Greenland, Ø Stervoldgade 10, DK 1350 Copenhagen, Denmark.

Precambrian Basement

The greater part of Greenland forms the extension of the Precambrian shield of North America (Fig. 2). Along its northern and eastern margins this shield area is bounded by Paleozoic fold belts (see below). Field observations, supported by isotopic age determinations, have shown that the Precambrian basement consists of three major units. On either side of a central Archean block, two younger

nant rocks are granulite or amphibolite facies gneisses of granodioritic to quartz dioritic composition. The central block also contains remnants of several successions of metasedimentary and metavolcanic rocks and plagioclase-rich basic intrusive complexes, which locally display conspicuous chromite layering. The basic intrusives and the metavolcanic units in the gneisses can be used as marker horizons; the structural interference patterns suggest that the central block has passed through complex stages of development. Incorporated metavolcanics and metasediments are occasionally well preserved and have been interpreted as low-grade metamorphosed remnants of supracrustal rock overlying an older, reworked basement which has been included in the regional low-pressure granulite facies metamorphism, approximately 3000 m yr ago. Within the high-grade metamorphic complex, there are enclaves of amphibolite facies gneisses which apparently have escaped the granulite facies metamorphism; these have been given radiometric ages of over 3750 m yr; that is, they are among the oldest known rocks on earth.

The central basement is cut by several swarms of basic dikes which have been ascribed to a period of volcanic activity some 2000–2600 m yr ago. The dikes are quite fresh and undeformed in the central block, but become metamorphosed and deformed by boudinage and folding in the areas affected by younger movements both to the N and to the S. The area N of the central basement is known as the *Nagssugtoqidian mobile belt*. Deformation most probably began during the period of dike intrusion—i.e., about 2500 m yr ago in this area—and continued presumably intermittently until 1700 m yr ago. Most of the area of the Nagssugtoqidian mobile belt is occupied by older gneisses which have been reworked by these later movements. However, between 71 and 72°30'N there is a well-preserved cover of geosynclinal metasediments up to 8000 m thick. The age of deposition of this sequence with respect to the earliest deformation in the southern part of the Nagssugtoqidian mobile belt is as yet unknown. The generation of extensive domes in the underlying, older gneisses has initiated gravity folding in the metagraywackes of the cover. Both the metasediments and the reworked gneisses have a K–Ar age of 1700–1800 m yr.

At their southern margin the gneisses of the central basement are unconformably overlain by a thick sequence of virtually unmetamorphosed sedimentary and volcanic rock suits, deposited at some time between 2000 and 2600 m yr ago. Traced southward the metamorphic grade of this supracrustal cover increases until the rocks are indistinguishable from the supracrustals of the central block. The southern gneiss area is known as the *Ketilidian mobile belt*.

To the N of Julianehåb (61°30'N) the gneisses and the supracrustals are cut off by an extensive area of granite and granodiorite which was emplaced during several different periods, most probably beginning around 2600 m yr ago. The active period ended with the intrusion of postkinematic calc-alkaline rock suites, including norites, pyroxene diorites, hypersthene monzonites, and rapakivi granites. The latest of these were presumably intruded 1780 m yr ago.

South of the 61st parallel the basement rocks are either metasediments or gneisses derived from them, intersected by a variety of synkinematic and postkinematic plutons. The supracrustal units, which locally have reached a granulite facies metamorphic grade, cannot be correlated with any radiometric unit north of the granite area. A radiometric age of 1830 m yr indicates the age of the last major metamorphism in this area.

After the plutonic events around 1700 m yr ago, southern Greenland was deeply eroded before the beginning of cratogenic Gardar sedimentation and magmatism. The *Gardar sediments* are continental sandstones and conglomerates which are intercalated with normal to alkaline basalt flows. Only a limited area of these surficial rocks is preserved today. Dikes and intrusive complexes from this period—comprising olivine gabbro, anorthosite, augite syenite, nepheline syenite, including peralkaline types, and alkaline granites—occur over a wider area. Radiometric dating of the late Gardar plutonic phase has given ages of about 1150 m yr (Table 1).

In northeastern Greenland the *Carolinidan fold belt* has been recognized, forming the trunk of the northern branch of the Caledonian orogeny.

Late Precambrian and Paleozoic: The Caledonian Orogeny

In northern and eastern Greenland crystalline rocks are overlain by late Precambrian and early Paleozoic sediments of continental or epicontinental origin. The basement is only exposed locally at the edge of the Inland Ice. Farther away from the margin of the Precambrian shield, the foreland cover strata merge into the thick geosynclinal deposits of the Paleozoic fold belts.

East Greenland Fold Belt. The fold belt of eastern Greenland represents the western part of the Caledonian system, which is found on both sides of the North Atlantic Ocean (Fig. 3). In eastern Greenland it flanks the Precambrian

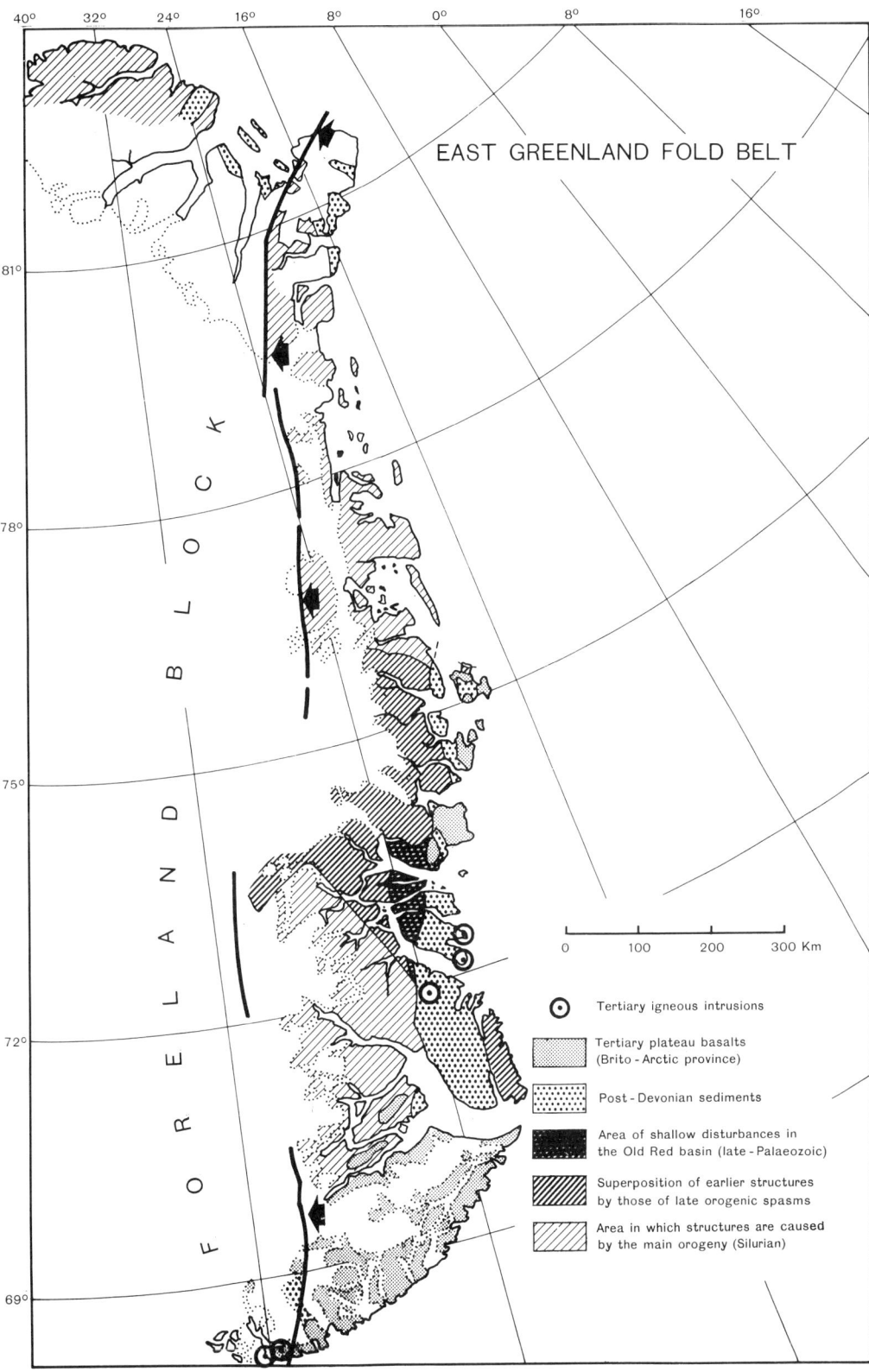

FIGURE 3. Sketch map of the East Greenland Fold Belt.

TABLE 1. Pre-Quaternary: Stratigraphy and Major Geological Events in South Greenland

Time	Rocks		Deformation and Crustal Movements
	Supracrustal	Infracrustal	
Late Phanerozoic		Dolerite dikes	
Gardar (ab. 1275–1000 m yr)	Eriksfjord Formation Continental sandstones, alternating with lava flows	Dikes and intrusive centers of the Gardar alkaline province	Important faulting (wrench type common) throughout the period
Ketilidian	Andesitic volcanism ?		
1780 m yr		Postkinematic granites, Rapakivi granites, appinites Granites, migmatites, gneisses, and schists (culmination of regional metamorphism)	Repeated folding
1830 m yr			
	Qipisarqo Group (2000 m) Agglomerates, semipelitic sediments, conglomerate		
	Sortis Group (>2500 m) Pillow lavas, pyroclastics and some sediments	Gabbro	
	Vallen Group (ca. 1703 m) Eugeosynclinal sediments, miogeosynclinal at base, and basal conglomerate		Development of sedimentary basins
Pre-Ketilidian ab. 2600 m yr		Dolerites (Igavik dikes) (major swarms)	Faulting
		Migmatites, gneisses, and schists	Repeated folding
	Tartoq Group (at least 2000 m) Pillow lavas, pyroclastics, and sediments		
		Metamorphites?	
		Gabbro anorthosites	

shield between the 70th and the 82nd parallel.

Sedimentation in the Pre-Caledonian trough was influenced by the extension of the Carolinidian basement. A central ridge of folded gneisses at about 77°N divides the sedimentary basin into a southern and a northern region. South of that barrier the central eastern Greenland geosyncline gradually developed. The accumulated sediments reach a thickness of 16,000 m, predominantly deposited in the late Precambrian. Four cycles of sedimentation can be distinguished:

1. The *Lower Eleonore Bay Group,* up to 10,000 m thick, mainly arenaceous and argillaceous sediments and including ophiolites and other characteristic features that indicate eugeosynclinal conditions.

2. The *Upper Eleonora Bay Group,* 4000 m thick, of pelitic and calcareous sediments.

3. The *Tillite Group,* 200 to 1000 m thick, generally of miogeosynclinal character and including a tillite horizon.

4. *Cambro-Ordovician* rocks, 3000 m thick, consisting mainly of pelitic sediments; Silurian strata have not been observed in the southern region. Devonian clastics overlie the Ordovician with an angular unconformity.

In the northern region, basin-and-swell structures control sedimentation. The main elements are a marginal trough in the eastern part of Kronprins Christian Land (81°N) and a shallow basin between the truncated Carolinidian Fold Belt and the shield platform to the west. The late Precambrian sediments of both areas can probably be correlated with the Tillite Group of the southern region.

Unfolded sediments adjacent to the Caledonian mobile belt also include a distinct cycle of Paleozoic sedimentation, starting with a transgressive Lower Cambrian series, followed by Ordovician to Silurian limestones, about 2500 m thick, and topped by graptolitic shales. These shales were found to underlie an overthrust on Kronprins Christian Land, and thus provide the oldest possible age for the beginning of orogenic movement in northeastern Greenland.

The main Caledonian orogeny, during the Silurian, affected the whole central geosynclinal trough. Overthrusting toward the western shield area took place at that time. In addition to the metamorphosed equivalents of the Eleonore Bay Group of sediments, the thrust blocks include appreciable quantities of reactivated basement gneisses. According to plate tectonic concepts, the orogeny resulted from a collision of the East European shield with the North American one, joining the two northern land masses together until the Cenozoic rifting.

South of the 76th parallel late-orogenic movements occurred during the Devonian in the intramontane basin in which continental sediments of Old Red sandstone facies accumulated (see below). During this stage, superficial folding was accompanied by minor thrusting and by intrusion of granites. Before the close of the Paleozoic era, the southern part of the continental basin was affected by minor disturbances and faulting, which concluded the orogenic cycle.

The structural development in the two regions distinguished above took a different course, apparently related to the magnitude of the sedimentary sequences. In central eastern Greenland, a major part of the accumulated sediments underwent metamorphism and granitization, including the underlying basement gneisses. Large-scale folding was accompanied by migmatization and gave rise to a core of granitoid rocks enveloped by relatively brittle, low-grade metamorphosed schists. North of the Carolinidian barrier, where the strata were not so thick, tangential thrusting predominated. Migmatization affected the reworked basement gneisses almost completely.

In central eastern Greenland a 7000-m-thick pile of continental molasse accumulated in a graben-like intramontane basin that developed at the end of the main orogeny. They show a similarity to the Old Red sandstones of the British Isles. The principal groups of lower vertebrates are represented, and indicate a Middle to Upper Devonian age. The eastern Greenland Devonian is acknowledged as the only locality to yield *Ichthyostega,* the oldest known tetrapod.

Post-Devonian faulting parallel to the present coastline affected much of the outer fjord region. During the Carboniferous and Lower Permian, continental sedimentation continued. Transgression across the truncated Caledonian Fold Belt began in northeastern Greenland in the Upper Carboniferous, gradually progressing to the south, where it was completed during the Upper Permian. This marked the beginning of a cycle of predominantly marine sedimentation that persisted through the Mesozoic.

North Greenland Fold Belt. The North Greenland Fold Belt forms the eastern continuation of the Innutian orogenic system of the northern Canadian Arctic Archipelago, a complex of folded Paleozoic rocks deposited in the Franklinian geosyncline and of slightly deformed Upper Paleozoic strata. The mobile belt runs parallel to the northern coast of Greenland. In northern Peary Land (83°N) the local northern boundary is a thrust that separates Lower Paleozoic metasediments from the Kap Washington Group of presumed Tertiary volcanics. To the south the folded rocks pass into an extensive platform of late Precambrian to Silurian strata which unconformably overlie the Precambrian crystalline basement, exposed in the inner Victoria Fjord only. In the Thule area (77°N) in the western part of northern Greenland, a separate basin is found, in which Precambrian sandstones rest upon the basement.

In both trough and shelf zones of the geosyncline, rather complete sections from Cambrian to Silurian (Wenlock-Ludlow) have been observed. Early Devonian sediments exist in the trough. The strata of the platform and the shelf area are dominated by calcareous rocks, particularly in the Ordovician and Silurian sections. The rock suites of the trough zone are mainly arenaceous and argillaceous. Reef development is common in the shelf zone, especially during the Silurian, where complex facies changes occur between reefal calcareous rocks and off-reef argillites.

In eastern Peary Land, marine Pennsylvanian and Permian strata unconformably overlie the folded Lower Paleozoic rocks. Further outliers occur to the southeast, particularly in eastern Kronprins Christian Land, where deposits of Mississippian age form the base of the succession.

Paleozoic diastrophism, occurring in Devonian time, resulted in polyphase deformation and accompanying metamorphism. Three phases of more-or-less coaxial parallel folds have been recognized, the majority of which are overturned to the N toward the assumed central region of the orogen. The metamorphic grade increases in a northern direction. Along the northern coast of Peary Land, the metasediments reach high-pressure amphibolite facies conditions.

Late Phanerozoic (Cretaceous-Tertiary) diastrophism is marked by regional faulting, some folding of the Upper Paleozoic-Mesozoic-Tertiary rock suites in eastern Peary Land, by thrusting and, finally, by regional metamorphism. The major thrust in northern Peary Land pushed Lower Paleozoic phyllites and schists northward, covering the Kap Washington Group of volcanics.

Unfolded Mesozoic and Tertiary

Eastern Greenland. Mesozoic sediments are widely distributed east of the Caledonian mountain range, between the 70th and 75th parallel. In the central area Triassic strata overlie the Permian beds concordantly, grading from a lower marine into an upper continental series. The rich ammonite fauna found in the marine sediments shows affinity to that of the Lower Triassic of Canada and Spitsbergen. Vertebrates are represented by species of sharks, ganoids, crossopterygia, and stegocephalia. The continental series is correlated with the Muschelkalk and Keuper of central Europe.

In the Scoresbysund area (70°N), beds of limnic Rhaeto-Liassic up to 400 m thick contain a flora rich in species similar to those found in the Upper Triassic to Lower Jurassic of central Europe. A nearly continuous succession of predominantly marine Jurassic sediments rest conformably on the Rhaeto-Liassic sequence. Its ammonite fauna is related to the Jurassic ammonites of northern Europe. Marine Cretaceous sedimentation continued uninterrupted until the Campanian; its fauna was basically similar to that of the northern European Cretaceous. Tertiary sediments (Eocene, Oligocene, and Miocene) are only found in some isolated localities in an area that is otherwise covered by pre-Eocene basalt flows resting directly on the basement. These basalts, determined radiometrically as Paleocene (c. 60 m yr), appear to date the initiation of the first phase of rifting and sea-floor spreading in the Labrador and Norwegian seas (Vogt *et al.,* 1969, 1972).

Plateau (flood) basalts belonging to the *Brito-Arctic province* cover the region between 68 and 70°N, attaining a maximum thickness of 4000 m in the coastal area and thinning out toward the ice cap.

Syenitic and granitic intrusives of Tertiary age penetrate the Paleozoic and Mesozoic sediments north of Scoresbysund. Mineralized veins and ore shoots, carrying galena, sphalerite, chalcopyrite, and molybdenite, are associated with this intrusive period. A Tertiary syenogabbroic intrusive complex, among which is the well-known Skaergaard intrusion, is found in the Precambrian gneiss area immediately S of the plateau basalts.

Western Greenland. In western Greenland, between 69 and 72°N, Cretaceous to early Tertiary sediments are preserved below a basalt cover. The sedimentary sequence, about 2000 m thick, rests on Precambrian basement. In the lower lacustrine beds, two distinct floral groups have been recognized: the Kome flora, of Gault age, and the Atane flora, which is presumed to be Cenomanian. In the overlying predominantly marine sequence, a third floral group, the Pautût flora, of Senonian age, is found together with *Inoceramus*. On the Nûgssuaq peninsula (71°N) Senonian is unconformably overlain by a Dano-Paleocene series. Conglomerates, sandstones, and bituminous shales contain a molluscan fauna which is closely related to that of the

Europian Danian and Montian. Intercalated tuff horizons date the initiation of volcanic activity as Danian.

The Montian is also represented by a lacustrine series on the nearby island of Disko, which contains the Upper Atanikerdluk flora. The species distribution of the successive floras indicates a gradual lowering of the mean annual temperature in this region toward the end of the Tertiary.

Volcanic activity was preceded by faulting. After extrusion of pillow lavas, the whole region was gradually covered by a succession of plateau basalt flows as in eastern Greenland. The lower half are olivine basalts and picrites; after a relatively short break in the extrusive sequence, differentiation of the basalts took place toward a tholeiitic composition. Volcanic activity was concluded by the extrusion of trachyandesitic flows.

Northern Greenland. The Upper Permian sandstone section in eastern Peary Land passes upward into Lower Triassic sandstones and shales containing a vertebrate fauna of Upper Scythian age and a Middle Triassic sequence containing ammonites of Anisian age; they have affinities with the Triassic fauna of North America, Spitsbergen, and Siberia. The lacustrine beds in eastern Peary Land with plant remains and the sandstones of Kronprins Christian Land are of Cretaceous to Tertiary age.

Cretaceous-Tertiary volcanic activity is indicated by swarms of dolerite dikes in the metasediments of Peary Land and by a sequence of flat-lying, nonmetamorphosed, bedded rhyolitic and andesitic lavas and tuffs at the margin of the Arctic Ocean, bordering northern Peary Land. They have been called the *Kap Washington* Group of volcanics.

Quaternary Glaciation

The central ice cap has a lenticular shape and covers by far the greatest part of Greenland, reaching a maximum thickness of about 3000 m at "Eismitte," the winter quarters and last resting place of Alfred Wegener, the founder of the continental drift theory.

In the interior the substratum of the Inland Ice is bowl-shaped and situated approximately at present sea level. The formation of the ice cap took place between 2 and 4 m yr ago, at or before the beginning of the Quaternary. Presumably piedmont-type glaciers were gradually formed in the central lowland areas, being nourished from firn accumulation in the high mountain ranges of the coastal area toward the east. The separated glacial lobes finally merged into the ultimate ice cap, aided by a progressive change in albedo caused by the increasing areal expansion of the glaciers. Eventually equilibrium was reached between accumulation and ablation of firn ice, as is the case at the present time. It is highly probable that the ice cover existed during the whole Quaternary period. However, there are some indications that the areal extension of the Inland Ice was at times slightly less than it is today. During the Wisconsin stage, however, the coastal plateau of western Greenland was almost completely glaciated, as deduced from morphological features and from observations on postglacial uplift. Despite a spread of measured values, based on the position of upper marine limits, the accumulated data show a uniform trend of isostatic compensation, amounting to an uplift of 110 to 150 m above present sea level for the main land mass of the coastal fringe.

Measurements of the postglacial uplift also demonstrate that the present distribution of the central ice mass has remained essentially the same since Boreal time. During the climatic optimum in Atlantic and Subboreal time, the extension of the Inland Ice advanced over northern Greenland; its margin has been retreating again in Subatlantic time.

A general minor advance of coastal glaciers was initiated in medieval time; around the turn of the present century, however, this trend was reversed and the glacial lobes are again retreating. In recent years the study of drill cores extracted a thickness of ice down to 1390 m in the central area of the ice cap which has yielded data as to the long-term variations of the oxygen isotope composition of the ice; such data reflect climatic changes during the past nearly 100,000 years. Climatic oscillations with periods of 120, 940, and 13,000 years have been deduced.

JAN BONDAM

References

Allaart, J. H., Bridgwater, D., and Henriksen, N., 1969. "Pre-Quaternary geology of Southwest Greenland and its bearing on North Atlantic correlation problems," *in* M. Kay, ed., *North Atlantic–Geology and Continental Drift. A Symposium* (Mem. 12). Tulsa: Am. Assoc. Petrol. Geologists, 859–882.

Berthelsen, A., and Noe-Nygaard, A., 1965. "The Precambrian of Greenland," *in* K. Rankama, ed., *The Precambrian,* vol. 2, 113–262.

Dansgaard, W., Johnsen, S. J., Clausen, H. B., and Langway, C. C., 1971. "Climatic record revealed by the Camp Century ice core," *in* K. K. Turekian, ed.,

The Cenozoic Glacial Ages. New Haven: Yale Univ. Press, 37–56.
Dawes, P. R., 1971. "The North Greenland fold belt and environs," *Bull. Geol. Soc. Denmark,* **20,** 197–239.
____, and Soper, N. J., 1973. "Pre-Quaternary history of North Greenland," *Am. Assoc. Petr. Geol.,* Mem. **19,** 117–134.
Flinn, D., 1971. "On the fit of Greenland and Northwest Europe before continental drifting," *Proc. Geol. Assoc. Lond.,* **82**(4), 469–472.
Geological Survey of Greenland, 1970. *Tectonic–Geological Map of Greenland.* Geol. Surv. Greenland, 1 : 2,500,000, compiled by A. E. Escher.
____, 1971. *Quaternary Map of Greenland.* Geol. Surv. Greenland, 1 : 2,500,000, compiled by A. Weidick. (Also, Weidick, A. 1971. "Short explanation to the Quaternary Map of Greenland," *Geol. Surv. Greenland Rept. 36.*)
Haller, J., 1970. "Tectonic map of east Greenland (1:500,000). An account of tectonism, plutonism, and volcanism in East Greenland," *Medd. Groenland,* **143**(5).
____, 1971. *Geology of the East Greenland Caledonides.* New York: Wiley-Interscience, 413p.
Le Pichon, X., Hyndman, R. D., and Pantot, G., 1971. "Geophysical study of the opening of the Labrador Sea," *J. Geophys Res.,* **76,** 4724–4743. (Also, Vogt, P. R., "Letters," *ibid.,* **77**(26), 5054–5057.)
Pulvertaft, T. C. R., 1968. "The Precambrian stratigraphy of western Greenland," *Proc. Rep. 23d Intern. Geol. Congr.* Czechoslovakia, **4,** 89–107.
Raasch, G. O., ed., 1961. *Geology of the Arctic.* Toronto: Toronto Univ. Press, 1196p.
Rosenkrantz, A., and Pulvertaft, T. C. R., 1969. "Cretaceous–Tertiary stratigraphy and tectonics in northern West Greenland," *in* M. Kay, ed., *North Atlantic–Geology and Continental Drift: A Symposium* (Mem. 12). Tulsa: Am. Assoc. Petrol. Geologists, 883–898.
Tarling, D. H., and Otulana, H. I., 1972. "The paleomagnetism of some Tertiary igneous rocks from Ubekendt Ejland, West Greenland," *Bull. Geol. Soc. Denmark,* **21,** 395–406.
Vogt, P. R., *et al.,* 1969. "Discontinuities in sea-floor spreading," *Tectonophysics,* **8,** 285–317.
____, *et al.,* 1972. "Discussion," *J. Geophys. Res.,* 77(26), 5054–5057.
Weidick, A., 1968. "Observations on some Holocene glacier fluctuations in West Greenland," *Medd. Groenland,* **165**(6), 202p.

Cross-references: *Canada; Canada–Arctic Archipelago; North America.*

GRENADA

The southernmost of the British Windward Islands in the Lesser Antilles, Grenada is situated at about 12°N and 62°W. It was formerly part of the British West Indies, becoming independent in 1974. The island was discovered by Columbus in 1498 and originally named "Concepción." It was settled by the French in 1650 but occupied by Britain after 1783. The capital is St. George's, where there is a good harbor.

Climatically the islands lie in the NE trade wind belt, but are rarely struck by hurricanes. The precipitation is very variable, with a maximum on the windward side, ranging there to over 4000 mm, dropping to only 1000 mm in the S. The rainy season is from May to December. The greatest heat occurs from July to September, the maximum being 32.4°C and the minimum 19.9°C.

The coastline is strongly cliffed, especially on the western side, but the southern coast is more gentle, dissected into 20 long, drowned valleys (up to 1 km across and as much as 200 m deep).

Part of the inner volcanic arc of the Caribbean belt, Grenada is a spectacular volcanic island. Measuring 33 by 19 km, and covering 280 km^2 (125 sq mi), it has a mountainous spine of heavily forested ridges and gorges, rising to 838 m. There are two crater lakes; one of them, Grand Etang, in the center of the island, is at 530 m, and overflows into a stream. The other, 2 m below sea level, occurs in a 95-m-deep crater and is 0.5 km across.

The basement of the island consists of a folded Eocene to Oligocene volcano-sedimentary succession. The post-Oligocene rocks include rare limestones but are mainly volcanic. Two suites of volcanic rocks are recognized (Sigurdsson et al., 1973): (1) a strongly undersaturated suite of olivine-rich alkali picrites, basanites, and alkali basalts, occurring mainly as lava flows with associated explosion craters; and (2) a suite of calc-alkaline andesites. Mount St. Catherine (840 m), at the northern end of Grenada, is the youngest volcanic structure on the island. It probably formed during the Pleistocene. No eruptions have been recorded in historic times, but thermal springs occur at six localities around the flanks of the mountain (Robson and Tomblin, 1966).

Grenada itself is the largest and most southerly island in the *Grenadine Group,* which consists of 8 larger and 125 smaller islands and rocks, covering in all about 100 km^2. They stretch for 95 km in a SSW-NNE trend between Grenada and *Saint Vincent* (q.v.) and rise from a submarine bank that is mainly less than 50 m in depth. The southern Grenadines are administered by Grenada; here the largest island is Carriacou. The northern Grenadines come under Saint Vincent. The Grenadines are mainly volcanic, partly coral reefs; there are both cliffed islands and some with sandy beaches, because of their generally low relief (max. elevation 335 m).

On *Carriacou,* which is 15 km long and

covers 34 km² with a maximum elevation 297 m, there are reef limestones of Middle to Upper Oligocene age overlying and overlain by tuffs and agglomerates. These are followed by the "Carriacou Limestone" of Aquitanian age. After a Lower Miocene orogenic phase, there followed the Miocene "Grande Baie Beds," deepwater shales and tuffs.

RHODES W. FAIRBRIDGE
WILLIAM J. REA

References

Anglejan, B. d', and Mountjoy, E. W., 1973. "Submerged reefs of the Eastern Grenadines shelf margin," *Bull. Geol. Soc. Am.,* 84, 2445–2454.
Davis, W. M., 1926. "The Lesser Antilles," *Am. Geogr. Soc., Publ.* 2, 207p.
Earle, K. W., 1924. *Geological Survey of Grenada and the Grenadines.* Grenada: Govt. Printing Off.
Rea, W. J., 1970. "Andesites of the Lesser Antilles," *Proc. Geol. Soc. Lond.* (1662), 39–46.
Robson, G. R., and Tomblin, T. F., 1966. "West Indies," *Catalogue of the Active Volcanoes of the World including Solfataric Fields.* Naples: Intern. Assoc. Volcanol., pt. 20, 56p.
Robinson, E., and Jung, P., 1972. "Stratigraphy and age of marine rocks, Carriacou, West Indies," *Bull. Am. Assoc. Petrol. Geologists,* 56(1), 114–127.
Sapper, K., 1903. "Besuch der Insel St. Vincent," *Centralbl. Geol. Min.,* 248–258.
Schuchert, C., 1929. "Geological history of the Antillean region," *Bull. Geol. Soc. Am.,* 40, 327–360.
Sigurdsson, H., Tomblin, J. F., Brown, G. M., Holland, J. G., and Arculus, R. J., 1973. "Strongly undersaturated magmas in the Lesser Antilles Island Arc," *Earth Planet. Sci. Lett.,* 18, 285–293.
Trechmann, C. T., 1935. "Geology and fossils of Carriacou, W. I.," *Geol. Mag.,* 72, 529–555.

Cross-references: *St. Vincent; West Indies; Windward Islands.*

GUADELOUPE

(Including La Désirade, Marie Galante, Isles des Saintes)

Covering an area of 1780 km² (687 sq mi), Guadeloupe is an overseas department of France. It is actually two islands, Basse-Terre and Grande-Terre, separated by a low-lying mangrove swamp and narrow tidal channel. Its dependencies include nearby Marie Galante, La Désirade, Isles des Saintes, and Iles de la Petite Terre, and 220 km (135 mi) to the northwest St. Barthélemy (St. Bartholomew) and St. Martin. All are islands situated in the *Leeward Islands* (q.v.), which comprise the northern half of the Lesser Antilles Island Arc.

The Lesser Antilles arc ridge widens abruptly in the Guadeloupe region, owing to the separation of the older and younger axes of volcanic activity north of Dominica. This separation of volcanic centers is expressed in the double islands of Guadeloupe, Basse-Terre to the west and Grande-Terre to the east. The youth of Basse-Terre is reflected in its mountainous terrain, whereas the morphology of Grande-Terre is more flat-lying, owing to the carbonate bank veneer which has covered the truncated basal volcanics. Isles des Saintes have a geological history similar to that of Basse-Terre, and the geology of Marie Galante and Iles de la Petite Terre corresponds to that of Grande-Terre. La Désirade is unique in the Lesser Antilles in being the only island with rocks representative of the earliest phase in the growth and development of the island arc.

Island Geology of the Guadeloupe Region

Grande-Terre (Fig. 1) consists of limestones, with the exception of the Abymes area, where thin laterites overlie a lower Miocene limestone conglomerate containing volcanic boulders. Underlying the conglomerate, unconformably, are strongly weathered volcanogenic units which can be seen in the more deeply carved valleys and on the sides of the higher hills near Abymes. The three stratigraphic units recognized on Grande-Terre are (1) the pre-Miocene volcanic substratum; (2) the sediments of the lower Miocene transgression; (3) the Quaternary deposits.

The structural deformation of Grande-Terre is minor, consisting mainly of normal faulting which has compartmented the island into several blocks, which were gently tilted toward the ESE. Slight folding has produced a recognizable WNW-ESE fold trend. The geology of Petite Terre and Marie Galante is identical to that of Grande-Terre.

Basse-Terre (Fig. 1) is the western member of Guadeloupe and is almost completely volcanic. This island is quite mountainous, with the highest elevation in the Lesser Antilles represented by the active Soufrière volcano (1467 m), which erupted in 1797, 1843, and 1956.

The volcanic rocks of Basse-Terre are all considered to be younger than the Burdigalian transgressive limestone sequence of Grande-Terre. Limited outcrops of these limestones occur in extreme northeast Basse-Terre and on both sides of the Blondeau River valley on the southern tip of the island. The volcanics of Basse-Terre were erupted through this time horizon.

On the basis of superposition and degree of

FIGURE 1. Geologic map of islands of Guadeloupe region.

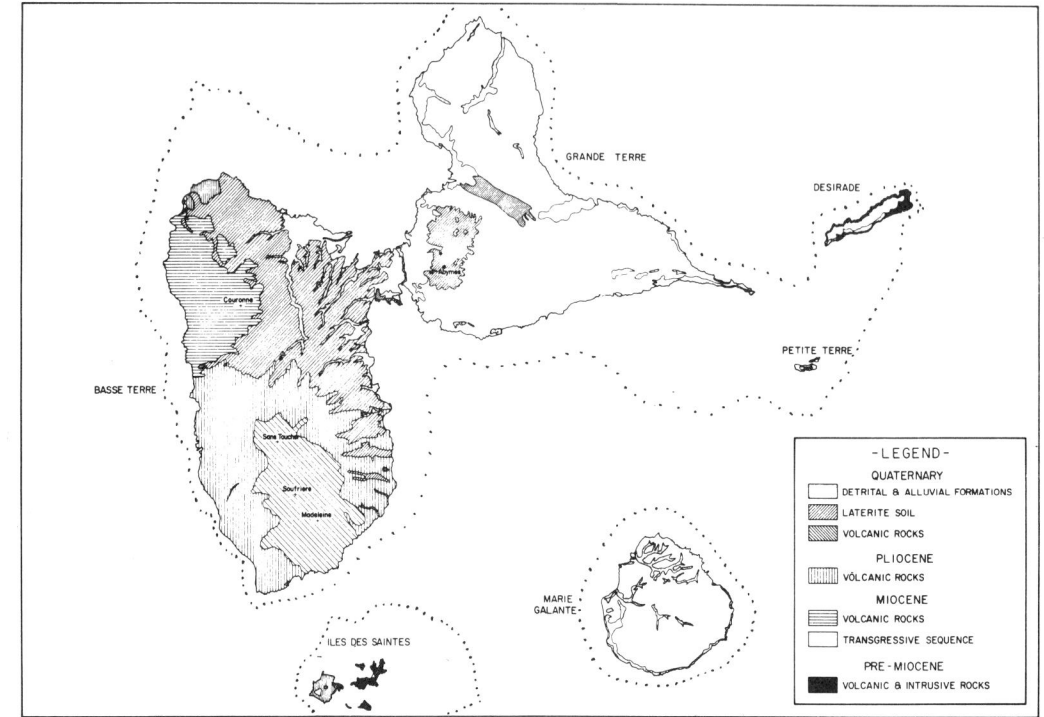

erosion, there are three recognizable phases of volcanism on Basse-Terre. The first phase, following the Burdigalian transgression, began in middle or late Miocene. The center of activity was located at Couronne (Fig. 1). This was followed by a Pliocene phase, which is considered to be the main phase of volcanic activity on Basse-Terre on the basis of the alignment and distribution of eroded strato-volcano remnants.

The final phase of volcanism, Quaternary to Recent in age, is represented by La Soufrière and La Madeleine in the southern and highest part of the island.

The three volcanic phases of Basse-Terre have produced both flows and nuée ardente–type deposits, with ash flows and explosion breccias predominating over lavas. The volcanic products are typical of the basalt-andesite-dacite series of the calc-alkaline suite. The main rock type is a labradorite-bytownite andesite.

The Isles des Saintes (Fig. 1) are a small group of entirely volcanic islands. Terre-de-Bas, to the W, consists almost entirely of labradorite andesite flows and was a Pliocene center. To the E, Terre-de-Haut consists of andesites and dacites of late Miocene age.

The geologic evolution of Basse-Terre and Isles des Saintes can be summarized as follows:

1. Initial eruptions along linear volcanic zones located on the western flank of the pre-Miocene volcanic ridge during late Miocene. The volcanic vents breached the early Miocene shallow-water limestones and underlying volcaniclastics associated with the older ridge. Volcanic centers were Couronnes and perhaps the two Mamelles.

2. Next there was general quiescence followed by renewal of activity during the Pliocene, which resulted in building the main axis of the island of Basse-Terre.

3. Again there was general quiescence, this time followed by the final Quaternary phase concentrated in the southern half of the island to build the well-preserved strato-volcanoes and endogenous domes. The Soufrière represents a still-active center of volcanism.

The island of La Désirade is the highest part of the southern scarp of a major radial fault zone which is expressed by the Desirade sea trough (Fig. 2). The proximal location to a major fault is reflected in the sheared aspect of many of the outcrops along the northern coast.

The volcanogenic rocks of La Désirade have been eroded to form a flat platform upon which a cap of Aquitanian limestones, averaging 50 m in thickness, has been deposited. The significance of La Désirade lies in the age and

FIGURE 2. Geologic map of La Desirade and location map.

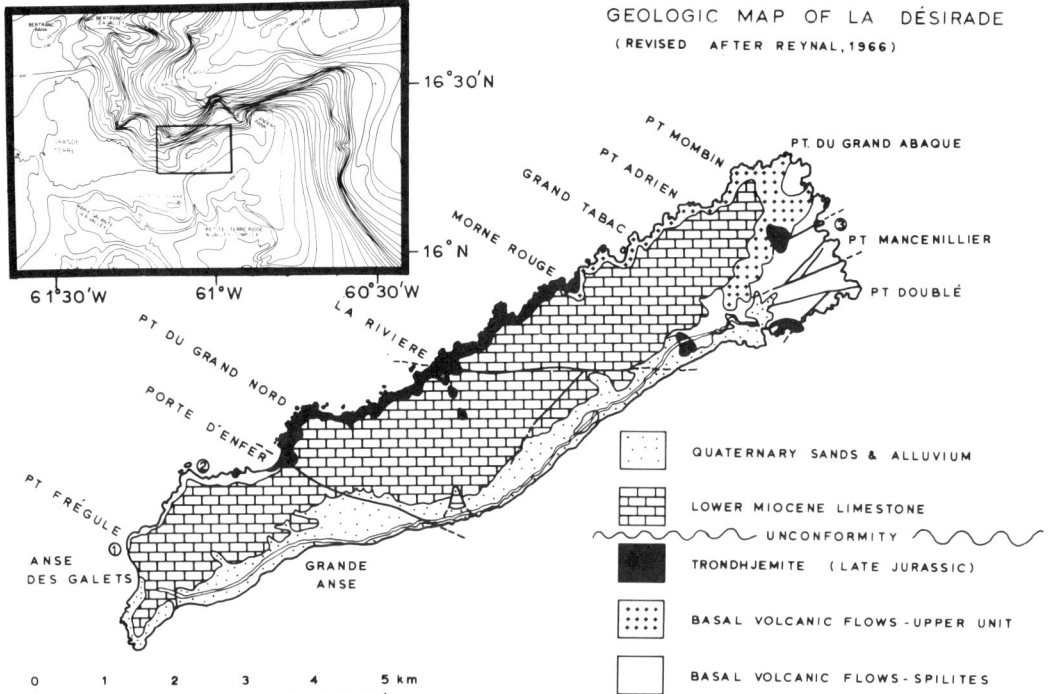

composition of the igneous basement, which constitutes an assemblage unique to the Lesser Antilles. The volcanic rocks belong to the spilite-keratophyre assemblage and are associated with a trondhjemitic hypabyssal pluton which has yielded radiometric minimum ages ranging from 71 to 150×10^6 years.

The spilites occur as a series of pillowed horizons with interbedded cherts, including jasper, and hyaloclastites. The mineralogy of the lavas is unique in the Lesser Antilles, consisting of secondary, low-temperature, unzoned albite (An_{5-8}), calcite, chlorite, epidote, some augite, and prehnite in the spilites, and unzoned albite, subhedral to euhedral porphyritic quartz, chlorite, and epidote in the quartz keratophyres. Epidote occurs as patch alterations in many of the pillow lavas, while hematite, calcite, chlorite, and prehnite fill amygdules and veins.

Chemical analyses of La Désirade samples rule out calc-alkaline affinities. They are characterized by extremely low K_2O, low Rb and Sr, high to extreme Na_2O/K_2O values, high K/Rb and low Rb/Sr ratios, and magnesium enrichment at the expense of total iron.

The fundamental question is whether the igneous complex of La Désirade represents oceanic tholeiite or island arc tholeiite. The rocks may be autochthonous oceanic crust on which the basal part of a volcanic ridge was constructed early in the development of the Lesser Antilles, then uplifted by post-Early Miocene faulting into its present position. A second possibility is that it is allochthonous oceanic crust obducted into its present position by sea-floor spreading. The third alternative, favored by petrochemical data and marine geophysical evidence, is that the basement rocks of La Désirade represent the continuation of the Greater Antilles and are the products of the oldest phase of volcanism.

Geologic History of the Guadeloupe Region

The geological, geochemical, and geophysical features suggest a geologic history which can be summarized as follows:

1. During the middle to late Jurassic, volcanic activity along a curvilinear zone produced submarine extrusions of island arc tholeiite affinity which initiated the upgrowth of a submarine volcanic ridge on the sea floor.

2. The spilites and keratophyres were succeeded by volcanogenic rocks of the calc-alkaline series which reflected a change in the nature of volcanism to a more explosive variety

and now emanating from coalescing volcanic centers to create an arcuate row of strato-volcanoes along the crest of the older submarine ridge. This phase of volcanism continued until Eocene-Oligocene time.

3. By late Oligocene-early Miocene time, all volcanism ceased and erosion truncated the islands to form the base upon which the Lower Miocene transgressive limestones were deposited.

4. The next stage was eastward crustal translation of the northern segment of the arc ridge along major transverse structural features such as the Désirade Fault Zone and the inferred Dominica Fault.

5. Volcanic activity was renewed during the late Miocene along linear zones now located on the western flank of the pre-Miocene volcanic ridge.

6. This younger phase of volcanism has continued along the same axis with only minor interruptions until the present day.

L. KENNETH FINK, JR.

References

Barrabé, L., 1934a. "Rapport sur les résultats d'une mission pour la recherche du pétrole à la Guadeloupe," *Ann. Off. Comb. Liq.* (Paris), 625–661.

____, 1934b. "Sur l'affleurement du socle ancien des Petites Antilles dans l'île de la Désirade," *C. R. Acad. Sci.*, **198**, 487–489.

____, 1934c. "Sur la transgression tertiaire qui a recouvert la partie orientale de la Guadeloupe," *C. R. Acad. Sci.*, **198**, 758–759.

____, 1952. "Sur l'origine des quartz bipyramidés des latérites du versant oriental de la Guadeloupe proprement dite," *C. R. Somm. Soc. Géol. France*, **16**, 334–336.

____, 1954. "Observations sur la constitution géologique de la Désirade (G)," *Bull. Soc. Géol. France*, Ser. 6, 3(28), 613–626.

____, 1956. "L'éruption de la Soufrière de la Guadeloupe," *C. R. Somm. Soc. Géol. France*, **13**, 223.

____, and Jolivet, J., 1958. "Les récentes manifestations d'activité de la Guadeloupe (Petites Antilles)," *Bull. Volcanol.*, Ser. 2, **19**, 143–157.

Cavelier, C., 1968. "Sur la nature superficielle et récente du 'substratum volcanique antémiocène' de la Grande-Terre (Guadeloupe)," *Bull. Soc. Géol. France*, Ser. 3, 7(9), 450–454.

Dickinson, W. R., and Hatherton, T., 1967. "Andesitic volcanism and seismicity around the Pacific," *Science*, **157**, 801–803.

Donnelly, T. W., Rogers, J. J. W., Pushkar, P., and Armstrong, R. L., 1971. "Chemical evolution of the igneous rocks of the eastern West Indies: an investigation of thorium, uranium, and potassium distributions, and lead and strontium isotopic ratios," *Geol. Soc. Am. Mem.* **130**, 181–224.

Fink, L. K., Jr., 1968. "Marine geology of the Guadeloupe Region, Lesser Antilles Island Arc," thesis, Univ. Miami, 121p.

____, 1970. "Evidence for the antiquity of the Lesser Antilles Island Arc," *Trans. Am. Geophys. Union*, 51(4), 326.

____, 1972. "Bathymetric and geologic studies of the Guadeloupe region, Lesser Antilles Island Arc," *Marine Geol.*, **12**, 267–288.

____, Harper, C. T., Stipp, J. J., and Nagle, F., 1971. "Tectonic significance of La Désirade," *Proc. 6th Caribbean Geol. Conf.*, Margarita, Venezuela, Abstr.

Hatherton, T., and Dickinson, W. R., 1969. "The relationship between andesitic volcanism and seismicity in Indonesia, the Lesser Antilles, and other island arcs," *J. Geophys. Res.*, **74**, 5301–5309.

Jakes, P., and White, A. J. R., 1972. "Major and trace element abundances in volcanic rocks of orogenic areas," *Bull. Geol. Soc. Am.*, **83**, 29–40.

Lewis, J. F., 1971. "Composition, origin, and differentiation of basalt magma in the Lesser Antilles," *Geol. Soc. Am. Mem.* **130**, 159–180.

Lidiak, E. G., 1965. "Petrology of andesitic, spilitic, and keratophyric flow rock, north-central Puerto Rico," *Bull. Geol. Soc. Am.*, 76(1), 57–88.

Martin-Kaye, P. H. A., 1960. "The double arc of the Lesser Antilles," thesis, Univ. London, 78p.

____, 1969. "A summary of the geology of the Lesser Antilles," *Overseas Geol. Mineral Resources*, 10(2), 172–206.

Mitchell, A. H., and Reading, H. G., 1971. "Evolution of island arcs," *J. Geol.*, 79(3), 253–284.

Pushkar, P., 1968. "Strontium isotope ratios in volcanic rocks of three island arc areas," *J. Geophys. Res.*, **73**, 2701–2714.

Reynal, A. de, 1961. *Carte Géologique Détaille de la France, Départment de la Guadeloupe, 1:50,000 Terre et Notice Explicative.* Paris: Minist. de l'Industrie.

____, 1966a. *Carte Géologique Détaille de la France, Départment de la Guadeloupe, 1:50,000 Feuille de Basse-Terre et des Notice Explicative.* Paris: Minist. de l'Industrie.

____, 1966b. *Carte Géologique Détaille de la France, Départment de la Guadeloupe. 1:50,000 Feuille de Marie Galante, La Désirade, Iles de Petite-Terre et Notice Explicative.* Paris: Minist. de l'Industrie.

Robson, G. R., and Tomblin, J. F., 1966. "West Indies," *Catalogue of the Active Volcanoes of the World including Solfatara Fields.* Naples: Intern. Assoc. Volcanol. pt. 20, 56p.

Tomblin, J. F., 1968. "Geochemistry and genesis of Lesser Antillean volcanic rocks," preprint of paper read at 5th Caribbean Geol. Conf., St. Thomas, V.I.

Weyl, R., 1966. "Geologie der Antillen," *Regionale Geologie der Erde.* Berlin: Gebr. Borntraeger, 410p.

Cross-references: *Antigua; Barbados; Vol. I, Caribbean Sea; Dominica; Vol. III, Island Arcs; Martinique; West Indies.*

GUAM

The largest and southernmost of the Mariana Islands, Guam is administered as a United States territory. It lies at $13°28'$N and

144°45'E. It is 50 km long, 6–18 km wide, and 549 km^2 (212 sq mi) in area. It is 407 m (1334 ft) at its highest point (Mt. Lamlam). Fringing reefs surround most of the island. The northern part is an emerged reef limestone plateau tilted to the SW and cliffed by the sea. The southern part is a range of mountains (max. elevation 407 m), chiefly volcanic.

The oldest rock unit is the Alutom Formation of Tertiary *b* and *c* age and forms the central part of the island. It is 900 m thick and consists of water-laid tuffaceous shales, sandstone, and conglomerate; lava flows and blocky breccias; and reworked tuff-breccia and conglomerate. Most of the southern part of the island is formed of a volcanic sequence, the Umutac Formation of Tertiary *e* and early Miocene age. The Facpi volcanic member consists of 450 m of pillow lavas, flow breccias, and tuffaceous shale. The Bonya Limestone (Tertiary *f*), the Alifan Limestone (Tertiary *f*), Barrigada Limestone (Tertiary *g*), and the Janum globergerinid Limestone (Tertiary *g*) are differentiated. The Mariana Limestone of Pliocene-Pleistocene age is a massive reef limestone making up the northern plateau and the fringing limestone along the eastern coast of southern Guam.

The island was uplifted in the early Pleistocene and terraces were cut during the shifting sea levels of glacial times. The youngest veneering coral limestone, deposited during the 2-m stand, has a radiocarbon date of 3400 ± 250 years. The soils are lateritic, but no significant bauxite deposits have been found in them.

Guam is divided into three structural provinces. The northern part was tilted to the SW in the Pleistocene. The central part is highly faulted and gravity-folded Tertiary volcanic rocks derived from a volcano W of Guam. The southern block of Miocene volcanic rocks is derived from a second, later volcano SW of the island. Guam is subject to numerous strong earthquakes. Most of the volcanic rocks are andesitic or basaltic and show typical island arc affinities. The island has been generally emergent since early Cenozoic time.

The first geological map and report on Guam was made by the writer in 1937 (but was kept confidential: Stearns, 1937). A thorough review of the voluminous geologic literature on Guam is given by Cloud et al. (1956, p. 9–20). A continuing study of the coral reefs is being undertaken at the University of Guam Marine Laboratory (established in 1970).

HAROLD T. STEARNS

References

Cloud, P. E., Jr., Schmidt, R. G., and Burke, H. W., 1956. "Geology of Saipan, Marianas Islands. 1. General geology," *U.S. Geol. Surv. Prof. Pap. 280-A.*

Kobayashi, K., 1972. "Reconnaissance paleomagnetic anal rock-magnetic study of igneous rocks of Guam, Mariana, and related sites," *in* M. Hoshino and H. Aoki, eds., *Izu Peninsula.* Tokyo: Tokyo Univ. Press, 385–390.

Stearns, H. T., 1937. *Geology and Water Resources of the Island of Guam, Marianas Islands,* classified report to U.S. Navy (now declassified).

Tracey, J. I., Jr., Schlanger, S. O., Stark, J. T., Doan, D. B., and May, H. G., 1964. "General Geology of Guam, Mariana Islands," *U.S. Geol. Surv. Prof. Pap. 403-A,* A1–A104. (See also *Prof. Pap. 403-B, 403-C,* and *403-D.*)

Cross-references: *Caroline Islands; Mariana Islands; Pacific Islands Trust Territory; Saipan.*

GUATEMALA

Guatemala, with an area of 67,658 km^2 (42,042 sq mi), is the northernmost of the Central American republics. It is geologically a part of the North American continent. Structural trends and geological history can be tied to Mexico and the rest of North America. Paleozoic and Mesozoic rocks and morphotectonic trends strike SE from Mexico and form an arc in central Guatemala convex to the south before striking northeasterly into the Caribbean.

The oldest rocks of the northern Caribbean region crop out in Guatemala. Great fault zones representing the landward continuation of the Cayman Trough, large serpentinite belts, the offshore Pacific Mid-American Trench, active volcanoes, and seismic zones all make this country of critical importance in theories of large-scale earth tectonics.

The great geological variety and magnificent volcanic scenery, coupled with colorful, still prevalent Indian culture and Mayan archeology, make the country most attractive for the geologist and traveler. However, apart from the remarkable reconnaissance surveys of Dolfuss and Montserrat printed in 1868 and Sapper in 1937, very little was published on the geology of Guatemala until 1957.

The country is completely covered by aerial photographs as well as by topographic maps at scales of 1:50,000 and 1:250,000. These are available from the Instituto Geográfico Nacional in Guatemala City, which also publishes geologic maps at various scales, as well as a geologic map of the country at a scale of 1:500,000.

Physiography

The country may be divided into four main physiographic divisions. These are, from S to N: (1) Pacific coastal plain, about 50 km wide,

FIGURE 1. Magnificent basin of Lake Atitlan, Guatemala. (Photo: C. Sitler, Guatemala City.)

consisting of detritus eroded from the volcanic highlands; (2) Tertiary to Recent volcanic province; (3) Cordilleran region across the middle of the country, made of crystalline rocks and a folded sedimentary belt to the north; and (4) the Petén lowland, occupying the northern third of the country.

The continental divide runs NW along the volcanic highlands. Numerous short streams rush down the volcanic slopes to the Pacific. Several larger streams flow into the Caribbean, including the Motagua, Polochic, and Sarstún. The Rio de la Pasión and Rio Chixoy flow northward, joining to form the Rio Usumacinta, which flows through Mexico and into the Gulf of Mexico.

The Petén Basin is predominantly an uninhabited lowland rain forest containing several large savannas. It is covered mostly by low-dipping Upper Cretaceous carbonates and Tertiary clastic and evaporitic rocks. The entire Petén Basin may be considered a continuation of the Gulf Coast province of North America. Salt domes in the S indicate that the Gulf salt province extends to this region. An extensive karst has developed on the carbonates in the Petén Basin and in the fold belt to the S.

There are several lake-filled basins in the volcanic highlands, including the twin caldera of Lake Ayarza, the volcano-tectonic Lake Amatitlán, and Lake Atitlán, a cauldron subsidence of breath-taking beauty (Fig. 1; see also Vol. III).

Stratigraphy

A thick sequence of predominantly sedimentary rocks that may be as old as 345 m yr makes up a metamorphic series of schists, gneisses, marbles, and amphibolites uplifted between the Motagua and Polochic fault zones. South of the Motagua valley is a zone of weakly metamorphosed graywacke, diabase, basalt, and chert of probable Paleozoic age that may represent, in part, a eugeosynclinal equivalent of the strongly metamorphosed sediments to the north (Fig. 2).

Phyllites and slates make up much of the Maya Mountains and the prevolcanic surface in southern Guatemala. These are generally taken to be equivalent to the Pennsylvanian-Permian rocks of the sedimentary belt.

Plutonic rocks of various ages occur in Guatemala. The oldest granites, associated with the metamorphic complex, were probably intruded about 345 m yr ago. Plutons occurring in the volcanic zone present problems of dating because of probable reheating during Cenozoic volcanism. Many of these rocks, ranging in composition from granite to diorite, are of Cretaceous age. Others are associated with Tertiary volcanism, acting as feeders and cutting volcanic rocks in southeastern Guatemala.

Numerous discontinuous serpentinite bodies, some exceeding 1000 km^2 in area, extend along the Motagua and Polochic fault zones. These were possibly emplaced in Cretaceous time.

An aggregate thickness of 10,000 m of sediments is present in the Cordilleran Fold Belt and Petén Basin. The oldest sediments are 2000 m of Pennsylvanian and Permian shales, sandstones, and conglomerates. The coarser rocks are generally restricted to the lower part of the section. These are gradationally overlain by 1000 m of Permian limestone. There is a facies change northward from the limestone to shales. The limestones are unconformably overlain by Upper Jurassic to Lower Cretaceous red beds, with an uneven distribution and maximum thickness of 1000 m. The red beds grade up into 3000 m of Cretaceous carbonates. There is a thickening and facies change in the subsurface of the Petén Basin to evaporites, perhaps in excess of 3000 m thick. The Tertiary is represented by more than 1000 m of Lower Tertiary clastic rocks and Upper Tertiary sediments exceed 1000 m in thickness.

Cenozoic volcanic rocks and pumice cover much of the southern half of the country. Strong volcanic activity started in Miocene time, covering the region with a series of predominantly rhyodacitic tuffs and ignimbrites, and lesser andesitic and dacitic lavas. Associated with these are volcanic sediments and lahars. Quaternary times have been marked by the development of large andesitic cones and rhyodacitic pumice eruptions and flows. Basaltic cinder-cone fields and flows are concentrated in the southeastern part of the country.

There is an impressive row of major volcanoes paralleling the Pacific coast; among these there is Volcán Tajumulco (4211 m), which is the highest point in Central America. A number

FIGURE 2. Generalized geological map of Guatemala (Instituto Geográfico Nacional, 1972).

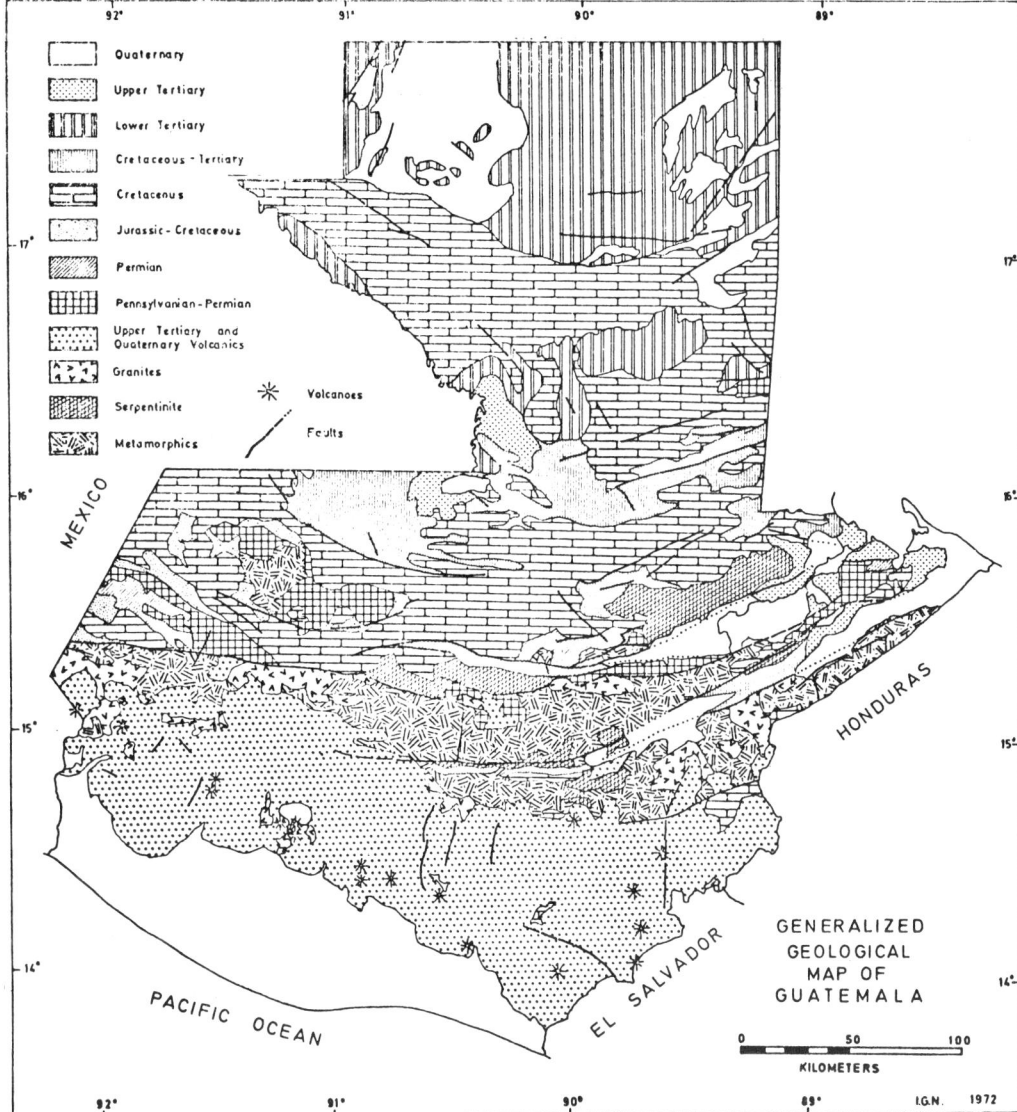

of the volcanoes are still active, especially V. Fuego, V. Pacaya, and Santiaguito. The latter is noteworthy as an especially active dacite dome complex which started growing in a recently (1902) blasted crater on the flanks of Volcán Santa María in 1922. It has been enlarging continuously with flows, small nuée ardentes, Pelean spines, ash eruptions, and the growth of coalescing dome units. The complex now measures 1 by 2 km, with a relief of greater than 450 m, and is still growing. Fumarole temperatures in excess of 800°C have been recorded.

Structure

A major orogenic revolution took place in mid-Paleozoic, probably Devonian time, with metamorphism, plutonism, and folding. Post-Permian activity resulted in some local angular unconformities, uplift, and a resultant hiatus in the sedimentary record from late Permian to late Jurassic time. Strong Laramide (Eocene) folding and faulting gave the region its present E-W structural grain. The NW cross-trending volcanic and structural system, including the Jalpatagua Fault, the Mid-American trench, and

N-S high-angle faulting, all of which are probably genetically related, are younger features superimposed upon the Laramide structures.

The Motagua and Polochic fault zones are parallel major fault systems lying along the landward extension of the Caribbean Cayman Trough, and as such make one of the major structural features of the earth's crust. It is probable that an ancestral Polochic fault zone existed at least as far back as Paleozoic time, and took on its present form possibly in Miocene time. The Polochic zone crosses the middle of the country in an E-W arc convex to the south that extends NW into Mexico. The Motagua zone, 50 km to the south, is lost under volcanic cover in the western part of the country.

These lengthy fault zones separate the sedimentary geologic province in the N from the uplifted crystalline basement lying between the faults. Large structures associated with the Cayman system include the Maya Mountains horst, the Lake Izabal graben, the Sierra de las Minas horst, and the uplifted sedimentary Cuchumatanes plateau.

Although large high-angle fault zones with kilometers of vertical displacement have been established, a large horizontal displacement might be expected in what appears to be major transcurrent faults but has not as yet been demonstrated in the field.

tropical weathering have resulted in commercial lateritic nickel deposits, especially N of Lake Izabal, where the International Nickel Company is scheduled to construct a large refinery complex. This will be the largest industry in Central America.

Small, high-grade vein copper deposits have been discovered associated with the ultramafic belt in Alta Verapaz.

The Petén Basin has potential as a possible future petroleum province. The first 16 wells produced only minor oil and gas shows, but in 1974 a significant oil discovery was made. The target has been Cretaceous carbonates. Salt dome structures in the southern part of the basin have had minor sulfur and petroleum shows. Poor accessibility has been a major deterrent. The gradual opening of this uninhabited region to colonization should bring more exploration and drilling. A sedimentary section under the volcanic detritus of the offshore Pacific coast is also an exploration target.

Hydroelectric potential has been exploited only in small part. Heavy rainfall and precipitous slopes are in part offset by strongly seasonal rainfall, limited potential storage reservoir capacity, and geology complicated by extensive faulting, hot springs, and karst. Geothermal energy has great promise, but has not yet been evaluated.

SAM BONIS

Economic Geology

There are numerous small, high-grade mineral prospects, but few productive mines in Guatemala. Poor accessibility and rugged terrain have hindered exploration. The rapid improvement of road facilities and the use of helicopters have accelerated exploration in the present decade.

In pre-Spanish time the Motagua valley was the center of jade production for the highly developed middle American Indian civilizations. At present the primary deposits appear exhausted, although low-grade alluvial boulders are still fairly common.

Small lead-zinc deposits have been mined in the Huehuetenango, Cobán, and Concepción Las Minas districts in the present century. A reported 40 million ounces of silver were taken from the Concepción district during the 19th century. The Huehuetenango and Cobán districts consist of replacement deposits in Cretaceous and Permian carbonates. Tertiary granodioritic intrusions and volcanics are responsible for mineralization in the Concepción district.

The large serpentinite masses subject to

References

Anderson, T. H., et al., 1973. "Geology of the western Altos Cuchumatanes, Northwestern Guatemala," *Bull. Geol. Soc. Am.*, **84**, 805–826.

Bonis, S., Bohnenberger, O., and Dengo, G., 1970. *Mapa Geológico de Guatemala*. Guatemala City: Inst. Geogr. Nacl., 1:500,000.

Clemons, R. E., et al., 1974. "Stratigraphic nomenclature of recognized Paleozoic and Mesozoic rocks of Western Guatemala," *Bull. Am. Assoc. Petrol. Geologists*, **58**(2), 313–320.

Dengo, G., 1968. *Estructura Geológica, Historia Tectónica y Morfología de América Central*. México: Centro Regional de Ayuda Técnica, 1–50.

Instituto Geográfico Nacional, 1961. *Diccionario Geográfico o de Guatemala*. Guatemala City: Inst. Geogr. Nacl., 2 vols.

Kesler, S. E., 1971. "Nature of ancestral orogenic zone in nuclear Central America," *Bull. Am. Assoc. Petrol. Geologists*, **55**(12), 2116–2129.

———, 1972. "Western extension of fault zones bounding the northern side of the Caribbean plate," *24th Intern. Geol. Congr.*, **3**, 238–244.

Roberts, R. J., and Irving, E. M., 1957. "Mineral deposits of Central America," *U.S. Geol. Surv. Bull.*, **1034**, 205p.

Rose, W. I., 1972. "Santiaguito Volcanic Dome, Guatemala," *Bull. Geol. Soc. Am.*, **83**, 1413–1434.

Vinson, G. L., 1962. "Upper Cretaceous and Tertiary stratigraphy of Guatemala," *Bull. Am. Assoc. Petrol. Geologists,* **44,** 1273–1315.
Weyl, R., 1961. "Die Geologie Mittelamerikas," *Regionale Geologie der Erde.* Berlin: Gebr. Borntraeger, 226p.
Williams, H., 1960. "Volcanic history of the Guatemalan Highlands," *Univ. Calif. Publ. Geol. Sci.,* **38**(1), 1–64.
Wilson, H. H., 1974. "Cretaceous sedimentation and orogeny in Nuclear Central America," *Bull. Am. Assoc. Petrol. Geologists,* **58,** 1348–1396.

Cross-references: *Belize; Central America—Regional Review; El Salvador; Honduras; Mexico; Nicaragua.*

GUIANA SHIELD—REGIONAL REVIEW

The Guiana Shield occupies a broad area in the NE of South America between the Atlantic Ocean, the Amazon, and the Orinoco rivers. Its ocean frontage, between these two rivers, is over 2000 km long, and its surface is 1,800,000 km^2. The shield straddles the equator and has a climate of equatorial type, with two dry and two rainy seasons. The heat is moderated by the trade winds blowing from the E and NE. High precipitation, diminishing from E to W varies from 2500 to 5000 mm (100–200 in) per year. Temperatures average around 26°C. The main part of the country is covered by continuous tropical forest. Narrow belts of savanna occur along the shore and in the center (the Rupununi and Boa Vista savannas).

Politically this wide zone is divided into several territories: the "three Guianas"—the Republic of Guyana—formerly British Guiana, (231,800 km^2); Surinam—Dutch Guiana (173,840 km^2); and French Guiana—La Guyane française (88,240 km^2), bordered by the Guayanas provinces of Venezuela to the W and Brazilian Guiana to the S and E, the last two covering 1,306,220 km^2 (see, respectively, *Guyana; Surinam; French Guiana; Venezuela; Brazil*).

Mining activities have everywhere preceded geological knowledge; the first navigators were led to these shores when searching for the legendary gold of El Dorado, and when they could not find it, the pioneers who gradually settled there took an interest in agriculture. The first gold nuggets were actually discovered in French Guiana in 1853, in Guyana in 1884, and in Surinam in 1885. In Guyana the first diamond was collected in 1887, followed in 1910 by the discovery of bauxite, the exploitation of which began 7 years later. In Surinam, bauxite was reported as early as 1903 by Du Bois, but the Moengo deposits began to be exploited only in 1927, and those of the Surinam River in 1939. In French Guiana, bauxite was recognized shortly before World War II and has been worked since 1949. Manganese was discovered in Guyana in 1903 and mined from 1960 to 1968, and an important mine was developed in the 1960s in the Amapá territory of Brazil.

The first geological data were reported in British Guiana by Barrington Brown and Sawkins in 1875, and by Harrison in 1908; in French Guiana by Ch. Valain (describing specimens collected by Crevaux) in 1881; in Surinam by Du Bois in 1901; and in Venezuela by Liddle, Newhouse, and Zuloaga. Systematic studies began in these countries only in 1945–1946.

Geological researches are at present conducted in the following institutions: Guyana: Geological Survey and Mines Department at Georgetown; Surinam: Geological and Mining Service (GMD) in Paramaribo; French Guiana: Carte géologique de la France (locally managed by the Bureau de Recherches géologiques et minières) Institut Français d'Amérique Tropicale; Venezuela: Direccion de Geología, Ministerio de Minas e Hydrocarburos.

Each of these institutes is engaged in preparing geological maps to various scales and also carries out mining, geophysical, and geochemical prospecting. Courses in geology are available at the university level in Venezuelan Guayana and teaching in the elements of geology for prospectors has started at the University of Guyana.

Geomorphology

Geologically the Guiana Shield is an ancient Precambrian land mass of long continental evolution made up of varied formations of sedimentary and igneous origin (volcanic and plutonic) that were metamorphosed and folded during at least two major cycles in Archean and Proterozoic time. They are eroded today into broad dissected pediplains consisting of immense forested plains, rarely exceeding 250 m in altitude, interspersed with rugged massifs rising to 500 and rarely 1000 m, and occasional tabular plateaus. The pediplains have been worn down into a pattern of water courses and secondary relief which enables a preliminary distinction of geological formations to be made from air photographs in spite of the heavy forest cover (Choubert, 1957).

The Guiana Shield is a broad uplift dipping away on all sides. The Precambrian basement dips to the W under youthful (Cenozoic) formations along the foot of the great Andean Cordillera (Llanos of Venezuela and Colombia). They

TABLE 1. Planation Surfaces

Name	Average Altitude	Probable Age
Not named	ca. 900 m	?
Kopinang	600–700 m	Late Cretaceous to early Tertiary
Kaieteur	400–450 m in Pakaraima Mts., 200–300 m near Bartica	Late Tertiary
Rupununi	100–140 m rising to 180 m plus at some distance from coast	End of Tertiary
Mazaruni	ca. 75 m (valley floor plains)	Quaternary, still active

disappear to the S under the marine Paleozoic trough of the Amazon Basin and to the N under the marine Quaterary and Tertiary, which form a fringe along the Atlantic shore.

Tabular postgranite detrital sediments of Proterozoic age overlie the old folded metamorphic and granitized formations in the central part of the shield. They form the high, cliff-bordered plateaus and mesas ("tepui") of the Pacaraima Mountains in Guyana, Brazil, and Venezuela, culminating in Mt. Roraima (ca. 2500 m), which marks the border between these three countries. East of a rift valley in the center of the shield this divide falls away from W to E in the underlying crystalline rocks, forming the Serra Acarai and Serra Tumucumaque (Tumuc Humac) on the border of Brazil with Guyana, Surinam, and French Guiana, where it does not exceed 700 m elevation.

The Guiana Shield falls into several natural regions:
1. Coastal Plain.
2. Central pediplains and rugged mountains of Precambrian metamorphic and granitic rocks.
3. High plateaus of the Pacaraima Mountains (Venezuela, Brazil and Guyana), formed by the tabular Roraima Formation of Proterozoic age.
4. Southern pediplains descending to the Rio Negro and the Amazon.

At least five erosion cycles with three major planation surfaces or stepped erosion bevels can be recognized in Guyana. These are interpreted as representing pediplains developed at successively lower levels during long periods of standstill which occurred during the periodic uplift of the Guiana Shield. These cyclical movements are correlated and dated by comparison with stratigraphic gaps in the Guyana coastal succession as established by palynological methods and with similar features in neighboring sedimentary basins. The observed planation surfaces (McConnell, 1968) are shown in Table 1.

The succession of planations is related to those described in French Guiana by B. Choubert, in Surinam by L. O'Herne, in eastern Brazil by L. C. King, and is compared with similar features in the shield areas of Africa; the surfaces are downwarped near the coast.

The main factors controlling the present-day drainage patterns of the major rivers of Guyana are in part rooted in the original consequent drainage of the Guiana Shield, in part controlled by lithology and structure, and in part due to fairly recent warping; river captures and diversions of drainage have also taken place.

Geological Formations

The geological formations of the Guianas consist chiefly of the folded Precambrian basement, with its capping of detrital Proterozoic and limited cover rocks of Tertiary and Quaternary age principally developed in the coastal plains and in the Takutu rift valley of Guyana (Fig. 1).

Folded Precambrian Basement. The Guiana Shield is not only largely covered with Amazonian rain forest under which there is a deep saprolite due to tropical weathering that renders difficult the identification of rock types, but it shares with other shield areas the intricate complications inherent in Precambrian geology. It is therefore a great tribute to the pioneers that geological mapping had progressed so far before the advent of air photography in the postwar years. With the new methods of photogeology great advances were made, particularly in French Guiana (Choubert, 1957), and a map correlating the geology of French Guiana, Surinam, and Guyana was produced at the International Geological Congress in 1952 by three of the pioneers of Guiana Shield geology (Choubert et al., 1954). Much more detailed geological mapping has been completed in these countries and tied into mapping in Venezuela and Brazil. Moreover, an

FIGURE 1. Outline geologic map of the Guiana Shield. 1, coastal formations and white sands; 2, Takutu Formation; 3, Paleozoic; 4, granite and gneiss, in part reactivated ca. 2000 m yr; 5, Imataca, Kanuku, Adampada-Falwatra, and Ile de Cayenne complexes; 6, Roraima Formation; 7, granites of trans-Amazonian cycle; 8, granites of trans-Amazonian cycle; migmatitic; 9, Pastora, Mazaruni, Armina, Orapu, and Bonidoro groups; 10, Carichapo, Barama, Paramaca, and Amapa groups; 11, southern limit of Guyana eugeosynclinal facies; 12, granophyric and porphyritic granites; 13, Cuchivero, Surumu, Iwokrama, Dalbana, and Kuyuwini igneous complexes; 14, Cinaruco, Muruwa, and Ston formations; 15, major dislocation zones.

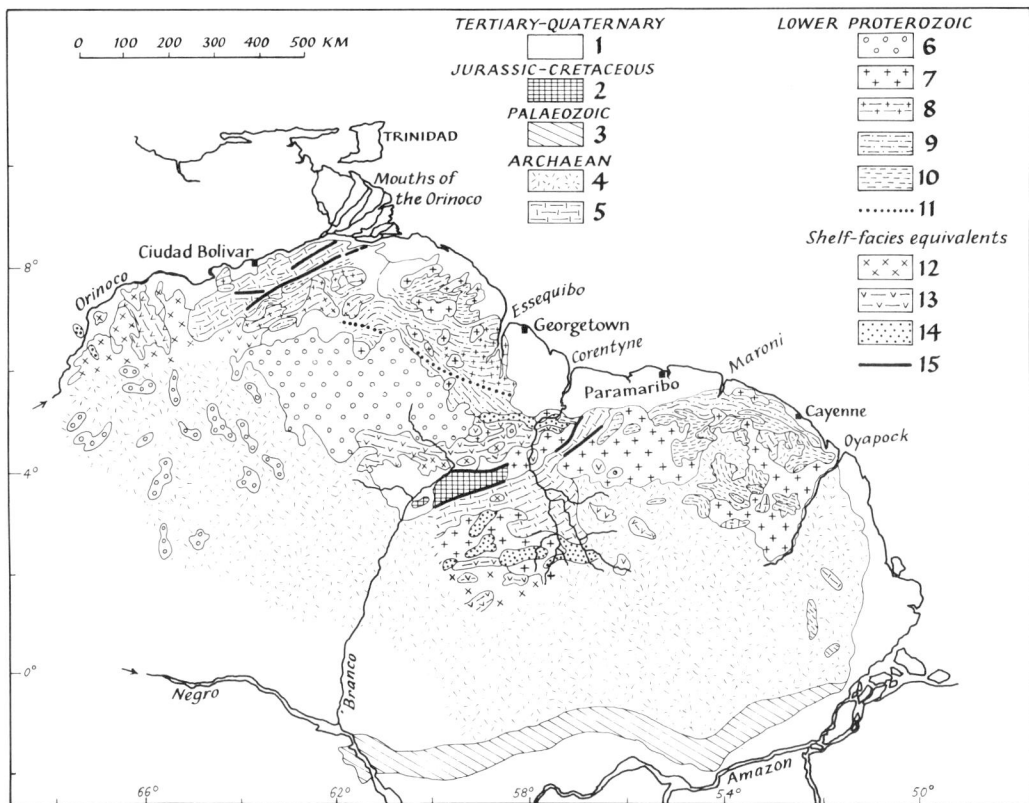

effective campaign of isotope dating has brought out a similar overall age pattern reaching from French Guiana through Surinam and Guyana to Venezuela. It is therefore possible to draw together some of the common geological features that have emerged and to establish a widely accepted correlation of the Precambrian formations of the shield. Present geological evidence and geochronological data indicate the following subdivision into three major units:

1. *Archean*, older than ca. 2500 m yr and consisting of high-grade gneisses with pyroxene granulites (cf. charnockitic gneisses) and itabirites, and of a widespread granitoid "basement." These rocks were folded and metamorphosed in an *Imatacan Orogeny* dated 3000–2700 m yr by Hurley and Rand (1971) and Hurley et al. (1972), which would correspond to the Hylean Cycle of Choubert (1969), and have been in part rejuvenated during later episodes.

2. *Lower Proterozoic* metasediments and metavolcanics, defined as younger than ca. 2500 m yr, folded, metamorphosed, and older than widely distributed granites of a *trans-Amazonian cycle* well dated as 2000–1800 m yr in age.

3. *Tabular Proterozoic* quartzitic sandstones and shales younger than the granites but older than basic intrusives dated about 1700 m yr old.

The use of the classical terms "Archean" and "Proterozoic," such as applied in the Canadian Shield, appears to be justified by the latest isotope dating of principal orogenic cycles in the Guiana Shield by Hurley and others.

Archean. Outstanding in the Archean subdivision are a number of strongly folded, high-grade metasedimentary and metavolcanic complexes characterized by banded ironstones (itabirites) and pyroxene granulites (Table 2). The *Imataca Complex* of eastern Venezuela

311

TABLE 2. Correlation Table

	French Guiana–Surinam Basin			Brazil
	French Guiana	Surinam		Brazil
Cenozoic	Alluvium and coastal sediments	Alluvium and coastal sediments		Alluvium and coastal sediments
Cretaceous				Nova Olinda Basalts Serra Tucano (Rio Branco)
Jurassic				
Permo-Triassic		Dolerites, ca. 220 m yr		
Paleozoic				Cambrian-Carboniferous
Proterozoic Middle and Upper, ca. 1200 m yr		*E. Province* Nickerie metamorphic episode	*W. Shelf Province*	
Trans-Amazonian orogeny, 2000–1800 m yr	Dolerites	Dolerites Roraima Formation, ca. 1600 m yr ?? *great unconformity*		Roraima Intrusives Roraima Formation *great unconformity*
Lower	Granite Galibi Granite Caraibe Granite Guyanais (rejuvenated)	Granites 2 and 3, ca. 2000 m yr	Acidic plutonic and volcanic suite, ca. 1810 m yr Dalbana Volc. Fm.	Granites Surumu acidic suite Rio Branco terr.)
	Serie de l'Orapu Serie de Bonidoro Serie de Paramaca Volcanic, sedimentary	Armina Series Rosebel Series Paramaka Fm. Tapaje Group (ibabirites in south?)	Ston Fm. Matapi Fm.	Amapá Series Serra do Navio Gp.
Archean Imalacan orogeny 3000–2700 m yr	*great unconformity* Granite Guyanasi (in part remobilized ca. 2000 m yr)	*great unconformity* ?Granite 2 in part?		Granitic basement (in part rejevenated ca. 2000 m yr?)
	Serie de l'ile de Cayenne	Adampada Falawatra Group (high-grade metamorphics) ?Tapaje Group?		?Falsino River Granulites? (Amapa territory)

(Bellizzia and Martin-Bellizzia, 1956; Short and Steenken, 1962; Kalliokoski, 1965) consists largely of granitoid gneiss and includes important ironstone formations as well as gneisses in the granulite facies of metamorphism; it has been well dated (Hurley et al., 1967; Hurley and Rand, 1971) as about 3000 m yr or older. In southern Guyana the *Kanuku Group* or *Complex* (Singh, 1962; Berrangé, 1972) of high-grade gneisses and pyroxene granulites of charnockitic type is regarded as Archean for structural, petrographical, and stratigraphical

for the Guiana Shield

Guyana-Venezuela Basin					
Guyana			Venezuela		
Alluvium and coastal sediments					
Maestrichtian-Apoteyi Volc. Fm.					
Takutu Fm.: Wealden Facies (Repununi District only)					
Minor Basic instrusives					
S. Province	*Shelf Province*	*N. Province*	*E. Province*		*Shelf Province*
Rigid block reaction			?Bolivar dislocation zone		
K'mudku mylonite episode			?Granites (?ca. 1500 m yr?)		
Roraima Intrusives, ca. 1700 m yr					
Roraima Formation					
great unconformity					*great unconformity*
Keyuwini GP.	Iwokrama igneous suite	Younger Granite Group, 2000 m yr			
Acidic plutonic and Volcanic suite		Bartica Gneiss Complex	Granites–Encrucijada Granite		Cuchivero igneous Suite
Southern Guyana Granite Complex, ca. 1845 m yr	Muruwa Fm.	Mazaruni Group	Supamo Complex Rejuvenated (ca. 2000 m yr)		Parguaza Granite
		Haimaraka Fm.			
		Cuyuni Fm.	Pastora Gp.		Cinaruco Fm.
Kwitaro Group		Barama Group	Carichapo Group		
great unconformity			*great unconformity*		
(?Southern Guyana?) Granite Complex remobilized 2000 m yr	Makarapan Granite	Basement unknown	Supamo Complex (Granitoid Gneiss) Prereactivation		
			Imataca Complex (ca. 3400–3000 m yr)		
Kanuku Group (high-grade metamorphics)			Itabirites, qz.-feldspar Gneisses, hornblende–pyroxene gneisses		

reasons (Williams *et al.*, 1967; McConnell and Williams, 1970). These rocks have been deeply affected by later tectonothermal episodes (Barron, 1962; Singh, 1966; Berrangé, 1972) and dates of 2100–2000 m yr (Snelling and Berrangé, 1970; Spooner *et al.*, 1971) are believed to indicate a downgrading of isotope ratios. Mapping indicates that the *Adampada-Falawatra and Coeroni Group* of Surinam (Janssen, 1966; Bosma and Groeneweg, 1970) are continuous with the Kanuku Group. In this subdivision would also come the *Ile de Cayenne Group* or Hylean System of French Guiana defined by Choubert (1965, 1969) and consisting of amphibolites, quartzites, paragneisses, and terminated by the emplacement of extensive massifs of quartz diorites and granodiorites.

In southern Guyana large areas were mapped by the Geological Survey as "uncorrelated" granite and gneiss (1962; Williams et al., 1967). Berrangé (1972) has now mapped a large part of these as granulite and migmatite of the Kanuku Complex; another part, including the former "South Savanna Granite" (Barron, 1962; McConnell, 1961), is termed the "Southern Guyana Granite Complex" (see below) and regarded mainly as autochthonous granite rejuvenated in the trans-Amazonian cycle (ca. 2000 m yr old). Farther to the E, in southern Surinam and French Guiana, these broad, forested, granitoid pediplains continue and have been mapped in French territory as the older *granites guyanais* and the younger *granites caraïbes*; and in Surinam as *Granite 2* and in part *Granite 3*, for both of which Priem has determined trans-Amazonian dates. Choubert (1969) has, however, concluded that the *granites guyanais* may be in part of an age equivalent to the Archean, although locally rejuvenated by the trans-Amazonian episode. In a later map Bosma and Groeneweg (1970) have joined the Surinam granites in one complex and Priem et al. (1971) described them as acidic-plutonic representatives of the trans-Amazonian basement (1810 m yr old). It may be suggested, however, that, as in the case of the *granites guyanais*, some may represent older granite reactivated. Farther to the W Bellizzia (1972) and Martin-Bellizzia (1972) regard the granitoid rocks of the Amazonas Territory as the *Supamo Complex* of Archean age where not reactivated, and Barbosa and Ramos (1961) suggest the same age for the granitoid basement of the Rio Branco territory of Brazil. Pyroxene granulites included in granodioritic and adamellitic gneisses and migmatites have also been described by Scarpelli (1966) in the Falsino River area of the Amapá territory of Brazil.

Lower Proterozoic. The major unit of the Guiana Shield, now defined as "Lower Proterozoic" in age, consists of geosynclinal, shelf, and continental assemblages terminated by an episode of folding, predominant epizonal metamorphism, anatexis, and the emplacement of synorogenic and post–tectonic granites. Numerous radiometric dates on the granites by Choubert (1964), Snelling (1964, 1965), Priem et al. (1971, 1973), Posadas and Kalliokoski (1967) indicate an age of about 2000–1800 m yr for this tectonothermal episode. It is known as the Caraïbe episode in French Guiana, the *Akawaian* in Guyana (Williams et al., 1967), and is now designated by Hurley et al., (1967) as the *trans-Amazonian orogenic cycle*, owing to its widespread extent in South America. The Lower Proterozoic rocks appear to have been deposited in two distinct geosynclinal basins, a *Guyana-Venezuela basin* to the W and a *French Guiana-Surinam basin* to the E, separated by a dorsal of Archean fringed by Proterozoic of shelf facies (McConnell and Williams, 1970).

The *Barama-Mazaruni Supergroup* (or *Assemblage*) (McConnell, 1958; Williams et al., 1967) of northern Guyana is taken as typical of the Lower Proterozoic. In central and northwestern Guyana it is eugeosynclinal in character, consisting of (1) the *Barama Group* (the base of which is not known) of chiefly pelitic metasediments with metamorphosed lavas and pyroclastics, characterized locally by gondites and manganiferous phyllites, and overlain with conformable passage by the *Mazaruni Group*, consisting of (2) the *Cuyuni Formation* of pebbly sandstones and intraformational conglomerates of the graywacke suite with intercalated acid to basic volcanics, and (3) the *Haimaraka Formation*, also conformable and indicating, with mudstones and graywackes, the return of pelitic sedimentation but unaccompanied by significant volcanism. The whole assemblage has been folded together, imprinted with a conspicuous near-vertical cleavage, and metamorphosed in the greenschist facies with the emplacement of granites and gneisses (Younger Granite Group) of the Akawaian tectonothermal episode, which locally form migmatite complexes. Orthoquartzites, volcanic agglomerates, and tuffs begin to appear in the Mazaruni Group S of latitude $5-6°N$, indicating a passage toward the S and E from eugeosynclinal to shelf conditions as the Archean foreland is approached. The Cuyuni and Haimaraka formations are now replaced in this direction by the orthoquartzitic Muruwa Formation, which is overlain by a thick sequence of rhyodacitic metavolcanics (ignimbrites?) associated with porphyritic and granophyric granites termed the "Iwokrama igneous suite" or group (Berrangé, 1972). These rocks can be followed to the W and are seen to be continuous with the Surumu Formation (Barbosa and Ramos, 1959, 1961) of the Rio Branco territory of Brazil. The contact between the Lower Proterozoic geosynclinal basin of northern Guyana and the Archean foreland is the locus of a very important dislocation zone marked locally by the Takutu rift valley in southern Guyana.

To the NW, the eugeosynclinal Barama-Mazaruni Supergroup passes in Venezuela to the Pastora-Carichapo Supergroup (Kalliokoski, 1965; McCandless, 1968; Bellizzia, 1972; Martin-Bellizzia, 1972). The *Pastora Group* corresponding roughly to the Mazaruni Group and the *Carichapo Group* to the Barama Group. The Younger Granites of Guyana are represented by the trans-Amazonian granites dated about 2000 m yr by Hurley et al. (1967). The proximity of an Archean foreland to the S and W, in the NW of the Guiana Shield, is indicated by the orthoquartzites of the *Cinaruco Formation* and the *Cuchivero igneous suite,* described by McCandless (1966, 1968) and Mendoza, (1972) and correlated, respectively, with the Muruwa and Iwokrama formations of Guyana and the Surumu Formation of the Rio Branco territory of Brazil to the SSE toward which they are striking (Fig. 1). In Surinam S of latitude 5°N the *Dalbana* and *Ston* formations, described by Loemban-Tobing (1966), continue to the E as the Iwokrama and Muruwa formations, respectively; these indicate the continuation of shelf conditions in the Lower Proterozoic, marginal to an Archean foreland represented by the Falawatra gneisses. However, the presence of a spilite-keratophyre *Matapi Formation* beneath the Ston suggests that the latter is transgressive over an equivalent of the Mazaruni Group (Bosma and Groeneweg, 1970).

The arc of shelf conditions in the Lower Proterozoic thus extends W from northwestern Surinam through central Guyana to the Rio Branco territory of Brazil, bending to the WNW to reach the Rio Caura area (7°N, 65°W) of Venezuela. This great arc thus forms a wide border to the area of eugeosynclinal facies defined in the Barama-Mazaruni Assemblage of northern Guyana (Fig. 1) and the Pastora-Carichapo of Venezuela. Thus a wide *Guyana-Venezuela basin* of Lower Proterozoic geosynclinal formations is defined, elongated WNW-ESE, and bisected to the N by the Atlantic coast. To the S and W a foreland is indicated comprising the Archean rocks previously described. The boundary between the Lower Proterozoic and Archean is marked by the great dislocation zones of southern Guyana, western Surinam, and the Estado Bolívar of Venezuela.

To the E, beyond the NE-SW-striking dorsal of the Archean Falawatra gneisses, lies a large area of granite and comagmatic acidic volcanics (Verhofstad, 1970; Bosma and Groeneweg, 1970) which is termed, by Priem et al. (1971), on the basis of a number of age determinations, the "Trans-Amazonian acidic plutonic-volcanic basement." Priem correlates these rocks with the Dalbana and with the Iwokrama igneous suite of Guyana, suggesting that all these igneous rocks represent the terminal phase of the Trans-Amazonian tectonothermal episode, and would thus be younger than the Mazaruni Group. The Kuyuwini Group of southern Guyana (Barron, 1962) has now been described by Berrangé (1972) as consisting of rhyodacitic lavas and tuffs with consanguineous intrusive porphyries, granophyres, and subvolcanic granitic rocks, provisionally correlated with the Iwokrama igneous suite, and hence with the trans-Amazonian acidic basement of Surinam. In southern Guyana, between the Kanuku Complex and the granito-volcanic Kuyuwini Group, lies the great Southern Guyana Granite Complex, including the former South Savanna Granite (Barron, 1962), which is marginally intrusive into the Kanuku Group (Singh, 1966). Berrangé (1972), however, regards it as largely autochthonous or parautochthonous, suggesting the possibility that it represents the Archean basement reactivated and now giving trans-Amazonian dates (Snelling and McConnell, 1969). Overlying, and intruded by this granite, lie five separate enclaves of metasedimentary formations, termed by Berrangé (1972) the "Kwitaro Group."

The relationship of the rocks of the Kwitaro Group to the trans-Amazonian tectonothermal episode indicates correlation with the Lower Proterozoic of the Guiana Shield, and they may be related to the rocks of the French Guiana-Surinam basin rather than to those of northern Guyana.

Outstanding pioneer work in French Guiana since 1948 by Choubert (1965, 1969), Barruol, Brouwer, and others has firmly established the stratigraphy of that country. The *série de Paramaca* appears to correspond to the Barama Group, and it also contains gondites and manganiferous phyllites. It is overlain unconformably (Barruol, 1961) by the chiefly detrital *série de Bonidoro*, which resembles the Cuyuni Formation, and is correlated with it by general consent. Overlying the Bonidoro, is the essentially argillaceous *série de l'Orapu*, which may represent the Haimaraka Formation. Volcanism, although important in the Cuyuni and present in the Haimaraka, is practically absent from the Bonidoro and Orapu, and their resemblance to a flysch facies of geosynclinal infilling is thus notable. Although separated by unconformities, the Paramaca, Bonidoro, and Orapu have been folded together during the same major tectonic cycle, and penetrated by the *Caraïbe* granites and gneisses, dated 2200–1800 m yr by Choubert (1964), which are thus of trans-Amazonian date. As already stated, some of the granites in French Guiana and Surinam may represent Archean basement reactivated in the trans-Amazonian episode, and this has been accepted by Choubert (1969).

The succession in eastern Surinam is much obscured by the trans-Amazonian granitization, but a *Paramaka Formation* is continuous with the Paramaca of French Guiana and is succeeded by the *Rosebel* and *Armina* formations, respectively, psammitic-psephitic and pelitic-psammitic (O'Herne, 1966), which correspond roughly to the Bonidoro and Orapu. All these formations have been grouped, together with the Ston, as the *Marowijne Group* by Bosma and Groeneweg (1970).

The *French Guiana-Surinam basin* described above appears to have been shallower than the Guyana Lower Proterozoic basin, as volcanic rocks are confined to the Paramaca. Moreover, the Bonidoro and Orapu are clearly unconformable, with basal conglomerates, whereas in Guyana the succession is conformable with intraformational turbidite-type conglomerates and spilite-keratophyre volcanism.

Owing to the metamorphism and the difficulty of establishing structure, no estimate of the thickness of the Lower Proterozoic formations in Guyana has been possible, but Choubert (1965) gives a figure of 4000 m for the lower sedimentary member of the Paramaca in French Guiana and estimates thicknesses of 2000 m each for the Bonidoro and Orapu.

Tabular Proterozoic. The vast, flat-lying Roraima Formation overlies unconformably the folded Lower Proterozoic and occupies an area of approximately 163,000 km^2 in Guiana, Venezuela, and Brazil, forming the Pakaraima Mountains in the center of the shield, with a small outlier in Surinam. The formation has been much reduced and dissected by erosion, and Gansser (1954) estimates that it originally covered an area of some 1,200,000 km^2. This formation consists of quartzitic sandstones and quartzites with some conglomerates, shales, and rare jaspers, and culminates in the high plateau of Mt. Roraima. It overlies unconformably the folded Lower Proterozoic and granites and is intruded by dolerites and quartz dolerites of the Roraima Intrusive Suite (Hawkes, 1966b), which Snelling (Snelling and McConnell, 1969) has dated as 1695 ± 66 m yr of age. It is therefore concluded that the Roraima Formation was deposited in the 1800–1700 m yr interval, soon after the uplift and erosion of the Lower Proterozoic mobile belt. Priem *et al.* (1973) has given numerous dates of around 1600 ± 50 m yr for the Tafelberg outlier in Surinam. Barbosa and Ramos (1961) have suggested that a Roraima Formation of Proterozoic age was overlain by a Permo-Carboniferous "Kaieteur" Formation occupying the highest summits. The Pakaraima Mountains have, however, been surveyed in some detail by the Geological Survey of Guyana (Bateson, 1966), and there is every indication that the Roraima Formation is a simple unit divided into three members with only local unconformities, and a date of 1570 m yr (Snelling and McConnell, 1969) has been given for the High Level Sill in Mt. Roraima. There is thus no doubt that the entire Roraima Formation is of Proterozoic age. Bateson estimates its thickness as about 2000 m.

Proterozoic Dolerites. Tholeiitic dolerites occurring as large bodies, major dikes, inclined sheets, and sills are found throughout the basement of the Guiana Shield. They are particularly well developed as sills up to 450 m in thickness cutting the Roraima Formation, and as dikes in the adjacent basement, where they have been termed the *Roraima intrusive suite* by Hawkes (1966a, 1966b). Bellizzia (1957) also studied these dolerites in Venezuela. An average date for the suite of 1675 m yr was established in Guyana (McConnell and Snelling, 1969), and a similar age has been determined in Surinam by Priem *et al.* (1971, 1973).

Late Proterozoic. The Kanuku Group and the Southern Guyana Granite Complex have been affected by belts of intense shearing, directed NE or ENE, with mylonitization and fusion, producing dense black pseudotachylites and even flaser gneisses. This is the *K'mudku mylonite episode* of Barron (1962), described in more detail by Berrangé (1972). Mica ages in adjacent trans-Amazonian granites have been affected by this episode, and the best estimate of its age is 1200 m yr, being the youngest K–Ar age of these micas (Snelling and McConnell, 1969). The same tectonothermal event has been detected in Surinam and termed the *Nickerie metamorphic episode,* dated also at 1200 m yr (Priem *et al.*, 1971).

A granite pod in one of the great faults of the Bolivar dislocation zone in Venezuelan Guayana has given K–Ar dates of 850 and 1070 m yr (Kalliokoski, 1965), and other low dates in the trans-Amazonian granites may indicate the effects of this same tectonic episode.

Paleozoic. A strip of cover rocks of Paleozoic age bounds the southern margin of the Guiana Shield in the Amazon River trough (see *Brazil*).

Mesozoic. Late Triassic dolerite dikes are widespread; in the W they trend mainly NE-NNE and in the E mainly NNW. The K–Ar dates are ca. 220 m yr. Mesozoic sediments occur in the Takutu rift valley in southern Guyana, and in the Rio Branco territory of Brazil (Barbosa and Ramos, 1961, p. 24; and "nota final") in the center of the shield. Marine Maestrichtian has also been recognized underlying Tertiary and Quaternary in boreholes in the coastal plain of Guyana. Mid-Cretaceous basalts are sporadically associated with the rifting and elsewhere.

Cenozoic. An important but rather thin wedge of Tertiary and Quaternary sediments forms a coastal fringe, narrowest in French Guiana and broadest at the Berbice-Correntyne river embayment (see separate articles, *Guyana, Surinam, French Guiana*).

Structure

Fundamentally the Guiana Shield is a great oval dome of folded metamorphic and crystalline Precambrian rocks capped by the tabular Proterozoic Roraima Formation. The Precambrian is strongly downwarped at the continental margin and overlain by the Mesozoic and Cenozoic formations of the coastal plain, these also penetrate the shield in a deep NE-SW-striking downwarp between the Berbice and Corentyne rivers.

Structures in the Archean are obscure, owing to metamorphism and granitization, but are steep and strike predominantly ENE-WSW to NE-SW. The Lower Proterozoic has been folded along axes directed chiefly WNW-ESE and is affected by a conspicuous subvertical cleavage that obscures the relatively open character of the folds (see structure section in Williams *et al.*, 1967, Fig. 3).

The shield has been affected by two very important dislocation zones. The *Bolívar dislocation zone* in Venezuelan Guayana (Fig. 1) extends ENE-WSW over 400 km and displays vertical mylonite zones up to 1 km wide with crushing extending over a width of up to 4 km (Short and Steenken, 1962; Kalliokoski, 1965). The movement is considered to be dominantly left-lateral transcurrent, but as the shears mark the limit between the Archean and Lower Proterozoic complexes, some vertical movement is also likely. The *Kanuku-Bakhuys Mountains dislocation zone* crosses the center of the shield between 3 and 4°N and also marks the limit between Archean and Lower Proterozoic complexes. In Guyana this zone is followed for 160 km by the Takutu rift valley, 50 km in width, which contains a thickness of about 2 km of Mesozoic sediments (see *Guyana*).

Mineral Resources

The Guiana Shield is famous for the high-grade bauxite deposits that occur in its coastal area, and for rich hematite iron ore mined in Venezuelan Guiana. Gold is everywhere associated with the Lower Proterozoic assemblages and occurs in peribatholithic quartz veins; some are very rich, and there are numerous small placer deposits, many of which are still worked in a small way. Diamonds occur chiefly in Guyana, Venezuela, and Brazil where they are recovered from river gravels; their immediate origin has been traced to conglomerates in the lower member of the Roraima Formation, but the pipes are not known. Small diamond workings are also associated with the Rosebel Formation in Surinam. Manganiferous phyllites and gondites are included in the lower pelitic-volcanic group of the Lower Proterozoic and have been extensively prospected; one mine was under production for 5 years in Guyana and an extensive deposit is being worked in the Amapá Province of Brazil.

Prospecting is difficult in all the territories of the Guiana Shield, owing to the extensive forest cover and thick saprolite. With the advent of aerial photography and sophisticated geophysical, geochemical, and scanning techniques, intensive mineral search campaigns are yielding indications of copper and other metals.

RICHARD B. McCONNELL
BORIS CHOUBERT

References

Barbosa, O., and Ramos, J. R. de A., 1959. "Territoria do Rio Branco," *Div. Geol. Mineral.*, (Brazil) **196**, 46p.

———, 1961. "Principal aspects of the geomorphology and geology in the Territory of Rio Branco, Brazil," *Proc. 5th Inter. Guiana Geol. Conf.*, Georgetown, 1959, 33–36.

Barron, C. N., 1962. "The geology of the South Savannas Degree Square," *Geol. Surv. Br. Guiana Bull.*, **33**, 29p.

Barruol, J., 1961. "Le Bonidoro en Guyane Française," *Proc. 5th Inter-Guiana Geol. Conf.*, Georgetown, 1959, 57–67.

Bateson, J. H., 1966. "Some aspects of the geology of the Roraima Formation in British Guiana," *Trans. 3rd Caribbean Geol. Conf.*, Jamaica, 1962, 144–150.

Bellizzia, A., 1957. "Consideraciones petrogeneticas de la provincia magmatica de Roraima (Guayana Venezolana)," *Bol. Geol.*, (Venezuela), **4**(9), 53–81.

———, 1972. "Presentación del mapa geológico de la Guayana Venezolana," *9th Inter-Guiana Geol. Conf.*, Ciudad Guayana, May, 1972.

———, and Martin-Bellizzia, C., 1956. "Imataca Series," *Stratigraphic Lexicon of Venezuela, Minist. Minas Hidrocarburas, Spec. Publ.* **1**, 254–256.

Berrangé, J. P., 1972. "The tectonic/geological map of Southern Guayana," *9th Inter-Guiana Geol. Conf.*, Ciudad Guayana, May, 1972.

Bosma, W., and Groeneweg, W., 1970. "Review of the stratigraphy of Suriname," *Proc. 8th Guiana Geol. Conf.*, Georgetown, 1969, **5**, 29p.

Choubert, B., 1957. "Essai sur la morphologie de la Guyane," *Mém. Carte Géol. Dét. de la France*, Paris: Dept. Guyane Fr., 43p.

_____, 1964. "Ages absolus du Precambrien guyanais," *C. R. Acad. Sci.,* **258,** 631–634.

_____, 1965. "Etat actuel de nos connaissances sur la géologie de la Guyane française," *Bull. Soc. Géol. France Ser.* 7, **7,** 129–135.

_____, 1969. "Les Guyano-Eburneides de l'Amerique du Sud et de l'Afrique Occidentale," *Bull. B.R.G.M.,* Ser. 2, **4**(4), 39–68.

_____, Schols, H., and Bracewell, S., 1954. "La carte géologique des trois Guyanes," *19th Intern. Geol. Conf.,* **14,** 371–377.

Damuth, J. E., and Fairbridge, R. W., 1970. "Equatorial Atlantic deep-sea arkosic sands and ice-age aridity in tropical South America," *Bull. Geol. Soc. Am.,* **81,** 189–206.

Gansser, A., 1954. "Observations on the Guiana Shield (S. America)," *Eclog. Geol. Helvet.,* **47,** 77–112.

Hammen, T. van der, and Burger, D., 1966. "Pollen flora and age of the Takutu Formation (Guyana)," *Leidse Geol. Medede.,* **38,** 173–180.

Hawkes, D. D., 1966a. "Differentiation of the Tumatumari-Kopinang dolerite intrusion, British Guiana," *Bull. Geol. Soc. Am.,* **77,** 1131–1158.

_____, 1966b. "The petrology of Guiana dolerites," *Geol. Mag.,* **103,** 320–335.

Hurley, P. M., and Rand, J. R., 1971. "Outline of Precambrian chronology in lands bordering the South Atlantic, exclusive of Brazil," Mass. Inst. Tech. Dept. Earth Planet. Sci., Progr. Rept., 23–35.

_____, et al., 1967. "Test of continental drift by comparison of radiometric ages," *Science,* **157,** 495–500.

_____, Melcher, C. C., Pinson, W. H., Jr., and Fairbairn, H. W., 1968. "Some orogenic episodes in South America by K-Ar and whole-rock Rb-Sr dating," *Caribbean J. Earth Sci.,* **5,** 633–638.

_____, Kalliokoski, J., Fairbairn, H. W., and Pinson, W. H., 1972. "Progress report on the age of granulite facies rocks in the Imataca Complex, Venezuela," *9th Inter-Guiana Geol. Conf.,* Ciudad Guayana, May 1972.

Janssen, J. J., 1966. "Bauxite in the Adampada-Kabalebo area, Surinam," *Proc. 6th Inter-Guiana Geol. Conf.,* Belem-Macapa.

Kalliokoski, J., 1965. "Geology of north-central Guayana Shield, Venezuela," *Bull. Geol. Soc. Am.,* **76,** 1027–1050.

McCandless, G. C., 1968. *Mapa Geologico de la Region Septentrional del Escudo de Guayana Venezuela.* Puerto Ordaz, Venezuela: Orinoco Mining Company, 1:500,000.

McConnell, R. B., 1958. "Provisional stratigraphical table for British Guiana," *Annual Report.* Geol. Surv. Br. Guiana, 35–53.

_____, 1961. *British Guiana, report on the Geological Survey Dept. for the year 1960.* Georgetown: Geol. Survey, 72 p.

_____, 1968. "Planation surfaces in Guyana," *Geogr. J.,* **134,** 506–520.

_____, and Snelling, N. J., 1969. "The geochronology of Guyana," *Geol. Mijnb.,* **48,** 201–213.

_____, and Williams, E., 1970. "Distribution and provisional correlation of the Precambrian of the Guiana Shield," *Proc. 8th Inter-Guiana Geol. Conf.,* Georgetown, **1,** 1–22.

Martin-Bellizzia, C., 1972. "Relaciones estratigráficas y paleotectónicas del Escudo Guayanés," *9th Inter-Guiana Geol. Conf.,* Ciudad Guayana.

Mendoza, V., 1972. "Geologia del area del Rio Suapure, NW del Escuda de Guayana, Estado Bolivar, Venezuela," *9th Inter-Guiana Geol. Conf.,* Ciudad Guayana.

O'Herne, L., 1966. "A short introduction to the geology of Surinam," *7th Guiana Geol. Conf.,* Paramaribo.

Posadas, V. G., and Kalliokoski, J., 1967. "Rb-Sr ages of the Encrucijada granite intrusive in the Imataca Comples, Venezuela," *Earth Planet Sci. Lett.,* **2,** 210–214.

Priem, H. N. A., Boelrijk, N. A. I. M., Hebeda, E. H., Verdurmen, E. A. T., and Verschure, R. H., 1971. "Isotopic ages of the trans-Amazonian acidic magmatism and the Nickerie Metamorphic Episode in the Precambrian basement of Suriname, South America," *Bull. Geol. Soc. Am.,* **82,** 1667–1680.

_____, et al., 1973. "Age of the Precambrian Roraima Formation in northeastern South America: evidence from isotopic dating of Roraima pyroclastic volcanic rocks in Suriname," *Bull. Geol. Soc. Am.,* **84,** 1677–1684.

Scarpelli, W., 1966. "Preliminary geological mapping of the Flasino River," *7th Guiana Geol. Conf.,* Surinam.

Short, K. C., and Steenken, W. F., 1962. "A reconnaissance of the Guayana Shield from Guasipati to the Rio Aro, Venezuela," *Bol. Inform., Assoc. Venezolana Geol., Mineral y Petrol.,* **5,** 189–221.

Singh, S., 1966. "Geology and petrology of part of the Guyana Shield in the south Savanna-Rewa area of Southern Guyana," *Geol. Surv. Guyana, Bull.,* **37,** 127p.

Snelling, N. J., 1964. "Age determination unit," *Annual Report 1963.* London: Overseas Geol. Surv., 36.

_____, 1965. "Age determination unit," *Annual Report 1964.* London: Overseas Geol. Surv., 36.

_____, and Berrangé, J. P., 1970. "The geochronology of Guyana II, results obtained in the period 1966–69," *Proc. 8th Guiana Geol. Conf.,* Georgetown, 1969.

_____, and McConnell, R. B., 1969. "The geochronology of Guyana," *Geol. Mijnb.,* **48,** 201–213.

Spooner, C. M., et al., 1971. "Rb-Sr whole-rock age of the Kanuku complex, Guiana," *Bull. Geol. Soc. Am.,* **82,** 207–210.

Verhofstad, J., 1970. "The geology of the Wilhelmina Mountains in Surinam, with special reference to the occurrence of Precambrian ash-flow tuffs," Ph.D. thesis, Univ. Amsterdam.

Williams, E., Cannon, R. T., and McConnell, R. B., 1967. "The folded Precambrian of northern Guyana related to the geology of the Guiana Shield, with age data by N. J. Snelling," *Geol. Surv. Guyana, Records,* **5,** 60p.

Cross-references: *Brazil; French Guiana; Guyana; South America; Surinam.*

GUYANA

The Republic of Guyana (formerly British Guiana), with an area of 231,800 km^2 (83,000 sq mi), is situated on the *Guiana Shield* (q.v.)

the northeastern coast of South America. Its capital is Georgetown, a port on the Demerara River. Geological surveying started in 1867 with the work of Brown and Sawkins (1875) and continued (Harrison, 1908) sporadically until the Geological Survey of British Guiana was formed in 1933 and laid the foundations for the present interpretation of the geology and structure of the country (Bracewell, 1947, 1956). Progress was difficult, owing to the cover of tropical rain forest and the deep surface layer of saprolite. Geological surveying was intensified in 1957 with the help of a grant from the British Government, and the establishment of a new geological map (McConnell and Dixon, 1960) served as basis for an extensive airborne prospecting campaign backed by the United Nations Special Fund. Following independence, the Geological Survey and Mines Department has pursued an active policy of mineral search. Guyana has become the third largest exporter of bauxite in the world and has been a steady producer of gold and diamonds.

Geomorphology

Guyana, elongated from N to S and occupying a central position in the *Guiana Shield* (q.v.), falls into three geomorphic areas governed by the geological constitution.

1. The *coastal plain,* underlain by Neogene sediments, forms a fertile strip 25–35 km wide, supporting flourishing cultivations of sugar and rice, in which 90% of the population lives.

2. Separated from the coastal strip by an infertile belt of (Quaternary) white sands, *Precambrian crystalline and metamorphic rocks* cover 60% of the country, forming extensive pediplains at a general altitude of 100–150 m with rugged highlands rising to 300–400 m. In the far S the general level rises gradually to some 180–220 m, and a number of ranges, such as the Kanuku, Marudi, Amuku, and Acarai mountains, reach elevations of 600–1000 m. The pediplains are covered with Amazonian rain forest, with the exception of the Rupununi savannas in the W between latitudes 2°30' and 4°N. The country is well watered, with great rivers, but navigation is impeded by many rapids.

3. In the NW the tabular *Proterozoic sediments* of the Roraima Formation form the Pakaraima Mountains, forested plateaus with an altitude of 600–1000 m rising from the plains with impressive scarps over which rivers cascade in spectacular falls, such as Kaieteur and Amaila. Several high mesa-like ("tepui") rise above the general level, culminating in Mt. Roraima (2730 m) at the point where Venezuela, Brazil, and Guyana meet.

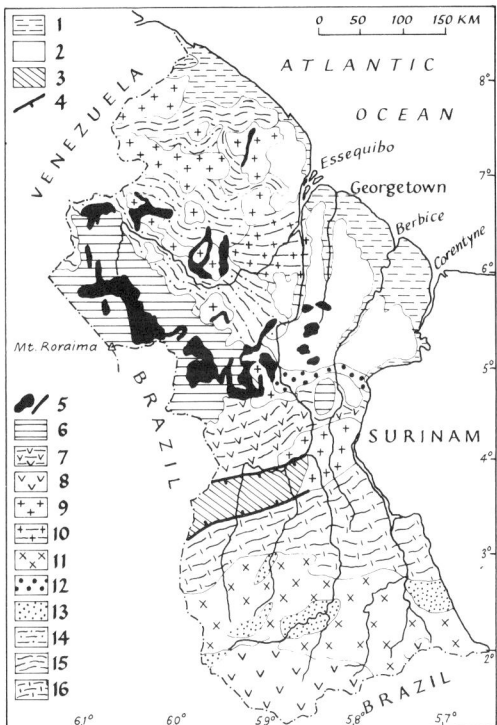

FIGURE 1. Geologic map of the Republic of Guyana. 1, Demarara Formation and alluvium; 2, white sands; 3, Takutu Formation; 4, rift faults; 5, Roraima intrusive suite; 6, Roraima Formation; 7, Iwokrama Formation; 8, Kuyuwini Formation; 9, Younger Granite Group; 10, Younger Granite Group, migmatitic; 11, southern Guyana granite complex; 12, Muruwa Formation; 13, Kwitaro Group; 14, Mazaruni Group; 15, Barama Group; 16, Kanuku Group.

Geological Formations

The geomorphic subdivisions reflect the geological framework of the country and emphasize the predominance of Precambrian formations (Table 1). The progress of geological mapping in Guyana, in conjunction with geochronological data for the whole of the shield, makes it convenient to divide the main assemblages of Precambrian rocks into Archean and Proterozoic, with the division at about 2500 m yr of age. The folded Proterozoic rocks lie in a northern geological province (Williams *et al.,* 1967), N of the Takutu rift valley (Fig. 1), whereas the southern province, S of this structural feature, is largely formed by Archean gneisses and a granite complex that appears to have resulted from the reactivation of a preexisting basement (McConnell and Williams, 1970; Berrangé, 1972). Shelf facies rocks mark the passage between the two provinces.

Archean. *Kanuku Complex.* These rocks, previously described by Barron (1962) and

TABLE 1. Stratigraphic Table for Guyana

	Southern Province	Shelf Province	Northern Province
Cenozoic	Alluvium and coastal sediments		
	Maestrichtian sands and clays		
Cretaceous	Apoteri Volcanic Formation (basaltic) ca. 115 m yr		
	Takutu Formation: Wealden Facies		
Jurassic			
Permo-Triassic	Minor basic intrusives		
Paleozoic			
Proterozoic			
Middle and Upper	Roraima intrusive suite (ca. 1700 m yr.)		
	Roraima Formation (ca. 1800–1700 m yr.)		
	Great unconformity		
Lower, trans-Amazonian orogenic cycle, ca. 2000–1800 m yr	Kuyuwini igneous suite	Iwokrama igneous suite	Younger Granite group
	Southern Guyana Granite complex		Granitization of Bartica complex
	Granitization and migmatization of Kanuku complex and Archean basement	Muruwa Formation	Mazaruni Group
			Haimaraka Fm.
			Cuyuni Formation
	Kwitaro Group		Barama Group
		Great unconformity	
Archean Imatacan orogeny (?ca. 3000 m yr)	Kanuku complex Granulitic gneissose Basement complex (largely reactivated ca. 2000 m yr)	Makarapan Gneiss	No basement visible

Singh (1966), have recently been mapped in detail by Berrangé and Johnson (Berrangé, 1972, 1973) and are considered to consist of high-grade metasediments and metavolcanics of an Archean "Proto-Kanuku Complex," migmatized and granitized during the trans-Amazonian orogenic cycle (Hurley et al., 1967) dated about 2000 m yr old. The complex is divided into migmatites and granulites. The *Kanuku Migmatites* form the bulk of the complex and are described as (1) paragneiss containing biotite, garnet, hypersthene, silimanite, and cordierite in varying proportions; (2) calcsilicate paragneiss; and (3) quartzite, banded ironstone, basic granulite, pyribolite, and amphibolite. The *Kanuku Granulites* consist of (1) biotite-hypersthene (hornblende) granulite; (2) biotite-(garnet, sillimanite, cordierite) granulite; (3) hypersthene-biotite enderbite; and (4) basic granulite, pyribolite, and amphibolite. Isotope dating (Snelling and Berrangé, 1970;

Spooner et al., 1971) shows that the Kanuku Complex was affected by the trans-Amazonian episode around 2100–2000 m yr ago. On the basis of lithology, structural position, polyphase metamorphism (McConnell and Williams, 1970), and the somewhat ambiguous isotope dating, the Kanuku Complex is correlated with the Imataca Complex of eastern Venezuela dated (Hurley et al., 1967) at about 3000 m yr.

Granite and Gneiss. The large granite and gneiss complex forming the nucleus of the *Guiana Shield* (q.v.), and appearing to be of Archean age, has been largely reactivated at about 2000 m yr, and will be described below as the Southern Guyana Granite Complex.

Proterozoic. *Lower Proterozoic.* The northern geological province of Guyana, N of the parallel 5°N, is formed predominantly by an assemblage of metasediments, metavolcanics folded and metamorphosed, mainly in the greenschist facies, with large granitic batholiths dated about 2000 m yr (Snelling and McConnell, 1969). Previously defined as an "assemblage" by McConnell (1959) and Williams et al. (1967) owing to the difficult problems of correlations, subsequent detailed mapping and geochronological dating now justify the designation "Barama-Mazaruni Supergroup." Clearly eugeosynclinal and of graywacke facies N of 6°N, where the Mazaruni Group has been shown to overlie the Barama Group conformably, to the S the sediments become coarser, passing to orthoquartzites of the shelf province, and shallow-water volcanic tuffs and agglomerates appear. It has been suggested that the Mazaruni Group overlaps the Barama Group to the S and W, and overlies it unconformably (Berrangé, 1972), but an acceptable unconformity has not yet been observed.

The *Barama Group* is developed in the northwest district of Guyana (Williams et al., 1967) and consists of chiefly pelitic metasediments and metamorphosed basic volcanic rocks, with gondites and manganiferous phyllites near the top of the succession.

The *Mazaruni Group* consists of the *Cuyuni Formation* of pebbly sandstones and intraformational conglomerates of the graywacke suite, with intercalated volcanics of spilite-keratophyre type, overlain conformably by the *Haimaraka Formation* of mudstones and graywackes, but without significant volcanism.

The whole Barama-Mazaruni Supergroup has been folded together on WNW-ESE axes, imprinted with a conspicuous subvertical cleavage with the same strike, and metamorphosed in the greenschist facies with the emplacement of the granites and gneisses of the *Younger Granite Group*, which locally forms migmatite complexes such as the Bartica Complex. This granitization has been well dated (Snelling and McConnell, 1969) at about 2000 m yr, and thus represents clearly the trans-Amazonian orogenic cycle (Hurley et al., 1967).

Proterozoic Shelf Facies. South of latitude 6°N the eugeosynclinal facies of the Mazaruni Group is replaced by indications of shallow-water deposition, such as the occurrence of volcanic agglomerates, now known farther to N (although previously reported in error) and orthoquartzitic intercalations. Acid volcanics also predominate, and the Mazaruni Group is replaced by (a) the *Muruwa Formation* of orthoquartzitic sandstones and conglomerates, and (b) the overlying *Iwokrama Igneous Complex*. Granite and doleritic intrusions obscure the relationship here, and these two formations may be unconformable over the Mazaruni Group, and have been united by Berrangé (1972) as the *Burro-burro Group*. The Iwokrama Igneous Complex (Carter, M. W., 1961) consists of metamorphosed acid-intermediate tuffs and lavas with genetically associated adamellite, granophyre, and feldspar porphyry dated 1915 m yr (Snelling and Berrangé, 1970). Similar rocks in Surinam have been dated 1810 ± 40 m yr by Priem et al., (1971) who propose to classify them as a high crustal level trans-Amazonian plutonic-volcanic suite, more or less contemporaneous with the deeper-level granites and gneisses of the same age emplaced in the Lower Proterozoic of the *Guiana Shield* (q.v.).

Southern Guyana Granite Complex. In the southern geological province, S of the Kanuku Group, a South Savanna Granite (Barron, 1962; Singh, 1966) was dated 1845 ± 130 m yr (Snelling and McConnell, 1969) and Berrangé (1972) has now extended its limits and renamed it the Southern Guyana Granite Complex, regarding it as a largely autochthonous and parautochthonous product of the granitization of a preexisting basement. He also recognizes intrusive phases, such as the Marudi Granodiorite, which are associated with gold deposits. It would thus be a katazonal representative of the trans-Amazonian tectonothermal cycle. Singh (1966) showed that this granite was, at least marginally, intrusive into Archean and Proterozoic formations.

Kwitaro Group. Berrangé (1972) has united under this title metasedimentary enclaves in the southern province (Barron, 1962, Singh, 1966), including the former Marudi Group. The most abundant rocks are pelitic and semipelitic metasediments, now much metamorphosed. They may represent Lower Proterozoic rocks of the French Guiana-Surinam basin (see *Guiana Shield* and *French Guiana*) developed E of the Kanuku basement dorsal.

Kuyuwini Group. Rhyodacitic lavas and tuffs with consanguineous intrusive porphyries, granophyres, and subvolcanic rocks are found in the extreme south of Guyana. The volcanics generally forming narrow, elongated E-W trending belts. Evidence indicates that this whole comagmatic series was intruded late in the trans-Amazonian cycle (Berrangé, 1972) and represents a late orogenic shelf facies. It may thus indicate the proximity of the southern boundary of the central Archean nucleus of the Guiana Shield.

Tabular Proterozoic. *Roraima Formation.* Overlying the Lower Proterozoic and the trans-Amazonian granites, the Roraima Formation (Martin-Kaye, 1952; Bateson, 1966) consists of some 2000 m of current-bedded sandstones and quartzites—pink, light gray, or red in color—interbedded with shales and jaspers. There are conglomerates near the base, which is unconformable but overlies a fairly hilly erosion surface. The Roraima sandstones have been silicified but not metamorphosed, and rise with a steep escarpment above the Precambrian basement. The age of the formation must lie between the minimum age of the trans-Amazonian granites, 1800 m yr and 1700 m yr, the approximate age of the dolerites described in the next paragraph.

Roraima Intrusive Suite. Tholeiitic dolerites intrude both the basement and the Roraima Formation. They occur as major dikes, inclined sheets, and sills. The present distribution of these features is not haphazard (Hawkes, 1966a, 1966b) but is related to the depth of erosion. The sills are restricted to the Roraima Formation, in which they attain up to 360 m in thickness, the inclined sheets occur in the basement areas near the escarpment, and the dikes are found only where erosion has reached deeper levels in the basement.

On the basis of a number of age determinations on these dolerites and on a contact mica hornfels an average date of 1700 m yr has been established, thus fixing a minimum date for the Roraima Formation. Priem et al. (1973) have fixed a comparable date for the Proterozoic dolerites of Surinam, and for a tuff in the Roraima Formation.

The Roraima Formation is not notably faulted or disturbed in Guyana, and detailed surveys have established that it is all a continuous sequence, not divided by any major unconformities.

Late Proterozoic. Rejuvenated mica ages have been found in the Kanuku and Southern Guyana Granite Complexes, and are associated with an extensive ENE-WSW to NE-SW system of shear zones of the *K'mudku Mylonite Episode* (Barron, 1962) dated 1200 m yr old (Snelling and McConnell, 1969); and wide crush zones associated with the mylonites give confusing K-Ar dates. A similar rejuvenation phase has been termed the "Nickerie Metamorphic Episode" in western Surinam by Priem et al. (1973) and also given a date of 1200 m yr. The effects of this episode appear to be limited to southern Guyana and western Surinam, but may appear again adjacent to the Bolivar dislocation zone in Venezuela (see *Guiana Shield*).

Paleozoic. A minor dike suite of dolerites of olivine-basalt type was distinguished in Guyana by Hawkes (1966a). A date of 450 m yr (Snelling and McConnell, 1969) may be applicable, but it appears more likely that these are Permo-Triassic dolerites, such as found in Surinam by Priem et al. (1973). No other Paleozoic rocks are known in Guyana.

Takutu Formation. The Takutu Rift Valley, referred to above, is a feature about 160 km in length directed ENE-WSW and 50 km wide. It lies on the dislocation zone separating the Archean Kanuku Complex from the Lower Proterozoic Iwokrama igneous suite at latitude 3–4°N, and has been proved by aeromagnetic and gravity surveys to be a steep-sided graben (McConnell et al., 1969), containing a thickness of some 3–4 km of sediments. These easily eroded sediments outcrop in river beds, beneath a Pleistocene cover, and consist of soft, red-brown, current-bedded sandstones and variegated shales. Originally thought to be of Permo-Triassic age on the basis of fragmentary fossils (McConnell, 1959), boreholes have disclosed that the top 100 m at least contains a Jurassic to Lower Cretaceous pollen flora, and, on the basis of lithology and the presence of ostracods, represent a "Wealden facies" (van der Hammen and Burger, 1966). The age of the deeper beds in the graben is unknown.

The Takutu Formation continues to the W into the Rio Branco territory of Brazil, where it forms the Serra Tucano, attributed (wrongly) by Barbosa and Ramos (1959, p. 24; 1961) to the Roraima Formation.

The *Apoteri Volcanic Formation,* associated with the Takutu Formation, consists of amygdaloidal tholeiitic basalt flows, locally pillowed, that are related to the faulting of the Takutu Rift Valley (Berrangé, 1972). A date (K–Ar) of 114 m yr is given by Snelling and Berrangé (1970). Basalt flows also occur near the Serra Tucano.

Superficial and Coastal Deposits. Very important features of the geology of Guyana are the Tertiary and Quaternary deposits that form the coastal plain because (1) they underlie the fertile coastal strip; (2) they contain an artesian aquifer (the "A sands") from which boreholes supply the coastal area with potable water; (3) the commercial bauxite deposits are associated with them; and (4) they continue to thicken to the N on the Guyana shelf and hold there the promise of oil or gas. Preliminary studies by Grantham and Paton (1938) were followed by Bleackley (1956). Finally, in order to assist prospecting for bauxite and oil and to establish more exactly the status of the artesian aquifer, a palynological investigation was commissioned by the Geological Survey and certain mining companies resulting in two fundamental reports (van der Hammen, 1963; van der Hammen and Wijmstra, 1964). It was found that the age of the deposits stretched from Holocene back to at least Upper Cretaceous.

An outstanding feature of Guyana (and Surinam) is the wide belt of comparatively barren *white sands* (up to 160 km wide) separating the fertile coastal plains from the Precambrian outcrops in the interior (Fig. 1). These Berbice Sands are a leached surface wash derived in part

FIGURE 2. Structural profile across Guyana from Mt. Roraima to the coast at the Waina River.

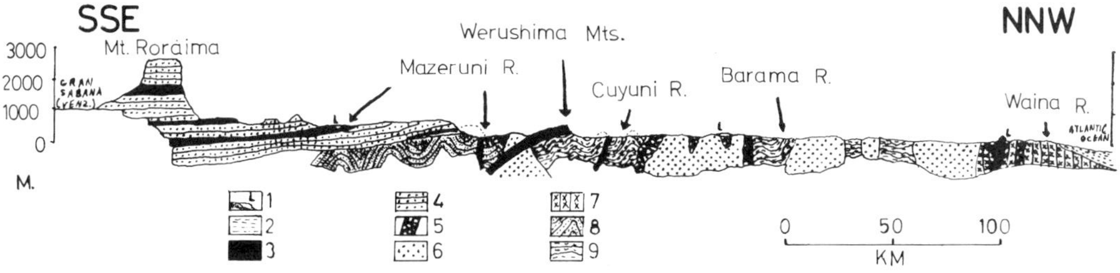

from Tertiary and Quaternary deltaic sands of the coastal deposits, and in part from the weathering of the Roraima and granitic formations. The sands are white, owing to the leaching of the normal yellow die of iron oxides by the high content of organic acids in the waters of the tropical forest environment.

Normally of shallow depth, the superficial and coastal deposits were found to be nearly 2000 m (6260 ft) thick in a borehole (Kugler et al., 1944) E of the Berbice River, in the center of a NE-SW-striking downwarp of basement between Georgetown and the Corentyne River (Fig. 2). This thickness is reduced to a 100 m or less in the "bauxite belt," which forms a wide arc, concave to the NE, between Georgetown and Paramaribo, reaching inland a maximum distance from the sea of 140 km on the Berbice River. The bauxite is found to represent (at least in part) a period of emergence and weathering under hot-wet, lacustrine or mangrove swamp conditions during the Middle and Upper Eocene, and to have formed during the breakdown of alumina-rich sediments (arkose?) or weathered basement (Valeton, 1971).

The coastal deposits thicken very considerably on the Guyana shelf offshore and are being explored for hydrocarbons.

Geomorphology

Geomorphology is of particular economic importance in Guyana as bauxite and aluminous laterite occur on relatively level surfaces that have been exposed to humid tropical weathering over long periods. Three such cyclic planation surfaces are recognized (Bleackley, 1964; McConnell, 1968), dated (1) late Cretaceous to early Tertiary, (2) late Tertiary, and (3) end-Tertiary. The first occurs at about 700 m in the Pakaraima Mountains, the second at about 450 m. The end-Tertiary surface is widespread at about 300 m in the interior of Guyana and elsewhere in the Guiana Shield, and the steepness of the inselbergs arising from this pediplain suggests periods of arid, desertic erosion during the Pleistocene. Berrangé (personal communication, 1969) has, in fact, found sand dunes in the central Rupununi savannas, and Damuth and Fairbridge (1970) record evidence of arkosic sediments in the deep Atlantic, indicating arid conditions in South America during Pleistocene glacial phases. All the cyclic surfaces are downwarped beneath the coastal plain but are represented by hiatuses in the succession (Fig. 26 in van der Hammen and Wijmstra, 1964).

During the Tertiary, humid-hot conditions alternated with tropical weathering, which produced saprolites and crusts at several stages. The early Quaternary, in contrast, is marked by a widespread residual leached sand (Berbice white sands of Guyana; Zanderij Formation of Surinam). This appears to be the result of the repeated deforestation of much of the hinterland during the arid (cool-dry) phases of the Pleistocene, when the sea level was progressively lowered. Extensive sheets of colluvial, alluvial, and deltaic fans covered the coastal plain and continental shelf. This process was repeated at each cold stage and alternated with renewed humid leaching (with reforestation) during each warm stage. Pollen analyses indicate that the cool-dry stages were marked by grassy savannas and semiarid conditions.

In the interior, notably on granitic rocks, the semiarid Pleistocene stripping of the old saprolite cover has left many bare bedrock surfaces. These are being currently attacked by chemical weathering, leaving hollows comparable to some Mediterranean "tafoni" (see *Encyclopedia of Geomorphology*). Often filled with mossy peat today, these are known in South America as "oricangas" (Bakker, 1957). The same semi-arid erosion history produced dells, partly comparable to those of periglacial landscapes in higher latitudes.

Mineral Resources

The *Guiana Shield* (q.v.) is a complex area of Precambrian rocks similar to the Canadian and

Brazilian shields, which are rich in minerals, but prospecting is difficult, owing to the cover of tropical forest and deep surface decomposition. A description of the mineral resources of Guyana was given by Stockley (1955), and since 1962 an active prospecting campaign by all modern methods has been carried out by the Geological Survey and Mines Department under the auspices of the United Nations, and many mineral indications are now being explored.

Bauxite is the principal mineral and Guyana is the third largest exporter in the world, 4,238,000 long tons of dried bauxite equivalent being produced in 1969, from which 296,115 tons of alumina were exported and the rest exported in various other forms. The present production is from a distinct bauxite belt in which the deposits are famous for their high quality. The country, however, with its great extent of pediplaned surfaces of crystalline and basic rocks and its humid tropical climate, is an ideal area for the formation of lateritic deposits (Bleackley, 1964; Valeton, 1971), and it contains many millions of tons of aluminous laterite, representing great reserves of both iron and alumina, not economically exploitable by present methods of extraction, but which may be developed in the future.

Gold has been exploited since 1884, reaching a peak production of 140,000 oz in 1893 but falling to 2102 oz by 1969. It comes mainly from small alluvial diggings, but rich quartz reefs were discovered and a number of small mines worked for several years.

Diamonds were first worked in 1887, the production rising to 220,000 carats in 1923; 56,486 carats were exported in 1969. The diamonds are derived from conglomerates near the base of the Roraima Formation and exploited in gravels at the base of the Pakaraima Mountains escarpment, and in the mountains themselves.

Manganese was discovered in commercial quantities by the Geological Survey and production began in 1960, rising to 180,000 long tons in 1966, but ceasing after 1968.

Indications of *chromium, copper,* and *molybdenite* have been explored but no commercial deposits discovered. Exploration has also followed up discoveries of radioactive minerals, but without commercial success as yet.

Active exploration for hydrocarbons is taking place in the coastal and offshore sedimentary formations.

RICHARD B. McCONNELL

References

Bakker, J. P., 1957. "Quelques aspects du probleme des sédiments correlatifs en climat tropical humide," *Z. Geomorphol.*, **1**, 3–43.

Barbosa, O., and Ramos, J. R. de A., 1959. "Territorio do Rio Branco," *Div. Geol. Mineral.*, (Brazil) **196**, 46p.

———, and Ramos, J. R. de A., 1961. "Principal aspects of the geomorphology and geology in the Territory of Rio Branco, Brazil," Proc. 5th Inter-Guiana Geol. Conf., 33–36.

Barron, C. N., 1962. "The geology of the South Savanna degree square, British Guiana," *Geol. Surv. Br. Guinea Bull.*, **33**, 29p.

Bateson, J. H., 1966. "Some aspects of the geology of the Roraima Formation in British Guiana," *Trans. Third Caribbean Geol. Conf.*, Jamaica, 144–150.

Berrangé, J. P., 1972. "The tectonic/geological map of southern Guyana," *9th Inter-Guiana Geol. Conf.*, Ciudad Guayana, May, 1972.

———, 1973. *Tectonic/Geological Map of Southern Guyana* (Map 1186). Inst. Geol. Sci. London, 1: 500,000.

———, in press. "The geology of southern Guyana, South America," *Inst. Geol. Sci. Lond., Overseas Div. Mem.* **4**.

Bleackley, D., 1956. "The geology of the superficial deposits and coastal sediments of British Guiana," *Geol. Surv. Br. Guiana Bull.*, **30**, 46p.

———, 1964. "The bauxites and laterites of British Guiana," *Geol. Surv. Brit. Guiana Bull.*, **34**, 156p.

Bracewell, S., 1941. "The geomorphology of British Guiana," *Geol. Mag.*, **78**, 463–469.

———, 1947. "The geology and mineral resources of British Guiana," *Bull. Imp. Inst., Lond.*, **45**(1), 47–69.

———, 1956. "British Guiana," *in* W. F. Jenks, ed., *Handbook of South American Geology.* New York: Geol. Soc. Am. Mem. **65**, 89–98.

Brinkmann, R., and Pons, L. J., 1968. "A pedo-geomorphological classification and map of the Holocene sediments in the coastal plain of the three Guianas," *Soil Surv. Pap.*, **4**, 40p.

Brown, C. B., and Sawkins, J. G., 1875. *Reports on the Physical, Descriptive, and Economic Geology of British Guiana.* London: Edward Stanford, 297p.

Carter, M. W., 1961. "The acid volcanic–plutonic relationship in the northern Rapununi," *Proc. 5th Inter-Guiana Geol. Conf.*, Georgetown, 1959, 129–134.

Damuth, J. E., and Fairbridge, R. W., 1970. "Equatorial Atlantic deep-sea arkosic sands and ice-age aridity in tropical South America," *Bull. Geol. Soc. Am.*, **81**, 189–206.

Dixon, G., and George, H. K., 1964. *Bibliography of the Geology and Mining of British Guiana.* Georgetown: Geol. Surv. Br. Guiana, 87p.

Grantham, D. H., and Paton, R. F., 1938. "Geology of the superficial deposits of British Guiana," *Geol. Surv. Br. Guiana Bull.*, **11**, 1–122.

Guyana Geological Survey, 1969. "A further guide to mineral exploration in Guyana," *Guyana Geol. Surv. Bull.*, **38**, Suppl., 47p.

Harder, E. C., 1936. "British Guiana and its bauxite resources," *Can. Min. and Met. Bull.*, **295**, 739–758.

Harrison, J. B., 1908. *The Geology of the Goldfields of British Guiana.* London: Dulaw.

Hawkes, D. D., 1966a. "Differentiation of the Tumatumari–Kopinang dolerite intrusion, British Guiana," *Bull. Geol. Soc. Am.*, **77**, 1131–1158.

———, 1966b. "The petrology of Guiana dolerites," *Geol. Mag.*, **103**, 320–323.
Hurley, P. M., *et al.*, 1967. "Test of continental drift by comparison of radiometric ages," *Science*, **157**, 495–500.
Kersen, J. F. van, 1955. "Bauxite deposits in Suriname and Demerara (British Guiana)," *Leidse Geol. Mededel.*, **21**, 247–375.
Kugler, G., Griffiths, J. C., Mackenzie, S. C., and Stainforth, R. M., 1944. "Report on exploration for oil in British Guiana," *Geol. Surv. Br. Guiana Bull.*, **20**, 78p.
McConnell, R. B., 1959. "The Takutu Formation in British Guiana and the probable age of the Roraima Formation," *Trans. 2nd Caribbean Geol. Conf.*, 163–170.
———, 1968. "Planation surfaces in Guyana," *J. Geogr.*, **134**, 506–520.
———, and Dixon, C. G., 1960. "A geological map of British Guiana," *Rept. 21st Intern. Geol. Congr.*, Copenhagen, **9**, 39–50.
———, and Williams, E., 1970. "Distribution and provisional correlation of the Precambrian of the Guiana Shield," *Proc. 8th Guiana Geol. Conf.*, Georgetown, **1**, 1–22.
———, Masson Smith, D., and Berrangé, J. P., 1969. "Geological and geophysical evidence for a rift valley in the Guiana Shield," *Geol. Mijnb.*, **48**, 189–199.
Martin-Kaye, P. H. A., 1952. "The Roraima Formation in British Guiana," *Geol. Surv. Br. Guiana Bull.*, **22**, 1–32.
Priem, H. N. A., Boelrijk, N. A. I. M., Hebeda, E. H., Verdurmen, E. A. T., and Verschure, R. H., 1971. "Isotopic ages of the trans-Amazonian acidic magmatism and the Nickerie metamorphic episode in the Precambrian basement of Suriname, South America," *Bull. Geol. Soc. Am.*, **82**, 1667–1680.
———, Boelrijk, N. A. I. M., Hebeda, E. H., Verdurmeu, E. A. T., and Verschure, R. H., 1973. "Age of the Precambrian Roraima Formation in northeastern South America: evidence from isotopic dating of Roraima pyroclastic volcanic rocks in Suriname," *Bull. Geol. Soc. Am.*, **84**, 1677–1684.
Singh, S., 1966. "Geology and petrology of part of the Guyana Shield in the south Savanna-Rewa area of southern Guyana," *Guyana Geol. Surv. Bull.*, **37**, 127p.
Snelling, N. J., and Berrangé, J. P., 1970. "The geochronology of Guyana II." *Proc. 8th Guiana Geol. Conf.*, Georgetown, **8**, 15p.
———, and McConnell, R. B., 1969. "The geochronology of Guyana," *Geol. Mijnb.*, **48**, 201–213.
Spooner, C. M., *et al.*, 1971. "Rb-Sr whole-rock age of the Kanuku compex, Guiana," *Bull. Geol. Soc. Am.*, **82**, 207–210.
Stockley, G. M., 1955. "The geology of British Guiana and the development of its mineral resources," *Geol. Surv. Br. Guiana Bull.* **25**, 102p.
Tate, G. H. H., 1930. "Notes on the Mount Roraima Region," *Geogr. Rev.*, **20**(1), 53–68.
Valeton, I., 1971. "Tubular fossils in the bauxites and the underlying sediments of Surinam and Guyana," *Geol. Mijnb.*, **50**(6), 733–741.
Hammen, T. van der, 1963. "A palynological study on the Quaternary of British Guiana," *Leidse Geol. Mededel.*, **29**, 125–180.
———, and Burger, D., 1966. "Pollen flora and age of the Takutu Formation (Guyana)," *Leidse Geol. Mededel.*, **38**, 173–180.
———, and Wijmstra, T. A., 1964. "A palynological study on the Tertiary and Upper Cretaceous of British Guiana," *Leidse Geol. Mededel.*, **30**, 183–242.
Williams, E., *et al.*, 1967. "The folded Precambrian of northern Guyana related to geology of the Guiana Shield, with age data by N. J. Snelling," *Guyana Geol. Surv. Records*, **5**, 60p.

Cross-references: *Brazil; French Guiana; Guiana Shield–Regional Review; South America; Surinam.*

H

HAITI

Occupying the western third of the island of Hispaniola in the West Indies, Haiti has an area of 27,750 km^2 (10,714 sq mi). It is bounded on the north by the Atlantic Ocean, on the east by the *Dominican Republic* (q.v.), on the south by the Caribbean Sea, and on the northwest by the Windward Passage, which separates it from *Cuba* (q.v.).

Nearly two-thirds of Haiti is covered by WNW-ESE and W-E trending mountain ranges, the principal ones being the Massif du Nord, Montagnes Noires, Massif de la Selle, and Massif de la Hotte. Fertile plains and valleys lie between these mountains, with luxuriant vegetation in the central part (sugar). The western part gives way to dry savannas underlain by ferruginous duricrust ("iron pan") and the easternmost part to dense xerophytic forest.

Bauxite and copper are mined and there are undeveloped deposits of gold, silver, antimony, tin, sulfur, nickel, gypsum, and coal.

The oldest fossiliferous rocks in Haiti are reported to be Lower Cretaceous (Hauterivian to Barremian), consisting of partly silicified pelagic limestones and radiolarites, interbedded with volcanics, mainly submarine basalts followed by andesites (Mitchell, 1953). In the northern regions the basalts are covered by andesites and dacites, unconformably (Kesler, 1968). It is claimed by some writers that these rocks rest on a basal complex of moderately to slightly metamorphosed rocks which may go back to Hercynian (Appalachian) time, but confirmation is lacking. This older history was once favored by Weyl thirty years ago and doubted by Butterlin (1954, 1960), who is responsible for much of the field work.

The Upper Cretaceous is marked by an increasing proportion of sediments, and the Senonian is marked by limestones and argillites, the *Trois Rivières Formation.* In southwestern Haiti, especially in La Hotte Mountains, there are thick limestones, containing a pelagic fauna, the *Macaya Formation,* also mainly Senonian. It is associated with extensive submarine basaltic lava flows. These formations are similar to the deposits encountered in the Venezuelan basin of the Caribbean Sea and indicate that this part of Haiti corresponds to a fault-bounded block, emerging from the sea (Edgar and Saunders in Anon., 1971).

Orogenic activity of the Laramide phase occurred at the end of the Cretaceous period, leading to much of the metamorphism of the *Basal Complex.* Folds are found running NW-SE, en échelon, accompanied by intrusions predominantly granodioritic. Ultrabasic material is represented by serpentinized peridotite, basics by dolerite, and acidic differentiated by quartz diorites.

A new transgression came with the Paleocene, and marine sedimentation continued up to the mid-Eocene. At first there are coarse conglomerates, sandy shales, calcareous sandstones, and clastic limestones (*Marigot Formation*). The Lower Eocene (*Abuillot Formation*) is also clastic in some areas, but is dominated by a pelagic bathyal facies with chalky limestones, passing to calcarenites near the shore. Particularly in the S and in the Barahona peninsula there was continued subsidence. In the Middle Eocene, notably in the Northwestern peninsula and in the Black Mountains, there was extensive subsidence comparable to that of the Sierra Maestra in eastern *Cuba* (q.v.), marked by a thick formation of basaltic and andesitic tuff with thin limestones (*Perodin Formation*). Elsewhere there are calcarenites of the "yellow limestone" type (see *Jamaica*), the *Plaisance Formation* (cf. *Hidalgo Formation,* in *Dominican Republic,* q.v.). Clastic facies reappear south of the central Cordillera following uplift there and renewed subsidence to the south.

Upper Eocene is marked by general regression, in connection with Upper Middle Eocene orogenic activity, except in the extreme north and south, where pelagic sedimentation with submarine lavas continued (*Ennery Formation* and *Neiba Formation,* respectively). Orogeny occurred in the central regions, marked by granodiorites and dolerites.

With the Lower Oligocene came further subsidence in the south with pelagic chalky limestones and chert. Middle Oligocene saw a general transgression over much of the north and center, at first marked by clastics followed by carbonates. The *La Crête Formation,* consisting of sandstones and limestones, is succeeded by the *Madame Joie Formation* of shales and limestones. Nepheline volcanics occur in the Trou d'Eau Mountains. Calcareous sedimentation continued into the Upper Oligocene, except for new areas of uplift.

Orogenic activity began again toward the close of the Oligocene and continued until modern times mainly in E-W trends. The folding is partly isoclinal to the north and partly to the south. Synorogenic flysch facies, with a great thickness of sandstones and shales of the *Thomonde Formation* and *Rivière Grise Formation,* are of Oligo-Miocene age. Troughs that developed during the uplift became filled with molasse clastics in considerable thickness—the *Las Cahobas Formation* and *Morne Delmas Formation* of Miocene age. Molasse conglomerates and other clastics in the Pliocene extended to the southern part of the country (*Rivière Gauche Formation*), but most of the region was emerged and undergoing erosion with peneplanation. It is marked by continental accumulations in places. Block faulting of these peneplaned surfaces occurred in late Pliocene and Quaternary times, in places passing to "fault folds" and, for some writers, northerly directed overthrusts (Mitchell, 1954; Meyerhoff, 1954); some were accompanied by eruptions of limburgite and nepheline basalt (MacDonald and Melson, 1969). Raised beaches and reefs of the Pleistocene show differential uplift as well as eustatic emergences, and in the Bombardopolis Plateau reach over 600 m elevation (300 m in the Sierra Maestra of Cuba).

From the geotectonic point of view, Haiti—as part of Hispaniola—occupies part of a "microplate," cut off by strike-slip planes from *Cuba* (q.v.) on the one hand and from *Puerto Rico* (q.v.) on the other. Partly on the basis of seismic data, it has been postulated by Bracey and Vogt (1970) that the Atlantic oceanic crust is today underthrusting Hispaniola island in a NNE to SSW direction, generating a short island arc structure.

JACQUES BUTTERLIN

References

Anon, 1971. *6th Conf. Geol. del Caribe, Resumenes,* 41p. plus suppl.
Bracey, D. R., and Vogt, P. R., 1970. "Plate tectonics in the Hispaniola area," *Bull. Geol. Soc. Am.,* 81, 2855–2860. (Also "Discussion," *Ibid.,* 82, 1123–1128.)
Butterlin, J., 1954. *La géologie de la République d'Haïti et ses rapports avec celle des régions voisines.* Port-au-Prince: Publ. Comm. 150th Anniv. Indep. Republ. Haiti, 446p.
———, 1960. "Géologie générale et régionale de la République d'Haïti," *Trav. et Mem. Inst. Htes Etudes Am. Latine* (Paris), 6, 194p.
Kesler, S. E., 1968. "Igneous rocks of the Terre Neuve Mountains and Massif du Nord, Haiti," *5th Caribbean Geol. Conf.,* Abstr.
———, 1971. "Petrology of the Terre-Neuve Igneous Province, northern Haiti," *Geol. Soc. Am. Mem. 130,* 119–137.
Logan, R. W., 1968. *Haiti and the Dominican Republic.* New York: Oxford Univ. Press, 228p.
MacDonald, W. D., and Melson, W. G., 1969. "A late cenozoic volcanic province in Hispaniola," *Caribbean J. Sci.,* 9(3–4), 81–91.
Meyerhoff, H. A., 1954. "Antillean tectonics," *Trans. N. Y. Acad. Sci.,* Ser. 2, 16(3), 149–155.
Mitchell, R. C., 1953. "New data regarding the dioritic rocks of the West Indies," *Geol. Mijnb.,* 15, 285–295.
———, 1954. "Hauteurs de profils agrandis et errements tectoniques dans les travaux sur les Antilles," *Cahier Geol. Thoiry,* 23, 201–205.
Van den Bold, W. A., 1974. "Neogene of central Haiti," *Am. Assoc. Petrol. Geologists Bull.,* 58(3), 533–539.
Woodring, W. P., Brown, J. S., and Burbank, W. S., 1924. *Geology of the Republic of Haiti.* Port-au-Prince: Dept. Public Works, 631p.

Cross-references: *Cuba; Dominican Republic; Puerto Rico; West Indies.*

HONDURAS

One of the Central American countries, the second largest, Honduras occupies an area of 112,090 km^2 (43,277 sq mi). It is bordered by the Caribbean Sea and the Gulf of Honduras (for about 600 km) on the N, Nicaragua on the E and S, the Gulf of Fonseca and the Pacific (for about 70 km) on the S, El Salvador on the SW, and Guatemala on the W. Its capital is Tegucigalpa, situated in the southern interior and its chief (North coast) harbor is Puerto Cortes. During the Mayan civilization, its chief city was Copan, where tremendous ruins still survive. It was on the north coast (Cape Honduras, near Trujillo) that Christopher Columbus made his first landfall on the actual continent of America in 1502.

Except for the coastal plains on the NE and S, Honduras is a mountainous, forested, and generally fertile country. Two-thirds of the country is over 300 m elevation. Nevertheless, the country lacks high peaks and the maximum elevation is only 2590 m (Cerros de Culmi).

The main divide (an extension of the Nicaragua Cordillera) lies in the south and the principal streams flow to the NE, cutting up the country into secondary NE-SW ranges (e.g., Sierra de la Esperanza, Montañas de Colon),

FIGURE 1. Geological sketch map of Honduras (based upon the "Geological Map of North America," U.S.G.S., 1965). Explanation: Qc, Quaternary continental; Qtv, Quaternary/Upper Tertiary volcanics; Tpv, Pliocene volcanics; Tki/Ki, Lower Tertiary and Cretaceous granitic intrusives; Ku/Kl, Upper and Lower Cretaceous; Pz, Paleozoic metamorphic rocks; pKm/Pzm, Pre-Cretaceous and Paleozoic metamorphics; pC, Precambrian crystalline rocks.

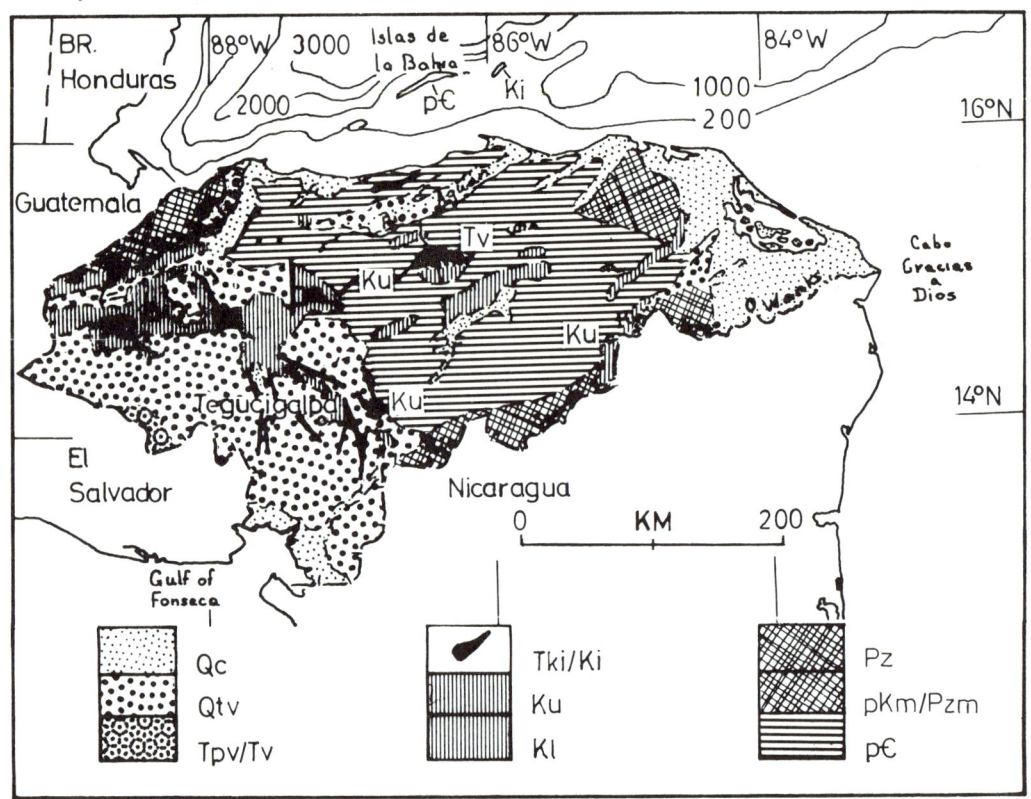

and subparallel basins (known as "bolsones" locally, although not exactly appropriate in the geological use of the term). Nevertheless, there are numerous high-level terraced plains suggestive of dry Pleistocene climates (Bengtson, 1926). These trends follow mainly structural lines with fault controls. From E to W, the main rivers are the Segovia (also called Coco, Wanks), Patuca, Sico, Aguan, Ulua, and Chamelecon. In the S the major streams are the Rio Choluteca and the Rio Goascorán, which flow into the Gulf of Fonseca. There is one major lake, the Lago de Yojoa (17 km long), the site of many archaeological finds.

The eastern part of the country slopes down gradually to the swampy 20,000-km² coastal plain of the *Mosquitia Coast*, which Honduras shares with Nicaragua. It is marked by extensive barrier beaches and lagoons (the largest: Laguna Caratasca, 80 km in length). Offshore, to the NE, is a broad shallow platform, the Nicaragua Rise, extensively capped by shallows and coral reefs (Banco Gorda) and small sand cays ("cayos"). The northern coast is more abrupt, but paralled by an ENE-WSW-trending belt of islands (Islas de la Bahia), marking the border of the Cayman Trough. An outlying member of this belt is the Swan Islands, a former U.S. possession. (*Note:* The Cayman Trough, or Trench, is referred to as the "Bartlett Trough" in some older publications.) The south coast, around the Gulf of Fonseca, is marked by extensive mangrove swamps.

There are abundant mineral resources but these are largely undeveloped, although lead, zinc, silver, and gold are mined. Placer gold has been exploited since antiquity. Other deposits are copper, iron, antimony, and coal. Seasalt is extracted by evaporation. The economy rests principally on bananas, coffee, and timber.

Most of the country is underlain by a relatively stable geotectonic block, the *Honduras Massif*, which consists of crystalline rocks, metasediments, and intrusives of pre-Mesozoic age. They may be Paleozoic or possibly in part Precambrian. The massif is extensively block-faulted, and split meridionally by the Honduras Graben, which is actually a series of discontinuous N-S graben better known collectively as the Honduras Depression, which extends from the Ulua Valley and Comayagua to the Gulf of Fonseca on the Pacific coast. This rift zone opened about early Pliocene to create a West Honduras block and an East Honduras block. The regional tilt is to the S, and thus the basement outcrops on the northern side, the southern portion being overlapped by Cretaceous marine sediments (limestones and tuffs), and by various sequences of Tertiary and Quaternary volcanics.

The southern and western margins of the Honduras Massif are classified by Moody (this volume: *Central America*) as the "Nicaragua Volcanic Upland," the westerly extension of which is seen in *El Salvador* (q.v.), where it terminates in the Lempa valley. In the Honduras Massif, apart from the N-S graben system, the principal block-fault component is E-W and the north coast (Gulf of Honduras) is downdropped with antithetic blocks along this trend so that the crystalline rocks crop out again in the Bay Islands. The east coast is cut off by another N-S fault line, but here, toward the S, there is a coastal plain where oil exploration wells disclose a thick section of Cretaceous and Tertiary.

In the region NE of Tegucigalpa, there is an important (1000 m) sequence of gray shales and sandstones said to contain late Triassic or early Jurassic flora, which is reminiscent of the almost global taphrogeny associated with the beginning of the Atlantic drifting. Although since the mid-Tertiary much of the Honduras Massif has remained above sea level and, unlike its neighbors, free from extensive volcanism except along the southern border, it has been widely blanketed by ash deposits.

Geological History

The crystalline foundation of the Honduras is part of "Nuclear Central America" (Vinson and Brineman, 1963), largely a metamorphosed Paleozoic block that until the late Mesozoic is believed to have been situated adjacent to the Guatemala complex (Dengo and Bohnenberger, 1969; Malfait and Dinkelman, 1972). The two blocks may have lain south of Mexico originally. E-W stretching began early, and the Triassic/Jurassic accumulations E of Tegucigalpa suggest that there was mid-Mesozoic crustal tension here, but little is known of this phase.

In late Cretaceous time the East Pacific Plate began underthrusting the North American Plate along the Southern Mexico Trench, which then extended around the border of southern Honduras, leading to subduction, volcanism, and granitization along this belt. Laramide intrusives are widespread throughout Honduras, in southern Guatemala, and northeastern Nicaragua (McBirney and Williams, 1965).

During the late Tertiary (Pliocene), this ancient crystalline block became coupled to the NE-moving East Pacific/Caribbean Plate and sheared off from southern Mexico-Yucatan along a major transcurrent fault system, the Polochic-Motagua-Cayman Trough line, with a left-lateral displacement. Sheared granodiorites have been dredged from the wall of the Cayman Trough (Fox *et al.,* 1971). This strike slip may have exceeded 400 km in total shift, moving the Honduras (with the Nicaragua Rise and Jamaica) to the ENE, initiating subduction in the Cuban Trench. Displacement continues along this line, as shown by current seismicity.

During the Mesozoic evolution, oceanic crust lay to the south and an open sea connection existed from southern Mexico to the opening Atlantic and Tethys. After the Laramide orogenic phase a geosynclinal belt and island arc began to develop to the south (see *Nicaragua*), and downwarps developed in the NW (Ulua basin) and E (Mosquitia basin). In the former some 600 m of Upper Cretaceous and Eocene red beds and limestones accumulated, but the warping then ceased.

In central Honduras, north of Tegucigalpa, the Lower Cretaceous is represented by the Yojoa Group of carbonates and these are overlain by 3000 m of red beds, the Valle de Angeles Group. This group is intercalated by a 470-m sequence of calcilutite, marlstone, and calcarenite, a shallow marine facies with mollusca and echinoids that suggest a Cenomanian age.

In the Mosquitia basin (which extends into northeastern Nicaragua) there was continued subsidence as rifting away of the western Caribbean began. Some 4000 m of Upper Cretaceous limestones and shales are reported and nearly 5000 m of Tertiary deltaic, fluvial, and shallow marine clastics today form the coastal plain and shelf region, which extends 200 km into the Caribbean.

Complex taphrogenic stresses were repeatedly applied to the Honduras Massif during the Tertiary. The opening of the Honduras Sphenochasm probably involved counterclockwise rotation of the Yucatan block, in view of the fact that the Polochic-Motagua-Cayman megashear is arcuate. There are thus the early Tertiary fractures that parallel this fault system (ENE-WSW in Honduras), and the NNE-SSW trends of the western Caribbean margin, which were followed in the post-late Miocene times by the N-S graben faults of the Ulua-Fonseca trend, which coincided with the maximum phase of movement along the Polochic-Motagua megashear, according to Dengo and Bohnenberger (1969).

RHODES W. FAIRBRIDGE

References

Bengtson, N. A., 1926. "Notes on the physiography of Honduras," *Geogr. Rev.,* **16,** 403–413.

Carpenter, R. H., 1954. "Geology and ore deposits of the Rosario mining district and the San Juancito Mountains, Honduras, Central America," *Bull. Geol. Soc. Am.,* **65**(1), 23–38.

Dengo, G., and Bohnenberger, O. H., 1969. "Structural development of northern Central America," *Am. Assoc. Petrol. Geologists,* **11,** 203–220.

Fox, P. J., Schreiber, E., and Heezen, B. C., 1971. "The geology of the Caribbean crust; Tertiary sediments, granitic and basic rocks from the Aves ridge," *Tectonophysics,* **12**(2), 89–109.

Helbig, K. M., 1959. "Die Landschaften von Nordost-Honduras," *Petermanns Geogr. Mitt.,* Gotha, Erg. **268,** 270p.

Hoffstetter, R., 1960. "Amerique Central," *Lexique Stratig. Intern.* (Paris), **5,** fasc. 2.

Horne, G. S., Atwood, M. G., and King, A. P., 1974. "Stratigraphy, sedimentology, and paleoenvironment of esquias formation of Honduras," *Am. Assoc. Petrol. Geologists,* **58**(2), 176–188.

Malfait, B. T., and Dinkelman, M. G., 1972. "Circum-Caribbean tectonic and igneous activity and the evolution of the Caribbean plate," *Bull. Geol. Soc. Am.,* **83,** 251–272.

McBirney, A. R., and Bass, M. N., 1969. "Geology of Bay Islands, Gulf of Honduras," *Am. Assoc. Petrol. Geologists Mem. 11,* 229–243.

_____, and Williams, H., 1965. "Volcanic history of Nicaragua," *Cal. Univ. Pubs. Geol. Sci.,* **38,** 177–242.

McGrew, P. O., 1942. "Field museum paleontological expedition to Honduras," *Science,* 96(2482), 85.

Maldonaldo-Koerdell, M., 1953. "Plantas del Retico-Liasico y otros fosiles triasicos de Honduras, C.A.," *Ciencia,* **12**(11–12), 294–296.

Melhado, A. R., and Moran, M. C., 1953. *Geografia General de la República de Honduras.* Tegucigalpa.

Mills, R. A., 1959. "A geologist discusses Honduras oil prospects," *Petroleo Interamericano,* **17**(5), 39–44.

_____, et al., 1967. "Mesozoic stratigraphy of Honduras," *Bull. Am. Assoc. Petrol. Geologists,* **51,** 1711–1786.

Olson, E. C., and McGrew, P. O., 1941. "Mammalian fauna from the Pliocene of Honduras," *Bull. Geol. Soc. Am.* **52**(8), 1219–1243.

Redfield, A. H., 1923. "The petroleum possibilities of Honduras," *Econ. Geol.,* **18**(5), 474–493.

Roberts, R. J., and Irving, E. M., 1957. "Mineral deposits of Central America," *U.S. Geol. Surv. Bull.,* **1034,** 205p.

Vinson, G. L., and Brineman, J. H., 1963. "Nuclear Central America, hub of Antillean transverse belt," in O. E. Childs, ed., *Backbone of the Americas.* Tulsa: Am. Assoc. Petrol. Geologists, 101–113.

Washington, H. S., 1921. "Obsidian from Copan (Honduras) and Chichen Itza (Yucatan)," *J. Wash. Acad. Sci.,* **11**(20), 481–487.

Williams, H., and McBirney, A. R., 1969. "Volcanic history of Honduras," *Cal. Univ. Pubs. Geol. Sci.,* **85,** 101p.

Wilson, H. H., 1974. "Cretaceous sedimentation and orogeny in nuclear Central America," *Am. Assoc. Petrol. Geologists,* **58**(7), 1348–1396.

Cross-references: *Belize; Cayman Islands; Central America–Regional Review; El Salvador; Guatemala; Jamaica; Mexico; Nicaragua.*

I

IZU-OGASAWARA-IWO ISLANDS (NAMPO-SHOTO)

The Izu-Ogasawara or Izu-Bonin Islands, which include the Iwo Islands, extend southward from Honshu (24–35°N, 139–143°E) for nearly 1500 km in the northwest Pacific. The climate is mild to subtropical. At 27°N, 142°E, the annual average temperature is 22°C; January, 17°C; July, 27°C; with 1600 mm mean precipitation. The islands are mostly well forested with valuable timber, but some of the northern islands have been so exploited that today they are mainly grass covered.

The *Izu Islands,* the northern group, are also called the *Shichito Islands.* They have long been a part of the same province as the Izu Peninsula of Honshu. The *Ogasawara Islands,* the middle group, were first discovered by a Samurai voyager, S. Ogasawara, in 1593, after whom the islands were named. They had been uninhabited until the beginning of the last century and called Munin, meaning "uninhabited" in Japanese, which became altered to "Bonin." The term *Bonin Islands* is thus a synonym of the Ogasawara Islands and became the one adopted in the scientific literature in English. However, it is never used in the literature in Japanese, since the islands are no longer uninhabited. They became in Japanese possession in 1875. The *Iwo Islands,* the southern group, were found by Captain J. Cook in 1779 and became in Japanese possession in 1891. They are also called the *Kazan Islands.* Iwo means "sulfur" and kazan means "volcano" in Japanese. The three groups of islands, including Minami-Tori or Marcus Island (24°N, 154°E) and Okino-Tori or Parece Vela Island (20°N, 136°E), under Japanese sovereignty, are officially known collectively as the *Nampo-Shoto,* which means the "southern islands" in Japanese. Altogether these islands cover about 400 km² (Fig. 1).

The Izu-Ogasawara Islands are the crestal part of the Izu-Ogasawara Ridge, which trends N-S and forms a typical island arc. (This is the "Izu-Bonin Ridge" of Minato et al., 1965). The highest point is the peak of Hachijo Island (854 m). The *Izu-Ogasawara Trench* lies to the E and is remarkably parallel to the ridge. The deepest point is at 29°N, 143°E, and it is 9656 m deep. The Izu-Ogasawara Ridge has two major chains, the east one called the "Ogasawara Ridge" (named by Yoshiwara, 1902), and the west one the "Shichito-Iwojima Ridge." A trough 4144 m deep lies between them. The Ogasawara Ridge, where the highest point is the peak of Chichi Island (462 m high), represents the outer arc, and the Shichito-Iwojima Ridge the inner or volcanic arc. The same double arc of the Izu-Ogasawara Ridge extends to the S as the *Mariana arc* (q.v.), which is fronted by the deep Mariana Trench. The Ogasawara plateau, the shallowest area being 774 m deep, is southeast of the Ogasawara Islands and stands in the way between the Mariana Trench and the Izu-Ogasawara Trench, which is, however, continuous with the Japan Trench farther north. Regardless of the discontinuity of the trenches, the Izu-Ogasawara arc and the Mariana arc are often called jointly the *Izu-Mariana arc* or the Shichito-Mariana arc, which usually includes

FIGURE 1. Sketch map showing locations of the Izu and Ogasawara islands.

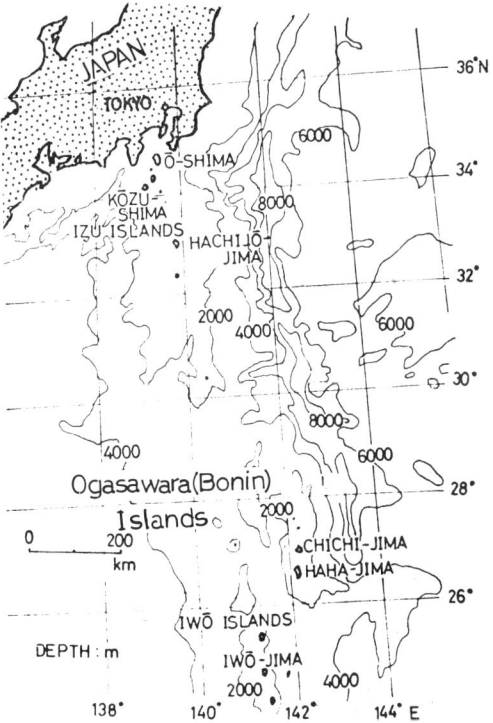

the Yap Ridge and the Palau Ridge to the SW of the Mariana Ridge (see *Caroline Islands*). The Izu-Mariana arc separates the Philippine Sea Basin from the West Pacific Basin.

The Izu-Ogasawara Trench is associated with a negative gravity anomaly belt and the negative value, reaches -252 mgal after free-air reduction. The outer arc is associated with a positive gravity anomaly belt and the positive value reaches +341 mgal of free-air anomaly at Hutami on Chichi Island, which is one of the highest values in the Japanese region. Heat flow has been measured as 1.5–2.0 HFU (10^{-6} cal/cm^2 sec) along the inner arc. Seismicity is fairly high, but no large earthquake with $M > 7.5$ has occurred. The reason may be that the crust is oceanic. Almost all seismic foci except those associated with volcanic activity are in a layer dipping about 45° from the trench toward the W. The deepest foci are deep in the mantle around 550 km deep. Travel times of P waves observed at the station in Chichi Island for shallow earthquakes along the Pacific coast of northeastern Japan are on the average 8% shorter than those for shallow earthquakes in southwestern Japan. The anomalously high velocity of seismic waves under the outer arc would indicate that the West Pacific plate with high Q and high V is actively underthrusting beneath the Izu-Ogasawara arc.

Izu Islands

The synonym "Shichito Islands" means "seven islands" in Japanese, which are O-shima, To-shima, Nii-jima, Kozu, Miyake, Mikura, and Hachijo islands. They are not the largest seven and there are three other major islands, named Shikine, Aoga-shima, and Tori. All of them are Pliocene to Quaternary volcanic islands (Table 1). Most of the basaltic rocks are tholeiitic. Submarine volcanic activities are reported between them. The one near Bayonnaise Rocks is notably active. The rocks form a part of a caldera rim and a postcaldera cone is called *Myojinsho*.

Ogasawara Islands

There are three clusters in the Ogasawara Islands: the northern or Muko ("son-in-law") group, with four islands; the middle or Chichi ("father") group, with seven islands; and the southern or Haha ("mother") group, with six islands. The islands expose andesitic lavas, partly with pillow texture, and pyroclastic rocks (all groups) interbedded with Eocene *nummulites* limestone (Haha Group). They are strongly folded and faulted, and covered by gently dipping Miocene *Lepidocyclina* limestone (Chichi Group), probably unconformably. There is a glassy hypersthene andesite called *boninite* in the volcanic complex (Chichi Group). Boninite was named by Petersen. Potassium-argon ages of 26 m yr (Chichi Group) and 40 m yr (Haha Group) were dated for the lower or volcanic complex. Coral reefs fringe some parts of the islands.

TABLE 1. Volcanoes of the Izu Islands

	Last Major Activity	Rock
O-shima	1951	Basalt
To-shima		Basalt
Nii-jima	886	Rhyolite
Shikine		Rhyolite
Kozu	838	Rhyolite
Miyake	1962	Basalt
Mikura		Basalt → andesite
Hachijo	1606	Basalt, andesite, dacite
Aoga-shima	1785	Basalt (rim), andesite (cone)
Myojinsho	1960	Basalt (rim), dacite (cone)
Tori	1939	Basalt → andesite

Iwo Islands

The major island of this group is Iwo Island, and its two cousins are Kita-Iwo Island and Minami-Iwo Island, respectively, to N and S. All three are Quaternary volcanoes, the last activity being in 1957 at Iwo and in 1889 at Kita-Iwo. Iwo Island consists of trachyandesite, but Kita-Iwo Island consists of tholeiitic basalt. Uplifted coral reefs are found on Iwo Island.

Nishino-shima Island

Nishino-shima Island (27°15'N, 140°53'E) is situated about 120 km west of the Ogasawara Islands and lies on the volcanic chain halfway between the Izu Islands and the Iwo Islands. The island is a part of an andesitic volcano. The volcano set up an intense submarine eruption and produced a new island during 1973.

ARATA SUGIMURA

References

Davis, W. M., 1928. "The coral reef problem," *Am. Geog. Soc. Spec. Publ. 9*.

Dietz, R. S., 1954. "Marine geology of northwestern Pacific; description of Japanese bathymetric chart 6901," *Bull. Geol. Soc. Am.*, 65, 1199–1224.

Fitch, T. J., 1972. "Plate convergence, transcurrent

faults, and internal deformation adjacent to Southeast Asia and the western Pacific," *J. Geophys. Res.*, 77(23), 4432–4460.

Hanzawa, S., 1925. "Foraminifera bearing rocks from Okinawa Island and the Ogasawara Islands," *J. Geol. Soc. Japan*, **32**, 461–484 (in Japanese, incl. detailed geologic map).

Heiskanen, W. H., 1945. "The gravity anomalies on the Japanese Islands and in the waters east of them," *Ann. Acad. Sci. Fennicae,* Ser. A., pt. 3, Geol.–Geogr., **8**, 22p.

Ichikawa, M., 1970. "Seismic activities at the junction of Izu-Mariana and southwestern Honshu arcs," *Geophys. Mag.*, **35**(1), 55–69.

Kaneoka, I., Isshiki, N., and Zashu, S., 1970. "K-Ar ages of the Izu-Bonin Islands," *Geochem. J.*, **4**, 53–60.

Katsumata, M., and Sykes, L. R., 1969. "Seismicity and tectonics of the western Pacific: Izu–Mariana–Caroline and Ryukyu–Taiwan Regions," *J. Geophys. Res.*, **74**, 5923–5948.

Kuno, H., 1959. "Origin of Cenozoic petrographic provinces of Japan and surrounding areas," *Bull. Volcanol.*, Ser. 2, **20**, 37–76.

_____, 1962. *Catalogue of the Active Volcanoes of the World, including Solfatara Fields.* Roma: Intern. Assoc. Volcanol., pt. 11.

Ludwig, W. J., Den, N., and Murauchi, S., 1973. "Seismic reflection measurements of Southwest Japan Margin," *J. Geophys. Res.*, 78(14), 2508–2516.

Matsuda, T., 1962. "Crustal deformation and igneous activity in the South Fossa Magna, Japan," *Am. Geophys. Union Monogr. 6*, 140–150.

Minato, M., *et al.*, 1965. *The Geologic Developments of the Japanese Islands.* Tokyo: Tsukiji Shokan Co., 442p.

Mogi, A., 1972. "Bathymetry of the Kuroshio Region," in *Kuroshio–Its Physical Aspects.* Tokyo: Univ. Tokyo Press, 53–80.

Otuka, Y., 1938. "A geologic interpretation on the underground structure of the Sititô–Mariana island arc in the Pacific Ocean," *Bull. Earthquake Res. Inst.*, **16**, 201–211.

Takai, F., *et al.*, 1964. *Geology of Japan.* Berkeley: Univ. Calif. Press, 279p.

Tsuya, H., 1937. "On the volcanism of the Huzi volcanic zone, with special reference to the geology and petrology of Idu (Izu) and the Southern Islands," *Bull. Earthquake Res. Inst.*, **15**(1), 215–357.

Utsu, T., 1971. "Seismological evidence for anomalous structure of island arcs with special reference to the Japanese region," *Rev. Geophys. Space Phys.*, 9(4), 839–890.

Watanabe, T., Epp, D., Uyeda, S., Langseth, M., and Yasui, M., 1970. "Heat flow in the Philippine Sea," *Tectonophysics*, **10**, 205–224.

Yoshiwara, S., 1902. "Geological age of the Ogasawara Group (Bonin Islands)," *Geol. Mag.*, **9**, 296–303.

Cross-reference: *Mariana Islands.*

J

JAMAICA

Lying in the Caribbean Sea about 145 km (90 miles) S of Cuba and 160 km (100 miles) W of Haiti, Jamaica with an area of 10,960 km^2 (4,232 sq mi) is the largest of the former British West Indies. It is 240 km long (E-W) and 60-80 km across. Physically it is (60%) a limestone plateau with extensive areas of "cockpit" and "cone" karst (and large poljes) in the W. The interior is mountainous, with WNW-ESE trends. The highest peak is in the Blue Mountains (E) rising to 2292 m (7042 ft). The western region is lower, and the south is marked by a series of four wide coastal plains, fronted in places by reefs with coral sand cays. The northern coast is largely steep-to, dropping away directly to the Cayman Trough, but the southern coast is marked by bays and much mangrove. Offshore to the S is the Caribbean shelf, which has an area of 5000 km^2 above the 200-m line. It connects to the Pedro Banks and central America (Fig. 1)

Jamaica is rich in bauxite and gypsum. The bauxite production (from lateritic clays) represents almost 25% of the world's output. Jamaica's ceramic clay, silica sand, and marble is also commercially productive. In spite of some exploratory drilling, no traces (or seeps) of hydrocarbons have been located. Since 1970, the annual value of mineral commodities produced in Jamaica has exceeded $200 million. Bauxite accounts for more than $75 million (7.5 million tons) and processed alumina $100 million. In addition, 7.5 million tons of cement and other building materials—gypsum, sand, etc.—are produced annually.

Lying in the belt of NE trade winds, Jamaica enjoys a mild to warm climate (average temperature in February 24.4°C, in July 27.5°C), reaching maxima on the southeastern coast with 32°C average maxima in the summer. February is the driest month, with 60-150 mm and October the wettest with 250-400 mm. The highest rainfall, in the interior at Mooretown, averages 5600 mm, but on the southern coast it falls to less than 1000 mm. In the three centuries 1655-1955, there have been 31 hurricanes. Vegetation is heavy in the mountains, with some dense forests giving way to savanna on parts of the southern coast. Rivers are relatively short and unimportant; in the karst country most of the drainage is underground, with many closed depressions.

Jamaica is located near an active seismic belt, and there have been numerous destructive earthquakes recorded since the original discovery by the Spaniards.

The geology consists of a series of mainly volcanic rocks of Cretaceous (Barremian to Maestrichtian) age, locally with interbedded rudistid limestones and mudstones (Fig. 2). The sequence is intruded by granodiorites, dated by K-Ar at 63-85 m yr (Lewis et al., 1973). Metamorphic rocks (amphibolites and greenschists), confined to the southwestern Blue Mountains, are probably the same orogenic age but may include pre-Mesozoic and Jurassic sediments. Dike intrusion occurred around 68 m yr, and general uplift, block faulting, and minor felsite intrusion followed at about 52-47 m yr.

The Cretaceous is unconformably overlain

FIGURE 1. Deeply dissected landscape of the Rio Grande in the John Crow Mountains of Jamaica, near Port Antonio. (Photo: Jamaica Government.)

FIGURE 2. Geological sketch map of Jamaica, showing principal sedimentary and igneous formations. (By courtesy of the Geological Survey Department, Kingston, Jamaica, from 1:750,000 sheet of 1966.) 1, alluvium (Quaternary); 2, coastal fms. (M. Miocene-Pleistocene); 3, White Limestone (M. Eocene-L. Miocene); 4, Yellow Limestone (M. Eocene); 5, Richmond Formation (L. Eocene); 6, Wagwater Formation (L. Eocene; 7, undifferentiated sediments and metamorphics (U. Cretaceous); A, Halberstadt Volcanics (L. Eocene); B, Newcastle Volcanics (L. Eocene); C. andesitic volcanics (U. Cretaceous); D, granodiorite (U. Cretaceous); E, serpentinite (Cretaceous).

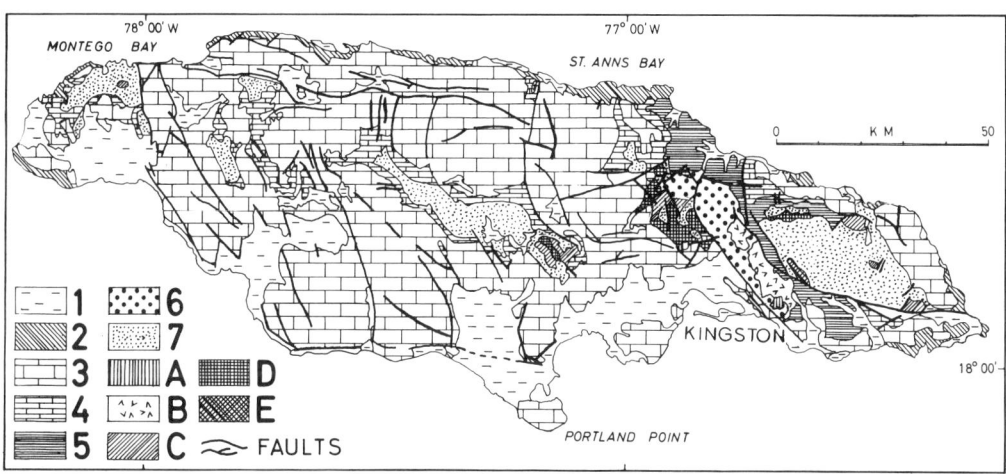

by a Lower Eocene clastic series (Wagwater, Richmond, and lower Chapelton formations) which passes upward by intercalation into carbonates (Yellow Limestone and White Limestone groups); these carbonates approach 2000 m in thickness. The Wagwater includes red beds, evaporites, and volcanic extrusives (porphyritic andesites and some dacites) confined to eastern Jamaica. The overlying Richmond is mainly a flysch facies with local volcanics. In central Jamaica the sometimes lignitic clastics and impure carbonates of the Chapelton Formation yield a variety of shallow-water fossils. The White Limestone (Middle Eocene to Middle Miocene), 1500–2000 m thick, includes a wide spectrum of carbonate types representing shallow and deep-water facies.

Over central Jamaica the White Limestone is subaerially weathered, producing spectacular karst scenery. Superficial deposits include locally abundant pockets of bauxite. Marginally the White Limestone is succeeded by Middle Miocene to Lower Pleistocene marls, clays, and limestones of the Coastal Group (Buff Bay, Bowden, August Town, and Manchioneal formations). Later Quaternary deposits include the partly submerged alluvial cone of the Liguanea Plain and other deposits along the south coast, locally exceeding 300 m thickness. A series of raised coral reefs of late Pleistocene age is developed mainly along the north coast. Elevations on these reefs vary and indicate continued warping and faulting into the Holocene.

EDWARD ROBINSON

References

Burke, K., 1967. "The Yallahs Basin: a sedimentary basin southeast of Kingston, Jamaica," *Marine Geol.,* 5(1), 45–50

Cant, R. V., 1973. "Jamaica's Pleistocene reef terraces," *Geologie en Mijnbouw,* 52, 157–160.

Chubb, L. J., 1955. "The Cretaceous succession in Jamaica," *Geol. Mag.,* 92(3), 177–195.

_____, and Burke, K., 1963. "Age of the Jamaican granodiorite," *Geol. Mag.,* 100, 524–532.

Daneš, J. V., 1914. "Karststudien in Jamaica," *Sitz. K. böhm. Gesell. Wiss. Jahrg., Math. Natl. Kl.,* 20, 1–72.

Goreau, T., and Burke, D., 1966. "Pleistocene and Holocene geology of the island shelf near Kingston, Jamaica," *Marine Geol.,* 4(3), 207–225.

Hill, R. T., 1899. "Geology of Jamaica," *Bull. Mus. Comp. Zool., Harvard,* 34, 1–256 (map).

Hose, H. R., and Versey, H. R., 1956. "Palaeontological and lithological divisions of the Lower Tertiary limestones of Jamaica," *Col. Geol. Mineral. Res.,* 6, 19–39.

Hughes, I. G., 1973. *The Mineral Resources of Jamaica* (Bull. 8). Kingston: Geol. Surv. Dept., 89p.

Lewis, J. F., *et al.,* 1973. "Potassium–argon retention ages of some Cretaceous rocks from Jamaica," *Bull. Geol. Soc. Am.,* 84, 335–340.

Matley, C. A., 1929. "The Basal Complex of Jamaica with petrographical notes by F. Higham," *Quart. J. Geol. Soc. Lond.,* **85**(4), 440–490, pls. 23–25.

Robinson, E., 1967. "Submarine Slides in White Limestone group, Jamaica," *Bull. Am. Assoc. Petrol. Geologists,* **51**(4), 569–578.

_____, 1969. "Geological field guide to Neogene sections in Jamaica," *West Indies J. Geol. Soc.* (Jamaica), **10**, 1–24.

_____, and Lewis, J. F., 1972. "Field guide to aspects of the geology of Jamaica," in T. W. Donnelly, ed., *International Field Institute Caribbean Field Guide, 1970.* Washington, D.C.: Am. Geol. Inst., 44p.

Steers, J. A., 1940. "The coral cays of Jamaica," *Geogr. J.,* **95**, 30–42 (4 pls.).

Sweeting, M. M., 1973. *Karst Landforms.* New York: Columbia Univ. Press, 362p.

Trechmann, C. T., 1922. "The Barrettia beds of Jamaica," *Geol. Mag.,* **59**(701), 501–514, pls. 18–20.

_____, 1923. "The yellow limestone of Jamaica and its mollusca," *Geol. Mag.,* **60**(710), 339–367, pls. 14–18.

_____, 1924. "The Cretaceous limestones of Jamaica and their mollusca," *Geol. Mag.,* **61**(723), 385–408, pls. 22–26.

_____, 1927. "The Cretaceous shales of Jamaica," *Geol. Mag.,* **64**(751, 752), 27–42, 49–65, pls. 1–4.

_____, 1936. "The basal complex question in Jamaica," *Geol. Mag.,* **73**, 251–287.

Zans, V. A., 1951. *Economic Geology and Mineral Resources of Jamaica* (Publ. 1). Kingston: Geol. Surv. Dept., 61p.

_____, Chubb, L. J., Versey, H. R., Williams, J. B., Robinson, E. and Cooke, D. L., 1963. *Synopsis of the Geology of Jamaica* (Bull. 4). Kingston: Geol. Surv. Dept., 72p.

Cross-references: *Cayman Islands; Cuba; Guatemala; West Indies.*

JUAN FERNÁNDEZ

Comprising a rugged and wooded volcanic island group, the Juan Fernández Islands lie in the Pacific Ocean, about 580 km (360 miles) W of Valparaiso, Chile. They are owned by Chile. The two chief islands, Más Afuera (Isla Robinson Crusoe) and Más a Tierra (Isla Alejandro Selkirk), are some 160 km (100 miles) apart, W and E, respectively. They have a combined area of about 185 km^2 (70 sq mi), Más Afuera is 7 by 10 km and 85 km^2, rising to 1650 m (6560 ft); Más a Tierra is 5 by 20 km and 95 km^2, rising to 916 m; a smaller island, Santa Clara, is only 5 km^2. They were discovered by Juan Fernández in 1572 and are famous notably for Daniel Defoe's classic book *Robinson Crusoe* (about Alexander Selkirk's adventures). Selkirk was a Scottish seaman who was shipwrecked and survived on Más a Tierra from 1704 to 1709. The islands were renamed by the Chilean authorities in honor of Selkirk and his storybook hero.

The islands lie at the intersection of an E-W submarine ridge and a N-S one that trends toward *San Ambrosio* and *San Felix* (q.v.). The topography is deeply dissected, and the coasts are cliffed and without coral reefs. Terraces have not been reported. The volcanic foundations are "presumably" of Tertiary age and consist of feldspathoidal basalts, basanites, picrites, and soda-trachytes. Quensel (1912) described the geology, and the paleogeography and biology have been studied by Skottsberg (1925). Of the once-favored "Pacific Continent," employed by biogeographers to explain biological distribution, there is, alas, no trace (see Quensel, 1952; Skottsberg, 1920, 1921, 1925). Sea-floor spreading data show that the oceanic crust hereabouts is probably of Eocene age.

RHODES W. FAIRBRIDGE

References

MacArthur, R. H., and Wilson, E. O., 1967. *The Theory of Island Biogeography.* Princeton, N.J.: Princeton Univ. Press, 203p.

Quensel, P. D., 1912. "Die Geologie der Juan Fernandez Inseln," *Geol. Inst. Upsala Bull.,* **11**, 240–290.

_____, 1952. *The Natural History of Juan Fernandez and Easter Island,* vol. 1. Upsala, Sweden: Almquist and Wiksell, 37–87.

Skottsberg, C., 1920. *The Natural History of Juan Fernández and Easter Islands.* Upsala, Sweden: Almquist and Wiksell, 3 vols.

_____, 1921. "Die Juan Fernandez-Inseln," *Z. Ges. Erdk.* (Berlin).

_____, 1925. "Juan Fernandez and Hawaii, a phytogeographic discussion," *Bernice P. Bishop Mus. Bull.,* **16**.

Cross-references: *Chile; Easter Island and Sala y Gómez; Galápagos Islands; South America.*

K

KERMADEC ISLANDS

The Kermadec Islands form part of a chain of widely spaced groups of emergent and submarine volcanoes extending SW from Tonga to New Zealand along an "island arc" ridge bordering the northwestern flank of the Tonga-Kermadec Trench.

The largest island, Raoul (or Sunday Island), about 1100 km NE of Auckland (N.Z.), was discovered by d'Estrecasteaux in 1793. It is approximately 8 km across, deeply dissected and rises to peaks of more than 500 m which ring a caldera up to 1.6 km across. Offshore islets and shoals and the steep cliffs show that the island is the summit remnant of a large and complex volcanic pile. Present activity is usuin the caldera and on sea cliffs to the west. But there have been numerous eruptions from both the crater and Denham Bay in the last 200 years, notably in 1847, 1853, and in 1872, when there was an eruption from Green Lake in the crater and an island was built up in Denham Bay (Wolverine Shoal). The last eruption, in 1964 (Healy et al., 1965) was mainly phreatic from a chain of craters on the caldera floor. New Zealand maintains a meteorological station on the island, and since 1957 a seismograph has been in operation. A soil survey was made by Wright and Metson (1959).

Raoul is essentially a stratovolcano, a volcanic complex built up by eruption from many vents. The rocks (Brothers and Searle, 1970) are typical of a tholeiitic suite ranging from olivine basalts to basaltic andesites and dacite; typical andesites are not represented. All but the most basic olivine-rich rocks in the oldest formation are quartz normative, and pigeonite and hypersthene are abundant in many of the basaltic andesites. The whole island has been mantled with pumice breccia, which is also abundant in an extensive fan-like terrace on the northern coast. The pumice is presumably related to explosive eruptions associated with formation of the caldera. Superficial and local coatings of ash and volcanic muds in and about the caldera indicate that later phases of activity were mainly phreatic.

Boulders of two categories of plutonic rocks are found on Raoul: holocrystalline rocks including gabbro, diorites, and quartz diorites; and *cummulate blocks* of gabbroic composition consisting of coarse crystalline phases (olivine, pyroxene, and calcic plagioclase) bonded by volcanic glass, sometimes as little as 5% by volume. Such rocks are so friable that they shatter with the tap of a hammer.

Macauley, the next largest island, is 108 km SW of Raoul (Brothers and Martin, 1970). It has been reduced by erosion to a residual little more than 2 km across. The sequence of lava flows and tuffs dip away from the northwestern side of the island where there is a small crater. The rocks consist of high alumina basalt, tuffs of rhyolitic pumice, and accidental blocks of basalt like those of Raoul.

The smaller islands are described by Thomson (1926) but as yet no petrological study of their rocks has been made.

A group of submarine volcanoes has recently been discovered on the submarine ridge midway between the Kermadecs and New Zealand. Of these the most active is Rumble III (Kibblewhite, 1966) about 250 km ENE of Auckland. It is built up from 2000 m to within 120 m of the surface. Brothers (1967) has identified a basaltic andesite here and reported (1970) on the petrochemical affinities along the whole arc.

E. J. SEARLE

References

Brothers, R. N., 1967. "Andesite from Rumble III volcano, Kermadec-Tonga ridge, Southwest Pacific," *Bull. Volcanol.*, 31, 17–19.

———, 1970. "Petrochemical affinites of volcanic rocks from the Tonga-Kermadec island arc, Southwest Pacific," *Bull. Volcanol.*, 34, 308–329.

———, and Martin, K. R., 1970. "The geology of Macauley Island, Kermadec Group, Southwest Pacific," *Bull. Volcanol.*, 34, 330–346.

———, and Searle, E. J., 1970. "The geology of Raoul Island, Kermadec Group, Southwest Pacific," *Bull. Volcanol.*, 34, 7–34.

Healy, J., Lloyd, E. F., Banwell, C. J., and Adams, R. D., 1965. "Volcanic eruption on Raoul Island, November, 1964," *Nature*, 205, 743–745.

Kibblewhite, A. C., 1966. "The acoustic detection on location of an underwater volcano," *N.Z. J. Sci.*, 9, 178–199.

Thomson, J. A., 1926. "Volcanoes of the New Zealand-Tonga volcanic zone," *N.Z. J. Sci. Tech.,* **8,** 345–371.

Wanoa, R. J., and Lewis, K. B., 1972. "Gazetteer of seafloor features in the New Zealand region," *N.Z. Oceanogr. Inst.,* **1**(5), 67–106.

Wright, A. C. S., and Metson, A. J., 1959. "Soils of Raoul (Sunday) Island, Kermadec Group," *N.Z. Soil Bur. Bull.,* **10,** 49p.

Cross-references: *Fiji; New Zealand; Tonga.*

L

LEEWARD ISLANDS

Situated in the northeastern part of the E-facing loop of the West Indies island arc, the Leeward Islands constitute the northerly and northwesterly chains of the *Lesser Antilles* (the southerly ones being the *Windward Islands,* q.v.). The *Virgin Islands* (q.v.), sometimes included with the Leeward Islands, lie to the W and are rather distinct geologically; they are treated here separately.

As described here the Leeward Islands are separated from the Virgin Islands platform by the deep Anegada (or Jungfern) Passage and extend from 15°00' to 18°35'N and 61°40' to 63°20'W. The term "Leeward Islands" (in French = Iles-du-vent) comes from the fact that the trade winds here are from the E to NE, and the Caribbean Current sets to the W, so that the northeastern islands in the Lesser Antilles are "up-wind;" in other languages the concept is reversed. (*Note:* There is a corresponding "Leeward" and "Windward" grouping in the Society Islands of French Polynesia.) Administratively the islands have long been colonies of France (*Guadeloupe*, q.v., with St. Martin and St. Barthélemy), the Netherlands (St. Maarten, the southeastern part of St. Martin; and Saba; see *Netherlands Antilles*), and Britain. The former British Leeward Islands are now in part independent. *St. Christopher-Nevis-Anguilla* (q.v.) constitute a "British Associated State;" likewise *Antigua* (q.v.) and *Dominica* (q.v.), while *Montserrat* (q.v.) remains a British colony.

The Leeward Islands are arranged in two distinct belts:

1. The *"inner, volcanic arc"* (part of the "Volcanic Caribbees," see *West Indies*), which includes Saba, St. Eustastius, St. Christopher (St. Kitts), Nevis, Redonda, and Montserrat. They are rather evenly spaced vents, more than 30 km apart, that continue southward in Guadeloupe and Dominica, followed by the Windward Islands. The volcanic activity is traced back to Eocene time and some continues today—e.g., Mt. Soufrière on Nevis and another of the same name on Guadeloupe. On St. Eustatius and St. Christopher the volcanics are interbedded with Pliocene and Pleistocene limestones (Godwin Gut and Brimstone Hill formations, respectively). On Montserrat the Roche Bluff formation (Pleistocene) includes interbedded tuffs and limestones. They lie in the trade wind belt and are thus subtropical; the higher islands receive an adequate rainfall.

2. The *outer, limestone arc* (part of the "Limestone Caribbees"), which includes Sombrero, Dog, Anguilla, St. Martin (Sint Maarten), St. Barthélemy (St. Bartholomew), Barbuda, and Antigua. The largest of the islands is Antigua, which is treated here separately. The eastern part of Guadeloupe actually belongs to this belt, and likewise its offshore islets of Désirade and Marie Galante. In places there is a sedimentary cover of late Oligocene to Miocene, mainly foraminiferal or oolitic limestones with tuffs, which was uplifted, somewhat tilted, and faulted in the late or early Pliocene, but which has been rather stable since late Pliocene times. Today these islands are deeply degraded

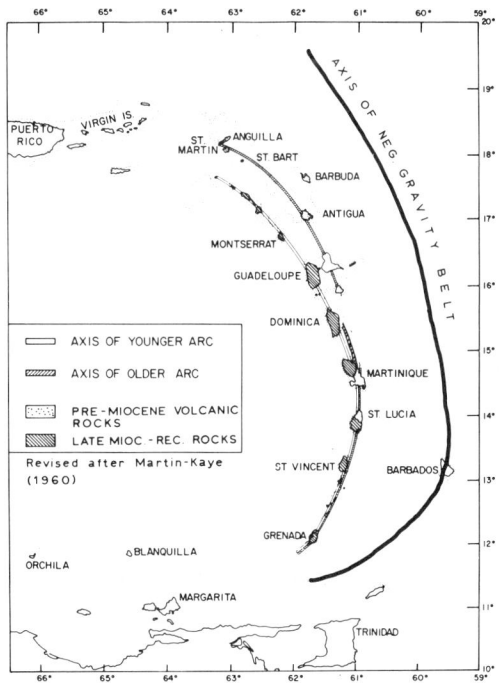

FIGURE 1. Lesser Antilles island arc with axes of older and younger arcs and axis of negative gravity anomaly belt indicated (revised after Martin-Kaye, 1960)

and have only low hills. There are Quaternary eustatic terraces. According to Davis (1924, 1926) Sombrero, Antigua, and Marie Galante are crowned by uplifted atolls. Owing to their low elevations the islands today present little obstruction to the trade winds (from E to NE) and are rather dry to semiarid.

Geological History

The Leeward Islands form the northern part of the emergent crest of the N-S submarine volcanic ridge produced by at least two distinct long-term volcanic phases. Figure 1 illustrates the older and younger axes connecting coeval volcanic centers which have been recognized in the Lesser Antilles. North of Dominica an offset of the older axis has resulted in a separation of the older and younger volcanic belts to form the double row of islands. From Dominica southward the two zones are superimposed, to form a single row. Historically, because of the dominant rock type exposed on the islands, as mentioned above, the inner row was referred to as the Volcanic Caribbees (from Saba to Grenada) and the outer, the Limestone Caribbees (from Sombrero to Marie Galante; see article: *West Indies*).

The islands of the older arc have foundations of andesitic and dacitic flows of calc-alkaline affinity, and are Eocene to Oligocene in age. The radiometrically determined late Jurassic-Cretaceous ages of the igneous basement of La Désirade, however, suggest that the Lesser Antilles ridge and the Greater Antilles are one continuous feature with similar pre-Miocene histories. An alternative interpretation offered by Fink *et al.* (1971) for the pillowed spilites, cherts, and quartz keratophyres associated with hypabyssal trondhjemite is that the La Désirade basement is part of an ophiolite suite and represents upfaulted or obducted Atlantic sea floor crust. The cessation of volcanism around middle Eocene-early Oligocene time was followed by erosion, partial truncation, minor hypabyssal intrusive activity, and capping of the volcanic complexes with a thin flat-lying transgressive sequence of late Oligocene-early Miocene age.

During this general cessation of volcanic activity, a major offset of the submarine ridge occurred along a line just N of Dominica. This was attributed to different rates of isostatic uplift caused by greater thicknesses of sediment in the southern end of the pre-Miocene Lesser Antillean Trench.

Renewed volcanism in the late Miocene and early Pliocene produced the younger, inner row of islands from Saba to Isles des Saintes. In the single row of islands from Dominica to Grenada, the late Tertiary to Holocene volcanics overlie unconformably the older volcanogenic and sedimentary series.

Structure and Petrology

Structurally, the double arc of the Leeward Islands consists of one broad asymmetric ridge associated with the older volcanic centers and the younger aligned volcanic centers which are perched on the western flank of the main ridge. A major transverse fault, just N of Dominica, was proposed to account for the axial offset of the older zone of volcanism which produced the abrupt widening of the arc ridge at that point.

The Leeward Islands conform chemically and petrologically to the circum-oceanic calc-alkaline volcanic suite. The general chemical features of the Lesser Antilles calc-alkaline assemblage are high alumina, low alkali, particularly low K_2O content, very low titanium, and general silica enrichment. The only petrochemical pattern revealed by several recent studies has been a heterogeneous distribution of rock types along the arc. Although there has been no agreement on a specific physiochemical model which might explain these variations, most workers have proposed some form of Upper Mantle-Lower Crust multiple-stage derivative process. Attempts to demonstrate the spatial and temporal trends of potash content and iron enrichment which are characteristic of other island arc regions have been unsuccessful.

L. KENNETH FINK, JR.
RHODES W. FAIRBRIDGE

References

Baker, P. E., 1968. "Comparative volcanology and petrology of the Atlantic Island Arcs," *Bull Volcanol.*, **32**, 189–206.

Bold, W. A. van den, 1970. "Ostracoda of the lower and middle Miocene of St. Croix, St. Martin and Anguilla," *Caribbean J. Sci.*, **10**(1–2), 35–61.

Bunce, E. T., Phillips, J. D., Chase, R. L., and Bowin, C. O., 1970. "The Lesser Antilles Arc and the eastern margin of the Caribbean Sea," *in* A. E. Maxwell, ed., *The Sea,* vol. 4. New York: Wiley-Interscience, 359–386.

Butterlin, J., 1959. "Microfaune et âge de deux formations calcaires de la Martinique," *C. R. Somm. Soc. Géol. France,* fasc. 2, 42–43.

Christman, R. A., 1953. "Geology of St. Martin, St. Bartholomew and Anguilla, Lesser Antilles," *Bull. Geol. Soc. Am.*, **64**, 65–96.

Cleve, P. T., 1871. "On the geology of the northeast-

ern West India Islands," *Kongl. Svenska Ak. Handlingar,* 9(12), 48p.
Davis, W. M., 1924. "A tilted-up, beveled-off atoll," *Science,* 60, 51–56.
———, 1926. "The Lesser Antilles," *Am. Geogr. Soc. Spec. Publ. 2,* 207p.
Dickinson, W. R., and Hatherton, T., 1967. "Andesitic volcanism and seismicity around the Pacific," *Science,* 157, 801–803.
Donnelly, T. W., 1964. "Evolution of Eastern Greater Antillean island arc," *Bull. Am. Assoc. Petrol. Geologists,* 48, 680–696.
———, Rogers, J. J. W., Pushkar, P., and Armstrong, R. L., 1971. "Chemical evolution of the igneous rocks of the eastern West Indies: an investigation of thorium, uranium, and potassium distributions, and lead and strontium isotopic ratios," *Geol. Soc. Am. Mem.* 130, 181–224.
Drooger, C. W., 1951. "Foraminifera from the Tertiary of Anguilla, St. Martin and Tintamarre (Leeward islands, W.I.)," *Konink. Ned. Akad. Wetensch.* (Amsterdam), Ser. B, 54, 54–65.
Fink, L. K., Jr., 1968. "Marine geology of the Guadeloupe Region, Lesser Antilles Island Arc," thesis, Univ. Miami, 121p.
———, 1970. "Evidence for the antiquity of the Lesser Antilles Island Arc," *Trans. Am. Geophys. Union,* 51(4), 326.
———, Harper, C. T., Stipp, J. J., and Nagle, F., 1971. "Tectonic significance of La Désirade," *Proc. 6th Caribbean Geol. Conf.,* Margarita, Venezuela, Abstr.
Hatherton, T., and Dickinson, W. R., 1969. "The relationship between andesitic volcanism and seismicity in Indonesia, the Lesser Antilles, and other island arcs," *J. Geophys. Res.,* 74, 5301–5309.
Jakes, P., and White, A. J. R., 1972. "Major and trace element abundances in volcanic rocks of orogenic areas," *Bull. Geol. Soc. Am.,* 83, 29–40.
Lewis, J. F., 1971. "Composition, origin, and differentiation of basalt magma in the Lesser Antilles," *Geol. Soc. Amer. Mem.* 130, 159–180.
Lidiak, E. G., 1965. "Petrology of andesitic, spilitic, and keratophyric flow rock, north-central Puerto Rico," *Bull. Geol. Soc. Am.,* 76(1), 57–88.
MacGregor, A. G., 1938. "The volcanic history and petrology of Montserrat, with observations on Mt. Pelée, in Martinique," *Phil. Trans. Roy. Soc. Lond.* Ser. B, 229, 90p.
Martin-Kaye, P. H. A., 1960. "The double arc of the Lesser Antilles," thesis, Univ. London, 78p.
———, 1969. "A summary of the geology of the Lesser Antilles," *Overseas Geol. Mineral Res.,* 10(2), 172–206.
Mitchell, A. H., and Reading, H. G., 1971. "Evolution of island arcs," *J. Geol.,* 79(3), 253–284.
Mitchell, R. C., 1953. "New data regarding the dioritic rocks of the West Indies," *Geol. Mijnb.,* 15, 291–294.
Molengraaff, G. J. H., 1931. "Saba, St. Eustatius and St. Martin," *Leidsch. Geol. Mededel.,* 5(458), 126p.
Molnar, P., and Sykes, L. R., 1969. "Tectonics of the Caribbean and Middle America regions with focal mechanisms and seismicity," *Bull. Geol. Soc. Am.,* 80(9), 1639–1684.
Perret, F. A., 1939. "The volcano-seismic crisis at Montserrat," *Carnegie Instit., Publ. 512,* 1–76.
Pushkar, P., 1968. "Strontium isotope ratios in volcanic rocks of three island arc areas," *J. Geophys. Res.,* 73, 2701–2714.
Robson, G. R., and Tomblin, J. F., 1966. "West Indies," *Catalogue of the Active Volcanoes of the World including Solfatara Fields.* Naples: Intern. Assoc. Volcanol., pt. 20, 56p.
Schuchert, I. C., 1935. *Historical Geology of the Antillean-Caribbean Region.* New York: Wiley.
Thomas, H. D., 1942. "On fossils from Antigua and the age of the Seaforth limestone," *Geol. Mag.,* 79(1), 49–61.
Tomblin, J. F., 1968. "Geochemistry and genesis of Lesser Antilles volcanic rocks," Reprint of paper read at 5th Caribbean Geol. Conf., St. Thomas, V.I.
Westermann, J. H., and Kiel, H., 1961. "The geology of Saba and St. Eustatius, with notes on the geology of St. Kitts, Nevis, and Montserrat," *Natuur. Stud. Surinam Publ. 24,* 175p.
Weyl, R., 1966. *Geologie der Antillen.* Berlin: Gebr. Borntraeger, 410p.

Cross-references: *Antigua; Barbados; Vol. I, Carribbean Sea; Dominica; Guadeloupe; Martinique; St. Christopher-Nevis-Anguilla; Virgin Islands; West Indies; Windward Islands.*

LINE ISLANDS

Lying in the central Pacific Ocean, the Line Islands consist of the Caroline Atoll, Christmas Island, Fanning Atoll, Flint Island, Jarvis Island, Malden Island, and Palmyra Island. They are NE of Fiji and Samoa, NW of Tahiti, and S of the Hawaiian Islands. Most of them are under the administration of the United Kingdom; Johnston, Palmyra, and Jarvis islands are possessions of the United States.

The Line Islands are all coral reefs, atolls, or reef limestone islands, that are on a discontinuous NW-SE line of oceanic volcanic cones extending for 3700 km (2300 miles), from the western limits of the Tuamotu group (*French Polynesia* q.v.) almost to the latitude of Hawaii. They range from $17°N$ to $12°S$, and $150-169°W$. The underlying oceanic crust here is of late Cretaceous age according to sea-floor spreading data.

Johnston Island, the most northerly, is sometimes treated separately from the Line Islands; it is the most isolated, although it lies in the same trend, separated from the nearest emerged reef by some 20 seamounts. It is a partly drowned atoll (Emery, 1965) that was artificially extended in 1940 to build an 1800-m (6000-ft) airstrip.

Next come Kingman Reef and *Palmyra Island* (an atoll, 8 km across), which share a common pedestal. *Washington Island* seems to be an elevated atoll; it is 5 km across and contains a freshwater lake and a swamp (described by Christophersen, 1927). *Fanning Island* is an atoll 17 km across, but with not over 4 m elevation; it has a good anchorage in the central lagoon and has been worked for guano.

Christmas Island (not to be confused with the phosphate island in the Indian Ocean, q.v.) is said to be the world's largest atoll, 25 km in diameter. (There are, of course, much larger ring-reef structures forming insular barrier reefs or near-atolls, e.g., in the East Indies, New Guinea, Truk, etc.). Most of the Christmas Island rim is no more than 3 m in elevation, but some dunes reach 20 m. There are also lakes or ponds within the rim belt, mostly brackish. The windward reef flat is 5 km wide.

Jarvis Island lies 400 km SW of Christmas Island, off the main trend of the Line Islands, and is an emerged atoll with a saucer-shaped middle, which is the former lagoon. The rim is about 6 m above sea level and surrounded by a fringing reef. The surface, dry and desert-like, has been worked for guano.

Some 700 km SE of Christmas Island comes *Malden Island,* another elevated atoll, 6 km across and up to 11 m in elevation, surrounded by a contemporary fringing reef. The central depression is swampy and has been worked for guano. *Starbuck* is also a guano island, slightly emerged (4.5 m) and 8 km across. *Caroline* (not to be confused with the group of this name in the western Pacific, q.v.) is a perfect example of an atoll that has also been worked for guano. *Vostok* is a small reef island only 600 m across. Finally, *Flint Island* is also coral and reported to be about 15 m in elevation.

No volcanic materials are visible on any of the islands.

RHODES W. FAIRBRIDGE

References

Christophersen, E., 1927. "Vegetation of Pacific Equatorial Islands," *Bernice P. Bishop Mus. Bull.,* **44,** 45–46, 1–79.
Davis, W. M., 1928. "The coral reef problem," *Am. Geog. Soc. Spec. Publ.* **9,** 1–596.
Emery, K. O., 1956. "Marine geology of Johnston Island and its surrounding shallows, central Pacific Ocean," *Bull. Geol. Soc. Am.,* **67,** 1505–1519.
Gregory, H. E., 1925. "Malden Island," *Bernice P. Bishop Mus. Bull.,* **21,** 30–32.
Hague, J. D., 1862. "On phosphatic guano islands of the Pacific Ocean," *Am. J. Sci.,* Ser. 2, **34,** 224–243.
Holden, E. S., et al., 1883. "Report on the eclipse expedition to Caroline Islands," *Mem. Nat. Acad. Sci.,* **2,** 1–146.
Hutchinson, G. E., 1950. "The biogeochemistry of vertebrate excretion," *Am. Mus. Nat. Hist. Bull.,* **96,** 1–554.
Malahoff, A., 1971. "Magnetic lineations over the Line Islands Ridge," *Bull. Geol. Soc. Am.,* **82,** 1977–1982.
Marshall, P., 1911. "Oceania," *Handbuch Reg. Geol.,* 7(2), 1–36.
Power, F. D., 1925. "Phosphatic deposits of the Pacific," *Econ. Geol.,* **20,** 266–281.
Ritchie, G. S., 1958. "Sounding profiles between Fiji, Christmas and Tahiti Islands," *Deep-Sea Res.,* **5,** 162–168.

Cross-references: *French Polynesia; Oceania; United States–Hawaii.*

LOYALTY ISLANDS

Situated in French Melanesia, the Loyalty Islands (Îles Loyauté) extend from 20°20'S to 22°38'S and 165–169°E, in a NW-SE trend. The islands are administered from New Caledonia (Noumea). They rise from a NW-SE ridge that is aligned to the east and parallel to that of New Caledonia, and separated from it by a trough over 2000 m in depth, and on the other side by the much deeper depression of the New Hebrides trench that reaches 7660 m in depth.

The Loyalty Islands are all late Tertiary coral reefs and ancient atolls, crowning volcanic foundations. From SE to NW they comprise the following islands.

Walpole Island is an emerged coral reef rising to 75 m elevation in places and containing pockets of phosphate (derived from guano) that has been mostly exploited. A small intermediate point, *Durand Reef,* occupies a position midway between Walpole and Maré.

Maré Island is an elevated atoll that has been tilted during uplift to reach 60 m in the NW and 138 m in the SE (the windward side, exposed to the SE trade wind). On the NW there are numerous ancient passes, the former lee side (showing that the wind conditions were the same at its time of growth). In detail the old reef rim consists of little table reefs or "faros" (see *Maldives*), with a former lagoon depression in the middle. The reef cap consists both of corals grown in situ and detrital accumulations. The former lagoon, at 45–70 m elevation, is a plateau tilted NW, formed of detrital limestones and dolomites, with pockets of phosphate. In the middle there are three small outcrops of an olivine basalt, the only surficial trace of the island's volcanic foundation.

The emerged atoll slopes of Maré are

notched by former littoral terraces, benches, and undercuts at 15 levels, evidence of former sea levels. Those at 12, 8–9, 5, 3–4, 2–2.6, and 1.5 m are horizontal and are evidently post-deformational and interpreted as eustatic. The older terraces are tectonically tilted NW and slightly warped.

The basalt is K-Ar dated as 29 ± 4 m yr, thus late Oligocene or early Miocene. The oldest reef limestones near the basalt outcrops have been assumed to be late Miocene or Pliocene (but without paleontological control). There is a narrow fringing reef living around Maré today.

Tiga Island is a small elevated atoll, similar to the others, and marked by four ancient terraces at 72–78, 60–70, 28–34, and 4–5 m.

Lifu (or Lifou) *Island* is likewise an ancient atoll emerged to an elevation of 60 m. The original reef rim, 200–2000 m wide, encircles a former lagoon, now a plateau gently tilted SW. The older limestones contain foraminifera of probable Miocene age, including *Operculina, Amphistegina, Gypsina, Planorbulina, Calcarina,* and *Globigerina.* There are also more recent assemblages of Pliocene to early Pleistocene age. Five ancient marine levels have been recognized.

Ouvea Island is an ancient atoll tilted W but mainly submerged today. The former reef rim is only visible in the E, where it reaches an elevation of 46 m. It is built of massive and detrital corals and there is also a small sector of the former lagoon floor above sea level today. Much of the former atoll today represents the foundation for living reefs around a modern lagoon to the N, S, and W, the floor tilting WNW. The older limestone contains foraminifera of possibly Miocene age. It is notched by four eustatic terraces.

Beautemps-Beaupré Island is a half atoll built over an ancient atoll stump and tilted NW. There is only a small outcrop of the older limestone (Eo), in the SE, at 3.5 m above low-tide level.

Astrolabe Reefs are half atolls without any surface trace of their older foundations. A quite isolated drowned atoll, the *Petrie Reef,* appears in the same trend 200 km NW.

In summary, the Loyalty Islands are an extremely interesting series of Miocene-Pliocene coral reefs, former atolls, now elevated and tilted, rising from a tectonic ridge related to the New Caledonian orogenic belt. During the early Pleistocene they were warped, block-faulted, and tilted, to be affected in late Quaternary times by multiple eustatic notching.

J.-P. CHEVALIER

References

Chevalier, J.-P., 1968. "Géomorphologie de l'Île Maré," *Expédition francaise sur les récifs coralliens de Nouvelle Calédonie,* vol. 3. Paris: Fondation Singer-Polignac, 1–50.

Davis, W. M., 1928. "The coral reef problem," *Am. Geogr. Soc. Spec. Publ. 9,* 596p.

Haeberle, F. R., 1952. "Coral reefs of the Loyalty Islands," *Am. J. Sci.,* **250,** 656–666.

Koch, P., 1958. "Hydrologéologie des îles Loyauté," *Bull. Géol. Nouvelle-Calédonie* (Nouméa), **1,** 135–185.

Launey, J., and Recy, J., 1970. "Nouvelles données sur une variation relative récente du niveau de la mer dans toute la région Nouvelle-Calédonie-Îles-Loyauté," *C. R. Acad. Sci.,* Ser. D, **270**(18), 2159–2161.

Sarasin, F., 1917. *Neu-Caledonien und die Loyalty-Inseln.* Basel, 281p. Trans. to French with Roux, J.

____, 1917. *La Nouvelle-Calédonie et les Iles Loyauté.* Paris: Fischbacher, 296p.

____, 1925. *Nova Caledonica,* vol. 4. Berlin: Kreidel, 177p.

Cross-references: *Fiji; French Polynesia; New Caledonia; New Guinea; New Hebrides; Solomon Islands.*

M

MARIANA ISLANDS

The Mariana Island group forms a double arcuate chain of some 15 volcanic and volcanic-limestone islands in the western Pacific, approximately 2500 km east of the Philippine archipelago. The islands form the eastern limit of the Philippine Sea. Originally settled by Micronesian peoples, the islands have since passed through Spanish, German, Japanese, and American hands. Ferdinand Magellan, in 1521, was the first European visitor to the Marianas, which he named Islas de los Ladrones. At present all the islands except Guam are part of the U.S. Pacific Islands Trust Territory (Micronesia), with the seat of government on Saipan. *Guam* (q.v.), the largest and southernmost island, is a separate territory of the United States. The following geologic review is based on Cloud *et al.* (1956), Tracey *et al.* (1964), and Karig (1971), except when explicitly noted.

The islands stretch over a length of 800 km in two chains that form parts of the Mariana island arc system (about 12°30' to 21°N, and 144° to 146°E). The six southern uplifted volcanic-limestone islands lie in the eastern chain. Guam is the largest and southernmost of these but the eastern chain continues about 50 km to the SW and carries Santa Rosa Reef and Galvez Bank at depths less than 30 m. Ferdinand de Medinilla, a small raised limestone platform, is the northernmost exposed part of the ridge which continues at depths of 1500–2000 m northward to join the Bonin (Izu-Ogasawara) Ridge. Other major islands in this group include Rota, Tinian, Aguijan, and Saipan.

The geologic histories read from the strata of the eastern chain islands are relatively similar, but because the ridge is broken by cross-trending normal faults, individual blocks, in some cases consisting of entire islands, have somewhat different stratigraphic sections. Guam, which consists of at least three such blocks, is a good example.

Sea-floor spreading suggests that the underlying crust east of the Marianas is as old as Jurassic. On Guam, Saipan, and Tinian the oldest exposed units consist of volcanic and volcaniclastic suites that may be late Eocene or older. These rocks range in composition from dacites and other silicic varieties on Saipan to tholeiitic basalts on Guam. Andesitic rocks are widespread. The volcanic activity appears to have been both submarine and subaerial. Overlying and interfingering with the volcanic rocks are late Eocene shallow-water calcareous sediments. The eastern ridge thus came into existence near sea level by late Eocene time. Volcanism appears to have ceased or waned by the early Oligocene.

A second major period of volcanism occurred approximately from the late Oligocene to late-middle Miocene, as determined by outcrops on Guam and Saipan and by the results of deep-sea drilling in the Philippine Sea to the west (Fischer *et al.*, 1971). Rock types during this episode were basaltic to andesitic in composition. Limestones containing evidence of subsidence were deposited during the late Miocene and into the early Pliocene.

Uplift and the formation of fringing terraces began, probably in middle or late Pliocene, and have continued to the present with interruptions and minor reversals. Terrace heights reach a maximum elevation of slightly over 450 m on Saipan and Rota, and Mio-Pliocene limestone forms the highest peaks on Guam (400 m).

A third pulse of volcanism began in the western (inner) chain approximately in late Pliocene time and continues to the present. However, this volcanic activity is reflected in the contemporaneous sediments of the eastern ridge only as a diagnostic mineral suite in the insoluble fraction of the limestones (Schlanger, 1964), and underlines one difficulty in interpreting the history of arc systems from ridge-top geology.

The eruptive centers during earlier pulses of volcanic activity are thought to have been several tens of kilometers west of the present eastern chain, but this area is now occupied by a large scarp system separating the ridge from the deep oceanic Mariana Basin to the west.

Tectonic deformation on the eastern ridge is limited to normal faulting and tilting of large blocks, which generally have a westward component of dip. The only mineral resources of note on this group of islands are minor phosphate and manganese deposits on Saipan and Rota. Construction materials and traces of bauxite, ochre, and iron ore have been noted.

The northern group of islands are isolated and seldom-visited volcanic cones, stretching from Farallon de Pajaros, or Uracas, in the north to Anatahan in the south. Pagan has been mapped in some detail (Corwin et al., 1957), but the other islands have received only brief examination. This chain of volcanoes includes at least one more submarine center to the north of the subaerial group and several more submarine cones to the south, as far as the latitude of Guam (Corwin and Tracey, 1965).

These volcanoes apparently began erupting in late Pliocene time; Farallon de Pajaros, Guguan, Asuncion Agrihan, Pagan, and Esmeralda Bank, opposite Saipan, have been active during the recorded history of the region (Kuno, 1962).

Although these island arc volcanoes have been commonly referred to as calc-alkaline and andesitic, they are strictly neither. The predominant lava erupted is a low-alkali tholeiitic basalt with 48-52% SiO_2, and is distinctly calcic. Such rocks are typical of tholeiites described from intraoceanic island arcs (Jakes and White, 1971). Very few examples of more silicic eruptions have been reported. However, explosive eruptions are a typical mode of eruption, and the abundant fine fragmental material may prove to be more silicic than the flows. Tectonic uplift of these volcanic islands has not been reported. They range in form from composite stratocones reaching elevations of 965 m on Agrihan to calderas breached by the sea as at Maug.

The volcanic (inner) chain lies 30 to 40 km W of the crest of the limestone-volcanic (outer) islands along the entire length of the Mariana island arc system, rather than converging to the north as suggested earlier. Seismic reflection profiles demonstrate that these volcanoes are located along or slightly west of the boundary scarps of the eastern chain.

Nearly every feature of Pacific island arc systems is classically developed in the Mariana region. The eastern Mariana chain is a frontal arc separated from its trench by a slope broken into a smooth, deeply sedimented upper slope and a steeper lower slope that probably reflects tectonic disruption. The Mariana Trench, 150 to 200 km east of the frontal arc, reaches the greatest known oceanic depth of 11,034 m S of Guam (Challenger Deep). Earthquakes, believed associated with the underthrusting (subduction) of ocean crust and upper mantle beneath the trench, reach depths of more than 600 km and define a steeply dipping, irregular surface that steepens to vertical beneath the islands. West of the volcanic chain is the Mariana Basin, a zone of active sea-floor spreading that seems to be related to underthrusting in the trench, but in a way not yet understood.

Pulses of sea-floor spreading initiated near the volcanic chain of the Mariana arc system appear to have occurred several times during the Tertiary, opening a series of deep marginal basins. The southern part of the "Iwo Jima Ridge" of Hess (1948), here identified as the West Mariana Ridge (and in the *Times Atlas* as the "South Honshu Ridge"), is one of several remnants of the Mariana Ridge which were left behind as the arc system extended internally.

DANIEL E. KARIG

References

Bracey, D. R., and Ogden, T. A., 1972. "Southern Mariana arc: geophysical observations and hypothesis of evolution," *Bull. Geol. Soc. Am.,* 83, 1509-1522.

Cloud, P. E., et al., 1956. "Geology of Saipan, Mariana Islands," *U.S. Geol. Surv. Prof. Pap. 280-A,* 126p.

Corwin, G., and Tracey, J. I., Jr., 1965. "Marine geologic investigations near the Palau Islands and Guam, Mariana Islands," *Cruise Narrative and Scientific Results,* vol. 1. Intern. Indian Ocean Exped., U.S.C. and G.S. Ship Pioneer 1964.

_____, Bonham, L. D., Terman, M. H., and Viele, G. W., 1957. "Military geology of Pagan, Mariana Islands," U.S. Army Corps of Engineers, intelligence dossier, 259p.

Fischer, A. G., et al., 1971. *Initial Reports of the Deep Sea Drilling Project,* vol. 6. Washington: U.S. Govt. Printing Off.

Hess, H. H., 1948. "Major structural features of the western north Pacific, an interpretation of H.O. 5485, Bathymetric Hart, Korea to New Guinea," *Bull. Geol. Soc. Am.,* 59, 417-446.

Jakes, P., and White, A. J. R., 1971. "Composition of island arcs and continental growth," *Earth Planet. Sci. Lett.,* 12, 224-230.

Karig, D. E., 1971. "Structural history of the Mariana Island arc system," *Bull. Geol. Soc. Am.,* 82, 323-344.

Kuno, H., 1962. "Japan, Taiwan, and Mariana," *Catalogue of the Active Volcanoes of the World.* Naples: Intern. Volcanol. Assoc., 11, 245-252.

Larson, R. L., and Chase, C. G., 1972. "Late Mesozoic evolution of the Western Pacific Ocean," *Bull. Geol. Soc. Am.,* 83, 3627-3644.

Schlanger, S. O., 1964. "Petrology of the limestones of Guam," *U.S. Geol. Surv. Prof. Pap. 403-D,* 52p.

Tracey, J. I., et al., 1964. "General geology of Guam," *U.S. Geol. Surv. Prof. Pap. 403-A,* 104p.

Cross-references: *Caroline Islands; Guam; Izu-Ogasawara-Iwo Islands; Oceania; Pacific Islands Trust Territory; Saipan.*

MARQUESAS ISLANDS (ISLES MARQUISES)

Lying in the South Pacific Ocean approximately 10°S, 140°W, some 1200 km (740 miles) NE of Tahiti, the Marquesas Islands consist of 11 islands. A part of *French Polynesia* (q.v.), the combined area is 1280 km² (492 sq mi). Nuku Hiva is the largest island and reaches an elevation of 1185 m (3887 ft). Other islands include Hiva Oa, the second largest, on which the highest peak is 1073 m (4520 ft). Ua Pu reaches 1232 m (4043 ft) and Fatu Hiva reaches 1200 m (3940 ft). The southern cluster of islands is sometimes called the Mendana Islands.

The islands rise from a 200-km-broad NW-SE-trending submarine ridge that rises from over 4000 m and parallels those of the Tuamotus and Society Islands (qq.v.).

The island foundations are unquestionably oceanic volcanic accumulations (olivine basalts), probably beginning with the Paleocene (from sea-floor spreading evidence). In a second stage there was explosive volcanism on Nuku Hiva and Fatu Hiva, producing calderas and large craters accompanied by trachyte flows and extensive pyroclastics. A third stage, probably Quaternary, was marked by dike and sill injection, the petrology indicating progressive differentiation to acid types. The rocks, according to Lacroix (1928), are essentially nepheline-free. In this respect Chubb (1930) points out that they form a province related to Easter, Pitcairn, and Galapagos islands.

Most of the islands are marked by an uplifted marine abrasion platform that converts the interiors into infertile plateaus at elevations up to 800 m. This surface has been deeply weathered and altered to laterite. On Nuku Hiva an erosional platform is reported at about 400 m elevation, associated with foraminiferal sands. There is no uplifted coral reef, but there are coral sands at Tahvata at about 80 m elevation and on Na Huka at 20 m (Chevalier, personal communication).

Motu Iki is small but up to 220 m high and is largely covered with guano. Most of the islands are marked by a late Quaternary, 2-3 m, shore platform. There are very few coral reefs today inspite of the low latitude. This may in part be due to the occasional incidence of an eddy from the cold Peru (Humboldt) Current, and the lack of spawn from other sources of coral populations. The existing reefs are chiefly in the bays, only about 15 species being represented (mainly *Porites, Pocillopora,* and *Millepora,* according to Chevalier, personal communication).

RHODES W. FAIRBRIDGE

References

Adamson, A. M., 1939. "Review of the faunas of the Marquesas Islands and discussion of its origin," *Bernice P. Bishop Mus. Bull.,* 159, 1-39.

Aubert de la Rüe, E., 1958. "Observations sur le volcanisme tertaire et quaternaire de quelques îles de le Polynésie Française," *Bull. Volcanol.,* Ser. 7, 19, 159-178.

———, 1959. "Etude géologique et prospection minière de la Polynesie francaise," *Recherche Géologique et Minerale en Polynesie Francaise.* Paris: L'Inspection Génerale des Mines Géologiques, 7-42.

Chubb, L. J., 1930. "Geology of the Marquesas Islands," *Bernice P. Bishop Mus. Bull.,* 68, 1-71.

Lacroix, A., 1928. "Nouvelles observations sur les laves des îles Marquises et de l'ile Tubuai," *C. R. Acad. Sci.,* 187, 364-369.

Cross-references: *Easter Island and Sala y Gómez; French Polynesia; Galápagos Islands; Oceania; Pitcairn Islands; Society Islands; Tahiti; Tuamotu Islands.*

MARSHALL ISLANDS

Part of Micronesia, and of the U.S. Trust Territory of the Pacific, the Marshalls are mainly atolls with a combined land area of 1813 km (700 sq mi). Lying in the west central Pacific Ocean, about 3700 km WSW of Hawaii and 5500 km N of Auckland, New Zealand, the Marshall Islands are composed of 34 atolls and coral islands, disposed mainly in two parallel belts running NW-SE, 230 km apart, the Ralik (Ralek) chain and the Ratak (Radak) chain, respectively, on the W and E. In the Ralik chain there are 6 atolls and one coral island; in the Ratak Chain 13 atolls and one coral island; the rest are on isolated seamounts. The principal atoll in the former is Kwajalein and in the latter Arno. Bikini and Eniwetok, involved in nuclear bomb testing program in the 1950s, are the two most isolated northerly representatives of the Ralik chain.

The annual mean temperature is 27°C (81°F) and the average annual rainfall is 4060 mm (160 in) in the southern atolls and 2030 mm (80 in) in the northern islands.

The structural line of the Marshalls is continued to the SSE in the *Gilbert and Ellice Islands* (q.v.) but is terminated to the NNW except perhaps for a culmination of the submarine Mid-Pacific Mountains, *Wake Island* (q.v.). There is also a wide gap to the west, separating them from the *Caroline Islands* (q.v.). Sea-floor spreading data show that the underlying oceanic crust in this part of the Pacific is of late Cretaceous age.

Because of the nuclear testing program the northern Marshalls have been subject to the most intensive geological, geophysical, and ecologic surveys of any isolated island group in the world. Most significant perhaps are two deep bores put down on Eniwetok atoll which encountered the basalt seamount basement at 1287 m (4222 ft) and 1411 m (4630 ft).

The principal formations penetrated were Miocene and Upper Eocene, the latter just above the basalt. The original volcanic cone is believed to have been planed off by sea-level erosion. During the progressive submergence of the cone, shallow-water limestones and reef facies accumulated. There were several interruptions, notably in the Oligocene, when complete emergence occurred, and a tropical paleosol developed, marked by terrestrial snails comparable to those of the high forested islands of this latitude today (Ladd, 1958).

The major problem of the submergence of these Mid-Pacific atolls has not yet been solved. Kuenen (1954) suggests that each seamount foundation subsided separately. Charles Darwin and later Tester (1950) and Ladd (1958) concluded from the pattern of submergence that the entire central Pacific was drowned by as much as 350 m. Menard (1964) proposed that this region represented a former mid-ocean rise, which he called the "Darwin Rise," and which is supposed to have slowly subsided since early Tertiary time. Fairbridge (1961) suggests that the relative motion of the North Pole, from the area of Kamchatka, away from the Pacific section over the last 50–100 m yr could cause a geodetic rise of sea level over the whole Mid-Pacific quadrant. The rate of crustal compensation was, of course, always slower than the sea-level adjustment for each polar shift (probably in brief accelerations), so that at no time was the shift so rapid as to "drown" all the corals. Each shift of the polar axis covered could only be a few minutes of arc or else the reefs would have been submerged too deeply for continued upward growth. The total emergences that occurred from time to time must represent either reverse polar shift (toward the Pacific quadrant) or tectono-eustatic drops, related to plate tectonics. There is no evidence of broad tectonic uplift or subsidence in the central Pacific area.

The Marshall Islands have featured in extended studies of late Quaternary eustatic sea-level changes. Shepard et al. (1967) do not find support here for the concept of a high mid-Holocene level, for which there is more evidence in other regions (Fairbridge, 1961).

RHODES W. FAIRBRIDGE

References

Cole, W. S., 1957. "Large Foraminifera from Eniwetok atoll drill holes," *U.S. Geol. Surv. Prof. Pap. 260-V,* 743–784.

Dobrin, M. B., et al, 1954. "Seismic studie of Bikini atoll," *U.S. Geol. Surv. Prof. Pap. 260-J,* 487–505.

Emery, K. O., Tracey, J. I., Jr., and Ladd, H. S., 1954. "Geology of Bikini and nearby atolls," *U.S. Geol. Surv. Prof. Pap. 260-A,* 1–265.

Fairbridge, R. W., 1961. "Eustatic changes of sea level," in *Physics and Chemistry of the Earth,* vol. 4. New York: Pergamon Press, 99–185.

Fosberg, F. R., 1954. "Soils of the northern Marshall atolls, with special reference to the Jemo series," *Soil Sci.,* 78, 99–107.

____, 1957. "Description and occurrence of atoll phosphate rock in Micronesia," *Am. J. Sci.,* 255, 584–592.

____, et al., 1956. "Military geography of the northern Marshalls," U.S. Army Chief Eng., Intelligence Div., USAFFE, 1–320.

Johnson, J. H., 1954. "Fossil calcareous algae from Bikini atoll," *U.S. Geol. Suv. Prof. Pap. 260-M,* 537–545.

Kuenen, P. H., 1954. "Eniwetok drilling results," *Deep-Sea Res.,* 1, 187.

Ladd, H. S., 1958. "Fossil land shells from western Pacific atolls," *J. Paleontol.,* 32, 183.

____, 1965. "Tertiary fresh-water mollusks from Pacific Islands," *Malacologia,* 2(2), 189–197.

____, 1973. "Bikini and Eniwetok atolls, Marshall Islands," in *Biology and Geology of Coral Reefs,* Vol. 1. New York: Academic Press, 93–112.

____, Ingerson, E., Townsend, R. C., Russell, M., and Stephenson, H. K., 1953. "Drilling on Eniwetok Atoll, Marshall Islands," *Bull. Am. Assoc. Petrol. Geologists,* 37, 2257–2280.

Leopold, E. B., 1970. "Miocene pollen and spore flora of Eniwetok Atoll, Marshall Islands," *U.S. Geol. Surv. Prof. Pap. 260-II,* 1133–1185.

Menard, H. W., 1964. *Marine Geology of the Pacific.* New York: McGraw-Hill, 271p.

Nugent, L. E., Jr., 1946. "Coral reefs in the Gilbert, Marshall, and Caroline Islands," *Bull. Geol. Soc. Am.,* 57, 735–780.

Raitt, R. W., 1957. "Seismic-refraction studies of Bikini and Kwajalein atolls," *U.S. Geol. Surv. Prof. Pap. 260-S,* 507–527.

Revelle, R., and Emery, K. O., 1957. "Chemical erosion of beach rock and exposed reef rock," *U.S. Geol. Surv. Prof. Pap. 260-T,* 699–709.

Schlanger, S. O., 1963. "Subsurface geology of Eniwetok Atoll," *U.S. Geol. Surv. Prof. Pap. 260-BB,* 991–1066.

Shepard, F. P., 1970. "Lagoonal topography of Caroline and Marshall Islands," *Bull. Geol. Soc. Am.,* 81(7), 1905–1914.

____, et al., 1967. "Holocene changes in sea level: evidence in Micronesia," *Science,* 157, 542–544.

Tayama, R., 1953. "Coral reefs in the South Seas," *Bull. Hydro. Office* (Tokyo), 2(941), 292p.

Tester, A. C., 1950. "Marine terraces of the Pacific Ocean area," *18th Intern. Geol. Congr. Rept.,* pt. 8, 72.

Thurber, D. L., Broecker, W. S., Blanchard, R. L., and Potratz, H. A., 1965. "Uranium-series ages of Pacific Atoll coral," *Science,* 149(3679), 55–58.

Cross-references: *Caroline Islands; Gilbert and Ellice Islands; Pacific Islands Trust Territory; Wake Island.*

MARTINIQUE

One of the islands of the West Indies and a French overseas "départment," Martinique lies between Dominica and St. Lucia, about 210 km (130 miles) south of Guadeloupe. Martinique has an area of 1115 km^2 (431 sq mi). The island is thickly vegetated and mountainous with a high, active volcano, Mt. Pelée (1463 m; 4428 ft) and a nearby inactive cone, Mt. Carbet (1207 m; 3960 ft).

The climate is tropical, hot and rainy, from July to October with winds from the S. Winds are easterly from March to June, whereas from November to February the air is cooled by northerly winds. Average precipitation is 2200 mm, but lower on the southern coast and higher in the mountains. Notable hurricanes occurred in 1767, 1839, 1891, 1903, and 1928.

Situated at 14°40'N, 61°00'W, it lies just S of latitude 15°N, which is commonly taken as the boundary between the Windward and the Leeward islands. Thus Martinique is classified here in the Windward Islands (Iles-au-vent), but others place it in the Leeward Islands. The island is roughly ovoid, trending NW-SE, 62 by 32 km. Unlike so many of the Lesser Antilles, there are several deep embayments both in the E (Baie du Galion and Havre du Robert) and in the W (Baie de Fort de France and Cul de sac du Marin). The northern part of the island consists of youthful volcanics and the coasts are steep and cliffed in places. The southern coast, in contrast, is marked by terraced limestones and low cliffs with numerous coral reefs, with another of limestone islets and sand cays.

There are numerous torrential streams, and the upper slopes of the mountains are densely forested. The lower slopes are under cultivation, mainly sugar.

Martinique suffered one of history's most catastrophic events, when in 1902 Mt. Pelée erupted ash, gas, and nuées ardentes, completely destroying St. Pierre, the former capital (at its southern foot), killing 40,000 people. The term *pelean*-type eruption has since become standard for this type of volcanic activity. The lavas are entirely andesites and dacites.

The oldest rocks identified on Martinique include a great thickness of tuffs and limestone. Grunevald (1964, 1965) believes they may be Lower Eocene or even late Cretaceous because they contain *Belosepia*. In the conglomerates there are boulders of quartz diorite, gabbro, and mica schist, which may belong to the late Cretaceous or early Tertiary "basement" observed on St. Bartholomew (see *Guadeloupe*). The alternating tuffs and lavas range from Lower Oligocene up to mid-Miocene (probably Helvetian). According to Barrabé (1955) the sequence is mostly marginal to a volcanic foundation. There was WSW-ENE-trending deformation in the post-Middle Miocene stage.

Pliocene and Pleistocene high sea levels left raised beach deposits and locally elevated coral reefs.

JACQUES BUTTERLIN

References

Anderson, T., 1903. "Recent volcanic eruptions in the West Indies, *Geogr. J.*, **21**.

Barrabé. L., 1955. "Contribution à l'étude stratigraphique et tectonique des formations sédimentaires de la Martinique," *Rev. Inst. Fr. Petrol. Ann. Comb. Liq.*, **10**, 295-308.

Butterlin, J., 1959. "Microfaune et âge de deux formations calcaires de la Martinique," *C. R. Somm. Soc. Géol., France*, fasc. 2, 42-43.

Cossman, M., 1913. "Etude comparative de fossiles miocéniques recueillis à la Martinique et à l'isthme de Panama," *Bull. Conchyologie*, **61**, 1-64.

Fink, L. K., Jr., 1968. "Marine geology of the Guadeloupe Region, Lesser Antilles Island Arc," Ph.D. thesis, Univ. Miami.

Grunevald, H., 1964. "Géologie de la Martinique," *Thèse Fac. Sci. Univ. Paris*, Ser. A (4238), 144p. (map).

____, 1965. "Géologie de la Martinique," *Mém. Carte Géol. Fr. Dépt. Martinique*, 144p. (with maps).

Gunn, B. M., Roobol, M. J., and Smith, A. L., 1974. "Petrochemistry of the Peléan-type volcanoes of Martinique," *Bull. Geol. Soc. Am.*, **85**, 1023-1030.

Heilprin, A. 1903. *Mont Pelée and the Tragedy of Martinique*. Philadelphia: Lippincott, 335p.

____, 1908. *The Eruption of Pelée: A Summary and Discussion of the Phenomena and Their Sequels*. Philadelphia: Lippincott, 72p.

Laborde, D., 1912. "Sur l'existence de blocs calcaires métamorphosés dans les tufs ponceux de la Montagne Pelée," *C. R. Acad. Sci.*, **154**, 824-826.

Lacroix, A., 1904. *La Montagne Pelée et Ses Éruptions*. Paris: Acad. Sci., 662p.

____, 1908. *La Montagne Pelée après Ses Éruptions*. Paris, 136p.

MacGregor, A. G., 1938. "The volcanic history and petrology of Montserrat, with observations on Mt. Pelée, in Martinique," *Phil. Trans. Roy. Soc. Lond.*, Ser. B, **229**, 90p.

Perret, F. A., 1935. "The eruption of Mt. Pelée 1929-1932," *Publ. Carnegie Inst. Washington*, **458**, 126p.

Revert, E., 1949. *La Martinique*. Paris: Nouvelles Editions Latines, 559p.

Cross-references: *Dominica; Guadeloupe; St. Lucia; West Indies.*

MEXICO

Mexico, with an area of 1,969,269 km^2 (758, 259 sq mi), is the smallest of the three countries that form the continental mass of the

North American continent. The country lies in the southern one-sixth of the continent and comprises mainland Mexico, the peninsulas of Baja (Lower) California and of Yucatán, and more than 250 islands. Physiographically speaking, the northern limit of Central America is formed by the Isthmus of Tehuantepec, but the political boundaries of Mexico extend far beyond it, so that in this sense Mexico also occupies the northern part of Central America. On the other hand, considering strictly geological provinces and major morphotectonic units, such as, for example, the Sierra Madre Oriental fold belt, the continuity of which can be identified in northern Guatemala, in Belize (British Honduras), and even in northern Honduras, one should really extend the southern limit of continental North America practically to latitude 15°N. It is clear, therefore, that no simple, universally acceptable boundary is available, and it is optional to consider that part of Mexico beyond the Isthmus of Tehuantepec either as part of continental North America or as part of the Central American isthmus. Furthermore, anthropological considerations, based on the distribution of the Maya Indians and their culture, which are more widespread in the northern Central American countries, tend to support a Central American classification of Mexico beyond the Isthmus of Tehuantepec, thus including also the Yucatán Peninsula.

Physiography

Several attempts to divide Mexico into physiographic provinces have been made, but the most realistic one is that of Raisz (1959), who first prepared a fairly detailed physiographic diagram from aerial photographs covering the entire country and, later, on the basis of these data, supplemented by published maps and articles, introduced the subdivision as shown in Fig. 1.

Baja California is a peninsula, about 1700 km long and about 50–100 km wide, which extends SE from the California border, between the Pacific Ocean on the west and the Gulf of California (Golfo de Cortés to the Mexicans) on the E. It is predominantly a desert to semi-desert country. Its mountain ranges are great granitic batholiths, which are flanked on the W by Mesozoic metamorphic and sedimentary rocks and elsewhere by Cenozoic volcanic and sedimentary rocks. The peninsula as a whole can be visualized as a great fault block that became separated from mainland Mexico during the last 50 m yr or so.

FIGURE 1. Physiographic provinces of Mexico (after Raisz, 1959).

FIGURE 2. View of a portion of the Sierra Madre Occidental, from the Durango-Mazatlán highway. (Photo: Zoltan de Cserna.)

The area of the *Buried Ranges* embraces primarily the Sonoran Desert, a region characterized by the presence of isolated and deeply dissected mountain ranges of granitic and metamorphic rocks, disposed in N-S-trending fault blocks separated by wide valleys or basins filled with late Cenozoic alluvium and locally even with wind-blown sand—an *Inselberg* landscape. This province resulted in part from the same processes that produced the Basin and Range Province of the southwestern United States and, in part, from the abundant debris supply from the adjacent Sierra Madre Occidental on the east.

The *Sierra Madre Occidental* is a medially dissected region of coalescing volcanic plateaus, made up primarily of rhyolitic and rhyodacitic pyroclastics flows, which are either horizontal or slightly tilted, with an average altitude of 2000 m above sea level (Fig. 2). As the principal rain systems come from the W and drop their moisture content by the time they reach the highest summits, the Basin and Range area lying to the E, as well as parts of the Central Mesa, are areas of rain shadow.

The *Basin and Range Province* is similar, although not identical, to that of the *Buried Ranges*. Its similarity consists in that it is also a southward extension of the Basin and Range Province of the southwestern United States into Mexico. It consists of NW-SE-trending isolated mountain ranges, which are separated by wide valleys or bolsones—mainly of interior drainage—that resulted from Tertiary block faulting. However, the mountain ranges are composed of strongly folded Mesozoic sedimentary rocks, principally Lower Cretaceous limestones. This province lies between the Sierra Madre Occidental on the W, the Sierra Madre Oriental on the E, and by the Cross Ranges of the Sierra Madre Oriental on the S. The average elevation of this province is approximately 1500 m above sea level.

The *Central Mesa* is the counterpart of the *Basin and Range Province,* S of the *Cross Ranges,* and extends practically to latitude $20°N$, to the northern limits of the *Neovolcanic Plateau.* The Cenozoic basin fill in this province is greater than in the Basin and Range area, which is primarily due to the entirely endorheic nature of its drainage. In later Cenozoic times the Neovolcanic Plateau acted as a huge E-W-trending dam, blocking the earlier southward drainage of the province. In the southern portion of this province lies the *Bajio,* which can be considered one of the bread baskets of Mexico.

The *Sierra Madre Oriental Province* represents the Mexican portion of the Mesozoic foreland fold and thrust belt of the North American Cordillera. It consists of NNW-SSE-trending anticlinal mountain ranges separated by synclinal valleys. Some of the mountain ranges reach altitudes of around 3000 m above sea level, although the average elevation of the belt is around 2000 m. The belt is developed on the site of the late Jurassic-early Cretaceous miogeosyncline next to the Paleozoic craton, which lay to the E and acted as a foreland during the early Eocene orogeny. The Sierra Madre Oriental extends from southwest Texas (Quitman Mountains) to the Guatemala border, and continues beyond that into northern Guatemala, southern British Honduras (Belize), and to northern Honduras. It is bordered on the E and N by the Gulf Coastal Plain and the Yucatán Peninsula. Its western and southern boundaries are the Basin and Range, the Central Mesa, and the Sierra Madre del Sur physiographic provinces.

With the exception of its northern part, which was filled chiefly by the Río Grande (Río Bravo) drainage system, the *Gulf Coastal Plain* is a lowland, formed by clastic Cenozoic sediments that were derived by erosion from the Sierra Madre Oriental. The Cenozoic sediments have a general gulfward dip and one approaches the Gulf of Mexico crossing over progressively younger rocks. Near the latitude of the Tropic of Cancer, broadly folded Mesozoic rocks stand out from the surrounding Cenozoic terrain (Sierra de San Carlos and Sierra de Tamaulipas), whereas farther S and SE, late Cenozoic volcanics, chiefly basalts, rest on the sediments (Tuxtla volcanoes). In the northern portion of the Isthmus of Tehuantepec, salt domes intruded the Cenozoic sediments, contributing important oil and mineral wealth in that region.

The *Neovolcanic Plateau* crosses Mexico in a roughly E-W direction from the Pacific Ocean to the Gulf of Mexico between latitudes 18 and $20°N$. The highest mountains of the country are in this province, such as the two magnificent stratovolcanoes, Orizaba (5747 m) and Popocatépetl (5452 m; Fig. 3). The province is a late Cenozoic feature; the earliest volcanic products consist mainly of rhyodacitic composition, whereas the lastest ones are of olivine basaltic composition. The volcanic pile, which consists of coalescing lava flows and associated volcaniclastics, has an average maximum thickness of over 1000 m, whereas the province as a whole has an average elevation of about 2000 m above sea level. The principal cities of Mexico, including the capital of the Republic, are located on this province.

The *Sierra Madre del Sur Province* embraces practically all of the remainder of Mexico S of the *Neovolcanic Plateau* and SW and S of the *Sierra Madre Oriental.* It is a maturely dissected

FIGURE 3. Popocatépetl volcano, as seen from the west. (Photo: Zoltan de Cserna.)

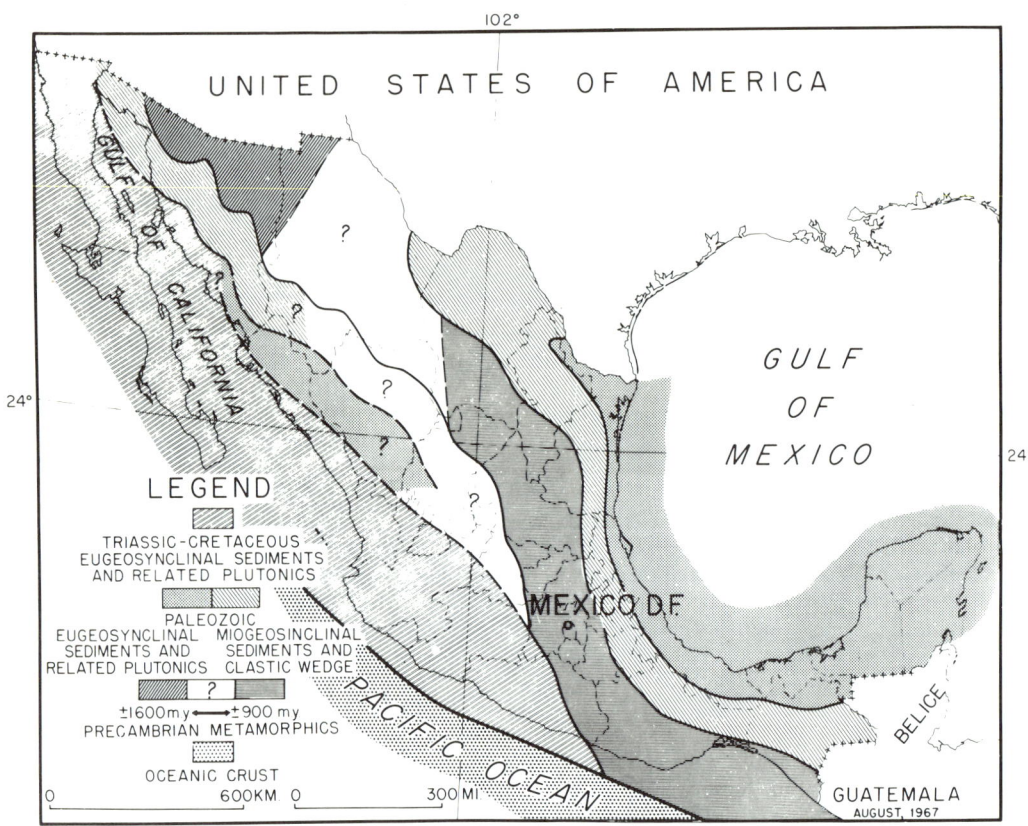

FIGURE 4. Precambrian-Mesozoic structural belts in Mexico (from de Cserna, 1969).

region drained by river systems going into the Pacific Ocean, such as the rivers Balsas and Atoyac. Were it not for the effusion of the Neovolcanic Plateau in later Cenozoic times, probably the Central Mesa S of the Cross Ranges would also have the same dissected topography as that of the present Sierra Madre del Sur. The rocks in this province consist of Precambrian and Lower Paleozoic metamorphic and granitic rocks, folded Mesozoic marine and continental sedimentary and volcanic rocks, Tertiary continental clastics, and chiefly acid volcanics. The picturesque southern coast (*costa brava*) of Mexico is essentially a drowned coastline, with relatively small stretches of locally developed coastal plain.

The *Yucatán Peninsula* is essentially a low platform, composed of flat-lying Miocene limestones, studded with sinkholes (*xenotes*) with the general physiographic characteristics and geologic history of the peninsula of Florida.

The two principal natural boundaries of Mexico, the Pacific Ocean on the W and S and the Gulf of Mexico on the E, have characteristics that are radically different. The continental shelf along the Pacific coast is virtually nonexistent, whereas along the Gulf of Mexico it is very wide and at places reaches more than 100 km in width.

Geologic Evolution

Precambrian. The oldest rocks that have been identified in Mexico are granitic and metamorphic and are well distributed in northwestern Sonora. The terrain of these rocks doubtless formed part of a medial Precambrian (±1700 m yr) craton that extended into Mexico from the southwestern United States. This craton developed from a eugeosynclinal belt that became consolidated around 1700 m yr ago, and during subsequent Precambrian times experienced erosion and peneplanation.

In the eastern half of Mexico, during later Precambrian times, another geosynclinal belt came into being, the *Oaxacan structural belt*. It received great volumes of clastics derived from the older Precambrian terrain to the W. About 150 km to the E of the margin of this older Precambrian terrain, active vulcanicity prevailed in a roughly N-S-trending belt, whose products, chiefly tuffs, also accumulated in this geosyncline, which probably was continuous with the Grenville geosyncline of southeastern Canada and northeastern United States. As there are no definite data through which this continuity could be conclusively proved, this paleogeographic interpretation is based on the petrology, structure, and radiometric age data gathered from the rocks. This Oaxacan structural belt was formed about 900 m yr ago and became welded to the W-lying older Precambrian terrain, thus enlarging the older craton, which together with it controlled sedimentation and tectonics during the succeeding Paleozoic era (Fig. 4).

Paleozoic. The Paleozoic sedimentary sequence actually started earlier in northwestern Mexico during latest Precambrian time with an accumulation, about 2000 m thick, of *Collenia*-bearing clastics and carbonates which pass upward without a break into the Lower Cambrian *Olenellus*-bearing carbonates. In that region, which comprised a platform area transitional to the W-lying miogeosyncline, the total aggregate thickness of the Paleozoic is about 2700 m. It consists chiefly of carbonates, with several major disconformities that came about primarily by temporary withdrawal of the seas.

The Paleozoic rocks along the present west coast of mainland Mexico have been identified in only a very few places where they are isolated remnants; their erosion and destruction was favored by the geologic events that took place in the region during the Mesozoic. Moreover the widespread Tertiary volcanic cover in the W, which makes up the Sierra Madre Occidental, further obliterated what was left by the end of the Mesozoic.

In northeastern and eastern Mexico the oldest Paleozoic rocks heretofore identified are of Tremadocian (Upper Cambrian-Lower Ordovician) age and are near the city of Oaxaca. Elsewhere the oldest sedimentary rocks are of Ordovician or Middle Silurian age. The lack of older sedimentary rocks is probably due to the topographically high position of the Mexican Precambrian craton in Cambrian times, and because the sea from the east did not reach W beyond the present Sierra Madre Oriental. In northern Chihuahua State sedimentation was practically continuous throughout the Paleozoic. The sediments are chiefly carbonates, about 1000 m thick, which were deposited in a shelf environment of the miogeosynclinal part of the Mexican Paleozoic Huasteca orthogeosyncline.

In east-central Mexico, near the latitude of the Tropic of Cancer, the Paleozoic sequence ranges from the Middle Silurian to Permian, with a total thickness of about 1500 m. The pre-Middle Silurian basement is formed by Precambrian gneiss of Grenville age, and by chlorite schists that have tectonic contacts with the gneiss and are tentatively considered to be Taconic. The upper part of the Silurian is composed of limestone and shale that grade into Devonian shale, novaculite, and sandstone, and finally into a Lower Mississippian flysch. The Pennsylvanian flysch was deposited over an erosion surface and grades upward into the Lower Permian flysch. In southeastern Puebla and northwestern Oaxaca, the Paleozoic sequence is composed of a Tremadocian calcareous shale (30 m), which is uncomformably overlain by about a 600-m-thick Devonian-Permian flysch. Near this area, however, the Precambrian basement gneiss of Grenville age is covered by an allochthonous mass of metasediments and metavolcanics, which in turn are overlain uncomformably by Permo-Carboniferous flysch that contains a few crinoid-bearing carbonate horizons. The radiometric dating gave an Ordovician age to the metavolcanics of the allochthon, indicating that it is Taconic. In southeasternmost Mexico, near the Guatemalan border, the oldest Paleozoic rocks are a Pennsylvanian (?) flysch, about 2000 m thick, which is overlain by a 1500-m-thick Lower Permian marine carbonate sequence.

The Paleozoic magmatic activity is evidenced by vulcanicity in the eugeosynclinal portion of the Huasteca orthogeosyncline during Cambrian (?) and Ordovician times, the rocks of which are preserved in the Taconic allochthons in Tamaulipas and adjacent Nuevo León and in southeastern Puebla. A late Ordovician granodiorite stock has been recently identified in southeastern Puebla, and in the states of Oaxaca and Chiapas several granitic plutons are suspected to be of Middle Paleozoic age. A similar mid-Paleozoic granite pluton is known in the vicinity of Manzanillo in Colima. Detrital zircon concentrate, from Upper Cretaceous sandstone in north-central Mexico, also yielded a Middle Paleozoic age, suggesting that these were derived from a Middle Paleozoic granitic source that is concealed today by Tertiary volcanics.

The stratigraphic record and local geological relations indicate that the first orogenic event of the Paleozoic took place in eastern Mexico, roughly between latitudes 18 and 24°N, along the present site of the Sierra Madre Oriental and adjacent Gulf Coastal Plain. Probably as a result of the formation of a welt in the eugeosyncline that had risen progressively during the Ordovician, this orogeny was marked by gravity sliding from E to W and permitted allochthonous masses to override the Precambrian craton. This tectonic event has taken place after the early

Ordovician and prior to the deposition of the Middle Silurian, correlating thus with the Taconic orogeny of the northeastern United States and Canada.

The second diastrophic event is recorded for the end of the Mississippian, for there is a pronounced unconformity between the Lower Mississippian and the Pennsylvanian. This phase, which consisted of folding, is comparable to the first deformational events of the late Paleozoic (Appalachian orogeny), also recorded for eastern and southeastern North America.

The third and most important deformation in eastern Mexico during the Paleozoic has taken place during latest Permian times, which probably also encompassed the early Triassic. This deformation, which is responsible for the final consolidation of the Huasteca structural belt, is called the *Coahuilian orogeny* and consisted of folding and thrusting from the E to the W.

The stratigraphic record for northwestern Mexico indicates three events that took place in the western geosynclinal belt during the Paleozoic. However, the existing stratigraphic gaps are not easy to evaluate in terms of tectonics. There is an unconformable relation of the Upper Devonian on top of the Middle Cambrian and here and there remnants of Lower Ordovician are in central Sonora State. It appears, therefore, that the region experienced tilting and large-scale erosion sometime during the Silurian and early Devonian. A second period of uplift and erosion took place at the end of the Mississippian and beginning of the Pennsylvanian; these were probably related to the emplacement of granitic stocks and batholiths. The last Paleozoic event took place at the very end of the Permian and consisted of folding and faulting; this deformation is referred to as the *Sonoran orogeny*.

Mesozoic. Sometime after the Permian and prior to the late Triassic, Mexico was divided into four segments by four WNW-ESE-trending transcurrent faults, the northernmost of these being the *Texas lineament,* a right-lateral fault system, the next to the S the left-lateral *Torreón-Monterrey fracture zone,* followed in the S by the right-lateral *Zacatecas fracture zone,* and, finally, the left-lateral *Mexico fracture zone.* Westward movement along the Texas lineament was the greatest amounting to about 400 km, which gradually decreased southward to about an estimated distance of less than 100 km along the Mexico fracture zone. These movements were doubtless related to the opening of the Atlantic Ocean and to the drifting of the Americas westward.

As a result of lateral displacements in the Precambrian-Paleozoic basement or craton, the Mesozoic framework of sedimentation became controlled by the new distribution of the positive or topographically high areas. Whereas the Paleozoic sedimentary source areas appear to have been developed along the E or present Gulf of Mexico side of the country and of lesser importance in the NW and W, the Mesozoic stratigraphic record indicates that geologic developments on the W controlled most of what happened over the rest of the country.

In northwestern Mexico a 1500-m-thick evaporite and coal-bearing clastic sequence of late Triassic age changes westward into a normal marine shale-carbonate sequence, about 1700 m thick, which accumulated in the miogeosyncline. Farther W, a eugeosynclinal belt bordered this miogeosyncline; this can be deduced from the presence of Upper Triassic quartzites, slates, and metavolcanics in Baja California, which in late Triassic times was adjacent to Sonora.

In central and east-central Mexico, on and right next to the fold and thrust belt produced by the Coahuilan orogeny and subsequently offset by the transcurrent faults, continental clastics in the form of red beds and associated volcanics accumulated in graben structures in late Triassic-early Jurassic times. On the site of the present Gulf Coastal Plain, black shales accumulated during the early Jurassic in an euxinic environment.

Toward the end of the early Jurassic epoch, a tectonic land started to form in the eugeosyncline resulting from metamorphism and anatexis. This land began to rise and enabled the gliding of one or of a series of allochthonous masses, composed of the western volcanic assemblage, to the east, over the Precambrian-Paleozoic craton. These allochthons did reach the central parts of Mexico, as they have been identified as far east as the cities of Zacatecas and Guanajuato. The eastward-moving slides or thrust plates also produced gentle folds in the red beds, volcanics, and black shales, which were in front of them to the E. This late early Jurassic-early middle Jurassic deformation is referred to as the *Zacatecas thrusting event* of the *Mexican geotectonic cycle.* During this cycle the Gulf of Mexico began to open. The Triassic red beds in graben fills in northeastern Mexico suggest analogies with the Newark Group of eastern North America. According to Moore and Castillo (1974), the Gulf is believed to have become about half-opened during the Jurassic. Ocean circulation became restricted and vast salt deposits accumulated, extending from Alabama to the southern Gulf.

Following this deformation a molasse sedimentation began in eastern Mexico, in a sinuous

FIGURE 5. Sierra de Minas Viejas, north of Monterrey, Nuevo León. In the foreground a diapiric mass of Upper Jurassic evaporite and associated red beds, forming the core of the anticline. The circular strike valley is developed on the Upper Jurassic and Lower Neocomian sediments. The prominent cliffs are formed by Barremian reef limestones (Cupido Limestone). (Photo: Zoltan de Cserna.)

and irregular basin, during the remainder of the Middle Jurassic with the accumulation of coal-bearing continental clastics, evaporites, and, finally, of normal marine sedimentation beginning with the Upper Jurassic (Oxfordian). This late Jurassic-Neocomian basin is generally called the *Mexican geosyncline,* an autogeosyncline bordered on the W by Precambrian and Upper Triassic-Lower Jurassic metamorphics, and on the E by deeply eroded and highly deformed Paleozoic miogeosynclinal sediments and flysch terrain. The maximum thickness of the nonmarine portion of this sequence (Middle Jurassic) is about 800 m, whereas the thickness of the evaporites and clastics that accumulated in northeastern Chihuahua, eastern Coahuila, southern Nuevo Leon, southern Veracruz, and in adjacent Tabasco is estimated to be around 3000 m (Fig. 5). The marine, mainly carbonate, sequence with a basal clastic blanket, representing the Upper Jurassic and Neocomian, reaches 1200 m in thickness and its Barremian portion locally contains important reef developments.

Starting in the late Jurassic in the west, a new eugeosynclinal regime came about that continued without interruption until the end of the early Cretaceous. This belt was W of the late early Jurassic-early Middle Jurassic tectonic land or welt in which argillites and volcanics accumulated with associated ophiolites. This sequence is more than 3000 m thick and is preserved metamorphosed in various parts in the western half of the Baja California peninsula.

As a result of a notable rise of the level of the sea toward the end of the early Cretaceous, during the Albian all of Mexico became flooded. The transgression was gradual and progressive, since the Aptian sediments around the former positive areas contain clastics, whereas over the positive areas there is a fairly thick succession of evaporites, the accumulation of which continued into early Albian. The Albian-lower Cenomanian sequence is essentially a carbonate (Fig. 6) about 600 m thick, containing three well-developed facies, which are (1) a rudistid reef facies on the edges of the former positive areas, (2) a miliolid-bearing lagoonal facies over the positive areas, and (3) a deep-water cherty limestone facies. The western boundary of this late early Cretaceous miogeosyncline was along a NW-trending line that would connect the junctions of the states of Oaxaca and Guerrero, and Sonora and Chihuahua.

FIGURE 6. Cañón del Sumidero with the river Grijalva in the bottom. This prominent canyon is near Tuxtla Gutiérrez, Chiapas, and is cut in the Albian limestones. (Photo: Zoltan de Cserna.)

About 100 m yr ago, just on the turn from early to late Cretaceous, the eugeosyncline underwent regional metamorphism and anatexis in the former site of Baja California. This process brought about a general rising of that region which progressed toward the E with time. As a result the clastics that were derived from the rising land were dumped eastward into the progressively sinking, and E-shifting trough or exogeosyncline, constituting a clastic or flysch wedge. The magnafacies of this flysch wedge coarsen upward and are overlapping eastward, covering all areas previously occupied by the miogeosyncline. The maximum thickness of this flysch wedge is more than 6000 m; it abruptly thins and diappears to the W and gradually thins to the E.

Cenozoic. Progressive uplift in the W is estimated to have attained an elevation about 3000 m above sea level by the end of the Paleocene. A corresponding sinking in the E produced a slope with an easterly dip. During early Eocene time, this slope became a glide plane over which the Upper Jurassic-Paleocene sequence slid from W to E and became folded and locally even thrusted as a result of resistance offered by the late Jurassic-Neocomian positive areas formed by Paleozoic deformed rocks, which acted as buttresses. This deformation took place during the early Eocene and constitutes the *Hidalgoan orogeny*. The foreland fold and thrust belt thus produced is the Sierra Madre Oriental. The tectonic style, observed today in this fold belt, is the result of several factors that intervened, which are (1) the overall distribution of the massifs; (2) the proximity of the massifs and hence crowding of the folds in front of them in the W and locally on the S (that is, the Cross Ranges); (3) the presence and distribution of the lowermost Upper Jurassic and of the Upper Aptian-Lower Albian evaporites; and (4) Upper Cretaceous granitic plutons, mainly in the Basin and Range Province and in the Central Mesa, the emplacement of which produced crustal folds (Fig. 7).

In western Baja California, the postmetamorphic and batholithic rocks consist of about 1000 m of Upper Cretaceous flat-lying conglomerates, sandstones, and shales, which are covered by Tertiary, also flat-lying, sediments.

Following the Hidalgoan orogeny the Sierra Madre Oriental fold belt gradually emerged and erosion set in (Fig. 8). A good part of the detritus was carried out to the Gulf of Mexico side of the fold belt and dumped into the Gulf, building up gradually what is today the Gulf Coastal Plain. Several major river systems contributed to the building of the coastal plain, such as the Río Grande in the north, the Pánuco and Tuxpan rivers farther south, the Papaloapan and Coatzacoalcos rivers in the Veracruz area, and the Grijalva and Usumacinta rivers to the E of the Isthmus of Tehuantepec. The coastal plain sediments constitute a molasse, as they grade from coarse to fine upward and at the same time become gradually more marine. The thickest Tertiary sedimentary sections in the Gulf Coastal Plain reach about 10,000 m; they are flat-lying with a regional Gulfward dip and are undeformed, except for local small high-angle faults or gentle uparching due to salt tectonics in the northern parts of the Isthmus of Tehuantepec. The oil and sulfur-producing salt domes in the isthmus began to form as a result of this sediment loading over the Upper Jurassic salt, which continued to rise, with some temporary halts, throughout the Cenozoic.

The southern portions of the Yucatán Penin-

FIGURE 7. Simplified tectonic map of Mexico.

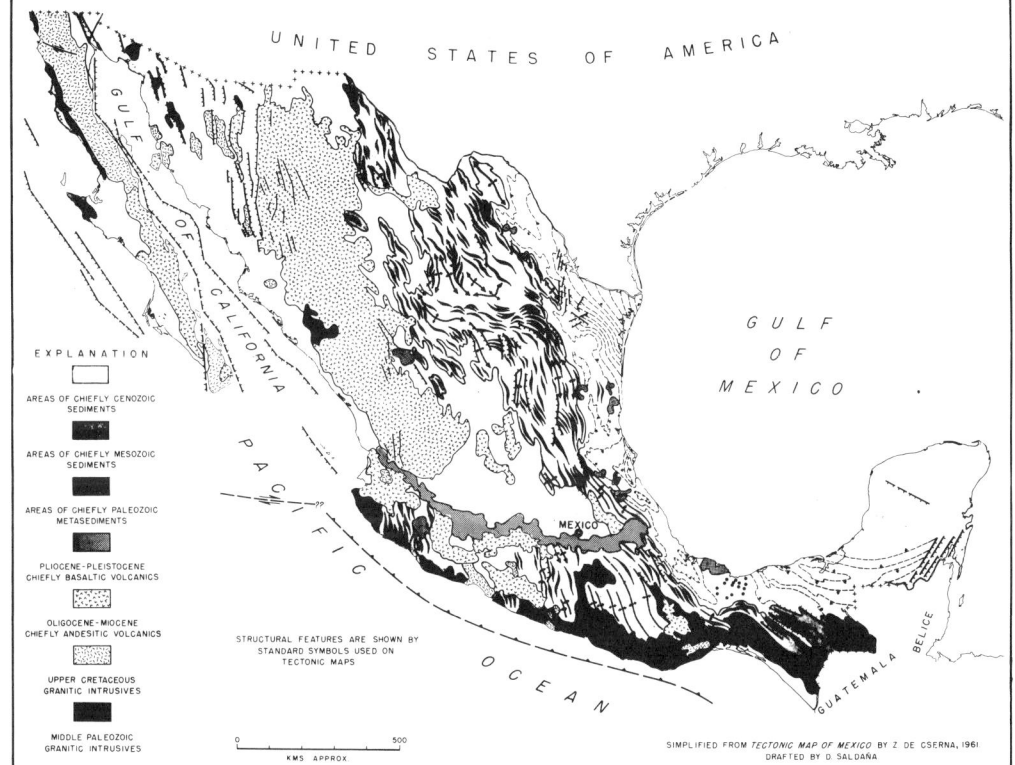

sula, most of which belongs to Guatemala and British Honduras (Belize), between the Hidalgoan fold belt and the present central part of the peninsula was the site of clastic sedimentation and evaporite accumulation during the Tertiary, prior to the Miocene. During the Miocene, the peninsula acted as a platform over which a 400-m-or-so-thick limestone blanket was deposited and which is today preserved undeformed. During the Pliocene and Pleistocene, this platform gradually emerged and attained its present elevation.

During the late Eocene and early Oligocene, central Mexico, comprising the Basin and Range Province, the Central Mesa, and the Balsas-Mexcala Basin, received the erosion débris of the fold structures that was not carried to the Gulf of Mexico, in endorheic basins that developed as a result of block faulting and attendant drainage blocking. Some of the grabens formed this way contain as much as 3000-m-thick continental red beds, horizontally or gently tilted; in places there are evaporites and associated diabasic, basaltic, or rhyolitic volcanics and small intrusives.

Vulcanicity became more important toward the end of the Oligocene and lasted throughout the Miocene. The most intense activity was centered in the western part of Mexico and diminished gradually toward the E. The products of this activity, about 2000 m thick, which are chiefly rhyolitic and rhyodacitic pyroclastic flows becoming andesitic toward the top of the sequence, built up what is known today as the *Sierra Madre Occidental*. Products of this vulcanicity did not reach farther E than the Sierra Madre Oriental. However, the Middle Tertiary sediments in the central parts of the Gulf Coastal Plain do contain some reworked tuffs.

At about this time the peninsula of Baja California began its journey northwestward, with the gradual opening of the Gulf of California. Although the exact dating of the opening of the Gulf of California is not yet established, the overall geological and geotectonic framework tend to indicate that most of the volcanics, both of the peninsula and of the islands of the gulf, are of middle Tertiary age and are quite similar to those that make up the Sierra Madre Occidental. Thus the present-day structure and tilted nature of Baja California must have come about after the Middle Terti-

FIGURE 8. Hausteca canyon, looking W (15 km W of Monterrey). Lower Cretaceous limestone exposed in frontal anticline of the Sierra Madre Oriental.

ary, as a result of movements along the southern continuation of the San Andreas fault system and the northern end of the East Pacific Rise.

At the north end of the East Pacific Rise a rather complex system of movements came about, which not only led to the opening of the Gulf of California through a series of transform faults, but also facilitated the formation of another transform fault with a major left-lateral component which constitutes the Jalisco-Nicoya fault and which actually is responsible for the present abrupt termination of Mexico against the Pacific Basin on the south.

Shortly after the formation of the Jalisco-Nicoya fault and as a result of renewed movements of Mexico in a southwesterly direction due to drifting, a subduction zone began to form with the development of the Mexico-Mesoamerica trench. In a line paralleling this trench, at a distance of about 250 km to the north, a series of volcanoes began to form during the Pliocene and continued their activity into modern times. These volcanoes and their associated lava flows and volcaniclastics built up the trans-Mexico volcanic belt (i.e., Neo-volcanic Plateau). The earliest products of this vulcanicity were rhyodacitic in composition and became gradually more basic, attaining olivine basaltic composition during the Pleistocene and Holocene. This line of volcanoes and related rocks constituted a huge dam across Mexico and blocked up the southward drainage of most of the central part of the country. Consequently, further filling of the Central Mesa by continental clastics took place during the Pliocene and Pleistocene, in contrast to the region south of this volcanic belt.

Since the end of the Miocene, the Yucatán Peninsula gradually emerged and attained its present configuration. Likewise, marine erosion and gradual filling along the southern coast of Mexico resulted in the development of narrow coastal plains from Jalisco to Chiapas. By the Pliocene, as the Gulf of California attained its approximate present configuration, headward erosion of the streams progressed eastward far enough to produce young dissection of the Sierra Madre Occidental and with their sediments to build up a narrow coastal plain in Sonora and Sinaloa.

During the Quaternary Mexico came under

the influence of the sharply contrasting, global climatic oscillations. During the cool maxima the snow line fell at least 1000 m and the highest mountains developed small glaciers; tills and moraines are found on Iztaccihuatl down to 2450 m and on Malinche to below 2630 m (Heine, 1973). Holocene "neoglacial" activity is reported on both at elevations around 4000–4400 m. Orizaba (5747 m) and Popocatepetl (5452 m) are regularly snowcapped today. At lower elevations, for example, in the central valley of Mexico (in the Neovolcanic Plateau), there was an alternation between aridity during glacial maxima and pluvial conditions during the solar radiation maxima. Large lakes formed and it is on poorly consolidated lake deposits that Mexico City is built, thereby posing extensive foundation problems. The wide extension of these lakes in mid-Holocene time was particularly favorable to the flowering of the Aztec and other cultures. Since that last pluvial climax there has been an oscillating regression of the lakes accompanied by expansion of the arid conditions in the northern part of the country.

Mineral Deposits

Mexico is a favored country as far as some minerals are concerned; however, it is very poor with respect to groundwater. The country has been the chief silver producer of the world for many years. Silver occurs chiefly with lead and zinc in manto-type deposits within the Upper Jurassic (Oxfordian) and Lower Cretaceous (Albian) limestones. However, the Pachuca district, which includes the biggest single silver mine of the country, is in Tertiary volcanics. The Albian limestones at several localities contain sizable iron deposits in conjunction with early late Cretaceous granitic plutons, although the most important production of iron ore has come from the apatite-bearing martite deposit of Cerro Mercado. The Upper Jurassic is noted for its large marine phosphorite reserves in north-central Mexico, whereas east-central Mexico contains a large manganese deposit. Mercury, fluorite, and tin are related to the Tertiary volcanic activity. Although some copper is recovered from the lead-zinc-silver ores, to date Mexico has only two disseminated copper deposits, which are in Sonora. One of these has been in production since the end of the last century (Cananea), whereas the other is in the process of development (Caridad).

The Gulf Coastal Plain is one of the principal hydrocarbon provinces of the world, including the classical "Golden Lane." Rich oil and gas production comes from Albian reef limestones, from Tertiary stratigraphic traps, and from Tertiary salt-dome-deformed sediments. Some of the salt domes, on the north side of the Isthmus of Tehuantepec, contain very important sulfur deposits.

Coal occurs in the Upper Triassic sediments in central Sonora and in the Middle Jurassic sediments of northwestern Oaxaca, but at present the production, utilized exclusively by the steel industry, comes entirely from the Upper Cretaceous bituminous coal basin of northern Coahuila.

ZOLTAN DE CSERNA*

*Publication authorized by the Director, Instituto de Geología, Universidad Nacional Autónoma de México.

References

Ballard, J. A., and Feden, R. H., 1970. "Diapiric structures on the Campeche shelf and slope, western Gulf of Mexico," *Bull. Geol. Soc. Am.*, 81, 505–512.

Benavides-Garcia, L., 1956. "Notas sobre la geologia petrolera de Mexico," in E. J. Guzman Jiménez, ed., *Symposium sobre Yacimientos de Petroleo y Gas*. Mexico, D. F.: Internat. Geol. Cong., 20,(3), 351–562.

Burckhardt, C., 1930. "Etude synthetique sur le Mesozoique Mexicain," *Soc. Paleont. Suisse, Mem. 49–50*, 280p.

Coogan, A. H., Bebout, D. G., and Maggio, C., 1972. "Depositional environments and geologic history of Golden Lane and Poza Rica Trend, Mexico: An alternative view," *Bull. Am. Assoc. Petrol. Geologists*, 56(8), 1419–1447.

Cornwall, I. W., 1971. "Geology and early man in Central Mexico," *Proc. Geol. Assoc. Lond.*, 82(3), 379–391.

Cserna, Z. de, 1960. "Orogenesis in time and space in Mexico," *Geol. Rundschau*, 50, 595–605.

———, 1961. *Tectonic Map of Mexico*. New York: Geol. Soc. Am., 1:2,500,000.

———, 1969. "Tectonic framework of southern Mexico and its bearing on the problem of continental drift," *Bol. Soc. Geol. Mexicana*, 30, 159–168.

———, 1971. "Precambrian sedimentation, tectonics and magmatism in Mexico," *Geol. Rundschau*, 60(4), 1488.

Curray, J. R., 1970. "Quaternary influence, coast and continental shelf of Western U.S.A. and Mexico," *Quaternaria*, 12, 19–34.

Gierloff-Emden, H. G., 1970. *Mexico–eine Landeskunde*. Berlin: de Gruyter, 634p.

Gonzalez-Reyna, J., 1956. "Riqueza minera y yacimientos minerales de Mexico," *Congr. Geol. Intern.* (Mexico), 20, monogr., 497p.

Guzman, E. J., and Cserna, Z. de, 1963. "Tectonic history of Mexico," *Am. Assoc. Petrol. Geologists Mem. 2*, 113–129.

Heine, K., 1973. "Zur glazialmorphologie und präkeramischen Archäologie des Mexikanischen Hochlandes während des spätglazials (Wisconsin) und Holozäns," *Erdkunde,* 27(3), 161–180.

Hernandez-Sanchez, M. S., ed., 1968. *Carta Geologica de la Republica Mexicana.* Mexico, D.F.: Comite de la Carta Geologica de Mexico, 1:2,000,000.

Imlay, R. W., 1943. "Jurassic formations of Gulf region," *Bull. Am. Assoc. Petrol. Geologists,* 27, 1407–1533.

———, 1944. "Cretaceous formations of Central America and Mexico," *Bull. Am. Assoc. Petrol. Geologists,* 28, 1077–1195.

Kesler, S. E., 1973. "Basement rock structural trends in southern Mexico," *Bull. Geol. Soc. Am.,* 84, 1059–1064.

King, P. B., 1969. "The tectonics of North America—a discussion to accompany the tectonic map of North America, 1:5,000,000," *U.S. Geol. Surv. Prof. Pap. 628,* 94p.

Lomnitz, C., Mooser, F., Allen, C. R., Brune, J. N., and Thatcher, W., 1970. "Seismicity and tectonics of the northern Gulf of California region, Mexico. Preliminary results," *Geofis. Intern.,* 10(2), 37–48.

Moore, G. W., and Castillo, L. del, 1974. "Tectonic evolution of the southern Gulf of Mexico," *Bull. Geol. Soc. Am.,* 85, 607–618.

Mooser, F., *et al.,* 1974. "Paleomagnetic investigations of the Tertiary and Quaternary igneous rocks: VIII," *Geol. Rundsch.,* 63(2), 451–483.

Orme, A. R., 1972. "Quaternary deformation of Western Baja California, Mexico, as indicated by marine terraces and associated deposits," *24th Intern. Geol. Conf.,* 3, 627–634.

Raisz, E., 1959. *Landforms of Mexico.* Cambridge, Mass. (map with text, 1:3,000,000). Priv. publ.

Salas, G. P., 1970. "Los Recursos naturales no renovables de Mexico," *Cons. Rec. Nat. no Renovables* (Mexico), 73, 111p.

Tamayo, J. L., 1962. *Geografia General de Mexico.* Mexico, D.F.: Inst. Mex. Investigaciones Economicas, 4 vols., 2620p.

Tardy, M., 1973. "Les Phases tectoniques du secteur tranverse de Parras Sierra Madre Orientale (Mexique)," *Bull. Soc. Geol. Fr.,* Ser. 7, 15(3–4), 362–366.

Wilson, J. L., Ward, W. C., and Brady, M. J., 1970. "Northeastern Yucatan, Mexico—a new area opens for study of carbonate-evaporite sediments," *J. Sed. Petrology,* 40(2), 745–749.

Cross-references: *Belize; Central America–Regional Review; Guatemala; North America; Revillagigedo Islands; United States; West Indies.*

MONTSERRAT

Situated at 16°45'N 62°12'W, Montserrat has an area of about 100 km^2 (39 sq mi). One of the Leeward islands of the West Indies, it lies 420 km (260 miles) SE of Puerto Rico and N of Guadeloupe. Montserrat (sometimes spelled Monserrat) was discovered in 1493 by Columbus, who named it after *Monserrado* (now called "Montserrat") a jagged mountain in Spain which has a celebrated monastery. The island, which was originally colonized by Irish settlers in 1632, was occupied by the French between 1664 and 1668 and again in 1782. Montserrat was finally ceded to Britain by the Treaty of Versailles (1783).

The climate of the island is governed by the NE trade winds, which bring heavy precipitation from July to November, reaching 4000 mm in the highest places, and with less than 1000 mm in the northern part of the island and along the south coast.

Montserrat is a mountainous island made up of the products of five volcanic centers, of Pliocene to Holocene age. The northern part of the island consists of andesitic lavas and pyroclastics erupted from the three older centers—the Harris-Bugby Centre (active about 4.2 m yr ago), Silver Hill (active until about 1.6 m yr ago), and the Centre Hills, which is a younger center than Silver Hill.

In the southern part of the island, South Soufrière Hill, which rises to 755 m (2479 ft), was active at least 1.6 m yr ago, but the most recent eruptions are considerably younger than this (Fig. 1). The earlier eruptions produced basaltic lava flows and a series of pyroclastic falls, which range in composition from basalt to andesite and which contain ultrabasic cumulate blocks. These eruptions probably built up a cone-shaped strato-volcano with a large central crater (estimated diameter about 1500 m). Late in the history of the volcano a composite dome of basaltic andesite was extruded within the crater, destroying the eastern part of the crater wall and much of the eastern flank of the volcano.

The blocking of the South Soufrière Hill vent may have been responsible for the shift of activity to the nearby Soufrière Hills (summit 915 m above MSL), where the early eruptions produced four andesite domes. Pyroclastic deposits are associated with these domes and charcoal from one of the later pyroclast flows gave a radiocarbon date of about 24,000 years (Shotton *et al.,* 1968). The andesite domes are truncated by a crater, English's Crater (a horseshoe-shaped crater, open on the E, diameter approximately 800 m), the formation of which appears to have been accompanied by the violent eruption of andesite pumice flows. A radiocarbon date of about 18,500 years (Shotton *et al.,* 1970) obtained on charcoal from one of the pumice flows may therefore give the age of formation of the crater. Late-stage events at the Soufrière Hills include the extrusion of the Castle Peak acid andesite dome within the crater,

FIGURE 1. Geologic sketch map of Montserrat.

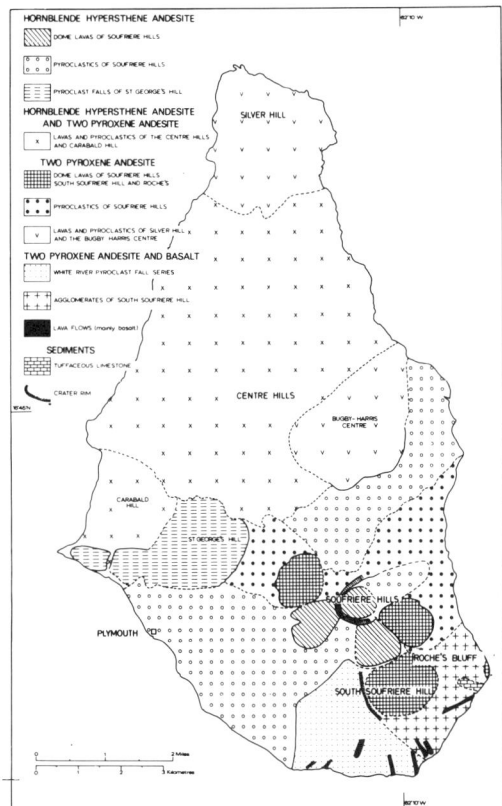

and this event may have been responsible for the breaching of the east wall of the crater. No historic eruptions have been reported at the Soufrière Hills, but seven active sulfur springs occur on the flanks of the volcano, and three periods of unusually intense solfataric activity accompanied by intense shallow focus earthquakes have occurred in the last 75 years (1897–1898, 1933–1936, and 1966–1967; see Robson and Tomblin, 1966; Shepherd et al., 1971). These "seismic crises" are almost certainly related to the movement of magma beneath the Soufrière Hills.

WILLIAM J. REA

References

MacGregor, A. G., 1938. "The volcanic history and petrology of Montserrat, with observations on Mt. Pélé, in Martinique," *Phil. Trans. Roy. Soc. Lond.*, Ser. B., **229**, 1–90.

Rea, W. J., 1970. "The geology of Montserrat, British West Indies," Ph.D. thesis, Oxford Univ.

———, 1974. "The volcanic geology and petrology of Montserrat, West Indies," *J. Geol. Soc. Lond.*, **130**(4), 341–366.

Robson, G. R., and Tomblin, J. F., 1966. "West Indies," *Catalogue of the Active Volcanoes of the World including Solfataric Fields*. Naples: Intern. Assoc. Volcanol., pt. 20, 56p.

Sapper, K., 1903. "Ein Besuch der Insel Montserrat (Westindien)," *Zbl. Min. Geol. Palaeontol.*, 279–287.

Shepherd, J. B., Tomblin, J. F., and Woo, D. F., 1971. "Volcano-seismic crisis on Montserrat, West Indies, 1966–67," *Bull. Volcanol.*, **35**, 143–163.

Shotton, F. W., Blundell, D. J., and Williams, R. E. G., 1968. "Birmingham University ratio carbon dates II," *Radiocarbon*, **10**, 200–206.

———, Blundell, D. J., and Williams, R. E. G., 1970. "Birmingham University radio carbon date IV," *Radiocarbon*, **12**, 385–399.

Cross-references: *Guadeloupe; Leeward Islands; St. Christopher-Nevis-Anguilla; West Indies.*

N

NAURU

Lying in an isolated position in the central Pacific between the Solomon and Gilbert islands and just south of the equator ($0°31S$ and $166°56E$), about 2000 km (1300 miles) northeast of Australia, Nauru Island has been an independent republic since 1968. It was discovered in 1798 by Captain Fearn, who named it "Pleasant Island." From 1888 until World War I it was part of the German colonial empire and before independence was an Australian mandate.

Nauru has a narrow coastal strip backed by coral cliffs and is completely surrounded by a fringing coral reef. The interior is a barren plateau with extensive deposits of phosphate, with reserves amounting to about 70 million tons, of which about 1.5 million tons are exported each year; this brings in an annual revenue of more than $10 million, enabling the inhabitants to enjoy one of the highest per capita incomes in this part of the world.

Nauru possesses no natural ports. Her shores descend almost vertically into deep waters, necessitating the building of special moorings for ships. The trade winds from the east blow for 9–11 months each year and there is little rain.

The island of Nauru is an uplifted atoll with a rim ("makatea") reaching 60–70 m (213 ft), which is terraced and contains many karst caves and sinkholes. The interior depression carries a small freshwater lake. The age of the limestone is not known, but the analogous formation on Ocean Island (see under *Gilbert and Ellice Islands*), 250 km to the east, is paleontologically determined as late Tertiary. The underlying oceanic crust is of Cretaceous age according to evidence of sea-floor spreading. Somewhat comparable phosphate islands are found in the *Carolines* (q.v.), the *Line Islands* (q.v.), and in the Indian Ocean (*Christmas Island*, q.v.).

RHODES W. FAIRBRIDGE

References

Ellis, A. F., 1936. *Ocean Island and Nauru*. Sydney.
Hambruch, P., 1912. "Entstehung, Bildung und Lagerung der Phosphate auf Nauru," *Z. Ges. Erdk.* (Berlin), 671–680.
———, 1914. *Nauru, Ergebnisse der Südsee Expedition, 1908–1910*. Hamburg.
Hutchinson, G. E., 1950. "The biogeochemistry of vertebrate excretion," *Am. Mus. Nat. Hist. Bull.*, **96**, 1–556.
Nugent, L. E., Jr., 1948. "Elevated phosphate islands in Micronesia," *Bull. Geol. Soc. Am.*, **59**, 977–994.
Power, F. D., 1925. "Phosphatic deposits of the Pacific," *Econ. Geol.*, **20**, 266–281.

Cross-references: *Caroline Islands; Gilbert and Ellice Islands; Line Islands.*

NETHERLANDS ANTILLES

Once known as the "Dutch West Indies," the Netherlands Antilles cover a total area of 1004 km^2 (390 sq mi) and consist of two groups of islands in the Caribbean. The leeward group (Netherlands Windward Islands)—St. Maarten (partly French), Saba, and St. Eustatius are situated east of the Virgin Islands and form part of the east Caribbean complex (see *West Indies*). The windward group (Netherlands Leeward Islands)—Aruba, Bonaire, and Curaçao at more than 900 km SW of the leeward group—forms the western most part of an E-W row of islands in front of the north coast of Venezuela. Both groups belong to the Lesser Antilles. (N.B.: The terms "windward" and "leeward" are used in opposite sense in English and other European languages.)

The climate, dominated by NE trade winds, is semiarid tropical. The generally low windward group, with a desert vegetation, differs from the more rugged leeward group, with a mixed vegetation due to microclimates.

Mineral resources are very limited. Phosphate is exploited on Curaçao; gold exploitation on Aruba was abandoned in 1916. Gypsum and pumicestone deposits on St. Eustatius could be economically important in the future. Curaçao and Aruba are important refinery centers for the crude oil from Venezuela.

Windward Group

The basic geology of the "A.B.C." islands was worked out by Rutten (1932), Molengraaff

(1929), Pijpers (1933), and Westermann (1932). These islands, forming part of the Aruba-Orchilla geanticline, have a geological history closely related to the Greater Antilles and different from the Venezuela coastal area.

The basement is a Laramide folded series of submarine basic lavas and tuffs ("Diabase formation" on Curaçao and Aruba; the lower part of the "Washikemba Formation" on Bonaire) of Cretaceous age and deposited in a trough situated at the present location of the islands. This complex was intruded by granodioritic and, subordinately, grabboic rocks in the Upper Cretaceous, probably Campanian time (Priem et al., 1966, 1967). Young, post-Danian intrusions are known on Curaçao.

Curaçao (444 km^2). The Knip Group of Campanian-Maestrichtian age, composed of chert-rich sediments, conglomerates, limestones, and tuffs, rests unconformably on the diabase basement with a sedimentation hiatus at the base. The basal part of this group, the Serou Teintje Formation, is composed of limestones with reef corals, *Lithothamnium*, rudists and conglomerates. Tuffs occur in the upper part of the group. Tectonic movements during the period resulted in changes in facies and thickness and the presence of turbidites and conglomerates. The Midden-Curaçao Group, a flysch-type deposit of Danian age and conglomerates, in the basal part overlies unconformably the former group. The main folding took place at the end of the Cretaceous and the beginning of the Tertiary, whereas these events had already affected the Venezuela coastal area in the Middle Cretaceous.

Bonaire (288 km^2). The more-than-5000-m-thick Washikemba Formation, a volcanic series with intercalations of cherts and radiolarites, can roughly be correlated with the diabase basement and the Knip Group of Curaçao. Ignimbrites in the middle part could coincide with the sedimentation hiatus (Beets and Lodder, 1967). The Rincon Formation, shallow-water deposits of Upper Campanian-Lower Meastrichtian age, overlies these; the Soebi Blanco Formation, coarse-grained conglomerates, could be correlated with the Midden-Curaçao Group on Curaçao. This whole complex was slightly folded and intruded by porphyrites of Campanian or older times.

Aruba (193 km^2). On Aruba the basement, diabase and tuffs, has no sedimentary cover of Upper Cretaceous−Lower Tertiary age as occurs on Curaçao and Bonaire. Batholiths of granodiorites, hooibergites, and some gabbros occupy an important part of this island.

After a post-Danian period of deformation and metamorphism, uplift and denudation affected the three islands and Tertiary deposits are sporadically preserved. Upper Eocene limestones with large foraminifera and marls, the Serou di Cueba Formation, occur on Curaçao and Bonaire; marine Miocene deposits, the Oranje Stad Formation, were found in a borehole on Aruba. The development of a deep-sea basin in this area which is still seismically active started probably at the end of the Eocene (Lagaay, 1969).

There is a gap during the rest of the Tertiary and the islands are largely coated by Quaternary formations; coral reef limestones and terraces occurr up to 210 m above sea level. According to older ideas of Gabb, Molengraaff, and Rutten, these islands were arched up and the reefs isolated by erosion. All the hills along the southwest coast of Curaçao, Aruba, and Bonaire are strikingly asymmetric, owing to the 25°SW dip of these reef caps. Schaub (1948) believed the reef limestones are relatively undeformed and built up during eustatic changes of the sea level. Zonneveld (1960) found that the steeply dipping sectors are in situ, off-reef talus slopes, and these in turn were truncated by flat terraces which correspond to stillstands in the eustatic oscillations during the Quaternary uplift. Lagaay (1969) explained the circumstance that older terraces are always in a progressively higher position than the younger ones due to gradual uplift of the islands, which started probably in the Lower Miocene, superimposed on eustatic sea-level changes.

Steeply W-dipping eolianites indicate strong E winds during periods of regression and identify each sea-level stand. Dry climate phases alternated with wet ones during which karst caves were developed. The cave deposits include rodents, the giant tortoise (*Geochelone*), and the ground sloth (*Pauloenus*), which suggest former land connections with South America at times of possibly intermediate humidity. Holocene evolution of the coast has led to cliff, barrier spit, and lagoon development. On the salt flats of Bonaire, white carbonate muds contain very recent dolomite, C^{14} dated at 1500−2200 B.P. (Deffeyes et al., 1964).

Leeward Group

Also included administratively in the Netherlands Antilles are the three islands in the east Caribbean area: St. Maarten, Saba, and St. Eustatius. St. Maarten forms part of the Limestone Antilles; Saba and St. Eustatius are situated in the volcanic inner belt of the Lesser Antilles.

St. Maarten, situated in the northern part of the Lesser Antilles, forms part of the nonvolcanic arc of the Limestone Antilles east of the volcanic inner arc. It covers a total area of 85 km², of which 37 belong to the Netherlands Antilles, the rest to France. Two hill systems cross the island, reaching 424 m, Paradise Peak, in the West and 311 m in the East.

Basic geology was worked out by Molengraaff (1929) and Westermann (1949). The basement is formed by the Point Blanche Formation, composed of tuffs with coarse-grained pyroclastics and inclusions of carbonate rocks of the Upper Eocene. Granodioritic intrusions in the Oligocene, joined by embryonic volcanic activities, provide a close geological relation with the Greater Antilles. Andesitic and basaltic fault intrusions represent the late- or postmagmatic phase.

The Lowlands Formation—limestones and marls of Upper Oligocene to Lower Miocene age (Drooger, 1951)—rests unconformably on an eroded surface.

Tectonic activity occurred successively in the Oligocene, the Lower Miocene, and in the younger Quaternary. Block faulting in MioPliocene resulted in a WNW tilting. Pleistocene coral reefs and coastal terraces, left by the oscillating sea, occur at 5–6 m above sea level with, locally (Point Blanche) with dips of 30° (Weyl, 1966).

Saba (13 km²) is formed by a simple andesitic volcanic cone situated in the inner volcanic belt of the Lesser Antilles. It reaches an elevation 869 m above sea level and approximately 1500 m above the sea floor. The cone of this extinct volcano of Middle or Upper Pleistocene age is well preserved and the local settlement appropriately named "The Bottom" is located in the middle of the crater. Sulfur and gypsum deposits occur as products of late volcanic activities. Hot, sometimes sulfurous springs have disappeared in recent times.

St. Eustatius (21 km²) is situated SE of Saba near *St. Christopher* (q.v.) and like Saba is a youthful volcanic island in the inner belt. The northwestern part is denuded and hilly. It forms the oldest, nonactive volcanic center, composed of tuffs, breccia, and andesitic lavas. The southeastern part is dominated by a younger, symmetric andesitic volcanic cone with an elevation of 600 m. The crater floor, the Quill, is at 273 m above sea level. This stratovolcano is surrounded by tuffs, breccia, and pumicestone layers. Interesting is the presence of the Middle Pleistocene White Wall Formation on the southern flank of the Quill volcano, composed of coral, molluscan, and *Lithothamnium* limestones, conglomerates, tuffs, and pumicestone with a 40–45°S dip. This formation represents a mixed, volcanic and shallow-marine deposit, formed after the Upper Pliocene volcanic activity in the northwestern part of the island; it was deformed and intersected by the younger Quaternary volcanic activities in the southeast. This area is still seismically active.

GERRIT C. BROUWER

References

Alexander, C. S., 1961. "The marine terraces of Aruba, Bonaire and Curaçao, Netherlands Antilles," *Ann. Assoc. Am. Geogr.*, **51**(1), 102–103.

Beets, D. J., and Lodder, W., 1967. "Indications for the presence of ignimbrites in the Cretaceous Washikemba Formation of the Isle of Bonaire, Netherlands Antilles," *Proc. Kon. Ned. Akad. Wetensch.*, (Amsterdam), Ser. B, **70**, 5p.

Bucher, W. H., 1952, "Geologic structure and orogenic history of Venezuela," *Geol. Soc. Am. Mem. 49*, 113p.

DeBuissonjé, P. H., 1964. "Marine terraces and subaeric sediments in the Netherlands Leeward Islands, Curaçao, Bonaire, and Aruba, as indications of Quaternary changes in sea level and climate," *Proc. Kon. Ned. Akad. Wetensch.* (Amsterdam), Ser. B, **67**(1), 60–79.

Deffeyes, K. S., *et al.*, 1964. "Dolomitization: observations on the island of Bonaire, Netherlands Antilles," *Science*, **143**, 678–679.

———, Lucia, F. J., and Weyl, P. K., 1965. "Dolomitization of recent and Plio-Pleistocene sediments by marine evaporite waters on Bonaire, Netherlands Antilles," *Soc. Econ. Paleontol. Mineral. Spec. Pap. 13*, 71–88.

Drooger, C. W., 1951. "Upper Cretaceous foraminifera of the Midden Curaçao beds near Hato, Curaçao (N.W.I.)," *Proc. Kon. Akad. Wetensch.* (Amsterdam), Ser. B, **54**(1), 66–72.

Hedberg, H. D., 1942. "Mesozoic stratigraphy of northern South America," *Proc. 8th Am. Sci. Congr.*, **4**, 195–227.

Lagaay, R. A., 1969. "Geophysical investigations of the Netherlands Leeward Antilles," *Verh. Kon. Ned. Akad. Wetensch. Afd. Natuurkunde*, Ser. 1, **25**(2), 86p.

Lucia, F. J., 1968. "Recent sediments and diagenesis of South Bonaire, Netherlands Antilles," *J. Sed. Petrology*, **38**, 845–858.

Martin-Kaye, P. H. A., 1969. "A summary of the geology of the Lesser Antilles," *Overseas Geol. Mineral Res.*, **10**(2), 172–206.

Molengraaff, G. J. H., 1929. *Geologie en hydrologie van het eiland Curaçao*. Delft: J. Waltman, thesis, 126p.

Pijpers, P. J., 1933. *Geology and Paleontology of Bonaire*. Utrecht: Geogr. Med., thesis, 103p.

Priem, H. N. A., Boelrijk, N. A. I. M., Verschure, R. H., Hebeda, E. H., and Lagaay, R. A., 1966. "Isotopic age of the quartz–diorite batholith on the island of Aruba, Netherlands Antilles," *Geol. Mijnb.* **45**, 3p.

———, *et al.*, 1967. "Isotopic age determinations on Surinam rocks," *Geol. Mijnb.*, **46**(1), 26–31.

Rutten, L. M. R., 1932, "De geologische geschiedenis der drie Nederlandse Benedenwindse Eilanden," *De West Indische Gids*, (Netherlands), 13-9.

Schaub, H. P., 1948. "Geological observations on Curaçao, N.W.I.," *Bull. Am. Assoc. Petrol. Geologists*, **32**, 1275–1291.

Westermann, J. H., 1932. *The Geology of Aruba.* Utrecht: Min. Geol. Inst. Rijksuniv., 129p.

———, 1949. "Overzicht van de geologische en mijnbouwkundige kennis der Nederlandse Antillen," *Roy. Inst. for the Indies, Mededel.,* 85(35), 169p.

———, and Kiel, H., 1961. "The geology of Saba and St. Eustatius, with notes on the geology of St. Kitts, Nevis, and Montserrat," *Natuur. Stud. Surinam Publ. 24,* 175p.

Weyl, R., 1966. "Geologie der Antillen," *Region. Geol. der Erde,* vol. 4. Berlin: Gebr. Borntraeger, 410p.

Zonneveld, J. I. S., 1960. "An aerial-photographic research in Curaçao, Aruba, and Bonaire," *Tijdschr., Kon. Ned. Aardr. Gen.,* 72(4), 389–400 (Dutch with Engl. abstr.).

———, 1968. "Quaternary climatic changes in the Caribbean and N. South America," *Eisz. Gegenw.,* 19, 203–208.

Cross-references: *Barbados; Colombia; Leeward Islands; South America; West Indies.*

NEW CALEDONIA

New Caledonia is a French overseas possession in the Southwestern Pacific. It is a long, narrow, mountainous island with a mean altitude of 600 to 1000 m, part of an archipelago of approximately 20,000 km^2, located between 17 and 23°S and between 163 and 167°E midway between New Guinea (1755 km to NW) and New Zealand (1555 km to the SSE). It lies on the "Melanesian" island arcs, which were folded in Cretaceous to Tertiary time (Alpine) and which parallel the northeastern coasts of the Australian continent (Fig. 1). About 50 km wide and 400 km long, the main island is extended for nearly 1000 km by shallow shelf platforms bordered by coral reefs and islets. Belep and Huon islands lie to the N and Pines Island to the S. The *Loyalty Islands* (q.v.)— Uvea, Lifou, Maré, a parallel chain of coral islands covering 1743 km^2—lie 100 km to the E, each with a core of volcanic rocks.

The Neocaledonian archipelago is located about midway between Australia and the "andesite line" of today, which separates the Melanesian "quasicratonic" area from the completely "oceanic" area of the Pacific plate. The Melanesian area is subdivided today into smaller, "microplates" (see *Fiji, New Guinea, New Hebrides, Tonga*).

From the *geomorphologic* point of view, New Caledonia appears as a structurally controlled, elongate island corresponding to the fold trends. It has been eroded into two slopes, eastern and western, separated in places by more or less flat plateau areas (peneplains) but frequently by a sharp, irregular crest, the "central chain," with a maximum altitude of 1650 m. The eastern and western slopes are asymmetric. The western slopes consist of undulating plains between low chains, dominated by tabular massifs of peridotite. In contrast the eastern coast is mountainous, with steep cliffs; it is intersected by the alluvial plains of several large rivers.

New Caledonia is thus a remnant of an earlier submergence and is bordered by a submarine shelf surrounded by one of the most beautiful barrier reefs in the world, which encloses a lagoon that reaches a width of more than 15 km to the west.

Discovered by Cook in 1774, New Caledonia was first studied geologically in 1863 by Deslongchamps (1866), who found Triassic fossils. Several descriptions followed, principally by mining engineers such as Garnier (1867–1871), who discovered nickel silicate ores, the principal mineral of which has since been named *garnierite.* Later came Glasser (1903–1904), whose work remained a standard for nearly 50 years. The first general stratigraphic investigation was by Piroutet (1917). There is also the work of Benson, who made a comparative geological study survey with New Zealand, and the great geographer W. M. Davis (1925), who described the majestic coasts and coral reefs. Several paleontologists studied the fauna, including Crie, Kilian, Heim, Heinz, and Ratte. The petrography of the crystalline rocks was described in two memoirs by the famous mineralogist Lacroix (1941–1942).

After World War II, a French government geological mission was sent to New Caledonia in 1946–1949, consisting of Avias, Arnould, and Routhier, the director. They made the first large-scale geological map of the island (Arnould et al., 1953) (1:100,000 scale in 10 sheets, each with explanatory notes), and generated three doctoral theses (Avias, 1953; Routhier, 1953; Arnould, 1958). The work of Avias was devoted to the stratigraphy and paleontology of pre-Cretaceous formations (Jurassic-Permian). That of Routheir is a monograph of the northwestern part of the island concerning principally Tertiary formations. That of Arnould specialized in the metamorphic formations in the N and NE. Further, Avias has published on the ultrabasis rocks, nickel deposits, and coastal swamps. Routhier published further on the iron, manganese, and chromium deposits.

In addition, during this period two missions studied the petroleum possibilities and published important results, especially on Tertiary microfaunas and the tectonics. Since 1957, various works, mainly of applied economic geology, were carried out by geologists from various organizations such as BUMIFOM, BRGM, and

FIGURE 1. The geotectonic position of New Caledonia in the Australasian region (adapted from Fairbridge, 1950).

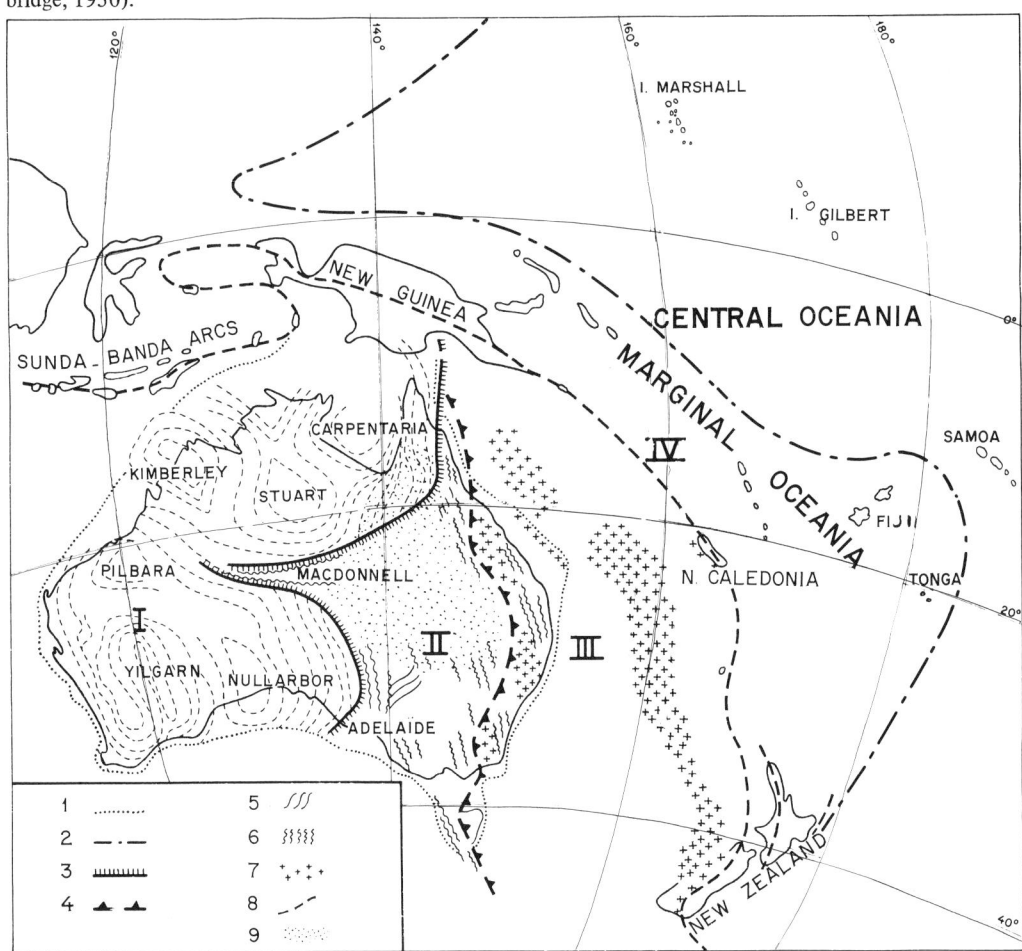

the "Service des Mines et de la Géologie de la Nouvelle Calédonie" (Tissot and Noesmoen, 1958). A thesis by Espirat (1963) dealt with the stratigraphy and petrography of the area (metamorphic) to the N and NE.

From 1960 to 1966 various missions studied the coral reefs, organized by the Museum National de'Histoire Naturelle of Paris and the Fondation Singer Polignac (Avias, Guilcher, Lucas, Doumenge, Chevalier, and others). Lillie and Brothers (1970), from New Zealand, and their pupils have also undertaken important studies comparing New Caledonian formations and those of New Zealand.

Starting in 1965, in addition to work by the Bureau de Recherches Géologiques et Minières and the "Le Nickel" Company, a geological mission has been organized by the University of Montpellier (with Avias, Coudray, and Gonord). Further, geologists and pedologists from the Institut Francais d'Océanie have been studying the ultrabasic rocks under the direction of Routhier (Guillon and Routhier, 1971).

Geology

New Caledonia forms a highly folded and faulted mountain chain (Eocene–Oligocene Alpine tectonics) with mild metamorphism (except in the NE) and ultrabasic rocks (peridotites), which form much of the southeastern (massif Minier du Sud) and western slopes N of Bourail (Fig. 2).

Paleozoic. The oldest dated rocks are the "multicolored tuff formation" (Avias, 1949) formed of rhyolitic and dacitic tuffs and breccia flows alternating with red and green mudstones and argillites. The upper part (Avias, 1953) contains *Maitaia* (= *Atomodesma*) *trechmanni*, a fossil characteristic of the Lower Permian (Artinskian) of New Zealand. Elsewhere

FIGURE 2. Geological and mineral map of New Caledonia (Routhier).

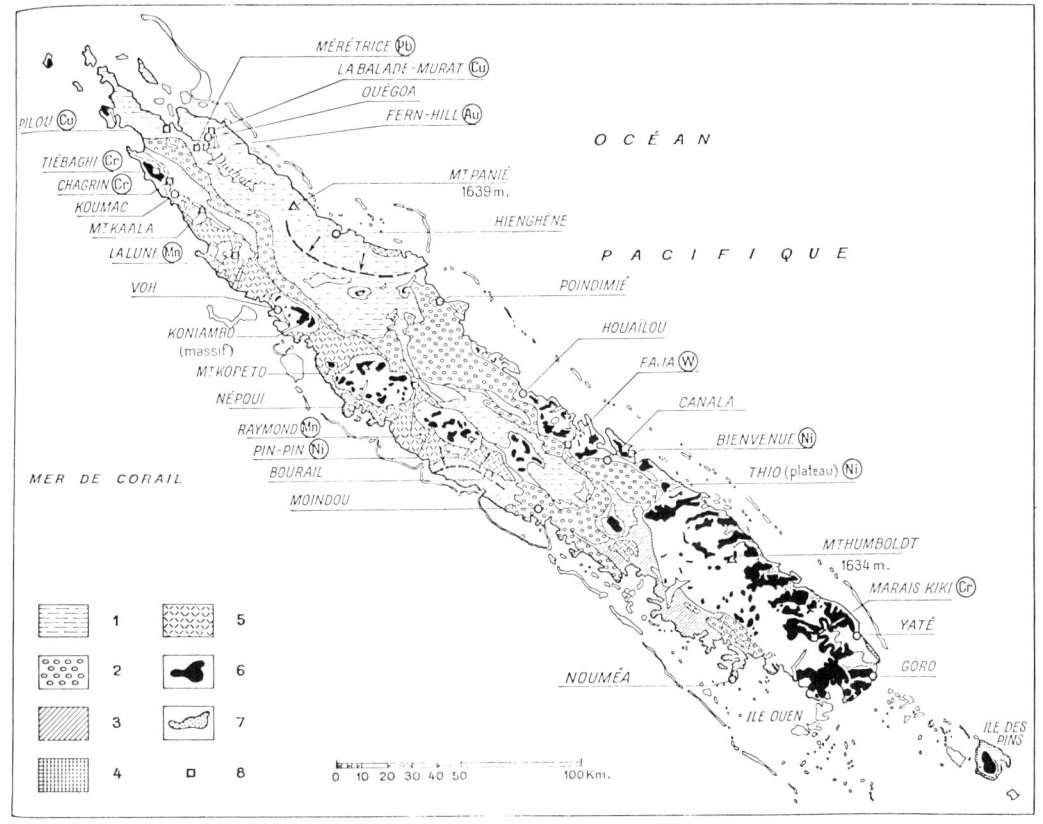

the facies are analogous to the Devonian–Permian sequence of eastern Australia. Cephalopod faunas, formerly referred to *Popanoceras* and *Wagenoceras*, are linked to *Cyclolobus* (Avias and Guerin, 1957), Upper Permian to Lower Triassic.

Mesozoic. Over the multicolored tuffs lies unconformably the Graywacke formation that represents Triassic and Jurassic. This thick sequence resulted from the erosion of Hercynian chains that were uplifted in the W in the area which then lay between New Caledonia and Australia known as "Tasmantis" or "Tasmantia." Because of drifting and plate motions, these two areas were then much closer. Essentially the rocks consist of graywackes, locally fossiliferous and calcareous. From a paleontologic point of view the following stages may be identified:

Trias:

Lower Trias (Werfenian) with ammonites, *Flemingites*

Middle and Upper Trias (Ladinian to Carnien) containing *Spiriferina Kaihikuana* and *Halobia*, with a zone of *Mytilus problematicus* and *Minetrigonia* (Carnian)

Upper Trias (Norian) with graywackes often passing to shell beds containing Monotidae, including (below) a zone of *Monotis richmondiana* and *Stenarcestes arnouldi*, and (above) zone containing *Monotis gigantea* and *Monotis routhieri* indicating the passage from Norian to Rhaetian

Rhaetian containing *Monotis* aff. *calvata* and *Anodontophora* at the bottom, and *clavigera bisulcata* at the top

Jurassic:

Lias (Lower Jurassic), includes:

Hettangien, containing ammonites (Psiloceratidae and Schlotheimidae), *Waehneroceras,* and lamellibranchs (*Otapiria marshalli*)

Sinemurien, containing *Arnioceras*

Middle and Upper Lias, contains *Pseudaucella marshalli*

Middle Jurassic seems to be missing and is represented by volcanic activity on the west coast and possibly the "breccia from Sarramea," a curious bed of consolidated volcanic sands containing titanomagnetite, which is also present in the same stratigraphic position in New Zealand

Late Middle Jurassic is marked by green graywackes with belemnites (west coast) plus black shales containing *Inoceramus* and *Aucella* (*Buchia*) on the east coast (Portlandian, previously reported, has not been confirmed)

An important gap occurred in the Lower and Middle Cretaceous and was followed by sandstones and carbonaceous shales, constituting the *"formation à charbon"* (of Avias and Routhier).

The faunas (collected by Piroutet, Avias, Routhier, Noesmoen, Lormand, and Koch) are exclusively Senonian (Upper Cretaceous) and include *Kosmaticeras, Baculites, Pacitrigonia, Lahillia,* and *Acanthocardia,* analogous to those of the Antarctic-Indo-Pacific province.

Throughout the Permian, Triassic, Jurassic, and Cretaceous formations of the west coast, numerous pieces of silicified wood are found, corresponding to floated logs coming from Tasmantia. Besides araucarioxylon and cedroxylon, there are *?Sahnioxylon aviasii* and *S. australe,* some Triassic species that are among the oldest known in the world (Boureau, 1954; Colonna-Salard, 1964), which are of interest in the problem of the origin of angiosperms.

In summary, the Mesozoic of New Caledonia was marked by sedimentation in the "Australo-Melanesian" trough (Triassic and Jurassic), the coastal facies of which was particularly fossiliferous on the west coast (the edge of "Tasmantia"), whereas the facies of the central chain and coast was more fine grained and often unfossiliferous. For the late Cretaceous period, the western zone ("external" zone) is missing, and the only deposits of this age are found in the central and eastern or "internal" zone.

Older Tertiary (Paleogene). In the same way there is a west coastal external Eocene, lying transgressively on an eroded Permo-Triassic basement, whereas a more complete internal Eocene occurs in the E; this zone is intensely folded and tectonized, and its facies are quite different from those of the external zone.

In the *external zone* the stratigraphy is relatively simple, and subsidence permitted the steady accumulation of detrital sediments to several thousands meters in thickness. It seems that they were derived from the erosion of cordilleras of the internal zone (*Ed. note:* thus an "exogeosyncline" in Kay's terminology). At the base are conglomerates with Permo-Triassic pebbles, calcareous lenses with *Orthophragmina, Discocyclina,* and algae, passing up to a very thick deposit (3000 m) of marly flysch, fine at the bottom, and sandy, breccia rich, or tuffaceous at the top.

In the *internal zone* the Tertiary sedimentation starts with conglomerates containing calcareous lenses with *Globigerina,* which pass up without a break from the Senonian. They are covered by tuffaceous shales and calc-schists, then by thick *volcanosedimentary formations* formed by submarine flows, andesitic basalt, dolerites, associated tuffs with radiolarite, and red claystone lenses, often interbedded with ultrabasic rocks (see further).

Younger Tertiary (Neogene) and Quaternary. On the early formations, now often highly folded and tectonized, are found horizontal argillaceous and calcareous *Miocene formations,* containing isolated corals and characteristic foraminifera.

These formations contain traces of ferruginous crusts (ferricrete), indicating that by late Oligocene time ultrabasic formations were being lateritized on the newly emerged mountain chain. This *carapace* continued to form during Pliocene and Quaternary times, but a major portion of it has now been removed by erosion. At present the primary carapace survives only on the residual plateaus or in the faulted sectors.

At the same period of this carapace formation, *coral reefs* were developing on the New Caledonian coasts for the first time as fringing reefs. These are rather striking features on account of high sea levels during the interglacial warm stages and notably during the last eustatic rise (Flandrian transgression). The present complex includes both living reefs (barrier reef and fringing reef) and dead reefs, eroded and drowned, or even uplifted, with calcarenitic lagoonal formations and old dune systems (*eolianites*), now highly eroded and partly submerged (Avias and Coudray, 1965). On some islands (Huon and d'Entrecasteaux) the calcarenites are superficially phosphatized by the guano of sea birds.

Besides the reef complex, Pliocene and Quaternary eluvial, fluvial, and coastal formations are also present. Today some of them are at relatively high levels, such as montmorillonitic clay with gypsum from Oua Tom or Tontouta ($H = +18m$). In the mangrove swamp muds of today can be seen gypsum crystallization, silicification, and calcareous concretions (Avias, 1960). At the mouths of rivers draining off the peridotite massifs there are ferruginous and metalliferous sediments being laid down in the modern lagoons.

Crystalline Rocks

Volcanic Rocks. During the geological history of the island, several volcanic cycles have occurred (rhyolites in the Devonian-Permian, titanomagnetitic sands in the Jurassic, trachytes and rhyolites in the Cretaceous, basaltic andesites and dolerites in the vulcanosedimentary Paleogene). In contrast, no volcanic formations of Neogene or Quaternary age are known.

Metamorphic and Plutonic Rocks. From a general metamorphic point of view, New Caledonia can be divided into three distinct sectors: a nonmetamorphic sector on the west coast, a low metamorphic sector in the central chain and east coast, and a high metamorphic sector to the extreme NE of the island.

In the second sector, metamorphism is expressed mainly by sericitization and chloritization; in the third sector it has produced micaschists, albitic gneiss, granites, and amphibolites (notably garnetiferous glaucophanites). Epidote and lawsonite are particularly abundant to the NE.

In the N three subzones may be distinguished: (a) a subzone of schists containing pumpellyite, (b) a subzone of glaucophane lawsonite, and (c) a subzone containing epidote glaucophane; these last two zones correspond to the higher micaschist zone in the sense of Jung and Roques.

From a chronological point of view, a low-grade metamorphism seems to have developed in the New Caledonian trough under the increasing overburden with "premonitory" movements at the end of the Jurassic, and this may explain the presence of metamorphic pebbles in the conglomerates of the Congo River, which were probably Cretaceous. Major metamorphism was related to folding at the time of the principal orogenesis (Alpine), and is cited by Avias to explain the formation of ultrabasic rocks and of their association of basic vulcanosedimentary rocks. These became chloritized and serpentinized in a first phase, and became sericitized and "peridotized" in a second phase.

Acid crystalline rocks include granites but are limited to a few places. (a) The double Massif of Koun in the "Grand Massif minier du Sud" is a dioritic granite containing biotite and amphibole. It is surrounded by a serpentinized peridotite containing folia of talc. (b) The Saint Louis Massif is intrusive into Cretaceo-Eocene formations adjacent to ultrabasic rocks. This is very heterogenous calc-alkaline granite and appears to be a result of granitization of the formations it intrudes and is younger than the ultrabasic rocks.

Besides granite, there are also some diorites and granodiorites, notably near the pass at Amieu, associated with andesitic vlows.

Most important there are the remarkable outcrops of ultrabasic rocks (in total, more than 6000 km^2), principally harzburgites, saxonites, and dunites. There are commonly also inclusions of chromite, pyroxene, or amphibole gabbros, and ouenites (gabbros containing green pigeonite and olivine); the fine-grained varieties of the latter were used by the natives to make their huge stone axes. There are also rare feldspathic rocks (ferromagnesian plagioclasolites, diorites, and even pegmatites).

The borders of the big ultrabasic massifs are frequently marked by serpentine, containing inclusions of the country rocks.

Tectonics

A residual Devono-Permian foreland on the eastern border of "Tasmantia" testifies to a late Hercynian orogenesis, cropping out here and there between Moindou and the Baie de Saint Vincent. This slightly folded but mostly faulted complex is mainly of graywacke. Against this foreland abut the various fold belts of the orogenic New Caledonian geosyncline. Here a relatively weak first phase of folding, accompanied by emergence and basic volcanism, seems to have taken place toward the end of the Jurassic. Later, in the same trough during Paleogene sedimentation there was uplift of cordilleran folds followed by erosion. Toward the end of this period very important submarine volcanic flows were emplaced before the paroxismal phase of folding (Oligocene). This last phase culminated with the formation of ultrabasic complexes which, according to some writers, were fissure eruptions (Routhier, 1953); or to others (Avias, 1949, 1967) might have resulted from the metasomatism of basic volcanosedimentary masses, which was then followed by a tectonic emplacement during an E-W thrusting. In any case, sheets of ultrabasic rocks are overthrust onto the remainder of the New Caledonian belt and involved in the general overturning toward the W, accompanied by folding and imbrication. At the same time, this generated breccias and fractures, as can be observed at the bottom or in the middle of the ultrabasites. According to observations by Gonord the orogeny can be subdivided into at least three subphases on the basis of microtectonic studies.

During, or just after this movement, there was a development of major fractures, trending N-S, which led to the fragmentation of this chain into sections and the scalloping of corresponding folds into arcs. Some differential movements of these sections would have finally caused their uplift in the extreme south to the S of Ounia, or in the Ile des Pins, etc. These block phenomena have controlled the growth of coastal reefs, subject also to eustatic variation of sea level.

In general, the tectonic picture is outwardly simple, chiefly anticlinorial in the central chain, monoclinal toward the E and W. In detail, however, it is complex because of overthrusting and block faulting. The history relates to that of the whole Melanesian region, which in terms of plate tectonics reflects the interactions of the Australian and Pacific plates, but is complicated by six smaller plates, undergoing progressive counterclockwise rotation (Van der Linden, 1969).

After the younger differential movements, the whole zone became remarkably stable, as shown by the uniform levels of the coral reefs. New Caledonia is thus in contrast to the particularly unstable zone (with frequent earthquakes and frequent volcanic activity) of the neighboring belt of the New Hebrides (q.v.). Several terraces connected with the eustatic variations of the sea level, and well known in Australia, for example, are found also in New Caledonia, at least in some sectors ("Abrolhos" and "Peron" terraces of Fairbridge). The coasts are in general *submerged coasts,* owing to the rise of sea level at the time of the Flandrian glacioeustatic transgression. There is also some tectonic warping of the Pleistocene terraces (Launay and Recy, 1972).

Mineralization and Economic Geology

Relative to its small size, New Caledonia is one of the most richly mineralized countries in the world. Besides *copper* in the Permian and *coal* lenses in the Upper Cretaceous, there are lenses of *manganese* ore, often associated with radiolarites in the vulcanosedimentary Paleogene. Also there are masses of *gypsum* in Pliocene and Quaternary formations. Hydrocarbon traces are noted in the Eocene near Koumac. The most important deposits, however, are those connected with metamorphic formations, especially in the northeastern part of the island associated with ultrabasic rocks where nickel, cobalt, chromium, and iron are in evidence.

In the first group the *copper* occurs as oxidized copper sulfides, in veins or "impregnations" in schists; about 10,000 tons of copper metal have already been mined. *Lead* and *zinc* occur as carbonates and sulfates; 3000 tons of lead and 2000 tons of zinc have been obtained so far. *Gold* is found as inclusions in pyritic quartz veins and as placers. One hundred kilograms has been extracted, mainly from the Fern Hill mine. *Mercury, tungsten, titanium, molybdenum,* and *barite* are only known in traces.

In the second group come the mineralizations that bring New Caledonia into world prominence and underly the whole economy.

Nickel. The long history of peneplanation and lateritization of massifs of ultrabasic rocks persisted probably since the Oligocene, or at least since the Miocene. This led to an exceptional concentration in *nickel,* which accumulated mainly at the base of the lateritic carapace in contact with the unaltered bedrock (this duricurst is otherwise almost exclusively made of iron oxides). The nickeliferous minerals are hydrated silicates of nickelferous antigorite type (*garnierites*), most often of a green color and variable appearance (pulverulent as nepouite or in concretions as noumeite). These minerals are mainly in fissures and crush zones near the bedrock–laterite contact. They may also impregnate altered peridotites. Some nickel can be found scattered in goethite. The average percentage of nickel and cobalt in the laterites is 1.20%, but may vary from 0.4 to more than 5%. The Ni^+Co content of intermediate ores between the bottom of laterites and the unaltered bedrock may exceed 14%. Nickel sulfides and arsenides are also known in the peridotites, but only as curiosities, e.g., orcellite.

Geomorphologic rules of location of mineable ores have been defined by Avias (1953). The ore can be found only at the peneplanation surface of Cycle 1, where it is intersected by younger erosion surfaces. At such places the subsequent erosion has been able to concentrate the thick ferruginous lateritic layer (up to 30 m thick), leaving the basal mineralized zone. Exploited by the "Le Nickel" Company with a treatment plant at Nouméa, the nickel ore extracted since its discovery in 1867 amounts to several tens of millions tons, with a percentage of nickel of more than 3%. The New Caledonia reserves are the most important in the world and production of the metal is over 30,000 tons per year.

Cobalt. A companion to nickel, cobalt is found in lower proportions in ferruginous laterites, especially as small concretions of bluish-black kidney ore.

Chromium. All the ultrabasic rocks carry chromite, which may sometimes be concentrated in masses or lenses (e.g., Tiebaghi mine, for a long time the most important chromium mine in the world).

Iron. The ferruginous laterites constitute a gigantic reserve of iron ore, consisting of more than 15 billion tons (Routhier, 1953), mainly goethite in the "earth" and hematite in the crust, which usually varies from 30 up to more than 65%. This ore invariably contains more than 1% chromium and reduces its interest from a metallurgical point of view.

JACQUES AVIAS
HENRI GONORD

References

Arnould, A., 1958. "Etude géologique de la partie Nord-Est de la Nouvelle Calédonie," thesis, Paris.
———, Avias, J., and Routhier, P., 1953 et seq. *Carte Géologique Coloriée au 1/100.000 de la Nouvelle Calédonie* (en 10 feuilles). Paris: O.R.S.T.O.M. (avec notices).
Avias, J., 1949. "Note préliminaire sur quelques observations et interprétations nouvelles concernant les péridotites et les serpentines de Nouvelle Calédonie (secteur central)," *Bull. Soc. Geol. France* Ser. 5, 19, 439–452.
———, 1953. "Contribution à l'étude stratigraphique et paléontologique des formations antécrétacées de la Nouvelle Calédonie centrale," *Sci. de la Terre* (Nancy), 1, 1–276.
———, 1958. "Sur l'existence d'une phase tectonique hercynienne tardive ayant affecté les formations antétriasiques de la côte Ouest de la Nouvelle Calédonie," *C. R. Acad. Sci.*, 246, 136–137.
———, 1959. Les récifs coraliens de la Nouvelle Calédonie et quelques uns de leurs problèmes," *Bull. Soc. Geol. France*, Ser. 7, 1, 424–430.
———, 1960. "A propos des vases bariolées gypsifères actuelles de la Nouvelle Calédonie et sur la genèse des marnes bariolées salifères du Trias," *Mem. B.R.G.M. Paris*, no. 15, 615–622.
———, 1967. "Overthrust structure of the main ultrabasic New Caledonia massives," *Tectonophysics*, 4(4–6), 531–542.
———, and Coudray, J., 1965. "Sur la présence d'éolianites en Nouvelle Calédonie," *C. R. Soc. Geol. France*, 327–329.
———, and Guerin, S., 1957. "Contribution à l'étude des faunes de céphalopodes permotriasiques de Nouvelle Calédonie Nautiloïdes et ammonoïdes du Permien et du Trias Inférieur," *Bull. Serv. Mines, Geol. Nouvelle Calédonie*, 1, 117–133.
Boureau, E., 154. "Découverte du genre Homoxylon Sahni dans les terrains secondaires de la Nouvelle Calédonie," *Mem. Mus. Hist. Nat.* (Paris), Ser. C, 3(2), 129–143.
Coleman, R. G., 1967. "Glaucophane schists from California and New Caledonia," *Tectonophysics*, 4(4–6), 479–498.
Colonna-Salard, M., 1964. "Contribution à l'étude de la Flore fossile de la Nouvelle Calédonie d'aprés des échantillons de bois minéralisés permotriasiques," Thèse, Paris (in *Paleontographica*).
Combes, P. J., 1963. "A propos du nickel dans les latérites nickélifères de la Nouvelle Calédonie," *C. R. Acad. Sci.*, 256, 211–212.
Davis, W. M., 1925. "Les côtes et les récifs coralliens de la Nouvelle Calédonie," *Ann. Géog.*, 34, 224–269, 332–359, 423–441, 521–558.
Deslongchamp, E., 1866. Documents sur la géologie de la Nouvelle Calédonie," *J. Conchyl.*, Ser. 3, 6, 384.
Fairbridge, R. W., 1950. "Problems of Australian geotectonics," *Scope* (Univ. West. Austral.), 1(5), 22–28.
Garnier, J., 1867. "Essai sur la géologie et les ressources minérales de la Nouvelle Calédonie," *Ann. Mines* (Paris) Ser. 6, 12, 1–92.
Glasser, E., 1903–1904. "Rapport sur les richesses minérales de la Nouvelle Calédonie," *Ann. Mines* (Paris), Ser 10, 5, 503–520, 623–693.
Gonord, H., and Coudray, J., 1966. "Précisions sur la position de la 'formation de la cathédrale' dans la presqu'île de Nouméa et les îles avoisinantes (côte sud-ouest de la Nouvelle Calédonie)" *C. R. Soc. Geol. France*, fasc. 2, 89.
Gubler, Y., and Pomeyrol, R., 1948. "Existence du Néogène marin en Nouvelle Calédonie," *C. R. Acad. Sci.*, 226(16), 1292–1293.
Guillon, J-H., and Routhier, P., 1971. "Les stades d'évolution et de mise en place des massifs ultramafiques de Nouvelle-Calédonie," *Bull. B.R.G.M.*, Ser. 2, 4, 5–37.
Koch, P., 1958. "Introduction a la géologie de la Nouvelle-Calédonie et dépendances. Notice explicative sur la carte géologique au 1/4,000,000 eme," *Bull. Geol. Nouvelle Caledonia*, 1, 9–22.
Locroix, A., 1942. "Les péridotites de la Nouvelle' Calédonie, levrs serpentines et leurs gîtes de nickel et de cobalt. Les gabbros qui les accompagnant," *Mém. Acad. Sci.* (Paris), Ser. 2, 66, 1–143.
Launay, J., and Recy, J., 1972. "Variations relatives du niveau de la mer et néo-tectonique en Nouvelle-Calédonie au Pléistocène supérieur et à l'Holocene," *Rev. Géogr. Phys. Géol. Dyn.*, Ser. 2, 14(1), 47–66.
Lillie, A. R., and Brothers, R. N., 1970. "The Geology of New Caledonia," *N.Z.J. Geol. Geophys.*, 13(1), 145–183.
Linden, W. J. M. van der, 1969. "Rotation of the Melanesian complex and of West Antarctica—a key to the configuration of Gondwana?" *Palaeogeogr., Palaeoclim., Palaeoecol.*, 6, 37–44.
Orloff, O., 1968. "Étude géologique et géomorphologique des massifs d'ultrabasites compris entre Houailou et Canala," thesis, Univ. Montpellier.
Piroutet, M., 1917. *Etude Stratigraphique sur la Nouvelle Calédonie*, Thèse, Paris; Mâcon, 313p.
Routhier, P., 1953. "Etude géologique du versant occidental de la Nouvelle Calédonie entre le col de Boghen et la pointe d'Arama," Thèse, *Mem. Soc. Géol. France* (67).
———, and Arnould, A., 1956. "Les gîtes de manganèse 'vulcanosédimentaires' de la Nouvelle Calédonie," *Symposium sur le Manganèse, 20 Cong. Géol. Intern.*, Mexico, 2, 313–330.
———, Caillère, S., and Kraut, F., 1956. "Etude géologique, minéralogique et structurale des gisements et minerajs de chrome de Tiébaghi (Nouvelle Calédonie)," *Bull. Soc. Géol. France*, Ser. 6, 6, 169–188.
Stanton, R. L., 1958. "Etude microscopique des chromites de nouvelle Calédonie," *Bull. Géol. Nouvelle Calédonia*, 1, 51–94.
Stearns, H. T., 1945. "Eustatic shorelines in the Pacific," *Bull. Geol. Soc. Am.*, 56, 1071–1078.
Tissot, B., and Noesmoen, A., 1958. "Les bassins de Nouméa et de Bourail (Nouvelle Calédonie)," *Inst. Fr. Petrol. Rev.*, 13(5), 739–759.

Waterhouse, J. B., 1956. "Recent French contributions to the geology of New Caledonia," *N.Z. J. Sci. Tech.*, 37, 587–97.

Cross-references: *Australasia–Regional Review; Fiji; New Guinea; New Hebrides; New Zealand; Norfolk Island.*

NEW GUINEA

The island of New Guinea ("Niugini" in the new phonetic style) consists politically of two parts, Irian Jaya or West Irian to the west and Papua New Guinea to the east. Irian Jaya (416,000 km^2, 160,618 sq mi) was formerly part of the Netherlands East Indies but is now an integral part of Indonesia. Papua New Guinea (474,650 km^2, 183,000 sq mi) is a new independent state (since 1975) succeeding the former Australian administered Territories of Papua and New Guinea. Papua New Guinea consists of the former British (and, since 1906, Australian) dependency of Papua (234,650 km^2, 90,600 sq mi) in the S and the former German colony (Kaiser-Wilhelm-land), later UN Mandate, the Territory of New Guinea (240,000 km^2, 93,000 sq mi) in the N. This latter area not only includes the northeastern part of mainland New Guinea but also the islands of New Britain (formerly Neupommern), New Ireland (formerly Neumecklenburg), which together with the Admiralty Islands constitute the Bismarck Archipelago, and finally the two westerly islands of the Solomon Chain (Buka and Bougainville). In this *Encyclopedia* the *Bismarck Archipelago* (q.v.) is treated separately and the *Solomon Islands* (q.v.) as a unit.

New Guinea is the second largest island in the world; it measures 2400 km (1500 miles) in length, with outlying islands covering about 805,000 km^2. The forbidding nature of the coasts of this jungle-covered island, its primitive inhabitants, and its position far from the main trade routes at first offered few incentives for European colonialization.

During the three centuries prior to 1884 only the coasts were charted, and from that year the reconnaissance of the interior began but made slow progress. The first find of gold in Papua in 1887 stimulated exploration both by government and private parties. Oil seepages were first reported in Papua in 1911, resulting in some exploration by various companies. The joint, large-scale effort of the Australasian Petroleum Company and Island Exploration Company lasted from 1937 to 1961. Some gas, economically without value, was the only discovery. The findings of the oil companies mentioned above have been published (Australasian Petroleum Company, 1961. The discovery of gold in commercial quantities, in 1898 in Papua and in 1921 in the Mandated Territory, gave the eastern part of the island a great advantage in economic development.

In West Irian the Netherlands Military Exploration group (1907–1915) mapped the larger part of the interior, and here the discovery of oil seepages, both on the northern coastal area and in the western peninsulas, led to geological mapping by the Netherlands Indies Mining Department (1917–1922). Between 1935 and 1960 exploration for oil was carried out by the Nederlands Nieuw Guinee Petroleum Maatschappij, first in the western peninsulas and around Geelvink Bay, and after 1955 in the areas to the E, by aerial mapping, geological and geophysical field work, and deep drilling. The only results were three small oil fields discovered in the western peninsulas in 1939–1940 (Visser and Hermes, 1962). The present paper is based largely on the published work of oil companies in both parts of New Guinea.

In Papua New Guinea and in several offshore blocks search operations for oil by several companies continue at the present time. Geological mapping is being carried out by the Australian Bureau of Mineral Resources. A 1:1 million map and a series of 1:250,000 geological maps have been issued or are in preparation.* During the last decade the establishment of the University of New Guinea in Port Moresby, with excellent library facilities and a small department of geology, is revolutionizing the scientific picture. A turning point in the history of the territory was marked by the meeting of the Australian and New Zealand Association for the Advancement of Science in Port Moresby in 1970 (see special issue of *Search,* 1, no. 5).

In the western part of the island the geological mapping by the government and the mineral exploration by a Netherlands Government-sponsored foundation was terminated in 1962 (d'Audretsch *et al.*, 1966). There was little further activity until 1967, when foreign exploration began again, including interest in offshore oil.

Geomorphic and Geotectonic Divisions

New Guinea and Australia are united by a common shelf, the Arafura Sea and Torres Strait—being shallower than 200 m. The region is enclosed by the 1000-m isobath that outlines the Australian continental block. To the W the Banda Island arcs belong to an intercontinental orthogeosyncline between Australia and Asia

FIGURE 1. Structural map of New Guinea and adjacent parts of Australia and the East Indies.

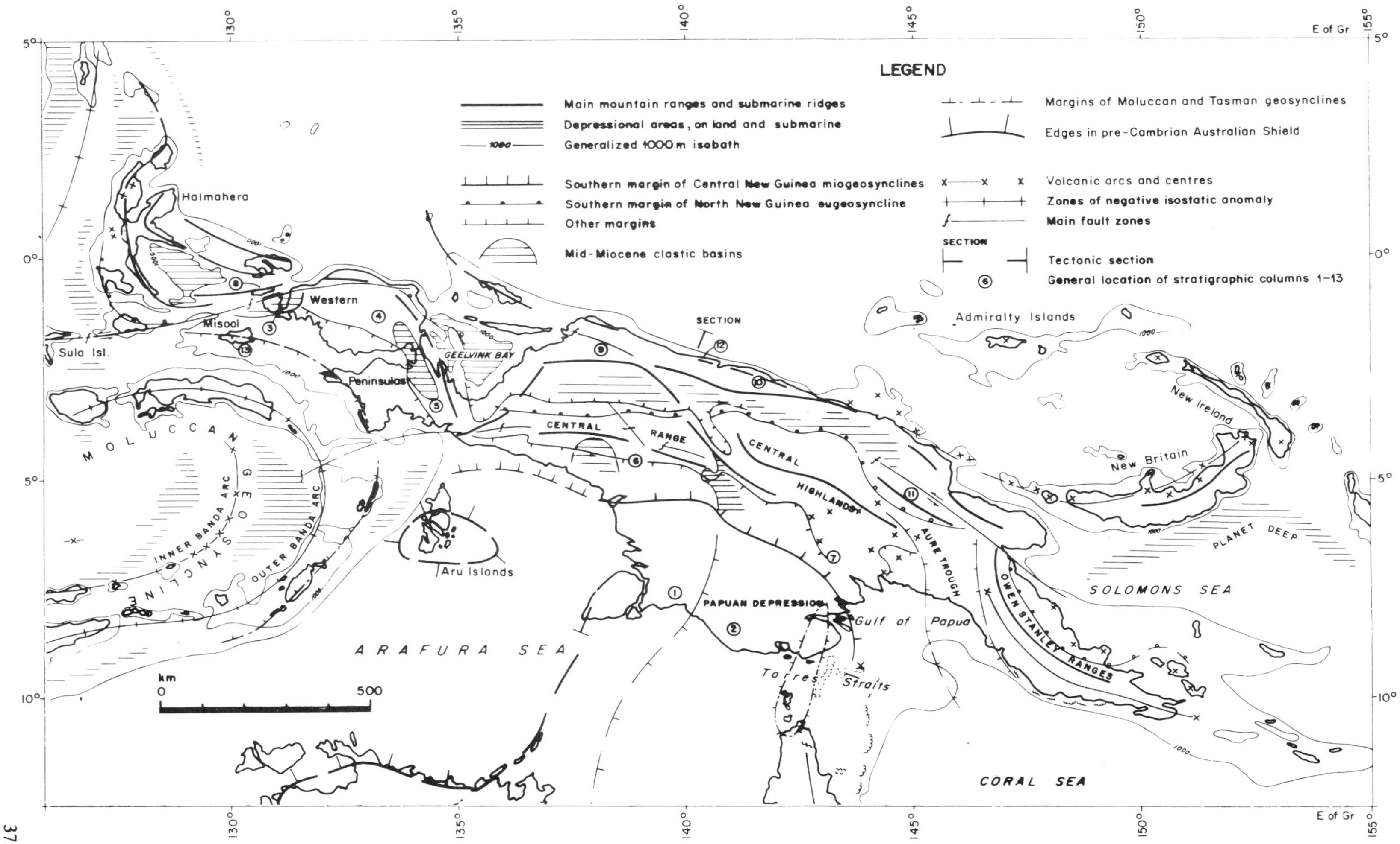

(Moluccan geosyncline). To the E lie the deep seas and island arcs of the Melanesian region. New Guinea lies in the mobile northern rim of the Australian craton. As the orogenesis is very youthful there exists a straightforward correlation between geomorphologic and tectonic elements, and the trends of coasts and mountain ranges follow the main structural trends.

The following **geotectonic units** are distinguished, from S to N (see map, Fig. 1, and tectonic section, Fig. 2):

1. Granite, presumably belonging to the *Australian Shield*, is probably close to the surface in the Aru Islands. In the southernmost tip of New Guinea and in the islands of the Torres Strait there is Upper Carboniferous granite and Carboniferous volcanics are part of the Tasman orogenic belt of eastern Australia.

2. The *stable shelf of southern New Guinea* extends from the Aru Islands to the Gulf of Papua. Southern New Guinea is covered mainly by lowlands and swamps. To the south a gentle E-W upwarp (the Oriomo trend) creates low plateaus and ridges with relics of a former piedmont alluvial plain and fluvial terrace deposits. The stable shelf was subjected to epeirogenic warping movements only (as in Aru), resulting in an almost horizontal incomplete stratigraphic sequence (Fig. 3, strat. cols. 1, 2). The sedimentary cover increases in thickness toward the E with a gentle slope into the Papuan depression, which is related to Australia's Great Artesian Basin.

3. Toward the N the stable shelf grades into the *miogeosyncline of central New Guinea*, where prolonged subsidence and deposition was followed by orogeny during the Cenozoic. This belt contains the highest mountain ranges of the island, rising to nearly 5000 m in the Nassau Range. The highest peak is Mt. Carstensz, at 4884 m. Initial orogenic movements took place in the Oligocene and early Miocene; the final phase was late Plio-Pleistocene and continues in places into the Holocene.

Five structural provinces are distinguished within the miogeosynclinal belt, from W to E.

(a) In the *western peninsulas* of West Irian the trends are parallel to the arcs of the Moluccan geosyncline. There is a pronounced relief that is expressed in a step-by-step rise from the Miocene and Plio-Pleistocene covered lowlands in the SW to the lofty mountain ranges (up to 2600 m) in the N and E. In the highlands strongly folded Silurian is unconformably overlain by a Permo-Carboniferous to Miocene sequence (strat. cols. 4, 5). Epeirogenic movements took place in the late Mesozoic (strat. cols. 3, 4).

(b) Orographic and structural trends of the *Central Range* are mainly E-W. The crestal part is situated toward the southern side of the range. The highest peaks are covered above 4350 m by firn, and small glaciers totalling about 10.5 km^2 in extent (Dozy, 1938; Verstappen, 1964; Peterson and Hope, 1972). During the latest Pleistocene cold period the snow line dropped by an average of 1000–1100 m, locally even more. A moraine formed by an ice tongue from the Carstensz glacier extended as low as 1700 m above mean sea level (dated by radiocarbon 10,100–10,500 B.P.). The southern slopes of these mountains drop down from nearly 5000 m within 30 km to the foothills, not more than 1000 m high, which pass, in turn, abruptly into the plains. In the southern slopes traces of old Paleozoics were found that are apparently conformably overlain by a complete sequence of Permo-Carboniferous to Miocene (strat. col. 6). In the crestal area (Nassau Range) there is often an unconformity between the Mesozoic and the Tertiary sequence (Dow, 1968; Van der Wegen, 1966). The uplift of the Central Range commenced in the N early in the Miocene and progressed S; it was accompanied by minor igneous activity, of intermediate to acid character. During the Plio-Pleistocene a periorogenic trough in the present foothill belts and plains to the S was filled with clastics

FIGURE 2. Structural profile through the central parts of New Guinea.

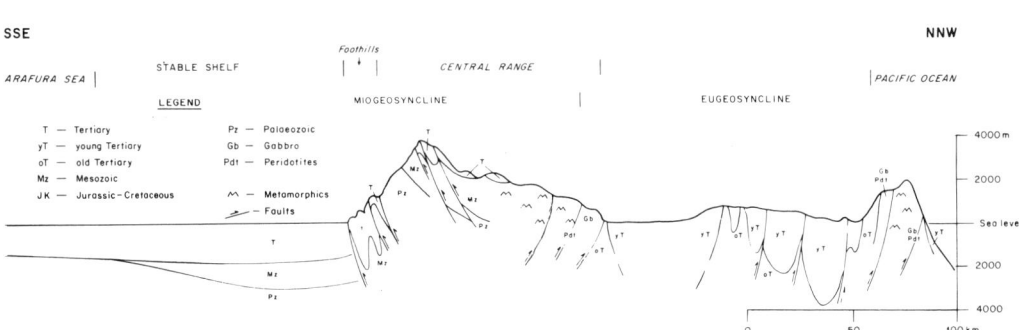

FIGURE 3. Stratigraphic columns (1–7) through the shelf and miogeosynclinal sectors of New Guinea. (For an explanation, see the text.)

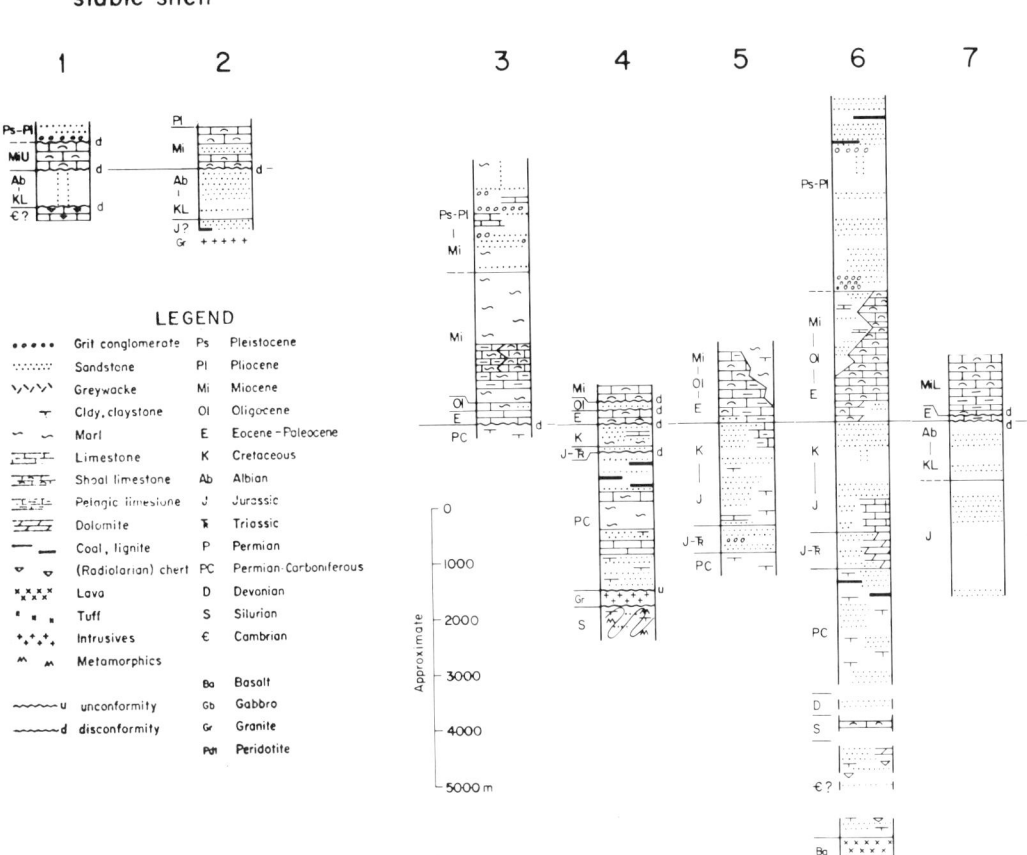

("molasse") derived from the rising range (strat. col. 6).

(c) A NW-SE-trending depression separates the Central Range province from the Central Highlands to the E. Here the highest peaks rise over 4000 m. In the tectonically highest raised northern part, metamorphics and granite are overlain by remnants of Mesozoics; in the NE in the transition to the eugeosyncline, some Permian and Traissic overlie granite (strat. col. 11). The Central Highlands represent basically a stable crustal block which is regarded as the northernmost extension of the Palaeozoic metamorphics and igneous rocks overlain by several thousand meters of Mesozoic and Tertiary sediments. These sediments are only gently folded, this being expressed in broad folds such as the Kubor and Müller anticlines. The northern boundary of the crustal block is formed by a westerly to northwesterly trending fault zone in the east, and the Lagiap fault zone in the west, roughly coinciding with the main drainage divide. Late Cretaceous and early Tertiary epeirogenic movements occurred in the southern part, where there is a wide foothill belt with a sequence beginning with Jurassic (strat. col. 7). The main orogenic movements are of Plio-Pleistocene age; they were followed by the ejection of large masses of basic to intermediate lavas and tuffs from large strato-volcanoes in the southern and eastern border zones. They are now extinct.

(d) The *Aure trough* or *Papuan Geosyncline* extends between the stable shelf to the S and the Central Highlands to the N and constitutes an area of prolonged subsidence and deposition followed by intensive orogeny manifested in tight compressional folding and high-angled strike faulting. The main structural trend is NW to WNW. A total thickness of probably over

FIGURE 4. Stratigraphic columns (8–13) through the eugeosynclinal sectors of New Guinea. (For an explanation, see the text.)

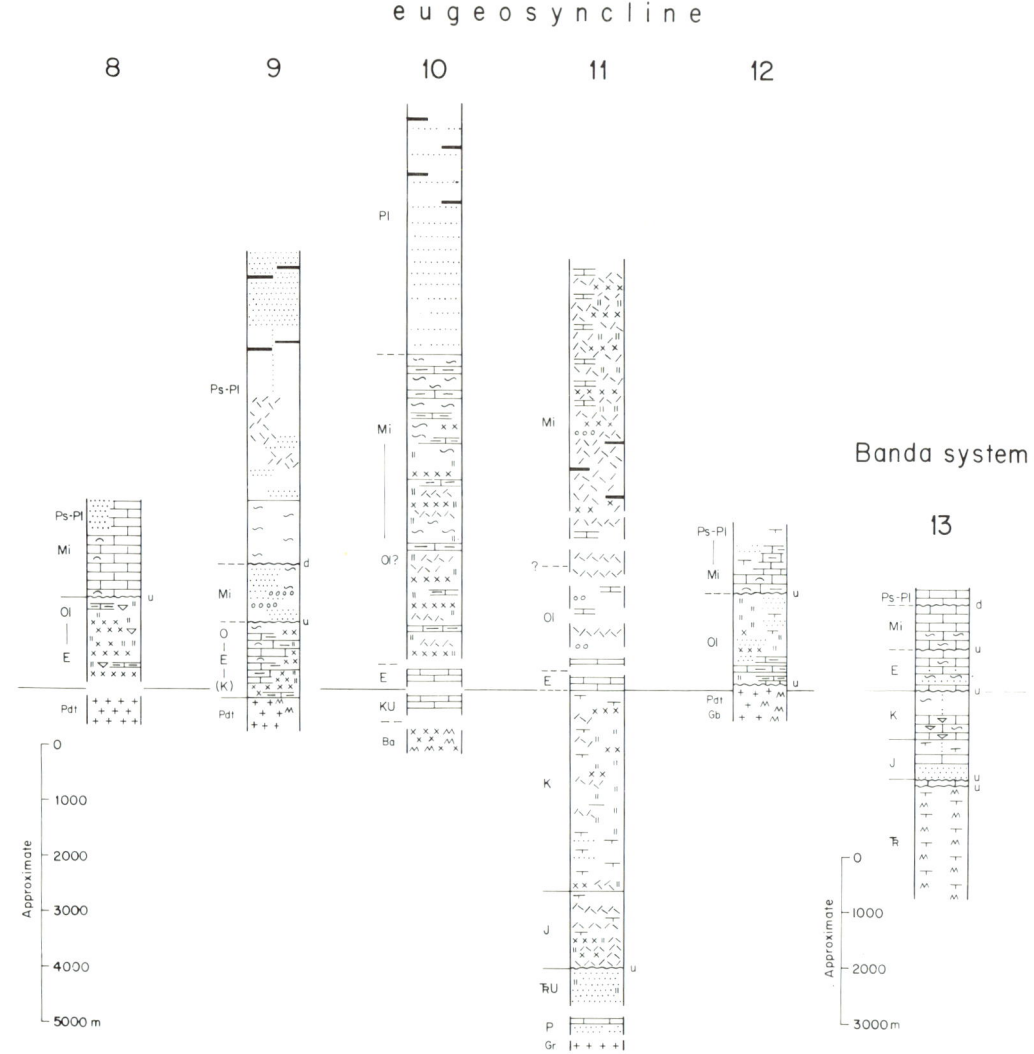

16,000 m of sediments, of which at least 15,000 m are of Miocene to Pliocene age, was accumulated in this trough. A great variety of sedimentary rocks is present, the most common of which are limestone, mudstone, siltstone, graywacke, sandstone, and shale. Plutonic igneous rocks and metamorphic rocks are notably absent, but volcanic cones of basic to intermediate composition occur in the central and eastern parts, forming spectacular landmarks. To the SE the Aure trough extends into the wide and deep (4650 m) depression of the Coral Sea, which is fed by several important submarine canyons (Winterer, 1970).

(e) The *Owen Stanley Ranges,* the rugged backbone of eastern Papua New Guinea, is separated from the Central Highlands by the eastern part of the Aure trough. The Owen Stanley Ranges, rising to over 4000 m in places, are a relatively uniform structural unit. They consist of a core of mainly low-grade Mesozoic (Jurassic) metamorphics with only few Tertiary intrusions of granodiorite, diorite, and gabbro. To the SW the metamorphics are flanked by a series of mainly basic lavas, agglomerates, conglomerates, and pyroclastics. To the NE the metamorphics are bounded by a prominent fault zone forming a sharp boundary with the ultrabasic belt.

4. To the N of the miogeosyncline, a mobile belt, the *eugeosyncline of northern New Guinea,* occupies the region between the north-

ern part of the central mountain ranges and the Pacific Ocean, extending from the island of Halmahera in the northwest to the Owen Stanley fault trough and the island of New Britain in the east. Intense igneous activity during the late Mesozoic and early Tertiary was accompanied and followed by strong subsidence and rapid deposition. Three provinces are distinguished in the eugeosynclinal belt from west to east:

(a) A number of arcuate *submarine ridges* connect Halmahera (*Indonesia,* q.v.) with northwestern New Guinea. In Halmahera an Upper Cretaceous to Eocene volcanic-sedimentary series overlies ultrabasic igneous rocks; a similar sequence is found in New Guinea, where it is unconformably overlain by younger sediments (Fig. 4, strat. col. 8). Compressional movement probably occurred at the end of the Oligocene; during the Plio-Pleistocene, block faulting and strong differential movements gave the area its present shape. An important W-E-directed fault zone (Sorong Fault Zone) with a large left-lateral component forms the southern border of this province and separates it from the miogeosyncline to the S.

(b) The *north coastal region* of New Guinea between Geelvink Bay and the Solomon Sea consists of low hills and elongate mountain ranges, in the W rising to over 2000 m, and in the E to over 4000 m, that enclose extensive swampy lowlands. The northern slopes of the Central Range and Highlands are formed of Jurassic and Cretaceous metamorphics, predominantly of low-grade ultrabasic and basic intrusives and early Tertiary basic volcanics, marking the southern eugeosynclinal rim. Its northern rim lies along the Pacific Ocean and is locally deformed into a high and narrow mountain chain (strat. col. 12). In between there is a deep trough that was formed after an orogenic phase. It received large thicknesses of Miocene and Plio-Pleistocene, mainly clastic sediments derived from the rising mountain ranges in the S (strat. cols. 9–11). During the Plio-Pleistocene, large differential vertical movements and block faulting gave rise to the present topography. In the E the major Ramu-Markham Valley fault zone, presumably with a left-lateral displacement, strikes into the Solomon Sea.

(c) In the region next to the Solomon Sea, the ranges northeast of the Owen Stanley Ranges are composed of a belt of early Tertiary ultrabasic to intermediate intrusions and Eocene sediments separated from the miogeosynclinal area to the SW by a narrow thrust-fault zone, the Owen Stanley fault, which forms a distinct topographic lineament. Young Tertiary sediments and young Tertiary to Holocene volcanics forming spectacular strato-volcanoes occupy the coastal stretches. In the mountainous island of New Britain presumably Paleozoic to Mesozoic metamorphics and granites are overlain by Tertiary sediments. It is suggested that the island is a disrupted part of the northern New Guinea eugeosyncline. The high seismicity of New Britain, with shallow, intermediate, and deep shocks, stands in strong contrast to the seismically quiet central and southern parts of the island of New Guinea. An active volcanic belt lies along the north coast of New Britain and extends to the W some distance off New Guinea's northern shore. There is an important caldera at Talasea, where the last eruption was 80 years ago (Lowder and Carmichael, 1970). The main part of the Solomon Sea is some 5000 m deep; an arcuate deep-sea trench (over 7000 m) parallels the southern coast of New Britain, evidently an area of contemporary subduction, with its "Benioff zone" dipping to the WNW.

5. To both W and E, "exotic" island arcs are welded onto New Guinea. In Misool and the low ranges along the southwestern shore of the western peninsulas, the sedimentary sequence is transitional to that of the Moluccan geosyncline (Fig. 4, strat. col. 13). In Misool there is evidence of a late Triassic orogeny; in the Moluccan geosyncline, sedimentation, partly bathyal, commenced in the early Permian and an ophiolitic complex was formed possibly during late Mesozoic to early Tertiary. Orogenesis took place in mid-Tertiary and mid-Quaternary.

In the E the Admiralty Islands (Bismarck Archipelago) and New Ireland belong to a young festoon of island arcs that extends along the Pacific border via the Solomons and New Hebrides to Fiji. The islands contain some schists and ultrabasics, older Tertiary basalts, and Miocene clastics and volcanics. Extinct and active volcanoes, high seismicity, and deep-sea trenches demonstrate the instability of this part of the earth's crust.

The structural development of New Guinea is the result of the interaction of the Australian and Pacific plates. As the Australian plate is now moving N and the Pacific plate moving W, the area is somewhat complicated (Carey, 1970). Strike-slip movements and rotation or differential motions of smaller lithospheric plates are predictable results. Seismic zones show that the present main contact between the Australian and Pacific plates is situated along the north coast of New Guinea and south coast of New Britain and the Solomons Islands; smaller lithospheric plates in the Bismarck and Solomon seas are indicated by lesser seismic zones (Denham, 1969). The relative movement between the Australian and Pacific plates at $3°S142°E$ is a compression of 10.7 cm/yr (Le Pichon, 1970). According to Carey (1970),

FIGURE 5. Left-lateral (sinistral) shears transecting New Guinea along E-W lines and relating to a "Melanesian magashear" (Carey, 1970). A-B, repeated, represents total displacement (1300 km). It should be stressed that this is a highly speculative interpretation, hardly supported by field evidence.

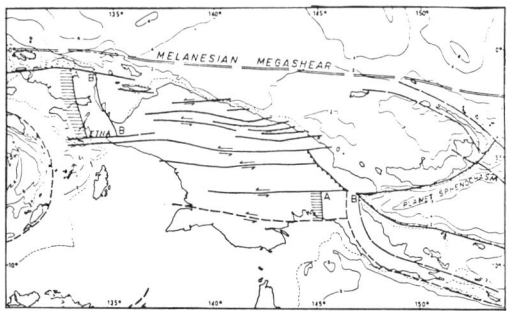

Davies and Smith (1971), Krause (1965, 1966), and others, this results in left-lateral (sinistral) shear along the line of the Solomon Islands, New Ireland, and Manus Island (Fig. 5), together with a possible rotation of smaller lithospheric plates in the Bismarck and Solomon seas (see *Bismarck Archipelago,* Fig. 1), the sinking and subduction of the Australian plate into the trench S of New Britian and the Solomon chain, and the buckling and overthrusting of lithosphere in the central New Guinea highlands.

Stratigraphic History

Lower Paleozoic. Sediments assumed to be deposited during the Cambrian, mainly because of lithologic affinities with rocks of that age in northern Australia, were found in boreholes in the western part of the stable shelf and in the southwestern part of the Central Range. They consist of a rapid alternation of fine clastics and carbonates, some 1600 m thick. They are underlain by basalts and intruded by quartz diabase (Fig. 3, strat. cols. 1, 6).

The only known *Silurian* in situ occurs in the northern mountainous area of the western peninsulas. The strongly folded slates and subordinate quartzites contain graptolites. They are several thousands of meters thick and are associated with granites (strat. col. 4).

In the southern slopes of the Central Range, river pebbles containing *Silurian* and *Devonian* fossils have been found, presumably derived from nearby outcrops (strat. col. 5). Whereas in the western peninsulas the Silurian is unconformably overlain by Permo-Carboniferous, photo and field geology indicate that the southern slopes are built up of a conformable N-dipping sequence. Hence in the Central Range we have no evidence of a Paleozoic orogeny, and there is no support for the assumption that the Central Range would form the connection between the Paleozoic orogen of the western peninsulas and the Tasman geosyncline in eastern Australia.

Upper Paleozoic. *Carboniferous* and *Permian* were found in a narrow belt that extends over almost the entire length of the miogeosyncline in the western part of the island. The Permo-Carboniferous commenced with a terrestrial sedimentation of quartzites with fossiliferous sandy shale intercalations, followed by a shallow marine sequence of slates, sandstones, and limestones, which finally becomes paralic in the upper part, as shown by the intercalation of coal seams with well-preserved plant remains (strat. cols. 4, 6). Fossils indicating Upper Carboniferous, Permian, and doubtful Lower Triassic occur. The sequence has a thickness on the order of 2450 m. In the Central Highlands in Papua New Guinea, marine limestone and sandstone with Permian brachiopods and foraminifera overlie granite (strat. col. 11). Isotopic dating of granites intruding the Torres Strait volcanics have given an age of 295 m yr or Upper Carboniferous (Richards and Willmott, 1970).

Triassic. Triassic rocks proved by fossils are known only from the eastern Central Highlands and from the island of Misool in Indonesia. In the former locality some 1500 m of graywackes, sandstones, siltstones, and some tuffs contain a shallow-marine fossiliferous association (strat. col. 11). In Misool there are more than 2000 m of shallow-marine, strongly folded clastics of Lower Carnian and possibly Ladinian age, separated by an angular unconformity from the thin (max. 100 m) Norian carbonates and clastics (strat. col. 13).

In the miogeosyncline in West Irian up to some 500 m of red and mottled sandstones, shales, and claystones—unfossiliferous and locally conglomeratic—are conformably intercalated between proved Permo-Carboniferous and Middle Jurassic sediments. Toward the E, dolomitic limestones, shales, and dolomites become predominant. The beds are for the most part paralic or terrestrial and have been assigned to the Triassic to Jurassic.

Jurassic and Cretaceous. In the Papuan depression, sandstones with minor lignites of presumably Jurassic age and shallow-marine Lower Cretaceous and Albian sandstones and mudstones, some 1500 m thick, overlie a granitic basement (strat. col. 2). In the western part of the stable shelf Barremian to Albian glauconitic sandstones and claystones (0–600 m preserved) disconformably overlie presumed Cambrian (strat. col. 1).

Jurassic and Cretaceous marine sediments are widespread in the miogeosyncline. The transgression is of Bajocian to Oxfordian age in the west conformably over the terrestrial Triassic and Jurassic; in the E over granite or metamorphics. In the W three major facies types are distinguished:

(a) A glauconitic sandstone-claystone facies with arenaceous foraminifera is found nearest the stable shelf (strat. cols. 5, 6).

(b) Another, shaly, slaty, phyllitic facies with ammonites and with subordinate arenites and limestones occurs toward the euogeosyncline in the N. Jurassic in a comparable argillaceous facies occurs as far to the W as the Sula Islands. The transition between these facies closely follows the present sinuous trends from the N of the western peninsulas through the isthmus W of the Geelvink Bay into the Central Range. In the N of the Central Range the sediments are mildly metamorphosed. The sediments range from the Bajocian into the Paleocene and are up to 2000 m thick.

(c) A third facies occurs in the mountains along the west coast of Geelvink Bay and consists of some 1000 m of fine-grained limestones and some marls with abundant pelagic foraminifera of Cenomanian and younger age (strat. col. 5).

In the Central Highlands similar facies changes occur. In the NE (strat. col. 11) some 5000 m of Jurassic and Cretaceous unconformably overlie the Triassic and Permian. Large masses of granodiorite have been intruded possibly during the late Triassic to early Jurassic movements. This sequence lies in the transition from the miogeosyncline to the eugeosyncline and is composed of marine shales, graywackes, and siltstones with considerable quantities of tuffs, basaltic agglomerates, and minor basalt flows.

An open-marine, partly bathyal, Jurassic-Cretaceous sequence is described from the island of Misool in the rim of the Moluccan geosyncline. About 1450 m of clastics, radiolarites, and pelagic limestones overlie the Triassic with a slight unconformity (strat. col. 13).

The facies distribution of the miogeosynclinal Jurassic and Cretaceous suggests a southern supply area, i.e., parts of the Australian continent; no indications of a northern land mass have been obtained.

Paleocene to Early Miocene. In the ranges along Geelvink Bay's west coast and in the Central Range the Tertiary overlies the Cretaceous without a break (strat. cols. 5, 6); elsewhere, however, movements of an epeirogenic nature took place, and early or middle Tertiary sediments disconformably overlie older formations (strat. cols. 1–4, 7).

Around the Mesozoic-Tertiary boundary, the supply of clastic material stopped and the sediments became largely calcareous. When clastic sedimentation was resumed later, the material was derived not from the S but from rising land masses in New Guinea itself.

It should be noted that the chronology of the Tertiary is based on planktonic and larger benthonic foraminifera, the correlation of which with the European classification is approximate and used here only for convenience.

Early Tertiary is absent in the stable shelf and is thin and incomplete in the larger part of the Central Highlands. In the western part of the miogeosyncline the early Tertiary consists mainly of shallow-water limestones with a tendency to grade N into pelagic ones (strat. cols. 3, 7). Their thickness is about 1000 m. A clastic development during the Oligocene in the northern part of the western peninsulas indicates the initial rise of the northern coastal mountain ranges (strat. cols. 3, 4).

In the eugeosyncline there are at least 4000 m of basic extrusives (basalt and spilites, volcanic breccias, tuffs, and tuffites) intercalated with pelagic and in some places shallow-water limestones, and in the E with graywackes (strat. cols. 8–12). In the ranges along the central part of the northern coast this sequence overlies a series composed of metamorphics, basic and ultrabasic intrusives, and some granite and diorite. The early Tertiary volcanics may be genetically related to the ultrabasics; the latter have possibly been emplaced during or rather before the early Tertiary. Early Miocene tectonic movements, presumably of a compressional nature, caused a general shallowing of the trough and were accompanied by the intrusion of gabbroic and dioritic masses.

In the southwestern coastal ranges of West Irian the early Tertiary is developed in shallow-water limestone facies, which grades into a pelagic facies toward the Moluccan geosyncline (strat. col. 13).

Middle Miocene to Late Miocene. This was an epoch of widespread transgression, during which the stable shelf of southern New Guinea was again invaded by the sea, and a thin sequence of shallow-water limestones was laid down (Fig. 3, strat. cols. 1, 2).

In the miogeosyncline some 600 m of shallow-water and reef limestones were deposited. However, on the southern side of the present mountain ranges and locally in the Central Range (Dow, 1968), the calcareous shelf was warped by basinal depressions in which some 2000–3000 m of clastics and locally marls, pelagic, and shallow-water limestones were deposited (strat. cols. 5, 6). The clastics were presumably derived from the rising ranges to the N. Open connections to the ocean, however, must have existed, especially in the eastern and western extremities, to account for the abundance of planktonic foraminifera. In the Aure trough over 6000 m of Miocene, mainly graywackes and mudstones, were deposited. In this trough the astonishing thickness of 12,000 m accumulated from late Oligocene to Pliocene time (Davies and Smith, 1971). A mid-Miocene reef limestone on the hinge line of the trough is over 1000 m thick. Strong volcanism, lasting into the Pleistocene, occurred on its northeastern edge.

In the eugeosyncline after shallowing at the end of the preceding epoch, the rate of subsidence increased again and some 2000 m or more of graywackes and shales were deposited with an admixture of volcanic and calcareous material (strat. cols. 9–11). Indications of major slumping are widespread. The clastic components show affinities to the rocks exposed in the present mountain ranges, and the implication is that the source of the material was situated in the northern parts of the ranges, the southern parts still being depositional areas.

Late Miocene and Pliocene. During this epoch the mountain ranges became fully emerged and contributed large amounts of material to the subsiding troughs adjoining to the N and S. Along the southern side of the central mountain ranges, a continuous exogeosynclinal trough, in which the separate basins of the preceding epoch had merged, received up to 5000 m of sands, clays, conglomerates, and lignites (strat. cols. 3, 6). To the N of the ranges, the sediments may reach a thickness of some 7000 m; here sedimentation commenced with *Globigerina* marls, followed by graywackes and subgraywackes, claystones, and siltstones with lignites; the latter become ever more prominent in the upper members of the formations, indicating a gradual silting up of the basin (strat. cols. 9, 10).

Vertical movements that brought into being the orographic features we see today occurred mainly along faults, in places accompanied by magmatic activity. In the crestal zone of the Central Range only a few bodies of quartz diorite were emplaced; large bodies of granite were intruded in the coastal ranges along the west coast of Geelvink Bay.

Pleistocene and Holocene. The orography of New Guinea is young, the rise of the mountain ranges lasting from young Tertiary into possibly Holocene times. The stable shelf of southern .New Guinea became an area of swampy plains, caused by eustatic oscillation and sedimentation, without tectonic deformation taking place. River terraces with well-rounded pebbles testify to a greater runoff and loading, or more pronounced topographic relief than now, possibly during glacial times. Eustatic rise after the last glaciation caused the inundation of the Arafura Sea and adjacent shallow shelf areas. Patterns of Pleistocene stream channels are still readily determined from the bathymetry, and the peculiar "sungeis" (drowned river channels) of Aru are evidence of youthful warping parallel to the New Guinea trend (Fairbridge, 1953).

In the Central Range of New Guinea positive isostatic anomalies are found, even outside the belt of ultrabasics. Hence it would seem that the young Tertiary to Quaternary uplift of the range cannot be ascribed to isostatic readjustment of a mass deficiency, but it could be due to a still active force that caused thrusting over the low block to the N (Fig. 2 tectonic section II). Uplift commenced in the N early in the Miocene and may have taken place along S-dipping thrusts. At later stages of the orogeny, first the southern part of the Central Range rose at the beginning of the Plio-Pleistocene, and afterward (late Plio-Pleistocene) the present foothills rose relative to the southern plains. Whereas both crestal parts and foothills are intensively folded and faulted parallel to the main trend of the range, the southern slopes show only N-dipping beds. It seems that in the southern parts stresses were relieved in two zones of dislocation, which for a large part could be N-dipping thrusts.

In the northern parts of the Central Highlands of eastern New Guinea, faulting seems to have played a dominant role. In the southern parts compression diminishes toward the stable shelf; an imbricate zone is followed by a strongly folded belt, a gently folded zone adjoins the plains. The imbrications are N dipping; the anticlines are asymmetric and steeper toward the S. The orogeny is Plio-Pleistocene. Piedmont deposits along the northern slopes today are marked by arkosic quartz-poor facies (Ruxton, 1970). There is evidence at high elevations of small Pleistocene glaciers and there is some periglacial activity (Löffler, 1970, 1972; Williams et al., 1972). Traces of postorogenic, now extinct, volcanoes dominate the landscape. The lava flows, tuffs, and agglomerates are of andesitic composition.

In the Aure trough, between the Central Highlands and the Owen Stanley Ranges, the strata have been subjected to strong compression causing numerous tight, steeply flanked and asymmetric folds. The region to the N of the Central Mountain Ranges was deformed mainly along the faults running parallel to the strike of the island. Here mountain chains rose, separating low swampy plains in intramontane depressions. In the low hilly areas there are narrow-crested and steep anticlines separated by wide and flat bowl-shaped synclines. Gravitational gliding and collapse of surficial formations over buried faults appear to have played a major role in shaping the structure (Jenkins, 1974).

In the northern region the orogeny is also of Plio-Pleistocene age, whereas raised Pleistocene to Holocene coral reefs, in some cases situated quite far inland, testify to the extreme youth of the latest movements. Radiometric dating shows uplift averaging 1 m/1000 years (Veeh and Chappell, 1970). Their work shows a sequence of reefs on the Huon Peninsula to be disposed in groups, evidently discontinuously raised [complex I, 6000 B.P.; II and III (10–30 m), 30,000–50,000 B.P.; IV (110 m), 60,000–74,000 B.P.; V (200 m), 116,000–140,000 B.P.; VI, 180,000–190,000 B.P.; VII, 210,000–250,000; and VIII, older and higher]. Post-Miocene uplift in southeastern Papua was in two main stages and may exceed 3000–4000 m (Smith, 1970).

In the main part of New Guinea between Geelvink Bay and the Solomon Sea, the mountain ranges show both E and N trends, forming together a huge trapezoid pattern that is reminiscent of similar structural outlines in the Australian craton. In strong contrast to this rectilinear pattern stands the sinuosity of the trends in both the Solomon Sea area and the western peninsulas. The trends in the latter are parallel to the Banda Island arcs of the Moluccan geosyncline, but the stratigraphic development shows them to be part of New Guinea and hence of Australia, the Moluccan geosyncline at present abutting them as an "exotic" element. Although in the western peninsulas and Sula Islands the facies distribution is similar to that of the regions to the E, an Australian source area is obviously obstructed by the Moluccan geosyncline. A solution might be found in the assumption that in the pre-Tertiary the south Moluccan and Misool region occupied a site farther to the W, and that the arcs of southern Moluccas and western peninsulas acquired their present relative positions only during the Tertiary, the eastward movement having been effected mainly along the Sorong fault zone. The sinuosity would then appear to be an acquired property.

In the Solomon Sea area the sea-facing arc of New Britain strikes topographically into the coastal mountain ranges of northeastern New Guinea, and although the distribution of seismic foci and the volcanic belt along the northern coast also suggest a structural connection, the geology of central New Britain seems more akin to the Owen Stanley Ranges than to northeastern New Guinea. To the NE, New Britain abuts against New Ireland, but the deep-sea trench to the S has a curvature conforming to the trends of both islands (with a "dog-leg" shape). Here also strike-slip movements of crustal blocks may have taken place, e.g., formation of the Solomon Sea by N drift of New Britain and left lateral drift of the Admiralty-New Ireland island arc relative to New Guinea. Then the volcanism and seismicity would suggest continuing movements.

Pacific-ward drift of Australia-New Guinea, as indicated by paleomagnetic measurements, has probably been the underlying geotectonic tendency since Permian times when Australia lay in the vicinity of the South Pole. A relatively stable central part forming the front of the advancing continent is adjoined by more mobile zones in the NE, N, and E. The structure of the western peninsulas of New Guinea was conditioned by interactions with neighboring Southeast Asia; in the region of the Solomon Sea the inferred movements of crustal plates may be due to interaction of the Pacific plate and the more mobile "microplates" of Melanesia (q.v. *Solomon Islands; New Caledonia; Fiji*).

Mineral Deposits

New Guinea was until recently considered relatively poor in mineral resources. A little gold and also silver in the E are the main precious metals.

The gold is mainly associated with Mesozoic and Tertiary granodioritic and dioritic intrusions in the northeastern coastal ranges and in the northeastern Central Highlands; it is also present in the northern and northeastern parts of the Owen Stanley Ranges. Economically unattractive occurrences are reported from terraces in the Eilanden River area in the southeastern foothills of the Central Range (Van der Wegen, 1966). The metal is concentrated mainly in placer deposits, which are the main producers. Some gold and copper deposits are known from the small islands off the "tail" of New Guinea (Louisiade Archipelago, Misima, and Tagula).

Oil production in West Irian dropped from over $\frac{1}{2}$ million m^3 to one-tenth this figure from 1955 to 1965. Gold production in the E dropped from about 2300 to 1200 oz in the same period. Important oil discoveries in the E from 1970 onward constitute the major economic potential.

Although the existence of copper deposits has been known for some time (Dozy, 1939), extensive exploration has only been started recently. One of the main obstacles to mining is poor accessibility. Mining of a large porphyry copper ore deposit of approximately 900 million tons of ore containing 0.5% copper on Bougainville Island (politically part of Papua New Guinea) started in 1972. This site will be among the largest open-cut copper mines in the world. On the mainland of New Guinea copper ore deposits of 2.5-3% have been proved in the Ertsberg in West Irian and in the eastern Star Mountains in Papua New Guinea and are being developed by Freeport Sulphur and Kennicot Copper, respectively. Radiometric dating of the gold and porphyry copper mineralization in the highlands ranges from late Miocene (Yanderra), Pliocene (Morabe) to mid-Pleistocenc (Mt. Fubilan), according to Page and McDougall (1972).

Nickel-cobalt ores of low grade but in large quantities have been reported from West Irian on Waigeo Island and in the Cyclops Mountains near Djajapura, and prospecting has been resumed recently. Nickel also occurs in the ultrabasic belt of eastern Papua but concentrations are low.

W. A. VISSER
ERNST LÖFFLER

*The Division of Land Use Research of the Commonwealth Scientific and Industrial Research Organization has undertaken land resources surveys over the last two decades and contributed to the knowledge on the geomorphology of the area. A geomorphological map with explanatory notes summarizes the information (Löffler, 1974).

References

Audretsch, F. C. d', Kluiving, R. B., and Oudemans, W., 1966. "Report on an economic geological investigation of NE Vogelkop (Western New Guinea)," *Verh. Kon. Ned. Geol. Mijnbk. Gen., Geol. Ser.*, 23, 151p.

Australasian Petroleum Company Pty, Lt., 1961. "Geological results of petroleum exploration in Western Papua, 1937-1961," *J. Geol. Soc. Austral.*, vol. 8, 1-133.

Bamford, R. W., 1972. "The Mount Fubilan (Ok Tedi) porphyry copper deposit ..." *Econ. Geol.*, 67, 1019-1033.

Blake, D. H., and Löffler, E., 1971. "Volcanic and glacial landforms on Mount Giluwe, Territory of Papua and New Guinea," *Bull. Geol. Soc. Am.*, 82(6), 1605-1614.

Bureau of Mineral Resources (Australia), 1952. *Geological Map of Australia and New Guinea.* Canberra: Bur. Mineral Res., 1:6,336,000.

Bureau of Mineral Resources (Australia), 1972. *Geology of Papua New Guinea.* Canberra: Bur. Mineral Res., 1:1,000,000.

Carey, S. W., 1970. "Australia, New Guinea, and Melanesia in the current revolution in concepts of the evolution of the earth." *Search*, 1(5), 178-189.

Chappell, J., 1974. "Geology of Coral Terraces, Huon Peninsula, New Guinea: A Study of Quaternary Tectonic Movements and Sea-Lvel Changes," *Bull. Geol. Soc. Am.*, 85, 553-570.

Coneybeare, C. E. B., and Jessup, R. G. C., 1972. "Exploration for oil-bearing sand trends in the Fly River area, western Papua," *APEA J.*, 12(1), 69-73.

Curtis, J. W., 1973a. "Plate Tectonics and the Papua-New Guinea-Solomon Islands region," *J. Geol. Soc. Austral.*, 20(1), 21-36.

———, 1973b. "The spatial seismicity of Papua New Guinea and the Solomon Islands." *J. Geol. Soc. Austral.*, 20(1), 1-20.

David., T. W. E., and Browne, W. R., 1970. *The Geology of the Commonwealth of Australia.* London: E. Arnold, 3 vols, 747p., 618p., and atlas.

Davies, H. L., and Smith I. E., 1971. "Geology of East Papua," *Bull. Geol. Soc. Am.*, 82, 3299-3312. See also Rod, E., 1974. "Geology of eastern Papua: discussion," *Bull. Geol. Soc. Am.*, 85, 653-658.

Denham, D., 1969. "Distribution of earthquakes in the New Guinea-Solomon Island region," *J. Geophys. Res.*, 74, 4290-4299.

Dow, D. B., 1968. "A geological reconnaissance in the Nassau Range, West New Guinea," *Geol. Mijnb.*, 47, 37-46.

Dozy, J. J., 1938. "Eine Gletscherwelt in Nied erländisch Neuguinea," *Z. Gletscherk.*, 26, 45-51.

———, 1939. "Geological results of the Carstensz Expedition, 1936," *Leidse Geol. Mededel.*, 11, 68-131.

Fairbridge, R. W., 1953. "The Sahul Shelf, Northern Australia, its structure and geological relationships," *J. Roy. Soc. W. Austral.*, 37(1), 1-33.

Finlayson, D. M., and Cull, J. P., 1973. "Structural profiles in the New Britain/New Ireland region," *J. Geol. Soc. Austral.*, 20(1), 37-48.

Glaessner, M. F., 1950. "Geotectonic position of New Guinea," *Bull. Am. Assoc. Petrol. Geologists*, 34, 856-881.

Haantjens, H. A., 1970. "New Guinea soils: their formation, nature and distribution," *Search*, 1(5), 233-238.

Heming, R. F., 1974. "Geology and Petrology of Raboul Caldera, Papua New Guinea," *Bull. Geol. Soc. Am.*, 85, 1253-1264.

Hermes, J. J., 1968. "The Papuan geosyncline and the concept of geosyncline," *Geol. en Mijnbouw*, 47(2), 81-97.

Jenkins, D. A. L., 1974. "Detachment tectonics in western Papua New-Guinea," *Bull. Geol. Soc. Am.*, 85, 533-548.

Katili, J. A., 1971. "A review of the geotectonic theories and tectonic maps of Indonesia," *Earth Sci. Rev.*, 7, 143-163.

Krause, D. C., 1965. "Submarine geology north of New Guinea," *Bull. Geol. Soc. Am.*, 76, 27-42.

———, 1966. "Tectonics, marine geology, and bathymetry of the Celebes Sea-Sulu Sea region," *Bull. Geol. Soc. Am.*, 77, 813-832.

Le Pichon, X., 1970. Correction to paper by Le Pichon "Sea floor spreading and continental drift," *J. Geophys. Res.*, 75, 2793.

Löffler, E., 1970. "Evidence of Pleistocene Glaciation in East Papua," *Austral. Geogr. Stud.*, 8, 16-26.

———, 1972. "Pleistocene glaciation in Papua and New Guinea," *Z. Geomorphol.*, Suppl. 13, 46-72.

———, 1974. "Geomorphological map of Papua New

Guinea, with explanatory notes," *Austral. Land Res.* Ser. 33.
Lowder, G. G., and Carmichael, I. S. E., 1970. "The volcanoes and caldera of Talasea, New Britain: geology and petrology," *Bull. Geol. Soc. Am.,* 81, 17–39.
Luyendyk, B. P., MacDonald, K. C., and Bryan, W. B., 1973. "Rifting history of the Woodlark Basin in the Southwest Pacific," *Bull. Geol. Soc. Am.,* 84, 1125–1134.
Milsom, J., 1973, "Papuan ultramafic belt: gravity anomalies and the emplacement of ophiolites," *Bull. Geol. Soc. Am.,* 84, 2243–2258.
Page, R. W., and McDougall, I., 1972. "Ages of mineralization of gold and porphyry copper deposits in the New Guinea Highlands," *Econ. Geol.,* 67, 1034–1048.
Peterman, Z. E., et al., 1970. "Sr^{87}/Sr^{86} ratios of the Talasea Series, New Britain, territory of New Guinea," *Bull. Geol. Soc. Am.,* 81, 39–40.
Peterson, J. A., and Hope, G. S., 1972. "A lower limit and maximum age for the last major advance of the Carstensz glacier, West Irian," *Nature,* 240(5375), 36–37.
Richards, J. R., and Willmott, W. F., 1970. "K-Ar age of biotites from Torres Strait," *Austral. J. Sci.,* 32, 369–370.
Ripper, I. D., 1970. "Global tectonics and the New Guinea–Solomon Islands region," *Search,* 1(5), 226–232.
Ruxton, B. P., 1970. "Labile quartz-poor sediments from young mountain ranges in Northeast Papua," *J. Sed. Petrology,* 40(4), 1262–1270.
Smith, I. E., 1970. "Late Cainozoic uplift and geomorphology in southeastern Papua," *Search,* 1(5), 222–225.
Thompson, J. E., and Fisher, N. H., 1965. "Mineral deposits of New Guinea and Papua and their tectonic setting," *Proc. 8th Comm. Min. Met. Congr., Australia–New Zealand,* 6, 115–148.
Tjia, H. D., 1973. "Displacement patterns of strike-slip faults in Malaysia–Indonesia–Philippines," *Geol. Mijnb.,* 52, 21–30.
Veeh, H. H., and Chappell, J., 1970. "Astronomical theory of climatic change: support from New Guinea," *Science,* 167, 862–865.
Verstappen, H. T., 1964. "Geomorphology of the Star Mountains," *Nova Guinea,* (Sci. Res. Netherlands N.G. Exped., 1959), 5, 101–158.
Visser, W. A., and Hermes, J. J., 1962. "Geological results of the exploration for oil in Netherlands New Guinea," *Verh. Kon. Ned. Geol. Mijnb. Gen., Geol. Ser.,* 20, 265p.
Wegen, G. van der, 1966. "Contribution of the Bureau of Mines to the geology of the Central Mountains of West New Guinea," *Geol. Mijnb.,* 45, 249–261.
Williams, P. W., et al., 1972. "Aspects of the Quaternary geology of the Tari-Koroba Area, Papua," *J. Geol. Soc. Austral.,* 18(4), 333–347.
Winterer, E. L., 1970. "Submarine valley systems around the Coral Sea Basin (Australia)," *Marine Geol.,* 8, 229–244.

Cross-references: *Australasia–Regional Review: Australia; Bismarck Archipelago; Solomon Islands.*

NEW HEBRIDES

The New Hebrides archipelago is aligned NNW-SSE, extending from 13 to 20°S, and approximately bisected by the 168°E meridian. The land area is 14,760 km^2 (5,700 sq mi). The territory (capital Port Vila) is administered as an Anglo-French condominium. The Santa Cruz Islands (see *Solomon Islands*) rise from the same submarine ridge system and are geologically part of the New Hebrides arc as are the isolated volcanic isles of Matthew and Hunter in the south.

The New Hebrides have been visited on occasions by natural scientists, including geologists, for nearly a century; the first substantial contribution on the geology of the group was made by Mawson (1905). However, extensive geological reconnaissance mapping was not begun until 1956 when the geologists of Compagnie Française des Phosphates d' Oceanie, the Bureau de Recherches Géologique et Minières, and Société Le Nickel investigated the mineral potential. Lacroix had made some important petrological observations earlier in 1939. Manganese was discovered on Efate Island and mined from 1962 to 1968.

A geological survey, part of the British service, was inaugurated in 1959, and has since completed the systematic geological mapping of most islands. Research projects including gravimetric, aeromagnetic, and vulcanological investigations were also undertaken by the survey in conjunction with various universities, which contributed several research students. The work of the Geological Survey is coordinated with that of the Condominium Mines Department which maintains stations in Vila and on Santo and Pentecost islands as part of the French seismological network controlled from New Caledonia.

Morphotectonics and Bathymetry

The New Hebrides island arc is part of the outer Melanesian arc system that extends from the northern coast of New Guinea through the Solomons to Fiji and the Tonga-Kermadec chain.

There are three island belts in the New Hebrides; a western belt comprising Santo and Malekula, the two largest islands; an eastern belt consisting of the islands of Maewo and Pentecost; and a longer central chain containing all but one of the recently active and extinct volcanoes.

Opinions differ on the nature of the connection between the New Hebrides and Tongan chains, expressed by the Hunter Ridge which swings from south of the New Hebrides NE through Fiji to Tonga. Carey (see references in Mitchell and Warden, 1971) regarded this feature as a major sigmoidal flexure—the "Fiji orocline." Others infer that a transcurrent fault spans the gap between the two ridges and suggested that the oppositely directed Tonga and New Hebrides island arcs were once continuous and were subsequently displaced by left-lateral movements.

The trench to the west of the New Hebrides is divided into northern and southern segments by a number of submarine ridges and troughs at the eastern end of the D'Entrecasteaux fracture zone (opposite Malekula and Espiritu Santo islands). The southern segment (7570 m) swings NE at its southern end; the northern Torres Trench truncates the San Cristobal trough at right angles and in turn abuts against the Johnson Deep which trends NE from Malaita in the Solomons. To the east, the comparatively shallow Fiji plateau (less than 3000 m) bounded on the south by the Hunter Ridge, separates the New Hebrides from Fiji.

Vulcanism and Seismicity

Nearly all the active and recently extinct volcanoes of the New Hebrides are confined to the central chain (Fig. 1), which is flanked by the two shorter belts of islands consisting mainly of older Neogene rocks. The islands of the central chain range in size from small central volcanoes 4-5 km in diameter, or even temporarily emerging submarine shoals such as Karua, to large caldera volcanoes such as Ambrym and Aoba with maximum dimensions of up to 30 km. Submarine silicic and alkalic lavas produced during an earlier (?) Pliocene volcanic cycle occur on some islands. Tholeiites and high alumina basalts with minor amounts of andesite and acidic lava characterize the younger Quaternary vulcanism (Colley, 1970). Ankaramites and picrite basalts bordering on an ultramafic composition are mainly restricted to Aoba. Mildly alkalic basalts (similar to those produced by typical oceanic volcanoes such as those in Hawaii) have been recorded on Aoba, Ambrym, and Erromango islands. Yasour volcano on Tanna Island currently produces trachyandesite, whereas Lopevi in the center of the chain and the Banks volcanoes in the north produce a high alumina basalt and tholeiitic andesite lava association. No consistent zonal variations in magma composition suggestive of derivation at different depths within the Benioff zone are evident in the New Hebrides. However, the composition of the lava erupted has varied with time at some volcanoes.

The New Hebrides arc is part of a belt of intense seismic activity. Most earthquakes originate within an E-dipping Benioff zone which approaches the surface in the vicinity of the western submarine trenches. There is a predominance of shallow foci within the region, though intermediate shocks are frequent and a cluster of hypocenters occurs at depths of 600-650 km. The penetration of the seismic plane quoted is consistent with rates of underthrusting of lithosphere beneath the arc: 8 cm/year according to Le Pichon. The New Hebrides seismic zone differs from that beneath most island arcs and has several anomalies. In terms of plate tectonics they can be explained by contortions in a descending lithospheric slab.

A relationship between deep earthquake foci, later shocks originating at intermediate to shallow depths, and ensuing eruptions of Lopevi and Gaua was established by Blot in 1966 but has not yet been demonstrated for the other active volcanoes.

Part of the same region of positive gravity anomaly system as the Solomons, the Bouguer gravity anomalies are relatively low over the western belt (+60 mgal over northwestern Santo) and high over the eastern belt (+210 mgal over Maewo) where they are associated with ultramafic intrusions (see Malahoff and Woollard as quoted by Mitchell and Warden, 1971).

The crust beneath the New Hebrides is thinner than that of typical continents and there is no evidence that fragments of an old continent are present. The crust probably consists of a thick accumulation of volcanic, intrusive, and sedimentary rocks overlying a faulted and depressed layer of normal oceanic crust.

Stratigraphy and Structure

Western Belt. Except for a narrow zone of pelagic red mudstones, the oldest rocks are clastic volcanics, limestones, and lavas of Lower Miocene age which total at least 4000 m on Malekula (Mitchell, 1966) and 3000 m on Espiritu Santo (Robinson, 1969). Volcanic conglomerates and breccias predominate, though calcirudites also occur. Quartz-free sandstones and calcarenites with Lower Miocene foraminifera and mudstones are interbedded with the volcanics. Much of the sediment was probably transported by submarine mass-flow movements, though an autoclastic origin is inferred for some submarine lavas (Mitchell, 1966; Robinson, 1969). Similar andesite breccias with

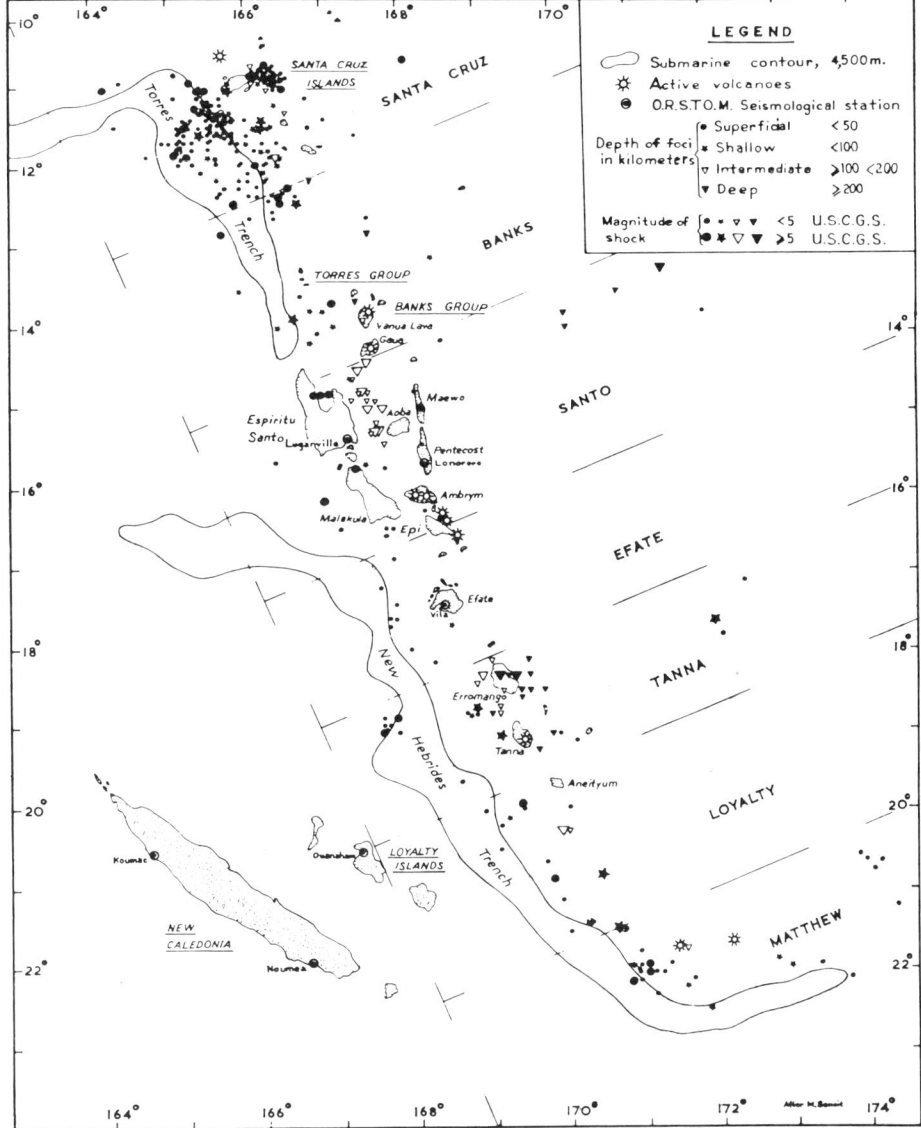

FIGURE 1. Seismic map of the New Hebrides.

marine tuffs, again with Lower Miocene foraminifera, form the basement of the Torres Islands north of Santo (Figs. 2 and 3).

Minor intrusions of microgabbro and microdiorite, and a gabbrodiorite complex cut the succession on Malekula and Santo islands, respectively. The volcanic association is predominantly andesitic and calc-alkaline, but basalts are more abundant on Santo than Malekula.

The Middle and possibly Upper Miocene sequence of volcanics, tuffs, mudstones, and carbonates with predominant sandstone facies alternating with mudstones, calcarenite, and calcilutites are at least 1500 m thick on Malekula and 3000 m on Santo (?mid-Miocene only). They are generally less indurated, faulted, and steeply dipping than the Lower Miocene rocks that they overlie with probable unconformity.

Eastern Belt. The oldest rocks are reworked Upper Eocene limestone clasts associated with dacite and basalt boulders in basal conglomerates of early Miocene age on Maewo Island (Liggett, Coleman, quoted in Mitchell and War-

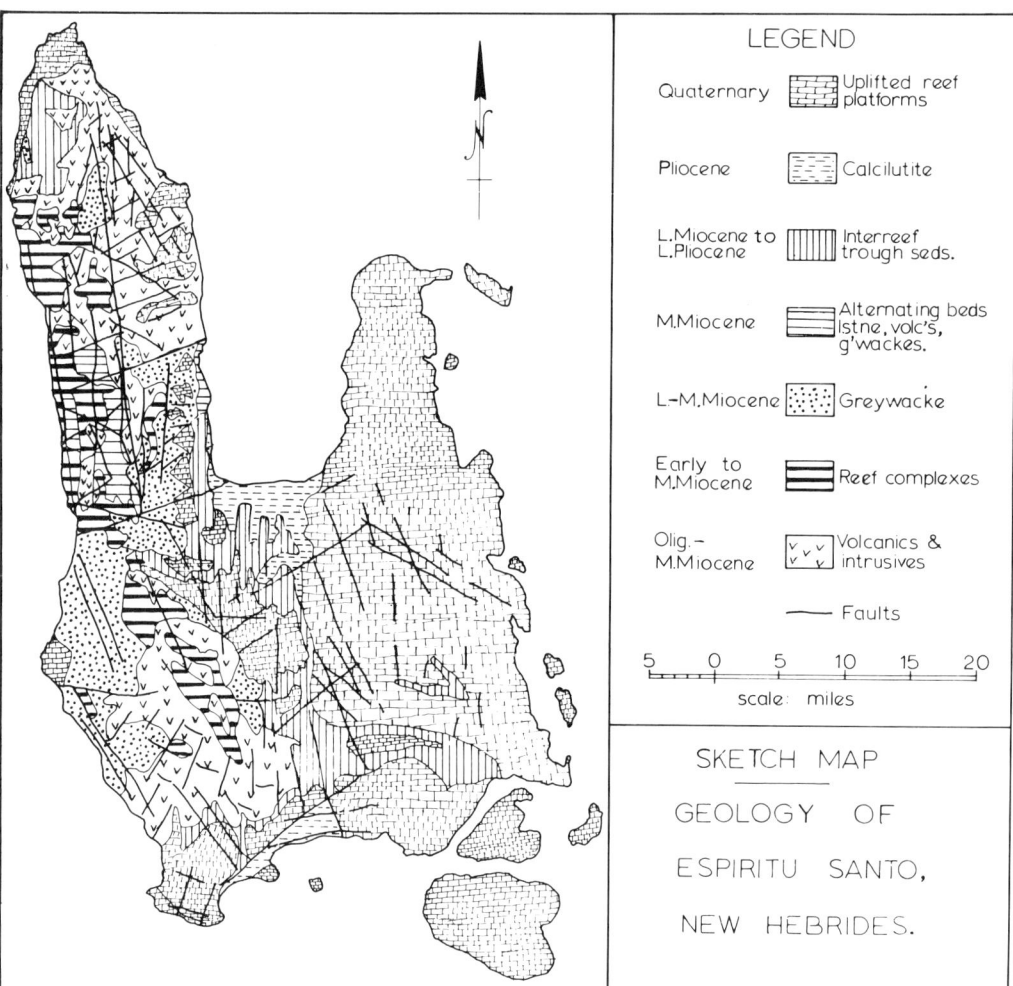

FIGURE 2. Sketch map of the geology of Espiritu Santo, New Hebrides.

den, 1971). A thick succession of submarine andesites and basalts of Oligocene-Lower Miocene age are apparently the oldest layered rocks on Pentecost Island. Basaltic lavas and volcaniclastic rocks (Maewo) but mainly basaltic rocks (Pentecost) accumulated during the Middle Miocene. Turbiditic volcanic sandstones and mudstones are interbedded with lava breccias in the upper part of the sequence toward the top of which sediments become increasingly abundant. On Maewo they contain reef limestone clasts and are overlain by calcarenites and calcilutites. On Pentecost there are interspersed beds of calcilutites rich in planktonic foraminifera (amounting to pelagic oozes) with little volcanic detritus. The bulk of these varied sediments appears to be post-Middle Miocene in age. The volcanic rocks belong to a suite of relatively primitive tholeiites.

Except for rare dikes on Maewo Island, intrusions in the eastern belt are confined to Pentecost Island. The latter include the only serpentinized ultrabasic rocks in the group which were probably tectonically emplaced, and gabbros and norites of probable pre-Middle Miocene age.

Central Chain. Rocks forming the islands of the central chain are mainly primary and reworked volcanics mostly erupted above sea level. Submarine lava breccias and lithic or pumiceous tuffs locally interbedded with pillow lavas on Aneityum, Erromango, Tanna, Efate, Western Epi, and Vanua Lava islands are possibly of Pliocene age and are the oldest exposed

FIGURE 3. Sketch map of the geology of Aoba.

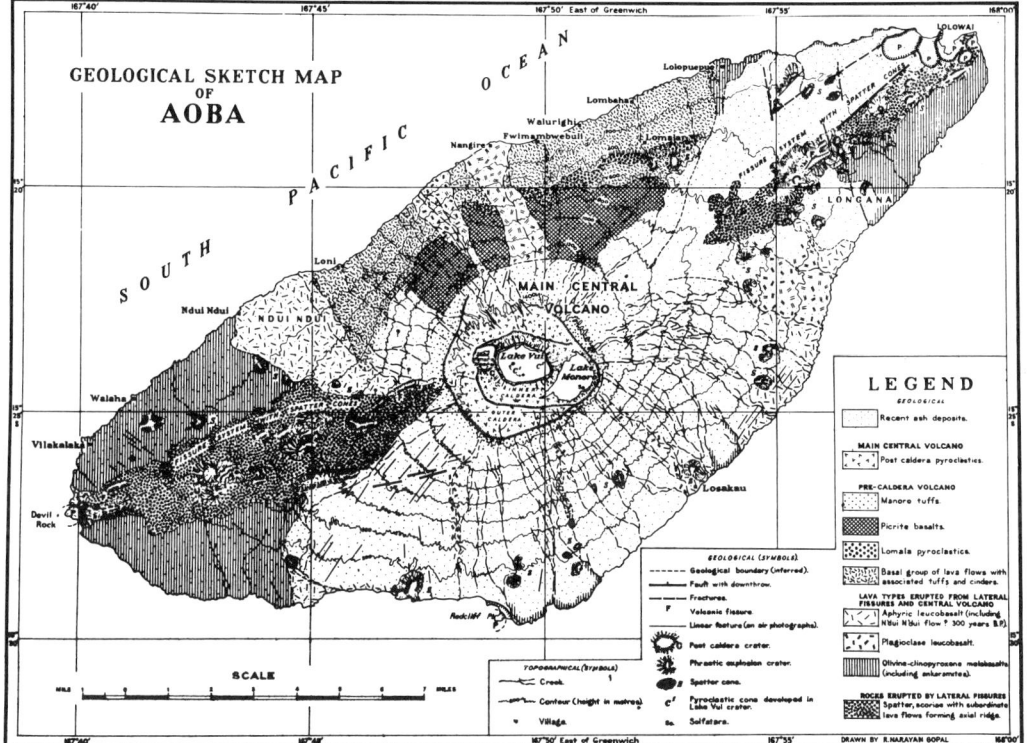

rocks in the central chain. More reworking occurred within the upper part of the succession where cross-bedded volcanic sandstones and calcarenites are common. They contain interbedded lenses of massive limestone sometimes emplaced as gravity slides. These submarine volcanics are often capped or surrounded by reef limestones and terraces. Similar limestone plateaus and terraces also overlie the older Neogene rocks of the eastern and western belts.

Volcanic growth is largely or entirely responsible for the formation of the other islands in the central chain. Islands such as Gaua, Aoba (Fig. 4), and Ambrym and the coalescing volcanoes of eastern Epi and most of the islets of the Shepherds have risen solely by volcanic growth. Most of the active and extinct stratiform volcanoes consist predominantly of pyroclastic material, though piles of subaerial basalt flows are common near the exposed base of many of the subaerial volcanoes and so suggest a shield-forming stage.

Minor intrusions, mainly dikes, are abundant on some small islands like Paama and Euvose, but generally they form only a small proportion of the volcanic structures. No plutons are exposed but magnetic data suggest the presence of high-level magma chambers perched beneath some of the volcanoes (Malahoff, 1970).

Structure

The structure of most islands in the group can be related to faulting and there is no evidence of regional folding or low-angle thrusting. Faulting and uplift occurred throughout much of the western belt late in the Lower Miocene, and again in the Upper Miocene and Pliocene. Fault graben and tilted terraces and plateaus of Pliocene-Quaternary age indicate post-Pliocene block faulting (Mitchell and Warden, 1971). Amphibolites on Pentecost Island suggest a phase of probable Miocene deformation and metamorphism. The earliest recognizable fault movements imparted a steep dip to pre-Middle Miocene rocks. On Maewo there is evidence of late Miocene to early Pliocene uplift and faulting; as in the western belt and some of the islands of the central chain, uplifted and tilted Pleistocene limestones reflect vertical movements during the Quaternary. Maximum uplift

FIGURE 4. Sketch map of the geology of South Malekula.

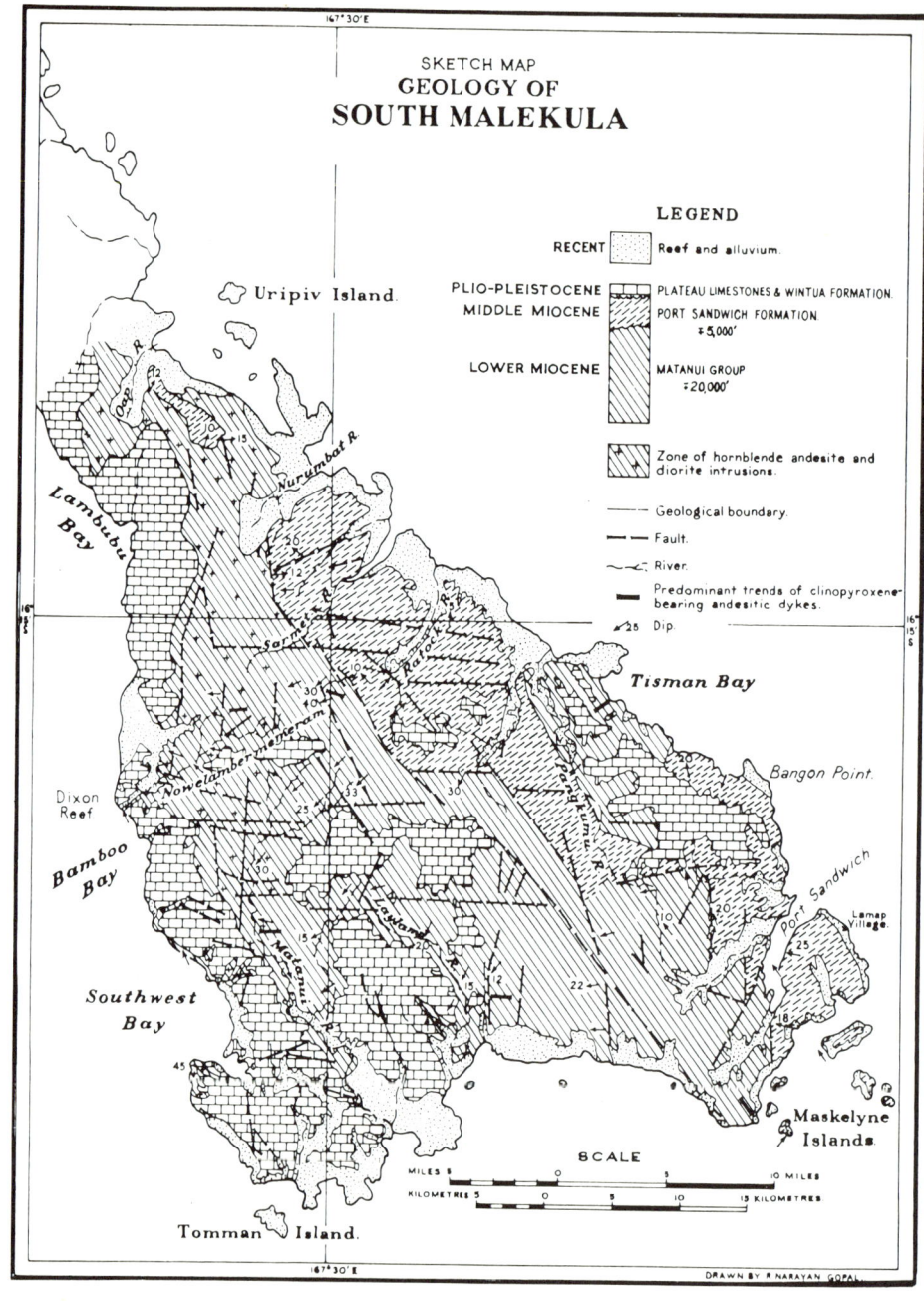

on the trench side is indicated by eastward tilted terraces in the north of Malekula, western Santo and Efate (Fig. 5).

Volcanic centers are aligned parallel to, or almost at right angles to, the axis of the active arc (Warden, 1967). Eruptive centers generally become progressively older away from the volcanic axis. Evidence of the tectonic control of volcanic activity within these two well-defined trends are the bipole aeromagnetic anomalies of up to 1000 γ recorded over all the principal volcanoes of the central chain with positive Bouguer gravity anomalies of up to 160 mgal (Malahoff, 1970). Major volcanic centers are concentrated at the intersection of the two trends, e.g., Ambrym and Aoba which possess strong E-W linear components.

The western belt was the site of an early

FIGURE 5. Diagrammatic cross section through South Malekula, illustrating probable geological history.

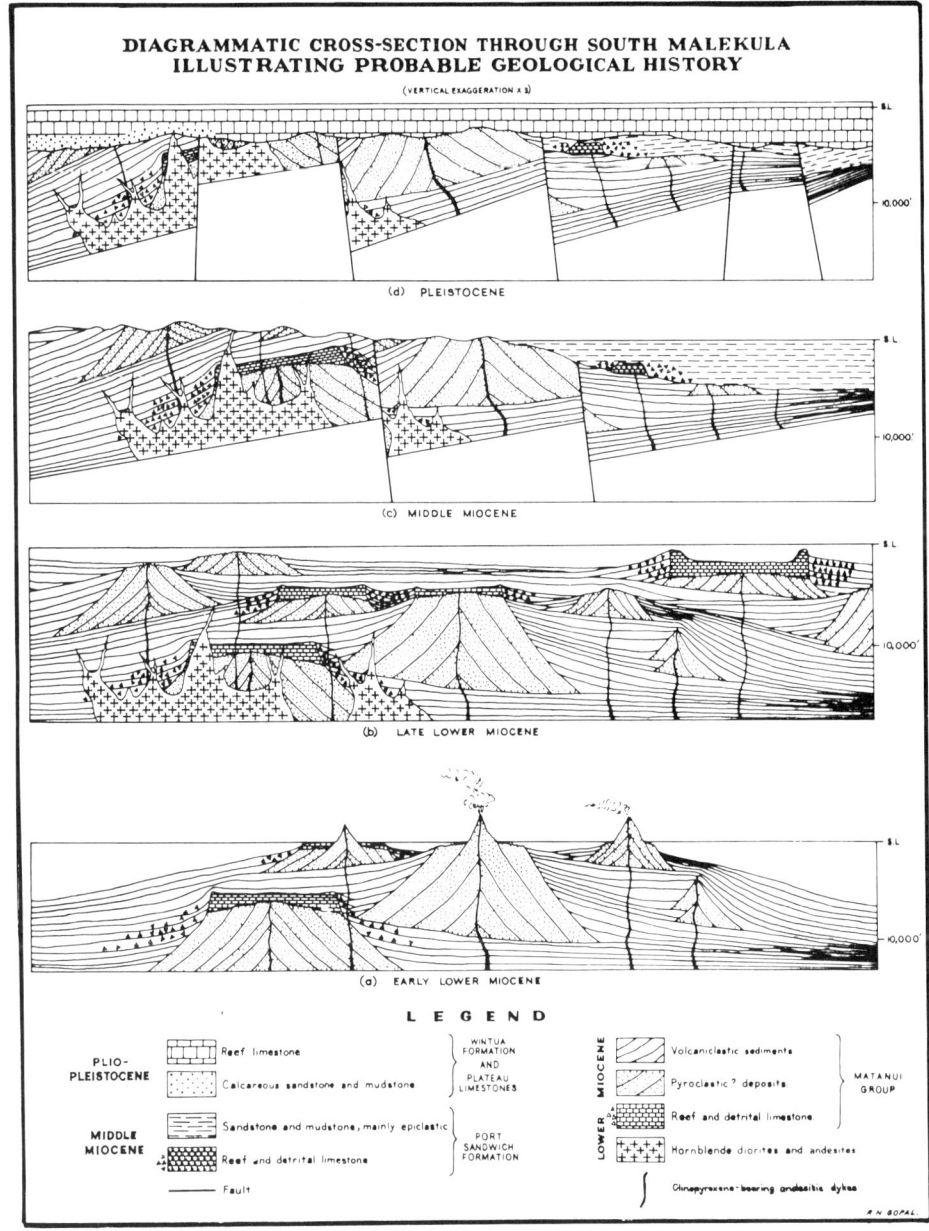

Miocene volcanic arc with a trench possibly located over the area of what is now the eastern belt. A change in the sea-floor spreading pattern in the Pliocene possibly initiated the present (apparently reversed) trench-arc system. Here also may lie the explanation of the contrasting volcanic suites of the eastern and western belts. The volcanoes of the central chain, tapping the existing Benioff zone, produce suites which, as in most other Quaternary island arcs, are predominantly tholeiitic.

Economic Geology

Manganese oxides occur on several islands and were mined at Forari on Efate Island between 1962 and 1968. Derived by the leaching of volcanic rocks, the manganese has been extensively redistributed and deposited at limestone contacts, and under conditions of lateritic weathering, as at Forari.

A rapid geochemical drainage reconnaissance of some islands by Conzinc Riotinto of Australia for potential occurrences of porphyry copper yielded negative results. However, traces of chalcopyrite and sulfides have been recorded in microdiorite on Malekula during the Geological Survey's mapping program. Traces of secondary copper ores are widespread in some Plio-Quaternary volcanic sequences; malachite is locally concentrated within vesicular zones in lava flows.

Alluvial gold has been reported from south Malekula, and chrome from south Pentecost, and sphalerite was recorded from both these islands. Native sulfur occurs in a small solfatara field on Vanua Lava in the Banks Islands.

ARTHUR J. WARDEN
PATRICK J. COLEMAN

References

Aubert de la Rüe, E., 1937. "Le Volcanisme des Nouvelles-Hébrides (Mélanésie)," *Bull. Volcanol.*, Ser. 2, **2**, 79–142.
Chapman, F., 1905. "Notes on the older Tertiary foraminiferal rocks on the west coast of Santo, New Hebrides," *Linn. Soc. N.S.W., Proc.*, **30**, 261–274.
Coleman, P. J., 1970. "Geology of the Solomon and New Hebrides Islands," *Pacific Sci.*, **24**, 289–314.
Colley, H., 1970. "Andesitic volcanism in the New Hebrides," *Proc. Geol. Soc. Lond.*, **1662**, 46–51.
Luyendyk, B. P., et al., 1974. "Shallow structure of the New Hebrides Island Arc," *Bull. Geol. Soc. Am.*, **85**, 1287–1300.
Malahoff, A., 1970. "Gravity and magnetic studies of the New Hebrides Island Arc," *New Hebrides Geol. Surv. Rept.* (Vila), 67p.
Mawson, D., 1905. "The geology of the New Hebrides," *Linn Soc. N.S.W., Proc.*, **30**, 400–484.
Mitchell, A. H. G., 1966. "Geology of South Malekula," *New Hebrides Geol. Surv. Rept.* (Vila), **3**, 42p.
———, and Warden, A. J., 1971. "Geological evolution of the New Hebrides Island Arc," *J. Geol. Soc. Lond.*, **127**, 501–529.
Obellianne, J. M., 1961. "Contribution à la connaissance géologique de l'archipel des Nouvelles-Hébrides," *Sci. de la Terre* (Nancy), 6(1958), 139–368.
Robinson, G. P., 1969. "The geology of North Santo," *New Hebrides Geol. Surv. Rept.* (Vila), 77p.
Sagatzky, J., 1959. "Contribution à l' étude géologique de l'ile Espiritu Santo (Nouvelles Hébrides). 1. Roches sédimentaires, *Bull. Soc. Géol. France*, Ser. 6, 8(1958), 501–509.
———, 1960. "Contributions a l'étude géologique de l'ile Espiritu Santo. 2. Roches volcaniques," *Bull. Soc. Geol. France*, Ser. 7, 1(1959), 588–593.
Shutler, R., Jr., 1970. "A radiocarbon chronology for the New Hebrides," *Proc. 8th Intern. Congr. Anthro. and Ethno. Sci.*, 1968, **3**, 135–137.
Warden, A. J., 1967. "The geology of the Central Islands," *New Hebrides Geol. Surv. Rept.* (Vila), **5**.

Cross-references: *Australasia–Regional Review; Fiji; New Caledonia; Solomon Islands; Tonga.*

NEW ZEALAND

New Zealand has an area of 268,675 km^2 and comprises two major islands—North Island (114,687 km^2 or 44,281 sq mi) and South Island (150,460 km^2 or 58,093 sq mi) separated by Cook Strait, 20 km wide—and outlying islands. Stewart Island, the largest (670 sq mi) of the outlying islands, is composed of granites and metamorphics; the Chatham Islands (372 sq mi) are composed of schist, Cretaceous and Tertiary sediments, and volcanics; Great Barrier Island (176 sq mi) is composed of volcanics, late Tertiary and Mesozoic graywacke; and numerous small islands are predominantly volcanic (*the Kermadecs* (q.v.), Little Barrier Island, White Island, Mayor Island, Three King's Island, Solander Island, The Snares, Campbell, Auckland, Bounty and Antipodes Islands). New Zealand also administers the *Tokelau Islands, Niue,* and *Cook Island* in the South Pacific, and the Ross Dependency (414,400 km^2) in Antarctica.

The landscapes of both North Island and South Island are dominated by axial ranges, of Upper Paleozoic and Mesozoic graywackes in the North Island, and graywackes with a western belt of schist in the South Island. Volcanic landscapes feature prominently in the North Island whereas the rugged highlands of the South Island are dominated by Paleozoic or older sediments, together with igneous and metamorphic rocks in the South Island (Figs. 1 and 2).

Geological Exploration of New Zealand dates from the pioneer work of Hochstetter (1858), Hector, von Haast, and Crawford (1860–1861), although scattered observations had been made before, notably by Dieffenbach (1843) and the surveyor Heaphy (1854). The New Zealand Geological Survey with Hector as director was constituted in 1865 and much valuable work was accomplished in its first 30 years, notably by Hector, Cox, and McKay. The Geological Survey now has about 50 specialists, and its work ranges from Pleistocene geology to glaciology, volcanology, mineral resources, water supply, engineering geology, paleontology, and geochemistry, as well as regional geological mapping, on a scale of 4 miles to 1 inch and 1 mile to 1 inch.

Early in the present century when Bartrum, Cotton, Speight, and Park were appointed to the universities at Auckland, Wellington, Christ-

FIGURE 1. Geological map of New Zealand North Island. Data supplied by N.Z. Geol. Sur. Adapted from *New Zealand Official Yearbook, 1963.*

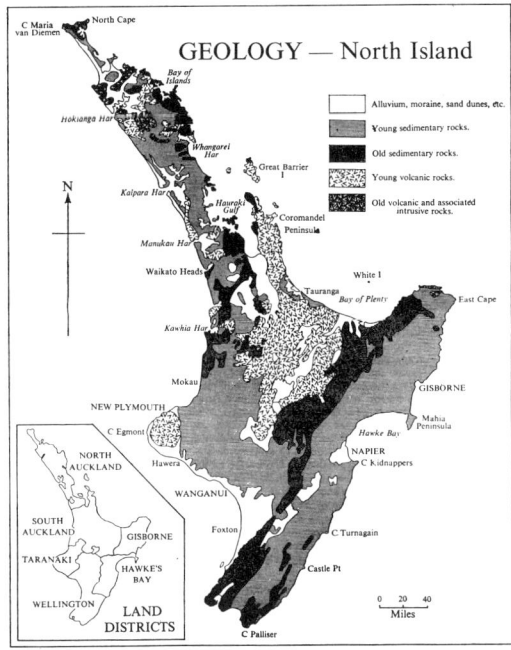

church, and Dunedin, respectively, an important phase of university geological research began, characterized by many noteworthy studies.

Mining Activity

The alluvial gold rushes of the 1860s (Otago, 1861; Westland, 1864) were of considerable importance in New Zealand's economic development. Not only did they lead to an increase in population and wealth, but also to the rapid exploration of remote country. Auriferous quartz lodes were also worked (Coromandel area, with silver; also in West Nelson and Otago). Over 27 million ounces of gold have been exported. Peak years were 1866 and 1906.

Small amounts of lead and zinc have been found in the Coromandel area and antimony and tungsten ores are found in the South Island schists. Erosion of andesite and granite has produced extensive coastal black (iron)sand deposits in Taranaki and South Auckland (andesitic origin) and W. Nelson and Westland (granitic origin) and these are being actively worked, mainly for use in the Japanese iron and steel industry.

Subtropical weathering of basalts in Northland has led to the formation of bauxite deposits. Cinnabar, the mercury ore, has been obtained from hot spring areas in Northland. Small manganese and copper deposits are found associated with volcanics in Northland and chromite has been mined from serpentine rocks in Nelson. Uranium was found in the Buller Gorge in 1955, and in the Auckland Islands, 1500 km south of South Island.

Industrialization has accelerated the utilization of nonmetallic minerals (aggregates, building stones, sands, limestones, bentonite, dolomite, magnesite, perlite, etc.), mineral fuels, geothermal steam, and ground water.

New Zealand coals range in age from Cretaceous (Westland) to Tertiary (Waikato, Southland, etc.). Most North Island coals are subbituminous and the main coalfields are in the Wakato Basin and in northern Taranaki. Subbituminous coalfields also occur in Nelson, Westland, Otago, and Southland in the South Island. The west coast of South Island has the only deposits of bituminous coal, which are worked at Puponga, Reefton, Buller, and Greymouth. Anthracite occurs locally in Canterbury (lignites altered by a volcanic intrusion) and in southern Westland.

Surface oil seeps are known throughout the large Tertiary basins of New Zealand, but drilling has been largely unsuccessful. A small oil and gas field near New Plymouth has been

FIGURE 2. Geological map of New Zealand South Island. Data supplied by N.Z. Geol. Sur. Adapted from *New Zealand Official Yearbook, 1963.*

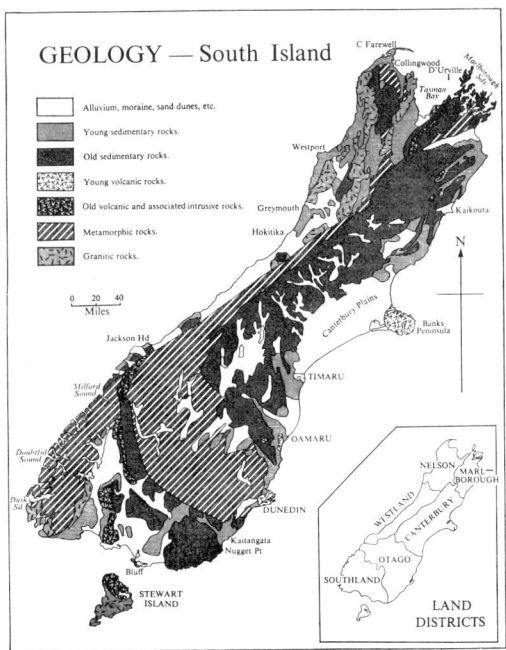

worked since early this century. In 1959 a gas-condensate field was discovered at Kapuni, Taranaki, at a depth of about 4000 m, associated with Eocene coal measures. Reserves have been estimated as 655 billion cu ft of gas and 26 million bbl of condensate. In 1969 another large gas-condensate field, called Maui, was discovered at a comparable stratigraphic horizon offshore from southern Taranaki. Reserves have been estimated as 5 trillion cu ft of gas and 75 million bbl of condensate.

Oil shales are known from the North and South islands, and may have economic possibilities at Orepuki and Nevis where up to 2000 million tons of shales are estimated to average 12.6 gal of oil and 14.6 lb of ammonium sulfate per ton.

Structural Regions

Fiordland. The Fiordland region includes the southern part of Stewart Island as well as the southwestern part of South Island (Fiordland area). The mountain summits here rise northward along the main divide from 1000 to 3000 m in the Darran Range. The main rocks are metamorphosed quartzite, graywacke, tuffaceous graywacke, and limestone, 10-15 km thick and ranging from Cambrian to Devonian, or possibly Carboniferous, and affected by late Paleozoic or early Mesozoic orogeny; the most altered rocks are now sillimanite gneisses and garnet-hypersthene gneisses. In a coastal strip in the extreme northwest of Southland near Milford Sound, a narrow belt has been intensely deformed and mylonitized by movements along the Alpine Fault.

In the late Paleozoic and early Mesozoic the Fiordland rocks were invaded by a sequence of syenitic and granitic dikes and batholiths. The oldest fossiliferous rocks are Ordovician graptolite slates at Preservation Inlet, but some of the adjacent metamorphic rocks are probably Cambrian.

Fiordland's ancient dome originally had radial drainage; Pleistocene glaciation has etched out a complicated rectangular network, which includes high-level lakes (e.g., Lake Te Anau) gouged out in preexisting valleys. Further there is a magnificant forested fiord terrain extending for some 250 km round the southwestern coast; and drowned U-shaped valleys extending 15-30 km inland, carved in the resistant rocks, have retained their precipitous sides, frequently plunging 1000 m or so into the sea (Fig. 3).

The Alpine Fault forms the northern boundary of the region, extending from Westland into Fiordland to Milford Sound and then prob-

FIGURE 3. Mitre Peak, and Milford Sound, in Fiordland. (Photo: N.Z. Nat. Publ. Studios.)

ably running just offshore, accounting for the spectacularly cliffed coast and the deep water close inshore.

The older rocks are separated by a major unconformity from remnants of Tertiary rocks formerly more extensive but now confined to the southern and eastern parts of Fiordland.

West Coast of South Island. This region includes that part of Westland west of the Alpine Fault and western Nelson. The landscape ranges from forested mountains (peaks up to 1900 m) in western Nelson to moraine-covered coastal plains in Westland.

Most of the country's pre-Carboniferous rocks are found within this region. They range in age from presumed Precambrian and contain Cambrian, Ordovician, Silurian, Devonian and, at one locality, Permian fossils. In western Nelson these rocks are intruded by some of the largest granite masses known in New Zealand, disposed in N-S-trending belts. A Mesozoic peneplain was cut over the whole region before Tertiary sedimentation started and its exhumed surface is now well exposed on the flanks of the mountains in the less deformed regions. In northern Westland Lower Tertiary coal measures were laid down in fault depressions.

The Alpine Fault forms the western boundary of this region in Westland; it separates the low-lying Cretaceous to Pleistocene rocks, resting on a basement of Precambrian and Paleozoic, from the high schist and graywacke ranges of the Southern Alps. The scarp of the Alpine Fault forms a natural western boundary to the Alps and its course is marked in many places by a prominent fault-line valley. In places the precipitous western downthrow side falls 1600 m, forming a massive natural rampart. Both dextral strike slip and vertical movements have occurred along the fault; it has been suggested that a total horizontal shift of 480 km has occurred, probably starting in the Late Jurassic-Early Cretaceous Rangitata Orogeny. Horizontal movements continued on the fault in the

Kaikoura Orogeny and upward displacement of over 18,000 m occurred simultaneously, forming the Southern Alps as they are known today.

Southland Syncline. This prominent structural feature appears at the Alpine Fault in northwestern Otago province and curves in an arc across Southland Province (see Figs 1 and 2). The oldest sediments in the Southland Syncline are Permian tuffaceous graywackes marking the northern margin of the syncline. They flank the Otago schists and extend across Southland as a broad SE-curving belt from the Humboldt Mountains through the Thomson Mountains to the Waikaka area (and thence through Otago to the Kaitangata coast). These sediments become more strongly folded and partly schistose as they approach the Otago schists which are, at least in part, their metamorphosed equivalents.

Rocks of similar age form the other margin of the Southland Syncline flanking Fiordland and extending from the Hollyford Valley southward and southeastward through the Takitimu Mountains to Riverton, Bluff, Ruapuke Island, and the northern part of Stewart Island (where they flank Fiordland-type granites to the south). These rocks are dominantly basalts, tuffs and agglomerates, but with thin sediments, and attain a thickness of about 23,000 m, attesting to the greatest volcanicity known in the New Zealand region. Linear intrusions of ultrabasic and basic rocks are associated with them (e.g., Longwood Range, Bluff Peninsula, Ruapuke Island, and the northern part of Stewart Island) and along the northern margin of the syncline (e.g., Livingstone Mountains, Otama Hills, and the Clinton-Waipahi area of Otago). Triassic and Jurassic rocks form the core of the syncline and strike across eastern Southland to the southeastern Otago coast.

Nelson Syncline. Upper Paleozoic and Triassic sedimentary and volcanic rocks have been folded into this structural feature between the plutonics and metamorphics of West Nelson (to the W) and the schist belt of Marlborough (to the E). The rocks can be matched with similar rocks in the Southland Syncline and it has been suggested that both synclines were formerly continuous and have been displaced horizontally by the Alpine Fault. This also implies a horizontal movement of 480 km; further support for this theory stems from the fact that the West Nelson plutonics and metamorphics can be matched with those of Fiordland and the Marlborough Schist matched with the Otago Schist.

Otago and Alpine Schist. Schist forming the undermass of this region was produced by the regional metamorphism of up to 15,000 m of graywackes and argillites of the New Zealand geosyncline, probably Upper Paleozoic or Triassic in age. During the early stages, they were intensely folded on a grand scale into a series of recumbent nappe folds. This gave rise to subhorizontal schistosity over a large area, and steeply dipping schistosity near the edges. This deformation is thought to have occurred during the Rangitata Orogeny (late Jurassic-early Cretaceous). Erosion exposed the schist toward the middle of the Cretaceous, and a peneplain was achieved during the Tertiary; this is now exposed as an exhumed surface over much of Otago.

Along the Southern Alps the chlorite schists grade west into biotite, garnet, and oligoclase schists and gneisses. The schistosity planes strike NNE at an acute angle to the metamorphic zonal boundaries and the Alpine fault and three stages of folding and metamorphism are recognized.

Marlborough Schist. Typical Otago chlorite schist reappears in Marlborough on the western side of the Alpine Fault and its main continuation, the Wairau Fault, and it is thought that it has been displaced from the Otago area by transcurrent movement along with the Nelson and Southland synclines and the basement rocks of West Nelson and Fiordland.

Alpine Graywacke. Complexly folded and faulted geosynclinal graywackes and argillites make up the rugged mountain ranges of the Southern Alps, attaining altitudes up to nearly 4000 m, e.g., Mt. Tasman (11,475 ft) and Mt. Cook (12,349 ft). The same rocks also continue NE to form the Spenser Mountains of Marlborough and extend E in Canterbury to form the foothills of the Southern Alps and probably to form the basement under the Canterbury Plains. Toward their boundary with the Otago schist in the south, the graywackes and argillites become even more sharply folded, with axial planes that probably dip uniformly towards the schist. The oldest rocks are thought to occur in some of the ranges marginal to the Alps, in south Canterbury (e.g., Kirkliston Range and Hunters Hills), and these are locally schistose and thought to be of Permian and Lower Triassic age. Elsewhere in Canterbury and in Marlborough the rocks contain Upper Triassic fossils and toward the northern Canterbury coast in the Cheviot area, Upper Jurassic fossils.

Canterbury Plains and Banks Peninsula. The Canterbury Plains, about 60 km wide and rising from sea level to 500 m, consist of huge fans of gravel and sand built up by the main rivers at a time when glaciers extended well out from the higher part of the Alps, probably at about the culmination of one of the main stages of the last glaciation. The fans have since been cut into by the rivers and the sea, but the underlying rocks, the continuation of those in the region to the west, are exposed only near the Alpine foothills.

Two early to mid-Pleistocene volcanoes composed of interbedded flows of basalt and andesite form Banks Peninsula (rising to nearly 1000 m above sea level). Magnetic anomalies show that similar basic rocks underlie the Plains to the southwest and outcrop still farther southwest, at Timaru and Geraldine. Unfossiliferous graywacke is exposed within the eastern volcano of Banks Peninsula, many miles from the nearest of the old rocks.

Eastern Marlborough. This region represents the southern end of a late Mesozoic geosyncline that extends along the eastern coast of North Island. The oldest rocks are redeposited graywackes similar to the Alpine Graywackes but younger. At the axis, which lies close to the coast, they are followed conformably by redeposited sandstones and siltstones with bands of redeposited conglomerate. Eastward these beds grade laterally into thin shelf deposits that rest unconformably on the graywacke. Rare fossils in the graywacke range from Upper Jurassic to Albian. The shelf deposits are more fossiliferous, all Cretaceous stages from Aptian to Maestrichtian being represented.

In this region the Alpine Fault splits into a number of branches, including the Awatere, Kekerengu, and Hope-Kaikoura faults. These continue across Cook Strait and may be correlated with North Island faults. The resultant fault blocks include the Seaward and Inland Kaikoura ranges, culminating in the peak Tapuaenuku (2890 m).

East Coast Ranges. The east coast ranges are the northern continuation of the eastern Marlborough region and are composed of similar rocks. The oldest are Upper Jurassic graywackes in southern Hawkes Bay province. The Lower Cretaceous is confined to the south and consists of crushed dark mudstone and massive tuffaceous graywacke sandstone with bands of redeposited conglomerate. The graywacke sandstone forms conspicuous craggy hills. All the Upper Cretaceous stages are represented and show the same upward facies change as in eastern Marlborough, from redeposited through to shelf sediments. Sedimentation continued throughout the greater part of the Tertiary but in the late Tertiary the region was strongly folded along NE-SW axes and the folds were cut across by numerous transcurrent faults, some of which are still active.

Raukumara Peninsula. The main structural feature of the Raukumara Peninsula is the southern end of a Cretaceous geosyncline. The axis plunges north and corresponds to the central highlands of the region. The margin of the geosyncline is fairly well defined on the southeast, south, and southwest, Cretaceous beds thinning and changing in facies much the same as in eastern Marlborough. These rocks form the main divide between the Bay of Plenty and the east coast. The Raukumara Range reaches over 1500 m and the Huiarau Range over 1200 m, forming a series of massive ridges, densely forested and deeply dissected by precipitous gorges and forming some of the most inaccessible country in North Island.

Late Mesozoic volcanics form a complex mass (Matakaoa Volcanics) in the extreme northeastern section of the region.

Western Northland. The oldest rocks here are the Tangihua volcanics; they are geosynclinal and extremely thick without interbedded sediments other than tuff and small limestone lenses. They closely resemble the Matakaoa volcanics of the Raukumara Peninsula and are thought to be of the same age. Fossils show that full Upper Cretaceous and Lower Tertiary sequences are probably present but the prevalence of slumping and the great depth of weathering of the rocks makes deciphering the stratigraphy and structure very difficult. Marine andesites of Miocene age form a fairly continuous belt along the west side of the region. Basalt flows occurred in the Pliocene and Pleistocene and these now form prominent plateaus.

Eastern Northland. The pre-Miocene sequence and structure of eastern Northland show few points of resemblance with those of western Northland. The oldest rocks are steeply dipping graywackes with interbedded basalts and limestone lenses from which Permian foraminifera and corals have been obtained. Mesozoic beds are apparently absent and the next younger beds—Eocene coal measures and marine mudstone, and Oligocene limestone—rest on a peneplaned graywacke surface and are now preserved only on the western side of the region. Post-Oligocene rocks are mainly represented by volcanics. Miocene andesites and agglomerates form jagged peaks on the coast near Whangarei and Plio-Pleistocene basalt plateaus and scoria cones are present around the Bay of Islands.

Southeast Auckland. The northern part of the southeastern Auckland region consists of Coromandel Peninsula to the east and lowlands to the west. The Coromandel Peninsula and its southern continuation the Kaimai Range are separated from the lowlands (Firth of Thames and Hauraki Plains) by a prominent fault scarp. In turn the Hauraki Plains are separated from the Waikato Basin by graywacke hills with Upper Jurassic fossils. Similar hills almost completely enclose the Waikato Basin and the Waikato River cuts through them at Karapiro and Taupiri Gorge.

Coromandel Peninsula, Great Barrier Island,

and Kaimai Range are composed of Miocene andesites and Pliocene ignimbrites and rhyolites probably floored by Jurassic graywackes, which outcrop at several places in the peninsula. Silver-gold lodes in the Coromandel volcanics have been extensively mined.

The older rocks in the Waikato Basin and Hauraki Plains are hidden by Late Pleistocene pumice sands—outwash from the volcanic belt to the south—and thicknesses of Tertiary sediments are probably also present.

The southern part of the region is mainly overlain by ignimbrite from the volcanic belt and slopes to the north. Graywacke hills project through the ignimbrite in the west and in the east there is a narrow coastal belt of Pliocene freshwater beds that interfinger with marine beds at the coastal end of the volcanic belt.

Southwest Auckland. The older rocks of the southwestern Auckland region, fossiliferous Triassic and Jurassic shelf sediments, are warped into a regional syncline (Kawhia Syncline), which was developed during the Upper Jurassic-Lower Cretaceous Rangitata Orogeny. Although at times hidden by younger rocks, this syncline can be traced for almost the length of the region and it undoubtedly extends south beneath the Tertiary sediments of Taranaki and across Cook Strait to join the Nelson Syncline. Fossils are more abundant on the western than on the eastern side of the syncline, the oldest being Upper Triassic in age. They suggest that the sequence is divided by a break in the mid-Jurassic.

The lower part—tuffaceous graywacke, sandstone, and siltstone—is similar in thickness and fossil content to sediments of the same age in the Nelson Syncline and Southland Syncline. Together the sediments of these three synclines represent shelf and transitional beds that were deposited on the western side of the New Zealand Geosyncline. They extended north to the western side of New Caledonia. The upper part of the sequence consists of shallow-water marine and freshwater deposits that extend up into the uppermost Jurassic. Cretaceous beds are absent and erosion has removed the cover of Tertiary strata from much of the region. The deposits that survive, mainly in faulted depressions, average more than 300 m in thickness in few areas. In the northern part of the region Upper Eocene coal measures, with subbituminous coal of considerable economic importance, rest on deeply leached graywacke.

Large areas are covered by Quaternary basalts and andesites. The large cones of Pirongia and Karioi are conspicuous at the southern end of the region as are the many small but younger and less-dissected cones near Auckland at the northern end.

Taranaki. Of all the New Zealand structural regions none has a simpler surface geology than Taranaki. In the north the Mesozoic strata of the Kawhia Syncline disappear beneath Tertiary marine beds that extend southward to form wide lowlands pierced in the west by a line of volcanic vents; a chain of andesite volcanoes has grown along these vents culminating in the giant cone of Mt. Egmont (2500 m), with its secondary cone Fantham's Peak on its eastern flank. These Tertiary rocks, eroded to form plains, underlie Taranaki's farmland, and successive eruptions from the volcanoes have mantled them with thick deposits of the fertile ash that forms the basis of its agricultural productivity.

Tertiary sedimentation began in Taranaki in the early Eocene with deposition of coal measures—with seams of subbituminous coal of some economic importance—followed by limestone and other relatively thin sediments in the Oligocene. During early Miocene times the Taranaki area became a broad, rapidly sinking basin in which more than 5000 m of marine sedimentary rocks accumulated.

These rocks began to emerge in the late Tertiary and Quaternary, first in the north, then successively southward. As a result the older beds are progressively more deeply buried toward the south and progressively younger marine beds are encountered southward at the surface. The Tertiary beds extend beneath the volcanics near New Plymouth, on the north side of Mt. Egmont, and have been penetrated by numerous bores in New Zealand's only oil field. Since 1959, a number of bores reaching depths as much as 4000 m at Kapuni, south of Mt. Egmont, have also revealed natural gas below the thick cover of Tertiary marine strata.

The volcanics of Taranaki are of two ages. The northern and eastern groups, comprising the Sugar Loaves, Whareorino, and Pehimatea, are of early Pleistocene age and dacitic composition. The central Taranaki volcanoes, andesitic in composition, are three vents aligned NW-SE, all of Upper Pleistocene and Holocene age; Kaitake is the oldest, then Pouakai, followed by Egmont, the youngest. Egmont was last active less than 360 years ago as shown by radiocarbon dating of charcoal from a Maori oven covered by the latest volcanic ash from Egmont.

Rangitikei Basin. The entire region of the Rangitikei Basin is covered by fresh-water and richly fossiliferous marine sands, silts, and coquina limestone of early Quaternary age. They were tilted seaward while being deposited and now have a roughly synclinal structure nearly parallel to the graywacke range to the east. Upper Tertiary marine sediments dip under from the north and probably rest on a graywacke basement. The unusually rapid Quaternary warping is possibly related to the major gravity anomaly of New Zealand. This is oval in shape and extends across Cook Strait to the north end of the Marlborough Sound and the Bouguer values reach minus 160 mgal in the Rangitikei area. The anomaly must be one either to extremely thick crustal rocks, possibly graywacke or schist, or to deep-seated crustal variations, related to the line of the two major New Zealand tectonic features—the Alpine fault of the South Island and the volcanic belt of the North Island. The Quaternary sediments are broken by normal faults probably related to those of the volcanic belt and contrasting with the transcurrent or transcurrent-reverse faults in the regions to the east.

Wairarapa and Hawke's Bay Depression. Early Quaternary marine and freshwater sediments similar to those of the Rangitikei Basin cover most of the region. The basal beds contain thick bands of conglomerate derived from the axial range, and date the major uplift of that region. To the east they rest on the Upper Tertiary beds of the Wairarapa Range, and to the west and north either directly on the graywacke of the axial range or on Upper Tertiary beds. The underlying rocks are graywackes, probably Jurassic on the east and Triassic on the west.

Several major clockwise transcurrent faults, continuations of South Island faults, cut through the region and have displaced late Pleistocene terraces.

Axial Range of North Island. The region is composed entirely of graywacke, except for one small outlier of Lower Tertiary marine sediments, several small outliers of Upper Tertiary marine sediments, and in the north a thin covering of Quaternary pumice and ignimbrite. In the south near Wellington the graywacke is of Upper Triassic age, and in the north near Whakatane it is probably Jurassic. Gradual increase in degree of metamorphism toward the west in the center of the North Island, reaching subschist rank in the Kaimanawa Range, may indicate increasing age. The structure is probably complex throughout.

Volcanic Belt. The presently active volcanoes of New Zealand—Ruapehu, Ngauruhoe, Tongariro, Tarawera, and White Island—are confined to the volcanic belt and are on a line that strikes toward the volcanoes of the Kermadec Islands, Tonga, and Samoa (Fig. 4). The line is of major importance and has been the site of persistent volcanism since the early Quaternary. The volcanoes outside the belt, such as Mt. Egmont and some of the Auckland cones, have erupted during the last 2000 years and therefore should still be regarded as active, but with the exception of Mt. Egmont they were probably active for short periods only.

FIGURE 4. Mt. Ngauruhoe seen in eruption, Tongariro National Park. (Photo: N.Z. Inform. Serv.)

The volcanic sequence is 1000 m or more thick in the central part of the belt and contains a variety of basaltic, andesitic, and rhyolitic volcanoes. Most of the stratified rocks are acid tuffs extending considerable distances from vents that are now hidden. Ignimbrites, probably formed from nuée ardentes, and pumice shower deposits are the most important. The volcanic belt is cut by numerous NE-SW predominantly normal faults, contrasting with the transcurrent nature of the faulting in adjacent regions to the east and southeast.

Geological History

There is no firm evidence of Precambrian events in New Zealand, but the Greenland and Waiuta groups of Westland and South West Nelson, unfossiliferous lightly metamorphosed graywackes and argillites, have been mapped as Precambrian. Older Paleozoic rocks are preserved only in small areas of Fiordland and Northern West Nelson, but they are important in showing that even in these early times New Zealand was the site of geosynclinal deposition, although with the supply of sediment slackening in the Lower Devonian. Their faunas show cosmopolitan affinities at first but an austral element appears in the Lower Devonian. Middle and Upper Devonian rocks are not definitely known and an orogenic phase, the *Tuhua Orogeny*, spanned this time and a large part of the Carboniferous, during which occurred a fundamental reorganization of New Zealand structure and pattern of sedimentation occurred during this period.

The next sediments are extremely thick deposits accumulated east of a foreland of older rocks (of which Fiordland, coastal Westland,

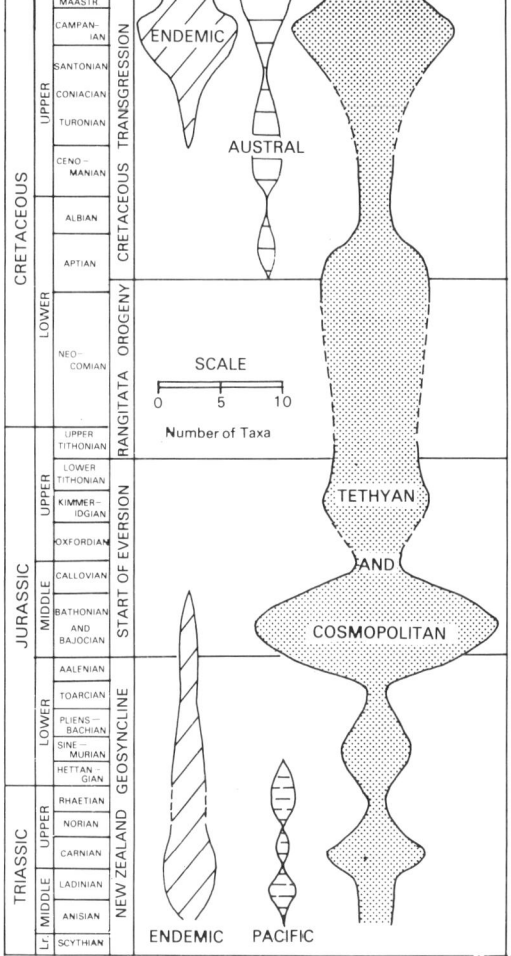

FIGURE 5. Biogeographic elements represented among incoming taxa of bivalvia and gastropoda in the Mesozoic of New Zealand (after Fleming, 1967).

the geosyncline. The rocks of the marginal facies have now been folded but can be traced, as the Southland Syncline, from the southeastern coast of Otago across Southland to the Alpine Fault. They are absent for 500 km along the Alpine Fault, but reappear as the Nelson Syncline and are presumably continuous northward, passing under Cook Strait and the Tertiary rocks of Taranaki, to reappear on the western coast of South Auckland. The axial sediments were strongly folded and faulted and now form the axial ranges of both North Island and South Island as well as the basement rock over much of the remaining country. Beyond the main islands, the course of the geosyncline is speculative. The western foreland can be traced through Stewart Island and The Snares to the Auckland Islands. The geosynclinal sediments reappear at New Caledonia to the north northwest and in Chatham Island to the east.

Affinities of the faunas populating the New Zealand Geosyncline range from predominantly Tethyan or Indo-Pacific in the Permian (with a cool-water "austral" element), Middle and Upper Jurassic and Lower Cretaceous to strongly endemic in the Triassic and Lower Jurassic (Figs. 5 and 6).

The first shallowings in the New Zealand and North West Nelson are the emergent parts), in a rapidly subsiding geosyncline ("New Zealand Geosyncline") embracing the remaining parts of South Island and all of North Island. The oldest rocks in the New Zealand Geosyncline are the unfossiliferous schists of Otago, Westland, Marlborough, and a narrow strip in the Kaimanawa Range (North Island), perhaps locally as old as Carboniferous but of uncertain age because of their great alteration.

In the Permian, Triassic, Jurassic, and Lower Cretaceous a marginal facies of fossiliferous rocks deposited in shelf conditions on the western side of the geosyncline can be distinguished, and an axial facies of contemporary but poorly fossiliferous graywackes and argillites, more indurated and folded, was deposited near the axis of most rapid sinking of

FIGURE 6. Faunal changes related to the orogenic events in the Mesozoic of New Zealand. The diagram shows percentages of the various biogeographical elements among new bivalve and gastropod arrivals in each stage, somewhat generalized. The increase in Tethyan-Cosmopolitan elements in the Middle to Upper Jurassic corresponds with the climax of the Rangitata orogeny that temporarily ended isolation of the southwest Pacific by establishing shallow-water routes to the northwest. The renewed development of endemic elements in the Upper Cretaceous reflects the increasing isolation of New Zealand. Appearance of Austral elements in the Cretaceous suggests that the migration paths provided by the west wind drift in the southern Ocean in modern times first became established at this time (after Fleming, 1967).

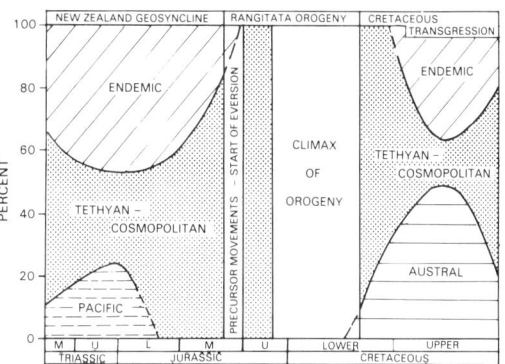

Geosyncline occurred in the Middle and Upper Jurassic and these were followed by the *Rangitata Orogeny,* probably in Upper Tithonian and Neocomian times, for no fossils of these ages have been found. By Aptian times the geosynclinal area was restricted to the east coast. The folded geosynclinal belt was at first mountainous but was worn down to low relief before the end of the Cretaceous, when marine transgression ensued. The land was progressively submerged until in the Oligocene the site of modern New Zealand was occupied by an extensive archipelago (Fig. 7).

The New Zealand Geosyncline was originally straight or gently curved and current theories consider it to have been a linear trough marginal to Gondwanaland; the Lord Howe Rise, the Campbell Plateau and westernmost parts of the South Island constituted remnants of the foreland (Fleming, 1970; Griffiths, 1971). The New Zealand Geosyncline was

FIGURE 7. New Zealand coastline throughout late Paleozoic-Quaternary time. The distribution of land is shown by the diagonal line pattern and to assist in orientation each diagram is superimposed on an outline of the modern New Zealand land mass. Pre-Cretaceous transcurrent movement is shown along the Alpine fault.

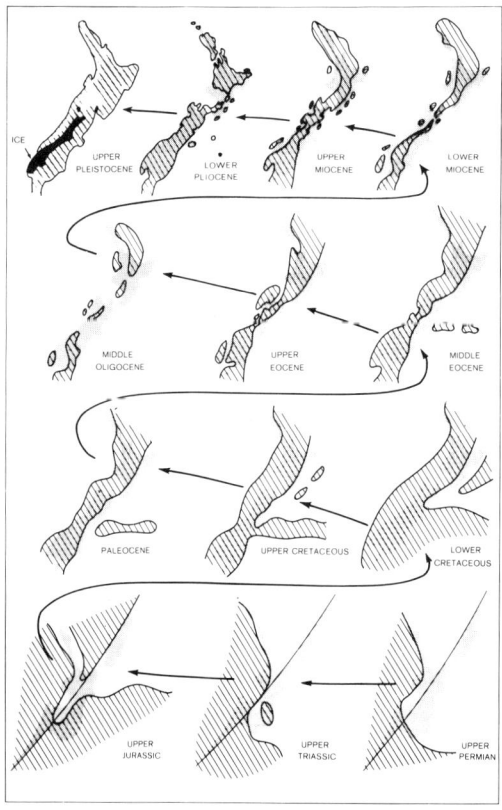

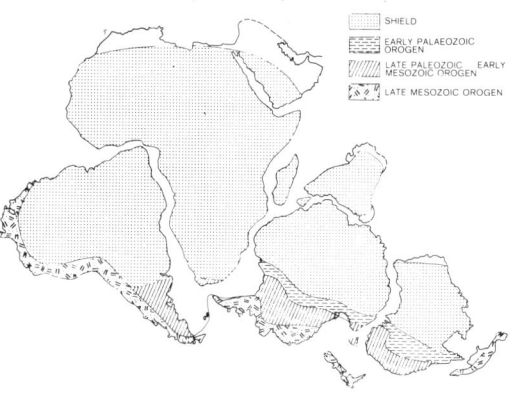

FIGURE 8. Relationship of the major structures of New Zealand to those of the remainder of Gondwanaland. Only the present-day land masses are shown; for details, see: Griffiths (1971) and Suggate (1971).

distorted and ruptured during the Rangitata Orogeny, which coincided with the main dispersal movements of the Gondwanaland continents.

Prior to the disruption of Gondwanaland the New Zealand Geosyncline was probably continuous with the Papuan Geosyncline (New Guinea, Indonesia) and the West Antarctic-Andean Geosyncline (Fig. 8). New Zealand and the West Antarctica-South American region are now separated by an extensive gap formed by the production of new ocean floor generated from the South Pacific Rise.

Tethyan faunal affinities continued into the Middle Cretaceous but in the Late Senonian and Maestrichtian strong Austral elements appear, the first to do so since the Permian (Figs. 5 and 6). The Paleocene and Eocene are characterized by strong endemic as well as Tethyan groups. Influx of warm-water Indo-Pacific forms typifies the Oligocene and reaches a climax in the Lower Miocene when *Cocos* groves grew in Northland.

The New Zealand Geanticline, the eroded stump of the mountains formed in the Rangitata Orogeny, started to rise in Miocene times and sedimentary basins formed to either side and large-scale submarine slumping was common. In Middle and Upper Miocene successive marine benthic faunas record progressive cooling of the seas around New Zealand by the extinction of subtropical genera that had marked the mid-Tertiary thermal maximum.

Earth movements became even more intense during the Pliocene, approaching the climax of the *Kaikoura Orogeny* to which New Zealand owes its present geography. Differential movements were most acute in a mobile belt trending from Raukumara Peninsula obliquely

SW through the Southern Alps to Fiordland and decreased on either side toward the more stable districts of Northland and Otago-Stewart Island.

Progressive extinction of warm-water genera continued throughout the Pliocene, but even so the temperatures were still probably warmer than at present.

Four marine stages—two glacial, two interglacial—are recognized in the Lower and Middle Pleistocene. Owing to earth movements in central New Zealand, Upper Pleistocene sediments are clearly separated from those of the Lower and Middle Pleistocene.

Four glaciations and three interglacial intervals have been recognized in the New Zealand Upper Pleistocene. These deposits are either coastal terraces formed during interglacial stages of high sea level or alluvial deposits of river beds aggraded during glacial stages of low sea level and of rapid erosion due to periodic deforestation of the mountains.

During the maximum phase of the last glaciation (Otiran Stage) permanent snow level was at least 1100 m lower than at present and ice tongues carved glacial valleys on either side of the main divide from Fiordland to North West Nelson and to a minor extent in the northern mountains (Tararua Range, Egmont, National Park). Glaciers extended below present sea level on the southwestern coast of South Island and in some eastern lake basins. In Westland ice tongues joined to form a continuous piedmont glacier. Fell-field or tundra occupied vast areas in south and central North Island. Periglacial conditions extended below sea level at Wellington and solifluxion produced vast quantities of coarse debris. Extensive outwash plains were formed and from these wind picked up dust to form loess deposits on lowlands as far north as central North Island, there interfingering with ash from the active volcanoes. The coastline followed the present-day continental shelf edge, linking North and South island, and most outlying islets.

The Flandrian transgression drowned low-lying coastal areas, forming the intricately embayed coastlines characteristic of the more stable parts of New Zealand (e.g., Northland). Numerous Holocene deposits have been confirmed by radiocarbon dating.

GRAEME R. STEVENS

FIGURE 9. Large-scale submarine slumping in the Lower Miocene Waitemata Group, exposed in the Whangaparaoa Peninsula, north of Auckland. Note the sharp contacts above and below the slumped beds in A; detail in B. (Photos: courtesy M. R. Gregory.)

References

Andrews, J. E., and Eade, J. V., 1973. "Structure of the Western Continental Margin, New Zealand, and Challenger Plateau, Eastern Tasman Sea," *Bull. Geol. Soc. Am.*, **84**, 3093–3100.

Austin, P. M., Sprigg, R. C., and Braithwaite, J. C., 1973. "Structure and petroleum potential of eastern Chatham Rise, New Zealand," *Bull. Am. Assoc. Petrol. Geologists*, **57**(3), 477–497.

Christoffel, D. A., 1971. "Motion of the New Zealand Alpine Fault deduced from the pattern of sea-floor spreading," *Roy. Soc. N.Z. Bull.*, **9**, 25–30.

Cotton, C. A., 1958. *Geomorphology*, 7th ed. Wellington: Whitcombe & Tombs, 505p.

Cullen, D. J., 1970. "Tectonic map of the south-west Pacific 1:10,000,000," *N.Z. Oceanogr. Inst. Chart, Misc. Ser. 20.*

Fleming, C. A., 1967. "Biogeographic change related to Mesozoic orogenic history in the southwest Pacific," *Tectonophysics*, **4**, 419–427.

———, 1970. "The Mesozoic of New Zealand: chapters in the history of the Circum-Pacific mobile belt," *Quart. J. Geol. Soc. Lond.*, **125**, 25–170.

Gregory, M. R., 1969. "Sedimentary features and penecontemporaneous slumping in the Waitemata Group, Whangaparaoa Peninsula, North Auckland, New Zealand," *N.Z. J. Geol. Geophys.*, **12**(1), 248–282.

Griffiths, J. R., 1971. "Reconstruction of the southwest Pacific margin of Gondwanaland," *Nature*, **234**, 203–207.

Grindley, G. W., Harrington, H. J., and Wood, B. L., 1959. "The geological map of New Zealand 1:2,000,000," *N.Z. Geol. Surv. Bull.*, **66**.

Hay, R. F., et al., 1970. "Geology of the Chatham Islands," *N.Z. Geol. Surv. Bull.*, **83**, 86p.

Healy, J., 1962. "Structure and volcanism in the

Taupo Volcanic Zone, New Zealand," *The Crust of the Pacific Basin* (Monogr. 6). Washington, D.C.: Am. Geophys. Union, 151–157.

Hornibrook, N. de B., 1968. "Handbook of New Zealand microfossils (Foraminifera and Ostracoda)," *N.Z. Geol. Surv. Info. Ser.* **62**, 136p.

Landis, C. A., and Bishop, D. G., 1972. "Plate tectonics and regional stratigraphic metamorphic relations in the southern part of the New Zealand geosyncline," *Bull. Geol. Soc. Am.,* **83**, 2267–2284.

McLintock, A. H., ed., 1966. *An Encyclopaedia of New Zealand.* Wellington: N. Z. Govt. Printer, 3 vols.

Scholz, C. H., et al., 1973. "Detailed Seismicity of the Alpine Fault Zone and Fiordland Region, New Zealand," *Bull. Geol. Soc. Am.,* **84**, 3297–3316.

Stevens, G. R., 1974. *Rugged Landscape: The Geology of Central New Zealand.* Wellington: A. H. & A. W. Reed, 286p.

Suggate, R. P., 1971. "Mesozoic–Cenozoic development of the New Zealand region," *Pacific Geol.,* **4**, 113–120.

———, Stevens, G. R., TePunga, M. T. (eds.), 1975. *The Geology of New Zealand.* Wellington: N. Z. Gov't. Printer, 2 vols.

Maps:

Geological maps, geological and paleontological bulletins issued by N.Z. Geol. Surv. Dept. of Scientific and Industrial Res.

Geological and paleontological articles in the *N.Z. J. Sci. Tech.,* and *Trans. Roy. Soc. N.Z.* (now *J. Roy. Soc. N.Z.*).

Cross-references: *Antarctica; Australasia–Regional Review; Kermadec Islands; New Caldeonia; New Guinea; Oceania; Tonga.*

NICARAGUA

With an area of 130,000 km² (50,193 sq mi), Nicaragua is the largest Central American republic. Nicaragua adjoins Honduras on the north and Costa Rica on the south, with its western coastline along the Pacific Ocean and its eastern coastline along the Atlantic. The eastern coastal area is low, hot, and swampy where the annual rainfall sometimes reaches 760 cm (300 in.). The central region is a plateau with dense forests and rolling hills cut by rivers. About 300 cm (120 in.) of rain falls here annually. The western region is mountainous with many lakes where the temperatures are moderate, ranging from 4.1 to 35°C with about 150 cm of rainfall during the wet season from May to December. The principal river is the Rio Coco (Segovia), which is common also to Honduras.

The Cordillera range of mountains runs northwest to southeast through the center of Nicaragua. Between these mountains and a

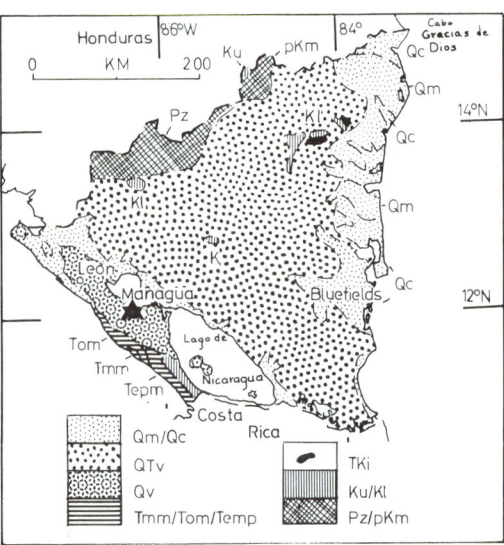

FIGURE 1. Geological outline map of Nicaragua (based upon the "Geological Map of North America," U.S.G.S., 1965). Explanation: Qm, Quaternary marine; Qc, Q continental; Qv, Q volcanics; Tmm/Tom/Temp, Tertiary/Miocene/Oligocene/Eocene/Paleocene, marine; Tki, Tertiary and Cretaceous intrusives; Ku/Kl, Upper and lower Cretaceous; Pz/pKm, Paleozoic and pre-Cretaceous metamorphics.

range of volcanic peaks to the west, lie Lake Managua and the 160-km (100-mile)-long Lake Nicaragua. Lake Managua is surrounded by volcanic mountains, including the nearly mile-high Mt. Momotombo on the lake's northern shore. Another range of volcanic mountains, the Cordillera Isabelia, stretches along the northern border from east to west. The highest peak is the Cerro Chachagón, 1804 m (6300 ft).

In 1931 the capital city of Managua was partly destroyed by earthquake and fire, and again in 1972.

Gold, silver, copper, tungsten, precious gems, and gypsum are mined.

Geologically the country appears, at least partly, to be underlain by ancient crustal blocks (Paleozoic), continuous with that exposed in the Honduras Massif, but extensively covered by Oligocene-Miocene volcanics and dropped down by faulting on the southwest. This is Moody's "Nicaragua Volcanic Upland" (this volume; see *Central America–Regional Review*). It is highest to the west (1500 m or more) and slopes down eastward toward the coastal plain on the Caribbean side.

The downfaulted sector to the southwest is occupied by the Lake Nicaragua and Lake Managua depression, associated with which there is a chain of Pleistocene-Holocene volcanoes, partly active, trending NW-SE, reaching their highest culmination in Volcan de Viejo. This belt is cut off near the Costa Rica border by an E-W strike-slip fracture zone, a continuation of the eastern Pacific Clipperton Zone. Emerging in the coastal hills beneath the younger volcanics is a belt known as the Rivas Arch, consisting of block-faulted Upper Cretaceous to Miocene sediments, mainly clastics, tuffs and lavas.

In the Caribbean (Mosquito) coastal plain, up to 100 km wide, there is a section of at least 4500 m of Tertiary clastics, over perhaps 4000 m of Upper Cretaceous limestones and shales. An exploration well here, offshore, drilled on Touche Island in 1957 to 14,997 ft, encountered high-pressure gas (Karim et al., 1966).

Structural Evolution

The oldest basement rocks in the country are the Paleozoic crystalline blocks that belong to the "Nuclear Central America" complex of Guatemala and Honduras. Its southern margin, according to Malfait and Dinkelman (1972), became involved in subduction by the East Pacific Plate in late Mesozoic/early Tertiary time and a belt of Laramide volcanics and granitic intrusions were introduced along the southern "leading edge."

In the Paleocene a south-facing island arc developed here and by early Oligocene the subduction zone extended to the SE, toward that of Colombia in South America (the Bolivar Trench). From this stage on, an "isthmian link" of volcanic islands progressively built up toward the SE, eventually (about 4 m yr ago) to connect through Panama with the South American continent.

Offshore the Mid-America Trench continues to this day to be a site of progressive subduction, as evidenced also by the high seismicity and explosive volcanism along the Pacific margin.

RHODES W. FAIRBRIDGE

References

Brown, R. D., Jr., Ward, P. L., and Plafker, G., 1973. "Geologic and seismologic aspects of the Managua, Nicaragua," Earthquakes of December 23, 1972, *U.S. Geol. Surv. Prof. Pap. 838*, 34p.

Dengo, G., Bohnenberger, O., and Bonis, M., 1970. "Tectonics and volcanism along the Pacific Marginal Zone of Central America," *Geol. Rundschau*, **59**(3), 1215–1232.

Hoffstetter, R., 1960. "Amerique Central," *Lexique Stratig. Intern.* (Paris), **5**, fasc. 2.

Karim, M., Chilingar, G. V., and Hoylman, H. W., 1966. "Northeast Nicaragua has gas and oil indications," *World Oil*, **162**(4), 84–96.

Lloyd, J. J., 1963. "Tectonic history of the south Central American orogen," *Backbone of the Americas*. Tulsa: Am. Assoc. Petrol. Geologists, 88–101.

Malfait, B. T., and Dinkelman, M. G., 1972. "Circum-Caribbean tectonic and igneous activity and the evolution of the Caribbean Plate," *Bull. Geol. Soc. Am.*, **83**, 251–272.

Mills, R. A., and Hugh, K. E., 1974. "Reconnaissance Geologic Map of Mosquitia Region, Honduras and Nicaragua Caribbean Coast," *Bull. Am. Assoc. Petrol. Geologists*, **58**(2), 189–207.

Roberts, R. J., and Irving, E. M., 1957. "Mineral deposits of Central America," *U.S. Geol. Surv. Bull.*, **1034**, 205.

Weyl, R., 1961. "Die Geologie Mittelamerikas," *Regionalen Geologie der Erde*. Berlin: Gebr. Borntraeger.

———, 1969–1970. "Geologische Bilder aus Mittelamerika," *Natur und Museum*, **99**, 415–423, 559–570; **100**, 120–128, 269–278, 362–370.

Cross-references: *Central America–Regional Review; Costa Rica; El Salvador; Guatemala; Honduras; Panama.*

NIUE ISLAND

Lying in the South Pacific, some 2400 km (1500 miles) NE of New Zealand, 450 km E of the Tonga Islands, at 19.02°S and 169.55°W, the uplifted coral atoll island of Niue has a warm climate (av. 24.8°C) and is buffeted occasionally by severe hurricanes. The rainfall is about 2000 mm, but greater on the east (windward) side in the southern summer; the rest of the year is rather dry under the influence of the SE trade winds and there may be serious droughts. Its area is 259 km^2 (about 100 sq mi). Its soil is sparse but fertile (Wright and Westerndorp, 1965).

The first European discovery of Niue was recorded in 1774 by Captain Cook, who called it "Savage Island." In the present century it came under New Zealand responsibility as part of the *Cook Islands* (q.v.) dependency, and since 1962 has been semi-independent. The island exports copra and bananas. Some 20% of the island is forested. The principal town and port facility is Alofi.

The emerged atoll rises from a drowned volcanic cone. The atoll is about 17 km in diameter and rises to about 65 m, the former lagoon floor forming an interior plain where there is a water problem because of karst drainage. The former reef rim rises in steep cliffs from a narrow contemporary fringing reef with a notable [nontilted] terrace at 27 m and several minor benches (Schofield, 1959). Much of the limestone has been dolomitized (Schofield pers. obs.) and some soils are abnormally radioactive

(Fieldes et al., 1960; Schofield, 1967). From the evidence of sea-floor spreading the underlying oceanic crust is Cretaceous in age.

J. C. SCHOFIELD

References

Anon., 1968. *Reports on Niue and the Tokelau Islands.* Wellington: Dept. of Maori and Islands Affairs.

Fieldes, M., Bearling, G., Claridge, G. G. C., Wells, N., and Taylor, N. H., 1960. "Mineralogy and radioactivity of Niue Island soils," *N.Z. J. Sci,* **3,** 658–675.

Kennedy, T. F., 1966. *A Descriptive Atlas of the Pacific Islands.* Wellington.

Schofield, J. C., 1959. "The geology and hydrology of Niue Island, South Pacific," *N.Z. Geol. Surv. Bull.,* **62,** 1–28.

———, 1967. "Origin of radioactivity at Niue Island," *N.Z. J. Geol. Geophys.,* **10,** 1362–1371.

Wright, A. C. S., and Westerndorp, F. J. van, 1965. "Soils and agriculture of Niue Island," *N.Z. Soil Bureau Bull.,* **17.**

Cross-references: *Cook Islands; New Zealand; Tonga.*

NORFOLK ISLAND

Norfolk Island and its companion islets of Philip and Nepean lie midway between New Zealand and New Caledonia on the Norfolk Island Ridge at 29°05′S and 167°57′E. The island is approximately 5 km by 8 km, is bounded by cliffs 50 m to 90 m high except for three shallow bays, and rises to 315 m in two peaks. Its soil is fertile, and its climate is mild, with an annual rainfall of 1100 mm.

Captain Cook discovered the island in 1774. Up until 1855 it was used as a penal settlement, and in the following year the descendants of the *Bounty* mutiny were transferred here from Pitcairn Island. Since 1913 Norfolk Island has been a Territory of the Commonwealth of Australia. It is connected by an air service to Sydney and Auckland, and enjoys a modest tourist trade.

Following brief early reports on the geology by Laing (1912), Speight (1912), and Hutton and Stephens (1956), recent work (McDougall and Aziz-ur-Rahman, 1972; Jones and McDougall, 1973) confirms that the bulk of the island consists of olivine basaltic lavas and interbedded tuffs. The oldest rocks form sequences of as many as ten almost flat-lying flows exposed on the E coast. K-Ar dates of these lavas range from 3.01 to 3.08 m.yr. (Late Pliocene). The greater part of Norfolk Island comprises lavas and less common tuffs erupted mainly from the vicinity of the summits of Mount Pitt and Mount Bates during the period 2.75 to 2.35 m.yr. Quench zones in the basalt indicate that the sealevel during eruption stood at about its present level, reflecting the stability of the Norfolk Ridge at this latitude for the past 3 m.yr.

Coordinated K-Ar and paleomagnetic measurements on the basalts of Norfolk Island provide a means of refining the age estimate of the boundary between the Gauss normal and the Matuyama reversed polarity epochs, and help refine the ages of the Mammoth and Kaena polarity events.

Quaternary sediments fringe the S coast of Norfolk Island and form the nearby Nepean Island. Late Pleistocene calcareous eolianite is overlain by a complex of Holocene lagoonal black mud (with fossil logs of the Norfolk pine, *Araucaria heterophylla*) and beachrock calcarenite.

Philip Island lies 6 km south of Norfolk Island, and is 2 km by 2 km in plan, and 280 m high. Owing to early overstocking with rabbits and goats, the island is almost completely barren, and its ochres, reds and browns contrast with the greens of Norfolk Island in mute testimony to past ecological sins. Philip Island's volcanic history is part of that of Norfolk Island, but additionally contains evidence of a previous phase in the form of inclusions in tuff of Lower Miocene reef limestone (Coleman and Veevers, 1971).

JOHN J. VEEVERS

References

Coleman, P. J., and Veevers, J. J., 1971. "Microfossils from Philip Island indicate a minimum age of Lower Miocene for the Norfolk Ridge, South-west Pacific," *Search,* **2**(8), 289.

Green, T. H., 1973. "Petrology and geochemistry of basalts from Norfolk Island," *J. Geol. Soc. Aust.,* **20**(3), 259–272.

Hutton, J. T., and Stephens, C. G., 1956. "The palaeopedology of Norfolk Island," *J. Soil Sci.,* **7,** 255–269.

Jones, J. G., and McDougall, I., 1973. "Geological history of Norfolk and Philip Islands, southwest Pacific Ocean," *J. Geol. Soc. Austral.*

Laing, R. M., 1912. "Notes on the chief physiographic features of Norfolk Island," *Trans. N.Z. Inst.,* **45,** 323–326.

Linden, W. J. H. van der, 1971. "Western Tasman Sea geomagnetic anomalies," *N.Z. Oceanograph. Inst. Misc. Ser. 19,* 1:2,000,000.

McDougall, I., and Aziz-ur-Rahman, 1972. "Age of the Gauss-Matuyama boundary and of the Kaena and Mammoth events," *Earth Planet. Sci. Lett.,* **14,** 367–380.

Shor, G. G., Kirk, H. K., and Menard, H. W., 1971. "Crustal structure of the Melanesian area," *J. Geophys. Res.,* 76(11), 2562–2586.

Speight, R., 1912. "On a collection of rocks from Norfolk Island," *Trans. N.Z. Inst.,* 45, 326–331.

Cross-references: *Australasia–Regional Review; New Caledonia; New Guinea; New Zealand.*

NORTH AMERICA

North America comprises the United States, Canada, Greenland, Mexico, and the tiny French colony of St. Pierre and Miquelon. Each is also treated separately in this volume. The total area of the continent is approximately 24 million km^2.

Precambrian Tectonic Divisions

By means of several hundred isotope age dates made during the past few years, three major divisions of Precambrian orogeny of North America have been delineated and a fourth is questionable. They are shown on Fig. 1. The oldest division of igneous activity and metamorphism is called the *Kenoran,* and the time of orogeny was about 2500 m yr ago. It is probably the core region of the continent about which younger belts of orogeny have developed. To the north and northeast is a belt of orogeny, known as the *Hudsonian,* that is 1700 m yr old. A small secondary nucleus exists in the Great Slave and Great Bear lakes region of Kenoran age. It is surrounded by Hudsonian dates. South of the Kenoran in the United States is a vast region with dates mostly of 1250 to 1400 m yr, but some geochemists believe these dates are minimal and that the true age of orogeny here is about 1700 m yr, like that of the Hudsonian. In this sense the Hudsonian belt more clearly surrounds the Kenoran core.

The Grenvillian belt of orogeny, about 1000 m yr old, is classically developed in southern Ontario and Quebec, but it is now known to comprise the crystalline piedmont of the eastern margin of the United States and exposures of Precambrian rock in Texas. It probably extends to Greenland, as shown on Fig. 1.

The region of Manitoba, the Dakotas, Montana, Wyoming, Minnesota, and northern Michigan has yielded mixed dates of the Kenoran and Hudsonian, and it is believed that here the younger Hudsonian is superposed on the older Kenoran. In the Great Lakes region additional Grenvillian dates occur and probably indicate the superposition of a third episode of orogeny.

The Beltian trough or geosyncline rests across and on top of the Hudsonian belt and is clearly younger. Its sediments have been folded in places, eroded, and covered by the thick Purcell strata. These also were somewhat folded and eroded before burial by Cambrian strata. The age of the Beltian strata is not yet well determined, but it must be somewhere between 1000 and 800 m yr.

Early and Mid-Paleozoic Evolution

Dominant in early and mid-Paleozoic times were the great marginal geosynclines (see Fig. 2). Both the eastern and western geosynclines were composed of two divisions, the outer eugeosyncline and the inner miogeosyncline. Feldspathic sedimentary rocks interbedded with andesitic and basaltic volcanic rock are characteristic of the eugeosyncline. Fossil-bearing marine shales and limestones also occur, especially in the western eugeosyncline. The miogeosynclines are replete with shales, sandstones, limestones, and dolomites, mostly marine and fossiliferous.

The eugeosynclines at the same time having much volcanism were beset with orogeny, meta-

FIGURE 1. Major divisions of Precambrian orogeny.

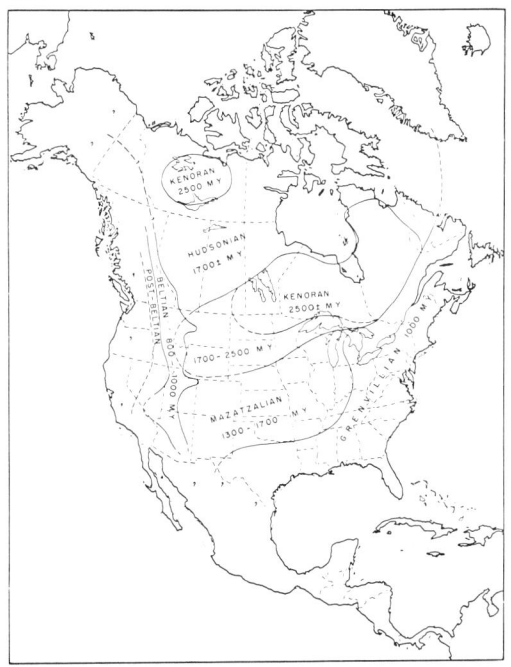

FIGURE 2. Early and mid-Paleozoic North America.

Late Paleozoic Evolution

Nearly all the northern part of the continent had become emergent by the Pennsylvanian period as shown by the map of Fig. 3. The fold belt of the Canadian Arctic islands had subsided and become invaded by Pennsylvanian seas, and a blanket of sediment was spread across the eroded folds. The Pennsylvanian seas invaded a small area in northern Yukon Territory and in northeastern Alaska.

During one part or another of the Pennsylvanian, most of the central United States was covered with seas and in this region, especially from Kansas westward, a number of major uplifts occurred. The uplifts in New Mexico, Colorado, Utah, and Arizona are known as the *Ancestral Rockies.* They rose sharply from the surrounding seas and shed much clastic sediment around them like skirting fans. As the sediments accumulated, the uplifts were gradually covered. Burial became complete during the Permian for most of them, but the higher parts of the two large ones in Colorado stood as islands until the late Mesozoic.

Other uplifts related to the Ancestral Rockies rose in Nebraska, Kansas, Oklahoma, and Texas during the Early Pennsylvanian but be-

morphism, erosion of uplifted belts, and sedimentation in narrow troughs. In the site of the eastern eugeosyncline the Taconian orogeny in Late Ordovician time was pronounced from northeastern Pennsylvania through western New England, New Brunswick, and Newfoundland. Sediments were shed eastward across New England and westward across Pennsylvania, West Virginia and Ohio. The Caledonian orogeny of Late Silurian time is noted in central Newfoundland, eastern Greenland, and in the Canadian Arctic archipelago. During Late Devonian, the major orogeny along the Atlantic margin occurred, the Acadian, in which vast batholiths were intruded and the strata intensely folded and generally metamorphosed. The rising chains of mountains supplied much sediment to an adjacent interior basin in Pennsylvania, West Virginia, and Ohio.

The Canadian Shield, as shown on Fig. 2, was probably transgressed by seas at one time or another more than shown but was generally emergent. A long broad extension to the southwest is known as the *Transcontinental Arch.* Other arches, several basins, and extensive shelf regions also characterized the development of the interior stable part of the continent during early and mid-Paleozoic times.

FIGURE 3. Pennsylvanian period.

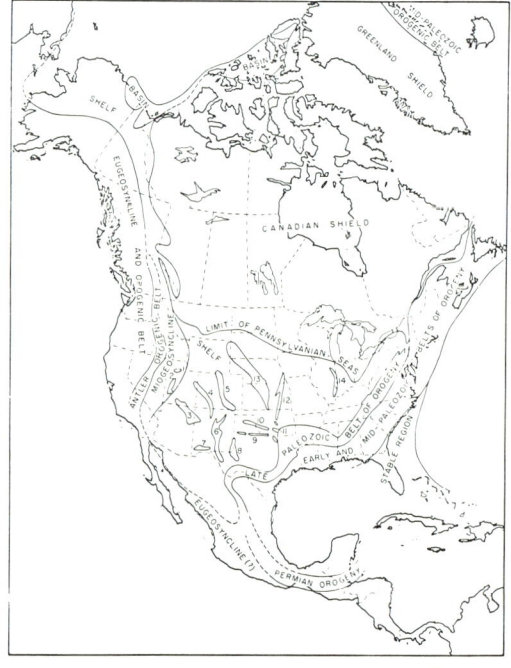

404

came completely buried by Late Pennsylvanian.

The most momentous event was the development of the Appalachian Mountains and their associates, the Ouachita Mountains in Arkansas and Oklahoma and the Marathon Mountains in west Texas. The Appalachians came into existence when the thick sediments of the miogeosyncline were folded and thrust faulted. The thrust sheets were moved toward the interior of the continent. The Ouachitas and Marathons exhibit similar structures to the Appalachians. The older Acadian and Taconic orogenic belts lie to the east of the newer Appalachians of Pennsylvanian age, which now constitute the Valley and Ridge province of physiographers and geographers.

Subsidence of the Acadian belt and the spread of seaways over it occurred in parts of New England, the Maritime Provinces, and Newfoundland. This resulted in the deposition of Mississippian and Pennsylvanian sediments unconformably on the older Acadian structural complex.

Crustal unrest evolved new structures in central Nevada, where the Antler orogenic belt had developed. It became mountainous in places and shed coarse clastic deposits on either side.

The geology of Mexico and Central America is shown for the first time, but not with much confidence. It is known that an orogenic belt of Permian age formed in southern Mexico, Guatemala, and Nicaragua, but its connection with the Appalachian belt through the Marathons or with the Antler orogenic belt is in question.

A zone of rift faulting occurred in the late Triassic that extended at least from the Bay of Fundy to South Carolina. Graben and downtilted blocks accompanied by basic igneous eruptions and sill intrusions are characteristic and pervade the Acadian orogenic belt. The basins created were the sites of continental sedimentation.

Late Mesozoic and Early Cenozoic Evolution

During the Jurassic and Cretaceous, a great inland seaway spread from the Gulf of Mexico to the Arctic Ocean and at times probably separated the continent into two isolated regions—the Canadian Shield principally on the east and the great new Cordilleran region on the west. After the Acadian and Appalachian orogenies, the eastern margin of the continent was left free of further mountain-making disturbances, but the western margin became engrossed in profound orogeny. The tectonic map of Fig. 4 spans the time from mid-Cretaceous to the Eocene epoch of the Tertiary.

As shown on the map, seas encroached over the margin of the continent from New York southward to Florida and all around the Gulf of Mexico. Florida and the Bahama Islands were part of an extensive shallow platform on which various thick calcareous sediments accumulated.

The Atlantic Ocean basin, the Gulf of Mexico basin (Mexican basin), and the Caribbean basin (the western part is known as the Colombian basin) had definitely come into existence. There is considerable evidence to indicate that Europe and Africa lay in contact with or close proximity to North America and South America during the Paleozoic and that a great breach occurred between them with the continents of the western hemisphere drifting westward. The drifting apart occurred during the Mesozoic and Cenozoic to open up the gap between and to create the Atlantic Ocean. The central zone of spreading is marked by the Mid-Atlantic Ridge. It is possible that South America was closer to North America then than now and that as the two continents moved apart the Mexican and Caribbean basins were created. Hence on previous tectonic maps neither the Atlantic Ocean or Gulf of Mexico are labeled, but from mid-Mesozoic onward we can be confident of their existence.

FIGURE 4. Tectonic map from mid-Cretaceous to the Eocene.

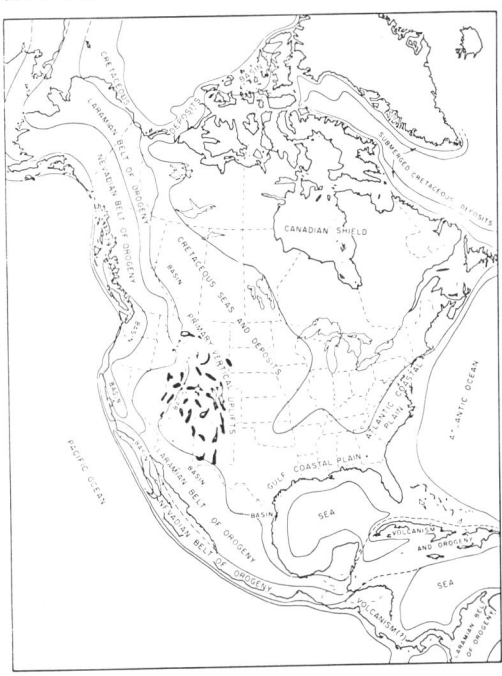

Concerning the development of the long and broad western Cordillera we should note first the Nevadian belt of orogeny. It is the most profound belt of disturbance in the west and somewhat analogous to the older Acadian belt on the eastern margin. It developed in the thick eugeosynclinal sediments, which were tightly folded, widely metamorphosed, and then injected by tremendous batholiths. The batholithic intrusions culminated a long build-up of deformational and volcanic events. The extensive granite terrane of the Sierra Nevada of California is one of the immense batholiths.

Almost all the large batholiths were intruded in Mid-Cretaceous time, which by isotope dating methods, is about 100 m yr before the present. This was a time of head-on collision of the westward advancing North American continent and the eastern Pacific plate. A deep zone of subduction had developed, which is described in detail in other articles in this encyclopedia.

Adjacent on the east to the Nevadian orogenic belt is the long, and in places narrow, Laramian orogenic belt. It is the region of the Rocky Mountains proper. It may be seen to extend from Alaska to the Greater Antilles, and a similar belt extends southward from Venezuela and Colombia to form the Andes of South America. The Laramian belt has a number of divisions along its lengthy course, each with its characteristic structures. In large part it consists of folded and thrust-faulted miogeosynclinal sediments. The well-known Lewis overthrust of Glacier National Park at the international border between Canada and the United States is an example of one of the great faults. There the Precambrian Beltian strata have been pushed 15 to 75 km eastward over the Cretaceous shales on a nearly horizontal thrust surface.

One of the most important structural divisions of the Laramian belt is the group of large anticlinal uplifts extending from Montana through Wyoming, Colorado, Utah, and New Mexico. Such ranges as the Wind River and Big Horn in Wyoming, Uinta in Utah, Front and Sawatch in Colorado, and Sangre de Cristo and Zuni in New Mexico are examples. They are all wonderfully scenic ranges sculptured by erosion from the large oval-shaped anticlines. Some of the anticlines were asymmetrical and even faulted along one or both sides, and by now erosion has exposed the Precambrian rocks in the center of most of them. The Rush Memorial in the Black Hills is a megalithic monument sculptured in a Precambrian granite in the core of the Black Hills anticlinal uplift.

Wide arid and semiarid basins separate the uplifts. These include the Powder River Basin between the Black Hills and the Big Horn uplifts, and the Green River south of the Wind River Range and north of the Uinta Range. They trapped much sediment as lake and river flood-plain deposits during the Tertiary and in these strata are the favorite mammalian fossil hunting grounds of the geologist.

Some narrow basins along the Pacific Coast sank in Late Cretaceous time within the Nevadian orogenic belt. They sank to great depth and received an equally great thickness of clastic sediments from both land areas on the east and on the west.

Tertiary Evolution

The seas of the Gulf and Atlantic margin of the continent south of New York retreated gradually from Late Cretaceous as the land rose until they finally reached their present shore line (Fig. 5). In fact, the entire southern half of the continent, including Mexico, Central America, and the Antilles, has generally risen about 100 m, whereas the northern half has tilted downward. From Chesapeake Bay to Labrador, Hudson Bay, and the Arctic islands on the east and from Puget Sound to Alaska and the Arctic Ocean on the west, the sea level has risen with

FIGURE 5. Tertiary evolution in North America.

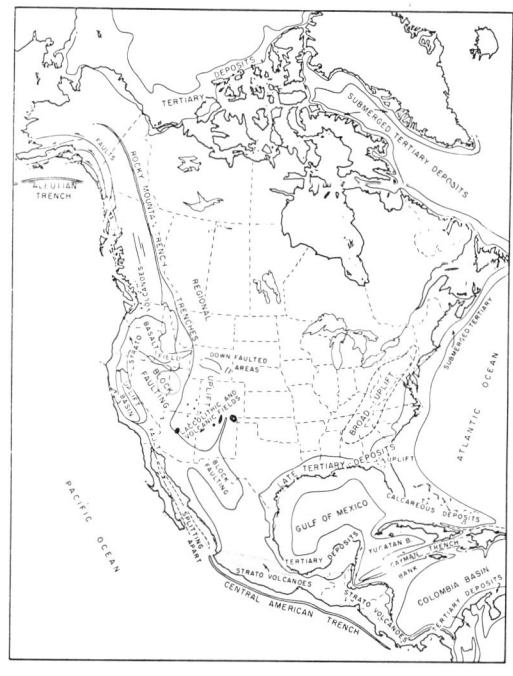

considerable marginal submergence, e.g., the drowned river valleys and fiords on any detailed map. This appears as a continent-wide tilting in which north of a hinge line from Chesapeake Bay to Puget Sound the continent appears to have tilted downward, and south of it the continent has risen. Superposed on the continent-wide tilting is a broad arching of the Rocky Mountain region from the Missouri to the Rockies and the Colorado Plateau in which 1500–2500 m of uplift has occurred. This broad arching has affected the Great Plains, Laramian ranges, intermontane basins, and plateaus equally. It is probably a matter of expansion of the upper mantle causing the broad uplift of the overlying crust.

Besides the broad arching of the Cordilleran region, there was a tremendous outbreaking of vulcanism. Numerous volcanic fields from the Aleutians and Bering Strait to South America were built during the Tertiary. Volcanic eruptions in the western United States and Mexico were particularly intense, voluminous, and long continued. Fissure eruptions of incandescent heavy clouds of volcanic dust were emitted that rolled over the countryside for hundreds of square kilometers, annihilating and burying all life. These were particularly marked in western Utah and Nevada and occurred mostly in the early Oligocene. Later great fissure basalt flows were extruded in Oregon and Washington to build the extensive Columbia lava field in Miocene time. It ranges between 500 and 1000 meters thick. The Sierra Madre Occidental is a most extensive volcanic field in western Mexico of the mid-Tertiary. The numerous high volcanic cones of southern Mexico and Central America are mostly Quaternary and many are still active. The High Cascades of Oregon and Washington are also lofty and extremely scenic Quaternary cones. The Aleutian Islands have had a long Cenozoic volcanic history, and its most impressive exhibit is a long arcuate row of exquisite Quaternary cones.

Equally impressive is the breaking up of the complex Mesozoic and early Tertiary Cordillera by faulting. An extensive system of generally N-S faults developed in the Oligocene and later and resulted in uplifted, dropped, and tilted blocks, and long narrow trenches. Adjustments along some faults are still taking place. The Great Basin province from the Sierra Nevada to the Wasatch Range is a series of fault blocks of this orogeny. Some blocks have been elevated and others depressed, but the bounding blocks, the Sierra Nevada and the Wasatch, are conspicuously uplifted beyond the rest. Each uplifted block generally defines a modern range.

The faulting of the Basin and Range province is part of a breaking up or fragmentation of the entire western margin of North America in the late Cenozoic. Fragmentation appears to have followed the episode of subduction in which the Cordilleran mountain belts were chiefly created. The regimen of great colliding plates underwent a reorganization and, instead of subduction, a NW translation regimen developed, along with extension and the development of the Basin and Range system and other rift systems.

As the map shows (Fig. 5) the block faulting extends southward into Mexico. The slice or sliver of continental margin west of the San Andreas fault including Baja California has moved progressively northwestward from its original adjacent position connected to the Sierra Madre del Sur, and as a result the Gulf of California has opened up.

From western Montana northward through British Columbia and the Yukon a long single trench extends from the Basin and Range province. It is most surely of downfaulted origin and is called the *Rocky Mountain Trench.*

The broad arching or uplift of the middle and late Cenozoic of the Cordilleran region, the widespread vulcanism, and the profound block faulting are all believed to be the result of deep-seated upper-mantle expansion and evolution of basaltic magma.

A narrow belt of deformation, involving folding and faulting along the narrow Pacific margin of California, Oregon, and Washington has resulted in the Coast Ranges. The Great Valley between the Sierra Nevada and the Coast Ranges is a result of the uptilting of the Sierra Nevada block on the east and the building of the Coast Ranges on the west in Late Cenozoic time. The Great Valley is mostly underlain by Late Cretaceous and Cenozoic sediments.

The Greater Antilles was a belt of volcanism and deformation in the Cretaceous and until well into the Tertiary was mostly submerged. It gradually rose in the late Cenozoic to its partially emergent condition. A profound trench between Cuba and Jamaica and extending to Guatemala has settled along faults some 5000–7000 m in the late Cenozoic. This is the Cayman Trench. North of this is Yucatan, another downfaulted basin. Possibly the Mexican and Colombia basins have widened during the Cenozoic as North and South America have drifted somewhat more apart.

Figure 5 shows the submergent condition of the Gulf and Atlantic coastal plains in mid-Tertiary and the emergent Arctic land areas. Since then the southern coasts have risen and became emergent and the northern coasts have generally subsided and become drowned. These activities seem to be continuing. The extensive ice caps of the northern part of the continent depressed the crust under them, and since the melting a maximum response in uplift of at least 400 m has occurred.

A. J. EARDLEY*

*Armand J. Eardley died Nov. 7, 1972, in Salt Lake City.

References

Adkinson, W. L., ed., 1966. "Stratigraphic cross-section of Paleozoic rocks, Colorado to New York," *Am. Assoc. Petrol. Geologists Cross Sec. 4.*
American Commission on Stratigraphic Nomenclature, 1961. "Code of stratigraphic nomenclature," *Bull. Am. Assoc. Petrol. Geologists,* 45(5), 645–665 (also 1970, reprint, rev.).
Andrews, J. T., 1970. "Present and post-glacial rates of uplift for glaciated northern and eastern North America," *Can. J. Earth Sci.,* 7, 703–715.
Anon., 1962. *Tectonic Map of the United States.* Tulsa: Am. Assoc. Petrol. Geologists (jointly with U.S. Geol. Surv.), 2 sheets, 1:2,500,000 (rev. of 1944 ed.).
———, in prep. Geological Highway Maps. Tulsa: Am. Assoc. Petrol. Geologists. 1. Mid-Continent; 2. Southern Rockies; 3. Pacific Southwest; 4. Mid-Atlantic Region; 5. Northern Rockies; 7. Texas; and others (Pacific NW, Northern Great Plains, Great Lakes, NE Region, SE region).
Childs, O. E., and Beebe, B. W. eds., 1963. *Backbone of the Americas* (Mem. 2). Tulsa: Am. Assoc. Petrol. Geologists.
Chipping, D. H., 1971. "Palaeoenvironment significance of chert in the Franciscan formation of Western California," *Bull. Geol. Soc. Am.,* 82(6), 1707–1711.
Clark, T. H., and Stearn, C. W., 1960. *The Geological Evolution of North America.* New York: Ronald Press, 434p. (2nd ed., 1968, 570p.).
Cloud, P., 1971. *Precambrian of North America, Geotimes,* 3, 13–18.
Cram, I. H., ed., 1971. *Future Petroleum Provinces of the United States—Their Geology and Potential* (Mem. 15). Tulsa: Am. Assoc. Petrol. Geologists, 2 vols., 15, 803p., 692p.
Douglas, R. J. W., ed., 1970. *Geology and Economic Minerals of Canada.* (Econ. Geol. Ser. 1, 5th ed.). Ottawa: Geol. Surv. Can.
Dunbar, C. O., and Waage, K. M., 1969. *Historical Geology,* 3rd ed. New York: Wiley, 556p.
Eardley, A. J., 1962. *Structural Geology of North America,* 2nd ed. New York: Harper & Row, 743p. (1st ed., 1951).
———, 1968. "Bonneville chronology: correlation between the exposed stratigraphic record and the subsurface sedimentary succession," *Bull. Geol. Soc. Am.,* 78, 907–909.
Fillon, R. H., 1971. "Possible causes of the variability of post-glacial uplift in North America," *Quat. Res.,* 1(4), 522–531.
Fisher, G. W., et al., eds., 1970. *Studies of Appalachian Geology, Central and Southern.* New York: Wiley, 460p.

Flawn, P. T., chm., 1967. *Basement Map of North America.* Washington, D.C.: U.S. Geol. Surv., Am. Assoc. Petrol. Geologists Basement Rock Project Committee, 1:5,000,000.
Flint, R. F., 1959. *Glacial Map of the United States East of the Rocky Mountains.* New York: Geol. Soc. Am.
———, 1971. *Glacial and Quaternary Geology.* New York: Wiley, 892p.
Fox, F. G., 1969. "Some principles governing interpretation of structure in the Rocky Mountain orogenic belt," *Time and Place in Orogeny.* London: Geol. Soc., 23–41.
Gilluly, J., 1969. "Geological perspective and the completeness of the geological record," *Bull. Geol. Soc. Am.,* 80, 2303–2312.
Halbouty, M. T., ed., 1970. *Geol. of Giant Petroleum Fields.* (Mem. 14). Tulsa: Am. Assoc. Petrol. Geologists, 575p.
Heyl, A. V., 1972. "The 38th parallel lineament and its relationship to ore deposits," *Econ. Geol.,* 67, 879–894.
Higgs, D. V., 1949. "Quantitative aerial geology of the United States," *Am. J. Sci.,* 247(8), 575–583.
Holland, C. H., ed., 1971. *Cambrian of the New World. 1. Lower Palaeozoic Rocks of the World.* New York: Wiley-Interscience, 456p.
Hopkins, D. M., ed., 1967. *The Bering Land Bridge.* Stanford: Stanford Univ. Press, 495p.
Hough, J. L., 1958. *Geology of the Great Lakes.* Urbana: Univ. Illinois Press, 313p.
———, 1963. "The pre-historic Great Lakes of North America," *Am. Scientist,* 51, 84–109.
James, H. L., 1972. "Stratigraphic commission: note 40–subdivision of Precambrian: an interim scheme to be used by U.S. Geological Survey," *Bull. Am. Assoc. Petrol. Geologists,* 56(5), 1128–1133.
Kay, M., 1951. "North American geosynclines," *Geol. Soc. Am. Mem. 48,* 143p.
———, 1969. "North Atlantic—geology and continental drift," Tulsa: Am. Assoc. Petrol. Geologists *Mem. 12,* 1082p.
———, and Colbert, E. H., 1965. *Stratigraphy and Life History.* New York: Wiley, 736p.
King, E. R., and Zietz, I., 1971. "Aeromagnetic study of the midcontinent gravity high of the Central United States," *Bull. Geol. Soc. Am.,* 82, 2187–2208.
King, P. B., 1959. *The Evolution of North America.* Princeton, N.J.: Princeton Univ. Press, 190p.
———, 1969. "The tectonics of North America—a discussion to accompany the tectonic map of North America, scale 1:5 million," *U.S. Geol. Surv. Prof. Pap. 628,* 95p.
———, 1970. "The Precambrian of the United States of America; southeastern United States," *The Precambrian,* vol. 4. New York: Wiley, 1771.
King, R. E., ed., 1972. *Stratigraphic Oil and Gas Fields—Classification, Exploration Methods, and Case Histories* (Mem. 16). Tulsa: Am. Assoc. Petrol., 696p.
Kistler, R. W., 1974. "Phanerozoic batholiths in western North America: a summary of some recent work on variations in time, space, chemistry and isotopic compositions," in: F. A. Donath, ed., *Annual Review of Earth and Planetary Sciences,* vol. 2. Palo Alto: Annual Reviews, Inc., 478p.

Kummel, B., 1970. *History of the Earth.* San Francisco: W. H. Freeman, 707p.

Landes, K. K., 1970. *Petroleum Geology of the United States.* New York: Wiley, 571p.

Lanphere, M. A., and Reed, B. L., 1973. "Timing of Mesozoic and Cenozoic Plutonic Events in Circum-Pacific North America," *Bull. Geol. Soc. Am.,* 84, 3773–3782.

Larson, R. L., 1972. "Bathymetry, magnetic anomalies, and plate tectonic history of the mouth of the Gulf of California," *Bull. Geol. Soc. Am.,* 83, 3345–3360.

Leech, G. B., Lowden, J. A., Stockwell, C. H., and Wanless, R. K., 1963. *Age Determinations and Geological Studies* (Rept. 4, Pap. 63–17). Ottawa: Geol. Surv. Can.

LePichon, X., Hyndman, R., and Pautot, G., 1971. "Geophysical study of the opening of the Labrador sea," *J. Geophys. Res.,* 76(20), 4724–4743.

Lipman, P. W., Prostka, H. J., and Christiansen, R. L., 1972. "Cenozoic volcanism and plate-tectonic evolution of the western United States. 1. Early and middle Cenozoic," *Phil. Trans. Roy. Soc. Lond.,* Ser. A, 271, 217–248.

Locke, D. H., n.d. (c. 1967). *Selected Guides for Geologic Field Study in Canada and the United States of America* (ESCP Ser. RS-9). Englewood Cliffs. N.J.: Prentice-Hall, 56p.

McKee, E. D., et al., 1956. "Paleotectonic maps of the Jurassic System," *U.S. Geol. Surv. Misc. Geol. Inv. Map I-175,* 6p.

———, et al., 1959. "Paleotectonic maps of the Triassic System," *U.S. Geol. Surv. Misc. Geol. Inv. Map I-300,* 33p.

———, et al., 1967. "Paleotectonic maps of the Permian System," *U.S. Geol. Surv. Misc. Geol. Inv. Map I-450,* 164p.

Maher, J. C., 1960. "Stratigraphic cross-section of Paleozoic rocks, West Texas to Northern Montana," *Am. Assoc. Petrol. Geologists,* 15p.

———, 1965. "Correlations of subsurface Mesozoic and Cenozoic rocks along the Atlantic Coast," *Am. Assoc. Petrol. Geologists,* 18p.

———, and Applin, G. R., 1968. "Correlation of subsurface and Mesozoic rocks along the eastern Gulf Coast," *Am. Assoc. Petrol. Geologists Cross Sec. 6.*

Merrill, G. P., 1924. *The First One Hundred Years of American Geology.* New York: Hafner 773p. (reprint, 1969).

Meyer, R. F., 1968. *Geological Provinces Map.* Tulsa: Am. Assoc. Petrol. Geologists.

Murray, G. E., 1961. *Atlantic and Gulf Coastal Provinces of North America.* New York: Harper & Row, 692p.

Pitcher, Max G., ed., 1973. *Arctic Geology.* Tulsa: AAPG, Mem. 19, 747p.

Rankama, K., ed., 1970. *The Geologic Systems.* Vol. 4. *The Precambrian.* New York: Wiley-Interscience, 288p.

Ridge, J. D., 1972. "Annotated bibliographies of mineral deposits in the Western Hemisphere," *Geol. Soc. Am. Mem. 131,* 681p.

Roeder, D. H., 1967. *Rocky Mountains.* Berlin: Gebr. Borntraeger, 318p.

Ronov, A. B., and Migdisof, A. A., 1971. "Geochemical history of the crystalline basement and the sedimentary cover of the Russian and North American platforms," *Sedimentology,* 16, 137–185.

Ruedemann, R., and Balk, R. eds., 1939. *Geology of North America.* Berlin: Gebr. Borntraeger.

Sbar, M. L., and Sykes, L. R., 1973. "Contemporary compressive stress and seismicity in Eastern North America: An example of intra-plate tectonics," *Bull. Geol. Soc. Am.,* 84, 1861–1882.

Scholz, C. H., Barazangi, M., and Sbar, M. L., 1971. "Late Cenozoic evolution of the Great Basin, western United States," *Bull. Geol. Soc. Am.,* 82(11), 2979–2990.

Schöpf, J. D., 1787 (1972). *Geology of Eastern North America.* New York: Hafner, 384p., trans. from German by E. M. Spieker.

Schwab, F. L., 1972. "Geochemical history of the crystalline basement and the sedimentary cover of the Russian and North American Platforms—a discussion," *Sedimentology,* 19, 299–302.

Smith, R., 1967. "Stratigraphic cross-section of Paleozoic rocks, Oklahoma to Saskachewan," *Am. Assoc. Petrol. Geologists, Cross Sec. 5.*

Stevenson, J. S., ed., 1952. *The Tectonics of the Canadian Shield* (Roy. Soc. Canada Spec. Publ. 4). Toronto: Univ. Toronto Press, 180p.

Stewart, J. H. 1971. "Basin and range structure; a system of horsts and grabens produced by deep-seated extension," *Bull. Geol. Soc. Am.,* 82(4), 1019–1043.

Stille, H, 1940. *Einführung in den Bau Amerikas.* Berlin: Gebr. Borntraeger, 717p.

Summerson, C. H., and Swann, D. H., 1970. "Patterns of Devonian sand on the North American craton and their interpretation," *Bull. Geol. Soc. Am.,* 81, 469–490.

Thornbury, W. D., 1965. *Regional Geomorphology of the United States.* New York: Wiley, 609p.

Ulrych, T. J., and Reynolds, P. H., 1966. "Whole-rock and mineral leads from the Llano Uplift, Texas," *J. Geophys. Res.,* 71(12), 3089–3094.

Wheeler, H. E., 1963. "Post-Sauk and pre-Absoroka Paleozoic stratigraphic patterns in North America," *Bull. Am. Assoc. Petrol. Geologists.,* 47(8), 1497–1526.

Zen, E., 1972. "The Taconide Zone and the Taconic Orogeny in the western part of the Northern Appalachian Orogen," *Geol. Soc. Am. Spec. Pap. 135,* 72p.

Cross-references: *Canada; Greenland; Mexico; Saint-Pierre and Miquelon; United States–General; West Indies.*

O

OCEANIA

"Oceania" is a rather loose term roughly synonymous with "South Sea Islands," of which there are perhaps 20,000 in number. Sometimes the term is taken to include Australasia, but this is not recommended; it is usually employed to embrace those regions of the Pacific Ocean, basically inhabited by the *Micronesians* in the northwest, the *Polynesians* in the northeast, center, and south, and the *Melanesians* in the southwest.

All these people are believed to have stemmed from Southeast Asia, coming in successive waves, but mainly in small groups. The Polynesians were the farthest travelers, reaching Hawaii, the Marquesas, Easter Island, and New Zealand over a period of about 500–3000 years ago, according to evidence of C^{14} dating. Contacts with the Americas and their aboriginal inhabitants were undoubtedly made from time to time, so that there is support for some of Heyerdahl's arguments for cultural ties; however, no large-scale east to west central Pacific migrations seem at all likely. An apparent correlation between the pictograph script of Easter Island with that of Mehendjo-Daro in the Indus valley is still enigmatic.

The anthropological boundaries of the Pacific do not match the geological limits. In the southwest the whole of Melanesia lies on the Australasian side of the Andesite Line (which passes around the outside of the Solomons, the Melanesian Border plateau, Fiji, Tonga, and so to New Zealand). Further, the Tongans and New Zealand Maoris are classified as Polynesian. Remaining in the typically "South Seas" environment are most of the Micronesian islands and the rest of the Polynesian outposts (notably, in Hawaii, Samoa, the Cook Islands, and the Tahitian chain).

Political Units

Within the main Pacific basin, as defined by the Andesite Line, the political limits are briefly as follows:

1. To Japan: Nampo-Shoto including Bonin Island (Ogasawara-gunto), Izu, and Kazan groups; Daito Island; in part, U.S. military administration.

2. To United States: The Hawaiian chain, constituting the State of Hawaii; American Samoa (Tutuila, with former naval base at Pago Pago), a "territory." Also scattered islets in the west and northeast Pacific (Guam, Wake I., Johnson I., and divided control, with U.K. or N.Z., of the Phoenix Is., the Line Is. and the Tokelau Is.).

To United States under United Nations Trusteeship: Some 2000 islands, being most of Micronesia, and embracing the Caroline Is., Marshall Is., and Mariana Is., as the "Trust Territory of the Pacific Islands" under administration of Department of the Interior, Washington; before World War I under German colonial occupation, mandated to Japan in 1922, and occupied by U.S. since 1944.

3. To France: The Society Is., the Tuamotus, the Marquesas, the Gambier Is., the Tubuai Is., and Clipperton I. (see *French Polynesia*)

4. To United Kingdom: Gilbert Is., Ellice Is., Ocean I., Pitcairn, Henderson, and Ducie Is.; divided control, with U.S., of the Phoenix Is., and Line Is.

5. To New Zealand: Cook Is., Niue I., divided control, with U.S., of Tokelau (Union) Is.

6. To Chile: Easter I. (Isla de Pascua), Sala y Gomez; Juan Fernández; San Félix, San Ambrosio.

7. To Ecuador: Galápagos Is.

8. To Colombia: I. de Malpelo.

9. To Costa Rica: I. del Coco.

10. To Mexico: Is. Revillagigedo, Is. Guadelupe.

11. Independent: Western Samoa (German colony till 1902; N.Z. Trust Territory after 1920; independent 1962.) Also: Nauru, independent since 1969.

A number of Pacific territories were, as of 1975, working toward self-determination: the Micronesian area (Marianas, Carolines, and Marshalls) in the U.S. sphere; Fiji in the British sphere; and the Cook Islands in the New Zealand area of responsibility.

Scientific and Educational Establishments

Because of its several trust territories in Oceania, and owing to the undeveloped state or economic difficulties of many of the islands, the United Nations, particularly through its dependent group UNESCO (Hq., Place Fontenoy, Paris), has a sustained interest in Oceania and supports various projects. An International Symposium on the Oceanography of the South Pacific was organized in 1972 by the New Zealand National Commission for UNESCO in cooperation with its Royal Society of N.Z.

In 1947 Australia, France, the Netherlands, New Zealand, the United Kingdom and the United States met in Canberra to establish a *South Pacific Commission* dedicated to the scientific, educational, health, and economic interests of the region. With its principal offices or agencies in Noumea (New Caledonia) or in Suva (Fiji), this commission has provided continuing advisory services which have already contributed a good deal to the general advancement of the region.

Important educational and scientific establishments include the *University of Honolulu* in Hawaii, the *University of Guam,* and now the *University of the South Pacific* near Suva in Fiji. Anthropological and natural history museums are in Honolulu (Bernice P. Bishop Museum), Tahiti (Papeete), Cook Is. (Rarotonga), in Fiji, and in New Zealand. A *Directory of Asian-Pacific Museums* is issued by the Bishop Museum Press.

Local geological surveys exist only in Fiji, the New Hebrides, and the Solomons, but national surveys (the U.S.G.S., the New Zealand Geological Survey, the New Zealand Geophysics Division, the British Institute of Geological Sciences and the French B.R.G.M.) have carried out important missions, and in Hawaii there is a branch office of the U.S.G.S., as well as a volcanological service. There is an Oceanographic Institute in New Zealand that publishes oceanographic maps and scientific reports on the southwest Pacific, and at the University of Hawaii there is also a Geophysical Institute.

Pacific Science Association. Under the auspices of a political body, the Pan-Pacific Union, a first Pan-Pacific Science Conference was organized in Honolulu, Hawaii, in 1920. Under Herbert E. Gregory, then director of the Bernice P. Bishop Museum in Honolulu, this group evolved into the *Pacific Science Association,* which has played a major role in the scientific development of Oceania and its borderlands. It organizes conferences and maintains a Pacific Science Council. Publications of its conferences (Proceedings) and committees are of outstanding importance. (Many are still in print and available, usually through the national academies of the host institutions.) The successive congresses have been as follows:

First: Honolulu, Hawaii, 1920
Second: Melbourne and Sydney, Australia, 1923
Third: Tokyo, Japan, 1926
Fourth: Batavia and Bandoeng, Java, 1929
Fifth: Victoria and Vancouver, B.C., Canada, 1933
Sixth: Berkeley, Stanford, and San Francisco, Calif., 1939
Seventh: Auckland and Cristchurch, N.Z., 1949
Eighth: Diliman, Quezon City, Philippines, 1953
Ninth: Bangkok, Thailand, 1957
Tenth: Honolulu, Hawaii, 1961
Eleventh: Tokyo, Japan, 1966
Twelfth: Canberra, Australia, 1971

In addition to its regular congresses this association maintains standing committees on the various disciplines, those of special interest at the present time being: Solid Earth Sciences, Geography, Marine Sciences, Fresh-water Sciences, Pacific Islands Ecosystems, Museums in Pacific Research, and Conservation and Environmental Protection. Various intercongress activities include the holding of symposia and cooperating with interunion meetings, e.g., in August, 1971—"Regional Symposium on Conservation of Nature: Reefs and Lagoons" (Noumea).

Other Unions. The International Geological Congress (now the International Union of Geological Sciences), the International Union of Geodesy and Geophysics, and other unions have from time to time organized special commission activities dedicated to Oceania and Pacific problems. The INQUA, or International Union for Quaternary Research, had its 9th meeting in New Zealand in 1973, the first time in the Pacific region.

Publications. Numerous scientific works on the earth sciences of Oceania are diffused through the major national journals, and to some extent "buried" in the weighty reports of exploring and scientific expeditions. The material is thus extremely scattered, and, from the point of view of field scientists, largely inaccessible. Thus, major research is usually undertaken from external centers with good libraries such as Washington, London, Paris, Tokyo and Sydney. Honolulu and Wellington contain the best equipped library facilities within the area itself.

Specialized journals of continuing interest include the monthly Pacific Science Association *Information Bulletin* (reaching vol. 27 in 1975), *Pacific Science* (Honolulu), the *Atoll Research Bulletin* (published first by the Pacific Science Board of the National Academy of Sciences in Washington, later by the Smithsonian Institution; it reached bulletin no. 175 by 1975), the *Cahiers du Pacifique* (issued by the Foundation Singer-Polignac in Paris, no. 18 by 1975; it carries particularly useful abstracts of new literature), the *Journal de la Société des Océanistes* (published by the Musée de l'Homme, Paris), *Pacific Geology* (published in Tokyo by Tsukiji Shokan Publ. Co. since 1968), and above all, various serials (bulletins, etc.) published by the *Bernice P. Bishop Museum* in Honolulu, founded in 1889. G. K. Hall and Co. (Boston) has published in 9 volumes a facsimile of 143,600 cards of the "Dictionary Catalog of the Bernice P. Bishop Museum." An annotated bibliography of French Polynesia was published in 1966 (*Cahiers du Pacifique*, no. 9).

A particularly helpful source of literature information is *"Island Bibliographies"* (Marie-Hélène Sachet and F. Raymond Fosberg: 1955, Washington, D.C., N.A.S.; out of print, but reproductions available from National Technical Information Service, Springfield, Va. 22151); a *"Supplement"* (448 p.) was issued in 1971.

Geology

Geologically, Oceania largely corresponds to the main Pacific basin, that is, inside the "Andesite Line" (of Marshall, 1916) wherein the islands are exclusively volcanic or coral (see discussion in Schmidt, 1957). Continental islands and island arc belts are extensively distributed around the western periphery of the main Pacific basin, partly isolating it from series of marginal seas—Bering Sea, Japan Sea, Philippine Sea, the East Indian basins, Coral Sea, etc., from the landmasses of Asia and Australia. The western boundary of the main Pacific basin is geologically dictated by a belt of deep-sea trenches, island arcs, high seismicity, numerous active volcanoes, etc.

This dynamic western border of the main Pacific basin is the *"Girdle of Fire,"* a discontinuous zone of active subduction, where the oceanic crust is sliding under the borders of the Asiatic and Australian plates at rates of up to 10 cm per year, to the accompaniment of shallow-, medium-, and deep-focus earthquakes down to depths of about 700 km. Igneous rocks in this zone are marked by the calc-alkaline "Pacific" suite, typically but not exclusively andesitic. The volcanics of the main basin are primarily in the alkaline "Atlantic" or oceanic type, typically basaltic and directly related to the upper mantle, but some of the more evolved groups also develop andesites. The boundary is thus called the *Andesite Line* (q.v., in Vol. V). Since Marshall's term was nonexclusive, others have preferred to designate the boundary in another way, such as "Marshall Line" (e.g., Macpherson, 1946).

Whereas the framework of the Pacific has, at least during Phanerozoic time, been more or less continuous though elastic, the ocean floor has been in a semicontinuous state of growth and overturn. Under sea-floor spreading, no part of the main basin seems to be older than the early Mesozoic (in the region of the Mariana Islands). In contrast, however, to the other oceans this basin has not closed during this time, a fact which justifies Stille's description of the Pacific as the "Urozean." Gregory's picture of former Pacific land masses (Gregory, 1930) is now rejected. For the geological history reference should be made to the article "Pacific Ocean" in *Encyclopedia of Oceanography* (see also Menard, 1964). The key references on the geology of Oceania are collected at the end of this entry.

Briefly, it can be concluded, from recent studies of sea-flooring spreading and paleomagnetics, that since mid-Cretaceous times the Mid-Pacific crust has shifted about $30°$ of arc to the north (Winterer, 1973). Soon after their formation as mid-oceanic seamounts many Mesozoic volcanoes became planed off by wave erosion to form guyots or atoll foundations. Progressive submergence of these foundations, as deduced by Charles Darwin, can now be explained in terms of a geoidal rise of sea level which has accompanied the secular displacement of the what was formerly the south Pacific crust into more and more equatorial latitudes (Fairbridge, 1961, p. 110). As sea-level rose here so did the atoll-building coral reefs, but in the north Pacific, former (Cretaceous) reefs were carried progressively into more northern and colder waters, so that only guyots remain today, still capped by remains of the ancient reefs.

An important geotectonic feature of the Pacific, and one which distinguishes it from all other oceans is that no part of the central basin (i.e., Oceania) contains fragments of ancient continental origin (such as the "micro-continent" of Seychelles in the Indian Ocean). On the contrary, the islands are almost exclusively of volcanic origin, derived from eruptions from beneath youthful oceanic crust, or of coral growth which has been superimposed on those volcanic foundations.

Because of the progressive nature of the growth of the Pacific oceanic crust, it is apparent that no island can be older than the crust on which it rests. A list of these maximum basement ages is given on Table 1. Whereas many of the volcanic islands only disclose rocks

TABLE 1. Maximum Possible "Basement Ages" for the Island Groups of Oceania, Deduced on the Basis of Paleomagnetic Analyses of the Pacific Oceanic Crust

Jurassic 190–135 m yr	Mariana Is., Wake I.
Lower Cretaceous 135–110 m yr	Marshall Is., Phoenix Is., Line Is., Gilbert and Ellice Is.
Middle to Upper Cretaceous 110–65 m yr	Nauru, Samoa, Niue, Cook Is., Society Is., Tubuai Is., Tuamotu Is.
Paleocene 65–53 m yr	Marquesas Is.
Eocene 53–37 m yr	Juan Fernández Is.
Oligocene 37–23 m yr	Gambier Is., Pitcairn Is.
Miocene 23–5 m yr	Clipperton Is., Revillagigedo Is., Galápagos Is.
Pliocene 5–2 m yr	Easter Is.

Courtesy Roger L. Larson.

of the last few million years, numbers of atolls are represented only by reef materials dating from the last 6000 years (when sea level reached its present stand approximately, after its negative oscillations during the last glacial period). It has been proposed (by the writer) that about 6000 years ago mean sea level rose to about 3m above its present datum and that it has oscillated since then, with important intermediate stands at about 1.5 m and 0.5 m around 2500 years B.P. and 1000 years B.P. (Fairbridge, 1961). Those high sea level stages alternated with emergences of up to 3–4 m.

The question is still controversial, but large numbers of radiocarbon dates appear to confirm the idea. If these dates are in any way correct, it is evident that during the high sea level stands many of the atolls of Oceania (which are often less than 3 m above MSL), would have been uninhabitable to the early Polynesians, and could only have been practicable for colonization during the relatively low sea level stages.

The structural pattern of the Pacific crust involves two principal contrasting trends. E-W fracture zones (becoming ENE in the north, and ESE in the south), and NW SE ridges with intervening depressions. The fracture zones are well understood as transform fault systems, while the ridges constitute growth belts of volcanoes.

From the viewpoint of vertical motion, it is recognized that the oldest part of the Pacific Ocean crust is also the deepest, while the newest, the active sea floor spreading ridge crest is the shallowest part of the crust. Superimposed on this basic growth trend there is a curious "ground-swell" of undulations that can be best understood from Fig. 1 (from Chubb, 1957). Studies of coral reefs, raised or depressed, disclose that certain ridges are now rising and others are sinking. This sinking evidence was once thought (by Menard) to be limited to an area of the western Pacific, his "Darwin Rise," but it is widespread, and alternates with uplift as recognized more than a century before by Darwin. Chubb (1957) established a definite temporal oscillation.

Biogeography of Oceania

Oceania has long presented an enigma to the biogeographers, because of the problem of explaining the terrestrial populations, faunal and

FIGURE 1. Diagrammatic sections, not to scale, showing the movements of the Tuamotu, Society, and Austral islands. Volcanic Islands dotted; coral reefs black (Chubb, 1957).

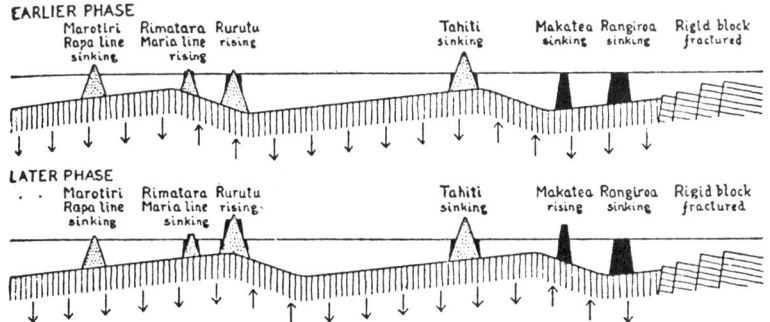

floral, of these widely distributed islands. For a long time a certain school of biogeography assumed the existence of an ancient "Pacific continent." In a quite fanciful way Gregory (1930) presented paleogeographic maps illustrating stages of earth history with narrow geosynclines crossing the Pacific, while much of the area was assumed to be continental.

Such reconstructions are widely regarded as impossible, if one considers the geophysical character of the earth's crust; and unnecessary, if one considers the wide dispersal of organisms that is possible over very extended periods of time by means of the "normal accidents" of winds, waves and currents, i.e. flotsam and jetsam. The best that can be allowed is the concept of island "stepping stones" first considered by Wallace (1881), and summarized by Ladd (1960), and further by Menard and Hamilton (1963). The widespread distribution of volcanic seamounts rendered the region far more like an archipelago during the late Cretaceous. On the other hand, many of the biogeographic discussions involved the margins of Oceania, e.g., the Melanesian Borderland. In such quasicratonic belts, true "land bridges" or "isthmian links" have existed from time to time.

RHODES W. FAIRBRIDGE

References

Adamson, A. M., 1939. "Review of the faunas of the Marquesas Islands and discussion of its origin," *Bernice P. Bishop Mus. Bull.*, 159, 1–39.

Andel, T. H. van, and Bukry, D., 1973. "Basement ages and basement depths in the eastern equatorial Pacific from deep sea drilling project legs 5, 8, 9, and 16," *Bull. Geol. Soc. Am.*, 84, 2361–2370.

Aubert de la Rüe, E., 1958. "Observation sur le volcanisme tertaire et quaternaire de quesques iles de la Polynesie Française," *Bull. Volcanol.*, Ser. 2, 19, 159–178.

———, 1959. "Etude géologique et prospection minière de la Polynesie française," *Recherche Geólogique et Minerale en Polynesie Française*. Paris: L'Inspection Generale des Mines Géologiques, 7–42.

Avias, J., et al., 1956. "Oceanie proprement dite," *Lexique Stratig. Intern.* (Paris) 6, fasc. 2, 286p.

Bender, M. L., 1973. "Helium-uranium dating of corals," *Geochim Cosmochim. Acta*, 37, 1229–1247.

Brigham, W. T., 1900. "An index to the islands of the Pacific," *Mem. Bernice P. Bishop Mus.*, 1.

Catala, R. L. A., 1957. "Report on the Gilbert Islands: some aspects of human ecology," *Atoll Res. Bull.*, 59, 1–187.

Chubb, L. J., 1927. "Mangaia and Rurutu; a comparison between two Pacific Islands," *Geol. Mag.*, 64, 518–522.

———, 1934. "The structure of the Pacific basin," *Geol. Mag.*, 71, 289–302.

———, 1957. "The pattern of some Pacific Island chains," *Geol. Mag.*, 94(3), 221–228. (Discussion, *Ibid.*, 1961, 98(2), 170–171; 1962, 99(4), 279–283).

Clague, D. A., and Jarrard, R. D., 1973. "Tertiary Pacific plate motion deduced from the Hawaiian-Emperor Chain," *Bull. Geol. Soc. Am.*, 84, 1135–1154.

Coleman, P. J., ed., 1973. *The Western Pacific. Island Arcs, Marginal Seas, Geochemistry*. Nedlands: Univ. of West Aust. Press, 676p.

Cotton, C. A., 1969. "The pedestals of oceanic volcanic islands," *Bull. Geol. Soc. Am.*, 80, 749–760.

Crossland, C., 1928. "Coral reefs of Tahiti, Moorea, and Raratonga," *Linn. Soc. Lond. J. Zool.*, 36, 577–620.

David, T. W. E., and Sweet, G., 1904. "The geology of Funafuti," in W. J. Sollas *et al.*, *The Atoll of Funafuti*. London: Roy. Soc. Lond., Coral Reef Comm.

Davis, W. M., 1928. "The coral reef problem," *Am. Geogr. Soc. Spec. Publ. 9*.

Dietz, R. S., 1954. "Marine geology of northwestern Pacific; description of Japanese bathymetric chart 6901," *Bull. Geol. Soc. Am.*, 65, 1199–1224.

Dobrin, M. B., Perkins, B., Jr., and Snavely, B. L., 1949. "Subsurface constitution of Bikini Atoll as indicated by a seismic refraction survey," *Bull. Geol. Soc. Am.*, 60, 807–828.

Emery, K. O., Tracey, J. I., Jr., and Ladd, H. S., 1954. Geology of Bikini and nearby atolls," *U.S. Geol. Surv. Prof. Pap. 260-A*, 265p.

Fairbridge, R. W., 1950. "Landslide patterns on oceanic volcanoes and atolls," *Geogr. J.*, 105, 84–88.

———, 1961. "Eustatic changes in sea level," *Physics and Chemistry of the Earth*, vol. 4. New York: Pergamon, 99–185.

———, and Stewart, H. B., Jr., 1960. "Alexa Bank, a drowned atoll on the Melanesian Border Plateau," *Deep-Sea Res.*, 7, 100–116.

Furon, R., 1959. *La Paléogéographie*, 2nd ed. Paris: Payot, 405p.

Gregory, J. W., 1930. "The geological history of the Pacific Ocean," *Quart. J. Geol. Soc. Lond.*, 86, 72–136.

Guilcher, A., 1970. "Les variations relatives du niveau de la mer au Quaternaire en Melanesie et en Polynesie," *Quaternaire*, 12, 137–143.

Hamilton, E. L., 1950. "Sunken islands of the Mid-Pacific Mountains," *Geol. Soc. Am. Mem. 64*, 42–52.

———, and Rex, R. W., 1959. "Lower Eocene phosphatized globigerine ooze from Sylvania guyot," *U.S. Geol. Surv. Prof. Pap. 260-W*.

Hess, H. H., 1946. "Drowned ancient islands of the Pacific basin," *Am. J. Sci.*, 244, 772–779.

Heyerdahl, T., 1968. "Sea routes to Polynesia," London: Allen & Unwin, 238p.

Hsü, K. J., and Schlanger, S. O., 1968. "Thermal history of the upper mantle and its relation to crustal history in the Pacific basin," *23rd Intern. Geol. Congr.*, Prague, 1, 91–105.

Hutchinson, G. E., 1950. "The biogeochemistry of vertebrate excretion," *Am. Mus. Nat. Hist. Bull.*, 96, 1–556.

Johnson, D. A., and Parker, F. L., 1972. "Tertiary radiolaria and foraminifera from the equatorial Pacific," *Micropaleontology*, 18(2) 129–143.

Joleaud, L., 1934. "Paléogéographie de l'Océan Pacifique," *Mem. Soc. Biogéographie,* **4,** 9–40.

Karig, D. E., 1974. "Evolution of arc systems in the western Pacific," in F. A. Donath, ed., *Annual Review of Earth and Planetary Sciences,* vol. 2. Palo Alto: Annual Reviews, Inc., 478p.

Keller, F., Jr., Meusakka, J. L., and Alldredge, L. R., 1954. "Airomagnetic surveys in the Aleutian, Marshall and Bermuda Islands," *Trans. Am. Geophys. Union,* **35,** 558.

Lacroix, A., 1927a. "La constitution lithologique des iles volcaniques de Polynesie australe," *Mem. Acad. Sci. Paris,* **59,** 1–80.

———, 1927b. "La constitution lithologique des volcans du Pacific central australe," *Bull. Volcanol.,* **4,** 218–231.

Ladd, H. S., 1960. "Origin of the Pacific Island molluscan Fauna," *Am. J. Sci.,* Bradley Vol., **258-A,** 137–150.

———, Tracey, J. I., Jr., and Gross, M. G., 1967. "Drilling on Midway Atoll, Hawaii," *Science,* **156,** 1088–1094.

———, et al., 1970. "Deep drilling on Midway Atoll," *U.S. Geol. Surv. Prof. Pap. 680-A.*

Larson, R. L., Smith, S. M., and Chase, C. G., 1972. "Magnetic lineations of early Cretaceous age in the western equitorial Pacific Ocean," *Earth Planet. Sci. Lett.,* **15**(3), 315–319.

Leeson, I., 1954. *A Bibliography of bibliographies of the South Pacific.* New York: Oxford Univ. Press (for S. Pacific Comm.), 61p.

MacDonald, C. A., 1949. "Hawaiian petrographic province," *Bull. Geol. Soc. Am.,* **60,** 1541–1596.

McKee, E. D., Chronic, J., and Leopold, E. B., 1959. "Sedimentary belts in lagoon of Kapingamarangi Atoll," *Bull. Am. Assoc. Petrol. Geologists,* **43,** 501–562.

Macpherson, E. O., 1946. "Cretaceous and Tertiary diastrophism in New Zealand," *N.Z. Rept. Sci. Ind. Res. Geol. Mem. 6.*

Marshall, P., 1916. "Oceania," *Handbuch Reg. Geol.,* 7(9), 36p.

Menard, H. W., 1964. *Marine Geology of the Pacific.* New York: McGraw-Hill, 271p.

———, 1973. "Depth anomalies and the bobbing motion of drifting islands," *J. Geophys. Res.,* **73**(23), 5128–5137.

———, and Hamilton, E. L., 1963. "Paleogeography of the tropical Pacific," in J. L. Gressitt, ed., *Pacific Basin Biogeography,* Pacific Science Congress, 10th, Honolulu, p. 193–218.

Newell, N. D., 1954. "Expedition to Raroia, Tuamotus," *Atoll Res. Bull.,* **31,** 1–21.

Power, F. D., 1925. "Phosphatic deposits of the Pacific," *Econ. Geol.,* **20,** 266–281.

Schmidt, R. G., 1957. "Petrology of the volcanic rocks; Geology of Saipan, Mariana Islands, pt. 2," *U.S. Geol. Surv. Prof. Pap. 280-B,* 127–176.

Sclater, J. G., and Detrick, R., 1973. "Elevation of midocean ridges and the basement age of JOIDES deep sea drilling sites," *Bull. Geol. Soc. Am.,* **84,** 1547–1554.

Shepard, F. P., et al., 1967. "Holocene changes in sea level," *Science,* **157,** 542–544.

Stark, J. T., and Howland, A. L., 1941. "Geology of Borabora, Society Islands," *Bernice P. Bishop Mus. Bull.,* **169,** 1–63.

Stearns, H. T., 1945. "Eustatic shore lines in the Pacific," *Bull. Geol. Soc. Am.,* **56,** 1071–1078.

———, 1946. "An integration of coral reef hypotheses," *Atoll Res. Bull.,* **22,** 1–6.

Stone, E. L., Jr., 1953. "Summary of information of atoll soils," *Atoll Res. Bull.,* **22,** 1–6.

Swartz, J. H., 1957. "Geothermal measurements on Eniwetok and Bikini atolls," *U.S. Geol. Surv. Prof. Pap. 260-U.*

Tayama, R., 1953. "Coral reefs in the South Sea," *Bull. Hydro. Office* (Tokyo), 2nd publ., 292p.

Tester, A. C., 1950. "Marine terraces of the Pacific Ocean Area," *18th Intern. Geol. Congr. Rept.,* 8, 72, Abstr.

Tracey, J. I., Jr., Ladd, H. S., and Hoffmeister, J. E., 1948. "Reefs of Bikini, Marshall Islands," *Bull. Geol. Soc. Am.,* **59,** 861–878.

Trichet, J., 1969. "Quelques aspects de la sédimentation calcaire sur les parties emergées, de l'atoll Muraroa," *Cahiers du Pacifique,* **13,** 1–14.

U.S. Office of Geography, 1967. *South Pacific; Official Standard Names Approved by the United States Board in Geographic Names.* Washington: Govt. Printing Off., 68p.

Veeh, H. H., 1966. "Th230 and U234-U238 Ages of Pleistocene high sea level stand," *J. Geophys. Res.,* 71(14), 3379–3386.

Wallace, A. R., 1881. *Island Life.* New York: Harper Bros.

Wentworth, C. K., and Ladd, H. S., 1931. "Pacific Island sediments," *Univ. Iowa Stud. Nat. Hist.,* **13**(2), 1–47.

Wharton, W. J. L., 1883. "Mangrove as a destructive agent," *Nature,* **29,** 76–77.

Wiens, H. J., 1962. *Atoll Environment and Ecology.* New Haven: Yale Univ. Press, 532p.

Williams, H., 1930. "Geology of Tahiti, Moorea, and Maiao," *Bernice P. Bishop Mus. Bull.,* **105,** 1–90.

Winterer, E. L., 1973. "Sedimentary facies and plate tectonics of equatorial Pacific," *Bull. Am. Assoc. Petrol. Geologists,* **57**(2), 265–282.

Wood, B. L., 1967. "Geology of the Cook Islands," *N.Z. J. Geol. Geophys.,* **10**(6), 1429–1445.

Cross-references: *American Samoa; Caroline Islands; Cook Islands; Easter Island and Sala y Gómez; French Polynesia; Galápagos Islands; Gambier Islands; Guam; Juan Fernández; Line Islands; Mariana Islands; Marquesas Islands; Marshall Islands; Nauru; Phoenix Islands; Pitcairn Islands; Society Islands; Tahiti; Tokelau Islands; Tuamotu Islands; United States–Hawaii; Western Samoa.*

P

PACIFIC ISLANDS TRUST TERRITORY (U.S.)

The "Trust Territory of the Pacific Islands," held by the United States under a United Nations Mandate since World War II, is essentially coextensive with Micronesia, an ethnic bloc. It is mainly populated by the Micronesians but includes others, such as Philippinos and Chamorros (pop. 91,448 in 1967). It embraces the *Marshall Islands* (q.v.), *Caroline Islands* (q.v.), and *Mariana Islands* (q.v.) except for *Guam* (q.v.), which has a separate administration. The government center of the Trust Territory is *Saipan* (q.v.) in the Marianas. A home-rule proposal was made in 1971.

RHODES W. FAIRBRIDGE

References

U.S. Navy Department, 1948. *Handbook on the Trust Territory of the Pacific Islands.* Washington, D.C.: U.S. Navy.

Cross-references: *Caroline Islands; Guam; Mariana Islands; Marshall Islands; Saipan.*

PANAMA

The Republic of Panama lies in the tropical "isthmian link" region, connecting Central and South America, between 7°00' and 9°35'N, and between 77°15' and 83°00'W. The land is sigmoidally shaped with a length of approximately 684 km (425 mi) along the curve, from Colombia on the east to Costa Rica on the west. The country varies in width from 45 km to about 193 km. To the north is the Carribean Sea and to the south is the Pacific Ocean. The area is 75,928 km² (29,306 sq mi). The land is very mountainous with high volcanic peaks; the highest mountain is El Baru (3469 m, 11,382 ft), near the Costa Rican border. Close to the Columbian border, the mountains rise to about 1700 m (to join the Andean chain) and the lowest part of this mountain belt is in the area where the Panama Canal was constructed (Fig. 1).

The country is particularly interesting from a geological viewpoint because it is a narrow neck of land that connects continental land masses; i.e., it is the land bridge between North and South America. The only other biogeographically comparable area is the strip of land at the north end of the Red Sea that connects Asia and Africa, but this is structurally quite different.

Major Geological Studies

A study was made by Comdr. Selfridge of the U.S. Navy in 1871–1873, of proposed interoceanic canal routes in present-day Panama and Columbia. Further investigations were made by Hersey in 1901 (a regional study); the Canal Zone was reexamined by Howe in 1907 and by MacDonald between 1914 and 1942. From 1939 to 1946 geological studies were made in connection with the construction of a third set of locks, and in 1947 and 1948 studies were made for construction of a sea-level canal at the present site (Woodring, 1957). Further studies in both Panama and Columbia have been authorized by the U.S. Congress to determine the feasibility and the best methods of constructing a sea-level canal. The Republic of Panama has also begun a natural resources study.

Geomorphic Divisions

The greater part of the country is in the mature stage of erosion with steep-sided hills and valleys. Few high-level erosion features are left between the various drainage basins. In the Darien area, the Tuira, Chucunaqui, and Bayano rivers are of old age and have been superimposed on relatively soft Tertiary bedrock structurally modified by the variable lithologies. The relief is low to moderate.

In the southern part of the central provinces near Aguadulce, a large peneplaned area has developed on pre-Tertiary volcanics and intrusives. The intrusives form large monadnocks. The soil developed on this surface is a typical laterite with concretionary limonite. Scattered on this old peneplain are old river gravels of

FIGURE 1. Political divisions of Panama.

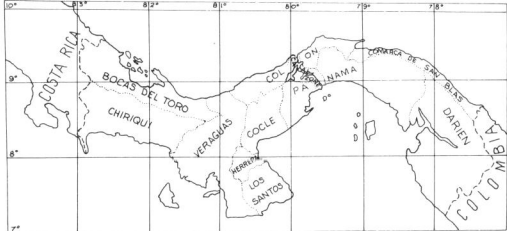

quartz, agate, petrified wood, and granodiorite.

Two Pleistocene volcanoes, El Baru in the west in Chiriqui province and El Valle de Anton in the central part in Cocle province have deposited extensive detrital fans, sands, tuffs, and volcanic conglomerates. These plains are drained by youthful streams flowing in steep-sided box canyons that radiate outward from their respective volcanic centers. Active hot springs in the vicinity of both craters indicate that volcanic activity has not been long quiescent.

In other parts of the country there are other craters that have shown very little erosion since their last eruption. The scoriaceous material shows little weathering. Some have flows with very thin soil cover in which only small shrubs have gained a foothold. They also have active hot springs and mineral springs.

Structural Framework

The basic structure of the isthmus consists of two large, arcuate anticlinal structures which appear to be overturned toward the north. These northerly convex structures, to the east and west, are cut by faults that radiate toward the north. These arcs are connected by a southerly, convex structure in the central provinces. This arc is correspondingly cut by faults radiating toward the south (Fig. 2).

The Azuero peninsula is block faulted with horst and graben structures. This peninsula is structurally severed from the rest of the country by a transpeninsular series of parallel faults that show scarps of late Pleistocene to Holocene age. The southern part of the peninsula contains several NW-trending faults that show evidence of recent movement. The Tonosi valley is a fault-bounded graben which contains rocks of middle to upper Eocene age.

The Bayano-Chucunaqui-Tuira River valley is formed in the broad geosyncline which continues eastward into Columbia where it joins the Bolivar geosyncline. The Tertiary sediments in the Panama portion of this geosyncline may be as much as 5000 m thick.

Geologic History

The geologic history of the Isthmus of Panama begins in Pre-Tertiary time, possibly Early Cretaceous. There are published reports of Cretaceous foraminifera and diatoms.

The early rocks are primarily marine calcarcous sediments and lie unconformably on older volcanics. In places they grade into terrestrial deposits containing volcanic tuffs, conglomerates, and basalt flows. Near the Canal Zone some of these pre-Tertiary volcanic tuffs have been metamorphosed to hornfels.

The early marine and terresterial sediments were buried deeply and somewhat tectonized by two periods of deformation in pre-Tertiary times. The strike of the folds of the first period is due N-S. Those of the second period, more intense deformation, is due E-W, partly isoclinal and overturned to the north. The axes of the folds do not intersect at right angles, but can be seen weaving in and out. In the core of the isthmus the pre-Tertiary folded sediments were intruded by plutons of granite, syenite, monzonite, gabbro, and diorite of coarse texture. Most of the pre-Tertiary rocks are found on the northern flanks of the northerly arcuate structures and on the southern flank of the southerly arcuate structure.

A third period of deformation occurred during the Tertiary after a long period of intensive erosion. This time the rocks were not so deeply buried and yielded by fracture. Rocks with abundant calcite were recemented but the others have remained uncemented.

From Middle to Upper Cretaceous times the Isthmus of Panama appears to have been part of a land mass that was above water. Intensive volcanism occurred in eastern Panama in late Middle Eocene. In late Middle to Upper Eocene the Bayano-Chucunaqui-Tuira geosyncline and

FIGURE 2. Structural patterns of Panama.

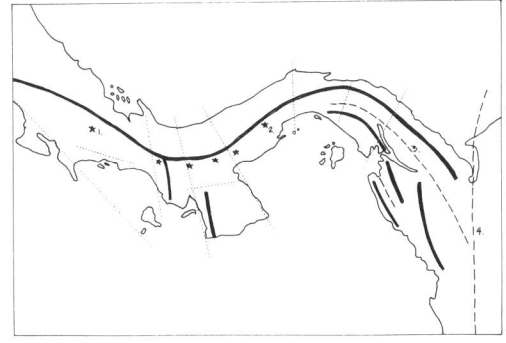

the Bolivar geosyncline began to form and the Atlantic and Pacific Oceans were joined. The islands of pre-Tertiary rocks that were remnants of the earlier land mass remained above water. Coral reefs were abundant around these islands.

The Oligocene period began with an outburst of volcanic activity which continued into early Miocene times. Vast deposits of volcanic conglomerate and tuff accumulated, interfingering with marine sandstones and siltstones. At this time the land mass of Panama apparently was connected to the mainland of North America, as indicated by a new find (by the writer, in 1965) of lower Miocene vertebrate fossils on the banks of the Panama Canal. Panama was still separated from South America by the deep Bolivar geosyncline to the east through which, from Paleocene or early Eocene to early Pliocene times, the Atlantic and Pacific marine fauna intermixed; at the same time the vertebrates in South America were effectively prevented from entering Panama.

By the late Pliocene, tectonic activity raised the entire isthmus above sea level where it has remained until the present. This made it possible for the vertebrates from both continents to mingle. At the same time an oceanographic barrier to warm E-W currents and faunal exchange was elevated (dated by the foraminiferal record in deep-sea cores at 5.7 m yr ago). This barrier, along with the comparable one in the Middle East, played a key role in initiating the Quaternary ice age.

In the Pleistocene, when from time to time the continent to the north was partly covered with glaciers, the flora and fauna were driven south in successive waves by the harsh climate. The isthmian migration route was aided by the lowered sea level in Panama which at times was at least 100 m lower than at present.

As the ice melted during the early Holocene, the sea level on both coasts rose, drowning valleys and lowlands developed during late Miocene or Pliocene time. These channels became gradually filled with up to 50 m of organic clays and peat known as "muck," the Atlantic-Pacific Muck Formation.

There is some evidence that the sea level may have reached as much as 1.5 m higher on the Atlantic coast than at present but this is open to debate.

Mineral Deposits

The history of mining in Panama began long before the Spaniards arrived. The Indians used crude, hand-carved wooden bateas, or basins, to wash alluvial gold from small rich placer deposits. When the Spanish came in 1510, they found numbers of gold ornaments and utensils among the Indians. Additional quantities of alluvial gold were recovered by the Spaniards by using crude ground-sluicing methods. The remains of their previous workings can still be found throughout the jungle. A few rich lode mines were also worked, such as at Campana and several in the central provinces. The most famous gold mine was El Espirito Santo at Cana where the Spanish found the largest gold-producing area in the Western Hemisphere until the gold rush in California in 1849. The mines at Cana were closed down in 1727 after an Indian uprising. These mines were reopened by an English company in 1893 and were operated until 1910. In the central provinces the old Spanish mine, El Gallo, is said to have been found by Columbus' men on his third voyage. It was worked again from 1912 until 1929, at which time the rich oxidized ores of the gossan ran out, and the sulfide body was apparently never touched.

Manganese was first mined on the north coast of Panama from 1894 to 1903, and later during World War I. Beach sands containing heavy minerals such as magnetite and ilmenite are found on the Pacific coast a few miles west of the Panama Canal. Low-grade lignite and subbituminous coal have been found. There has been exploration for oil but only small shows were found in Garachine in the Darien, as well as near the western border in Costa Rica. Porphyry copper deposits were found near the Colombian border in 1953 and in Chiriqui Province in 1957 by the author, and on the Atlantic slope west of Colon by the United Nations in 1967. These deposits are being explored at present under the auspices of the Panamanian government.

With the finding of large areas containing rocks of pre-Tertiary age and evidence of pre-Tertiary igneous activity, a greater interest in mining exploration is expected. Surface prospecting has been done in various parts of the country but lack of transportation and great distances without roads have made it very difficult.

R. H. STEWART

References

Andel, T. H. van, et al., 1971. "Tectonics of the Panama Basin, Eastern Equatorial Pacific," *Bull. Geol. Soc. Am.,* **82,** 1489–1580.

Bandy, O. L., and Casey, R. E., 1973. "Reflector Horizons and Paleobathymetric History, Eastern Panama," *Bull. Geol. Soc. Am.,* **84,** 3081–3086.

Case, J. E., 1974. Oceanic crust forms basement of Eastern Panamá," *Bull. Geol. Soc. Am.,* **85,** 645–652.

Fisher, S. P., and Pessagno, E. A., 1965. "Upper Cretaceous strata of northwestern Panama," *Am. Assoc. Petrol. Geologists Bull.,* **49**(4), 433–444.

Hershey, O. H., 1901. "The geology of the central portion of the Isthmus of Panama," *Univ. Calif. Dept. Geol. Bull.,* **2**(8), 231–267.

Howe, E., 1907. "Geology and the Panama Canal," *Econ. Geol.,* **2**(7), 145–281.

MacDonald, D. F., 1915. "Some engineering problems of the Panama Canal," *Bull. U.S. Bur. Mines,* **86,** 88p.

———, 1919. "Sedimentary formations of the Panama Canal Zone with specific reference to the stratigraphic relations of the fossiliferous beds," *Bull. U.S. Nat. Mus.,* **103.**

MacIlvaine, J. C., and Ross, D. A., 1973. "Surface sediments of the Gulf of Panama," *J. Sed. Petrology,* **43**(1), 215–223.

Malfait, B. T., and Dinkelman, M. G., 1972. "Circum-Caribbean tectonic and igneous activity and the evolution of the Caribbean Plate," *Bull. Geol. Soc. Am.,* **83,** 251–272.

Olsson, A. A., 1942. "Tertiary deposits of northwestern South America and Panama," *8th Am. Sci. Congr., Proc. Geol. Sci.,* Washington, 1940, **4,** 231–287.

Roberts, R. J., Irving, E. M., and Simmons, F. S., 1957. "Mineral deposits of Central America," *U.S. Geol. Surv. Bull.,* **1034**(with maps).

Schuchert, C., 1935. *Historical Geology of the Antillean–Caribbean Region.* New York: Wiley.

Selfridge, T. O., 1874. Reports of explorations and surveys to ascertain the practicability of a ship canal between the Atlantic and Pacific Oceans by way of the Isthmus of Darien," Washington, D.C.

Terry, R. A., 1956. "A geological reconnaissance of Panama," *Calif. Acad. Sci. Occ. Pap. 23,* 91p.

Woodring, W. P., 1957. "Geology and paleontology of Canal Zone and adjoining parts of Panama," *U.S. Geol. Surv. Prof. Pap. 306-A* (with maps).

Cross-references: *Central America–Regional Review; Colombia; Costa Rica; South America; West Indies.*

PARAGUAY

The Republic of Paraguay covers an area of 406,752 km² (157,047 sq mi) in the interior of South America, situated between Brazil, Bolivia, and Argentina. Its principal rivers are the south-flowing Paraná and the Rio Paraguay, tributaries of the Rio de la Plata, and between them live most of the population; here the rich soil is known as *tierra colorada.* Elsewhere the country is very sparsely populated. The climate is of subtropical savanna type, wet in the southern summer and dry in winter. In the east the precipitation of 1800–2000 mm supports a tropical rainforest (selva) in hill and plateau country. In the central part it is less humid with 1400–1600 mm, and in the west (Campos) there is a park savanna, passing northwestward into the dry savanna of the Chaco Boreal with scattered palms and thorns and 500–800 mm of rain.

The oldest geographical descriptions, with notes on geological features, were given by Schmidl (1567) and two Jesuit priests, Sepp (1716) and Stoecklin (1728). The first mineralogical and geological accounts are thanks to de Mersay (1860) and du Graty (1862). Useful details are found in the petrographic descriptions by Hibsch (1891) and Goldschlag (1913), as well as in the geomorphologic and geological studies by Carnier (1911, 1913), Bertoni (1913, 1921 first systematic field work), Schuster (1929) and Kanter (1930). Kanter published the most comprehensive work about the Gran Chaco. Beder (1918, 1923) and Boettner (1945) have described Paleozoic fauna.

Harrington has presented a detailed regional geology of the eastern part (1950), followed by a general review of the whole of Paraguay (1956); he, too, has studied the Devonian fossils of two deep boreholes in the Chaco. In 1959 Eckel published his reconnaissance of the geology and mineral resources, with considerable petrographic research (Eckel, 1959). The latest and most complete work on the regional geology and minerals is the author's book (Putzer, 1962). The revision of older Paleozoic faunas by Wolfart (1961) was very important; Harrington's (1950) alleged "Devonian" in eastern Paraguay belongs to the Silurian but both lower Devonian and Silurian are proved by fossils in the Chaco. Cretaceous alkalic rocks in both parts of the country were investigated by Putzer and van den Boom; Wilhelmy and Rohmeder (1963) provide many notes on morphology and geology. Menendez and Pöthe de Baldie (1967) have described Devonian spores from Chaco drill holes. Padula and Mingramm (1963, 1968) studied the Red Beds in the frontier area to the SW and interpreted their upper part as Cretaceous; the chances for oil are considered favorable in the western Chaco Boreal. Vivar and Morinigo (1969) prepared a general review on Paraguay geology, with new observations, for the *Geological Map of the World.*

Geological Maps of Paraguay. *Geological Map,* colored, 1:1,000,000 (Eckel, 1959, two sheets)

Soil Map, colored, 1:2,000,000 (Eckel, 1959)

Geological Map, colored, 1:3,000,000 (Putzer, 1962)

Structure

The general geotectonic structure of Paraguay is rather simple, consisting of three units

FIGURE 1. General geological map of Paraguay by H. Putzer.

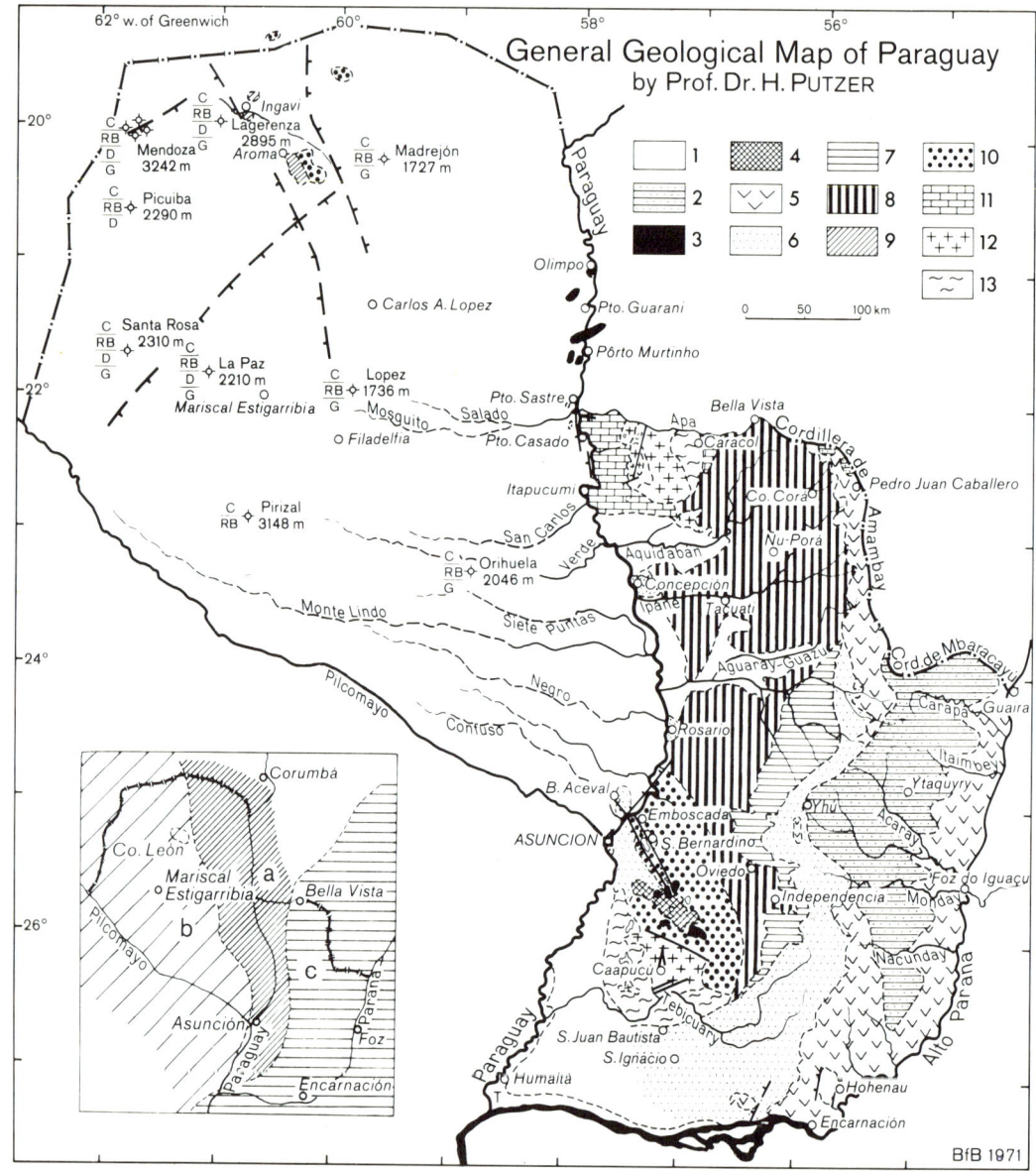

(Fig. 1b). The *Central Paraguayan Arch,* made of Precambrian rocks, separates two younger areas of subsidence: in the *Chaco Trough,* part of the belt of pericratonic basins between the Brazilian Shield and the Andine Geosyncline, and in the epicontinental Paraná Basin, that extends from Brazil (q.v.). The Precambrian Arch is a spur of the Central Brazilian Shield running N-S; its rocks are gneisses, crystalline schists, phyllites, quartzites, folded and faulted, intruded by granitic rocks and, locally (near Caapucú), covered by porphyries. The Caapucú anorogenic granite has an age of 468 m yr; the Bernardino granodiorite, 786 m yr; and the amphibolites of the Rio Apo, 1250–1056 m yr. The last folding was Assyntic. In the N a sequence of metasediments (Itapucumi Series) rests unconformably upon the folded Precambrian and is considered to be youngest Precambrian.

The *Chaco Trough* is an unstable depression filled with more than 2000 m of marine Paleozoic sediments and up to more than 2500 m of continental clastic sediments of Mesozoic and

FIGURE 2. Section across NW Chaco Boreal. 1, Cenozoic sediments; 2, red beds (Triassic-?Miocene); 3, Devonian; 4, Silurian; 5, Precambrian.

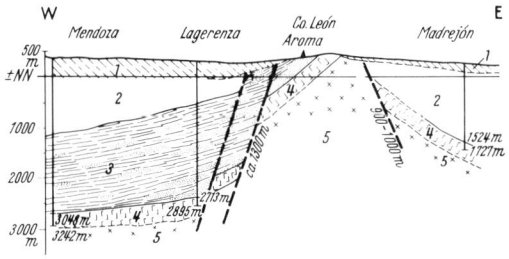

Cenozoic age. These sediments were not folded, but faulted and tilted, probably in the latest Mesozoic and Tertiary. The existence of small anticlinal structures is expected in the extreme W of the Chaco Boreal. Important faults are inferred with NW and NE trends (Fig. 1a). Cerro León and adjacent areas in the northern Chaco form a horst with a downthrow of some 1100 m to the W and of more than 2000 m to the E (Fig. 2). Along the Alto Paraguay N-S faults have downthrown the west limb of the Itapucumi Limestones. Other major faults may control the course of the river Paraguay from Rosario to Humaitá. The epeirogenetic uplift during the Pleistocene provoked a NW-SE and E-W trending system of fractures in the plateau basalts, used by the right tributaries of the Paraná, whose course follows a tectonic valley. For this reason the rivers Piratí, Carapá, Itaimbey, Acaray, Monday, and Ñacunday form cataracts of 50–30 m near their junctions; these falls represent a valuable resource for hydroelectric energy.

Alcalic intrusions between Fortin Olimpo and Puerto Sastre (syenites with foyaites) are related to the Mesozoic fracturing and to the alkalic province of the margin of the Paraná Basin; they are considered as Mesozoic, the phonolite at Pão de Açucar on the Brazilian border is Triassic, dated 239–209 m yr, according to Comte and Hasui (1971).

The greater part of eastern Paraguay is the southwest region of the *Paraná Basin,* an oval depression whose NNE striking axis is marked by the course of the Paraná river. About 2300 m sediments were deposited over a peneplain, now in the form of a giant homoclinal structure; the beds dip very gently to the E, NE, and SE. Sedimentation began with marine sandstones (Ordovician?). In Triassic times subsidence stopped, and continental sands rest unconformably upon older sediments and even directly on the Precambrian Arch. Some 140–120 m yr ago the basin suffered extensive fracturing and faulting; repeated lava flows were spread over the whole basin, sills and dikes intruding older sediments.

Basaltic lavas near Sapucai erupted over a long period of time from Jurassic until Cretaceous (178–109 m yr). The alkaline intrusions, essexites, and related rocks between Paraguari-Ybycui-Aguapety in eastern Paraguay are Early Cretaceous (149–129 m yr), according to Comte and Hasui (1971).

The important fracture zone of Lake Ypacaraí, a NW-SE-striking step fault zone of more than 100 km length with a downthrow of some 600 m to W, is of pre-Tertiary age. The basalt plugs near Asunción are Cenozoic (46 m yr), according to Comte and Hasui (1971).

Historical Geology

Precambrian rocks include mainly metasediments and igneous rocks. Granites and porphyries belong to the late Precambrian. Outcrops are known in the N, in the Apa Mountains between Caracol-San Luis-Machuca. This unit is related to the Algonkian Cuiabá Series in Mato Grosso, Brazil. In the eastern Chaco, 40 km W of Puerto Casado, gneisses, and granite were found in drillholes at 108 to 115 m. In the S from San Miguel to Quiindy and E of Quyquyó there is porphyry, and an isolated, interrupted band occurs between San Bernardino (red granodiorite), Paraguari (cherts, quartzite), and Escobar (Escobar Series, Karpoff, 1965).

Upper Precambrian (Proterozoic). In the Apa Mountains, a sequence of unmetamorphosed beds rests unconformably upon the folded early Precambrian. This Itapucumí Series, consisting of dolomitic, oolitic, and conglomeratic limestones, red and green marls, and shales is seen in many outcrops. The limestones form the cliff on the left bank of the Paraguay River between Vallemí and San Salvador, and some isolated hills in the eastern Chaco. The faulted limestones show intense karst development. The clastic basal San Luis Member crops out at Cerro Paiva. The Itapucumí Beds are gently folded and faulted; they are compared with the Corumbá Series in Mato Grosso, now generally considered Proterozoic. No fossils are known. The limestones supply the cement plant at Vallemí.

Ordovician. In eastern Paraguay a 1000-m sequence of clastic rocks has covered the folded Proterozoic of the Central Paraguay Arch between Emboscada-Iturbe and Carapeguá (formerly placed in the Devonian) and is Silurian and probably Ordovician (Wolfart, 1961). In the type section from Paraguarí to Itacurubí de la Sierra the oldest sediments are coarse conglomerates, up to 20 m thick, whose components are quartz, quartzite, porphyry, and cherts, from the nearby Precambrian, probably littoral facies, indicating an Ordovician transgression. It is overlain by 700-800 m of brown arkosic sandstones and white sugarey sandstone, without any fossils.

Silurian. Some 300-400 m of fossiliferous sandstones and shales follow conformably with dips to the E and NE; they belong to the marine Llandovery. The lower part is the Eusebio Ayala Sandstone, 200-250 m thick, whitish to brown at the top, very fine grained, and micaceous. It has furnished the oldest fossils from Paraguay: *Skolithos*, *Dalmanites* sp., *Trimerus* sp., *Proetus* sp., *Diplograptus* sp., and *Tentaculites trombetensis* (Clarke). The next member is the key of the sequence, the white kaolinic Vargas-Peña Shales, 10-30 m thick, with a rich austral marine fauna: *Anabaia paraia* (Clarke), *Calymene boettneri* (Harr.), *Eophacops* n.sp. A (Wolf.), *Dalmanites* sp., *Climacograptus innotus brasiliensis* (Rued.), *Tentaculites*, etc.

The uppermost Silurian consists of 100 m of soft, micaceous thin-bedded violet and green "Cerro Perro Sandstone." It has furnished *Australostrophia conradii* (Harr.), *Tentaculites trombetensis* (Wolf.), *Orthoceras* sp., *Calymene* ex aff. *boettneri* (Harr.), *Diacalymene* cf. *crassa* (Shir.), *Trimerus* n. sp. A (Wolf.), *Eophacops* n. sp. A (Wolf.), *Camarotoechia* sp., *Crinodea* sp. In the Chaco Boreal (Figs. 2 and 3) the slope of the inselberg Cerro León is made of gray quartzite and micaceous sandstones, dipping WSW.

FIGURE 3. Sections of five wildcat borings in the Chaco Boreal.

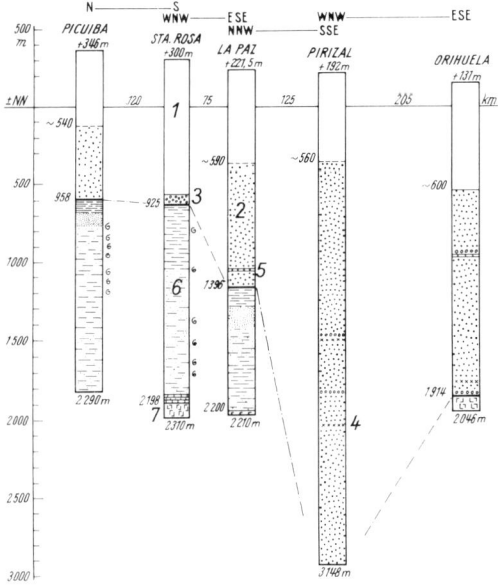

There, *Arthrophycus* and some brachiopods were found by Putzer and Wolfart. Similar white quatzites are reported from some wildcat drillholes (Eckel, 1959): Santa Rosa (2198-2310 m), La Paz (2220-2210 m), Orihuela (1914-2046 m), Lagerenza (2713-2895 m), Mendoza (3048-3242 m), Lopez and Madrejón (1524-1727 m). There is no doubt that the Chaco Boreal was invaded by the Silurian ocean. In eastern Paraguay the Llandovery was neritic.

Devonian. No Devonian exists in eastern Paraguay. At the very beginning of the Devonian the ocean withdrew into the Chaco Trough, probably due to uplift of the Central Paraguayan Arch, whereas subsidence continued in the Chaco Trough. Several 100 m of dark sandy marine shales and finegrained lenticular sandstones are found (Fig. 3) in the Chaco wildcats: Santa Rosa (925-2198 m), Picuiba (958-2290 m), La Paz (1396-2200 m), Mendoza (1490-3048 m), and Lagerenza (1200-2713). The facies of the Devonian is very similar to the oil-bearing Monos Shales in sub-Andine Bolivia. Surface outcrops are known at Ingavi and from the eastern part of Cerro León. The outcrops have furnished *Notiochonetes falklandicus* M. and Sh., *Australocoelia tourteloti* B. and C., and *Tentaculites stuebeli* (Clarke). Harrington (1950) has determined *Australospirifer* cf. *antarcticus* (Ulr.), *Pustulacia* sp. *Australostrophia arcei* (Ulr.), *Notiochonetes skottsbergi* (Clarke), *Tentaculites crotalinus* (Salt.), *T. jaculus* (Clarke). This austral fauna indicates a lower Devonian age. Menendez and Pöthe de Baldie (1967) list the palynological microfossils from Picuiba well (samples between 1160 and 1400 m) considering this part as middle Devonian. During the Devonian the direct connection between the ocean of the Andine geosyncline and the center of the Paraná Basin (where marine lower Devonian is proved) was interrupted.

Carboniferous. Continental beds of Pennsylvanian age rest with angular unconformity upon the Silurian in central and eastern Paraguay, and upon Precambrian in the N. The Upper Carboniferous sediments are glacial (tillites and varvites) and postglacial (fine-grained sandstone) and correspond closely to the Tubarão Series of the Brazilian Gondwana in the Paraná Basin (Fig. 4). In the N the sequence consists of preglacial sandstones, 2-3 tillites, separated by interglacial varvites, and sandstones, and to the top of the postglacial Upper Aquidauana Sandstone. This formation is identical with that known in Mato Grosso. The sandstones are rich in feldspar and kaolinite. It was Boettner (1947) who had first recognized the glacial origin of the Paraguayan Carboniferous in the N.

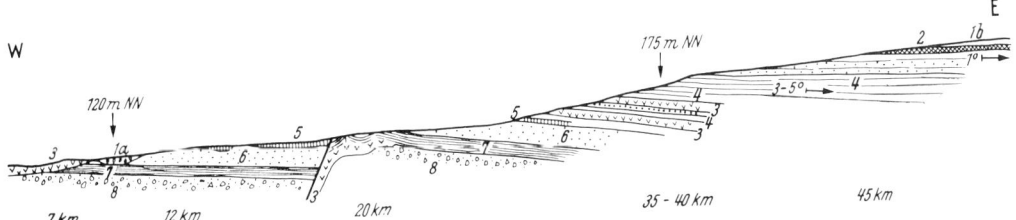

FIGURE 4. Lower Gondwana (U. Carboniferous).

The Paxixi member about 100 m thick is an alternating sequence of white and red varites with silty sandstones and brick-red siltstones; they are of glacial lacustrine and fluvioglacial origin. In the youngest, the Nioaque tillite near Itapopó, we found very coarse pebbles of crystalline rocks, Ordovician quartzite and much porphyry with hematite, proving transport from S.

Permian. A 300–450 m thick sequence of sandstones, passing up into sandstones alternating with shales is related to the Permian Passa Dois Series of the Paraná Basin in Brazil. The sequence is exposed in the Cordillera de Ybytyruzú to the S of Independencia. Shales and marls, observed near Independencia, are considered as equivalent of the lower Permian Iratí Formation by Vivar and Morinigo (1969). In 1923 Beder found, near Villarica, some ribs and a skull of *Mesosaurus,* the key fossil of the Brazilian Irati. The next member is an arkosic sandstone (Independencia facies, corresponding to the Estrada Nova Formation), containing: *Pinzonella neotropica* (Reed), *P.* cf. *illusea* (Reed), *Pyramus anceps* (Reed), *Leinzia similis* (Holdr.), and silicified *Dadoxylon* sp. The top part passes to argillaceous sandstones alternating with green, reddish, violet shales and siltstones, and correspond to an identical facies in the Rio de Rasto Formation (middle Permian).

Triassic. During the Upper Triassic the Paraná Basin was occupied by one of the most extensive ancient deserts of the world. Soft reddish sandstones of the Botucatú Formation, the Misiones Formation in Paraguay, form the lower part of the São Bento Series. Their vast extent suggests that the fluviatile and eolian sands passed beyond the margins of the Paraná Basin and reached (at least) the Central Paraguayan Arch.

The Misiones Sandstone is partly well stratified and partly thin-bedded with plenty of mica on the bedding planes, indicating subaquatic sedimentation, with basal conglomerates, dark red shales with signs of reworking; this facies was laid down by rivers and lakes. The sandstones are partly unstratified or cross-bedded and of eolian origin. The thickness varies from less than 50 m (Caaguzú) up to more than 250 m (Asunción). No fossils are known. The bedded facies serves as an excellent material for construction and is quarried at Hohenau and San Juan Bautista. The magnificent ruins of churches of the Jesuitic settlement "Guayra" are built of it.

Wildcat drillholes in the Chaco Boreal have found a thick series of soft red beds of continental origin, with some intercalations of conglomerates, gypsum, anhydrite and very little limestone, up to 2588 m thick (Pirizal well). Gypsum and anhydrite support the assumption of arid conditions. The formation of the red beds started in the Triassic but quite probably rather similar conditions of climate, erosion, transport, and deposition continued until Tertiary times. In 1961 Mädler (see Putzer, 1962) found palynologic proofs for the Triassic age in samples from Mendoza well from 919 to 948 m and from 999 to 1043 m: *Botryococcus* sp., *Benettiteae, Cycladeae.* In the neighboring northern Argentina from seismic measurements Padula and Mingramm (1968) have deduced an angular unconformity within the red beds and suggest that the lower part is Upper Triassic (Buena Vista Formation) and the upper part is the Lower Cretaceous San Cristobal Formation (maximum thickness 800 m). The last is oil-bearing in northern Argentina. The phonolite at Pão de Açucar has given K/Ar ages of 239 m yr (on biotite) and 207 m yr (on feldspar) according to Amaral (1967) and 209 m yr on whole rock samples (Comte and Hasui, 1971), and is thus Middle Triassic.

Jurassic-Cretaceous. The upper part of the São Bento Series of the Paraná Basin is the Serra Geral Formation, consisting of giant sheets of tholeiitic and doleritic plateau basalts in the overlying Tubarão Series, and in the Misiones Sandstone mostly above an erosional unconformity. In Brazil the absolute age of these lava sheets is established as 140–120 m yr, having flowed out over a long period.

FIGURE 5. Plain of the Chaco Boreal, Cenozoic sediments (oblique air photo: author).

Amigdaloidal facies (today with amethysts, agates, quartz) is typical of the top of each particular flow.

Plutonic bosses of alkali magmas occur in the eastern Chaco, in the southern part of the Central Paraguayan Arch and at the western margin of the Parana Basin; nepheline syenites with foyaites occur as isolated hills, and essexites with shonkinite have intruded the Precambrian, Ordovician, Pennsylvanian, and Triassic in the hilly country between Asunción and Independenia (Putzer and van den Boom, 1962). These alkali rocks do not show any tectonic stress, and appear to be products of differentiation of the parent magma of the plateau basalts; they certainly belong to the Mesozoic.

Eolian sand and poorly consolidated sandstone of very fine grain, light brown up to yellow, cover the plateau basalts in places. The same facies is found in Brazil (Caiuá Formation) and considered as Cretaceous. Outwash pebbles of conglomeratic Baurú Sandstone of Upper Cretaceous age are found upon the lava in the Cordillera de Amambay.

Upper Tertiary. Yellow and gray, fine- to medium-grained sands and gravel of the Argentine Late Tertiary Ituzaingo Formation are observed in the valley of the lower Paraguay up to its junction with the Paraná River.

Quaternary. The most extensive Quaternary formations are in the Chaco Boreal, where unconsolidated sandy beds reddish, green and yellow clays, with gypsum and loess of Pleistocene age represent the continuation of the Argentine Pampeano Formation. The sequence may reach some 100 m in thickness and is certainly of continental origin. The greater part is fluvio-lacustrine. The facies passes up into the Holocene. Volcanic ash, produced by young volcanoes of the Chilean-Bolivian cordillera, is mixed with the loose clastic material. This eolian transport has continued up till the present. Mammalian fossils such as *Megatherium, Glyptodon clavipes, Toxodon, Macrauchenia boliviensis,* and *Cuvieronius andium* were found south of Asunción (Bertoni, 1940) and in the valley of the Pilcomayo River (Vellard, 1934). In northeastern Paraguay the Xaraiés Member is a thin-bedded travertine (up to 8 m) mixed with a conglomerate of carbonate-cemented pebbles of Itapucumí Limestone. These rocks suggest alternating wet and dry tropical climates.

Lateritic soils and in places rich in limonitic concretions ("tacurú"), partly as hard, porous rock of goethite and alumina ("canga"), or as lateritic gravel ("itacurubí") occur particularly upon weathered plateau basalts.

Holocene sediments are found in the Chaco as eolian and fluviatile sands of very fine grain, with loam (loess) and clay. These youngest sediments are mostly rich in chlorides and sulfates (by evaporation) resulting in salt pans ("salitrales"). Peat is reported in the Paraná valley near Huamitá; dunes and blown sand are widespread over the highlands and cover some cliffs of the Alto Paraná; they and sandy valley fill are the youngest sediments in eastern Paraguay.

Paleogeography

During Precambrian times Paraguay was mostly occupied by geosynclinal seas; its argillaceous and sandy sediments were folded by the Assyntic orogeny (latest Precambrian). Later a gulf of the youngest Precambrian (or Cambrian?) ocean transgressed from the N, and the Itapucumí Series was deposited. At this time the Brazilian Shield, including Paraguay, became merged with the great continent of Gondwanaland.

During the early Paleozoic era the old mountains were first eroded and peneplaned; then, in *Ordovician* and *Silurian* times, the basins of the Chaco and Parana troughs began their long subsidence. The oldest marine fossils (Llandovery) in the Eusebio Ayala Sandstone prove a connection with an epicontinental ocean, probably through the Amazon Basin. In the Chaco Trough coarse sands mark the transgression of the Silurian.

At the beginning of the *Devonian* the Central Paraguayan Arch and the Paraguayan part of the Parana Basin became continental, but in the Chaco Trough subsidence continued and was intensified. In its eastern part clastic sediments are dominant, furnished from the arch, whereas in the western part bituminous shales and fine-grained sandstones were deposited. The marine fauna has a distinct austral aspect, so-called "malvino-kaffric" fauna. After the Devonian the ocean never reached Paraguay again.

Tillites, varvites, and sandstones in eastern Paraguay prove a glacial epoch in the Upper Carboniferous of the Paraná Basin, followed by fluviatile and lacustrine sandstones and siltites of the *Permian*.

Red fluviatile and eolian sandstones of the *Triassic* covered the Paraná Basin, reached the Chaco Trough, and merged there with red bed facies. The Triassic indicates a semi-desertic climate. The red beds, more than 3000 m thick in the Chaco Trough, prove subsidence of a very long period with a continuous transport of fine clastic material from former mountains. Their upper part is now placed in the *Lower Cretaceous*. Possibly the same facies continued into the Lower Tertiary.

At the end of the Gondwana epoch the Paraná Basin and adjacent areas suffered tensional stress, leading to fractures and the eruption of enormous basaltic-tholeiitic lava sheets. Sedimentation of fine-grained reddish sands continued after the effusion of the plateau basalts (140–120 m yr). These are *Cretaceous*. Fluviatile sands and gravels along the Paraná River indicate *Upper Tertiary* sedimentation, and likewise in the Chaco Trough. Fluvio-lacustrine and eolian sedimentation continued in *Pleistocene* times when, during the cold phases, a vast desertic plain was developed in the Chaco Boreal. Locally fanglomerates and travertine were deposited and lateritic soils formed in the subtropical highlands of eastern Paraguay.

Mineral Resources

Metallic ore deposits of great extent are not yet known. Hematite and magnetite veins of high quality occur in porphyry rocks at Caapucú, Paso Pindó, San Miguel, and Mburicacy. Small production took place during the war from 1864 to 1870; the reserves are certainly small. Lateritic iron, forming a crust of 1–2 m upon the basalt flows, occur between Encarnación and Puerto Presidente Stroessner. The mean grade is 35% iron; 2–5 million tons were measured and over 250 million tons inferred reserves (Alvarado, 1970). Manganese ores are mined on a small scale at Emboscada; pyrolusite has mineralized a fracture zone in (?)Ordovician sandstones. Rich ores have 46–49% manganese, but reserves are small. An interesting deposit of hypergene copper ores occurs near Villa Florida; malachite and azurite have mineralized a breccia of quartz porphyry. Mica, feldspar, and some beryl were mined in pegmatites of the Apa Mountains. Talc veins, 1–3 m thick, occur near Villa Florida; a large deposit of pyrophyllite, up to 5 m thick, a product of hydrothermal alteration in porphyry and tuffs, is known in northwest Caapucú.

The Silurian in the valley of the Peribebuy River yields sedimentary white kaolinite, mined at several places. Barite veins are observed at Fuerte Olimpo (Chaco); agates and amethysts, originally from the plateau basalts, are found in alluvial deposits of the Paraná River and its tributaries. Limestones and some marble crop out in the Apa Mountains; they are mined at Vallemi for portland cement. Bauxite may occur upon weathered alkalic eruptives.

Oil prospects in the western Chaco Boreal appear favorable; gas and oil shows are recorded in some of the 12 wildcats (Fig. 3), in a thick sequence of Devonian shales interbedded with fine-grained sandstone. The upper part of the red beds in the Chaco is now considered as the continuation of the oil-producing Lower Cretaceous on the Argentine Caimancito basin.

Copious *groundwater* is found in eastern Paraguay, where rainfall is between 1400 and 1900 mm/yr. Important aquifers are known in the Paleozoic sandstones, the Misiones Sandstone, the amigdaloidal facies of the plateau basalts, and in alluvial valley sediments. The hydrogeological situation in the Chaco Boreal is more limited. Rainfall decreases from 1300 to less than 600 mm/yr. Small aquifers occur at shallow depths (up to 10 m) but may be salty. An extensive second, deeper aquifer in Cenozoic sands has only highly mineralized groundwater; water supply in this region depends largely on stored surface water.

HANNFRIT PUTZER

References

Alvarado, B., 1970. "Iron ore deposits of South America," *Survey of World Iron Ore Resources*. New York: United Nations, 302–380.

Amaral, G., et al., 1967. "Potassium–argon ages of alkaline rocks from southern Brazil," *Geochim. Cosm. Acta*, 31, 117–142.

Beder, R., 1923. "Sobre un hallazgo de fosiles pérmicos en Villarrica (República del Paraguay)," *Bol. Acad. Nacl. Cien. Cordoba*, 27.

____, and Windhausen, A., 1918. "Sobre la presencia del Devónico en la parte media de la República del Paraguay," *Bol. Acad. Nacl. Cien. Cordoba*, 33.

Bertoni, G. T., 1940. "Constitución geologica, clima y producciones minerales del Paraguay," *Geog. Econ. Nacl.* (Asunción), 130–157, 189–212.

Beurlen, K., 1970. "Geologie von Brasilien," *Regionalen Geologie der Erde*, vol. 9. Berlin: Gebr. Borntraeger, 444p.

Boettner, R., 1947. "Estudio geológico desde Puerto Fonciere hasta Toldo-Cué," *Rev. Fac. Quim. Farm., Univ. Asunción*, (6–7).

Comte, D., and Hasui, Y., 1971. "Geochronology of eastern Paraguay by the potassium-argon method," *Rev. Brasileira Geosci.* (Sao Paulo), **1**(1), 33–42.

Eckel, E. B., 1959. "Geology and mineral resources of Paraguay, a reconnaissance," *U.S. Geol. Surv. Prof. Pap. 327.*

Goldschlag, M., 1913. "Beitrage zur Kenntnis der Petrographie Paraguays und des angrenzenden Gebiets von Matto-Grasso," thesis, Jena, 59p. (Summary in *Mitt. Geogr. Gesell. München,* **8**(3), 293–301.)

Harrington, H. J., 1950. "Geologia del Paraguay oriental," *Univ. Buenos Aires, Contr. Cient.,* Ser. E, **1**, 82p.

―――, 1956. "Paraguay," *Geol. Soc. Am. Mem.* **65**, 103–114.

Hopkins, E. A., Crist, E., and Snow, W. P., 1968. *Paraguay 1852–1968.* New York: Am. Geogr. Soc., 64p.

Karpoff, M. R., 1965. "Observations géologiques du SE de Asuncion, Paraguay," *C. R. Acad. Sci.,* **261**(25), 5558–5560.

Menendez, C. A., and Pöthe de Baldie, E. D., 1967. "Devonian spores from Paraguay," *Rev. Palaeobot. Palynology,* **1**(1–4), 161–172.

Milch, L., 1894. "Über Gesteine aus Paraguay," *Tscherm. Min. Petr. Mitt.,* **14**, 383.

Morinigo, G. V., and Vivar, D. V. D., 1969. "Données sur la tectonique de la Republique du Paraguay," *Comm. Carte Geol. Monde* (Paris), **9**, 58–65.

Padula, E., and Mingramm, A., 1963. "The fundamental geological pattern of the Chaco-Paraná basin (Argentina) in relation to its oil possibilities." *6. World Petrol. Congr.,* Frankfurt, sect. 1, paper 1, 18p.

―――, and Mingramm, A., 1968. "Estratigrafia, distribución y cuadro geotectonico-sedimentario del 'Triasico' en el subsuelo de la llanura chacoparanaense," *Actas Terc. J. Geol. Argent.* (Buenos Aires), **1**, 291–331.

Putzer, H., 1962. "Die Geologie von Paraguay," *Regionalen Geologie der Erde,* vol. 2. Berlin: 183p.

―――, 1968. "Überblick über die geologische Entwicklung Südamerikas," in E. J. Fittkau, ed., *Biogeography and Ecology in South America,* vol. 2. The Hague: Hillary, 1–24.

―――, and Van den Boom, G., 1962. "Über einige Vorkommen von Alkaligesteinen in Paraguay," *Geol. Jb.,* Hannover, **79**, 423–444.

Vellard, J., vol. 1. 1934. "Sur quelque fossiles du Paraguay," *Mus. Nat. Hist. Nat.,* B.s. 2, **6**(1), 150–152.

Vivar, V. D., and Morinigo, G. V., 1969. "La géologie du Paraguay," *Comm. Carte Géol. Monde* (Paris), **9**, 65–74.

Wilhelmy, H., and Rohmeder, W., 1963. "Die La Plata-Länder (Argentinien-Paraguay-Uruguay)," Braunschweig: G. Westermann, 584p.

Wolfart, R., 1961. "Stratigraphie und fauna des alteren Paläozoikums (Silur, Devon) in Paraguay," *Geol. Jahrb.* (Hannover), **78**, 29–102.

Cross-references: *Argentina; Bolivia; Brazil; South America; Uruguay.*

PERU

Within its 1,420,000 km^2 (496,225 sq mi), extending from the equator to 18°S latitude, the Republic of Peru embraces an impressive variety of physiographic, tectonic, and climatic regions. Its capital is Lima and principal port Callao. Since the travels of Antonio Raimondi, which began in 1851, the accumulation of geologic knowledge about Peru has been considerable, although large areas remain comparatively unknown; only a small part of the country has been mapped on a scale of 1:200,000 or larger. A landmark in the development of geologic knowledge of Peru was Steinmann's *Geologie von Peru* (1929). Although mining and petroleum companies have done exploration and mapping, only a few regional studies have resulted from this work (e.g. McLaughlin, 1924; Petersen, 1965; early references—see recent papers). Many geological reports were published in the *Boletines* of the Cuerpo de Ingenieros de Minas del Perú and the Instituto Geológico del Perú. The Instituto Nacional de Investigación y Fomento Mineros continued the work of the earlier government organizations, chiefly in connec-

FIGURE 1. Map showing geomorphic regions of Peru and approximate boundaries of principal Triassic and Jurassic volcanic activity.

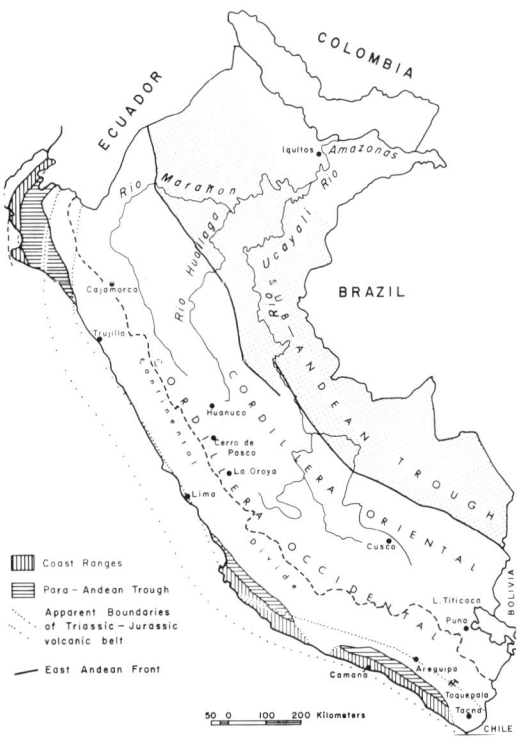

tion with mineral resources. An active mapping program formerly conducted by the Comisión Carta Geológica Nacional, now the Servicio Nacional de Geologia y Mineria, has produced (by April, 1973) 37 colored geological quadrangle maps (scale 1:100,000, cf., for example, Wilson et al., 1967). The *Boletin* of the Sociedad Geológica del Perú, published irregularly, contains many informative articles.

The latest geologic map of Peru, on a scale of 1:2,000,000, was published in 1956 by the Sociedad Geológica del Peru, and a 1:1,000,000 geological map serves as a base for the excellent metallogenic map published in 1969 by the Sociedad Nacional de Mineria y Petroleo. A recent summary of the geology of Peru was provided by Bellido (1969).

The principal centers of geologic training in Peru are the Universidad Nacional de Ingenieria and Universidad Mayor de San Marcos in Lima and the Universidad Nacional de San Agustin in Arequipa.

Geomorphic Divisions

Three regions dominate the morphotectonics of Peru: the desert coastal lowlands (below 2000 m), the Andes, and the jungle-covered lowland (below 2000 m) of the upper Amazon basin. Each of these regions can be further subdivided, but systematic work is still needed (Fig. 1).

The *Coastal Lowlands* may be divided into four types of terrain: The Andean Coastal Slope, the Para-Andean Depression, the Coast Range, and narrow segments of Coastal Plain. The Andean Coastal Slope is that portion of the lower Pacific slope of the Andes, with complex geology, which extends SW or W at a lower gradient than the average steep Pacific slope of the Andes. This province reaches the coast between Camaná and Atico, where it shows evidence of marine planation.

The Para-Andean depression (Fischer, 1956), between the Andes and the Coast Range, is a structural feature filled with Tertiary continental clastics in the south and with largely marine sediments elsewhere, particularly in northwestern Peru, where the Sechura Desert is more than 100 km wide.

In southern Peru the Coast Range borders the sea or is separated from it only by narrow remnants of elevated, wave-eroded coastal terraces. The SW slopes are steep, but the NE slopes are mostly buried beneath Tertiary clastics of the Para-Andean depression. The Coast Range decreases in elevation to the NW and the boundary between the Para-Andean depression and the Coast Range is cut off obliquely by the coastline near Camaná. The Coast Range consists mostly of Precambrian gneisses and intrusive rocks, with some locally fossiliferous upper Paleozoic strata, cut by Mesozoic intrusives and, like the Para-Andean depression, strongly affected by Cenozoic normal faulting. The equivalent geotectonic unit rises out of the Pacific again in northwestern Peru, first as offshore islands, then, with north trend, the Cerros de Illescas (Fischer, 1956). The whole west coast of Peru, together with basement rocks under the relatively narrow shelf, all strongly faulted, is the leading edge of the continent as it approaches the Peru-Chile trench, parallel to the coast some 150 km off shore. The Cerros de Amotape have a NE trend into Ecuador. These northern mountains consist largely of metamorphic and igneous rocks of Paleozoic age.

Because of the instability of the Pacific border throughout much of Cenozoic time, there are only small areas of true Coastal Plain. The largest is in the northwest, extending NE from the oil fields of Talara toward Ecuador, NW of the Cerros de Amotape. Between Lima and the Paracas Peninsula a narrow strip of coastal plain has been elevated and faces the sea with a wave-cut cliff. In southern Peru a limited coastal plain extends for a short distance NW of Tacna.

The *Andes* stretch from one end of Peru to the other, forming a high and rugged barrier between the coastal lowlands on the southwest and the jungle lowland on the northeast. Customarily in physiographic studies the range has been separated into three parts, the Cordillera Occidental (Western), Cordillera Oriental (Eastern), and the less distinctive intervening Cordillera Central of northern Peru. The Cordillera Occidental is the most continuous, forming the continental divide from Ecuador to southern Peru, where it separates the Titicaca depression from the Pacific drainage. The most spectacular glaciated peaks, culminating in Nevado Huascarán (elevation 6768 m, 22,195 ft), are in the Cordillera Blanca, some 300 km N of Lima. This Western Cordillera consists of folded and faulted Paleozoic and Mesozoic sequences cut by Upper Cretaceous and Tertiary intermediate intrusives and covered, especially in southern Peru, by Tertiary volcanics of great thickness. Several high, wide plains are located between the Cordilleras Occidental and Oriental. Especially notable are the Pampa de Junin in central Peru and the Lake Titicaca basin, continuous with the Poopo basin of Bolivia. Such basins appear to be of tectonic origin largely due to normal faulting; their relatively level plains are partly filled by young sediments and are thus not erosional, but rather aggradational features, formed in late Tertiary and Pleistocene time.

The Cordilleras Central and Oriental are less well defined as topographic units. The basement is a very great thickness of metasedimentary rock of mainly early Paleozoic age recently investigated by French geologists (Megard, 1967; Megard et al., 1971). Intrusives of late Paleozoic age cut these older rocks. In central and northern Peru younger Paleozoic and Mesozoic sequences rest unconformably on the older Paleozoics and have in turn been folded and faulted. The eastern slope of the Andes drops abruptly through the Montaña or high jungle country to the jungle lowlands (the Selva). Great tributaries of the Amazon have cut through the Andean front to form a series of gorges and rapids. Upstream the same rivers and their tributaries have cut long, deep subsequent valleys parallel to Andean trends.

East of the Andes, extending the length of Peru, is the *Subandean Foreland*. According to Ham and Herrera (1963), a persistent system of steep faults separates the Andes from the Subandean foreland trough. Foreland folds, commonly faulted along their eastern side, decrease in amplitude toward the Brazilian Shield. The break between the eastern slopes of the Andes and the Subandean Foreland is thus geomorphically and geologically sharp in most places. The complex structure of the Andes is in striking contrast to the vast region of Subandean Tertiary continental clastic rocks which rest on a sequence of marine Paleozoic and marine and continental Mesozoic strata.

Geologic History

An elongated depressed zone at the western margin of the Precambrian Brazilian Shield began to form and receive sediments early in the Paleozoic. Remnants of late Precambrian crystalline rocks (a K/Ar date of 642 + 16 m yr obtained near Arequipa by Stewart and Snelling is reported by Cobbing and Pitcher, 1972b) and possibly early Paleozoic rocks are preserved along the coast in southern Peru, and in isolated areas along the central and eastern Cordilleras. The oldest known Paleozoic sediments are Ordovician, a very thick sequence of argillaceous and sandy sediments, now dark slates, phyllites, and schists preserved in a belt extending from Bolivia northwest to the upper Marañon River in northern Peru (Fig. 2). These geosynclinal sediments may represent a time span from Cambrian through Devonian, but no Cambrian fossils have been found thus far in Peru.

In the Lake Titicaca region shales, siltstones

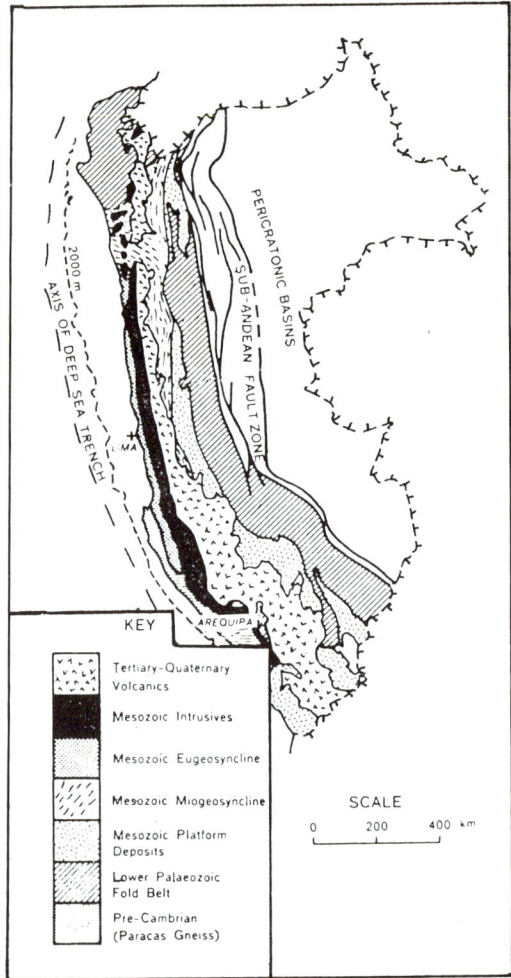

FIGURE 2. Generalized geological sketch map of Peru, compiled from the geological map of Peru at 1:4,000,000 (1969) by the Servicio Nacional de Geolgia y Mineria (Cobbing, 1972).

and sandstones some 3000 m. thick are largely of Early and Middle Devonian age (Newell, 1949), but include Silurian and possibly Ordovician strata (Boucot and Megard, 1972). Near Camaná the Devonian rests on Precambrian gneiss. Middle and Lower Devonian strata probably extended, with considerable thickness, along the length of the Andes in Peru. The section apparently thinned rapidly both to the east and west. An important early *Hercynian orogeny* along the whole Andean axis in Peru is indicated by tight folding, local metamorphism to green schist facies of lower Paleozoic strata, and the absence of upper Devonian rocks. Post-tectonic coarse to fine, locally tuff-bearing clastics of Mississippian age are widely developed. Concordantly above these rocks is a middle

FIGURE 3. Diagrammatic SW-NE stratigraphic section of the Mesozoic strata of Peru.

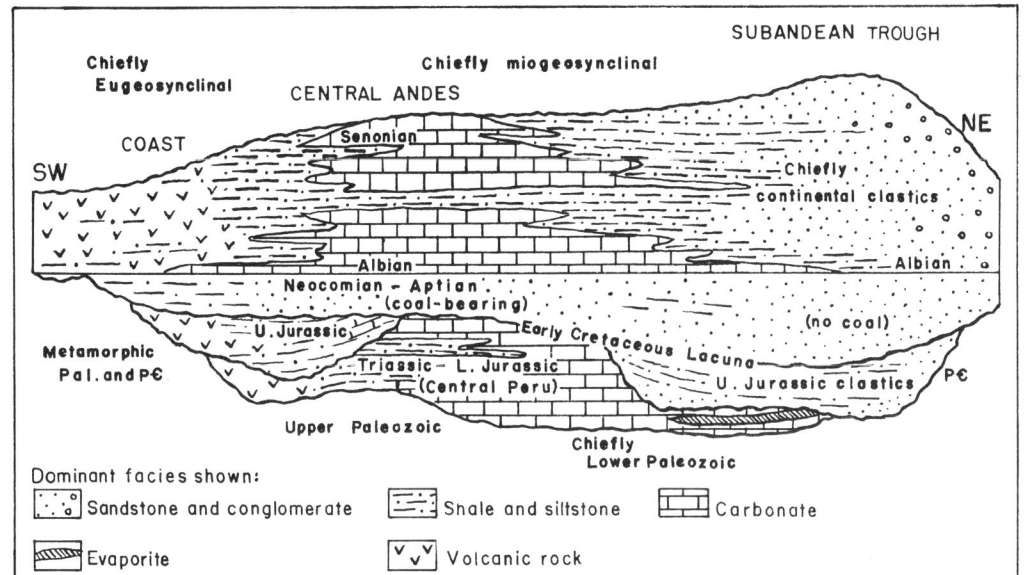

Pennsylvanian chiefly marine sequence, followed with little discordance by Lower to Middle Permian shale and limestone, especially in southern Peru (Newell et al., 1953).

A late Hercynian disturbance is marked by andesitic to rhyolitic flows and volcaniclastic rocks of middle to late Permian age. Thick, red, coarse to fine post-orogenic clastic rocks (molasse) at the top of the Permian sequence are widespread in central and southern Peru, and recognized in many places farther north. Permian granites occur in NW and central Peru and perhaps elsewhere.

Since Lower and Middle Triassic strata are missing, evidently much of Andean Peru was land from the later Permian to the beginning of the Upper Triassic. During this interval, however, salt beds, which later gave rise to diapiric structures, were formed in the Subandean region (Huallaga Valley).

During late Triassic and early Jurassic time a carbonate sequence with a thickness of more than 2000 m accumulated in a long and relatively narrow shelf area east of the present continental divide of northern and central Peru. Concurrently, thick andesitic volcanic rocks in large part submarine, with interbeds of dark shale and impure limestone, accumulated to the west. Still farther west, according to Fischer (1956), both Triassic and Jurassic rocks are missing, so that Albian and younger Cretaceous rocks rest directly on Paleozoic metamorphic rocks (Fig. 3). The lower Mesozoic volcanic belt is recognized along the coast in southern Peru, and inland as far as Arequipa and Moquegua. Near Arequipa, Liassic andesitic flows and tuffs, contain bioherms with an abundant coral fauna. From late Triassic into middle Jurassic time, volcanic activity prevailed along and offshore from the present coast of Peru and joined with a similar zone in Chile (Fig. 1). This is the first well defined volcanic island arc recorded in the geologic history of the country. No plutons of this age are yet known in this belt.

Just to the east of this Triassic/Jurassic volcanic belt in southern Peru a parallel narrow foreland belt received a thick accumulation of flysch-like clastic sediments and minor carbonates. This accumulation continued through Callovian time. The apparent northeastern limit of Jurassic sedimentation in southern Peru is some 30 km west of Lake Titicaca.

Volcanism in a western archipelago was a precursor to crustal disturbance. Upper Jurassic strata (Tithonian) appear to rest on disturbed Middle Jurassic beds south of Nazca (Rüegg, 1956). The next significant event is the very widespread accumulation of sandstones, largely terrestrial, with coal beds, but with some marine intercalations, during early Cretaceous time. The unconformity below these clastics is significant, most pronounced in central and northern Peru, but also evident in the south from structural relations. In central Peru the coal-bearing Neocomian to Aptian Goyllarisquisga group rests on Jurassic or Triassic limestones; near Puno, in the south, Neocomian or Aptian sandstone rests on Upper Jurassic sediments or overlaps onto Devonian strata; in eastern Peru a Neocomian sandstone rests disconformably on Upper Jurassic to Ordovician rocks.

Transgressively above the nonmarine Lower Cretaceous clastics, carbonate accumulation occured close to but slightly to the west of the earlier Mesozoic shelf sediments (Fig. 3). Benavides (1956), on the basis of ammonite zones, distinguished limestones or marls of Albian to Santonian age. Coney (1971) mapped thick Albian to Turonian limestones in the Cordillera de Huayhuash, north-central Peru. Although marine sedimentation was essentially continuous in central and northern Peru through the Upper Cretaceous, the equivalent zone was above sea level in the south after the deposition of the Cenomanian Ayavacas limestone.

In the Lake Titicaca region thick continental clastics, in part red beds, are the highest Cretaceous strata, mostly trapped in broad, low-level valleys as a result of vertical block movements. In the Subandean foreland, however, Lower and Middle Cretaceous strata include marine and fluvio-lacustrine sediments, probably in large part derived from the east. These give way to a white quartz sandstone, a littoral marine facies, during Late Cretaceous time.

To the west, not far from the present coastline, more than 2000 m of volcanic rocks accumulated in Late Cretaceous to early Tertiary time. These volcanics, dominantly andesites but ranging to rhyolites, accompanied the Andean batholith. The plutonic rocks of the batholith extend more or less continuously along the Andean coastal slope and the para-Andean depression (see Geologic Map of Peru, 1956). Emplacement of the multiple intrusions appears to have begun in Late Cretaceous time and continued well into the Tertiary. A radiometric date of 58.7 ± 1.8 m yr was obtained on diorite at Toquepala (Laughlin et al., 1968). Cobbing and Pitcher (1972a) report ages of 76 ± 3 m yr on early tonalite and 33 ± 1 m yr on a late adamellite north of Lima. In southern Peru a basal Tertiary conglomerate contains boulders from the batholith. The average composition of the batholith is granodiorite or tonalite. Various phases of intrusions have been described, ranging from gabbro to granite. Early phases were syntectonic. Zones of contact metamorphism around the batholith are extensive.

The end of the Cretaceous and the emplacement of the Andean batholith marked a distinct change in Andean history. In the vicinity of Talara, in the NW part of Peru, Tertiary marine clastics including numerous turbidites accumulated in an active tectonic zone characterized by much normal faulting. Dominantly red continental clastics and lake beds were deposited in isolated elongate basins parallel to the present trend of the Andes between the Cordilleras Occidental and Oriental. In central and southern Peru the clastics reached great thickness, but it is not yet possible to make specific correlations between one area and the next. During early Tertiary time there was renewed andesitic and basaltic volcanism along a line just to the west of the principal Tertiary basins. Sedimentary and volcanic rocks interfinger along this line, which passes to the west of Lake Titicaca and along the continental divide northwest past the latitude of Cerro de Pasco.

During the Tertiary, east of the already positive Andean area dominantly continental clastics accumulated to great thicknesses (up to 6100 m in Madre de Dios); they are highly varied, fine to coarse, and irregularly distributed. Only during the Oligocene was there one final marine invasion into the Subandean area, probably through a strait from the Pacific in northern Peru.

The late Tertiary was marked in the Andean belt by normal and reverse faulting, folding, tilting, intrusion of stocks, sills and dikes, and continued volcanic activity. Although the folding in many places appears to have been intense and isoclinal, it was probably superficial. Overturning and thrusting toward the northeast in the Andes of central Peru may be an indication of decollement due to vertical movements. Tertiary strata with structure more complicated than that of the underlying Paleozoics and Mesozoics are not unusual in the central part of the country. Development of the modern Andes began with uplift in the late Tertiary. Normal faulting has been of major importance. Intermittent periods of stand-still allowed the formation of several widely recognized erosion surfaces, some of which are now conspicuously tilted. Rapid uplift has resulted in very deep dissection (see, for example, Fig. 4).

In southern Peru volcanic activity has continued until the present time. Rhyolitic pyroclastic flows accompanied many eruptions but the dominant rock type is andesite. The famous volcano El Misti, near Arequipa, is a beautiful composite volcano (Fig. 5).

Shallow earthquakes are frequent on and offshore along the coast of Peru; intermediate depth shocks are farther east, and deep-focus earthquakes have their foci below the Subandean region of the margin of the shield. According to James (1971) the depth to the Moho discontinuity reaches a maximum of about 70 km under the Cordillera Occidental of southern Peru (Fig. 6). The Pacific Nazca plate appears to be underthrusting the South America plate

FIGURE 4. Celebrated ruin of Machu Picchu, in the Peruvian Andes. (Courtesy: Braniff International.)

FIGURE 6. Contour map of depth (in kilometers) to the M discontinuity beneath the central Andes (from James, 1971b).

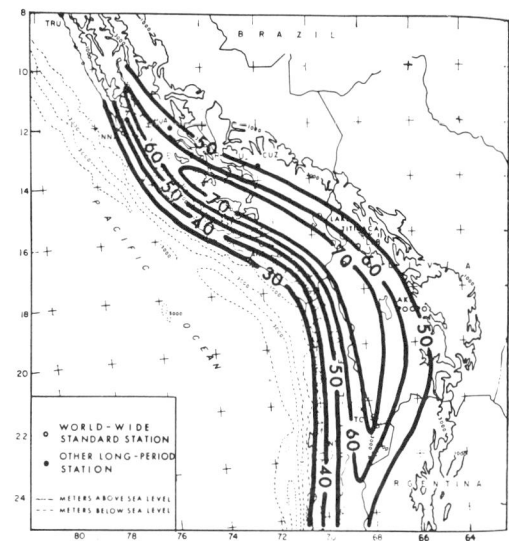

FIGURE 5. Composite volcano Misti, as seen from near Yanahuara. (Photo Arequipa Casa de la Cultura.)

TABLE 1. Metal Production, 1969

Metal	Production 1969 (metric tons)
Copper	206,100
Zinc	314,700
Lead	162,900
Iron ore	9,620,000
Silver	34,147,000 troy oz
Gold	127,722 troy oz

From *Minerals Yearbook*.

at a low angle, in combination with deep normal faulting (Abe, 1972).

Mineral Deposits

Peru is a producer of a variety of metals, but only copper, zinc, lead, gold, silver and iron ore are mined in quantity (see Table 1).

The present major copper producer is the Toquepala Mine, some 50 km from the coast near Tacna in southern Peru. This is a disseminated, low-grade sulfide deposit in breccia pipes and intrusives cutting volcanic rocks of probable Cretaceous age. Open pit mining started in 1959. Reserves were then estimated at 400,000,000 tons with 1% copper.

In the Andes of central Peru is a major base and precious metal mining region. Cerro de Pasco, Morococha and Casapalca are famous, old, but still active mining camps. More recently developed mines of importance are at Atacocha, Yauricocha and Cobriza. Cerro de Pasco, which has a history dating from the seventeenth century, has been an important producer of silver, copper, zinc and lead. High-grade oxidized or enriched silver-copper ores at the surface gave way below to enargite-rich veins and irregular orebodies cutting a monzonite stock and a large pyrite-silica body. Pyrite and silica form a massive replacement of Triassic limestone in a mile-long crescent on the east contact of the stock. Within the pyrite-silica mass are major bodies of lead and zinc ore.

The Oroya smelter of the Cerro de Pasco Corporation in 1968 produced 65,873 metric tons of refined zinc, 86,346 metric tons of lead, 53,210 metric tons of copper, 43,160 ounces of gold, and 20,371,000 ounces of silver.

Most of Peru's iron ore production comes from Marcona, near the coast SE of Lima.

Petroleum production in Peru is chiefly from the Talara and adjacent oil fields along the northwestern coast with some offshore production. The oil and gas are in Tertiary clastics, largely in stratigraphic traps. Exploration in the Subandean region of eastern Peru has led to the discovery of several fields producing from Cretaceous sandstones. Production of oil in 1970 was over 26 million barrels. On October 4, 1968, the Peruvian government expropriated the holdings of the International Petroleum Company. Petroleum exploration, production, and refining is now carried on almost entirely by Petroperu, formerly the Empresa Petrolera Fiscal, or by private companies under agreement with the government monopoly.

WILLIAM F. JENKS

References

Abe, K., 1972. "Mechanisms and tectonic implications of the 1966 and 1970 Peru earthquakes," *Phys. Earth Planet. Interiors*, 5, 367–379.

Bellido, B. E., 1969. "Sinopsis de la geologia del Peru," *Serv. Nacl. Geol. Mineral.*, 22, 1–54 (with 1:4,000,000 geol. map).

———, et al., 1956. *Mapa Geologico del Peru*, Lima: Soc. Geol. Peru, 1:2,000,000.

Benavides C., V., 1956. "Cretaceous system in northern Peru," *Bull. Am. Mus. Nat. Hist.*, 108, 357–493.

———, 1962. "Estratigrafia pre-Tertiaria de la region de Arequipa," *Soc. Geol. Peru*, 38, 5–63.

———, 1968. "Saline deposits of South America," *Geol. Soc. Am. Spec. Pap. 88*, 249–290.

Bogdanoff, A. A., 1970. "The making of international geologic maps of South America," *Geotectonics, Acad. Sci. USSR.*, 1, 1–7.

Boucot, A. J., and Megard, F., 1972. "Silurian of Peru," *Geol. Soc. Am. Spec. Pap. 133*, 51.

Cobbing, E. J., 1972. "Tectonic elements of Peru and the evolution of the Andes," 24th *Intern. Geol. Congr.*, 3, 306–314.

———, and Pitcher, W. S., 1972a. "The coastal batholith of central Peru," *J. Geol. Soc. Lond.*, 128, 421–460.

———, and Pitcher, W. S., 1972b. "Plate tectonics and the Peruvian Andes," *Nature*, 240, 51–53.

Coney, P. J., 1971. "Structural evolution of the Cordillera Huayhuash, Peru," *Bull. Geol. Soc. Am.*, 82, 1863–1884.

Dalmayrac, B., 1970. "Mise en evidence d'une chaine ante-ordovicienne et probablement précambrienne dans la Cordillère orientale du Pérou Central (région de Huanuco)," *C. R. Acad. Sci.*, Ser. D, 270, 1088–1091.

Fischer, A. G., 1956. "Desarollo geológico de noroeste Peruano durante el Mesozoico," *Soc. Geol. Peru*, 30, 177–190.

Giletti, B. J., and Day, H. W., 1968. "Potassium–argon ages of igneous rocks in Peru," *Nature*, 220(5167), 570–572.

Ham, C. K., and Herrera, L. J., Jr., 1963. "Role of subandean fault system in tectonics of eastern Peru and Ecuador," *Am. Assoc. Petrol. Geologists* Mem. 2, 47–61.

Hamilton, W., 1969. "The volcanic central Andes - a modern model for the Cretaceous batholiths and tectonics of western North America," *Proc. Andesite Conf., Bull. Dept. Geol. Mineral. Ind. St. Oregon*, 65, 175–184.

Hillebrandt, V. A. von, 1970. "Die Kreide in der Zentralkordillere ostlich von Lima (Peru, Sudamerika)," *Geol. Rundschau*, **59**(3), 1180–1203.

James, D. E., 1971a. "Plate tectonic model for the evolution of the central Andes," *Bull. Geol. Soc. Am.*, **82**, 3325–3346.

———, 1971b. "Andean crustal and upper mantle structure," *J. Geophy. Res.*, **76**, 3246–3271.

Jenks, W. F., 1956. "Peru," *Geol. Soc. Am.* Mem. **65**, 217–247.

Koch, E., 1959. "Geology of the Maguia oilfield in eastern Peru and its regional setting," *5th World Petrol. Congr.*, Sec. 1, **32**, 1–10.

———, 1962. "Die Tektonik im Subandin des Mittel-Ucayali-Gebietes, Ostperu," *Geotekt. Forch.* (Stuttgart), **15**, 1–67.

———, 1968. "Ein tektonischer Schnitt durch Mittelperu," *Geol. Rundschau.*, **57**, 615–621.

Laughlin, A. W., Damon, P. E., and Watson, B. N., 1968. "Potassium–argon dates from Toquepala and Michiquillay, Peru," *Econ. Geol.*, **63**, 166–168.

Megard, F., 1967. "Commentaire d'une coupe schématique à travers les Andes Centrales du Pérou," *Rev. Géogr. Phys. Géol. Dyn.*, **9**(4), 335–346.

———, Dalmayrac, B., Laubacher, G., Marocco, R., Martinez, C., Paredes, J., and Tomasi, P., 1971. "La chaine hercynienne au Pérou et en Bolivie, premiers resultats," *Cahier ORSTROM*, Ser. Géol. 3, **1**, 5–44.

Newell, N. D., 1949. "Geology of the Lake Titicaca region Peru and Bolivia," *Geol. Soc. Am. Mem.* **36**, 111p.

———, Chronic, J., and Roberts, T. G., 1953. "Upper Paleozoic of Peru," *Geol. Soc. Am. Mem.* **58**, 276p.

Perales-Calderon, F., 1970. "Glosario y tabla de correlacion de las unidades estratigraficas del Peru," *Primer Congr. Latinoamericano Geol.*, Lima.

Petersen, U., 1965. "Regional geology and major ore deposits of Central Peru," *Econ. Geol.*, **60**, 407–476.

Portugal, J. A., 1974. "Mesozoic and Cenozoic stratigraphy and Tectonic events of Puno-Santa Lucia Area, Dept. of Puno, Peru," *Bull. Am. Assoc. Petrol. Geologists*, **58**(6), 982–999.

Rüegg, W., 1956. "Geologie zwischen Canete-San Juan, 13°00'–15°24' Sudperu," *Geol. Rundschau*, **56**, 775–858.

Steinmann, G., 1929. *Geologie von Peru*. Heidelberg: Winter, 448p. (Spanish translation, 1930, Heidelberg).

Stewart, J. W., et al., 1974. "Age determinations from Andean Peru: a reconnaissance survey," *Bull. Geol. Soc. Am.*, **85**, 1107–1116.

Travis, R. B., 1953. "La Brea-Parinas oil field, northwestern Peru," *Bull. Am. Assoc. Petrol. Geologists*, **37**(9), 2093–2118.

Wilson, J. J., 1963. "Cretaceous stratigraphy of central Andes of Peru," *Bull. Am. Assoc. Petrol. Geologists*, **47**(1), 1–34.

———, Reyes, L., and Garaya, J., 1967. "Geológia de los Cuadrangulos de Mollebama, Tayabamba, Huaylas, Pomabamba, Carhuaz y Huari," *Serv. Geol. Minearl.*, **16**, 1–95.

Cross-references: *Bolivia; Chile; Colombia; Ecuador; South America.*

PHOENIX ISLANDS

Situated in the central Pacific midway between Fiji and Hawaii, the eight Phoenix Islands are unlike most of the other oceanic island groups in being, not in a linear trend, but in a ring, as seamounts surmounting a low dome. The group's location is 2 to 5°S, and 170 to 175°E. On a similar dome some 500 km NW are the islands of Howland and Baker, also treated here. All ten islands are of coral limestone, but only Canton is an atoll, most of the islands having been partly uplifted. According to their position, the sea-floor spreading history would suggest a late Cretaceous age for the basement.

Canton Atoll has a broad, circular lagoon with a U-shaped rim island 32 km long, that has been used to build a major airfield, at one time important (in the prejet era) on the U.S.-Australia commercial route. The climate is arid and there is no fresh water. The other seven islands of the Phoenix group (*sensu stricto*)— Enderby, Phoenix, Birnie, Sydney, Hull, Gardner, and McKean—are all small uplifted atolls reaching elevations of 10–15 m.

Howland Island (0°48'N, 176°38W) and *Baker Island* (0°12'N, 176°29'W) are both uplifted atolls, utilized for airstrips; having no water, they are uninhabited, however, though they have furnished some guano in the past.

The southerly Phoenix Islands are administered by the United Kingdom, Howland and Baker islands by the United States, and Canton Island has a joint administration.

RHODES W. FAIRBRIDGE

References

Davis, W. M., 1928. "The coral reef problem," *Am. Geogr. Soc. Spec. Publ.*, **9**.

Gregory, H. E., 1925. "Trip B–Baker and Howland Islands," *Bernice P. Bishop Mus. Bull.*, **21**, 25–28.

Hague, J. D., 1862. "On phosphatic guano islands of the Pacific Ocean," *Am. J. Sci.*, Ser. 2, **34**, 224–243.

Hutchinson, G. E., 1950. "The biogeochemistry of vertebrate excretion," *Am. Mus. Nat. Hist. Bull.*, **96**, 1–556.

Power, F. D., 1925. "Phosphatic deposits of the Pacific," *Econ. Geol.*, **20**, 266–281.

Cross-references: *Cook Islands; Gilbert and Ellice Islands; Line Islands; Tokelau Islands.*

PITCAIRN ISLANDS

A group of four islands in the southeastern Pacific, administered by Britain under the Governor of Fiji. The area of the colony is 48 km² (18.5 sq mi). Pitcairn proper lies at 25.04S and 130.06W, 6800 km (4200 mi) southwest of Panama and 5100 km (3200 mi) northeast of New Zealand, Pitcairn proper (area of 2 sq mi; 5.2 km²) is the only inhabited island of this group that includes Oeno, Henderson, and Ducie Islands. Its people are descendants of mutineers from the *H.M.S. Bounty* (1790) and Polynesian women from Tahiti. Some migrated to *Norfolk Island* (q.v.), though many returned.

The climate is warm all year, mostly under the influence of the SE Trades except for brief reversals during the southern summer. There is a good rainfall and much of the island is cultivated.

Only one of the islands is volcanic, Pitcairn (5 km across and 333 m elevation), a silica-rich basalt with some hypersthene andesite, deeply weathered and cliffed without fringing reefs. *Oeno* is an atoll 3 km in diameter with a shallow lagoon. *Henderson* is an uplifted atoll, the plateau 15 m high and 7 km wide, bounded by steep cliffs and fringing reefs. *Ducie* is an atoll, low and 2 km across; it is the most southerly atoll of the central Pacific. The islands represent a trend connecting Tahiti, the Gambiers, and Easter Island. The underlying oceanic crust is of Oligocene age according to plate-tectonic evidence.

RHODES W. FAIRBRIDGE

References

Davis, W. M., 1928. "The coral reef problem," *Am. Geog. Soc. Spec. Publ.,* 9.

Marshall, P., 1918. "Notes on the geology of the Tubuai Islands and of Pitcairn," *N.Z. Inst. Trans. Proc.,* 1, 278–279.

Cross-references: *Easter Island and Sala y Gómez; Gambier Islands; Oceania; Tahiti.*

PUERTO RICO

The Commonwealth of Puerto Rico, the easternmost and smallest of the Greater Antilles, is in United States territory, 8896 km² (3435 sq mi) in area. It is roughly rectangular in shape and approximately 170 km E-W in length and 60 km in width. It lies between latitudes 17°55' and 18°30'N and longitudes 65°37' and 67°15'W.

The island is surrounded by a narrow insular shelf that drops to oceanic depths reaching 2000 and 3000 m in some places within 3 km of the shore.

The small island of Mona, in the middle of the passage separating Puerto Rico from the Dominican Republic to the west, belongs to Puerto Rico politically, as do the islands of Vieques and Culebra to the east.

The central backbone of Puerto Rico, the Cordillera Central, is a rugged terrane of steep-sided mountains reaching an elevation of 1338 m at their highest point, Cerro de Punta.

The main watershed is displaced somewhat to the south of the center line of the island and the major streams flow northward to the Atlantic. The almost continuous air flow of the NE trade winds results in the rainfall being concentrated in the northern parts (over 5000 mm annually in El Yunque), and the south, and particularly the southwest, being in a semiarid rainshadow (about 750 mm annually). Most of the southward-flowing streams are intermittent.

There are low-lying plains separated by ridges; four are found along the west coast and two on the east. The country supports a dense vegetation except in the south and southwesterly areas. Coffee and tobacco plantations are extensive in the mountain areas, and sugar is grown in the plains. Agriculture in the island has suffered a considerable decline in recent years. Remnants of tropical rain forest are found on Cerro de Punta and on El Yunque in the northeast. In the southwest, cactus and thornbrush prevail.

Geological Work

The first geological work in Puerto Rico was a reconnaissance survey by the Scientific Survey of Puerto Rico and the Virgin Islands, directed by Berkey, a joint endeavor of the New York Academy of Science and Columbia University in the 1920s. The results were summarized by Meyerhoff in *The Geology of Puerto Rico* (1933). After that nothing was done until 1953 when the Commonwealth Government contracted with the U.S. Geological Survey for a geological mapping program that is now more than two-thirds completed. The Water Resources Division of the U.S. Geological Survey also started investigations at about the same time. During the 1950s graduate students from Princeton University carried out some mapping projects in connection with the Caribbean Research Project headed by the late H. H. Hess. Geologists have been employed by various government departments during the last 15 years and a geological service in connection with environmental protection is planned. A geochemical laboratory has been established within the Department of Public Works. A Department of Geology has been developed in the University of Puerto Rico at Mayaguez and a

FIGURE 1. Geological map of Puerto Rico, generalized from Briggs (1964). 1, Serpentinite complex–serpentinites, amphibolite, and cherts; 2, Upper Cretaceous-Eocene–mainly volcanics; pyroclastics, lavas and volcanically derived sediments; some limestones; 3, Oligocene-Miocene–mainly white or buff limestones, marls, some clastics; 4, Quaternary–alluvium, beach deposits; 5, intrusives–chiefly tonalite and granodiorite; 6, faults (highly generalized).

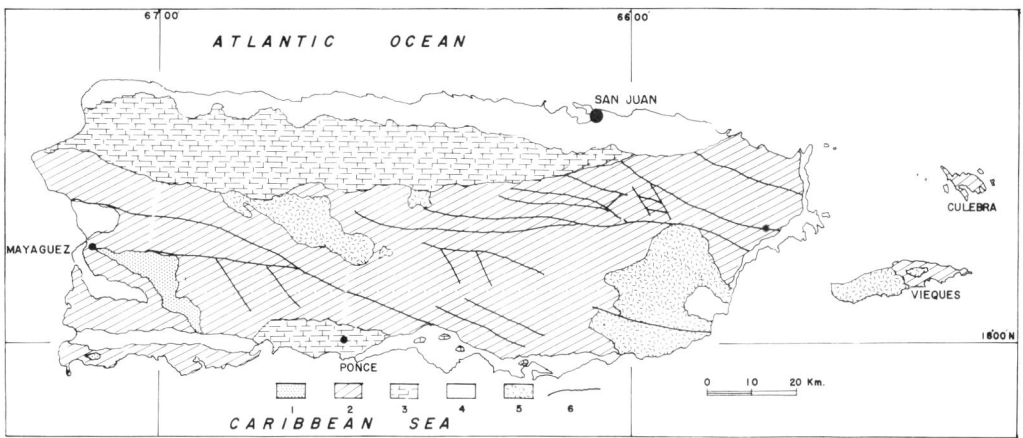

Geomorphic and Structural Divisions

The most striking structural feature is the contrast between the strongly faulted and folded volcanogenic Cretaceous-Eocene rocks, which form the central part of the island and the relatively undeformed mid-Tertiary limestones that overlap them to the north and south with a marked angular unconformity. A steep escarpment forms the southern edge of the northern limestone area. For convenience the Cretaceous-Eocene rocks are commonly referred to as the "Older" series, and the mid-Tertiary limestones as the "Younger" series.

The "Older" series are notably faulted. Two major fault zones with left-lateral strike-slip displacement divide them into three blocks, each of which is intensely faulted and folded (Fig. 1). The dominant structural trend is WNW-ESE, although minor faults trend NE, N, and NW. The dominant WNW trend is clearly seen in the topographic lineaments.

In the southwestern block the rocks are folded into a series of broad anticlines and synclines somewhat asymmetric to the south. An anticlinal ridge following the structural trend from Mayaguez on the west coast toward the ESE is underlain by serpentinite which appears to have been emplaced tectonically. Other smaller serpentinite bodies occur to the southwest. They are associated with amphibolites, spilitic rocks, and cherts.

In the central block, two plutonic intrusives, predominantly of tonalite and granodiorite, are emplaced in the "Older" series. A number of smaller intrusions are also found. Intrusion seems to have taken place in Late Cretaceous and Eocene time. It was accompanied or followed by extensive hydrothermal activity. Rocks at the contracts with the intrusives have been converted to hornfelses. The plutonic rocks are more susceptible to erosion and form lower, more subdued topography than the surrounding hornfels zone and other country rocks. Away from the intrusives, the "Older" series generally show a low level of metamorphism (zeolite facies). Limestones locally interbedded with the volcanics give rise to karstic features–sinkholes, caves, lapiés. In the southern part of the central block, N-directed gravity slides have affected the Eocene rocks.

In the northeastern block fold trends tend to be NE-SW in contrast to the WNW trends of the central block.

The overlying rocks of the "Younger" series, Oligocene to Miocene in age, are predominantly white or buff limestones and practically undeformed. They give rise to a distinctive topography, particularly in the northern outcrop, where beautifully developed karst features are found. Sinkholes, caves, underground drainage, and steep-sided residual hills (known variously as "mogotes," "pepinos," or "haystack hills") have resulted from the solution of the limestones. The escarpment at the southern edge of the northern outcrop is surmounted along much of its length by jagged peaks ("tower karst"). These rocks dip gently about 4 or 5°

toward the north and their E-W strike contrasts with the WNW strike of the older rocks.

The southern outcrop of mid-Tertiary limestones is not so extensive as the northern. Faulting has affected them markedly. Karstic features are present but not pronounced.

Erosion surfaces are found at average elevations of 780, 620, 470, 330, 270, and 180 m. Most rivers are terraced, but no studies have been made of them. Submarine terraces or scarps have been noted at various depths around the island, those at about 18 and 6.5 m being the more prominent.

Around the coast, Quaternary raised beaches, raised reefs, and marine terraces are found but are not widely developed. Fossil sand dunes (eolianites) are found in the central part of the northern coastal areas.

Stratigraphic History

The oldest rocks dated by fossils are radiolarian cherts of late Jurassic-early Cretaceous age associated with the serpentinite complex in the southwest. Amphibolite in the complex has yielded a radiogenic age of 110m yr (low Cretaceous), but this may only represent the date of the latest metamorphism. It may possibly be as old as Paleozoic. The relationship of the serpentinite to surrounding rocks is obscure, but it has been suggested that it forms a northward-directed nappe structure. It is thought by some to be part of the old ocean floor, tectonically emplaced near a former plate boundary and in parts remobilized. Detrital serpentinite breccias apparently overlying the massive serpentinites suggest submarine mudflow deposits eroded from a serpentinite extrusion on the sea floor.

The rocks of Upper Cretaceous to Lower Eocene age, which form the main mass of the island, are mainly volcanic in origin, consisting of lava flows, pyroclastics, and volcanically derived sediments that show considerable lateral and vertical variation. Lenticular masses of limestones are interbedded. Extensive faulting renders correlation difficult. In particular the horizontal displacements, measurable in tens of kilometers, across the two major zones of strike-slip faulting divide the island into blocks, so that correlations between them are only generalized. A large number of formations have been named whose complex relationships are typified by the diagram (Fig. 2) illustrating the stratigraphy of an area in the center of the island.

The Cretaceous-Eocene rocks are divided at disconformities and unconformities into five sequences, the major unconformities occurring in the upper part of the Cretaceous.

The volcanic rocks are chiefly andesitic lavas, breccias, and tuffs, though compositions vary from basalt to dacite. Lavas in the middle sequence show alkaline affinities whereas the younger lavas are calc-alkaline. The older flows are commonly pillow lavas and it is evident that the early volcanic activity was submarine. Subaerial eruptions occur increasingly in the younger rocks. The total thickness of these rocks is probably of the order of 6000–8000 m.

Fossils are quite abundant in places. Most valuable for correlation purposes are the foraminifera; but ammonites, rudistids, and other mollusca, corals, and algae are found locally.

Plutonic intrusions were emplaced in stages during the latter part of the Cretaceous and the Eocene. They are predominantly tonalitic and granodioritic in composition.

The rocks of the "Younger" series overlie the older rocks with a strong angular unconformity. They are relatively unaffected by the deformation that is so prominent in the older rocks and which therefore must have occurred toward the end of the Cretaceous and into the Eocene. The younger rocks consist of detrital conglomerates and clays at the base followed by predominantly limestones with a thickness in all of about 1500–1800 m. They range in age from Oligocene to Miocene (or possibly Pliocene). They are quite fossiliferous, the commonest fossils being corals, molluscs, echino-

FIGURE 2. Stratigraphic relations in a part of central Puerto Rico; typical of the Upper Cretaceous stratigraphy. 1, Robler Fm.; 2, Malo Breccia; 3, Vista Alegre Fm.; 4, Tetuán Fm.; 5, Cotorra Tuff; 6, Cariblanco Fm.; 7, conglomerate; 8, lavas. (Adapted from Briggs, 1966.)

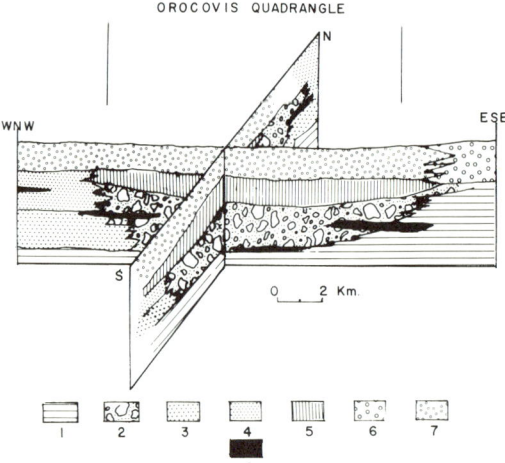

derms, though again foraminifera are the most important for correlation purposes. Some vertebrate bones (sea cow) have been found.

Correlation between the northern and southern outcrops has not been established and it does not seem that there was any direct communication between the two depositional areas. The higher erosion surfaces, mentioned earlier, were probably developed during the late Miocene. Quaternary deposits—beach sands, sand dunes, raised reef deposits, and marine terraces are found at various elevations around the coastal areas, particularly in the north and west.

The island of Mona consists of limestone of Miocene age capped with a lithologically similar limestone of Pleistocene age.

The parallelism between the E-W trends in the Oligo-Miocene rocks and the line of the Puerto Rico Trench 140 km to the N has suggested that the trench had come into existence by that time. Deep-sea sediments suggest it began to form in the late Eocene as a left-lateral strike-slip developing a long, narrow graben. Tectonic activity still persists there as evidenced by the numerous earthquakes.

Economic Geology

A small amount of mining for copper, gold, manganese, zinc, lead, and platinum was carried out during Spanish colonial times. A magnetite mine was operated for 2 years in the early 1950s and then abandoned.

In the late 1950s prospecting with the employment of geochemical and geophysical methods disclosed two low-grade porphyry copper deposits associated with the pluton in the center of the island and development is planned.

The major mineral exploitation is for cement making, utilizing the mid-Tertiary limestones. The volcanic rocks and older limestones are used for constructional purposes. Some of the older limestones are used for ornamental stone. Pleistocene quartz sands are employed for glass making and some clays are used for ceramics. Three test wells have been drilled in the mid-Tertiary limestone areas for petroleum without favorable indications.

JOHN D. WEAVER

References

Bowin, C., 1972. "Puerto Rico Trench negative gravity anomaly belt," *Geol. Soc. Am. Mem. 132*, 339–350.

Briggs, R. P., 1964. "Provisional geological map of Puerto Rico and adjacent islands," *U.S. Geol. Surv., Misc. Geol. Inv. Map 1*, 392p.

____, 1966. "The Malo Breccia and Cotorra Treff in the Cretaceous of central Puerto Rico," *Bull. U.S. Geol. Surv.*, 1254-A, 23–29.

Galloway, J. J., and Heminway, C. E., 1941. "The Tertiary foraminifera of Porto Rico," *N.Y. Acad. Sci., Sci. Surv. of Porto Rico*, 3(4), 275–491.

Glover, L., III, 1971. "Geology of the Coamo area, Puerto Rico, and its relation to the volcanic arc-trench association," *U.S. Geol. Surv. Prof. Pap. 636*, 102p.

Hooker, M., 1969. "Bibliography and index of the geology of Puerto Rico and vicinity 1866–1968," San Juan: Geol. Soc. Puerto Rico, 53p.

Kaye, C. A., 1957. "Notes on the structural geology of Puerto Rico," *Bull. Geol. Soc. Am.*, 68(1), 103–118.

____, 1959a. "Coastal geology of Puerto Rico, geology of Mona and notes on the age of Mona Passage," *U.S. Geol. Surv. Prof. Pap. 317-C*.

____, 1959b. "Shoreline features and Quaternary shoreline changes, Puerto Rico," *U.S. Geol. Surv. Prof. Pap. 317-B*, 49–139.

____, 1960. "Geology of the San Juan metropolitan area, Puerto Rico," *U.S. Geol. Surv. Prof. Pap. 317-A*, 1–48.

Mattson, P. H., 1966. "Geological characteristics of Puerto Rico," *Continental Margins and Island Arcs, Geol. Surv. Can. Pap. 66-15*, 124–138.

____, 1973. "Middle Cretaceous nappe structures in Puerto Rican ophiolites and their relation to the tectonic history of the Greater Antilles," *Bull. Geol. Soc. Am.*, 84, 21–38.

____, and Pessagno, E. A., 1974. "Tectonic significance of Late Jurassic-Early Cretaceous Radiolarian chert from Puerto Rico ophiolite," *Geol. Soc. Am.*, abs. with Programme, 1974 Ann. Mtgs., 859.

Mitchell, R. C., 1954. "A survey of the geology of Puerto Rico," *Univ. Puerto Rico Agric. Exp. Station, Tech., Pap. 13* (Rio Piedras), 187p.

Molnar, P. H., and Sykes, L. R., 1969. "Tectonics of the Caribbean and Middle American regions from focal mechanisms and seismicity," *Bull. Geol. Soc. Am.*, 80, 1639–1684.

Moussa, M. T., and Seiglie, G. A., 1970. "Revision of mid-Tertiary stratigraphy of southwestern Puerto Rico," *Bull. Am. Assoc. Petrol. Geologists*, 54(10), 1887–1998.

Seiglie, G. A., 1973. "Revision of mid-Teritary stratigraphy of southwestern Puerto Rico," *Am. Assoc. Petrol. Geologists*, 57, 405–406.

Weaver, J. D., 1961a. "Erosion surfaces in the Caribbean and their significance," *Nature*, 190(4782), 1186–1187.

____, 1961b. "Institute of Caribbean Studies field excursion to Isla Mona," *Caribbean J. Sci.*, 1(1), 30–32.

____, 1962. "The nature of the 'Nipe Clay' on Las Mesas, western Puerto Rico," *Z. Geomorphol.*, 6(2), 213–232.

Weyl, R., 1966. "Geologie der Antillen," *Region. Geol. der Erde*, vol. 4. Berlin: Gebr. Borntraeger, 410p.

Cross-references: *Dominican Republic; Haiti; Leeward Islands; Virgin Islands; West Indies.*

R

REVILLAGIGEDO ISLANDS

The Islas Revillagidego, belonging to Mexico, are volcanic, four in number, and situated 400 km (260 miles) S of the tip of Baja California, about 19°N and 111–115°W. One volcano is active. They are situated along an E-W line, corresponding to the deep-sea Clarion fracture zone, west of the mid-ocean ridge. Sea-floor spreading information suggests a Miocene date for the oceanic crust here.

Isla Clarion lies in the west, I. Roca Partida in the middle, and I. San Benedicto and I. Socorro (see Fig. 1) in a pair (N-S) in the east. They have been visited often but rarely described. There are geological notes by Mooser and Maldonado-Koerdall (1961), A. F. Richards (1958), and Emery (1948).

RHODES W. FAIRBRIDGE

References

Emery, K. O., 1948. "Submarine geology of Ranger Bank, Mexico," *Bull. Geol. Soc. Am.,* **59**, 790–805.

Mooser, F., and Maldonado-Koerdall, M., 1961. "Pene-contemporaneous tectonics along the median Pacific Ocean coast," *Geogisica Intern.,* **1**(3).

Richards, A. F., 1958. "Transpacific distribution of floating pumice from Isla San Benedicto, Mexico," *Deep-Sea Res.,* **5**, 29–35.

———, 1960. "Rates of marine erosion of tephra and lava at Isla San Benedicto, Mexico," *Proc. 21st Intern. Geol. Congr.,* Copenhagen, pt. 10, 59–64.

Cross-references: *Easter Island and Sala y Gómez; Galápagos Islands.*

FIGURE 1. Oblique air photograph of Socorro Island (elev. 1130 m), with recent lava stream from parasitic cone, on the right. (Photo: U.S. Navy, Aug. 6, 1954.)

S

ST. CHRISTOPHER-NEVIS-ANGUILLA

The islands of St. Christopher (St. Kitts), Nevis, and Anguilla comprise a "British Associated State" in the Leeward Islands (q.v.) of the Lesser Antilles Island Arc. St. Kitts and Nevis are the southernmost of three islands that form the emergent crest of Statia Bank which lies along the axis of the young phase of volcanic activity in the Lesser Antilles arc. Anguilla is the northernmost of the islands of the Anguilla Bank which was formed by coalescing, coeval volcanic centers which formed along the axis of the pre-Miocene phase of volcanism.

St. Kitts

Morphologically, St. Kitts is elongated in a NW-SE direction and has an area of 176 km^2 (68 sq mi). It is 36 km long with a low neck in the SE. Almost entirely volcanic, four geological units (see Fig. 1) were distinguished by Martin-Kaye (1959) and are listed below in order of decreasing age:

1. Salt Pond Peninsula and Canada Hills
2. The South East Range
3. The Middle Range
4. Mt. Misery

The stratigraphic order is deduced in part on the superposition and in part on the degree of dissection, due to the paucity of paleontological control. These units correspond to the main volcanic centers, all of Plio-Pleistocene age. The ages of the older units, the Salt Pond Peninsula Unit and the South East Range, indicate that the volcanics are all younger than 7×10^6 yr B.P. K-Ar age determinations on lavas from these units yielded the oldest age of about 7 ×

FIGURE 1. Geologic map of St. Kitts (after Martin-Kaye, 1959; Baker, 1968a).

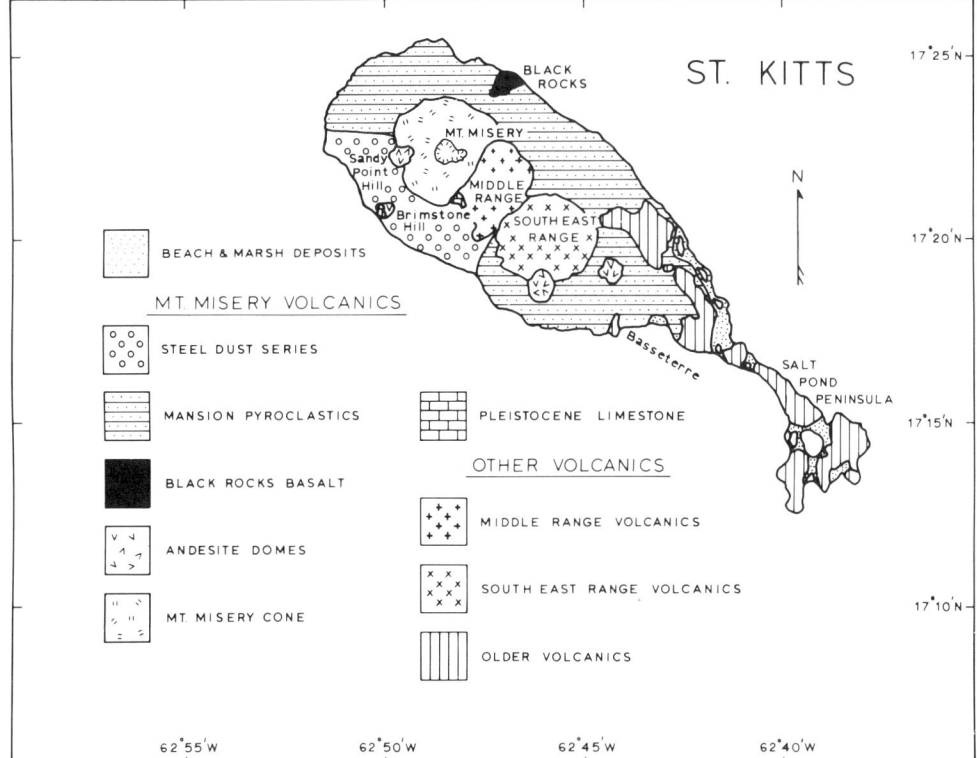

10^6 yr B.P. for a lava flow near the base on the south side of the main peak of the South East Range and $2-2.5 \times 10^6$ yr B.P. for the Salt Pond Peninsula volcanic rocks.

Associated with the main volcanic centers are several lava domes such as Brimstone and Sandy Point Hills, Mt. Misery Peak, Ottley's Mt., and Monkey Hill (Fig. 1).

Salt Pond Peninsula and Canada Hills Volcanics. The rocks of this southeastern part of St. Kitts are the last representatives of the oldest volcanic centers and are considered to be the basement of St. Kitts by Martin-Kaye (1959). The volcanics consist mainly of coarse-block agglomerates, minor amounts of tuff, and some intrusive andesites. Two main volcanic centers for this unit are suggested, one in the vicinity of Basse-Terre and the other marked by the present Great Salt Pond.

Petrographic descriptions for various rocks in this unit are as follows:

Blocks in agglomerates and intrusive andesites. Pyroxene andesites usually carrying hypersthene and diopsidic augite and occasionally a little basaltic hornblende. The ferromagnesian component is variable and usually minor. 2V determinations on the hypersthenes indicates them as fairly iron rich. The plagioclases range between labradorite and andesine.

Needsmust Quarry andesite. Hypersthene phenocrysts up to 5 mm are set in a weakly flow-oriented pilotaxitic groundmass of feldspar and granules of pyroxene and opaque minerals. The plagioclases, normally zoned labradorite, commonly carry oriented inclusions.

Sir Timothy's Hill andesite. Ragged and often angular phenocrysts of andesine-labradorite, diopsidic augite and hypersthene.

The South East Range. The South East Range volcanics consist of andesite lavas and pyroclastics, apparently from more than one phase of volcanism and several ill-defined centers of activity. Of interest are the two lava domes, Monkey Hill and Ottley's Mountain (Fig. 1). One petrographic analysis is reported for an andesite from Ottley's Mt. as follows; 2-pyroxene andesite: plagioclase, hypersthene, clinopyroxene, amphibole, and quartz phenocrysts in a fine-grained groundmass of plagioclase, pyroxene, cristobalite, and a little glass.

The Middle Range. Although well developed as a definite volcanic center and similar in composition to the South East Range, the Middle Range is smaller than the adjacent centers. Outcrops are limited but are mainly andesitic flows and agglomerates. At 330 m elevation on the southwestern side of the Middle Range, in the vicinity of Godwin Gut, is an exposure of limestone, apparently Plio-Pleistocene in age based on a well-preserved specimen of *Montastrea annularis*.

Mt. Misery. Mt. Misery (1156 m, 3711 ft) is the youngest volcanic center on St. Kitts and is typical of strato-volcanoes which characterize volcanism in the Lesser Antilles. Of particular interest on St. Kitts are the secondary volcanic centers associated with Mt. Misery. Brimstone Hill, Farm Flat, and Sandy Point Hill are andesite domes located on the flanks of the main cone. Sandy Point Hill and Farm Flat are composed of two-pyroxene (usually hypersthene and augite) andesites. Brimstone Hill, in contrast, is made up of hypersthene andesite and is the only occurrence of this on Mt. Misery. During the protrusion of Brimstone Hill, Pleistocene limestone beds were carried up on the flanks. A radiocarbon dated coral from the limestone indicated an age of 44,000 ± 1200 yr B.P. (N.B.: a "minimum" age, for such samples. *Ed.*)

Coarse grained ejectamenta are known from Mt. Misery and were described as hornblende-eucrites, anorthite-bearing blocks (similar to those from St. Vincent), and gabbroic blocks (including olivine gabbros, quartz gabbros, and norites).

Nevis

Situated only 4 km SE of St. Kitts, Nevis is a circular conical island rising to 985 m, and covering 130 km^2 (50 sq mi). In spite of its small size, seven volcanic centers have been recognized on Nevis by Hutton (1965). There were three phases of volcanism of different ages associated with these centers. On the basis of apparent age, the volcanogenic rocks on Nevis can be divided into three main units, which are listed below in order of decreasing age. This apparent age classification is based on degree of dissection and superposition of units. Since there is no modern geological map of Nevis, only the eruptive centers are depicted in Fig. 2.

Basal Volcanic Unit. This unit is represented by identified relict volcanic centers at Windy Hill, Mt. Lily (Hurricane Hill and Round Hill, also), Saddle Hill, Red Cliff (Round Hill in south), and Cades Point where any pre-existing volcaniclastic units have been stripped away by erosion to expose residual central volcanic domes, often massive and well-jointed.

Butler's Mountain Dome. The dissected remnant central dome of Butler's Mountain overlies the older volcanic basement and projects through the youngest ejectamenta from Nevis Peak. Butler's Mountain is considered intermediate in age because of its state of dissection and overlying contact with the Basal Unit.

Nevis Peak Volcano. Nevis Peak is the youngest volcanic center on Nevis and is a typical strato-volcano which is comprised mainly of

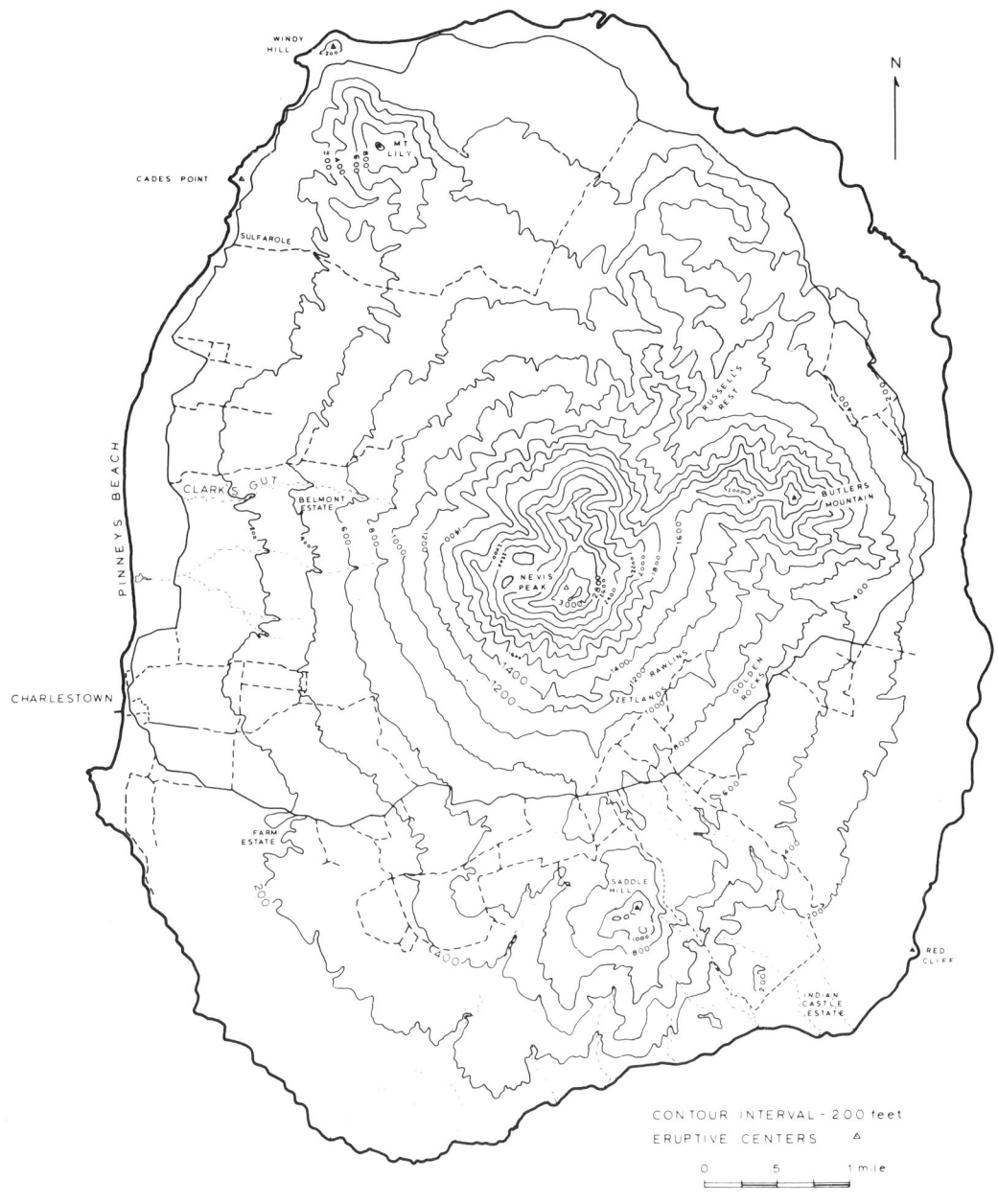

FIGURE 2. Map of Nevis, showing eruptive centers.

pyroclastics (tuffs and agglomerates) which are widely distributed over the island. The structure of Nevis Peak is somewhat complicated by the fact that it developed on the remains of the older volcano whose center was Butler's Mountain. These older units project through the flank on the southern and northeast sides.

Of interest on Nevis is the extensive region in the Belmont Estate area (Fig. 2), where recent fumarolic activity has produced alunite-opalcristobalite assemblages. The following minerals have been identified in the Belmont area: alunite, natroalunite, halloysite, carphosiderite, jarosite, natrojarosite, halite, coquimbite, tamarugite, and occasionally kaolinite, alunogen, gypsum, and chalcanthite. Fumarolic activity is still present in the vicinity of Farm's Estate and near Cades Bay (Fig. 2).

In spite of the general similarity of the rock types associated with the various volcanic centers, several distinctions can be made. C. O. Hutton made a detailed petrological study of the rocks on Nevis. Generally, the chemical analyses indicate a trend following closely that of Daly's basalt-andesite-dacite-rhyolite series.

The petrographic features of the main centers were described as follows (see Fig. 2):

Basal Volcanic Unit. Porphyritic hypersthene augite dacites with much tridymite, rare biotite and varying amounts of glass; hyalopilitic two-pyroxene dacites with common tridymite; and glassy dacites with microspherulitic textures common; plentiful and often coarsely crystalline hornblende and rare relict olivine.

Intermediate Age Unit. Holocrystalline feldspar-rich dacites with opacite and tridymite.

Nevis Peak Volcano. Strongly porphyritic hornblende, slightly glassy dacites with much tridymite, quartz "xenocrysts" present here; and vitrophyric hornblende, hypersthene dacites, some devitrification.

Major element analyses of Nevis volcanic rocks yield three main results.

1. FeO/MgO ratios are relatively low when compared to tholeiitic magma fractionation products.
2. Al_2O_3 content is high.
3. K_2O content is relatively low.

All three results have been shown by Tomblin (1968) to be typical for rocks in the Lesser Antilles.

Anguilla

Anguilla has an area of 70 km^2 (35 sq mi) and is physiographically quite distinct from St. Kitts and Nevis. Although politically placed with the islands of the Miocene-Recent volcanic arc, its geology and position 100 km northeast of St. Kitts and Nevis, place it with the outer pre-Miocene arc. It is a long, narrow and flat-lying island comprised of a Lower Miocene transgressive sequence overlying truncated volcaniclastics. Due to the proximity of St. Martin and St. Bartholomew, the volcanics are believed to be the age equivalent of the Pointe Blanche formation, which is Eocene-Oligocene.

Sombrero (5 km^2), the site of a lighthouse station; only 12 m in elevation, it is situated on an isolated point of uplift of the Anguilla Bank about 20 km NW of Anguilla and consists of Lower Miocene limestones. It was once (erroneously) reported to be an uplifted atoll. The island has a wide wave-cut platform all around, giving it the appearance from the sea of a Mexican hat, or hence its name.

L. KENNETH FINK, JR.

References

Baker, P. E. 1968a. "Guide for the excursion to St. Kitts," *Guidebook for 5th Caribbean Geol. Conf.,* St. Thomas, V.I., 34p.

———, 1968b. "Petrology of Mt. Misery Volcano, St. Kitts, West Indies," *Lithos,* **1,** 124–150.

———, 1970. "The geology of Mt. Misery Volcano, St. Kitts," *Trans. 4th Caribbean Geol. Conf.,* Trinidad, W.I., 361–365.

Christman, R. A., 1953. "Geology of St. Bartholomew, St. Martin, and Anguilla, Lesser Antilles," *Bull. Geol. Soc. Am.,* **64**(1), 65–96.

Cope, E. D., 1883. "On the contents of a bone cave in the island of Anguilla," *Smithsonian Contrib.,* **25,** 30p.

Drooger, C. W., 1951 "Fóraminifera from the Tertiary of Anguilla, St. Martin and Tintamarre (Leeward Islands, W.I.)," *Proc. Kon., Ned. Akad. Wetensch.* (Amsterdam), Ser. B, **54,** 54–65.

Eames, F. E., Banner, F. T., Blow, W. H., and Clark, W. J., 1962. *Fundamentals of Mid-Tertiary Stratigraphical Correlation.* New York: Cambridge Univ. Press, 163 p.

Earle, K. W., 1925. *Report on the Geology of St. Kitts-Nevis (B.W.I.) and the Geology of Anguilla (B.W.I.)* London: Crown Agts. Colon., 50p.

Fink, L. K., Jr., 1970a. "Field guide to Nevis, Lesser Antilles," in T. W. Donnelly, ed., *International Field Institute Guidebook to the Caribbean Island Arc System.* Washington, D.C.: Am. Geol. Inst. 11p.

———, 1970b. "Field guide to St. Kitts, Lesser Antilles," in T. W. Donnelly, ed., *International Field Institute Guidebook to the Caribbean Island Arc System.* Washington, D.C.: Am. Geol. Inst., 13p.

Hutton, C. O., 1965. "The mineralogy and petrology of Nevis, Leeward Islands, British West Indies," *4th Caribbean Geol. Conf.,* Trinidad, W.I., 383–388.

Julien, A. A., 1867. "On the geology of the key of Sombrero, W.I.," *Ann. Lyceum Nat. Hist. N.Y.,* **8,** 251–278.

Lewis, J. F., 1964. "Mineralogical and petrological studies of plutonic blocks from the Soufriere Volcano, St. Vincent, B.W.I," Ph.D. thesis, Oxford Univ., 270p.

Martin-Kaye, P. H. A., 1959. *Reports of the geology of the Leeward and British Virgin Islands.* Castries, St. Lucia: Voice Publ. Co., 129p.

———, 1969. "A summary of the geology of the Lesser Antilles," *Overseas Geol. Mineral Res.,* **10**(2), 172–206.

Rittman, A., 1962. *Volcanoes and Their Activity.* New York: Wiley, 305p.

Sapper, K., 1903. "Ein Besuch der Inseln Nevis und St. Kitts," *Centralbl. Min.,* 284–287.

Spencer, J. W., 1901. "On the geological and physical development of St. Christopher chain and Saba banks," *Quart. J. Geol. Soc. Lond.,* **57,** 534–544.

Tomblin, J. F., 1968. "Geochemistry and genesis of Lesser Antillean volcanic rocks," *5th Caribbean Geol. Conf.,* St. Thomas, V.I., 153.

Trechmann, C. T., 1932. "Notes on Brimstone hill, St. Kitts," *Geol. Mag.,* **69**(6), 241–264.

Westermann, J. H., and Kiel, H., 1961. "The geology of Saba and St. Eustatius, with notes on the geology of St. Kitts, Nevis, and Montserrat," *Surinam Natur. Stud. Publ. 24,* 175p.

Weyl, R., 1966. "Geologie der Antillen," *Regionalen Geologie der Erde,* vol. 4. Berlin; Gebr. Borntraeger, 410p.

Cross-references: *Antigua; Leeward Islands; Virgin Islands; West Indies.*

ST. LUCIA

The largest of the British Windward Islands, in the Lesser Antilles volcanic island arc, St. Lucia is situated S of Martinique and N of St. Vincent, at $13°54'$N and $60°59'$W. The island is a former colony, now a British Associated State. It covers 603 km^2 (238 sq mi), and measures 43 X 22 km. It is a deeply dissected, very mountainous island, and one of the most beautiful in the West Indies. There is a deep bay and good harbor at Castries. The highest point is Morne Ginne, 951 m (3145 ft), and there are two former volcanic necks, the Pitons (Fig. 1), which rise to over 800 m. There is a boiling sulfur spring at Ventine in a former crater near the Petit Piton. The southern part of the island consists only of hills interspersed with broad alluvial fans. There are many coral reefs here. Davis (1926) illustrates a delta plain filling a drowned valley embayment on the west coast, clear evidence of recent tectonic stability and the deep Pleistocene dissection.

The volcanic rocks appear to be late Tertiary (Miocene) and include large amounts of andesite and dacite with minor basalt flows. The andesites are unusually rich in potash.

RHODES W. FAIRBRIDGE

References

Davis, W. M., 1926. "The Lesser Antilles," *Am. Geogr. Soc. Spec. Publ. 2,* 207p.

Martin-Kaye, P. H. A., 1969. "A summary of the geology of the Lesser Antilles," *Overseas Geol. Mineral Res.,* 10(2), 172–206.

Rea, W. J., 1970. "Andesites of the Lesser Antilles," *Proc. Geol. Soc. Lond.,* (1662), 39–46.

Sapper, K., 1903. "Zur Kenntnis der Insel St. Lucia in Westindien," *Centralbl. Min.,* 273–278.

Spencer, J. W., 1902. "On the geological development of Dominica, with notes on Martinique, St. Lucia, St. Vincent and the Grenadines," *Quart. J. Geol. Soc. Lond.,* 58, 341–353.

Weyl, R., 1966. "Geologie der Antillen," Berlin: Gebr. Borntraeger, *Region. Geol. der Erde,* vol. 4. 410p.

Cross references: *Grenada; Martinique; St. Vincent; West Indies; Windward Islands.*

SAINT-PIERRE AND MIQUELON

The sole-remaining (and oldest) French possession in North America, the group of small islands known as Saint-Pierre and Miquelon, is situated about 25 km off the coast of Newfoundland (W of Burin Peninsula). The islands cover only 241 km^2 (93 sq mi), including inland waters. The group is intersected by the coordinates $47°00'$N and $56°28'$W.

Apart from rocky islets, there are really four islands, Le Cap–Miquelon–Langlade, which are joined by sand and gravel bars, and Saint-Pierre, which is isolated (Fig. 1a). There is an extraordinary variety of geologic, geomorphic, and mineral attractions, so much so that Aubert de la Rüe (1932, 1944, 1951), who made the most detailed studies here, described the group as "un véritable musée géologique naturel" (1948).

FIGURE 1. The Pitons, St. Lucia, a pair of dacite plugs, stripped of their pyroclastic surroundings by erosion. (Photo: R. Weyl.)

Geology

The oldest rocks, assigned by Aubert de la Rüe (1951) to the Upper Precambrian, occur in the northernmost of the islands, Le Cap, and also on the adjacent part of Miquelon. These are highly metamorphosed sediments, basic intrusive rocks and lavas: mica-schists, spotted (pinite) schists, quartzo-phyllites (very fine-grained amphibolitic quartzites with conspicuous bands of magnetite), paragneisses containing cordierite and more rarely tourmaline, and orthoamphibolites. The whole sequence has been tectonized, and intruded by granite, diorite and numerous acid dikes (microgranite, aplite, pegmatite) representing consolidated residual magma associated with the granite intrusion, as well as by basic dikes (basalt and dolerite). Some of the latter were apparently intruded contemporaneously with the acid dikes whereas others belong to a younger intrusive episode recognized in the other islands of the group.

Emplacement of the granite led to migmatization of the metamorphic rocks of the eastern part of Le Cap (lit-par-lit injection). At the western end, however, where the rocks appear less severely metamorphosed, a clear metamorphic aureole is evident around the granite intrusion of Cap Blanc. The Cap Blanc granite (biotite monzonite), although exposed over only a relatively small area at the western end of Le Cap, belongs to a much larger body at depth as shown not only by the migmatization of the eastern part of Le Cap, where the aplite and pegmatite dikes are especially abundant, but also by the reappearance of this granite in the islets called the Veaux Marins about 12 km southwest of Cap Blanc. Numerous quartz veins, many probably representing the extreme ends of pegmatite dikes, are associated with the rocks of Le Cap. Epidote is abundant in some of these veins; red feldspar, stilbite, prehnite and copper minerals have also been noted.

The age of the granite has not been established. It postdates the metamorphism of the Precambrian sediments and associated volcanic rocks of Le Cap and pre-dates the rhyolites of the other islands which are Upper Cambrian or younger (for evidence see Saint-Pierre below). The Cap Blanc granite might, therefore, be Paleozoic but it could equally well be Hadrynian as granites of this age are known in southeastern Newfoundland.

Miquelon, apart from the narrow belt of metamorphic rocks like those of Le Cap bordering the northwest coast of the island, consists exclusively of volcanic rocks, a thick sequence of rhyolitic and less widely distributed, possibly earlier, andesitic flows and breccias. They are much altered and commonly epidotized. Basalt flows present between the metamorphic rocks to the north and the rhyolites to the south may be related to the former as they are, in part, similarly metamorphosed. They are clearly unrelated to the numerous younger basalt (dolerite) dikes that cut these flows as well as the neighbouring rhyolites and andesites. The basalt flows are also intruded by the rhyolite dikes and it seems likely, therefore, that they belong to an older volcanic episode. Quartz veins, rare in the basalts, are ubiquitous in the rocks of the rest of the island.

The third island in the connected group is Langlade. In contrast to the others, this consists almost entirely of unmetamorphosed sedimentary formations, folded and steeply dipping, arranged in parallel belts of unequal width striking NE-SW across the island. Similarly aligned faults have resulted in the same succession reappearing several times across the width of the island (Fig. 1b).

Four formations were distinguished by Aubert de la Rüe (1951) and assigned to the Lower Paleozoic; he considered three of them conformable but was less certain of the stratigraphic relationship of the fourth, a series of fine to coarse predominantly red clastics (conglomerates, arkoses, sandstones sometimes micaceous and cross-bedded, shales) whose contacts with the other formations are nearly everywhere obscured by Pleistocene glacial deposits or faulting. These red sediments are restricted to the eastern side and the north coast of the island (Fig. 1b). The conformable formations are (1) a sequence of fissile black shales with occasional thin, irregular and discontinuous interbeds of limestone, locally nodular, that also contains large concretions showing cone-in-cone structure, and has yielded trilobite fragments (only *Paradoxides davidis* has, so far, been identified); (2) a series of less fissile green shales (some red or gray) with intercalations of micaceous sandstone, quartzite and lenticular beds of limestone containing unidentifiable fossil fragments; (3) gray to white quartzites in beds up to 40 cm thick frequently showing cross-bedding, interbedded with greenish micaceous sandstones and with red and green shales similar to those of (2). The black shale sequence containing *Paradoxides davidis* has the characteristics of the medial to late Middle Cambrian Manuels River Formation in southeastern Newfoundland (Hutchinson, 1962).

The order of deposition of the sedimentary formations according to Aubert de la Rüe (1951) was (1) black Middle Cambrian shales, (2) green shales, (3) quartzites, and (4) red clastics [possibly conformable with (3) in a river section in northeastern Langlade, Fig. 1b]. The beds overlying the Cambrian shales would then be Upper Cambrian to possibly Ordovician age. However, quartzites very similar to (3) typically underlie Cambrian strata in southeastern and southern Newfoundland (Random Formation; Blue Pinion Formation, Widmer, 1953), Cape Breton Island, Nova Scotia (Quartzite member at the top of the Morrison River Formation, Weeks, 1954) and the St. John region of New Brunswick (Glen Falls Formation, Alcock, 1938). Furthermore, in several areas they overlie red-bed sequences, probably of continental origin, resembling the red clastic formation of Langlade. These quartzite occurrences are at different stratigraphic levels—Late Precambrian or basal Cambrian in Trinity Bay, southeastern

FIGURE 1. a, The main islands of the Saint-Pierre and Miquelon group; b, geological map of Langlade after Aubert de la Rüe (1951); the stratigraphic succession is a reinterpretation.

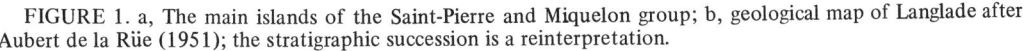

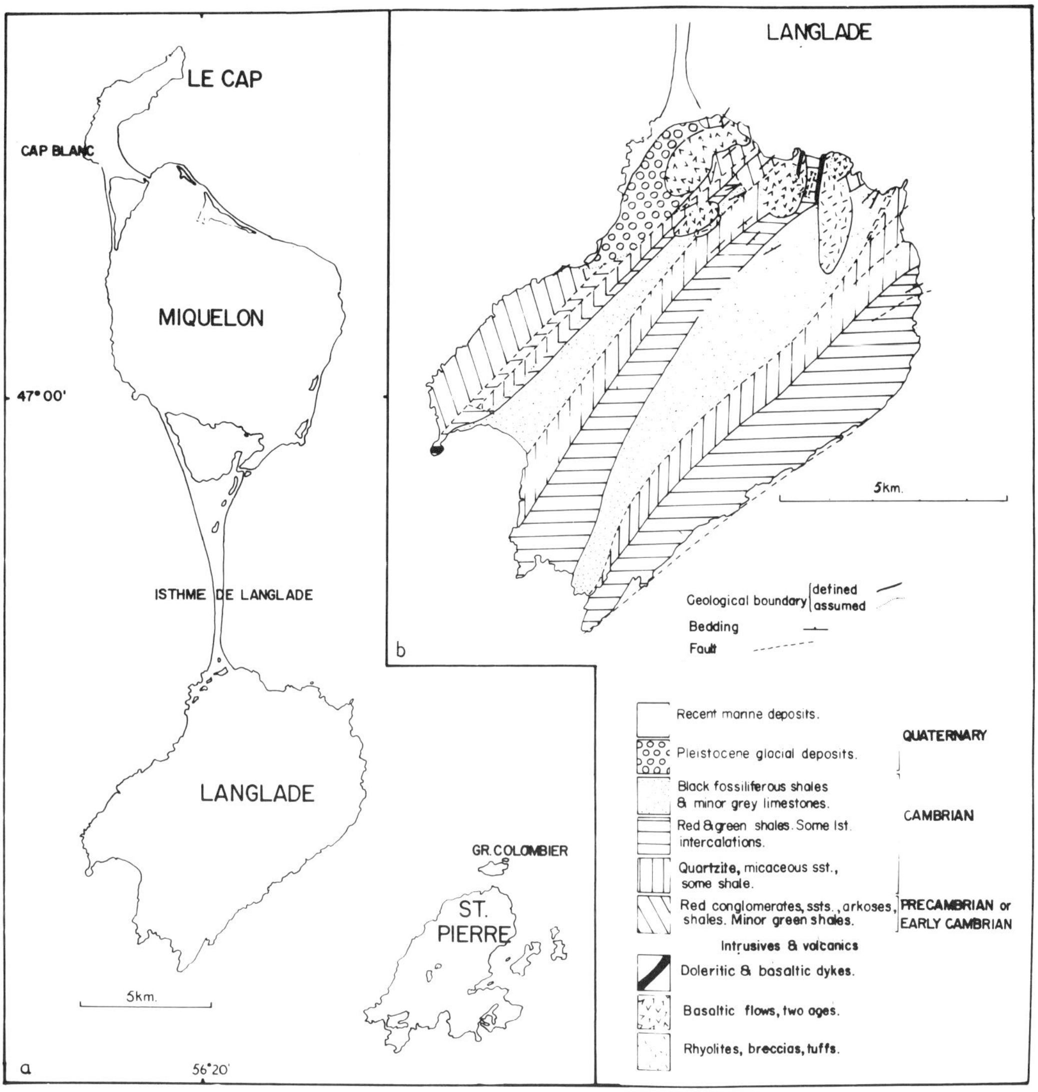

Newfoundland, Lower Cambrian in Nova Scotia and New Brunswick, Middle Cambrian in Hermitage Bay (Blue Pinion Formation), southern Newfoundland—which indicates that the quartzite formation is diachronous and probably represents reworked coastal sediments along the shoreline of an advancing sea during the course of its transgression. Apart from interbedded sandstones, the green shale sequence is remarkably like the Chamberlains Brook Formation conformably underlying the Manuels River Formation in southeastern Newfoundland. The presence of the sandstones can be explained as resulting from deposition of the green shale sequence not far from the shoreline, which would be in keeping with the transgression of the sea into the area only at the beginning of the Middle Cambrian. In view of the similarities referred to between the Langlade sedimentary formations and Cambrian successions elsewhere in the region, it is considered that the order of deposition was: (1) red clastics (Hadrynian or Lower Cambrian), (2) quartzites (late Lower Cambrian or earliest Middle Cambrian), (3) green shales (early Middle Cambrian), and (4) black shales (medial to late Middle Cambrian).

Volcanic rocks occupy only a tenth of Langlade. These are basalt flows, younger rhyolites and more recent basalt (dolerite) dikes. There are four principal masses of basalt in northern Langlade. Two of them are associated with the red clastic formation, and their extrusion was contemporaneous with the deposition of the sediments; the other two masses appear to be younger than the Cambrian formations. Three zones of rhyolite adjoin the basalt flows. Numerous basalt and dolerite dikes intrude the older volcanic rocks as well as the sedimentary formations.

Saint-Pierre and nearby islets are entirely volcanic and although formed dominantly of a thick accumulation of calc-alkaline (rhyolitic) flows, breccias and tuffs, andesites are also present in the extreme northern and southern parts of the island. These gray to green andesites are fine grained or porphyritic and locally slightly vesicular.

The variously colored rhyolites, commonly pale red to purple, form thick flows, glassy to porphyritic in texture; some of the finer-grained varieties show flow structure and some contain spherulites up to 6 cm across. The majority of the rhyolite breccias are flows but some are explosion breccias. The latter, like the tuffs, are usually well-bedded. The angular fragments within the breccias are mostly rhyolitic, similar in composition to the matrix, less commonly andesitic and more rarely of shale and quartzite torn from the underlying sediments, apparently like those of Langlade, that are nowhere exposed on Saint-Pierre or neighbouring islets; two xenoliths of granite similar to that of Le Cap have also been found. The tuffs have been strongly silicified and, in places, even altered to jasper.

Epidotization is widespread in the rhyolites and responsible for the green color of some of the flows. Pale red rhyolites owe their color to a pink manganiferous epidote, close to piedmontite, either disseminated or in small, visible carmine crystals. The manganiferous epidote has resulted from alteration of an amphibole now rarely found in the fresh state.

The rhyolite flows, breccias, and tuffs of Saint-Pierre, Langlade, and Miquelon are similar to one another and belong to the same post Middle Cambrian volcanic episode.

Basalt dikes (basaltic to doleritic) intruding the rhyolites and andesites are common and widespread and rarely exceed a metre or two in width; two of them, however, striking NE-SW in the western half of the island, are noteworthy in being up to 40 m across and traceable for much of the length of the island. Some circular basalt outcrops in the southwestern corner of the island are thought to be volcanic necks. These basic intrusives are the most recent rocks exposed on the islands apart from unconsolidated Pleistocene to Recent deposits.

Structurally the islands are dominated by NE-SW trends corresponding to those of adjacent Newfoundland, the rocks being heavily folded and faulted along these lines. These early Appalachian tectonics are discussed in more detail in other entries (e.g., *Canada–Atlantic Province*). Jointing is conspicuous everywhere and some zones are so highly fractured that they appear crushed; in places there is evidence of intense tangential stress.

Geomorphology

The islands preserve traces of long (presumably Mesozoic/Cenozoic) peneplanation like the rest of the Appalachian belt. The peneplain remnants are found at about 130 m above MSL. The most extensive of them forms a plateau extending over most of Langlade; it is marginally dissected and ends abruptly at steep coastal cliffs. Smaller areas at about the same level are present on Miquelon and in southeastern Saint-Pierre but the remainder of these islands and Le Cap show a more varied and rugged relief, the highest hills and peaks reaching 200 m on Miquelon and 250 m on Saint-Pierre. The surface of the islands is largely covered by moorland, peatbogs and swamps with numerous small ponds, and many areas are covered by dense thickets of wind-deformed bushes and dwarf trees.

The archipelago was completely ice-covered during one or more of the Pleistocene glaciations. At those times the sea was at a lower level, the islands were most likely joined to Newfoundland and the ice sheet moved across the islands from Newfoundland. The mantle of till thus includes some unusually large erratics as well as a great many smaller rock fragments of Newfoundland origin. The clasts derived from Newfoundland are generally rounded whereas those of local origin are nearly always angular. The principal striations on smoothed and polished bedrock surfaces are NW-SE and the southeasterly direction of ice movement is confirmed by the various erratics from Langlade that are found in the moraines of Saint-Pierre. Marine erosion of coastal drift sequences has produced large sand and gravel spits and bars. Their development led to Miquelon becoming connected by a double tombolo to Le Cap and likewise to Langlade, both enclosing large lagoons at the Miquelon shore. The second, the Langlade isthmus, is a particularly fine example 12 km long. Beach and bay-mouth bars fringe the greater part of the northern and eastern coasts of Miquelon.

The original bays are now lagoons; some are saline or brackish but where fed by streams they contain fresh water. The spits and bars are locally called "dunes," although they are not of eolian origin and are, in fact, constructed largely of pebbles. True dunes occur locally on Langlade isthmus and on the west coast of Langlade. The former are live dunes up to 50 m high, whereas the latter are now fixed and grass-covered.

Economic Resources

Numerous minerals have been found in trace or minor amounts on the islands but so far no deposits of economic value have been discovered in spite of extensive exploration. The latter has been chiefly for sources of copper and iron. Signs of copper are widespread in the metamorphic rocks of Le Cap and in the red clastic formation and associated basalt flows of Langlade (bornite, chalcocite, chalcopyrite). Hematite (of secondary origin) associated with fractured rhyolite breccias, is sufficiently abundant on the large islet of Grand Colombier, just north of Saint-Pierre (Fig. 1a), to give rise to magnetic anomalies in the area. Although there are apparently no workable surface or near-surface concentrations of copper or iron ore, they may exist at depth.

In the peatbogs there are considerable quantities of diatomite and peat. Estimates of the quantity of diatomite in the ponds on Saint-Pierre and on Langlade show that at least 100,000 tons is present on each of these islands and much of it is of good quality and exploitable.

The islands have, in the past, owed their economic viability to the French fishing industry but in recent years the development of the port of Saint-Pierre as a supply and transshipment center and the growth of tourism has considerably expanded the economy.

MICHAEL M. ANDERSON

References

Alcock, F. J., 1938. "Geology of Saint John region, New Brunswick," *Geol. Surv. Can. Mem. 216,* 65p.
Arnold, F. K., 1941. "Islands adrift: St. Pierre and Miquelon," *Nat. Geogr. Mag.* (Washington), 80(6), 743–768.
Aubert de la Rüe, E., 1932. "Premiers résultats d'une mission géologique aux îles Saint-Pierre et Miquelon," *Rev. Géogr. Phys. Geol. Dyn.,* 5(4), 417–456.
____, 1944. *Saint-Pierre et Miquelon.* Montreal: L'Arbre (Coll. "France Forever"), 261p.
____, 1948. "Un musée géologique naturel: Saint-Pierre et Miquelon," *Rev. Gen. Sci,* 4(1), 5–10.
____, 1951. *Recherches géologiques et minières aux îles Saint-Pierre et Miquelon.* Paris: Office de la Recherche Scientifique Outre-Mer, 75p.
Hutchinson, R. D., 1962. "Cambrian stratigraphy and trilobite faunas of Southeastern Newfoundland," *Geol. Surv. Can. Bull.,* 88, 148p.
Weeks, L. J., 1954. "Southeast Cape Breton Island, Nova Scotia," *Geol. Surv. Can. Mem. 277,* 112p.
Widmer, K., 1953. "The geology of the Hermitage Bay area, Newfoundland," Ph.D. thesis, Princeton Univ., 459p.

Cross-references: *Canada–Atlantic Provinces; United States–New England Region.*

ST. VINCENT

A volcanic island in the inner, volcanic arc of the Lesser Antilles, St. Vincent belongs to the Windward group, located between St. Lucia (q.v.) and Grenada (q.v.). The smaller Grenadine Islands are administratively divided by St. Vincent and Grenada. They include Bequia, Union, Caouan, Mustique, Mayero, Baliceaux, and Ile de Quatre. A British West Indian colony, St. Vincent is, like the other Windwards populated by a mixed race who, besides English, speak an Afro-French patois, a reflection of early French missionary education. The island was discovered by Columbus in 1498 on St. Vincent's day (January 22).

Situated at $13°09'N$, $61°14'W$, St. Vincent is 30×18 km and covers 389 km^2 (133 sq mi); the Grenadine dependencies embrace another 44 km^2 (17 sq mi). Consisting of heavily wooded hills with deeply dissected valleys, the island is crowned by the volcano La Soufrière (1165 m; 3822 ft) at the northern end. It erupted in 1718, 1812, 1902, and 1904 (at the same time as Martinique's Mt. Pelée). It has a crater 1.5 km across and 500 m deep, today occupied by a crater lake. The oldest rocks are augite and hypersthene andesites, both lavas and tuffs. In the older agglomerates metamorphic limestones and vein quartz are incorporated. The younger lavas are basaltic. There are also ultrabasic blocks rich in olivine.

There are many small streams but they are likely to run dry in winter. The west coast is steep and cliffed, but with a good harbor at St. Georges. The east coast is gentler and even has a narrow coastal plain to the N. There are four Quaternary terraces here. There is also some artificial terracing, much of the country being cultivated for sugar. It was to St. Vincent that the admirable Captain Bligh, R. N., introduced the breadfruit tree from Tahiti after the famous mutiny in the Tonga Islands (1789).

RHODES W. FAIRBRIDGE

References

Anderson, T., and Flett, J. S., 1903. "Report on the eruption of the Soufrière in St. Vincent in 1902 and on a visit to Mtgne Pelée in Martinique," *Phil. Trans. Roy. Soc. Lond.*, Ser. A, **200**, 353–553.

André, E., 1902. "The volcanic eruption at St. Vincent," *Geogr. J.*, **20**, 60–68.

Aspinall, W. P., Sigurdsson, H., and Shepherd, J. B., 1973. "Eruption of Soufrière Volcano on St. Vincent Island, 1971–1972," *Science*, **181**(4095), 117–124.

Earle, K. W., 1947. "The geology of St. Vincent and the neighboring Grenadines," in B. Gibbs, *A Plan of Development for the Colony of St. Vincent*. Port-of-Spain.

Flett, J. S., 1908. "Petrographical notes on the products of the eruptions of May, 1902, at the Soufriere in St. Vincent," *Phil. Trans. Roy. Soc. Lond.*, Ser. A, **200**, 305–332.

Hay, R. L., 1959a. "Formation of the crystal-rich glowing avalanche deposits of St. Vincent," *B.W.I. J. Geol.*, **67**, 540–562.

———, 1959b. "Origin and weathering of late-Pleistocene ash deposits on St. Vincent," *B.W.I. J. Geol.*, **67**, 65–87.

Hill, O. M., 1959. "St. Vincent in the Windwards," *Can. Geogr. J.*, **59**(5), 162–167.

Hovey, E. O., 1902. "Martinique and St. Vincent; a preliminary report upon the eruptions of 1902," *Am. Mus. Nat. Hist.*, **16**, 333–372.

Lewis, J. F., 1964. "Mineralogical and petrological studies of plutonic blocks from the Soufrière Volcano, St. Vincent, B.W.I.," Ph.D. thesis, Oxford Univ., 270p.

Pinchow, R. P., 1961. "Description de l'Ile de Saint-Vincent," *Bull. Soc. Hist. Martinique* (Fort-de-France).

Robson, G. R., 1968. "Excursion guides; field trip guide, St. Vincent, *Trans. 4th Caribbean Geol. Conf.*, 454–457.

Wager, L. R., 1962. "Igneous cumulates from the 1902 eruption of Soufrière St. Vincent," *Bull. Volcanol.*, **24**, 93–99.

Cross-references: *Grenada; St. Lucia; West Indies; Windward Islands.*

SAIPAN

Saipan lies about 15°N and 146°E, 250 km NNE of *Guam* (q.v.) in the southern part of the *Mariana Islands* (q.v.), an island-arc belt of 15 small islands strung along a convex submarine ridge more than 700 km long (N-S) midway between Japan and New Guinea and 2000 km E of the Philippines.

Saipan is second in size in the group, 20 × 9 km, covers 124 km^2 (48 sq mi), and is hilly, up to 474 m (1555 ft) high (Mt. Topotchau). A succession of terraces, mainly erosional, step down to the sea sloping away from a strongly dissected central volcanic area in this N-S-oriented island. The oldest rocks are subaerial dacitic pyroclastic and flow rocks of (?)Eocene age called the Sankakuyama Formation. Over and around this formation are the andesitic pyroclastic and subordinate flow rocks of the Hagman Formation. These rocks were reworked by the sea to form the Densinyama Formation, consisting mainly of conglomerate and sandstone. These grade laterally and upward into a 160 m succession of warm-water bank limestones known as the Matansa Formation. The Hagman, Densinyama, and Matansa are upper Eocene (Tertiary *b*). The Mariana Ridge was in existence, near sea level by early Tertiary time. The Fina-sisu Formation consists of Oligocene (?) andesitic flows and marine tuffs.

The core of the island consists of 1500 m of these rocks, mainly volcanic of Eocene and Oligocene age. They are overlain by 400 m of bioclastic and coral-algal limestones and unconsolidated sediments of Miocene, possibly Pliocene, and Pleistocene to Holocene age. The early Miocene Tagpochau Limestone accumulated at depths too deep for coral to grow. The Donni member of tuffaceous sandstone accumulated downslope from the Tagpochau Limestone. This member contains Tertiary zone *e* fossils. Thin terrace sands and gravels of (?)Pliocene age lie on benches truncating the Miocene strata. The Mariana Limestone, probably of early Pleistocene age and reaching 150 m above sea level, is composed of lithified reef complex. The Tanapag Formation is a late Pleistocene emerged fringing reef complex. The writer observed that coral reef limestones on Saipan, as on many other islands in the Pacific, formed late in the history of an island indicating that conditions were not conducive to massive coral reef growth until late Pliocene and the Pleistocene. Massive reef growth requires either tectonic or glacial eustatic fluctuations of sea level. Between 12 and 25 marine platforms or terraces cut across the older rocks on Saipan, or consist of reef deposits of Pleistocene limestones. The last two still-stands of the sea were at the 2 and 1 m worldwide eustatic levels.

The dominant faults trend N-NE and dip W. Minor local folding was noted. The oldest dated faults are post-early Miocene. Earthquakes are common. There are minor deposits of manganese and phosphate.

HAROLD T. STEARNS

References

Cloud, P. E., Jr., Schmidt, R. G., and Burke, H. W., 1956. "Geology of Saipan, Mariana Islands," *U.S. Geol. Surv. Prof. Pap. 280-A*, 126p.

Cole, W. S., and Bridge, J., 1953. "Geology and larger foraminifera of Saipan Island," *U.S. Geol. Surv. Prof. Pap. 253*, 45p.

Ladd, H. S., 1966. "Chitons and Gastropods (Haliotidae through Adeorbidae) from the Western Pacific Islands," *U.S. Geol. Surv. Prof. Pap. 531*, 98p.

Tayama, R., 1938. "Topography, geology and coral reefs of Saipan Island (in Japanese)," *Palau Tropical Indus. Bull.*, 1, 1–62, English trans. in files of U.S. Geol. Surv.

Cross-references: *Caroline Islands; Guam; Mariana Islands; Pacific Islands Trust Territory.*

SAN AMBROSIO AND SAN FÉLIX ISLANDS

Lying in the South Pacific, at 26°21'S and 79°54'W and 26°17'S and 80°7'W, respectively, some 960 km (600 miles) W of Chile, San Ambrosio and San Félix are two small, barren, uninhabited islands discovered by the Spanish navigator Juan Fernández in 1574. They are administered by Chile. San Ambrosio rises to above 480 m (1570 ft) and is 3 km in diameter. San Félix is about 180 m (600 ft) high and 2 km across. They are situated about the intersection of two major submarine structural trends, the Juan Fernández Ridge, and the Easter Island Fracture Zone. The oceanic basement here evolved as part of the Nazca Plate, and the island foundations cannot be older than Eocene (Herron, 1972).

The two islets are the tops of volcanic seamounts. They rise from twin platforms 3–6 km wide and truncated at 140 m depth. The islets have cliffed coasts, without coral reefs or terraces. Neither volcano is active. The geology has been observed by Bailey Willis and later reported on by Willis and Washington (1924). The Tertiary or early Quaternary volcanics of San Ambrosio are basanites associated with some nepheline-rich lavas. San Félix appears to be somewhat younger, with very fresh fragments of basanite glass (very hydrated), together with blocks of phonolitic trachyte and obsidian.

San Félix and San Ambrosio are also known collectively as the Islas do los Desventurados (Misfortune).

RHODES W. FAIRBRIDGE

References

Herron, E., 1972. "Sea-floor spreading and the Cenozoic history of the East-Central Pacific," *Bull. Geol. Soc. Am.*, 83, 1671–1692.

Willis, B., and Washington, H. S., 1924. "San Felix and San Ambrosio, their geology and petrology," *Bull. Geol. Soc. Am.*, 35, 365–384.

Cross-references: *Easter Island and Sala y Gómez; Galápagos Islands; Juan Fernández; Oceania.*

SOCIETY ISLANDS

Lying in the South Pacific, some 6500 km (4200 miles) SW of San Francisco, the Society Islands cover an area of ca. 1500 km^2 (560 sq mi). They are administered by France. They represent a 700 km (450 miles) chain consisting of two clusters of volcanic and coral islands. None of the volcanoes is active. The Windward group ("Iles du Vent") includes Tahiti (the largest), Moorea, Maiao, Mehetia, and Tetiaroa; the Leeward group ("Iles Sous le Vent") includes Raiatea and Bora-Bora. (N.B.: In the West Indies, the English language use of the names Windward and Leeward islands is in the opposite sense.)

Great geological interest attaches to these islands because they contain many nepheline-bearing rocks, and Chubb (1927) designated a specific nepheline province in this part of the Pacific. This phenomenon is now recognized as being related to progressive evolution under sea-floor spreading. In the Society chain, which trends NW-SE, there is a graded sequence of 15 volcanic centers, ranging from the youngest, Mehitia, in the E to the highest cone in Tahiti-Nui (2237 m) and then dropping off in size to the NW, apparently matching an increasing maturity and degree of erosion until at the extreme NW there are three atolls. The last, like other mid-Pacific seamounts, may have their foundations going back to Upper Cretaceous, although these examples have not been drilled.

Mehetia is a 455-m (1494-ft) youthful cinder cone with many recent lava flows, in places burying Holocane coral reefs.

Tahiti consists of two basaltic cones connected by a narrow isthmus. Both are well developed but deeply gullied (see *Tahiti*).

Moorea has its volcanic plug well preserved, rising to 1207 m (3960 ft), but the flanks are much more deeply eroded than in Tahiti. Faulting on the northern side has produced high cliffs and the island is tilted over to the SW. The old lavas, basalts, trachytes, phonolites,

FIGURE 1. Bora-Bora, in the Society Islands, a partially drowned volcanic island, showing development of extensive barrier reef that represents upgrowth during submergence. Note the drowned embayments of the volcanic stump, proving former deep subaerial erosion to a much lower base level.

and minor pyroclastics are extensively affected by laterization.

Tubuai-Manu (also called Maiao or Moruiti) has a 160-m high volcanic ridge, a remnant of a very deeply dissected cone. Thick residual laterites and clays cover the olivine basalt flows. The volcanic island is ringed by an unusual double barrier reef with twin lagoons. (This island should not be confused with the Tubuai Islands, q.v.).

Tetiaroa lies off to the N of Moorea, quite separate from the main trend of islands. It is an atoll with islets on the rim.

Among the "Iles Sous le Vent," the most easterly is *Huahine,* a double volcanic island, divided by a narrow strait, but united by a single encircling reef, 15 X 10 km; the larger cone reaches 680 m (2233 ft). The dominant rock is a labradorite basalt, with minor trachytes and phonolites. The lavas are deeply weathered to laterite and kaolinite.

Raiatea and *Tahoa* are a similar pair of cones, surrounded by a single reef, 21 X 14 km. Raiatea reaches 1033 m (3380 ft). The lavas consist of olivine basalts and agglomerates, with trachytes and phonolites.

Tahoa is a circular island, 12 km in diameter, and reaches 590 m (1934 ft). It is mainly olivine basalt extensively intruded by dikes and minor gabbro.

Bora-Bora (Fig. 1) includes a deeply eroded but jagged cone, with secondary centers (including Toopua), surrounded by wide fringing reefs, a lagoon, and barrier reef (16 km in diameter). It is a classic "Darwin-type" volcanic island and barrier. The neck reaches 590 m (1937 ft) elevation. There are hundreds of radial dikes. The lavas are olivine basalts with very few pyroclastics, evidently lavas of very great fluidity, showing dips as low as 3°. The cone is now deeply dissected, embayed, and weathered to laterites and clays.

Motu-Iti (or Tubai) lies NW of Bora-Bora and is an atoll 10 km in diameter. Due W is Maupiti (or Marua), another deeply eroded stump of a basalt cone surrounded by a wide coral reef. It reaches 379 m elevation. There is a gabbroic dike. *Motu One* (Mohelia, Mopeha), *Fenua Ura* (or Scilly I.), and *Bellingshausen* are situated at the extreme NW of the Society chain and are isolated atolls.

J.-P. CHEVALIER

References

Aubert de la Rüe, E., 1958. "Observations sur le volcanism tertiaire et quaternaire de quelques îles de la Polynésie Française," *Bull. Volcanol.,* Ser. 2, **19,** 159–178.

Becker, M., *et al.,* 1974. "Phases d'érosion–comblement de la vallée de la Papenoo et volcanisme subrécent à Tahiti, en relation avec l'évolution des îles de la Société (Pacifique Sud)," *Mar. Geol.,* **16,** 71–77.

Chubb, L. J., 1927. "Mangaia and Rurutu; A comparison between two Pacific Islands," *Geol. Mag.,* **64,** 518–522.

Crossland, C., 1928. Coral reefs of Tahiti, Moorea and Rarotonga, *Linn. Soc. Lond. J. Zool.,* **36,** 577–620.

Guilcher, A., Berthois, L., Doumenge, F., *et al.,* 1969. Les récifs et lagons coralliens de Mopelia et de Bora-Bora (Iles de la Société) et quelques autres récifs et lagons de comparaison (Tahiti, Scilly, Tuamotu occidentales); morphologie, sédimentologie, fonctionnement hydrologique. *Fr. Off. Rech. Sci. Tech. Outre-Mer. Mém. 38,* 103p.

Iddings, J. P., and Morley, E. W., 1918. "A contribution to the petrography of the south sea islands," *Proc. Nat. Acad. Sci.,* **4,** 110–117.

Lacroix, A., 1910. "Les roches alcalines de Tahiti," *Bull. Soc. Géol. France,* Sér. 4, **10,** 91–124.

———, 1923. "Archipel de la Société (Tahiti, etc.)," *Minéralogie de Madagascar* (Paris), **3**, 279–289.

———, 1927. "La constitution lithologique des îles volcaniques de la Polynésie australe," *Mém. Acad. Sci. Paris*, **59**, 1–82.

———, 1933. "Recent observations on the mineralogical and chemical constitution of the intrapacific lavas (South Central Pacific)," *Proc. 5th Sci. Congr.*, **3**, 2438–2542.

Marshall, P., 1915. "The geology of Tahiti," *N.Z. Inst. Trans. Proc.*, **47**, 361–376.

Obellianne, J. M., 1955. "Contribution à l'étude géologique des îles des Établissements français de l'Océanie," *Sci. de la Terre* (Nancy), **3**(3), 1–146.

Stark, J. T., and Howland, A. L., 1941. "Geology of Bora-bora, Society Islands," *Bernice P. Bishop Mus. Bull.*, **169**, 1–63.

Williams, H., 1933. "Geology of Tahiti, Moorea and Maiao," *Bernice P. Bishop Mus. Bull.*, **105**, 1–90.

Cross-references: *Cook Islands; French Polynesia; Oceania; Tahiti; Tuamotu Islands; Tubuai Islands.*

SOLOMON ISLANDS

The Solomon Islands occupy some 1200 km of the southwestern border of the Pacific Ocean basin, from 3 to 12°S and 154 to 163°E. Although many of their features are not fully understood, they are acknowledged as excellent examples of fractured island arcs in which faulting is the dominant structural feature. The Solomons offer data that is essential to an understanding of the structure and genesis of the southwest Pacific (Fig. 1). They have particular relevance to the "Tasmantis" concept of foundered land areas (termed "quasicraton" by Stille) E of Australia; a concept that has been discussed by a number of authors, including Avias, Benson, Fairbridge, and Glaessner, and is now an important aspect of the application of plate tectonic theory in this region. To the NW the Solomon Group is connected to the swirl of the *Bismarck Archipelago* (q.v., including New Britain and New Ireland) and so to *New Guinea* (q.v.) by the Lihir Group of small basaltic volcanic islands; in a geological sense these are a part of the Solomon Islands. At their southeastern end they are separated by an abrupt gap from the New Hebrides which are joined to Fiji by an arcuate E-swinging bathymetric ridge, the Hunter Island Rise (for geographic data refer to *The Times Atlas of the World*, volume 1, plates 9, 10, and 15.)

The Solomon Islands proper are a chain of islands E of New Guinea which extend from latitude 5°S, longitude 154°E to latitude 12°S, longitude 162½°E. They have a population of about 150,000 and a land area of about 38,850 km^2 (approximately 15,000 sq mi). Most of the islands are mountainous, jungle covered, and sparsely inhabited. The mountains are, for the most part, block faulted (Mt. Popamanisiu on Guadalcanal is about 2455 m high; 8005 ft). There are also many volcanoes, only a few of which are currently active. The rivers are short and torrential with many cataracts and waterfalls; their courses are commonly controlled by faults.

Although the group was discovered four centuries ago, in 1568, by Mendaña, the Spanish mariner, they were then lost to sight and rediscovered some 200 years later by Louis de Bougainville.

Although Mendaña's men reported gold, hence the name anticipating the "wealth of Solomon," the islands were never a serious target for itinerant prospectors: as a result of the depredations of "blackbirders" (a euphemism for forced labor recruiters), the inhabitants were justifiably hostile to outsiders.

The islands are divided administratively under an agreement reached originally by Britain and Germany in 1899. After World War I, Buka and Bougainville came under the Australian mandate (Trust Territory of New Guinea) and are now part of the recently independent Niu Guini. The other islands, the "British Solomons," remain under the protection of the United Kingdom.

Only in 1950 were systematic geological studies first begun by members of a geological expedition from the University of Sydney in conjunction with the then newly established geological survey (headquarters at Honiara, the administrative capital on Guadalcanal Island). Prior to 1950, H. B. Guppy had made scattered but acute observations on the coastal geology which were published in 1887.

Since 1950 the local geological survey has been most active, assisted by personnel from Australian universities and Imperial College, London, and all the islands have now been covered by general surveys. Mapping on a scale of 1:50,000 is well advanced. The Geophysical and Polar Research Institute, University of Wisconsin, completed an initial gravity survey of the land areas; a United Nations special development project followed with extensive geophysical and geochemical surveys. Magnetic and sea-gravity work has been carried out in the region by Hawaii Institute of Geophysics.

The following account is based on the results of these institutions and the work of survey personnel (in particular that of J. C. Grover, R. B. Thompson, P. Pudsey-Dawson, and B. D. Hackman), R. L. Stanton (University of New England), F. K. Rickwood (British Petroleum Company), and that of the writer. References are given in Coleman (1970).

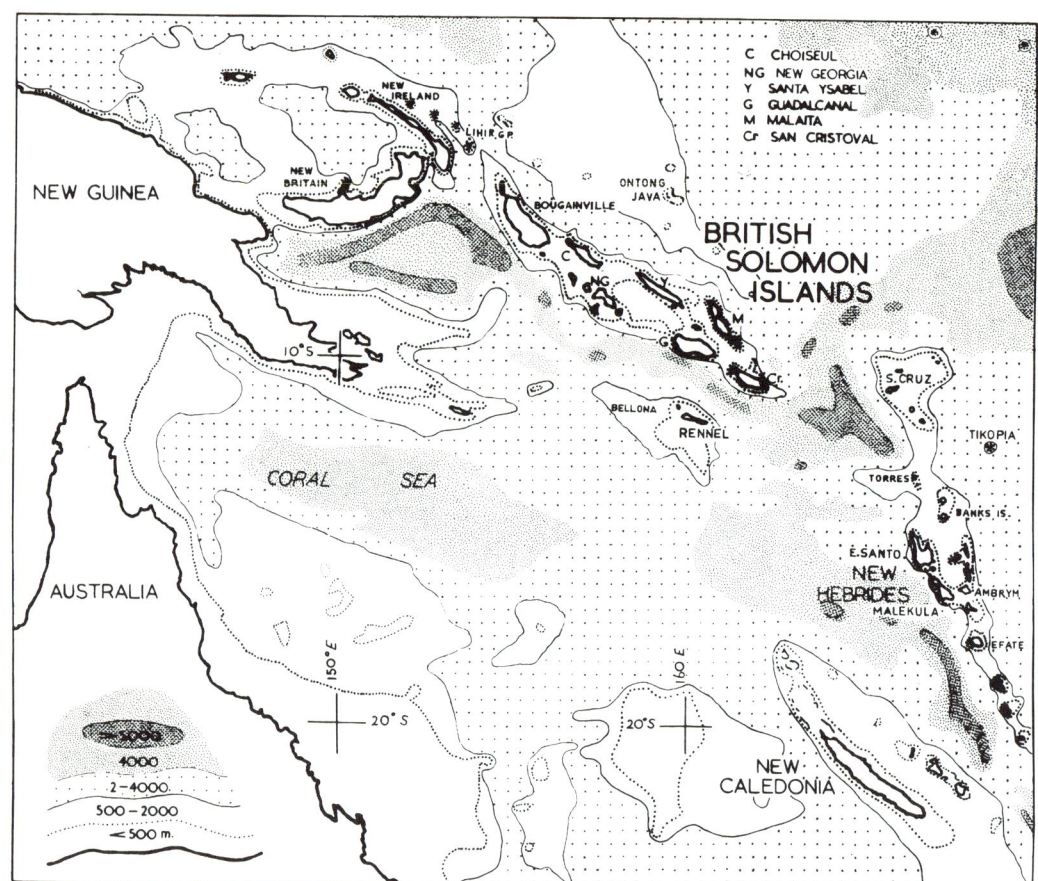

FIGURE 1. Southwest Pacific, showing location of the Solomon Islands.

Morphotectonics and Bathymetry

The Solomon Islands are essentially a double chain (Fig. 2) closed at the northern end by Bougainville (6250 km^2) and at the southern end by San Cristobal (4500 km^2). Choiseul (2450 km^2), Santa Ysabel (4500 km^2), and Malaita (3750 km^2) on the Pacific side are separated by a narrow stretch of sea (The Slot or New Georgia Sound) from the Shortland Islands, New Georgia (3250 km^2), Russell Islands, Guadalcanal, and San Cristobal forming the southwestern limb. They stand on a submarine ridge, the axis of which is slightly sigmoidal and over which the water is usually less than 500 m deep, between New Georgia and Santa Ysabel. On its southwestern side this high is bordered by the Solomon Trench (convexly arcuate towards Australia) which is generally more than 4000 m deep and includes the Planet Deep (9140 m), an unnamed deep of over 5000 m off southwestern Guadalcanal and the San Cristobal Trench (about 7000 m). It is interrupted in the middle by the Woodlark Ridge and appears to be terminated transversely by the Torres Trench, SW of the Santa Cruz group, some 200 miles E of San Cristobal. To the NE the Solomons are separated by a minor linear deep from the Ontong Java Plateau, a platform that supports Ontong Java and other prominent atolls (Kapingamaringi, etc.) and extend to the eastern *Carolines* (q.v.).

Vulcanism and Seismicity

Only a few volcanoes are at present active in the region (notably Mt. Bagana on Bougainville and a submarine volcano in the New Georgia group), but there are at least 30 extinct volcanoes with well-preserved cones and craters, the most spectacular of which is Kolombangara in the New Georgia group. There are still others that are older and have suffered a good deal of

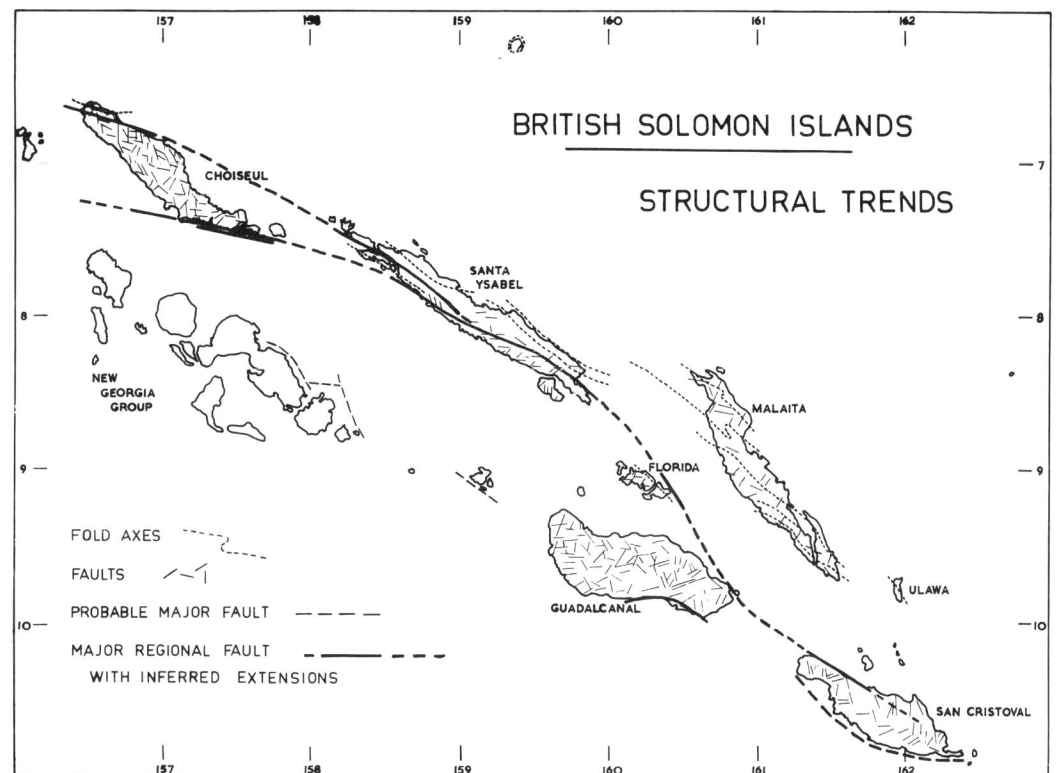

FIGURE 2. British Solomon Islands, structural trends.

erosion. Most of these foci are confined to an arcuate belt, convex to Australia, which stretches from the northwestern tip of Bougainville to the western end of Guadalcanal. The bulk of the extrusives are basalts, andesites or basaltic andesites, the older of them tending to be more basic.

Seismicity in the Solomon Islands is very high. Most of the earthquake shocks are shallow with foci about 50 km and are located along a belt adjacent to the northeastern side of the Solomon trough. Intermediate- and deep-focus earthquakes are comparatively rare, except in the area of the Planet Deep. Denham (1969) suggests that the seismic plane is steeply inclined toward the Pacific in the NW but near vertical over the remaining SE. Seismic evidence indicates a variable crustal thickness in the region from about 15 to 30 km. The recent gravity data support this figure; they also show the existence of very large positive anomalies associated with remarkable gradients. The Solomons are part of a regionally positive system extending from New Guinea to Fiji, first detected by the perturbations of orbiting satellites.

Stratigraphy and Structure

In both the distribution of rock types and in their structure the Solomon Islands can be considered as made up of three provinces: the Central, Volcanic, and Pacific provinces (Fig. 3). The *Central province* includes part of Bougainville, Choiseul, the southeastern half of Santa Ysabel, Florida, Guadalcanal, and most of San Cristobal. These areas have extensive exposures of a Mesozoic igneous and metamorphic "Basement complex" which includes lavas, greenschists, amphibolitic schists, serpentinized ultramafics, gabbroic, and granitoid rocks. Potassium/Argon age determinations indicate that the schists on Choiseul were metamorphosed during Eocene time.

The "Basement complex" is composed essentially of a pile of lavas predominantly basaltic, although basaltic pillow lavas and andesites are also common. In some areas these lavas may be over 1000 m thick; they were in place before the end of Oligocene time. They are overlain by Miocene sediments that begin with reef limestones and foraminiferal and algal calcarenites representing a quiescent stage in the

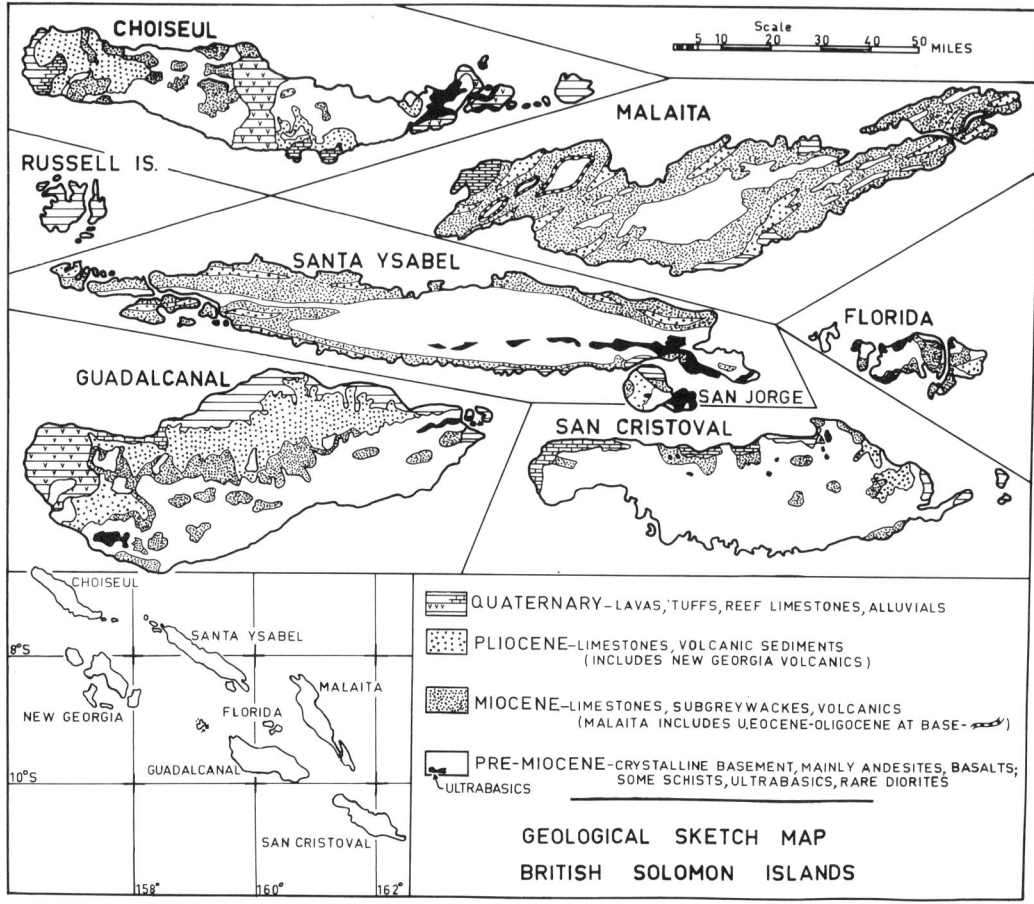

FIGURE 3. Geological sketch maps of individual islands in the Solomons.

regional tectonics, followed by reworked tuffs, volcanic sandstones, and subgraywacke. A similar and equally thick layer of Pliocene sediments completes the Tertiary succession. On Guadalcanal this Miocene-Pliocene succession is over 4600 m thick, but on the other islands of the province is much less than this. The sediments were deposited in subsiding fault-bounded troughs.

Quaternary rocks are represented mainly by coastal fringes of uplifted reef masses (now up to 360 m above sea level) and flood-plain conglomerates and alluvium; only small areas are blanketed by Holocene volcanics (parts of central and eastern Choiseul and western Guadalcanal).

The Central province is shaped and shattered by faulting; folds are rare. The faults are steeply dipping and usually in two sets, one trending 80–100°, the other at 145–180°, resulting in a reticulate pattern. A third trend, 45–225°, is, however, structurally important. On Guadalcanal, movement along the faults has proceeded in an orderly fashion to produce an asymmetric spine of step-faulted mountains, the Kavo Ranges, which fall off steeply southward.

The *Volcanic province* includes much of Bougainville, the Shortland Islands, the New Georgia group, Borokua, Russell Islands, Savo Island (a Mt. Pelée type of volcano), and westernmost Guadalcanal. With the exception of a few inactive volcanoes on Choiseul, it contains all the recognizable volcanic cones and the currently active ones. The New Georgia group is essentially an assemblage of cones linked by their own extrusives. The lavas and pyroclastics are mainly basaltic andesites, and probably no older than Pliocene. Faulting is intense and many of the faults are active. Many of these islands are fringed by uplifted reefs; out to sea, growing reef is splendidly developed. The Volcanic province may be regarded as a youthful extension of the Central province.

The *Pacific province* includes Malaita, Ulawa, and, in a marginal sense, the northeastern flank of Santa Ysabel and eastern San Cristobal. Here the succession begins with basaltic lavas of over a thousand meters thick. Some of these basalts are very basic and include alnoitic types. On Malaita the lavas are overlain by 600 m of organically derived, deep-water, calcareous silts with hardly any terrigenous material. The oldest are Late Cretaceous with many chert layers. Coarser grained sediments with an increasing amount of terrigenous material then follow (Late Miocene). The sediments range through to the Quaternary. The whole succession of about 1200 m is folded with moderate intensity and in contrast to the Central province faulting is incidental to folding of the "cascade" type. The marginal areas of Santa Ysabel and San Cristobal have a much higher terrigenous content and are not so thick. Basement complex rocks similar to those of the Central province do not occur. It is likely that Malaita, especially, is geologically part of the Ontong Java Plateau to the NW and has been caught up, as it were, by collision of the Solomons with that great platform in the Late Tertiary.

The Pacific province is separated from the Central province by a major shear system, expressed most emphatically by a great thrust that runs through the southwestern flank of Santa Ysabel, roughly parallel to the axis of that island. This shear system is also the locus of a serpentine belt. There are a few other ultrabasic areas, notably on Guadalcanal, which are associated with other large faults: Marau to the E, Suta in the center (SW-trending) and Gausava to the SW. The ultrabasics of the region began to be emplaced probably in Late Oligocene time.

Isolated Atolls North of the Solomons

Located on the Ontong Java Plateau a series of giant atolls are also administered with the Solomons (British and Australian) in two groups. They lie to the N of the Solomon Islands and are roughly parallel to them. From W to E they include Kilinailau (about $4°50'S, 155°30'E$), Tanu (about $4°50'S, 157°17'E$), Ontong Java or Lord Howe Atoll ($5°20'S, 159½°E$), Nukumanu (104 km N of Ontong Java), and Sikiana or Stewart Islands ($8°28'S, 162°34'E$). Of these only Ontong Java and Sikiana belong with the British Solomons. Parts of these atolls have been stabilized to form vegetated sand cays only a few feet above sea level. None of them appear to have recent volcanic associations.

Isolated Atolls South of the Solomons

Rennell and Bellona islands (centered about $11½°S, 160°E$) are fine examples of uplifted atolls. Rennell especially is magnificently preserved and all the fundamental features of atoll form can still be recognized; its rim is now 110 m above sea level. Both islands lie on a bathymetric rise (Rennell Ridge) which connects with westernmost New Guinea by way of the Louisiade Archipelago and Pocklington Reef. In the past, this ridge had also a connection with New Caledonia to the SE.

Isolated Islands in the New Hebrides Region

Also administered by the British Solomons are numbers of islands to the SE that are structurally quite separate from the Solomons and distributed along the New Hebrides Ridge. Largest are the *Santa Cruz Islands* some of which are active volcanoes. The largest island is Ndeni (about 550 m), the geology of which is similar to that of the islands of the western belt (e.g., Espiritu Santo) of the New Hebrides (q.v.).

The *Duff Islands* are a linear group of small volcanic basaltic islands, trending NW from latitude $10°S$, about 80 km NE of the Santa Cruz Islands.

Cherry and *Mitre* islands (centered about $12°S, 170°E$) and *Tikopia* (about $12°S, 169°E$) lie E of the main body of the Santa Cruz Islands. They are small basaltic islands rising steeply from the sea floor, which in the Vityaz Trench is 5084 m deep. This is the structural margin of the "Melanesian border plateau" and appears to have been involved in complex left-lateral strike-slip faulting. Tikopia is a fine example of an explosively breached cone, but shows few signs of volcanic activity at present.

Economic Geology

Although an established mining industry in the Solomon Islands is still in its infancy, the Geological Survey has found a number of potentially valuable ore deposits, some of which have also been investigated by private companies. Gold occurs at Gold Ridge in central Guadalcanal but is widely disseminated in quartz stringers intrusive into Upper Miocene-Pliocene sediments from underlying andesitic lavas. Alluvial gold from this source occurs in several rivers nearby. These occurrences are potentially exploitable. Nickel ores are found in association with ultrabasic bodies on western Choiseul and Santa Ysabel. Copper is prominent in dioritic "porphyry copper" bodies in central western Guadalcanal. The largest of these, the Koloula Diorite, on the southern coast, is now being evaluated as a commercial proposition. Metallic sulfides are present in great quantity in heavily mineralized areas in many parts of central Guadalcanal. A small but commercial deposit of manganese ore occurs on Hanesavo Island in the Florida group. Bauxite is mined on Rennell, but the available tonnage is not large by world standards. Although the area available for testing is small, the thick sedimentary succession in eastern Guadalcanal offers chances of petroleum accumulations. The largest rivers on Guadalcanal have some hydroelectric potential. Phosphate deposits occur on Bellona Island, probably in commercial quantity. Bougainville, in the New Guinea Territory, possesses an extremely large porphyry copper deposit (at Panguna) which is now being mined; some 30 million tons of reserves are estimated.

PATRICK J. COLEMAN

References

Coleman, P. J., 1970. "Geology of the Solomon and New Hebrides Islands, as part of the Melanesian Re-entrant, Southwest Pacific," *Pacific Sci.,* 24(3), 289–314.

_____, ed., 1973. *The Western Pacific: island arcs, marginal seas, geochemistry.* Nedlands: Western Australia Press, 675p.

Colley, H., 1970. "Andesitic vulcanism in the New Hebrides," *Proc. Geol. Soc. Lond.,* 1662, 46–51.

Curtis, J. W., 1973a. "Plate tectonics and the Papua–New Guinea–Solomon Islands region," *J. Geol. Soc. Austral.,* 20(1), 21–36.

_____, 1973b. "The spatial seismicity of Papua New Guinea and the Solomon Islands," *J. Geol. Soc. Austral.,* 20(1), 1–20.

Denham, D., 1969. "Distribution of earthquakes in the New Guinea–Solomon Islands region," *J. Geophys. Res.* 74(17), 4290–4299.

Fountain, R. J., 1972. "Geological relationships in the Panguna porphyry copper deposit, Bougainville Island, New Guinea," *Econ. Geol.,* 67, 1049–1064.

Grover, J. C., 1968. "The British Solomon Islands: some geological implications of the gravity data, 1966," in L. Knopoff, C. L. Drake, and P. J. Hart, eds., *The Crust and Upper Mantle of the Pacific Area* (Geophys. Monogr., 12). Washington D.C: Am. Geophys. Union, 296–306.

Halunen, A. J., Jr., 1973. "Heat flow in the western equatorial Pacific Ocean," *J. Geophys. Res.,* 78(23), 5195–5208.

_____, and Herzen, R. P. von, 1973. "Heat flow in the western equatorial Pacific Ocean," *EOS,* 54(7), 740, Abstr.

Page, R. W., and McDougal, I., 1972. "Geochronology of the Panguna porphyry copper deposit, Bougainville Island, New Guinea," *Econ. Geol.,* 67, 1065–1074.

Rose, J. C., Woollard, G. P., and Malahoff, A., 1968. "Marine gravity and magnetic studies of the Solomon Islands," in L. Knopoff, C. L. Drake, and P. J. Hart, eds., *The Crust and Upper Mantle of the Pacific Area,* Washington D.C: Am. Geophys. Union, 379–410.

Stanton, R. L., and Bell, J. D., 1969. "Volcanic and associated rocks of the New Georgia Group, British Solomon Islands Protectorate," *Overseas Geol. Mineral Res.,* 10(2), 113–145.

van Deventer, J., and Postuma, J. A., 1973. "Early Cenomanian to Pliocene Deep-Marine Sediments from North Malaita, Solomon Islands," *J. Geol. Soc. Aust.,* 20(2), 145–150.

Cross-references: *Australasia–Regional Review; Bismarck Archipelago; Fiji; New Caledonia; New Guinea; New Hebrides; Oceania.*

SOUTH AMERICA

South America covers an area of about 17,783,000 km^2 (7,335,000 sq mi). It is shaped roughly like an acute triangle, with the longest side on its western border and the acute angle at its southern tip. It extends from about 12°N to 56°S, and from about 35°W to 81°W, being bounded on the E by the Atlantic Ocean, and on the W by the Pacific Ocean. The northernmost shores of the continent, from Trinidad to western Colombia, are bathed by the Caribbean Sea.

Geological exploration of South America began during the last years of the Spanish colonial period with the travels of A. von Humboldt who, between 1799 and 1803, visited extensive areas of what are now Venezuela, Colombia, Ecuador, and Peru. His travels were followed, in the 19th century, by the work of W. L. Eschwege, P. Lund, A. d'Orbigny, Charles Darwin, I. Domeyko, F. de Castelnau, H. Karsten, V. von Helmreichen, D. Forbes, L. Agassiz, H. Gorceix, J. Heusser, C. Hartt, O. Derby, J. Branner, J. Sawkins, C. Velain, A. Stelzner, L. Brackebusch, T. Wolf, A. Hettner, and J. Evans, to mention only the most important pioneers.

By the end of the 19th century the main geological features of the continent were broadly known, and semidetailed work was already being carried out by the geological surveys that had been organized in a few countries. Great advances have been made during the last 30 years, thanks not only to the work of the different geological surveys and teaching centers by now well established in all countries, but also to important contributions from the oil and mining industries.

Mining in South America was already practiced by the Indians, from Colombia to Bolivia, in pre-Columbian days. The Spanish conquest brought about intensive mining of gold and silver in early colonial times. In 1537 gold was discovered in Colombia and from that date to its independence early in the 19th century, that country produced more than 1000 tons. The famous silver mines of Potosí in Bolivia and of Cerro de Pasco in Peru were opened in 1545 and 1630, respectively. The Potosí mine, now exhausted, produced more than 1.5 billion dollars worth of silver during 300 years of intensive operation, whereas the Cerro de Pasco mine yielded, between 1630 and 1886, about 17,000 tons. Though it became essentially a copper mine in the late 19th century, Cerro de Pasco still produces considerable amounts of silver. Mercury was mined in Colombia during colonial days, and in 1538 the Spaniards opened the first emerald mine there. To this day, Colombia maintains almost a world monopoly in the production of emeralds. Gold and diamonds were discovered in Minas Gerais, Brazil, in 1693 and 1710, respectively. In 1783 platinum was found in Colombia and until 1819 it was the only country that produced this metal. At that time, however, platinum had little value and was used mostly to counterfeit silver coins; hence its original Spanish name "platina," a diminutive of "plata" (silver).

Late in the 19th century, gold and silver mining lost their preeminence, whereas the exploitation of copper, lead, zinc, tin, manganese, iron, and other ores gained in importance. Early in this century, coal began to be exploited, and subsequently oil, discovered in the early twenties, rose to become the leading energy source of the continent.

Geomorphology

South America can be subdivided into more than a score of geomorphic provinces, but here only its salient characteristics will be summarized.

The outstanding feature of the continent is the high Andean belt, running close to its western border from eastern Venezuela to southern Chile. This complex cordillera, often branching into several subparallel ranges, forms the divide between the Pacific-Caribbean and Atlantic watersheds. Its average altitude ranges between 3000 and 5000 m (10,000 to 17,000 ft), but between Ecuador and central Chile long mountain ridges rise well above 5000 m, culminating at Cerro Aconcagua in western Argentina (7005 m; 23,129 ft), the highest peak of the Western Hemisphere.

East of the Andes, a wide belt of low plains extends from the Llanos of Venezuela to the Pampas of Argentina and expands eastward along the wide valley of the Amazon River. For the most part these plains rise very gradually from sea level to about 200 m, but in the Beni and northern Chaco areas of Bolivia and Paraguay they reach altitudes of 500 m. East of the plains, as well as N and S of the Amazon valley, large plateaus and rolling highlands form most of the Guianas and Brazil. Extra-Andean Patagonia is characterized by a similar relief. Most of the highlands attain only 500 to 700 m above sea level, but the central Guiana plateau rises abruptly to 2800 m and comparable elevations characterize the rugged highlands of eastern Minas Gerais in Brazil.

In accordance with the westerly location of the Andean Cordillera, most of the continent has an Atlantic drainage, involving three main river systems. The largest is the Amazon system, which drains an area of about 7,050,000 km^2 (2,700,000 sq mi) covering large parts of the Guianas, Colombia, Ecuador, Peru, Bolivia, and central Brazil. The Amazon River (Amazonas in local languages) itself, born in the high Andes of Peru, is more than 6500 km (4000 mi) long. It has more than 200 tributaries, 100 of which are navigable, 17 of them being 1600 to 3700 km long. The second largest system is that of the Paraná-Paraguay rivers, which drains an area extending over southern Brazil and Bolivia, Paraguay, Uruguay, and northern Argentina. The Paraná River itself, born near the Brazilian coast in Minas Gerais, is 4400 km long. Its main tributary, the Paraguay River, is 2000 km long. Near its mouth, the Paraná receives the waters of the Uruguay River, jointly forming the short but extremely wide Río de la Plata (River Plate). The third system is that of the Orinoco, which drains most of Venezuela and parts of Colombia. The main river is 2400 km long and it is actually connected with the Amazon system by the Cassiquiere branch of the Río Negro, a tributary of the Amazon, and by the Pimichin and Termi branches of the Atabapo, a tributary of the Orinoco.

Geologic History

Precambrian. Radiometric age determinations of South American Precambrian rocks began in the early 1960s. Dates are still scarce in relation to the huge areas, but they make it possible to give a generalized picture of the development of the continent during Precambrian time. It should be borne in mind, however, that this provisional picture, very different to that of a few years ago, will need modifications as new information is obtained.

Lower Precambrian. Five cratonic nuclei have been recognized in the northern, central, and eastern parts of the continent. These are the Guiana Shield, the São Luiz Cratonic area, the Central Brazilian Shield, the São Francisco

Shield, and the Uruguay-Tandilia Cratonic area, all of which were affected by early Precambrian orogenic episodes.

1. *Gurian Orogenic Cycle.* The oldest South American rocks are exposed in the Guiana Shield, forming an E-W elongated belt in the northern part of the Venezuelan Guiana and in neighboring areas of Guyana. They consist of a great thickness of high-grade metamorphic gneisses, granulites, and charnockites, known as the Imataca complex, which have yielded ages between 3000 and 3400 m yr. The tectonothermal episode, termed "Gurian orogenic cycle," took place about 2700 m yr ago. In southern Guyana and Suriname, similar high-grade metamorphic gneisses, granulites, and charnockites (Kanuku and Falawatra groups, respectively) have furnished a few age determinations of about 2600 m yr.

Outside the Guiana Shield, metamorphics originated by the Gurian orogenic cycle have been found in a few Brazilian localities, such as the Itacaiúnas River of Pará, the upper basin of the Tocantins River in Goiás, the area W of Goiania, the Rio das Velhas in Minas Gerais, and the lower Paramirím River S of Barras in Bahia. In all these areas, ages of more than 2700 m yr have been determined.

2. *Trans-Amazonian Orogenic Cycle.* The five cratonic nuclei of the continent were strongly affected by a well-defined orogenic event, called "Trans-Amazonian orogenic cycle," which took place between 1700 and 2200 m yr ago, reaching maximum intensity about 2000 m yr ago. This cycle affected most of the Guiana Shield, where gneisses and other metamorphics intruded by acid rocks, have yielded ages ranging between 1800 and 2200 m yr.

The metamorphic rocks of the small São Luiz Cratonic area of coastal Brazil have given concordant Rb/Sr and Ar/K ages of about 2000 m yr. This area is separated from the central Brazilian and São Francisco shields by a much younger Upper Precambrian-Lower Paleozoic orogenic belt. In a predrift assemblage of Africa and South America, it forms the southernmost tip of the large West African Shield, affected by the Eburnean orogenic cycle 2000 m yr ago.

The central Brazilian Shield has been subdivided into an eastern half, called "Tocantins-Tapajós cratonic area," and a western half, termed "Rondonia Cratonic area." In the eastern part of the shield, metamorphic rocks affected by the Trans-Amazonian orogenic cycle of 2000 m yr are widespread. In the western, or Rondonia part, the few radiometric age determinations carried out so far have failed to find rocks of this age, but have revealed the existence of anorogenic Upper Precambrian tin-bearing granites, 900 to 1200 m yr old, which probably represent intrusions within the old cratonic basement.

Most of the high-grade metamorphics and intrusives of the São Francisco Shield—which extends over parts of Goiás, Minas Gerais, and Bahia along the São Francisco River, reaching the Atlantic coast in the Salvador area—have ages of 1800 to 2000 m yr, corresponding to the Trans-Amazonian orogenic cycle.

This cycle also affected the Uruguay-Tandilia cratonic area. In southern Uruguay, from a little E of Montevideo to the mouth of the Uruguay River, granites and granodiorites have Rb/Sr ages between 1700 and 2000 m yr, whereas in the Northern Hills of Buenos Aires—the "Tandilia"—granitic rocks, gneisses, and mylonites have yielded ages between 1800 and 2200 m yr.

Upper Precambrian

3. *Espinhaçian Cycle.* The metamorphic rocks of the Espinhaço group of Minas Gerais and Bahia, exposed along a narrow belt E of the São Francisco River following the Serra do Espinhaço, have given Rb/Sr and K/Ar ages ranging between 1300 and 1800 m yr. It is difficult to ascertain whether these ages represent a true orogenic cycle or a partial rejuvenation of the older rocks of the São Francisco Shield.

4. *Roraima Sedimentary Episode and Basic Intrusives.* Large parts of the Guiana Shield are covered by the thick, flat-lying Roraima formation, consisting chiefly of cross-laminated reddish sandstones with minor arkoses and shales, conglomerates, and thin ignimbritic layers resting with sharp discordance on the old basement rocks. The Roraima formation is intruded by doleritic sills, the age of which has been tentatively established as between 1550 and 1650 m yr. Although the time of sedimentation of the Roraima formation is still somewhat uncertain, it is believed that the beds were deposited about 1600 to 1650 m yr ago. It should be emphasized that the Roraima formation—long regarded as Mesozoic—represents the oldest known platform cover of South America.

5. *Nickerie Metamorphic Episode.* The southern part of Guyana and its extension into southwestern Suriname, was affected by a tectonothermal event about 1200 m yr ago, called "Nickerie Metamorphic episode," related to the development of mylonitic belts and zones of shearing. It seems likely that a wide zone of transcurrent movement, which overprinted ages of about 1200 m yr, runs approximately ENE-WSW along the Guiana Shield between latitudes 3 and 4°N, taking a course more to the NE in southwestern Suriname.

6. *Uruaçuian-Minas Cycle.* The sedimentary-tectomagmatic-metamorphic-local graniti-

zation cycle represented by the Araxá group of Goiás, and the Andrêlandia and Minas groups of Minas Gerais, has been termed "Uruaçuian-Minas cycle." The Araxá group is older than 980 m yr and younger than 1400 m yr, whereas the Minas group seems older than 850 m yr and apparently younger than 1300 m yr. The cycle, therefore, is regarded as covering the interval between 900 and 1300 m yr B.P. The belt of nickeliferous peridotites of Goiás and Minas Gerais belong to this cycle, and apparently also the tin-bearing anorogenic granites of Rondonia, 900–1200 m yr old. The existence of an orogenic belt belonging to this cycle in southern Mato Grosso and northeastern Paraguay seems likely, as pegmatites and amphibolites exposed near the Río Apa have given K/Ar ages of 1250 and 1050 m yr, respectively.

7. *Brazilian Cycle.* The term "Brazilian cycle" will be here restricted to designate the sedimentary-tectomagmatic-metamorphic cycle represented by such groups as the Macaúbas, Itacolomí, Canastra, and Tocantins, which form a belt around the southern half of the São Francisco Shield. Though radiometric age determinations are scarce, it seems that this cycle covers the interval 700–900 m yr B.P. The Cuiabá Group of southern Mato Grosso seems to belong to this cycle, and apparently also some microdiorites and granadiorites of southeastern Paraguay which have furnished a few K/Ar ages ranging between 786 and 960 m yr; these are rather questionable, however, because of their very large plus or minor errors.

8. *Bambuí Sedimentary Episode.* Most of the São Francisco Shield, particularly the area of western Bahia, eastern Goiás, and northern Minas Gerais between the Serra do Espinhaço in the E and the Tocantins River in the W, is covered by the thick Bambuí Group, consisting of sandstones, arkoses, limestones, and dolomites, practically devoid of folding and metamorphism. These rocks, traditionally regarded as Silurian, were found to be 600 m yr old by K/Ar and Rb/Sr determinations carried out in 1967. The Araras and Corumbá groups of southern Mato Grosso, and the Itapucumí group of northeastern Paraguay, characterized by a predominance of limestones and dolomites, seem to represent contemporaneous platform covers of the central Brazilian Shield. It should be mentioned that the Bambuí Group, as well as the Araras-Corumbá-Itapucumí sequences, were tectonized along theirs borders by late movements related to the last orogenic cycle that affected the cratonic areas.

9. *Pan-American Cycle.* The cratonic nuclei of South America are flanked by sweeping orogenic belts formed of Upper Precambrian to Lower Paleozoic metamorphics and intrusives. In all these belts Rb/Sr and K/Ar measurements have given ages ranging between 500 and 700 m yr, with average values of 600 to 640 m yr. Though three such belts have been recognized—the Cariri, the Paraíba, and the Paraguay-Araguaia—the "western belt" of the continent should be added. As all of them correspond to a single geologic cycle, the name "Pan-American" cycle is proposed here to distinguish it. In the following, the individual belts will be briefly described.

The *Cariri orogenic belt,* a wide belt of metamorphic and intrusive rocks belonging mainly to the Ceará group, is found in northeastern Brazil, N of the São Francisco River and E of the Parnaíba Basin. It trends NW in the N, swinging W in the S where it flanks the northern edge of the São Francisco Shield Rb/Sr whole rock determinations, confirmed by K/Ar analyses, indicate that the main phase of the cycle took place 640 m yr ago. However, K/Ar measurements of the granitic pluton of the Serra de Meruoca and of volcanics of the Jaibara Basin have given minimum ages of 420–430 m yr, indicating that these are posttectonic events.

The Cariri belt probably continues W below the sedimentites of the Parnaíba Basin, as rocks between 485 and 540 m yr old are found along the Southern edge of the São Luiz Cratonic area. In a predrift assemblage of Africa and South America, this belt forms the continuation of the Pan-African orogenic belt of Nigeria.

The *Paraíba orogenic belt* is a long metamorphic belt, extending along coastal Brazil from southern Bahia to Rio Grande do Sul and southeastern Uruguay. It comprises the Canudos, Jangada, Lavras, and Jequitaí groups of Bahia, and the Serra dos Orgãos, São Roque, Açunguí, Brusque, Porongos, and Lavalleja groups of São Paulo, Paraná, Santa Catarina, Rio Grande do Sul, and Uruguay. The main orogenic phase seems to have occurred about 600 m yr ago, but in Rio Grande do Sul the slightly folded metasedimentites and associated acid and basic intrusives of the Porongos Group are about 500 m yr old. In southeastern Uruguay, granites, granodiorites, and migmatites have given Rb/Sr ages ranging between 500 and 600 m yr.

In southern Minas Gerais this belt bifurcates. The main branch continues along the coast to the SE in Uruguay, but another sweeps W around the Southern border of the São Francisco Shield and then bends to the N, following the course of the Tocantins River.

Very few radiometric age determinations have been made in the *Paraguay-Araguaia orogenic belt,* which seems to extend from the upper Araguaia River in Goiás S into central

Paraguay, following a sinuous course along the southeastern border of the central Brazilian Shield. A few K/Ar determinations on micas from southern Mato Grosso have given ages of about 500 m yr, whereas a single analysis on potash feldspar of the nonmetamorphic Caacupú granite of eastern Paraguay yielded an age of 468 ± 25 m yr.

In all likelihood Precambrian rocks exist all along the *Andean belt* of South America and in the *Pampean Ranges* and *northern Patagonian Massif* of Argentina, and possibly in the deep subsurface of the Deseado Massif of southern Patagonia. In many extensive areas, however, these rocks are now masked by Paleozoic and younger deposits and by intrusives.

Except for a couple of analyses along the southeastern foot of the Andes de Mérida of Venezuela, which gave ages of about 600 m yr, practically no radiometric age determinations of possible Precambrian rocks have been carried out in the Andean belt. In spite of this fact, it is almost certain that in addition to the high-grade metamorphics of the Andes de Mérida, the metamorphic rocks of the Perijá, Santander, Quetame, and Garzón massifs of the eastern Cordillera of Colombia, the Mazamorras-Sibundoy complex of the central Cordillera of Colombia and its southern continuation into the Cordillera Real of Ecuador, and the metagraywackes and slates of the eastern Cordilleras of northern Argentina have a late Precambrian age. The last-named rocks (Puncoviscana formation) are intruded by anorogenic granites and granodiorites which have given early Cambrian K/Ar ages.

The eastern Cordilleras of northern Argentina grade S into the Pampean Ranges Massif, a vast spindle-shaped basin and range region formed chiefly of high metamorphics and acid intrusives with a patchy covering of continental Upper Paleozoic and younger deposits. Radiometric age determinations have been carried out in rocks of the central part of the massif (Córdoba and San Luis). Except for a tonalite from Córdoba, which gave 735 ± 75 m yr, the rocks have early Paleozoic ages: the gneisses are about 540 m yr old, and the granitic intrusives range between 465 and 515 m yr old. This seems to indicate that the rocks of the Pampean Ranges Massif belong in the Pan-American orogenic cycle.

As regards the northern Patagonian Massif, the few Rb/Sr age determinations carried out so far on whole rock samples of granites, monzonites, quartz diorites, and adamellites have given late Permian ages of about 230 m yr. However, these rocks intrude undated metamorphics, which are probably Precambrian. This age of the "basement" is suggested by the

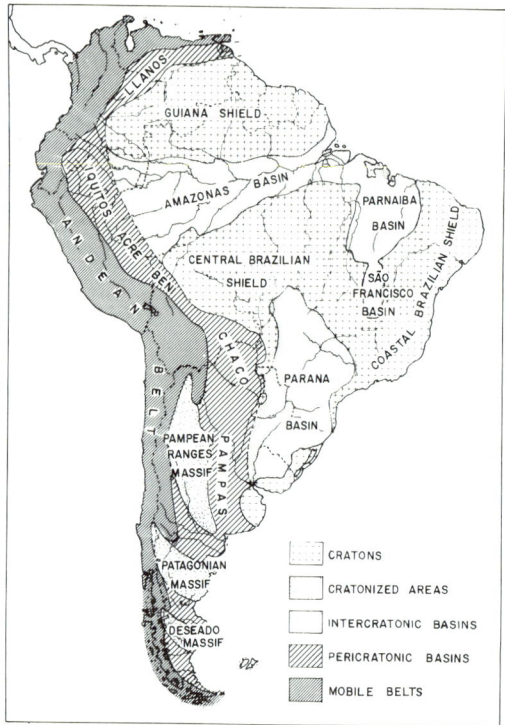

FIGURE 1. Phanerozoic geotectonic elements of South America.

presence of nonmetamorphic fossiliferous Silurian clastics in the area, and very likely also of Ordovician limestones.

Phanerozoic. No attempt will be made here to explain the Phanerozoic evolution of the continent in terms of plate tectonics, as any model, or models, that could be proposed at present must be based, perforce, on hazardous oversimplifications and *ad hoc* supplementary hypotheses. On the other hand geosynclinal terminology will be avoided, as it has lately become evident that the so-called geosynclinal deposits of the Andean belt are, basically, continental shelf sediments. It is clear, however, that since early Paleozoic time the geologic history of South America has been controlled by the location and interplay of the different geotectonic elements that form the structural framework of the continent. These elements are shown sketchily in Fig. 1.

Paleozoic. *Lower Paleozoic (Cambrian to Devonian).* Early in the Cambrian, the western belt of the continent developed negative tendencies and was invaded by the sea. Lower, Middle, and Upper Cambrian fossiliferous limestones were deposited in Western Argentina, Middle Cambrian clastics in Northern Argentina and Southern Bolivia, and Middle Cambrian limestones in the eastern Cordillera of Colombia. At the close of the Cambrian, Sardian movements gently warped and uplifted these beds (Fig. 2).

After a short erosion period, the western belt subsided once again and a very thick sequence of marine Ordovician beds accumulated. Fossiliferous sediments are known from Venezuela to Western Argentina. In Middle Tremadocian time, a short-lived, highly localized, Alpine-type glaciation occurred in northern Argentina. The Ordovician invasion reached a maximum during the Llanvirnian, when a shallow marine transgression flooded the Amazon Basin and a narrow seaway, interposed between the Pampean Ranges and northern Patagonian Massif, which connected the Precordilleran trough of western Argentina with the southern proto-Atlantic Ocean. By Llandeilian time the sea was already regressing and the Amazon Basin was abandoned. At the close of the Ordovician, Taconian movements gently warped, tilted, and uplifted the thick sedimentary prisms. These movements, as well as the Sardian, were not attended by intrusions in the mobile belt. On the other hand, igneous rocks ranging from early Cambrian to late Ordovician age, related to the last phase of the Pan-American orogenic cycle, intruded vast areas of the Pampean Ranges Massif and of the Cariri, Paraíba, and Paraguay-Araguaia belts.

During the early Silurian (Llandoverian), the western belt of the continent, the Amazon Basin, the Iquitos-Acre-Beni-Chaco region, eastern Paraguay, and the seaway connecting the Precordillera with the southern proto-Atlantic Ocean were invaded by the sea. In addition, a thick clastic sequence accumulated in the aulacogene of the Southern Hills of Buenos Aires in eastern Argentina. In the eastern Cordilleras of Bolivia, the granulometric gradient plainly shows that most of the Silurian sediments were derived from a western source area, probably formed of uplifted Cambrian and Ordovician rocks. In Middle Silurian (Wenlockian) time the sea abandoned the Amazon basin and eastern Paraguay, but sedimentation continued in the western troughs during the Ludlovian. The end of the Silurian was marked by extremely mild Caledonian movements which gently warped, tilted, and uplifted the sedimentary rocks. A most striking and baffling exception are the gneisses of the southern coast of Peru and of the neighboring Arequipa area, which were metamorphosed 400 m yr ago, i.e., at the close of the Silurian. Near Camaná, the gneisses are overlain by fossiliferous Devonian beds.

In early Devonian (Coblentzian) time renewed subsidence of the Andean troughs, from Colombia to southern Chile, brought about a new marine invasion that followed the pattern set up in the Silurian, but the sea was able to overflow the western troughs and the largest of all marine transgressions that ever affected the continent invaded the Amazon and Parnaíba basins, the Iquitos-Acre-Beni-Chaco region, the Paraná Basin, the Southern Hills aulacogene, the seaway between the Pampean Ranges and northern Patagonian massifs, and the Falkland Islands. In Colombia the Devonian trough had a "Mediterranean" character, interposed between the Guiana Shield in the E and the then uplifted central Cordillera-Santa Marta Hills in the W, which was the main source area of the sediments. In late Eifelian time the Paraná Basin was abandoned by the sea and in the late Givetian the same thing happened with the Amazon and Parnaíba basins, but in many parts of the western troughs mixed marine and continental sedimentation continued well into the Frasnian.

At the close of the Devonian, or shortly after, the western belt of the continent was affected by Acadian or Bretonian movements. They were very intense in Colombia, Ecuador, central Peru, and western Argentina, where the

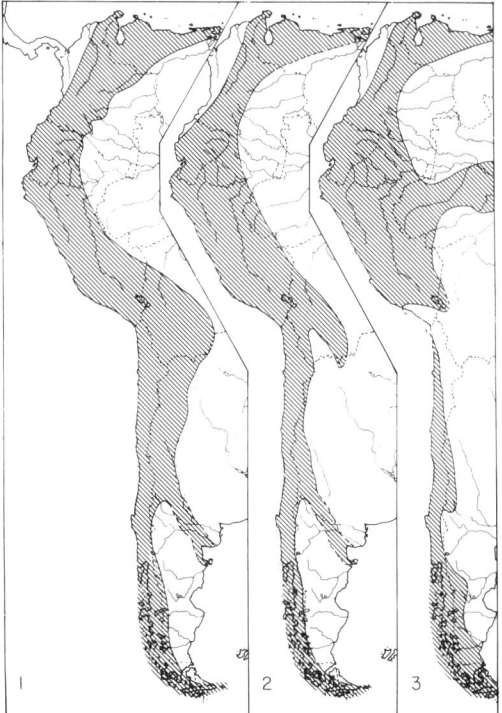

FIGURE 2. Three stages in the development of the Paleozoic marine western troughs of South America: 1, Middle Ordovician (Llanvirnian); 2, late Middle Devonian (Upper Givetian); 3, Middle Pennsylvanian (Desmoinesian).

Devonian and older sediments were strongly folded and in places metamorphosed, but they were very gentle to practically imperceptible in southern Peru, Bolivia, and northern Argentina. These were the first truly tecto-orogenic movements to affect long stretches of the Andean belt and cause the uplift of important mountain ranges.

Upper Paleozoic (Mississippian to Permian).
In Mississippian time only the Andean areas of Colombia and Chile were invaded by the sea. In western Argentina, interfingering marine and continental beds were deposited, whereas in western Peru, southern Bolivia, and northern Argentina only continental beds accumulated. Localized acid volcanic activity was felt in Peru and Argentina, and several small Alpine-type glacial centers appeared between southern Bolivia and western Argentina. In the extra-Andean region only the Parnaíba Basin received mixed marine and continental sediments. At the close of the Mississippian, or early in the Pennsylvanian, the first of a series of Hercynian movements affected the Andean belt. They were intense in western Argentina, but gentle to very weak elsewhere.

In Middle Pennsylvanian time the sea once again invaded the Andean area, from Venezuela to Southern Chile, and transgressed over the Amazon and Parnaíba basins. Conditions in Western Argentina and conterminal areas of Chile were similar to those prevailing in the Mississippian. In Late Pennsylvanian time the sea abandoned the extra-Andean basins, and a glaciation of continental proportions spread from several centers over large areas of southern Brazil, Uruguay, Paraguay, southern Bolivia, and Argentina. In the Paraná Basin and in the Southern Hills aulacogene, shallow-water marine transgressions alternated with the glacial episodes. The glaciation of South America was adjacent in a predrift assemblage to that of South Africa and other parts of the "Gondwana" continent. Sometime near the close of the Pennsylvanian, moderate tecto-orogenic movements and uplifts affected the Andean belt between Colombia and Western Argentina.

In early Permian time the Andean troughs from Colombia to southern Chile were invaded by the sea. At the same time a general amelioration of the climate brought about the end of the Late Pennsylvanian glaciation and continental beds accumulated in the Parnaíba and Paraná basins, and in the Chaco-Pampas region. In the aulacogene of the Southern Hills of Buenos Aires, continental beds bearing *Glossopteris* flora remains were deposited, alternating with shallow marine clastics bearing an Australian *Eurydesma* fauna.

Strong to moderate movements at the end of the early Permian affected the Andean belt, which was abandoned by the sea. These movements were very intense in the Concepción area of coastal Chile and in the Pincipal Cordillera of western Argentina, where the Upper Paleozoic rocks were folded and metamorphosed into micaschists between 270 and 245 m yr ago. During late Permian time, only continental red beds and vulcanites accumulated in Peru, from Leimebamba to Lake Titicaca, whereas eolian and fluviatile sediments were deposited in the Paraná Basin, the Chaco-Pampas region, and the Southern Hills aulacogene. At the close of the Permian, renewed tecto-orogenic movements were felt in the Andean belt, whereas epeiorogenic uplifts affected the extra-Andean regions of the continent. The several phases of the Hercynian orogenic cycle were attended by acid to mesosilicic intrusions and extrusions which in Chile, the Principal Cordillera of western Argentina, and the northern Patagonian Massif have Rb/Sr and K/Ar ages ranging between 270 and 225 m yr. The few radiometric age determinations carried out so far in the northern part of the Andean Belt have not detected comparable rocks, in spite of the fact that in all likelihood they are also present.

Mesozoic. Except for marine and continental clastics with intercalated vulcanites in central Chile, and andesitic flows and tuffs in western Argentina succeeded by thick continental deposits, no Lower or Middle Triassic rocks are known in South America. The Scythian and Anisian beds of Argentina contain reptilian and amphibian faunas closely comparable to those of South Africa. An acid-mesosilicic volcanic cycle began in the Ladinian, climaxed in the Karnian, and dwindled in the Norian, affecting some areas of Colombia and large regions between northern Chile and southern Patagonia. During the Upper Triassic, continental beds continued to accumulate in western Argentina, bearing a rich *Dicroidium* flora and reptilian remains, and some basic volcanism was felt. Early in the Norian the sea invaded the northern Andean Belt, from western Colombia to southern Bolivia, whereas eolian and fluviatile sediments were deposited in the Paraná, Parnaíba, and Amazon basins in central Venezuela and in northeastern Colombia. In the Late Triassic "germanotype" movements affected some western areas and sedimentation ceased, whereas granitic intrusions occurred in the Principal Cordillera of western Argentina and in the eastern Cordilleras of northern Bolivia (Fig. 3).

During Jurassic and Cretaceous times, three sedimentary troughs appeared *en échelon* along the western border of the continent, separated by two subpositive areas. They may be termed the "Venezuelan-Peruvian," "Chilean-Argen-

FIGURE 3. Three stages in the development of the Mesozoic marine western troughs of South America: 1, Upper Triassic (Carnian); 2, Lower Jurassic (Liassic); 3, Lower Cretaceous (Neocomian).

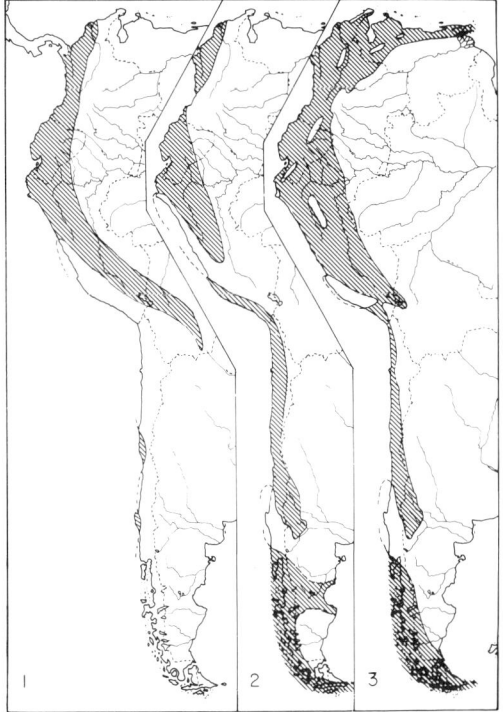

tine," and "Patagonian" troughs, respectively. Each basin had its own complex and distinctive geologic history, characterized by dissimilar cycles of marine and continental deposition and erosion. However, all of them were affected by moderately strong Nevadan differential uplifts, succeeded by very intense volcanic activity during the Kimmeridgian. It is likely that the extreme folding of the Southern Hills of Buenos Aires aulacogene, which was not attended by intrusions or extrusions, took place in late Jurassic time.

No Jurassic sediments are known in the extra-Andean areas, but some basic lavas were extruded in the Paraná Basin. These were succeeded in Early Cretaceous time by a sequence of basaltic flows that spread over the entire basin covering an area in excess of 1,000,000 km^2 (385,000 sq mi) and reaching a maximum thickness of about 1000 m. At the same time, basic sills intruded the sedimentites of the Parnaíba and Amazon basins and of the Pampas region. Basic volcanic activity was also felt in central and western Argentina. Immediately after this major volcanic episode, the Paraná, São Francisco, Parnaíba, and Amazon basins were the site of continental deposition. Early in the Neocomian, and even maybe in the Late Jurassic, several taphrogenic troughs began to subside along coastal Brazil and Argentina, receiving great thicknesses of continental to brackish water sediments. Marine beds were accumulated in some of them from the Albian-Aptian onward.

During intra-Senonian time, fundamental changes were introduced by Subhercynian tecto-orogenic movements which affected the three western troughs. These movements, which were preceded by intense volcanic activity and attended by the intrusion of huge granodioritic batholiths, strongly folded the Mesozoic and older rocks and brought about the initial uplift of the Principal Cordillera, from southern Argentina to Colombia. As a result the Venezuelan-Peruvian basin between northern Colombia and Ecuador was split longitudinally into two troughs (E and W), the Chilean-Argentine basin disappeared, and the Patagonian trough was considerably reduced. Continental beds accumulated in different extra-Andean regions.

Cenozoic. In Danian time an ephemeral Atlantic transgression flooded the subnegative areas of Patagonia and the taphrogenic troughs of coastal Buenos Aires.

During the Tertiary not less than four orogenic movements, each one consisting of several phases, intermittently uplifted the western belt of the continent into what is now the complex Andean Cordilleras. The movements, mainly vertical but occasionally strongly tangential, were attended by acid intrusions and volcanic activity.

As the different mountain units were intermittently uplifted, more or less continuous sedimentation persisted in the intermontane basins and along the foot of the rising ranges, the resulting piedmont deposits becoming involved in subsequent orogenic movements. Marine to brackish water accumulation prevailed in the Venezuelan basins and in the coastal belt of Colombia, Ecuador, and Peru until Middle Miocene time, but the Patagonian trough was abandoned by the sea in the Oligocene with the uplift of the Coast and Patagonian Cordilleras. East of the rising mountains, from northern Colombia to southern Patagonia, great thicknesses of continental beds accumulated during different Tertiary intervals. In late Miocene to late Pliocene times, most of the Andean ranges were strongly uplifted, whereas the Pampas Ranges Massif was broken up into numerous fault blocks forming a complex basin and range system. These movements, which were attended by intense volcanic activity in Colombia, Ecuador, Peru, Bolivia, Argentina, and Chile, were felt all over the continent and were responsible for the uplift of the Brazilian and Guianan highlands.

During Pleistocene time there was further rapid uplift in the Andes and intense volcanism. The orogenic and volcanic activity persists to this day, as witnessed by the strong earthquakes that frequently affect the Andean belt, and by the great active volcanoes of Colombia, Ecuador, and southern Chile. Numerous Alpine-type glacial centers appeared along the high Cordilleras between northern Colombia and western Argentina. South of that latitude a true Pleistocene continental glaciation covered most of Patagonia. Three or four periods of advance and retreat of the ice have been reported from different areas. At present extensive ice fields exist only in the Patagonian-Fueguian Cordillera, but small glaciers persist further N, mainly in the high Principal Cordillera of Chile and Argentina and in the Cordillera Blanca of Peru. During the cold phases the tropical regions became extensively dessicated, resulting in pedimentation and valley filling, whereas extensive dune fields developed in Argentina and southern Brazil.

Mineral Deposits

South America has a wealth of mineral deposits, both metallic and nonmetallic. Some of them such as the large nickel deposits of central Brazil, the rich iron-manganese ores along the Brazilian-Bolivian border, the huge tin reserves of Rondonia in western Brazil, and the important porphyry copper deposits of western Argentina—are not yet exploited. Here only the principal mineral production of South America, including fuels, will be listed according to the different producing countries, mentioned in alphabetical order. *Argentina:* lead, zinc, iron, tin, tungsten, uranium, oil, coal. *Bolivia:* tin, silver, tungsten, copper, lead, zinc, antimony, bismuth, gold, oil. *Brazil:* iron, manganese, rock crystal, diamonds, tungsten, beryllium, tantalum, niobium, thorium, mica, zircon, coal, oil. *Chile:* copper, lead, zinc, gold, silver, molybdenum, bismuth, mercury, tungsten, niobium, iron, manganese, sodium saltpeter, iodine, coal, oil. *Colombia:* platinum, emeralds, gold, coal, oil. *Ecuador:* gold, silver, oil. *Guianas:* bauxite, gold, manganese, diamonds. *Peru:* copper, vanadium, bismuth, lead, zinc, silver, tin, antimony, molybdenum, tungsten, iron, selenium, sulfur, coal, oil. *Venezuela:* iron, gold, diamonds, oil.

HORACIO J. HARRINGTON*

*Deceased, Dec. 20, 1973.

References

Ahlfeld, F., 1970. "Zur Tektonik des andinen Bolivien," *Geol. Rundschau,* 59(3), 1124–1140.

*Almeida, F. F. M. de, 1971. "Geochronological division of the Precambrian of South America," *Rev. Brasil. Geociencias* (Sao Paulo), 1(1), 13–21.

___, 1972. "Tectono-magmatic activation of the South American platform and associated mineralization," *24th Intern. Geol. Congr.,* Canada, 3, 339–346.

Aubouin, J., 1972. "Chaines liminaires (Andines) et chaines géosynclinales (Alpines)," *24th Intern. Geol. Congr.,* Canada, 3, 438–456.

Bürgl, H., 1967. "The orogenesis in the Andean system of Colombia," *Tectonophysics,* 4(4–6), 429–443.

Casertano, L., 1963. "General characteristics of active Andean volcanoes and a summary of their activities during recent centuries," *Seismol. Soc. Am. Bull.,* 53, 1415–1433.

Childs, O. E., and Beebe, B. W., eds., 1963. *Backbone of the Americas* (Mem. 2). Tulsa: Am. Assoc. Pet. Geologists.

Cobbing, E. J., 1972. "Tectonic elements of Peru and the evolution of the Andes," *24th Intern. Geol. Congr.,* Canada, 3, 306–315.

Commission de la Carte Géologique de Monde, 1964. *Carte Géologique de l'Amérique du Sud.* Rio de Janeiro, 1:5,000,000.

Coney, P. J., 1971. "Structural evolution of the Cordillera Huayhuash, Andes of Peru," *Bull. Geol. Soc. Am.,* 82, 1863–1884.

Creer, K. M., 1970. "Review and interpretation of palaeomagnetic data from the Gondwanic continents," *Second Gondwana Symposium, South Africa, Proc. and Pap.,* Pretoria: CSIR, Scientia, 55–72.

___, 1972. "A palaeomagnetic survey of South American rock formations," *Phil. Trans. Roy. Soc. Lond.,* 267 (1183), 98p.

Dalziel, I. W. D., and Cortes, R., 1972. "Tectonic style of the southernmost Andes and the Antarctandes," *24th Intern. Geol. Congr.,* Canada, 3, 316–327.

Francheteau, J., and Le Pichon, X., 1972. "Marginal fracture zones as structural framework of continental margins in south Atlantic Ocean," *Bull. Am. Assoc. Petrol. Geologists,* 56, 991–1007.

Gansser, A., 1973. "Facts and theories on the Andes," *J. Geol. Soc. Lond.,* 129(2), 93–132.

*Gerth, H., 1932–1941. "Geologie Sudamerikas," *Regionalen Geologie der Erde.* Berlin: Gebr. Borntraeger, 3 vols.

___, 1955. "Der geologische Bau der südamerikanischen Kordillera," *Regionalen Geologie der Erde.* Berlin: Gebr. Borntraeger, 266p.

Ham, C. K., and Herrera, L. J., Jr., 1963. "Role of Subandean fault system in tectonics of eastern Peru and Ecuador," in O. E. Childs and B. W. Beebe, eds., *Backbone of the Andes* (Mem. 2). Tulsa: Am. Assoc. Petrol. Geologists, 47–61.

Hamilton, W., 1969. "The volcanic central Andes—a modern model for the Cretaceous batholiths and tectonics of western North America," *Proc. Andesite Conference, Oregon, 1968, Bull. Dept. Geol. Min. Indus.,* 65, 175–184.

*Harrington, H. J., 1962. "Paleogeographic development of South America," *Bull. Am. Assoc. Petrol. Geologists*, 46, 1773–1814.

———, 1963. "Deep focus earthquakes in South America and their possible relation to continental drift," *Polar Wandering and Continental Drift, S.E.P.M. Spec. Publ. 10*, 55–73.

Herrero-Ducloux, A., 1963. "The Andes of western Argentina," in O. E. Childs and B. W. Beebe, eds., *Backbone of the Americas* (Mem. 2). Tulsa: Am. Assoc. Petrol. Geologists, 16–28.

James, D. E., 1971. "Plate tectonic model for the evolution of the central Andes," *Bull. Geol. Soc. Am.*, 82, 3325–3346. (Discussion in Helwig, J., 1973. *Ibid.*, 84, 1493–1500.)

Jenks, W. F., ed., 1956. *Handbook of South American Geology* (Mem. 65). New York: Geol. Soc. Am., 378p.

Lohmann, H. H., 1970. "Outline of tectonic history of Bolivian Andes," *Bull. Am. Assoc. Petrol. Geologists*, 54, 735–757.

Lomnitz, C., 1962. "On Andean structure," *J. Geophys. Res.*, 76, 351–363.

Miller, H., 1971. "Das Problem des hypothetischen 'Pazifischen Kontinentes' gesehen von der chilenischen Pazifikküste," *Geol. Rundschau*, 60, 927–938.

Morrison, R. P., 1973. *Geological Structure of South America*. Essex, England: Longman, 550p.

Petersen, U., 1965. "Regional geology and major ore deposits of Central Peru," *Econ. Geol.*, 60(3), 407–476.

Radelli, L., 1967. "Géologie des Andes Colombiennes," *Trav. Lab. Géol. Fac. Sci. Grenoble*, 6, 470p.

Sauer, W., 1971. "Geologie von Ecuador," *Regionalen Geologie der Erde*, vol. 11. Berlin: Gebr. Borntraeger, 316p.

*Singewald, J. T., 1943. "Bibliography of economic geology of South America," *Geol. Soc. Am. Spec. Pap. 50*, 159p.

*Stille, H., 1940. "Einführung in den Bau Amerikas," *Regionalen Geologie der Erde*. Berlin: Gebr. Borntraeger, 717p.

Vicente, J-C., 1972. "Aperçu sur l'organisation et l'évolution des Andes argentino-chiliennes centrales au parallele de l'Aconcagua," *24th Intern. Geol. Congr.*, Canada, 3, 423–436.

Vuilleumier, B. S., 1971. "Pleistocene changes in the fauna and flora of South America," *Science*, 173(3999), 771–779.

*Weeks, L. G., 1947. "Paleogeography of South America," *Bull. Am. Assoc. Petrol. Geologists*, 31, 1194–1241.

Zeil, W., 1964. "Geologie von Chile," *Regionalen Geologie der Erde*, vol. 3. Berlin: Gebr. Borntraeger, 233p.

Cross-references: *Antarctica; Argentina; Brazil; Central America–Regional Review; Chile; Colombia; Easter Island and Sala y Gómez; Ecuador; Falkland Islands; French Guiana; Guiana Shield–Regional Review; Guyana; Juan Fernández; Netherlands Antilles; Panama; Paraguay; Peru; San Ambrosio and San Félix Islands; Sub-Antarctic Islands; Surinam; Uruguay; West Indies.*

*Extensive bibliographies.

SUB-ANTARCTIC ISLANDS

The geology of those islands south of 35°S (latitude of the Cape of Good Hope) and relatively remote from continental crustal masses (Fig. 1) is described here. They fall into four groups: (1) islands on the Scotia Ridge (Scotia Arc) that bathymetrically joins South America to the Antarctic Peninsula; (2) islands on seismically active ridges of the southern Atlantic, Indian, and Pacific Oceans; (3) islands on aseismic plateaux in the southern Indian and Pacific Oceans; and (4) islands on the continental shelf of Antarctica.

Almost all these sub-Antarctic islands have an unpleasant climate, ranging from temperate maritime for the more northerly ones, to polar maritime for those nearer the Antarctic continent. Vegetation is sparse and very specific, e.g., *Acaena ascendens, Springlea antiscorbutica, Poa cooki*, and *Azorella selago* on the islands between 40 and 50°S. Those islands south of 50°S consist almost entirely of bare rock, talus, and beach deposits, partially covered by an ice cap with glaciers flowing off into the ocean. Only the northernmost of the islands considered here, Tristan da Cunha on the Mid-Atlantic Ridge (Fig. 1), is inhabited other than by the personnel of scientific bases. The area S of 60°S is covered by the Antarctic Treaty, the islands N of this latitude are claimed by various nations.

Islands of the Scotia Arc

The South American and West Antarctic segments of the circum-Pacific mobile belt are joined physiographically by the discontinuous submarine Scotia Ridge that forms a loop extending for some 1500 km eastward into the South Atlantic Ocean basin (Fig. 2). It is locally emergent, forming small, but extremely rugged and heavily glaciated islands (Figs. 2, 3, and 5). The E-trending north and South Scotia Ridges are composed of continental rocks that can be correlated with those in both southernmost South America and the Antarctic Peninsula: (1) metamorphic complexes of probable pre-late Paleozoic age; (2) ?late Paleozoic marine sediments, mainly graywacke and shale; (3) Middle to Late Jurassic and Cretaceous calc-alkaline volcanics and sediments; (4) a late Mesozoic to earliest Cenozoic batholithic intru-

FIGURE 1. Map showing the location and geotectonic setting of the sub-Antarctic islands (modified after Girod and Nougier, 1972).

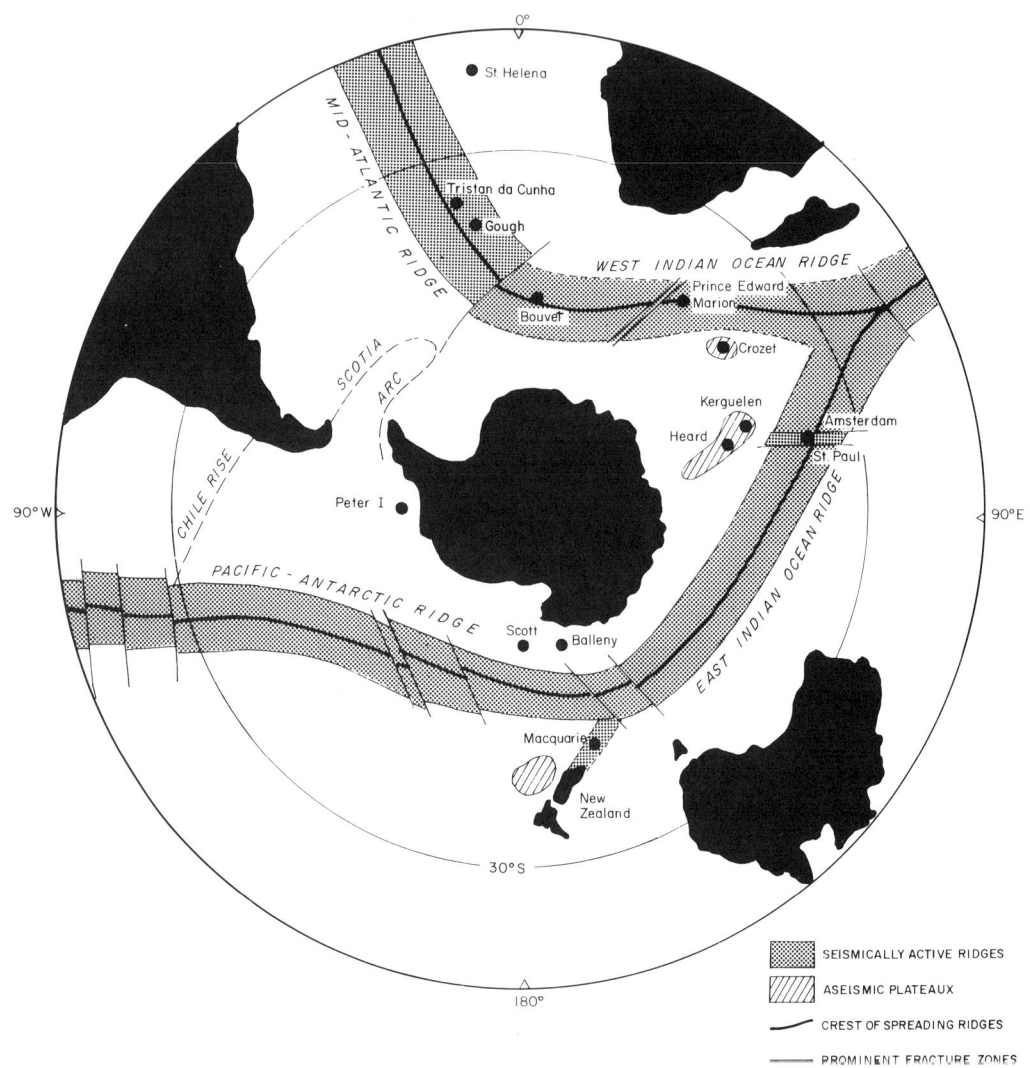

sive complex; and (5) Cenozoic sediments and volcanics (see review by, and references in, Dalziel and Elliot, 1973). Active volcanoes occur at the west end of the South Scotia Ridge. The South Sandwich Islands that close off the Scotia Arc in the E (Figs. 2 and 4) are analogous to the Lesser Antilles in the Caribbean, and consist wholly of recent volcanic deposits. Most of them have active fumaroles, and volcanic eruptions have occurred on several in historic times (Baker, references in Dalziel and Elliot, 1973).

The metamorphic complexes were affected by polyphase deformation and broadly synchronous low- to medium-grade regional metamorphism, probably before the late Paleozoic. The ?late Paleozoic sediments were strongly folded, but little metamorphosed, in the early Mesozoic Gondwanian orogeny, and the Mesozoic and earlier rocks were all deformed during the late Mesozoic-early Tertiary Andean orogeny. It was believed that an almost straight cordillera joined the Andes to the Antarctic Peninsula until the latest Mesozoic, being bent, disrupted, and fragmented during the Cenozoic to form the North and South Scotia Ridges (Dalziel and Elliot, 1973). However, a more complex picture is now emerging (Dalziel, 1974). The South Sandwich volcanic arc developed after this event in the late Cenozoic.

South Georgia (British; Claimed by Argentina). The island of South Georgia on the

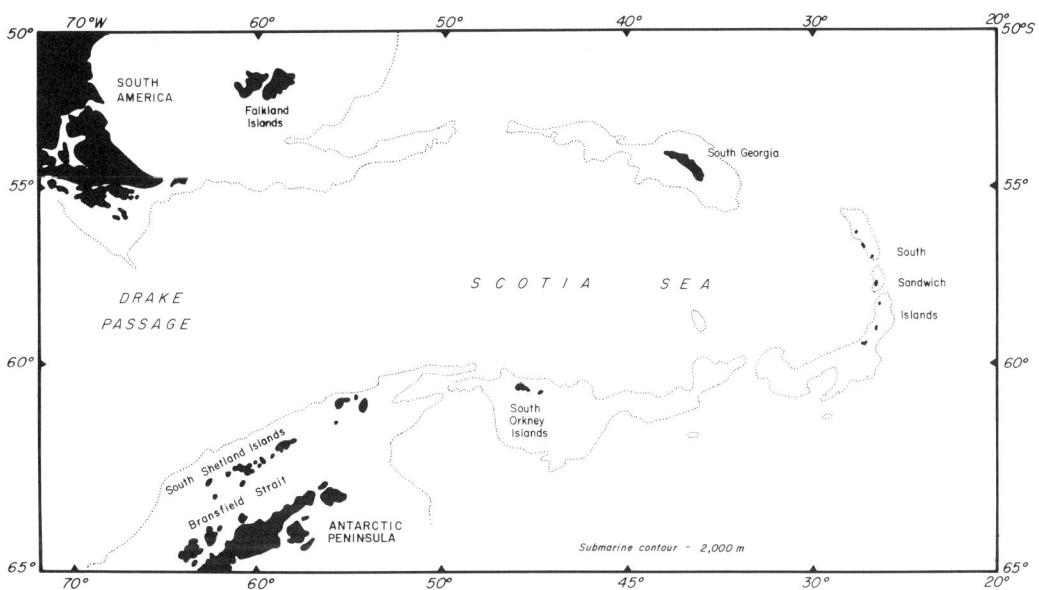

FIGURE 2. Scotia Arc.

North Scotia Ridge (54–55°S, 36–38°W) is over 170 km long and has an average width of less than 32 km. It is elongated in a WNW-ESE direction and has a mountainous spine (the Allardyce Range) rising to over 2800 m. Captain James Cook was first to land there in 1775, and H. Will of the German International Polar Year Expedition (1882–1883) made the first significant geological observation that, unlike most oceanic islands, South Georgia is composed mainly of sedimentary and metamorphic, rather than igneous, rocks. Almost all recent work has been undertaken by the British Antarctic Survey.

Most of South Georgia is underlain by the Cumberland Bay sedimentary sequence derived from a volcanic terrain and interbedded with spilitic lavas. These rocks are probably all early Cretaceous. The sequence is highly folded about axes parallel to the length of South Georgia and overturned to the NNW (Fig. 3).

On the NE there are quartz-rich graywackes and shales known as the Sandebugten sequence, again highly deformed but virtually unmetamorphosed. On lithic grounds this has been correlated with the sediments of ?late Paleozoic age around the Scotia Arc, but it is more likely to be a facies equivalent of the Cumberland Bay (see Dalziel, 1974).

An igneous complex (?"Andean") cuts the

FIGURE 3. Folding in the Cumberland Bay sequence, South Georgia, looking SE (from Trendall, 1953, see Dalziel and Elliot, 1973: British Antarctic Survey photograph).

FIGURE 4. South Sandwich Islands (from Kemp and Nelson, 1931, see Adie, 1964).

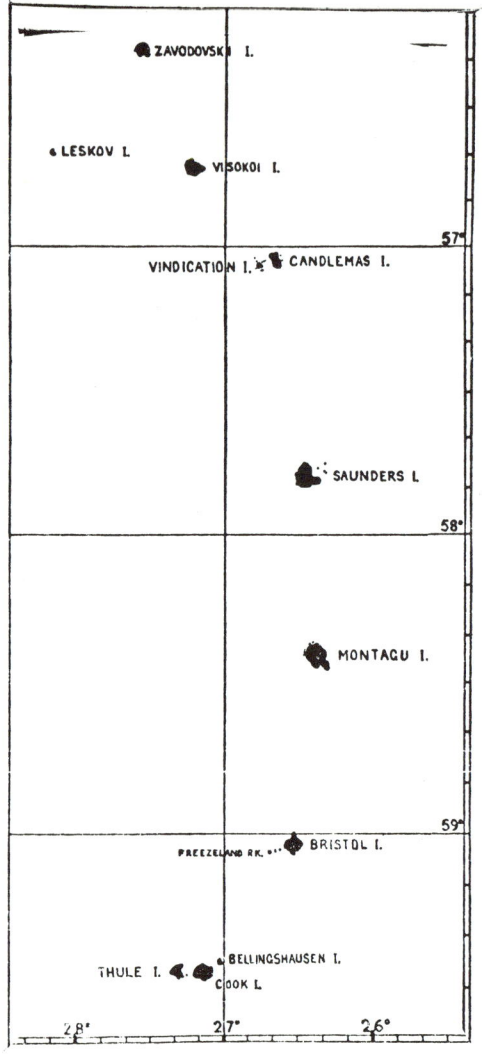

Cumberland Bay rocks at the southeastern tip of the island. The igneous complex and the Cumberland Bay sequence are both cut by posttectonic basic dykes (Adie and Trendall, references in Dalziel and Elliot, 1973).

South Sandwich Islands (British). The 11 islands of the South Sandwich group form a gentle arc (approximately 300 km long and concave to the W) located at 56–59°S, 26–28°W (Figs. 2 and 4). The smallest island, Leskov, is situated 50 km W of the main arc and has a circumference of only 2.5 km. The largest, Montagu, 40 km in circumference, is surmounted by the highest peak in the group, Mt. Belinda (1395 m). Captain Cook discovered the South Sandwich Islands in 1775. The first geologic specimens from the islands were collected by Captain C. A. Larsen of S. S. *Undine* in 1908. Recent work has been led by P. E. Baker of the University of Leeds.

The islands are Cenozoic volcanoes. Eight of the eleven are known to have been active in this century. The rocks are mostly effusive. The oldest dated are 4 m yr, although there is reason to believe that volcanicity was initiated as early as 8 m yr ago (Dalziel and Elliot, 1973). The lavas are considered to have been derived from primary tholeiitic basalt (Baker, see Dalziel and Elliot, 1973). They are estimated to consist of basalt ($<54\%$ SiO_2), 67%; andesite (54–63% SiO_2), 28%; dacite (63–70% SiO_2), 4%; rhyolite ($>70\%$ SiO_2), $<1\%$. The only known rhyolite was erupted from a seamount 56 km NW of Zavadovski Island (Figs. 4 and 5) in 1962.

The South Sandwich Islands are the locus of shallow and intermediate depth earthquakes (Barazangi and Dorman, Isacks and Molnar, see Dalziel and Elliot, 1973).

South Orkney Islands (Antarctic Treaty Area). The South Orkney Islands, on the central part of the South Scotia Ridge at 60°40'S between 44°30'W and 47°W (Figs. 2 and 6), were discovered by George Powell and Nathaniel Palmer in December 1821. The first geologic work of consequence was undertaken by J. Harvey Pirie, the geologist of the Scottish National Antarctic Expedition of 1902–1903 that wintered over in a base on Laurie Island. Work in the western islands has been carried out by the British Antarctic Survey since the mid-1940s, and recently on Laurie, Fredriksen, Powell, and Signy islands by the author.

Coronation, Signy, and Powell islands (Fig. 6) are largely formed of metamorphic rocks (Adie, 1964; Dalziel, see Dalziel and Elliot, 1973). Garnets and fragments of schist have been reported (by Tilley, see Dalziel and Elliot, 1973) in the Graywacke-Shale Formation of Laurie and Fredriksen islands (Fig. 6) that are lithically similar to the ?late Paleozoic sediments elsewhere in the Scotia Arc region. Hence the metamorphics are assigned to the pre-late Paleozoic "basement" complex. Undeformed conglomerates, believed to be Creta-

FIGURE 5. Zavadovski Island, South Sandwich Islands (from Kemp and Nelson, 1931, see Adie, 1964).

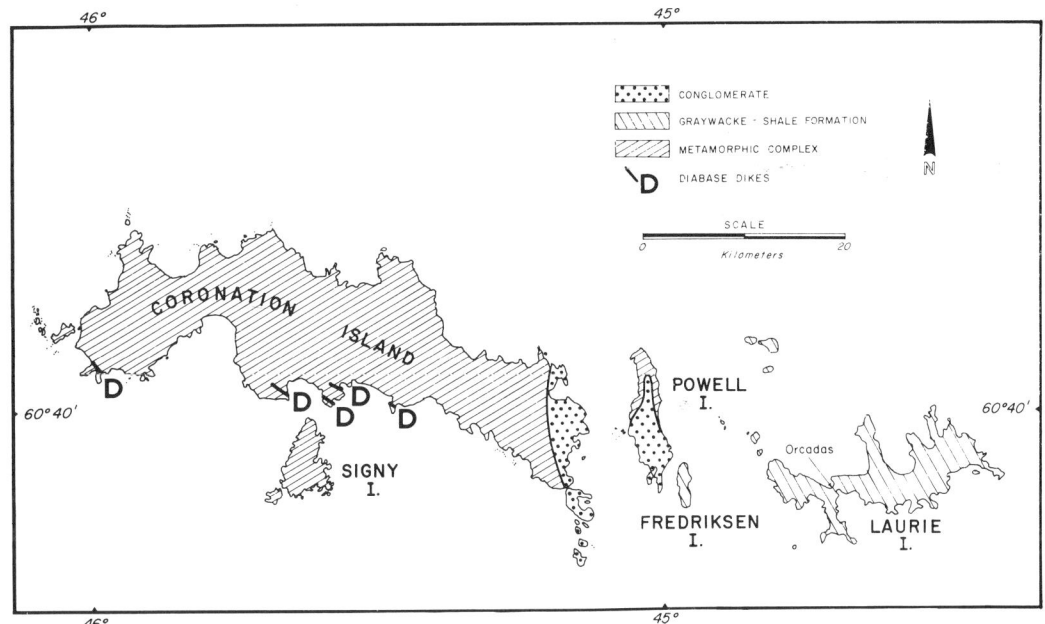

FIGURE 6. Geology of the South Orkney Islands (from Dalziel, 1971, see Dalziel and Elliot, 1973).

ceous, unconformably overlie the metamorphic rocks and the Graywacke-Shale Formation (Adie, 1964).

Posttectonic basic dikes of unknown age cut the basement complex on Coronation Island (Fig. 6), and also reportedly the conglomerates (see Harrington *et al.,* in Adie, 1972).

South Shetland Islands (Antarctic Treaty Area). At the eastern end of the South Scotia Ridge, the narrow but relatively deep Bransfield Strait separates the NE-SW trending chain of the South Shetland Islands from the Antarctic Peninsula (Fig. 2). The South Shetland group lies between 61 and 63°30'S and between 54 and 62°45'W. It was first sighted by William Smith in February 1819. Some of the islands have little relief (e.g., Low and Snow islands), others are extremely rugged. Smith Island rises to over 2000 m. Many early visitors reported volcanism on Deception and Bridgeman islands (Fig. 7). Recent work in the group has been undertaken by the British Antarctic Survey, The Instituto Antarctico Argentino, the Instituto Antarctico Chileno, the Soviet Institute of the Geology of the Arctic, and U.S. scientists, including the author.

The isolated Elephant Island subgroup in the NE, where Sir Ernest Shackleton's expedition took refuge after the loss of the *Endurance* in 1916 (Wordie, 1921), consists of metamorphic rocks (Fig. 7) comparable to those in the South Orkneys (Tyrrell, Dalziel *et al.,* and Dalziel, see Dalziel and Elliot, 1973), and (on Gibbs Island) a dunite-serpentinite complex. The age of the metamorphics is uncertain, but like those in the South Orkneys they are probably pre-late Paleozoic (see Dalziel and Elliot, 1973), and possibly Precambrian (Iltchenko, in Adie, 1972).

Rocks lithically comparable to the ?late Paleozoic sediments of the region occur on Livingston Island (Fig. 7), the second largest of the group. Once again they are unmetamorphosed but highly deformed, being folded (on axes parallel to the South Shetlands chain) probably in the early Mesozoic (Dalziel, in Adie, 1972).

Late Jurassic and Cretaceous sedimentary and volcanic rocks occur on Livingston Island (González-Ferrán *et al.,* see Dalziel and Elliot, 1973), Snow Island, and also on King George Island (largest of the group).

There are a few small "Andean" granitic intrusives, and Cenozoic volcanicity (with a mildly alkaline cast) was widespread. Eruptions of similar material occurred on Deception Island in 1967, 1969, and 1970 after a (historically) long dormant period (Baker *et al.,* Baker and McReath, Schultz, Gonzáles-Ferrán *et al.,* and Orheim, see references in Dalziel and Elliot, 1973).

FIGURE 7. Geology of the South Shetland Islands (modified after Adie, see Dalziel and Elliot, 1973).

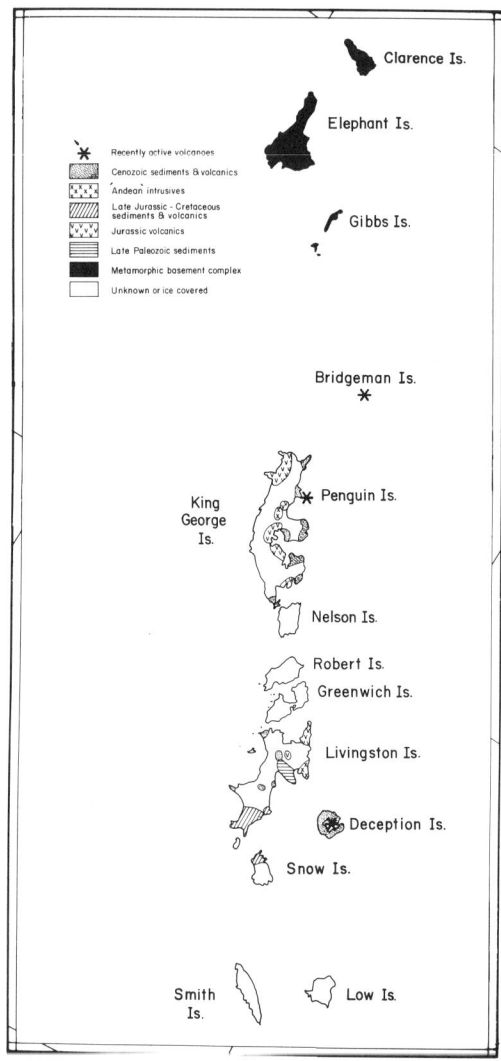

Islands on Seismic Ridges

All but one of the sub-Antarctic Islands in this group are situated on mid-oceanic ridges (Fig. 1) known to be characterized by extensional stress and normal faulting. The one exception, on the complex Macquarie Ridge that bathymetrically joins New Zealand to the Pacific-Antarctic mid-oceanic ridge, is located where there is a component of extension normal to the trend of the ridge (Hayes et al., in Adie, 1972). Hence it is not surprising to find that all these islands consist almost entirely of Cenozoic volcanic rocks.

Tristan da Cunha group (British). Tristan (as it is known), the only sub-Antarctic island now inhabited by a nonscientific community (Edinburgh settlement), is located on the eastern margin of the Mid-Atlantic Ridge at 37°05'S, 12°17'W. It was discovered by the Portuguese admiral Tristão d'Acunha in 1506. The main island, 12 km in diameter, is a near perfect volcanic cone (Fig. 8) rising to an elevation of 2060 m from the slopes of the Mid-Atlantic Ridge 3700 m below sea level (Fig. 9). Inaccessible and Nightingale islands are smaller, the former measuring 4.8 X 4 km and rising to nearly 600 m, the latter 2.5 km^2 in area with a relief of 400 m. They are now inactive.

Isolated rock specimens were collected by various expeditions including the *Challenger* in 1885, H.M.S. *Odin* in 1904, and the M.V. *Quest* in 1921–1922. The first systematic geologic study was undertaken by Dunne of the Norwegian Antarctic Expedition in 1937–1938. Following an eruption that forced the temporary evacuation of the islanders in 1961 a Royal Society of London expedition carried out an extensive study of the group in 1962 (Baker *et al.*, 1964, see Girod and Nougier, 1972).

The main island (Fig. 8) consists of subhorizontal basalt flows intercalated with red tuffs, the amount of pyroclastics decreasing outward from the main vent where there is a radial initial dip. There are buried parasitic cones and dykes. Inaccessible Island is similar, but has more small intrusives. Nightingale Island consists of porphyritic trachyte flows and intrusives into yellow ash, and agglomerate.

Gough Island (Formerly Gonçalo Alvarez– British). At 40°19'S, 9°56'W there is another island on the eastern flank of the Mid-Atlantic Ridge (Fig. 1). Gough Island measures 14 km X 6 km. It is rugged and deeply dissected, has two summits of approximately 900 m, and was discovered by Gonçalo Alvarez (another Portuguese seaman) about 1506. In 1731 it was rediscovered by Gough on board the *Richmond*. Specimens collected by a New London, Connecticut, whaling captain were described in 1893 by L. V. Pirsson (see reference by Girod and Nougier, 1972). Important recent work was undertaken by R. W. Le Maitre in 1955–1956 and published in 1960 and 1962 (see also reference list of Girod and Nougier, 1972).

Five phases of volcanicity can be recognized (see section, Fig. 10). From oldest to youngest they are basaltic flows in the E (Lower Basalts); trachytic flows in the W (Lower Trachytes); basaltic flows (Middle Basalts); trachytic domes (Upper Trachytes); recent basaltic flows (Upper Basalts).

Bouvet Island (Also Commonly Known as

FIGURE 8. Aerial photograph of Tristan da Cunha taken by the Royal Air Force on 3 April 1961 from 42,000 ft. (from Baker et al., 1964, see Girod and Nougier, 1972).

Bouvetøya–Norwegian). The island of Bouvet is a flattened stratovolcano located at 54°26'S, 3°24'E atop the southernmost part of the Mid-Atlantic Ridge (Fig. 1). It is 9.5 km (E-W) × 7 km (N-S), and is surmounted by an ice-covered plateau reaching 935 m (Fig. 11). Although discovered in 1739 by French navigator J. F. C. Bouvet de Lozier, it was 1898 before its position was correctly charted by the *Valdivia* cruise, and 1929 before an account of the geology was published by Holtedahl based on offshore dredging by the Deutsche Tiefsee-Expedition (1898), and landings by the two 1927–1929 Norwegian Antarctic Expeditions (Broch, 1946, see reference list of Girod and Nougier, 1972).

The geology as a whole is poorly known. The central vent of the volcano has produced alternately lavas, tuffs, and pyroclastics. A benmoreite eruption on the northwestern coast between February 1955 and January 1958 was studied by Baker and Tomblin (see Girod and Nougier, 1972).

Marion Island (South African). A shield volcano, Marion Island, is situated at 46°54'S, 37°45'E near the axis of the western arm of the Mid-Indian Ocean Ridge (Fig. 1). It rises from a depth of 3700 m to over 1200 m above sea level. The island is oval, 24 km (E-W) × 16 km (N-S). Short visits were made by the *Challenger* Expedition in 1875, and by Jeannel in 1938. Verwoerd and other members of the 1965 South African Expedition (see Girod and Nougier, 1972) have undertaken the only extensive geologic study.

A succession of gray lavas, yellow tuffs, and tillites is overlain by black pahoehoe flows erupted from numerous vents and covering most of the island (Fig. 12).

Prince Edward Island (South African). Only 22 km from Marion Island at 46°38'S and 37°57'E lies Prince Edward Island; also on the flank of the western branch of the Mid-Indian Ocean Ridge. Prince Edward covers 52 km² and reaches a height of 675 m. The South African Expedition of 1965 (Verwoerd, see Girod and Nougier, 1972) carried out the first geologic work there. The island is probably a compound subsidiary vent of the volcano forming Marion Island. The same two episodes of volcanicity (older gray lavas and younger black lavas) can be recognized.

Amsterdam Island (French). Located at 37°52'S, 77°30'E, Amsterdam Island lies very near the axis of the eastern branch of the Mid-Indian Ocean Ridge. It was discovered by Sebastian del Cano, a member of Ferdinand Magellan's 1522 expedition. The geology was first studied by the Austrian *Novara* expedition in 1857, and later in more detail by the *Gauss* Expedition in 1901–1903. The stratovolcano that forms the island has parasitic cones radially arranged with respect to a central group of vents and is cut by several faults on the southwestern flank (for references see Girod and Nougier, 1972).

St. Paul (French). First shown on a chart in 1559, St. Paul Island (38°43'S, 77°30'E) lies 80 km S of Amsterdam. (On ancient maps the names of St. Paul and (New) Amsterdam are

FIGURE 9. Bathymetry of the Tristan da Cunha group (from Baker *et al.*, 1964, see Girod and Nougier, 1972).

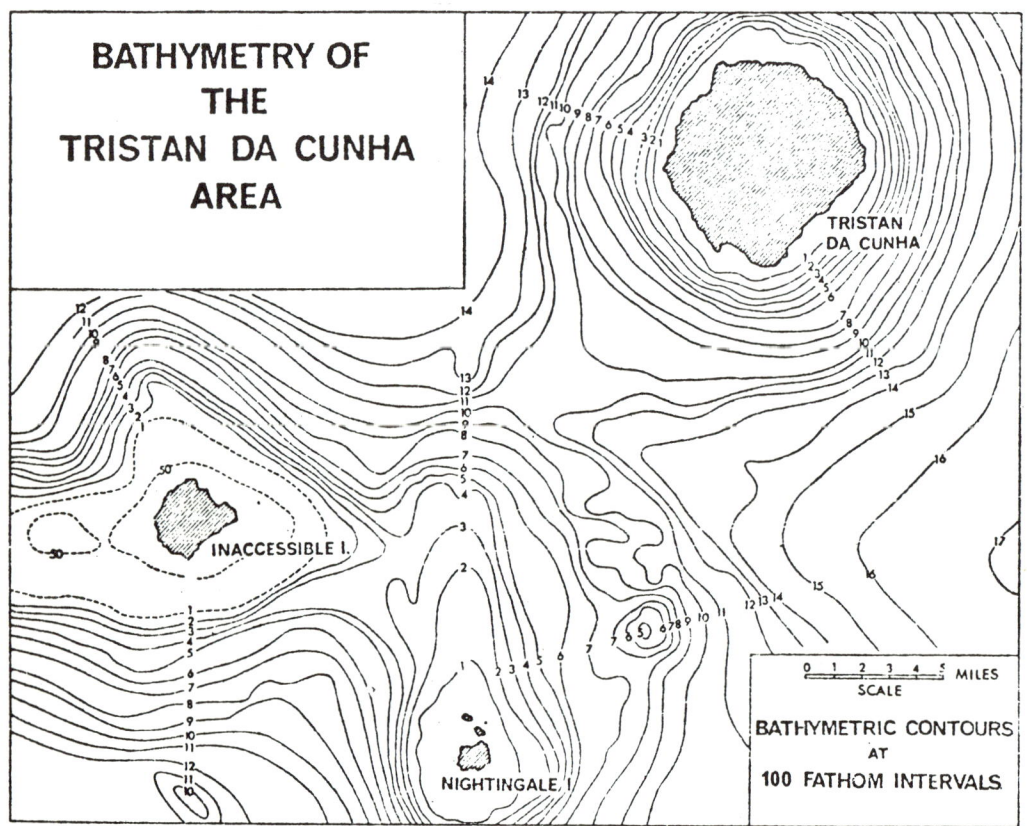

FIGURE 10. Geologic cross section of Gough Island (after Le Maitre 1968, see Girod and Nougier, 1972).

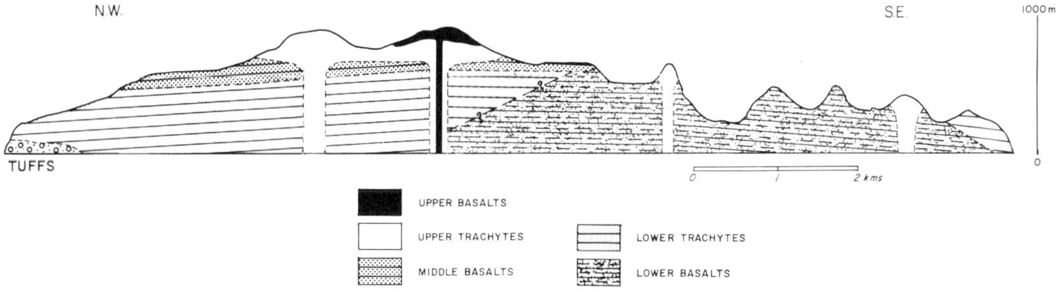

frequently reversed.) It once covered approximately 20 km², but is now less extensive due to the volcanic cone 270 m high being cut by a vast central crater breached by the sea (Fig. 13). Early work there was undertaken by the *Novara* expedition (1857-1859), and by Charles Vélain during the French expedition to observe the transit of Venus (1874-1875). The geology was later studied briefly by E. Aubert de la Rue. Activity in the past was both effusive and explosive. Small cones occur on radial fissures. The deposits consist of thin tholeiitic basalt flows, tuff, and rhyolite flows (for references see Girod and Nougier, 1972).

Macquarie Island (Australian). The Macquarie Ridge joins New Zealand bathymetrically to the Pacific-Antarctic Ridge (Fig. 1). Macquarie Island is located on a short zone with a significant component of extensional stress acting normal to it (Hayes *et al.*, in Adie, 1972). This zone is flanked by others with compressional components as well as the transform (transcurrent) fault motion that characterizes the ridge as a whole. Thus the tectonic situation of Macquarie Island is essentially the same as that of the sub-Antarctic islands on mid-oceanic spreading ridges. It differs only in lying on a shorter spreading ridge segment.

The coordinates of the island are 54°30'S–54°45'S, 158°57'E. It is situated on an extensive submerged platform, measures 38 km (N-S) by 5 km, and consists of a plateau 200-250 m high with central hills in the S rising to 425 m. Captain Fred Hasselborough out of Sydney, New South Wales, reported the island in April 1810, and the first extensive geologic work was undertaken by L. R. Blake during the Australian Antarctic Expedition of 1911 and reported by D. Mawson (see Girod and Nougier, 1972). H. T. Ferrar (see also Girod and Nougier, 1972) earlier collected specimens there.

Macquarie Island is formed mainly of basaltic volcanic rocks, dolerite dike swarms, gabbro, and serpentinized peridotite masses (Mawson, see Girod and Nougier, 1972; Varne and Rubenach, see Fig. 14). According to Varne and Rubenach an upper layer of unmetamorphosed and greenschist facies basaltic extrusive rocks may have overlain amphibolite facies dike swarms, gabbros, and serpentinized peridotites (Fig. 14). They believe that the rocks could represent a section through the crust of an oceanic-spreading ridge. This appears to be consistent with the tectonic setting of the island outlined above.

Islands on Aseismic Ridges

Two main plateaus in the southern Indian Ocean between the east and west branches of the Mid-Indian Ocean Ridge emerge to form the island groups of Crozet, Kerguelen, and Heard (Fig. 1). The nature of these aseismic ridges is not known for certain; some of them may represent microcontinents.

Crozet Islands (French). On a plateau at the edge of the west Mid-Indian Ocean Ridge, the Crozet Islands consist of two groups: Ile aux Cochons, Ile aux Pingouins, and Les Apôtres at 46°S, 50°E; and Ile de l'Est and Ile de la Possession that lie 16 km to the E. The islands were discovered in 1772 by Nicholas Marion-Dufresne. His second in command, Crozet, made a landing. Reference to early geologic work there by the German *Gauss* Expedition

FIGURE 11. Bouvet Island from the E (from Holtedahl, 1929).

FIGURE 12. Geologic map of Marion Island (from Verwoerd and Langenegger 1967, see Girod and Nougier, 1972).

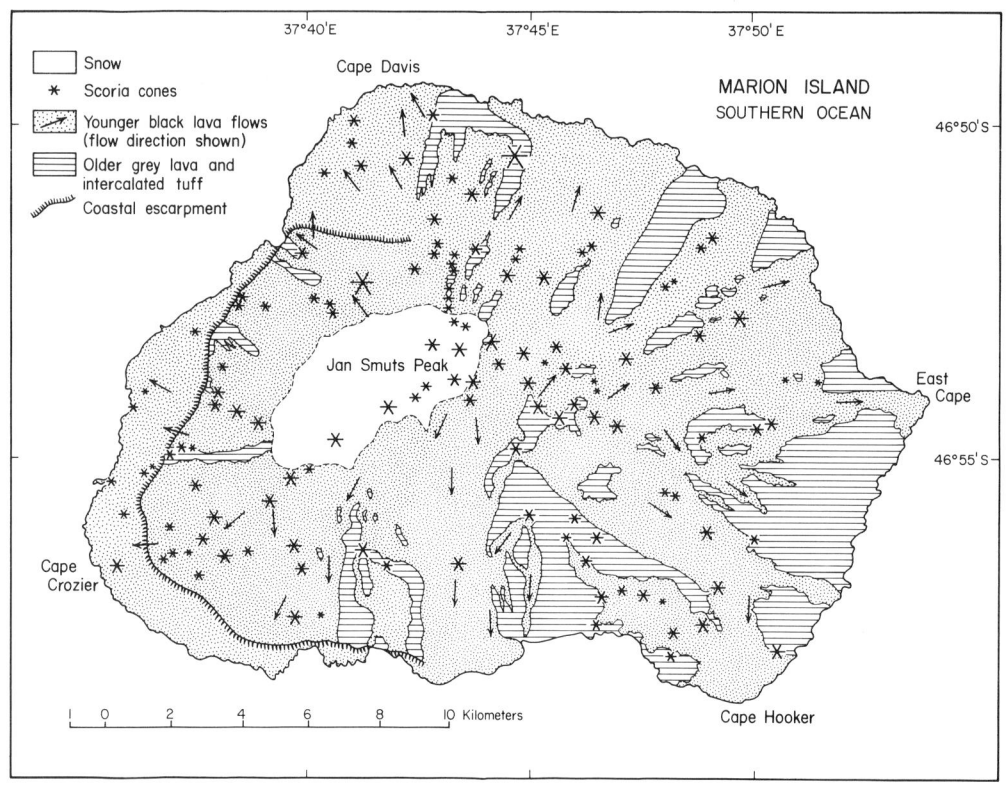

(1901–1903) is made by Girod and Nougier (1972).

Ile aux Cochons is a volcanic cone rising to 600 m and consisting of alternating pyroclastics and flows with marginal scoriaceous cones. Ile de la Possession and Ile de l'Est are each 120–130 km^2 in area. The former rises to 934 m, and the latter to 1100 m. Ile de la Possession has a NW-trending ridge from which 5–20 m thick basalts erupted. These are cut by intrusions of monzonite and syenite. Subsidence preceded the formation of reddish-colored scoriaceous cones. Ile de l'Est is a deeply dissected shield volcano consisting of two basaltic units: a basal one cut by dikes, and an upper complex overlain by layers of pyroclastics and thin basalt flows.

Kerguelen Islands (French). The Kerguelen Islands (48°33'S, 68°30'E) cover 7000 km^2 (Fig. 15). They consist of Ile de Kerguelen (a highly dissected main island) with more than 300 smaller islands and islets including Iles Nuageuses, Ile de l'Ouest, Ile Roche, and Roche Salamanca. The group was discovered by Yves de Kerguelen in 1772. Geologic exploration was conducted in 1929–1931 by E. Aubert de la Rüe, and since 1961 by J. Nougier.

The volcanic complex of the Kerguelen Islands is made up of over 1000 m of subhorizontal plateau basalt flows of Lower Tertiary-Pliocene age erupted from several fissures and shield volcanoes. In the N the plateau basalts are cut by trachyte and rhyolite intrusives; in the SE, flows, sills, domes, and spines of trachyphonolite and phonolite occur in the basalts. A syenite complex covering 350 km^2 in the SW has cut, uplifted, and metamorphosed the basalts. More localized stratovolcanic basalts and trachyphonolites also occur, being cut by Quaternary scoriaceous cones and very recent pumice and trachytic tuffs. There is a long history of faulting (see references in Girod and Nougier, 1972).

Heard Islands (Australian). Heard, Shag, and McDonald islands (52°05'S, 73°30'E) occur on the plateau joining the Kerguelen Islands to East Antarctica (Kerguelen-Heard plateau). According to Ewing and others, and Schlich and others (see Girod and Nougier, 1972) the plateau is covered by layers of sediments more

FIGURE 13. Inside the volcanic caldera of St. Paul Island showing the installations erected by the French expedition to observe the transit of Venus in 1874-1875 (from Velain 1875, see Girod and Nougier, 1972).

than 2000 m thick. Heard is 2785 m high and is a volcano 21 km in diameter at sea level with a secondary crater in the NW (Fig. 16). Early work was undertaken by the *Gauss* Expedition (1901–1903), by Aubert de la Rüe (1929); recent studies by Stephenson (for references see Girod and Nougier, 1972).

A paleocaldera 5 km in diameter forms a plateau 2300 m high topped by the actual summit (Mt. Mawson) of the main volcano (Big Ben) that has steep sides terminating in cliffs and is covered by an ice cap.

The oldest known rocks are early Tertiary pelagic limestones intruded by hypabyssal rocks. Deposition of glacial sediments and submarine lavas (? in the late Tertiary or Pleistocene) followed a period of erosion. After a

FIGURE 14. Geologic map of Macquarie Island (by R. Varne and M. J. Rubenach, Department of Geology, University of Tasmania). Crosses, serpentinized peridotite and gabbro.

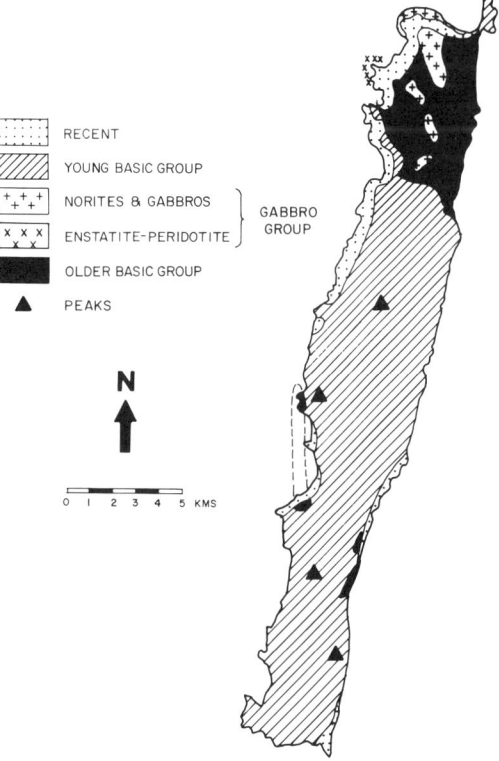

FIGURE 15. Schematic geologic map of the Kerguelen Islands (from Nougier, in Adie, 1972): 1, alkali basalts; 2, syenitic volcano-pluton; 3, alkali trachytes and rhyolites intrusive province; 4, trachyphonolitic and phonolitic intrusive province; 5, Pliocene stratovolcanoes; 6, Pliocene volcanism of limburgites and basanites; 7, small even-grained massifs (nepheline-syenite, diorite, gabbro); 8, Quaternary Vulcanian-type cones.

further short period of erosion in the Big Ben volcano was built up.

The northwest volcanic center (20 km^2) is formed by pyroclastic and partly submarine lavas. Recent volcanicity has resulted in the formation of cinder cones and flows.

Rocks of the Islands on Seismic and Aseismic Ridges (Excluding the Scotia Arc)

Almost certainly basalts predominate over all other volcanic rocks on these sub-Antarctic islands (Table 1). Otherwise the relative abundance of rock types is not reliably known, but Girod and Nougier (1972) have recorded the occurrences (Table 1). Most of the following information is taken from their summary.

Numerous varieties of basaltic rocks have been recorded. Ankaramites and oceanites are known on Tristan, Gough, the Crozet, and Kerguelen islands. Tachylites (like those dredged from the Atlantic Ocean) occur on Macquarie Island. Most of the basaltic lavas are clearly alkaline, the few found to have a tholeiitic composition appear to have been extensively altered.

Most oceanic basalts contain more than 2% TiO_2. The basalts from the islands on mid-

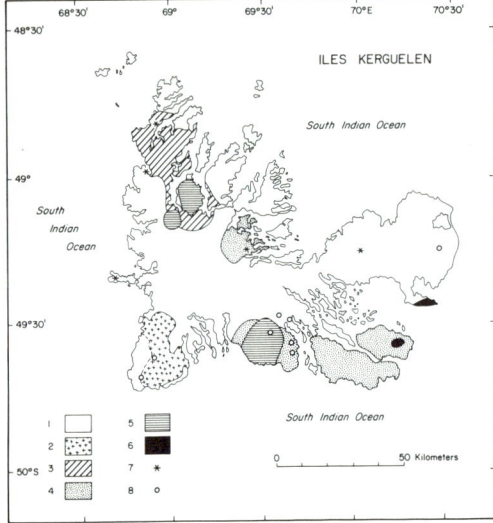

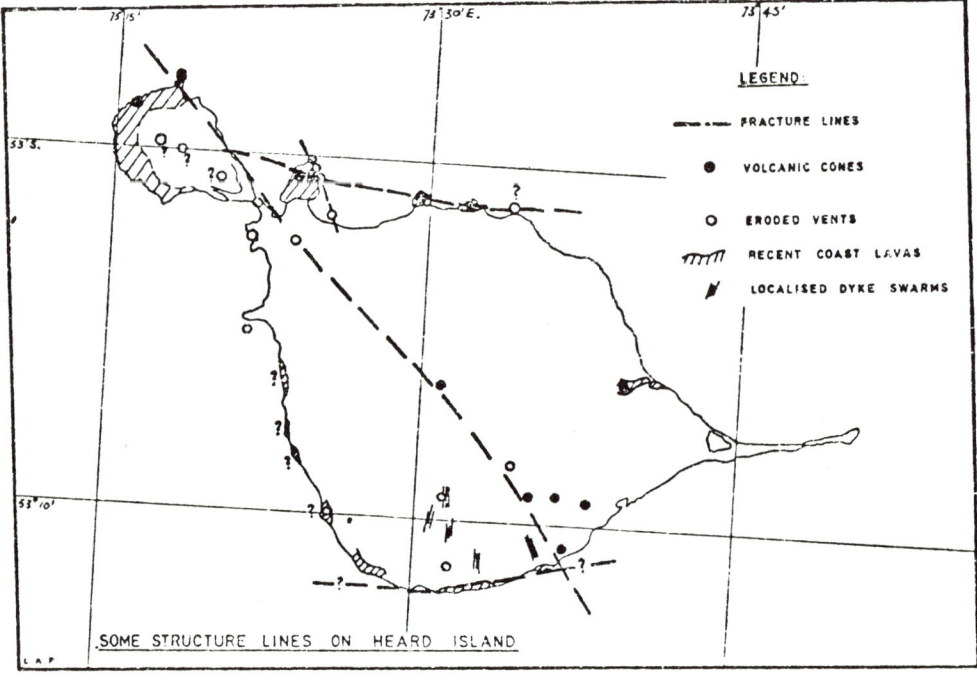

FIGURE 16. Principal structural elements of Heard Island (from Stephenson 1964, see Girod and Nougier, 1972).

Fig. 8. Map showing some of the structure lines on Heard Island.

TABLE 1. Rock Types of the Sub-Antarctic Islands on Seismic and Aseismic Oceanic Ridges

	Tristan da Cunha	Gough	Bouvet	Marion Prince Edward	Amsterdam	St. Paul	Macquarie	Chozet	Kerguelen	Heard
Lavas										
Basaltic	x	x	x	x	x	x	x	x	x	x
Intermediate	x	x	—	?	?	—	x	?	x	?
Trachyte	x	x	x	—	—	—	—	—	x	x
Phonolite	x	—	—	—	—	—	—	—	x	—
Ryolite	—	—	x	—	—	x	—	—	x	—
Other rocks										
Xenoliths	Gabbro	Gabbro	—	—	—	—	—	—	Pyroxenite, peridotite	Dolerite
Intrusive rocks	—	—	—	—	—	—	Gabbro, horite, eucrite, allivalite, harzburgite, dunite	Monzonite, syenite, dicrite?	Monzonite, micromonzonite, granite, syenite	Micromonzonite

Source: After Girod and Nougier (1972).

oceanic ridges conform to this pattern, but some from the Crozet and Kerguelen islands contain less than 2% TiO_2, being in this respect more like basalts erupted on oceanic margins.

The restricted number of lavas of intermediate composition that have been reported from sub-Antarctic islands is possibly the effect of incomplete study. They have been reported from Tristan, Gough, Bouvet, Marion, and Kerguelen.

Trachytes are known on Tristan, Gough, Bouvet, St. Paul, Kerguelen, and Heard islands. Phonolites occur on Kerguelen, and rhyolites on Bouvet and Kerguelen. There seems to be a relationship (according to Baker and others, quoted by Girod and Nougier, 1972) between lava composition and position with respect to a mid-oceanic ridge. Bouvet, situated on a ridge crest, has a strongly differentiated rhyolitic sequence, Tristan and Gough (on ridge flanks) have differentiated trachytic sequences. The farther from the ridge crest, the more undersaturated are the lavas with respect to silica.

A variety of plutonic rocks occurs. Gabbroic xenoliths are common; ultramafic xenoliths are relatively rare. The Crozet Islands, and the Kerguelen and Heard islands, contain monzonites, diorites, syenites, and granites (Kerguelen). A singular igneous suite occurs on Macquarie Island: gabbro, norite, eucrite, allivalite, harzburgite, and diorite. A large syenitic volcano-pluton has recently been discovered in the Kerguelen Islands.

Sedimentary rocks are confined to islands on the aseismic plateaus: Miocene globigerinal oozes on Macquarie, and early Tertiary limestones on Heard Island.

The chronology of volcanic activity in the sub-Antarctic islands is as yet incomplete (Table 2). The aseismic plateaus of Crozet and Kerguelen-Heard (and also Macquarie) may have existed throughout the Tertiary at least. The oldest rocks dated (by K/Ar whole rock method) from the Crozet Islands at 25 m yr, and those from Kerguelen-Heard at 30–25 m yr. Perhaps the initial great outflow of lavas on these plateaus took place in the Lower Tertiary. Subaerial volcanism at Kerguelen continued during the Tertiary. The earliest eruptions on islands on the mid-oceanic ridges took place at this time (18 m yr in the Tristan group). In the Upper Miocene (8.5 m yr) the plutonic complexes of Kerguelen were emplaced. Gough Island was built up in the Pliocene (6 ± 2 m yr) and the stratovolcanoes of Kerguelen 2 m yr B.P.

Volcanicity was common in the recent Quaternary: limburgites and basanites on Kerguelen and Heard; basalts on Crozet, Bouvet, Marion, Prince Edward, St. Paul, and Amsterdam. Present activity is largely confined to fumaroles More significant recent extrusive activity occurred in 1961 on Tristan, and also on the NW of Bouvet Island (Westwind Beach) sometime between February 1955 and January 1958. Fumarolic activity on St. Paul volcano and on the otherwise seismically inactive islands of the Kerguelen-Heard plateau, is particularly interesting.

TABLE 2. Known Geochronologic and Paleomagnetic History of the Oceanic Sub-Antarctic Islands, Excluding the Scotia Arc

After Girod and Nougier (1972).

Geochronologic measurement only, filled triangle; geochronologic and paleomagnetic measurements, open circle for normal magnetism, closed circle for reversed magnetism; estimated chronology only, open triangle. The horizontal lines represent present-day sea level.

Islands off the Continental Shelf of Antarctica

Balleny Islands (Antarctic Treaty area). Some 240 km N of the coast of East Antarctica at 66°–67°S and 162°–165°E lie the Balleny Islands. They consist of three main islands: Young, Buckle, and Sturge, and three smaller ones. Young Island (the largest) is 35 km × 9 km, and rises to over 1000 m.

The Balleny Islands are all composed of volcanic rocks of late Tertiary to Recent age, apparently related to the McMurdo Volcanic Group that occurs on the Antarctic mainland in the vicinity of the Ross Sea. The petrographic province is characterized by the association of hornblende and olivine basalts with phonolitic trachytes and phonolites as well as alkaline basalts.

Ferrar (see reference in Anderson, 1965) records that the volcano forming Buckle Island was active in 1901. Olivine trachybasalt, basalt (of various textures, effusive and pyroclastic), and a coarsely crystalline epidote rock have been reported from the islands (Mawson, 1950, reference in Anderson, 1965).

Scott Island (Antarctic Treaty area). Scott Island lies approximately 300 km NE of Cape Adare in Antarctica (at 67°24'S, 179°55'W). It is only about 1200 m long and 600 m wide. Like the Balleny Islands it is composed of Cenozoic volcanics similar to those of the McMurdo Volcanic Group. Trachyte specimens were collected there by Ferrar (see Anderson, 1965).

Peter I Island (Antarctic Treaty area). Peter I is located in the Bellingshausen Sea off West Antarctica at 68°47'S, 90°35'W. It was discovered by Bellingshausen in 1821, is elongated N-S, and consists of Miocene volcanics some of which have been dated at 13 m yr by the K/Ar whole rock method (Craddock, 1970). The *Odd I* Expedition of Consul Lars Christensen determined the volcanic nature of the island as early as 1927 by dredging off the west coast. Basalt, andesite, and trachyandesite are known. The rocks are gray and reddish colored. The bedding is horizontal.

Sea-Level Changes

The effects of marine erosion are striking in all the sub-Antarctic islands. Yet the numerous signs of relative change in sea level are difficult to interpret.

Around the Scotia Arc, Holtedahl (1929) noted submarine platforms at 100–200 m around South Georgia, 270 m around the South Orkneys, and 245–285 m around the South Shetlands. "Drowned" coast lines are characteristic of all these islands (e.g., Laurie Island, Fig. 6), and hence considerable submergence is indicated. Conversely, the concordant summits of the mountainous spine of South Georgia can be interpreted as remnants of a substantially uplifted and deeply dissected peneplain (Adie, 1964). Raised beaches have been recorded as

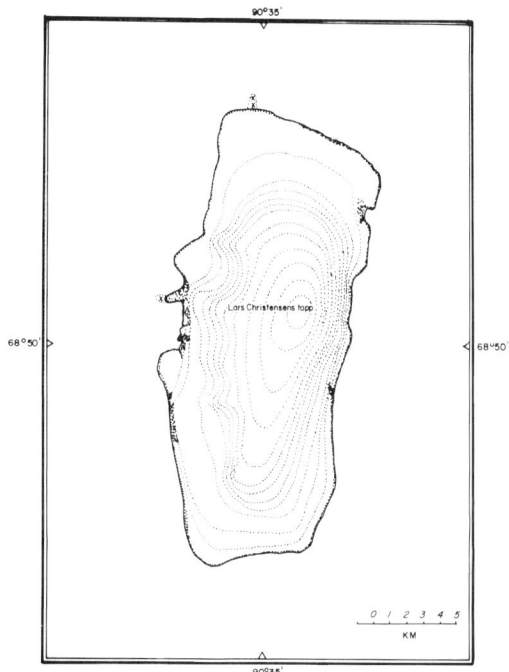

FIGURE 17. Peter I Island (after Holtedahl, 1929).

follows: 2.1–3.4 m and 7.3–8.0 m on South Georgia (with wave-cut platforms up to 50 m); 20–21.5 m and four other levels in the South Sandwich Islands (see Fig. 5); 3.0–4.5 m, 6.0–7.5 m, 9–12 m, and 21–30 m in the South Orkneys (with wave-cut terraces to 40–45 m); 10 levels, ranging from 1.5–2.4 m to 73–88 m on King George Island in the South Shetlands (Adie, 1964).

The various levels are usually ascribed to isostatic adjustments and eustatic changes, but there is also definite evidence of (?longer-term) vertical tectonic movements affecting Pliocene-Pleistocene deposits (Adie, 1964).

The youthfulness of the main island of Tristan da Cunha is shown by the lack of subaerial erosion, the steep cliffs up to 1000 m high on the windward side, and the lack of a submerged platform. The extinct cones of Nightingale and Inaccessible islands are surrounded by extensive platforms less than 185 m deep (Fig. 9), presumably reflecting little recent change in sea level, though the narrow beaches of Tristan itself (Fig. 8) do suggest a slight recent drop. The Crozet Islands do not show any effect of recent (Riss or Würm) glaciation. Raised beaches and wave-cut notches 3–4 m above sea level (locally covered by volcanic deposits) are thought to reflect a eustatic change in sea level at 5500 B.P. (Bellair 1963, see Girod and Nougier, 1972).

Holtedahl (1929) interpreted the gently W-sloping surface of Peter I Island with its line of cliffs (Fig. 17) as an abraded marine platform with wave-cut notches.

There is similar evidence of relative change in sea level on most of the other islands. Much more study is required, however, before any reliable synthesis can be attempted.

The author's work in the Scotia Arc region was supported by the National Science Foundation, Office of Polar Programs. Helpful criticism was provided by Peter E. Baker of the University of Leeds and Jacques Nougier of the Université de Paris.

I. W. D. DALZIEL

References

Adie, R. J., 1964. "Geological history," in R. Priestley, R. J. Adie, and G. Robin de Q, eds., *Antarctic Research*. London: Butterworth, 118–162.

____, ed., 1972. *Antarctic Geology and Geophysics*, Oslo: Universitetsforlaget, 876p.

Anderson, J. J., 1965. "Bedrock geology of Antarctica; a summary of exploration 1831–1962," *Am. Geophys. Union Antarctic Res. Ser. 6*, 1–70.

Baker, P. E., and Griffiths, D. H., 1972. "The evolution of the Scotia Ridge and Scotia Sea," in Sutton, J., et al., *A Discussion on Volcanism and Structure of the Earth, Phil. Trans. Roy. Soc. Lond.*, Ser. A, **271**, 151–185.

Craddock, C., 1970. "Radiometric age map of Antarctica," *Am. Geog. Soc. Antarctic Map Folio 12*, Geol. plates XIX.

Dalziel, I. W. D., 1972. "Large-scale folding in the Scotia Arc," in R. J. Adie, ed., *Antarctic Geology and Geophysics*. Oslo; Universitetsforlaget, 47–55.

____, 1974. "The margins of the Scotia Sea," in C. A. Burk and C. L. Drake, eds., *The Geology of Continental Margins*. New York: Springer-Verlag, 567–579.

____, and Elliot, D. H., 1973. "The Scotia Arc and Antarctic Margin," in F. G. Stehli and A. E. M. Nairn, eds., *The Ocean Basins and Margins. 1. The South Atlantic*. New York: Plenum, 171–246.

Girod, M., and Nougier, J., 1972. "Le volcanisme de iles sub-Antarctiques," in R. J. Adie, ed., *Antarctic Geology and Geophysics*. Oslo: Universitetsforlaget, 777–788.

Holtedahl, O., 1929. "On the geology and physiography of some Antarctic and sub-Antarctic islands," *Scientific Results of the Norwegian Antarctic Expeditions 1927–28 and 1928–29*, vol. 3. Oslo: Dybwad, 172p.

John, B. S., and Sugden, D. E., 1971. "Raised marine features and phases of glaciation in the South Shetland Islands," *Br. Antarct. Surv. Bull.*, **24**, 45–111.

Nougier, J., 1970. Contribution à l'Etude Géologique et Géomorphologique des Iles Kerguelen. Paris:

Comité Nat. Fr. Rech. Antarctiques, Inst. Géogr. Nat., 2 vols., 440p. 255p.

Quilty, P. G., Rubenach, M., and Wilcoxon, J. A., 1973. "Miocene ooze from Macquarie Island," *Search*, **4**(5), 163–164.

Van Zinderen Bakker, E. M., et al., eds., 1972. *Marion and Prince Edward Islands*. Cape Town: A. A. Balkema, 427p.

Watkins, N. D., et al., 1974. "Kerguelen: Continental Fragment or Oceanic Island?", *Bull. Geol. Soc. Am.*, **85**(2), 201–212.

Will, H., 1884. "Das Exkursionsgebiet der Deutschen Polarstation auf Sud-Georgien in geognostischer, floristischer und faunistischer Beziehung," *Dtsch. Geogr. Bl.* (Bremen), **7**, 116–144.

Wordie, J. M., 1921. "Shackleton Antarctic Expedition 1914–1917: geological observations in the Weddell Sea area," *Trans. Roy. Soc. Edinburgh*, **53**, pt. 1.

Cross-references: *Antarctica; Argentina; Australia; Chile; Falkland Islands; Vol. VIII, Pt. 2, St. Helena; Vol. VIII, Pt. 2, South Africa.*

SURINAM

Surinam, with an area of 163,800 km², is situated on the NE coast of South America, bordered in the west by Guyana, in the east by French Guiana, and in the south by Brazil. It has a humid tropical climate with a mean annual temperature of 27.3°C and an average annual rainfall of 220 cm. The greater part of the population lives in the coastal area, where mining and agriculture represent the basic economy.

The Geologisch Mijnbouwkundige Dienst (Geological and Mining Service) was established in 1943, although there was much earlier geological work. Gold mining started in the 1850s, the first bauxite was exported in 1922, and a large volume on the geology of the country appeared in 1931 (Ijzerman). The Geological and Mining Service has had considerable growth and now employs approximately 500 men. Geologists and mining engineers have been trained abroad. Since 1971 courses are given to train geological and geophysical surveyors and drillers.

Knowledge of the geology is obtained by the study of air photographs, geological fieldwork, aerogeophysical surveys and ground follow-up, geochemical surveys, and drilling. At present about 75% of the surface has been covered by geological mapping. Maps have been issued with scales of 1:200,000 and 1:100,000. A photo-geological map covering the whole country and showing a large variety of landscapes has been

TABLE 1. Stratigraphy of Surinam

#	Unit	Age	Description
1	CORANTIJN GROUP	Cretaceous (not cropping out) to Recent	Unconsolidated sands and clays mainly
2	APATOE DOLERITE	Permo-Triassic 227 ± 10 m yr	Pigeonite dolerite dikes
	Shearing and mylonitization of Precambrian Nickerie Metamorphic Episode, 1200 ± 100 m yr		
	AVANAVERO DOLERITE	Precambrian 1500–1800 m yr	Sills and dikes of hypersthene bearing pigeonite gabbro and dolerite
3	RORAIMA FORMATION	Precambrian 1600–1650 m yr	Subhorizontal quartzitic sandstones and conglomerates, with local tuff intercalations
	Acidic plutonic-volcanic magmatism of Trans-Amazonian Orogenic Cycle:		
4	GRANITOID ROCKS		Granites, granodiorites and quartzdiorites, with local leucogranites
5	DALBANA RHYOLITE	Precambrian 1810 ± 40 m yr	Slightly metamorphosed rhyolitic-dacitic lavas and tuffs including ash-flow tuffs
	Folding and metamorphism of geosynclinal stage of Trans-Amazonian Orogenic Cycle		
6	MAROWIJNE GROUP	Precambrian	Geosynclinal metasediments and metavolcanics
	Armina and Rosebel Fmns. (E. Sur.) and Ston Formation (W. Sur.)		Flysch–(Armina) and molasse-type (Rosebel, Ston) metasediments including schists, phyllites, quartzites, graywackes, subgraywackes, and conglomerates
	De Goeje Gabbro(?)	Relative age and origin uncertain	Older gabbroic (and ultrabasic?) intrusives
	Paramaka Formation (E. Sur.) and Matapi Spilite (W. Sur.)	1890 ± 90 m yr	Basic and acid extrusives, and clastic and chemical metasediments, including spilites, basalts, and basic tuffs; rhyolites and dacites; and schists, phyllites, and quartzites
7	FALAWATRA GROUP and COEROENI GROUP	Precambrian Relative age and origin uncertain (older basement?)	Charnockitic granulites, migmatic gneisses, and cordierite-sillimanite gneisses (associated with younger mylonitic granite gneisses and basic intrusives)

Source: Bosma and Oosterbaan (1972). Numbers refer to legend of Figure 1.

prepared by O'Herne (1969a, 1969b) with scales of 1:1,000,000 and 1:500,000.

Geological data are exchanged with neighboring countries at the Inter-Guiana Geological Conferences which are held about every three years in one of the countries of the shield.

Geology

The greater part of Surinam consists of Precambrian rocks which form part of the Guiana Shield. In the north the shield is overlain by Cretaceous to Holocene sediments increasing in thickness in a WNW direction. Near the mouth of the Corantijn River, on the axis of the "Guiana Basin", a thickness of 2000 m is reached.

Table 1 shows the present picture of the stratigraphy of Surinam (mainly after Bosma and Groeneweg, 1970). A simplified geological map is shown in Fig. 1 (Bosma and Oosterbaan, 1972). The white spaces on this map represent areas not yet covered by fieldwork.

Falawatra Group. The highly metamorphosed rocks of this group, including charnockitic granulites and sillimanite gneisses, can be correlated with similar rocks in Guyana, French Guiana, and Venezuela (Table 1 in *Guiana Shield*). The main occurrences are in western Surinam, e.g., the Bakhuis Mountains, a horst of SW-NE trend, and the Coeroeni area, to the south of it. Preliminary dating points to ages of 1900 to 2500 m yr. Large bauxite deposits occur on parts of the Bakhuis Mountains.

Marowijne Group. This group consists of several formations, forming part of a huge geosyncline, the Lower Proterozoic Mobile Belt that was folded and metamorphosed during the

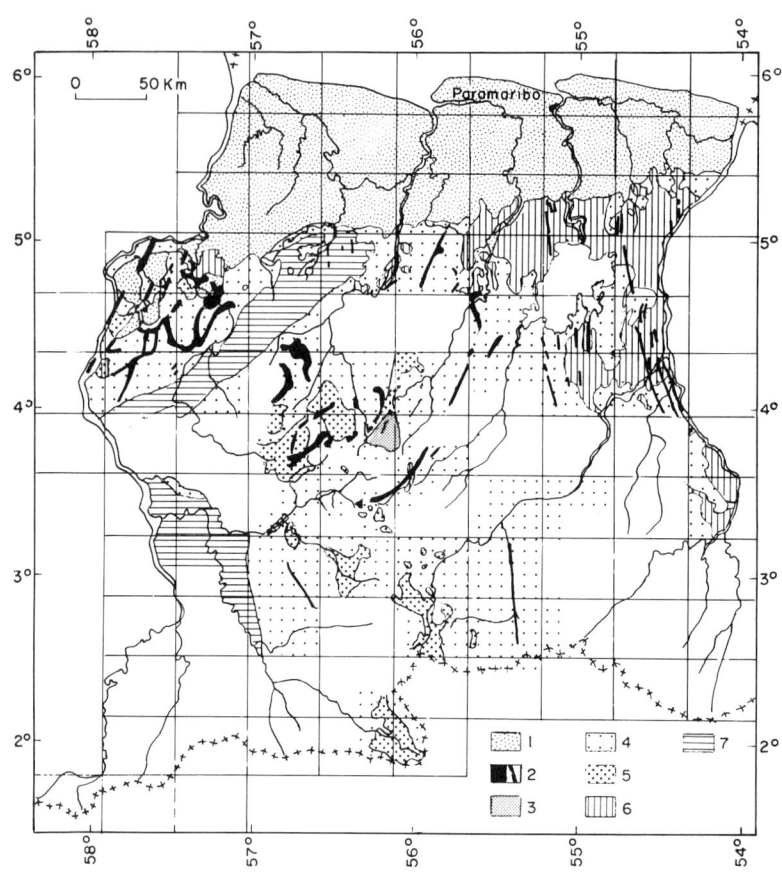

FIGURE 1. Geological sketch map of Surinam (after Bosma and Oosterbaan, 1972). For legend, see corresponding numbers in Table 1.

Trans-Amazonian Orogeny (McConnell and Williams, 1970).

The Paramaka Formation consists of metasediments, metavolcanics, and metamorphic basic intrusive rocks. The formation is found mainly in the eastern part of the country. It is often covered by a hard cap of laterite or bauxite. Most of the gold occurrences are in this formation. The Matapi Spilite of western Surinam may be of the same age as the Paramaka Formation.

The stratigraphical place of the De Goeie Gabbro or Older Basic Intrusives, which are found at several localities in eastern Surinam, is still uncertain. The rocks, comprising gabbro, norite, and metagabbro, were originally recognized as part of the Paramaka Formation.

The Armina Formation shows metasediments from conglomerates to phyllites of which the argillaceous deposits predominate, giving it a flysch-like appearance. The Rosebel Formation has a more arenaceous development and is a molasse-type sediment. The succession of the latter two formations has not been established with certainty, but Surinam geologists usually take Armina as the older, in contrast with most students of the geology of French Guiana. The Ston Formation in western Surinam can probably be correlated with Rosebel.

Granitoid Rocks and Dalbana Rhyolite. During the Trans-Amazonian Orogeny large intrusions of granitic rocks (granite to quartz-diorite) occurred, about contemporaneous with extrusions of rhyolites and dacites (Dalbana Rhyolite). In the Wilhelmina Mountains and the Dalbana Creek area (Corantijn) there are welded tuffs. All granites and rhyolites in Surinam appear to be of the same age (1810 ± 40 m yr, Priem *et al.,* 1971). Granites and acid volcanics probably belong to the same magmatic suite. The theory of rejuvenation (see McConnell and Williams, 1970) finds little support in Surinam (see *Guiana Shield*).

Roraima Formation. After the Trans-Amazonian Orogeny the basic pattern of the Guiana Shield was more or less completed. Erosion, local sedimentation, minor basic intrusions and faulting were the agents that gave it its final appearance.

The shield must have been partly peneplaned before the Roraima Formation was deposited on it. This formation, consisting of subhorizontal sandstones and conglomerates with some thin ignimbrite and tuff intercalations, covers large parts of Guyana, Venezuela, and Brazil. In Surinam it is only found on and near the Tafelberg in the center of the country. The most recent dating of the tuff yielded an age of 1599 ± 18 m yr (Priem et al., 1973).

Avanavero Dolerite. Both before and after deposition of the Roaraima Formation the Guiana Shield was intruded by gabbros and dolerites. The dolerites can probably be divided in two groups, one with ages of 1550–1650 m yr and one of about 1750 m yr. Palaeomagnetic research has shown that the two groups have different paleomagnetic poles (Veldkamp et al., 1971).

Nickerie Metamorphic Episode. At about 1200 m yr ago a tectonothermal event occurred in western Surinam and adjacent Guyana during which large SW-NE shear zones were formed. The Bakhuis zone originated at this time but was subsequently rejuvenated probably up to the Tertiary.

Apatoe Dolerite. In the eastern part of the country N-S to NNW-SSE directed dolerite dikes occur which are of Permo-Triassic age (Priem et al., 1968). They probably point to stress in the shield prior to the breaking up of the South American and African continents (see also May, 1971).

The Coastal Area. The Upper Cretaceous to Holocene deposits form the Corantijn Group. Figure 2 shows a simplified N-S profile through the coastal area near Paramaribo. The stratigraphy is mainly after Montagne (1964) and the ages have been determined by palynological studies (Van der Hammen and Wijmstra, 1964; Wijmstra, 1969, 1971). Marine fossils have only been found in the Paleocene and Holocene. The generalized stratigraphy of the Guiana Basin is given in Table 2. For practical reasons a quite different approach to the stratigraphy was given by Noorthoorn van de Kruijff (1970); see Fig. 2. It was based on two sudden increases of density found in oil exploration holes. These "density shifts" were used for correlation purposes. A density shift is an indication of the denudation of a considerable thickness of formerly overlying layers, after which younger sediments were deposited on top. One density shift occurs between Upper Cretaceous and Paleocene, the other in the mid-Lower Miocene. The first suggests the erosion of about 150 m of sediments, the latter of about 60 m. The "bauxite hiatus" (Upper Eocene–Lower Miocene) does not show a density shift.

It is clear from the profile that the Paleocene and Eocene on the landward side have been protected against erosion where capped by bauxite. This has been covered by younger de-

FIGURE 2. Simplified N-S section of the coastal area near Paramaribo (Krook and Mulders, 1971).

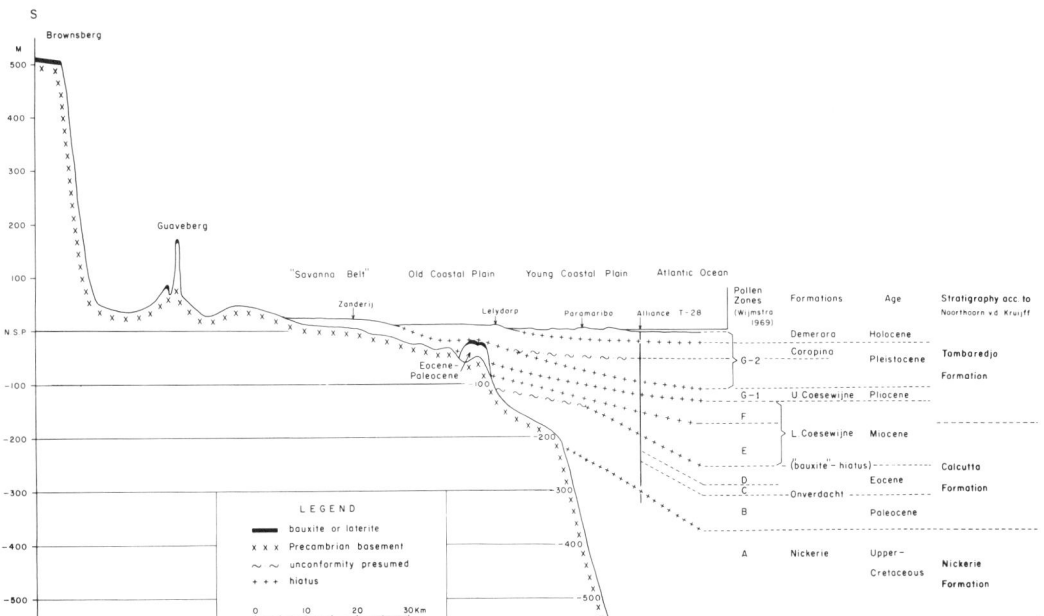

TABLE 2. Stratigraphic Sequence in the Coastal Area in Relation to Climate and Sea-Level Movements

Age		Nature of Deposits	Stratigraphy of the Guiana Basin	Climate	Transgressions and Regressions
Holocene		Mainly clay, some sand (beaches and ridges)	Upper clays	Humid	Transgression
Pleistocene	Glacials	Some clay; erosion and soil formation; sand on continental shelf		Dry	Regressions
	Interglacials	Mainly clays, some sand; clay on shelf		Humid	Transgressions
Pliocene		Coarse unsorted sands and kaolinitic clays; mainly very coarse unsorted sands (Savanna Belt)	Upper sands	Very dry Dry	Regression
Miocene	Upper	Missing (hiatus)		Dry?	Large regression
	Middle	Mainly clays	Intermediate clays	Humid	Transgression
	Early	Missing (hiatus)		Humid	Transgression
		Mainly clays		Dry?	Large regression
		Mainly clays		Dry and humid seasons	Regression
		Coarse sands	A-sands		Transgression
Oligocene? (no pollen)				Dry?	Regression
		Bauxite	Bauxite	Dry and humid seasons	Regression
Eocene to Paleocene		Arkosic sand (weathered to bauxite) Kaolinite clays (bauxite area) Sands	Alternating sands and clays	Dry	Regression

Nature of deposits mainly as inferred from drillholes.

FIGURE 3. Coastal area of Surinam (from Zonneveld, 1968a).

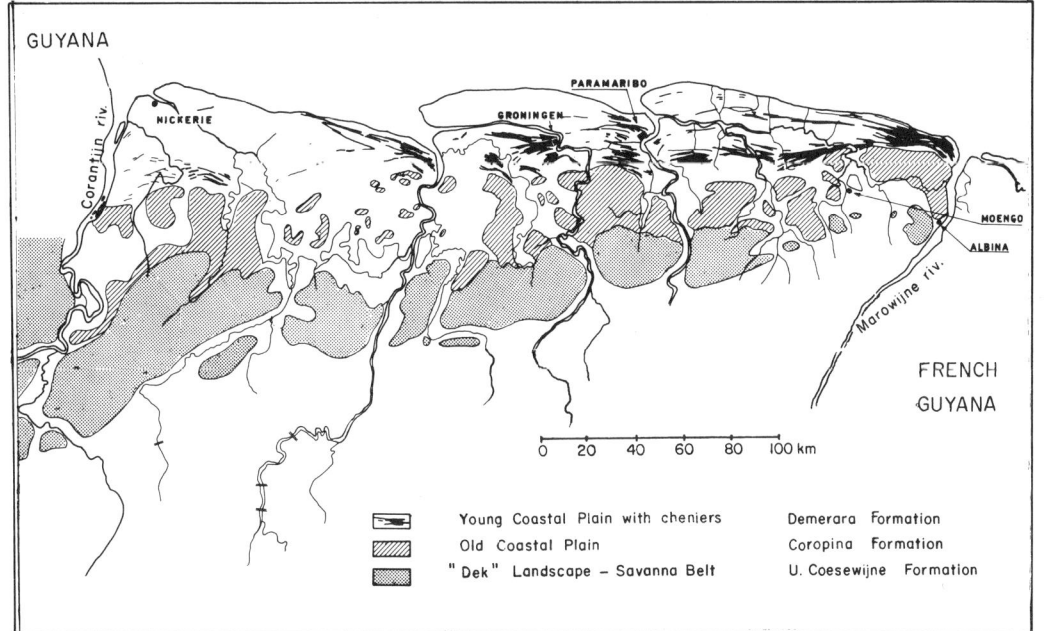

posits. The principal outcropping deposits are the Upper Coesewijne Formation, the Coropina Formation and the Demerara Formation (Fig. 3). In addition, in the Moengo area the Onverdacht Formation, capped by bauxite, is also found at the surface.

During the Pliocene the Upper Coesewijne Formation was deposited, mainly coarse sands, but kaolinitic clays and lignites occur as well. Heavy mineral investigations showed that the uppermost part was of very nearby origin, probably supplied by short braided rivers or even alluvial fans in a relatively dry climate. Under a savanna vegetation during the dry (glacial) stages in the Pleistocene part of these deposits were leached to a high degree in the short rainy seasons, resulting in the creation of the white quartz sands, nowadays still covered with savannas or savanna forests. The more loamy brownish sands support a high rain forest. The E-trending Coesewijne Formation is known as the Savanna Belt.

During the Pleistocene transgressions coastal plains were formed although on the landward sides the formations were almost entirely removed by erosion during the subsequent regressions. In the drill holes in the coastal plain, however, the whole succession can still be found (Wijmstra, 1969, 1971). Only the Upper Pleistocene deposits still crop out at present, the Coropina Formation. The Lower Coropina Formation (possibly of Mindel-Riss interglacial age) is found only very locally. The Upper Coropina (Riss-Würm Interglacial) consists of a northern sandy part (Lelydorp Sands) formed as barrier islands and a southern part of silty clays (Para Clays or Old Sea Clays), see Fig. 4 (after Veen, 1970). The Coropina deposits form the Old Coastal Plain. The present Old Coastal Plain consists of hundreds of large and small erosion remnants, originated during the low sea level of the Würm regression. During the Holocene sea level rise the creeks and rivers were filled up with marine clays and peat.

The Young Coastal Plain consists of marine with swampy terrain interrupted by "cheniers" or "ridges" (for detailed description of the Y.C.P., see Brinkman and Pons, 1968). Most ridge complexes occur west of the main rivers, which suggested formation by sand supply from local rivers. Heavy mineral investigations, however, showed that the fine sandy ridges west of

FIGURE 4. Relation of landscape and stratigraphy in the coastal plain of Surinam (after Veen, 1970).

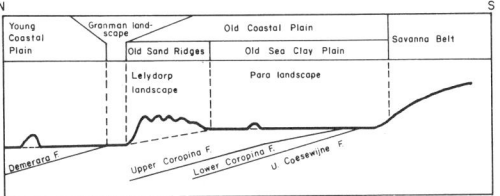

Paramaribo all share one particular mineral association consisting mainly of staurolite, epidote and zircon. This mineral association does not resemble the associations of the rivers. The unweathered fine sands of the ridges contain green pellets consisting partly of chamosite. Since large areas with the same kind of sand have been found on an ancient delta off the Marowijne River it is assumed that this sand is the source of the beachridges. Submarine currents transported the sand landward. Most ridges are found west of the rivermouths due to the fact that here the concentrating effect of the waves is greatest, being under the influence of the NE trade winds. The sandy ridges east of Paramaribo contain predominating staurolite. These mostly coarse sands have been mainly derived from French Guiana and the Marowijne River. Staurolite is a typical mineral of the Armina Formation which occurs especially in eastern Surinam and adjacent French Guiana. The sands have been transported along the coast by beach drift.

Originally all rivers in the coastal plain discharged straight into the Atlantic Ocean. In later times the smaller rivers became gradually deflected to the west due to the supply of large quantities of sand and Amazon mud (Zonneveld, 1968a). At present there is hardly any accretion of the land. At some places erosion predominates, at other places sedimentation.

An excellent description of the coastal area in relation to the landscape was given by Van der Eyk (1957).

The Continental Shelf and the Guiana Marginal Plateau. Near the relatively shallow edge (100 m) of the wide continental shelf of the Guianas remnants of upper Pleistocene coral reefs occur. Sediments on the shelf vary from coarse sands on the outer part to fine sands on the central part to pelitic near the present coast (Nota, 1958, 1969). The sandy deposits are of Pleistocene age. The nearshore pelitic deposits are of Holocene age and have been supplied by the Amazon and the Caribbean Current. Some 85—90% of these sediments are from the Andes which explains the clay mineral composition of kaolinite, illite and montmorillonite, and contributes to the fertility of the Young Coastal Plain.

To the NE of the Continental Shelf a borderland plateau feature of much larger dimensions is found, the Demerara Rise or Guiana Marginal Plateau (Fig. 5), at a depth of about 1500—2000 m. Figure 6 shows a section through this plateau with data from several sources (Colette et al., 1971; Gillmann et al., 1972; Hayes et al., 1971; Fox and Heezen, 1970). The oldest rock, dredged on the northern flank of the plateau, is a sandstone of

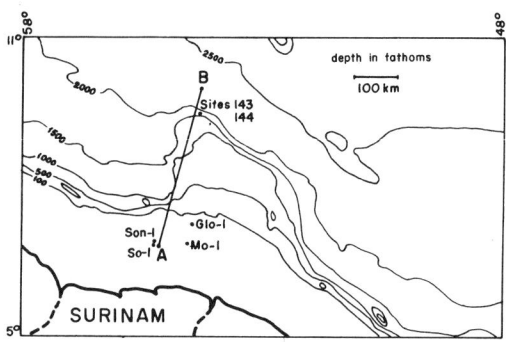

FIGURE 5. Continental shelf and the Guiana Marginal Plateau, with drill sites (after Colette et al., 1971).

Upper Jurassic age. A nearby drill hole (144) showed an unconformity between Upper Cretaceous and Paleocene. The total thickness of the sediments of the Guiana Plateau is as yet unknown but estimates of the southern part of more than 10 km have been made. The plateau is asymmetric. To the N and NE the continental slope is steep, to the west a gently shelving slope is found. On the plateau several deep submarine canyons occur. The seismic surveys, however, were not detailed enough to show the pattern of these canyons.

Cretaceous pollen from drill hole GLO-1 showed the same successions and associations as in West Africa. During the Tertiary the Guiana flora diverged progressively more and more from the African flora.

Paleoclimatology

During the Tertiary and the Quaternary Surinam saw a succession of dry and humid climatic phases. Indications of these climatic oscillations are given by palynological studies of the Cenozoic sediments (Wijmstra, 1971). Climatic influences are also reflected by the geomorphology and the nature of the sediments in the coastal plain and along the rivers (Krook, 1970). Table 2 has been prepared with data from various sources. It is, of course, of a very generalized nature. Only during the mid-Miocene (zone F, fig. 2) three transgressions and regressions have been distinguished, but the zone as a whole is transgressive. It is apparent that relatively dry phases are usually contemporaneous with regressions at which times the sediments are often coarse grained. This is particularly clear in the Pleistocene. The glacial phases caused both regressions of the sea and dry phases in the tropics. The dry and somewhat cooler periods caused less chemical weathering, the disappear-

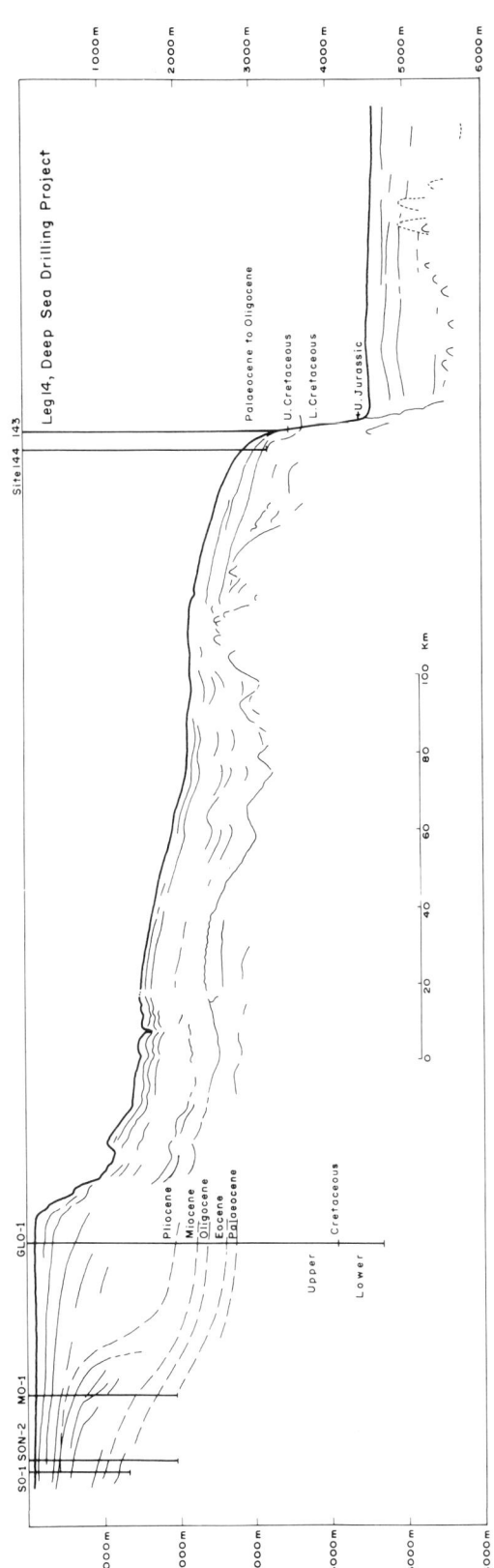

FIGURE 6. Section A-B through the Guiana Marginal Plateau (after Colette *et al.*, 1971).

ance of most of the rainforest (which was replaced by savannas) and severe erosion of the original weathered layer (saprolite) and the underlying bedrock. Thus sandy sediments originated, often of an arkosic nature (Damuth and Fairbridge, 1970). The above is valid for the whole of tropical South America. The Amazon supplied less clay during the dry periods which made it favorable for coral growth on the edge of the continental shelf of the Guianas.

The influence of the eustatic changes may be traced back into the Tertiary. There were major regressions during the Upper Eocene-Oligocene and the Upper Miocene that may have been related to ice formation in Antarctica and this in turn to periods of worldwide increased activity of plate tectonics (see Frerichs, 1970).

Geomorphology

Studies of the morphology of the interior and sediments in the coastal area have shown that unconformities in the Cenozoic deposits can be correlated with certain erosion levels in the hinterland (King, 1964; in Guyana see McConnell, 1966). This shows a cyclic development of erosional landscape formation and contemporaneous sedimentation by intermittent upwarping of the interior and downwarping of the coastal belt.

King distinguished the following levels:

Early Tertiary Surface (Paleocene-Eocene). This surface has only been preserved where laterite or bauxite crusts protected it against erosion. On the corresponding deposits in the coastal plain an Upper Eocene to Oligocene hiatus is found, the "bauxite hiatus." The formation of the Early Tertiary Surface occurred mainly by pedimentation under arid to semi-arid climatic conditions. This resulted in the deposition of arkosic sands in the coastal area. Bauxites were formed both on the erosion surface and on the sandy deposits, now distinguished, respectively, as plateau bauxite and coastal plain bauxites. Most of the latter have been buried below younger sediments. In the Moengo area in NE Surinam, however, they form conspicuous, rather low, plateaus in the coastal plain.

Late Tertiary I Surface (Oligocene-Lower Miocene). This is a generally rolling landscape, in contrast to Guyana where it is a more conspicuous surface. It correlates with the Upper Miocene hiatus (Messinian in the Mediterranean).

FIGURE 7. Bonita Peak with the "Devils Egg" (air photo, C.B.L., from Zonneveld, 1968b), dissected margin of early Tertiary peneplain surface of ca. 1000–1050 m elevation. Note "pseudo-karst" rills on granitic slopes, common in the warm-wet tropics.

Late Tertiary II Surface (Pliocene). A flat landscape covering the greater part of Surinam. Its erosion products form the Upper Coesewijne Formation of the Savanna Belt. During the Quaternary renewed incision occurred in the fluvial phases (interglacials and postglacial), interrupted by pedimentation in the dry phases (glacials).

Remnants of the Early Tertiary Surface can now be found as flat-topped hills and mountains capped by laterite or bauxite with heights varying from 300 m (N part of Bakhuis Mountains) to 700 m (Lely Mountains in eastern Surinam). In some parts of the country, however, considerably higher mountains are found (highest summit: Juliana Peak, 1180 m), usually on grainte or acid volcanic rocks, partly on the Roraima sandstone (Tafelberg). In the Wilhelmina Mountains a summit level can be observed at about 1000 m. This may be a remnant of the Early Tertiary Surface (Fig. 7). Its distinctive height is probably due to general uplift of the central part of the country. The Wilhelmina Mountains consist of large granite massifs which arise abruptly in the pediplain of the Upper Tertiary II Surface (Verhofstad, 1969).

Other remnants of the E.T.S. are probably the highest parts of the Acarai Mountains (SW Surinam and adjacent Guyana), the Oranje Mountains, and the Tumac Humac Mountains (SE Surinam). Several large inselbergs have about the same height, which shows their old age.

The morphology of the Bakhuis Mountains in western Surinam bears witness to a succession of monsoonal, semi-arid and humid tropical climates. This morphology has been preserved by lateritization and bauxitization. The plateaus (E.T.S.) are often bordered by gently sloping pediments. These are separated from lower lying pediments by rather steep slopes. The lowest pediments are just above the present river level which means they are of relatively recent formation. Plateaus and pediments have hard bauxite and laterite caps; the slopes (20—25°) show soft bauxite and lateritic colluvium which covers bauxites originated in situ. At a few places bauxititized alluvial fans are encountered, also witnesses of past arid climates.

Several minor laterite plateaus all over the country may belong to the Upper Tertiary I Surface but the occurrence of more than three levels but must not be excluded. So far, however, little detailed geomorphologic fieldwork has been done.

The Upper Tertiary II Surface, which covers the greater part of Surinam, has been subject to cyclic erosion during the Pleistocene. In the dry phases pediments and gravelly alluvial fans were formed along the rivers. They are locally covered with loamy terrace material indicating humid consitions. In the later stages of the pluvials incision occurred (de Boer, 1972, Zonneveld, 1969).

Erosion during the dry phases was rather intense due to the deforestation of large parts of the shield. The older weathering mantle was removed to such a degree that many creeks and rivers still follow the directions of the faults and fault zones, often SW-NE and NW-SE.

During the Pleistocene many minor and major river captures took place on the flat Upper Tertiary II Surface. Even at present the process is still at work in the flat interfluvial areas between large river systems.

In the rivers on the shield there are many rapids, called "Sulas" in Surinam. Partly these rapids are due to hard formations, e.g. dolerite dikes. Others show intricate braided patterns (Fig. 8), originated by a change from a dry to a

FIGURE 8 Sula complex in the Corantijn River (vertical air picture, from Zonneveld, 1968b, width of picture 4.4 km).

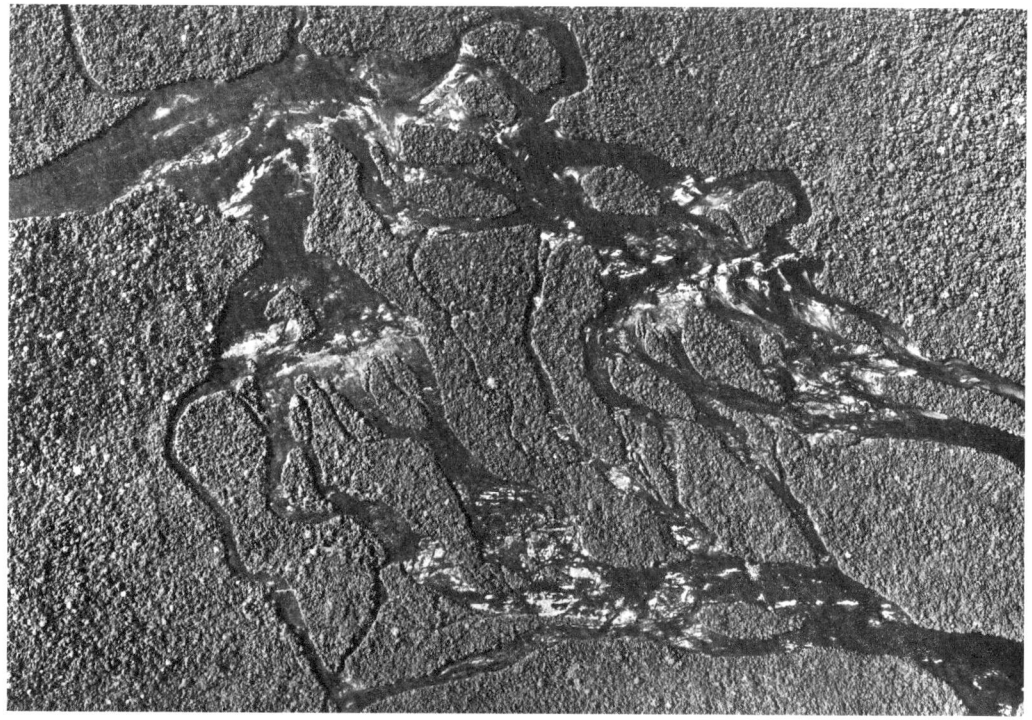

humid climate when the river had to make its way across a stripped planation surface (Garner, 1967).

Mineral Resources

The mineral occurrences and indications of Surinam are summarized in Fig. 9. Bauxite is Surinam's most important mining product. So far only the coastal plain bauxite has been mined. Export from the Moengo area began in 1922 and from the Onverdacht area (S of Paramaribo) in 1941. Since 1965 alumina is produced with the Bayer process. Part of this is reduced to aluminum with the Hall-Herault process, using power of a hydroelectric station in Afabaka (100 km S of Paramaribo), where a dam was built in the Surinam River (for site of lake, see Fig. 9). The production in 1971 of bauxite, alumina, and aluminum was 3,697,000, 1,276,000, and 54,000 tons, respectively. The potential coastal plain bauxite zone occurs in an E-W belt (Fig. 10) where exploration still proceeds. Large occurrences of plateau bauxite are found south of the coastal belt, and are being explored at present for mining in the near future.

Gold has been mined since the 1850s, reaching a peak in 1907 with a production of 1200 kg. Production declined since then to a few tens of kg annually at present. The primary gold is associated with quartz near granite contacts in the metamorphic rocks in eastern Surinam.

Most gold has been found in alluvial deposits. Locally some platinum has been found, associated with the gold.

Ore bearing pegmatites occur scattered in the NE of the country. Two deposits have been mined. One produced some 1700 tons of amblygonite and 20 tons of cassiterite and tantalite, the other 9 tons of beryl. Both deposits have been abandoned but prospection is still proceeding.

Iron ore is found mainly in the form of laterites in many plateaus in the interior. Estimates of a total of several billion tons with an average content of 35% have been made. Lateritic manganese ores occur in several parts of the country. Nickel and chromium have been indicated locally in ultrabasic rocks. Some copper has been found which was associated with granitic rocks.

For building purposes crushed stone, gravel, sand, and clay are of importance. The white quartz sands of the Upper Coesewijne Formation are suitable for glass production. The kaolinite deposits underlying the coastal plain bauxites have not been utilized as yet.

A kyanite quartzite with a possible reserve of 3 million tons occurs NE of the Afobaka Dam. Some diamonds have been found in alluvial gravels. They were possibly derived from kimberlites associated with Apatoe dolerites.

Most fresh groundwater in the coastal plain is obtained by drilling. The water occurs is aquifers of Coesewijne age which are recharged in the Savanna Belt (see Fig. 2).

Hydrocarbons have been explored mainly since 1965, both on and off shore. Some oil was found in the coastal plain at shallow depths (180–300 m) but this proved of no commercial importance. Two wells drilled on the continental shelf showed minor indications. Nothing is known so far about possible oil reserves in deposits underlying the Guiana Marginal Plateau. However, black shale was reported in the Upper Cretaceous on the N flank of the plateau (drillhole site 144).

A summary on the mineral deposits has been given by Bosma *et al.* (1972).

LEENDERT KROOK

FIGURE 9. Major mineral indications in Surinam (after Bosma and Oosterbaan, 1972).

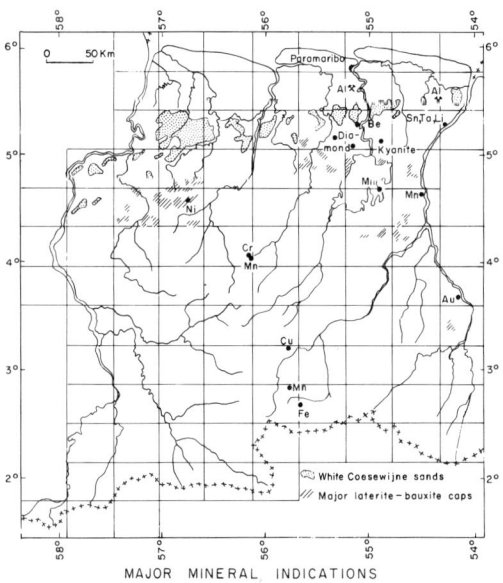

MAJOR MINERAL INDICATIONS

References

Boer, M. W. H. de, 1972. "Landforms and soils in eastern Surinam," *Wageningen, Agricultural Res. Rept. 771.* 169p.

Bosma, W., and Groeneweg, W., 1970. "Review of the stratigraphy of Surinam," *Proc. 8th Guiana Geol. Conf.,* Georgetown, Guyana, August 1969.

____, and Oosterbaan, W. E., 1972. *30 Years Geologi-*

FIGURE 10. Bauxite belt in the Guiana coastal plain (from Krook, 1969).

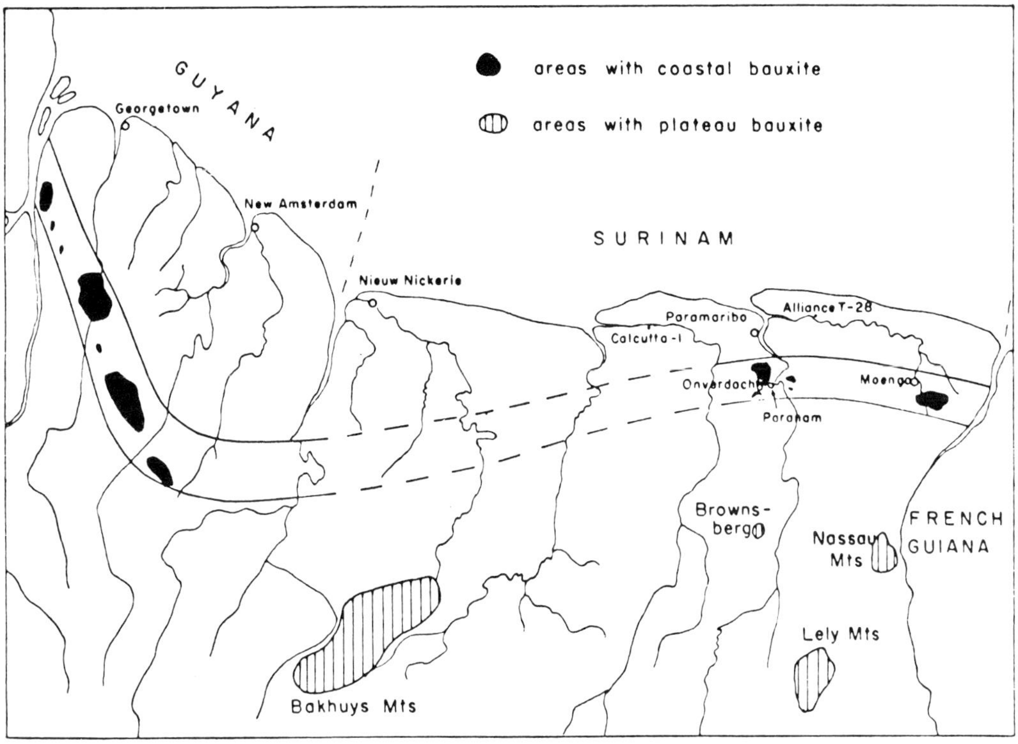

cal and Mining Service of Suriname, 1973. Suriname: Geol. Mijnb. Dienst, 31p.

———, Ho Len Fat, A. G., and Welter, C. C., 1972. "Minerals and mining in Suriname," paper presented at 9th Guiana Geol. Conf., Puerta Ordaz, Venezuela, May 1972.

Brinkman, R., and Pons, L. J., 1968. "A pedogeomorphological classification and map of the Holocene sediments in the coastal plain of the three Guianas," Soil Surv. Pap. 4, 40p.

Colette, B. J., Schouten, J. A., Rutten, K. W., Doornbos, D. J., and Staverman, W. H., 1971. "Geophysical investigations on the shelf of Surinam, H.NL.M.S. Luymes, 1969," Hydrographic Newsletter, Spec. Publ. 6, 17–24.

Damuth, J. E., and Fairbridge, R. W., 1970. "Equatorial Atlantic deep-sea arkosic sands and ice-age aridity in tropical South America," Bull. Geol. Soc. Am., 81, 189–206.

Eyk, J. J. van der, 1957. "Reconnaissance soil survey in northern Surinam," thesis, Agricultural Univ., Wageningen, 99p.

Fox, P. J., and Heezen, B. C., 1970. "Jurassic sandstone from the tropical Atlantic," Science, 170, 1402–1404.

Frerichs, W. E., 1970. "Paleobathymetry, paleotemperature, and tectonism," Bull. Geol. Soc. Am., 81, 3445–3452.

Garner, H. F., 1967. "Rivers in the making," Sci. Am., 216, 84–94.

Gillmann, M., Jardine, S., Belsky, C. Y., Cassan, J. P.,

Durif, P., Grodidier, E., and Prestat, B., 1972. "Etude stratigraphique et sédimentologique d'une coupe du Crétacé-Tertiaire au large du Surinam," 9th Inter-Guiana Geol. Conf., Puerta Ordaz, Venezuela, 20p.

Hammen, T. van der, and Wijmstra, T. A., 1964. "A palynological study of the Tertiary and Upper Cretaceous of British Guiana," Leidse Geol. Mededel., 30, 183–241.

Hayes, D. E., Pimm, A. C., Benson, W. E., Berger, W. H., Rad, U. von, Supko, P. R., Beckman, J. P., Roth, P. H., and Musich, L. F., 1971. "Deep sea drilling project, leg 14," Geotimes, Feb. 1971, 14–17.

Ijzerman, R., 1931. "Outline of the geology and petrology of Surinam," thesis, Utrecht, The Hague, 519p.

King, L. C., 1964. "Cyclic denudation in Surinam," internal report, Geol. Mining Serv. Surinam.

Krook, L., 1969. "The origin of bauxite in the coastal plain of Surinam and Guyana," Geol. Mijnb. Dienst, Mededel., 20, 173–180.

———, 1970. "Climate and sedimentation in the Guianas during the last glacial and the Holocene," Proc. 9th Inter. Guiana Geol. Conf., Georgetown, Guyana, August 1969.

———, and Mulders, M. A., 1971. "Geological and related pedological aspects of the Upper Coesewijne Formation," Geol. Mijnb. Dienst, Mededel., 21, 183–208.

McConnell, R. B., 1966. "Notes on the erosion bevels

and geomorphology of British Guiana," *Trans. 3rd Caribbean Geol. Conf.,* Jamaica, April 1962, 115–159.

____, and Williams, E., 1970. "Distribution and provisional correlation of the Precambrian formations of the Guiana Shield," *Proc. 8th Inter-Guiana Geol. Conf.,* Georgetown, Guyana, August 1969.

May, P. R., 1971. "Pattern of Triassic-Jurassic diabase dikes around the North Atlantic in the context of predrift position of the continents," *Bull. Geol. Soc. Am.,* 82, 1285–1292.

Montagne, D. G., 1964. "New facts on the geology of the "young" unconsolidated sediments of northern Surinam," *Geol. Mijnb.,* 43, 499–515.

Noorthoorn van de Kruijff, J. F., 1970. "Surinam onshore exploration 1968/1970. Summary, petroleum engineering and geology," Shell Suriname Exploratie- en Produktie Maatschappij N.V., internal report, Geol. Mijnbouwk. Dienst.

Nota, D. J. G., 1958. "Sediments of the western Guiana Shelf," Reports on the Orinoco Shelf Expedition, II, Thesis, Utrecht, 98p.

____, 1969. "Geomorphology and sediments of western Surinam Shelf; a preliminary note," *Geol. Mijnb.,* 48, 185–188.

O'Herne, L., 1969a. "A photogeological study of the basal complex of Surinam," *Geol. Mijnbouwk. Dienst, Mededel.,* 20, 53–149.

____, 1969b. "Presentation of the photogeological map of Surinam, 1:1,000,000," *Proc. 7th Intern. Guiana Geol. Conf.,* Paramaribo, Nov. 1966. (*Verh. K.N.G.M.G.,* 27, 49–52.)

Priem, H. N. A., Hebeda, E. H., Boelrijk, N. A. I. M., and Verschure, R. H., 1968. "Isotopic age determinations on Surinam rocks. 3. Proterozoic and Permo-Triassic Basalt magmatism in the Guiana Shield," *Geol. Mijnb.,* 47, 17–20.

____, Boelrijk, N. A. I. M., Hebeda, E. H., Verdurmen, E. A. T., and Verschure, R. H., 1971. "Isotopic ages of the trans-Amazonian acidic magmatism and the Nickerie metamorphic episode in the Precambrian basement of Suriname, South America," *Bull. Geol. Soc. Am.,* 82, 1667–1680.

____, Boelrijk, N. A. I. M., Hebeda, E. H., Verdurmen, E. A. T., and Verschure, R. H., 1973. "Age of the Precambrian Roraima Formation in northeastern South America: evidence from isotopic dating of Roraima pyroclastic volcanic rocks in Suriname," *Bull. Geol. Soc. Am.* 84, 1677–1684.

Schönberger, J. M. H., and de Roever, E. W. F., 1975. "The possible origin of diamond in the Guiana Shield," *Geology,* 3.

Veen, A. W. L., 1970. "On geogenesis and pedogenesis in the Old Coastal Plain of Surinam," *Publ. Fysisch Geogr. en Bodemk. Lab. Univer. Amsterdam,* 14, 176p.

Veldkamp, J., Mulder, F. G., and Zijderveld, J. D. A., 1971. "Palaeomagnetism of Suriname dolerites," *Phys. Earth Planet. Interiors,* 4, 370–380.

Verhofstad, J., 1969. "Geological conclusions from aerial photograph analysis of the Wilhelmina Mountains," *Geol. Mijnbouwk. Dienst, Meded.* 20, 151–164.

Wijmstra, T. A., 1969. "Palynology of the Alliance well," *Geol. Mijnb.,* 48, 125–133.

____, 1971. "The palynology of the Guiana Coastal Basin," thesis, Amsterdam, 62p.

Zonneveld, J. I. S., 1968a. "The evolution of the coastal area of Suriname (South America)," *Means of Correlation of Quaternary Successions, Proc. 7th Congr. INQUA.* Salt Lake City: Univ. Utah Press, 577–589.

____, 1968b. "Some aerial photographs from the Surinam jungle," *Geografisch Tijdschrift,* N.R. 2(5) 528–535.

____, 1969. "Preliminary remarks on summit levels and the evolution of the relief in Surinam (S. America)," *Proc. 7th Inter-Guiana Geol. Conf.,* Paramaribo, Nov. 1966. (*Verh. K.N.G.M.G.,* 27, 53–60.)

Cross-references: *Brazil; French Guiana; Guiana Shield–Regional Review; Guyana; South America.*

TAHITI

One of the Windward group of the Society Islands (q.v.), the mountainous island of Tahiti lies in the South Pacific at 17°37'S and 149°26'W. Its area is 1045 km² (402 sq mi). Papeete, on the north coast, has a major airport and is the administrative center of *French Polynesia* (q.v.). There are four major peaks, the highest being Mt. Orohena (Fig. 1), which rises to a height of 2237 m (7340 ft).

The island is built of a double cone of olivine basalts, connected by a low isthmus and surrounded by a single barrier reef. The larger cone, Tahiti-Nui, lies to the NW, the smaller, Tahiti-Iki or Taiarapu, to the SE. Both cones preserve their general initial shape, but are deeply gullied, the precipitous gorges and valleys being graded to Pleistocene low sea levels and the mouths are now dammed back by reefs and fluvial sedimentation.

In Tahiti-Nui the core contains a 1 km wide plutonic plug of nepheline monzonite, theralite, and syenite, which was discovered by Marshall (1915). It evidently crystallized about 2000 m below the level of the original crater. Its existence gave rise to certain references to "granitic" and "continental or sialic" rocks in Tahiti, but it is clearly an oceanic basaltic differentiate. The lavas are dominantly alkaline basalts, basanites, with some ankaramites and oceanites, with minor tahitites, trachytes, and phonolites. The lavas are highly fluid and for the most part deficient in pyroclastics. Williams (1933) noted about 80 m of tuffs and breccias. There are five parasitic cones. There are thin covers of laterite and some red paleosol layers. Little work has been done on absolute chronology, but the physiography speaks for a late Tertiary date for the main eruptions, and sea-floor spreading data suggest a Cretaceous date for the underlying oceanic crust.

Around the coast there are "fossil" cliffs up to 300 m high in places, fronted by a narrow Holocene coastal plain of raised reefs and littoral sands. Modern beach sands are partly coralline, but there are also black lithic sands (basalt). There is a narrow carbonate-filled lagoon and barrier reef, partly missing in the NE.

J.-P. CHEVALIER

References

Becker, M., et al., 1974. "Phases d'erosion–comblement de la vallée de la Papenoo et volcanisme subrécent à Tahiti, en relation avec l'évolution des îles de la Société (Pacifique Sud), *Marine Geol.,* **16,** 71–77.

Crossland, C., 1928a. "Coral reefs of Tahiti, Moorea and Rarotonga," *Linn. Soc. Lond. J. Zool.,* **36,** 577–620.

———, 1928b. "The island of Tahiti," *Geog. J.,* **71,** 561–585.

Dana, J. D., 1886. "A dissected volcanic mountain (Tahiti) some of its revelations," *Am. J. Sci.,* Ser. 3, **32,** 247–255.

Lacroix, A., 1910. "Les Roches alcalines de Tahiti," *Bull. Soc. Géol. France,* Ser. 4., **10,** 91–124.

Marshall, P., 1915. "The geology of Tahiti," *N.Z. Inst. Trans. Proc.,* **47,** 361–376.

Obellianne, J. M., 1955. "Contribution à l'étude géologique des îles des Établissements français de l'Océanie," *Sci. de la Terre* (Nancy), 3(3), 1–146.

Williams, H., 1933. "Geology of Tahiti, Moorea, and Maiao," *Bernice P. Bishop Mus. Bull.,* **105,** 1–90.

Cross-references: *French Polynesia; Oceania; Society Islands.*

FIGURE 1. Deeply dissected central cone of Tahiti, Mt. Orohena (2237 m), showing knife-edge ridges and erosion to base level ca. 100 m below present.

TOKELAU ISLANDS (UNION ISLANDS)

A short chain of atolls with a land area of only 10 km² (4 sq mi), these islands became officially the Tokelau Islands in 1944, being earlier referred to as the "Union Islands." They are administered by New Zealand. A British protectorate from 1877, they were grouped in with the Gilbert and Ellice islands (q.v.) in 1916, but taken over by New Zealand in 1926. They are situated at 8–10°S, 171–173°S. (The nearby atoll of Swains Island is administered with *American Samoa*, q.v.)

Lying in the South Pacific Ocean, some 485 km N of Western Samoa, 3380 km NE of New Zealand, the Tokelau Islands consist of three atolls: Atafu, Fakaofo, and Nukunonu. They extend over a distance of 160 km in a NW-SE trend. Each atoll contains a central lagoon, surrounded by a ring reef on which are situated a large number of islets, the maximum elevation not exceeding 4 to 5 m.

Most of the year the islands lie in the belt of the SE trade winds, but in the southern summer (December–March) they come within the Intertropical Convergence zone, marked by heavy rain (av. 2900 mm; exceptionally up to 4500 mm). Mean annual temperature is 27.8°C. There is a small trade in copra and matting.

Geologically the islands are exclusively Quaternary coral limestone (slightly emerged reefs) and calcarenites (Holocene beach rock and loose sand). The geological history has probably been comparable to that of the Gilbert and Ellice Islands, involving upgrowth of coral on a basement of oceanic basaltic cones. Sea-floor spreading history suggests that the underlying crust is of Cretaceous age.

RHODES W. FAIRBRIDGE

References

Anon., 1968. *Reports on Niue and the Tokelau Islands.* Wellington, N.Z.: Dept. Maori and Islands Affairs.

Brigham, W. T., 1900. "An index to the islands of the Pacific," *Mem. Bernice P. Bishop Mus.,* 1.

Davis, W. M., 1928. "The coral reef problem," *Am. Geog. Soc. Spec. Publ. 9.*

Marshall, P., 1912. "Oceania," *Handbuch Reg. Geol.,* 7(2), 1.

Cross-references: *Gilbert and Ellice Islands; New Zealand; Oceania; Western Samoa.*

TONGA

Lying in the southwest Pacific Ocean, some 645 km east of Fiji, the kingdom of Tonga (also known as the "Friendly Islands" after visits by Captain Cook in 1773 and 1777) consists of an archipelago of about 200 islands, most of them uninhabited, low-lying, and of coral formation (but some of volcanic origin). The total land area is about 670 km^2. The chain trends NNE-SSW, 15–22°S, 173–177°W. By convention Tonga is taken to lie W of the Date Line. A great number of the low islands have no running water, the only supply coming from wells and stored rain water. The average rainfall is 1675 mm, falling between May and November. Hurricanes are frequent. The climate is relatively cool for this latitude.

The capital town is Nukualofa on Tongatabu (Tongatapu). Apart from limestones there are no mineral resources, but the volcanic ash makes a good soil and exports include copra and bananas.

Structural Position and the Volcanic Arc

Geotectonically Tonga lies along the eastern edge of the Melanesian province, and is therefore within the "Andesite Line." It embraces twin "island arcs" (here, rectilinear), an outer non-volcanic belt, marked by limestones islands and coral reefs (notably Tongatabu, Ha'apai, Vavau), and an inner volcanic belt (Tofua, Kao, Late, Falcon I). Submarine topography is well shown on a series of N.Z. Oceanographic Institute 1:200,000 charts (e.g., Eade, 1972). Niuafo'ou is isolated and lies to the west of the main belt; its crater contains a freshwater lake and a lava flow of historic date (Somerville, 1896). Its lavas are distinctive, basaltic, and its position has been discussed by Fairbridge and Stewart (1960) and others. The other volcanoes are mainly andesitic, basalt-andesitic and dacitic and quiescent, but Falcon Island is relatively active (dacitic). It periodically erupts, emerges as an island and disappears once more. Eruptions along the inner arc average once in every four years, and there are other minor submarine events. The volcanoes coincide with en échelon fractures along the Tonga Ridge trending NE-SW, in such a way as to open under the right-lateral shear associated with the whole New Zealand Alpine Fault/Tonga trend, but they would tend to close under the E-W compression of Pacific crustal subduction ("West Drift"). The volcanic belt lies 200 km W of the Tonga trench axis (Fig. 1).

The earliest volcanic rocks of the Tonga are pre-late Eocene basalt and uralitized gabbro on Eua, in the extreme E side of the chain, i.e., closest to the trench. It seems that the site of eruptions has shifted progressively westward. (The concept of the "andesite line" must therefore be treated as a dynamic feature, shifting through time: Karig, 1970). The lavas are exceptionally poor in potash and other alkalis,

FIGURE 1. Map of Tonga showing active volcanoes and submarine contours (after Schofield, 1967); with suggested transform faults and former sea-floor spreading axes in the Lau Basin (postulated by Sclater *et al.*, 1972).

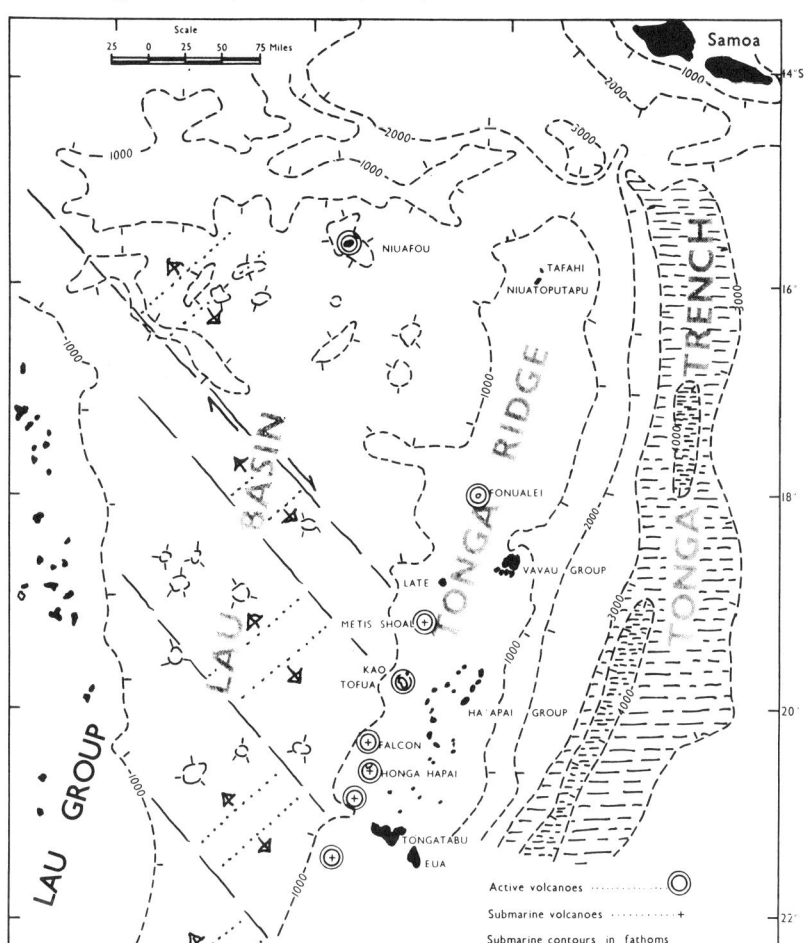

with enrichment of iron (Bryan *et al.*, 1972). The Lau Ridge (Fiji) is of similar nature.

A closely planned seismic triangulation system of stations on Tongatapu, Fiji, Samoa, and Rarotonga (Cook Islands) have disclosed the nature of one of the planet's major subduction zones along the Tonga Trench (incl. "Horizon Deep" 10,881 m., 5950 fm), and a key area in the elucidation in the "new global tectonics" (Fig. 2, Barazangi *et al.*, 1972).

The crust W of Tonga, and as far as the Australian continent, is what geotectonicists call "quasicratonic" (following Stille), i.e., partly oceanic, partly sialic, with consolidated belts of intermediate thickness dating back to the Paleozoic, e.g., in New Caledonia, which alternate with sectors of new oceanic crust. The crustal sector immediately W of Tonga is the Lau Basin, an acute triangular shaped depression which separates Tonga from the Lau Ridge.

The Lau Basin is marked by numerous diagonal minor ridges and troughs, in the S part trending NE-SW, swinging around to the NW in the N, between Niua-Fo'ou and Fiji, where the principal feature is known as Peggy Ridge. Karig (1970) suggested that the basin was extensional, filled by oceanic crust during the last 10 m yr. Detailed surveying by Sclater *et al.* (1972) led to the interpretation of the echelon trends of the Lau Basin as a series of NE-SW sea-floor spreading axes offset by five NW-SE transform faults, the largest and most active being Peggy Ridge.

FIGURE 2. Schematic cross section perpendicular to the Tonga arc, showing lithospheric plates (dotted) and the high- and low-attenuation zones (Q) as inferred from a study of seismic behavior (Barazangi et al., 1971).

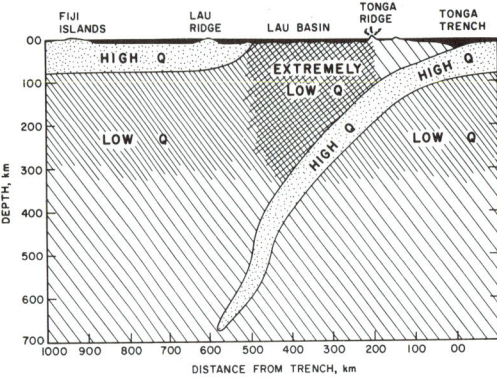

Geology of the Nonvolcanic Islands

Tongatabu is the largest island in the southern group of the Tonga chain. It is 40 x 16 km and has an area of 427 km² (165 sq mi). The highest point is 68 m (225 ft) above sea level. The best description is by Lister (1891), and there is a recent discussion by Schofield (1967). It is essentially an emerged coral island with 6–8 m of palagonitized fire-fountain debris unconformably overlying the coral in places, indicating nearby volcanic eruptions during the Pleistocene. Today this tuff is weathered to a rich and fertile soil, unusual for coral islands.

The limestones are typical coral reef rock of Plio-Pleistocene age slightly tilted to N. Large wave eroded notches are common along the coast at the 1.5–4.5 m levels, and benches cut into the limestone occur at 30, 45, and 60 m, particularly on the southeast coast. The entire south coast is unprotected by any barrier reef and drops into fairly deep water from the channelled ramp at the outer edge of the narrow reef flat. The flat is 30 m wide at Nokolo and is covered with encrusting corals and algal rimmed terraces. It is kept wet by spectacular blowholes of karst origin. Well-defined sea caves occur toward the south shore at 5 and 30 m. In spite of numerous earthquakes, the island must have been fairly stable for much of the late Pleistocene because most of the emerged shorelines and marine benches appear to be at world-wide glacioeustatic levels.

Eua is 20 by 7 km, north to south, and 350 m high. Ohonua is the largest village. Eua is unique for this region because of its plutonic rocks and also for a bed of well preserved marine Eocene fossils, which are separated easily from their tuffaceous matrix. Eua began in the Eocene as a submarine volcano which deposited andesitic, rhyolitic and dacitic tuffs and lavas below sea level. Most of the rock is composed of dense lavas, but ash and pillow lavas occur in small amounts. A dike complex is exposed above sea level. Three ages of pre-Pliocene marine tuffs and three ages of limestone are recognized, all limestones being of foraminiferal and algal origin. The island was uplifted about 350 m above sea level in the Pliocene when the first coral reef limestone was deposited.

Tectonic and glacial eustasy, with uplift as well has left clearly defined emerged and tilted reefs on Eua, indicating an epoch of resubmergence during the late Pliocene and early Pleistocene. World-wide eustatic shoreline levels are found at 30 m, 8 m, 3.5–1.5 m, and at 0.5 m, at progressively younger stages. Tilted terraces may be traced as high as 230 m (Hoffmeister, 1932). Stream erosion has been negligible because of karst sink holes in the limestone. Marine erosion, probably along an ancient fault scarp, has been dominant in forming the giant sea cliff, more than 300 m high along the east coast. Faults are numerous. Folding is common in the weak tuffs.

In the central group of the Tonga chain, the *Ha'apai Group*, apart from the volcanic arc, the only islands are low, emerged coral limestones, presumably late Pleistocene (from personal reconnaissance, R.W.F.).

The northern or *Vavau Group* possesses older, tilted coral limestones. At Talau there are tilted terraces and around 127, 85, and 45 m, with almost the same sequence at Mo'unga Lafa, 157, 109, and 45 m.

HAROLD T. STEARNS
RHODES W. FAIRBRIDGE

References

Anon., 1968. *Pacific Island Yearbook*, Sydney.
Baker, P. E., et al., 1971. "The geology of Tofua Island, Tonga," *Cook Bicentenary Expedition in the Southwest Pacific, Bull. Roy. Soc. N.Z.*, 8, 67–69.
Barazangi, M. et al., 1971. "Lateral variations of seismic wave attenuation in the upper mantle above the inclined earthquake zone of the Tonga Island arc," *J. Geophys. Res.*, 76(35), 8493–8516.
Bauer, G. R., 1970. "The geology of Tofua Island, Tonga," *Pacific Sci.*, 24(3) 333–350.
Brodie, J. W., 1970. "Notes on volcanic activity at Fonualei, Tonga," *N.Z. J. Geol. Geophys.*, 13(1), 30–38.
Bryan, W. B., et al., 1972. "Geology, petrography, and

geochemistry of the volcanic islands of Tonga," *J. Geophys. Res.*, 77(8), 1566–1585.
Daly, R. A., 1916. "Petrography of the Pacific Islands," *Bull. Geol. Soc. Am.*, 27, 325–344.
Eade, J. V., 1971. "Tonga bathymetry," *N.Z. Oceanogr. Inst.*, Ocean. Chart Ser., 1:1,000,000.
———, 1972. "Ha'apai bathymetry," *N.Z. Oceanogr. Inst.* Chart., Island Ser., 1:200,000.
Ewart, A., and Bryan, W. B., 1972. "Petrography and geochemistry of the igneous rocks from Eua, Tongan Islands," *Bull. Geol. Soc. Am.*, 83, 3281–3298.
Fairbridge, R. W., and Stewart, H. B., Jr., 1960. "Alexa Bank, a drowned atoll on the Melanesian Border Plateau," *Deep-Sea Res.*, 7, 100–116.
Guest, N. J., 1959. "Geologic mapping, Tonga," *Annual Report for 1958* (Council pap. 17). Suva: Fiji Geol. Surv. Dept., 3.
Hoffmeister, J. E., 1932. "Geology of Eua, Tonga." *Bernice P. Bishop Mus. Bull.*, 96, 93p. Contains chapters by Alling and Whipple.
Jaggar, T. A., 1930. "Volcanoes of New Zealand–Tonga Belt," *Volcanol. Lett.*, 265, 2–4.
———, 1931. "Geology and geography of Niuafoou Volcano," *Volcanol. Lett.*, 318, 1–3.
Karig, D. E., 1970. "Ridges and basins of the Tonga-Kermadec island arc system," *J. Geophys. Res.*, 75, 239–254.
Ladd, H. E., 1970. "Eocene mollusks from Eua, Tonga," *U.S. Geol. Surv. Prof. Pap. 640-C*, 11p.
Lister, J. J., 1891. "Notes on the geology of the Tonga Islands," *Quart. J. Geol. Soc. Lond.*, 47, 590–617.
Orbell, G. E., 1971. "Soil surveys–Vava'u and adjacent islands, Tonga Islands," *Cook Bicentenary Expedition in the Southwest Pacific, Bull. Roy. Soc. N.Z.*, 8, 125–130.
Ostergaard, J. M., 1935. "Recent and fossil marine mollusca of Tongatabu," *Bernice P. Bishop Mus. Bull.*, 131.
Raitt, R. W., et al., 1955, "Tonga Trench," *Geol. Soc. Am. Spec. Pap.* 62, 237–254.
Schofield, J. C., 1967. "Notes on the geology of the Tongan Islands," *N.Z. J. Geol. Geophys.*, 10(6), 1424–1428.
Sclater, J. G., et al., 1972. "Crustal extension between the Tonga and Lau Ridges: petrologic and geophysical evidence," *Bull. Geol. Soc. Am.*, 83, 505–518.
Snow, P. A., 1969. *Bibliography of Fiji, Tonga and Rotuma.* Miami: Univ. Miami Press.
Somerville, B. T., 1896. "Account of a visit to Niuafou, South Pacific," *Geogr. J.*, 7, 65–71.
Stearns, H. T., 1971. "Geologic setting of an Eocene fossil deposit on Eua Island, Tonga," *Bull. Geol. Soc. Am.*, 82, 2541–2552.
Talwani, M. I., et al., 1961. "Gravity anomalies and crustal section across the Tonga Trench," *J. Geophys. Res.*, 66(4), 1265–1278.
Tarling, D. H., 1966. "The palaeomagnetism of the Samoan and Tongan Islands," *Geophys. J. Roy. Astron. Soc.*, 10(5), 497–513.
Thomson, J. A., 1926. "Volcanoes of the New Zealand–Tonga volcanic zone," *N.Z. J. Sci. Tech.*, 8, 354–371.

Cross-references: *Fiji; Kermadec Islands; New Hebrides; New Zealand; Western Samoa.*

TRINIDAD AND TOBAGO

Lying 11 km (7 miles) NE of the coast of Venezuela, the islands of Trinidad and Tobago are the southernmost of the Lesser Antilles of the West Indies. They have a combined area of 5128 km^2 (1980 sq mi). Between Trinidad (4828 km^2; 1864 sq mi), which is closest to Venezuela, and Tobago (300 km^2; 116 sq mi) to the NE, is a 30-km channel. Trinidad is crossed from E to W by two narrow mountain ranges, up to 1000 m, following the north and south coasts, with a low central range. South of it the land is rolling, whereas the land between the central and northern ranges is flat. A ridge of ancient volcanic origin runs down the middle of Tobago.

Petroleum, coal, lignite, iron, clay, limestone, and gypsum are found on the islands. Crude oil production is not so high as in former times, but there have been recent discoveries of offshore accumulations. The celebrated Pitch Lake, 114 acres, contains one of the world's largest supplies of natural asphalt. Oil and gas fields are limited to the southern third of the island. The offshore discoveries are in Miocene clastics to the SE of the island.

Structurally Trinidad's history is divided into a highly dynamic Cretaceous and younger sedimentary megacycle, and a pre-Cretaceous cycle of intense orogeny, producing the metamorphic rocks of the Northern Range and overthrusting them to the S. The Cretaceous was folded along NE-SW trends and the Tertiary formations were laid down over this irregular and unstable basement. Continued folding during sedimentation occurred along E-W axes, producing five distinctive anticlinal belts. The youthful folding mobilized water, oil, and gas-saturated formations, producing complex sedimentary diapir structures, including mud volcanoes, some of large size that are still active. The island and its faults are still fairly seismic.

The oldest rocks, collectively known as the "Caribbean" Series, are metasediments, micaschists, crystalline limestones, quartzites, etc. Tithonian (Upper Jurassic) ammonites have been identified by L. F. Spath. To the S and in the Central Range there is the Cuche Shale and Maridale Marl with ammonites and rudistids indicating a Lower Cretaceous age, followed by siltstones and shales, rich in foraminifera, mollusca, and corals, of Upper Cretaceous age. Remarkable slump blocks and chaos formations are found in these formations.

Eocene is represented by the Point à Pierre Formation, a thick sequence of fossiliferous shales, a wildflysch facies. It is followed by thick Oligocene silts and clays, which pass up into Lower Miocene, represented by globigerina marls, limestones, pebbly clays and sandstone, the St. Croix Formation, which in places is 1000 m thick. Progressive folding led to the development of many partly independent troughs in the Middle and Upper Miocene. The stratigraphy, because of tectonics and gravity sliding, is highly complex, both on and offshore.

The Upper Miocene contains the famous LaBrea oil sands. Miocene, Pliocene, and Quaternary are represented by raised beach deposits and a small Pleistocene flora has been described.

The island of Tobago rises from an extension of the same continental shelf as Trinidad. It forms the easterly termination of the northern or coastal range of Venezuela. Its core rocks are Mesozoic metasediments and basic igneous rocks, just as in Trinidad's Northern Range. Overlapping these in the SW there is a flat-lying platform rich in Upper Miocene fossils, notably *Balanus*. In turn this is overlain by a Pleistocene raised coral limestone, which in places reaches 35 m above sea level.

RHODES W. FAIRBRIDGE

References

Andel, T. H. van, and Sachs, P. L., 1964. "Sedimentation in the gulf of Paria during the Holocene Transgression, a sub-surface acoustic reflection study," *J. Marine Res.*, 22, 30–50.

Anon., 1953. *Caribbean Atlas*. New York: Macmillan, 16p.

Barr, K. W., 1952. "Limestone blocks in the Lower Cretaceous Cuche formation of the central range, Trinidad, B.W.I.," *Geol. Mag.*, 89(6), 417–425.

____, and Saunders, J. B., 1968. "An outline of the geology of Trinidad," *Trans. 4th Caribbean Geol. Conf.*, 1–10.

Berry, W., 1925. "Tertiary flora of the island of Trinidad," *Johns Hopkins Univ. Stud. Geol.* 6, 71–160.

____, 1926. "A Pleistocene flora from the island of Trinidad," *Proc. U.S. Nat. Mus.*, 66(2558), Art. 21, 1–9.

Illing, V. C., 1928. "Geology of the Naparima region of Trinidad," *Quart. J. Geol. Soc. Lond.*, 84, 1–56.

Kugler, H. G., 1936. "Summary digest of geology of Trinidad," *Bull. Am. Assoc. Petrol. Geologists*, 20, 1439–1453.

____, 1956. "Trinidad," *Handbook of South American Geology* (Mem. 65). New York: Geol. Soc. Am., 351–365.

____, 1959. *Geological Map and Cross-sections of Trinidad.* Zürich: O. Füssli; London: E. Stanford, 1:100,000.

Liddle, R. A., 1928. *The geology of Venezuela and Trinidad.* Forth Worth, Tex.: 552p.

Maury, C. J., 1935. "The Soldado rock section," *Science*, 82(2122), 192–193.

Maxwell, J. C., 1948. "Geology of Tobago, B.W.I.," *Bull. Geol. Soc. Am.*, 59(8), 801–854.

Nugent, N., 1811. "The Pitch-Lake of Trinidad," *Trans. Geol. Soc.*, 1, 63–96.

Peckham, S. F., 1895. "The Pitch-Lake of Trinidad," *Proc. Geol. Soc.*, 16, 460–470.

Potter, H. C., 1968. "A preliminary account of the stratigraphy and structure of the eastern part of the Northern Range, Trinidad," *Trans. 4th Caribbean Geol. Conf.*, 15–20.

____, 1973. "The overturned anticline of the Northern Range of Trinidad near Port-of-Spain," *J. Geol. Soc. Lond.*, 129(2), 133–138.

Renz, H. H., 1942. "Stratigraphy of northern South America, Trinidad and Barbados," *Proc. 8th Am. Geol. Soc. Congr.*, 4, 513–571.

Rutsch, R., 1939. "Entwicklung tropischamerikanischer Tertiärfaunen und Kontinental-verschiebung Hypothesis," *Geol. Rundschau*, 30, 362–372.

Suter, H. H., 1951. "The general and economic geology of Trinidad, B. W. I.," *Colon. Geol. Mineral Res.*, 2(3–4); 3(1), 134 p. (rev. 2nd ed., 1960).

Tomblin, J. F., ed., 1970. "Field guides to the geology of Trinidad," *Intern. Field Inst. Guidebook.* Washington, D.C.: A.G.I., 18p.

Trechmann, C. T., 1934. "Tertiary and Quaternary beds of Tobago," *Geol. Mag.*, 71, 481–493.

____, 1935. "Fossils from the Northern Range of Trinidad," *Geol. Mag.*, 72, 166–175.

Vaughan, T. W., and Cole, W. S., 1941. "Preliminary report on the Cretaceous and larger Tertiary Foraminifera of Trinidad," *Geol. Soc. Am. Spec. Pap. 30.*

Waring, G. A., and Harris, G. D., 1926. *Geology of the Island of Trinidad.* Baltimore: Stud. in Geol. No. 7. Baltimore: Johns Hopkins Univ., 170p. (with maps and plates).

Wilson, C. C., 1958. "The Los Bajos Fault and its relation to Trinidad's oilfield structures," *J. Inst. Petrol.*, 44, 124–236.

Cross-references: *Barbados; West Indies.*

TUAMOTU ISLANDS

The Tuamotu, also once called the Paumotu or Low Archipelago, in the South Pacific represents the largest group of atolls in the world. Under the administration of *French Polynesia* (q.v.), these islands and reefs extend from 14 to 22°S and 135 to 148°W. They are situated along NW-SE trends between the Society Islands and the Marquesas. There are 78 atolls and one elevated coral island. The underlying oceanic crust in this region is of Cretaceous age.

The largest atoll is *Rangiroa (or Rairoa)*, 105 km (66 miles) long, which is the administrative center. The raft *Kon Tiki* (Heyerdahl) landed here on its voyage from Peru. Most of the atolls have multiple rim islands or islets. The more mature ones have eroded limestone, but the younger ones are only sandy cays that are liable to change each time there is a hurricane. The area lies within the belt of the SE trade winds, but northern cyclones are liable to occur in the southern, summer months (December–March). Rangiroa is reported to have coral limestone up to 5 m above present sea level.

One island, *Makatea*, the one closest to the Society Islands, is an uplifted atoll, 70 m in elevation, with major phosphate deposits. The limestone is said to be Eocene in age (Repelin, 1919). Its shores are steeply cliffed and terraced.

Deep borings put down on Mururoa in 1964–1965 disclosed volcanic basement at 415 and 438 m, capped by coral debris, reef limestone, or sand, with the top 100 m poorly consolidated (cf. Funafuti, in *Gilbert and Ellice Islands;* also *Marshall Islands*). Atomic testing by the French government on Mururoa in recent years has directed considerable scientific attention to it (Chevalier et al., 1968; Deneufbourg, 1969; etc.).

RHODES W. FAIRBRIDGE
J.-P. CHEVALIER

References

Agassiz, A., 1906. "The eastern tropical Pacific...," *Mem. Mus. Comp. Zool.,* 33, 1–75.

Aubert de la Rüe, E., 1959. "Etude géologique et prospection minière de la Polynésie Française," *Recherche Géologique et Minérale en Polynésie Française.* Paris: Inspection Générale des Mines et de la Géologie, 7–42.

Chauveau, J., Deneufbourg, G., and Sarcia, J., 1967. "Observations sur l'infrastructure de l'atoll de Mururoa (Archipel des Tuamotu, Pacifique Sud)," *C. R. Acad. Sci.,* 265(16), 1113–1116.

Chevalier, J.-P., Denizot, M., Mougin, J. L., Plessis, Y., and Salvat, B., 1968. "Etude géomorphologique et bionomique de l'atoll de Mururoa (Tuamotu)," *Cahiers du Pacifique,* 12, 1–144.

Crossland, C., 1928. "Coral reefs of Tahiti, Moorea and Rarotonga," *Linn. Soc. Lond. J. Zool.,* 36, 577–620.

Deneufbourg, G., 1969. "Les forages de Mururoa," *Cahiers du Pacifique,* 13, 47–58.

Emery, D. P., 1932. "Tuamotu Survey," *Bernice P. Bishop Mus. Bull.,* 94, 40–50.

Newell, N. D., 1954a. "Expedition to Raroia," *Atoll Res. Bull.,* Washington, D. C., 31, 2–23.

———, 1954b. "Reefs and sedimentary processes of Raroia," *Atoll Res. Bull.,* 36, 1–35.

———, 1956. "Geological reconnaissance of Raroia (Kon Tiki) atoll, Tuamotu archipelago," *Am. Mus. Nat. Hist. Bull.,* 109, 313–372.

Ranson, G., 1957. "Observation sur les îles basses des Tuamotu," *8th Pacific Sci. Cong., Philippines,* 3A, 989–1008.

Repelin, J., 1919. "Sur un point de l'histoire de l'océan Pacifique," *C. R. Acad. Sci.,* 168, 237–239.

Trichet, J., 1969. "Quelques aspects de la sédimentation calcaire sur les parties émergées de l'atoll Mururoa," *Cahiers de Pacifique,* 13, 1–14.

Wiens, H. J., 1962. *Atoll Environment and Ecology.* New Haven: Yale Univ. Press.

Cross-references: *Gilbert and Ellice Islands; Marquesas Islands; Marshall Islands; Oceania; Society Islands.*

TUBUAI (AUSTRAL) ISLANDS

A chain of islands in the South Pacific, occupying the most southerly row of NW-SE trends in *French Polynesia* (q.v.), the Tubuai or Austral group consists of one atoll and six volcanic islands, four with coral reefs. There are also two major reef-capped seamounts in the chain, President Thiers Bank (−33 m) and Neilson (Lancaster) Reef (−5 m). Neilson is the most southerly, purely coral reef in the Pacific. The chain extends from 22-28°S and 143-155°W. The same trend continues to the NW in the northerly Cook Island chain.

The most northerly of the Tubuais is an atoll, Maria (Sand) Islands, two rim islets on a triangular reef 5 km across. Next, in sequence, follow the volcanic islands: Rimatara, Rurutu, Tubuai, Raivavae (Vavitu), Rapa, and the Bass Islets (Morotiri). Rimatara has an eroded volcanic cone surrounded by a raised reef rim, now 7–8 m above sea level, stepping down to a ring lagoon and a contemporary barrier reef, 10 km in diameter. Rurutu is similar, but without a barrier reef, having only a fringing reef, and the old reef is tilted and elevated to 60–80 m with a patch possibly as high as 150 m; it is about 11 km wide, topped by a plateau at 240 m elevation. Chubb (1927b) has compared Rurutu with Mangaia in the Cook Islands (q.v.). The limestone is partly covered by basalt lapilli and palagonite tuff. There are intermediate erosion benches or emerged reefs at 180, 80, 2, and 0.5 m. It has been claimed that there was schist on Rurutu, but it turns out to be a platey phonolite (Jérémine, 1959). Little is known of the chronology of the islands, but sea-floor spreading suggests that the underlying crust is Cretaceous.

Tubuai and Raivavae are degraded volcanic cones with contemporary barrier and fringing reefs with no evidence of elevation. The lagoons are shallow and probably partly infilled. Rapa and Morotiri are more isolated and are the most southerly islands of the entire mid-Pacific; they have little living coral, perhaps because of the latitude (28°S) and cool currents in winter. Rapa has limestone uplifted to 30 m and the usual eustatic bench at 2–3 m. Morotiri consists of four volcanic islets, spread over 3 km and rising to 106 m. Rapa and Morotiri are remnants of caldera explosions. Both are deeply eroded and cliffed.

It has been claimed that, like the Hawaiian Islands, the Tubuais become progressively younger from NW to SE, beginning with a submerged stump (now crowned by an atoll), ending with a younger volcanic remnant. However, the analogy should not be pushed too far, because there is no really youthful or contemporary eruptive center at the southeastern end of the chain.

RHODES W. FAIRBRIDGE
J.-P. CHEVALIER

References

Aubert de la Rüe, E., 1956. "Sur la présence du manganèse à Rurutu," *20th Intern. Geol. Congr., Manganese Symp.*, **4**, 373.

———, 1958. "Observations sur le volcanisme tertiaire et quaternaire de quelques îles de la Polynésie Française," *Bull. Volcanol.* Sér. 2, **19**, 159–178.

———, 1959. "Etude géologique et prospection minière de la Polynésie française," *Recherche Géologique et Minéralogique en Polynésie Française.* Paris: Inspection Génerale des Mines et de la GéOlogie. 7–42.

Chubb, L. J., 1927a. "The geology of the Austral or Tubuai Islands," *Quart. J. Geol. Soc. Lond.*, **83**, 291–316.

———, 1927b. "Mangaia and Rurutu; a comparison between two Pacific Islands," *Geol. Mag.*, **64**, 518–522.

Davis, W. M., 1928. "The coral reef problem," *Am. Geog. Soc. spec. Publ.*, 9.

Jérémine, E., 1959. "Sur quelques laves de l'île Raivavae," *Bull. Volcanol.*, Sér. 2, **21**, 117–126.

Lacroix, A., 1927. "La constitution lithologique des volcans du Pacifique central austral," *Bull. Volcanol.*, **4**, 218–231.

———, 1928. "Nouvelles observations sur les laves des îles Marquises et de l'île Tubuai," *C. R. Acad. Sci.*, **187**, 364–369.

Marshall, P., 1918. "Notes on the geology of the Tubuai Islands and of Pitcairn," *N.Z. Inst. Trans. Proc.*, **1**, 278–279.

Smith, W. C., and Chubb, L. J., 1927. "The petrography of the Austral or Tubuai Islands," *Quart. J. Geol. Soc. Lond.*, **83**, 317–341.

Cross-references: *Cook Islands; French Polynesia; Oceania; Society Islands.*

TURKS AND CAICOS ISLANDS

Situated in the western Atlantic, to the SE of the Bahama Islands and N of Hispaniola, the Turks and Caicos Islands are a British colony, formerly a dependency of Jamaica. They lie between 21° and 22°N and 71° and 72°30'W and have an estimated land area of 430 km^2 (166 sq mi).

The climate is characteristic of the high-pressure NE trade-wind belt, maintaining a mild to warm climate the year round, but with rather frequent hurricanes in early and late summer. Those in 1945 and 1960 caused great damage. The rainfall for Grand Turk averaged about 685 mm between 1907 and 1971 but over the last 10 years this has fallen to about 584 mm. There are no rivers or streams and water supply is limited to brackish well-water and rainwater catchments.

Of the eight main islands, most of the population is centered on Grand Turk, South Caicos, and North Caicos with a scattering on Middle Caicos, Providenciales, and Salt Cay. East Caicos and West Caicos are at present uninhabited. There are rolled limestone airstrips on all the inhabited islands and paved international airfields on Grand Turk and South Caicos.

The islands are geologically similar to the Bahamas and are formed of Quaternary marine limestones partially mantled by eolianites and beach ridges. There is extensive karst dissection which developed during the low sea-level stages of the Pleistocene and the Atlantic-facing shorelines are paralleled by extensive coral reefs. A radiocarbon date of an emerged reef off South Caicos may indicate a mid-Holocene high sea level of about 1 m.

Natural resources that have been exploited in the past include guano and solar evaporated salt. The extraction of guano became uneconomic at the beginning of this century but the salt industry continued as the main employer until 1964 when operations at Grand Turk and South Caicos closed down. At the present time salt raking continues only on Salt Cay.

JOHN D. MATHER

References

Agassiz, A., 1893. "A reconnaissance of the Bahamas and of the elevated reefs of Cuba in the Stream Yacht 'Wild Duck'," *Bull. Mus' Comp. Zool.*, **26**.

Anon., 1966. *Turks and Caicos islands, Report for the Years 1963 and 1964.* London: HMSO, 52p.

Booy, T. de, 1918. "The Turks and Caicos Islands," *Geogr. Rev.*, **6**, 37–51.

Doran, E., Jr., 1955. *Land-forms of the Southeast Bahamas* (Publ. 5509). Austin, Tex.: Univ. Texas, 38p. (includes the Turks and Caicos Islands).

Klein, H., Hoy, N. D., and Sherwood, C. B., 1958. *Geology and Groundwater Resources in the Vicin-*

ity of the Auxiliary Air Force Bases, British West Indies. Tallahassee, Florida: U.S. Geol. Surv., open file rept., 142p. (containing discussion of Grand Turk).

Ray, C., and Sprunt, A. N., 1971. *Parks and Conservation in the Turks and Caicos Islands*. Washington, D. C.: U.S. Govt. Print. Off., 45p.

Richards, H. G., 1966. "Dates of some Pleistocene coral reefs in the West Indies," *Geol. Soc. Am., Spec. Pap. 101,* 373.

Sharples, S. P., 1883. "Turks Island and the guano caves of the Caicos Islands," *Proc. Boston Soc. Nat. Hist.,* 77, 242–252.

Cross-references: *Bahamas; Bermuda; Dominican Republic; West Indies.*

U

UNITED STATES OF AMERICA—GENERAL

Sharing the continent of North America with Canada, Mexico, and Greenland, the United States of America (usually abbreviated U.S., or U.S.A.) comprises 50 states, 48 of which are "coterminous," whereas two, Alaska and Hawaii, are separated by land or water. The capital, Washington, occupies a small federal enclave, the District of Columbia. Other areas under U.S. sovereignty include: *Puerto Rico, Virgin Islands* (U.S.), *American Samoa, Guam, Wake Island,* and other minor islands (see separate treatments in this volume). Together the "outlying areas" cover about 1550 km² (6000 sq mi), providing a total U.S. area of 9.52 million km². *The United States Trust Territory of the Pacific Islands* (Micronesia) are under U.N. mandate (21,974 km²). Panama Canal Zone, under U.S. jurisdiction, is treated here under *Panama* (q.v.).

The central reference point of the coterminous United States, adopted for the triangulation network, is at Meades Ranch, Kansas, 39°13′26.686″N and 98°32′30.506″W. The *area* of the coterminous United States (including lakes) amounts to 7.98 million km² (3,026,789 sq mi), with Alaska (1.52 million km², 586,400 sq mi), and Hawaii (16,700 km², 6406 sq mi). The *coast line* (measured to the nearest 30″ of latitude on a 1:1,200,000 chart) of the coterminous United States is 19,926 km (12,383 statute miles), of Alaska and Hawaii 8035 km (4993 mi). In contast, the *shore line* (measured in detail to tide height and including islands) is, respectively, 142,639 km (88,633 mi) and 86,390 km (53,677 mi). The coterminous U.S. extends from West Quoddy Head, Maine (66°57′W) to Cape Alava, Washington (124°44′W), and from Lake of the Woods (49°23′N) to Cape Sable, Florida (25°07′N). The highest mountain is Mt. McKinley (Alaska) with 7061 m (20,320 ft), or, in the coterminous U.S., Mt. Whitney (California) with 4418 m (14,494 ft). The longest river is the Mississippi-Missouri with 6020 km (3741 mi).

Geologic and Physiographic Regions

The United States has many diverse geologic and physiographic regions. Fenneman numbered 25 provinces (and many subprovinces) within the coterminous states. For convenience and simplicity these regions are now grouped into a dozen major provinces or regions, and discussed in separate articles in this volume. By way of introduction to these regional treatments, the basic characteristics of each province are outlined below.

The continent *North America* (q.v.) consists of a central shield and platform flanked by mountain belts to E and W. The United States includes major parts of each of these three divisions. This seemingly simple pattern permits the recognition of three major geotectonic and physiographic or morphotectonic regions (Fig. 1) which in turn are subdivided into many smaller units (Fig. 2). The major regions have no formal names and are considered here as (1) the *North American Platform* (shield, craton, lowland); (2) the *Appalachian Region,* which includes the Appalachian Mountains proper, together with Gulf and Atlantic Coastal Plains; and (3) the *Cordilleran Region* and associated western provinces, including the intermontane plateaus. *Alaska* and *Hawaii* (q.v.) are considered separately.

The *North American Platform* consists of a shield or central craton in which Precambrian crystalline rocks form the basement. In the United States these rocks are exposed in northern Minnesota and Wisconsin as part of the *Laurentian Upland province* which in turn is part of the *Canadian Shield* (q.v. under *Canada*).

FIGURE 1. Major geotectonic regions of the United States: (1) the North American platform, (2) the Appalachian region, and (3) the Cordilleran region.

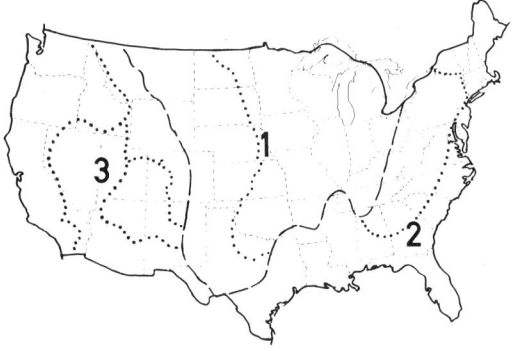

FIGURE 2. Major physiographic regions of the United States.

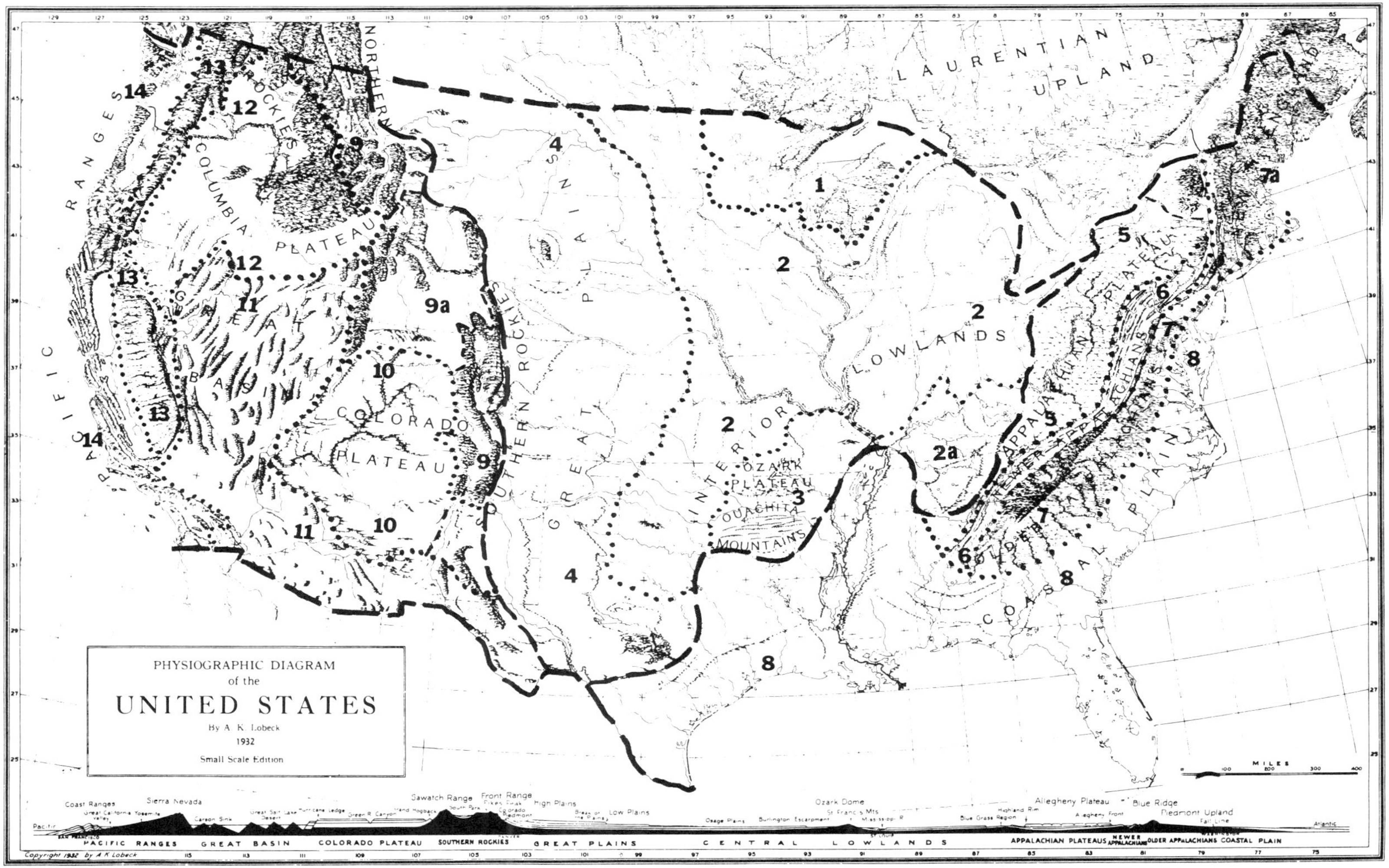

Peripheral to this nucleus or core region (and dipping away gently to the S, W, and E) the basement is covered by Paleozoic rocks, mainly little disturbed marine sedimentary facies, constituting the *Midwestern Region* (q.v.). As used in this volume the Midwestern province includes the Laurentian or Superior Upland, the Central Lowland, the Interior Low Plateaus, the Ozark Plateau, and the Ouachita Mountains (Fig. 1).

The Paleozoic rocks of the craton generally form a lowland, plains, or low plateaus, structurally characterized by broadly warped basins and domes. Plateau and fold-belt topography is largely limited to the southern-most regions. Much of the northern and central Midwestern province has been strongly affected by glacial and periglacial processes in the Quaternary.

The *Great Plains Province* (q.v.) is tectonically part of the Central Platform, with some deformation near its western limits related to the relatively recent orogenic events in the Rocky Mountains. During the Cenozoic uplift of these mountains a piedmont blanket of continental sediments was deposited over the plains, so that the Paleozoic (and Mesozoic) rocks and structures are largely buried here.

The *Appalachian Region* (q.v.) has a classic orogenic belt of mountains that dominates the eastern United States from Alabama to the Canadian border (q.v. *Canada: Atlantic Province* for the northeastern extension of this belt). The western portion of the Appalachians consists of Paleozoic sedimentary rocks that have undergone moderate deformation (mostly open folding of "Juratype") but no appreciable metamorphism. The sediments represent a typical miogeosynclinal facies (early Paleozoic), passing to an exogeosynclinal assemblage (late Paleozoic). Recent plate tectonic theories would suggest deposition of parts of the column in a continental terrace environment (paraliageosyncline). Thus Cambrian and some of the Ordovician layers have a source from the craton and were laid down in relatively shallow water on a terrace that faced the then open proto-Atlantic Ocean. Later sediments reflect the orogenic and mountain-building events associated with ocean closing and continental plate collisions, so that much of the upper Paleozoic material had a source to the E, and is partly continental (i.e., proto-Eur-African).

The rocks are nearly horizontal in the region of the Appalachian Plateau but are strongly folded and thrust faulted in the E, where there is a fold-mountain topography designated as the "Newer Appalachians" (Fig. 2). The eastern portion of the Appalachians ("Older Appalachians") consists of metamorphic rocks, largely lower Paleozoic and Precambrian metasediments, with many intrusive bodies. These crystalline rocks are dominated by a high rugged mountain belt, the Blue Ridge, and an adjacent region of hills, a dissected plateau, called the "Piedmont province." The northern extension of the Appalachians is termed the *New England Region* (q.v.), which lacks the younger Appalachian folds and suffered more or less intense glaciation during the Quaternary period.

The *Coastal Plain (Atlantic Coastal Province, Gulf Coastal Province,* q.q.v.) occurs to the E and S of the Central Platform. The Coastal Plain consists of a series of seaward dipping sediments, late Mesozoic and Cenozoic in age, with some continuing deposition on the continental shelf today.

The *Pacific Cordilleran Region* west of the Central Platform, along a more or less N-S boundary, is found a complex series of mountain belts and intermontane plateaus, with many important tectonic and physiographic units. In this volume these units are discussed as *Rocky Mountains Province, Colorado Plateau Province, Basin and Range Province, and Pacific Cordilleran Region* (q.q.v.).

The *Rocky Mountain Province* (q.v.) includes Precambrian crystallines as well as sedimentary rocks of all periods. The mountain belt has had a complicated history and, in fact, contains areas with quite diverse early histories. The entire region can be treated as a unit because the present-day landscape and the most obvious structures result from diastrophism (and subsequent erosion) in late Mesozoic and early Cenozoic time. This "Laramide" phase resulted in massive block faulting, intense folding (in narrow belts), and some important volcanism. Uplift of the Rockies resulted in the shedding of sediment over the Paleozoic platform to the E, thus producing the detrital cover of the Great Plains province.

Along the Pacific Coast is found the *Pacific Cordilleran Region* (q.v.) which consists of several different orogenic belts, containing intensely deformed rocks ranging from Paleozoic to Cenozoic in age. Extensive areas of igneous intrusions, mostly Mesozoic, are also found. Cenozoic deformation and volcanism have determined the contemporary tectonic fabric and physiographic character of the region. Physiographically there are two main mountainous zones, a rugged and seismically active belt, the Coast Ranges, and an interior high relief belt which includes the Sierra Nevada and Cascade mountains. Between the two belts there are lowland troughs filled with late Cenozoic sediments or lavas. The Coast Ranges are split longitudinally by the celebrated San Andreas fault system, which is a dextral strike-slip with highly active contemporary seismicity.

Between the Rocky Mountains and the Pacific ranges are found the intermontane provinces: Colorado Plateau; Columbia Plateau; and Basin and Range.

The *Colorado Plateau* (q.v.) consists mainly of Paleozoic and younger sedimentary rocks that are nearly horizontal, but nevertheless have been uplifted, faulted, and in places subjected to monoclinal folding above the basement faults. The Colorado River system has deeply dissected this high-altitude region resulting in a network of scenic canyons and gorges amidst mesas and plateaus.

The *Columbia Plateau* is a distinctive area, discussed in this volume as part of the *Pacific Cordilleran Region* (q.v.). Cenozoic lava flows and continental sediments blanket a diverse bedrock which includes elements of orogenic belts exposed elsewhere in the Pacific Cordilleran belt.

The *Basin and Range Province* (q.v.) is characterized by Cenozoic block faulting along a NNW-SSE trend, resulting in a series of uplifted mountains and downfaulted basins. Bedrock includes the basic units of Paleozoic and Mesozoic orogenic belts which are no longer coherent tectonic entities. A considerable thickness of Cenozoic fill is found in the basins, along with much volcanic material, especially in the W and N of the province. Geophysical evidence suggests that the region is largely under the E-W tension (with minor strike-slip components), with related crustal thinning, 29–30 km, in contrast to 43–45 km under the surrounding territory (Prodehl, 1970). It is marked by high heat flow and much seismicity.

Regional Physiography

North American Platform. As indicated in the introduction, the central United States between the Appalachians and the Rocky Mountains is mostly a broad lowland or plain with a few regions of plateaus, hills, and low mountains. In this volume it is convenient to treat it in two divisions. The *Midwestern Region* (q.v.) of this Central Platform is the largest part of the region and is marked by the lowest relative relief. In the N it includes the Superior Upland, which is structurally an extension of the Laurentian Upland province of the Canadian Shield. Its southern parts embrace the Interior Low Plateaus, Ozark Plateaus, and Ouachita Mountains. The core of the Midwest Province is the Central Lowland, also known as the "Interior Lowland." The second division is the higher western part of the central United States, the *Great Plains Province* (q.v.), which extends to the E of the Rocky Mountains from the Canadian border to the Rio Grande.

Central Lowland. The central lowland covers over 15% of the United States. The area has a low altitude and slight local relief, with a mantle of glacial or eolian deposits in most areas that smooth the ground surface and conceal the underlying bedrock. In the N the Pleistocene cover is largely morainic (till and deglacial deposits), whereas in the S and SE, notably E of the Mississippi, there is extensive loess. The area has a continental climate, abundant water in the form of lakes and rivers, and a rich soil, supporting a diverse economy.

Bedrock consists mostly of Paleozoic marine sedimentary rocks of platform facies usually rather thin, and broadly deformed into very large basins and domes, the warping in part being contemporaneous with deposition.

The area of the Great Lakes was formed by Pleistocene glacial lobes modifying preglacial drainage patterns. The Michigan basin forms the geologic nucleus of this area which has few topographic features older than 15,000 years. The northwestern part of the central lowland (e.g., Minnesota) is a low-relief plain characterized by vast numbers of small lakes developed in glacial drift that masks the preglacial topography. Between the two lake regions occurs the Wisconsin "Driftless Area" where Quaternary modification of pre-Pleistocene topography has been relatively slight, and is mostly limited to periglacial processes.

South of the lake regions occurs a vast till and loess plain, which corresponds to the fertile farmland known as the "Corn Belt" (Nebraska-Ohio). The youngest parts of the plain are rather flat, but to the S and W older tills have been dissected to produce an increasingly hilly terrain. Still farther S is the Osage section (Kansas-Texas) where a Pleistocene cover is lacking. Here the gently dipping strata have been somewhat beveled by erosion, resulting in a landscape with low hills and cuestas.

Superior Upland (or Laurentian Upland, Canadian Shield). The Superior Upland is an extension of the Laurentian Upland of Canada into the United States, as are the Adirondack Mountains (included here, for convenience, within the *New England Region,* q.v.). Precambrian crystalline rocks and metasediments, modified by glacial erosion and deposition, typify the region. In some areas the topography has low relief, reflecting a Precambrian planation surface ("paleoplain") which continues as an unconformity beneath the Paleozoic sediments to the S. Elsewhere broad undulating structures are common, such as the synclinal basin forming Lake Superior. Rich copper and iron ore bodies are found in this region.

Interior Low Plateaus. This plateau region extends from Ohio through central Kentucky to Tennessee. The terrain is underlain by a broad domal structure in the Precambrian basement identified as the Cincinnati Arch. A subdued low plateau topography, mostly developed on limestone, provides a landscape that is very distinctive from the glacially modified terrain to the N and the more rugged terrain of the Appalachian Plateaus to the E. Escarpments, cuestas, small plains, and karst features are common. Much of the land is used in agriculture, though the soil is relatively poor.

Ozark Plateau. The Ozarks are a hilly upland formed by domelike plateaus of gently tilted Paleozoic limestone, surrounding a nucleus of exposed Precambrian crystalline basement rocks (e.g., St. Francis Mountains). Dissection of the layers has produced hills and rounded scarps. The soils are rather poor for the most part, but valuable stratiform lead and coal are mined, notably in the "Tri-State Area" of Missouri, Kansas, and Oklahoma.

Ouachita Mountains. The Ouachita Mountains and the nearby Arbuckle Mountains of Arkansas and Oklahoma represent a E-W late Paleozoic fold belt related to the Appalachians far to the E, the connections being hidden beneath the Mississippi embayment. The Arkansas Valley is located in a parallel structural depression associated with the regional fold pattern. Locally faults and massive intrusions of igneous rocks occur. Thus geotectonically this region correlates with the Appalachians, but because of its isolation and physiographic continuity with the country to the N, it is more appropriate to group this landscape along with the Central Platform.

Great Plains. Geologically the Great Plains are part of the Central Platform of the United States, but the essential terrain features of the region are related to events associated with the development of the Rocky Mountains. Above its basement of Precambrian age, there is a cover of Paleozoic and Mesozoic marine sedimentary rocks found today forming a very broad N-S trending depression, with the deepest layers occurring near the Rocky Mountain front.

The Late Cretaceous to early Tertiary Laramide events led to extensive erosion from the uplifted Rockies. Much of this material was deposited as great sheets of piedmont alluvium, which covered the gently dipping sedimentary rocks of the central platform area E of the mountains. These extensive piedmont deposits almost buried the Rockies in their own debris and their postdepositional (late Cenozoic) emergence and dissection, aided by Quaternary climatic changes, which is responsible for much of the present mountainous terrain.

The piedmont alluvial sheet E of the Rockies is still partly intact (e.g., in Wyoming) and in places slopes steadily up to the mountain "divide"; elsewhere the front is more dissected and abrupt. The topography is thus quite diverse. The "high plains" area represents the original undissected alluvial surface, and has a monotonous gentle slope from the mountain front toward the central lowland. Elsewhere dissection has produced badlands (e.g., in North and South Dakota), or scarps (e.g., central Kansas, east-central Colorado, the Pecos Valley, and parts of Texas). Glaciation has masked the piedmont deposits in much of the Dakotas, whereas loess and sand dunes are a factor in the formerly periglacial regions, notably in Nebraska and Kansas. The Black Hills and central Texas regions expose Precambrian crystalline cores with annular hogbacks or cuestas of Paleozoic sediments, whereas in southern Colorado and northern New Mexico the terrain is greatly modified by extensive Quaternary volcanic activity. The Great Plains represent an area of vast agricultural resources, especially grazing.

Appalachian Region—Mountains and Coastal Plains. The second great morphotectonic division of the United States contains the folded Paleozoic belt to the E of the shield and central platform and the wedges of younger rocks forming the coastal plains to the E and S of the folded Paleozoic belt. Geosynclinal sediments of Paleozoic age have been extensively deformed to produce the *Appalachian Mountains* (q.v.) which have a complex history including folding, faulting, intrusions, and metamorphism, mainly in the early, middle, and late Paleozoic. The sediments from erosion of these mountains have been shed onto the *Atlantic Coastal Plain* (q.v.) and part of the *Gulf Coastal Plain* (q.v.) which additionally contains sediments derived from the Central Platform. The Coastal Plain sediments are largely unconsolidated, dip gently seaward, and are late Mesozoic and Cenozoic in age.

Newer Appalachians ("Valley and Ridge" Region). This distinctive belt of highly linear ridges and parallel valleys is built largely of miogeosynclinal and (later) exogeosynclinal sediments of the Paleozoic Appalachian geosyncline. They were subjected to décollement (disharmonic uncoupling) and crumpling during several orogenic episodes, particularly during the late Paleozoic (approximately mid-Devonian, Late Pennsylvanian). The resulting thrust faults and overturned folds have their relative movement toward the NW. Bedding-plane thrusts occur extensively, but are most common in the S (e.g., Tennessee), whereas folds are more prominent in the NE (e.g., Pennsylvania); the bedding plane thrusts come to the surface, in part along the Allegheny front.

Plate tectonic theory suggests that the classic Appalachian geosyncline developed as a rift between the North American craton and that of

Eur-Africa in earliest Paleozoic time. This rift repeatedly opened and closed in later periods. The earliest (Cambro-Ordovician) miogeosynclinal rocks may represent a continental rise and terrace wedge ("miogeocline") built W into the proto-Atlantic ocean. Subsequent continental collisions brought about early deformation of this material, and erosion of the resulting mountain belts caused shedding of deltaic materials on top of the continental terrace accumulations. Décollement, faulting, and overturned folding was followed by a long period of planation during the Permian and Triassic. The axial belt was then shattered by development of a late Triassic rift system, which is traced from Georgia to northeastern Canada (see also across the Atlantic: cf. *Portugal*).

Postdeformation erosion, in several cycles during the Mesozoic and Tertiary, has produced a characteristic ridge and valley topography in which structure and land forms are intimately related. Synclinal and anticlinal ridges of sandstone and quartzite underlie the rugged forested ridges. Valleys occur characteristically on limestone and shale, and include a continuous low region extending from Alabama to Canada. This "Great Valley" corresponds to the boundary of the Older Appalachians. the Newer Appalachian belt is a variegated region, alternating between poor, barren ridge country and valuable farm lands, with some industry. There are extensive mineral resources, notably hard coal and base metals.

Blue Ridge. The Blue Ridge is the structural backbone of the present-day Appalachian Mountains. Bedrock is primarily Precambrian crystallines that have been dissected to produce a rugged terrain. The Blue Ridge includes the Great Smokies which contain the highest mountain summit (6617 ft) in the eastern United States. Historically these mountains have been a major barrier to transportation, but now the Blue Ridge is used for recreation and forestry.

Older Appalachians (Piedmont). According to classic theory the early Paleozoic eugeosynclinal facies of the Appalachian geosyncline have been intensely deformed to produce the metasediments found in the Piedmont region E of the Blue Ridge. Igneous intrusions (Paleozoic) are also common. Plate tectonic theory ascribes the origin of some of these rocks to a combination of oceanic crust (ophiolite sequence) generated at some ancient oceanic ridge, and deep-sea sediments piled atop this crust (e.g., turbidites, pelagic muds). Plate collisions are held responsible for the complicated structures now seen in the area as well as most of the metamorphism and igneous activity.

A Triassic rift zone with block faulting and volcanic activity can now be seen in a series of lowlands within the Piedmont. The Piedmont region has been extensively eroded to produce a terrain of low, rolling hills, often with fairly flat, peneplaned upland surfaces. It carries important forests and development is extensive.

New England. The crystalline Older Appalachians extend into New England, where both the Blue Ridge and Piedmont types of terrain have been upwarped and modified by Pleistocene glacial erosion (especially uplands) and deposition (especially lowlands and coastal regions). Hence the soils are rockier, the terrain more irregular, and lakes more common than to the S. Economic activity in New England thus has been based more on industrial development than on agriculture. In this volume the article *New England Region* (q.v.), includes discussion of the Adirondack Mountains, a domed upland region that is geologically allied with the Canadian Shield (Laurentian Upland).

Atlantic Coastal Plain. The Coastal Plain of the United States extends from Long Island (or Cape Cod, strictly speaking) to the Rio Grande and continues into Mexico (and on to Yucatan). Very flat coastal regions of low-lying early and late Cenozoic sedimentary deposits overlap inland older deposits (as old as mid-Cretaceous) of slightly greater relief. The seaward dip found throughout the region results in a cuesta-form topography, with low scarps directed toward the interior. The basement rocks are related to the Appalachian-Piedmont section, but include buried grabens of late Triassic age. Jurassic is completely missing, and plate tectonic data suggest that the adjacent Atlantic rift of the present cycle did not open here until the mid-Cretaceous. The Atlantic Coastal Plain is subdivided into arcuate divisions by transverse warping, some still active. Especially in its northern area, there are numerous embayments representing submerged river valleys. Barrier beaches with interior lagoons are found all along the Atlantic coast. Glacial topography is a factor in the far N (Long Island-Cape Cod-Nantucket), whereas raised Pleistocene terraces and shallow elliptical depressions, termed "Carolina Bays" (q.v. Vol. III, *Encyclopedia of Geomorphology*), are common in the S. Peninsular Florida is a limestone region, marked by extensive karst topography, a highly distinctive landscape that lacks the cuestas typical of the remaining Coastal Plain areas. The Atlantic Coastal Plain is dominantly agricultural with poor forested land ("sand barrens"). Commerce and industry are associated with the seaports.

Gulf Coastal Plain. Facing the Gulf of Mexico, the Gulf Coastal Plain resembles the Atlantic Coastal Plain in many features. However, the deposition of postorogenic sediments began as early as the Jurassic along the Gulf, and the total thickness of the sediment wedge is very large, in places exceeding 10 km. The extensive deposition is due to regional subsidence and to a continuous supply from the vast source region of the continental interior, Appalachian Mountains, and Rocky Mountains, almost all of which is funneled into the Gulf through the Mississippi. Its great delta and alluvial embayment is one manifestation of this sediment supply, but large fractions move along the shelf or out into the Gulf depression, in part by turbidity flows. Away from the delta, barrier beaches and cuestas are the typical topography. Agriculture and commerce are important, but there are also vast geologic resources related to deposits of petroleum, salt, and sulfur.

Cordilleran Region and Associated Western Provinces

The western United States has been the site of diverse geologic events from Precambrian times to the present. A Paleozoic geosyncline, with eugeosynclinal, miogeosynclinal, and platform facies, occupied a N-S belt in middle and late Paleozoic, and was subject to progressive deformation (mid-Paleozoic Antler orogeny, among others). Through most of subsequent geologic times, the region was marked by changeable archipelagoes, and orthogeosynclinal to zeugogeosynclinal facies. Further sediments were deposited in Mesozoic seaways, with marine deposition terminated by orogenic events in late Mesozoic and early Tertiary (to Eocene) times. These events spanning an extended sequence of phases, are termed "Laramide" in the Rocky Mountain area, and "Nevadan" in the Pacific coastal region, formerly thought to be quite distinctive (respectively, late Cretaceous and late Jurassic). Absolute dating suggests some overlap. These younger deformations, with subsequent erosional, depositional, volcanic, and tectonic events, have determined the visible physiographic pattern of the western United States and have obscured much of the earlier history.

Rocky Mountains. The Rocky Mountains in the United States can be subdivided into four provinces: Northern Rockies, Middle Rockies, Wyoming Basin, and Southern Rockies. Their eastern aspect, abutting the Great Plains, is often termed the "Rocky Mountain front," usually a major fault or overthrust structure.

The Southern Rocky Mountains extend from northern New Mexico to southern Wyoming, and produce the mountainous terrain typical of central Colorado. Precambrian crystalline rocks have been arched up, usually along normal and thrust faults, to form the core of the mountains. The overlying sedimentary rocks, mostly Paleozoic and Mesozoic, have been deformed by the uplift, and subsequent dissection has formed spectacular cuestas and hogbacks marginal to the crystalline cores. Broad basins or "parks" occur between many of the mountain areas and are partly filled with Cenozoic sediments shed from the adjacent massifs since the onset of the Laramide orogeny. Volcanic activity is largely limited to southwestern Colorado where an extensive blanket of lava and tuff is found in the San Juan Mountains; scattered intrusions occur elsewhere. The Southern Rockies form an effective topographic barrier, with the "continental divide" at elevations not lower than 6000 ft. There are diverse and rich metallic deposits, extensive exploitable forests, and a numerous recreational attractions.

The Wyoming Basin is an area where post-Laramide deposition has largely blanketed the basement structure, so that the topography consists of broad alluvial and lacustrine basins with isolated areas of folded and faulted mountain ranges. Pleistocene dissection of the fill has caused resurrection of some buried topographic features and has also led to widespread superposition of drainage. Soft coal deposits, evaporites, and valuable accumulations of petroleum are found.

The Middle Rockies actually involve three different mountain belts, with intervening basins similar to the Wyoming Basin. Central Wyoming consists of features very similar to the Southern Rockies, having Precambrian crystallines at the core of the ranges, with deformed sedimentary rocks at the margins. Western Wyoming, eastern Idaho, and parts of adjacent Montana have small ranges and intermediate basins somewhat like those found in the Basin and Range province, except that much of the faulting has been along thrust planes. The Absaroka Mountains and Yellowstone Plateau are areas where Cenozoic volcanic rocks have covered the earlier rocks and structures. The Middle Rockies have seismically active areas, and show evidence of deformation throughout geologic time up to the present. The area has extensive grazing (cattle and sheep), irrigation agriculture, and considerable oil production.

The Northern Rockies are a varied and complex region, extending from Idaho and Montana into western Canada. At least three distinctive mountain types can be recognized. Many ranges in the S of the Northern Rockies are faulted blocks of Paleozoic sedimentary rocks, not unlike the Basin and Range structures further south. South-central Idaho, and several other areas, are the sites of vast Cretaceous intrusions, e.g., grantite batholiths, which produce a homogenous rugged terrain. In northern Montana the Laramide deformation of Paleozoic rocks produced extensive thrust sheets directed toward the E and a markedly linear terrain of N-S trending ridges and trenches. The Northern Rockies have rich mineral deposits, and also provide forest and recreational resources.

Colorado Plateau. This scenic province consists of extensive areas of nearly horizontal sedimentary rocks, mostly Paleozoic and Mesozoic, which have been differentially upwarped along faults and fault-related monoclinal flexures. The high plateaus have been deeply incised by an integrated drainage system, thus producing steep-walled canyons in which the relationship between structure and topography is expressed with a clarity not surpassed in many other parts of the world. This clear control of land forms by geology, enhanced by the arid climate and sparse vegetation, has made the Colorado Plateau an area where many important geologic and geomorphic principles have been derived, especially after the classic surveys made in the late-nineteenth century. Igneous structures are common, including volcanoes, volcanic necks, lava-capped mesas, and small intrustions (e.g. laccoliths).

Subdivisions within the Colorado Plateau are related to various structural flexures and to volcanic activity. Volcanic materials are particularly common at the margins of the province, whereas structural upwarps occur scattered throughout. For example, the Grand Canyon area is structurally very high, hence the deep dissection that has produced the striking canyons. Grazing, forestry (at high altitudes), and tourism are important on the Colorado Plateau, and appreciable oil resources have been developed.

Basin and Range Province. The bedrock of this province has undergone a complex history, including Paleozoic deformation (Antler orogeny) of an early Paleozoic geosyncline, as well as Mesozoic geosynclinal deposition and extensive deformation during the Nevadan and Laramide orogenies. However, the entire terrain reflects Cenozoic geologic processes, particularly normal faulting, with a NNW-SSE trend, along a very broad rift zone between the Rockies and Sierra Nevada. Range erosion and sedimentation has supplied basin fills, particularly fan and playa lake deposits, and volcanic activity has been abundant, particularly in the N and W of the province.

The Great Basin, mostly in Nevada, has youthfully revived mountainous terrain, and many closed basins with interior ("endorheic") drainage. Northward in Oregon volcanic outpourings of the Columbia Plateau partly obscure the fault pattern and are themselves deformed by quite recent activity. Southward in Arizona and part of New Mexico the fault structures tend to be older (e.g., early Tertiary) and the mountains are more subdued. Drainage integration has occurred and the area drains to the sea ("exorheic"). In southeastern New Mexico and western Texas the terrain is actually somewhat transitional to the Colorado Plateau, and is grouped with the Basin and Range partly because it is contiguous to that province.

Oasis settlements occur in the Basin and Range, mostly in Arizona, whereas mineral deposits are abundant throughout the region, which includes one of the world's richest copper mines.

Pacific Cordillera. The Columbia Plateau is often considered as an intermontane province, but the basement structure of this region is intimately related with that of the Sierra Nevada and Cascade mountains, and the province is here considered part of the Pacific Cordillera. Huge outpourings of lava, flood basalts, extruded throughout the Cenozoic, cover basement structures. The lava is interbedded with river and lake deposits and the overall terrain is a plateau, with only moderate structural control of terrain (eg., escarpments), due to the limited lithologic contrast available.

Subdivisions of the Columbia Plateau are often based on differing ages of surface lavas (hence subsequent length of erosion and modification). Local structural peculiarities are also important. Block faulting, related to the Basin and Range, is important in southeastern Oregon, whereas folding is important in parts of Washington. Locally the basement rises above the sea of extrusive rocks to form a mountainous massif (e.g., Blue Mountains, Oregon). Pleistocene drainage features, some related to catastrophic flooding, are a unique terrain feature of the Columbia Plateau, and are well expressed in the "scablands" and "coulees" of Washington. The Columbia Plateau is primarily an area of grazing and agriculture.

The Sierra Nevada Mountains are a large block-faulted range, tilted to the W, consisting mainly of massive granite batholiths related to the Nevadan orogeny (late Mesozoic), although postdating much of the orogenic deformation. Tertiary lava flows are found with increasing frequency northward, until one reaches the southern Cascade Mountains, where volcanoes and lavas dominate the terrain.

The Pacific Coast Ranges include a variety of mountain types. The Klamath Mountains are closely related to the Sierra Nevada in lithology and history, but without the effects of block faulting. The remaining ranges consist of strongly deformed sedimentary rocks and metasediments. Folding of geosynclinal sediments is important in most ranges but some areas, particularly central California, have faulting of Tertiary marine beds as the dominant structural pattern.

Much of the Coast Range structure is now being interpreted as the product of ancient plate collisions and subduction. The more recent deformation is tied to further oblique collisions that have produced a number of shear zones, related mainly to the active dextral motion of the San Andreas strike-slip system.

Alaska and Hawaii

Alaska. *Alaska* (q.v.) is a state with highly varied geology and terrain, including mountains of several types, coastal plain areas, and an extensively glaciated landscape. The mountains include areas analogous to the Rockies and Pacific Coast Ranges, as well as a modern island arc, the Aleutian arc, which is overriding the Pacific plate from N to S. Structural valleys, trending roughly E-W, occur between many of the major ranges, and the most extensive coastal plain is the oil-rich Arctic slope, bordering the Arctic Ocean.

Alaska is an area of immense natural resource wealth, as yet little developed due to low population pressure, adverse climate, and extensive development of permanently frozen ground.

Hawaii. *Hawaii* (q.v.) is a chain of Cenozoic basaltic volcanoes trending WNW-ESE which get younger to the E, with the "big island" of Hawaii itself containing many famous active craters. A tropical climate and tropical agriculture dominate the life and human development in this oceanic region.

LEE WILSON

Appendix: Sources of Further Information

The primary source for further information about United States geology is the U.S. Geological Survey. Main survey offices are found in Washington, D.C.; Denver, Colorado; and Menlo Park, California. Additionally there are offices in most state capitals, which contain information about local geology and water resources.

The American Association of Petroleum Geology, P.O. Box 979, Tulsa, Oklahoma 74101, has also published much information about U.S. geology, and has an excellent series of small-scale highway maps that contain valuable generalized data.

Additionally the following list of American state geologists may be consulted. A letter to a state office should specify what information is needed (e.g., list of published maps, bulletins, etc.; or information on a particular area).

Alabama
Philip E. LaMoreaux, State Geologist
Geological Survey of Alabama
P. O. Drawer O
University, AL 35486

Alaska
Ross G. Schaff, State Geologist
Division of Geological Survey
3001 Porcupine Drive
Anchorage, AK 99501

Arizona
William H. Dresher, Acting Director
Arizona Bureau of Mines
University of Arizona
Tucson, AZ 85721

Arkansas
Norman F. Williams, State Geologist
Arkansas Geological Commission
446 State Capitol
Little Rock, AR 72201

California
James E. Slossen, Chief
California Division of Mines and Geology
1416 9th Street, Room 1341
Sacramento, CA 95814

Colorado
John W. Rold, Director
Colorado Geological Survey
254 Columbine Bldg.
1845 Sherman Street
Denver, CO 80203

Connecticut
Hugo F. Thomas, State Geologist
Connecticut Geological and Natural
   History Survey
Department of Environmental Protection
Staff Office Building
Hartford, CT 06115

Delaware
Robert R. Jordan, State Geologist
Delaware Geological Survey
University of Delaware
Newark, DE 19711

Florida
Charles W. Hendry, Jr., Chief
Florida Department of Natural Resources
Bureau of Geology
403 West Tennessee Street
Tallahassee, FL 32304

Georgia
Sam M. Pickering, Jr., Director
Department of Natural Resources
19 Hunter Street
Atlanta, GA 30334

UNITED STATES OF AMERICA – GENERAL

Hawaii
Robert T. Chuck
Division of Water and Land Development
Department of Land and Natural Resources
P. O. Box 373
Honolulu, HI 96809

Idaho
John G. Bond, Chief
Idaho Bureau of Mines and Geology
Moscow, ID 83843

Illinois
Jack A. Simon, Acting Chief
Illinois State Geological Survey
Natural Resources Building
Peabody East of South Sixth
Urbana, IL 61801

Indiana
John B. Patton, State Geologist
Indiana Department of Natural Resources
Geological Survey
611 North Walnut Grove
Bloomington, IN 47401

Iowa
Iowa Geological Survey
123 North Capitol Street
Iowa City, IA 52242

Kansas
William W. Hambleton, State Geologist
and Director
Kansas Geological Survey
1930 Avenue A, Campus West
University of Kansas
Lawrence, KS 66044

Kentucky
Wallace W. Hagan, State Geologist
and Director
Kentucky Geological Survey
University of Kentucky
307 Mineral Industries Building
Lexington, KY 40506

Louisiana
Leo W. Hough, State Geologist
Louisiana Geological Survey
Box G, University Station
Baton Rouge, LA 70803

Maine
Robert G. Doyle, Director
Bureau of Geology
Department of Conservation
Augusta, ME 04330

Maryland
Kenneth N. Weaver, Director
Maryland Geological Survey
214 Latrobe Hall
The Johns Hopkins University
Baltimore, MD 21218

Massachusetts
W. Richard Boehmer, Acting Director
Division of Mineral Resources
State Office Building, Government Center
100 Cambridge Street
Boston, MA 02202

Michigan
Arthur E. Slaughter, State Geologist
Geological Survey Division
Department of Natural Resources
Michigan Department of Conservation
Lansing, MI 48926

Minnesota
Matt Walton, Director
Minnesota Geological Survey
1633 Eustis Street
St. Paul, MN 55108

Mississippi
William H. Moore, State Geologist
and Director
Mississippi Geological Survey
P. O. Box 4915
Jackson, MS 39216

Missouri
Wallace B. Howe, State Geologist
and Director
Missouri Geological Survey and
Water Resources
P. O. Box 250, Buehler Park
Rolla, MO 65401

Montana
S. L. Groff, State Geologist
and Director
Montana Bureau of Mines and Geology
Montana College of Mineral Science
and Technology
Butte, MT 59701

Nebraska
Vincent H. Dreeszen, Acting Director
Conservation and Survey Division
113 Nebraska Hall
Lincoln, NB 68508

Nevada
John H. Schilling, Director
Nevada Bureau of Mines and Geology
University of Nevada
Reno, NV 89507

UNITED STATES OF AMERICA – GENERAL

New Hampshire
Glenn W. Stewart, State Geologist
Department of Resources and Economic
 Development
Office of State Geologist
James Hall, Room 117
University of New Hampshire
Durham, NH 03824

New Jersey
Kemble Widmer, State Geologist
New Jersey Bureau of Geology and
 Topography
P.O. Box 2809
Trenton, NJ 08625

New Mexico
Frank E. Kottlowski, Director
New Mexico Bureau of Mines
 and Mineral Resources
Campus Station
Socorro, NM 87801

New York
James F. Davis, Acting State Geologist
Geological Survey
New York State Museum and Science Service
Room 973
Albany, NY 12224

North Carolina
Stephen G. Conrad, Director and
 State Geologist
Department of Natural and
 Economic Resources
Division of Resource Planning
 and Evaluation
P. O. Box 27687
Raleigh, NC 27611

North Dakota
Edwin A. Noble, State Geologist
North Dakota Geological Survey
University Station
Grand Forks, ND 58201

Ohio
Horace R. Collins, Division Chief and
 State Geologist
Division of Geological Survey
Department of Natural Resources
Fountain Sq., Bldg. B
Columbus, OH 43224

Oklahoma
Charles J. Mankin, Director
Oklahoma Geological Survey
830 Van Vleet Oval, Room 163
Norman, OK 73069

Oregon
R. E. Corcoran, State Geologist
State Department of Geology and
 Mineral Industries
1069 State Office Building
Portland, OR 97201

Pennsylvania
Arthur A. Socolow, State Geologist
Bureau of Topographic and
 Geological Survey
Dept. of Environmental Resources
Harrisburg, PA 17120

Puerto Rico
Eduardo Aguilar-Cortes, Director
 and State Geologist
Program of Geology and Mineral Resources
Industrial Research Department
Economic Development Administration of
 Puerto Rico
G.P.O. Box 3088
San Juan, Puerto Rico 00936

South Carolina
Norman K. Olson, State Geologist
Division of Geology
South Carolina State Development Board
Harbison Forest Road
Columbia, SC 29210

South Dakota
Duncan J. McGregor, State Geologist
South Dakota Geological Survey
Science Center, University
Vermillion, SD 57069

Tennessee
Robert E. Hershey, State Geologist
Department of Conservation
Division of Geology
G-5 State Office Building
Nashville, TN 37219

Texas
W. L. Fisher, Director
Bureau of Economic Geology
The University of Texas
University Station, Box X
Austin, TX 78712

Utah
Donald T. McMillan
Utah Geological and Mineralogical Survey
103 Utah Geological Survey Building
University of Utah
Salt Lake City, UT 84112

Vermont
Charles G. Doll, State Geologist
Vermont Geological Survey
University of Vermont
Burlington, VT 05401

Virginia
James L. Calver, State Geologist
Virginia Division of Mineral Resources
P. O. Box 3667
Charlottesville, VA 22903

Washington
Vaughan E. Livingston, Jr.
Division of Geology and Earth Resources
Department of Natural Resources
14th and Jefferson
Olympia, WA 98504

West Virginia
Robert B. Erwin, Director and
 State Geologist
West Virginia Geological and
 Economic Survey
Box 879
Morgantown, WV 26505

Wisconsin
Meredith E. Ostrom, State Geologist
 and Director
Wisconsin Geological and Natural
 History Survey
1815 University Avenue
Madison, WI 53706

Wyoming
Daniel N. Miller, Jr., State Geologist
Geological Survey of Wyoming
P. O. Box 3008
University Station
Laramie, WY 82071

References

American Geological Institute, 1972. *Directory of Geoscience Departments.* Washington, D.C.: Am. Geol. Inst., 214p.

Anon., 1957. *National Atlas of the United States.* Washington, D.C.: U.S. Geological Survey.

Brobst, D. A., and Pratt, W. P., eds., 1973. "United States mineral resources," *U.S. Geol. Surv. Prof. Pap. 820.*

Clark, T. H., and Stearn, C. W., 1968. *The Geological Evolution of North America,* 2nd ed. New York: Ronald Press, 570p.

Collins, R. S., Geraghty, J. J., and Miller, D. W., 1973. *Water Atlas of the United States,* 2nd ed. Port Washington, N.Y.: Water Information Center, 200p.

Cram, I. H., ed., 1971. *Future Petroleum Provinces of the United States–Their Geology and Potential* (Mem. 15, no. 2). Tulsa: Am. Assoc. Petrol. Geologists, 1–803, 805–1496.

Dickinson, W. R., ed., 1974. "Tectonics and Sedimentation," *S.E.P.M.,* Sp. Publ. 22, 204p.

Douglas, E. M., 1930. "Boundaries, areas, geographical centers, and altitudes of the U.S. and the several states," *U.S. Geol. Surv. Bull., 817.*

Fenneman, N. M., 1931. *Physiography of Western United States.* New York: McGraw-Hill, 534p.

———, 1938. *Physiography of Eastern United States.* New York: McGraw-Hill, 714p.

Gilluly, J., Reed, J. C., Jr., and Cady, W. M., 1970. "Sedimentary volumes and their significance," *Bull. Geol. Soc. Am.,* 81, 353–376.

Holland, C. H., ed., 1971. *Cambrian of the New World.* Vol. 1. *Lower Palaeozoic Rocks of the World.* New York: Wiley-Interscience, 456p.

Hunt, C. B., 1973. *Natural Regions of the United States and Canada.* San Francisco: W. H. Freeman, 725p.

King, P. B., 1959. *The Evolution of North America.* Princeton, N.J.: Princeton Univ. Press.

Küchler, A. W., 1964. "Potential natural vegetation of the coterminous United States," *Am. Geogr. Soc. Spec. Publ. 36.*

Landes, K. K., 1970. *Petroleum Geology of the United States.* New York: Wiley-Interscience, 571p.

Lobeck, A. K., 1948. *Physiographic Provinces of North America.* New York: Geographic Press (Hammond), map, notes.

Locke, D. H., n.d. (c. 1967). *Selected Guides for Geologic Field Study in Canada and the United States of America.* (ESCP Ser. RS-9), Englewood Cliffs, N.J.: Prentice-Hall, 56p.

Long, H. K., 1971. *A Bibliography of Earth Science Bibliographies of the United States of America.* Washington, D.C.: Am. Geol. Inst., 19p. State-by-state listing.

Matthews, W. H., III, 1965. *Selected Maps and Earth Science Publications for the States and Provinces of North America,* (ESCP Ser. RS-4). Englewood Cliffs, N.J.: Prentice-Hall, 42p.

Proctor, P. D., et al., 1968. "A coast to coast tectonic study of the United States," *Univ. Mo. Res. J.,* 1, 1–156.

Prodehl, C., 1970. "Seismic refraction study of crustal structure in the western United States," *Bull. Geol. Soc. Am.,* 81, 2629–2646.

Ridge, J. D., 1972. "Annotated bibliographies of mineral deposits in the Western Hemisphere," *Geol. Soc. Am., Mem. 131,* 681p.

Thornbury, W. D., 1965. *Regional Geomorphology of the United States.* New York: Wiley, 609p.

Visher, S. S., 1954. *Climatic Atlas of the United States.* Cambridge: Harvard University Press, 403p.

Wright, H. E., Jr., and Frey, D. G., 1965. *The Quaternary of the United States.* Princeton, N.J.: Princeton Univ. Press, 922p.

Cross-references: *American Samoa; Bahamas; Canada; Cuba; Mexico; North America; Pacific Island Trust Territory; Puerto Rico; United States–Alaska, Appalachian Region, Atlantic Coastal Province, Basin and Range Province, Colorado Plateau Province, Great Plains Province, Gulf Coastal Province, Hawaii, Midwestern Region, New England Region, Pacific Cordilleran Region, Rocky Mountain Province; Virgin Islands; West Indies.*

UNITED STATES–ALASKA

Alaska is the northwestern extremity of the continent of North America, W. of the Yukon district of Canada (long 140°W), with a projection along the Pacific coast to 130°W. It is a peninsula bordered by the Pacific Ocean, the Bering Sea, and the Arctic Ocean, and is separated from Asiatic U.S.S.R. by the narrow, shallow Bering Strait. Roughly 1½ million km^2

(586,000 sq mi) of Alaska lie above sea level, and it possesses a continental shelf almost as great (primarily in the Bering Sea). Alaska extends through 20° of latitude and across 58° of longitude.

The climate, geology, and resources of the state are as varied as its size suggests. In climate the range is from temperate rain forest in the SE to full arctic conditions along the north coast. The principal physiographic and tectonic elements of western North America are all to be found in Alaska, including extensions of the Rocky Mountains, the intermontane plateaus, and the Pacific Coast Ranges (Fig. 1). Some of these features can be traced even farther W into Siberia. The southwestern extremity of Alaska includes the Aleutian Range of the Alaska Peninsula and the Aleutian Islands, the only true island arc system on the North American coast.

Although mineral exploitation is a major industry in Alaska, much of its geologic study is at no more than a reconnaissance level and large areas are unmapped. The U.S. Geological Survey and the Alaska Department of Natural Resources have been systematically exploring, usually regions of known economic importance. Historically gold and copper, among the metals, have been significant; currently gold is still produced but at low levels, and there is small scale production of copper, platinum, antimony, mercury, and barite. There are widespread deposits of relatively low-rank coal, developed only for local consumption.

Petroleum, first discovered on a significant scale in 1957 near Cook Inlet in southern Alaska, is now by far the most valuable mineral resource, especially since the discovery in 1968 of large reserves in the North Slope region. Large unexplored potential petroleum areas, particularly under the continental shelf, still remain to be prospected.

Geologic information is so sparse for most of Alaska that there are few comprehensive reviews. Most useful are summaries of the physiography by Wahrhaftig (1965), the geologic and tectonic history by Gates and Gryc (1963) and Miller et al. (1959), the Quaternary geology by Péwé et al. (1965), and the special problems of the Bering Land Bridge by Hopkins (1967) and Nelson et al. (1972). The U.S. Geological Survey has published a useful review of the economic geology (U.S. Geological Survey, 1964) and an excellent reconnaissance map of the state (Dutro and Payne, 1957). The Geological Survey, the Coast and Geodetic Survey, and the National Academy of Sciences have each reported extensively on the mechanics and consequences of the 1964 earthquake. Stoneley (1971) has made a useful preliminary synthesis of the structural evolution of Alaska in the light of crustal plate tectonic concepts (see also Grow and Atwater, 1970; Marlow et al., 1973; Moore, 1972, 1973a, 1973b; and others).

In petrology the name of the state is honored in the term *alaskite* (proposed by Spurr in 1900 for an alkali granite, a light-colored dike or plutonic rock essentially of alkali-feldspar and quartz).

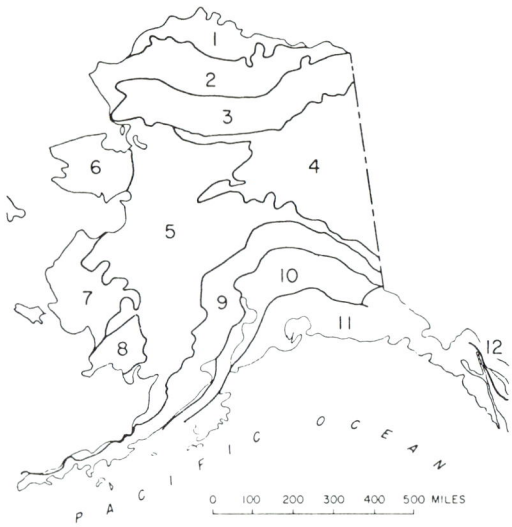

FIGURE 1. Geomorphic provinces of Alaska (after Wahrhaftig, 1965). Explanation: 1, Arctic Coastal Plain province; 2, Arctic Foothills province; 3, Arctic Mountains province; 4, Northern Plateaus province; 5, Western Alaska province; 6, Seward Peninsula province; 7, Yukon-Kuskokwim Coastal Lowland province; 8, Ahklun Mountains province; 9, Alaska-Aleutian province; 10, Coastal Trough province; 11, Pacific Border Ranges province; 12, Coast Mountains province.

Climatology and Biogeography

The present climate of Alaska is extremely varied. The southeastern panhandle and most of the Pacific coastal zone and the Aleutian chain have an extremely humid cool, maritime climate. This permits the simultaneous development in southeastern Alaska of magnificent temperate rain forests dominated by western hemlock and Sitka spruce next to extensive ice fields in the coastal mountains, many of which with glaciers extending to sea level. Precipitation is strongly influenced by local topography, amounting to over 550 cm (220 in.) in some parts of the southeast. Temperature in the coastal areas, thanks to the moderating influence of the Pacific, shows very little variation from the mean annual value of 5°C (40°F).

Much of the coastal area and continental shelf of southern Alaska was completely covered by glaciers during Pleistocene glacial maxima (Hopkins, 1967), as well as the Brooks Range in the N, but most of interior Alaska was never glaciated (Fig. 2). Reforestation since the last glacial retreat has been rapid. The present tree line along the Pacific coast of Alaska is controlled by both temperature and wind regime. It is now moving both E and S in response to post-Pleistocene amelioration of climate. In historic times the tree line has moved past the Kodiak city area, treeless when settled by the Russians in the late 18th century but now with well-developed forest at lower elevations. The Aleutian Island chain is unforested at present, but trees can survive if protected from winds. The entire southwestern coast is subject to frequent storms and much fog and though the temperatures are relatively mild, the climate is severe and much of the biota is similar to that of higher latitudes.

The coastal mountain ranges are among the highest mountains in North America and show an extensive contemporary development of glaciers—a function of temperature, elevation, and an abundance of precipitation (Péwé and Reger, 1972). The Bering and Malaspina glaciers of the central Gulf of Alaska coastline are the sole remaining examples in the world of a common Pleistocene phenomenon, piedmont glaciers—alpine glaciers that have coalesced to spread across a coastal plain.

Alaska's interior has a high-latitude continental climate characterized by extreme temperatures. Fairbanks, with a mean annual temperature of −4°C (25°F), has a recorded high of 38°C (99°F) and a low of −55°C (−66°F). Periods of several weeks below −40°C (−40°F) are not uncommon in the winter, whereas summer temperatures are often in the 20–25°C (68–75°F) range. Precipitation averages 35 cm (14 in.), but evaporation rates are low and the interior is relatively well watered. Snow covers much of the area for 8 months of the year (October to May). Subarctic *taiga* forest, dominated by spruce, covers much of the interior, punctuated by a variety of willows, birch, alder, and aspen, particularly along rivers, on uplands, and on recently burned areas. *Muskeg,* or bog develops in poorly drained areas, and *permafrost*—permanently frozen soil often with distinct ice wedges—left over from the last glacial stage, is scattered throughout. Because of the low precipitation, much of interior Alaska has never been actually glaciated. Glacial and periglacial features are found nevertheless throughout the state, and as much as 25% of the area is covered with unconsolidated Quaternary and Holocene deposits.

Tundra, the typical arctic and alpine flora of mosses, lichens, grasses, sedges, and miniature flowering plants, is widely scattered throughout Alaska but is dominant N of the Brooks Range. This region, the North Slope, is under true arctic conditions. Extremes in temperature are not as great as in the interior, but the mean is around −12°C (10°F). Precipitation is very low, averaging 15 cm (6 in.), but evaporation rates are low and poor drainage produces many small lakes. There are frequent severe storms but, because so much of the Arctic Ocean is ice covered, little moisture is available.

Physiography and Geomorphology

Most of Alaska is mountainous, punctuated by the flood plains of the major rivers. It is usually divided into physiographic provinces comparable to those of the western United States and Canada (Fig. 1). The Pacific mountain system, including the coastal mountains, an intervening lowland belt, and the Alaska Range, corresponds to the Coast Ranges, lowland, and the Sierra Nevada in the continental United States. The Brooks Range has been compared to the Rocky Mountains system, but here the analogy breaks down, and it may be more closely related to the circumarctic tectonic province including the Canadian Arctic Archipelago and Wrangel Island (off the Siberian coast).

The Pacific mountain system in Alaska bifurcates near the Canadian border to form two arcs, separated by a belt of lowlands. The southern arc includes the Fairweather, Chugach, and Kenai ranges and continues SW to Kodiak Island; it borders the Gulf of Alaska with a rather narrow coastal plain and continental shelf in most areas. The northern arc, consisting of the St. Elias and Wrangell mountains and the Alaska Range, leads into the largely volcanic Aleutian Range. It is separated from the Kenai-Chugach Mountains by the Copper River and Cook Inlet lowlands. It has a number of active volcanoes, especially toward the W, and includes the highest peaks in North America. The Aleutian Range and Islands contain at least 76 volcanoes, of which 36 have had activity recorded since 1760. Contemporary alpine glaciation is widespread in the Pacific mountain system, particularly the maritime ranges to the S and E, which are blanketed by extensive ice fields. Such conditions, varying only in intensity, have persisted for the past 3 m yr. At the present time the firn line ranges from about 1000 m on the south coast to nearly 3000 m on the north face of the mountains.

UNITED STATES–ALASKA

FIGURE 2. Bathymetric sketch map of the Bering Sea and Aleutian Arc showing the contrasting structural (fault) trends of western Alaska and the easternmost USSR (Burk, 1965). The Bering Seas area is divided into two: the shallow Bering Shelf ("Beringia") that became exposed during the last glacial maxima, and the deep Bering Basin in the southwest; the boundary is believed to be the site of a Mesozoic island arc.

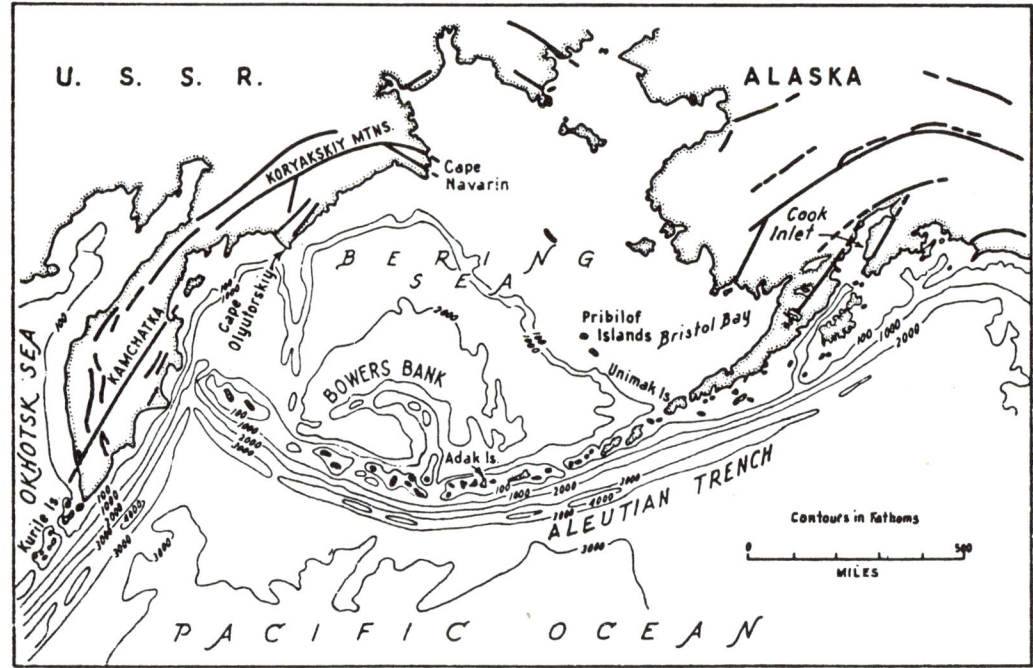

The interior of Alaska is characterized by a number of broad valleys, dominated by the Yukon and its major tributaries, the Tanana and the Koyukuk. Just to the N of the western Alaska Range lies the Kuskokwim which almost joins the Yukon to form an extensive delta in the Bering Sea. Separating the major valleys are many intermediate-scale mountain ranges and uplands. The mountains of the interior have no contemporary glaciation but show signs of sporadic ice action during Pleistocene maxima. Loess is widespread on the uplands and may be tens of meters thick. Pingoes are observed in many parts. During all later Cenozoic time precipitation apparently was blocked by the mountain barriers both to N and S, so that continental glaciers could never develop in the Alaskan interior. This ice-free area may have been a major migration route for primitive men, as well as arctic mammals, between Asia and North America when the Bering Strait was exposed by lowered sea level.

The broad continental shelf of the Bering Sea is a continuation of the Interior Province (Fig. 2). The scattered islands of the E constitute the only relief, and in the N the shelf is

FIGURE 3. Existing and Pleistocene glaciation of Alaska (adapted from Wahrhaftig, 1965). Explanation: A, belt of continuous permafrost; B, discontinuous permafrost; C, sporadic permafrost; D, no permafrost.

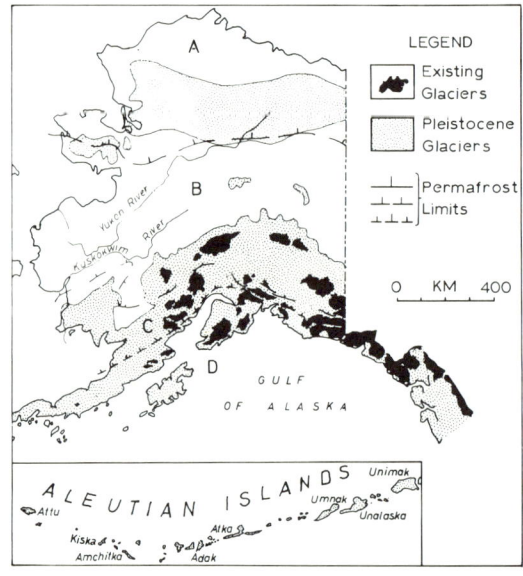

516

continuous with Siberia to form *Beringia,* an extensive land-bridge area that connected Asia and North America during glacioeustatic low sea-level stands in the Pleistocene (Hopkins, 1967). The Yukon and Kuskokwim rivers transport large volumes of sediment to the shelf, much of which is moved by prevailing currents to deposition sites N of the Bering Strait. At times of low sea level these sediments were either deposited in basins within the shelf area or carried S to the abbysal areas of the Bering Sea. A well-developed submarine canyon system with coalescing deep-sea fans testifies to the importance of this drainage during glacial maxima.

The Brooks Range and its western extensions, the DeLong and Baird mountains, form a rugged barrier between the interior and the North Slope. Elevations are lower than in the Alaska Range; direct glacial effects are relatively sparse, reflecting the low precipitation. Contemporary ice action is restricted to small alpine glaciers in the eastern Brooks Range. Periglacial phenomena are prominent, and permafrost underlies virtually the entire area (Fig. 3). Such features as pingoes are often seen in alluvial areas.

Similarly on the Arctic coastal plain the permafrost is sometimes hundreds of meters thick. Because summer thaw of the soil layer rarely exceeds 1 m, drainage is exceedingly poor and lakes and marshy tundra are widespread. Thaw lakes are oriented parallel to prevailing winds, and there are extensive networks of ice-wedge polygons that cover much of the coastal plain. Relief of the North Slope is very slight and in some areas pingoes form the only noticeable hills. The rivers, all of which head in the Brooks Range, tend to be braided meandering streams, briefly carrying large volumes of material during spring break-up floods, but at other times are sluggish.

Tectonic Framework

Since the Lower Paleozoic, orogenic activity in southern Alaska has been confined to a series of belts parallel to the Pacific continental margin (Fig. 4). These can be interpreted to represent a history of continental growth by the addition of successively developed eugeosynclines. The pattern has been complicated by local intrusions and volcanic activity, and by the superposition, in some areas, of several generations of tectonism. In the SE of interior Alaska there are thoroughly metamorphosed rocks, which are derived from Lower Paleozoic sediments and volcanics. Several cycles of tectonism are inferred, the final major activity being in late Mississippian or Pennsylvanian. To the S, a Permian-Triassic geosyncline developed in the present site of the Alaska Range and was metamorphosed in the early Jurassic. Still farther S, in the region of the Chugach-St. Elias Mountains, a Jurassic geosyncline developed but was tectonized in the Cretaceous (Fig. 5). In the outer part of the Chugach Range and (probably) beneath Prince William Sound there is a Cretaceous geosyncline, deformed during the early Tertiary. A Tertiary geosyncline that persisted until the early Quaternary lies beneath the coastal plain and continental shelf of the central Gulf of Alaska. Tertiary successor basins are also found beneath parts of the Bering Sea continental shelf and in some of the neighboring lowlands. The Aleutian Trench, S of the Alaska Peninsula and the Aleutian Islands, can be regarded as a contemporary eugeosyncline.

The most prominent feature of the structure of southern Alaska is the pronounced bend in the Alaska Range that occurs just NE of Cook Inlet. To the E, the ranges strike NW; W of the oroclinal fold, the strike is to the SW, a rotation of at least $65°$. The major fault systems to the S and in interior Alaska also show this configuration, and several of the larger systems show dextral strike-slip displacement, as if there had been repeated counterclockwise rotation of this province with respect to northern Alaska. The Denali fault system, lying just N of the Alaska Range, has been traced from the fjord country of southeast Alaska, W to Bristol Bay in the southern Bering Sea. On the Chugach-St. Elias system, the northern boundary of the Gulf of Alaska coastal plain, movement appears to have been dominantly vertical.

Northern Alaska has a simpler tectonic pattern than the southern sections. Miogeosynclinal conditions persisted through much of the Paleozoic and Mesozoic, ending with orogenic activity during the Jurassic. During this time there was a source of clastic sediments to the N beyond the present shore line. The axis of Jurassic uplift corresponded to the axis of the Brooks Range (Fig. 6), and a flysch basin developed to the N. In the Brooks Range to the W there was extensive N thrusting and gravity sliding of nappes of the older sediments during the early Cenozoic.

Stratigraphy

Representatives of each geologic period appears to be present (Dutro and Payne, 1957). Metamorphic rocks of Precambrian and lower Paleozoic age are believed to underlie much of

FIGURE 4. Morphotectonic units of Alaska (adapted from Gates and Gryc, 1963). Explanation: 1, Tertiary (and Quaternary) igneous and pyroclastic rocks; 2, Tertiary sediments; 3, late Mesozoic igneous rocks; 4, late Mesozoic geosynclinal rocks; 5, late Mesozoic geanticlinal areas; 6, major faults.

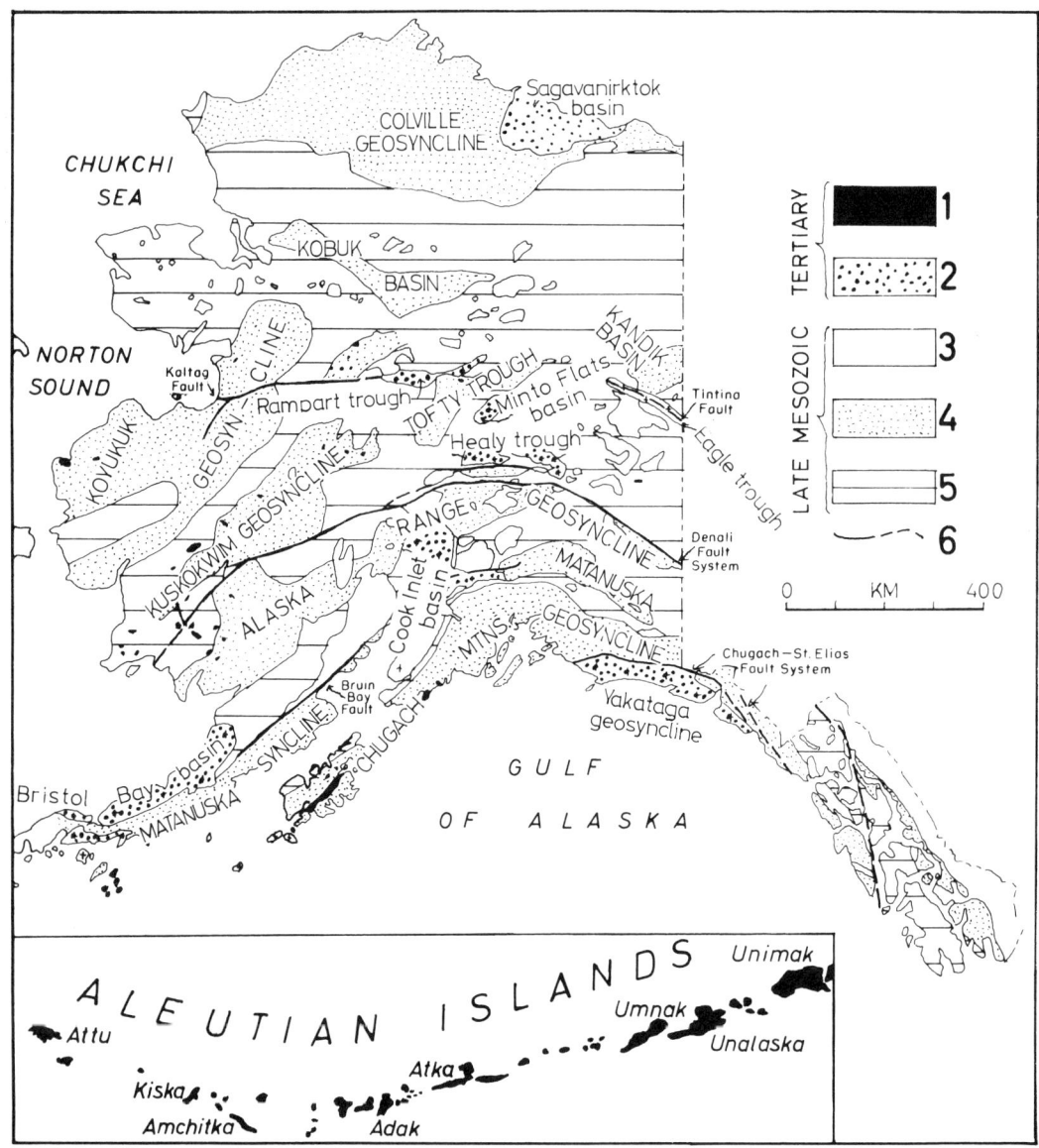

the state. They are exposed in southeastern, central, and western Alaska. The largest area of outcrop is the Birch Creek Schists, which form the Yukon-Tanana Upland, a sequence of highly metamorphosed eugeosynclinal sediments and volcanics of probable lower Paleozoic age. Unmetamorphosed Precambrian marine sediments are found just to the N, in the Yukon-Porcupine area. Possible Precambrian metasediments are also reported in the Ahklun Mountains region of western Alaska.

Paleozoic and Mesozoic sediments are present throughout Alaska. In the S they are an eugeosynclinal assemblage of graywackes, arkose, and shales with abundant volcanics in places. The miogeosynclinal environment of northern Alaska produced a very considerable thickness of marine limestones, sands, and shales which has recently been demonstrated to be highly petroliferous. These are widely exposed in the Brooks Range and occur in the subsurface beneath the North Slope, where the

FIGURE 5. Pillow lavas on the northwest shore of Ingot Island, Prince William Sound, Alaska. (Photo: U. S. Geological Survey Bull. 989-E, 1954.)

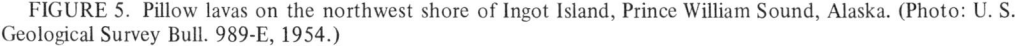

Mississippian particularly has excellent reservoir possibilities. Facies relationships within the North Slope Paleozoic and Mesozoic strata strongly suggest a hinterland lying N of the present continental margin. Another great accumulation of Mesozoic detrital rocks (graywacke) was in the Kuskokwim area near the Bering Sea. Just to the N of this region, in the Yukon-Koyukuk province, there is a considerable development of a Cretaceous volcaniclastic sequence grading upward into nonmarine facies.

The southern Coast Ranges show a particularly well-developed Cenozoic sequence, and local basins of Tertiary and Quaternary accumulation are found scattered throughout the state. In particular, the Tertiary of the Cook Inlet basin has proven to be an important source of petroleum. Tertiary sediments underlie much of the Alaska continental shelf and may also prove to be petroliferous. Glacial conditions developed locally in Alaska as early as Miocene time, and both Pliocene and Quaternary tills, loess, and glacial-marine sediments are found throughout Alaska.

Mesozoic intrusive and extrusive rocks are found in many parts of Alaska. They can be divided roughly by age into middle Mesozoic (Nevadan), largely found in the Alaska Range area, and late Mesozoic (Laramide) material, N of the Yukon. Some Cenozoic intrusive are exposed in the Coast Ranges, and volcanic activity has been significant in various parts of southern Alaska throughout much of the Cenozoic. Presently active volcanoes are found as far E as the Wrangell Mountains, and volcanism comes to dominate all other geologic processes in the Alaska Peninsula and Aleutian Arc.

FIGURE 6. Oblique air photo of the Brooks Range and Fulton River, Alaska. (Courtesy: British Petroleum Co., ref. 9194/28.)

Geologic History

Alaska's geology is not sufficiently well known to permit a comprehensive synthesis, but modern tectonic theories permit a general reconstruction. Through Paleozoic and much of Mesozoic time, Alaska shared the history of western North America. The southern part of the state was quite likely a continental margin, a eugeosynclinal zone. This zone appears to have gone through as many as five distinct cycles which suggest subduction in late Paleozoic, the early Jurassic, the late Jurassic, the early Tertiary, and the late Tertiary. The Aleutian Arc and Trench appears to be a contemporary eugeosyncline.

In northern Alaska during much of the Paleozoic and Mesozoic there was a northern source of sediments located offshore. In the middle Jurassic it became miogeosynclinal facing N.

Plate tectonic theory suggests that Alaska may have rotated into its present position since mid-Cretaceous times. One hypothesis assumes that northeastern Siberia and North America were part of a single crustal plate which became bent by sea-floor spreading within the Arctic Ocean basin. Rotation of Alaska into line with the North American Cordillera could place the Canadian Arctic Archipelago against the north coast of Alaska, providing the offshore sedimentary source area suggested in the stratigraphy of the North Slope. Simultaneously the Pacific crustal plate has been moving toward the NW, with the Aleutian Arc system representing a crustal subduction zone. It is marked by high seismic and tectonic activity with a well-developed Benioff zone dipping N.

Mineral Resources

Petroleum is by far the most valuable resource (Cram, 1971; Morgridge and Smith, 1972). Commercial production was first entirely from the Cook Inlet area, but vast reserves were discovered on the North Slope, between the Brooks Range and the Arctic Ocean, with estimates of recoverable oil from 5 to 10 billion barrels. Other areas in Alaska, especially beneath the continental shelves, may contain important quantities of petroleum. Oil and gas seeps are well known on the Alaska Peninsula and in the Gulf of Alaska Tertiary Province. Considerable reserves of coal are present in several areas, usually as bituminous or lower grade, and as are yet very little developed. Coking-grade coal has been reported from the Kukpowruk River area in extreme northwestern Alaska.

Alaska was famous for gold production in the past, but in recent decades production has dropped owing to the exhaustion of accessible placers. Production of gold up to 1960 totaled almost 30 million fine ounces. Important gold districts included the stream placers of the interior (Fairbanks, Iditarod, Chisana, and Livengood), the famous beach placers of Nome, and few lode mines in the Juneau area and elsewhere.

Platinum has been produced from placers in the Goodnews Bay area since the 1930s. Copper mining has been important in the Ketchikan area in the SE and in the upper Copper River region in the Alaska Range. Deposits near Bornite in the NW are currently under development. There are significant reserves of mercury in the Kuskokwim district. Barite is produced on a small scale near Petersburg in the SE, and minor amounts of jade and soapstone are mined.

FREDERICK F. WRIGHT

References

Brosge, W. P., and Tailleur, I. L., 1971. "Northern Alaska petroleum province," in I. H. Cram, ed., *Future Petroleum Provinces of the United States* (Mem. 15, vol. 1). Tulsa: Am. Assoc. Petrol. Geologists, 68–99.

Burk, C. A., 1965, "Geology of the Alaska Peninsula–island arc and continental margin," *Geol. Soc. Am. Mem. 99*, 250p.

Cameron, C. P., and Stone, D. P., 1970. "Outline geology of the Aleutian islands with paleomagnetic data from Shemya and Adak islands," *Alaska Univ., Geophys. Inst. Rept. R-213*, 153p.

Churkin, M., Jr., Carter, C., and Eberlein, G. D., 1971. "Graptolite succession across the Ordovician-Silurian boundary in south-eastern Alaska," *Quart. J. Geol. Soc. Lond.*, **126**, 319–330.

Cram, I. H., 1971. "Future petroleum provinces of the United States; their geology and potential," in I. H. Cram, ed., *Future Petroleum Provinces of the United States* (Mem. 15). Tulsa: Am. Assoc. Petrol. Geologists, **1**, 1–803; **2**, 805–1496.

Davies, W. E., 1972. "The Tintina Trench and its reflection in the structure of the Circle area, Yukon-Tanana Upland, Alaska," *24th Intern. Geol. Congr.*, **Canada, 3**, 211–216.

Dutro, J. T., Jr., and Payne, T. G., 1957. *Geologic Map of Alaska*. Washington, D.C.: U.S. Geol. Surv., 1:2,500,000.

Forbes, R. B., and Lanphere, M. A., 1973. "Tectonic significance of mineral ages of blueschists near Seldovia, Alaska," *J. Geophys. Res.*, **78**(8), 1383–1384.

Freeland, G. L., 1973. *Rotation History of Alaskan Tectonic Blocks*. Miami: NOAA.

Fritts, C. E., and Tuell, E. J., 1972. *Bibliography of Alaskan geology, 1965–1968*. College, Alaska: Alaska Geological Survey, 112p.

Gates, G. O., and Gryc, G., 1963. "Structure and tectonic history of Alaska, in O. E. Childs and B. W. Beebe, eds., *The Backbone of the Americas* (Mem. 2). Tulsa: Am. Assoc. Petrol. Geologists, 264–277.

Grow, J. A., and Atwater, T., 1970. "Mid-Tertiary tectonic transition in the Aleutian Arc," *Geophys. Soc. Am. Bull.*, **81**, 3715–3722.

Gryc, G., 1971. "Summary of potential petroleum resources of Region 1-Alaska," in I. H. Cram, ed., *Future Petroleum Provinces of the United States*. (Mem. 15). Tulsa: Am. Assoc. Petrol. Geologists, 55–67.

Hopkins, D. M., 1967. "The Cenozoic history of Beringia–a synthesis," in D. M. Hopkins, ed., *The Bering Land Bridge, 7th Congr. INQUA*. Stanford: Stanford Univ. Press, 451–484.

———, Rowland, R. W., and Patton, W. W., Jr., 1972. "Middle Pleistocene mollusks from St. Lawrence Island and their significance for the paleo-oceanography of the Bering Sea," *Quat. Res.* **2**, 119–134.

Irving, W. N., and Harrington, C. R., 1973. "Upper Pleistocene radiocarbon-dated artifacts from the northern Yukon, *Science*, **179**(4071), 335–340.

Johnson, P. R., and Hartman, C. W., 1969. *Environmental Atlas of Alaska*. Univ. of Alaska, Inst. of Arctic Environmental Engineering, 111p.

Jones, J. G., 1971. "Aleutian enigma: a clue to transformation in time," *Nature*, **229**(5284), 400–403.

Karlstrom, T. N. V., and Ball, G. E., eds., 1969. *The Kodiak Island Refugium: Its Geology, Flora, Fauna and History*. Toronto: Ryerson Press, 262p.

Knebel, H. J., and Creager, J. S., 1973. "Yukon River: evidence for extensive migration during the Holocene transgression," *Science*, **179**, 1230–1232.

McKenzie, G. D., and Goldthwait, R. P., 1971. "Glacial history of the last eleven thousand years in Adams Inlet, southeastern Alaska," *Bull. Geol. Soc. Am.*, **82**, 1767–1782.

Marlow, M. S., Scholl, D. W., Buffington, E. C., and Alpha, T. R., 1973. "Tectonic history of the central Aleutian arc," *Bull. Geol. Soc. Am.*, **84**, 1555–1574.

Martin, A. J., 1970. "Structure and tectonic history of the western Brooks Range, De Long Mountains and Lisburne Hills, northern Alaska," *Bull. Geol. Soc. Am.*, **81**, 3605–3622.

Miller, D. J., Payne, T. G., and Gryc, G., 1959. "Geology of possible petroleum provinces in Alaska, with an annotated bibliography by E. H. Cobb," *U.S. Geol. Surv. Bull.*, **1094**, 131p.

Moore, J. C., 1972. "Uplifted trench sediments: southwestern Alaska-Bering shelf edge," *Science*, **175**, 1103–1105.

———, 1973a. "Cretaceous continental margin sedimentation, southwestern Alaska," *Bull. Geol. Soc. Am., 84,* 595-614.

———, 1973b. "Complex deformation of cretaceous trench deposits, southwestern Alaska," *Bull. Geol. Soc. Am., 84,* 2005-2020.

Morgridge, D. L., and Smith, W. B., Jr., 1972. "Geology and discovery of Prudhoe Bay field, eastern Arctic slope, Alaska," *Stratigraphic Oil and Gas Fields* (Mem. 16). Tulsa: Am. Assoc. Petrol. Geologists, 489-501.

Murray, C. G., 1972. "Zoned ultramafic complexes of the Alaskan type: feeder pipes of andesitic volcanoes," *Geol. Soc. Am. Mem. 132,* 313-338.

Nelson, C. H., Hopkins, D. M., and Scholl, D. W., 1974. "Cenozoic sedimentary and tectonic history of the Bering Sea," in D. W. Hood, ed., *Intern. Symp. on Bering Sea Studies.* Univ. Alaska Press.

North, F. K., 1971. "The Cambrian of Canada and Alaska," in C. H. Holland, ed., *Cambrian of the New World.* New York: Wiley-Interscience, 219-324.

Ovenshine, A. T., and Brew, D. A., 1972. "Separation and history of the Chatham Strait Fault, Southeast Alaska, North America. *24th Intern. Geol. Congr., 3,* 245-254.

Page, R. A., Jr., and Lahr, J., 1971. "Measurements for fault slip on the Denali, Fairweather, and Castle Mountain faults, Alaska," *J. Geophys. Res., 76*(35), 8534-8543.

Péwé, T. L., ed., 1965. "Central and southern Alaska," *INQUA Guidebook F.* Lincoln: Neb. Acad. Sci., 141p.

———, and Reger, R. D., 1972. "Modern and Wisconsinan snowlines in Alaska," *24th Intern. Geol. Congr., Canada, 12,* 187-197.

———, Hopkins, D. M., and Giddings, J. L., 1965. "The Quaternary geology and archeology of Alaska," in H. E. Wright, Jr. and D. G. Frey, eds., *The Quaternary of the United States.* Princeton, N.J.: Princeton Univ. Press, 355-374.

Pitcher, M. G., ed., 1973. *Arctic Geology.* Tulsa, Oklahoma: *Am. Assoc. Petrol. Geologists Mem. 19.*

Plafker, G., 1969. "Tectonics of the March 27, 1964, Alaska earthquake," *U.S. Geol. Surv. Prof. Pap., 543-I,* 11-174.

Proctor, P. D., and Carlile, R. E., 1972. "Alaska–its mineral potentials and environmental challenges," *Bull. Am. Assoc. Petrol. Geologists,* 56(1), 171-178.

Reed, B. L., and Lanphere, M. A., 1973. "Alaska-Aleutian Range batholith: geochronology, chemistry, and relation to circum-Pacific plutonism," *Bull. Geol. Soc. Am., 84,* 2583-2610.

Richards, H. G., 1974. "Tectonic Evolution of Alaska," *Bull. Am. Assoc. Petrol. Geologists,* 58(1), 79-105.

Richter, D. H., and Matson, N. A., 1971. "Quaternary faulting in the eastern Alaska Range," *Bull. Geol. Soc. Am.,* 82(6), 1529-1540.

Rickwood, F. K., 1970. "The Prudhoe Bay field," *Geological seminar on the North Slope of Alaska, Proc.,* Los Angeles: Am. Assoc. Petrol. Geologists, L1-L5.

Schmoll, H. R., Szabo, B. J., Rubin, M., and Dobrovolny, E., 1972. "Radiometric dating of marine shells from the Bootlegger Cove Clay, Anchorage area, Alaska," *Bull. Geol. Soc. Am., 83,* 1107-1114.

Smith, P. S., 1939. "Areal geology of Alaska," *U.S. Geol. Surv. Prof. Pap., 192.*

Spurr, J. E., 1900. "Classification of igneous rocks according to composition," *Am. Geologist,* 25, 210-234.

Stoneley, R., 1967. "The structural development of the Gulf of Alaska sedimentary province in southern Alaska," *Quart. J. Geol. Soc. Lond.,* 123, 25-57.

———, 1971. "A note on the structural evolution of Alaska," *J. Geol. Soc. Lond.,* 127, 623-628.

Tocher, D., 1972. "General introduction," *The Great Alaska Earthquake of 1964.* Washington, D.C.: Nat. Acad. Sci., 1-7.

U.S. Geological Survey, 1964. *Mineral and Water Resources of Alaska.* U.S. 88th Cong., 2nd Sess., Comm. on Interior and Insular Affairs, Comm. Print, 179p.

von Huene, R., 1972. "Structure of the continental margin and tectonism at the eastern Aleutian Trench," *Bull. Geol. Soc. Am., 83,* 3613-3626.

Wahrhaftig, C., 1965. "Physiographic divisions of Alaska," *U.S. Geol. Surv. Prof. Pap. 482,* 52p.

Cross-references: *Canada–Cordilleran Region; United States–General; Vol. VIII, pt. 2, U.S.S.R.*

UNITED STATES–APPALACHIAN REGION

The Appalachian region is in the southeastern United States, and extends from New York City 1400 km SW to Alabama, with a width of up to 500 km. It is the exposed part of a segment of the Appalachian orogenic belt that is commonly known as the Appalachian Mountains or Appalachian Highlands. The Appalachian region is composed of rocks that were variably deformed and afterward subjected to prolonged erosion. This erosion has carved the rocks into mountains, plateaus, and lowlands, arranged in NE-SW trending belts or provinces that follow the grain of the rock structures (see Fig. 1). From NW to SE, these provinces are as follows.

The *Appalachian Plateaus* (Allegheny and Cumberland Plateaus), nearest the continental interior, are composed of little-deformed strata that have been cut into a maze of ridges and valleys. Relief of the plateaus increases to the SE away from the interior lowlands, and they attain heights of up to 1500 m in eastern West Virginia. Their southeastern border is a well-defined, linear escarpment.

In the *Valley and Ridge province* the strata are more strongly folded and faulted, and have been etched into subparallel ridges and valleys

FIGURE 1. Map of Appalachian region showing principal geographic features. *Explanation of symbols:* 1, plains; 2, low dissected terrain; 3, plateaus, with bordering escarpments; 4, strike ridges and valleys; 5, mountains.

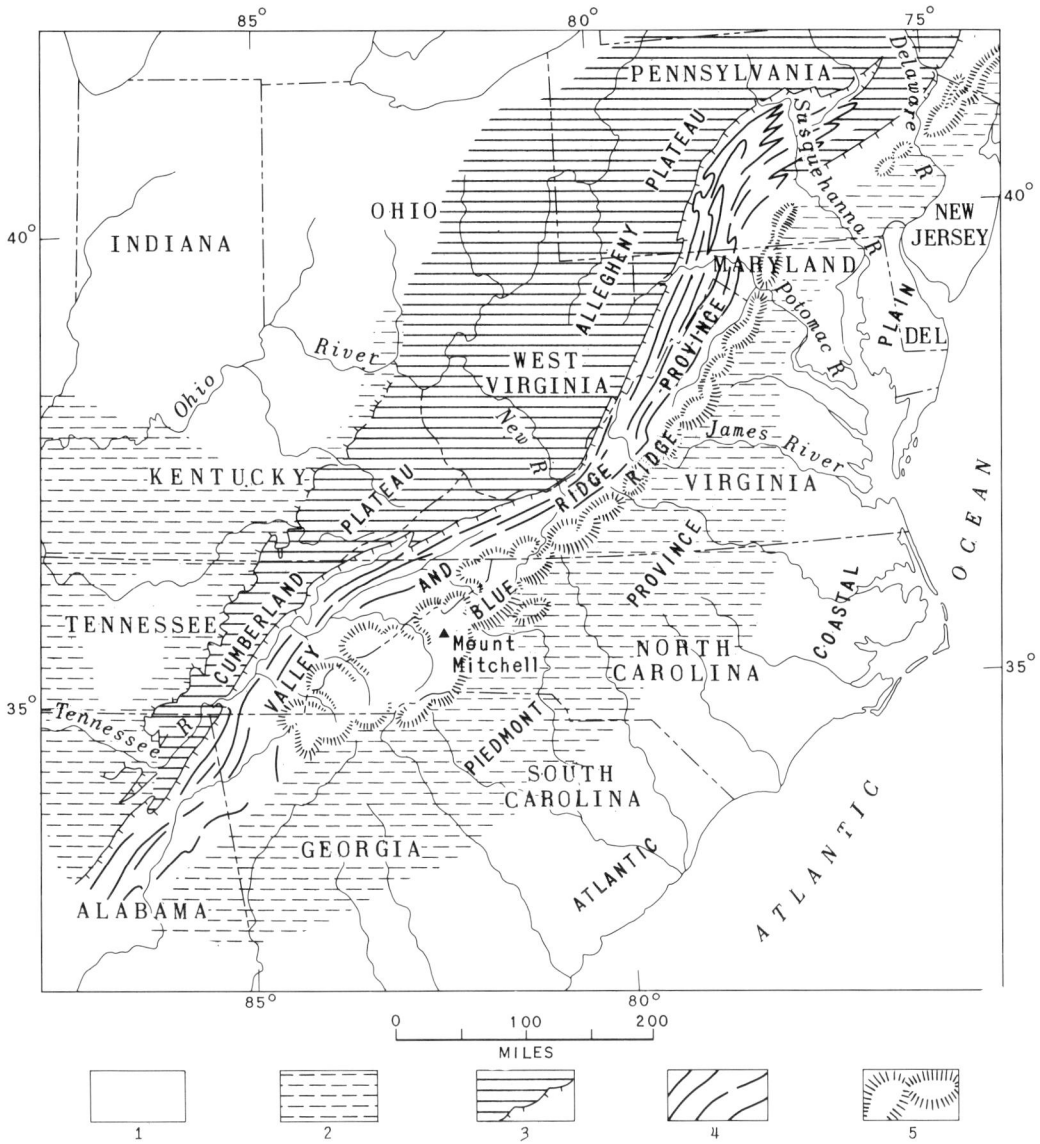

according to their resistance to erosion. Most of the ridges are formed on sandstones and quartzites, and most of the valleys on carbonate rocks.

The *Blue Ridge province,* or central mountain axis, is composed of more homogeneous, dominantly crystalline rocks. It extends from southern Pennsylvania to northern Georgia, beginning toward the NE as a narrow chain, but widening to the SW into a massive highland that includes such high summits as Mt. Mitchell, North Carolina (altitude 2027 m, 6684 ft).

The *Piedmont province,* or easternmost unit of the Appalachian Region, is composed mainly of metamorphic and plutonic rocks that have been worn down to rolling hills and occasional residual mountains. Along its southeastern edge, the rocks of the orogenic belt pass beneath younger deposits of the Atlantic Coastal Plain. Gradients of streams that flow from the Piedmont into the coastal plain change abruptly at the edge, producing a "fall zone." The Atlantic Coastal Plain merges to the SW with the similar Gulf Coastal Plain; the latter extends entirely across the southwestern prolongation of the orogenic belt, terminating the Appalachian region in this direction.

From Pennsylvania to central Virginia, the Appalachian region drains SE into the Atlantic Ocean by the Susquehanna, Potomac, James, and other rivers, which head in the plateaus on the NW and cross the ridges farther to the SE in a succession of water gaps. From central Virginia SW the Atlantic drainage heads on the southeastern slope of the Blue Ridge, and the remainder of this part of the Appalachian region drains NW into the Ohio River, through the New (Kanawha) River, the Tennessee River, and its tributaries.

The landforms of the Appalachian region have been developed by normal processes of weathering and stream erosion, but in the extreme north (New York and Pennsylvania) they were modified by glacial sculpturing during the Pleistocene. Stone rings, boulder trains, and other periglacial phenomena mark many upland areas. The bedrock surfaces have weathered into residual clay and soil, which to the S attain an appreciable thickness on the rocks most susceptible to decay (the carbonate, metamorphic, and plutonic rocks). Most of the highland areas are covered by dense hardwood and conifer forests of great variety.

Geologic Investigations

The Appalachian region was explored early by American geologists, and their observations produced many classic concepts. H. D. Rogers and W. B. Rogers, who were State Geologists of Pennsylvania and Virginia in the 1830s to 1840s, obtained a clear conception of the folding and faulting of the rocks of the Valley and Ridge Province. This was amplified by the work of other state surveys, and during the latter part of the 19th century by the U.S. Geological Survey, which produced an extensive set of geologic folios. Bailey Willis, who participated in the folio work, analyzed the mechanics of Appalachian structure. At about the same time W. M. Davis studied the ridges and valleys of Pennsylvania and New Jersey and developed concepts of multiple erosion cycles and of adjustment of streams to structure. (Bauxite deposits to the S appear to be related to an early Tertiary peneplain.) I. C. White formulated the anticlinal theory of petroleum accumulation during work in the oil and gas fields in Pennsylvania and adjacent states.

The Appalachian region continues to be a fruitful field for study by geologists. The surface features of the Valley and Ridge province and the plateaus to the NW are now well apprehended, but new data and new concepts regarding the deeper structure are being obtained from drilling and geophysical surveys. Knowledge of the Piedmont province has lagged behind that of the rest of the region, but progress has been made in the last few decades by means of detailed mapping, supplemented by precise structural observations and isotopic dating.

Stratigraphy

The stratigraphic record in the Appalachian region falls into three parts: (1) That on the NW, in the Valley and Ridge province and adjacent plateaus, where there is a well-known Paleozoic sequence of miogeosynclinal facies, merging into a platform facies toward the continental interior. (2) That of the central axis, or the Blue Ridge province, where Precambrian rocks of several ages emerge from the Paleozoic cover. (3) That on the SE, in the Piedmont province, where there is a metamorphosed sequence of rocks of eugeosynclinal facies, Precambrian to early Paleozoic in age, invaded by large volumes of plutonic rocks.

Valley and Ridge Province and Appalachian Plateau Area. The Valley and Ridge province contains a nearly conformable sequence of sedimentary rocks up to 10,000 m thick, which includes representatives of all the Paleozoic systems. This thins and becomes less complete toward the continental interior. Drilling in the plateau area indicates that much of the Cambrian, an important constituent of the sequence in the Valley and Ridge province, wedges out to the NW by overlap on the basement surface.

The sequence has few volcanic components. Some beds of lava occur in the basal Cambrian of Tennessee and Virginia and the Middle Ordovician of Pennsylvania. Thin layers of volcanic ash (metabentonite) are widespread at several levels in the Ordovician.

Precambrian rocks at the base of the Paleozoic sequence emerge in the Blue Ridge to the SE, but are not exposed farther to the NW. Much of the surface of the Valley and Ridge province is formed of Cambrian to Devonian rocks, although the Mississippian is preserved in many downfolds. The Pennsylvanian forms most of the surface of the plateaus to the NW, where it is capped by remnants of Lower Permian. The Pennsylvanian is also infolded in a few places to the SE, as in the Anthracite and Broadtop basins at Pennsylvania, and near Birmingham, Alabama.

Along the western edge of the Blue Ridge the Precambrian is overlain by Lower Cambrian clastic rocks (arkose, quartzite, and argillite). These are followed in the Valley and Ridge by a thick carbonate body (limestone, dolomite, and some units of calcareous shale) that includes the remainder of the Cambrian, the Lower Ordovician, and in places the Middle Ordovician. Higher parts of the sequence are more varied. Carbonate units are thin and discontinuous and most of the rocks are clastic—marine below, nonmarine above. Marine clastics begin in the Middle Ordovician in places, at higher levels in others. Nonmarine clastics appear from one place to another in the Upper Devonian or at various levels in the Mississippian; they dominate the Pennsylvanian and also the Lower Permian. Coal measures are prominent in the Pennsylvanian, but also occur in the Mississippian and earliest Permian. The Silurian and Mississippian contain some evaporite beds. The clastic rocks of Ordovician and younger ages were derived from the SE; to the NW they not only thin but become finer textured. According to plate tectonic theory, this source lay in what is today northwest Africa.

Blue Ridge Province. The basement rocks of the Appalachians emerge in the Blue Ridge the core of which, from Maryland through North Carolina, is formed of granite, migmatite, and orthogneiss which have yielded isotopic ages of about 1000 m yr or of Middle Proterozoic age. Similar rocks, yielding similar ages, have been penetrated by drilling beneath the plateaus on the NW. They also emerge to the SE in domes near Baltimore, Maryland, but are exposed in a few other places in the Piedmont region.

The basement rocks of the Blue Ridge province are overlain with profound unconformity by the next succeeding strata. In places such strata are the Lower Cambrian clastics, but in others they are an intervening sequence of late Proterozoic age. In the NE of the Blue Ridge the late Proterozoic is chiefly mafic to felsic lavas with ages of about 800 m. yr., but between them and the basement is generally a thin layer of nonvolcanic sedimentary rocks; this layer is the tapering edge of much thicker units in the Piedmont (see below). The late Proterozoic of the southwestern part of the Blue Ridge is all nonvolcanic clastic sedimentary rocks, forming the Ocoee Series which is more than 8000 m thick.

Piedmont Province. The Piedmont province SE of the Blue Ridge, like the Valley and Ridge province NW of it, contains a sequence of supracrustal rocks younger than the Precambrian basement, but its stratigraphy is less certain because of metamorphism, interruption by plutonic rocks, and general absence of fossils.

The lower part of the sequence seemingly dominates the NW of the province, where it includes the Glenarm Series of Maryland and Pennsylvania, the Evington Group and Lynchburg Formation of Virginia, and unnamed paraschists and paragneisses elsewhere. All these units include very thick masses of fine to coarse clastics, many parts with graded bedding and other features of turbidites. They are probably of late Proterozoic and of Cambrian age, the Proterozoic part being an extension of the Ocoee Series and comparable nonvolcanic sedimentary rocks of the Blue Ridge.

Somewhat younger strata of Paleozoic age are infolded in the synclines to the NW in the Piedmont and are even more extensive to the SE, especially in the Carolina Slate Belt (see Fig. 2). A few Cambrian and Ordovician fossils have been found at one place or another, and isotopic dates of early Paleozoic age have been obtained here and there. No evidence is available for any supracrustal rocks younger than Ordovician in the Piedmont, except in Alabama in the northwestern belts close to the Coastal Plain overlap. In the Carolina Slate Belt the rocks are little metamorphosed except near granite plutons, but equivalents of the Slate Belt rocks farther NW are more metamorphosed.

The supracrustal rocks of the Piedmont are invaded by varied plutonic rocks. Gabbros and diorites occur in Maryland and North Carolina and are seemingly of early Paleozoic or latest Proterozoic age. A younger, more extensive suite of granitic rocks includes foliated, concordant varieties, and more massive cross-cutting varieties. The granitic rocks have yielded isotopic dates of 500–300 m. yr., and were evidently emplaced during a wide span of Paleozoic time, during various stages of the deformation and metamorphism of the supracrustal rocks.

Extending through the Piedmont region from Pennsylvania to North Carolina are strips and patches of Upper Triassic rocks. They lie unconformably on the metamorphic and plutonic rocks, and in Pennsylvania overlap the edge of the Paleozoic sedimentary rocks of the Valley and Ridge province; they were laid down in fault depressions after the main deformation in the Appalachians. They are nonmarine "taphrogeosynclinal" deposits, partly red, including conglomerate, arkose, shale, and some coal. In places they are invaded by diabase dikes and sills; related mafic dikes extend long distances into the surrounding older rocks. Comparable facies may be traced through the Canadian Atlantic provinces and match up with similar deposits in westernmost Europe.

FIGURE 2. Map of Appalachian region showing geologic and tectonic features. Explanation of symbols: 1, basement rocks (Middle Proterozoic); 2, metamorphic rocks of Piedmont Province (late Proterozoic and early Paleozoic); 3, weakly metamorphosed rocks of Carolina slate belt (early Paleozoic); 4, granitic plutonic rocks (Paleozoic); 5, miogeosynclinal and platform deposits (Paleozoic); 6, Triassic rocks; 7, edge of coastal plain deposits (Cretaceous and Tertiary); 8, thrust fault (barbs on upthrown side); 9, normal fault (hachures on downthrown side); 10, transcurrent fault; 11, anticlines.

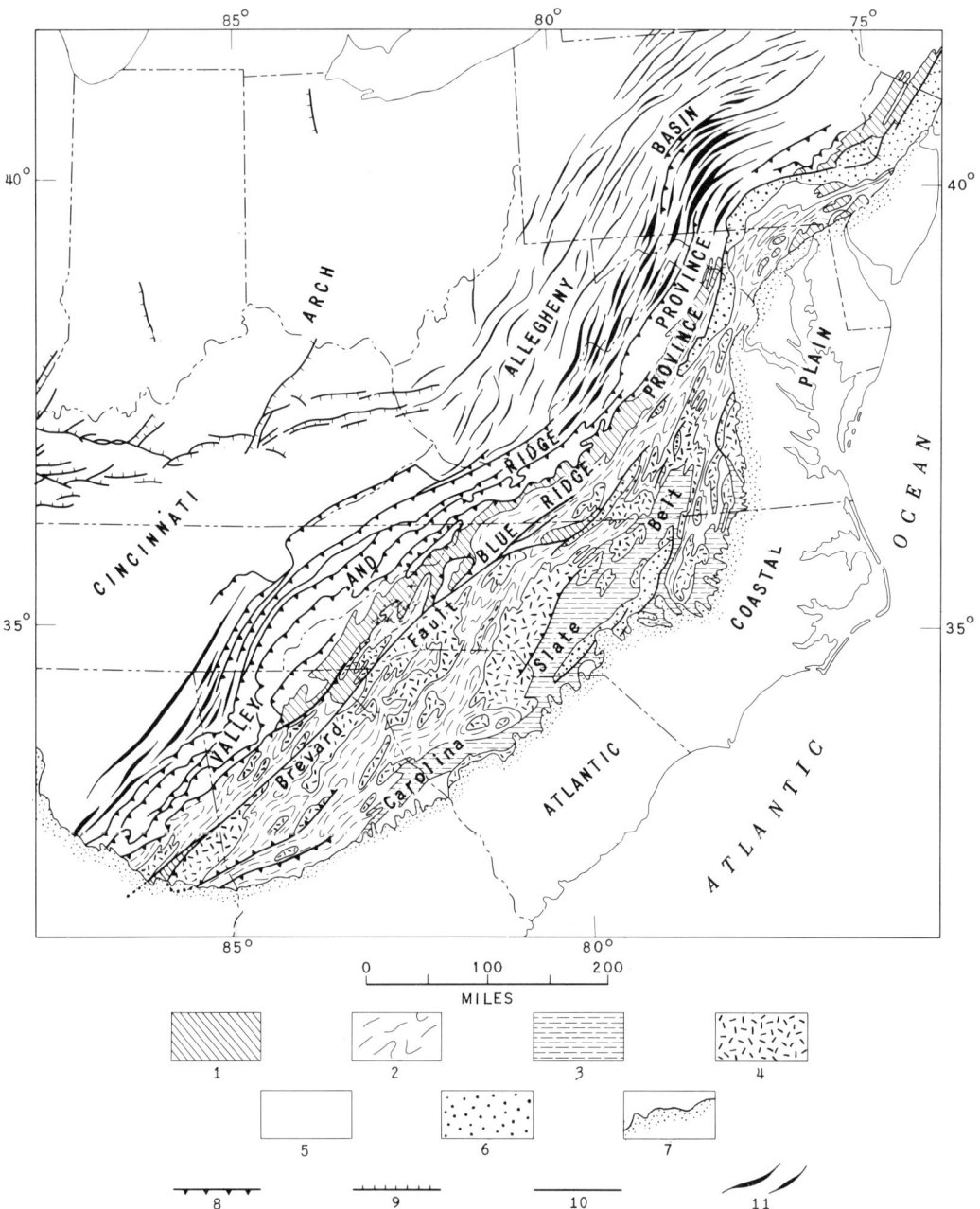

Tectonics

The rocks of the Allegheny and Cumberland plateaus are today downwarped into a broad synclinorium, lying between the Cincinnati Arch of the continental interior and the more deformed rocks to the SE, which border them abruptly along a structural front (see Fig. 2). The rocks of the plateaus dip gently, but are warped into low folds parallel to those farther to the SE. For some distance NW of the structural front they are also transversed by low-angle thrusts of small to moderate displacement, which follow the weaker strata.

The rocks of the Valley and Ridge province are thrown into long, closely spaced folds, whose northwest sides are commonly overturned or broken by thrust faults. These faults increase in number and magnitude to the SW and in Tennessee dominate the folds. Many of the faults probably branch from major sole faults at depth, formed along one or more weak layers in the sedimentary sequence. Much of the deformation of the province is by flexure; slaty cleavage occurs only in the weaker strata and there is little metamorphism. The basement beneath the sedimentary rocks probably participated very little in the deformation.

The southeastern edge of the Valley and Ridge province is a zone of strong uplift, which brings the Proterozoic basement to view in the Blue Ridge. From Pennsylvania to central Virginia the Blue Ridge is an anticlinorium, strongly overfolded to the NW. In central Virginia low-angle thrusts appear along the overfolded edge, and increase in magnitude to the SW. Large displacements on the thrusts are attested by windows of overridden rocks in Tennessee and North Carolina, some of them 10 to 50 km behind the leading edges. The structural style of the Blue Ridge differs from that of the Valley and Ridge province; deformation is more by shear than by flexure, and involves not only the sedimentary cover but the basement; the basement is also involved in the low-angle thrusting. Cleavage strengthens in the Blue Ridge, and progressively higher-grade metamorphic minerals appear to the SE.

From Virginia NE the boundary between the Blue Ridge and the Piedmont is largely homoclinal, younger rocks succeeding older toward the SE. From Virginia SW to the Coastal Plain border the boundary is the remarkably straight *Brevard zone*, marked throughout by faulting and cataclasis, which juxtaposes contrasting rocks and structures on the two sides. The nature and age of the Brevard zone are still incompletely understood. Strikeslip displacement undoubtedly occurred upon it, but this probably was superposed on earlier structures of quite different character; possibly latest movements occurred toward the end of the Paleozoic, but there is some indication of movements as early as the Devonian.

Structure of the Piedmont province is still incompletely known. Gross anticlinoria and synclinoria are apparent from the patterns of the metamorphic and plutonic rocks, but these are superposed on earlier, highly complex plastic folds, some of which, at least, are recumbent nappes. Metamorphic grade increases to the SE from the Blue Ridge, but in the North Carolina-South Carolina segment it reaches a climax well to the NW of the edge of the coastal plain. Farther SE, as in the Carolina slate belt, the rocks are again weakly metamorphosed.

The taphrogenic Triassic in the Piedmont consists of remnants of tilted rocks that are generally downfaulted on one or both sides, with trends that nearly follow the grain of the older rocks. Two main belts occur. One is on the NW close to the Blue Ridge, extending from Pennsylvania to North Carolina, and is downfaulted on the northwest side. Another on the SE is close to the coastal plain, extending from Virginia across North Carolina, and is downfaulted on the southeast side. In North Carolina the two belts are symmetrically placed on opposite sides of the metamorphic climax in the older rocks.

All the deformed rocks of the Appalachian orogenic belt—Precambrian, Paleozoic, and Triassic—are truncated at the edge of the Atlantic and Gulf Coastal Plains, where they are covered unconformably by gently tilted Cretaceous and Tertiary deposits. They extend to the SE and SW beneath this cover, where they have been penetrated in places by drilling and are indicated elsewhere by geophysical surveys. The geophysical surveys suggest that metamorphism of the rocks beneath the coastal plain deposits may decrease to the SE, as they have lower velocity properties toward the coast and onto the continental shelf than farther to the NW. Drilling in southern Georgia and northern Florida, much farther south, has disclosed nearly flatlying, unmetamorphosed, fossiliferous Ordovician to Devonian rocks beneath the coastal plain cover. These lie well to the SE of the trend of the Appalachian orogenic belt, and their relations to the Appalachian structures are uncertain. Paleontologically they correlate with West Africa.

Geophysical Data

In the Appalachian region, according to explosion seismology, the crust is 32 to 36 km thick, and thickens to a root of about 45 km in the mountains of the Blue Ridge Province of western North Carolina. The crust of the plateau and cratonic area to the W is 37 to 43 km thick, or somewhat thicker on the average than that of the Appalachian region.

Bouger gravity anomalies are weakly positive in the Piedmont province and Atlantic Coastal Plain, and moderately to strongly negative in the Valley and Ridge province and plateau area, the two regions being separated by an abrupt gradient (from + 20 milligals to less than −100 milligals, locally). The slope of this gravity gradient indicates that it reflects density differences in the shallow part of the crust, but the amplitude is such that the gradient is probably not caused by density contrasts between crystalline and sedimentary rocks. Toward the NE, where the Blue Ridge is an autochthonous uplift, the gradient is on its northwestern side. Farther SW, where the Blue Ridge rocks have been transported on low-angle thrusts, the gradient lies increasingly to the SE until it is 80 km SE of the Blue Ridge in Georgia.

Heat-flow measurements by Diment indicate a value of approximately 1 cal/cm^2 sec for the Appalachian region. This figure is significantly lower than the worldwide average and is comparable to values measured in shield areas.

Recorded earthquake epicenters in the eastern United States are significantly concentrated along a belt that follows the trend of the Appalachians, near or a little SE of the Blue Ridge, with much less seismic belts on either side. However, clusters of epicenters along this belt have no obvious relation to other tectonic or geophysical features—especially a cluster that extends SE from western North Carolina to the coast at Charleston, South Carolina, which was the site of a major earthquake in 1886.

Tectonic history

Appalachian structures evolved through much of Paleozoic time, with early phases in the Proterozoic and continuing instability through the Triassic. The early phases involved mainly geosynclinal sedimentation, but orogenic events began early toward the SE, on the site of the Piedmont province, and overlapped the geosynclinal events to such an extent that this area was nearly consolidated by mid-Paleozoic time. Most of the orogenic events on the site of the Valley and Ridge Province toward the NW came later, after the geosynclinal events and late in Paleozoic time. The tectonic history of the Appalachian region, as now understood, thus differs from the traditional concept that most of its structures were created by an Appalachian revolution near the end of Paleozoic time, and are younger than most of the structures in the New England region; histories of the Piedmont and New England region are more nearly comparable than was formerly supposed.

On the site of the Piedmont province and parts of the Blue Ridge a geosyncline formed in late Proterozoic time over the previously consolidated older Precambrian basement. The deposits were largely nonvolcanic, except at the northwestern edge. Although the deposits were laid down in an unstable environment, they were not appreciably deformed or metamorphosed until later. During at least the first half of Paleozoic time, eugeosynclinal deposits were spread over the late Proterozoic deposits, especially toward the SE.

In the Piedmont province, plutonic, metamorphic, and probably orogenic events are indicated by isotopic dates. Significant clusters of dates occur near 450 to 500 m yr, 350 m yr, and 250 m yr, the first being most abundant toward the NW, the last toward the SE. These dates invite comparison with known orogenic events in the New England region—the early Paleozoic *Taconic orogeny,* the mid-Paleozoic *Acadian orogeny,* and later Paleozoic orogenies (notably the *Allegheny orogeny*—see below). Various considerations, including the record of clastic deposition in the Valley and Ridge Province, suggest that the earlier isotopic dates express a greater orogenic event than the later ones, which may be due merely to "reheating." The effects of this orogeny extended into the Valley and Ridge Province in eastern Pennsylvania, where there is an angular unconformity at the top of the Ordovician, but the unconformity fades out to the SW in that province.

The clastic rocks of the middle and upper parts of the miogeosynclinal sequence in the Valley and Ridge province were derived from areas in the Piedmont province that were in the process of uplift and deformation. Nevertheless (except in a few places, as in eastern Pennsylvania) this sequence is nearly conformable, indicating that its rocks were themselves not appreciably deformed until after the youngest strata were deposited there. Part of the deformation was later than the Lower Permian rocks, which are involved in the broad folds of the plateau area; this has been called the *Allegheny orogeny.* Earlier deformations may have occurred in parts of the Valley and Ridge province, where Mississippian and even older rocks are the youngest preserved. Successive events can be inferred here from superposed structures; for example, low-angle thrust faults have been folded, and the folds have been broken by later faults. However, no method for dating these events is available.

Continuing instability of the Appalachian region is recorded by the block-faulted structures of the Triassic sedimentary rocks and mafic intrusives ("Palisadian" disturbance). These structures are younger than the deformational climax, thus postorogenic. They were produced merely by taphrogeny—tilting and warping—and by blockfaulting and dike formation of tensional origin.

The region became much more stable later in Mesozoic time, as shown by the truncation of the Triassic and older rocks by the little disturbed Cretaceous coastal plain deposits. From then until the present the region has been affected only by epeirogenic upwarping in the axial region and downwarping toward the margins. Upwarping and erosion produced the Appalachian region described here; downwarping and deposition concealed other parts of the Appalachian orogenic belt beneath the Atlantic and Gulf Coastal Plains.

Erosion in the upwarped areas sculptured the rocks according to their resistance to erosion, producing the varied mountains, plateaus, and lowlands of the Appalachian region. This warping occurred during successive episodes that were separated by lengthy periods of relative stability. Remnants of several accordant planation surfaces have been recognized in the mountains and valleys of the region, which many geologists believe were leveled by erosion during the times of stillstand and dissected during the times of uplift; this interpretation has been questioned by certain geomorphologists. Downwarping in the coastal plains was clearly episodic, as the deposits laid down there record many times of transgression and regression. The Oligocene, for example, is completely absent. The occurrence of earthquakes (admittedly infrequent) during modern times indicates that crustal movements in the Appalachian region have not entirely ceased.

Economic Mineral Deposits

The Appalachian region contains many economically useful mineral deposits, the exploitation of which has had a major influence on American history. The Appalachian Plateaus on the NW contain large amounts of mineral fuels, whereas the other provinces to the SE contain workable deposits of metallic and nonmetallic minerals.

Mineral Fuels. The occurrence of coal in the Appalachian Plateaus has long been known, but the coal was not fully exploited until the industrializiation of the middle-eastern states in the last half of the 19th century; its exploitation has contributed to the growth of industrial complexes such as that around Pittsburgh, Pennsylvania. Most of the coal beds are in rocks of the Pennsylvanian System, but a few occur in strata immediately below and above; they are principally in Pennsylvania and West Virginia, but extend into adjacent states, as well as along the trend of the province to the SW to Alabama. About 60 coal beds occur in Pennsylvania, of which 10 are mineable; one of them, the Pittsburgh Coal, is as much as 4 m (13 ft) thick, and contains more than 22 billion tons of coal. Throughout the Appalachian Plateaus, the coal is of bituminous grade, but where the Pennsylvanian rocks have been synclinally downfolded in eastern Pennsylvania, it has been altered to anthracite. The anthracite fields have an area of only 1200 km^2, but have produced more than an eighth of the total output of coal in North America. Besides the Paleozoic coal of the Appalachian Plateaus, minor coal beds also occur in the Triassic rocks of parts of the Piedmont province.

The strata beneath the coal beds in the Appalachian Plateaus contain oil and gas and were the first in North America to be exploited for these fuels, beginning with their discovery in the Drake Well of northwestern Pennsylvania in 1859. Subsequent exploitation resulted in the creation of the Standard Oil Company and other corporate giants. As late as the end of the 19th century the annual oil production of the province was more than half of that in the United States, but it has since declined whereas production of provinces elsewhere has increased. Oil produced in the province is notable for the high quality and quantity of its lubricating fraction, and it still contributes about 10% of the lubricants of the United States. Productive horizons occur in all the systems from the Ordovician to the Pennsylvanian, but the principal horizons are in the Devonian and Mississippian, which include the prolific Oriskany and Berea Sands.

Metallic and Nonmetallic Mineral Deposits. Weathering of carbonate rocks and other bedrock of the Valley and Ridge province has created residual accumulations of brown iron ore, which were extensively used by settlers during the early days of the republic for the manufacture of ironware. At one time the brown ores accounted for much of the iron production of the country, but they were largely mined out by the beginning of the 20th century. Some of the residual accumulations also contain manganese oxides which were worked during World War I and II. Iron mining still continues in the stratified red iron ores of the Silurian System; it once accounted for 10% of the annual production of the United States, but has now greatly declined. Principal workings are near Birmingham, Alabama, where iron ore, coal beds, and limestone for flux are favorably juxtaposed.

Metallic sulfide deposits are more sporadically distributed than the deposits of iron and mineral fuels, but are locally important. Zinc and lead sulfides occur in the carbonate rocks of the Valley and Ridge province of southwestern Virginia and eastern Tennessee. The lead was mined as early as the Revolutionary War, but exploitation of the zinc began only in the latter part of the 19th century; nearly a dozen mines are now operating, and economically valuable deposits continue to be discovered. Barite occurs in similar situations and is worked in many places. Copper sulfides were discovered early in the 19th century in the metamorphic rocks at Ducktown, Tennessee, and have been mined there from 1845 to the present, although now mainly for their sulfur rather than for their copper. Similar but lesser copper sulfide deposits occur elsewhere in the metamorphic rocks along the strike of the Blue Ridge province.

Gold occurs in the metamorphic and plutonic rocks of the Piedmont Province, and is also redistributed in the adjacent stream alluvium. It aroused the interest of the original Spanish explorers, and of the settlers during the early days of the republic, and resulted in a "gold fever" in the 1830s (preceding by several decades the California Gold Rush), which was largely responsible for the removal of the Cherokee Nation from its gold-bearing lands to Indian Territory W of the Mississippi. Placer mining, followed by lode mining, was extensive in many of the Piedmont states during the 19th century, with total production of about 50 million dollars, but has long been moribund.

Nevertheless the metamorphic and plutonic rocks of the Piedmont and Blue Ridge provinces have continued to yield many useful economic products, some of which are unique in the economy of the country—marble from northern Georgia; titanium minerals from Virginia; tungsten from a district athwart the Virginia-North Carolina boundary; mica and feldspar from many districts, but especially from Spruce Pine, North Carolina; kyanite from the metamorphic rocks of many areas; and a host of lesser mineral products.

PHILIP B. KING

References

Bentley, R. D., and Neathery, T. L., 1970. *Geology of the Brevard Fault Zone and Related Rocks of the Inner Piedmont of Alabama.* Univ. Alabama, Ala. Geol. Soc., 119p.

Berkland, J. O., and Raymond, L. A., 1973. "Pleistocene glaciation in the Blue Ridge Province, southern Appalachian Mountains, North Carolina," *Science,* 181, 651–653.

Carpenter, R. H., Fagan, J. M., and Wedow, H., Jr., 1971. "Evidence on the age of barite, zinc, and iron mineralization in the Paleozoic rocks of east Tennessee," *A Paleoaquifer and Its Relation to Economic Mineral Deposits, Econ. Geol.,* 66, 792–798.

Cloos, E., 1947. "Oolite deformation in the South Mountain fold, Maryland," *Bull. Geol. Soc. Am.,* 58(9), 843–918.

Coates, D. R., Landry, S. O., and Lipe, W. D., 1971. "Mastodon bone age and geomorphic relations in the Susquehanna Valley," *Bull. Geol. Soc. Am.,* 82(7), 2005–2009.

Cooper, B. N., 1964. "Relation of stratigraphy to structure in the southern Appalachians," in W. D. Lowry, ed., *Tectonics of the Southern Appalachians* (Mem. 1). Virginia Polytechnic Inst., Dept. Geol. Sci., 81–114.

_____, 1968. "Profile of the folded Appalachians of Western Virginia," *Univ. Mo. Res. J.,* 1(1), 27–64.

Epstein, J. B., and Epstein, A. G., 1972. "The Shawangunk Formation (Upper Ordovician (?) to Middle Silvrian) in eastern Pennsylvanian," *U.S. Geol. Surv. Prof. Pap. 744,* 43p.

Faill, R. T., 1973. "Tectonic development of the Triassic Newark–Gettysburg Basin in Pennsylvania," *Bull. Geol. Soc. Am.,* 84, 725–740.

Fisher, G. W., Jr., Pettijohn, F. J., Reed, J. C., Jr., and Weaver, K. N., 1970. *Studies of Appalachian Geology: Central and Southern.* New York: Wiley-Interscience, 460p.

Frey, M. G., 1973. "Influence of Salina salt on structure in New York-Pennsylvania part of Appalachian Plateau," *Bull. Am. Assoc. Petrol. Geologist.,* 57(6), 1027–1037.

Gwinn, V. E., 1964. "Thin-skinned tectonics in the plateau and northwestern Valley and Ridge province of the central Appalachians," *Bull. Geol. Soc. Am.,* 75(9), 863–900.

_____, 1970. "Kinematic patterns and estimates of lateral shortening, Valley and Ridge and Great Valley provinces, south-central Pennsylvania," in G. W. Fisher, *et al.,* eds., *Studies of Appalachian Geology: Central and Southern.* New York: Wiley, 127–146.

Hadley, J. B., 1964. "Correlation of isotopic ages, crustal heating and sedimentation of the Appalachian region," in W. D. Lowry, ed., *Tectonics of the Southern Appalachians* (Mem. 1). Virginia Polytechnic Inst., Dept. Geol. Sci., 33–45.

Harris, L. D., 1970. "Details of thin-skinned tectonics in parts of Valley and Ridge and Cumberland Plateau provinces of the southern Appalachians," in G. W. Fisher, *et al.,* eds., *Studies of Appalachian Geology: Central and Southern.* New York: Wiley, 161–173.

Hatcher, R. D., Jr., 1971. "Stratigraphic, petrologic, and structural evidence favoring a thrust solution of the Brevard problem," *Am. J. Sci.,* 270(3), 177–202.

_____, 1972. "Developmental model for the southern Appalachians," *Bull. Geol. Soc. Am.,* 83, 2735–2760.

Hill, W. T., McCormick, J. E., and Wedow, H., Jr.,

1971. "Problems on the origin of ore deposits in the lower Ordovician formations of east Tennessee," *A Paleoaquifer and Its Relation to Economic Mineral Deposits, Econ. Geol.,* **66**(5), 700–804.

Hopson, C. A., 1964. "The crystalline rocks of Howard and Montgomery Counties," in E. Cloos, G. W. Fisher, and C. A. Hopson, *The Geology of Howard and Montgomery Counties, Maryland,* Maryland Geol. Surv., 27–336.

King, P. B., 1950. "Tectonic framework of southeastern United States," *Bull. Am. Assoc. Petrol. Geologists,* **34**(4), 635–671.

———, 1970. "The Precambrian of the United States of America," in K. Rankama, ed., *The Precambrian,* vol. 4. New York: Wiley, 1–71.

Krinsley, D. H., 1973. "Age of the Mount Laurel and Navesink formations at Marlboro, New Jersey, from K-Ar measurement of glauconite," *Bull. Geol. Soc. Am.,* **84**, 2143–2146.

Kulp, J. L., and Eckelman, F. D., 1961. "Potassium-argon isotopic ages on micas from the southern Appalachians," *N.Y. Acad. Sci. Ann.,* **91**, 408–419.

Laurence, R. A., 1968. "Ore deposits of the southern Appalachian region," in J. D. Ridge, ed., *Ore Deposits of the United States, 1933-1967* (Grafton-Sales Vol. 1). Am. Inst. Mining, Met. Petrol. Engs., 155–168.

Lovlie, R., and Opdyke, N. D., 1974. "Rock Magnetism and Paleomagnetism of some intrusions from Virginia," *J. Geophy. Res.,* **79**(2), 343–349.

Lowry, W. D., and Cooper, B. N., 1970. "Penecontemporaneous downdip slump structures in Middle Ordovician limestone, Harrisonburg, Virginia," *Bull. Am. Assoc. Petrol. Geologists,* **54**, 1938–1945.

Maxey, L. R., 1973. "Dolerite dikes of the New Jersey Highlands: probable comagmatic relation with the Mesozoic Palisades Sill and the dolerite dikes of eastern United States," *Bull. Geol. Soc. Am.,* **84**, 1081–1086.

Maxwell, J. A., and Davis, M. B., 1972. "Pollen evidence of Pleistocene and Holocene vegetation on the Allegheny Plateau, Maryland," *Quat. Res.,* **2**, 506–530.

Meyerhoff, H. A., 1972. "Postorogenic development of the Appalachians," *Bull. Geol. Soc. Am.,* **83**, 1709–1728.

Ratcliffe, N. M., 1971. "The Ramapo fault system in New York and adjacent northern New Jersey: a case of tectonic heredity," *Bull. Geol. Soc. Am.,* **82**, 125–142.

Rich, J. L., 1934. "Mechanics of low-angle overthrust faulting as illustrated by the Cumberland thrust block, Virginia, Kentucky, and Tennessee," *Bull. Am. Assoc. Petrol. Geologists,* **18**(12), 1584–1596.

Rodgers, J., 1949. "Evolution of thought on structure of middle and southern Appalachians," *Bull. Am. Assoc. Petrol. Geologists,* **33**(10), 1643–1654.

———, 1970. *The Tectonics of the Appalachians.* New York: Wiley, 12–65, 164–202.

———, 1972. "Latest Precambrian (Post-Grenville) rocks of the Appalachian region," *Am. J. Sci.,* **272**, 507–520.

Stearns, R. G., 1955. "Low-angle overthrusting in the central Cumberland Plateau, Tennessee," *Bull. Geol. Soc. Am.,* **66**(6), 615–628.

Stirewalt, G. I., and Dunn, D. E., 1973. "Mesoscopic fabric and structural history of Brevard Zone and adjacent rocks, North Carolina," *Bull. Geol. Soc. Am.,* **84**, 1629–1650.

Tobisch, O. T., and Glover, L., III, 1971. "Nappe formation in part of the southern Appalachian Piedmont," *Bull. Geol. Soc. Am.,* **82**, 2209–2230.

Turner, S., 1970. "Timing of the Appalachian/Caledonian orogen contraction," *Nature,* **227**(5253), 90.

Walker, R. G., 1971. "Nondeltaic depositional environments in the clastic wedge (upper Devonian) of central Pennsylvania," *Bull. Geol. Soc. Am.,* **82**(5), 1305–1326.

Webb, E. J., 1969. "Geologic history of the Cambrian system in the Appalachian basin," *Ky. Geol. Surv. Spec. Publ. Ser. 10* (18), 7–15.

Woollard, G. P., 1958. "Areas of tectonic activity in the United States as indicated by earthquake epicenters," *Trans. Am. Geophys. Union,* **39**(6), 1135–1150.

Young, D. A., 1971. "Precambrian rocks of the Lake Hopatcong area, New Jersey," *Bull. Geol. Soc. Am.,* **82**, 143–158.

Cross-references: *Canada–Atlantic Provinces; North America; United States–Atlantic Coastal Province, Gulf Coastal Province, Midwestern Region, New England Region.*

UNITED STATES–ATLANTIC COASTAL PROVINCE

The Atlantic Coastal Plain extends along the east coast of the United States from Long Island to Florida. A submarine extension of the coastal plain is represented by the continental shelf. The relative width of these two units varies. On Long Island, the emerged part is very narrow, whereas the shelf is almost 160 km wide. The continental shelf is even wider off the coast of New England and eastern Canada.

Both coastal plain and shelf consist primarily of unconsolidated Cretaceous and Tertiary sediments frequently capped with thin deposits of Quaternary age. To the W of the coastal plain is the Piedmont province (see *Appalachian Region*) made up mainly of crystalline rocks of Paleozoic age or older. The line between the coastal plain and the Piedmont is known as the "Fall Line" (or "Fall Zone" when there is not a sharp line).

Triassic basins (grabens) underlie parts of the coastal plain of Maryland, Virginia, the Carolinas, and Georgia, and wells in these basins have revealed rocks similar to those of the Newark group of Pennsylvania and New Jersey.

Structural Features

Early workers regarded the Atlantic Coastal Plain as a simple monocline with rocks dipping gently seaward. More detailed mapping, both on the surface and in the subsurface, has revealed some complexity. The Atlantic Coastal province between New Jersey and Georgia has now been divided into nine major structural features. These were summarized by Richards (1967). Additional structural features are also known from Florida.

Cape May Slope. This name was proposed for the area between the outcropping basement rocks near New York City and the mouth of Delaware Bay (Richards, 1967). Because of the scarcity of data on the subsurface geology, not much is known about the detailed structure of the region. Undoubtedly there is considerable relief on the basement as shown by detailed mapping in the area of Camden, New Jersey (Barksdale *et al.*, 1958).

Several geophysical surveys have been made, but very little has been released for publication. The slope of the basement rock is about 12 m/km (65 ft/mi) increasing to 18 m/km (100 ft/mile) near the coast and beyond, i.e., from 1 in 85 to 1 in 55. Several minor structures have been recognized (Woollard, 1941).

Salisbury Embayment. Magnetic work by Balsey *et al.* (1946) suggested that there were irregularities in the basement in the vicinity of Salisbury, Maryland. Three unsuccessful oil tests were drilled in this area, the first being that of the Ohio Oil Company near Salisbury. A thicker section of Cretaceous sediments was found than expected (1206 m, 3980 ft); below that was a 51 m (169 ft) section of shale and sandstone, correlated with the Newark deposits of Triassic age. Basement rock was encountered at 1675 m (5529 ft). The other oil tests, drilled near Berlin and Ocean City, Maryland, showed that the formations become deeper toward the E, forming what is called the *Salisbury Embayment* (Richards, 1948, 1949). This embayment extends to the W of the District of Columbia line.

Although the Triassic rocks encountered were all nonmarine, it is significant that the Cretaceous sediments become more marine toward the present coast. Eventually drilling on the shelf off Maryland may reveal a further extension of the Salisbury Embayment with a thicker section, and possibly lead to the discovery of oil.

Fort Monroe High. The basement rock rises from the Salisbury Embayment to the vicinity of Fort Monroe, Virginia, where it has been encountered at a depth of only 680 m (2246 ft). This structural region has been called the *Fort Monroe High* (Richards and Straley, 1953, p. 105). Some of the sediments of Upper Cretaceous age (Monmouth-Matawan) pinch out, whereas those of basal Upper Cretaceous age (Raritan) are reduced in thickness and occur only in the subsurface. On the other hand there is a thickening of deposits of Early Cretaceous age. Cederstrom (1945) cites geological and geophysical evidence to indicate a major fault in the vicinity of the James River.

Hatteras Low. This structural feature has also been called the *Albemarle Basin* and the *Pamlico Basin*. In this zone the basement rock drops sharply from the Fort Monroe High to a low near Cape Hatteras, North Carolina, where it was encountered in an unsuccessful oil test drilled by Esso Standard Oil Company in 1946 at a depth of 3042 m (9975 ft). The slope of the basement is probably much more complex than shown in the diagram (Fig. 1) and several structural irregularities have been reported (Johnson and Straley, 1953). Other examples of the irregularity of the basement rock in this area are granite hills near Fountain and Smithfield, North Carolina, which are surrounded by unconsolidated coastal plain sediments.

Many of the Cretaceous and Tertiary sediments pinch out toward the N against the Fort Monroe High.

Cape Fear Arch or "Great Carolina Ridge". Basement rock rises from the Hatteras Low to only 336 m (1109 ft) beneath the surface near Wilmington, North Carolina. McCarthy and Straley (1937) conducted geomagnetic investigations on this "ridge," and observed a series of anomalies subparallel to Appalachian tectonic trends. These were attributed to structural conditions, lithologic variations, and topography of the pre-Mesozoic basement.

Beaufort Basin. The basement again drops S of the Great Carolina Ridge forming the Beaufort Basin. Deep drilling and geophysical studies have revealed numerous structural features (Richards and Straley, 1953), including at least two buried Triassic basins. Movement along a fault at the edge of the Summerville Basin may possibly have caused the serious earthquake in Charleston, South Carolina, in 1886. According to some authorities there is a deep-seated left-lateral strike-slip fault paralleling the coast here (see Shaler's Line, *below*).

Okefenokee Embayment. The basement continues to dip along the Atlantic coast to the S, although relatively little is known about its exact configuration. The basin is bounded on the W by the Central Georgia Uplift. The Okefenokee Embayment is sometimes called the *Southeast Georgia Basin* or the *Savannah Basin*.

Central Georgia Uplift. On the W, Georgia is bordered by the Decatur Arch that follows

FIGURE 1. Schematic diagram of main structural features of coastal plain between Long Island and South Carolina (From Richards, 1967.)

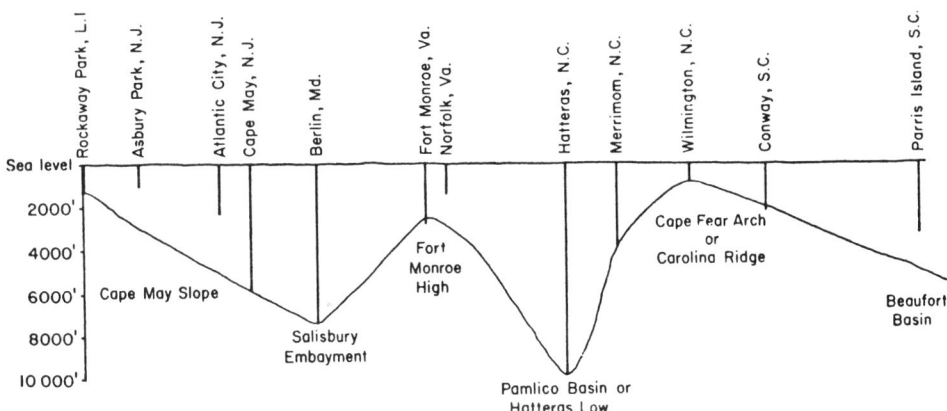

the trend of the Chattahoochee River. The central Georgia uplift, is related to the Ocala uplift of Florida and may be the northern extension of one of its branches. This upwarp separates the Atlantic Coastal province from the eastern Gulf Coastal province. A physiographic feature, Trail Ridge, trends roughly N-S and is thought by some to represent a Pleistocene offshore bar; others regard it as older.

Suwannee River Basin. This name was proposed by Braunstein (1958) for the Paleozoic basin below Cretaceous deposits in parts of Georgia, Florida, and Alabama. Wells drilled into this basin have yielded fossils of ages varying from Ordovician to Pennsylvanian.

Structural Features in Florida. A NE-trending negative structural feature in southern Georgia and northeastern Florida is called the *Suwannee Strait.* It corresponds to a lineament extending up to the Carolina coast, a feature first recognized by Shaler and called "Shaler's Line" (Husted, 1972). It has been active, on and off, since the late Cretaceous. On that structure, Cretaceous rocks are very thin or missing. South of the Suwannee Strait is the Peninsular Arch, a major positive structural feature that is mainly in Florida. Upper Cretaceous rocks are also very thin, suggesting uplift. The general area of the Peninsular Arch was uplifted again as the Ocala uplift in Tertiary time.

Geological History

Paleozoic. Little is known about the geological history of the Atlantic Coastal Plain during the Paleozoic era since rocks of that age are deeply buried. Samples encountered at the bottom of deep wells from New Jersey to the Carolinas are crystalline rocks of unknown age. Some unmetamorphosed sedimentary rocks of Paleozoic age with marine fossils have been found under coastal plain sediments in Georgia and Florida, suggesting that at that time there was a close relationship with west Africa and Spain. Palynological material suggests that at least for the Silurian to Devonian (Llandoverian to Gedinnian) the two regions were connected by shallow epeiric seas with numerous shoals (Cramer, 1971).

Triassic. A series of taphrogenic, fault basins or grabens was formed during Triassic time. As far as is known, these were filled with fresh water or salinas, although at least some of them—for example, the Salisbury Embayment—may have had connections to the sea. The climate was evidently subtropical and the sediments typical redbeds.

Jurassic. No marine deposits of Jurassic age are known from the Atlantic Coastal Plain except in the deep well at Cape Hatteras, North Carolina, where Swain (1962) found some Jurassic ostracodes between depths of 2603 and 3020 m (8500 and 9878 ft) lying directly on the basement. The rest of the Jurassic shoreline was apparently far E of that of the present day. A gently sloping land surface, called the *Fall Zone Peneplain,* extended E from the eroded Appalachian Mountains to the Jurassic shore line. Rivers flowing E across this peneplain depositing sediment on what was then the floor of a narrow ancestral Atlantic Ocean. Someday when deep drilling is attempted offshore, near the outer margin of the continental rise, marine fossils of Jurassic age should be encountered, as may be predicted from the paleo magnetic patterns.

Early Cretaceous. At the end of Jurassic

time, or at least the very beginning of the Cretaceous, the Fall Zone Peneplain was uplifted and the gravels that had been deposited on its surface were eroded and redeposited in deltas in eastern Pennsylvania, New Jersey, Delaware, and Maryland. One of the early Cretaceous rivers apparently flowed E across Maryland, approximately along the course of the present Potomac River, and then across the "Eastern Shore" along the line of the Salisbury Embayment, which probably started to evolve during Triassic time. Similar rivers were present both N and S of this ancient Potomac. Deposits formed by these rivers consisted of alternating sands and clays, with frequent zones of lignite. A few fossil freshwater molluscs have been found in these deposits and some fragmental bones of dinosaurs have been found near Baltimore and Washington. These sediments are known as the *Potomac Formation* (or Group). Near Baltimore, this formation reaches a thickness of 150 m and has been subdivided into three units.

At the same time that these rivers were transporting alluvial material E toward the sea, there was a progressive encroachment by the sea itself—part of the world-wide Cretaceous transgression. Therefore the subsurface (downdip) Potomac beds consist of interfingering marine and alluvial deposits. Marine fossils, probably of Early Cretaceous age, have recently been found in a well near Cape May, New Jersey (Shapiro and Richards, 1969).

The dividing line between the Lower and Upper Cretaceous is very difficult to recognize on the Atlantic Coastal Plain since deposition was probably continuous. For example, Jordan (1962) uses the term *Potomac Formation* for deposits of both Early and Late Cretaceous age in the state of Delaware.

Late Cretaceous. As Cretaceous time progressed, the sea encroached more and more on the land with the result that the Raritan Formation of New Jersey and Long Island contains alternating marine and freshwater beds. The Magothy Formation overlies the Raritan and sometimes the two units cannot be separated. Like the Raritan, the Magothy is partly marine and partly nonmarine, again suggesting alternating submergence and emergence. The Raritan Formation is probably equivalent to the Tuscaloosa Formation of the South Atlantic and Gulf Coastal Plains. Like the Raritan, this formation is largely nonmarine in outcrop, but contains marine elements in the subsurface. In North Carolina, the term *Middendorf Formation* is used for deposits at least partly equivalent to the Tuscaloosa (Heron and Wheeler, 1959, 1965).

The sea encroached more and more on the land as late Cretaceous time progressed, but probably did not extend across the Fall Line. In New Jersey and Delaware the post-Magothy deposits are divisible into two main units—the Matawan and Monmouth Groups—each of which is subdivided into several distinct formations (Table 1). The deposits suggest alternating shallow and relatively deep water with possibly one or more intervals of complete withdrawal, especially in Delaware. A recent summary of the stratigraphy of New Jersey and Delaware has been given by Owens *et al.* (1970).

The late Cretaceous deposits pinch out against the Fort Monroe High, near the Virginia capes, no deposits equivalent to the Matawan or Monmouth being known from the surface or subsurface of Virginia. In North Carolina the two distinct marine units are again recognized, the deposits being known as the *Black Creek Formation* and *Peedee Formation*. It has, however, not been possible to recognize the distinct units as in New Jersey and Delaware. The marine phases interfinger with nonmarine phases along part of the outcrop. In Georgia several units, probably equivalent to the Black Creek and Peedee, are recognized (see Table 1); in Florida, however, all deposits of Cretaceous age are deeply buried.

Paleocene-Eocene. There was probably a withdrawal of the sea at the end of Cretaceous time, for there is generally a disconformity below the Paleocene. In the subsurface, however, they may be continuous.

Along the Gulf Coastal Province (q.v.), where much more work has been done because of oil drilling, the Paleocene-Eocene has been divided into four main units as follows: 4, Jackson (Upper Eocene); 3, Clayton (Middle Eocene); 2, Wilcox (Lower Eocene); 1, Clayton (Paleocene). For convenience, these terms will be used in discussing the units of the Atlantic Coastal Plain.

The earliest Tertiary shorelines of New England probably lay E of those of the present day, for no marine deposits of this age are known N of New Jersey except on the continental shelf of New England and Nova Scotia. Much of the Coastal Plain of New Jersey was submerged during the Paleocene and early Eocene and the highly fossiliferous Hornerstown and Vincentown formations of that state date from these epochs. The exact time equivalents of these formations have been matters of speculation for many years. The Hornerstown, with its highly glauconitic sand is generally regarded as Paleocene (Clayton), whereas the Vincentown with its extensive fauna of foraminifera, bryozoa, echinoids, etc., is thought to be part Paleocene and part Lower Eocene (Clayton-Wilcox) in age.

TABLE 1. Correlation of Cretaceous and Tertiary Formations of Atlantic Coastal Plain

Age	Long Island	New Jersey	Delaware	Maryland	Virginia	North Carolina	South Carolina	Georgia
Pleistocene				Only briefly discussed in this paper				
Pliocene	Manetto	Beacon Hill	Bryn Mawr	Brandywine?	Sedley Kilby (?)	Croatan Waccamaw James City	Waccamaw	Charlton
Miocene	—	Cohansey Kirkwood { St. Marys Calvert	Cohansey Chesapeake Group	Cohansey North Keyes Chesapeake Group	Yorktown St. Marys Choptank Calvert	Duplin Yorktown St. Marys Calvert Trent	Duplin Raysor Hawthorn Tampa	Duplin Hawthorn Tampa
Oligocene	—	—	—	—	—	—	Flint River Cooper	Flint River Suwanee Cooper
Eocene	Shark River Manasquan	Piney Point Vincentown Hornerstown	Piney Point Rancocas	Piney Point Chickahominy Nanjemoy Aquia Brightseat	Piney Point Chickahominy Nanjemoy Aquia Mattaponi	Castle Hayne	Barnwell Castle Hayne McBean Santee Warley Hill Congaree Black Mingo	Barnwell Ocala McBean Wilcox
Paleocene						Beaufort		Clayton
Late Cretaceous Post-Raritan		Monmouth Group { Tinton Red Bank Navesink Mt. laurel ; Matawan Group { Wenonah Marshalltown Englishtown Woodbury Merchantville	{ Red Bank Mt. Laurel-Navesink ; { Wenonah — — Merchantville	Monmouth Matawan Magothy		Peedee (incl. Snow Black Creek Bladen Member Eutaw Middendorf	Peedee Black Creek Middendorf = Tuscaloosa	Providence Ripley Hill) Cusseta Blufftown Eutaw Tuscaloosa
Early Cretaceous Raritan	Raritan	Magothy Raritan Potomac	Magothy Potomac	Raritan Potomac Group { Patapsco Arundel Patuxent	Raritan Potomac Group { Patapsco Arundel Patuxent	Cape Fear	1?	1

Source: Richards (1967).

The Vincentown Formation is world famous for its fossils and has been correlated with the Landenian of Europe. A paleoecological study of some elements of its macrofauna has recently been made.

The exact equivalents of the Hornerstown and Vincentown formations S of New Jersey are somewhat uncertain because of major differences in their faunas which can probably be accounted for by different ecological conditions during deposition. The Brightseat Formation of Maryland, exposed on the outskirts of Washington, D. C., is regarded as basal Paleocene, equivalent to the Hornerstown of New Jersey or possibly slightly older. Elsewhere in Maryland as well as in Virginia, the Aquia Formation, also highly glauconitic, is either of Midway or Wilcox age, or possibly both. The type locality of the formation at Aquia Creek, near Fredericksburg, Virginia, near the Fall Line, with its large number of marine fossils, especially *Turritella mortoni,* suggests a fairly extensive submergence of the Virginia coastal plain during this time interval (Fig. 2a).

Only a relatively slight submergence of North Carolina is indicated for Midway-Wilcox time. Fossils of this age are known only from the subsurface in the Cape Hatteras well and in the Beaufort Formation in the subsurface of Pitt County (Brown, 1958, 1963).

South of the Cape Fear Arch, however, there was marine submergence during these epochs, various formations containing marine fossils having been laid down (see Table 1). The entire state of Florida was probably submerged during the Paleocene and Lower Eocene, but fossiliferous deposits of these ages are known only from deep wells, located especially in the southern part of the state.

A few fossils of Claiborne age found on Martha's Vineyard (in beds upthrust by glacial tectonics) indicate a transgression also in the Cape Cod region. In New Jersey, the Manasquan-Shark River formations and in Maryland-Virginia, the Nanjemoy Formation, indicate a partial submergence of the coastal plain, but probably not as extensive as that of the earlier episodes of the Paleocene-Eocene.

On the other hand, it is probable that the Claiborne seas extended far inland in North Carolina and actually crossed the Fall Line. Marine fossils of middle or late Eocene age have been found at isolated localities near Clayton, Lillington, and between Raleigh and Durham, in the Piedmont province (Richards, 1950).

No actual outcrops of late Eocene (Jackson) age are known N of North Carolina, although fossils of this age have been found in wells near the present shoreline of New Jersey, Delaware, Maryland, and Virginia.

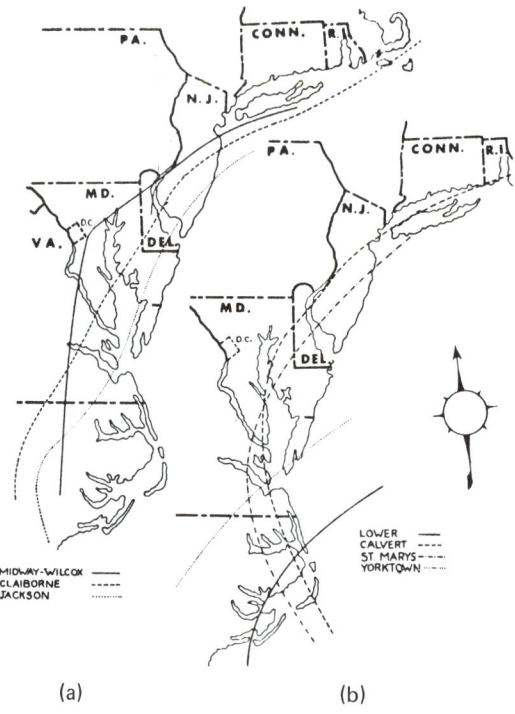

FIGURE 2. (a) Approximate position of Eocene shorelines between Cape Cod and Cape Hatteras (from Richards, 1968); (b) approximate position of Miocene shorelines between Cape Cod and Cape Hatteras (from Richards, 1968).

At least half of the coastal plain of North Carolina was submerged by seas of Jackson age as indicated by the highly fossiliferous deposits of the Castle Hayne Formation. There was probably a conspicuous promontory near present Cape Fear in southern North Carolina. This might have been the beginning of the structural feature known as the "Cape Fear Arch" mentioned earlier. In any case there was extensive marine submergence both N and S of this promontory during Jackson time.

Much of the surface of northern Florida is made up of the Ocala Limestone of Jackson age. This is one of the important aquifers of the region. Just below the Ocala Limestone in parts of Florida is another limestone, containing unusual fossils (Tethyan fauna) more related to those of southern Europe than to other Eocene localities in America. The exact age of these fossils is unknown; they may be of Jackson age, or late Claiborne (Richards and Palmer, 1954).

Oligocene. The shoreline of Oligocene age probably lay E of the present shore N of North Carolina as inferred by the lack of fossils of this age in outcrop or the subsurface. Microfossils of Oligocene age have been reported from wells in eastern North Carolina and outcrops in the "Trent Formation" near Silverdale and Belgrade are regarded as Oligocene by some workers (Lawrence, 1966) and as Lower Miocene by others (Richards, 1950). The formation may be of both ages.

Marine deposits of Oligocene age are known from outcrops in South Carolina, Georgia, and Florida (see table 1), but much of Oligocene time seems to have been regressive in many parts of the world.

Miocene. The Cape Fear Arch continued to exist into early Miocene time, only slight submergence, if any, having occurred on the coastal plain during this time interval. The Trent Formation of North Carolina and the Tampa Limestone of Florida were formerly dated from the early Miocene (Kellum, 1926; Richards, 1943, 1950). However, as stated above, more recent work by Lawrence and others has suggested that at least some of these deposits should be regarded as Oligocene.

As Miocene time progressed, there was increased submergence of the Atlantic Coastal Plain. During middle Miocene time the shoreline probably extended across New Jersey from Asbury Park to the vicinity of Salem and then across Delaware, Maryland, and Virginia to the vicinity of Petersburg, Virginia. Deposits made by this sea are called the *Kirkwood Formation* (in part) in New Jersey and the *Calvert Formation* farther S. Fossils are abundant, especially in the famous Calvert Cliffs on the western shore of Chesapeake Bay in Maryland (Fig. 2b).

The Calvert Formation disappears from the surface near the Virginia-North Carolina line, and it is probable that the depression of the land did not extend much farther S and that the shoreline turned abruptly toward the present coast. During Calvert time, the Cape Fear Arch probably extended far out into the sea.

The Choptank Formation was laid down at the close of Calvert time, and frequently the two formations are regarded as a single unit. In any case, the water was more shallow and probably a little cooler. There was probably a regression and erosion following the deposition of the Calvert-Choptank formations.

A similar early Miocene submergence occurred S of the Cape Fear Arch making up the Hawthorne Formation of South Carolina and Georgia and the Choctawhatchee Formation of Florida.

The next advance of the sea was marked by the deposition of the St. Mary's Formation. The maximum depression of the land was slightly farther S than it had been during Calvert time, and the St. Mary's shoreline probably crossed New Jersey from the vicinity of Atlantic City to Millville and then across Delaware, Maryland, and Virginia, but did not reach as far inland as had the Calvert Sea. Near Richmond, Virginia, the St. Mary's shoreline probably overlapped that of the Calvert, causing the maximum advance of the sea to be somewhere near Richmond. Not much is known about deposits of this age in North Carolina being probably buried by younger accumulations. The Cape Fear Arch was most likely still above water.

In late Miocene time there was certainly a major regression followed by another marine invasion in places accompanied by the sinking of the land. This sea deposited the Yorktown Formation and did not extend as far N as New Jersey, but crossed the present coast not far N of Yorktown, Virginia. It overlapped the Calvert and St. Mary's shorelines near Richmond, Virginia, and extended a few miles over the Piedmont rocks. By this time the Cape Fear Arch was submerged with the result that the late Miocene sea covered much of the coastal plain from Virginia to Florida. The Duplin Formation of North Carolina, with its extensive fauna of warm-water mollusks, is probably equivalent to the Yorktown Formation.

New Jersey, Delaware, and most of Maryland were above water during the late Miocene, for no marine fossils of that age are found in those regions. The Cohansey Sand of New Jersey and the North Creek Sand of Maryland (Hack, 1955) may have been deposited at the same time as the marine Yorktown and Duplin formations farther S. Large parts of Florida, especially in the panhandle, were submerged by the late Miocene sea.

Pliocene. The history of the Pliocene epoch along the Atlantic Coastal Plain is somewhat uncertain. Probably the sea withdrew at the end of the Miocene, but when it advanced again is not precisely known. No Pliocene marine deposits are found N of North Carolina, the only deposits possibly of that age having been described are clearly nonmarine (such as the Beacon Hill Gravels in New Jersey—and these may be early Pleistocene). Extensive marine deposits containing abundant fossils make up the Croatan and Waccamaw formations of the Carolinas and the Caloosahatchee Formation of Florida. Until very recently these have been regarded as Pliocene in age, mainly because of the large number of extinct species of mollusks. However, recent work of DuBar (1958, 1959) has suggested that these formations are of early Pleistocene age. Further work, including some absolute geochronological dating, will be necessary before the exact age of these formations can be determined.

Pleistocene. Although the Pleistocene glaciers did not reach the Atlantic Coastal Plain S of Long Island and central New Jersey, they did have indirect influence on the geological history of the region. Pleistocene shorelines were affected in three ways: (1) eustasy, changing sea level due to the advance and melting of the ice; (2) isostasy, the oscillations of the crust due to ice loading and unloading; and (3) tectonics, movements of the land irrespective of the ice. The first factor was world-wide; the second factor affected mainly the glaciated areas, but may have had secondary repercussions along the New Jersey coast and farther S (according to Walcott). The third factor, tectonic movement, probably has had some influence on the coastal plain, but much work remains to be done.

The older Pleistocene deposits from New Jersey to Virginia or North Carolina appear to be mainly nonmarine. In any case no marine fossils have been found, nor is there positive evidence that the relative level of Pleistocene seas were higher than about 10 m above present sea level in this sector. In South Carolina and Georgia, in contrast, there is paleontological evidence of an early Pleistocene shoreline at an elevation of about 20 m (Colquhoun et al., 1968). Physiographic evidence for higher Pleistocene shore lines is debatable.

For the late Pleistocene there is both physiographic and paleontological evidence of a shoreline at an elevation of about 8—10 m all the way from New Jersey to Florida (Fig. 3). It is believed that this shoreline dates from the last interglacial stage (Sangamon). In New Jersey the deposits laid down in this sea make up part of the Cape May Formation. Farther S the formation has various names, but the name used in most reports is the Pamlico Formation (Richards, 1962, 1970).

As stated under the discussion of the Pliocene, DuBar regards the Croatan and Waccamaw formations of the Carolinas and the Calloosahatchee Formation of Florida as early Pleistocene rather than Pliocene as generally supposed.

Paleontology

Space will not permit a complete discussion of the fossils found on the Atlantic Coastal Plain, nor a list of all the published works on the subject. Many of the publications already cited contain information on fossils and localities. In this section will be cited merely a few of the more extensive works on Cretaceous, Tertiary, and Quaternary fossils in the area.

Cretaceous fossils are numerous, but with a few exceptions, poorly preserved. In New Jersey, some 500 species of invertebrate fossils

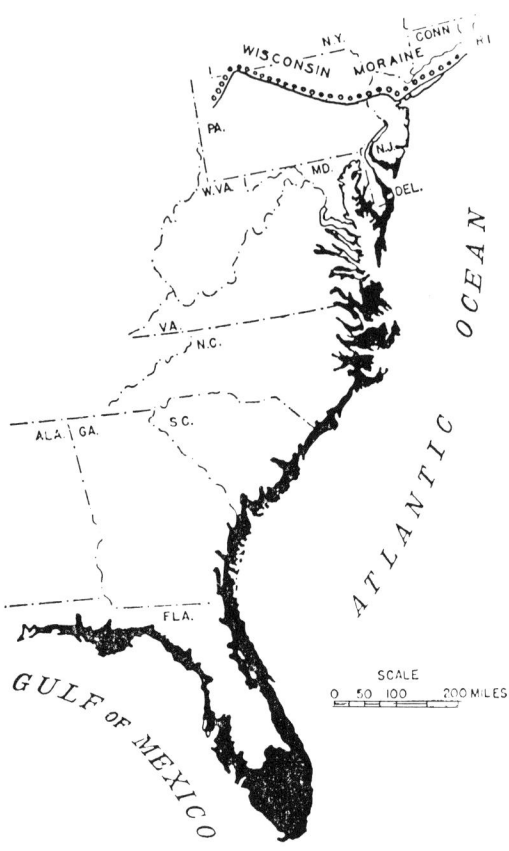

FIGURE 3. Sketch map showing position of Sangamon (Pamlico) shoreline along the Atlantic Coastal Plain. It corresponds to a world sea level ("Tyrrhenian") 7—8 m above that of today (Richards).

have been recorded from various localities, many of which are no longer accessible. These were completely described and figured by Weller (1907) and updated by Richards et al. (1958—1962). Among the vertebrate fossils of Cretaceous age (listed in Richards et al., 1962) should be mentioned *Hadrosaurus foulkii,* from Haddonfield, New Jersey, the first dinosaur to be found in America.

Many of the Cretaceous fossils from Delaware are included in the New Jersey reports; some 114 species from excavations for the Chesapeake and Delaware Canal and figured by Groot et al. (1954). Cretaceous invertebrates from Maryland and North Carolina are discussed by Clark et al. (1916) and Stephenson and Rathbun (1923), respectively.

At one time most of the Paleocene-Eocene formations of New Jersey were regarded as Cretaceous, and consequently their faunas are covered in the work of Weller (1907). The bryozoa of the Vincentown Formation of New Jersey are fully treated in a work by Canu and Bassler (1933). The richer Eocene faunas of Maryland, including many Virginia species, are described by Clark et al. (1901). The Eocene fossils of the southern Atlantic Coastal Plain have not been monographed but are discussed in a large number of shorter publications.

Miocene invertebrate fossils are generally better preserved than those of the Cretaceous and Paleocene-Eocene. Several monographs cover these faunas: New Jersey (Richards and Harbison, 1942), Maryland, Virginia and North Carolina (Gardner, 1943) and Florida (Gardner, 1926–1937).

The extensive Pliocene molluscan fauna of Florida has been discussed by Dall (1890-1905) and more recently by Olsson and Harbison (1953). The smaller number of species of the Waccamaw and Croatan formations have not been monographed, but are covered in a number of shorter publications.

The marine Pleistocene faunas are essentially the same as that of the modern seas in the region. For descriptions and figures, see Shattuck et al. (1906) for Maryland, and Richards (1962) for a general review.

Economic Geology

Although the Atlantic Coastal Plain is not rich in economic resources, it does contain certain deposits that play important parts in the economy of the region. Among the most obvious resources are the deposits of sand, gravel, and clay, the latter perhaps being most significant in New Jersey and Georgia. Limestone is extensively quarried in Florida and locally farther N. Phosphate is also mined in Florida; it occurs also to the N but is no longer worked. Glauconite was formerly extensively dug as a fertilizer or for use as a water softener, and is still obtained from one or two pits in New Jersey.

Thus far the only commercial petroleum from the coastal plain has come from several fields in Florida. However, considerable exploration has been carried out all along the coast offshore as far N as New Jersey.

Groundwater is an important resource of many coastal plain formations, and will become even more significant as the region increases in population. The U.S. Geological Survey and the various state surveys devote considerable attention to this problem.

HORACE G. RICHARDS

References

*Indicates references with extensive bibliographies.

Balsey, J. R., et al., 1946. "Magnetic maps of Worcester County and parts of Wicomico County, Maryland," *U.S. Geol. Surv. Oil and Gas Invt. Prelim. Map 46.*

Barksdale, H., et al., 1958. "Groundwater resources of the tri-state region adjacent to the lower Delaware River," *N.J. Div. Water Policy and Supply Spec. Rept. 13,* 190p.

Bass, M. N., 1969. "Petrography and ages of crystalline basement rocks of Florida–some extrapolations," *Am. Assoc. Petrol. Geologists,* **11,** 283–310.

Braunstein, J., 1958. "Habitat of oil in Eastern Gulf Coast," *Habitat of Oil,* Tulsa; *Am. Assoc. Petrol. Geologists,* 511–522.

Brown, P. M., 1958. "Well logs from the Coastal Plain of North Carolina," *N.C. Div. Mineral Res. Bull.,* **72,** 100p.

____, 1963. "The geology of northeastern North Carolina," *Atlantic Coastal Plain Geological Association Guidebook, 4th Field Conf.,* 44p.

Canu, F., and Bassler, R. S., 1933. "The Bryozoan fauna of the Vincentown Limesand," *U.S. Nat. Hist. Mus. Bull.,* **165,** 108p.

Cederstrom, D. J., 1945. "Geology and groundwater resources of the Coastal Plain of southeastern Virginia," *Va. Geol. Surv. Bull.,* **63,** 384p.

Clark, W. B., et al., 1901. *The Eocene Deposits of Maryland.* Md. Geol. Surv.

____, et al., 1916. *The Upper Cretaceous of Maryland.* Md. Geol. Surv.

Colquhoun, D. J., 1971. "Glacio-eustatic sea level fluctuation of the Middle and Lower Coastal Plain, South Carolina," *Quaternaria,* **15,** 19–34.

____, Herrick, S. M., and Richards, H. G., 1968. "A fossil assemblage from the Wicomico Formation in Berkeley County, South Carolina," *Bull. Geol. Soc. Am.,* **79,** 1211–1220.

Cooke, C. W., 1971. "American emerged shoreline compared to levels of Australian marine terraces," *Bull. Geol. Soc. Am.,* **82,** 3231–3234 (with discussion by Ward, W. T., Ross, P. J., and Colquhoun, D. J., 1973. *Ibid.,* **84,** 1835–1838).

Cramer, F. H., 1971. "Position of the North Florida Lower Paleozoic block in Silurian time: phytoplankton evidence," *J. Geophys. Res.,* **76** in (20). 4754.

Dall, W. H., 1890–1905. "Contributions to the Tertiary fauna of Florida," *Trans. Wagner Free Inst. Sci.,* **3.**

Dodd, J. R., and Siemers, C. T., 1971. "Effect of late Pleistocene karst topography on Holocene sedimentation and biota, lower Florida Keys," *Bull. Geol. Soc. Am.,* **82,** 211–218.

DuBar, J., 1958. "Stratigraphy and paleontology of the late Neogene strata of the Caloosahatchee River area of southern Florida," *Fla. Geol. Surv. Bull.,* **40.**

____, 1959. "The Waccamaw and Croatan formations of the Carolinas," *S.C. Develop. Board, Geol. Notes,* **3,** 1–9.

____, 1972. *Neogene Stratigraphy of the Lower*

Coastal Plain of the Carolinas (F.G.-2). S.C. Develop. Board, 128p.

Emery, K. O., and Milliman, J. D., 1970. "Quarternary sediments of the Atlantic continental shelf of the United States," *Quarternaria*, 12, 3–18.

Foord, E. E., Parrott, W. R., and Ritter, D. F., 1970. "Definition of possible stratigraphic units in north central Long Island, New York, based on detailed examination of selected well cores," *J. Sed. Petrology*, 40(1), 194–204.

Gardner, J. 1926–1937. "The molluscan fauna of the Alum Bluff Group of Florida," *U.S. Geol. Surv. Prof. Pap.* 142-A-F.

____, 1943. "Mollusca from the Miocene and Lower Pliocene of Virginia and North Carolina," *U.S. Geol. Surv. Prof. Pap.*, 199.

Gibson, T. G., 1970. "Late Mesozoic-Cenozoic tectonic aspects of the Atlantic coastal margin," *Bull. Geol. Soc. Am.*, 81, 1813–1822.

Gilluly, J., 1969. "Geological perspectives and the completeness of the geological record," *Bull. Geol. Soc. Am.*, 80, 2303–2312.

Groot, J. J., Organist, D., and Richards, H. G., 1954. "Marine Upper Cretaceous formations of the Chesapeake and Delaware Canal," *Del. Geol. Surv. Bull.*, 3.

Hack, J. T., 1955. "Geology of the Brandywine area and origin of the upland of southern Maryland," *U.S. Geol. Surv. Prof. Pap.*, 267-A, 21–43.

Heezen, B. C., and Sheridan, R. E., 1966. "Lower Cretaceous rocks (Neocomian-Albian) from the submarine Blake Escarpment, western North Atlantic, *Science*, 154, 1644–1647.

Heron, S. D., and Wheeler, W. H., 1959. *Guidebook for Coastal Plain Field Trip featuring Basal Cretaceous Sediments of the Fayetteville area, North Carolina.* New York: Geol. Soc. Am., SE Sect. Coastal Plain Field Trip, 44p.

____, and Wheeler, W. H., 1965. "The Cretaceous formations the Cape Fear River," *Atlantic Coastal Plain Geol. Assn., Guidebook, 5th Field Conf.*, 55p.

Hoyt, J. H., and Henry, V. J., Jr., 1971. "Origin of capes and shoals along the southeastern coast of the United States," *Bull. Geol. Soc. Am.*, 82, 59–66.

Husted, J. E., 1972. "Shaler's Line and Suwannee Strait, Florida and Georgia," *Bull. Am. Assoc. Petrol. Geologists*, 56, 1557–1560.

Johnson, M. E., and Richards, H. G., 1952. "Stratigraphy of Coastal Plain of New Jersey," *Bull. Am. Assoc. Petrol. Geologists*, 36, 2150–2160.

Johnson, W. R., and Straley, H. W., 1953. "Geomagnetics of North Carolina Coastal Plain," *Ga. Geol. Surv. Bull.*, 60, 132–135.

Jordan, R. R., 1962. "Stratigraphy of the sedimentary rocks of Delaware," *Del. Geol. Surv. Bull.*, 9, 51p.

Kellum, L. B., 1926. "Paleontology and stratigraphy of the Trent Marl in North Carolina," *U.S. Geol. Surv. Prof. Pap.* 143.

Lawrence, D., 1966. *Paleoecology of an Oligocene Oyster Deposit at Belgrade, North Carolina.* New York: Geol. Soc. Am., SE Sec. Program 32, Abstr.

*LeGrand, H. E., 1961. "Summary of geology of Atlantic Coastal Plain," *Bull. Am. Assoc. Petrol. Geologists*, 45, 1557–1571.

Marine, I. W., and Siple, G. E., 1974. "Buried Triassic Basin in the central Savannah River area, South Carolina and Georgia," *Bull. Geol. Soc. Am.*, 85, 311–320.

Mattick, R. E., et al., 1974. "Structural framework of the United States Atlantic Outer Continental Shelf north of Cape Hatteras," *Bull. Am. Assoc. Petrol. Geologists*, 58(6), Part II, 1179–1190.

McCarthy, G. R., and Straley, H. W., 1937. "Magnetic anomalies near Wilmington, North Carolina," *Science*, 85, 362–364.

*Murray, G. E., 1961. *Geology of the Atlantic and Gulf Coastal Province of North America.* New York: Harper & Row, 692p.

Oaks, R. Q., Jr., and DuBar, J., eds. 1974. *Post-Miocene Stratigraphy Central and Southern Atlantic coastal plain.* Logan, Utah: Utah State Univ. Press.

Olsson, A., and Harbison, A., 1953. "Pliocene mollusca of southern Florida," *Acad. Nat. Sci. Phila. Monogr.* 8.

Owens, J., et al., 1970. "Stratigraphy of the outcropping post-Magothy formations in southern New Jersey and northern Delmarva Peninsula, Delaware and Maryland," *U.S. Geol. Surv. Prof. Pap.* 674.

Richards, H. G., 1943. "Additions to the fauna of the Trent Marl of North Carolina," *J. Paleontol.*, 17, 518–526.

____, 1948. "Studies on the subsurface geology of the Atlantic Coastal Plain," *Proc. Acad. Nat. Sci. Phila.*, 100, 39–76.

____, 1949. "The occurrence of Triassic rocks in the subsurface of the Atlantic Coastal Plain," *Proc. Penn. Acad. Sci.*, 23, 45–48.

____, 1950. "Geology of the coastal plain of North Carolina," *Trans. Am. Phil. Soc.*, 40(1), 1–83.

____, 1962. "Studies on the marine Pleistocene," *Trans. Am. Phil. Soc.*, 52(3), 1–141.

____, ed., 1965. "Central Atlantic Coastal Plain," *INQUA Guidebook B-1.* Lincoln: Neb. Acad. Sci., 28p.

____, 1967. "Stratigraphy of Atlantic Coastal Plain between Long Island and Georgia: review," *Bull. Am. Assoc. Petrol. Geologists*, 51, 2400–2429.

____, 1968. "The Tertiary history of the Atlantic coast between Cape Cod and Cape Hatteras," *Palaeogeogr., Palaeoclimatol., Palaeoecol.*, 5, 95–104.

____, 1970. "Changes in shoreline during the past million years," *Proc. Am. Phil. Soc.*, 114, 198–204.

____, and Harbison, A., 1942. "Miocene invertebrate fauna of New Jersey," *Proc. Acad. Nat. Sci. Phila.*, 94, 167–250.

____, and Judson, S., 1965. "The Atlantic Coastal Plain and the Appalachian Highlands in the Quaternary," in H. E. Wright, Jr., and D. G. Frey, eds., *The Quaternary of the United States.* Princeton: Princeton Univ. Press, 129–136.

____, and Palmer, K. V. W., 1954. "Old World affinities of some Eocene mollusks from Florida," *19th Intern. Geol. Congr.*, Algiers, 19, 35–38.

____, and Straley, H. W., 1953. "Geophysical and stratigraphic investigations in the Atlantic Coastal Plain," *Ga. Geol. Surv. Bull.*, 60, 101–115.

____, et al., 1958–1962. "The Cretaceous fossils of New Jersey," *N.J. Geol. Surv. Bull.*, 61, 2 vols.

Shapiro, E., and Richards, H. G., 1969. "Lower Cretaceous (?) mollusks from a deep well near Cape May,

New Jersey," *Acad. Nat. Sci. Phila., Notulae Naturae*, **418**, 9p.

Shattuck, G. B., et al., 1906. *The Pliocene and Pleistocene of Maryland*. Md. Geol. Surv.

Shideler, G. L., Swift, D. J. P., Johnson, G. H., and Holliday, B. W., 1972. "Late Quaternary stratigraphy of the inner Virginia Continental shelf: a proposed standard section," *Bull. Geol. Soc. Am.*, **83**, 1781–1804.

Stephenson, L. B., and Rathbun, M. J., 1923. "Cretaceous formations of North Carolina," *N.C. Geol. and Econ. Surv.*, **5**.

Swain, F., 1962. "Ostracoda from wells in North Carolina: 2. Mesozoic Ostracoda," *U.S. Geol. Surv. Prof. Pap. 234-B*, 59–93.

Weller, S., 1907. A report on the Cretaceous paleontology of New Jersey," *N.J. Geol. Surv. Paleontol. Ser.*, **4**.

Woollard, G. P., 1941. "Geophysical methods of exploration and their application to geological problems in New Jersey," *N.J. Geol. Surv. Bull.*, **54**, 89p.

Cross-references: *Canada–Atlantic Provinces; North America; United States–Appalachian Region, Gulf Coastal Province, New England Region.*

UNITED STATES–BASIN AND RANGE PROVINCE

The term "Basin and Range Province" carries two connotations, one pertaining to regional physiography, the other to structural geology. The geographic boundaries of the province are defined on the basis of physiography, but "Basin and Range structure," although a dominant characteristic of the province, is not confined to the province.

Basin and Range Physiographic Province

Definition and Areal Extent. The Basin and Range physiographic province refers to the arid to semiarid region in the western United States characterized by a landscape of quasi-parallel elongate ranges alternating with valleys floored by detritus. The province includes most of Nevada, and parts of Utah, Arizona, California, Oregon, Idaho, New Mexico, and Texas (see Fig. 1). It is bounded on the E by the Wasatch Mountains, the Colorado Plateau, and the western edge of the Great Plains of Texas and New Mexico; [and] on the W by the Cascade, Sierra Nevada, and Tehachapi mountains and the Transverse and Coast ranges. The Northern boundary is less distinct; Fenneman (1946) considers the divide N of which drainage is to the Snake River as the northern limit in Idaho and southeastern Oregon, but through central Oregon he somewhat arbitrarily selects a line near the 43d parallel. The province extends S to about 25.5°N latitude in Mexico (q.v.), and includes parts of Sonora, Chihuahua, Coahuila, Durango, and Baja California Norte (Raisz, 1959).

Related Terms. *Basin Range province* was used by Powell, King, and Gilbert (see references in Nolan, 1943); Fenneman added "and" to avoid the implication that "ranges" predominate in the region. J. H. Mackin (1960) suggests the use of *basin-range* as an adjective, e.g., basin-range topography. The *Great Basin* (see Fig. 1), named by Capt. J. C. Fremont and now incorporated as a subdivision of the Basin and Range province, is characterized by interior ("endorheic") drainage. Other subdivisions of the province recognized by Fenneman (see Fig. 1) include the *Sonoran Desert, Salton Trough, Mexican Highland*, and *Sacramento Sections*.

Physiographic Features. Isolated, elongate, subparallel mountain ranges separate broad, nearly flat-floored valleys or basins, many of which have no external ("exorheic") drainage. Ranges are commonly 60–140 km long, are from 5 to 25 km wide at their bases, and are spaced at about 25–40 km from crest to crest. Many ranges have a well-dissected, mature topography of V-shaped valleys and narrow divides, and some are notably asymmetrical,

FIGURE 1. Extent of Basin and Range physiographic province and its subdivisions. (After Fenneman, 1946, and Raisz, 1959.)

having one steep escarpment and a gentle backslope, particularly in Nevada and Utah. In southwestern Arizona, basin deposits have encroached upon the ranges and erosion (mainly by pedimentation) has reduced many of them to small spiny remnants of once larger and more extensive mountain masses.

Altitudes range from 85 m (280 ft) below sea level in Death Valley to 4,356 m (14,246 ft) above sea level at White Mountain Peak near the California-Nevada boundary. Altitudes of valleys and basins rise generally from the SW of the province to the N and NE; for example, Mohawk Valley in southwestern Arizona is 150 m, and Ruby Valley, in eastern Nevada, is 1700 m above sea level.

In the northern half of the province, fault scarps in different stages of dissection commonly lie along one or more flanks of the ranges. Scarps may be relatively undissected linear escarpments 1–7 m high, cutting Holocene alluvial fans at the base of the ranges, or imposing mountain fronts rising 1000 m or more above the valley floor. Other features indicative of faulting include truncation of internal structure of the mountain at the range front, linear or broadly arcuate plan of range front, hour-glass valleys, and faceted spurs. In the S of the province, range-bounding faults are not reflected so noticeably in the physiography.

Alluvial fans spread basinward from the ranges, in many places coalescing to form a bajada (see Vol. III, *Encyclopedia of Geomorphology*), but in places what appears to be an alluvial fan is a veneer of gravel from zero to 100 m or more thick covering a pediment cut in rock or older basin-filling deposits.

Shallow playa lakes intermittently pond in the undrained basins during rare periods of heavy rain. Flat pavement-like surfaces of clay and silt commonly form after the lakes evaporate, and in the lowest part of the playa a salt-encrusted salina may develop. In some basins deflation by wind action removes the alluvial deposits faster than they are deposited. Yardangs are preserved in places. Wind has spread a thin veneer of loess over extensive areas, and in some places dunes occur on the lee side of the basins and migrate upslope. A relatively flat surface armored by pebbles, a desert pavement ("reg"), is commonly produced by wind erosion in such areas.

Rivers and Lakes. The Humboldt River flows W across much of the Great Basin, rising in the Ruby Mountains and flowing W to the Humboldt and Carson Sinks, where the water sinks in or evaporates. From the Wasatch and other ranges to the E the Bear and Weber rivers drain into Great Salt Lake, Provo River drains into Utah Lake, and the Sevier River drains into Sevier Lake.

From the Sierra Nevada on the W, the Truckee, Carson, and Walker rivers flow E into Pyramid Lake, Carson Sink, and Walker Lake, respectively. Pyramid and Walker lakes are remnants of Pleistocene *Lake Lahontan,* which covered large areas of western Nevada in Wisconsin time. In the eastern Great Basin, Great Salt Lake, and Utah Lakes are corresponding remnants of Pleistocene *Lake Bonneville* (Eardley et al., 1973). Both Lake Lahontan and Lake Bonneville had many undulations of lake levels, very likely synchronous to a degree, and each reached two major high-level stands; Lake Lahontan at about 1330 m (4380 ft) above sea level and Lake Bonneville at about 1560 m (5150 ft) above sea level. Former shorelines of these and other smaller Pleistocene lakes are still conspicuous in the Great Basin.

The southern part of the province is drained by major rivers that rise in the Rocky Mountains or along the rim of the Colorado Plateaus. Among these are the Colorado River, Gila River, and the Rio Grande. Major tributaries of the Gila, including the Salt, San Pedro, and Santa Cruz rivers, drain parts of Arizona and New Mexico.

Climate. The province has an arid to semiarid climate. Precipitation in the basin areas is generally less than 250 mm/year (e.g., Yuma, Arizona, 78 mm) although in many ranges precipitation is as much as 350 mm/year. As the province spans more than $10°$ of latitude, temperatures in the N and S differ. At Phoenix, Arizona, the midsummer daily maxima usually exceed $38°C$ ($100°F$) whereas in northern Nevada temperatures in excess of $32°C$ ($90°F$) are uncommon. Temperatures also vary greatly with altitude; it is not uncommon for patches of snow to linger above altitudes of 3000 m on the Ruby Range throughout the summer, and there is an active rock glacier in the Snake Range.

Deserts. The province is the locale of North America's major deserts. The Sonoran Desert section of the province includes the Mojave Desert of California and the Desert Region of southwestern Arizona. Other desert areas include the Salton Trough and Death, Panamint, Searles, and Saline valleys of California; the Black Rock, Smoke Creek, and Amargosa deserts and the Humboldt and Carson sinks of Nevada; and the Great Salt Lake and Sevier deserts of Utah.

Climatic Influence on Land Forms. The land forms of the province are the product of an interaction of arid and semiarid climate, rock type, and geologic structures. Weathering is characterized more by mechanical changes resulting from great diurnal fluctuations in temperature than by chemical changes that accompany water-saturated, vegetation-covered conditions. Erosion processes are dominated by the intermittent, catastrophic erosion by flash floods, which transport great quantities of rock waste in a few minutes or hours, followed by intervals of years during which erosion is negligible. Mud flows are common and ephemeral stream beds are commonly braided or dissected by gullys ("arroyos").

Basin and Range Structure

Definition and Areal Extent. *Basin and Range structure* refers to late Cenozoic structures that control in large measure the topography of the Basin and Range physiographic province. The term *fault block,* introduced by W. M. Davis (see Gilbert, 1928, p. 1), and particularly the related term *fault-block structure,* in reference to range blocks dominated by great normal faults of late Cenozoic age, are commonly equated with the term *Basin and Range structure.* The normal faults can be referred to as *Basin and Range faults.* However, other types of late Cenozoic deformation, such as folding, strike-slip faulting and thrust faulting, are also recognized and locally are dominant.

Large late Cenozoic normal faults in the western United States, although concentrated in the Basin and Range physiographic province, are not restricted to that province, but extend into the Rocky Mountains, the Colorado and Columbia plateaus, and the ranges of California (see Fig. 2). The Sierra Nevada and Wasatch mountains, although not included in the province, are also in fact great tilted blocks, bounded on one flank by faults, different only in size from many smaller ranges within the province. Thus it is not possible to define a "structural province" having clearly prescribed boundaries.

Geologic History. A complex series of structural events differing from place to place preceded the development of Basin and Range structures. These earlier structures exerted little control on the later block faulting.

Pre-Cenozoic. Gneiss, schist, and granite of early Precambrian age are overlain unconformably by extensive deposits of Precambrian and Lower Cambrian quartzose sediments. At least three widespread episodes of metamorphism and deformation are recorded in the Precambrian rocks. The oldest, called the *Mazatzal*

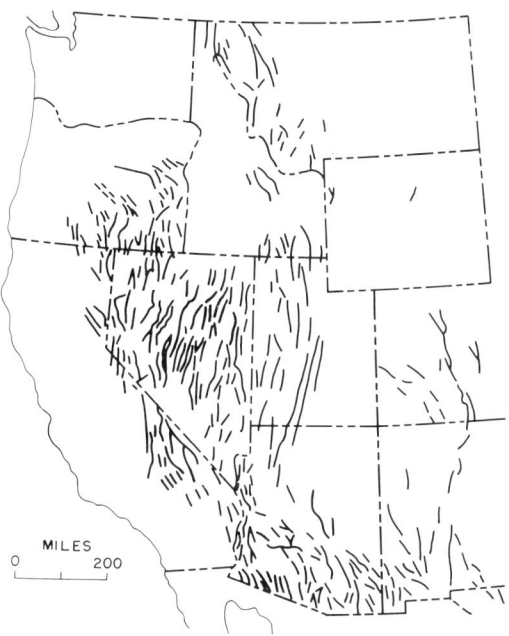

FIGURE 2. Distribution of late Cenozoic Basin and Range faults in the western United States (after Gilluly, 1963).

Revolution, occurred about 1.7 billion years ago, a second unnamed episode occurred about 1.4 billion years ago, and a third, called the *Grand Canyon disturbance,* took place about 0.8 billion years ago.

During Paleozoic time the *Cordilleran geosyncline* accumulated shale, chert, and quartzite in western Nevada and carbonate rocks in western Utah and eastern Nevada. The Wasatch line trending NE through central Utah marks the boundary between geosynclinal and cratonal deposition typical of most of the province in Arizona, New Mexico, and Texas. The *Antler orogeny* of middle Paleozoic (Upper Devonian–Lower Mississippian) time, characterized by the Roberts Mountains thrust, and the *Sonoma orogeny* and the *Golconda thrust* of late Paleozoic (Upper Permian) time represent pulses of great eastward tectonic transport and thrusting in Nevada.

During early Mesozoic time seas again transgressed large parts of the province. In Middle Jurassic and Early Cretaceous time the western part of the Great Basin section of the province was intensely deformed and the major episode of batholithic intrusion since Precambrian time took place. Little sedimentary record represents Cretaceous time, except in the S of the province. Some volcanic rocks were extruded and additional intrusive rocks were emplaced in Cretaceous time.

In the last part of the Mesozoic and the early part of the Cenozoic, during *Laramide orogeny*, thrust faults developed in a belt that extends from southern Nevada through western Utah. Here are found the Willard, Nebo, Charleston, Muddy Mountain, Keystone, and other thrusts along which plates moved E, in some places tens of miles. To some degree in one part or another of the province, orogeny may have been almost continuous from middle Paleozoic through early Tertiary time.

Cenozoic. Following Laramide orogeny early Tertiary time is poorly represented except by local terrestrial deposits and eruptive volcanic rocks. A great surge of andesitic volcanism in middle Tertiary (latest Eocene) time seems to indicate a fundamental change in the upper mantle beneath the province. This widespread volcanic activity persisted for the following 20 million years, from late Eocene to early Miocene, ending abruptly about 18 million years ago. A second, but petrochemically different basaltic volcanism, started about 17 million years ago over large parts of the region and heralded the beginning of the most recent (taphrogenic) stage of structural development, namely, the fragmentation of the crust into blocks that were to become Basin and Range block mountains and basins. In most parts of the province the present major ranges and basins had taken shape by early Pliocene time. Normal faults also developed during Oligocene and Miocene time, but these earlier faults commonly have trends different from those of Pliocene and later faults.

Structural Features. Blocks bounded by faults have been raised, dropped down, or tilted; the relatively high parts have been eroded and sculptured to form the ranges, and eroded debris has been deposited on adjacent downdropped blocks or on the lower flanks of tilted blocks to form relatively flat-floored basins of the province.

Examples of relatively raised blocks or horsts are the Humboldt, Ruby, and Toiyabe ranges. Tilted blocks are exemplified by the West Humboldt Range, northern Shoshone Mountains, and Cortez Mountains. Moore (1960, p. B409) (see Fig. 3) suggests that many of the tilted blocks are bounded by fault surfaces that are crudely spoon shaped and concave toward the downdropped block. These blocks are commonly tilted toward regional topographic highs. The geometry of such block faulting requires expansion of the upper surface or crust of the earth, and estimates suggest crustal extension on the order of 100 km for

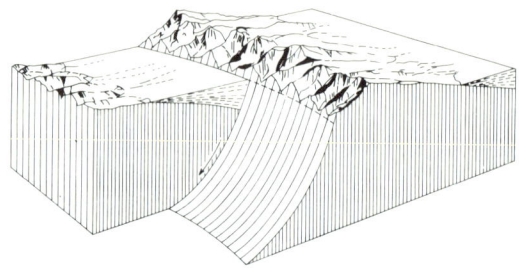

FIGURE 3. Block diagram showing typical Basin and Range fault block structure (after Moore, 1960).

the entire Great Basin (Thompson and Burke, 1974).

In parts of Nevada, California, and Utah, block faulting still continues; for example, blocks have moved in historic time with accompanying earthquakes and surface rupture in Owens Valley, California (1872) and in Pleasant Valley (1915), Dixie Valley (1954), and Gabbs Valley, Nevada (1932). Scarps that formed along blocks which moved during the recent past and have been eroded by different amounts occur along many range fronts. In Dixie Valley the spreading rate during the late Quaternary has been approximately 1 mm/yr (Thompson and Burke, 1973).

Gravity studies indicate that many basins are filled to depths of a 1000 m or more by poorly consolidated sediments, probably mostly of Cenozoic age. The bedrock floor under these poorly consolidated sediments is below sea level in some basins, in contrast to the summit altitudes of 3000 m or more above sea level of some ranges. Altitudes for some basin floors and for the bedrock surface below the basin floors follow: Death Valley, California (−85 m −3000 m); Long Valley, California (+2000m, −1500 m); Buena Vista Valley, Nevada (+1200 m, −1200m); Maricopa area between Sacaton and Palo Verde Mountains, Arizona (360 m, −1100 m).

In addition to ranges and basins controlled by relative movement of fault-bounded blocks, some topographic relief has also resulted from gentle folding in late geologic time. For example, the hills around the SW of Big Smoky Valley in Esmeralda County, Nevada, and the Hot Springs Range and adjacent Eden Valley in Humboldt County, Nevada, owe their general configuration to folding as much as to faulting.

Over much of the province, adjacent fault blocks appear to have moved primarily vertically with respect to each other; but lateral movement, including strike-slip faulting and oroclinal warping, is especially pronounced in the southwestern Great Basin section and in the Mojave Desert, and may be more prominent elsewhere within the province than has yet been recognized. The San Andreas fault bounds the Mojave Desert on the SW and is known to have had right-lateral strike slip of many kilometers, possibly more than 500 km. The Garlock fault bounds the Mojave Desert on the N and NW and is believed to have had at least 60 km of left-lateral strike slip. In the *Las Vegas shear zone,* and the related *Walker Lane* of southwestern Nevada, there are major lineaments with right-lateral strain. Both display right-lateral strike-slip faulting associated with right-lateral oroclinal bending. Relatively few strike-slip faults are known in the S of the province, although lineaments that trend NW-SE across Arizona are suggestive of a tectonic style similar to that in the Walker Lane. The latter may pass into the important *Texas Lineament,* which, like several others, has great importance in the economic field (Wertz, 1970; Roberts, 1966).

Geophysical investigations show that the Basin and Range province has certain characteristics that sharply distinguish it from the adjoining provinces. It is a region of anomalously high heat flow (Sass *et al.,* 1964) and of high seismicity. Earthquake focal mechanism studies clearly define a consistent E-W direction of Basin and Range extension along NW-SE-trending normal faults (Scholz *et al.,* 1971). The crust beneath the province is about 29 km, which is relatively thin (Tatel and Tuve, 1955; Pakiser, 1963; Prodehl, 1970) and it is seismically layered. The lower "mafic" layer of higher velocity (6.6–7.0 km/sec) is separated by a discontinuity or narrow transition zone from the upper "silicic" layer of lower velocity ($\sim$6.0 km/sec). Typical continental crust exhibited by the Colorado Pleateau to the E is about 40 km thick and has a broad gradiation zone between its lower and upper parts. In addition to the thin and layered crust, the upper part of the mantle beneath the Basin and Range has an anomalously low Pn velocity of about 7.8 km/sec (normal upper mantle velocities are about 8.2 km/sec). Considering the seismic data on the crust and upper mantle and gravity measurements, Thompson and Talwani (1964) calculate that the anomalous upper mantle is at least 20 km thick. This low-density, low-velocity upper mantle must supply isostatic support for the elevated Basin and Range region.

Origin of Basin and Range Structure. That these structures represent dominantly vertical displacements of comparatively rigid blocks has been a conclusion accepted by most students of the province since its proposal by Gilbert and his colleagues in 1874, but the absolute sense of this vertical movement and the cause of the movements are still unresolved.

Regional extension is inferred from the preponderance of normal faults and from an interpretation of gravity and seismic data (Thompson, 1959). Other geologic and geophysical evidence, such as basaltic volcanism associated with rift faulting, high regional heat flow, thin crust, and low velocity upper mantle, emphasize the similarities between the Basin and Range province and other regions of crustal expansion (e.g., East African Rift, Rhine graben of Europe, Lake Baikal depression in USSR). Some geologists, however, argue that the extension is a component of regional compressional or rotational strain. Strike-slip faulting is indeed characteristic of parts of the province, and many range-front faults display oblique slip rather than simple dip slip.

Among many ideas that have been proposed to explain the origin of the structures are those of Le Conte (1889, p. 260–262) who postulated that the region between the Wasatch and Sierra Nevada was arched by intumescent lava, and that the arch then collapsed (see Fig. 5a–c). He likened the resulting arrangement of the collapsed blocks to rhomboidal blocks of wood floating on water (see Fig. 4a). Mackin (1960, p. 127) suggested that block movements "repre-

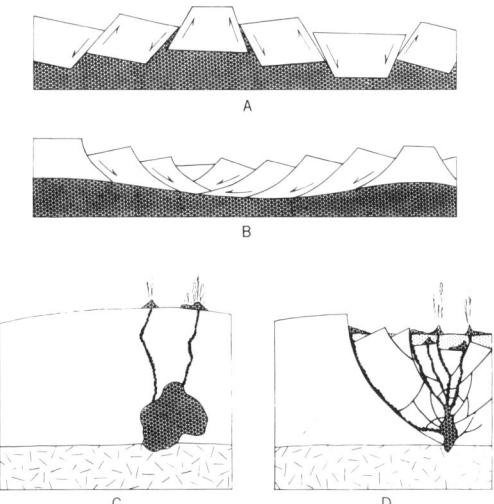

FIGURE 4. Diagrams showing some interpretations of origin of Basin and Range fault block structure: A, analogy of floating blocks; B, gravity sliding; C and D, collapse accompanying extrusion of volcanic material.

FIGURE 5. Diagrammatic representation of one concept of history of Basin and Range province. A, arching of crust over hot expanding zone (stippled)—the locus of magma generation; B, gravity sliding of sheets away from arch and thrusting accompanying lateral expansion; C, collapse of crust accompanied by frothing eruptions and development of block faulting.

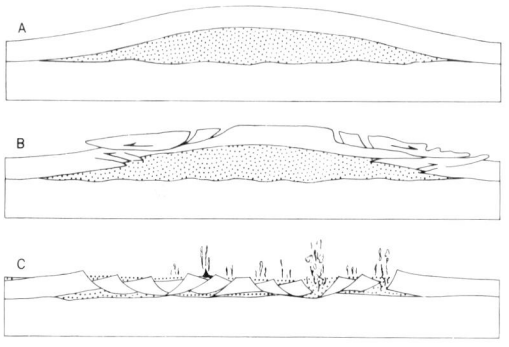

sent failures of the crust due to withdrawal of lateral support caused by frothing eruption of many thousands of cubic miles of magma formed by fusion of crustal rocks" (see Fig. 4c and d). Moore (1960) has suggested that gravity sliding has played an important part, and that range blocks have slid away from regional topographic highs, tilting, in landslide-block fashion, back toward the high away from which they moved (see Fig. 4b).

The concept of arching followed by collapse is probably most applicable to the Great Basin section. A great mass of terrigenous clastic rocks, including the Wasatch and Green River formations of early Tertiary age in central and eastern Utah and Wyoming, attests to a highland or arch to the W in the area that is now the Basin and Range province. Low-angle faults which in part may reflect E gravity sliding in late Cretaceous and early Tertiary time off this highland or arch include the Snake Range, Nebo, Charleston, and Willard thrusts. Clearly there has been great differential vertical movement approximately along the line of the Wasatch Range, which changed the land form from a high E-facing escarpment in early Tertiary time to the existing high W-facing Wasatch Range escarpment.

A regional structural arch characterizes the Mexican Highland section of central Arizona, for the top of the Precambrian in this section is in general 1000 m or more higher than in the Colorado Plateaus province or in southwestern Arizona. Perhaps here collapse of the arch is at a less advanced evolutionary stage than in the Great Basin section, although in general Quaternary block faulting is less evident than in the Great Basin section.

Recognition and distinction of first-, second-, third-, and higher-order structures present problems. For example, blocks on a scale of 100 m to a few kilometers are known to be deformed by landslide mechanisms, and Longwell (1945, Fig. 12) illustrates the collapse of an anticline, 35 km from flank to flank, on low-angle normal faults. However, the fundamental mechanism that produced the characteristics of the province still remains debatable. Various tectonic models have been proposed to explain the unusual volcanic and tectonic features of the province, especially the Great Basin part of it. These include speculation that the East Pacific Rise extends northward beneath the province (Menard, 1964; Wilson, 1970; McKee, 1971) or that a subduction zone with a Cenozoic oceanic plate was subducted beneath the region (Atwater, 1970; McKee, 1971; Christiansen and Lipman, 1972; Lipman et al., 1972; Noble, 1972), or that a moving mantle diapir or hot spot existed beneath it during the latter part of the Cenozoic (Cook, 1962, 1968; Armstrong et al., 1969; Scholz et al., 1971). These models are subject to intensive testing.

ROBERT E. WALLACE*

*Publication authorized by the Director, U.S. Geological Survey.

References

Anderson, R. E., 1971. "Thin skin distension in Tertiary rocks of southeastern Nevada," *Bull. Geol. Soc. Am.*, **82**, 43–58.

____, 1973. Large magnitude Late Tertiary strike-slip faulting north of Lake Mead, Nev. *U.S. Geol. Surv. Prof. Pap. 794*, 18p.

____, et al., 1972. "Significance of K-Ar ages of Tertiary rocks from the Lake Mead Region, Nevada–Arizona," *Bull. Geol. Soc. Am.*, **83**, 273–288.

Armstrong, R. L., et al., 1969. "Space-time relations of Cenozoic silicic volcanism in the Great Basin of the western United States," *Am. J. Science*, **207**, 478–490.

Atwater, Tanya, 1970. "Implications of plate tectonics for the Cenozoic tectonic evolution of western North America," *Bull. Geol. Soc. Am.*, **81**, 3513–3536.

Burchfiel, B. C., Pelton, P. J., and Sutter, J., 1970. "An early Mesozoic deformation belt in south-central Nevada–southeastern California," *Bull. Geol. Soc. Am.*, **81**, 211–215.

Christiansen, R. L., and Lipman, P. W., 1972. Cenozoic volcanism and plate-tectonic evolution of the western United States. 2. Late Cenozoic," *Phil. Trans. Roy. Soc. Lond.*, Ser. A, **271**, 249–284.

Cook, K. L., 1962. "The problem of the mantle-crust mix-lateral inhomogeneity in the uppermost part of the Earth's mantle," in H. E. Landsberg and J. Van

Mieghem, eds., *Advances in Geophysics*, Vol. 9. New York: Academic Press, 295–360.

―――, 1966. "Rift system in the Basin and Range Province," *The World Rift System, Geol. Surv. Can. Pap. 66-14,* 246–277.

―――, 1968. "Evidence for the Eastern Pacific Rise and mantle convection currents under western North America," *Geol. Soc. Am. Spec. Pap. 101,* p. 43.

Corbitt, L. L., and Woodward, L. A., 1973. "Tectonic framework of Cordilleran Foldbelt in southwestern New Mexico," *Am. Assoc. Petrol. Geologists Bull.,* 57(11), 2207–2216.

Crawford, A. L., 1963. "Surface, structure and stratigraphy of Utah," *Utah Geol. Mineral. Surv. Bull.,* 54a, 175p.

Eardley, A. J., et al., 1973. "Lake cycles in the Bonneville Basin, Utah," *Bull. Geol. Soc. Am.,* 84, 211–216.

Eaton, G. P., 1972. "Deformation of Quaternary deposits in two intermontane basins of southern Arizona, U.S.A.," *24th Intern. Geol. Congr.,* Canada, 3, 607–616.

Elston, W. E., et al., 1973. "Tertiary volcanic rocks, Mogollon-Datil Province, New Mexico, and surrounding region: K-Ar dates, patterns of eruption and periods of mineralization," *Bull. Geol. Soc. Am.,* 84, 2259–2274.

Fenneman, N. M., 1931. *Physiography of Western United States.* New York: McGraw-Hill, 534p.

―――, 1946. *Physical Divisions of the United States.* Washington, D.C.: U.S. Geol. Surv. map.

Fleck, R. J., 1970. Tectonic style, magnitude, and age of deformation in the Sevier orogenic belt in southern Nevada and eastern California," *Bull. Geol. Soc. Am.,* 81, 1705–1720.

Gilbert, C. M., and Reynolds, M. W., 1973. "Character and chronology of basin development, western margin of the Basin and Range Province," *Bull. Geol. Soc. Am.,* 84, 2489–2510.

Gilbert, G. K., 1928. Studies of Basin Range structure," *U.S. Geol. Surv. Prof. Pap.,* 153, 92p.

Gilluly, J., 1963. "The tectonic evolution of the western United States," *Quart. J. Geol. Soc. Lond.,* 119, 133–174.

Hamilton, W., and Myers, W. B., 1966. "Cenozoic tectonics of the western United States," *Rev. Geophys.,* 4, 509–549.

Hayes, P. T., 1970. "Cretaceous paleogeography of southeastern Arizona and adjacent areas," *U.S. Geol. Surv. Prof. Pap., 658-B,* 42p.

Johnson, A. J., and Nairn, A. E. M., 1972. "Jurassic palaeo-magnetism," *Nature,* 240(5383), 551–552.

Kamp, R. C. van de, 1973. "Holocene continental sedimentation in the Salton Basin, California: a reconnaissance," *Bull. Geol. Soc. Am.,* 84, 827–848.

Le Conte, J., 1889. "On the origin of normal faults and of the structure of the Basin region," *Am. J. Sci.,* Ser. 3, 38, 257–263.

Lipman, P. W., et al., 1972. "Cenozoic volcanism and plate-tectonic evolution of the western United States, I. Early and Middle Cenozoic," *Phil. Trans. Roy. Soc. London,* 271, 217–248.

Longwell, C. R., 1945. "Low-angle faults in the Basin and Range Province," *Trans. Am. Geophys. Union,* 26(1), 107–118.

Lucchita, I., 1972. "Early history of the Colorado River in the Basin and Range Province," *Bull. Geol. Soc. Am.,* 83, 1933–1948.

McKee, E. H., 1971. "Tertiary igneous chronology of the Great Basin of western United States—implications for tectonic models," *Bull. Geol. Soc. Am.,* 82, 3497–3502.

―――, and Burke, D. B., 1972. "Fission–track age bearing on the Permian–Triassic boundary and time of the Sonoma orogeny in north-central Nevada," *Bull. Geol. Soc. Am.,* 83, 1949–1952.

―――, and Silberman, M. L., 1970. "Geochronology of Tertiary igneous rocks in central Nevada," *Bull. Geol. Soc. Am.,* 81, 2317–2328.

―――, et al., 1970. "Middle Miocene hiatus in volcanic activity in the Great Basin area of the western United States," *Earth Planet. Sci. Letters,* 8, 93–96.

Mackin, J. H., 1960. "Structural significance of Tertiary volcanic rocks in southwestern Utah," *Am. J. Sci.,* 258, 81–131 (contains excellent bibliography).

Menard, H. W., 1964. *Marine geology of the Pacific.* New York: McGraw-Hill, 271p.

Moore, J. G., 1960. "Curvature of normal faults in the Basin and Range Province of the western United States," *U.S. Geol. Surv. Prof. Pap. 400-B,* Art. 188, B409–B411.

Noble, D. C., 1972. "Some observations on the Cenozoic volcano-tectonic evolution of the Great Basin, western United States," *Earth Planet. Sci. Letters,* 17, 142–150.

Nolan, T. B., 1943. "The Basin and Range Province in Utah, Nevada, and California," *U.S. Geol. Surv. Prof. Pap., 197-D,* 141–196 (contains excellent bibliography).

Ohlen, H. R., and McIntyre, L. B., 1965. "Stratigraphy and tectonic features of Paradox Basin, Four Corners area," *Am. Assoc. Petrol. Geologists Bull.,* 49(11), 2020–2040.

Pakiser, L. C., 1963. "Structure of the crust and upper mantle in the western United States," *J. Geophys. Res.,* 68(20), 5747–5756.

Palmer, A. R., 1971. "The Cambrian of the Great Basin and adjacent areas, western U.S.," *Lower Paleozoic Rocks of the World. Vol. 1. Cambrian of the New World.* New York: Wiley-Interscience, 1–78.

Peterson, D. N., and Nairn, A. E. M., 1971. "Palaeomagnetism of Permian redbeds from the southwestern United States," *Geophys. J. Roy. Astron. Soc.,* 23, 191–205.

Prodehl, C., 1970. "Seismic refraction study of crustal structure in the western United States," *Bull. Geol. Soc. Am.,* 81, 2629–2646.

Raisz, E., 1959. *Landforms of Mexico.* Cambridge, Mass.: Geography Branch, Off. Naval Res.

Riva, J., 1970. "Thrusted Paleozoic rocks in the northern and central HD Range, northeastern Nevada," *Bull. Geol. Soc. Am.,* 81, 2689–2716.

Roberts, R. J., 1966. "Metallogenic provinces and mineral belts in Nevada," *Nevada Bur. Mines Rept.,* 13A, 47–72.

———, 1968. "Tectonic framework of the Great Basin," *Univ. Mo. Res. J.* Ser. 1, **1**, 101–119.
Ross, R. J., Jr., 1970. "Ordovician brachiopods, trilobites, and stratigraphy in eastern and central Nevada," *U.S. Geol. Surv. Prof. Pap. 639*, 103p.
Sass, J. H., et al., 1971. "Heat flow in the western United States," *J. Geophys. Res.*, **76**, 6376–6413.
Sbar, M. L., et al., 1972. "Tectonics of the Intermountain Seismic Belt, western United States: microearthquake seismicity and composite fault plane solutions," *Bull. Geol. Soc. Am.*, **83**, 13–28.
Scholz, C. H., Barazangi, M., and Sbar, M. L., 1971. "Late Cenozoic evolution of the Great Basin, western United States, as an ensialic interarc basin," *Bull. Geol. Soc. Am.*, **82**, 2979–2990.
Seager, W. R., 1970. "Low-angle gravity glide structures in the northern Virgin Mountains, Nevada and Arizona," *Bull. Geol. Soc. Am.*, **81**, 1517–1538.
Shaw, D. R., 1965. "Strike-slip control of Basin–Range Structure indicated by historical faults in western Nevada," *Bull. Geol. Soc. Am.*, **76**, 1361–1378.
Shurbet, D. H., and Cebull, S. E., 1971. "Crustal low-velocity layer and regional extension in Basin and Range Province," *Bull. Geol. Soc. Am.*, **82**, 3241–3244.
Smith, P. B., 1970. "New evidence for a Pliocene marine embayment along the lower Colorado River area, California and Arizona," *Bull. Geol. Soc. Am.*, **81**, 1411–1420.
Stewart, J. H., 1971. "Basin and Range structure: A system of horsts and grabens produced by deep-seated extension," *Bull. Geol. Soc. Am.*, **82**, 1019–1044.
Stormer, J. C., Jr., 1972. "Ages and nature of volcanic activity on the Southern High Plains, New Mexico and Colorado," *Bull. Geol. Soc. Am.*, **83**, 2443–2448.
Tatel, H. E., and Tuve, M. A., 1955. "Seismic exploration of a continental crust," *Geol. Soc. Am. Spec. Pap. 62*, 35–50.
Thompson, G. A., 1959. "Gravity measurements between Hazen and Austin, Nevada: a study of Basin–Range structure," *J. Geophys. Res.*, **64**(2), 217–229.
———, 1973. "Rate and direction of spreading in Dixie Valley, Basin and Range Province, Nevada," *Bull. Geol. Soc. Am.*, **84**, 627–632.
———, and Burke, D. B., 1972. "Cenozoic Basin Range tectonism in relation to deep structure," *24th Intern. Geol. Congr.*, Canada, 84–90.
———, and Burke, D. B., 1974. "Regional geophysics of the Basin and Range province," *Ann. Rev. Earth Planet. Sci.*, **2**, 213–238.
———, and Talwani, M., 1964. "Crustal structure from Pacific basin to central Nevada," *J. Geophys. Res.*, **68**, 4813–4837.
Wertz, J. B., 1970. "The Texas Lineament and its economic significance in southeast Arizona," *Econ. Geol.*, **65**, 166–181.
Wilson, E. D., 1962. "A resume of the geology of Arizona," *Ariz. Bur. Mines Bull.*, **171**, 140p.
Wilson, J. T., 1970. "Some possible effects if North America has overridden part of the East Pacific rise," *Geol. Soc. Am. Abs.*, **2**(7), 722.

Cross-references: Mexico; North America; United States–Colorado Plateau Province, Pacific Cordilleran Region, Rocky Mountain Province.

UNITED STATES–COLORADO PLATEAU PROVINCE

The Colorado Plateau province lies within the four states of Arizona, Utah, Colorado, and New Mexico, between the Rocky Mountains and the Basin and Range province, reaching the Great Plains in the SE (Fig. 1). The Colorado Plateau proper occupies an area of about 338,000 km^2 (130,000 sq mi). The southeastern part in the New Mexico lobe is least typical and referred to here as the "southeastern salient"; It adds another 130,000 km^2 (50,000 sq mi), making a total of 468,000 km^2 (180,000 sq mi). This is an area about twice that of Great Britain. It is a relatively high country, a land of individual plateaus rising from 1500–3000 m (5000–10,000 ft) above sea level, separated by scarps ("cuestas") formed in part by canyon cutting and stripping back of less resistant layers, in part by displacement along major faults. There are local mountain areas over 3300 m (max 3927 m, 13,089 ft in the La Sal Mountains). Low elevations are found only in deep valleys, notably the canyon of the Colorado River, where at the western edge of the plateau, a narrow gorge is cut to 300 m in a tableland 1800 m high.

FIGURE 1. Index map of Colorado Plateau Province.

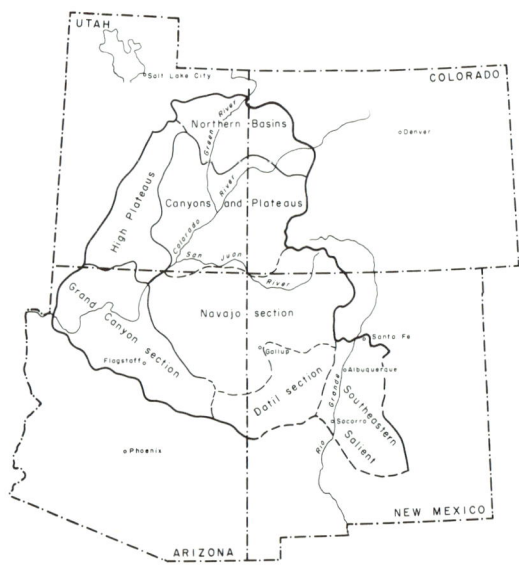

Exploration and Mining

One of the earliest geologists to make observations in this region was Jules Marcou, a Swiss-French geologist attached to Lt. A. W. Whipple on an exploration of the 35th parallel in search for a railroad route. This survey was made in 1853–1856 and led across the southern part of the province. In 1857–1859, the geologist J. S. Newberry was a member first of Lt. J. C. Ives' expedition up the Colorado and then of Capt. J. N. Macomb's across the southern and eastern part of the plateau. John Wesley Powell made his first trip down the Green and Colorado rivers in 1869, and his second in 1871. In the following decade, members of the Powell survey and of two other early Federal surveys investigated various parts and aspects of the region. W. H. Holmes of the Hayden survey mapped in the Mesa Verde area in 1874, and he and A. C. Peale investigated the La Sal Mountains in 1875, recognizing the intrusive character of the rocks forming the mountains, foreshadowing the work of G. K. Gilbert on the Henry Mountains. Clarence E. Dutton, of the Ordnance Department, U.S. Army, working on the Powell survey, studied the high plateaus of Utah in 1874–76, and later, in the next decade, the Mt. Taylor and Zuni plateaus. Like Powell, he was impressed with the amount of erosion, and applied the concept of isostasy to the elevated position of the plateau country. G. K. Gilbert, first attached to the Wheeler survey, later to the Powell survey, made a classical study on the Henry Mountains, clarifying concepts already suggested by Peale and Holmes and introducing the term "laccolite" (later "laccolith") into geologic literature. H. E. Gregory made very broad and useful studies of many parts of the plateau country in the early third of the present century. With the search for oil and gas beginning in the 1920s and the search for uranium in the 1940s, details of geology have become well known.

Mining has been largely restricted to coal, which has been mined since the 1880s, and to vanadium and uranium, which have been mined since the same decade. It was not until the 1940s and 1950s, however, that uranium became important.

Centers of Learning and Scientific Research

The area is one of the least populated in North America. The University of New Mexico is situated in Albuquerque, in the southeastern salient; the New Mexico Institute of Mining and Technology is in Socorro, New Mexico; Northern Arizona University is in Flagstaff, Arizona; and Fort Lewis College is in Durango, Colorado. Collections are included in the Museum of Northern Arizona at Flagstaff, the Utah Field House of Natural History at Vernal, the museum at the Dinosaur National Monument near Vernal, and the geology museum at the University of New Mexico. The U. S. Geological Survey carries on work in various parts of the plateau, and several geological societies are active in the area. The latter include the Arizona Geological Society, the Four Corners Geological Society, the Intermountain Association of Petroleum Geologists, the New Mexico Geological Society, and the Utah Geological Society.

Structural Framework and Physiographic Divisions

The Colorado Plateau is an area of relatively simple structure, becoming somewhat more complicated along the margins, particularly along the eastern margin and in the southeastern salient. Flat-lying or relatively gently dipping strata, with weak strata stripped away and resistant strata capping erosional remnants, form the typical landscape pattern. Warping is generally broad. Sharp flexures or monoclines occur along major faults. Along the margins, fault blocks produce distinct basins and ranges, transitional features of the Basin and Range Province. Since Precambrian time, the province has been chiefly a stable block, resisting the forces that have produced extensive deformation on all sides. The western margin is still active today, part of the "Intermountain Seismic Belt" that forms the eastern limit of the Basin and Range Province and extends N to the Rocky Mountain Trench. Moderate earthquake shocks are quite frequent along this line (Sbar et al., 1972).

The major units of the province are the northern basin section, high plateaus of Utah, Grand Canyon section, canyon and plateau section along the upper Colorado River, Navajo section, Datil section, and southeastern salient. The general structural grain is N, but locally, particularly along the eastern and northeastern margin, trends swing to the NW (the Paradox fold and fault belt) and even to the W (the Uinta Basin).

Grand Canyon Section. The section of greatest uplift and of greatest denudation is the Grand Canyon. It is characterized by plateau features on late Paleozoic rocks, chiefly Permian, modified locally by Tertiary volcanic features, e.g., the San Francisco Peaks (up to 1500 m above the plateau) in the Flagstaff area (Fig. 2). Elsewhere, S of the Grand Canyon, the

FIGURE 2. Schematic E-W profile across Grand Canyon and Navajo Sections of the Colorado Plateau. pЄ, Precambrian; LMP, Lower and Middle Paleozoic; UP, Upper Paleozoic; Tr–J, Triassic and Jurassic; K, Cretaceous; T, Tertiary. Length of section: 640 km (400 miles), vertical scale schematic. Structural notations above; geographic notations below.

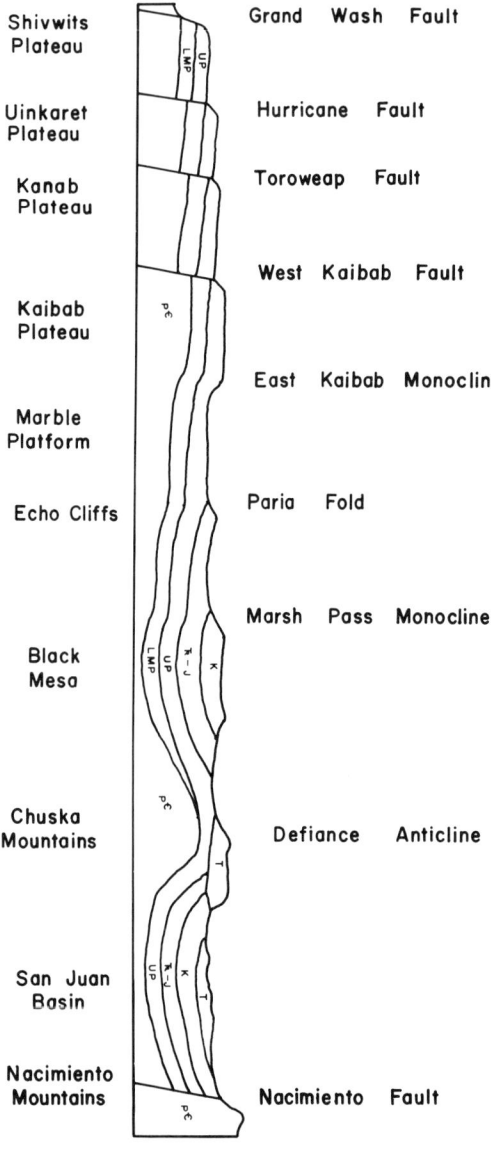

ern margin to the central part of the Grand Canyon section; farther E, the Permian drops down again in steps but is obscured by younger rocks (Mesozoic). The canyon of the Colorado River displays a classical section: Precambrian in the inner gorge, Paleozoic in the upper cliffs and benches.

High Plateaus of Utah. Northward from the Grand Canyon, the strata dip gently N with steps preserved on the more resistant Mesozoic rocks. The culmination of this stairway to the N is found in Tertiary shales and sandstones, and lavas cap the highest levels (3300 m). Faults, many continuing from the Grand Canyon region, separate N-trending plateaus, giving rise to linear valleys. In Zion Canyon and Bryce Canyon, erosion of the margins of these steps has produced spectacular scenery.

Northern Basin Section. Eastward from the high plateaus is a broad structural downwarp, topographically appearing as a dissected plateau surface, lapping onto the Uinta Mountains to the N, turning up against the Rocky Mountains

FIGURE 3. Sketch map of the Four Corners region (of Utah/Colorado/Arizona/New Mexico) with its regional basins and uplifts. A notable feature of the Paradox Basin was the deposition of evaporites in the mid-Pennsylvanian (Paradox evaporites, Desmoinesian), which have been mobilized along NW-trending anticlines (Ohlen and McIntyre, 1965).

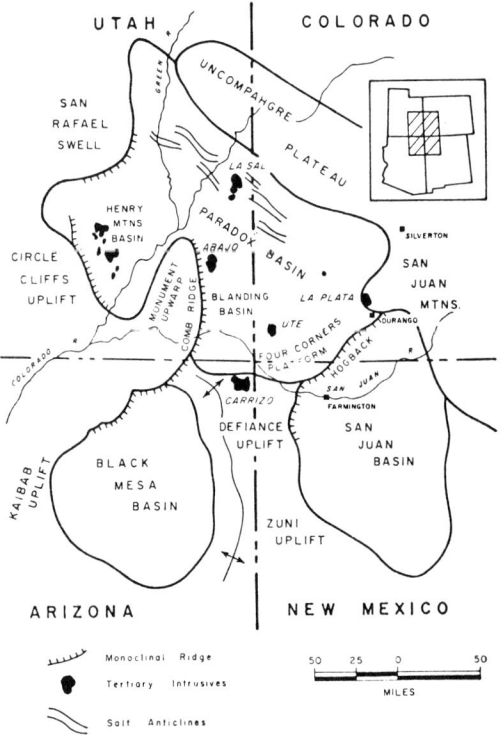

surface is fairly uniform on the surface of the Permian Kaibab limestone; the southern margin of the province is generally marked by a pronounced escarpment ("Mogollon Rim"). North of the Grand Canyon, separated by a number of major N-trending faults and monoclines, there are a series of giant steps rising from the west-

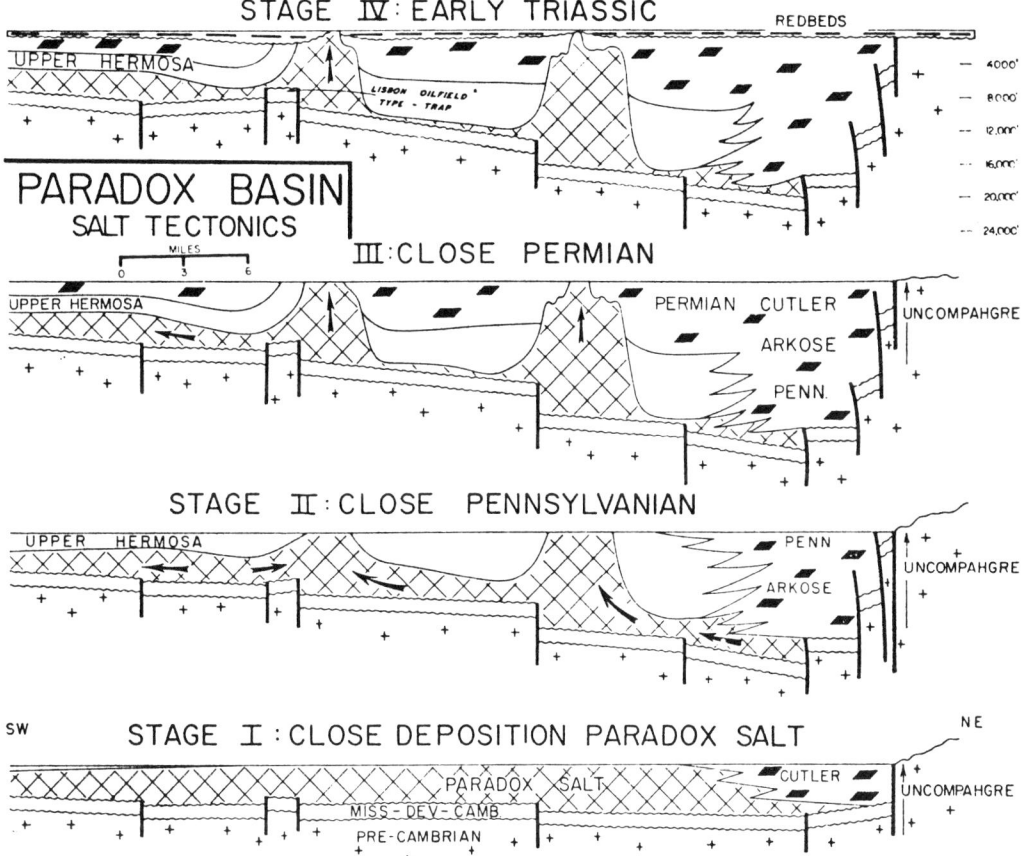

FIGURE 4. Four stages in the diapiric growth of salt anticlines ("halokinesis") in the Paradox Basin: I, mid-Pennsylvanian; II, latest Pennsylvanian; III, latest Permian; IV, early Triassic. Note, the progressive development of basement faults, especially bordering the Uncompahgre uplift (Ohlen and McIntyre, 1965).

on the E in the Grand Hogback, and overlooking the lower lands of the upper Colorado River to the S from the heights of the Book and Roan Cliffs. The rocks are chiefly Tertiary, with Cretaceous along the margins. South of the Uinta Mountains this section is represented by the Uinta Basin, but E, separated by a relatively narrow arch, is the Piceance Basin, which ends in uplifts against the Rocky Mountains.

Canyon and Plateau Section of the Upper Colorado River. South of the Uinta Basin, E of the southern high plateaus, lies a region principally of Mesozoic strata, cut by canyons, from which weak layers have been stripped back, producing an extensive area of buttes, mesas, and plateaus. Broad domal structures are local features here, with distinctive peripheral cliffs. Laccoliths form prominent mountainous areas: the Henry Mountains, La Sal Mountains, Abajo Mountains, Carrizo Mountains, and the Ute Mountains. This section is crossed by the NW-SE Paradox fold and fault belt, characterized by salt intrusions in the anticlines.

Navajo Country. Lying S of the canyon lands and E of the Grand Canyon, the Navajo country is mainly Mesozoic with Tertiary sandstones and shales in places, notably within the San Juan Basin of northwestern New Mexico. The region is marked by broad downwarps, or basins, and smaller upwarps, with intermediate platforms and saddles. The youngest rocks, Cretaceous and Tertiary, appear in the central parts of the downwarps; Jurassic and Triassic rocks appear on the margins and in the upwarps. Physiographically, this section resembles the canyon lands, but the canyons are less abundant; mesas and plateaus are more extensive.

Datil Section. Characterized by great thicknesses of lavas and tuffs, the Datil section is also marked by plateaus, but sedimentary rocks

FIGURE 5. Typical butte isolated by erosion in a broad upwarped area of southern Utah; here in Monument Valley a resistant Triassic conglomerate (Shinarump) caps the softer red-brown Triassic (Moenkopi) shales and sandstones. (Photo: Rhodes Fairbridge.)

are generally buried beneath the volcanic rocks. Surfaces are therefore more irregular and differences in erosional resistance are slight, so that bench and slope topography disappears. Near the southern edge of the Colorado Plateau Province, this section is greatly dissected along this margin. Volcanic necks, mostly basalt, are a distinctive feature of the northeastern edge.

Southeastern Salient. Least typical of the entire province, the southeastern salient has features of both Basin and Range structure and Rocky Mountain structure. It is a region of fault-block mountains and intermontane basins. The Rio Grande trough occupies a series of such basins, some filled with 1000 m or more of Tertiary and Quaternary alluvial sediments,

FIGURE 6. The "Goosenecks," an incised meander sector of the San Juan River, in southeast Utah. The river here crosses the gently folded east flank of the Monument upwarp. Steady downcutting during uplift has incised the canyon 260 m through the Permian Cutler Formation into the Pennsylvanian Hermosa Formation. (Photo: Rhodes Fairbridge.)

TABLE 1. Generalized Stratigraphic Sequence for the Four Corners Area of the Colorado Plateau

Period	Group	Formation	Thickness
Tertiary		Wasatch-San Jose (ss, sh)	75–600 m
		Animas (ss) Nacimiento (sh)	50–700 m
Cretaceous		McDermott (cgl)	10–90 m
		Kirtland (sh)	190–510 m
		Fruitland (ss-sh)	69–90 m
		Pictured Cliffs (fm)	10–85 m
		Lewis (sh)	22–510 m
	Mesaverde Group	Cliff House (ss)	0–75 m
		Menefee (coaly ss, sh)	240–480 m
		Pt. Lookout (ss)	30–90 m
	Mancos Group	Hosta (ss)	30–60 m
		Mullato (sh)	75–120 m
		Gallup (ss)–Tocito (ss)	55–75 m
		L. Mancos (sh)	200–650 m
		Greenhorn (ls)	10–20 m
		Graneros (sh)	15–45 m
		Dakota (ss)	30–80 m
		Burro Canyon (ss)	
Jurassic	Morrison Group	Brushy Basin (sh)	30–90 m
		Westwater (ss)	0–90 m
		Recapture (sh)	0–180 m
		Salt Wash (ss)	30–105 m
	San Rafael Group	Bluff (ss)	5–75 m
		Summerville (sh)	5–60 m
		Pony Express (ls)	0–22 m
		Entrada (ss)	20–135 m
		Carmel (sh)	10–200 m
	Glen Canyon Group	Navajo (ss)	0–180 m
		Kayenta (sh)	5–95 m
———		Moenave (ss)	30–105 m
Triassic		Wingate (ss)	30–180 m
		Chinle (sh)	125–300 m
		Shinarump (cgl)	0–35 m
		Moenkopi (sh)	0–180 m
Permian		Kaibab (ls) Hoskinnini (sh)	0–250 m
		De Chelly (ss) 0–180 m Cutler (ark, ss)	300–525 m
Pennsylvanian		Supai (sh) 0–150 m Hermosa (ls)	240–600 m
		Molas (sh) 0–60 m Naco (ls)	500–600 m
Mississippian		Leadville (ls) 20–30 m Redwall (ls)	2–200 m
Devonian		Ouray (ls)	0–90 m
		Elbert (ls)	15–90 m
		Aneth (sh, dol)	0–55 m
Cambrian		Ignacio (qtzt)	0–35 m
		Tonto	0–240 m
Precambrian		Grand Canyon Series (ss, ls, sh, qtzt)	
		Vishnu (schists, granite)	

ss, sandstone; sh, shale; ls, limestone; cgl, conglomerate; qtzt, quartzite; dol, dolomite; ark, arkose; ---, unconformity.

TABLE 2. Generalized List of Formations of the Southeastern Salient of the Colorado Plateau Province

Age	Group	Formation	Thickness
Tertiary		Santa Fe (ss, silt)	300 m+
Cretaceous		Mesaverde (ss)	750–1050 m
		Mancos (sh)	105–510 m
		Dakota (ss)	30–75 m
Triassic		Dockum (sh)	300–360 m
Permian	Manzano Group	San Andres (ls)	25–300 m
		Yeso (ss, silt, ls)	120–225 m
		Abo (ss)	165–330 m
Pennsylvanian	Magdalena Group	Madera (ls, sh)	320–600 m
		Sandia (ss, ls)	30–150 m
Mississippian		Helms (ls)	0–20 m
		Rancheria (ls)	0–90 m
		Lake Valley (ls)	10–20 m
		Caballero (ls)	5–20 m
Devonian		Percha (sh)	5–30 m
		Sly Gap (sh, ls)	0–15 m
		Onate (silt)	0–10 m
Silurian		Fusselman (dol)	5–15 m
Ordovician	Montoya Group	Cutter (ls) 15–50 m ... Valmont (ls)	45–65 m
		Aleman (ls)	35–55 m
		Upham (ls)	5–25 m
		Cable Canyon (ss)	5–10 m
		El Paso (ls)	105–155 m
Cambrian		Bliss (ss)	35–50 m
Precambrian		Granite, gneiss, schist	

but draining from one basin to the next. Other basins not in the Rio Grande system are endorheic. The ranges generally expose Precambrian rocks at the base, Pennsylvanian or Permian rocks at the crest, and younger rocks on the back slopes.

Stratigraphy, Orogeny, and Igneous Activity

The Colorado Plateau province is deservedly celebrated as the land of the red beds: sandstones and shales in various shades of red, generally Permian to Jurassic in age (see Table 1).

The complete sequence is not found in any one place. Cambrian rocks are mainly sandstones, the rest of the lower Paleozoic predominantly carbonates. Shale and sandstone become more abundant in the upper Paleozoic, Mesozoic, and Tertiary.

Two tectonic periods are recorded in Precambrian rocks: the earliest marked by folding, metamorphism, granites, and pegmatites; the second resulting in great fault movements, but accompanied by much tilting of the individual blocks. Each episode was followed by extensive and profound erosion. Renewed diastrophism developed toward the close of the Mesozoic, but it was only a mild taphrogeny, reflected also indirectly from activity in neighboring areas. The rising Rocky Mountains furnished great quantities of clastic sediments, which formed an extensive mantle of Tertiary sands, silts, and muds. The region was a more or less stable block resisting the deformation that affected the surrounding regions, yielding only along major faults and in broad warps.

Igneous activity is recorded in Precambrian rocks—flows, dikes, sills, and larger intrusions of mafic and silicic rocks. The long-ensuing calm interval saw no vulcanism in the province, until the Tertiary. This late activity includes laccolithic intrusions of intermediate rocks (diorites and monzonites), which date from the early Miocene (22-25 m. yr.) according to Stormer (1972). Basaltic eruptions of both central and fissure type followed, with great accumulations of intermediate and silicic tuffs and flows (Datil section), and local plugs and necks of minette, trachybasalt, and monchiquite (Hopi Buttes, Shiprock, and associated features). This group of alkalic igneous rocks is found mainly along the Arizona—New Mexico line.

Radiometric dating shows the principal basalts to be late Miocene (14.8—10.1 m. yr.). These were followed by the major phase of upwarp and faulting from 10—6 m. yr., succeeded in turn by some late basalts (McKee and Anderson, 1971), which continued up until the Holocene (Stormer, 1972). The basalts are olivine-rich of the tholeiitic type found elsewhere in the western United States at this stage. According to the interpretation of plate tectonics, these volcanics are related to spreading phenomena centered under the Basin and Range Province (q.v.).

The area of the *southeastern salient* underwent a somewhat different history (see Table 2). The Precambrian record here is similar to that of the rest of the province, but the Paleozoic history was marked by the emergence of the central part during most of the early Paleozoic; furthermore, the southern part was submerged during every period, and the entire salient became covered during the late Paleozoic, so that carbonate accumulations continued to dominate almost to the end of the era.

Continental conditions returned in places during the Permian, and became general in the Mesozoic when conditions in this salient were similar to the rest of the Colorado Plateau Province. With the Laramide, disturbances at the salient were more severely deformed, developing fault-block mountains and intermontane basins.

Special Localities and Features

Localities supplying rare specimens can be found in the Colorado Plateau province, but they are not particularly abundant. Secondary uranium and vanadium minerals are found notably in the Grants district in New Mexico, the Uravan district of Colorado, and in southeastern Utah. Fine specimens of copper, copper-lead, copper-zinc, and zinc minerals have been found in the Hansonburg and Magdalena districts near Socorro, New Mexico. Gilsonite is found in the Uinta Basin. Petrified wood, much of it highly colored, is common in the Chinle Formation.

Invertebrate fossils are to be found in many limestones, but the most distinctive feature of the Colorado Plateau province is its accumulation of vertebrate remains. These are found in rocks ranging from Permian to Tertiary. The Jemez Mountains at the eastern edge of the San Juan Basin are known for reptiles and mammals. The first Eocene mammal discoveries in the Southwest were made in the Rio Grande Valley. During the extensive mining of uranium, many dinosaur bones, replaced by uranium minerals, have been uncovered, particularly in the Grants area. Perhaps the most spectacular dinosaur locality is that now set aside as Dinosaur National Monument in Utah and Colorado, where there are displays of skeletons in place.

Features of spectacular geologic interest are the *Grand Canyon*—profound stream erosion and bench and slope topography; *Zion Park*—huge monoliths of white cross-bedded Jurassic sandstone; *Bryce Canyon and Cedar Breaks*—badland topography with bizarre spires in pink, early Tertiary fluvial and lacustrine rocks; *Capitol Reefs Monument*—erosional monoliths of the Glen Canyon group; *Dinosaur National Monument*—large features of stream erosion and dinosaur fossils in place; *Arches National Monument*—spectacular erosional features in red beds, chiefly the Entrada and Carmel formations; *Goosenecks of the San Juan River*—deeply incised meanders in southeastern Utah; *Natural Bridges National Monument*—majestic spans of massive sandstone, mainly erosion products of incised meanders; *Mesa Verde Park*—high mesa on Cretaceous rocks, the site of early Indian cliff dwellers; *Sunset Crater National Monument*—excellent and easily accessible view of recent cinder cone; *Canyon de Chelly*—steep-walled canyons in massive sandstone and remarkable Indian ruins; *Painted Desert*—beautiful varicolored landscape on Triassic strata; *Petrified Forest*—Triassic forest preserved in agate.

Mineral Resources

The Colorado Plateau is the leading producer of uranium in the United States and one of the largest producers in the world. The four states containing the Colorado Plateau are all among the five leading uranium-producing states. Production is chiefly from the eastern edge: the southeastern border of the San Juan Basin and the area along the southern part of the Utah-Colorado line. This has also been the general region of vanadium production in the United States since the 1890s. Ore horizons are principally in Jurassic rocks, to a lesser degree in Cretaceous and Triassic rocks.

Oil and gas have been widely discovered since the 1920s. The Uinta Basin on the N (mainly oil) and the San Juan Basin on the S (mainly gas) have been the principal producing areas. Production has come chiefly from Pennsylvanian and Cretaceous rocks and in the Uinta Basin, Eocene rocks. Gilsonite and other solid hydrocarbons are also found and mined in the Uinta Basin, and tremendous reserves of oil shale characterize the eastern part of this section. Helium is produced at two plants, located near the Four Corners area. Coal, chiefly bituminous to subbituminous is present in considerable reserves in the Cretaceous rocks to the E of the plateau province (50 billion tons in northwestern New Mexico alone).

J. PAUL FITZSIMMONS

References

Anderson, R. Y., Dean, W. E., Jr., Kirkland, D. W., et al., 1972. "Permian Castile varved evaporite sequence, West Texas and New Mexico," *Bull. Geol. Soc. Am.,* **83**(1), 59–85.

Babenroth, D. L., and Strahler, A. N., 1945. "Geomorphology and structure of the East Kaibab monocline, Arizona, and Utah," *Bull. Geol. Soc. Am.,* **56**, 107–150.

Baldwin, E. J., 1973. "The Moenkopi Formation of north-central Arizona: an interpretation of ancient environments based upon sedimentary structures and stratification types," *J. Sed. Petrology,* **43**(1), 92–106.

Beal, M. D., 1967. *Grand Canyon.* Flagstaff, Ariz.: Falkington, 38p.

Best, M. G., and Hamblin, W. K., 1970. "Implications of tectonism and volcanism in the western Grand Canyon," *Utah Geol. Soc. Guidebook,* **23**, 75–79.

Cater, F. W., and Craig, L. C., 1970. "Geology of the salt anticline region in southwestern Colorado," *U.S. Geol. Surv. Prof. Pap.* **637**, 80p.

Cram, I. H., ed., 1971. "Future petroleum provinces of the United States—their geology and potential," *Am. Assoc. Petrol. Geologists Mem.* **15**, 2 vols., 803p., 692p.

Eardley, A. J., 1951. *Structural Geology of North America.* New York: Harper & Row, 386–409.

Fenneman, N. M., 1931. *Physiography of Western United States.* New York: McGraw-Hill, 274–325, 385–395.

Ford, T. D., and Breed, W. J., 1973. "Late Precambrian Chuar Group, Grand Canyon, Arizona," *Bull. Geol. Soc. Am.,* **84**, 1243–1260.

____, Breed, W. J., and Mitchell, J. S., 1972. "Name and age of the Upper Precambrian basalts in the eastern Grand Canyon," *Bull. Geol. Soc. Am.,* **83**, 223–226.

Gilluly, J., 1963. "The tectonic evolution of the western United States," *Quart. J. Geol. Soc. Lond.,* **119**, 133–174.

Gregory, H. E., 1950. "Geology and geography of the Zion Park Region, Utah and Arizona," *U.S. Geol. Surv. Prof. Pap.,* **220**, 200p.

Hack, J. T., 1942. "Sedimentation and volcanism in the Hopi Buttes, Arizona," *Bull. Geol. Soc. Am.,* **53**, 335–372.

Hunt, C. B., 1956. "Cenozoic geology of the Colorado Plateau," *U.S. Geol. Surv. Prof. Pap.,* **279**, 99p.

____, 1967. *Physiography of the United States.* San Francisco: W. H. Freeman, 277–307, 320–323.

Kelley, V. C., 1955a. "Monoclines of the Colorado Plateau," *Bull. Geol. Soc. Am.,* **66**, 789–804.

____, 1955b. "Regional tectonics of the Colorado Plateau and relationship to origin and distribution of uranium," *Univ. New Mexico Publ. Geol.* **5**, 120p.

____, and Clinton, N. J., 1960. *Fracture Systems and Tectonic Elements of the Colorado Plateau.* Albuquerque: Univ. New Mexico Press, 104p.

Keyes, C. R., 1938. "Basement complex of the Grand Canyon," *Pan-Amer. Geol.,* **70**(2), 91–116.

Kottlowski, F. E., Cooley, F. E., and Ruhe, R. V., 1965. "Quaternary geology of the southwest," in H. E. Wright, Jr., and D. G. Frey, eds., *The Quaternary of the United States.* Princeton, N.J.: Princeton Univ. Press, 287–298.

Kurie, A. E., 1966. "Recurrent structural disturbance of Colorado Plateau margin near Zion National Park, Utah," *Bull. Geol. Soc. Am.,* **77**(8), 867–871.

Lochman-Balk, C., 1971. "The Cambrian of the craton of the United States," *Lower Palaeozoic Rocks of the World. Vol. 1. Cambrian of the New World.* New York: Wiley-Interscience, 79–167.

McKee, E. D., 1969. "Paleozoic rocks of Grand Canyon," *Grand Canyon Guidebook.* Four Corners Geol. Soc., 78–90.

____, and McKee, E. H., 1972. "Pliocene uplift of the Grand Canyon region—time of drainage adjustment," *Bull. Geol. Soc. Am.,* **83**, 1923–1932.

McKee, E. H., and Anderson, C. A., 1971. "Age and chemistry of Tertiary volcanic rocks in north-central Arizona and relation of the rocks to the Colorado Plateaus," *Bull. Geol. Soc. Am.,* **82**, 2767–2782.

McKnight, E. T., 1940. "Geology of area between Green and Colorado Rivers, Grand and San Juan counties, Utah," *U.S. Geol. Surv. Bull.,* **908**, 147p.

Martin, P. S., and Mehringer, P. J., Jr., 1965. "Pleistocene pollen analysis and biogeography of the Southwest," in H. E. Wright, Jr., and D. G. Frey, eds., *The Quaternary of the United States.* Princeton, N.J.: Princeton Univ. Press, 433–451.

Merrill, R. K., and Péwé, T. L., 1972. "Late Quaternary glacial chronology of the White Mountains, east-central Arizona," *J. Geol.,* **80**, 493–501.

Molenaar, C. M., 1969. "Lexicon of stratigraphic names used in northern Arizona and southern Utah east of the Paleozoic hinge line," *Grand Canyon Guidebook.* Four Corners Geol. Soc., 68–77.

Ohlen, H. R., and McIntyre, L. B., 1965. "Stratigraphy and tectonic features of Paradox basin, Four Corners area," *Am. Assoc. Petrol. Geologists,* **49**, 2020–2040.

Pakiser, L. C., 1963. "Structure of the crust and upper

mantle in the western United States," *J. Geophys. Res.,* **68**(20), 5747–5756.

Palmer, A. R., 1971. "The Cambrian of the Appalachian and eastern New England regions, eastern United States," in C. H. Holland, ed., *Cambrian of the New World.* New York: Wiley-Interscience, 169–217.

Peirce, H. W., and Scurlock, J. R., 1972. "Arizona well information," *Ariz. Bur. Mines Bull.,* **185**, 195p.

Péwé, T. L., 1968. *Colorado River Guidebook, Lees Ferry to Phantom Ranch.* Tempe, Ariz.: T. L. Péwé, 78p.

____, and Updike, R. G., 1970. "Guidebook to the geology of the San Francisco Peaks, Arizona," *Plateau,* **43**(2), 45–102.

Reger, R. D., and Batchelder, G. L., 1970. "Late Pleistocene molluscs and a minimum age of Meteor Crater, Arizona," *Ariz. Acad. Sci. J.,* **6**(3), 190–195.

Sbar, M. L., et al., 1972. "Tectonics of the Intermountain Seismic Belt, Western United States: microearthquake seismicity and composite fault plane solutions," *Bull. Geol. Soc. Am.,* **83**, 13–28.

Schneider, R. C., et al., 1971. "Petroleum potential of Paradox Region," *Future Petroleum Provinces of the United States; their Geology and Potential* (Mem. 15, vol. 1). Am. Assoc. Petrol. Geologists, 470–488.

Sharp, R. P., 1940. "A Cambrian slide breccia, Grand Canyon, Arizona," *Am. J. Sci.,* **283**, 668–672.

Shoemaker, E. M., Case, J. E., and Elston, D. P., 1958. "Salt anticlines of the Paradox basin," *Guidebook to the Geology of the Paradox Basin.* Intermountain Assoc. Petrol. Geol., 39–59.

Spieker, E. M., and Billings, M. P., 1940. "Glaciation in the Wasatch Plateau, Utah," *Bull. Geol. Soc. Am.,* **51**, 1173–1198.

Stokes, W. L., 1948. "Geology of the Utah–Colorado salt dome region with emphasis upon Gypsum Valley, Colorado," *Guidebook to the Geology of Utah.* Utah Geol. Soc., **3**, 3–40.

Stormer, J. C., Jr., 1972. "Ages and nature of volcanic activity on the southern high plains, New Mexico and Colorado," *Bull. Geol. Soc. Am.,* **83**, 2443–2448.

Strahler, A. N., 1944. "Valleys and parks of the Kaibab and Coconino plateaus, Arizona," *J. Geol.,* **52**, 361–387.

____, 1948. "Geomorphology and structure of the West Kaibab fault zone and Kaibab Plateau," *Bull. Geol. Soc. Am.,* **59**, 513–540.

Thornbury, W. D., 1965. *Regional Geomorphology of the United States.* New York: Wiley, 405–441, 499–505.

Visher, G. S., 1971. "Depositional processes and the Navajo Sandstone," *Bull. Geol. Soc. Am.,* **82**, 1421–1424.

Walcott, C. D., 1895. "Algonkian rocks of the Grand Canyon," *J. Geol.,* **3**, 312–330.

Williams, H., 1936. "Pliocene volcanos of the Navajo–Hopi country," *Bull. Geol. Soc. Am.,* **47**, 111–172.

Wilson, E. D., 1962. "A resume of the geology of Arizona," *Ariz. Bur. Mines Bull.,* 140p.

Much recent and extensive information on the geology of this province is to be found in the guidebooks and symposia of the Utah Geological Society, the Intermountain Association of Petroleum Geologists, the New Mexico Geological Society, and the Four Corners Geological Society. All contain good bibliographies.

Cross-references: *Mexico; North America; United States–Basin and Range Province, Rocky Mountain Province.*

UNITED STATES–GREAT PLAINS PROVINCE

The Great Plains physiographic province, one of the major great land forms of North America, extends from the Peace River plains in northwestern Canada S and SE to terminate in the Edwards Plateau, a limestone upland without Tertiary mantling sediments, in central Texas. This is a distance of about 3700 km or through about $32°$ of latitude, from $29°N$ to $61°N$. It is located wholly E of the Rocky Mountains.

The width of the Great Plains in western Canada ranges from about 300–550 km; and from the Canadian–United States border its width (the central Great Plains) is about 650 km N to central southern Kansas. Farther S, it narrows (300-550 km). The elevation at the W from the eastern front of the Canadian Rockies S across Montana, eastern Wyoming, Colorado, and New Mexico averages about 1500 m above sea level, and its eastern margin approximates the 300 m contour line throughout its entire length. Physiographically, it is bounded on the W by the Rocky Mountains, and on the E it merges into the Interior Lowlands of the great interior composite Mississippi-Missouri-Ohio river drainage basin.

The surface gradient, where the original High Plains surface still is almost completely intact in the "Gangplank" area of southeastern Wyoming into western Nebraska, slopes steeply from the Rocky Mountain front at the W, more gently across western Nebraska, and continues to decrease and flattens out in eastern Nebraska. The total area of the North American Great Plains is about $11,250 \text{ km}^2$ (700,000 sq mi) with three-fifths of this area in the United States.

General Structure

Structurally the Great Plains is a wide belt of shallow synclinal structures with the sedimentary formations from late Cambrian to Cretaceous age generally dipping toward the central axis from both E and W. The deepest part of this belt above the Cretaceous Pierre shale appears to be in central and western Nebraska, where it resembles an E-tipping, oval bowl. Here is found the thickest and most complete sequence of continental Tertiary sediments from Oligocene to Pliocene E of the Rocky Mountains, the maximum thickness exceeding 600 m.

Climate

The Great Plains lie mainly in a semiarid rainfall belt with about 500 mm of average annual precipitation at the E and but little more than 300 mm at the west. In the SE it may exceed 600 mm.

The topography and climate are fairly typical of a midlatitude steppe plain, although in northern Canada it merges in the lower Peace River plain into the zone of permafrost, whereas at the southern extremity, around Austin and San Antonio in Texas, there are mean annual temperatures of $20°C$.

Soil, Vegetation, and Agriculture

At the far N some underdeveloped podsols are present with some swampy muskeg areas. In general, however, gray chernozem and reddish-brown (chestnut) soils predominate to the central Great Plains, and most reddish and brown soils, characteristic of warm to cool and dry climate, are found farther S. All are high-lime pedocals.

The natural vegetation is mostly steppe and prairie grasses, usually the short types, but in the Sand Hills of Nebraska some tall grasses enhance the range. Conifers, pine, and cedar, and mixed types of forest such as elm, hackberry, cottonwood, willow, and others and at higher elevations poplar and aspen thrive in the larger valleys, where springs and seepage provide more soil moisture than is available on the tablelands.

Winter wheat is the most widely grown crop in the southern and central plains as far N as Central Nebraska. Farther N spring wheat is grown in the Dakotas, Montana, and N to the Peace River plains. Barley and rye are also grown. Cattle grazing and ranching are also important.

Alfalfa and sugar-beet farming occur in the major irrigated valleys in Nebraska, Colorado, and Kansas. Great expansion in agriculture began from Nebraska to Texas in the middle 1930s, when it became feasible to use large underground supplies of water from 60 to 120 m depth to irrigate formerly dry prairie lands.

Corn and vegetable crops had not been generally cultivated except on valley bottom lands and then mostly in the eastern parts of the Great Plains in Nebraska and Kansas, but when large-scale irrigation became feasible on large acrages of formerly dry lands, corn and alfalfa plantings were increased many times over. More than 25,000 irrigation wells are in use in Nebraska alone.

Until about a century ago, the Great Plains were regarded as and termed the "Great American Desert." Many early explorers reported the region to be unfit for human habitation. The Spanish explorers had to use celestial navigation and artificial towers ("stakes") to find their way around on the featureless plain of the Texas panhandle and adjoining New Mexico, the least eroded area. The U.S. Army at an early date regarded the southern Great Plains as so inhospitable that it imported camels for use in that area; and many years later when some of the bones of these camels were discovered, it caused a mild paleontological uproar. It was not until the period of transcontinental railroad building that the Great Plains began to interest permanent settlers.

The Great Rivers

Several large rivers have headwaters in the intermontane basins in the Rocky Mountains and flow E across the Great Plains through well-developed, mature valleys. These are the Missouri River, including the Yellowstone at the N, the North Platte and the South Platte, the Arkansas, mainly across Colorado and Kansas, the Canadian, and the Pecos River to the southwest. These rivers have had long histories which began with widespread constructional sedimentation over the central and southern Great Plains as far back as early Tertiary time; later they underwent complicated Pleistocene erosional and depositional histories in the quite long records of terrace cuts and fills with extensive deposits of gravels, sands, and silts dating from the earliest Pleistocene time to the present. Some of these deposits are important aquifers.

A number of lesser rivers, such as the Little Missouri, Cheyenne, White, Niobrara, Republican, Cimarron, Red (Oklahoma and Texas), Brazos, the Colorado (Texas), and a number of others head in the central and western parts of the Great Plains; for the most part they have only a Pleistocene history, but some in the southern plains appear to have originated in the very late Pliocene (Ogallala), or in earliest preglacial Pleistocene time.

FIGURE 1A. Physiographic sketch map of the northern Great Plains (slightly modified after Fenneman, 1931). Note position of Devil's Tower (Fig. 4). Buried valleys show that the ancestral Yellowstone and other rivers flowed NE into Canada prior to the Quaternary glaciation. They were deflected to the SE and today the principal end moraines are closely paralleled by the course of the Missouri River.

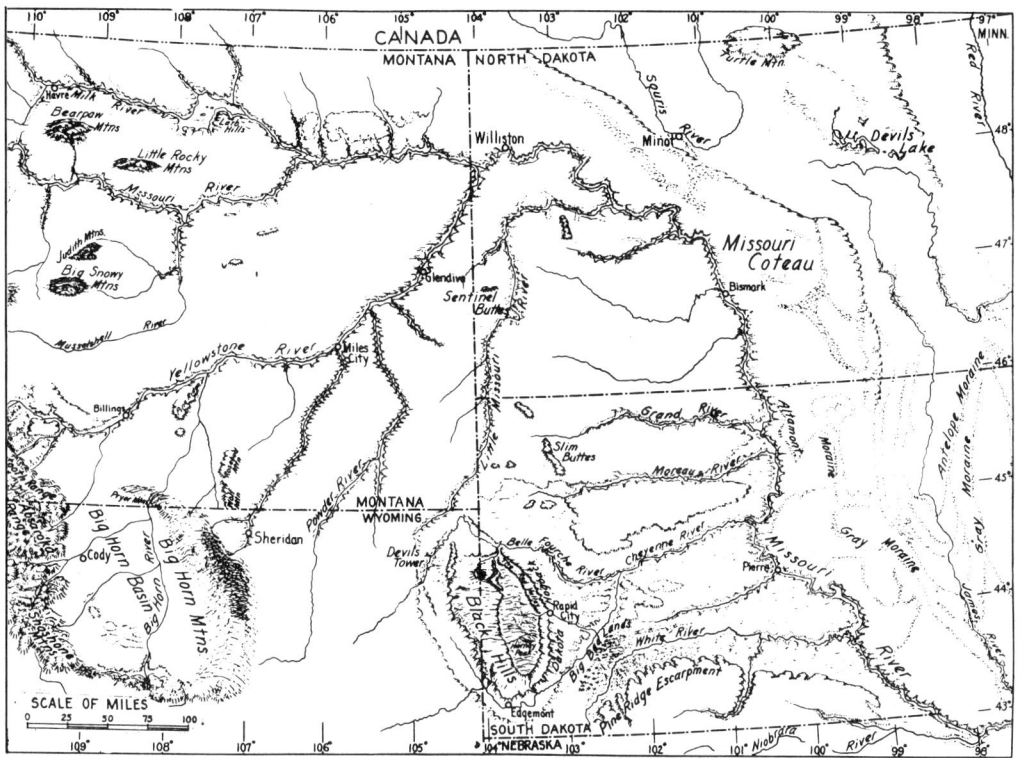

Origin of the Great Plains

The Rocky Mountains and the Great Plains, like fraternal twins, were born out of the same great Cretaceous geosyncline during the Laramide revolution, which began in late Cretaceous, Pierre time, perhaps almost a 100 m yr ago. The western half of this ancient pre-Cenozoic sea floor became the lofty, highly folded and faulted Rocky Mountains.

Most of the eastern half of the Cretaceous seaway was elevated gently and, as the waters drained away, the almost featureless surface came to be exposed as a very wide, slightly concave plain sloping very gently E; from time to time it was elevated higher and higher at the W and tilted more steeply to the E, but sagged down somewhat along an ill-defined N-S axis. Faulting developed along the Rocky Mountain front. Part of the old sea floor at the extreme E is classified physiographically as a part of the "Interior Lowlands," from Minnesota and the eastern Dakotas to central Texas.

The "High Plains" came into existence as a major geomorphic form by constructional, fluvial sedimentation beginning at the time of the Cretaceous-Tertiary transition, in response to the uplift of most of the Rocky Mountain system, which turned out to be a very long and complicated process. This was not completed until the highest Ogallala beds were in place at the end of Pliocene times in the central and southern Great Plains.

The uplift of all of the tectonic elements of the Rocky Mountains did not take place at one time therefore and changes continued through the Paleocene and Eocene epochs. Subsequently several additional uplifts occurred during the later Tertiary and Pleistocene times. The construction of the High Plains was well stated by W. D. Johnson (1901, p. 631). "Hence, with a few small exceptions at slightly lower levels, the High Plains are Tertiary in the date of their building. They were blocked out in plateau fragments in the Pleistocene; but it would seem there can be little question that these surfaces

FIGURE 1B. Physiographic sketch map of the central Great Plains (from Fenneman, 1931). Comparison with a structure contour map of the Denver Basin (Fig. 11) discloses that the present High Plains cover what has been a long-subsiding structural basin (see also Fig. 2).

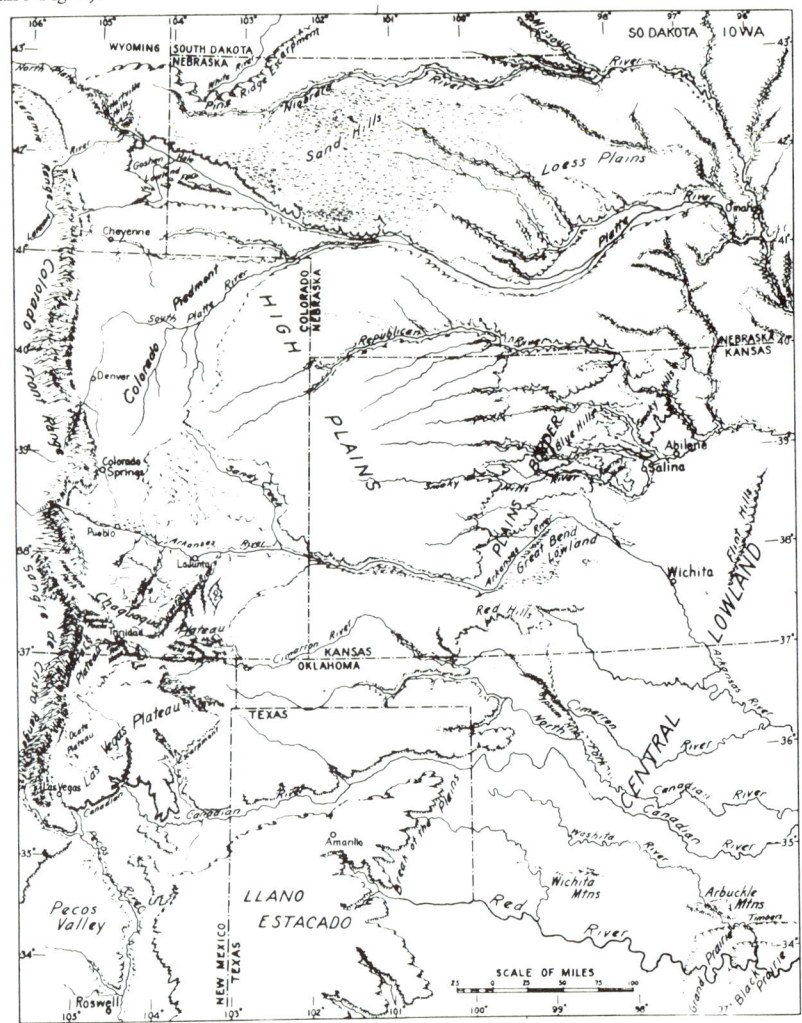

are survivals, without sensible change, from as long ago as the beginning of the Pleistocene." This is especially true of the Ogallala High Plains plateau areas, extending from the North Platte River to Texas. At first, after uplift was initiated during Paleocene and Eocene, sedimentation was confined to the complimentary intermontane basins or valleys, which, except in the northern Great Plains, are restricted to the Rocky Mountain region. The mountain areas continued to rise, tilting the Great Plains to the E. This caused the E-flowing rivers to cross out of the intermontane basins onto the adjacent Great Plains. Alluvial sheets of early Tertiary age were thus accumulated E of the mountains.

As the Rocky Mountains were progressively uplifted, the sediments were spread farther and farther E. In this way a great detrital apron came to be deposited progressively farther to the E, covering eventually most of the central and southern Great Plains. By the Oligocene (White River time), the sands and silts reached into western Nebraska. Later uplifts followed by extensive erosion, resulted in the removal of much of the Oligocene in certain areas, e.g., in Montana, North and South Dakota, and in Colorado, where Cretaceous shales are now widely exposed. By Pliocene time (Ogallala Group), the sheet of Tertiary sediments covered much of Nebraska, half of Kansas, the pan-

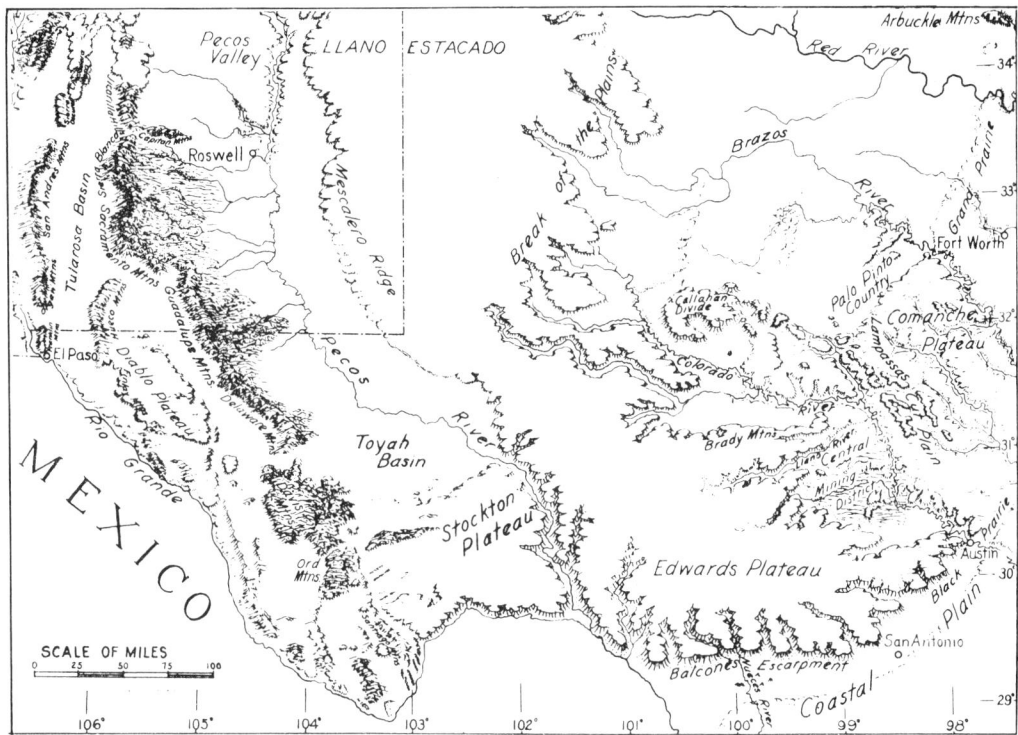

FIGURE 1C. Physiographic sketch map of the southern Great Plains (modified after Fenneman, 1931). The southwestern limits of this province follows a line west of the Pecos Valley down to the Big Bend of the Rio Grande, and the southeastern margin is marked by the Balcones Escarpment of the Edwards Plateau.

handle of Oklahoma, all of eastern Colorado, and eastern New Mexico from the Rocky Mountain front E across most of the panhandle of Texas.

The streams and rivers responsible for this sedimentation were broad, and shallow, of braided type and sediment clogged. The beds ranged up to 1 km wide, with the channels 1 to 10 m deep, comparable to the channels of the present rivers. There were also "lagoons," cutoff lakes, and other areas of slack water, where fine deposits accumulated; but probably no extensive freshwater lakes ever existed. Between the streams and along their courses, there were grass and vegetation-covered swamps and lowland flats. From time to time eolian sand and silt drifted into dunes and hummocks on the dry flats. The coarser gravels and sands fill the old channels; the silts and clays settled in the quieter backwaters and over the floodplains, or were shifted about by the wind. In early Tertiary time, deposits of gypsum and other evaporites represent periods of desiccation, whereas in later times (Tertiary) dry seasons are represented by zones of hard caliche. During every episode of extensive alluviation, however, the supply of water in the streams had to be adequate to maintain deep saturation and a high groundwater table in the thick sediments already deposited to maintain full channels, competency, and capacity of the rivers to transport sediment. In arid times the streams may have been dry for long periods, the water table far below the surface, and wind action could have been vigorous for long intervals; during the Pleistocene with its cold phases and arid climaxes, much loess might have been transported out of the region toward the E.

Paleozoic and Mesozoic Subsurface

Northern Great Plains. *Lower Paleozoic.* The Upper Cambrian rests on Precambrian crystalline metamorphic and igneous rocks. It contains abundant sandstones and some limestones. The Ordovician includes thick shales with some pervious sandstone and carbonates. The Silurian is similar and up to 335 m thick in the Williston Basin.

Upper Paleozoic. The Devonian system, 600–900 m thick in places, contains shales and

FIGURE 2. Three E-W profiles through the Great Plains' Above: from the Black Hills to the E of the Missouri River, passing through the Badlands of South Dakota (Darton, 1919: U.S.G.S. Bull. 691). Middle: The "Gangplank" sector of Wyoming and Nebraska. Below: The southern plains from the Pecos Valley to the E along latitude 34°45'N (U.S.G.S. Water Supply Paper 154).

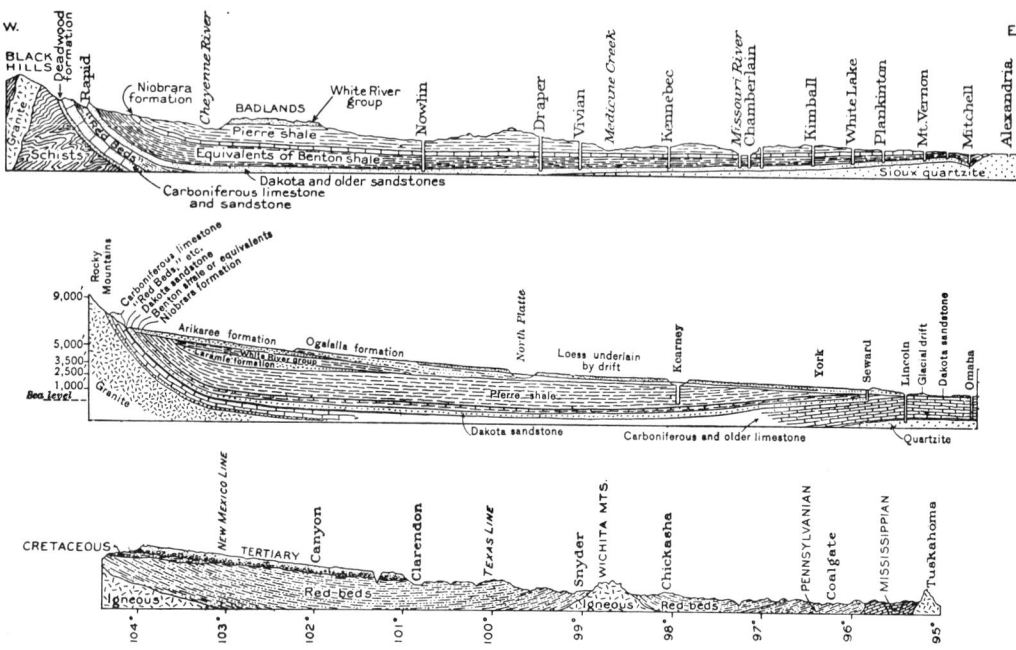

carbonates. Exposed Upper Devonian reefs have long been known in the Canadian Rocky Mountains, but it was not until about 1947 that Devonian reefs were discovered from oil drilling in the northern Great Plains, mainly in the Williston Basin.

The Mississippian includes carbonaceous dark shales and thick, massive limestones, such as the Madison and the Pahasappa. The Pennsylvanian and Permian systems contain large amounts of sandstones and sandy shales, and also some extensive limestones, mostly rather impure but with extensive stratigraphic continuity, as the Minnelusa (Pennsylvanian) and the Minnekahta (Permian).

Triassic System. The Triassic system is thin or absent in the northern Great Plains region. It is present and thickens in the geosyncline to the NW of the Williston Basin in Saskatchewan. Whereas some of the Triassic is marine to the E, these beds interfinger with red beds and evaporites.

Jurassic System. The Jurassic system under the northern Great Plains is generally about 150–300 m thick. It consists of sandstone, shale, and limestone, and the colors are quite varied. It thickens to 2500 m in the deeper part of the geosyncline to the W and NW, and much of it is a gray marine shale and limestone, with a number of clean, persistent sandstone formations deposited in the "Sundance Sea" (Callovian and Oxfordian stages).

The Upper Jurassic beds from the central Great Plains in Colorado and Wyoming, in the Black Hills of South Dakota, and in Montana represent a region consisting of mainly nonmarine, continental facies, which were deposited in shallow bodies of water, marshes, and swamps. These constitute the Morrison Formation. The Morrison is most noted for the great variety and abundance of its dinosaur remains. The great reptiles had appeared in the Triassic, but they attained their greatest development in the Jurassic, and continued to be dominant in the Cretaceous.

Central and Southern Great Plains. *Paleozoic Systems.* Every Paleozoic system is represented under the central and southern Great Plains from Nebraska and Wyoming to Texas, with much the same facies and structures, as in the northern Great Plains. An area of unusual structural and economic interest includes a little of southwestern Kansas, most of the Oklahoma panhandle, the northern panhandle of Texas, and a little of southwestern Oklahoma. It lies almost entirely within the High Plains. There there is a buried E-W faulted belt, the Amarillo mountain range, which was more or

FIGURE 3. Generalized profile through the Zuni sequence across the Central Rockies and northern Great Plains. Wavy boundaries indicate earlier sequence divisions (Sloss, 1963).

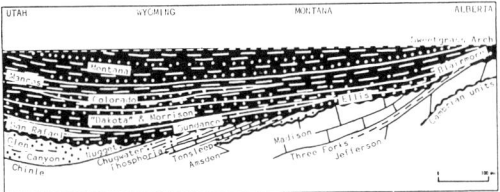

less contemporaneous in uplift (Pennsylvanian) with the Ouachita belt in Oklahoma, with which it is lineated. This long feature is in structural continuity with the earlier (Mississippian) "Ancestral Rocky Mountains" in central Colorado. This "mountain range" is such only in geological sense, inasmuch as it is completely buried under thick Triassic, Cretaceous, and Tertiary sediments. It is of great economic importance.

The southern extremity of the Great Plains covers the West Texas Permian Basin. This region has the most complete stratigraphic section of Permian formations in North America, mainly developed as marine shales and limestones.

In the central and southern Great Plains, the Upper Pennsylvanian and Permian formations consist of alternating, rather thin marine beds of limestone and shale, which usually have great stratigraphic continuity and maintain uniform thicknesses for great distances. The sandstones are commonly lenticular and more uneven. This lithology persists in the north and westward under thick younger deposits, to the Highwood Mountains in eastern Wyoming. Farther S this lithology gives way to a great increase in red beds, more gypsum, anhydrite, and salt to the W, where the Permian red beds are difficult to distinguish from the overlying Triassic red beds. The last Permian marine sedimentation took place in three terminal marine basins in west Texas. They were the Midland, Delaware, and Marfa basins.

Triassic and Jurassic. Terrestrial red beds of Triassic age are widespread, consisting of red and maroon sandstone and bright red sandy shale, from 50 m thick to more than 300 m, with lenses of gypsum and commonly with a persistent bed of gypsum up to 5 to 10 m thick at the top. They are present in the subsurface in western South Dakota, western Nebraska, eastern Wyoming and Colorado, and underly the High Plains Tertiary beds in eastern New Mexico and a large part of the Texas panhandle. These red beds are exposed in the Black Hills, the red valley of the "Red Race Track." They are known as the Spearfish Formation in the Black Hills and eastern Wyoming, the Chugwater farther W in Wyoming, and the Lykens Formation in eastern Colorado; the last is mainly in the subsurface but exposed in Colorado to the SE. The Dockum Group is present in eastern New Mexico and E into Texas, where the Tertiary Ogallala rests unconformably on it.

Some of the Jurassic formations, for example the Sundance marine beds and the continental Morrison, which have been noted in the northern Great Plains, are also present in the subsurface of western South Dakota and western Nebraska. They are exposed in the Black Hills, and are widespread in eastern Colorado and in northeastern New Mexico. Some Jurassic continental sandstones and sandy light-colored to red shales occur in some other places in the southern Great Plains.

Cretaceous System of the Great Plains

The Cretaceous rocks were deposited in a very wide, generally shallow, geosyncline (or epicontinental sea), which at its greatest width extended far into Iowa and Minnesota to the E and more than halfway across Utah to the W, about 1200 km. N-S it reached from the Gulf of Mexico to the Arctic Ocean. The facies consist largely of shales and limestones. The lower or basal beds are sandstones with some shales.

Volcanic activity, associated with the Laramide movements, especially during later Cretaceous time, provided voluminous ash falls, which repeatedly settled through the marine waters over large areas, and accumulated in thick beds, which now constitute important layers of very pure bentonite. Such beds occur widely in the Pierre Shale, and thinner beds are present in other older formations, serving as ideal time markers for stratigraphic correlation.

The Fox Hills, Lance, and Hell Creek beds total 300–450 m or more in thickness of shales and sandstones are in the main marine, but to some extent are transitional from the Cretaceous marine environment to the more terrestrial Tertiary environment. They are well developed in certain areas in northwestern South Dakota, NE into North Dakota, and in Montana. The Lance is thickest in the Powder River Basin, followed by thick early Tertiary deposits.

Tertiary of the Northern Great Plains

Paleocene. Almost all of the northern Great Plains surface in Canada is underlain by Cretaceous formations of widely ranging ages within that system. They are followed by the Paleocene Fort Union beds, and these also cover large areas in Montana and North Dakota. It would seem likely that the Fort Union must have been eroded extensively over a large area in this U.S.–Canadian region. Whatever Tertiary cover this large northern area of the Great Plains may have had, it appears to have been almost completely stripped away, and the region may never have had any Miocene cover. It is doubtful if Eocene sediments were ever very extensive here.

The Paleocene Fort Union Formation extended over a large area in eastern Montana, western North Dakota, and northwestern South Dakota. The basal beds are considered to be the earliest truly continental Tertiary deposits, and their accumulation followed the latest transition from true Cretaceous beds, the Fox Hills, Lance, and Hell Creek beds in undisturbed sequence over much the same areas.

To the E, however, in the vicinity of Bismark, North Dakota, the Lebo Shale Member of the Fort Union Formation grades into the Ludlow Shale, which in turn undergoes a facies change into the Cannonball Member (with gray, marine shales, 90 m thick). This is the last vestige of any marine sedimentation in the interior of North America. An earlier view regarded the Cannonball member as a facies of the older Hell Creek Formation of the Lance (Cretaceous).

In general the Fort Union Formation consists of varicolored layers of whitish, yellowish to yellow-brown, interbedded mudstones, sands, sandstones, siltstones, and clay shales, in some places with considerable amounts of lignite, especially in the Ludlow member. It also contains some beds of bentonite. Facies changes in the formation are common and numerous. It ranges widely in thickness, but reaches 425–460 m.

Eocene. The Eocene Golden Valley Formation lies unconformably on the Fort Union. It consists of micaceous sands, silts, and white and yellow clays. It contains some coal seams, and it is about 30 m thick. It apparently does not occur in Montana, but it caps many small buttes and mesas in western North Dakota, and it underlies Oligocene White River beds in the Killdeer Mountains in Dunn County, North Dakota.

Eocene (40.5 m yr B.P.; Bassett, 1961) volcanic activity may have been responsible for the formation of the Devil's Tower, Wyoming, as its distinctive columnar jointing is characteristic of cooling in a crater, classifying it as a volcanic neck. Some writers, however, have prefered to view the Devil's Tower as an eroded laccolith

FIGURE 4. Devil's Tower (or Mato Tepee), an eroded pluton, situated in the northern plains (northeastern Wyoming). Rising 250 m above the plain, it consists of tinguaite or phonolite porphyry. Established as a U.S. National Monument in 1906. Photo: A. L. Lugn.

due partly to the presence in the area of a number of laccolith domes (for a discussion see Vol. III of *Encyclopedia of Geomorphology*).

Oligocene. Oligocene White River beds in the northern Great Plains rest unconformably on the Paleocene Fort Union Formation or on the Eocene Golden Valley Formation. White River deposits (Chadron and Brule formations in southwestern South Dakota and Nebraska) in this area are up to 60 m thick, and consist of sands, silts, silty clays, some hard, nodular calcareous beds, and whitish volcanic ash layers. The lower part (Chadron) is gravelly and conglomeratic and tends to be more grayish in color. The upper part (Brule) is finer in texture and pinkish to reddish in color.

The Cypress Hills Plateau, in the SW of Saskatchewan, is capped with a thick bed of coarse gravel and conglomerate not unlike the Chadron Formation of northwestern Nebraska. The Cypress Hills Formation has been dated as of Oligocene age on the basis of good paleontological evidence.

It appears that there had been widespread deposition of the Oligocene White River beds in eastern Montana, western North Dakota, and adjacent Saskatchewan; they were very probably in geographical continuity with the more extensive and thicker deposits to the S in the central Great Plains. Many White River Group erosion remnants are present on the outer slopes of the Black Hills. Gravels on an old erosion surface (Mountain Meadow surface) in the Black Hills have been correlated with the basal Chadron gravel in South Dakota and Nebraska.

No Miocene deposits appear to have been identified in the northern Great Plains, and the region seems to have undergone erosion through the Miocene and most of the Pliocene to the time of the deposition of the Flaxville Formation (late Pliocene).

Late Pliocene (Flaxville Formation). The Flaxville Gravel is a valley-fill deposit, a cut-and-fill terrace in the wide, relatively shallow valleys of the Missouri, the Yellowstone, and the Little Missouri rivers in Pliocene time. These major streams probably flowed at about 150 m or more below the Cypress Hills level, that is, on the assumption that the Rimroad gravel in northeastern Montana is correlative with the upper member of the Cypress Hills gravel. By the end of Flaxville time, the major streams were flowing at 175 to 200 m above the present valley floors.

Calcareous sandstone or "mortar beds" overlie the thick gravels and are a part of the Flaxville Formation. They are almost identical lithologically to the Kimball Formation in Nebraska and Kansas, which includes the "Sidney" gravels at the base. Further, these beds also contain fairly abundantly a small variety of fossil seeds of *Celtis* sp. (hackberry), which are common in the Kimball Formation.

Subsequent to Flaxville deposition, erosion dissected the Flaxville plain into many large and small plateaus, for which the formation forms the caprock.

It seems that there may never have been any physical or geographic continuity between the Flaxville Formation in Montana and the Kimball Formation in Nebraska and Kansas; but there is a time, climatic, and environmental correlation between them.

Tertiary of the Central and Southern Great Plains

To the W of this general area, the oldest Tertiary deposits are mostly confined to intermontane basins of the Rocky Mountain region. At the foot of the Rocky Mountains in the basin E of Denver, an area of Eocene beds of about 10,000 km^2 (4000 sq m) rests on Lance deposits. A second area of Eocene beds is present in the Spanish Peaks area farther S, and this extends into New Mexico.

White River Group. The Oligocene White River beds are the most extensive Tertiary deposits in the central Great Plains, being present in a considerable area of northeastern Colorado, underlie about 30% of Nebraska, and occur extensively in southeastern Wyoming, southwestern South Dakota, and on the outer slopes of the Black Hills. Many outliers of the White River known in the NW of South Dakota, in western North Dakota, and in eastern Montana have already been noted. The former extent of the White River beds in the central and northern Great Plains may once have been nearly double their present known geographic occurrence.

White River deposits rest unconformably on the Pierre shale or older beds. In northwestern Nebraska in the subsurface (over the Chadron Dome), the White River (Chadron Formation) rests on Jurassic Morrison beds and a succession of older Cretaceous formations. Farther to the SW in Nebraska, the White River beds lie on Pierre shale, "transition beds," and Fox Hills and Lance beds.

The lower part of the White River group, the Chadron Formation, can be seen in a conspicuous landmark, Sugar Loaf Butte, a short distance NE of Orella, Nebraska. Here, the basal 1–4 m is the Chadron gravel or conglomerate. Above it there is about 40 m of grayish to greenish silty clay, with some sandy beds, and some harder calcareous layers, causing the butte to stand out prominantly as a whitish, monolithic pillar on the level landscape. The famous Toadstool Park locality only 5 km S of Orella contains fine exposures of the Brule Clay 161 m thick. This is one of the most typical known exposures of the Brule Formation. It is made up of massive silty clay, nodular and blocky

FIGURE 5. Differential erosion in the Oligocene Chadron Formation, at Toadstool Park, NW of Crawford, Nebraska, with "Metamynodon Channels," a cross-bedded sandstone. (Photo: A. L. Lugn.)

FIGURE 6. Toadstool Park Badlands, near Crawford, Nebraska, eroded in the Oligocene White River beds; Chadron Formation below and Brule Clay above. Note scale of man, center. (Photo: A. L. Lugn.)

sandstone layers, and contains several beds of whitish volcanic ash. At the base of the lower or "A" zone of the Orella Member is the famous "Toadstool" sandstone, 2.5–7 m thick, containing the fossiliferous "Metamynodon channels" (Fig. 5). It has been suggested that some beds of the Brule clay may have been eolian.

Toadstool Park is in the area of the Little Bad Lands of Nebraska, N of the town of Crawford. The Big Bad Lands of South Dakota are located about 65–125 km SE of Rapid City, S.D., where the White River formations are thick and dissection is generally intricate. The forms produced in the Chadron are generally rounded, with smooth outwardly convex surfaces. Bad Lands features developed on the Brule Clay have high, sharp, narrow ridges and sharp pinacles, with outwardly concave surfaces (Fig. 6). Also, the gray and grayish-green colors of the Chadron beds and the varied pink to reddish colors of the Brule Clay results is striking color contrasts. This is greatly enhanced if seen shortly after a shower of rain.

The White River beds contain the fossil remains (bones) of a great variety of extinct mammals. The Bad Lands have been important collecting grounds of paleontologists for more than 100 years.

Miocene Arikaree Group and Hemingford Group (Table 1). The Miocene Arikaree Group is restricted to western Nebraska and small areas adjacent in South Dakota, Wyoming, and northeastern Colorado, where it rests in the main on Oligocene White River beds. In this group of formations the Gering, Monroe Creek, and Harrison are the best exposed, in the steep walls of the many canyons of the "Pine Ridge" across Niobrara County, Wyoming; also in northern Sioux, Dawes, and northern Sheridan counties, Nebraska, and in Shannon County and adjacent areas in South Dakota; and also in the North Platte River valley from Goshen County, Wyoming; and extensively in Scotts Buff, Morrill, and Banner counties, Nebraska. Exposures are especially good in the Scotts Bluff butte and to the SE in "Wild Cat Ridge."

The famous Agate Springs fossil quarries, where an abundance of many kinds of vertebrate remains have been collected from the Harrison beds for about 80 years, is located in central Sioux County about 37 km S of Harrison, Nebraska. *Daemonelix* is a celebrated feature in the Harrison (Fig. 7).

The Miocene Hemingford group of formations (Marsland and Sheep Creek) occurs in a more restricted area in Sioux, Dawes, Box Butte, and Sheridan counties, Nebraska.

Pliocene (Ogallala Group). The Pliocene Ogallala Group is present in a fairly large area of south-central South Dakota, underlies much of Nebraska, about one-third of Kansas, large areas in eastern Colorado, the panhandle of Oklahoma, and most of the Llano Estacado of Texas and New Mexico. The Ogallala beds lie on the White River deposits, Arikaree formations, and Pierre and Niobrara shales in South Dakota and Nebraska, and on Pierre shale and older Cretaceous formations in Kansas, Colorado, Oklahoma, Texas, and New Mexico. The geographic area of Ogallala occurrence almost delineates the original area of the Tertiary (Pliocene) "High Plains" region of the Great Plains, with the exception of the very large areas E of the Rocky Mountains that have been stripped of most of the Tertiary cover.

FIGURE 7. Trace fossil *Daemonelix* ("Devil's corkscrew") in the Miocene Harrison Formation from near Harrison, Nebraska (Photo: Larry Rider.)

TABLE 1. Tertiary Formation of Nebraska
(South Dakota and Nebraska to Texas and New Mexico: Lugn,
1939, 1956)

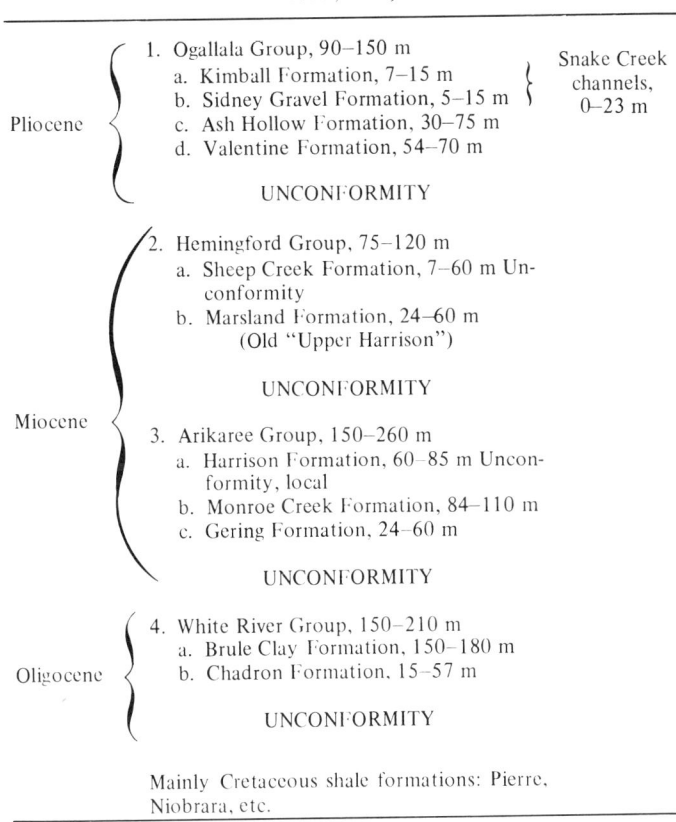

Pliocene	1. Ogallala Group, 90–150 m 　a. Kimball Formation, 7–15 m ⎫ Snake Creek 　b. Sidney Gravel Formation, 5–15 m ⎬ channels, 　c. Ash Hollow Formation, 30–75 m ⎭ 0–23 m 　d. Valentine Formation, 54–70 m UNCONFORMITY
Miocene	2. Hemingford Group, 75–120 m 　a. Sheep Creek Formation, 7–60 m Unconformity 　b. Marsland Formation, 24–60 m 　　(Old "Upper Harrison") UNCONFORMITY 3. Arikaree Group, 150–260 m 　a. Harrison Formation, 60–85 m Unconformity, local 　b. Monroe Creek Formation, 84–110 m 　c. Gering Formation, 24–60 m UNCONFORMITY
Oligocene	4. White River Group, 150–210 m 　a. Brule Clay Formation, 150–180 m 　b. Chadron Formation, 15–57 m UNCONFORMITY
	Mainly Cretaceous shale formations: Pierre, Niobrara, etc.

The Ogallala Group of formations (up to 175 m thick) consists of typical fluvial deposits of coarse sands and gravel channel fills with beds and lenses of floodplain silts and clays. Almost all the Ogallala beds consist of hard and soft zones; and the hard beds, for the most part, are cemented with caliche ('mortar beds"), which contain calcareous plant and root tubules (rhizo-concretions), and many zones contain stratigraphically significant fossil seeds of prairie grasses and other herbs. The fossil seeds are useful as local, zonal index fossils. Vertebrate fossils are present at almost all levels in many localities. Bone beds are most likely to be found where the animals may have been trapped in quick sands.

There is some evidence that some of the Ogallala deposits may have extended farther N, perhaps across northwestern Nebraska and eastern Wyoming, to the western side of the Black Hills. The Ogallala deposits maintain their lithologic and stratigraphic characteristics S into Texas and New Mexico.

Pleistocene of the Great Plains

All of the Great Plains area in Canada was glaciated by the continental glaciers, which at times merged at the front of the Rocky Mountains, with ice from the Cordilleran glaciers. Continental glaciation extended into northern Montana as far as 200 km S of the Canadian boundary along the Rocky Mountain front reaching E into North Dakota. Parts of the Great Plains in North Dakota and South Dakota were also glaciated. Actual glaciation affected but little of the Great Plains in the United States, but there was much periglacial activity.

The high Ogallala plain at the end of Pliocene time in all probability extended W intact to the Rocky Mountain front, where it extended and had physical continuity with deposits in mountain feeder valleys. Profound erosion by water and wind has stripped most of the Tertiary cover from large areas to the E of the mountains in eastern Colorado and New Mexico. The major through-flowing rivers developed mature wide valleys across the Great Plains from W to E. Across the central and southern Great Plains are the South Platte, the Arkansas, the North Canadian, and the Canadian. The headward erosion and development of the Pecos River and its many tributaries have stripped not only the Tertiary cover but extensively exposed the underlying Cretaceous and Triassic beds down to the Permian. This stripping of the Tertiary cover, especially of large areas of Pliocene Ogallala beds, from the western Great Plains (High Plains) began very early in the preglacial Pleistocene.

The Pleistocene gravel and sands, valley-fill deposits, are notable especially in Nebraska and Kansas. In southern Nebraska three large, wide valleys were filled with a succession of sand and gravel during the Nebraskan glaciation (David City Gravel and Holdrege Formation); it was followed in Aftonian interglacial time by an extensive layer of sandy silts and clay (Fullerton Formation). With the second or Kansan glaciation, a more extensive sheet of sand and gravel was spread over the Fullerton Formation.

The correlation of Pleistocene volcanic ash deposits in recent years has led to a new time classification of the beds, particularly the Pearlette ash, and of their point of origin (Izett et al., 1972). Late Kansan or early Yarmouth time is represented by extensive volcanic ash deposits over most of the central Great Plains, the Pearlette ash, type "O" (approximately 0.6 m yr old). This ash bed is associated with several layers of fluvial silt and clay (comprising the Sappa Formation), not unlike the earlier Fullerton bed. The similar-appearing but stratigraphically distinct Pearlette ash, type "S" (approximately 1.2 m yr old), is especially widespread in both Nebraska and Kansas, as well as in some adjacent areas. It has been suggested that both the Pearlette ash type "O" and type "S," as well as the early Pleistocene type "B" (2.0 m yr old), may have been distributed from eruptions occurring in the Yellowstone Park, Wyoming, area (Izett et al., 1972).

It would appear that this wide spread in the dating of these ash falls might very well indicate at least two specific ages within the Pleistocene. There can be little doubt that the 0.6 m yr old ash, type "O," is the *one true Pearlette ash.* The older or type "S" of more than a half million years earlier (the Coleridge ash in Nebraska), dated at about 1.2 m yr old, is not Pearlette at all, but is an ash fall of late Nebraskan or Aftonian age. The type "B" ash may be a still earlier Nebraskan ash, which has not yet been recognized in Nebraska. The above suggestion indicates another important dating; that is, the duration of the Kansan-Yarmouth epochs appears to have totaled 600,000 to 800,000 years.

FIGURE 8. *Archidiskodon imperator maibeni* (Barbour), the largest recorded elephant, 4 m high at the shoulder, found in the Pleistocene Loveland soil (over the Loveland Loess) in Lincoln County, Nebraska; exhibited at the University of Nebraska Museum. (Photo: A. L. Lugn.)

Farther W in the North Platte River valley in Nebraska, there is an almost identical development of high-level fluvial deposits. The Kansan deposits are capped by younger accumulations of loess. The dating of these terrace beds can be correlated by means of early Pleistocene vertebrate fossils (Fig. 8). A succession of lower terraces, mostly strath or cut-terraces with only thin layers of sand and gravel (1–10 m thick), correlate with later glacial advances and retreats. Terrace successions can be duplicated in most of the through-flowing rivers in the Great Plains from Nebraska and Wyoming S to Texas and New Mexico.

The high-terrace Blanco Formation in the panhandle of Texas, the Rexroad beds in SW Kansas, and the lower member of the Broadwater Formation in western Nebraska ("T^5" or "Terrace Hills"), and including also the Holdrege and David City deposits in south-central Nebraska are believed to be correlative and Nebraskan in age. If a sufficiently long span of time is accepted for the development of the wide, mature, pre-glacial Pleistocene valleys (several kilometers wide and 60–120 m deep) below the original "High Plains" level, which contain the above sediments, then the North American Pleistocene can quite properly be correlated with the Calabrian and the Villafranchian deposits of S Europe.

Later Pleistocene Events and the Loess

During the Pleistocene further uplift occurred in the Rocky Mountains which reached maximum elevations in Colorado and Wyoming, by Illinoian time. As a result, the central Great Plains area, mostly in eastern Wyoming, eastern Colorado, Nebraska, and Kansas came to have a very dry climate during the glacial phases, with long periods of drought and dust-bowl conditions occurring repeatedly but which became most significant in mid to late Wisconsin time. The extensive and thick accumulation of loess deposits in Nebraska and Kansas of Illinoian and Wisconsinan ages is one of the most interesting features of this period (Fig. 9).

The thickest loess deposits anywhere in North America (Loveland: up to 30 m; Peorian and Bignell: 60–75 m) occur in the Loess Hills region, an area to the SE of the Sand Hills region, to the N of the Platte-Loup River valley in central Nebraska, and also in the western part of the extensive Loess Plain between the Platte and Republican rivers in south-central Nebraska. The thick loess extends SW into northwestern Kansas, where thicknesses range up to 30 m. The thickest loess zone extends from NE Nebraska to the SW for 500 km into western Kansas, and it ranges in width from 40 to 120 km. To the SE across Nebraska and Kansas the loess decreases in thickness and is finer in texture, which suggests source areas to the NW.

These source areas were in north-central Nebraska, the great Sand Hills Region, which covers 50,000 km (20,000 sq mi). These sand dunes overly older Ogallala beds (Fig. 10). Other areas of sand dunes are present in Nebraska, S of the Platte River, notably in Lincoln County and adjoining areas into northeastern Colorado, all lying to the NW of the thickest loess.

The Sand Hills Region noted above is still almost completely underlain by middle and lower Ogallala sediments (mainly Ash Hollow and Valentine formations). It has been eroded

FIGURE 9. West-facing loess bluffs in central Lincoln County, Nebraska, rising 60 m above the very thinly loess-covered Tertiary plain, top of Kimball Formation of Ogallala Group in foreground. (Photo: A. L. Lugn.)

FIGURE 10. Sand Hills region in Hooker County, Nebraska. (Photo: A. L. Lugn.)

and dissected by fluvial erosion before the ages of principal eolian action. Early Pleistocene drainage courses had been established, and these early Pleistocene valleys had become in part filled with early and middle Pleistocene deposits, largely if not entirely before Loveland (late Illinoian) time. It appears likely that about 50% of the Tertiary beds (Ogallala) in the Sand Hills region has been stripped off.

The large admixture of very fine sand and silt in the thick loess across Nebraska and western Kansas can be attributed to the abundance of this material in the Ogallala of the Sand Hills Region, which was then being stripped by water and wind erosion. Erosion and reworking of both the dune sand and the *in situ* Ogallala beds by pluvial and fluvial processes between periods of eolian action must have contributed materially to the development of the Sand Hills Region and the removal of loessic materials and sands from the area. The silt and clay were simply winnowed out across the surface by wind action, which left the sand and coarser materials behind as residuals to form the sand dunes. The fines became loess deposits to leeward of the Sand Hills Region source area.

It appears from the field evidence that the sandy nature of the loess of the central Great Plains is directly related to composition of the Ogallala beds, which are almost ideally constituted to have yielded large quantities of loessic material, and also to leave behind large amounts of residual dune sand.

There is also evidence that the stripping of large areas to the W of the Great Plains during the Pleistocene contributed much eolian silt and clay to the loess deposits in the Midwest, even to the Mississippi valley. Most of the central Great Plains loess thus had a distinctly "Regional Source" or a "Desert Source" (see Lugn, 1968, for discussions on the eolian origin of the Loveland and Peorian loesses).

FIGURE 11. Structure contour map of the Denver Basin (center) and Raton Basin (south), drawn on the pre-Pennsylvanian surface (from Volk, in A.A.P.G. Mem. 15, 1971). Contour interval = 1000 ft and in part 500 ft. Strippled area in the west is the Rocky Mountain Front Range.

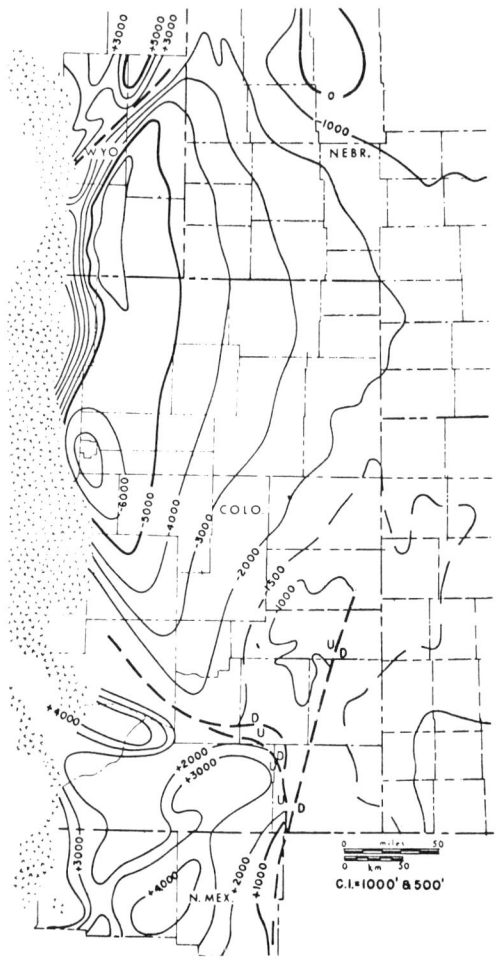

Economic Geology

It seems rather ironic that for about 100 years, the Great Plains Region in the United States was so little understood, went unappreciated, and was held in such low regard that it was called the "Great American Desert." It is perhaps not surprising that its great wealth of underground natural resources was never suspected.

The natural resources in the subsurface are primarily its natural gas and petroleum, trapped in thick Paleozoic and Mesozoic formations. These older sediments were deposited in a number of different cratonic basins and shelf areas over a period of about 400 m yr, e.g., Williston, Denver, and West Texas basins (Figs. 11 and 12).

In the northern Great Plains the Cambrian has limited shows of oil and seems to lack source beds; it contains potential reservoir rocks to which there has been migration in certain structural settings. In contrast, the Ordovician contains good source beds, as well as sandstone and carbonate reservoirs. The Silurian, on the other hand, has less potential.

The famous Devonian reef formations of western Canada have been traced also into the northern Great Plains, notably in the Williston Basin, which has been outstandingly prolific. Reef traps also occur in the Mississippian. In the northern plains region the Pennsylvanian and Permian are less promising.

In the central and southern Great Plains oil and gas have been produced from rocks of every Paleozoic period, although the facies are much the same as in the north. Of great importance is the E-W fracture belt that overlies the buried "Amarillo Mountains," situated in southwestern Kansas, western Oklahoma, and northwestern Texas. This district was once one of the greatest natural gas fields in the world, and is still producing. In addition to considerable petroleum, natural gasoline, and natural gas, which is carried by pipelines to the north-central and eastern areas of the United States, carbon black is processed from the gas, and most of the country's helium (extracted from natural gas) has been produced from a few wells in this area. The unusualness of this region is due to subsurface structural conditions arising from the existence of an old buried range of

FIGURE 12. Diagrammatic E-W section through North Dakota, showing part of the Williston Basin and its wedging out to the NE toward the Canadian Shield (from Ashmore, in A.A.P.G. Mem. 15, 1971).

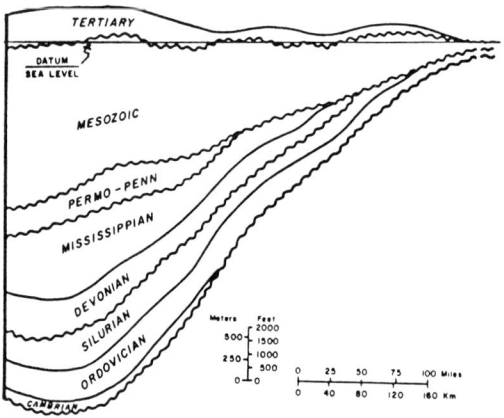

mountains, the Amarillo Mountains, which are now buried under a thick cover of Permian, Triassic, Cretaceous, and Tertiary sediments.

To the south again lies the West Texas Permian Basin, which has produced phenomenal quantities of oil and gas, mainly from Middle and Upper Permian reservoirs, but additional production is obtained from all Paleozoic systems.

In addition to oil and gas in the Paleozoic, widespread use is made of the gypsum, anhydrite, and salt deposits of the Permian from Kansas to Texas.

In the Mesozoic, the largely continental Triassic and Jurassic are generally unfavorable for the generation and trapping of hydrocarbons, but the Cretaceous is mainly marine and the basal sandstones (e.g., Dakota Sandstone) commonly truncate Paleozoic oil-bearing formations ("oil-bleeders"), which furnish hydrocarbons that are trapped in the porous sandstones. They are particularly important in the 150,000-km^2 Denver Basin of northeastern Colorado, southwestern Nebraska, and southeastern Wyoming. Jurassic and Cretaceous coals are widespread in Montana and there are also Tertiary coals in the northern plains.

Another product of great economic value in the Mesozoic is the bentonite (altered ash) found in remarkably continuous, though thin, layers, notably in the Upper Cretaceous Pierre Formation.

Cenozoic formations are essentially continental and lack valuable economic mineral resources, with the notable exception of subsurface water. As indicated earlier, the discovery of extensive artesian and fluvial gravel water bodies revolutionized the economy of the Great Plains. The Quaternary sands and gravels are also of great value in the construction industry, while the clays and loess are of value in brick and tile manufacture. Much of the most valuable agricultural land is underlain by loessic soils (28 million ha in the central plains and adjacent upper Mississippi states).

Raw materials, such as limestone, clay or shale, and siliceous sand or sandstone for the manufacture of Portland cement, are generally available to abundant in areas where a good market exists, usually near large cities.

ALVIN L. LUGN

References

Agnew, A. F., and Tychsen, P. C., 1965. "A guide to the stratigraphy of South Dakota," *S.D. State Geol. Surv.,* **14**, 195p.

Alden, W. C., 1932. "Physiographic and glacial geology of eastern Montana and adjacent areas," *U.S. Geol. Surv., Prof. Pap. 174,* 3–18.

Bassett, W. A., 1961. "Potassium-argon age of Devil's Tower, Wyoming," *Science,* **134**(3487), p. 1373.

Bayne, C. K., Davis, S. N., and Howe, W. B., 1971. "Regional Pleistocene stratigraphy," *Kans. Geol. Surv., Spec. Distrib. Publ. 53,* 5–8.

Carlson, M. P., 1970. "Distribution and subdivision of Precambrian and lower and middle Paleozoic rocks in the subsurface of Nebraska," *Neb. Geol. Surv., Rep. Invest 3,* 20p.

Carlson, C. G., and Anderson, S. B., 1965. "Sedimentary and tectonic history of North Dakota part of Williston Basin," *Bull. Am. Assoc. Petrol. Geologists,* **49**(11), 1833–1846.

____, and Eastwood, W. P., 1962. "Upper Ordovician and Silurian rocks of North Dakota," *N.D. Geol. Surv.,* **38**, 52p.

Collier, A. J., and Thom, W. T., Jr. 1918. "The Flaxville Gravel and its relation to other terrace gravels of the Northern Great Plains," *U.S. Geol. Surv. Prof. Pap. 108-J,* 179–184.

Condra, G. E., and Reed, E. C., 1959. "The geological section of Nebraska," *Neb. Geol. Surv.,* **14A**, 82p.

Cox, E. J., et al., 1962. "Geology of selected highway strips in South Dakota," *S.D. State Geol. Surv.,* **93**, 184p.

Crandell, D. R., "Geology of the Pierre area, South Dakota," *U.S. Geol. Surv. Prof. Pap. 307,* 83p.

Dort, W., Jr. and Jones, J. K., Jr. eds., 1970. "Pleistocene and recent environments of the central Great Plains," *Kans. Univ. Dep. Geol. Spec. Publ. 3,* 433p.

Elias, M. K., 1942. "Tertiary prairie grasses and other herbs from the high plains," *Geol. Soc. Am. Spec. Pap. 41,* 176p.

Fenneman, N. M., 1931. *Physiography of Western United States.* New York: McGraw-Hill, 534p.

Flint, R. F., 1955. "Pleistocene geology of eastern South Dakota," *U.S. Geol. Surv., Prof. Pap.* **262**, 173p.

Frye, J. C., and Leonard, A. B., 1952. "Pleistocene geology of Kansas," *State Geol. Surv. Kans. Bull.* **99**, 230p.

Fuller, J. G. C. M., and Porter, J. W., 1962. "Cambrian, Ordovician and Silurian formations of the northern Great Plains, and their regional connections," *J. Alberta Soc. Petrol. Geologists,* **10**(8), 455–485.

Hainer, J. L., 1956. "The geology of North Dakota," *N.D. Geol. Surv.,* **31**, 46p.

Hattin, D. E., 1965. "Upper Cretaceous stratigraphy, paleontology, and paleoecology of western Kansas (Field conference guidebook for annual meeting, Geol. Soc. Am. and associated societies, Kansas City, Missouri, 1965)," *State Geol. Surv. Kans.,* 69p.

Howard, A. D., Jr., 1957. "Cenozoic history of northeastern Montana and northwestern North Dakota, with emphasis on the Pleistocene," *U.S. Geol. Surv. Prof. Pap. 326,* 107p.

Izett, G. A., Wilcox, R. E., and Borchardt, G. A., 1972. "Correlation of a volcanic ash bed in Pleistocene deposits near Mount Blanco, Texas, with the Gnaje Pumice Bed of the Jemez Mountains, New Mexico," *Quat. Res.,* **2**, 554–578.

Johnson, W. D., 1901–1902. "The High Plains and their utilization," *U.S. Geol. Surv., 21st Ann. Rept., pt. 4,* 601–741; *ibid., 22nd Ann. Rept., pt. 4,* 631–669.

Landes, K. K., 1970. "Petroleum geology of the United States," New York: Wiley-Interscience, 571p.

Lueninghoener, G. C., 1947. "The post-Kansan geologic history of the Lower Platte Valley area," *Univ. Neb. Studies*, **2**, 82p.

Lugn, A. L., 1935. "The Pleistocene geology of Nebraska," *Neb. Geol. Surv.*, **10**(2), 223p.

——, 1939. "Classification of the Tertiary system in Nebraska," *Bull. Geol. Soc. Am.*, **50**, 1245–1276.

——, 1968. "The origin of loesses and their relation to the Great Plains in North America," *Proc. VIIth Congr. INQUA*, **12**, 139–182.

Maher, J. C., ed., 1960. "Stratigraphic cross section of Paleozoic rocks—west Texas to northern Montana," *Am. Assoc. Petrol. Geologists*, 18p.

Merriam, D. F., 1963. "The geologic history of Kansas," *Kans. State Geol. Surv. Bull.*, **162**, 317p.

Moore, R. C., and Merriam, D. F., 1959. "Guidebook, 23rd field conference," *Kans. Geol. Soc. and State Geol. Surv. Kans.*, 51p.

Naeser, C. W., Izett, G. A., and Wilcox, R. E., 1973. "Zircon fission-track ages of Pearlette Family ash beds in Meade County, Kansas," *Geology*, **1**(2), 93–95.

Pettijohn, W. A., 1966. "Eocene paleosol in northern Great Plains," *U.S. Geol. Surv. Prof. Pap. 550-C*, C61–C65.

Schultz, C. B., 1959. "The camel story," *Neb. State Museum Mus. Notes*, **8**, 4p.

Schultz, L. G., 1965. "Mineralogy and stratigraphy of the lower part of the Pierre Shale, South Dakota, and Nebraska," *U.S. Geol. Surv. Prof. Pap. 392-B*, 19p.

Simpson, H. E., 1960. "Geology of the Yankton area, South Dakota and Nebraska," *U.S. Geol. Surv. Prof. Pap. 328*, 124p.

Sloss, L. L., 1963. "Sequences in the cratonic interior of North America," *Bull. Geol. Soc. Am.*, **74**, 93–113.

Smith, R., and Burchett, R., 1967. "Stratigraphic cross section of Paleozoic rocks of Nebraska," *Am. Assoc. Petrol. Geologists Cross Sec. Publ.* **5**, 9p.

Stout, T. M., et al., 1971. "Guidebook to the late Pliocene and early Pleistocene of Nebraska," Lincoln: *Nebr. Univ., Conserv. Surv., Div.-Nebr. Geol. Surv.*, 109p.

Weimer, R. J., and Haun, J. D., 1960. "Guide to the geology of Colorado," Denver: *Rocky Mtn. Assoc. Geol.*, 303p.

Wulf, G. R., ed. 1968. "Guidebook, Black Hills Area South Dakota, Montana, Wyoming," *Wy. Geol. Assoc. 20th Field Conf.*, 243p.

Cross-references: *Canada–Northern Great Plains Province; North America: United States–Gulf Coast Province, Midwestern Province, Rocky Mountain Province.*

UNITED STATES–GULF COASTAL PROVINCE

More attention has been given to the Gulf coastal province of North and Central America as an oil and gas producing region than to any other area in the western hemisphere. This effort has been rewarded by an approximate 60 billion barrels of discovered oil and abundant data regarding its habitat. The province has also attracted attention as a youthful (Cretaceous and Tertiary) geosyncline whose sediment thickness exceeds 15,000 m in its deepest part. Much of the recent sediment has been transported by the Mississippi River which carries an estimated 730 million tons of solid and dissolved material into the Gulf of Mexico each year. These and other aspects of the Gulf coastal province make it a classic area for studies in Holocene sedimentation and stratigraphy, and these have provided conceptual keys for recognition of analogous deltaic systems in older rocks.

Location and Area

Included within the Gulf coastal province are parts or all of 11 states in the United States and 14 states in Mexico, parts of British Hounduras and Guatemala, and an extensive area of continental shelf in the Gulf of Mexico (Fig. 1). The total area of the province is nearly 1,500,000 km² (about 500,000 sq mi), of which approximately 500,000 km² (200,000 sq mi) is continental shelf. The province ranges in width from less than 100 km near Vera Cruz, Mexico, to over 1000 km at the Mississippi River embayment.

The inner margin of the coastal province

FIGURE 1. Structural map showing Gulf coastal province of United States and Mexico.

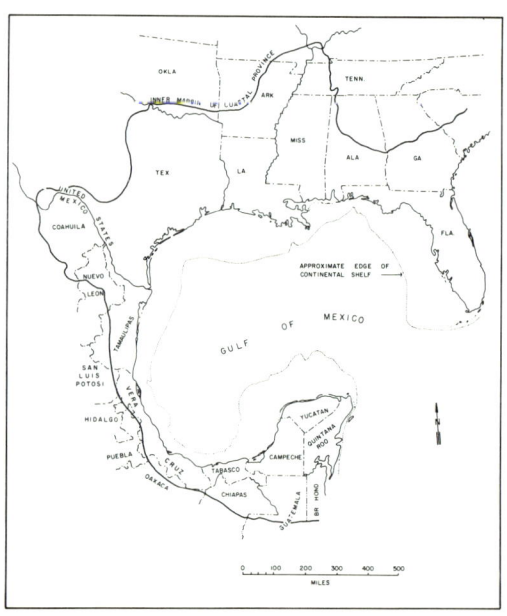

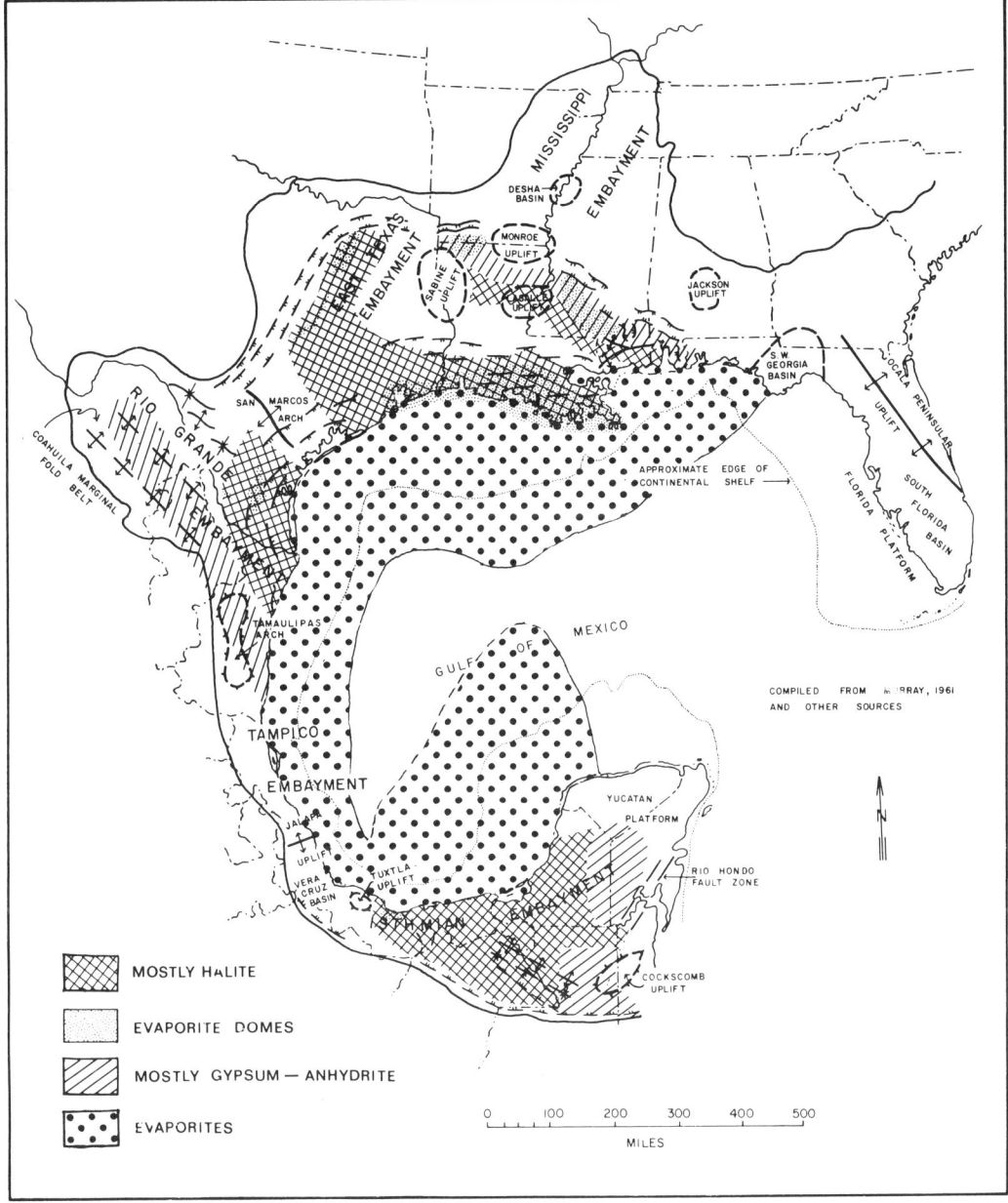

FIGURE 2. Evaporite areas of Gulf coastal province and Mexico.

forms an irregular boundary with the central craton of the United States to the N; and its marginal orogenic belts of Paleozoic and older rocks to the W and S it is bounded by the mountainous volcanic provinces and fold belts of Mexico that contain rocks of Precambrian to Tertiary age. The seaward boundary is arbitrarily placed at the edge of the continental shelf, beyond which is the deeper water of the Gulf of Mexico. The water reaches a maximum depth of about 4000 m in the Sigsbee Deep in the south-central Gulf basin.

Geomorphic Divisions

Rimming the Gulf of Mexico, the Gulf coastal province is a single physiographic unit comprised of the coastal plain and the adjoining continental shelf. The Ocala peninsular uplift of

TABLE 1. Stratigraphic Table for the Gulf Coastal Province

System	Series	Stage or Group
QUATERNARY	RECENT	
	PLEISTOCENE	HOUSTON
TERTIARY	PLIOCENE	CITRONELLE
	MIOCENE	FLEMING
	OLIGOCENE	CATAHOULA VICKSBURG
	EOCENE	JACKSON CLAIBORNE WILCOX
	PALEOCENE	MIDWAY
CRETACEOUS	GULFIAN	NAVARRO TAYLOR AUSTIN EAGLEFORD WOODBINE (TUSCALOOSA)
	COMANCHEAN	WASHITA FREDERICKSBURG TRINITY
	COAHUILAN	NUEVO LEON & DURANGO GROUPS of MEXICO
JURASSIC (UPPER)		COTTON VALLEY HAYNESVILLE SMACKOVER } LOUARK GROUP
JURASSIC (LOWER ? and MIDDLE)		LOUANN SALT WERNER FORMATION
PRE - JURASSIC		MOREHOUSE - EAGLE MILLS FORMATIONS

Source: Modified from Murray (1961).

central Georgia and Florida separates this province from the physiographically similar Atlantic coastal province (q.v.) to the NE (Fig. 1).

Coastal Plain. The coastal plain, or emergent segment of the Gulf coastal province, extends from the Florida peninsula in an arcuate trend around the Yucatan peninsula. Three features characterize the coastal plain: (1) deltaic and coastwise plains adjacent to the shoreline; (2) a band of Pleistocene steplike depositional terraces parallel to the coast; and (3) an inner region of belted topography developed by differential erosion of Tertiary strata. In addition, five major physiographic embayments occur on the coastal plain (see Fig. 2): the Mississippi embayment, the east Texas embayment, the Rio Grande embayment, the Tampico embayment, and the Isthmian embayment (Isthmus of Tehuantepec).

Generally the coastal plain has a subdued topography formed on sedimentary rocks that dip to the Gulf at 1–3°. However, two areas in Mexico, included in the plain, are dominated by mountainous topography; these are the Coahuila highlands and the Chiapas highland. Within these two highlands youthful folded strata are common, in contrast to other areas of the coastal plain. With the exception of these two highland areas, the plain has an average slope to the Gulf of 1:1000. In the United States section the coastal plain width ranges from 250–500 km; in Mexico it narrows generally to the S (Fig. 1).

Continental Shelf. The continental shelf is a seaward extension of the coastal plain but there are several differences—a slightly greater surface slope, an absence of drainage pattern, and a general lack of soil profiles. The continental shelf is widest (300 km) off the Florida and Yucatan peninsulas and is narrowest (35 km) along the west side of the Gulf (Fig. 1). The slope of the shelf averages about 1:400, although it has a more gentle slope to about the 50-fathom line whereupon it steepens slightly to a depth of about 70 fathoms. At this depth, which is the approximate edge of the continental shelf, the slope increases markedly to 1:8 (600 ft/mi) down the continental slope into the central Gulf basin. The shelf edge is marked by growth faults.

Although generally flat, the continental shelf has over 200 domelike topographic prominences, nearly all of which are interpreted as diapiric salt structures. A concentration of these features occurs along the edge of the shelf off the coast of Mexico. Their relief averages over 30 m and a few exceed 200 m (Ewing and Antoine, 1966).

Several large rivers most important of which is the Mississippi, are continually supplying sediments to be deposited on the continental shelf and slope, and thus gradually extending the continental margin.

Structural Features

Gulf of Mexico Basin. Mesozoic to Holocene strata within the Gulf coastal province form a thick sedimentary accumulation in the Gulf of Mexico basin. Centers of maximum sedimentation lie in the northern and southern Gulf of Mexico and exceed 15,000 m in thickness (Fig. 2). The basin probably began to form following late Paleozoic deformation of the bordering Appalachian, Ouachita, and eastern Mexican systems. It was certainly receiving sediment by the middle Jurassic (see Table 1). Its floor has a crust of normal occanic type (6.6 km/sec) and a corresponding upper mantle (8.3 km/sec). There is little magnetic expression over the Sigsbee knolls, Sigsbee escarpment, or other submarine structural features except at the oceanic crust-continental crust transition (Heirtzler et al., 1966). This is due in part to its great thickness of sedimentary fill. "The Gulf of Mexico basin appears to be a fragment of main ocean basin that happens to be an effective sediment trap" expresses a traditional view (Menard, 1967). Several different explanation of the opening of the basin have been offered in terms of plate tectonics and "oceanization."

Miogeoclinal accumulation of sediments has resulted in seaward building of the coastal plain into the Gulf of Mexico basin through the Mesozoic and Cenozoic eras. The continental margins were depressed as deposition continued

FIGURE 3. Diagrammatic structure map showing fault trends and salt structures of the Gulf Coast geosyncline (from Lehner, 1969, Fig. 3).

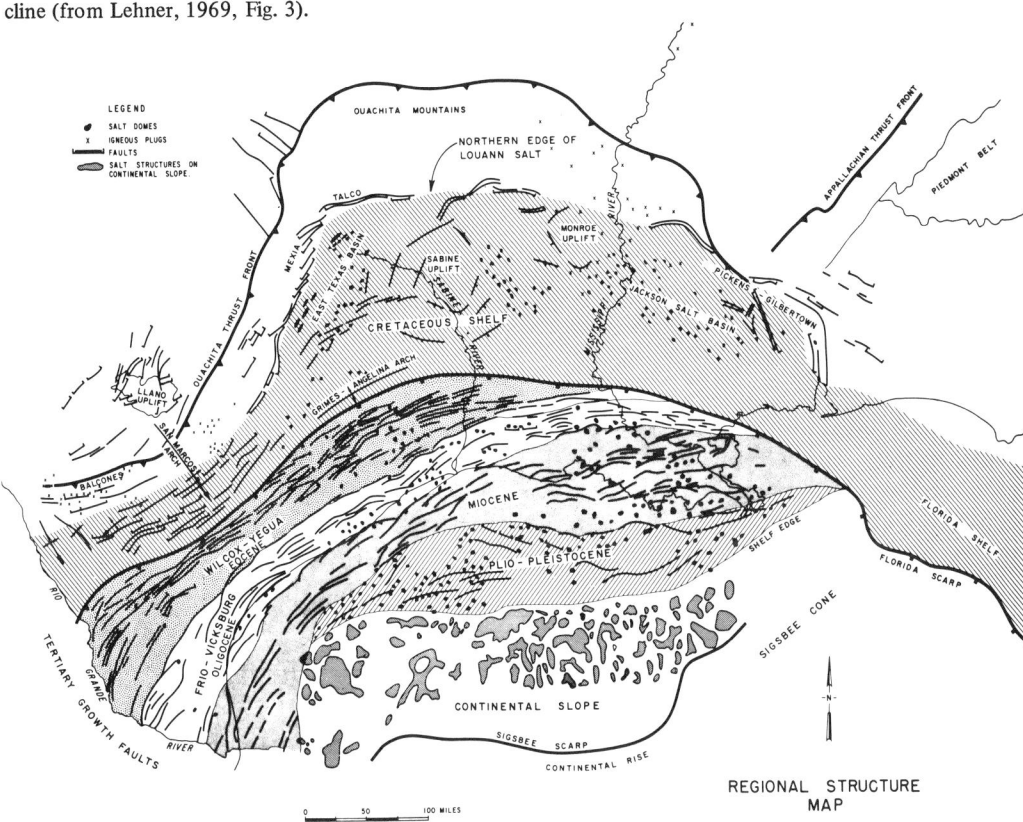

and now the sediments dip basinward. Sedimentation was often concentrated in large deltaic accumulations; indeed, the rivers on the coastal plain of Texas and Louisiana have maintained their present position and intermittently produced deltas at least since the early Eocene. Centers of maximum deposition shifted from river to river through time with source of supply; broadly speaking, this shift was from the eastern United States during the Mesozoic to the western United States through the Cenozoic. Broad structural arches and embayments formed on the edge of the continent as sedimentation amplified the salients and recessions in the underlying late Paleozoic fold belt.

Positive regional elements that characterize the Gulf coast geosyncline are classified as uplifts (essentially equidimensional) and arches (elongate) as shown in Figs. 2 and 3. Only the Jackson uplift (dome) is generally considered to result, at least partially, from igneous intrusion, although Bornhauser (1958) suggested that igneous activity is responsible for the La Salle, Monroe, and Sabine uplifts as well. Elongation of the arches suggests an underlying structural control; for example, the San Marcos arch, which lies along a transverse axis of the Ouachita fold belt.

Salt Structures. More than 300 salt ("halokinetic", diapiric) structures occur onshore in the Gulf Coastal province and are associated with sections of unusually thick sediments, or in some cases with fault zones or flexures (Murray, 1966; Fig. 3). Offshore, salt ridges and stocks are ubiquitous in the Gulf of Mexico from Tallahasse, Florida, around to Yucatan, Mexico, and extend across the continental shelf and rise out to about the 3000 m isobath. Slower seismic velocities indicated that the evaporites from the Rio Grande embayment to Yucatan are predominantly anhydrite in contrast to the more usually occurring halite. Coring has subsequently confirmed this interpretation (Fig. 2).

Salt structures display a variety of features dependent on the thickness of the source bed, its depth below the surface, and the relation of the salt to adjacent sedimentary rocks. A gener-

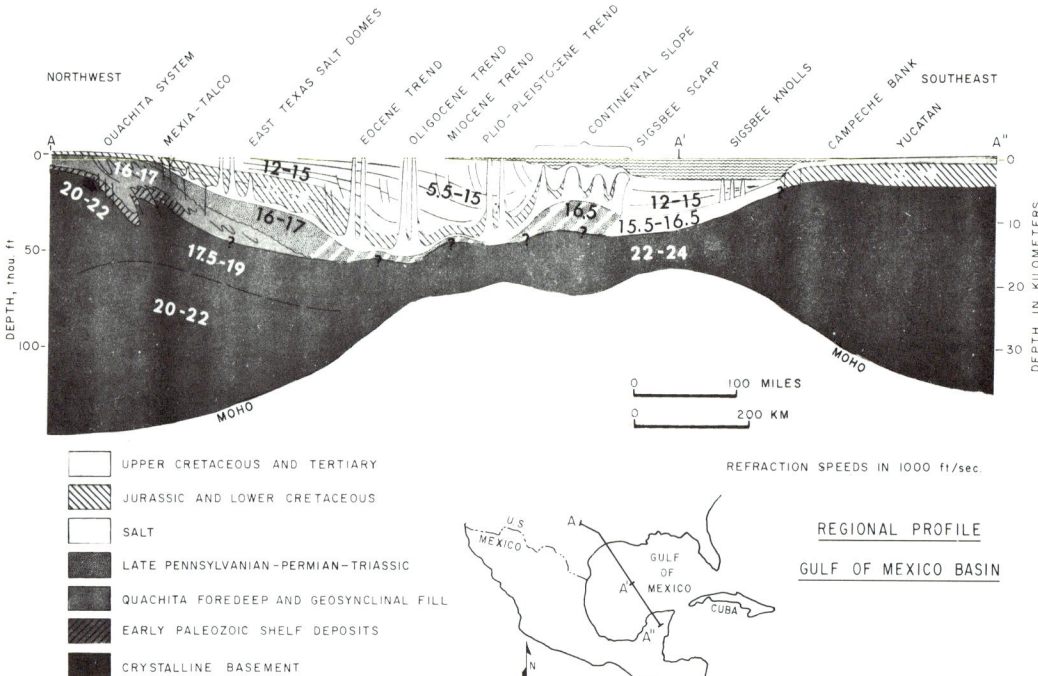

FIGURE 4. Cross section through Gulf of Mexico from Texas to Yucatan (from Lehner, 1969, Fig. 43).

alized spectrum of salt structures evolves basinward in response to these factors from broad, low pillows upslope, to ridges of increasing relief, to ridges of greater relief with stocks rising from them, to salt spines to greater relief, to detached diapirs (Murray, 1966). This appears to be true along any single local line of section or basin wide. Salt diapirs with a deep source characterize the northern Gulf of Mexico basin, whereas salt ridges are present in the western Gulf and low ridges with a shallow source in the southern Gulf. This pattern of occurrence suggests a sequential development of salt structures from N to S in response to loading (Figs. 3 and 4).

The configuration of individual salt bodies varies markedly. Some domes are flat topped, others have irregular spines and knobs of salt projecting into overlying strata. Individual salt stocks may have straight, nearly vertical walls or they may be inclined or have a mushroom shape. In some cases the mass of rising salt has become detached from the source bed and moved upward in a bubblelike fashion; other domes widen at depth and are connected to salt layers at their base.

A "cap rock" of distinctive lithology occurs on the top of many salt domes. It is characterized by a sequence (from top to bottom) of calcite, sulfur, gypsum, and anhydrite, although one or more of the three upper units are commonly absent. Cap rock forms from the solution of salt by ground water, and bacterial alteration of the residual anhydrite to yield calcite (with dolomite crystals), sulfur, and gypsum, and commonly attain a thickness of 300 m or more.

Structural features characteristically related to salt domes include beds thinning across the domes; local angular unconformities; rim synclines; and subsidence (normal) faults (Bornhauser, 1958). Local angular unconformities attest to the progressive upward movement of the salt and its domal effect on overlying sediments; strata that thin over the dome result from contemporaneous upward movement and sedimentation. Rim synclines on the dome flanks result from flow of the salt toward the dome from the periphery; thicker beds in the syncline result from concurrent sedimentation in the deepening syncline. Normal (gravity) faults are commonly related to and contemporaneous with salt doming; a major graben structure may develop or a complex pattern of faults may occur (Fig. 5).

Movement of salt or "halokinogenesis" occurs by gravity flow under conditions in which the salt behaves as a viscous fluid. In

FIGURE 5. Typical salt dome structures in (A) map and (B) section views.

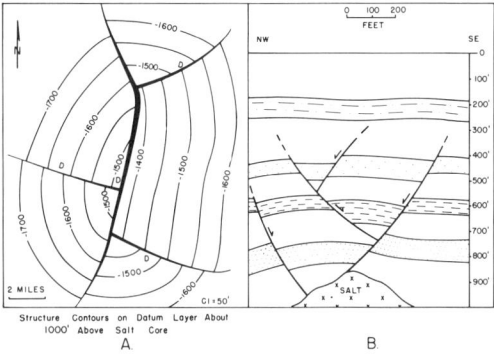

order for salt to flow upward, the density of overlying sediments must exceed that of the salt. The density of salt remains constant at 2.2 gm/cc whereas that of sediments increases with depth, surpassing that of salt at a depth of 1000 m or so. The resulting density inversion produces salt flowage upward ("diapirism") through the overlying strata.

Upward flowage of the salt mass commonly occurs in irregularly distributed, localized centers, with differential vertical advancement in spinelike fashion. Detailed mapping of salt layers exposed in mines within salt domes and subsurface studies of sediments surrounding salt cores indicate this movement developed along fracture zones produced in the roof rocks as the salt intrudes. Atwater and Forman (1959), in southern Louisiana, recognized abundant diapiric shale in association with salt doming. They suggested that the shale behaves similarly to the salt, and has breached overlying sediments and intruded upward rather than being dragged along by the salt cores. Their data indicate a complex growth history of these domes, and they postulate local concentrations of large thicknesses of salt within the mother salt bed prior to upward migration.

Downslope flowage of salt into the Gulf of Mexico basin is volumetrically and structurally more important than vertical flowage. Gravitational movement downslope combined with movement from higher toward lower overburden pressures accounts grossly for the bathymetry of the Gulf of Mexico basin. Initial depositional dips basinward, combined with deltaic sedimentation building the continental margin basinward, produces a gravity head and a basinward pressure gradient in the salt. The result is basinward salt flowage that tends to produce ridges or anticlines parallel to the sedimentary front which could trap and retain the subsequent influx of sediments from the continent, preventing their dispersal as turbidites on the abyssal plain (see Figs. 3 and 4). Continued accumulation of sediments behind and onlapping the ridge of salt would induce renewed salt flowage toward the basin and the growth of a new ridge just basinward of the previous one (Ewing and Antoine, 1966).

Two ridgelike systems of aligned salt domes, the Eocene and Oligocene trends, occur in the northwestern Gulf Coastal Plain. A third system, the Plio-Pleistocene trend, occurs halfway out the continental shelf. The base of each of these systems lies buried between 10,000 and 15,000 m. The hummocky bathymetry of the continental slope and the steep Sigsbee escarpment reflect a fourth system that underlies it. This system is a geanticlinal salt ridge with rounded anticlines protruding from it. The base of this system lies at about 8000 m (see Fig. 4; Lehner, 1969).

Faults. Faults are widely distributed over the Gulf Coastal Plain and occur in patterns of local or regional extent (Figs. 2 and 3). Normal faults with the same general characteristics predominate. Dips range from about 35 to 70° with a tendency of faults to flatten with depth (Fig. 6). Displacement on faults ranges from a few meters to more than 1500 m. Recurrent movement of a fault or system of faults is common. It is recognizable at the surface by offset Pleistocene and Holocene terraces, and in the subsurface by increasing displacement with depth (Fig. 6). It is probable that tectonic movement on many faults continues to aseismically recur at rates slower than those recorded

FIGURE 6. Generalized section view of coastal sediments showing normal faulting with contemporaneous deposition.

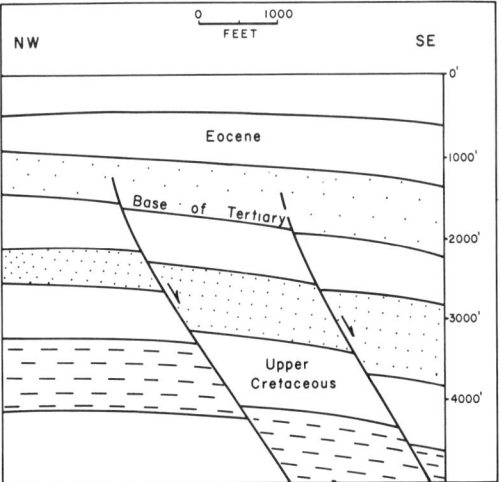

on faults near depressured reservoirs and aquifers (Reid, 1973). Penecontemporaneous (depositional; growth) faults are recognized by increased strata thickness from the upthrown to the downthrown sides (Fig. 6). Some faults display normal drag on the upthrown side; reverse drag is common on the downthrown blocks. Both are most commonly found on faults that flatten with depth, and generally are accentuated by sedimentation contemporaneous with movement of the fault. Strikes of faults closely parallel the regional trends of sedimentary units, with the exception of faults related to localized domal structures. Although some individual faults occur, they generally form in a parallel or an echelon pattern.

Two main zones of faulting are observed in the United States sector of the Gulf basin: an inland zone of older faults; and a wider, younger system along the edge of the coastal plain. The inland zone parallels both the regional strike of Cretaceous rocks and the updip limit of the underlying Jurassic salt, and delineates the subsurface position of the buried late Paleozoic Ouachita fold belt (Fig. 2). Faulting is apparently related to the pull-apart effect in the sediments above the salt due to its flow to the Gulf and to subsidence of the continental margin along and to the S of the Ouachita orogenic belt. The inland zone is characterized by a series of grabens: the Balcones, Luling, and Mexia-Talco systems in central to northeastern Texas; the south Arkansas; and the Pickens-Gilbertown in Mississippi and Alabama. Comparable inner-boundary faults are present in the southern Gulf in Vera Cruz, Chiapas, Guatemala, and British Honduras (Fig. 2).

The youngest systems are closely related to centers of concentrated deltaic sediment deposition. These faults commonly reflect a history of recurrent movement during deposition; increased thicknesses of stratigraphic units on the downthrown blocks may be up to seven times the thickness of units on the upthrown side. Such penecontemporaneous faults are more prevalent in post-Eocene sediments and they increase in abundance in younger strata. This increase is probably largely due to the steadily increasing rate of sedimentation in the Gulf since the Cretaceous, reflecting uplift in the Rocky Mountains and Appalachians, as well as opening of the Gulf basin.

These faults commonly dip toward the Gulf (down-to-the-coast faults), but they may be paired to form grabens. The prevalence of Gulfward-dipping faults may be attributed to their association with gravity flow folding (Bornhauser, 1958) or gravity sliding (Kehle, 1970). As sediments are deposited, underlying less competent beds tend to flow downdip faster than the superincumbent more competent beds causing the overlying strata to stretch and form normal faults or grabens at the updip end of the slide and to be compressed into folds at the downdip end of the slide. Flowage continues in response to gravity, sedimentation, and obstruction at the downdip end, and movement of pull-apart normal faults varies accordingly.

Regional Stratigraphy

General. Mesozoic and Cenozoic sediments accumulated in arcuate belts in concentrations which Murray (1961) has likened to lenticularly flattened link sausages. Localized centers of deposition may exceed 800 m in thickness. Centers for each successively younger stratigraphic unit generally occur seaward (downdip) from the thickest deposit known of the previous older unit.

Cyclic depositional sequences appear to constitute natural divisions of the Mesozoic and Cenozoic strata in several regions of the province. These cyclic sequences coincide with well-developed deltaic complexes.

The generalized geologic column (Table 1) reveals a fairly complete depositional sequence from upper Jurassic to the present time. Little is known about the pre-Jurassic stratigraphy of

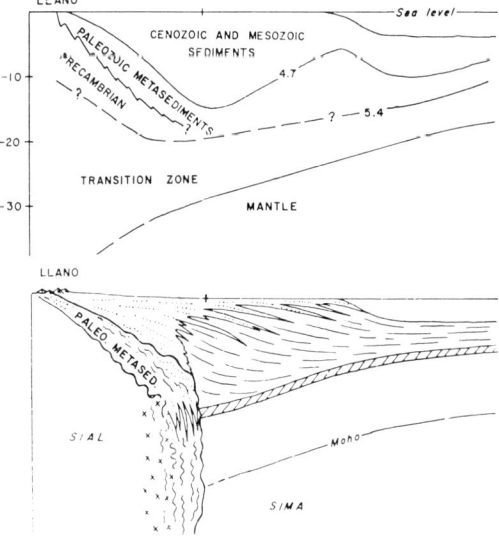

FIGURE 7. Interpretation of the Gulf Coast geosyncline (Dietz, 1964). *Above:* generalized profile from the Llano uplift to the central Gulf; *below:* interpretation offered by Dietz. The diagonal ruled layer is the Louann (Jurassic) salt; the question as to whether it was formed in deep water, or by normal evaporation, and whether the Gulf was landlocked, or whether it has flowed laterally, is still controversial.

the Gulf Coastal Plain; the *Eagle Mills* and *Morehouse* formations may be equivalent facies.

Pre-Cretaceous. In the northern Gulf the oldest sediments that are part of the province occur as a thick sequence of red sandstone and shale referred to as "Eagle Mills" strata. They are overlain by the *Werner Formation* (sandy conglomerate below anhydrite) and the *Louann salt* (Fig. 7). The Louann, which has an estimated maximum thickness of at least 1500 m, is the source for the salt domes. The Werner-Louann deposits characteristically have a shoreward (coarse clastic) facies that grades offshore (downdip) into darker and finer clastics, carbonates, and evaporites. Rocks equivalent to the Eagle Mills-Werner-Louann sequence are believed to be present in the deep subsurface of south Texas and northeastern Mexico, and the eastern half of the Yucatan peninsula. Farther S along the Gulf edge in Mexico these strata crop out in the marginal mountainous regions and are known in the subsurface.

Age of the Eagle Mills-Werner-Louann sequence is not definitely known, but is generally considered to be Jurassic. This sequence overlies rocks of Permian age and is overlain by Upper Jurassic strata (Louark Group), so the possible range is late Permian to Middle or late Jurassic.

Upper Jurassic strata are of widespread occurrence in the subsurface of the Gulf area. These beds have been divided into two principal sequences: the *Louark Group* and the *Cotton Valley Group*. Both sequences are characterized by a wide range of lithologies that include updip (near shore) red sandy sediments that grade basinward into dark-colored clayey and calcareous facies.

Cretaceous. Oldest (Lower) Cretaceous strata in the Gulf Coastal Plain are commonly placed in the *Coahuila series*—subdivided into the Durango and Nuevo Leon Groups. The *Comanchean Series* (Trinity, Fredericksburg and Washita) in general conformably and transitionally overlies the Coahuila. These Lower Cretaceous rocks are mainly red sandy and shaly strata at the outcrop in the shelf area and separated by the Stuart City barrier reef from the deeper facies, which change downdip into predominantly dark, clayey calcareous deposits (Fig. 8). These units thicken markedly progressing downdip in the subsurface to a maximum of 2500 m.

Gulfian (Upper Cretaceous) deposits unconformably overlie Comanchean and older rocks in updip areas of the Gulf coast. Seaward the boundary generally marks the position of a pronounced unconformity. The Gulf Series is subdivided into five smaller rock units. The

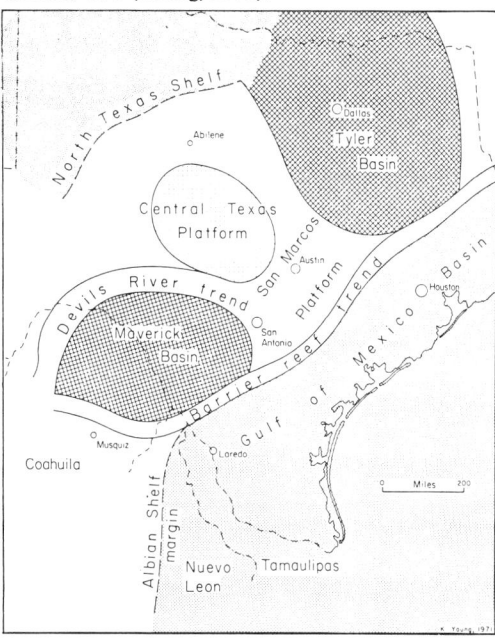

FIGURE 8. Position of barrier reef on the Texas shelf margin during much of Comanchean (Lower Cretaceous) time (Young, 1972).

older of these units (Woodbine and Eagleford) are principally sandstone and shale; younger Gulfian units (Austin, Taylor, and Navarro) are predominantly calcareous rocks: chalks, limestones, and marls (soft shaly limestones). However, in Alabama and Georgia the younger Gulfian rocks are mainly sandstone and shale. Also present in the Gulfian sequence are reefs (Tayloran (?) and Navarroan) on top of several uplifts in the northern Gulf; water-laid volcanic materials; and both intrusive and extrusive igneous rocks.

Tertiary. Tertiary strata of the province are generally grouped into two major divisions on the basis of a significant break in deposition at the close of the Oligocene. Older Tertiary deposits contain a wide variety of rocks deposited in marine, continental, and marginal shoreline environments, largely deltaic and characterized by cyclic repetion of rock types. Paleocene and Lower Eocene (*Wilcox*) strata include chalks and marls, prodeltaic shales and interbedded deltaic sands and shales. Carbonaceous clays and lignite are locally important adjuncts. Cyclic deposition is prominent in the Eocene *Claiborne;* calcareous strata grade upward into nearshore deltaic and fluviatile (river) sediments including lignites. Upper Eocene (*Jackson*) and Oligocene (*Vicksburg* and *Catahoula*) strata consist chiefly of marine, largely calcareous deposits, but include sands, clays, and volcanic-ash deposits, particularly in the updip regions of the central-northern Gulf coast.

Neogene (post-Oligocene) deltaic sediments exceed 6000 m in thickness. They are mainly fluviatile and deltaic sand and clay accumulations in the updip facies. Characteristically the units thicken markedly Gulfward as they grade into marine clays and marls. Sedimentary units include the *Fleming* (Miocene), *Citronelle* (Pliocene), and *Catahoula;* the latter is herein considered Oligocene (See Table 1), although it is possibly Miocene in age.

Pleistocene (*Houston*) and Holocene sediments are of fluviatile, marginal shoreline, and marine deltaic types. Terrace and valley-filling alluvial deposits characterize the fluviatile units. These materials grade into marginal shoreline deposits that parallel the present coastline; offshore they grade into marine sediments.

Marine Resources

Oil and Gas. Hydrocarbon accumulations are widely distributed in the United States and Mexico from the inner boundary of the province to far out on the continental shelf. Discoveries are recorded in strata of Late Jurassic to Holocene age, as well as in underlying Paleozoic rocks. Oil and gas occur from depths of 6000 m to within 15 m of the surface. In the U.S. part of the province, approximately 40 billion barrels of oil and 125,000 billion cu ft of natural gas have been produced.

Major production in the United States is almost entirely from sandy reservoirs; in Mexico the larger fields occur in carbonate strata. There are 77 major fields (those with 100 million barrels of discovered oil) in the United States part of the province. Almost two-fifths of the total production comes from the 10 largest fields. In the Mexican part of the province, seven major fields produce approximately half of the estimated 12 billion barrels of discovered hydrocarbons. Smaller fields in the United States and Mexico produce from sand, carbonate, and cap-rock reservoirs.

In addition to lithology, the porosity and permeability of the materials, as well as local and regional structural features, contribute to the development of oil and gas accumulations. Concentration of hydrocarbons commonly occurs in traps developed by normal faults, related often to salt doming.

Lignite and Coal. Commercial deposits of lignite, bituminous coal, and cannel coal occur in sedimentary strata of Late Cretaceous, Paleocene, and Eocene age in Texas. Strip mining is used almost exclusively to remove the material. Average annual production of lignite and coal is approximately 2.5 million tons.

Evaporite Deposits. Gypsum, salt, and related sulfur deposits are important industrial minerals found in the Gulf Coastal Plain. Gypsum and associated anhydrite occur both as cap rock on salt domes and as basin sediments. The deposits are of Jurassic and Cretaceous age.

Salt is present in domal and anticlinal structures and as bedded deposits (mainly of Jurassic age). Production is primarily from domes in Louisiana, Texas, and Mexico; conventional underground methods of mining are used for shallow domes, whereas the solution process is utilized for deeper structures. Annual production of salt is approximately 27 million tons.

Native sulfur occurs as disseminated deposits in the cap rock of salt domes. Most important commercial deposits are in Louisiana, Texas, and Vera Cruz, where the sulfur is extracted by the Frasch solution process. This process is currently restricted to deposits from depths of less than 800 m. Production of sulfur is in excess of 8 million tons annually.

Groundwater. Groundwater, an essential natural resource of the Gulf Coastal Plain, is used for agricultural and domestic and industrial purposes. Water-bearing strata, which range in age from Jurassic to Holocene, crop out in belts approximately parallel to the inner margin of the province and dip Gulfward at a low angle.

In the United States part of the province, sand and gravel deposits are most productive, although water is derived from carbonate rocks and, in some places, from fractured shales. In contrast, carbonate rocks are the most important water reservoirs in the Mexican Gulf Coastal Plain, with significant supplies in Quaternary alluvial deposits and valley fills.

Miscellaneous Materials. A variety of industrial materials, primarily nonmetals, are utilized for numerous purposes in industry and construction. Sand and gravel are important for construction, serving in concrete, road aggregate, and foundation material. Clay has numerous uses such as structural clay products, ceramic materials, filter in chemical processes, and in drilling mud. Bentonite, derived from alteration of volcanic ash, possesses properties suitable for drilling mud and oil filtering and decolorizing. Chalks, limestone, marl, and shell deposits are used as agricultural lime, for road metal, and in the manufacture of cement. Limestone is widely employed as building stone in Yucatan, British Honduras, and Guatemala.

Bauxite, iron, and uranium are commercially exploited from Tertiary units of the northern Gulf province. Various ores including copper, gold, lead, manganese, mercury, and silver have been mined in the western province of Coahuila, Nuevo Leon, Tamaulipas, and Vera Cruz.

WILLIAM M. REID
ROBERT E. BOYER

References

Anon., 1972a. *Tectonic map of Gulf Coast Region, U.S.A., 1:1,000,000.* Tulsa: Gulf Coast Assoc. Geol. Soc. and Am. Assoc. Petrol Geologists.

Anon., 1972b. *Texas* (Geol. Highway Map 7). Tulsa: Am. Assoc. Petrol. Geologists.

Atwater, G. I., and Forman, M. J., 1959. "Nature of growth of southern Louisiana salt domes and its effect on petroleum accumulation," *Bull. Am. Assoc. Petrol. Geologists* **43**, 2592–2622.

Barnes, E., and Eifler, G. K., Jr., 1972. "Cartography of Quaternary deposits of Texas," *Etudes Quat. Monde, 8th Congr. INQUA,* Paris, 1969.

Bernard, H. A., and LeBlanc, R. J., 1965. "Resume of Quaternary geology of the northwestern Gulf of Mexico Province," in H. E. Wright and D. G. Frey, eds., *Quaternary of the United States.* Princeton, N.J.: Princeton Univ. Press, 137–185.

Bornhauser, M., 1958. "Gulf coast tectonics," *Bull. Am. Assoc. Petrol. Geologists,* **42**, 339–370.

*Braunstein, J., ed., 1970. *Bibliography of Gulf Coast Geology* (Spec. Publ. 1). New Orleans, La.: Gulf Coast Assoc. Geol. Soc., 1045p.

Crosby, G. W., 1971. "Gravity and mechanical study of the Great Bend in the Mexia-Talco Fault Zone, Texas," *J. Geophys. Res.,* **76**(11), 2690–2705.

Dietz, R. S., 1964. "Wave-base, marine profile of equilibrium, and wave-built terraces: reply," *Bull. Geol. Soc. Am.,* **75**, 1275–1282.

Ewing, M. and Antoine, J., 1966. "New seismic data concerning sediments and diapiric structures in Sigsbee deep and upper continental slope, Gulf of Mexico," *Bull. Am. Assoc. Petrol. Geologists,* **50**, 479–504.

Fisk, H. N., 1944. *Geological Investigation of the Alluvial Valley of the Lower Mississippi River.* Vicksburg, Miss.: U.S. Army Corps Engineers, Miss. River Comm., 78p.

*Halbouty, M. T., 1967. *Salt Domes Gulf Region, United States and Mexico.* Houston, Tex.: Gulf Publ. Co., 425p.

Heirtzler, J. R., Burckle, L. H., and Peter, G., 1966. "Magnetic anomalies in the Gulf of Mexico," *J. Geophys. Res.,* **71**, 519–526.

Humphrey, W. E., 1956. "Tectonic framework of northeastern Mexico," *Trans. Gulf Coast Assoc. Geol. Soc.,* **6**, 25–35.

Kehle, R. O., 1970. "Analysis of gravity sliding and orogenic translation," *Bull. Geol. Soc. Am.* **81**, 1641–1664.

Kirkland, D. W., and Gerhard, J. E., 1971. "Jurassic salt, central Gulf of Mexico, and its temporal relation to circum-Gulf evaporites," *Bull. Am. Assoc. Petrol. Geologists,* **55**(5), 680–686.

Lankform, R. R., and Rogers, J. W., 1969. *Holocene Geology of the Galveston Bay Area.* Houston, Tex.: Houston Geol. Soc., 134p.

Lehner, P., 1969. "Salt tectonics and Pleistocene stratigraphy on continental slope of northern Gulf of Mexico," *Bull. Am. Assoc. Petrol. Geologists,* **53**, 2431–2479.

Lozo, F. E., and Smith, C. I., 1964. "Revision of Comanche Cretaceous stratigraphic nomenclature, southern Edwards Plateau, southwest Texas," *Trans. Gulf Coast Assoc. Geol. Soc.,* **14**, 285–307.

May, J. P., 1972. "Geology and history of the Gulf of Mexico: a discussion of the Late Neogene deposits of some of the coastal regions," *Bull. Geol. Soc. Am.,* **83**,(10), 3155–3156.

Menard, H. W., 1967. "Transitional types of crust under small ocean basins," *J. Geophys. Res.,* **72**, 3061–3073.

*Murray, G. E., 1961. *Geology of the Atlantic and Gulf Coastal Province of North America.* New York: Harper & Row, 692p.

*____, 1966. "Salt structures of Gulf of Mexico basin—a review," *Bull. Am. Assoc. Petrol. Geologists,* **50**, 439–478.

Rainwater, E. H., 1964. "Transgressions and regressions in the Gulf Coast Tertiary," *Trans. Gulf Coast Assoc. Geol. Soc.,* **14**, 217–230.

Reid, W. M., 1973. *Active Faults in Houston, Texas.* Ph.D. dissertation, Univ. Texas at Austin, 122p.

Rose, P. R., 1972. "Edwards Group, surface and subsurface, Central Texas," *Univ. Texas, Bur. Econ. Geol. Rept. Inv. 74,* 198p.

Russell, R. J., 1971. "The coast of Louisiana," *Applied Coastal Geomorphology.* New York: Macmillan, 98–115.

Tucker, D. R., 1962. "Subsurface Lower Cretaceous, central Texas," in W. L. Stapp, ed., *Contributions to the Geology of South Texas.* San Antonio: South Texas Geol. Soc., 117–216.

Vernon, R. O., and Puri, H. S., 1964. "Geologic map of Florida." *Florida Div. Geol. Map Ser. 18.*

Walthall, B. H., and Walper, J. L., 1967. "Peripheral Gulf rifting in northeast Texas," *Bull. Am. Assoc. Petrol. Geologists* **51**(1), 102–110.

Wilhelm, O., and Ewing, M., 1972. "Geology and history of the Gulf of Mexico," *Bull. Geol. Soc. Am.,* **83**, 575–600.

Williamson, J. D. M., 1959. "Gulf coast Cenozoic history," *Gulf Coast Assoc. Geol. Soc.,* **9**, 14–29.

Young, K., 1972. *Cretaceous Paleogeography: Implications of Endemic Ammonite Faunas* (Geol. Circ. 72-2). Austin: Univ. Texas, 13p.

*Items starred contain long bibliographies.

Cross-references: *Mexico; North America; United States–Appalachian Region, Atlantic Coastal Province, Great Plains Province, Midwestern Region.*

UNITED STATES–HAWAII

The Hawaiian Archipelago consists of the islands and reefs, strung out NW-SE for 2500

FIGURE 1. Map of the main Hawaiian Islands with inset showing their position in the Pacific Ocean.

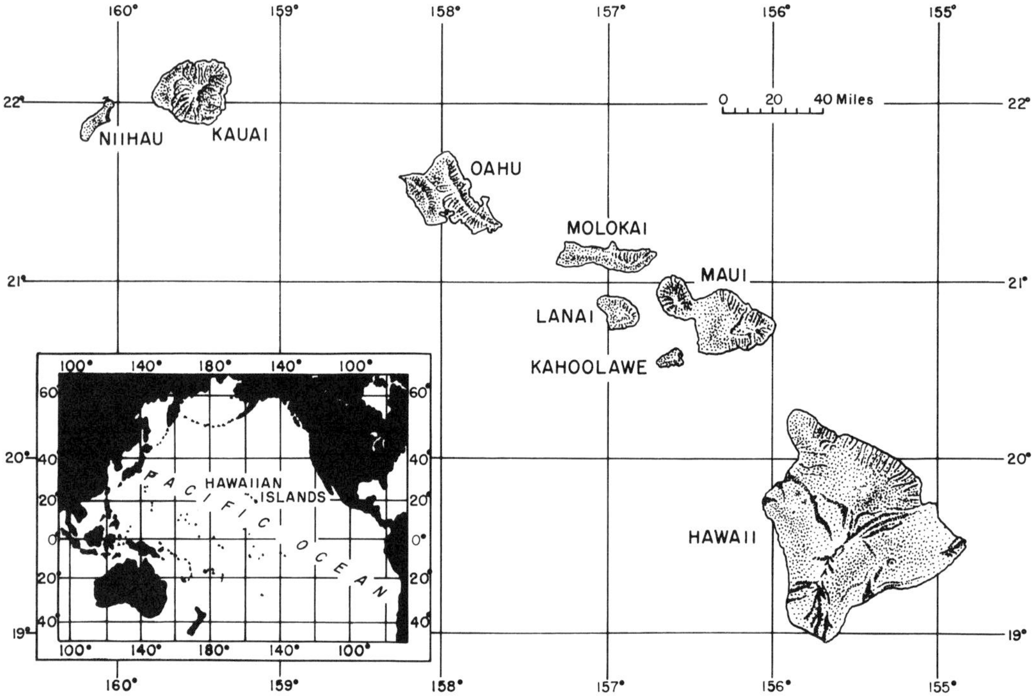

km, between 18°54′ to 28°15′N and 154°40′ and 171°75′W. (Figs. 1 and 2). Hawaii itself (the "Big Island") is the largest and southeasternmost island and has the only active volcanoes. The archipelago was first settled by Polynesians about 1200 years ago, acceded to the United States in 1898 and became the 50th state in 1959.

Although the islands are at the N margin of the tropics, they have a subtropical climate because currents from the Bering Sea cool the region. The temperature of the surrounding waters is about 5°C lower than the average for this latitude, resulting in the poor development of coral reefs. The ocean at Waikiki, Oahu, ranges from 70° to 85°F.

The Hawaiian Islands lie in the belt of NE trade winds which persist throughout the year but are occasionally interrupted during the winter by southerly or "kona" winds which blow for a few days at a time. Relatively low islands such as Kahoolawe and Lanai are sheltered from the trade winds by other islands and consequently are very dry.

The northeastern sides of the mountains are usually wettest because of the prevailing wind. Maximum precipitation occurs between 600 and 2000 m, depending upon the form and height of each island. Heaviest rainfall comes from southerly storms; on Jan. 25, 1956, about 1000 mm (40 in) fell in 24 hours at Kilauea Plantation, Kauai. Above 2000 m the precipitation decreases, and the peaks are semiarid. As the winds descend the lee slopes, they become warmer and drier, causing arid and semiarid climates on the leeward sides of the islands. On the "big island" of Hawaii, however, eddies result in prevailing southwest winds on the lee side so that the climate in the leeward districts is fairly wet. The annual rainfall in the islands ranges from 250 mm or less on the lee coasts to nearly 12,000 mm (450 in.) in the wettest belts. In one year 624 in. of rain was recorded on the summit of Kauai at an altitude of 5,170 ft (Table 1).

Previous Work

The earliest geologic work was done by J. D. Dana during the famous Wilkes Expedition in 1838–1842. He introduced the Hawaiian words "aa" and "pahoehoe" into the geologic literature for the two types of basaltic lava flows, and suggested that the volcanoes become progressively younger from NW to SE. Next came the volcanologist, T. A. Jagger, who founded the Hawaiian Volcano Observatory at Kilauea

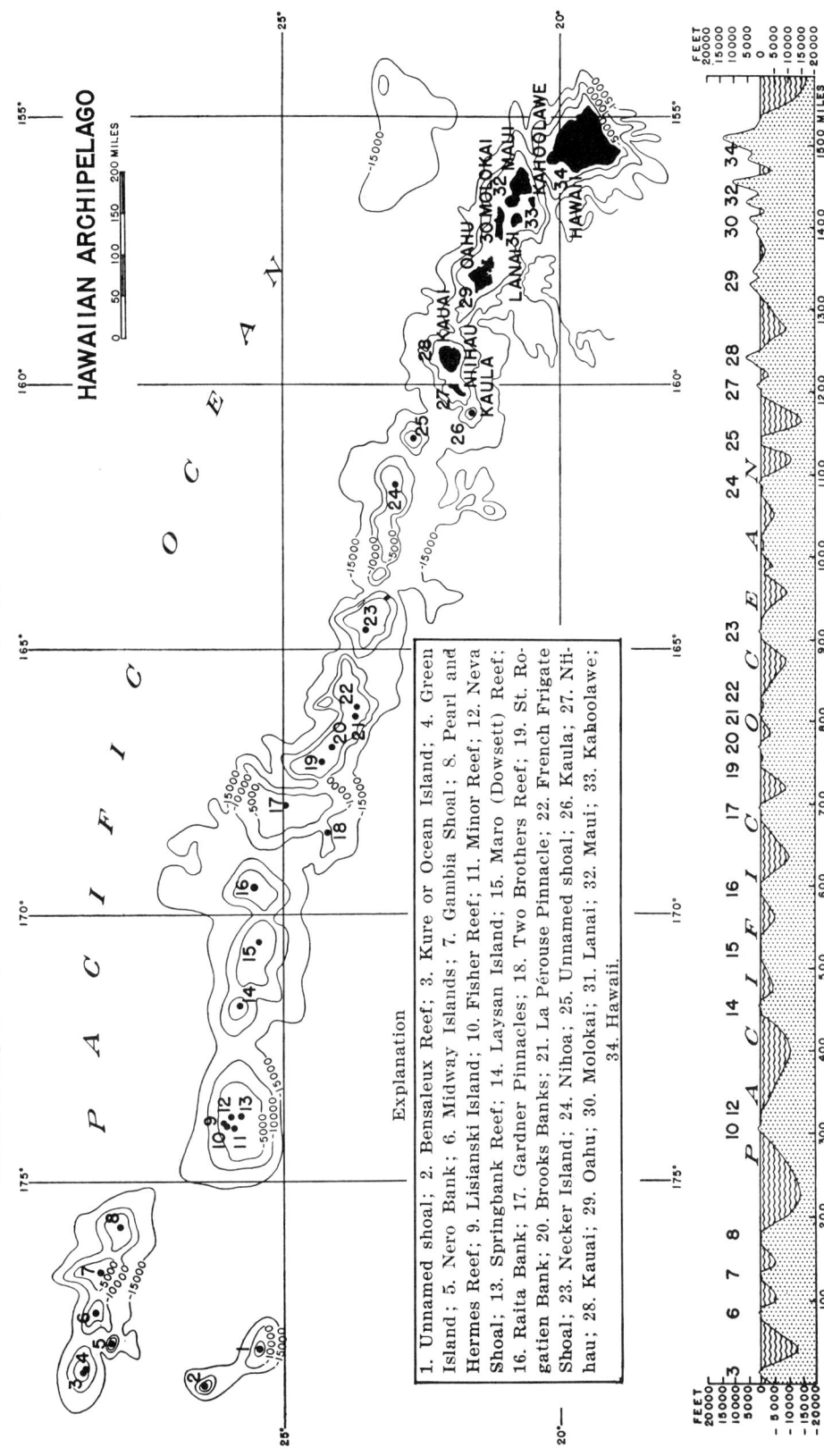

FIGURE 2. Map and longitudinal section of the Hawaiian Archipelago, showing submarine contours in feet.

UNITED STATES-HAWAII

TABLE 1. Area, Altitude, and Diameter of the Larger Islands

Island	Area (km²)	Altitude (m)	Maximum Diameter (mi)
Hawaii	10,436	4,201	87.3
Maui	1,883	3,055	38.4
Oahu	1,563	1,228	40.0
Kauai	1,435	1,574	29.9
Molokai	672	1,514	37.0
Lanai	363	1,035	13.3
Niihau	186	388	9.7
Kahoolawe	115	447	10.9
Total	16,638		

Volcano in 1911, which continues today under the U.S. Geological Survey. In 1924 a systematic study, with publication of geologic maps of each island, was started by the U.S. Geological Survey in cooperation with the territory, and thirteen volumes have been printed. The Department of Geology of the University of Hawaii has existed since 1921. Since 1963, the geophysical side has been advanced by the Hawaiian Institute of Geophysics. The Bernice P. Bishop Museum has also furthered geological work through its publications, collections, and fellowships.

Geomorphology

The geomorphology of the islands is very diverse (Fig. 3). The young volcanoes are constructional geomorphic forms in places with surfaces of historic lava flows. These volcanoes stand in contrast to the older islands where chemical weathering has penetrated up to 50 meters in places and some of the extinct volcanoes are so deeply eroded that they are nearly buried in their own waste. In places on the coasts (e.g., Kauai), sheer cliffs up to 1000 m high exist as a result of faulting and marine erosion. There are numerous terraces caused by emergence of coral reefs. "Deserts" with drifting dunes exist within a short distance of swamps fed by tropical rainfall. An offshore sounding 65 km SW of Mauna Kea (elevation 13,784 ft) of 3040 fathoms indicates the total height of Mauna Kea above its base to be 9800 m (32,024 ft) (Dietz and Menard, 1953, p. 103).

The chief geomorphic land forms are:

1. Constructional forms built by lava flows:
   a. Shield volcanoes, such as the active volcanoes of Mauna Loa and Kilauea, Hawaii, with slopes of 2 to 10°.
   b. Plateaus or isthmuses built by the lava flows of one volcano banking against a preexisting volcano.
   c. Flat-floored valleys and plains caused by lava flows partly filling former V-shaped valleys or burying several low interstream divides.
2. Constructional forms around secondary vents (Fig. 4):
   a. Pumice, cinder, spatter, and lava cones.
   b. Tuff cones built by phreatomagmatic explosions caused by magma erupting through coral reefs or under the sea. Diamond Head on Oahu is this type.
   c. Bulbous domes caused by viscous trachyte lavas piling steeply around their vents.
3. Volcanic forms due to collapse:
   a. Calderas due to collapse of the summit of the volcanoes, exemplified by Halemaumau Caldera on Kilauea. Ancient calderas may be low areas now, due to erosion, as, for example, the lowlands around Kailua on Oahu. The calderas in Hawaii are listed on Table 2.

FIGURE 4. Profiles of secondary cones. Ash cones after consolidation are called tuff cones. (There is no vertical exaggeration of scale.)

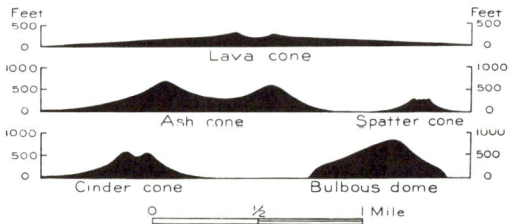

FIGURE 3. Comparative profiles of the mountains of the Hawaiian Islands. (Vertical scale is exaggerated about four times.)

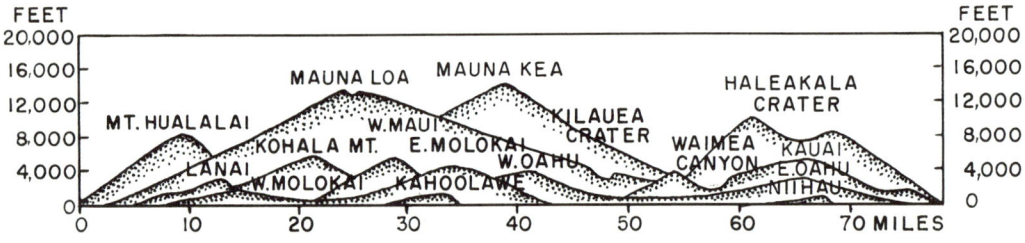

TABLE 2. Calderas in the Hawaiian Islands

Island	Mountain or Name	Diameter (max.) km	mi
Kauai	Mt. Waialeale	21	13
Kauai	Lihue depression[a]	18	11
East Maui	Haleakala[b]	12	7.5
Oahu	Koolau	10	6
Oahu	Waianae	8	5
Lanai	Palawai Basin	6	4
Molokai	East Molokai	7	4.5
Hawaii	Mauna Loa	6	3.7
Kahoolawe	No name	5	3
Hawaii	Kilauea	4.5	2.9
Kauai	Haupu caldera	4	2.5
Maui	Iao Valley	3	2

[a]Proof of a caldera in this depression is not certain.
[b]Believed to be chiefly due to erosion, perhaps guided by summit faults. If so, it is not a caldera.

    b. Pit craters—collapse forms smaller in area than calderas but of similar origin, such as the Chain of Craters on Kilauea Volcano.
    c. Grabens along rift zones, such as the Makaweli graben on Kauai.
4. Fault scarps some of which are more than 500 m high, such as those on Hawaii and Kauai Islands.
5. Forms due to stream erosion:
    a. Amphitheater-headed valleys, characteristic of basaltic volcanoes in regions of high rainfall. Their origin is complex but important factors causing them are illustrated in Fig. 5. Their headwalls merge to form extensive cliffs as the "Pali" on Oahu where several of these valleys are cut into the ancient Koolau caldera.
    b. Corrugated slopes where the shield-shaped lava domes are cut by numerous, nearly parallel valleys as on the leeward slopes of northwest Oahu.
6. Forms due to stream deposition:
    a. Extensive fans and deltas. Some of the deposits are derived from mudflows.
    b. Flat-floored valleys due to alluviation of large valleys by 400 m or more.
7. Forms due to marine erosion:
    a. Sea cliffs.
    b. Emerged terraces chiefly cut into alluvium.
    c. Submerged shelves formed during lower stands of the sea.
8. Forms due to marine and wind deposition and coral growth:
    a. Fringing coral reefs.
    b. Beach and beach dune deposits, including black glassy sand beaches resulting from the comminution of lava flows entering the sea.
    c. Emerged reef deposits and emerged marine plains.
    d. Eolianite or ancient lithified dunes.
    e. Recent dunes such as the "Barking Sands" of Kauai (see "Singing Sands" in *Encyclopedia of Geomorphology*)
    f. Tsunami deposits. The tsunamis of 1865, 1946, and 1960 left huge boulders along the shelving coasts.
    g. Lagoonal deposits.
9. Swamps: the most extensive are on the mountains in areas of high rainfall and poor drainage, such as the sphagnum swamp on Kohala Mountain, Hawaii, and the Waialeale Swamp on the top of Kauai and Eke Swamp on the top of West Maui. Other large swamps exist where drowned valleys are blocked by beach ridges, such as Kawainui Swamp on Oahu.
10. Planation surfaces: these occur where interstream divides have been cut low and alluviation has leveled out the intervening areas. The Kaneohe area on Oahu is nearly a peneplain or pediplain.
11. Glacial moraine: one on Mauna Kea was left by a Wisconsin glacier.
12. Landslide and soil avalanches: these occur following heavy rains and earthquakes. Huge slides on Maui followed the earthquake of 1938 and a disasterous one on Hawaii followed an earthquake in 1865. Huge undersea landslides and slumps may have produced some of the coastal faults such as those on the south slope of Kilauea and the ancient cliffs on Mauna Loa near Naalehu. These have been named "seaward slip faults" (Stearns and Clark, p. 85, 1930). The subaerial lavas may slide on saturated tuffs formed during the early submarine phase of the volcano, a condition noted by Fairbridge (1950).
13. Submarine delta cones: fanning out from the submarine canyons at depths ranging from 2000 to 3000 m off the windward coast of Oahu are extensive submarine deltas where the waste of the great valleys have been dumped (Hamilton, 1957).
14. Talus deposits: large talus deposits skirt the high

FIGURE 5. Formation of amphitheater-headed valleys. A. Youthful dome with radial drainage. B. Details of stream piracy that lead to a master stream, the precursor of an amphitheater-headed valley. C. Waterfalls caused by alternating resistant and nonresistant beds of basalt. D. Dome with amphitheater-headed valleys formed by the master streams. Repeated eustatic lowering of sea level has led to frequent rejuvenation of streams.

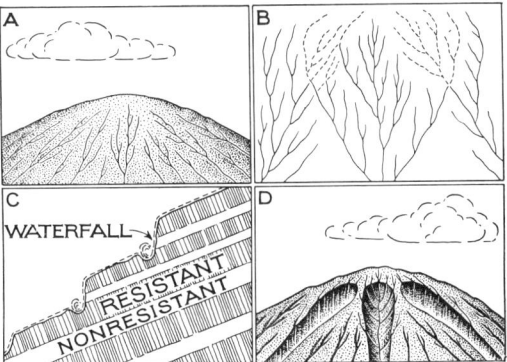

TABLE 3. Ancient Shorelines in the Hawaiian Islands

Approximate Altitude (ft)	(m)	Shelf or Terrace (name)	Age (B.P.)	Type Locality (Island)
0		Present	Present	–
5	1.5	Kapapa	4000 ±	Oahu
−15	−4.5	Koko	Late Holocene	Oahu
−350	−106	Mamala	12,000 ±	Oahu
−60	−18	Makai Range	Late Wisconsinan	Oahu
−120	−36	Makapuu	Late Wisconsinan	Oahu
−150	−45	Lahaina Roads[a]	Wisconsinan	Maui
−80	−24	Kaneohe[a]	Wisconsinan	Oahu
−185	−56	Penguin Bank[b]	Wisconsinan	Molokai
2±	0.6±	Leahi	115,000 ±	Oahu
−300(?)	−90	Kawela	Early Wisconsinan	
25	7.5	Waimanalo[c]	125,000 ±	Oahu
12	3.5	Kailua	Late Sangamon	Oahu
−30	9	Olomanu	Sangamon	Oahu
−350(?)	−106	Waipio	Illinoian	Oahu
−240±	−73±	Makua[a]	Illinoian(?)	Oahu
45	14	Waialae	Early Illinoian(?)	Oahu
70	21	Laie	Early Illinoian(?)	Oahu
95	28	Kaena	Yarmouth	Oahu
−350(?)	−106(?)	Kahipa	Kansan	Oahu
−205	−62	Ewa	Kansan	Oahu
25	7.5	PCA	Kansan	Oahu
55	16.5	Kahuku	Kansan	Oahu
250±	76±	Olowalu	Aftonian(?)	Maui
325±	99±	–	–	Lanai
375±	114±	–	–	Lanai
560	170	Manele	Early Pleistocene(?)	Lanai
625	190	Kaluakapo	Early Pleistocene(?)	Lanai
1200	366	Mahana	Early Pleistocene(?)	Lanai
−1200 to −1800	−548	Lualualei	Early Pleistocene(?)	Oahu
−3000 to −3600	−1100	Waho	Early Plesitocene(?)	Oahu

[a]Position in sequence uncertain.
[b]Penguin Bank is too extensive to have been formed during this short-lived stand of the sea.
[c]Two shorelines 22 and 27 ft (5.7, 8.2 m) above mean sea level.

canyon walls and the fault scarps. The older deposits of this sort are so decomposed so that they can be cut like cheese. These are thus in striking contrast to the talus deposits in the semi-arid southwestern United States where chemical weathering is of minor importance.

Ancient Shore Lines

The highest fossiliferous marine deposit is at 1069 ft on Lanai. The stripping of soil above this level is evidence of submergence to about 366 m. Numerous emerged shorelines and submerged shelves exist on the islands. Their elevations and relative ages are given in Table 3 and Fig. 6. Coral limestone occurs at 318 m (1050 ft) below sea level in a well near Diamond Head, Oahu. Because of their worldwide occurrence it seems likely that oscillations from about minus 135 m to plus 75 m are glacio-eustatic. There are also tectono-custatic factors related to changes in the floors of ocean basins and tectonic factors related in part to the slow uplift and subsidence of the Hawaiian Ridge. Miocene shallow-water fossil corals were dredged from a broad shelf about 550 m below sea level 10 km SW of Oahu indicating a submergence of that amount since the Miocene (Menard et al., 1962). This age assignment is now in doubt (Stearns, 1974).

From the dating of fossil coral on Oahu it appears that the widespread Waimanalo "25-ft" (8-m) shoreline, the maximum stand of the sea during the Sangamon interglacial, occurred about 120,000 years ago (Veeh, 1966; Ku et

FIGURE 6. Map of Oahu showing emerged reefs, fringing reef and type localities of emerged and submerged shorelines and shelves.

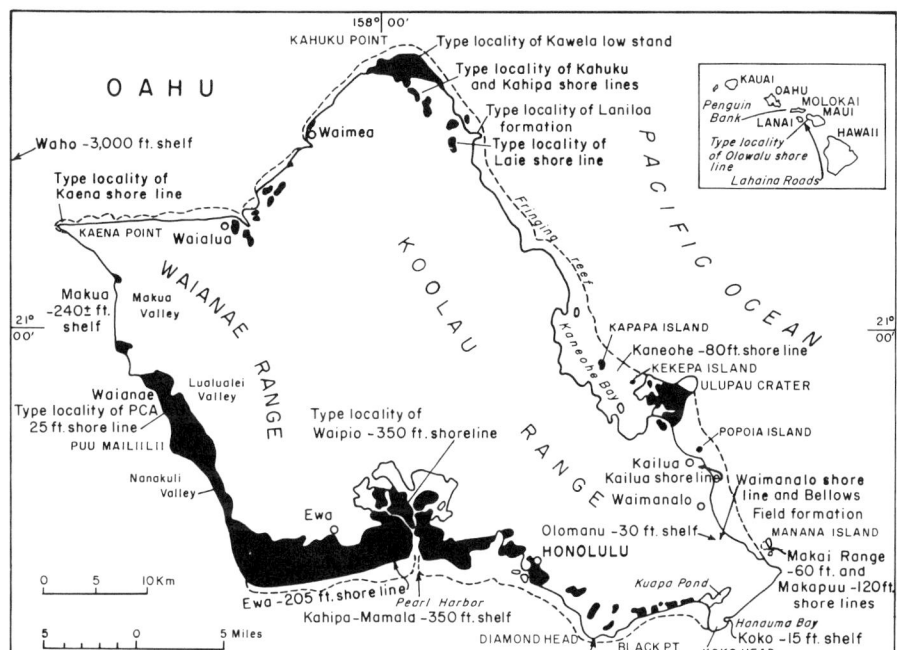

al., 1974), which may be compared with similar dates on Barbados and elsewhere. On Oahu the sea has stood higher by about 60 cm (2 ft) and 1.5 m (5 ft) above present sea level since that time (Stearns, 1974). Uranium series dating of shells from the "modern" beach rock at Kahuku Point, Oahu, indicate a Wisconsin age.

General Geology

The Hawaiian Islands are a chain of shield-type basaltic domes built over a fissure zone 2500 km long. This tectonic feature has existed since at least early Tertiary. The lava has risen along tension cracks bounding lozenge-shaped blocks strung out linearly from SE to NW. The vents are spaced about 45 km apart. Aeromagnetic surveys by Malahoff and Woollard (1966) show that the volcanic vents and rift zones are aligned along fissures parallel to ENE-WSW Molokai Fracture Zone which extends from the North American Continent.

The Hawaiian Ridge is bordered by a "moat" or trough about 5500 m deep which is bounded on the outside by a low rise, the broad Hawaiian Arch which slopes up to about 600 m above the depression. The arch loops around the island of Hawaii and dies out on the northwest side (Dietz and Menard, 1953). The trough and arch are thought to have been caused by the downbowing of the crust as a result of plastic and/or elastic response to the weight of the islands. Detailed soundings offshore have revealed numerous submarine volcanoes, mostly small cones, but one, however, is 2000 m high and 16 km across. Those adjacent to Oahu align with the NE-SW rifts of the secondary Quaternary cones on Oahu (Hamilton, 1957).

In 1963, Wilson set forth the hot spot and drifting plate theory for the origin of the Hawaiian Islands. Numerous articles have been written since to explain the linear arrangement of the Emperor and Hawaiian chains as a result of a Pacific plate sliding northwestward and northward over a hot spot in the earth's crust located under Kilauea and Mauna Loa volcanoes on the Island of Hawaii. The rate of movement of the plate has been given variously from 2 cm to 12.5 cm/yr (Claque and Jarrard, 1973b), based upon the fact that the islands increase progressively in K-Ar age northwestward from Kilauea. The hypothesis is based on the fact that the oldest seamount at the northern end of the Emperor chain was submerged about 70 m yr ago, and Midway Island, near the western end of the Hawaiian chain, is about 20 m yr old, whereas Kilauea, at the eastern end of the Hawaiian chain, is still erupting.

The bulk of the Hawaiian volcanic domes consist of lava flows dipping 3° to 10° away from their source and only rarely separated by thin soil beds, indicating rapid accumulation of flows such as is taking place now on Mauna Loa Volcano. Thin soils between flows in some volcanoes show that the time interval separating eruptions lengthened toward the close of the dome-building epoch. Many of these soil beds are decomposed vitric tuff, which during the early phase of eruptions is generally deposited in small quantities by lava fountains near the vents.

Fissure eruptions characterize Hawaiian volcanoes. The usual eruption is preceded by local earthquakes. The fissures are a few cm to a few m wide, and limited to definite rift zones. The widest dike known in Hawaii is 12 m across. Eruptions often begin with a lava fountain which is caused by frothing at the top of the lava column when pressure on the enclosed gases is released. Depending upon effervescence, these range from a line of spatter cones 1–20 m high to a line of cinder cones 20–100 m high. Rivers of pahoehoe pour from the fissure; but as it flows down the mountainside, the lava usually changes to aa type. Some of the lava flows entering the sea disrupt violently and build fragmental cones. Pillow lavas form offshore becoming larger progressively as the water deepens. Recorded eruptions have lasted from a few hours to years, and the flows have ranged in length from a few m to 50 km. The lava flows reach velocities in their main channels of 15–40 km per hour depending upon the steepness of slope. The discharge of (?)*juvenile water* increases in direct proportion to the amount of effervescence. Cognate secondary cones are superimposed on the great lava domes (Fig. 4).

Geologic History. Each of the Hawaiian volcanoes has passed through more or less similar stages in development (Fig. 7), as follows:

Stage 1. Building of a volcano from the ocean floor to sea level. During this submarine phase, the volcano lays down chiefly pillow lavas and at shallow depths produces large quantities of ash and pumice as a result of the contact of the magma with seawater. When a cone first rises above sea level, it is composed largely of weakly consolidated ash which is rapidly eroded by the sea. Lava flows soon veneer the cone, and the erosive effectiveness of wave action is decreased greatly.

Stage 2. Once the cone is above sea level, thin sheets of highly fluid olivine basalts (tholeiites) are rapidly poured out, usually from two major rifts and one minor one as well as at their intersection. Eventually a shield-shaped dome is built. In the dominantly olivine-bearing basalts, small feldspar phenocrysts are common and pyroxene phenocrysts are scarce. A few thin beds of lithic and vitric tuff may be deposited. Because the surface is highly porous and the time interval between eruptions is short, stream erosion is nonexistent. Mauna Loa between 1850 and 1950 liberated approximately 4 billion m^3 of lava (MacDonald *et al.*, 1960, p. 23).

Stage 3. The volcano gradually collapses over the vent areas, commonly forming a caldera on the summit and grabens along the major rifts. The composition of the lavas does not change appreciably, nor does the time interval between eruptions lengthen. Lavas ponded in closed fault basins are very massive, however, and in physical appearance differ greatly from the pre-caldera lavas. When eroded they form sheer cliffs which usually show columnar structure. Lithic and vitric tuff beds may be developed more frequently than in the first phase. The highest wall of the caldera usually bounds a segment between the two rift zones which intersect at an obtuse angle. Generally the sea-

FIGURE 7. Eight stages in the geologic history of the Hawaiian Chain.

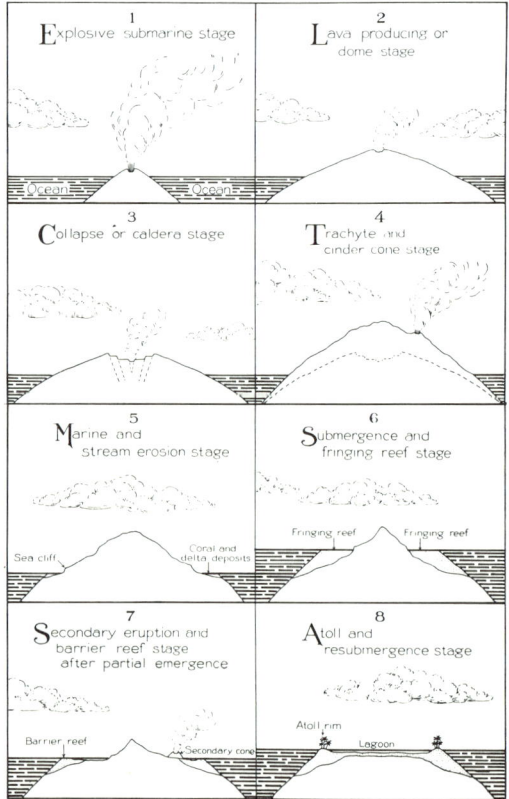

TABLE 4. Stratigraphic Rock Units in the Hawaiian Islands

Age	Niihau	Kauai	Oahu - Waianae Range	Oahu - Koolau Range	Molokai - West Molokai	Molokai - East Molokai	Lanai
Historic				Coral fill			
Recent	Younger alluvium, playa deposits, and unconsolidated calcareous beach and dune sand	Younger alluvium, unconsolidated calcareous beach and dune deposits and marly lagoon deposits.	Younger alluvium and unconsolidated calcareous beach and dune deposits	Younger alluvium, unconsolidated beach and dune deposits, and younger rocks in the Honolulu volcanic series	Younger alluvium and unconsolidated calcareous beach and dune deposits	Younger alluvium and unconsolidated calcareous beach deposits	Younger alluvium and unconsolidated calcareous beach and dune deposits
		~~Local unconformity~~~					
Pleistocene	Lithified calcareous dunes, emerged marine limestone, dunes of volcanic sand, older alluvium, and Kiekie volcanic series	Lithified calcareous dunes, older alluvium and rocks of the Koloa volcanic series, including the Palikea formation.	Lithified calcareous dunes, emerged marine limestone, older alluvium, and Kolekole volcanics	Lithified calcareous dunes, emerged marine limestone, older alluvium, and older rocks in the Honolulu volcanic series	Lithified calcareous dunes and older alluvium	Emerged marine limestone, older alluvium, Kalaupapa basalt, and Mokuhooniki tuff	Lithified calcareous dunes, emerged marine limestone, and older alluvium
	~~Great erosional unconformity~~~						
Pliocene and older	Paniau volcanic series	Waimea Canyon volcanic series: main caldera-filling (upper or Olokele formation), graben-filling (upper or Makaweli formation), secondary caldera-filling (Haupu formation) and extra-caldera (lower or Napili formation).	Waianae volcanic series; upper, lower, and middle members	Koolau and Kailua volcanic series	W. Molokai volcanic series	E. Molokai volcanic series; upper, lower, and caldera complex members	Lanai volcanic series

Age	Kahoolawe	Maui - West Maui	Maui - East Maui	Hawaii - Kohala Mountain	Hawaii - Mauna Kea	Hawaii - Mauna Loa Volcano	Hawaii - Hualalai Volcano	Hawaii - Kilauea Volcano
Historic			Volcanics of 1790(?)			Historic member of the Kau volcanic series and mudflow of 1868	Historic member of the Hualalai volcanic series (Volcanics of 1800–01)	Historic member of the Puna volcanic series
Recent	Red loess	Younger alluvium and unconsolidated calcareous beach and dune deposits	Younger alluvium, unconsolidated beach and dune deposits, and younger rocks in the Hana volcanic series	Younger alluvium, unconsolidated calcareous beach deposits, and black sand dunes	Younger alluvium, unconsolidated calcareous beach deposits, volcanic sand dunes, vitric ash deposits, and upper member of the Laupahoehoe volcanic series	Black sand dunes, and the prehistoric member of the Kau volcanic series	Unconsolidated calcareous beach deposits, and younger rocks of the prehistoric member of the Hualalai volcanic series	Black sand dunes, and prehistoric member of the Puna volcanic series
	~~Local unconformity~~~							
Pleistocene	Older alluvium	Emerged marine limestone, older alluvium, and Lahaina volcanic series	Kaupo mudflow, older alluvium, and older rocks in the Hana volcanic series including the Kipahulu member	Older alluvium, Pahala ash, and Hawi volcanic series	Glacial deposits, older alluvium, lower member of the Laupahoehoe volcanic series, Pahala ash, and younger rocks in the Hamakua volcanic series	Pahala ash and Kahuku volcanic series	Pahala ash and older rocks of the Hualalai volcanic series including the Waawaa volcanics	Pahala ash and Hilina volcanic series
Pliocene and older	Kanapou volcanic series: caldera and extra caldera members	Honolua volcanic series; Wailuku volcanic series	Kula volcanic series Honomanu volcanic series	Pololu volcanic series	Older unexposed rocks in the Hamakua volcanic series	Ninole volcanic series	Probably deep unexposed rocks in the Hualalai volcanic series	

ward slope of this high wall ceases to be flooded with lavas; and as a result, canyons are eroded into it, while lavas continue to veneer the other slopes. Sea cliffs are another characteristic feature of the high-walled slope. Two distinct physiographic stages may exist, therefore, on the same volcano.

Mauna Loa, Kilauea, and Lanai are now in this third volcanic phase. The other domes have passed through it. The West Molokai dome adjacent to the summit collapsed but apparently a caldera did not form.

Stage 4. When the volume of the lavas poured out exceeds the amount of the collapse, the caldera and grabens are partly or entirely obliterated. Time intervals between eruptions grow progressively longer, and the composition of the lavas may change gradually or abruptly to more alkalic types. Trachyte, andesite, or closely related rocks are laid down in thick sheets, chiefly as aa flows. In this phase, the more highly ferromagnesian lavas usually contain large phenocrysts of one or all the following minerals: pyroxene, olivine, and feldspar. Peridotitic and gabbroic cognate inclusions are common. High lava fountains which build large cinder cones characterize most eruptions, and bulbous domes may be formed. Vitric tuff beds increase in number and thickness. The profile of the dome steepens and becomes studded with cones. Some of the vents lie outside of the rift zones. Erosion has resulted in local unconformities between some of the flows. Hualalai, Mauna Kea, and Kohala domes are in this stage.

Volcanoes pass through this stage in either of two sequences. In West Maui, Kohala, and East Molokai the tholeiitic basalts are succeeded by (1) a thin incomplete veneer of olivine porphyritic basalts usually carrying augite phenocrysts, (2) a short pause, and (3) eruptions of the alkalic oligoclase basalt clan (hawaiite, mugearite) and trachytes which form an

TABLE 5. Leeward Islands of Hawaii

	Height	
	m	ft
Volcanic Islands		
17. Gardner (basalt flows and dikes)	58	190
21. La Pérouse Pinnacle in French Frigate Atoll (basalt flows)	37	122
23. Necker (basalt flows and dikes)	84	277
24. Nihoa (basalt flows and dikes)	277	910
26. Kaula (tuff)	167	550
Emerged Coral Atolls, Atolls or Near Atolls		
1. Unnamed	Breakers	
2. Bensaleux	Breakers	
3. Kure or Ocean (name of atoll)		
4. Green (sand island inside of Kure atoll)	6	20
6. Midway	13	43
8. Pearl and Hermes	3	12
9. Lisianski	13	44
10. Fisher	Breakers	
11. Minor	Breakers	
14. Laysan	17	56
15. Maro (Dowsett)	Breakers	
18. Two Brothers		?
22. French Frigate (excluding La Pérouse Pinnacle)	3	10
Shoals (Probably Submerged Atolls)		
5. Nero	−151	−492
7. Gambia	−25	−84
12. Neva	−5	−18
13. Springbank	−33	−108
16. Raita	−16	−54
19. St. Rogatien	−22	−72
20. Brooks	−25	−84
25. Unnamed	−58	−192

The numbers refer to those in Fig. 2.

incomplete veneer and add only a hundred m or so of lava to the summit of the volcano. The second sequence is typified by Haleakala and Mauna Kea in which no inactive period has occurred, trachytes and oligoclase basalts are rare or absent, and andesine basalts dominate but are interbedded with both olivine basalts and picrite basalts carrying large augite phenocrysts. These lavas may add 1000 m or more to the height of the volcano.

Stage 5. Marine and stream erosion partly destroy the volcanic dome. The effectiveness of these agents depends upon the height, which determines the rainfall, and upon the exposure, whether the island lies to the lee of another. Those which have been in the fifth stage for a long period develop fringing reefs, high sea cliffs, and deep valleys.

Stage 6. Deep submergence partly drowns the islands and extensive fringing reefs develop. Barrier reefs may develop or the growth of coral may be interrupted by secondary volcanic eruptions as shown in the next stage.

Stage 7. A rejuvenation of volcanicity with secondary eruptions occurs. These lavas, extruded during Pleistocene and Holocene times, commonly contain or lack pyroxene, feldspar, and olivine phenocrysts. Peridotitic and gabbroic cognate inclusions are usual. Vitric or crystal-lithic-vitric tuffs may be wide spread. The latter result chiefly from phreatomagmatic explosions. The lavas are unconformable upon extruded rocks of all preceding stages, indicating a long erosion interval. The domes of Haleakala, West Maui, East Molokai, Kahoolawe, Koolau, Waianae, Kauai, and Niihau are in this seventh volcanic phase. Many of these later vents show no close relationship to the ancient rift systems of the volcanoes on which they are formed. Many lie on N-S rifts, especially on Niihau, Kauai, and Oahu. Some volcanoes may never go through this stage.

Stage 8. If submergence continues or the island is planed off by marine and stream erosion, especially rapid during the fluctuating sea-levels of the Pleistocene, an atoll may develop on the eroded and submerged volcanic mass. The atolls in the leeward part of the archipelago were formed in this manner.

Stratigraphy. A correlation of the stratigraphic rock units in the Hawaii Islands is given in Table 4.

Offshore Islets

Thirty-nine islets lie close to the main islands. Seven islets are secondary tuff cones of late Pleistocene or Holocene age. Nineteen are remnants of lava flows isolated from the main islands by marine erosion. Two islets are part of the dike complex of the Koolau Volcano; three are lithified dunes of Pleistocene age; three are remnants of cinder cones; one is the remnant of a secondary nepheline basalt flow or crater fill; one is composed of reef limestone, Salt Lake tuff, and earthy sediments; two are reef limestone of Pleistocene age; and one is of unconsolidated sand.

Leeward Islands

Extending northwestward from Niihau are 26 islets (Table 5), reefs, and shoals known as the Leeward Islands. They mark the summits of submarine volcanoes (Fig. 2). West of Gardner Island (Fig. 2, no. 17), only low coral islands are found, whereas to the E many are remnants of basaltic cones.

Kaula is a secondary tuff cone built on a reef platform, whereas the other volcanic islands are probably eroded remnants of primary volcanic domes. The rocks are composed of andesine, picrite, olivine, and nepheline basalts. Andesine basalt is rare. A detailed study of the petrography and geochronology has been recently completed (Dalrymple et al., 1974).

Coralline algae are the principal constituent of the living and emerged reefs. At *Midway* and probably at some of the nearby islands, the reefs contain an unusually large number of barnacles. The emerged reef on Midway stands 1.5 m above sea level and has a carbon-14 age of 2400 yr (Ladd et al. 1970). Two holes drilled on Midway reached the basaltic foundation, one reached Miocene (Tertiary Zone g) at 445 ft and basalt at 516 ft. A second hole reached Miocene at 500 ft and basalt at 1216 ft. All the sediment below about 590 ft is Lower Miocene (Todd and Low, 1970).

Age of the Islands

The age of many of the superficial lavas has been determined by the K/Ar method (McDougall, 1963). The relative ages seem consistent with field evidence but the ages of Kohala and Ninole Volcanic Series appear inconsistent with the amount of erosion they have suffered (see Tables 6 and 7).

As remarked long ago by Charles Darwin, the geologic history of the Pacific as a whole is one of great submergence (Stearns, 1945), and may have a geodetic explanation (see discussion under: *Oceania*). Numerous submarine canyons off the windward coast of Oahu have their mouths 2000–3000 m below sea level (Hamilton, 1957). In the opinion of some geologists,

TABLE 6. Ages of Hawaiian Volcanoes

Island	Volcano	Range of K-Ar Ages (m yr)
Kauai	–	5.6–3.8
West Oahu	(Waianae Range)	3.4–2.7
East Oahu	(Koolau Range)	2.2–2.5
Niihau	(Paniau)	2.6±
West Molokai	–	1.8
East Molokai	–	1.5–1.3
West Maui	(Wailuku Mts.)	1.3–1.15
East Maui	(Haleakala)	0.8–
Hawaii	(Kohala Mt.)	Less than 1.0
	(Puu Waawaa)	0.4
	(Ninole V.S. Mauna Loa)	Less than 0.5
	(Laupahoehoe, Mauna Kea)	0.6

these canyons could hardly have been cut in hard lava rock by turbidity currents; hence they appear to indicate a submergence of this amount. Subsidence of this amount must in part predate the neritic Miocene fossils at 520 m below sea level. Miocene fossils were obtained from a core in the Hawaiian "moat" off Gardner Pinnacle (Shor, p. 567, 1960). Part of the submergence may well be isostatic, but submarine terracing and other evidence shows that it was highly episodic, and thus calls for another explanation.

Post-erosional Quaternary eruptions are found on every island except Lanai. Mauna Kea has lava flows resting on Wisconsin glacial debris. Haleakala erupted last about 1790 and Hualalai erupted last in 1801. Kilauea and Mauna Loa have erupted frequently in historic time.

Petrology

The main volume of lavas that built the Hawaiian volcanoes is tholeiitic basalt which is believed to have differentiated by partial melting and fractional crystallization of primary mantle material (Yoder and Tilley, 1962). They range from picrite basalts (rich in olivine phenocrysts) to fine-grained dense basalts with no olivines visible. In the waning phase when several hundred years or more intervene between eruptions, fractional crystallization led to the eruption of alkalic basaltic lavas (hawaiite), oligoclase basalts (mugearites) and finally even to trachytes. Trachytes are found on Hualalai volcano and Kohala Mt., West Maui Mt., the East Molokai volcano, and a rhyodacite in the Waianae Mt., on Oahu. The uranium content of the basalts varies from 0.21 to 1.8 ppm and increases in quantity in the late differentiated lavas. The proportion of alkalic rocks and rock types in the caps of the differentiated volcanoes is given in the Table 8.

TABLE 7. Average K/Ar Ages of Certain Lavas of the Pleistocene Honolulu Volcanic Series on Oahu

	Years B.P.
Koko Head vent on the Koko Fissure	40,000
Sugar Loaf vent on the Tantalus Fissure	67,000
Kaupo basalt	32,000
Kalama vent on Koko Fissure	34,000
Kaimuki basalt	282,000
Black Point basalt	297,000
Punchbowl	296,000
Nuuanu basalt	419,000
Salt Lake bombs	430,000
Kalihi basalt (upper valley)	460,000
Kaau Crater	647,000
Castle basalt	853,000

Source: Gramlick et al., 1971.

Mechanism of Eruption

Seismic evidence indicates that magma originates at a depth of about 60 km (37 mi) and finds its way to the surface through a complex but nearly vertical fracture system during the rapid building of the volcano, pausing at a shallow reservoir within the crust (Fig. 8). Tumescence measured by tilt meters at the summit of Kilauea accompanies the intrusion of the magma into the reservoir system. Measurable summit collapse usually follows the extrusion of the lava, or the movement of magma into the lateral rifts.

Swarms of shallow, local earthquakes sometimes numbered in the thousands per day with progressively higher epicenters indicate the rupturing of the edifice as the magma rises to the surface. These decrease and harmonic tremor begins when the lava is erupting. When the eruption ceases local earthquakes continue as

TABLE 8. Proportions of Rock Types in Hawaiian Volcanoes

Alkali Rocks	Volume of Volcano (%)
Kohala Mountain (Hawaii)	0.1
West Maui	3.0
East Maui (Haleakala)	0.8
Mauna Kea (Hawaii)	0.7
Waianae Range, Oahu	0.3

Rock Types in Caps of Hawaiian Volcanoes	
Alkalic basalt	41.0
Andesites (Hawaiite and mugearites)	56.5
Ankaramite	1.7
Trachyte	0.8

MacDonald and Abbott, 1970.

the summit of the volcano collapses over the deflated reservoir.

The reservoir is shown as a laccolith-like body by MacDonald (1963) from whose model Fig. 7 was partly constructed. However, the dike complexes in eroded cores of Hawaiian volcanoes have dikes closely spaced resembling hundreds of vertical parallel concrete walls which would structurally prevent emplacement of a horizontal or laccolith-shaped intrusion.

The magma reservoir is more likely to be bounded by nearly vertical walls, except locally where dikes have been removed by melting or stoping. A gravity study made by the Hawaii Institute of Geophysics indicates that rock in the eroded Koolau caldera is too dense, at a depth of 1300 m below sea level, to be basaltic.

Possibly olivines and pyroxenes have settled out during the long cooling period to form large bodies of olivine and clino-pyroxene holocrystalline rocks. The occurrence of large quantities of such xenoliths in many late lava flows, for example, the 1801 flow on Hualalai, supports this hypothesis (Stearns and MacDonald, 1946, p. 147; and Richter and Murata, 1961, p. B215).

Typical composition in volume percent of the magmatic gases as determined by Naughton and Barnes are (in volume percent): H_2O, 62.50; CO_2, 20.20; SO_2, 13.20; N_2, 1:29; H_2, 1.90; CO_2, 1.66; S_2, 0.42; Cl_2, 0.05; A, 0.04; and CuCl, trace. They may be all or in part elements reaching the surface of the earth for the first time ("juvenile"). However, some of the H_2O probably is derived from rainfall and ground water and absorbed by the magma on its way to the surface.

The basalt magma as it reaches the surface has a temperature of 1050–1200°C, a viscosity of $2-4 \times 10^3$ poises and a gas content ranging from 2% by weight at the beginning to 0.5% during the later stages. Below the level at which gas bubbles start to form in it, the magma has a specific gravity of about 2.73. Flows continue to move at temperatures far below the liquidus of similar dry melt in the laboratory, in some cases probably to less than 800°C (MacDonald, 1963).

Drill holes into the lava lakes formed in Kilaueaiki in 1959 and in Alae Crater in 1963 indicate that it takes 6 months to form a crust 6 m thick. The isotherm at the base of this crust was 1067°C (Peck et al., 1964).

During 1969–1973 Kilauea Volcano was nearly in constant eruption in the Alae Crater area, building a lava dome called Mauna Ulu.

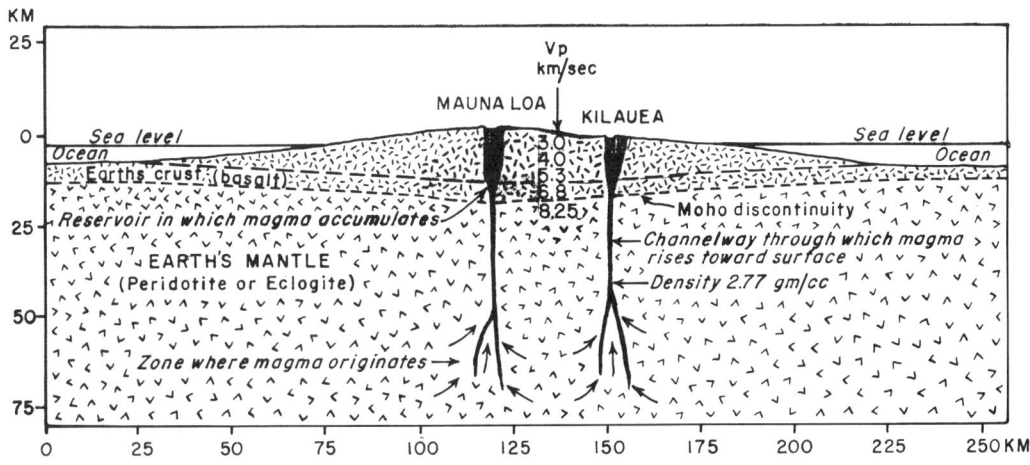

FIGURE 8. Schematic section of Mauna Loa and Kilauea volcanoes showing the source of the lava based on seismic data.

FIGURE 9. Air photograph of Kilauea crater with Mauna Loa in the distance.

Flows from the vent entered the ocean from 1969 onward, forming pillow lavas (Moore et al., 1973). This longest eruption in Kilauea's history has built a lava shield about 2 km long, 1 km wide, and 100 m high (Swanson et al., 1971) at a site actually predicted by the author (Stearns and Clark, 1930).

Economic Geology—Water and Nonmetallic Minerals

No metallic minerals are mined although extensive low-grade alumina (bauxite) deposits exist. The most valuable mineral resource is ground water. Large sums of money are spent to recover it. The U.S. Geological Survey reports that total ground water consumption on Oahu in 1970 was 689 million gallons per day. Besides drilled and dug wells, deep shafts and extensive tunnels are driven to recover perched, confined, and basal water (see Vol. IVA). Fresh basal ground water floats on seawater because it is lighter and, where covered with a caprock, is commonly artesian. A tunnel 10 km long recommended by the writer was completed on Molokai to recover water confined at high level in the dike complex.

Aggregate and sand (crushed basalt) are valuable nonmetallic resources. Dense rock satisfactory for aggregate is scarce on most of the islands. It is necessary to import rock to Oahu from Molokai. Coral beach sand is preferable to crushed basalt but is scarce on Oahu and much is hauled from Molokai on barges.

Cement is manufactured on Oahu using reef limestone as the main ingredient. The silica is imported.

Clay is mined locally for bricks but had not been used extensively in the past because of its high shrinkage. New sources and technology have reduced shrinkage so that good quality clay tile and brick have been produced on Oahu.

"Cinders" (volcanic ash) from Hawaii and Molokai are used extensively instead of soil for raising orchids because of their ability to hold moisture and their excellent drainage. They are also employed for lightweight aggregate.

Building stones in Hawaii consist of four types:

1. Reef and beach rock, a pleasant yellowish-white limestone easily worked.

2. Dense basalt for cut stone. Its use has declined in recent years as the old oriental workmen skilled in splitting the stone have died.

3. "Moss" rock is widely used. It is the weathered and pitted basaltic boulders from the arid areas of Oahu.

4. "Hilo" rock, a thin-layered pahoehoe quarried on Hawaii is in vogue for veneer because of its light weight, ease of cutting, and interesting texture.

HAROLD T. STEARNS

References

Claque, D. A., and Jarrard, R. D., 1973a. "Hot spots and Pacific plate motion," *Trans. Am. Geophys. Union,* **54,** 238.

____, 1973b. "Tertiary Pacific plate motion deduced from the Hawaiian-Emperor Chain," *Bull. Geol. Soc. Am.,* **84,** 1135–1154.

Cross, W., 1915. "Lavas of Hawaii and their relations," *U.S. Geol. Surv. Prof. Pap. 88,* 97p.

Dalrymple, G. B., Silver, R. A., and Jackson, E. D., 1973. "Origin of the Hawaiian Islands," *Amer. Scientist,* **6,** 294–308.

____, Lanphere, M. A., and Jackson, E. D., 1974. "Contributions to the petrography and geochronology of volcanic rocks from the Leeward Hawaiian Islands," *Bull. Geol. Soc. Am.,* **85,** 727–738.

Dana, J. D., 1849. "Geology", *U.S. Exploring Expedition, 1838–1842, under the Command of Charles Wilkes,* vol. 10. U.S. Navy, 756p.

Dietz, R. S., and Menard, H. W., 1953. "Hawaiian Swell, Arch, and Deep and subsidence of the Hawaiian Islands," *J. Geol.,* **61,** 99–113.

Doell, R. R., and Dalrymple, G. B., 1973. "Potassium–argon ages and paleomagnetism of the Waianae and Koolau volcanic series, Oahu, Hawaii," *Bull. Geol. Soc. Am.,* **84,** 1217–1242.

Eaton, J. P., 1962. "Crustal structure and volcanism," *Am. Geophys. Union Monogr. 6,* 13–29.

____, and Murata, K. I., 1960. "How volcanoes grow," *Science,* **132,** 925–938.

Fairbridge, R. W., 1950. "Landslide patterns on oceanic volcanoes and atolls," *Geog. J.,* **115,** 84–92.

Gramlick, J. W., Lewis, V. A., and Naughton, J. J., 1971. "Potassium dating of Holocene basalts of the Honolulu Volcanic Series," *Bull. Geol. Soc. Am.,* **82,** 1399–1404.

Gross, M. G., Milliman, J. D., Tracey, J. I., Jr., and Ladd, H. S., 1969. "Marine geology of Kure and Midway Atolls, Hawaii: a preliminary report," *Pacific Sci.,* **23**(1), 17–25.

Hamilton, E. L., 1957. "Marine geology of the southern Hawaiian Ridge," *Bull. Geol. Soc. Am.,* **68,** 1011–1026.

Heezen, B. C., and MacGregor, I. D., 1973. "The evolution of the Pacific," *Scientific American,* **229**(5), 102–112.

Jackson, E. D., Silver, E. A., and Dalrymple, G. B., 1972. "Hawaiian-Emperor Chain and its relation to Cenozoic circumpacific tectonics," *Bull. Geol. Soc. Am.,* **83,** 601–618.

Jagger, T. A., 1920. "Seismometric investigation of the Hawaiian lava column," *Seismol. Soc. Am. Bull.,* **10,** 155–175.

Ku, T-L., Kimmel, M. A., Easton, W. H., and O'Neil, T. J., 1974. "Eustatic sea level 120,000 years ago on Oahu, Hawaii," *Science,* **83,** 959–962.

Ladd, H. S., Tracey, J. I., and Gross, M. G., 1970. "Deep drilling on Midway Atoll," *U.S. Geol. Surv. Prof. Pap. 680-A,* A1–A22.

Lum, D., and Stearns, H. T., 1970. "Pleistocene stratigraphy and eustatic history based on cores at Waimanalo, Oahu, Hawaii," *Bull. Geol. Soc. Am.,* **81,** 1–16.

MacDonald, G. A., 1947. "Bibliography of the geology and water resources of the Island of Hawaii," *Hawaii Div. Hydrography Bull.,* **10,** 191p.

____, 1949. "Petrography of the Island of Hawaii," *U.S. Geol. Surv. Prof. Pap. 214-D,* 51–96.

____, 1963. "Physical properties of erupting Hawaii magmas," *Bull. Geol. Soc. Am.,* **74,** 1071–1077.

____, and Abbott, A. T., 1970. *Volcanoes in the Sea.* Honolulu: Univ. Hawaii Press, 441p.

____, Davis, D. A., and Cox, D. C., 1960. "Geology and groundwater resources of the Island of Kauai," *Hawaii Div. Hydrography Bull.,* **13,** 212p.

McDougall, I., 1964. "Potassium–argon ages from lavas of the Hawaiian Islands," *Bull. Geol. Soc. Am.,* **75,** 107–128.

Malahoff, A., and Woollard, G., 1966. "Magnetic measurements over the Hawaiian Ridge and their vulcanological implications," *Bull. Volcanol.,* **29,** 735–759.

Menard, H. W., Allison, E. C., and Durham, J. W., 1962. "A drowned Miocene terrace in the Hawaiian Islands," *Science,* **138**(3543), 896–897.

Moore, J. G., Phillips, R. L., Grigg, R. W., Peterson, D. W., and Swanson, D. A., 1973. "Flow of lava into the sea, 1969–1971, Kilauea Volcano, Hawaii," *Bull. Geol. Soc. Am.,* **84,** 537–546.

Peck, D. L., Moore, H. G., and Kojima, G., 1964. "Temperatures in the crust and melt of Alae lava lake, Hawaii after the August 1963 eruption of

Kilauea Volcano—a preliminary report," *U.S. Geol. Surv. Prof. Pap. 501-D*, D1–D7.
Porter, S. C., 1971. "Holocene eruptions of Mauna Kea volcano, Hawaii," *Science,* 172(3981), 375–377.
———, 1972. "Distribution, morphology, and size frequency of cinder cones on Mauna Kea Volcano, Hawaii," *Bull. Geol. Soc. Am.,* 83, 3607–3612.
Richter, D. H., and Murata, K. J., 1961. "Xenolithic nodules in the 1800–1801 Kaupulehu flow on Hualalai volcano," *U.S. Geol. Surv. Prof. Pap. 424-B,* 215–217.
Shaw, H. R., 1973. "Mantle convection and volcanic periodicity in the Pacific: evidence from Hawaii," *Bull. Geol. Soc. Am.,* 84, 1505–1526.
Shepard, F. P., 1963. "Thirty-five thousand years of sea level," *Essays in Marine Geology in Honor of K. O. Emery,* Los Angeles: Univ. S. Calif. Press, 1–10.
Shor, G. G., 1960. "Crustal structure of the Hawaiian Ridge near Gardner Pinnacles," *Seismol. Soc. Am. Bull.,* 50, 563–573.
Stearns, H. T., 1939. "Geologic map and guide of the Island of Oahu, Hawaii," *Hawaii Div. Hydrography Bull.,* 2, 75p.
———, 1940a. "Geology and ground-water resources of the Islands of Lanai and Kahoolawe, Hawaii," *Hawaii Div. Hydrography Bull.,* 6, 177p.
———, 1940b. "Supplement to the geology and ground-water resources of the Island of Oahu, Hawaii," *Hawaii Div. Hydrography Bull.,* 5, 164p.
———, 1945. "Late geologic history of the Pacific Basin," *Am. J. Sci.,* 243, 614–626.
———, 1961. "Eustatic shore lines on Pacific Islands," *Z. Geomorphol.,* Suppl. 3, 1–16.
———, 1966a. *Geology of the State of Hawaii.* Palo Alto, Calif: Pacific Books, 266p.
———, 1966b. *Road Guide to Points of Geologic Interest in the Hawaiian Islands.* Palo Alto, Calif: Pacific Books, 66p.
———, 1970. "Ages of dunes on Oahu, Hawaii," *Bernice P. Bishop Mus. Occ. Pap.,* 24(4), 49–72.
———, 1974. "Submerged shorelines and shelves in the Hawaiian Islands and a revision of some of the eustatic emerged shorelines," *Bull. Geol. Soc. Am.,* 85, 795–804.
———, and Clark, W. O., 1930. "Geology and water resources of the Kau District, Hawaii, with a chapter on ground water in the Hawaiian Islands by O. E. Meinzer," *U.S. Geol. Surv. Water-Supply Pap. 616,* 194p.
———, and MacDonald, G. E., 1942. "Geology and ground-water resources of the Island of Maui, Hawaii," *Hawaii Div. Hydrography Bull.,* 7, 344p.
———, and MacDonald, G. A., 1946. "Geology and ground-water resources of the Island of Hawaii," *Hawaii Div. Hydrography Bull.,* 9, 363p.
———, and Vaksvik, K. N., 1935. "Geology and ground-water resources of the Island of Oahu, Hawaii," *Hawaii Div. Hydrography Bull.,* 1, 479p.
Swanson, D. A., Jackson, D. B., Duffield, W. A., and Peterson, D. W., 1971. "Mauna Ulu eruption, Kilauea Volcano," *Geotimes,* 16, 12–16.
Todd, R., and Low, D., 1970. "Smaller Foraminifera from Midway drill holes," *U.S. Geol. Surv. Prof. Pap. 680-E,* E1–E49.

Veeh, H. H., 1966. "Th 230/U238 and U234/U238 ages of Pleistocene high sea level," *J. Geophys. Res.,* 71, 3379–3386.
Ward, W. T., 1973. "Correlation of Pleistocene shorelines in Gippsland, Australia and Oahu, Hawaii," *Bull. Geol. Soc. Am.,* 84, 3087–8092.
Wilson, J. T., 1963. "A possible origin of the Hawaiian Islands," *Canadian J. Physics,* 41, 863–870.
Woodward, P. W., 1972. "The natural history of Kure atoll, northwestern Hawaiian Islands," *Atoll Res. Bull.,* 164, 318p.
Yoder, H. S., Jr., and Tilley, C. E., 1962. "Origin of basalt magma," *J. Petrol.,* 3, 342–532.

Reference may also be made to *Publications of the Bernice P. Bishop Museum,* the *Hawaiian Volcano Observatory,* and the *Hawaii Institute of Geophysics.*

Cross-references: *American Samoa; Guam; Oceania; Pacific Islands Trust Territory; Society Islands; Tahiti; United States; Wake Island.*

UNITED STATES–MIDWESTERN REGION

The Midwestern region consists of about 1.5 million km^2 (600,000 sq mi) and includes parts or all of 15 states. It extends from Minnesota S 1700 km to Arkansas and from Ohio W 1200 km to Nebraska. Paleozoic sedimentary rocks are at the surface throughout the greater part of this region but disappear both to the W and S beneath overlapping Mesozoic strata of adjacent regions. The Paleozoic formations have been eroded away from a large Precambrian area near Lake Superior and much smaller ones in the low Ozark Mountains of Missouri and the Arbuckle and Wichita mountains of Oklahoma. The northern two-thirds or more of the province are mantled by Pleistocene glacial deposits (Fig. 1).

Geological Studies

No area of comparable size in the western hemisphere has been investigated more intensively by geologists than the Midwestern region. Field studies beginning here more than a century ago have contributed as greatly to an understanding of geologic history as have those in any other part of North America. Nowhere on this continent is the Precambrian better known than in the Lake Superior district. The Paleozoic sediments of the province provide a standard section second in importance only to that of the classic New York area. Pleistocene glacial history was originally worked out here

FIGURE 1. Morphotectonic divisions of the Midwest region.

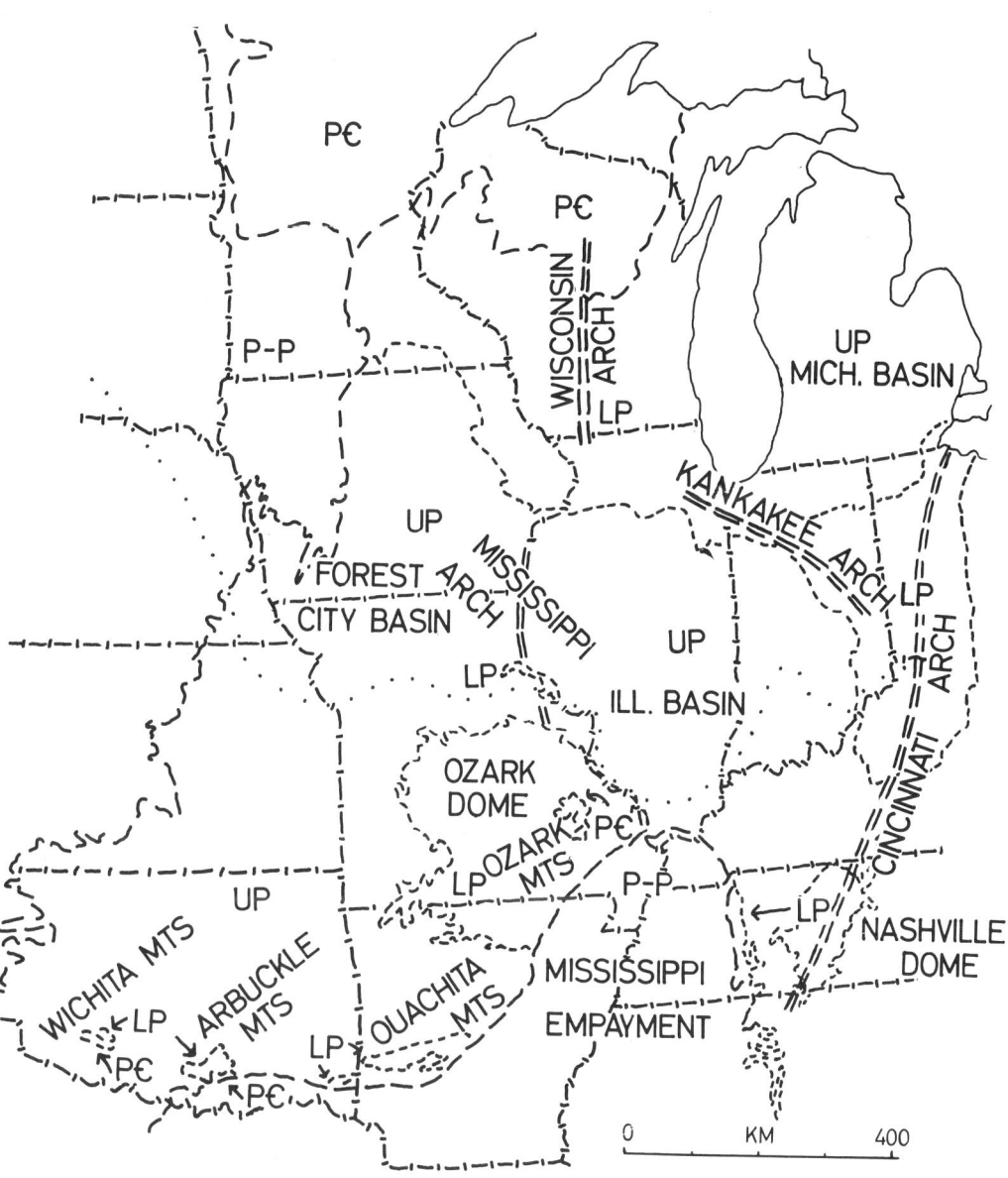

and is known in greater detail than in any other part of the United States or Canada.

Most of the states of the Midwestern region have active geologic surveys, that in Illinois being especially noteworthy. Most of the state universities and several well-known private institutions have strong geology departments. The one at the University of Chicago formerly was particularly outstanding. In more recent time the quality of others has rapidly improved and now little choice can be made between them. Geologic research continually pursued by persons connected with these surveys and universities has been supplemented importantly for many years by developments in a variety of mineral industries.

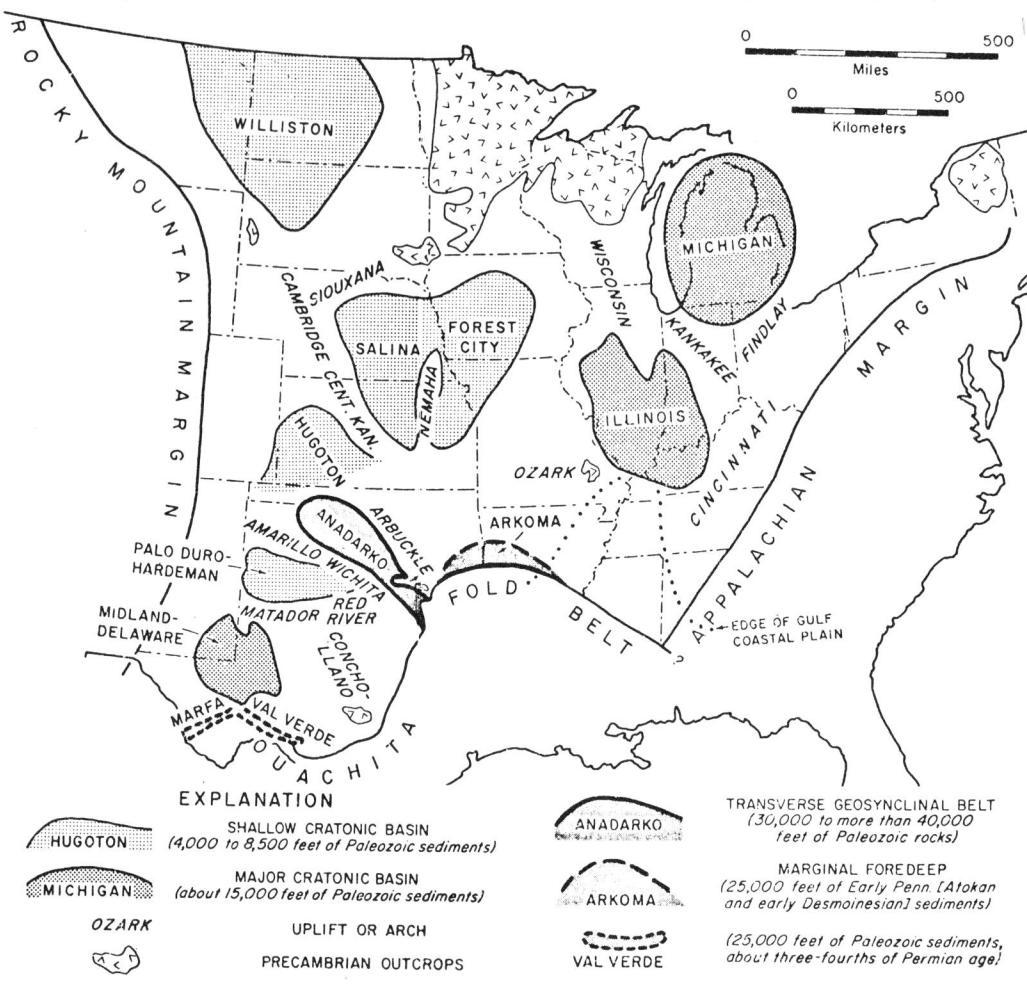

FIGURE 2. Principal structural features of the craton of the central United States. Names of features refer to basins or arches, according to the explanation below (after Ham and Wilson, 1967; from Dennis, 1972).

Structure

Lying parallel and W of the Appalachian belt, also trending NE-SW, there was a persistent positive swell, the *Cincinnati Arch,* and its northerly extension, the *Findlay Arch.* West of the Cincinnati Arch, which forms the eastern boundary of the Midwestern region, strata dip into the deep Illinois Basin (to W) and Michigan Basin (to N). These are separated by the NW-trending *Kankakee Arch* that merges with the gentle *Wisconsin Arch* plunging S from the Lake Superior Highlands and rising again as the *Mississippi Arch* to join the *Ozark Dome* in Missouri. Several less-defined structural basins lie W of the Mississippi River N and W of the Ozarks (notably the *Forest City Basin,* and beyond it, the *Salina Basin*). South of the Ozarks the strata dip into a belt of former deep geosynclinal troughs (notably *Arkansas Basin* and *Anadarko Basin*), where they have been much folded, overthrust N, and now rise in the Ouachita Mountains of Arkansas and Oklahoma (Fig. 2). A prominent zone of faulting produced by both tensional and compressional forces begins in Missouri on the NE flank of the Ozarks, trending ESE-WNW, crosses southern Illinois, extends into Kentucky and beyond. This zone (Heyl's "38th Parallel Lineament," 1972) is also the site of many "cryptovolcanic" structures (Bucher, 1936, 1963), sometimes (erroneously) classified as "meteorite craters."

Most of the structural features of the Midwestern region had their origin in the early Paleozoic, in part related to Precambrian basement features, and subsequently have been variously revived and accentuated. The greater part of this inherited structure, however, became further modified by earth movements at or near the close of the Paleozoic. Since then the region has been remarkably stable and was affected by little more than gentle warping. The region is essentially aseismic, but rare earthquakes occur along major basement faults (e.g., New Madrid) and in the former ice-covered area (still undergoing postglacial isostatic adjustments to removal of ice load).

Stratigraphy

Precambrian. Precambrian rocks in the Lake Superior district contain a measured sequence about 12,000 m thick and are structurally complex. They are divided into four great series and record three episodes of granitic intrusion, mountain building, and widespread erosion. The *Keewatin Series* of Archean age consists of repeated basaltic lava flows with subordinate sediments all much metamorphosed and intruded by Laurentian granite. Unconformably above these comes the *Knife Lake Series* of Aphebian (formerly "early Algonkian") age. It is composed principally of metamorphic slates, graywackes, and conglomerates with some thin zones of banded iron formation and, in the lower part, volcanic flows and tuffs. Most of these rocks are believed to have been deposited under nonmarine conditions. They are intruded by Algoman granite.

Next above an important unconformity is the *Huronian Series* of Paleohelikian (mid-

FIGURE 3. Time stratigraphic relationships of the Phanerozoic sequences in the North American craton. Black areas represent nondepositional hiatuses; white and stippled areas represent deposition. Stippling is used to differentiate successive depositional megacycles (Sloss, 1963).

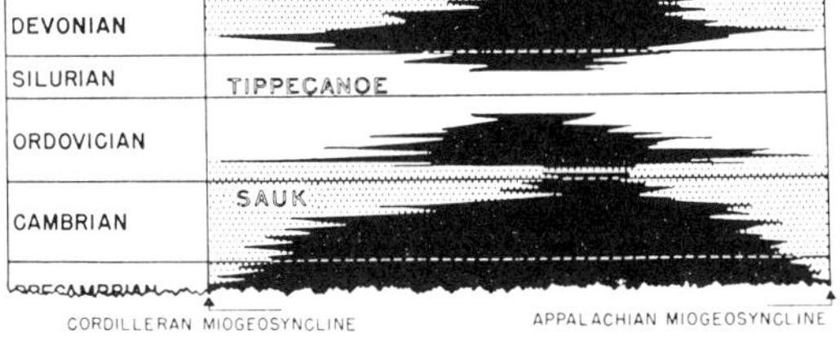

Algonkian) age. The rocks are moderately metamorphosed slates, quartzites, and dolomite and include the great iron formations, all interbedded with some basic flows and sills. They are believed to have accumulated in a marine environment. At the top of the succession and unconformably overlying older rocks is the *Keweenawan Series* (Neohelikian); about 8000 m thick. It consists mainly of sandstones, shales, and conglomerates with some lava flows and sills in the lower part. These beds are compact and hard but they are not metamorphosed except near intrusions of Killarney (Grenvillian) granite. Most of these rocks are believed to be nonmarine and to record a more or less arid climate. Boulder beds at several positions in the Precambrian have been interpreted as glacial tillites.

Precambrian rocks reach the surface in the eastern part of the Missouri Ozarks as mountain peaks reexcavated from beneath a former cover of Paleozoic sediments. They consist of rhyolitic flows and associated consolidated ash beds recording two episodes of volcanic action separated by the injection of intrusive granite. Precambrian granitic rocks also outcrop locally in Oklahoma.

Flat-lying red sandstone and conglomerate unconformably overlie the much disturbed Precambrian at several places near Lake Superior. Similar material up to 1000 m thick has been penetrated by deep wells on the flanks of the Michigan Basin and in northern Illinois. These beds of uncertain age have been referred to the Precambrian by some and to the Cambrian by other geologists.

Paleozoic Era. A nearly complete record of the Paleozoic is preserved in the Midwestern region (Table 1). The rocks representing this vast interval of time, more than 300 m yr, are much the thickest in the Ouachita Mountains of Arkansas and the nearby part of Oklahoma. Elsewhere they reach thicknesses of about 5000 m in some of the structural basins farther N. Many of the formations thicken inward toward the basin centers showing that these areas were sinking more or less continuously although irregularly during all this time.

Cambrian System. The Lower and Middle Cambrian are not represented in most of the Midwestern region. The Upper Cambrian, however, is widely found in the subsurface, but outcrops extensively only in the N from Wisconsin to Minnesota. Much smaller areas of outcrop occur in northern Illinois and in the several mountainous areas, previously mentioned, of Missouri, and Oklahoma. The formations in Wisconsin (*St. Croixan Series, Potsdamian*), are zoned on the basis of their trilobites, and provide the standard section for

TABLE 1. Generalized Composite Stratigraphic Section of Paleozoic Formations Based Mainly on Occurrences in Illinois[a]

Pennsylvanian
  McLeansboro Group
  Carbondale Group
  Tradewater Group
  Caseyville Group
	Devonian
	  Middle
	    Alto Limestone
	    Lingle Limestone
	    Grand Tower Limestone
	    Dutch Creek Sandstone

Mississippian
  Upper Chesterian
    Kinkaid Limestone
    Degonia Sandstone
    Clore Limestone
    Palestine Sandstone
    Menard Limestone
    Waltersburg Sandstone
    Vienna Limestone
    Tar Springs Sandstone
  Middle Chesterian
    Glen Dean Limestone
    Hardinsburg Sandstone
    Golconda Limestone
    Cypress Sandstone
  Lower Chesterian
    Paint Creek Limestone
    Bethel Sandstone
    Renault Limestone
    Aux Vases Sandstone
  Meramecian
    Ste. Genevieve Limestone
    St. Louis Limestone
    Salem Limestone
    Warsaw Limestone
  Osagean
    Keokuk Limestone
    Burlington Limestone
    Fern Glen Formation
  Kinderhookian
    Chouteau Limestone
    Prospect Hill Sandstone
    McCraney Limestone
    English River Sandstone
    Maple Mill Shale
    Louisiana Limestone
    Saverton Shale
    New Albany Shale
	Clear Creek Chert
	    Little Saline Limestone
	  Lower
	    Grassy Knob Chert
	    Bailey Limestone
	
	Silurian
	  Middle
	    Port Byron Dolomite
	    Racine Dolomite
	    Waukesha Dolomite
	  Lower
	    Kanakee Dolomite
	    Edgewood Dolomite
	
	Ordovician
	  Upper
	    Maquoketa Shale
	    Thebes Sandstone
	  Middle
	    Galena Dolomite
	    Decora Shale
	    Plattin Limestone
	    Joachim Dolomite
	    Dutchtown Dolomite
	    St. Peter Sandstone
	  Lower
	    Shakopee Dolomite
	    New Richmond Sandstone
	    Oneota Dolomite
	
	Cambrian
	  Upper
	    Jordan Sandstone
	    Trempealeau Dolomite
	    Frankonia Sandstone
	    Galesville Sandstone
	    Eau Claire Formation
	    Mt. Simon Sandstone
	
	Precambrian?
	  Fond du Lac Sandstone

[a]Other names and somewhat different classifications are recognized in other states.

North America (*Dresbachian, Franconian, Trempeleauan*). They are mostly sandstone, some shaly and some richly glauconitic, but include dolomitic strata particularly in the upper part. These beds seem to record two or more noteworthy fluctuations in the level of the late Cambrian sea. Southward in the Ozarks where the Upper Cambrian again rises to the surface, the basal part is sandstone again but most of the section consists of dolomite and limestone which evidently reflect offshore, warm, shallow environments. Much of this carbonate section is very cherty. Several of the Cambrian sandstones in northern Illinois are important freshwater aquifers.

Sloss (1963) has described this Midwest succession as a typical transgressive megacycle, his *Sauk Sequence,* which continued into the Lower Ordovician.

Ordovician System. Ordovician strata suc-

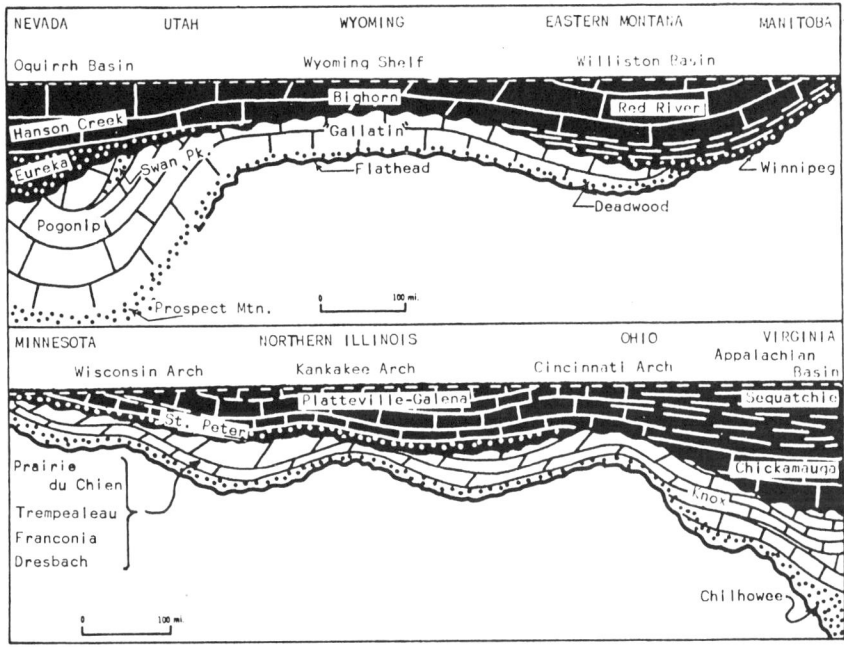

FIGURE 4. Diagrammatic cross sections of the Sauk Sequence (white) and the lower part of the Tippecanoe Sequence (black) (Sloss, 1963).

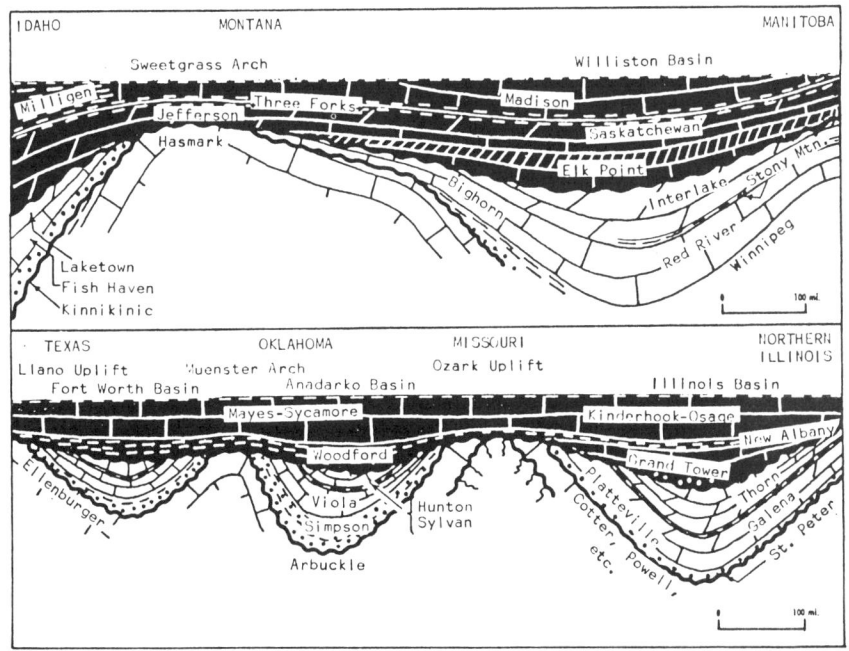

FIGURE 5. Diagrammatic cross sections of the Sauk and Tippecanoe sequences (white) and the basal part of the Kaskasia Sequence (black). Sequence boundaries are indicated by heavy wavy lines (Sloss, 1963).

ceed the Cambrian without notable stratigraphic discontinuity or lithologic change. The lower Ordovician (called *Canadian Series* after the Canadian River in Oklahoma) almost everywhere is dolomite with one or more sandy zones. It was terminated by a far-reaching discordance, the *Tippecanoe* (of Sloss) or *Owl Creek* (of Wheeler) regression. Much of the lower part of the Middle Ordovician (*Champlainian*) is remarkably pure quartz sandstone (*St. Peter*). This seems to be a relatively near-shore deposit of a sea that was slowly transgressing N. In northern Illinois it locally fills channels that cut downward into the Cambrian. The upper part of the Middle Ordovician (*Trentonian*) is somewhat cherty dolomite or nearly chert-free limestone. The Upper Ordovician (*Cincinnatian*) contains much shale (e.g., Maquoketa) with tongues of limestone in some areas and fine-grained sandstone E from the Ozark Dome, part of which may have been an emergent island. These strata probably are best known in southern Ohio and the neighboring part of Kentucky where they contain beautifully preserved fossils in great abundance. South of the Ozarks the Ordovician is mostly shale (e.g., Womble), sandstone (e.g., Blakely), and chert (e.g., Big Fork) contributed by the erosion of a land area lying somewhere farther in that direction.

The bulk of the Ordovician comprises a megacycle in its sedimentary evolution, part of the *Tippecanoe Sequence* of Sloss (or a shorter unit, the *Creek Sequence* of Wheeler).

Silurian System. The Silurian is mainly dolomite in most of the Midwestern region but includes limestone especially to the W and S. Much of it is very cherty. Impure reddish limestone and shale appear in the S, colored perhaps by residual subtropical soil washed into the sea from the Ozark Dome of Missouri and the Nashville Dome of Tennessee, parts of which may have been islands at this time. Sandstone occurs in Arkansas and is evidence of the erosion of a southern land mass. The Taconic orogeny of the Appalachian—New England belt may have reached its crescendo after the Lower Silurian (*Medinan*) and before the Middle Silurian (*Niagaran*). It was little felt in the Midwest, but Wheeler (1963) believes it should be designated as the upper limit for his Creek Sequence; to be followed by the *Tutelo Sequence,* which continued in a new cycle up into early Devonian.

The Upper Silurian (*Cayugan*) of Michigan contains salt in several layers reaching a total maximum thickness of more than 500 m. During the Silurian the central part of the Michigan Basin subsided about 1500 m relative to its borders. This is about one-third of the entire subsidence that has occurred there since the beginning of the Paleozoic. Similar but considerably less subsidence affected the S of what is now the Illinois Basin which during Paleozoic time was part of a much more extensive basin continuing an unknown distance farther S. A great system of organic reefs developed around the margins of the deeper basin and spread into the shallower waters to the N. Some of these reefs probably account for the isolation of the Michigan Basin and the precipitation of its salt. A few of the reefs may have risen 300 m above the subsiding sea floor.

Devonian System. The early Devonian sea was regressive and much restricted in central North America. Lower Devonian sediments (Helderbergian) seem to have accumulated in the Midwestern region only in its southern part within the deeper portions of the basin inherited from the Silurian. The rocks are mainly shale grading into siliceous limestone and chert.

A new sedimentary megacycle (the *Kaskaskia* of Sloss) began after the Helderbergian, beginning with Oriskany sandstone in the E, followed by shale. The pattern of land and sea evidently fluctuated somewhat at this time (in Europe, the Ardennic phase of the Caledonian orogeny). Middle Devonian strata are widely transgressive, mostly in limestone facies (e.g., Columbus, Wapsipinicon) except in the S of the region where chert is present. Rocks of Middle Devonian age overlap each other on the margins of the basins and overlie the Ordovician at many places. Part of the Ozark Dome probably existed as an island, and sand derived from it extends E as tongues within the limestone. The sea withdrew from much of the Midwestern region at about the end of Middle Devonian time. (This was the time of the Acadian orogeny in the Appalachians, or the Bretonic phase in northwestern Europe.) Normal faulting was active in southeastern Missouri and southern Illinois. The sea did not return until near the end of the Devonian except in northern Missouri and Iowa. There Upper Devonian shale and limestone record the existence of a marine embayment.

Much of the Midwestern region was subjected to erosion during considerable parts of the Devonian that varied from place to place. Some Middle Devonian sediments probably were deposited continuously across the major arches but later they were irregularly destroyed around the margins of the basins. Locally erosion cut deeply into older rocks and particularly where mild uparching had occurred, the entire Silurian section was removed. This erosion accounts not only for the unconformity regarded by some geologists as the Devonian-Mississippian boundary but also for part of the stratigraphic hiatus in many places where Pennsylvanian beds overlie pre-Mississippian formations.

Mississippian (Lower Carboniferous) System. The boundary between the Devonian and Mississippian systems is not agreed upon. Some geologists accept the unconformable base of the New Albany (Chattanooga) black shale formation that overlaps onto the Ordovician in a number of widely separated areas but others place it within the New Albany or higher stratigraphically. The New Albany black shale is remarkably persistent and uniform in its lithologic character throughout most of the Midwestern region. The greatest changes occur to the NW, where black color gives way to gray in northern Missouri and Iowa, and to the SW where, in Arkansas and eastern Oklahoma, shale grades into the upper part of thick, bedded chert whose lower part is Upper Devonian in age.

Succeeding Lower Mississippian (*Kinderkookian*), the strata show much variability around the Ozarks and N along the Mississippi River where limestone of several different types is well developed. Elsewhere they pass upward into the Middle Mississippian without much change. The lower half of the Middle Mississippian in the upper Mississippi Valley is a crystalline cherty limestone famous for the abundance and variety of its fossil crinoids. At some places near the Ozarks, the basal part is reddish suggesting the existence of a nearby island. Similar crinoidal limestone continues to the W and S. Widespread residual chert shows that some part of this formation once covered much if not all of the Ozark Dome. To the E, however, rapid change occurs and in Michigan this part of the section consists of silty shale and sandstone, and in Indiana contains bioherms. This transition is evidence for the presence of sediment producing land somewhere to the NE. From the Mississippi Valley SE the limestone also becomes increasingly siliceous and impure. The upper half of the Middle Mississippian is developed much more uniformly except to the NW where shale becomes increasingly conspicuous. Mostly these rocks are somewhat cherty limestones, finer grained, and a little darker colored than those below. Some of the beds are richly oolitic and others carry well-formed geodes.

The Upper Mississippian (*Chesterian*), which accounts for half the thickness of the system, consists of a succession of alternating sandstone and limestone-shale formations except in northern Arkansas where the strata grade upward from sandstone through shale to limestone without similar alternation and in Oklahoma where they are mostly shale.

The alternation of formations in the Upper Mississippian suggests a repeated cyclothemic pattern that became even more important in the following Pennsylvanian. Similarities include minor unconformities below the sandstones and thin local coal beds between the sandstones and marine limestone-shale formations. This general pattern begins near the top of the Middle Mississippian with two sandy zones in limestone and is repeated at least 10 times. It does not, however, duplicate the large number of members recognizable in well-developed Pennsylvanian cyclothems. This pattern in the Mississippian seems to be related to some contraction of the sea, lowering and fluctuating of its level, and deposition on and around a delta built by a large river flowing in from the NE.

Pennsylvanian (Upper Carboniferous) System. Except in the extreme S of the Midwestern region, Pennsylvanian strata are separated from older rocks by the most pronounced unconformity within the Paleozoic section of central North America. This is called the *Absaroka Sequence* by Sloss, and heralds the next sedimentational megacycle of the same name. In Illinois the Pennsylvanian occupies erosional channels up to 100 m deep, and from S to N it overlaps 1500 m of older beds as far down as the Middle Ordovician. Pre-Pennsylvanian faults are present locally in southern Illinois (the "38th Parallel Lineament," incorporating the Rough Creek fault zone). Relations similar in all these ways occur at other places. Lower Pennsylvanian (*Atokan*) strata are known only in northern Arkansas, where mostly they are lithologically similar to the underlying Upper Mississippian. To the S in the Ouachita Mountains 5000 m of sandstones and shales, overthrust N during the Appalachian orogeny, are of both Mississippian and Pennsylvanian ages. An abrupt facies change separates most of the Ouachita Paleozoic from that of its southern Appalachians (Thomas, 1972).

Everywhere in the Midwest Middle Pennsylvanian (*Desmoinesian*) beds occur at the base of the system. The lower part of this division is almost entirely nonmarine and consists of sandstone at the base, followed by coal with underclay below and a roof of marine fossil-bearing strata in several lithologically distinctive zones or members, and ending above with relatively thick unfossiliferous shale. Still higher in the section a larger-scale repetitive pattern becomes apparent with several differently developed cyclothems following each other in the same order. This latter type of cyclic repetition disappears in the uppermost Pennsylvanian.

The cyclic nature of the Pennsylvanian is best developed in the Illinois basin, where many of the cyclothems are nearly equally divisible

FIGURE 6. Diagrammatic cross sections showing relationships of lower part of Absaroka Sequence (black) to older units (white). Sequence boundaries are shown by heavy wavy lines (Sloss, 1963).

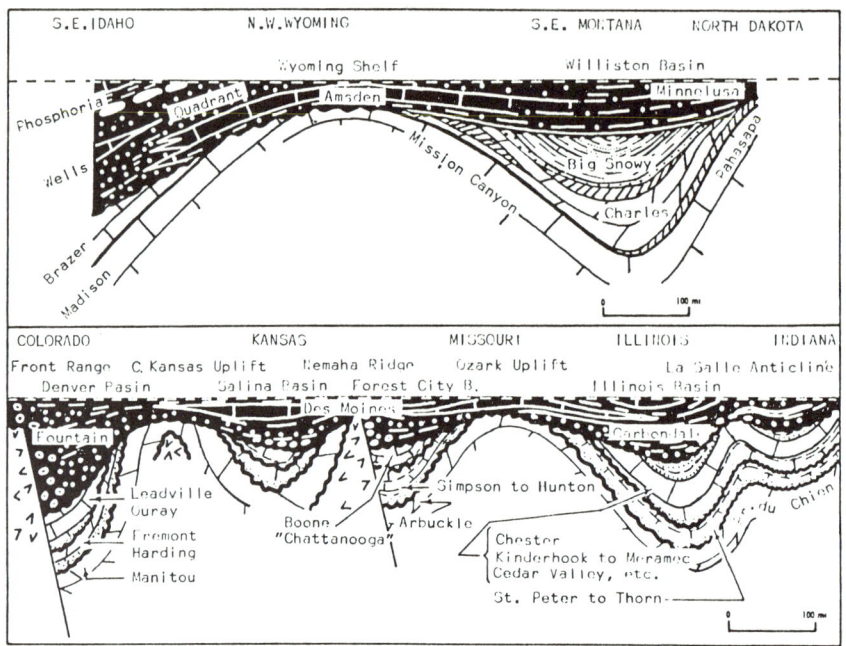

into a lower nonmarine half and an upper half, recording a brief shallow incursion of the sea. This balance is reduced to the W where the nonmarine members, particularly coals and sandstones in the lower half-cyclothems, become increasingly uncommon. This is evidence that the many transgressing marine embayments progressed across the Midwestern region from W to E. Similar reasoning based on the generally thicker, more numerous and coarser sandstones in the S and E of the province suggests that the land-derived sediments came from these directions.

Permian System. Permian strata, present only along the western border of the Midwestern region from Nebraska to Oklahoma, overlie the Pennsylvanian without physical discontinuity or noteworthy lithologic change. The lower beds (*Wolfcampian*) are mostly marine limestones and alternating shales in the pattern of incompletely developed cyclothems. Upward this pattern degenerates as the limestones disappear, the shales become increasingly red, and transition is made to a typical nonmarine redbeds succession with zones of gypsum and rock salt. Southward the red beds appear at lower and lower positions in the section and include sandstones, some which contain the fossil bones of terrestrial vertebrates.

Mesozoic Era. The Mesozoic is largely unrepresented by sedimentary deposits in the Midwestern region. This was mainly a long interval of time when much of the region stood as more or less low-lying land subject to almost continual erosion. Earth movements accentuated many of the older structures late in the Paleozoic or early in this interval and some new ones developed. The most important of the newer reactivations are in the folded and faulted Ouachita Mountains where there are some mid-Cretaceous granitoid and peridotite intrusions. There is also a zone of uplift cut by many faults and intruded by peridotitic lavas extending E from the Ozarks, and forming the southern boundary of the present Illinois Basin ("38th Parallel Lineament"). More than 1000 m of strata were eroded locally from southern Illinois and adjacent areas and Upper Cretaceous sediments of the Mississippi Embayment, some with marine fossils, here overlap onto the disturbed Paleozoic rocks.

Cenozoic Era. No noteworthy changes occurred in central North America at the transition from the Mesozoic to the Cenozoic except for withdrawal of the late Cretaceous marine embayments.

Tertiary Period. Much of the Midwestern region was reduced to a fairly level surface or peneplain in early Tertiary time. This was the result of a very long period of erosion that caused formation boundaries to retreat far from their original positions and produced the geologic pattern that exists today. Many of the Paleozoic rocks certainly were deposited across the Ozark Dome and the other great arches of this region but their former outer limits are not known. The peneplain S of the Midwestern region bears "lag deposits" of residual chert gravel and sand, mostly deeply iron stained (e.g., Lafayette gravels). Northward this surface was gently upwarped and such material, if it once extended farther in this direction, has been removed almost completely. A few patches of sand and gravel occur, however, adjacent to the upper Mississippi and lower Ohio rivers. These may be remnants of either late Cretaceous or early Tertiary deposits that formerly were more extensive.

Quaternary Period. The early Pleistocene was marked by increasing desiccation, with erosion, but left little trace in the stratigraphic record. More than 1,500,000 years ago the first (*Nebraskan*) of four great ice sheets formed in Canada and spread S to cover much of the Midwestern region. The second, or *Kansan,* reached farther SW. The third or *Illinoian,* was

FIGURE 7. Detail of the late Quaternary of the eastern Great Lakes. The Cary "Drift Border" represents the last readvance phase of the late Pleistocene (with moraines lying S of the Great Lakes), dated about 16,000 yr B.P. After rapid but cyclic retreat, the continental ice left successive moraine ridges. Glacial Lakes Whittlesey and Chicago formed between 14,000 and 13,000 yr B.P., the main spillways going to the SW through the Glacial Grand River and the Illinois River (Chicago Outlet). Isostatic tilting and further melting then permitted a northward shift of the lakes and, for a few centuries, the spillway shifted eastward to the Mohawk-Hudson, and eventually to the St. Lawrence Valley (see *Canada—Ontario Basin*).

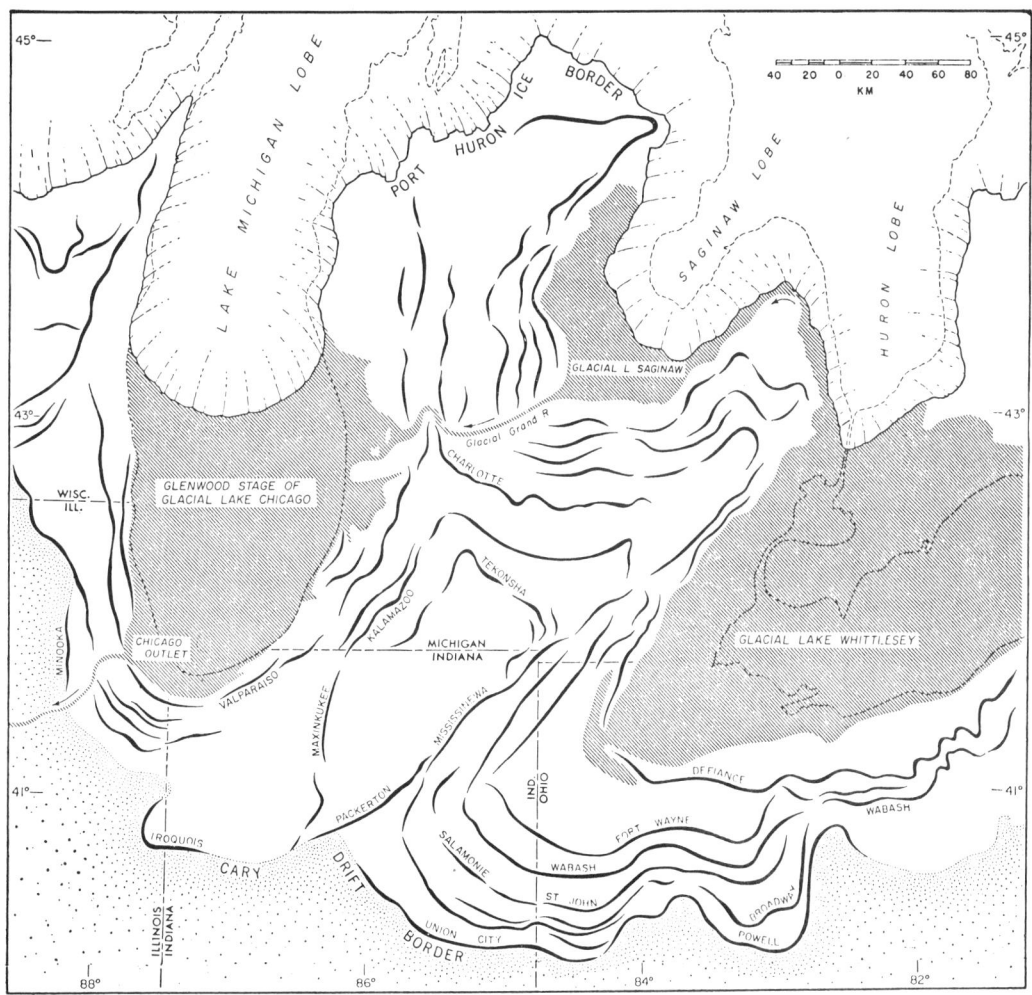

the most extensive and the outline of its farthest advance is marked approximately by the present courses of the Missouri, Mississippi, and Ohio rivers. These glaciers carried with them, and when they melted left behind, vast quantities of rock material, a distinctive till sheet of clay, pebbles, and giant erratics. This till (or "drift") covers the Paleozoic rocks with a thickness averaging 3–30 m, but as great as 400 m in parts of Michigan. Important "drift-free" areas occurred in Wisconsin and S Indiana (Schneider, 1970).

To the S of the ice fronts extensive areas of outwash accumulated, and during the cold maxima periglacial tundras developed but on a limited scale (Grüger, 1972). Widespread belts of loess were laid down under wind erosion during the arid conditions of each glacial crescendo and early retreat stage. Thickest accumulations are found E of the middle Mississippi Valley, while important gravels remained in the valley floor. Between the frigid epochs, an interglacial climate prevailed, at times warmer and wetter than today. On the loess or till surfaces forests grew and soils rapidly developed. The succession of tills, loess, and paleosols formed in this way reveals much of the history of the Pleistocene. The glaciers helped to scoop out the basins of the Great Lakes, diverted some rivers to the S, deeply channeled others as great floods accompanied the ice melting and isostatic tilting of lake basins. The glaciers left festoons along their outer limits and moraines where the retreating ice was halted temporarily. Much of the present physical geography of all but the S of the Midwestern region is the legacy of this glacial action. The last of the ice sheets (*Wisconsin*) began to melt back from this region about 15,000 years ago, and disappeared from the Great Lakes about 9000 years ago. The Holocene history has been marked by isostatic rebound in the N, as marked by numerous lake and river terraces, accompanied elsewhere by "interglacial" soil development.

Mineral Resources

The Midwestern region is richly endowed with mineral resources. The Illinois Basin, including parts of southeastern Indiana and western Kentucky, is one of the world's great Upper Carboniferous coal fields. Bituminous coal also occurs in lesser amounts in other parts of the region. Most of the iron mined in the United States has come from the Precambrian of Minnesota, Michigan, and Wisconsin. The highest-grade ore now is nearly exhausted but lower grades are being successfully and increasingly exploited. This province includes the world's most extensive and productive oil and gas fields of Paleozoic age in Illinois, Indiana, Michigan, Ohio, and Kentucky to the E of the Mississippi River, and in Oklahoma and Kansas to the W of it. Production is obtained from every geologic system from the Ordovician to the Pennsylvanian.

Much of the lead and zinc mined in the United States has come from three areas, the tristate district of Illinois, Iowa, and Wisconsin, the tristate district of Missouri, Kansas, and Oklahoma, and the Flat River district in southeastern Missouri. Production is from limestones of Cambrian, Ordovician, and Mississippian ages. Lead from Illinois once dominated world markets but now many of the deposits are essentially worked out. Barite is also mined.

Southern Illinois and western Kentucky have been the principal American source of fluorite although production in recent years has declined greatly. This mineral occurs in veins and bedded deposits associated with faults cutting Mississippian limestones. Southwestern Indiana is the site of the most important building stone industry in the United States. There Mississippian limestone is cut to accurate dimensions and shipped to all parts of this country and abroad. Salt is obtained by pumping brine from thick and extensive beds of Silurian rock salt far below the surface in the Michigan Basin and Permian salt is mined in Kansas. Copper is produced from the Precambrian in some of the world's deepest mines on the south shore of Lake Superior. Glass is manufactured from high silica sands mainly of Middle Ordovician age. Other products worthy of mention are cement, gypsum, high calcium chemical lime, rock wool, and of course great quantities of crushed stone, sand, and gravel used in construction and highway building.

J. MARVIN WELLER

References

Black, R. F., and Reed, E. C., 1965. "Upper Mississippi Valley," *INQUA Guidebook C.* Lincoln, Neb.: Neb. Acad. Sci., 126p.

_____, Goldthwait, R. P., and Willman, H. B., ed., 1973. *The Wisconsin Stage. Geol. Soc. Am. Mem. 136,* 334p.

Branson, E. B., 1944. "The geology of Missouri," *Univ. Mo. Stud.,* 19(3).

Bucher, W. H., 1936. "Cryptovolcanic structures in the United States," *Rept. 16th Intern. Geol. Congr.,* Washington, 1055–1084.

_____, 1963. "Cryptoexplosion structures caused from without or from within the earth ('astrobleme' or 'geobleme')?" *Am. J. Sci.,* 261, 597–649.

Carozzi, A. V., and Textoris, D. A., 1967. *Paleozoic Carbonate Microfacies of the Eastern Stable Interior.* Leiden: E. J. Brill, 42p.

Clark, T. H., and Stearn, C. W., 1968. *Geological Evolution of North America*, 2nd ed. New York: Ronald Press, 570p.

Cram, I., 1971. *Future Petroleum Provinces of the United States–Their Geology and Potential* (Mem. 15). Tulsa: Am. Assn. Petr. Geol., 2 vols., 1496p.

Croneis, C., 1930. "Geology of the Arkansas Paleozoic area," *Ark. Geol. Surv. Bull.,* **3**.

Damberger, H. H., 1971. "Coalification pattern of the Illinois Basin," *Econ. Geol.,* **66**(3), 488–494.

Dennis, J. G., 1972. *Structural Geology*. New York: Ronald Press, 532p.

Dott, R. H., Jr., and Dalziel, R. W. D., 1972. "Age and correlation of the Precambrian baraboo quartzite of Wisconsin," *J. Geol.,* **80**(5), 552–568.

Dury, D. H., and Knox, J. C., 1971. "Duricrusts and Deep-weathering profiles in southwestern Wisconsin," *Science,* **174**, 291–292.

Evenson, E. B., 1973. "Late Pleistocene shorelines and stratigraphic relations in the Lake Michigan Basin," *Bull. Geol. Soc. Am.,* **84**, 2281–2298. (Discussion: *ibid.,* **85**, 659–664.)

Flint, R. F., 1971. *Glacial and Quaternary Geology*. New York: Wiley, 892p.

Frye, J. C., 1973. "Pleistocene succession of the central interior United States," *Quaternary Res.,* **3**, 275–283.

———, Willman, H. B., and Black, R. F., 1965. "Outline of glacial geology of Illinois and Wisconsin," *The Quaternary of the U.S.* Princeton, N.J.: Princeton Univ. Press, 43–62.

Goldthwait, R. P., ed., 1965. "Great Lakes–Ohio River Valley," *INQUA Guidebook G*. Lincoln, Neb.: Neb. Acad. Sci., 110p.

Grüger, E., 1972. "Late Quaternary vegetation development in south-central Illinois," *Quaternary Res.,* **2**, 217–231.

Ham, W. E., and Wilson, J. L., 1967. "Paleozoic epeirogeny and orogeny in the central United States," *Am. J. Sci.,* **265**, 332–407.

Hare, M. G., and Morrow, E. H., 1973. *A Study of Paleozoic rocks in Arbuckle and western Ouachita Mountains of southern Oklahoma*. Louisiana: Shreveport Geol. Soc.

Heyl, A. V., 1972. "The 38th parallel lineament and its relationship to ore deposits," *Econ Geol.,* **67**, 879–894.

Hough, J. L., 1963. "The prehistoric Great Lakes of North America," *Am. Scientist,* **51**, 84–109.

Keller, G. R., and Cebull, S. E., 1973. "Plate tectonics and the Ouachita System in Texas, Oklahoma, and Arkansas," *Bull. Geol. Soc. Am.,* **83**, 1659–1666.

King, E. R., and Zietz, I., 1971. "Aeromagnetic study of the midcontinent gravity high of central United States," *Bull. Geol. Soc. Am.,* **82**, 2187–2208.

Leith, C. K., Lund, R. J., and Leith, A., 1935. "Precambrian rocks of the Lake Superior region," *U.S. Geol. Surv. Prof. Pap. 184*.

Lowenstam, H. A., 1950. "Niagaran reefs of the Great Lakes area," *J. Geol.,* **58**, 430–487.

Merriam, D. F., 1963. "The geologic history of Kansas," *State Geol. Surv. Kans. Bull.,* **162**.

———, ed., 1964. "Symposium on cyclic sedimentation," *State Geol. Surv. Kans., Bull.,* **169**, 2 vols.

Nicolaysen, L. O., 1972. "North American cryptoexplosion structures: interpreted as diapirs which obtain release from strong lateral confinement," *Geol. Soc. Am. Mem. 132*, 605–622.

Ocola, L. C., and Meyer, R. P., 1973. "Central North American rift system. 1. Structure of the axial zone from seismic and gravimetric data," *J. Geophys. Res.,* **78**(23), 5173–5194.

Pryor, W. A., and Amaral, E. J., 1971. "Large-scale cross-stratification in the St. Peter Sandstone, *Bull. Geol. Soc. Am.,* **82**, 239–244.

Ridge, D. J., 1971. *Annotated Bibliographies of Mineral Deposits in the Western Hemisphere* (Mem. 131–681). New York: Geol. Soc. Am.

Schneider, A. F., 1970. "A significant Pleistocene exposure in south-central Indiana," *Bull. Geol. Soc. Am.,* **81**, 3079–3084.

Sloss, L. L., 1963. "Sequences in the cratonic interior of North America," *Bull. Geol. Soc. Am.,* **74**, 93–114.

Snyder, F. G., 1968. "Tectonic history of Midcontinental United States," *Univ. Mo. Res.,* **1**(1), 65–77.

Thomas, W. A., 1972. "Regional Paleozoic stratigraphy in Mississippi between Ouachita and Appalachian Mountains, *Bull. Am. Assoc. Petrol. Geologists,* **56**(1), 81–106.

Trowbridge, A. C., and Atwater, G. I., 1934. "Stratigraphic problems in the upper Mississippi Valley," *Bull. Geol. Soc. Am.,* **45**, 21–81.

Wanless, H. R., 1950. "Late Paleozoic cycles of sedimentation in the United States," *18th Intern. Geol. Congr.,* London, 1948, pt. 4, 17–28.

———, and Shepard, F. P., 1936. "Sea level and climatic changes related to late Paleozoic cycles," *Bull. Geol. Soc. Am.,* **47**, 1177–1206.

Wayne, W. J., and Zumberge, J. H., 1965. "Pleistocene geology of Indiana and Michigan," *The Quaternary of the U.S.* Princeton, N.J.: Princeton Univ. Press, 63–84.

Webb, T., III, and Bryson, R. A., 1972. "Late- and postglacial climatic change in the northern Midwest, USA: quantitative estimates derived from fossil pollen spectra by multivariate statistical analysis," *Quat. Res.,* **2**, 70–115.

Weller, J. M., 1940. "Geology and oil possibilities of extreme southern Illinois," *Ill. State Geol. Surv., Rept. Inv. 71*.

———, 1961. "Pattern in Pennsylvanian cyclothems," *Third Conference on Origin and Constitution of Coal*. Nova Scotia: Dept. Mines., N.S. Research Found., 129–166.

———, and Sutton, A. H., 1940. "Mississippian border of Eastern Interior Basin," *Bull. Am. Assoc. Petrol. Geologists,* **24**, 765–858.

———, et al., 1944. "Symposium on Devonian stratigraphy," *Ill. State Geol. Surv. Bull.,* **68**, 89–222.

Weller, S., and Saint Clair, S., 1928. "Geology of Ste. Genevieve County, Missouri," *Mo. Bur. Geol. Mines*, Ser. 2, **22**.

Wheeler, H. E., 1963. "Post-Sauk and pre-Absaroka Paleozoic stratigraphic patterns in North America," *Bull. Am. Assoc. Petrol. Geologists,* **47**(8), 1407–1526.

White, W. S., 1966. "Tectonics of the Keeweenawan Basin, western Lake Superior region," *U.S. Geol. Surv. Prof. Pap. 524-E*, 23p.

Willman, H. B., and Frye, J. C., 1970. "Pleistocene stratigraphy of Illinois," *Ill. State Geol. Surv. Bull.,* **94**, 204p.

Cross-references: *Canada—St. Lawrence Lowlands of Quebec; North America; United States—Appalachian Region, Great Plains Province, Gulf Coastal Province.*

UNITED STATES—NEW ENGLAND REGION

The New England region usually constitutes the six northeasternmost states of the United States, but the geographic and geologic region described here also extends W across the valley of the Hudson River into the Adirondack area of New York State and NW to the St. Lawrence River in Quebec (Fig. 1). This larger region is a segment of the Appalachian orogenic belt and has a length of about 800 km and a width of about 300 km. The New England region is succeeded on the SW by another segment of the orogenic belt, the Appalachian region proper, but the connection is nearly broken near New York City, where the interior zones of the belt are overlapped by younger deposits or submerged. To the NE, the region extends directly into the Canadian Appalachians of Gaspé and the Maritime provinces (*see Canada—Atlantic Province*).

The New England region is composed of rocks that were variously deformed and later subjected to prolonged erosion, developing mountains, plateaus, and lowlands, which are less systematically arranged than in the Appalachian region farther to the SW.

The most persistent mountain axis extends from the Hudson Highlands N through the Berkshire Hills and Green Mountains in the W of the New England States, and thence NE into the Sutton Mountains of Quebec (Fig. 1). Like the Blue Ridge of the Appalachian region, these mountains are largely crystalline.

To the W of the mountain axis is a belt of lowlands and ranges cut on deformed sedimentary rocks that extends from the Valley and Ridge province in Pennsylvania N along the Hudson and Champlain valleys to the St. Lawrence River; implanted within it toward the S are the Taconic Mountains, formed mainly of metasediments. West of the lowlands are the Catskill Mountains, the NE terminus of the Appalachian Plateaus; and farther N the Adirondack Mountains, an outlier of the Laurentian Highlands or Canadian Shield.

To the E of the mountain axis, less regular mountain groups and knots are carved from the prevailing metamorphic and plutonic bedrock. The largest group is the White Mountains of New Hampshire, which contains the most rugged topography and the highest summits of the New England region, culminating in Mount Washington (altitude 1902 m, 6288 ft). Far to the E in central Maine is the smaller, isolated knot of Mount Katahdin. Between and around the mountains is dissected hill country, in many areas with nearly accordant summits that descend SE toward the coast. Along the Connecticut River a conspicuous belt of lowland extends N from Long Island Sound, following the structural grain of the rocks and terminating between the Green Mountains and White Mountains.

Along much of the Atlantic Coast of the New England province the metamorphic and plutonic bedrock extends directly beneath the water, with drowned embayments and outlying islands. The Atlantic Coastal Plain of Mesozoic and Cenozoic sediments (to SW) is here represented offshore, but fragments emerge on the southern offshore islands and on Cape Cod, although heavily masked by glacial terminal moraines (Fig. 2). To the NE of Cape Cod the extension of the coastal plain forms the submerged Georges Bank, lying between the Gulf of Maine and the edge of the continental shelf.

Most of the New England region drains to the S or SE into the Atlantic Ocean through the Hudson River, Connecticut River, and shorter streams farther NE, but the part NW drains into the St. Lawrence River (Fig. 1). Most of the St. Lawrence drainage area is N of the international boundary, but it extends well to the S along the W edge of Vermont to include the surroundings of Lake Champlain. The two lowest gaps in the whole Appalachian chain occur along the drainage divide on either side of the Adirondack Mountains—one between the Hudson River and Lake Champlain at an altitude of 44 m (145 ft) and one between the Mohawk River (the western branch of the Hudson) and the Great Lakes at an altitude of 190 m (430 ft).

The landforms of the New England region were profoundly modified during the Quaternary ice age, when even the highest peaks were eventually covered by continental glaciers. These removed the previous weathered cover, exposing bare rock in many areas, masking the intervening areas with drift and greatly disordering the drainage. Before and after continental glaciation, the rocks of the higher mountains were also sculptured by local cirque glaciers.

Before European settlement all the New England region was densely forested (except where the higher peaks extend above timber line), but until the middle of the 19th century it was being progressively cleared for farm land. Large parts of this farm land have since been abandoned, and have reverted to woodland. Nevertheless, northern Maine has always re-

FIGURE 1. Map of New England region, showing principal geographic features: 1, plains and lowlands; 2, low dissected terrain; 3, plateaus, with bordering escarpments; 4, mountains. Heights in feet.

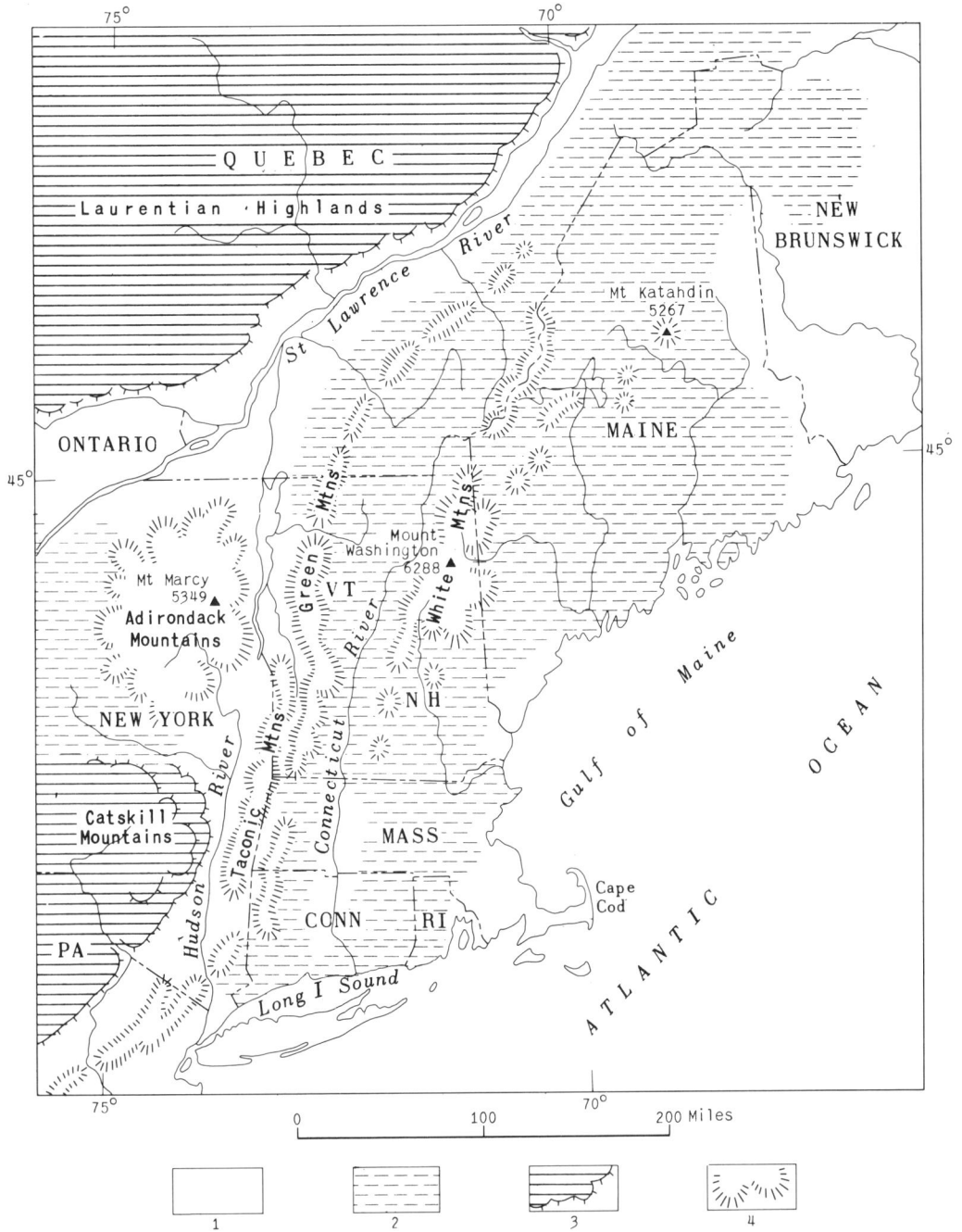

FIGURE 2. Late Pleistocene and moraines of southern New England, partially drowned by postglacial eustatic rise of sea level (Schafer and Hartshorn, 1965). Isostatic rise following deglaciation has been followed (in the last few thousand years) by a very gentle subsidence, causing further drowning.

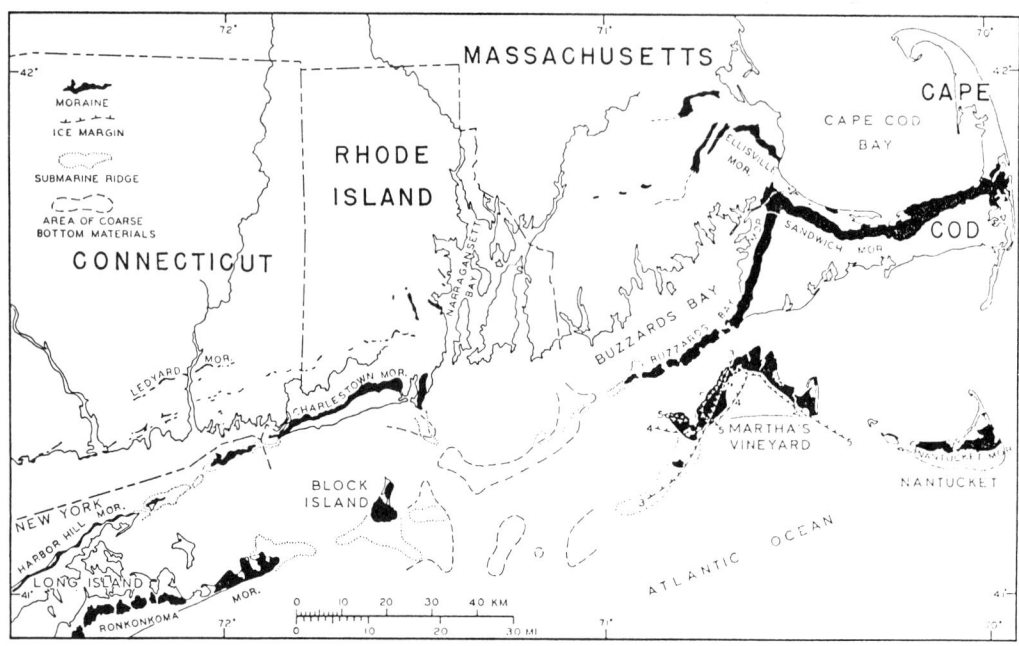

mained a wilderness, although it has been cut over many times for timber products. The S of the region now exhibits striking contrasts between densely populated urban and suburban areas and adjacent areas of forest.

Geological Investigations

The drift mantle of the New England region excited the interest of all the early geologists, but its true nature as a product of continental glaciation was not generally understood until Louis Agassiz became professor at Harvard in 1847, after having proposed the glacial interpretation in Europe some years before. The Triassic rocks of the Connecticut Valley also attracted attention, especially after Edward Hitchcock made known their remarkable fossil dinosaur footprints in 1836. The complex and obscure pre-Triassic rocks and structures of the New England region, however, long defied any convincing generalization—even though the region was the seat of many of the first learned institutions in the country (Harvard, Yale, Columbia), and the home of some of its leading geologists.

It was the gently tilted, fossiliferous strata immediately to the W that were ably deciphered by James Hall and his colleagues of the first New York survey between 1836 and 1843, making these strata the standard of reference for the Paleozoic of much of the eastern United States, and that provided Hall with the first intimations of the geosynclinal theory. The investigations by this survey in those parts of New York state E of the Hudson River produced only confusing results, and brought about the prolonged "Taconic controversy."

Some significant observations on the northwestern borders of the New England segment of the Appalachian orogenic belt were made by Sir William Logan after he became the first official Canadian geologist in 1842; he recognized especially the great structural discontinuity along its northwestern edge in Quebec, which has since been called *"Logan's Line."* Work by B. K. Emerson in Massachusetts, mainly in the latter part of the 19th century (summarized in a map published in 1917), produced stratigraphic interpretations well ahead of their time, with much of the pre-Triassic rocks assigned to various Paleozoic systems. Nevertheless, official maps published as late as 1932 continued to assign large parts of these rocks in New England to the Precambrian.

Modern work on the pre-Triassic rocks of the New England region began about 1930, under the leadership of M. P. Billings of Harvard University. His own surveys in New Hamp-

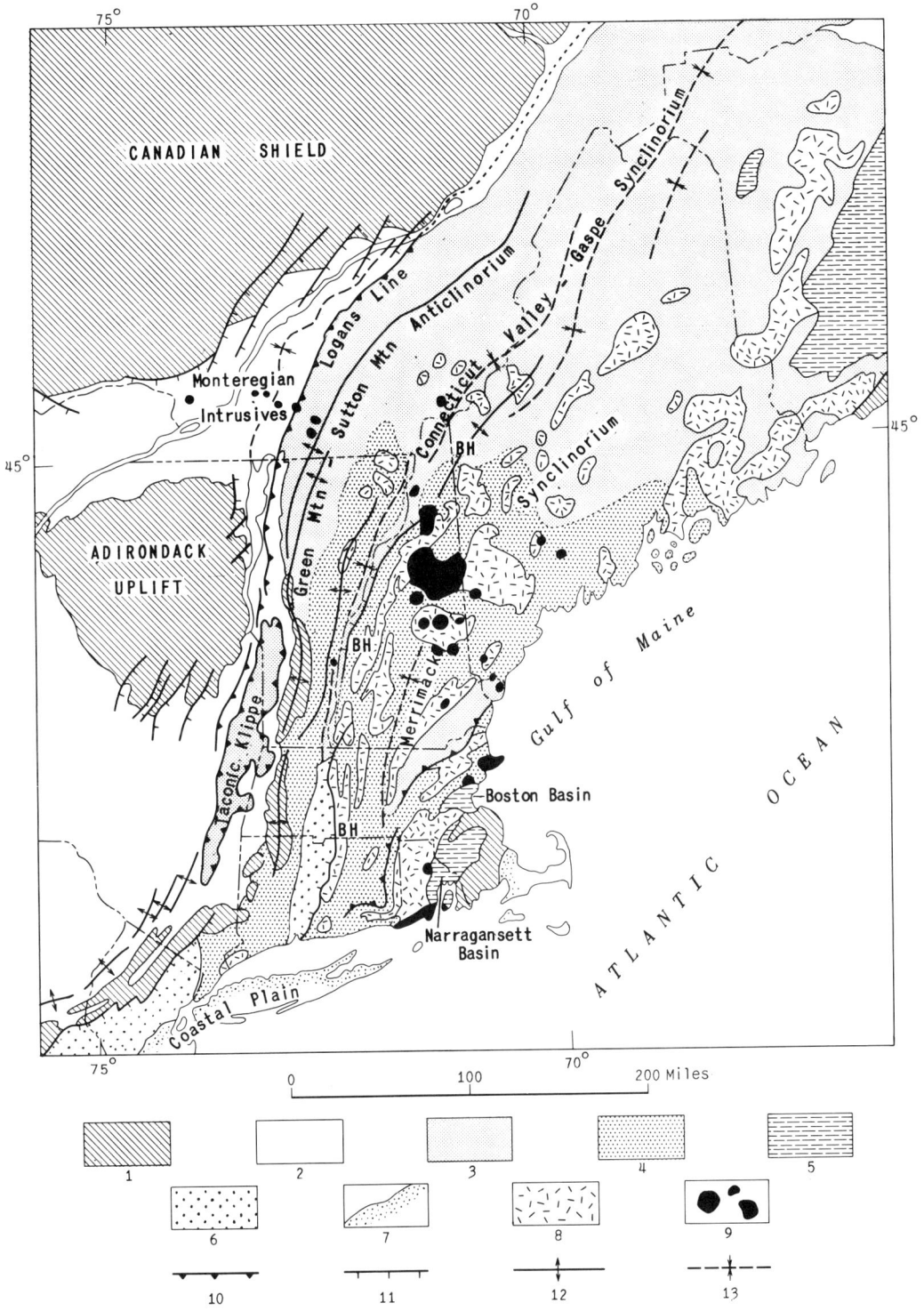

FIGURE 3. Map of New England region, showing geologic and tectonic features: 1, basement rocks (Proterozoic); 2, older Paleozoic miogeosynclinal and platform deposits (Cambrian to Devonian); 3, older Paleozoic eugeosynclinal deposits, unmetamorphosed or little metamorphosed (Cambrian to Devonian); 4, same, strongly metamorphosed (boundary with 3 is garnet isograd); 5, younger Paleozoic rocks (largely Mississippian and Pennsylvanian); 6, Triassic rocks; 7, edge of coastal plain deposits (Cretaceous and Tertiary); 8, granitic rocks, synorogenic to early postorogenic (Ordovician to Devonian); 9, Granitic and other intrusive rocks, postorogenic (late Paleozoic, early Mesozoic, and Cretaceous); 10, thrust fault (barbs on upthrown side); 11, normal fault (hachures on downthrown side); 12, anticline; 13, syncline; BH, Bronson Hill anticlinorium.

shire have been extended by his students and by the personnel of other New England universities. Detailed quadrangle mapping has resulted in the recognition of traceable formations, in the discovery of many fossils even in high-grade metamorphic rocks, and in an understanding of the sequence and interrelations of the plutonic rocks. Nevertheless, much remains to be done, especially in connection with the interpretation of the region in terms of plate tectonics (see, e.g., Bird and Dewey, 1970).

Stratigraphy

Most of the surface of the New England region is formed of older Paleozoic sedimentary and volcanic rocks, containing embedded plutonic rocks of several ages; the rest is formed of subjacent Precambrian basement and of superjacent younger Paleozoic and Triassic. Coastal-plain sediments (Cretaceous-Tertiary) cover the bedrock to the SE, and Quaternary glacial drift forms a discontinuous mantle. Most of the older Paleozoic rocks are of eugeosynclinal facies, but a miogeosynclinal facies forms a narrow belt on the W, merging with the platform deposits of the continental interior. Paleozoic history was diversified by three principal orogenies, but in the southern half of the region the historical record has been complicated by the strong regional metamorphism (Fig. 3).

Precambrian. Precambrian rocks are widely exposed NW of the New England region, in the Canadian Shield NW of the St. Lawrence River, and in the outlying Adirondack uplift (Fig. 2); these areas are part of the Grenville province, characterized by rather consistent Middle Proterozoic, or 1000-million-year-old isotopic dates. The rocks include paragneiss, paraschist, and marble of the Grenville Series, and small to large embedded masses of syenite, anorthosite, and other plutonic rocks.

Similar and broadly correlative metamorphic and plutonic rocks, reworked by Paleozoic orogenies, are upfolded or upfaulted near the northwestern margin of the Appalachian orogenic belt along the axis of the Hudson Highlands, the Berkshire Hills, and the Green Mountains; they also emerge in mantled gneiss domes nearby to the E.

In much of the E of the New England region the Precambrian is deeply buried, but it is exposed again in southeastern New Brunswick (east edge of Fig. 3), and in southeastern Massachusetts. These Precambrian rocks are younger than those of the Grenville province, and are part of the late Proterozoic Avalonian orogenic belt (see *Canada–Atlantic Province*); granodiorite in Massachusetts has been dated at 580 m yr (?reheated), yet is overlain by Lower Cambrian.

Miogeosynclinal and Related Older Paleozoic Rocks. Miogeosynclinal rocks are well displayed E of Lake Champlain in Vermont, along the edge of the orogenic belt, where they are about 2500 m thick. Overlying the Precambrian core of the Green Mountains are Lower Cambrian clastic rocks, including the prominent Cheshire Quartzite, which are followed by Lower Cambrian to Middle Ordovician carbonate rocks, topped in turn by Middle Ordovician shales. Rocks of similar kind and age extend S along the edge of the orogenic belt into Pennsylvania, but they wedge out to the N in Quebec, beyond which eugeosynclinal strata directly abut the platform area along Logan's Line. Former existence of a miogeosynclinal sequence here is attested by carbonate rocks with a wide variety of older Paleozoic ages, which occur as boulders in the eugeosynclinal sequence.

The miogeosynclinal sequence overlaps to the NW on the platform area. Around the Adirondack uplift, the Upper Cambrian Potsdam Sandstone is the basal deposit and the succeeding carbonate rocks have thinned to little more than 300 m. Farther onto the platform, near the edge of the Canadian Shield in Ontario and Quebec, the Cambrian and Lower Ordovician are largely missing and the Middle Ordovician commonly overlaps the Precambrian.

To the N from Albany, New York, Middle Ordovician shales are the highest strata preserved in either the miogeosynclinal or platform sequences, except for a downfolded remnant of Upper Ordovician clastics and red beds at the front of the orogenic belt in southern Quebec. Higher strata are preserved S of Albany and W of the Hudson River, where Silurian rocks succeed the Ordovician with a structural unconformity that is prominent into central Pennsylvania. The conformably overlying Devonian beds pass upward into continental clastic strata that form the Catskill Mountains. The unconformity below the Silurian records the Late Ordovician *Taconian orogeny,* and the folding and continental beds in the Devonian suggest the uplift due to a Middle to Late Devonian *Acadian orogeny* in the region to the E.

Implanted in, and overlying, the miogeosynclinal sequence in eastern New York and western New England is the celebrated "*Taconic Klippe,*" a eugeosynclinal sequence of shales and graywackes, whose fossils indicate that this sequence has the same age span as that of the miogeosyncline—from Lower Cambrian to Middle Ordovician. Apparently the Taconic rocks are an allochthonous mass, transported W into their present position; emplacement of the giant klippe occurred during deposition of the Ordovician sediments, which contain its debris, and it is overlain unconformably by the basal Devonian at Becraft Mountain S of Albany.

Eugeosynclinal Older Paleozoic Rocks. The older Paleozoic eugeosynclinal rocks are much thicker than the miogeosynclinal rocks to the W, and are largely slates, graywackes, and volcanics, rather than carbonate rocks and quartzites; they include not only Cambrian and Ordovician strata but also Silurian and Devonian. The eugeosynclinal and miogeosynclinal sequences contrast dramatically on the closely adjacent opposite flanks of the Green Mountain anticlinorium in Vermont. Great variations in stratigraphy of the eugeosynclinal rocks occur across the trends of the structures, and to a lesser extent along the trends—variations in amount of volcanic material, variations between deep-water and shallow-water sediments, etc.—implying the existence of belts of volcanism and of juxtaposed deep and shallow water at the time of deposition. Upon these depositional variations is superposed an increase in grade of metamorphism to the SW, so that rocks which are little altered or unaltered in Maine and nearby Canada are equivalent to high-grade schists and gneisses in southern New England; the boundary between little altered and extensively altered eugeosynclinal rocks is approximately along the garnet isograd, shown on Fig. 3.

Cambrian and Ordovician rocks form a belt along the whole NW of the eugeosynclinal area, which is as much as 80 km wide in Quebec but narrows to S across New England; Cambrian and Ordovician rocks reappear less extensively in anticlines and anticlinoria farther SE. Silurian and Devonian rocks form much of the central part of the eugeosynclinal area, especially in a wide belt along the Connecticut Valley-Gaspé synclinorium on the NW (Fig. 3) and in the Merrimack synclinorium that extends from southeastern Maine into eastern Connecticut.

In Quebec, at least, the Cambrian and Ordovician rocks of the belt to the NW were intensely folded (but little metamorphosed) during the Taconian orogeny, and parts of the belt may not have been much deformed since. On the SE of the belt the deformed rocks are overlapped by sandy and limy shelf deposits of Silurian age, but these pass within a short distance into thick eugeosynclinal deposits that extend up into the Devonian. A two-phase development of the eugeosyncline is thus indicated in this part of the region. The second phase terminates in continental plant-bearing beds of late Early Devonian age, which are preserved in some deeper downfolds from central Maine to the NE. Farther SE and S in the New England region, a double phase of eugeosynclinal development is not as clear. The Silurian is unconformable on the Ordovician in places, and in New Hampshire the Silurian even overlies the Late Ordovician Highlandcroft Plutonic Series; but elsewhere the two systems are seemingly conformable. In general, however, any Taconian deformation tends to be overwhelmed and obscured by the succeeding Acadian orogeny.

Younger Paleozoic Rocks. Younger Paleozoic rocks are preserved as remnants in the SE of the New England region (Fig. 3) and are much less deformed and altered than the older Paleozoic rocks, implying an intervening Acadian orogeny. They are largely continental deposits, including fanglomerates, red beds, coal, and volcanics, but marine layers and evaporites occur to the NE in New Brunswick. The oldest deposits, including the Perry and Mapleton Formations of easternmost Maine, are of Late Devonian age, but the greater bulk in New Brunswick is Mississippian and Pennsylvanian. These are topped by Permian red beds on Prince Edward Island (see *Canada–Atlantic Province*). The younger Paleozoic rocks of the Narragansett and Boston basins in Rhode Island and eastern Massachusetts (including the controversial Squantum "Tillite") may be largely Pennsylvanian, although fossil evidence for the ages of parts of them is not available.

Pennsylvanian red beds in the large area of central New Brunswick (Fig. 3) lie nearly flat, but the lower parts of even this sequence are more deformed. Deformation increases to the SW, indicating the effects of a late Paleozoic *Allegheny orogeny*. The rocks of the Boston and Narragansett basins are steeply folded, and the coal of the latter is largely altered to graphite; the younger Paleozoic rocks in southernmost Rhode Island are much metamorphosed, especially near Pennsylvanian or later granitic intrusives.

Triassic Rocks. Upper Triassic rocks of the Newark Group, comparable to those farther SW in the Appalachian region, form a downfaulted belt along the Connecticut Valley and some smaller outliers in southern New England, as well as another belt along the Bay of Fundy in Nova Scotia; the latter extends some distance to the SW beneath the Gulf of Maine. The Triassic rocks are continental red beds and arkoses, with a few finer-grained layers and several lava flows, as well as intrusive sheets and stocks. The Triassic rocks are wholly postorogenic, are unaltered and merely tilted, and are broken by small to large normal faults.

Plutonic Rocks. Ultramafic rocks form

small pods and lenses in the country rocks immediately SE of the Green Mountain–Sutton Mountain anticlinorium and attain their greatest proportion in southeastern Quebec. They are believed to be largely of Ordovician age.

Granitic rocks form extensive parts of the former eugeosynclinal area of the New England region. They have varied petrographic and structural characters and were emplaced over a prolonged period, extending from early Paleozoic into Mesozoic time.

The earliest granitic rocks are pre-Silurian and probably Ordovician in age, and their eroded surfaces are overlain in places by Silurian deposits. They include the Highlandcroft and Oliverian Plutonic Series of New Hampshire, several bodies in Maine, and others in Massachusetts and Connecticut. The Oliverian Series is notable for its distinctive structure; it forms a series of domes along the E of the Connecticut Valley from New Hampshire into Connecticut, in which it is nearly conformable with the superjacent rocks. Although its rocks are as granitic as any of the others, they may have formed in part by remobilization of felsic volcanic rocks that were originally parts of the supracrustal sequence.

Of much greater volume and extent throughout New England are granitic rocks of mid-Paleozoic age. They have yielded rather consistent 350 to 400 m yr isotopic dates, and are broadly contemporaneous with the *Acadian orogeny,* partly synorogenic, partly early postorogenic. They include both concordant sheets and throughbreaking bodies. In New Hampshire they are classed as the New Hampshire Plutonic Series.

Younger granitic rocks are wholly postorogenic and are generally more alkalic than the earlier ones. They are exemplified by the White Mountain Plutonic-Volcanic Series of New Hampshire, which includes a spectacular array of ring dikes, some of which enclose remnants of surficial volcanic rocks. The White Mountain Series itself is early Jurassic, with isotopic ages of about 185 m yr, but the alkalic rocks farther S in New England are late Paleozoic. The oldest precede the Pennsylvanian deposits of the Boston and Narragansett basins; the Pennsylvanian is itself invaded by nonalkalic granite at the south end of the Narragansett basin.

The youngest plutonic rocks of all, the *Monteregian intrusives* (Fig. 3), form a chain of stocks that extends E from Montreal (for which they are named) in the platform area, into the outer edge of the orogenic belt. Isotopic determinations indicate that these intrusives are of Cretaceous age.

Tectonics

The southern part of the New England region is bordered on the W by Paleozoic platform deposits that are warped into a synclinorium that plunges SW beneath the Appalachian Plateaus. Farther N, Precambrian basement rocks of the Adirondack uplift and Canadian Shield nearly impinge against the front of the orogenic belt, and areas of platform cover are less extensive (Fig. 3); the basement and its cover are broken in places by normal faults.

From Quebec into Vermont the front of the orogenic belt is the marked discontinuity of *Logan's Line,* a low-angle thrust fault of large displacement NW of which are Precambrian basement or gently tilted platform deposits, and SE of which are intensely folded and faulted Cambrian and Ordovician eugeosynclinal and miogeosynclinal rocks. Farther S, thrust faults are discontinuous along the front of the orogenic belt, but relations are complicated by the structural unconformity between the tilted Silurian and Devonian strata and the much more deformed Ordovician strata.

The rocks of the "Taconic Klippe" sequence which are implanted ("float") in the miogeosynclinal area a little back from the orogenic front are evidently an allochthonous mass that has been transported into the area along two or more low-angle thrust faults from a site farther E. The small thickness of the Taconic sequence and the incompetent nature of its materials suggest that it could not have been moved by lateral thrust, which lends plausibility to the concept that it is an enormous gravity slide.

A little back from the front of the orogenic belt, Precambrian rocks of the Grenville province have been raised in a chain of uplifts, beginning with the Hudson Highlands on the S and continuing to the N behind the Taconic mass in the Berkshire Hills and Green Mountains. Although the Precambrian plunges to N beneath Paleozoic cover in Vermont, the uplift extends far to the NE in the cover rocks into Quebec as the anticlinorium of the Sutton Mountains (Fig. 3).

The area of dominant eugeosynclinal rocks SE of this chain of uplifts is marked by intense deformation, and in the extremely metamorphosed S by large-scale plastic flowage. Nevertheless, little faulting accompanied the deformation and low-angle thrusts are lacking; the most prominent faults of the area may be of Triassic age, like those that border the basins of postorogenic Triassic strata. The eugeosynclinal rocks are thrown into a succession of gross synclinoria and anticlinoria, which extend across a width of about 300 km in the unmetamorphosed part to the NE but are compressed into about half this distance in the metamorphosed part to the SW.

Next SE of the chain of Precambrian uplifts and its extension in the Sutton Mountains is the Connecticut Valley—Gaspé synclinorium, which extends from the tip of Gaspé into Connecticut (Fig. 3), and encloses deep downfolds of strata as high as the Lower Devonian. In the metamorphic part of the synclinorium from southeastern Vermont S, its structure is confused by a chain of remarkable "mantled gneiss domes", in which remobilized basement rocks have risen by plastic flow through the supracrustal rocks.

The synclinorium is bordered to SE, that is, E of the Connecticut River, by the Bronson Hill anticlinorium (labeled BH on Fig. 3), which brings the lower part of the supracrustal sequence to the surface, but not its basement. Rather curiously, a narrow belt along its W flank through New Hampshire into Massachusetts has a metamorphic grade much lower than that on either side—which incidentally has greatly aided in deciphering the stratigraphy of this part of New England. Near the crest of the anticlinorium from New Hampshire to the S is a chain of mantled gneiss domes similar to those in the synclinorium farther W, but their cores are rocks of the Oliverian Plutonic Series rather than basement. The supracrustal rocks above the domes are thrown into recumbent nappes, up to several kilometers in breadth, which in southern New Hampshire are directed toward the W and in eastern Connecticut toward the E. Structure of the anticlinorium is further complicated by the large bodies of granitic rock of the New Hampshire Plutonic Series concentrated near its axis and by Triassic normal faults of large displacement, some of which juxtapose rocks of different metamorphic grade; these faults include the Ammonoosuc fault, which extends nearly the length of New Hampshire and was at first interpreted to be a high-angle thrust.

Beyond the Bronson Hill anticlinorium, nearer the coast, is the Merrimack synclinorium (Fig. 3), similar to the Connecticut Valley-Gaspé synclinorium farther NW but perhaps an even deeper downfold, again preserving thick bodies of supracrustal rocks as high as Devonian. In south-central Maine it forms a broad terrane of poorly differentiated slates. In New Hampshire, where the metamorphic grade is higher, equivalent rocks along the axis are sillimanite schists, but along the SE flank is another curious belt of low-grade metamorphic rocks like that west of the Bronson Hill anticlinorium. Structure of the supracrustal rocks in New Hampshire is confused by large plutonic bodies of the synorogenic New Hampshire Plutonic Series and of the postorogenic White Mountain Plutonic-Volcanic Series. In Massachusetts, much of the contents of the synclinorium were once assigned to the Pennsylvanian because of a reported occurrence of fossil plants near Worcester, but the great bulk of the rock is surely older. The synclinorium narrows and pinches out in Connecticut, where its structure is greatly confused by recumbent folds that originated in the Bronson Hill anticlinorium to the W.

Rhode Island and eastern Massachusetts are formed of a complex of older Paleozoic supracrustal rocks and of late Proterozoic (Avalonian) plutonic rocks, which are a fragment of another anticlinorium comparable to that along the same trend to the NE in southeastern New Brunswick; upon this complex are superposed the younger Paleozoic Boston and Narragansett basins. The steep folding of the rocks of these basins and the intrusion of their rocks by granite toward the S indicate that the earlier anticlinorial structure was much reworked by deformation late in Paleozoic time.

The Triassic rocks of the Connecticut Valley are tilted E toward a great border fault, part of whose displacement occurred during Triassic sedimentation. The structure of the Triassic rocks of the Connecticut Valley is a mirror image of that of the Triassic of New Jersey, lying *en échelon* to the W, where the rocks dip NW toward another border fault. The two faults may originally have been the edges of a continuous trough, a graben, of Triassic sediments whose center has been subsequently arched so that the connection has been removed by erosion; a small down-faulted block of this cover is preserved in SW Connecticut.

Geophysical Data

Bouguer gravity anomalies are weakly positive to weakly negative within the orogenic belt in the New England region, but there is an abrupt gravity gradient to the W and NW, a little behind the orogenic front, from around 0 milligals to −70 milligals. The crest of the gradient follows the Green Mountains and other Precambrian uplifts through western New England, but lies SE of the Precambrian of the Hudson Highlands in New Jersey and Pennsylvania. The western part of the Hudson Highlands, in fact, overlies an area of gravity minimum at the base of the gradient, as does the area of Taconic rocks farther N. The relation of the Taconic rocks to the gravity minimum supports the suggestion that they are an allochthonous mass, and implies that the Precambrian of the W of the Hudson Highlands may be allochthonous as well.

Epicenters of recorded earthquakes in the New England region are significantly concentrated in two zones, one that follows the central part of the orogenic belt from Connecticut to central Maine and another that follows the St. Lawrence River in Quebec; within these zones many epicenters are clustered in the White Mountains and on the N of the Adirondack Mountains. Many of the earthquakes of the region are minor, but several along the St. Lawrence (notably in 1663 and 1925) had magnitudes of 7 or 8, and were perceptible over much of eastern North America. The earthquakes of the region have sometimes been attributed to postglacial uplift, but distribution of their epicenters has little relation to glacial features and a much more obvious relation to structural trends in the bedrock. Postglacial isostatic rebound may have played a "trigger" action.

Tectonic History

Structures of the Appalachian orogenic belt in the New England region evolved through much of Paleozoic time, with continuing instability (taphrogeny) into the Triassic. The tectonic evolution was probably much the same as in the Appalachian region to the SW, but in the New England region details of the evolution are more plainly indicated by unconformities and sedimentary variations in the stratigraphic sequence, and by mutual relations of the different plutonic bodies, making it possible to distinguish several periods of orogeny.

Early phases of the tectonic evolution involved mainly geosynclinal sedimentation, the eugeosynclinal, miogeosynclinal, and platform areas being well defined by Early Cambrian time. So far as can be told from the exposed rocks, by far the larger part of the region was eugeosynclinal, both then and later, the miogeosynclinal area being only a narrow fringe at the edge of the continental platform.

The first strong deformation in the region occurred during the Taconian orogeny, as indicated by the structural unconformity between the Middle Ordovician and the Silurian along the orogenic front. The most spectacular effect of the orogeny was the emplacement of the allochthonous mass ("klippe") of Taconic rocks near the orogenic front. If this was emplaced as a gravity slide, there was probably an area of uplift in the source area to the E, possibly along the present Precambrian uplifts of the Green Mountains and Berkshire Hills. The whole belt NW of Cambrian and Ordovician rocks, from New England through Quebec to Gaspé, was evidently folded and faulted at about the same time. The effects of the Taconian orogeny farther SE are not as easily determined, partly because they have been obscured by later orogenies, but deformation at this time is indicated in places by unconformities, and plutonic rocks were emplaced, at least in New Hampshire and Maine.

The *Taconian orogeny* has generally been ascribed a Late Ordovician age, but was probably more prolonged. Earlier phases near the orogenic front are suggested by the Middle Ordovician shales and other clastic rocks that succeed the earlier carbonate rocks of the miogeosyncline. Farther back in the orogenic belt, as in Maine, unconformities have been observed within, or even at the base of, the Ordovician.

After the Taconian orogeny, the eugeosynclinal area assumed new and probably more restricted depositional patterns. Deep troughs received great thicknesses of Silurian and Devonian clastic rocks and volcanics, but the occurrence of shallow-water sediments in places suggests the existence of areas of submarine ridges. Occurrence of volcanic rocks is also variable, suggesting the existence of belts or areas of volcanic activity. Possibly areas of deep- and shallow-water sediments and areas of volcanism are related to the present gross synclinoria and anticlinoria of the region, although the evidence is inconclusive.

This second phase of eugeosynclinal development ended late in the Early Devonian, when eugeosynclinal deposits were covered, at least in places, by plant-bearing continental deposits. Presumably this marked the onset of the *Acadian orogeny* which thoroughly deformed much of the New England region. In the platform area coarse continental deposits were spread to the W during Middle and Late Devonian time, indicating vigorous erosion within the orogenic belt; moreover, the first postorogenic deposits within the orogenic belt are latest Devonian. Isotopic dating indicates that large volumes of granitic rocks, especially of the New Hampshire Plutonic Series, were also emplaced in the orogenic belt during Devonian and early Mississippian time.

Younger Paleozoic deposits occupy much smaller areas in the New England region than the older Paleozoic, and their original areas of deposition may not have been much larger than their present extent; it is unlikely that they ever formed a continuous deposit. Occurrence of fanglomerate in parts of the younger Paleozoic implies the existence of nearby areas of much relief.

The younger Paleozoic rocks in some areas toward the NE remain nearly flat, suggesting that little further deformation occurred here after the climactic Acadian orogeny. The younger Paleozoic of Rhode Island and Massachusetts, on the other hand, has been much folded, is partly metamorphosed, and is intruded by granite in places, indicating that to the SW the orogenic belt was still undergoing deformation late in Paleozoic time. This was broadly during the *Allegheny orogeny,* whose effects are prominent in the marginal part of the Appalachian region farther SW. Continuing plutonic activity is also manifest in the rather widespread postorogenic alkali granites, of which at least part (the White Mountain Plutonic-Volcanic Series) was emplaced as late as early Jurassic time.

Nevertheless, basal relations of the Upper Triassic rocks ("pre-Newark Surface") indicate that they were laid over a terrane that had already been thoroughly deformed, plutonized, and subsequently peneplaned. Continuing instability during and after the time when these rocks were deposited is shown by their tilting and faulting (Palisadian taphrogeny). By then truly compressional deformation had ceased.

The region became much more stable thereafter, and general peneplanation followed. From later Mesozoic time to the present it has been affected only by epeirogenic upwarping with marginal downwarping. Upwarping and erosion have produced the New England region described here and downwarping and deposition have produced the now mainly submerged coastal plain on the S and SE. The latest upwarping occurred during postglacial time, as a result of isostatic rebound following the removal of the continental glaciers that had covered the whole region as recently as 15,000 B.P.

Economic Mineral Deposits

The mineral resources of the New England region are very modest, especially when compared with the much greater mineral wealth of the Appalachian region, adjacent to the SW.

The region contains no significant mineral fuels. Coal in the Narragansett basin of Rhode Island is greatly metamorphosed and contains little combustible organic matter, but it has been mined from time to time for graphite. Showings of oil and gas have been encountered in the Paleozoic rocks between the Adirondack uplift and Canadian Shield in Ontario and Quebec, but there has been no commercial production.

Few deposits of metallic minerals are known. Sulfide mineralization occurs in eastern Vermont in a belt extending 30 km along the strike of the rocks, and has yielded about 120 million pounds of copper from several mines. Recent discoveries of commercial zinc and lead sulfide deposits in western New Brunswick has prompted a search for deposits in the comparable rocks and structures in Maine, but so far without significant results.

Nonmetallic mineral deposits are more important. Ultramafic rocks E of the Green Mountains in Vermont have long been mined for talc, and ultramafic rocks along the same trend in southeastern Quebec are the leading source of asbestos in North America. Pegmatite bodies, widely scattered in the metamorphic rocks of New England, have been worked for mica, feldspar, and other products, especially during World War II. Vermont has long been a leading producer of building stone—marble from the E of the miogeosynclinal belt, granite from plutonic bodies farther E, and verde antique from the ultramafic bodies. Slate has also been produced from the weakly metamorphosed shaly rocks of the miogeosynclinal and Taconic sequences in southwestern Vermont and in adjacent parts of New York State.

PHILIP B. KING*

*Publication authorized by the U.S. Geological Survey.

References

Ballard, R. D., and Uchupi, E., 1972. "Carboniferous and Triassic rifting: a preliminary outline of the tectonic history of the Gulf of Maine," *Bull. Geol. Soc. Am.,* 83, 2285–2302.

Billings, M. P., 1956. "Bedrock geology," *The Geology of New Hampshire.* Concord, N.H.: Plan. and Develop. Comm, 203p.

Bird, J. M., and Dewey, J. F., 1970. "Lithosphere plate-continental margin tectonics and the evolution of the Appalachian origen," *Bull. Geol. Soc. Am.,* 81, 1031–1060.

Boucot, A. J., 1961. "Stratigraphy of the Moose River Synclinorium, Maine," *U.S. Geol. Surv. Bull.,* 1111(E), 153–188.

Brookins, D. G., Berdan, J. M., and Stewart, D. P., 1973. "Isotopic and paleontologic evidence for correlating three volcanic sequences in the Maine Coastal Volcanic Belt," *Bull. Geol. Soc. Am.,* 84, 1619–1628.

Broughton, J. G., Fisher, D. W., Isachsen, Y. W., and Richard, L. V., 1961. "The geology of New York State," *Geologic Map of New York* (Map and Chart Ser. 3). N.Y. State Mus. and Sci. Service, 42p.

Cady, W. M., 1969. "Regional tectonic synthesis of northwestern New England and adjacent Quebec," *Geol. Soc. Am. Mem. 120,* 181p.

Diver, B. B. van, ed., 1971. *Geological Studies of the Northwest Adirondacks Region, Field Trip Guidebook.* Potsdam, N.Y.: N.Y. State Univ., Coll. Potsdam, Dept. Geol Sci., 164p.

Isachsen, Y. W., 1964. "Extent and configuration of

the Precambrian in northeastern United States," *N.Y. Acad. Sci. Trans.*, Ser. 2, **26**(7), 812–829.

Lindsay, J. F., Summerson, C. H., and Barrett, P. J., 1970. "A long-axis clast fabric comparison of the Squantum 'tillite', Massachusetts and the Gowganda Formation, Ontario," *J. Sed. Petrology*, **40**(1), 475–479.

Minard, J. P., et al., 1974. "Preliminary report on geology along Atlantic continental margin of northeastern United States," *Am. Assoc. Petrol. Geol. Bull.*, **11**(2), 1169–1178.

Newell, W. L., 1970. "Factors influencing the grain of the topography along the Willoughby arch in northeastern Vermont," *Geogr. Ann.*, Ser. A, **52**(2), 103–112.

Osborne, F. F., 1956. "Geology near Quebec City," *Naturaliste Canadien*, **83**(8–9), 157–223.

Pratt, R. M., and Schlee, J., 1969. "Glaciation on the continental margin off New England," *Bull. Geol. Soc. Am.*, **80**, 2335–2342.

Quinn, A. W., and Oliver, W. A., Jr., 1962. "Pennsylvanian rocks of New England," in C. C. Branson, ed., *Pennsylvanian System in the United States*. Tulsa: Am. Assoc. Petrol. Geologists, 60–73.

Rodgers, J., 1970. *The Tectonics of the Appalachians*. New York: Wiley, 66–147.

——, 1971. "The Taconic Orogeny," *Bull. Geol. Soc. Am.*, **82**, 1141–1178.

Rodgers, J., Gates, R. M., and Rosenfeld, J. L., 1959. "Explanatory text for preliminary geological map of Connecticut, 1956," *Conn. Geol. Nat. Hist. Surv. Bull.*, **84**, 64p.

Schafer, J. P., and Hartshorn, J. H., 1965. "The Quaternary of New England," in H. E. Wright, Jr., and D. G. Frey, eds., *The Quaternary of the United States*. Princeton, N.J.: Princeton Univ. Press, 113–127.

Spooner, C. M., and Fairbairn, H. W., 1970. "Relation of a radiometric age of granite rocks near Calais, Maine, to the time of Acadian orogeny," *Bull. Geol. Soc. Am.*, **81**, 3663–3670.

Stewart, D. P., 1971. "Pleistocene mountain glaciation, northern Vermont: discussion," *Bull. Geol. Soc. Am.*, **82**(6), 1759–1760.

Uchupi, E., 1966. "Structural framework of the Gulf of Maine," *J. Geophys. Res.*, **71**(12), 3010–3028.

Zartman, R. E., and Marvin, R. F., 1971. "Radiometric age (Late Ordovician) of the Quincy, Cape Ann, and Peabody granites from eastern Massachusetts," *Bull. Geol. Soc. Am.*, **82**, 937–958.

Zen, E., 1972. "Some revisions in the interpretations of the Taconic Allochthon in west-central Vermont," *Bull. Geol. Soc. Am.*, **83**, 2573–2588.

——, 1967. "Time and space relationships of the Taconic Allochthon and Autochthon," *Geol. Soc. Am. Spec. Pap. 97*, 107p.

——, White, W. S., Hadley, J. B., and Thompson, J. B., Jr., eds., 1968. *Studies of Appalachian Geology; Northern and Maritime*. New York: Wiley, 475p.

Cross-references: *Canada—Atlantic Provinces, St. Lawrence Lowlands of Quebec; North America; United States—Appalachian Region.*

UNITED STATES—PACIFIC CORDILLERAN REGION

Broadly speaking, the Cordillera encompasses the entire mountainous western quarter of the United States. However, as used here it is restricted geologically to Idaho N of the Snake River, all of Washington and Oregon, and California from the Sierra Nevada to the coast (Fig. 1). This great region, the Pacific Cordillera, presents its bold countenance to the Pacific Ocean for 3000 km—from Puget Sound to Baja California. Its average width is somewhat over 300 km and its area is approximately 1,000,000 km^2.

History of Geologic Activities

The geologic use of the name *Cordillera* goes back to Alexander von Humboldt. Walcott, in 1894, adopted it as a general term for the west. Schuchert (1910), through his paleogeographic studies, was responsible for establishing the

FIGURE 1. Geomorphic divisions of the Pacific Cordilleran region.

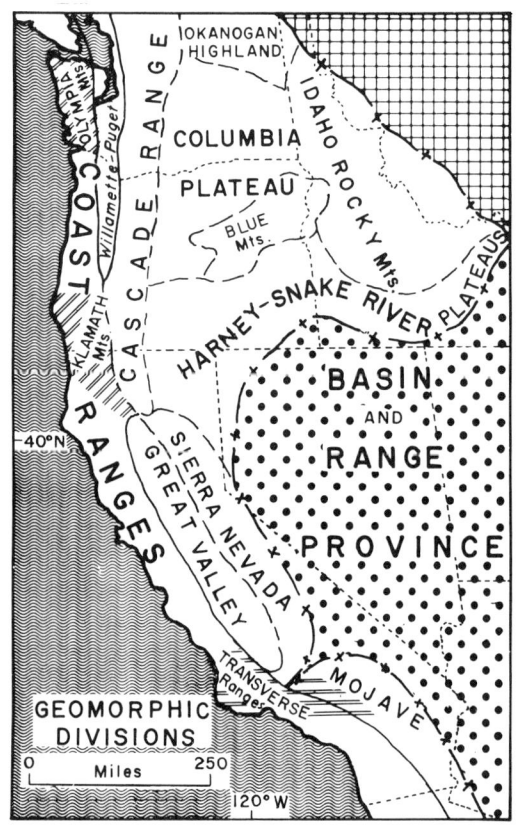

term *Cordilleran geosyncline* for this great belt of thick sedimentary and volcanic rocks that bordered the stable North American *craton* for nearly a billion years.

Gold was discovered in California in 1848 and rapidly thereafter all up and down the region. Serious geologic study was begun a decade later. It was first stimulated not by gold directly, but by exploration for the transcontinental railroads. The most famous early geologic reconnaissances (of King, Hayden, and Wheeler) did not extend into the Pacific states. California established its own geological survey under J. D. Whitney in 1860. Clarence King, organizer of the ambitious Fortieth Parallel Survey (1868-1873) and first director of the U.S. Geological Survey (1879-81), gained his initial experience under Whitney and named the region's highest peak in his honor.

Early reconnaissance of the Cordilleran province whetted the national appetite for the many great geologic treasures of "the West." Gold fever continued to dominate much of human activity here for half a century and the successive gold rushes motivated most of the geologic studies after 1879 by the newly formed U.S. Geological Survey. These excellent local studies contributed greatly to geologic thought in America, particularly to the formulation of W. Lindgren's unifying theories of ore deposits. Nevertheless for many years very little regional synthesis emerged. This only came much later as the national survey's scope of activities was broadened and partly as a by-product of the regionally oriented petroleum industry. State surveys and university work also contributed increasingly. In recent years this region has contributed much to major concepts of global tectonics, geosynclines and mountain building, volcanism, seismology, geomorphology, all branches of paleontology, and oceanography, to name but a few fields. It continues to yield an enormous mineral wealth.

The oldest and largest centers of geologic education in the west are in California. These include the University of California at Berkeley (established in 1873), University of Southern California (1880), Stanford University (1885), and the California Institute of Technology (1891). The University of Washington (1861), Oregon State University (1858), and University of Oregon (1872) also have contributed, as have such younger institutions as the University of California divisions at Los Angeles and La Jolla (the latter now including famous Scripps Institution of Oceanography). These centers have housed the activities of some of America's distinguished geologic pioneers, including J. LeConte, A. C. Lawson, J. C. Merriam, and (temporarily) G. K. Gilbert (all at Berkeley), Bailey Willis, J. Perrin Smith, and Eliot Blackwelder (at Stanford), and Thomas Condon (Oregon), to name but a few.

Geomorphic Divisions

Geomorphic character of the Cordilleran region is extremely varied, as witnessed by the fact that it contains the highest point in the contiguous 48 states, Mount Whitney (elevation 4419 m, 14,450 ft), located in the southern Sierra Nevada, and only 130 km W of Death Valley which contains the lowest point (85 m, 282 ft below sea level), located in the adjacent Basin and Range province (q.v.). The region contains eight national parks, numbers of national monuments or seashores, and national forest and state park areas of special geologic significance. The subdivisions generally listed from W to E, are as follows (see Fig. 1).

Coast Ranges. A complex belt of mountains rises abruptly from the Pacific to elevations up to 1200 m. Only in Puget Sound, southwestern Washington, around San Francisco Bay, and Los Angeles are there appreciable coastal lowlands. Within the Coast Ranges of California there are several prominent structurally controlled valleys. The southern Californian ranges are semiarid, whereas those N of San Francisco are very humid.

The Great Valley and the Willamette-Puget Depression. This discontinuous belt of lowlands lies E of the Coast Ranges. The Great Valley (720 by 80 km) is a flat alluvial plain that was an inland sea during most of Cenozoic time. It is very fertile where irrigated. This long, topographically depressed trend is interrupted at the middle of the province by the *Klamath Mountains,* which link the Coast Ranges with the Cascade and Sierra ranges (Fig. 1). To the N, the Willamette Valley, with its humid climate, is also fertile. The Puget Depression, though slightly larger, is extensively drowned by the sea, having been overridden and modified by Pleistocene ice sheets.

Cascade Range. Dominating the Pacific Northwest is the largely volcanic Cascade Range. It is a topographic composite of dissected upland (more or less the westernmost edge of the Columbia Lava Plateau) with elevations up to 1200 m, overshadowed by a N-S line of spectacular, snow-capped, conical volcanic peaks (including mounts Baker, 10,778 ft; Rainier, 14,410 ft; Hood, 11,245 ft; Shasta, 14,162 ft; and Lassen, 10,466 ft). Intensive Pleistocene glaciation of these high peaks and the entire N of the range is noteworthy, and small active glaciers still exist. The largest river of the province, the Columbia, is superimposed in a deep gorge across the middle of the Cascades and Coast Ranges.

Lava Plateaus of the Northwest. A broad, semiarid upland with average elevations of 500–1000 m—the *Columbia River Plateau*—extends for 300 km E from the Cascades, to northeastern Washington, eastern Oregon, and northern Idaho. In central Oregon the *Blue Mountains* complex interrupts the plateaus like islands in a sea, as do several of the northernmost ranges of the adjacent Basin and Range Province in the California-Oregon-Nevada border area. Summit elevations of these interior ranges average at least 2000 m; they are subhumid.

The Idaho-Northern Washington (Okanogan) Mountain Complex. On the E and N, the Columbia Plateau subprovince is bordered by a complex of mountains with maximum elevations in Idaho of over 3500 m, carved from complex pre-Tertiary rocks. These ranges merge with the Rocky Mountains province of British Columbia and Montana. They are separated from the Basin and Range province by the *Snake River plains,* an E crescentic extension of the southern Oregon (Harney) lava plateau, which reaches ultimately to Yellowstone Park. This arid expanse is crossed by the Snake River in a locally precipitous gorge.

Sierra Nevada. This is the single largest (800 by 80 km) range in the province, situated in eastern California. Peaks rise above 3000 m and have been intensely glaciated, producing fine scenery immortalized by the naturalist, John Muir. They are comparable only to the more glaciated Cascade and Olympic mountains of Washington. The east face of the tilted Sierra block is a precipitous, dissected fault scarp, whereas the western slope is comparatively gentle, though deeply dissected by canyons such as those followed by the transcontinental railroads through the N of the range. These same valleys and the Mother Lode foothills belt saw the famous Gold Rush of 1848–1849. The Basin and Range province adjoins the Sierras on the E, and the Mojave Desert, closely akin to that province, adjoins on the S. The SW of the Sierras adjoins the Coast Ranges at the S of the Great Valley.

Structural Divisions

Major structural contrasts across the Pacific Cordillera account for most geomorphic divisions. Several of the geomorphic provinces, however, cross structural ones (compare Figs. 1

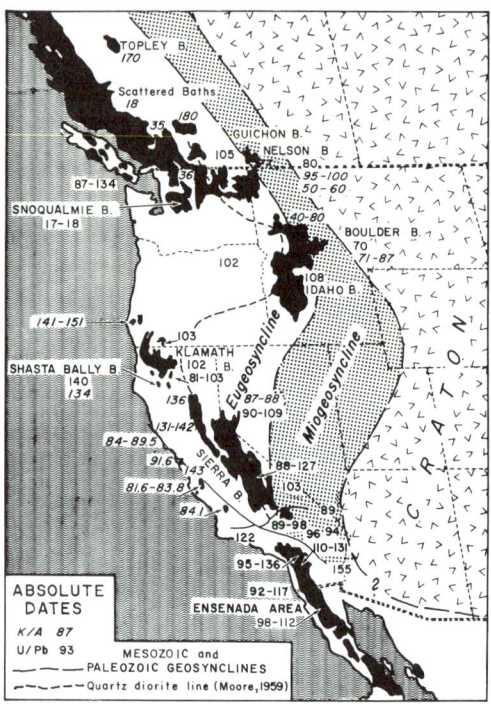

FIGURE 2. Absolute dates as of 1965 in relationship to the Mesozoic and Paleozoic geosynclines of the Pacific Cordilleran region.

and 2). Therefore, the latter are discussed below in a different order.

Pre-Late Cenozoic Orogenic Complex. An E-trending convex arc of deformed Paleozoic, Mesozoic, and early Cenozoic rocks extends from northern Washington into central Idaho, through scattered ranges in central Oregon to the Klamath Mountains subprovince on the coast, thence SE through the Sierra Nevada and, discontinuously, to Los Angeles and S into Baja California. This zone corresponds to the most intensely disturbed axis of the Cordilleran orogenic belt. Old sedimentary and volcanic rocks have been folded, faulted, and more or less metamorphosed along it. Large granitic batholiths are prominent. The most important ones are those of northern Washington, Idaho, Sierra Nevada, and southern California. There are smaller granitic plutons in the Klamath subprovince and the Blue Mountains of eastern Oregon (Fig. 2). It is largely because of the great batholiths and associated metamorphic rocks found in Idaho, that most of that state is included within the Pacific Cordilleran region.

Lava Plateaus of the Northwest. Immense volumes of late Cenozoic volcanic rocks and associated nonmarine sediments, locally more than 3000 m thick, partially mask the older orogenic belt. They include the Columbia River tholeiitic basalts (Oligocene-Miocene) and younger basalts (Miocene-Pleistocene), which continue to the S and to the SE under the downwarped Snake River plain. Though relatively simple structurally, this region nonetheless displays prominent faulting and minor folding. Very broad anticlinal structures closely similar in trend to the arcuate Mesozoic orogenic belt formed in the Columbia Plateau during late Miocene time. Later fault trends, particularly in southern Oregon, are generally sharply discordant with these fold patterns.

Cascade Range. The west edge of the interior plateaus terminates sharply at the line of basaltic, andesitic, and rhyolitic composite volcanic cones and calderas comprising the high Cascades. This volcanic "arc" is assumed to mark a great structural zone near the western margin of the lava plateaus.

Coast Range Upwarp of the Northwest. The Olympic Mountains of northwest Washington, are geologically related to the pre-Cenozoic belt as is the Klamath subprovince 500 km to the S. Between these sectors, however, the Coast Ranges are comprised solely of thick, dominantly marine Cenozoic clastic sediments and volcanics of the basalt-spilite-keratophyre association. They are gently folded along axes oblique to the N-S topographic axis.

The Willamette-Puget Depression. Though structural in the N, this feature is largely erosional in Oregon. The Willamette Valley follows approximately the westernmost edge of the lava plateau.

California Coast Range Province. This belt contains, in many ways, the most interesting and spectacular structures in the entire Pacific Cordilleran region, owing to both the intensity and youthfulness of shearing along a myriad of broad, NW-trending transcurrent fault zones. Its unique structural style can be traced from the southwestern Oregon coast to Baja California (Figs. 2 and 3).

The Coast Ranges also contain the famous Franciscan Complex, a very widespread, monotonous mixture of thick graywacke, basaltic-spilitic-keratophyric lavas, bedded chert, and their metamorphic equivalents. Because of lithologic homogeneity, complex structure, and sparsity of fossils, the Franciscan has thus far defied satisfactory subdivision. Such features can be understood in terms of assemblages of *mélange complexes,* comprising originally more or less separate rock suites deposited upon oceanic crust that are now intimately mixed and sheared together, presumably by the *subduction* of the Pacific oceanic plate beneath the continental margin during Mesozoic and early Cenozoic times. Also important in understanding Coast Range geology are the "blueschists" (containing blue amphiboles such as glaucophane and lawsonite) and ultramafic masses. Experimental studies have shown that blueschists reflect high pressure but low temperature metamorphism. Apparently underthrusting of an oceanic plate caused rapid increase of pressure, but then the rocks were elevated again before their temperature was raised significantly. Many of the ultramafic masses in the Coast Ranges appear to be sheets associated with major thrust zones. These sheets are thought to represent oceanic material that became structurally interleaved with geosynclinal rocks during plate interactions; the blueschists commonly are in contact with such sheets. The Franciscan Complex appears to represent a fossil subduction zone.

Most of the shearing seen in the Franciscan mélanges apparently predates the Cenozoic San Andreas fault system, a fact that was not widely appreciated until the late 1960s. That NW-trending fault system has been superimposed upon the Franciscan structures. The San Andreas fault itself, though the most famous, is but one of a family of related shear zones. The E-W Garlock shear, which lies outside and E of the Coast Ranges and marks the southern boundary of the Sierra Nevada, is apparently conjugate to the San Andreas system, as are the E-W trends within the Transverse Ranges of the southern coast (Fig. 2). This E-W trend appears again in the *Murray fracture zone* (see Vol. I: *Pacific Ocean*).

Complexity of deformation associated with the Coast Range shears reaches its extreme in the Transverse Ranges subprovince, which contains the intersection of the San Andreas and Garlock systems (Figs. 1 and 2); extensive late Cenozoic thrust as well as high-angle faults and tight folds are commonplace. The major shear zones have long been considered lateral or transcurrent faults, but controversy still exists over the total magnitude, antiquity, and ultimate cause of the movements. Some workers argue for as much as 600 km of total displacement over the past 130 m yr, but most recent studies indicate appreciable movement only since the late Oligocene or early Miocene. It was suggested long ago that the faulting was produced by a gigantic shear couple involving the entire Cordillera and the adjacent Pacific Basin. Such a "rotation of the Pacific" now is supported by evidence that the San Andreas is an immense *transform fault* connecting the northern terminus of the East Pacific Rise spreading axis located near Baja California with

UNITED STATES—PACIFIC CORDILLERAN REGION

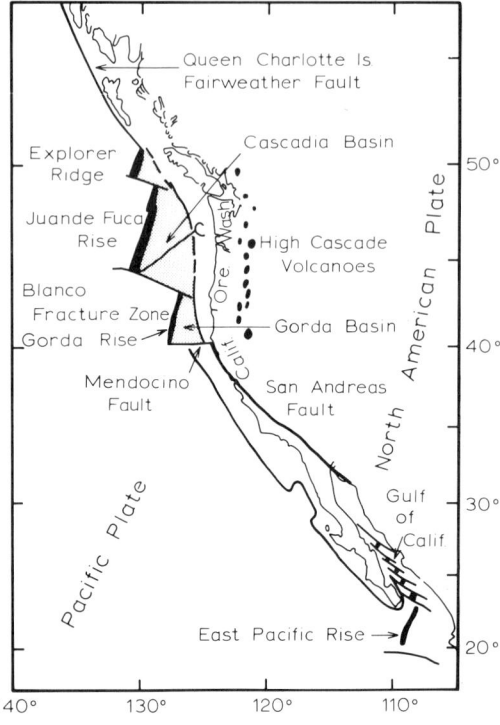

FIGURE 3. Location of tectonic features in the northeast Pacific and west coast of North America. Shaded region represents the Gorda plate (Silver, 1971a).

its continuation located W of Oregon and Washington (Fig. 3).

Great Valley. Between the Coast Ranges and the Sierra Nevada block lies the Great Valley, which is a deep downwarp that received great thicknesses of Cenozoic sediments. Here dominantly marine strata are but mildly deformed except adjacent to the Coast Range boundary shears.

Geophysical Character of the Cordilleran Crust

Crustal structure of the Pacific states is still poorly known, except in the broadest terms. A long-held interpretation of a narrow, 60-km-deep root beneath the Sierra batholith is no longer accepted; rather the Mohorovičić Discontinuity is now considered to be 45–50 km deep beneath both the Great Basin and the Sierras, rising gradually to 38 km near the coast in southern California. Off most of California, there appears to be a rather abrupt termination of the continental crust, whereas in the NW, petrologic and geophysical data suggest a more gradual transition from continental to oceanic crust. Off northern California there is magnetic evidence that oceanic crust lies beneath the continental shelf. There is a relatively high incidence of seismicity throughout the entire Pacific Cordillera and the adjacent Pacific Basin, but no deep-focus earthquakes at this time. Unusual crustal and upper-mantle seismic velocities, distribution of shallow seismicity, and limited heat-flow data have led to the speculation that the East Pacific Rise may in some way extend N beneath the Basin and Range province. Great active shear zones characterize the northeast Pacific sea floor much as they do the California Coast Ranges, though the E-W (Garlock) trend is dominant offshore. The oceanic fractures are expressed by bathymetry, seismicity, and magnetic anomaly patterns.

Geologic History

Stratigraphic Concepts of the Pacific Cordilleran Region. The Pacific Cordilleran region has clearly been a tectonically mobile (unstable) region throughout its known history, a portion of the western North American *Cordilleran orogenic belt,* stretching from Alaska to Panama, and continuing in South America. This entire belt is characterized stratigraphically by the presence of late Precambrian through Mesozoic strata roughly five times as thick as synchronous sequences of the stable interior (Fig. 2). After the geosynclinal concept was first developed in the Appalachian and Alpine regions, Schuchert in 1910 extended it to this western belt, recognizing it as the *Cordilleran geosyncline.* In keeping with the general thought of the times, he postulated a great persistent "borderland" of Precambrian crystalline rocks, named *Cascadia,* which was supposed to lie W of the geosyncline, off the present northwest coast, from which presumably most of the thick sediments were derived. At that time most metamorphics and large batholiths were assumed to be Precambrian, even though Whitney had discovered, as early as 1864, fossil evidence proving that the Sierra batholith must be post-Jurassic. Successively Daly in 1912, Crickmay in 1931, Wheeler in 1940, Kay in 1944, and Eardley in 1947 recognized that through most of the geosynclinal history, subsidence and sedimentary accumulations along approximately the western half of this great belt were accompanied by important volcanism. Throughout the Mesozoic, at least, this volcanism was nearly continuous (Gilluly, 1973). Stille in 1924 referred to the volcanic portion of the mobile belt as *eugeosynclinal,* distinct from the nonvolcanic *miogeosynclinal* portion (Fig. 2) located next to the stable *craton* (see Kay, 1951, for discussion and references). The concept of Precambrian borderlands was displaced in the 1940s by one of lands formed *within* the geosyncline. This modification of the geosynclinal theory was greatly dependent on the Pacific Coast region for documentation, for the evidence in the Appalachian and other well-known geosynclines is more obscure. As late as 1941 Schofield was interpreting evidence in terms of a Precambrian Cascadia borderland.

Kay (1951) stressed the fact that types of sediments deposited in the volcanic portion of the western geosyncline have varied tremendously in response to local and temporary circumstances, sometimes changing very rapidly. Only in a statistical sense, in contrast to other sedimentary environments, has the volcanic belt been dominated by accumulation of immature terrigenous clastic sediments ("graywackes") in an alternating turbidite/pelagic sequence ("flysch" facies). In fact, a complete spectrum of sedimentary rocks, including extensive limestones, is associated with the volcanic products, attesting to the extreme complexity and rapidity of change within mobile belts.

The Precambrian and the Early Continental Margin. Though Precambrian rocks are but sparsely known within the Pacific Cordillera itself, it is clear from adjacent areas that the western zone of continental crust had formed at least by late Precambrian time (about 1 b yr ago). Precambrian rocks are known in northern Idaho, the southern Klamath area, and in southern California. Late Precambrian sediments, just to the E of the province, were deposited unconformably upon an older, essentially continental (sialic) basement. Some of these basement rocks display structural trends markedly discordant with those of the Cordilleran mobile belt, thus indicating tectonic reworking of more ancient continental crust; complex overprinting of different orogenic regimes has occurred.

The presence of volcanic rocks in late Precambrian sequences in southeastern British Columbia and in the early Paleozoic of western Nevada—together with local unconformities and conglomerates—suggests that structural instability was an early characteristic of the Cordillera.

Early Paleozoic. Subsidence and thick accumulations marked the early Paleozoic with extensive, largely submarine, volcanic outpourings known during the Cambrian, Ordovician, and Silurian in nearby Nevada. Bedded chert and graptolitic black mudstones are associated with ellipsoidal basalts. In the southern Klamath Mountains, metavolcanics and metasediments probably represent synchronous accumulations. Silurian reef limestones in this area are the oldest well-dated fossiliferous strata in the Pacific Cordilleran region; similar rocks are known farther north in southeast Alaska.

Paleozoic Orogenesis. The plate tectonics theory suggests a tranquil Eocambrian-Devonian continental shelf in the eastern Cordillera (miogeosyncline) with abyssal oceanic crust beneath the Pacific Cordillera (eugeosyncline); subsequently the oceanic plate presumably began to be thrust E under the continental plate and caused orogenic disturbance. Beginning in late Devonian and early Mississippian time, the central part of the geosyncline in Nevada and Idaho suffered folding and thrust faulting (*Antler orogeny*), but these disturbances have not been recognized in California, Oregon, or Washington. This event does not appear to have been accompanied by metamorphism or plutonic activity. In the Klamath region the Devonian is represented by shales and reef limestones, whereas the Mississippian contains a monotonous sequence of alternating sandstone turbidites and shale. In central Oregon, carbonates are reported.

Late Paleozoic. Several pulses of Pennsylvanian and Permian diastrophism are recognized in Nevada and Idaho. Much coarse sedimentary debris was derived from erosion of elevated lands, including notably a Pennsylvanian chert gravel derived from the older bedded cherts. Farther W, finer terrigenous material and some extensive carbonate banks were the principal sedimentary facies. A distinctive Asiatic fusulinid fauna characterizes the Pacific Cordilleran belt in contrast to the typical American fauna of the miogeosyncline.

Permo-Triassic Volcanism and the Cassiar Orogeny. Widespread volcanism occurred in late Permian to early Triassic times, from Alaska to Mexico, the most widespread in the pre-Cenozoic history of the Cordilleran belt. Apparently it was a precursor to major orogenesis, the *Cassiar orogeny,* between medial Permian and Late Triassic time, an event recorded around the entire Pacific margin, e.g., the Akiyoshi orogeny of Japan. Recent speculation suggests the Cassiar event may represent the collision of an Asiatic plate with the old Cordilleran shelf.

Early Mesozoic. Later Triassic and Jurassic deposits are extremely varied and generously mixed with volcanics. Extensive graywackes, black mudstones, and conglomerates are typical, but important limestones, including organic reefs, also occur widely. Some of these have yielded famous fossil faunas, notably at Brock Mountain in the Klamath subprovince, at Taylorsville in the northern Sierras, and in the Wallowa Mountains of northeast Oregon. The paleogeography was rapidly changing during the early Mesozoic; a volcanic island arc and trench system persisted in the Pacific Cordillera with a shallow sea to the E.

Late Mesozoic Orogenesis. In northwest California, southwestern Oregon, and southwestern British Columbia, there is clear evidence of deformation, metamorphism, plutonic activity and extensive erosion near the end of the Jurassic time. Isotopic dating of diorites in the Klamath subprovince, northwest Sierras, and southernmost California indicates their crystallization about 135–155 m yr ago. This is the principal evidence of the most famous—but elusive—western tectonic event, the *Nevadan orogeny,* named for the Sierra Nevada. Its reality is now well established, but absolute dating also suggests that most of the great batholiths, long assumed to be "Nevadan," are in fact younger, chiefly Middle Cretaceous (80–110 m yr old, Fig. 2). N. L. Taliaferro long ago suggested that there were three orogenies in late Mesozoic: before the end of the Jurassic (*Nevadan*), at the beginning of Cretaceous (*Diablan*), and in the early late Cretaceous (*Santa Lucian*). In addition to evidence in different areas for disturbances, including emplacement of ultramafic and granitic plutons at each of these times, there is also evidence for activity at other times. Farther E the later of these diastrophic pulses seems to grade to the late Cretaceous-early Cenozoic *Laramide orogenesis* of the Rocky Mountain region. All of the Pacific Coast events now appear to be pulses of almost continuous disturbance culminating in the late Mesozoic orogenesis. This great tectonic revolution completely transformed the former geosyncline by early Cretaceous time, replacing it with a complex highland from which detritus was shed into a relict sea, a subsiding trough W of the main batholithic axis.

Early Cenozoic. Marine deposition persisted during Cretaceous and early Cenozoic times in much of the present California Coast Range area and in western Oregon and Washington. In the Northwest, volcanism was renewed on a grand scale, submarine (basaltic) to the W and subaerial (andesitic to rhyolitic) to the E. In the latter region during Oligocene and Miocene time, the great Columbia River flood basalts were erupted from fissures. Nonmarine sediments were also deposited, and many of these contain famous vertebrate and plant fossils. The region also contains many fine Cenozoic marine fossil localities; it is especially noted for studies of foraminifers.

Late Cenozoic and the Cascadan Orogeny. Early Cenozoic depositional and structural patterns strongly reflect the arcuate pattern of the Mesozoic orogenic system. Although the greatest diastrophism was pre-Tertiary, the same pattern of structural trends continued and was not changed until Miocene time and later. At about the same time, block faulting in the Basin and Range province became intense, the Sierra Nevada began to be tilted up, and the California Coast Range shear zones became extremely active, producing a complex series of rapidly changing depositional basins and elevated islands. Deep fissuring of the crust allowed basaltic volcanism to spread throughout much of the Pacific Cordilleran region. Emplacement of small, scattered plutons also occurred in the Cascade Range; the Snoqualmie batholith (Washington) is the largest of these, and has been dated as late Miocene to Pliocene in age (Fig. 2). All of these late Cenozoic phenomena have been referred to as the *Cascadan orogeny,* which is still in progress today.

An early and influential view held that the widespread Tertiary block faulting reflected tensional relaxation of strain within the crust after supposed great orogenic compression in the Mesozoic. This concept was regarded as a generalization for all orogenic belts, but is not supported by close analysis. According to plate tectonic theory, subduction of a Pacific plate beneath the continental one continued until the W-drifting plate overtook and collided with (and possibly overrode) the continuation of the East Pacific Rise spreading site in the present California region. Such a collision presumably caused the dramatic mid-Cenozoic change of tectonic behavior, which led to Basin-and-Range block faulting, San Andreas-type transform (strike-slip) faulting, and the Cascade and other volcanism. Off the northwest coast, where collision did not occur, subduction appears to be continuing (Fig. 3). If so, the deep trench that might be expected apparently is kept filled by rapid sedimentation.

The Present Scene. A major result of continued elevation of mountain masses, together with the general climatic deterioration of the late Cenozoic, has been the drying and cooling of the climate in much of the Pacific Cordillera. From a very mild, uniform subtropical climate with luxuriant vegetation over all of the land areas in the early Cenozoic, there has been an oscillatory but progressive retreat and isolation of more humid-climate (Arcto-Tertiary) floras as an arid-climate (Madro-Tertiary) flora spread N and W from northern Mexico. The restriction of the range of the redwood tree (*Sequoia* spp) from a nearly universal distribution to its two present (humid) habitats exemplifies these changes. The Klamath subprovince has been relatively stable during the late Cenozoic and has acted as a persistent relict floral center to which many of the older Arcto-Tertiary elements have retreated, but not without the intrusion of some Madro-Tertiary elements.

Quaternary (and earlier) glaciation has left significant marks on the Pacific Cordillera in the high mountains, as noted above, but also over most of northern Washington where continental ice sheets spread. Particularly interesting are the effects on drainage produced by temporary Pleistocene diversion of the Columbia River through Grand Coulee and the "channeled scablands" region where apparently sudden gigantic floods occurred as the contents of huge glacial lakes were decanted from the NE (see articles in Vol. III, *Encyclopedia of Geomorphology*). Both the glacial and erosional histories clearly have been cyclic or episodic, as in all of the mountainous west. Radiocarbon dating has added greatly to knowledge of the late Quaternary, and confirms a close correlation of glacial and postglacial events with those of the Upper Mississippi Valley region for the past 40,000 years. Older glacial events also are known, but with far less precision. Accurately dated volcanic ash layers that are petrologically traceable to specific Cordilleran eruptive centers are helping to delineate the Pleistocene history (Bishop Ash, Mazama Ash, etc.). Pleistocene periglacial phenomena are widely observed. Elevated coastal marine terraces are also numerous and widespread, attesting to profound recent changes of relative land and sea levels. In places, they are notably warped. In the Puget Sound area, glacioisostatic movements are still in progress.

The Pacific region is so youthful and restive that it provides some of the most graphic evidences of dynamic earth processes—literally natural experiments in progress. Numerous contributions to various modern plate tectonic interpretations are noted in the references. Glaciation is still active in the northern mountains, earthquakes and faulting are frequent, large areas continue to be elevated or depressed, and many of the volcanoes may only be resting for renewed outbursts.

Economic Mineral Deposits

Metals. Over the years the Pacific region has yielded immense wealth of metallic ores. The Mother Lode gold belt of the northern Sierras and the fabulous Comstock silver lode of nearby Nevada played a key historical role in white settlement and exploration. Gold districts (placers) were rapidly developed and depleted after 1849; a small amount of gold is still produced. Until about 1950, chrome was important in a coastal peridotite belt from northern California to Washington, and important nickel is still produced at Riddle in southwest Oregon. Manganese is present in the Olympic Mountains, but is not economically important. Silver, lead, and zinc are produced at Coeur d'Alene, northern Idaho, one of the largest contemporary mining districts. Tungsten, copper, and molybdenum are actively mined at Bishop, California, in the eastern Sierra foothills. Mercury is produced in a few districts, notably in the central California Coast Ranges. Uranium also has been exploited at scattered localities. The U.S. Geological Survey began its activities in the far west in the 1880s, and has played a valuable role in ore finding, but state, university, and, above all, individual investigators have led the way.

Nonmetals. Today nonmetallic deposits account for a greater dollar value than do metals. These include principally limestone for cement, high-aluminous clays, diatomite, building stone, and aggregate. High-quality limestone is at somewhat of a premium in the Pacific region because of the relative sparsity of extensive, pure carbonate rocks in very mobile tectonic provinces such as this. Nonetheless, important Paleozoic and Triassic limestones are useful, and some local Cretaceous and Tertiary limestones also are exploited in the Coast Ranges. Of the 12 western states, California ranks first in total metallic and nonmetallic mineral production according to the U.S. Bureau of Mines.

Fossil Fuels. Eocene subbituminous to bituminous coal was mined early in this century at Coos Bay, Oregon, and around Puget Sound, Washington, but has been abandoned. Oil, gas, sawdust, and hydroelectric power are the contemporary energy sources in the Pacific region.

Petroleum and natural gas are produced extensively only in California. Minor production exists in Washington, and active exploration, particularly offshore, continues in all three coastal states. California oil seeps were first discovered and tapped in the 1860s, and early wells were drilled next to seeps. After 1900 geology began to be used significantly in exploration, and its immediate success was phenomenal. No less than three major anticlinal accumulations were found in the single year, 1908. Owing to the youthfulness of southern California's geologic structures and to its climatic aridity, many of the greatest anticlinal traps were initially discovered almost solely by their topographic expression. Most of the production is in southern California, from Cenozoic rocks, whereas natural gas, first discovered in a water well in 1854, is produced chiefly from the Cretaceous of the northern Great Valley. Total production of one great district, the Midway-Sunset-Buena Vista field in the southern Great Valley, exceeds one million barrels to date. Total oil and natural gas production for the state has placed California among the top four or five producing states in the United States during the past half century.

R. H. DOTT, JR.

References

Addicott, W. O., 1970. "Tertiary paleoclimatic trends in the San Joaquin basin, California," *U.S. Geol. Surv. Prof. Pap. 644-D*, 9p.

Atwater, T., 1970. "Implications of plate tectonics for the Cenozoic tectonic evolution of western North America," *Bull. Geol. Soc. Am.*, 81, 3513–3536.

Bailey, E. H., ed., 1966. "Geology of northern California," *Calif. Div. Mines Geol. Bull.*, 190, 507p.

———, Irwin, W. P., and Jones, D. L., 1964. "Franciscan and related rocks and their significance in the geology of western California," *Calif. Div. Mines Geol. Bull.*, 183, 1–177.

Baksi, A. K., 1973. "Volcanic production rates: comparison of oceanic ridges, islands and the Columbian Plateau basalts," *Science*, 180, 493–495.

Baldwin, E. M., 1964. *Geology of Oregon*, 2nd ed., Ann Arbor, Mich.: Edwards Bros., 165p.

Bateman, P. C., 1968. "Geologic structure and history of the Sierra Nevada," *Univ. Mo. Res. J.*, 1(1), 121–131.

Berkland, J. O., 1972. "Paleogene 'frozen' subduction zone in the coast ranges of northern California," *24th Intern. Geol. Congr.*, Canada, 3, 99–105.

———, et al., 1972. "What is Franciscan?" *Bull. Am. Assoc. Petrol. Geologists*, 56(12), 2295–2302.

Chipping, D. H., 1972. "Early Tertiary paleogeography of central California," *Bull. Am. Assoc. Petrol. Geologists*, 56(3), 480–493.

Coleman, R. G., and Lanphere, M. A., 1971. "Distribution and age of high-grade blueschists, associated eclogites, and amphibolites from Oregon and California," *Bull. Geol. Soc. Am.*, 82, 2397–2412.

Crowell, J. C., 1968. "The California coast," *Univ. Mo. Res. J.*, 1(1), 133–156.

Davis, G. A., and Burchfiel, B. C., 1973. "Garlock Fault: an intracontinental transform structure, southern California," *Bull. Geol. Soc. Am.*, 84, 1407–1422.

Dott, R. H., Jr., 1965. "Mesozoic–Cenozoic tectonic history of the southwestern Oregon coast in relation to Cordilleran orogenesis," *J. Geophys. Res.*, 70(18), 4687–4707.

———, 1971. "Geology of the southwestern Oregon coast west of the 124th meridian," *Ore. Dept. Geol. Mines Indus. Bull.*, 69, 63p.

Easterbrook, D. J., 1969. "Pleistocene chronology of the Puget lowland and San Juan Islands," *Bull. Geol. Soc. Am.*, 80, 2273–2286.

Elders, W. A., et al., 1972. "Crustal spreading in southern California," *Science*, 178(4056), 15–24.

Ernst, W. G., 1970. "Tectonic contact between the Franciscan mélange and the Great Valley sequence–crustal expression of a late Mesozoic Benioff zone," *J. Geophys. Res.*, 75, 886–901.

Garfunkel, Z., 1973. "History of the San Andreas Fault as a plate boundary," *Bull. Geol. Soc. Am.*, 84, 2035–2042.

Garrison, R. E., 1972. "Inter- and intrapillow limestones of the Olympic Peninsula, Washington," *J. Geol.*, 80, 310–322.

Gastil, G., Phillips, R. P., and Rodriquez-Torres, R., 1972. "The reconstruction of Mesozoic California," *24th Intern. Geol. Congr.*, Canada, 3, 217–229.

Gilluly, J., 1963. "The tectonic evolution of the western United States," *Quart. J. Geol. Soc. Lond.*, 119, 133–174.

———, 1973. "Steady plate motion and episodic orogeny and magmatism," *Bull. Geol. Soc. Am.*, 84, 499–514.

Gresens, R. L., 1970." Serpentinites, blueschists, and tectonic continental margins," *Bull. Geol. Soc. Am.*, 81, 307–310.

Hamilton, W., 1969. "Mesozoic California and the underflow of Pacific mantle," *Bull. Geol. Soc. Am.*, 80, 2409–2430.

———, and Myers, W. B., 1974. "Nature of the Boulder Batholith of Montana," *Bull. Geol. Soc. Am.*, 85, 365–378.

Henyey, T. L., and Wasserburg, G. J., 1971. "Heat flow near major strike-slip faults in California," *J. Geophys. Res.*, 76(32), 7924–7946.

Hsü, K. J., 1968. "Principles of melanges and their bearing on the Franciscan-Knoxville paradox," *Bull. Geol. Soc. Am.*, 79, 1063–1074.

Huffman, O. F., 1972. "Lateral displacement of Upper Miocene rocks and the Neogene history of offset along the San Andreas fault in central California," *Bull. Geol. Soc. Am.*, 83, 2913–2946.

Hyne, N. J., et al., 1972. "Quaternary history of Lake Tahoe, California–Nevada," *Bull. Geol. Soc. Am.*, 83, 1435–1448.

Johns, R. H., ed., 1954. "Geology of southern California," *Calif. Div. Mines, Dept. Nat. Res. Bull.* (San Francisco), 170.

Jones, D. L., and Irwin, W. P., 1971. "Structural implications of an offset early Cretaceous shoreline in northern California," *Bull. Geol. Soc. Am.*, 82, 815–822.

Kay, M., 1951. "North American geosynclines," *Geol. Soc. Am. Mem. 48*, 143p.

Kays, M. A., 1970. "Mesozoic metamorphism, May Creek schist belt, Klamath Mountains, Oregon," *Bull. Geol. Soc. Am.*, 81, 2743–2758.

Kilkenny, J. E., 1971. "Future petroleum potential of region 2, Pacific coastal states and adjacent continental shelf and slope," *Future Petroleum Provinces of the United States; Their Geology and Potential* (Mem. 15, vol. 1). Tulsa: Am. Assoc. Petrol. Geologists, 170–177.

Kistler, R. W., et al., 1971. "Sierra Nevada plutonic cycle. 1. Origin of composite granitic batholiths," *Bull. Geol. Soc. Am.*, 82, 853–868.

LaFehr, T. R., 1965. "Gravity, isostasy, and crustal structure in the southern Cascade Range," *J. Geophys. Res.*, 70(22), 5581–5597.

Lanphere, M. A., 1971. "Age of the Mesozoic oceanic crust in the California Coast Ranges," *Bull. Geol. Soc. Am.*, 82, 3209–3212.

Lipman, P. W., Prostka, H. J., and Christiansen, R. L., 1972. "Cenozoic volcanism and plate-tectonic evolution of the western United States. 1. Early and

Middle Cenozoic," *Phil. Trans. Roy. Soc. Lond.,* Ser. A, **271**, 271–248.
Maxwell, J. C., 1974. "Anatomy of an Oregon," *Bull. Geol. Soc. Am.,* **85**, 1195–1204.
McKee, B., 1972. *The Geologic Evolution of the Pacific Northwest.* New York: McGraw-Hill, 394p.
McKenzie, D., and Julian, B., 1971. "Puget Sound, Washington, earthquake and the mantle structure beneath the northwestern United States," *Bull. Geol. Soc. Am.,* **82**, 3519–3524.
Menzer, F. J., Jr., 1970. "Geochronologic study of granitic rocks from the Okanogan Range, north-central Washington," *Bull. Geol. Soc. Am.,* **81**, 573–578.
Oakeshott, G. B., 1971. *California's Changing Landscapes; a Guide to the Geology of the State.* New York: McGraw-Hill, 388p.
Page, B. M., 1972. "Oceanic crust and mantle fragment in subduction complex near San Luis Obispo, California," *Bull. Geol. Soc. Am.,* **83**, 957–972.
Peterman, Z. E., et al., 1970. "Sr^{87}/Sr^{86} ratios of Quaternary lavas of the Cascade Range, northern California," *Bull. Geol. Soc. Am.,* **81**, 311–318.
Roberts, R. J., 1972. "Evolution of the Cordilleran fold belt," *Bull. Geol. Soc. Am.,* **83**, 1989–2004.
Ross, D. C., 1970. "Quartz gabbro and anorthositic gabbro: markers of offset along the San Andreas fault in the California coast ranges," *Bull. Geol. Soc. Am.,* **81**, 3647–3662.
Savage, J. C., and Burford, R. O., 1973. "Geodetic determination of relative plate motion in Central California," *J. Geophys. Res.,* **78**(5), 832–845.
Schmincke, H. U., 1967. "Stratigraphy and petrography of four upper Yakima Basalt flows in south-central Washington," *Bull. Geol. Soc. Am.,* **78**(11), 1385–1422.
Sharp, R. P., 1972. "Pleistocene glaciation, Bridgeport Basin, California," *Bull. Geol. Soc. Am.,* **83**, 2233–2260.
Shaw, H. R., et al., 1971. "Sierra Nevada plutonic cycle. 2. Tidal energy and a hypothesis for orogenic–epeirogenic periodicities," *Bull. Geol. Soc. Am.,* **82**, 869–896.
Silver, E. A., 1971a. "Tectonics of the Mendocino triple junction," *Bull. Geol. Soc. Am.,* **82**, 2965–2987.
_____, 1971b. "Transitional tectonics and late Cenozoic structure of the continental margin off northernmost California," *Bull. Geol. Soc. Am.,* **82**, 1–22.
_____, 1971c. "Pleistocene tectonic accretion of the continental slope off Washington," *Mar. Geol.,* **13**, 239–249.
Snavely, P. D., Jr., et al., 1973. "Miocene tholeiitic basalts of coastal Oregon and Washington and their relations to coeval basalts of the Columbia Plateau," *Bull. Geol. Soc. Am.,* **84**, 387–424.
Sumner, J. R., 1972. "Tectonic significance of gravity and aeromagnetic investigations at the head of the Gulf of California," *Bull. Geol. Soc. Am.,* **83**, 3103–3120.
Suppe, J., 1972. "Interrelationships of high-pressure metamorphism, deformation and sedimentation in Franciscan tectonics, U.S.A.," *24th Intern. Geol. Congr.,* **3**, 552–559.
Travers, W. B., 1972. "A trench off central California in late Eocene–early Oligocene time," *Geol. Soc. Am. Mem. 132,* 173–182.
Waters, A. C., 1962. "Basalt magma types and their tectonic associations: Pacific Northwest of the United States," *The Crust of the Pacific Basin* (Monogr. 6). Washington, D.C.: Am. Geophys. Union, 158–170.
Whetten, J. T., 1972. "Matrix-rich Pleistocene sediments from western Washington: incipient graywacke-type sedimentary rocks?" *Geol. Soc. Am. Mem. 132,* 573–584.
White, D. E., et al., 1973. "Thermal and mineral waters of nonmeteoric origin, California coast ranges," *Bull. Geol. Soc. Am.,* **84**, 547–560.

Cross-references: *Canada–Cordilleran Region; Mexico; North America; Vol. I, Pacific Ocean; Vol. III, Quaternary; United States–General, Alaska, Basin and Range Province.*

UNITED STATES–ROCKY MOUNTAIN PROVINCE

The Rocky Mountain Province, within the United States, is part of a larger Rocky Mountain belt that traverses western North America in a broadly N-S trend and in turn forms part of the 17,000-km "Backbone" of the Americas, which takes in the Andes as well, sometimes called the American *Cordilleran System* (following Alexander von Humboldt). Within this system the U.S. Rocky Mountain belt is restricted to a discontinuous mountain barrier that extends from the Canadian border through Montana, Idaho, Wyoming, and Colorado to northern New Mexico, where it dies out as a morphotectonic feature. Its eastern aspect forms a rugged mountain front that rises impressively above the Great Plains. Its crest often marks the continental divide. Its western border is usually transitional to the intermontane plateaus and basins (see *United States–Colorado Plateau Province; United States–Basin and Range Province;* also Columbia Plateau in *United States–Pacific Cordillera Region*).

Fenneman (1931) classified the United States Rocky Mountain system into four sub-provinces: the Northern Rocky Mountains, Middle Rocky Mountains, Wyoming Basin, and Southern Rocky Mountains. For simplicity, the second and third sectors are here treated together (Fig. 1). High peaks exceed 4300 m (14,200 ft) and the summit accordances are often 2000 m above the plains. Youthful reactivation of uplift has caused deep dissection and canyon cutting. Because of its spectacular scenery, many sections of this belt have been set aside as national parks and monuments (e.g., Glacier, Yellowstone, Grand Teton, and Rocky Mountain national parks and Dinosaur National

Monument). The belt covers 450,000 km² (180,000 sq mi).

Geological instruction is carried out at the principal state universities (e.g., notably Wyoming and Colorado), and each of the states has geological surveys and societies that issue geological maps, bulletins, and guidebooks. The Colorado School of Mines, established in 1874, is one of the country's leading schools of geological engineering and applied geology. A set of regional geological highway maps (and summary guides) is issued by the American Association of Petroleum Geologists; one map sheet treats the Northern and the Southern Rockies. The mountain state headquarters for the U.S. Geological Survey is situated at the Federal Center in Denver. The headquarters of the Geological Society of America are located at Boulder.

Regional morphotectonic summaries may be found in King (1959), Thornbury (1965), and Hunt (1967). Stratigraphic tables may be extracted from Kummel (1970) and the paleogeographic picture from Eardley (1962). An excellent symposium on Rocky Mountain geology was issued by the American Association of Petroleum Geologists in 1965 (Vol. 49, no. 11) and synthesized by Haun and Kent (1965). A magnificent "geologic atlas" of the region was issued in 1972 by the Rocky Mountain Association of Geologists.

Northern Rocky Mountains

A mountainous tract, rather than a distinct range, this subprovince consists primarily of an extensive segment of Precambrian crystalline and sedimentary rocks, tectonized before the Paleozoic, and subsequently disrupted in complex structures of late Mesozoic–Cenozoic origin, trending mainly NNE-SSW and giving rise to a trellis drainage pattern.

The principal rift is the *Rocky Mountain Trench*, which continues farther in a NNW trend along the length of the Canadian Rockies. Of outstanding interest in this sector is the younger Precambrian *Belt Series*, strata remarkable for their limited alteration and curious primitive fossils.

Marginal to this complex Precambrian massif there are scattered segments of Paleozoic geosynclinal rocks. In the NW, the Okanagan Highlands, rising to around 1500 m, are sculptured into rolling hills from heavily folded and altered Paleozoic sediments and Mesozoic to early Cenozoic granitic intrusives. To the SW the latter expand to tremendous proportions, the *Idaho Batholith*, which, in the Salmon River Mountains, is a dissected plateau measuring around 200 by 500 km (N-S). Its accordance of summit levels suggests a former peneplain to some observers. The gorge of the Salmon River is over 2000 m deep (exceeding those of the Grand Canyon and Hells Canyon). The batholith is bordered by metamorphosed Paleozoic geosynclinal rocks.

To the E and SE both Paleozoic and Mesozoic rocks are present in a thick miogeosynclinal sequence marked by complex folding and thrusting directed to the NE. Additional Mesozoic granite intrusions are found, for example, in the *Boulder batholith*, elevations in which reach nearly 4000 m.

In front of the eastern border of the Northern Rockies there is an important transitional zone—the "Disturbed Belt" or "Great Plains Features"—where the late Paleozoic–Mesozoic–early Tertiary rocks of the westernmost Great Plains are overturned and thrust by the E movement of the mountain structures (Fig. 2). In places there are low-angle thrust sheets that

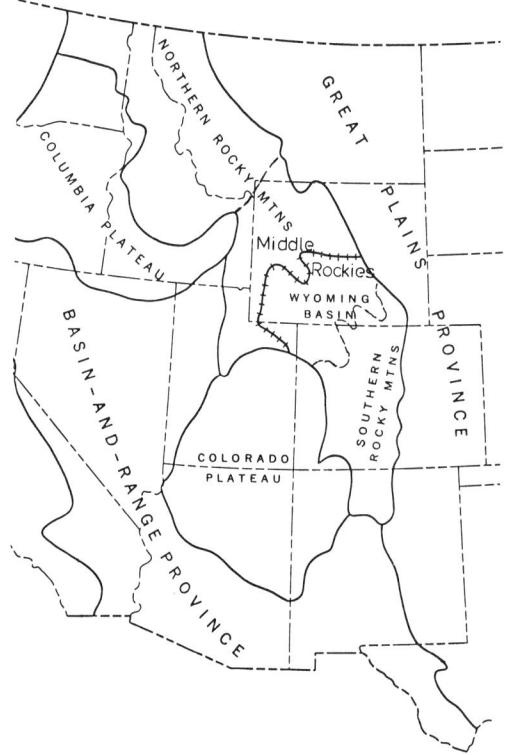

FIGURE 1. Principal physiographic divisions of the Rocky Mountain region. Some writers recognize only a Northern and Southern Rocky Mountain region, but in this encyclopedia entry a Middle Rocky sector is recognized (with Fenneman), south of the Montana Gap that separates the Wyoming-Utah structures from those to the north. The Wyoming Basin is treated with the Middle Rockies.

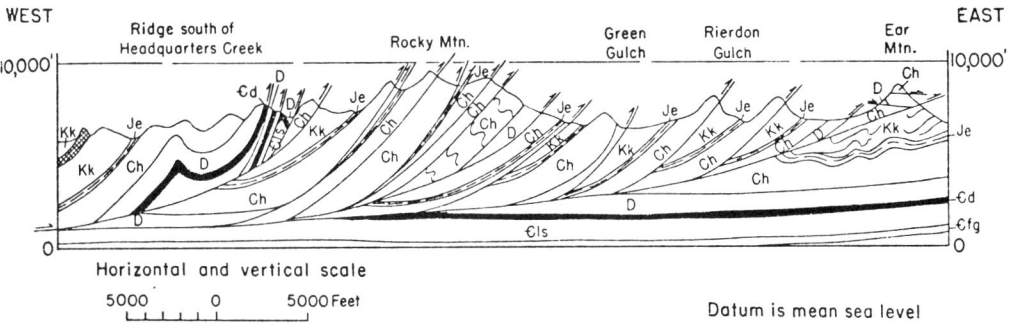

FIGURE 2. Eastward overthrust structures in the Sawtooth Range, Montana (Deiss, 1943b). Kk, Cretaceous Kootenai Formation, Je, Jurassic Ellis Formation; Ch, Mississippian Hannan Limestone; D, unnamed Devonian formation; C-d, Cambrian Devil's Glen Dolomite; C-ls, undifferentiated Cambrian limestone; C-fg, Cambrian Flathead Sandstone and Gordon Shale; Ba, Precambrian Belt Ahorn Quartzite; Bh, Precambrian Belt Hoadley Formation; cross-hatched pattern, diorite sill.

have moved eastward several tens of kilometers (e.g., Lewis Thrust). Precambrian to Mississippian thus are found superimposed over Cretaceous and younger rocks. An outstanding tectonic outlier of the Lewis thrust is Heart Mountain, a classic example of an erosional klippe. Numbers of these "thrusts" are demonstrably gravitational slides as shown first in the Bearpaw Mountains by Reeves (1946). Their major role was demonstrated by Pierce (1957), who called them "detachment thrusts." In the Alps they would be called "relief thrusts" or "nappes," because they have evidently slid over the existing relief. The "Disturbed Belt" structures describe a remarkable reversed S trending around the western end of the Crazy Mountain basin near Bozeman (the eastern end of the Montana Lineament, see below).

Between the ranges in the Northern Rocky Mountain province there are numerous linear troughs and grabens partially filled by late Tertiary and Quaternary continental deposits. The latter include considerable glacial debris. Elevations reach over 3000 m and in the N, at the Glacier National Park and in the Lewis and Clark ranges, there are numbers of small contemporary glaciers.

Quaternary continental glaciation came down the Canadian Rockies as far as Spokane in the SW and Missoula in the S. During deglaciation, a vast glacial lake, Lake Missoula, formed in western Montana, over 300 km long and in places 600 m deep. The opening of ice dams resulted in catastrophic flooding to the SW over the Columbia Plateau, responsible for the Spokane "scablands" (today, rough, dissected country), coulees (scoured channels), giant ripple structures, and so on. Small mountain glaciers formed farther S, leaving complex moraines and terraced valley fills.

Middle Rocky Mountains and Wyoming Basin

Separated almost completely from the Northern Rockies by the reentrant occupied by the Snake River plains in southwestern Idaho and by the line of the Yellowstone River in southwestern Montana, the principal overthrust belt of the Middle Rockies occupies a N-S zone extending along the western borders of Wyoming S into north-central Utah (Wasatch Mountains) and terminating in the E-W trend of the Uinta Mountains. In central Wyoming the Middle Rockies are represented by the Sweetwater Arch (Granite Mountains Uplift) and in northern Wyoming by the Beartooth and Bighorn Mountains and in the E by the Sherman Range.

The so-called Wyoming Basin occupies the southwestern and central parts of Wyoming and is partly subdivided into separate basins by fingers of the Middle Rockies, e.g. the Wind River Mountains (and by the Southern Rockies in the SE). The largest of these basins is the Green River Basin. It is convenient, therefore, to treat with the Middle Rockies also the various intermontane depressions. The basinlike character of the latter is partly offset by the fact that the mean elevation is often in excess of 2000 m, whereas the ranges generally do not exceed 3000 m. Geotectonically this belt corresponds to the "zwischengebirge" or median belts of Eurasia and the "altiplano" of the Andes.

The Middle Rockies are cut off, N and S, by two very remarkable transverse lineaments. The northern one, the *Montana Gap*, is a complex zone trending WNW-ESE associated with clusters of echelon faults (NE-SW), e.g., the Later Basin–Huntley Fault Zone, echelon folds (mainly NW-SE) that suggest a major basement lineament of left-lateral displacement. In the SE it is expressed in the Absaroka Fault and in the

NW by the Lewis and Clark faults and Osburn lineament of the Missoula area. The zone is straddled by the key mining centers of Butte and Helena and has been marked by serious earthquakes. This transverse axis also marks the crossover of the Cordilleran Geosynclinal Belt, with its Laramide overthrusts, that lies on the eastern front of the Northern Rockies, but on the western side of the Middle Rockies. The southern transverse lineament is marked by the southern border of the E-W Uinta Mountains, the *Cortez-Uinta Axis,* that projects also W into the Basin and Range Province and farther E veers to the SE and SSE in the echelon block faulting and torsional tectonics (counterclockwise twist) of the Southern Rockies.

Basement complexes in the Middle Rockies are of Precambrian rocks, mainly trend NE-SW to N-S and are dated in the 1300–1800 m yr range, but with some nuclei as old as 2.7 billion years. They are thus comparable to the Superior Province of the Canadian craton. This was the last general mobilization of most of this region, but reactivation in the form of broad warping and block faulting has repeatedly occurred during the Paleozoic, Mesozoic, and Cenozoic history. Thus the basins represent progressively reactivated downwarps, where remarkably complete sections are present in the middle of the basins and progressively more and more sequences are cut out along the steeply dipping borders. In places those boundaries are abrupt, faulted, and even somewhat overturned toward the basin, giving the uplift almost a diapiric form as in the Beartooth and Bighorn mountains. In the same way the Wind River and Uinta mountains are dominated by vertical tectonics, although more like draped anticlines with Precambrian cores.

As mentioned above, the western borders of the Middle Rockies contain geosynclinal Paleozoic and Mesozoic sequences that are rather typically "orogenic." They are strongly folded and, as in the Northern Rockies, overthrust from the W onto early Tertiary formations. Mansfield (1927) referred to this zone as the *Wyomide* orogenic belt.

Epeirogenic warping of the region set in as early as Oligocene time, but dominating all the topography of the Central Rockies is a major block faulting that occurred along N-S trends in late Cenozoic time. In this way the Teton Mountains of western Wyoming appear as giant W-tilted horst exposing a core of Precambrian, with an abrupt E-face, providing, from Jackson Hole, a scenery of Alpine type. The extraordinary E-facing fault may have a throw of over 6000 m extending N-S for about 60 km. In contrast, farther S, in the Wasatch Mountains of Utah, E of Great Salt Lake, the principal faults

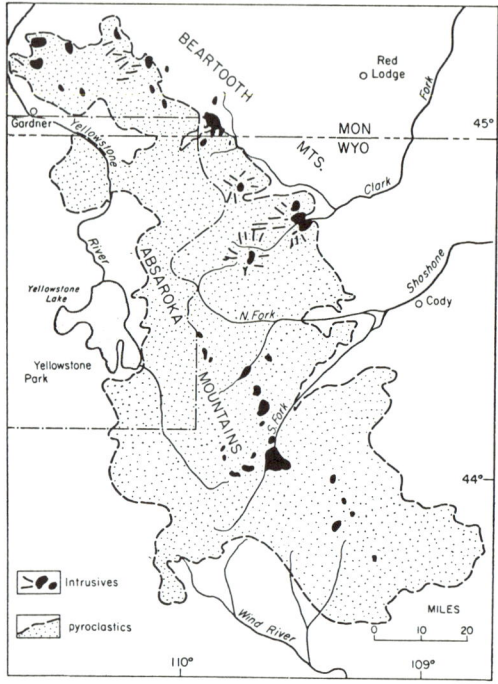

FIGURE 3. Sketch map of the late Cenozoic Absaroka Volcanic Field, in NW Wyoming and adjacent Montana. The Yellowstone National Park lies in the northwestern part. (Modified after W. H. Parsons, Billings Geol. Soc. Guidebook, 9th Conf.)

are on the western side and the range is tilted to the E. These faults represent the beginnings of the Basin and Range Province that lies to the W. The faulting is generally very youthful and according to Love (1954) the Tetons, for example, did not exist as mountains until after mid-Pliocene times.

Youthful volcanism played an important role in the northern part of the Middle Rockies. Here in the *Absaroka Volcanic Field* (Fig. 3), part of which is occupied by Yellowstone National Park, solfatara-stage volcanism is still in progress, and "Old Faithful" represents one of its impressive geysers. The late Tertiary to Quaternary eruptions are dominated by a giant caldera largely buried by flows, and ash blankets are believed to have erupted from numerous small vents, covering a landscape of considerable relief, with up to 500-m contrasts, and leaving it as a broad plateau, dissected by some deep canyons. Eruptions continued through the Pleistocene glaciations and ash interfingers with till. The Teton structures extend farther N but are buried beneath these younger extrusives.

Most of the Middle Rocky region was represented largely by open marine seaways during the Paleozoic and Mesozoic and the intervening depressions of western Wyoming only became isolated during the Cenozoic. Soft brown coals and lignites formed in places, and today these represent a vast energy reserve. In the Green River Basin (during the Eocene) some quite remarkable lacustrine formations accumulated that today contain extensive oil shales and a wide variety of curious evaporite minerals. Environmentally the site was comparable to some East African rift valley lakes today. The climate of the region was warm and partly humid, partly arid during much of the Tertiary, supporting a rich flora and fauna, notably mammals.

Many of the upland surfaces display traces of an undulating erosion surface that has been variously interpreted by geomorphologists. It could be of early Tertiary age and has been called the "Flattop" or "Summit" peneplain (see, e.g., Atwood and Atwood, 1938; an interpretation denied by Mackin, 1947, see below). Today it has considerable relief. The surface is said to be preserved beneath the mid- and late Cenozoic volcanics, and later it was appreciably warped and displaced. A younger erosion surface, partly a composite with the earlier one, had developed by the Pliocene and is known as the "Rocky Mountain," "Sherman," or "Sub-summit" peneplain. Some workers would classify it as a pediplain, especially in view of its great distance from the sea. The surface is continuous with the continental Pliocene Ogallala Formation of the Great Plains (q.v.). By this time the Rockies are visualized as having been reduced to "islandlike" ranges almost buried in their own debris. This younger and lower surface was interpreted by Mackin (1947) as a pediplain, but the higher ones he regarded as altiplanation features of Pleistocene age. It is true that altiplanation phenomena are present, but they appear to be merely secondary modeling.

Important and widespread regional uplift set in during the late Pliocene and continued on a reduced scale during the Quaternary. Furthermore, differential uplift occurred along existing warps and fault planes up to several thousand meters. This elevation not only accelerated erosion generally, but also materially increased the drop in temperatures during the Pleistocene. In the cold phases the snowline was probably 1000 m lower than today.

The Middle Rockies were extensively affected by glaciation during the Quaternary, and most of the mountains above 2500 m were affected by glaciers or strong periglacial action. In Wyoming these were largely Piedmont type in the mid-Pleistocene, with valley glaciers in Wisconsin times, according to observation by Blackwelder and later by Richmond. In the Tetons the valley glaciers descended to below 2000 m, and even lower in the Wasatch and Beartooth Mountains.

During the latest Tertiary and Quaternary, there has been a progressive lowering of base level to match the uplift. In part the latter has been a regional one, and the river gorges that were formerly regarded as antecedent are now generally felt to be superposed; that is, the rapid erosion of the soft alluvial and lacustrine basin fillings exhumed the old mountain "islands" and let down the drainage channels over them—e.g., the Green, Big Horn, Shoshone, Sweetwater, and other rivers.

Southern Rocky Mountains

The most uniform and clearly delineated of the Rocky Mountain provinces, the Southern Rockies, occupies a most characteristic N-S trend across central Colorado, dying out to the N in southeastern Wyoming, and to the S in north-central New Mexico. In the late Paleozoic "Ancestral Rockies" developed in this belt, but largely disappeared during the Mesozoic. Its main lines developed during the Laramide disturbance. The belt is not the product of E-W stresses but essentially vertical movements, modified as King (1959) suggested, by a counterclockwise torsion resulting in NNW-SSE en echelon structures. There is a major NE-SW shear and numerous related left-lateral strike-slip faults.

The Front Range occupies the principal N-S barrier against the Great Plains today, with vertical or slightly overthrust discontinuous border faults. To the N it merges with the Sherman Range of Wyoming, but it also bifurcates and the northwestern branch forms the Medicine Bow Mountains. To the W, but separated by the North Park Basin, is the Park Range uplift and the White River uplift. To the S of the Front Range and en echelon is the Sawatch uplift, continuing S from Colorado across into New Mexico, again en echelon, as the Sangre de Cristo Ranges. To the SW there is an exceptional feature, a broad swell, the San Juan uplift, which is probably a beveled Laramide uplift largely overlain by volcanic plateau deposits. Although most of the Northern and Middle Rockies drain W to the Pacific, two-thirds of the Southern Rockies drain E to the Gulf of Mexico (the North and South Platte, Arkansas, and Rio Grande rivers; the Colorado drains W).

As elsewhere in this province, the basement rocks are high-grade Precambrian metamorphics and granites dating around 1100–1300 m yr,

FIGURE 4. Structural features of the Southern Rockies (adapted from Eardley, 1968). Late Cretaceous and early Tertiary uplifts are indicated. Stippled areas are Precambrian rock exposures; ruled areas of the shelf (eastern Utah and Colorado) are exposures of Paleozoic rocks and are marked by: IP, mostly Pennsylvanian; P, Pennsylvanian and Permian. Ruled areas of the miogeosyncline (western Utah) are Cambrian exposures. K, major detached slide masses; KC, the Canyon Range slide mass; OR, Oquirrh Range; CU, Cottonwood uplift; M, Mesozoic sedimentary rocks; NFL, Newfoundland anticline; SBY, Stansbury Mountains. Both fold axes and thrusts faults in the Rocky Mountain Geosyncline are shown by bold lines.

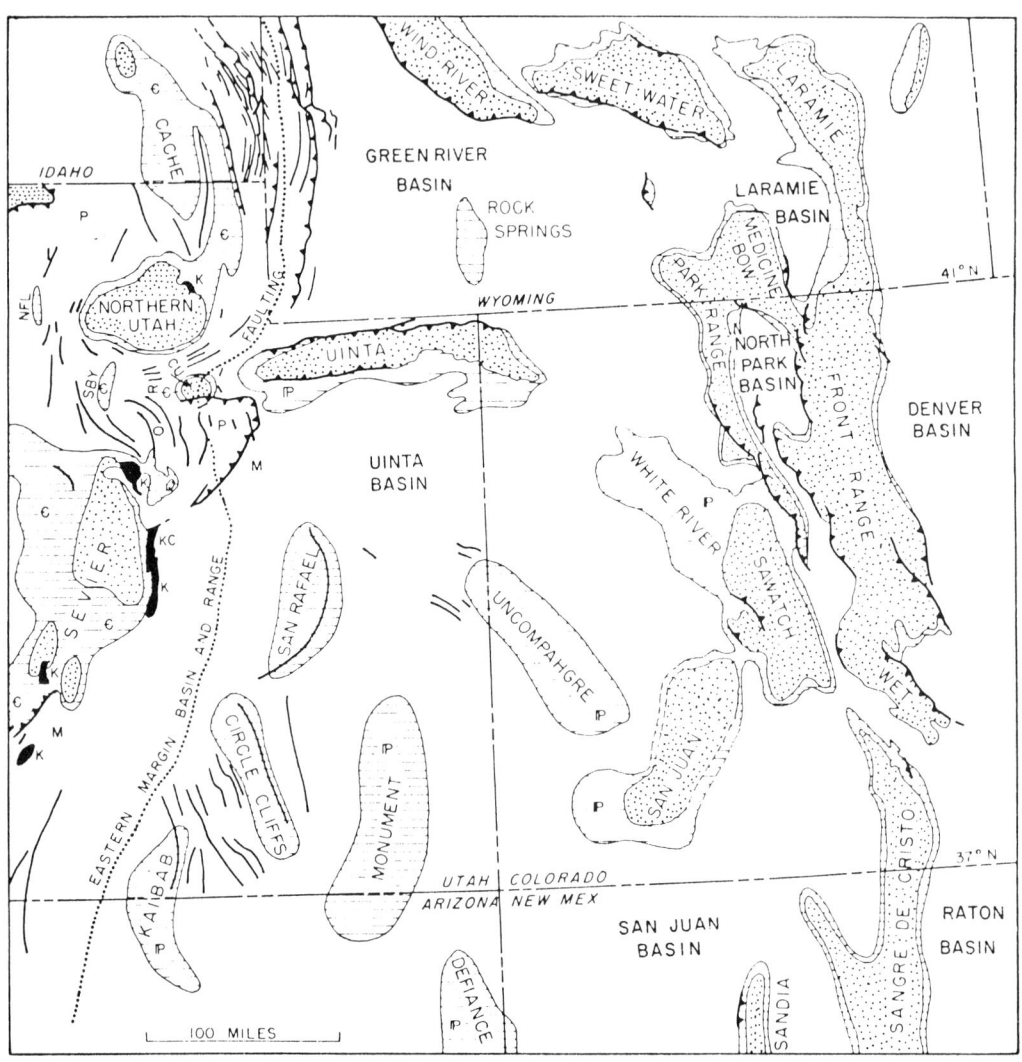

back to 2000 m yr and trending mainly along NE-SW lines. There is an interesting anorthosite E of Laramie; like the one in the Adirondacks it contains titaniferous magmatite pods. Less altered metasediments in the Medicine Bow Mountains contain magnificent specimens of stromatolitic algae.

Most of the Southern Rockies have Precambrian nuclei that have been progressively upwarped in broad anticlines throughout Paleozoic and Mesozoic times (Fig. 4). The vertical movements set in again during the Late Cretaceous and early Paleocene time, along new structural lines, with the *Laramide orogeny* (conventionally given as 65 m yr). The term can be misleading in some ways because the last true orogeny involving orthogeosynclinal sediments in the type area was in the middle Precambrian. In Laramide time there was extensive high-angle reverse faulting of a "fleur-de-lys" style, but most of the uplifts of the Phanerozoic were essentially epeirogenic and taphro-

FIGURE 5. Profile E-W across the Rocky Mountain Front Range to the Denver Basin (after Eardley, 1968, from the studies of Boos, Odiorne, and Evans).

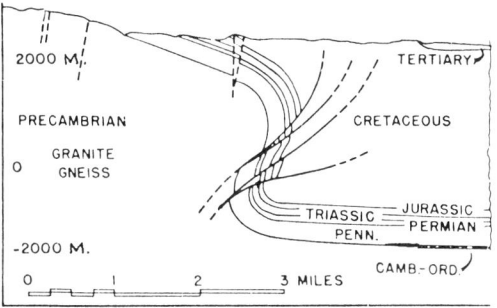

genic (block-faulted). By convention, nevertheless, the Laramide event has become an internationally recognized label for Late Cretaceous–early Tertiary diastrophism. It is commonly used in a rather loose way. In contrast to the Southern Rockies, the northern and western part of the Middle Rockies did indeed experience at this time an "alpinotype" orogeny with overthrusts, axial mobilization, and granite injection. Eardley (1963) has pointed out that in the southern Rocky Mountains the largest uplifts show border faults that droop over to the exterior and are thus gravitationally not compressionally, conditioned (Figs. 5 and 6). He postulated that they were raised by subcrustal differentiation.

In the Southern Rockies the Paleozoic-Mesozoic sequences (partly paraliageosynclinal or zeugogeosynclinal in the Kay terminology) were mainly marine and in midbasin sections several thousand meters of clastics and carbonates accumulated. Cambrian and Silurian are frequently missing, and commonly the younger Paleozoic rests directly on the Precambrian.

As in the northerly basins, the facies in the Southern Rockies changed to continental with the beginning of the Tertiary. Warm, humid climates led to deep chemical weathering and general reduction of relief, almost burying the giant anticlinal uplifts in their own debris. The Eocene Wasatch Formation, for example, is nearly 600 m thick, and in places there are over 2000 m of Tertiary basin fill, mainly of alluvial or lacustrine type.

In south-central Colorado there was an extensive development of Tertiary-Quaternary volcanics, the San Juan volcanic field being analogous in their way to the Absaroka field in the N. Smaller eruptive centers occur also in many areas, including the giant Valles Caldera in New Mexico. Extensive dike swarms are found in the Spanish Peaks area, in the Front Range, and in the Park Range. In the center of the Colorado ranges there are numerous stocks and larger intrusions such as the Mt. Princeton pluton.

Progressive regional uplift has brought about the exhumation of many of the ranges in the Southern Rockies, and as in the Central Rockies, there are numbers of superposed

FIGURE 6. Profiles across the Uinta and Wind River mountains (Middle Rockies) showing the nature of the border faults (Eardley, 1968).

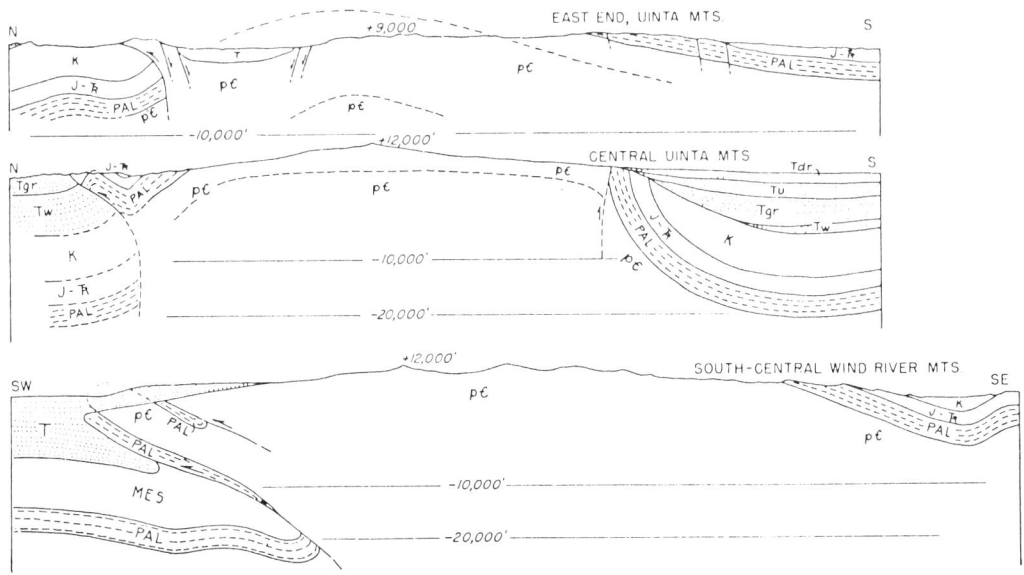

gorges, e.g., the Laramie and the Arkansas rivers in the Front Range and the upper Colorado in the Park Range. The Quaternary saw the development of many cirque glaciers and even large valley glaciers in the northern and higher mountains. Very widespread periglacial features developed. The vast talus streams and "rock glaciers" of the San Juan Mountains are classic examples.

Stratigraphic Summary

The oldest rocks in the Rocky Mountains are traditionally classified as *Archean*. Best known are those in the Beartooth and Bighorn mountains of Montana and Wyoming. Here there is evidence of a supracrustal sedimentary sequence that has gone through intense metamorphic recrystallization, and subsequently folded in several phases of decreasing intensity. It has successively been invaded by granites, the oldest showing metamorphism, the younger ones more mobility but less alteration. Radiogenic dates of the granite gneisses and pegmatites indicate the main metamorphism at 2.7 yr, for which the name *Beartooth orogeny* has been proposed (Skinner et al., 1969). The trend is NE-SW and comparable rocks occur in the Black Hills and in Manitoba, along the same orientation.

Middle Precambrian geosynclinal belts (1300–1800 m yr) probably crossed the area of the Middle Rocky Mountains of Colorado in a similar NE-SW trend, apparently continuing those of the SW of the Canadian Shield. In Wyoming the trend swings toward to the NW (Blackstone, 1963) and back to NE in the northwestern part of the state.

In eastern Idaho and western Montana a younger Precambrian miogeosyncline preserves a great thickness of shales (*Belt Supergroup*) that are in part fossiliferous, containing extremely well-preserved algae. The radiogenic dates suggest 1000–1100 m yr for the Beltian (Missoula Group). To the S there may be an analogous sedimentary sequence in the Uinta Mountain Group of Utah. Neither develop high-grade metamorphism. In most places in the province the Precambrian is truncated by a peneplaned surface ("Lipalian").

Cambrian seas extended in a broad shallow epicontinental to miogeosynclinal environment that deepened and thickened to the W (the *Cordilleran Geosyncline*). Strictly speaking, this is the "Meso-cordilleran Geosyncline," because to the W lay another, the Cordilleran Eugeosyncline. The geosynclinal boundary follows a more or less N-S trend, the "*Wasatch Line*," to the E of which lay the "Utah-Wyoming Shelf"

(Roberts et al., 1965). During the Paleozoic some 15,000 m of sediments filled this trough, detritals coming from both E and W. To the E of the shelf margin a comparable time span is represented by less than 5000 m (Armstrong and Oriel, 1965). There is a progressive overlap to the E of Cambrian shoreline facies (e.g., Sawatch Sandstone) through Lower Ordovician from the meridian of eastern Idaho and Utah to Colorado and the Dakotas. Thus nearshore clastics of early Cambrian age rest unconformably on the Precambrian basement in the W, and this facies shifts diachronously (Fig. 7) to late Cambrian in the E, with shales and carbonates succeeding them in the W (Lochman-Balk, 1956). In late Cambrian a southeastern sea may have joined the western one over central Colorado. The Lower Ordovician follows it concordantly in most areas with a widespread carbonate facies, thus terminating the first of the great Phanerozoic megacycles (Fig. 8), the *Sauk Sequence* (of Sloss, 1963).

The late Lower and Middle Ordovician initiates the *Tippecanoe Sequence*, which continued till late Silurian. Rocks of this sequence are unconformable in most places on Lower Ordovician to Precambrian. Basal clastics are followed by epicontinental limestones and sandstones in progressive overlap rather like that of the Cambrian. Although much of the record has been eroded away in a belt from northeastern Arizona to Wyoming, a few outliers remain to indicate continuous sedimentation during the transgression maxima. Of outstanding interest is the Harding Sandstone, which is rich in conodonts and also the first fish, the agnathids. The Fremont Limestone is marked by corals, brachiopods, and large mollusca.

Silurian was probably deposited without a

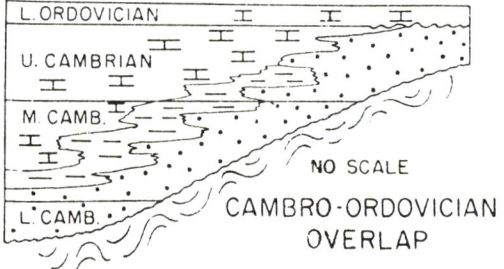

FIGURE 7. Diagrammatic cross section of the Cambrian-Ordovician overlap from west to east. Basal sands (dots) pass westward into shales (dashes) and limestones (blocks). From Haun and Kent (1965). Lithofacies are characteristically diachronous.

FIGURE 8. Megacycles of the North American craton with principal Rocky Mountain facies (from Haun and Kent, 1965). The sequence names are from Sloss (1963).

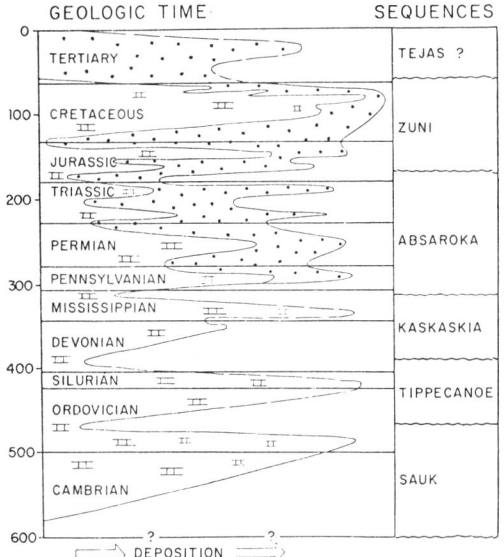

break over the Ordovician through most of the Rocky Mountain belt, but has suffered widespread erosion, so that the record is very spotty, mainly restricted to blocks in diatreme vents. The environment was that of a clear shallow sea over most of the region with mainly carbonate facies. Late Silurian regression was almost continent-wide (coinciding with the classical Caledonian climax, i.e., Ardennes phase, ca. 395 m yr).

The next megacycle, the *Kaskaskia Sequence,* was initiated after the Late Silurian disturbance and the Middle *Devonian* is transgressive over the extensively eroded craton. A *transcontinental arch* (Leverson, 1960), also known as *Souixia* in the midcontinent, was now a positive feature, extending in a NE-SW trend (the old Precambrian "grain") crossing eastern Colorado to New Mexico; Devonian sediments were thus restricted to the NW of it in the cratonic sector, mainly Wyoming, which was then an open epicontinental sea in continuity with the Cordilleran geosyncline along the western border of the province. Excellently preserved fossils in the Dyer Dolomite (brachiopods, bryozoa, and molluscs) are to be found in the White River Plateau area.

With the Lower Carboniferous (Mississippian) the marine sequence was renewed after a minor interruption on the craton (Fig. 9). To the W, however, in the Cordilleran geosyncline, there was a major orogenic disturbance at this time, the *Antler Orogeny.* During the course of the Mississippian the shelf–geosyncline boundary became shifted progressively to the E. Over the craton in some spots Mississippian rests unconformably on pre-Devonian rocks, and there seems to have been a general disconformity at its base. Minor uplift occurred in the region of the present Front Range and other elements of the "Ancestral Rockies," giving rise to littoral clastics in the basal Mississippian. The facies were carbonates for the most part (best known is the Leadville Limestone), but deepening of an E-W trough over Montana gave discontinuous access to the Williston Basin, where evaporites accumulated. There was a general regression in the Late Mississippian, accompanied by renewed uplift, expanding the Ancestral Front Range and Uncompahgre uplifts. Deep karst solution channels make the surface of the Mississippian in Wyoming and Colorado a classic paleogeomorphologic surface (with paleosols).

The next megacycle, the *Absaroka Sequence* (of Sloss), may have begun already in the latest

FIGURE 9. Mississippian isopach map of the Rocky Mountain region, indicating especially the Central Montana Trough, the Wyoming Shelf, and the Ancestral Rockies (Haun and Kent, 1965).

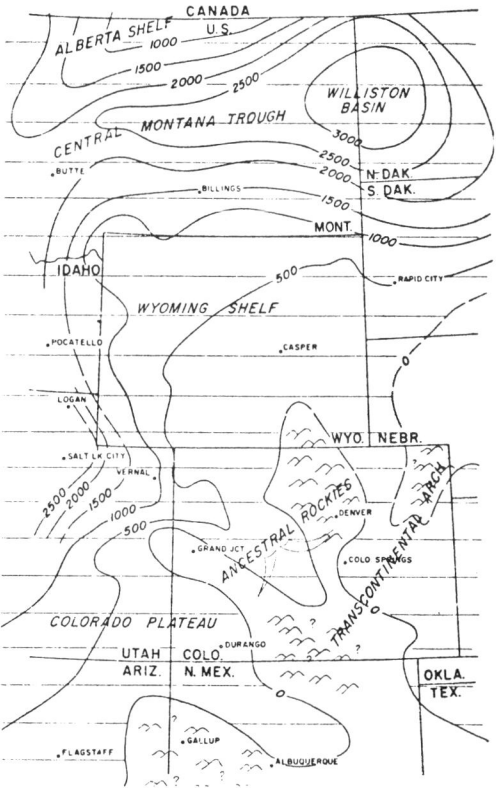

Mississippian. *Pennsylvanian* is again transgressive, in most places over an important unconformity. The Ancestral Front Range and Uncompahgre uplifts now shed vast quantities of clastics into the adjacent troughs. Along the eastern face of the Front Range, one of these formations, the Fountain Formation, forms spectacular "flatirons." Comparable formations in central and southwest Colorado are, respectively, the Minturn and the Hermosa. Here temporarily closed basins accumulated evaporites. To the NW, in central Wyoming and Idaho, there was still a fairly stable shelf environment marked by sandstones and carbonates.

The *Permian* boundary is generally a disconformity in Wyoming and the Cordilleran geosyncline belt expanded along an arc, bulging to the SE through Idaho–western Wyoming–northern Utah (Roberts et al., 1965). Dark shales (mudstones), phosphorites, and cherts accumulated (Park City Group, including Phosphoria Formation), and in eastern Wyoming pass to red beds and thin limestones (Gooseegg Formation, formerly Embar, etc.).

The Ancestral Rockies continued their positive movement in the Permian, shedding clastics so that much of Colorado was now emergent. Along the southwestern margin of the Uncompahgre Uplift, the accumulated relief permitted the accumulation of 5000 m of Permian and Pennsylvanian sediments. The coarse clastics were often arkosic, suggestive of the semiarid climates and the rapidity of the sediment burial. The borders of the uplift to the W and N were desertic with coastal barriers and lagoons. Dunes and offshore bars, spectacularly crossbedded and slumped in places, may be seen in the Lyons Sandstone. On the shelf, sedimentation was mainly carbonates and shale passing SE into evaporite and red-bed facies in cyclic oscillations. The red-bed environment suggests mudflat deposition during alternating wet and dry intervals (e.g., Maroon Formation, near Aspen). Primitive reptile tracks are found.

The Mesozoic transition is not marked by important events in the Rocky Mountain region, and the passage of the Permian into the Triassic in the Cordilleran miogeosynclinal belt is without a break. However, orogenic uplift developed on the western side of the trough in the late Triassic (an important worldwide event: the Palatine phase in Europe, the Akiyoshi phase in Japan), and further, in the Permian. As a result the miogeosynclinal phase now came to an end and the trough became gradually broken up into exogeosynclinal basins. Elsewhere on the cratonic region there is generally a disconformity, and there is a progressive onlap of Triassic from W to E onto the borders of the Ancestral Rockies—the Uncompahgre Uplift and northern Colorado Rockies being inundated by late Triassic, whereas the Front Range was greatly constricted. The facies, similar to those of the Permian, but more so, became dominated by red beds (red to pink, fine-grained sandstones and shales) with interfingering evaporites. A Sahara-type borderland is suggested, of coastal plains, mud flats, dunes, and deltas (e.g., the Chugwater Formation of Wyoming). The beds are almost totally devoid of fossils. Thin tongues of limestones, such as the Alcova in Wyoming, reached far to the E.

At the beginning of the Jurassic the Absaroka cycle was completed and the *Zuni Sequence* began. In the geosynclinal belt there was a progressive encroachment E onto the former shelf province. On the craton itself there was a progressive onlap from the W as the higher units transgressed the Triassic and older rocks. Most of Colorado remained emergent except for the NW and SW. Continued subsidence in the Cordilleran belt permitted a thick accumulation of limestone, shale, and coarser clastics in the Twin Creek. To the E there were still red beds as represented in the Sundance with shallow marine tongues (of boreal origin). This Middle to Late Jurassic Arctic gulf is sometimes called the "Sundance Sea." Nearshore bars and coastal plain dunes are represented by the Najavo Sandstone in the southwestern part of the province. The Entrada Sandstone in the northwestern part is a similar beach and dune sequence. In Late Jurassic the nonmarine mottled and variegated shales and sands of the Morrison became widespread. The Morrison marks a change to warm-humid climates, in contrast to the hot-dry pictures of the early Mesozoic. It is particularly celebrated for its fossil dinosaur deposits (including Como Bluff in Wyoming and Dinosaur National Monument in eastern Utah and northwestern Colorado).

Cretaceous is marked by a further stage of boreal invasions, with Arctic faunas, and there is generally an onlap of the Dakota Group from N to S. Eventually at the stage of the Skull Creek Shale (Santonian–Campanian), the Arctic seaway crossed the Transcontinental Arch to unite with the warm southern waters from the Mexican geosyncline. Late Jurassic and Early Cretaceous movements along the Cordilleran belt (Nevadan orogenic phase) produced extensive uplifts, and clastics were now shed from the W into narrow restricted basins. A notable recipient was the Green River Basin, which overlaps the Idaho–Wyoming border, accumulating 7000 m of mainly coarse clastics, a classical postorogenic molasse (Fig. 10), which interfinger with immense lacustrine oil shales of great economic potential.

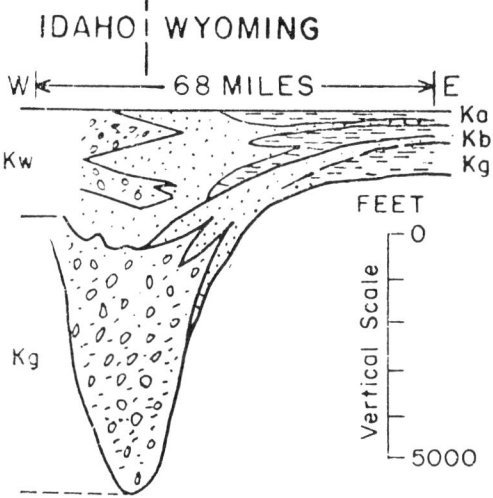

FIGURE 10. Diagrammatic cross section of the western part of the Middle Rockies, showing a classic example of molasse facies dumped into the western Green River Basin during the Early Cretaceous from the Idaho-Wyoming Thrust Belt, the newly emerged Wyomide (Nevadan) sector of the Cordilleran (orogenic belt) (Armstrong and Oriel, 1965). Abbreviations: Kw, Wayan Formation; Kg, Gannett Group; Ka, Aspen Formation; Kbr, Bear River Formation.

Important paleogeographic changes were now in progress and the entire Cordilleran belt became emergent, progressively raising also the western half of the province. Coastal plain and deltaic environments spread gradually eastwards, e.g., the fluvial Dakota Sandstone (in distinctive cyclical channel sequences), followed by the mostly continental Mesaverde Formation. Deepening sea in the late Cretaceous (Coniacian) brought in the Niobrara Limestone deposits, a characteristically white chalky facies, representing the contemporary worldwide population explosion of the calcareous pelagic marine nannoplankton (coccoliths and foraminifera). It passes up into the Pierre Shale (Santonian-Campanian), 1000 m or more, which contains concretions with exceptionally fine ammonites. The terminal retreat was marked in places by coal swamps, e.g., in the Laramie Formation.

During the Cretaceous evidence of the *Laramide Orogeny* became progressively more widespread, to reach a climax around the transition into the Tertiary (ca. 65 m yr). As indicated earlier, in its type area (the Colorado-Wyoming cratonic shelf), it was a "germanotype" disturbance, marked by block movements and not by general crustal mobilization or near-surface plate boundary effects. Along the miogeosynclinal front, from northern Montana to the Wasatch Line area of Utah, there was major thrust faulting and "alpinotype" structures, including (in Montana) nappe development and gravity slides. Along the Wasatch Line, the Idaho-Wyoming Thrust Belt about ten important imbricate slices have been mapped, with individual displacements (on planes dipping 15°W) of 15–20 km (Armstrong and Oriel, 1965). Some of the faults began moving from W to E in the earliest Cretaceous and some carried on into the Eocene. The degree to which giant low-angle nappes have participated in this is still somewhat controversial (Roberts et al., 1965). In the Rocky Mountain Province, its type area, the Laramide movement *sensu stricto,* is arbitrarily taken as having lasted from mid-Maestrichtian to late Eocene (Haun and Kent, 1965). An earlier movement, termed the *Cedar Hills Phase* (Eardley, 1962) in Utah occurred in the early Upper Cretaceous, and there were other minor phases. Each succeeding pulse shifted to a more easterly locus, which strongly suggests deep-seated subduction from the W.

With the termination of the Mesozoic came the end of the Zuni sequence as well as the *Tertiary* seas and also the beginning of the last or *Tejas Sequence.* The paleogeography of the Rocky Mountain province was now completely altered. Structural relief along the tectonized western margin of Middle Rockies and the eastern margin of the Northern Rockies developed to 5000 m or more, accompanied by massive block faulting, which spread generally over the province giving rise to immense volumes of postorogenic molasse-type clastics (Fig. 11). These piedmont facies spread out over the area and to the E of the Rockies (see *Great Plains Province*) and in scattered basins throughout the province. In central Colorado these basins are known as the Parks (North, South, and Middle). Up to 3000 m of sediment accumulated in the westernmost of these troughs. The depressions were partly fluvial and partly lacustrine. The extensive lake beds—for example, in the Green River Basin (Eocene)—accumulated potentially valuable oil shale deposits. Extensive volcanism developed, notably in two areas, the Absaroka (Yellowstone) field in the N and the San Juan volcanic field in the S (Fig. 12). Ash showers buried lake organisms and adjacent forests (including redwoods, poplar, and pine) leaving spectacular silicified fossils—e.g., in the Florissant Fossil Beds, National Monument, W of Colorado Springs. During late Cenozoic times, there has been progressive uplift and the bulk of sedimentation has shifted to the Great Plains province, whereas the upwarped Tertiary erosion surfaces, now preserved mainly as summit accordances, range up to 4300 m elevation.

Quaternary uplift of the Rocky Mountain

FIGURE 11. Early Tertiary basins in the Rocky Mountain Province (Haun and Kent, 1965). Heavy lines indicate the zero isopach for the Paleocene plus Eocene sediments. Isopach contour interval is 5000 ft. Eocene oil shales are shown within the dashed lines and high-grade oil shales are diagonally shaded. Basin abbreviations: Williston (WI), Bull Mountains (B), Crazy Mountains (CM), Powder River (PR), Big Horn (BH), Wind River (WR), Great Divide (GD), Green River (GR), Washakie (WA), Hanna (H), Laramie (L), Uinta (U), Piceance (P), North Park-Middle Park (NP), Denver (D), Raton (R), and San Juan (SJ).

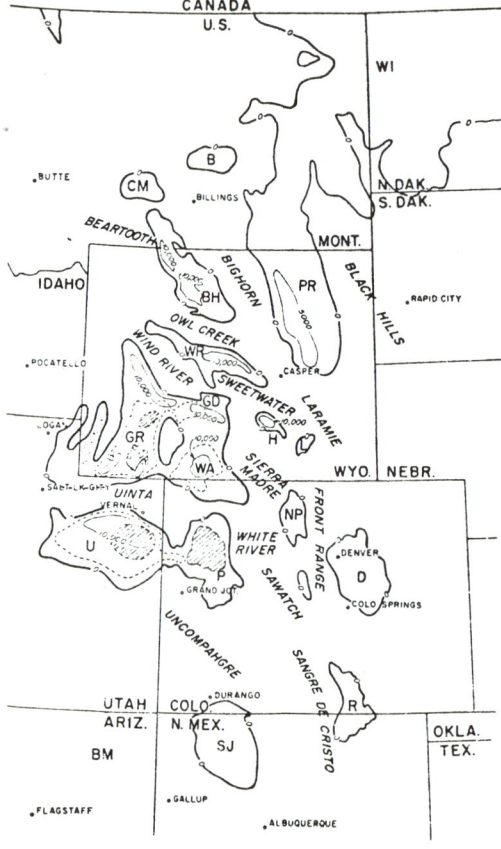

ments, and other data give evidence of multiple climatic oscillations.

Traces of ancestral man have not yet been identified prior to late Pleistocene. A famous spot, the Lindenmeier site, N of Fort Collins, Colorado, found in 1924, furnished a Folsom complex of arrowheads, scrapers, etc., in a large midden buried in an alluvial bank some 4 m below the surface. The site was comparable with the type example near Folsom, New Mexico, where the association with extinct bison (*Bison taylori*) proved its ambiguity. Elsewhere such points were found along with musk ox, mammoth, ground sloth, and other "ice age" mammals (Roberts, 1935). Radiocarbon dating indicates the range of this "Paleo-Indian" culture at around 11,000 to 8500 *B.P.* What makes the site of particular interest to Quaternary geology is its relatively high elevation, evidence of the very rapid amelioration of climate just before the Pleistocene–Holocene boundary (ca. 10,000 years B.P.). During the "hypsithermal"

FIGURE 12. Cenozoic igneous rocks: major intrusions (solid black), dikes (lines), and extrusives (Vs). Some of these areas became active during the Late Cretaceous. (From Haun and Kent, 1965.)

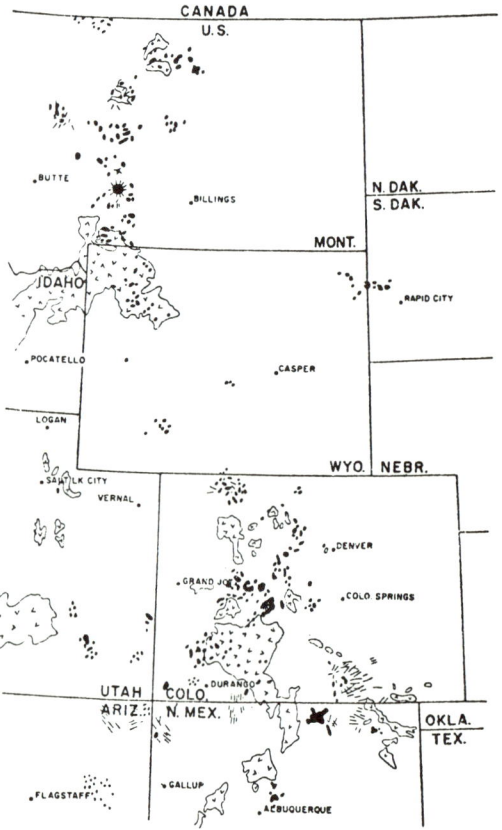

province led to deep dissection, further modified by the effects of repeated glaciation. In the northernmost parts of Idaho and northwestern Montana, the continuous cover of the Cordilleran ice sheet effectively buried even the highest mountains, but S of the Canadian border most of the glaciers were of the piedmont type, and to the SE of the Snake River plains only mountain glaciers were active. In the stratigraphic record only the last two glaciations are well documented by moraines, outwash and terraces, but scattered traces, incised pedi-

stage of the Holocene the Paleoindians were able to live in the high mountain valleys of Wyoming.

Seismicity

Late Cenozoic movements of faults, partly associated with volcanicity extending into the late Quaternary are widely in evidence in this province. A basalt dated as less than 27,000 B.P. is cut by faults along the Wasatch Line in southeastern Idaho. Neotectonics are disclosed also by tilted and dislocated youthful terrace deposits and erosion surfaces.

Earthquake shocks of low to moderate intensity are widespread within the province, but are not of high frequency. One severe shock, however, was recorded on June 27, 1925, E of Helena, Montana, triggering a major rockslide that blocked one of the main railroads for some weeks; the shock was felt over 750,000 km^2. The West Yellowstone quake of August 17, 1959, created a 6-m scarp, triggered landslides in the Madison Gorge that dammed the river and buried an occupied campground.

Mineral Deposits

The Rocky Mountain province, with its extensive Precambrian nuclear ranges, repeatedly reactivated movements, and numerous Phanerozoic volcanic and plutonic centers, contains many interesting minerals, including some of the world's great mineral deposits. Virginia City, Montana, was the scene of a gold rush in the 1850s. So was Colorado. Butte, Montana, is the site of one of the greatest copper concentrations and Climax, Colorado, is the world's biggest underground mine with a low-grade molybdenite deposit producing over $100 million of molybdenum annually.

In the San Juan Mountains there are numerous mineralizations of silver, lead, zinc, as well as copper and gold. First discoveries of gold were in placers leading to the Colorado gold rush period that began about 1859, but these were soon worked out. In a diagonal NE-SW transverse zone (from the San Juan Mountains to the northern part of the Front Range), the Colorado "Mineral Belt," a 70-km-wide shear zone, there are numerous gold and silver-lead mineralizations. Such centers as Central City, Leadville, Ouray, Telluride, and Silverton have long been famous but have greatly declined in recent decades. Cripple Creek (near Pikes Peak), discovered in 1890, alone produced 19 million oz of gold and 2 million oz of silver. The Schwartzwalder Mine, NW of Golden, is the principal source of pitchblende in the region today.

The deeper basins of Wyoming have furnished some prolific oil and gas fields, mainly from Cretaceous rocks, whereas the younger freshwater formations, notably of Green River (Eocene) age, contain immense oil shale deposits mainly in northwestern Colorado. Late Cretaceous and Paleocene soft coal and lignite are found extensively in Wyoming. The narrow structures of the Northern and Southern Rockies have been less fortunate with respect to sources of fossil fuels. In contrast, however, their metallic mineral resources are correspondingly large.

Uranium deposits in the Wyoming Basin make that state second only to New Mexico in U.S. reserves of radioactive fuels. According to some authorities, the uranium-rich solutions were leached downward from a Tertiary volcanic cover and accumulated in formations of favorable porosity.

There are several interesting areas of appreciable geothermal energy potential. Colorado alone possesses 113 thermal springs ranging from 21° to 84°C.

It is interesting, from the genetic point of view, that the really great mines are situated at key points in the geotectonic framework. The Leadville–Central City–Climax cluster in the "Colorado Mineral Belt" coincides with the intersection of the old Precambrian NE-SW grain of newer (Laramide) block faulting in NNW-SSE trends. Bingham (copper, lead, zinc) and Park City (lead, zinc, silver) in central Utah mark the intersection of the N-S Wasatch line and the E-W Uinta axis. Butte, Montana (copper, silver), lies near the junction of the Wasatch Line and a WNW-ESE shear zone.

RHODES W. FAIRBRIDGE

References

Alpha, A. G., 1958. "Tectonic history of Montana," in *Beartooth Uplift and Sunlight Basin Guidebook, Billings Geol. Soc.,* 9th Ann. Field Conf. 57–68.

Atwood, W. W., and Atwood, W. W., Jr., 1938. "Working hypothesis for the physiographic history of the Rocky Mountain region," *Bull. Geol. Soc. Am.,* 49(6), 957–980.

Armstrong, F. C., and Oriel, S. S., 1965. "Tectonic development of Idaho-Wyoming thrust belt," *Am. Assoc. Petrol. Geologists Bull.,* 49(11), 1847–1866.

Berg, R. R., 1960. "Cambrian and Ordovician history of Colorado," in *Guide to the Geology of Colorado,* Geol. Soc. Am., Rocky Mtn. Assoc. Geol., Colo. Sci. Soc., 10–17.

Blackstone, D. L., Jr., 1963. "Development of geologic structure in central Rocky Mountains," *Am. Assoc. Petrol. Geologists Mem.,* 2, 160–179.

_____, 1971. "Travelers guide to the geology of Wyoming," *Wyoming Geol. Surv.,* 90p.
Bradley, W. H., 1970. "Green River oil shale–concept of origin extended," *Bull. Geol. Soc. Am.,* **81,** 985–1000.
Christiansen, R. L., and Blank, H. R., Jr., 1972. "Volcanic stratigraphy of the Quaternary rhyolite plateau in Yellowstone National Park," *U.S. Geol. Surv. Prof. Pap. 729B,* 18p.
_____, and Lipman, P. W., 1972. "Cenozoic volcanism and plate-tectonic evolution of the western United States. II. Late Cenozoic," in Sutton, J., *et al.,* eds. "A discussion on volcanism and the structure of the earth," *Phil. Trans, Roy. Soc. Lond.,* **271A,** 249–285.
Chronic, J. H., 1972. "Prairie, peak and plateau: a guide to the geology of Colorado," *Colo. Geol. Sum. Bull.,* **32,** 126p.
Curry, W. H., 1971. "Summary of possible future peroleum potential, Region 4, Northern Rocky Mountains," in Future petroleum provinces of the United States; their geology and potential, Vol. 1. *Am Assoc. Pet. Geol. Mem.,* **15,** 538–546.
Deiss, C. F., 1935. "Cambrian-Algonkian unconformity in western Montana," *Bull. Geol. Soc. Am.,* **46,** 95–124.
_____, 1941. "Cambrian geography and sedimentation in the central Cordilleran region," *Bull. Geol. Soc. Am,* **52,** 205–262.
_____, 1943a. "Stratigraphy and structure of southwest Saypo quadrangle, Montana," *Bull. Geol. Soc. Am.,* **54**(2), 205–262.
_____, 1943b. "Structure of central part of Sawtooth Range, Montana," *Bull. Geol. Soc. Am,* **54**(8), 1123–1167.
Donnell, J. R., ed., 1960. "Geological road logs of Colorado," *Rocky Mt. Assoc. Geol.,* 90p.
Dort, W., Jr., 1962. "Glaciation of the Coeur d'Alene District, Idaho," *Bull. Geol. Soc. Am.,* **73,** 889–906.
Eardley, A. J., 1962. *Structural Geology of North America,* New York: Harper & Row, 2nd ed., 743p.
_____, 1963a. Structural evolution of Utah," *Utah Geol. Min. Surv. Bull.,* **54,** 19–29.
_____, 1963b. "Relation of uplifts to thrusts in Rocky Mountains," *Am. Assoc. Petrol. Geologists Mem.,* **2,** 209–219.
_____, 1968. "Major structures of the Rocky Mountains of Colorado and Utah," *J. Univ. Missouri,* V. H. McNutt Colloq., **1**(1), 79–99.
Engel, A. E. J., 1963. "Geologic evolution of North America," *Science,* **140,** 143–152.
Eugster, H. P., and Surdam, R. C., 1973. "Depositional environment of the Green River formation of Wyoming: a preliminary report," *Bull. Geol. Soc. Am.,* **84,** 1115–1120.
Fenneman, N. M., 1931. "Physiography of Western United States," New York: McGraw-Hill, 534p.
Gilluly, J., 1965. "Volcanism, tectonism and plutonism in the western United States," *Geol. Soc. Am Spec. Pap. 80,* 69p.
_____, 1970. "Crustal deformation in the Western United States," in Johnson, H., and Smith, B. L., eds., *The Megatectonics of Continents and Oceans.* New Brunswick, N. J.: Rutgers Univ. Press, 47–73.
Gussow, W. C., 1972. "The rate of mountain building," *24th Int. Geol. Congr.,* **3,** 355–362.

Harrison, J. E., 1972. "Precambrian belt basin of Northwestern United States: its geometry, sedimentation, and copper occurrences," *Bull. Geol. Soc. Am.,* **83,** 1215–1240.
Haun, J. D., and Kent, H. C., 1965. "Geologic history of Rocky Mountain Region," *Am. Assoc. Petrol. Geologists Bull.,* **49**(11), 1781–1800.
Hintze, L. F., 1959. "Ordovician regional relationships in north-central Utah and adjacent areas," in *Guidebook to the Geology of the Wasatch and Uinta Mountains Transition Area,* Intermtn. Assoc. Geol. 10th Ann. Field Conf., 46–53.
Hubert, J. F., Butera, J. G., and Rice, R. F., 1972. "Sedimentology of Upper Cretaceous Cody-Parkman Delta, southwestern Powder River Basin, Wyoming," *Bull. Geol. Soc. Am.,* **83,** 1649–1670.
Hunt, C. B., 1967. *Physiography of the United States.* San Francisco: W. H. Freeman, 480p.
King, P. B., 1959. *The Evolution of North America.* Princeton: Princeton Univ. Press, 190p.
Kummel, B., 1970. *History of the Earth,* 2nd ed. San Francisco: W. H. Freeman, 707p.
Leverson, A. I., 1960. *Paleogeologic Maps.* San Francisco: W. H. Freeman, 174p.
Lipman, P. W., Prostka, H. J., and Christiansen, R. L., 1972. "Cenozoic volcanism and plate-tectonic evolution of the Western United States. I. Early and Middle Cenozoic," in Sutton, J., *et al.,* eds., "A discussion on volcanism and the structure of the earth," *Phil. Trans. Roy. Soc. Lond.,* **271A,** 217–249.
Lochman-Balk, C., 1956. "The Cambrian of the Rocky Mountains and southwest deserts of the United States and adjoining Sonora Province, Mexico," in *El Sistema Cambrico: su Paleogeografia y el Problema de su Base,* Symposium, pt. II, 20th Internatl. Geol. Congr., Mexico, 529–661.
Love, J. D., 1954. "Periods of folding and faulting during late Cretaceous and Tertiary time in Wyoming," *Bull. Am. Assoc. Petrol. Geologists,* **38,** 1311–1312.
Lovering, T. S., and Goddard, E. N., 1950. "Geology and ore deposits of the Front Range, Colorado," *U.S. Geol. Surv. Prof. Pap. 223.,* 319p.
Mackin, J. H., 1947. "Altitude and local relief in the Bighorn area during the Cenozoic," *Wyo. Geol. Assoc. Field Conf.,* Bighorn Basin Guidebook, 103–120.
MacLachlan, J. C., 1961. "Cambrian, Ordovician and Devonian systems, eastern Colorado subsurface," in *Lower and Middle Paleozoic Rocks of Colorado,* Rocky Mtn. Assoc. Geol., 41–52.
McKee, E. D., 1956. "Paleotectonic maps of the Jurassic system," *U.S. Geol. Surv. Misc. Geol. Inv. Map I-175.*
_____, *et al.,* 1959. "Paleotectonic maps of the Triassic system," *U.S. Geol. Surv. Misc. Geol. Inv. Map I-300.*
_____, *et al.,* 1967a. "Paleotectonic maps of the Permian system," *U.S. Geol. Surv. Misc. Geol. Inv. Map I-450,* 164p.
_____, 1967b. "Paleotectonic investigations of the Permian system in the United States," *U.S. Geol. Surv. Prof. Pap. 515,* 271p.
Mansfield, G. R., 1927. "Geography, geology, and mineral resources of part of southeastern Idaho," *U.S. Geol. Surv. Prof. Pap. 152,* 409p.

Mudge, M. R., 1970. "Origin of the disturbed belt in northwestern Montana," *Geol. Soc. Am. Bull.,* **81,** 377–392. (Also: *ibid.,* **82,** 1139–1140.)

Muehlberger, W. R., 1965. "Late Paleozoic movement along the Texas lineament," *Trans. N.Y. Acad. Sci.,* **27**(4), 385–392.

Noble, J. A., 1970. "Metal provinces of the western United States," *Bull. Geol. Soc. Am.,* **81,** 1607–1624.

Perry, E. S., 1962. "Montana in the geologic past," *Mont. Bur. Mines Geol. Bull.,* **26,** 78p.

Peterson, J. A., 1957. "Marine Jurassic of northern Rocky Mountains and Williston basin," *Am. Assoc. Petrol. Geologists Bull.,* **41,** 399–440.

Pierce, W. G., 1957. "Heart Mountain and South Fork detachment thrusts of Wyoming," *Am. Assoc. Petrol. Geologists Bull.,* **41,** 591–626.

Price, R. A., 1971. "Gravitational sliding and the foreland thrust and fold belt of the North American Cordillera: discussion," *Bull. Geol. Soc. Am.,* **82,** 1133–1138 (see also: Mudge, M. R., 1970, *ibid.,* **81,** 377–392; reply: *ibid.,* **82,** 1139–1140).

Pride, D. E., and Hagner, A. F., 1972. "Geochemistry and origin of the Precambrian Iron Formation near Atlantic City, Fremont County, Wyoming," *Econ. Geol.,* **67,** *329*–338.

Prostka, H. J., 1973. "Hybrid origin of the absarokite-shoshonite-banakite series, Absaroka Volcanic Field, Wyoming," *Bull. Geol. Soc. Am.,* **84,** 697–702.

Reed, J. C., Jr., and Zartman, R. E., 1973. "Geochronology of Precambrian rocks of the Teton Range, Wyoming," *Bull. Geol. Soc. Am.,* **84,** 561–582.

Reeves, F., 1946. "Origin and mechanics of thrust faults adjacent to the Bearpaw Mountains, Montana," *Bull. Geol. Soc. Am.,* **57,** 1033–1047.

Richmond, G. M., 1965a. "Glaciation of the Rocky Mountains," in H. E. Wright and D. G. Frey, eds., *The Quaternary of the United States,* Princeton, N.J.: Princeton Univ. Press, 217–230.

———, 1965b. "Northern and middle Rocky Mountains," *Neb. Acad. Sci.,* INQUA Guidebook E, 129p.

Ridge, J. D., ed., 1968. "Ore deposits of the United States, 1933–1967," New York: Am. Inst. Min. Met. Pet. Eng., 2 vols., 1880p.

Rigby, J. K., 1963. "Devonian system of Utah," *Utah Geol. Min. Surv. Bull.,* **54,** 75–88.

Roberts, F. H. H., Jr., 1935. "A Folsom complex, preliminary report on investigations at the Lindenmeier site in northern Colorado," *Smithsonian Misc. Coll.,* **94**(4), 35p.

Roberts, R. J., *et al.,* 1965. "Pennsylvanian and Permian basins in northwestern Utah, northeastern Nevada and south-central Idaho," *Am. Assoc. Petrol. Geologists Bull.,* **49**(11), 1926–1956.

Rocky Mountain Association of Geologists, 1972. "Geologic atlas of the Rocky Mountain region," *Rocky Mtn. Assoc. Geol.,* 331p.

Roeder, D. H., 1967. "Rocky Mountains," Berlin: Gebr. Borntraeger, **5,** 318p.

Ross, C. P., 1956. "Belt series in relation to the problems of the base of the Cambrian system," in *El Sistema Cambrico: su Paleogeografla y el Problema de su Base,* Symposium, pt. II, 20th Internatl. Geol. Congr., Mexico, 683–699.

———, 1970. "The Precambrian of the United States of America: northwestern United States–the Belt Series," in Rankama, K., ed., *The Precambrian,* vol. 4. New York: Wiley-Interscience, 145–252.

Ross, S. H., and Savage, C. N., 1967. "Idaho earth science," *Idaho Bur. Min. Geol.,* 271p.

Rothrock, D. P., 1960. "Devonian and Mississippian systems in Colorado," in *Guide to the Geology of Colorado.* Denver: Rocky Mtn. Assoc. Geol., 17–22.

Sheldon, R. P., 1963. "Physical stratigraphy and mineral resources of Permian rocks in western Wyoming," *U.S. Geol. Surv. Prof. Pap. 313-B,* 49–273.

Skinner, W. R., Bowes, D. R., and Khoury, S. G., 1969. "Polyphase deformation in the Archean basement complex, Beartooth Mountains, Montana and Wyoming," *Bull. Geol. Soc. Am.,* **80,** 1053–1060.

Sloss, L. L., 1963. "Sequences in the cratonic interior of North America," *Bull. Geol. Soc. Am.,* **74,** 93–114.

Smedes, H. W., and Prostka, H. J., 1972. "Stratigraphic framework of the Absaroka volcanic supergroup in the Yellowstone National Park region," *U.S. Geol. Surv. Prof. Pap. 729-C,* 33p.

Smith, R. B., and Sbar, M. L., 1974. "Contemporary tectonics and seismicity of the western United States with emphasis on the Intermountain Seismic Belt," *Bull. Geol. Soc. Am.,* **85,** 1205–1218.

Steiner, M. B., and Helsley, C. E., 1972. "Jurassic polar movement relative to North America," *J. Geophys. Res.,* **77**(26), 4981–4993.

Stewart, J. H., 1972. "Initial deposits in the Cordilleran Geosyncline: evidence of a late Precambrian (850 m.y.) continental separation," *Bull. Geol. Soc. Am.,* **83,** 1345–1360.

Stille, H., 1940. *Einführung in den Bau Amerikas.* Berlin: Gebr. Borntraeger, 717p.

Stokes, W. L., 1968. "Multiple parallel-truncation bedding planes–a feature of wind-deposited sandstone formations," *J. Sed. Pet.,* **38**(2), 510–515.

Surdam, R. C., Eugster, H. P., Mariner, R. H., 1972. "Magadi-type chert in Jurassic and Eocene to Pleistocene rocks, Wyoming," *Bull. Geol. Soc. Am.,* **83,** 2261–2266.

Taubeneck, W. H., 1971. "Idaho Batholith and its southern extension," *Bull. Geol. Soc. Am.,* **82,** 1899–1928.

Thomas, H. D., 1949. "The geological history and geological structure of Wyoming," *Wyo. Geol. Surv. Bull.,* **42,** 28p.

Thornbury, W. D., 1965. *Regional Geomorphology of the United States.* New York: John Wiley, 609p.

Unterman, G. E., and Unterman, B. R., 1954. "Geology of Dinosaur National Monument and vicinity, Utah-Colorado," *Utah Geol. Min. Surv. Bull.,* **42,** 228p.

Weimer, R. J., and Haun, J. D., 1960. "Guide to the geology of Colorado," Denver: Rocky Mtn. Assoc. Geol., 303p.

Wertz, J. B., 1971. "Les linéaments et failles majeures de l'ouest des États-Unis leur importance au point de vue minier," *Rev. Géogr. Phys. Géol. Dyn.,* **13**(4), 315–326.

Cross-references: *Canada–Rocky Mountains and Eastern Cordilleran Region; United States–Basin and Range Province, Colorado Plateau Province, Great Plains Province.*

URUGUAY

Uruguay, with an area of 177,508 km^2 (72,172 sq mi) is characterized for the most part by a low, rolling terrain (highest hill, Sierra de las Animas, 340 m). It is generally monotonous and grass-covered; the coastal sector alternates between rocky headlands and pocket sand beaches.

The geology of Uruguay is quite well known in its major features. The geological map, 1:500,000 (1957), is a good representation but the geological literature on Uruguay is somewhat meager. The papers by Sellow (1827), d'Orbigny (1842), and Darwin (1846) called attention to the importance of the Pleistocene continental deposits; MacMillan (1933) and Walther (1948) worked on Precambrian rocks; Mendez Alzola (1938) on Devonian sediments and fossils; Falconer (1937) on Upper Paleozoic; Delaney (1967) on Cenozoic deposits; Butler (1970) and Leyden et al. (1971) on the structure of the continental margin. General summaries have been provided by Lambert (1941), Harrington (1956), Caorsi and Goñi (1958), and Bossi (1966).

Stratigraphy

Precambrian. The oldest rocks of Uruguay are exposed in the S of the country, especially along the coast of the Río de la Plata between Punta del Este and Colonia, and S of the Río Negro. But they also crop out in a narrow band in the NE between Tacuarembó and Aceguá.

Although the stratigraphy of the Precambrian has not yet been worked out, in general, the older crystalline rocks are divided into Lower Precambrian amphibolites, gneisses, and migmatites; Middle Precambrian dolomites, phyllites, schists, and migmatites (Lavalleja series, former Minas series); and Upper Precambrian phyllites, quartzites, and associated migmatites (Piedras de Afilar series). Very few radiometric datings have been carried out.

Lower Precambrian. These rocks crop out along the coast of the Río de la Plata, between Piriápolis and Colonia, especially in and around coastal headlands. The predominant rocks are granitoid, augen and foliated gneisses, but mica schists are also frequent. Pegmatite and aplite dikes are numerous.

The hill that forms the "cerro" of Montevideo is composed of amphibolite schists, dipping almost vertically, which according to Walther (1948) were formed by metamorphism of dioritic gabbro. These rocks also occur on the headlands at Juan Lacaza, Punta Artillero, and the vicinity of Colonia. West of Montevideo, at Playa Colorada, Punta Espinillo, and Punta Martín Chico, pink-colored migmatites crop out.

MacMillan (1933) has distinguished two generations of early Precambrian gneisses, the older rocks, with a NE-SW schistosity and dipping SE, and the younger with subvertical schistosity and striking E-W to ENE.

The metamorphic rocks are highly folded and contorted in a general NE direction, parallel to the Atlantic coast (NE-SW), with frequent local deviations roughly parallel to the Río de la Plata (NW-SE).

The predominant igneous rocks are granite (mainly biotite granite) and granodiorite, with ages that range between 2000 and 1700 m yr (Umpierre and Halpern, 1971), which correspond to the Trans-Amazonian orogenic cycle (Hurley et al., 1967).

Middle Precambrian. The Lavalleja series (former Minas series of Uruguay), a metamorphic complex, crops out in a long and narrow band of N-S trend, between the northern end of the Cuchilla Grande in the N and the town of Piriápolis in the S, and two or three isolated outcrops further W, one of which is 60 km NE of Colonia. North of Piriápolis it rests with marked angular unconformity on the early

FIGURE 1. Diagrammatic N-S profile from Montevideo to Rivera (based mainly on Harrington, 1956): p€, Precambrian (crystalline basement); D, Devonian sediments; P, Permian (Tres Islas, Frayle Muerto, Mangrullo, Paso Aguiar, Yaguarí, Buena Vista formations); Pd, Permian basaltic sills and dikes; Trs, Triassic sediments (Tacuarembó Sandstone); Trb or Kl, Lower Cretaceous basalts and related volcanics (Arapey lavas); Ku, Upper Cretaceous sediments.

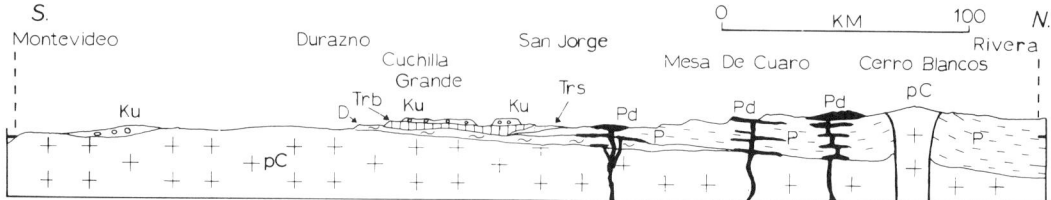

Precambrian gneisses. This unit is of undoubted sedimentary origin, consisting of whitish, massive quartzites almost devoid of stratification, dark gray to reddish slates, and phyllites grading into sericite schists, coarsely crystalline limestones, occasionally dolomitic, that occur as lenses up to 500 m wide and a few kilometers long, conglomerates with abundant roundstones of crystalline rocks and migmatites (augen-gneiss). These last crop out at Punta del Este, the most famous headland and tourist attraction on the Atlantic coast of Uruguay, where the weathered migmatites have been cut by numerous acidic dikes. These rocks show a strong lineation. Thick lava flows (rhyolites, trachytes) and volcanic breccias, together with talc and mica schists (La Pedrería and La Paloma) form intercalations in the metasedimentites. Scarce amygdaloidal basic rocks also occur.

This complex is highly folded and apparently attains a thickness of several thousand meters. The predominant strike varies between N and NE with a steep westerly dip. Rock fabric and joint analyses of these rocks show a superposition of the NE-SW over that of the NW-SE system (Delaney, 1967).

Upper Precambrian. The Lavalleja series is intruded by many "apotectonic" masses of granite (biotite granite, hornblende granite), granodiorite (Sauce batholith), with their train of pegmatites and aplites. Ten of these rocks have been dated radiometrically, with ages that range between 600 and 500 m yr (Umpierre and Halpern, 1971), which correspond to the Pan-American orogenic cycle (see article: *South America*).

In the NE of the Atlantic coastal plain younger Precambrian rocks occur, the Santa Teresa migmatites.

A somewhat younger complex, named "Lascano series" (former Aiguá series in part) by Caorsi and Goñi (1958), occurs in a wide belt of general NE trend between the towns of Lascano in the N and Minas in the S. They rest with sharp angular unconformity on the Lavalleja series. The dominant rocks are andesites and rhyolite porphyrys with scarce volcanic breccias. Highly altered amygdaloidal rocks are also occasionally observed.

The epimetamorphic Piedras de Afilar series crops out in the southern part of Canelones department, where it rests with angular unconformity on the older Precambrian biotite gneisses, granulites, and granites. The dominant rocks are silicified sandstones; arkoses, limestones, and mica schists also occur. The last two series are possibly also of late Precambrian age, although some authors believe them to be Lower Paleozoic.

In Piriápolis and its vicinity occur a series of alkali volcanics and hypabyssal rocks (arfvedsonite granite, nordmarkites, and porphyritic syenites) that trend NNW-SSE. Their age is not known, but since their trend is parallel with the La Plata tectonic direction and they are affected by faults of the NE-SW trending tectonic system (Delaney, 1967), they could possibly be of Upper Jurassic or Lower Cretaceous age.

Paleozoic. *Devonian.* Lower Devonian rocks are well developed in the central part of Uruguay where they form a narrow belt of general ENE trend, about 150 km long.

The sequence, which has been called "Durazno group," is about 280 m thick and begins with 150 m of unfossiliferous whitish arkose (Cerrezuelo formation), unconformably resting on Precambrian metamorphics. A basal conglomerate, not always present, grades upward into slightly coherent massive arkose which may display cross-lamination, the angular quartz grains of which are loosely cemented by a kaolinized feldspathic mass. This is succeeded by 95 m of gray-to-purple micaceous shales (Cordobés formation), occasionally somewhat silty or sandy, with thin hematitic intercalations. These last contain an abundant "austral" fauna of Lower Devonian fossils (about 70 species, Malvinokaffric province). The most characteristic species (identified by Mendez Alzola, 1938) are *Digonus noticus* (Clarke), *Australocoelia tourteloti* Boucout and Gill, *Lingula lepta* Clarke, *Derbyinia alta* Mendez Alzola, *Calmonia signifer* Clarke. The sequence ends with 35 m of reddish-violet laminated, shaly micaceous sandstone, bearing thin intercalations of calcareous sandstone (La Paloma formation) that have yielded two poorly preserved starfish.

The Lower Devonian sequence of Uruguay is similar to that of southern Brazil; the Cerrezuelo formation is the equivalent of the Furnas formation of Sao Paulo, and the Cordobés formation is equivalent to the Ponta Grossa beds.

Carboniferous. The San Gregorio formation crops out in west-central Uruguay up to the Brazilian border and corresponds to a glacial facies 250 m thick that rests unconformably on Devonian sediments or basement rocks. The formation consists of massive, poorly stratified tillite alternating irregularly with sandy tillite, well-stratified conglomerate, coarse-grained hard massive sandstone, banded shaly sandstone, and shales. Red, purplish, brown, and gray are the dominant colors. In some localities "roche mountonnée" surfaces carved on granite appear beneath the basal tillites. The glacial striae have dominant E-W direction. A marine intercalation is represented by bluish-to-yellowish plastic clays with phosphatic concretions that yield anaptychi, *Eoasianites* (*Glaphynites*) *rionegrensis* Closs of goniatites and orthocone cephalopods *Dolorthoceras chubutense* Closs (Closs, 1967), as well as fishes, arthropods, radiolarians, wood, and spores that indicate an Upper Carboniferous age. The glacial sediments are considered equivalent of the Itararé Formation of Brazil.

Permian. The Tres Islas formation, conformably succeeding the San Gregorio formation, consists of 100 m of gray, yellowish, brownish friable sandstones alternating with gray banded shales that occasionally include impure coal seams and a few conglomerates. The sandstones contain few and poorly preserved plant remains. The formation is assigned to the Lower Permian and is considered equivalent to the Río Bonito formation of southern Brazil.

Next come the sandy shales and sandstones of the Frayle Muerto formation. The main outcrops form a broad U-shaped band between the Yaguarón River in the E and Tacuarembó River in the W. The sequence, some 200 m thick, consists of gray to bluish-gray laminated and banded sandy shales and shaly sandstones. Thin layers and lenticular intercalations of impure gray limestone are quite frequent. The sandstones have furnished poorly preserved fish scales of *Acrolepis* and *Elomichthys* type and scarce plant remains have also been found. It is considered equivalent to the Palermo beds of Brazil.

The Mangrullo formation is formed of 70 m of black shales and gray limestones containing thin lenticular intercalations of black bituminous shales that have yielded remains of *Mesosaurus brazilienses* McGregor and fish scales with a few intercalations of hematitic bands and a local bed rich in pyrites. It is equivalent to the Iratí formation of southern Brazil.

The sequence of 50 m of grayish thinly bedded sandy shales and sandstones with few lenticular intercalations of limestone, which crops out in the upper reaches of the Río Negro, has been named "Paso Aguiar" formation, and has furnished abundant fragments of silicified wood, identified as *Dadoxylon* sp. The Frayle Muerto, Mangrullo, and Paso Aguiar formations are considered as of Middle Permian age.

The Yaguarí formation consists of purple, reddish, and gray shales that alternate with brick-red siltstones and gray and light green sandstones, which rest with a slight regional unconformity on the Middle Permian. The shales and sandstones contain concretionary balls of white, gray, or pink impure limestones, which may furnish fossils: *Ferrazia cardinalis* Reed and ?*Pyramus* sp. The sediments have yielded fossils "*Lucina*" oegra Cox, *Terraia altissima* (Holdhaus), ?*Pyramus falconeri* (Cox) in a few localities, which indicate an Upper Permian age and a marine to brackish water environment (Runnegar and Newell, 1971).

The Buena Vista formation is formed of soft, massive, thickly bedded, cross-laminated red sandstones, exposed between Melo and Cuchilla Grande. Some authors consider that Yaguarí and Buena Vista are facies of one and the same formation. The Yaguarí and Buena Vista formations are the equivalents of the Estrada Nova of southern Brazil. After the accumulation of the Buena Vista formation basaltic dikes and sills were extensively intruded into the Permian beds.

Mesozoic. *Triassic.* The Triassic is represented by the Tacuarembó sandstones that crop out in the north-central part of Uruguay, between Rivera in the N and Río Negro in the S. The sequence, some 400 m thick, rests with gentle unconformity over the underlying rocks, and it is formed of soft, friable, thinly bedded and often cross-laminated sandstones alternating with hard, compact, and more homogeneous beds. The soft, friable varieties predominate in the lower part, whereas the massive rocks predominate in the upper part, but both types alternate throughout the sequence. The colors vary; though dominantly brown and reddish, they are often white, gray, yellow, and pink.

To the W of Tacuarembó, the sandstones have furnished indeterminate gastropods, as well as fish remains identified by Walther (1933) as *Lepidotus* (? sp.). The Tacuarembó sandstones are dominantly eolian, but contain a few fluviatile and lacustrine intercalations. They are the equivalent to the Botucatú sandstones of Brazil.

Cretaceous. The northwestern part of Uruguay, N of Río Yi, is covered by the Arapey lavas, equivalent to the Serra Geral of Brazil. These are mainly tholeiitic basalts and attain a thickness of over 1000 m. The flows contain interstratified eolian sandy layers and rest on Tacuarembó sandstones. Radiometric age determinations show that most of the eruptions took place in the Lower Cretaceous (Umpierre and Halpern, 1971), with a few in the Jurassic, as in Brazil.

Unconformably resting on the Arapey lavas and older rocks is a sequence of continental sandstones and conglomerates of late Cretaceous age, that crops out in the W between Salto and Colonia, and in isolated patches in the central part of the country between Río Negro and Montevideo.

The basal part of the sequence is called "Guichón" formation and consists of 100 m of reddish, fine-grained, massive, poorly cemented, occasionally cross-laminated, shaly sandstones with rounded, unpolished quartz grains; they contain small white oolites, thin lenses of brownish-red clay shale, and occasionally impure limestone. These sandstones have yielded numerous crocodilian remains identified as *Uruguaysuchus aznarezi* Rusconi and *U. terrai* (Rusconi).

The Guichón sandstones grade upward into the 70 m thick sequence of strongly silicified, whitish to pinkish, occasionally reddish conglomerates and conglomeratic sandstones of the Mercedes formation. Much of the cement is calcareous, but it may be clayey. The upper part contains thin, impure limestone intercalations and occasionally manganese concretions. The unfossiliferous Mercedes beds cover a larger area than the Guichón sandstones.

The Asencio formation (former "Areniscas con Dinosaurios") consists of some 40 m of pale pink-to-white, fine-grained, poorly stratified, friable sandstones with rounded and polished quartz grains; the cement is clayey to calcareous. The upper part of the formation is dark red in color due to strong ferruginous cementation (Palacio member). These beds contain dinosaur remains identified by von Huene (1929) as *Titanosaurus australis* Lydekker, *Laplatasuarus araukanicus* von Huene, *Antarctosaurus wichmannianus* von Huene, and *Argyrosaurus superbus* Lydekker, which indicate a late Cretaceous (Senonian) age for the deposits.

The whole sequence is conformable, but the upper beds show a strong stratigraphic onlap. The sediments are evidently continental, but whereas the Guichón sandstones are mostly eolian, the Mercedes and Asencio formations are probably deltaic.

Cenozoic. The younger sediments of Uruguay consist of continental sandstones and shales, lacustrine limestones, coquina, marine beds, and terrace sands. The Tertiary stratigraphy is difficult to work out because of the discontinuity of the outcrops without any possible lateral correlation and, furthermore, there are no characteristic fossils.

Tertiary. On both margins of the Río Queguay and W of Durazno, a few small patches of Queguay limestone, unconformably resting on Upper Cretaceous sediments crop out. The sequence, 40 m thick, consisting of white-to-whitish massive, hard silicified limestone containing a little clastic quartz is of lacustrine origin. The beds have furnished scarce remains of gastropods identified as *Planorbis waltheri* von Ihering and ?*Bulimus* sp. The fossils do not permit an accurate age determination, but may be Oligocene(?).

East of the Río Uruguay, between Salto and Fray Bentos, continental silty sands are exposed in several large patches; similar beds crop out in an outlier a short distance NW of Montevideo. The sequence, Fray Bentos formation, some 80 m thick, rests on an erosion surface truncating Cretaceous and Precambrian rocks and consists of soft, friable, massive, light reddish silty sands devoid of stratification, with minor amounts of reddish montmorillonitic clays containing thin layers of volcanic ash. The beds have yielded gastropods, *Borus globosus, Strophocheilus lutescens* (King), and *Odontostamus dentatus* and mammal remains such as ?*Propachyrucos schiaffinoi* Kragl., *Palmiramys waltheri* Kragl. Similar sediments have been found in the northeastern corner of Uruguay, in the Chuy well at a depth of 150 m (Butler, 1970). The age of the Fray Bentos silty sands is still a moot question but according to Lambert they are Miocene.

A littoral facies of the Fray Bentos formation begins with green clay, followed by white, almost incoherent sandstone. The green clay bed, 1 m thick, has furnished remains of glyptodontids, *Stromaphoropsis scavonoi* Kragl., and rodents, *Cardiomys* sp., whereas the sandstone beds have yielded *Lingula bravardi* Doello Jurado. Some authors consider these beds as the base of the Camacho formation.

The Pliocene is represented by marine sandstones, Camacho formation (former Entre Ríos formation), exposed on the margin of the Río Uruguay S of Soriano. The sequence, 20 m thick, consists most commonly of whitish, coarse-grained, irregularly bedded, loosely cemented calcareous sandstone with scattered pebbles and very well-rounded quartz grains. The beds, highly fossiliferous (practically a coquina) have yielded *Megalonychops fontanoi* Kragl., *Ostrea patagonica* d'Orb., *O. puelchana* Borchert, *Venus munsteri* d'Orb., *Cardium robustrum* Phil., *Myochlamys paranensis* d'Orb., *Monophora darwini* Desh., and many others.

The Salto sandstones, exposed along the margins of the Río Uruguay N of Fray Bentos, are the continental equivalents of the marine Pliocene Camacho sandstones. The deposits consist of dark red and yellowish brown, soft, friable, fine-grained sandy silts very poorly cemented by clay and iron oxides, small lenticular intercalations of green clay, and conglomerates. They are regarded as partly fluviatile, and partly eolian sediments.

Quaternary. The Quaternary deposits extend in a discontinuous mantle a few meters thick throughout Uruguay. They are poorly stratified subaerial and fluviolacustrine sediments. Three principal lithological groups can be considered: (a) eolian and water-laid sands,

(b) shell beds forming cheniers, (c) peaty sands and marshy deposits.

The Azaratí formation consists of some 20 m of slightly coherent, yellow-to-reddish and brownish material that varies lithologically from conglomerate and silty clay to clayey and ash beds, with a preponderance of sand over both silt and clay. These materials are eolian in part, but not loess in the classical sense. Calcareous concretions, similar to loess kindchen, are frequent in certain layers.

The beds are highly fossiliferous and have furnished abundant remains of *Toxodon, Typotherium, Macrauchenia, Glyptodon, Panochtus, Smilodon, Equus,* etc. In addition, subaerial and fresh-water mollusks are frequent, such as *Bulimus, Borus, Planorbis,* and many others. Colluviation has played a very important part in the accumulation of these deposits.

The Chuy formation is exposed in northeastern coastal Uruguay from the Brazilian border to Santa Teresa and consists of well-sorted, predominantly quartzose sand, with well-rounded, polished, and iron-stained quartz grains.

The Vizcaíno formation (former Querandí formation) is a marine deposit left by the last (Holocene) transgression, exposed sporadically in numerous localities along the Atlantic coast as well as along the margins of the Ríos de la Plata and Uruguay. The beds are a few meters thick and consist of gray clays and subfossil shells, with a predominance of *Erodona mactroides* Daudin and *Mactra (Mactrotoma) isabellina* d'Orb. Three radiocarbon dates indicate 4460–5970 years B.P. (Delaney, 1967).

Gray-to-yellow clay beds with gypsum concretions are exposed near Bellaco and last, there are the sandy deposits, essentially quartzose, well-sorted sands. The principal peat deposit, 2.5 m thick, occurs in Laguna Negra.

Structure

The eastern and southern parts of Uruguay, an extension of the long coastal Brazilian Shield, are also formed mainly of Precambrian rocks, exposed almost continuously along the Atlantic seaboard and the Río de la Plata coast. The northern and western parts of the country are occupied by the southern extremity of the Paraná Basin of southern Brazil, northeastern Argentina, and eastern Paraguay, filled by Paleozoic and Mesozoic sediments and capped by sheets of Cretaceous basalt. The Pelotas Basin scarcely enters the northeastern corner of Uruguay. Each unit constitutes a geological province. The Pelotas and Paraná basins are separated by a divide formed by a narrow ridge of Precambrian rocks that connects the Brazilian Shield with the Uruguayan cratonic area. The plains of Uruguay are covered with Pleistocene deposits.

There has been no significant folding since Precambrian time, only faulting, oriented in preferential directions; faults or fault zones are superimposed one on another. Generally speaking, the major faults are aligned NE-SW and NW-SE. Apparently, the former is younger than the latter.

The early Precambrian rocks, highly folded and metamorphosed, have a general NE-SW schistosity, the middle Precambrian are N-NE, whereas the late Precambrian rocks are far less metamorphosed or disturbed and in some areas they are subhorizontal, but locally show a mild NW-NNW folding.

The Lower Devonian strata dip gently NW and their subsurface extent is quite limited.

The Upper Paleozoic and Mesozoic rocks, separated from the underlying rocks by a gentle regional unconformity, display rather simple structures, only locally complicated by moderate faulting. The Upper Paleozoic beds are either subhorizontal or very gently folded in open undulations except in the vicinity of faults. In the neighborhood of Aceguá on the Brazilian border several larger faults affect the Upper Paleozoic beds and expose patches of Precambrian rocks.

A slight regional unconformity occurs in the Permian sequence, at the base of the Yaguarí formation. The erosion surface cuts the Paso Aguiar, Mangrullo, Frayle Muerto, and Tres Islas deposits, from E to W. The relations of the Yaguarí formation are complicated by faulting, especially S of Tacuarembó. The folding and faulting of the Upper Paleozoic sediments, however, took place after the deposition of the Yaguarí and Buena Vista formations, and prior to the accumulation of the Tacuarembó sandstones, separated from the underlying deposits by a gentle angular unconformity. The Tacuarembó sandstones have a subhorizontal attitude or a very gentle regional dip to the W and WNW.

A gentle regional unconformity can be detected within the Cretaceous sequence. This is found at the base of the Guichón formation, which clearly rests on an erosion surface cutting the Arapey lavas in the N and the Precambrian rocks in the S.

At the base of the Queguay limestones and of the Quaternary deposits, other unconformities are recognizable.

In summary, the crystalline basement forms the fundamental geotectonic unit of Uruguay. It is a positive, heterogeneous but rigid crystalline mass, overlain by subhorizontal Upper Paleozoic beds. The western part of the cratonic area was affected by the Trans-Amazonian orogenic cycle, which took place in Uruguay between 2000 and 1700 m yr ago. Whereas the eastern part was affected by the Pan-American orogenic cycle, 600–500 m yr ago. The cratonic area became stabilized between the Cambrian and the Lower Silurian. Although it gives an impression of being an arched area, its margins, except the northern one, are faulted.

The Santa Lucía graben (former Canelones graben), with a depth of 1995 m (Castellanos well), lies N of Montevideo and has an ENE trend. It actually consists of two tectonic depressions, San Bautista to the N and El Sauce to the S, separated by a horst that crops out at Paso de los Francos. The basinfill consists of Arapey lavas at the base, Lower Cretaceous shales, and Upper Cretaceous red continental sediments. The total throw of the faults exceeds 2000 m. It is considered that the graben faulting of older Precambrian rocks and the forming of the Santa Lucía graben must have taken place since early Cretaceous onward.

Along the western border of the cratonic area lies the Paraná Basin, an intercratonic, epicontinental structure. It has been filled with Paleozoic and Mesozoic sedimentary rocks capped by Lower Cretaceous plateau basalts (Arapey lavas). This sequence rests on the Precambrian craton. Initial deposition was marked by a marine transgression during the Lower Devonian. Later, Gondwanic rocks were laid down in the basin between late Carboniferous and early Cretaceous times. Although the bulk of sedimentation took place between late Paleozoic and early Cretaceous, continental deposits of Tertiary and Quaternary ages also accumulated in the basin.

Broad, gentle warping occurred during the course of deposition of the Gondwanic sequence in the Paraná basin, especially during the Lower Triassic-Upper Jurassic. The area was characterized by fissure eruptions and continental sedimentation. During late Cretaceous time, epeiorogeny occurred. The formation of the Paraná Basin took place during the Middle Silurian to Middle Devonian.

The Pelotas Basin, a down-faulted coastal basin on the northeastern margin of the Uruguayan craton, is partially filled with sediments of Late Tertiary (marine) and Quaternary ages. It appears to have developed during Cretaceous-Tertiary time along preexisting trends in the Precambrian basement, truncating older structural trends. Off the Atlantic coast the basement occurs at shallow depths, about 250 to 400 m, according to reflection surveys (Butler, 1970; Leyden et al., 1971). The basement surface dips seaward.

JUAN CARLOS M. TURNER

References

Bossi, J., 1966. *Geología del Uruguay,* vol. 2. Montevideo: Univ. de la Rep., Dept. Publ., Col. Ciencias.

Butler, L. W., 1970. "Shallow structure of the continental margin, southern Brazil and Uruguay," *Bull. Geol. Soc. Am.,* 86(4), 1070–1096.

Caorsi, J. S., and Goñi, J. C., 1958. "Geología Uruguaya," *Inst. Geol. Uruguay,* 37.

Closs, D., 1967. "Upper Carboniferous anaptychi from Uruguay," *Assoc. Pal. Arg., Ameghiniana* (Buenos Aires), 5(4), 145–148.

Darwin, C., 1846. *Geological observations on the volcanic islands and parts of South America.* London, 647p.

Delaney, P. J. V., 1967. *Geomorphology and Quaternary Coastal Geology of Uruguay* (Coastal Stud. Ser.). Baton Rouge: Louisiana State Univ. Press.

Falconer, D. J., 1937. "La Formacion de Gondwana en el Nordeste del Uruguay, con referencia especial a los tenenos eogondwanicos," *Inst. Geol. del Uruguay,* 23b, 121p.

Goni, J. C., and Hoffstetter, R., 1964. "Uruguay," *Lexique Stratigr. Intern., Amérique Latine* (Paris), 5, pt. 9a, 202p.

Harrington, H. J., 1956. "Uruguay," in W. F. Jenks, ed., *Handbook of South American Geology* (Mem. 65). New York: Geol. Soc. Am., 115–128.

Huene, F. von, 1929. "Terrestrische Oberkriede in Uruguay," Centralblatt f. Min., etc., Abt. B, No. 4, 107–112.

Hurley, P. M., et al., 1967. "Test of Continental Drift by Comparison of Radiometric Ages," *Science,* 157, 3788, 495–500.

Lambert, R., 1939. "Biliographie géologique de la République Orientale de l'Uruguay," *Inst. Geol. del Uruguay,* 26, 70p.

———, 1941. "Estado actual de conocimientos sobre la geología de la Republica Oriental del Uruguay," *Inst. Geol. Uruguay,* 29(1940), 89p.

Leyden, R., Ludwig, W. J., and Ewing, M., 1971. "Structure of continental margin off Punta del Este, Uruguay, and Río de Janeiro, Brazil," *Bull. Am. Assoc. Petrol. Geologists,* 55(12), 2161–2173.

Macmillan, J. G., 1933. "Terrenos precámbricos del Uruguay," Inst. Geol. Uruguay, Bol. 18.

Mendez Alzola, R., 1938. "Fósiles devónicos del Uruguay," Inst. Geol. Uruguay, Bol. 24.

Orbigny, A. D. d', 1842. *Voyage dans l'Amérique Méridionale,* vol. III, part 3. Paris.

Parodiz, J. J., 1962. "Los Moluscos Marinos del Pleistocene Rioplatense," *Conm. Soc. Malacologica Uruguay,* 1(2), 29–46.

Runnegar, B., and Newell, N. D., 1971. "Caspian-like relict molluscan fauna in the South American Permian," *Bull. Am. Mus. Nat. Hist.,* 146(1).

Sellow, F., 1827. "Ueber das südlich Ende des gebirgszunges von Brasilien in der Provinz Sao Pedro do Sul und der Banda Oriental oder dem Staat Vom Montivideo," Berlin Akad. Wissens. Phys. Klasse, 217–293.

Umpierre, M., and Halpern, M., 1971. "Edades estroncio-rubidio en rocas cristalinas del sur de la República del Uruguay," *Asoc. Geol. Arg.,* (Buenos Aires), 26(2), 133–151.

Walther, K., 1933. "Restos de un pez ganoide de gran tamaño proveniente del Neogondwana uruguayo," Inst. Geol. Uruguay, Bol. 19.

———, 1948. "El basamento cristalino de Montevideo," Inst. Geol. Uruguay, Bol. 23.

Cross-references: *Argentina; Brazil; Paraguay; South America.*

V

VENEZUELA

Venezuela is situated on the north coast of South America. It has common boundaries, in clockwise direction, with Guyana, Brazil, and Colombia. The island of Trinidad is structurally an eastward prolongation of Venezuela, separated from it by the shallow Gulf of Paria. Off the northern coast are scattered islands partly Venezuelan, partly under the Dutch (see *Netherlands Antilles*).

Venezuela can be compared in shape to a stubby letter 'T', 1400 × 1150 km; its total area is 912,050 km^2.

The main wealth of the country lies in its prolific oilfields. The presence of oil (or pitch) was known in the days of the Spanish conquest, but exploitation began only after World War I. Metalliferous ores are mostly limited to small-scale operations, except for iron; high-grade, easily worked iron ores have been exploited since 1950. Placer deposits of gold have been worked by primitive methods since ancient times, and since the 1920s diamonds have been recovered from the same deposits. The primary source of each remains elusive.

An official geological survey has existed in Venezuela since 1936 (under various titles) and has become the present Dirección de Geología attached to the Ministerio de Minas e Hidrocarburos. The first training center for Venezuelan geologists was founded in 1938 in Caracas. It has evolved into the present Escuela de Geología, Minas y Metalurgia of the Universidad Central de Venezuela. Recently the Universidad de Oriente has started to organize a second school of geology in Ciudad Bolívar.

Geologic-Geomorphic Provinces

Venezuela displays physiographic extremes from coastal marshes to towering, snow-clad peaks; from tropical jungle to barren desert. Their configurations are closely governed by the geological factors (Figs. 1 and 2).

The backbone of the country is formed by the Venezuelan Andes and the Caribbean Mountains. The two ranges are approximately aligned though their tectonic characters are entirely distinct. The Andes are a block-faulted massif, whereas the Caribbean ranges are a compressive uplift, complexly folded and thrust-faulted. On the northwestern side of the Andes is the Maracaibo Basin, an extensive area largely covered by the shallow Lake Maracaibo. Its western boundary is the Perijá Range, and moderately high mountains related to the Eastern Cordillera of Colombia. Northeastward, the almost featureless Maracaibo Basin merges gradually into the hilly Falcón-northern Lara region. Although both areas are Tertiary depressions, the eastern part has been orogenically disturbed, whereas the western part has remained essentially stable.

Southeast of the Andes and south of the Caribbean Mountains are the vast plains or "llanos," which extend to the Orinoco River. They are traversed by innumerable tributaries of the Orinoco. Geologically this plains region is divided into two basins by the El Baúl Swell. West of the El Baúl Swell is the Barinas Basin, which is continuous with the Llanos Basin of Colombia. To the E is the Eastern Venezuela Basin. At the surface, the eastern basin slopes gently E to merge with the delta of the Orinoco, but geologically it extends beyond the delta and into southern Trinidad.

The Orinoco River follows the outcropping rim of the Guiana Shield. Consequently there is a sharp difference between the monotonous plains on the left bank and the mature uplift topography on the right bank. This Guiana region has a rolling landscape underlain by the Precambrian rocks of the shield. Scattered remnants of sandstone and conglomerate of the resistant Roraima formation, which lie horizontally on the Precambrian rocks, give rise to spectacular table mountains with immense vertical scarps from which cascade some of the world's highest waterfalls.

Three smaller but significant geomorphic features are (1) the uplifted peninsulas of Guajira (mostly in Colombian territory) and Paraguaná, both connected to the mainland by low-lying necks; (2) the small Tuy-Cariaco Basin, an elongate structural depression within the Caribbean Mountains; and (3) the Venezuelan Caribbean islands.

Geologic History

Precambrian rocks of the Guiana (Guyana) Shield occupy almost half of the surface of Venezuela. They also form the cores of the

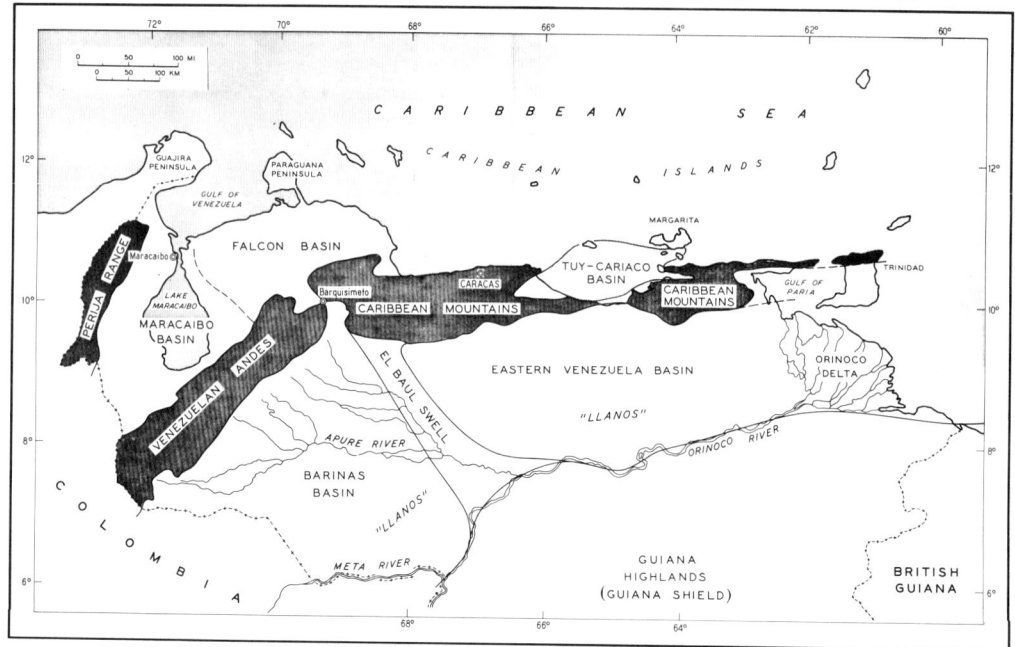

FIGURE 1. Morphotectonic provinces of the northern and central parts of Venezuela.

Venezuelan Andes and the Perijá Range and crop out locally in the El Baúl Swell and in the Caribbean Mountains. The Guiana Shield has remained a stable craton throughout most of geologic time.

The Paleozoic formations are too scattered and too restricted in outcrop area to allow a clear reconstruction of their geologic history. Cambrian is unknown. Ordovician and Silurian formations are limited to the southern flank of the Andes and to the El Baúl Swell. Devonian beds occur in the Perijá Range. Carboniferous and Permian units, mostly marine, are extensive in the Andes and appear to correlate with an equivalent sequence in the Perijá Range.

Possibly an early Paleozoic basin occupied the area between the shield and the eastern flank of the Andes, then a Devonian orogeny reversed the high and low areas and produced a basin extending from the present Andes across the Perijá Range and into Colombia. The Paleozoic ended with regional uplift, deep erosion, and peneplanation.

Extensive Triassic-Jurassic red beds initiate the long emergent phase that preceded the transgression of the Cretaceous seas toward the shield. Only in the Guajira Península are marine Triassic-Jurassic beds known.

The geologic history of Venezuela is better known from the Cretaceous onward, and it is convenient to treat the eastern and western parts of the country separately.

Eastern Venezuela in mid-Mesozoic was a broad gentle shelf stretching northward from the shield. The Cretaceous sea encroached southward and a wedge of sediments accumulated on the shelf, thickening into an offshore province. In the Upper Cretaceous a volcanic island arc formed along the present coastline, followed later by uplift, which brought into existence an asymmetric E-W-trending basin, a forerunner in the area of the present eastern Venezuela basin, which lies farther south. During the Tertiary the northern borderland grew by successive upthrusts along its southern margin, shifting the axis of the basin progressively southward. The main orogenic phases occurred during the Lower Miocene. Throughout the Tertiary the open sea was to the E. During evolution of this highly asymmetrical basin, littoral and inshore sands merged northward into deep marine shales with turbidites. By middle Miocene the orogeny had ceased, the basin filled with sediments, and the sea retreated eastward. Middle Miocene tilting permitted one final incursion of marine beds, but by the Pliocene, the basin was much as it is today.

Western Venezuela during the Cretaceous and early Tertiary was a stable emergent area (the Maracaibo Platform). As in the eastern part of the country, the Cretaceous transgression was from N and W and deep troughs developed around the platform, which continued its positive role. Cretaceous limestones and shales de-

FIGURE 2. Geologic map of Venezuela.

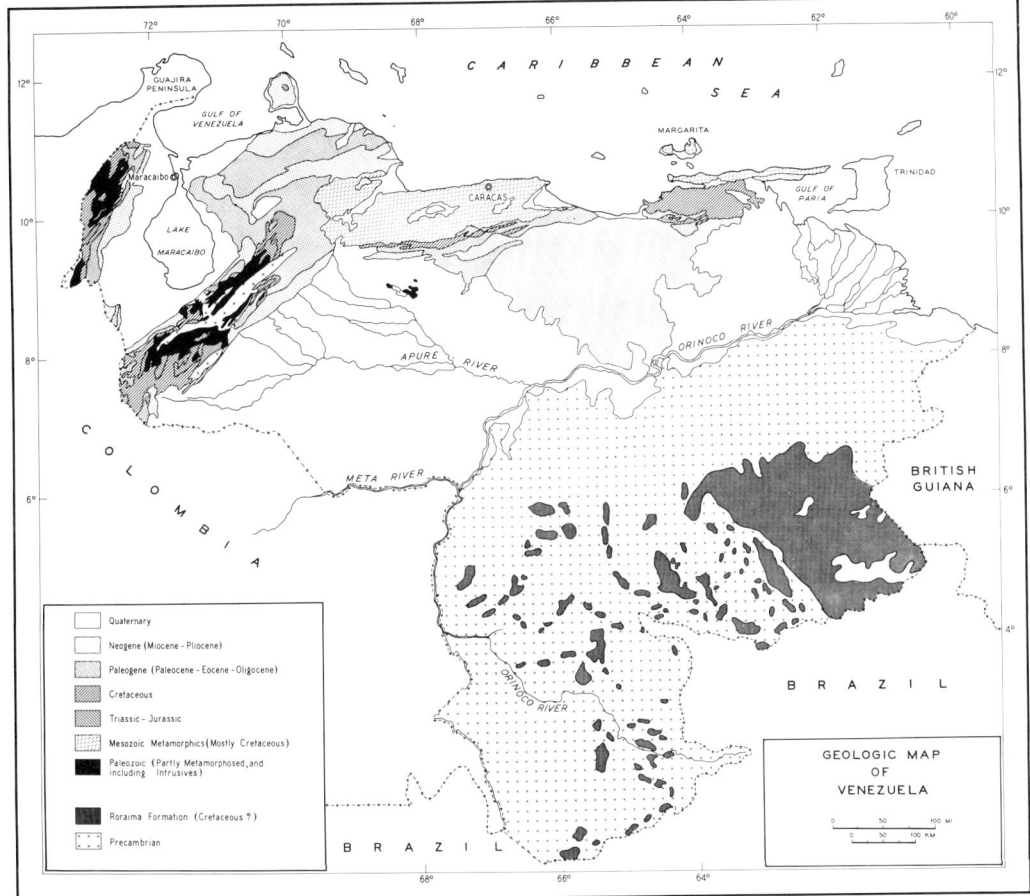

posited over the platform are much thinner than in the troughs. Paleocene and Eocene shallow-water sandstone and shale sequences also show the influence of the Maracaibo Platform. The deeper marine facies of the early Tertiary are found mostly E of Lake Maracaibo. The Maracaibo platform was not involved in the Tertiary orogeny. Instead, until Miocene times it suffered only epeirogenic movements, notably in the Paleocene and Oligocene.

Uplift of the Venezuelan Andes probably started in the late Eocene but important orogenic movements did not develop in western Venezuela until later in the Tertiary. During the Miocene the rising Andes and the Perijá Range produced thick piedmont deposits outlining the triangular depression of the Maracaibo Basin. The Andes are linear block-faulted uplifts, very distinct from the Caribbean Mountains with their asymmetric folds and low-angle thrusts.

Mineral Resources

Petroleum (see Fig. 3). The existence of petroleum has been known since the early days of exploration in the New World when pitch from seeps in various parts of the country was used for caulking ships and as medicine. A primitive kerosene still was put into operation in 1878 near San Cristóbal. By 1913 heavy oil was being produced from shallow wells drilled around the large asphalt lake at Guanoco, near the Gulf of Paria.

In 1914 the first commercial discovery was made in the Mene Grande field E of Lake Maracaibo. A 100,000 barrel per day blowout in 1922 attracted the attention of the oil world and began a succession of discoveries that eventually changed Venezuela from an agricultural country to the largest exporter of crude oil in the world. In 1926-1930 the Lagunillas, Tía

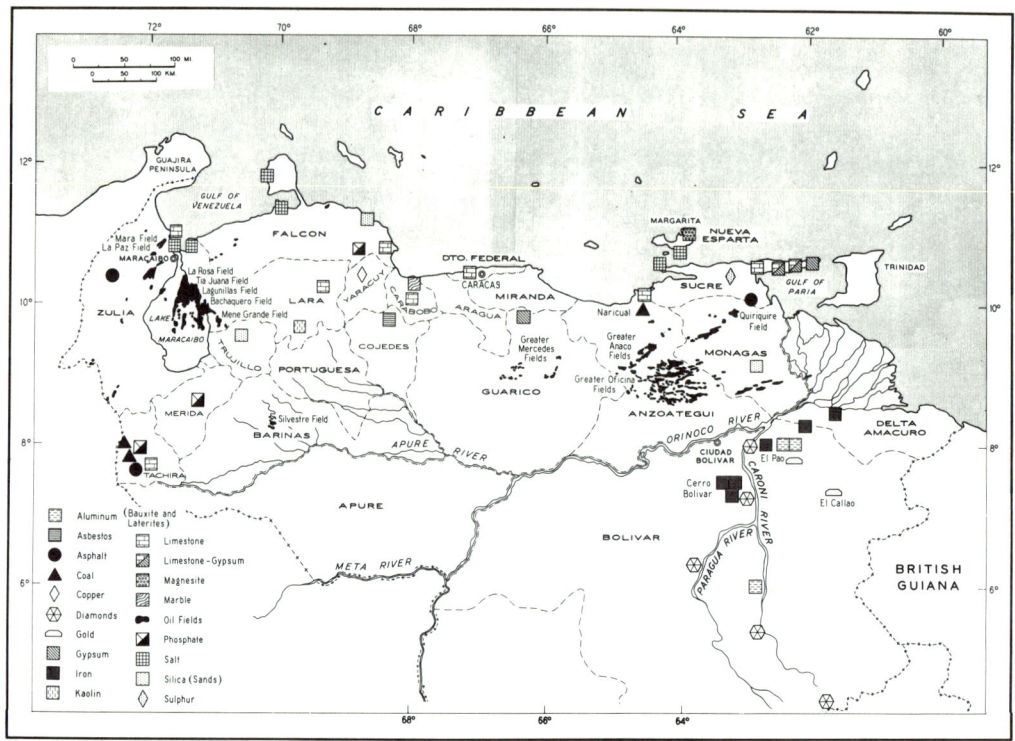

FIGURE 3. Mineral resources of Venezuela.

Juana, and Bachaquero fields were found in Lake Maracaibo; in 1928 Quiriquire, the first oilfield in eastern Venezuela followed, and in 1937, the Ofcina field. In 1941 the first of the Greater Mercedes fields was discovered, starting an exploratory rush in the "llanos" of central Venezuela; in 1945 Cretaceous production was obtained in the La Paz and Mara fields, west of Maracaibo; and in 1948 production was obtained in the Barinas basin. Following concessions granted in 1956 and 1957, the central part of Lake Maracaibo was developed with the discovery of the Lama, Lamar, Centro, Ceuta and other fields. Such important discoveries were not made without many failures. Costly but disappointing exploration programs have been carried out in the Falcón and Barinas basins, in Guárico, and in the Gulf of Paria.

Oil production has climbed from 121,116 barrels in 1917 to 1,353,400,000 barrels in 1970. Most of the oil today in Venezuela comes from the Miocene sandstones and Eocene sandstones of the Lake Maracaibo fields, which produce 81% of the total. Second in importance are the Greater Oficina and Greater Anaco areas, with a production of 14% of the total, virtually all from Oligo-Miocene sands.

Iron. Deposits of iron ore along a belt just S of the Orinoco River has long been known, and the remains exist of primitive smelters erected by Catalán Capuchins before 1750. After 50 years of exploration and concession activity, exploitation started during the 1940s, first at El Pao and then at the larger deposits of Cerro Bolívar, on opposite sides of the Caroní River, production of iron ore reaching a peak of 22 million metric tons in 1970. At first the entire production of iron ore was exported, principally to the United States; later, however, a proportion went to the state steel plant at the confluence of the Caroní River with the Orinoco.

Geologically the iron ores are secondary deposits enriched by meteoric leaching of highly ferruginous Precambrian quartzites. Solution of the quartz has left a residue of almost pure hematite and magnetite. The iron content of these high-grade ores is usually 64–68%. The quartzites and interbedded schists and gneisses are tightly folded and faulted. The ores are extracted by open-pit methods since the iron deposits form prominent hills standing up sharply above rolling country. Reserves of high-grade ore exceed 1.9 billion metric tons. In addition, four more iron deposits have been prospected and are now held as national reserves; they contain 54% of Venezuela's total iron ore reserves.

Gold. Gold has long been obtained in the Venezuelan Guiana by panning of placer deposits, but it contributes only a small part of the total production. Although an intensive search for the legendary golden city of El Dorado took place here during colonial days, it was not until the 1880s that the El Callao mines became a great world producer, reaching 8.2 million grams in 1885. The gold is found in quartz veins in hihgly metamorphosed Precambrian rocks. Later, production declined markedly, and in 1970 it was only 680,000 grams.

Diamonds. Both gem and industrial diamonds were recognized in the late 1920s in the placer deposits of the Venezuelan Guiana. Even though some rich finds have been made, including the 154-carat *Barrabas*, extraction has been sporadic, probably due to the small and scattered nature of the placer deposits. Production has fluctuated in the last 20 years from 50,000 carats to 512,000 carats (in 1970).

Phosphate. In the State of Falcón there are extensive deposits of phosphatized Miocene limestones in which collophane and dahlite are concentrated. About 10 million tons are held as national reserves.

Coal. Coal, ranging from lignitic to semianthracitic, is widely present in the Tertiary rocks, but only a few deposits are of commercial proportions. These are located in the state of Táchira and at Naricual, in northern Anzoátegui.

Other Mineral Deposits. The presence in neighboring Guyana of important deposits of bauxite and manganese ores suggests that they might extend into southeastern Venezuela. Some small deposits have been located but no development has been undertaken.

Local deposits of copper, nickel, mercury, magnesium, and other base metals, as well as asbestos, gypsum, and sulfur are known, and some have been mined on a small scale. Ample deposits of limestone have made possible an important cement industry.

AMOS SALVADOR

References

The geological literature is voluminous, especially on the petroliferous basins. Two fairly complete bibliographies have been published. The most important local geological journals are the *Boletín de Geología* and its predecessor, the *Boletín de Geología y Minería*, both published intermittently by the Government's Dirección de Geología, and the *Boletín Informativo,* published monthly since 1958 by the Asociación Venezolana de Geología, Minería y Petróleo. National congresses of geology or petroleum were held in 1937, 1938, 1951, 1959, 1962, and 1969, and their published proceedings provide a wealth of geologic information on the country. A stratigraphic lexicon was completed in 1956 and revised in 1970.

Maps

Mapa Geológico de la República de Venezuela, Ministerio de Minas e Hidrocarburos, Dirección de Geología, Scale 1:1,000,000, 1955. Same in Scale 1:2,000,000, 1959.

Mapa Geológico-Tectónico del Norte de Venezuela, Primer Congreso Venezolano de Petróleo, Scale 1:1,000,000, 1963 (1962).

Bibliographies

Hedberg, H. D., and Hedberg, F., 1945. Bibliografía e Indice de la Geología de Venezuela," *Revista de Fomento (Venezuela),* 7(58–59), 43–123.

Korol, B., and Forjonel, J., 1959. "Bibliografía e Indice de Geología, Minería y Petróleo de Venezuela, Primera Parte, 1950–1958," *Boletín de Geología (Venezuela),* 5(10) 121–211.

Bibliography of articles on Venezuela, Colombia, and Trinidad, published in the *Bull. Am. Assoc. Petrol. Geologists, Boletín Informativo,* A.V.G.M.P., 1(3), 101–107, 1958.

Bibliography of articles on Venezuela published in the *Geol. Soc. America Bull., Boletín Informativo,* A.V.G.M.P., 2(7), 181–182, 1959.

Publications

Bell, J. S., 1972. "Geotectonic evolution of the southern Caribbean Area," *Geol. Soc. America Mem. 132* (Hess vol.), 369–386.

Bellizzia, G. A., 1972. "Is the entire Caribbean Mountain belt of northern Venezuela allochthonous?, *Geol. Soc. America Mem. 132* (Hess vol.), 363–368.

Fiedler, V. G., 1970. "Die seismische Aktivität in Venezuela im Zusammenhang mit den wichtigsten tektonischen Bruchzonen," *Geol. Rundschau,* 59(3), 1203–1215.

Graf, C. H., 1969. "Estratigraphia del nordeste de Venezuela," *Assoc. Venez. Geol. Min. Pet.,* 12(11), 393–416.

Grauch, R. I., 1972. "Preliminary report of a late(?) Paleozoic metamorphic event in the Venezuelan Andes," *Geol. Soc. America Mem. 132* (Hess vol.), 465–473.

Lattimore, R. K., Weeks, L. A., and Mordock, L. W., 1971. "Marine geophysical reconnaissance of continental margin north of Paria Peninsula, Venezuela," *Bull. Am. Assoc. Petrol. Geologists,* 55(10), 1719–1729.

Maloney, N. J., 1966. "Geomorphology of continental margin of Venezuela, pt. 1, Cariaco Basin," Univ. Oriente, Inst. Oceanog. Bol., 5(12), 38–53.

———, 1967. "Geomorphology of continental margin of Venezuela, pt. 2, Continental terrace off Carupano," *Univ. Oriente, Inst. Oceanog. Bol.* 6(1), 147–155.
Maresch, W. V., 1972. "Eclogitic-amphibolitic rocks on Isla Margarita, Venezuela: a preliminary account," *Geol. Soc. America Mem. 132* (Hess vol.), 429–437.
Martinez, A. R., 1972. *Recursos de hidrocarburos de Venezuela*. Caracas: Edreca Editores, 151p.
Mencher, E., 1963. "Tectonic history of Venezuela," in "Backbone of the Americas," *Am. Assoc. Petrol. Geologists Mem. 2*, 73–87.
Miller, J. B., Edwards, K. L., Wolcott, P. P., Anisgard, H. W., Martin, R., and Anderegg, H., 1958. "Habitat of oil in the Maracaibo Basin, Venezuela," in L. G. Weeks, ed., *Habitat of Oil*, a symposium conducted by the Am. Assoc. Petrol. Geologists, 601–640.
Ministerio de Minas e Hidrocarburos, Dirección de Geología 1956. "Léxico estratigráfico de Venezuela," *Boletin de Geología* (Venezuela), Publ. Esp. 1, 728p. (also an English edition, "Stratigraphical Lexicon of Venezuela"). Contains extensive bibliography: 2nd ed., 1970, Publ. Esp. 4, 756p.
———, 1960–1961. Tercer Congreso Geológico Venezolano, Memoria, 4 vols., Boletin de Geología (Venezuela), Publ. Esp. 3, 1966p.
Renz, H. H., Alberding, H., Dalmus, K. F., Patterson, J. M., Robie, R. H., Weisbord, N. E., and MasVall, J., 1958. "The eastern Venezuelan Basin," in L. G. Weeks, ed., *Habitat of Oil*, a symposium conducted by the Am. Assoc. Petrol. Geologists, 551–600.
Santamaria, F., and Schubert, C., 1974. "Geochemistry and geochronology of the southern Caribbean-northern Venezuela plate boundary." *Geol. Soc. America Bull.*, 85(7), 1085–1098.
Schubert, C., 1970. "Glaciation of the Sierra de Santo Domingo, Venezuelan Andes," *Quaternaria*, 13, 225–246.
Shagam, R., 1972. "Andean research project, Venezuela: principal data and tectonic implications," *Geol. Soc. America Mem. 132* (Hess vol.), 449–463.
Sociedad Venezolana de Ingenieros de Petróleo, 1963. "Aspectos de la industria petrolera en Venezuela," 850p. (Contains the papers by Miller *et al.* and Renz *et al.* updated and translated into Spanish).
Weeks *et al.*, 1971. "Structural relations among Lesser Antilles, Venezuela, and Trinidad-Tobago," *Am. Assoc. Petrol. Geologists Bull.*, 55(10), 1741–1752.
Young, G. A., Bellizzia, A., Renz, H. H., Johnson, F. W., Robie, R. H., and MasVall, J., 1956. "Geología de las Cuencas Sedimentarias de Venezuela y de sus campos petrolíferos," *Boletín de Geología* (Venezuela), Publ. Esp. 2, 140p.

Cross-references: *Brazil; Colombia; Guiana Shield–Regional Review; Trinidad and Tobago.* See also: *Encyclopedia of Geomorphology, Lake Maracaibo.*

VIRGIN ISLANDS

The Virgin Islands are a group of small islands lying between latitudes $17°40'N$ and $18°50'N$ and longitudes $64°75'W$ and $65°10'W$ and altogether covering about 700 km^2. Politically they are divided into the British Virgin Islands (174 km^2; 59 sq mi) to the N and the American Virgin Islands (344 km^2; 133 sq mi) to the S. Tortola, Jost Van Dyke, Virgin Gorda, and Anegada are the principal islands of the British group; St. Thomas (27 sq mi), St. John (19 sq mi), and St. Croix (82 sq mi) belong to the United States group. There are over 50 smaller islets and cays most of which are uninhabited. (Vieques and Culebra belong to *Puerto Rico*, q.v.)

With the exception of St. Croix all the islands rise from the Virgin Island bank (av. depth 46 m) which is continuous with the insular shelf surrounding Puerto Rico. The edges of the bank drop away very steeply to oceanic depths. Between St. Croix and the other islands the depths exceed 3660 m. Structurally, therefore, it would seem that the Virgin Islands are closely related to Puerto Rico and form the eastern end of the Greater Antilles. About 150 km E is the northern end of the Lesser Antilles chain.

With the exception of Anegada, which is low and flat, the islands generally are rugged and with steep slopes. The highest point is Sage Mountain in Tortola with an elevation of 542 m. The southern part of Virgin Gorda, known as the "Valley," is relatively low and flat and St. Croix has a central lowland between the eastern and western ranges.

The climate tends to be rather dry with annual rainfall varying between 1000 and 1300 mm. Although in one or two parts there is quite dense tropical vegetation, a large part is covered with cactus and thornbrush. The islands have serious water supply problems. The economy rests almost entirely on tourism although there is a limited amount of truck and cattle farming and fishing.

Early geological work in the Virgin Islands was carried out by Cleve (1881) and Earle (1924). In the 1950s a survey of the British islands was made by Martin-Kaye, and in the 1960s work was done by graduate students of Princeton University as part of the Caribbean Research Project headed by the late H. H. Hess. The Water Resources branch of the U. S. Geological Survey, based in Puerto Rico, have carried out some investigations in the American islands.

Economic Geology

A small copper mine on Virgin Gorda was operated sporadically during the last century. The tailings also disclose molybdenite. Although at one time local building stones were exploited, they have now almost completely been replaced by concrete.

Stratigraphic History

The oldest rocks in the Virgin Islands are a series of submarine lava flows and breccias of spilitic and keratophyric composition, with some interbedded radiolarites, which are found in the southern parts of St. Thomas and St. John. They are thought to be of Early Cretaceous age but may be older. Radiolaria show some resemblance to Lower Cretaceous forms. The thickness is 4600 m or more.

Overlying them unconformably is a thick sequence composed mainly of augite andesite breccias and tuffs and totaling about 6000 m in thickness, including several limestone lenses.

This is followed unconformably by more than 1800 m of tuffaceous graywackes with a submarine slump breccia near the base containing large blocks of volcanics and of limestone. The limestone blocks contain molluscs, corals, echinoderms, and cephalopods suggesting an early Upper Cretaceous age.

Above the graywackes follows with probable unconformity a sequence of andesite breccias tuffs and volcanic sandstones more than 7600 m in thickness. Several limestone lenses are included that have yielded middle Eocene foraminifera.

Emplacement of a large batholith took place in middle or late Eocene time, which was probably also the time during which the major deformation and metamorphism took place. Volcanism still continued with the accumulation of a further 1800 m of tuffs and breccias. Fragments of the batholithic rocks are found in these later volcanics.

In St. Croix the succession is distinctive. It consists essentially of a great thickness of volcanically derived Upper Cretaceous sediments, turbidites, and tuffaceous sandstones amounting to nearly 10,000 m in thickness. The whole sequence appears to be of submarine origin and the presence of spilite and keratophyre fragments and indications of current directions in the turbidites suggest that the material was accumulated down a southward slope on the sea floor, probably from the region of the northern Virgin Islands; the tuffaceous sandstones were probably derived from airborne volcanic material from nearby vents. Microfossils are present in many of the rocks but are generally unidentifiable. However, foraminifera and a few macrofossils (rudistids) from one or two localities have indicated an Upper Cretaceous age.

After the deposition of these rocks, they were folded and faulted and intruded by stocks of gabbro and diorite. Slight metamorphism also occurred. A graben was formed in the central part of the island and this became filled with about 2300 m of sediment during late Oligocene and early Miocene time, although only the upper 180 m of marl and limestone containing foraminifera, molluscs, and corals is exposed at the surface. The lower part consisting mainly of mudstone is known from drilling.

Erosion surfaces, raised reefs, and submarine terraces indicate later fluctuations of sea level during the Pliocene and Quaternary and include limestone accumulations on Anegada.

Structural and Geomorphic Features

The greater part of the area is underlain by Cretaceous and Eocene rocks homoclinally dipping toward the N at angles averaging 40° in the American islands and steeper to overturned in the British. The strike is generally E-W, although cross-folding and faulting causes local variations.

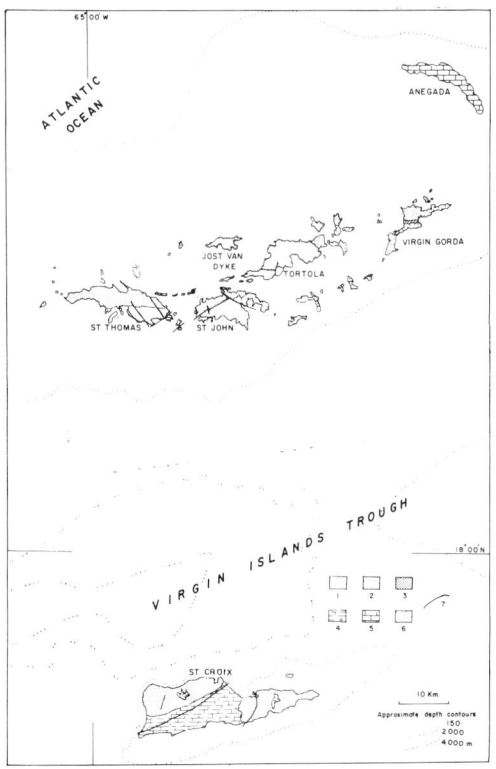

FIGURE 1. Geological map of the Virgin Islands (generalized from Donnelly, 1966; Helsley, 1960; Whetten, 1966). 1, ?Lower Cretaceous, mainly spilitic and keratophyric flows, some chert; 2, Upper Cretaceous, in the north mainly andesitic pyroclastics, in St. Croix, mainly volcanically derived sediments and tuffs; 3, Eocene, mainly pyroclastics; 4, Miocene, marls; 5, Pleistocene, limestone; 6, intrusive, mainly tonalite and granodiorite, some gabbro; 7, faults.

Two sets of faults affect St. Thomas and St. John oriented NW and NE, the former set having dextral strike-slip displacement and the latter sinistral. Between faults of the NE trending set, which also have a strong vertical component, a graben extends from western St. John to southeastern St. Thomas and has been traced to the SW as a submarine scarp that merges into the Virgin Island trough. The British islands are also affected by large numbers of small faults with sinistral and dextral displacements.

A large mid to late Eocene batholithic intrusion occupies most of Virgin Gorda and extends into Tortola, St. John, and intervening islets. It is largely of tonalite and granodiorite although compositions vary from gabbro (layered in places) to granitic pegmatites and include an orbicular diorite. A regional low-grade metamorphism has affected all the rocks and in the vicinity of the batholith a pyroxene hornfels facies is found. Numerous andesitic and basaltic dikes are both earlier and later than the intrusion of the batholith.

The rocks of the batholith have weathered into huge boulders which are now found piled up in places to form a rather bizarre landscape. The small islet of Fallen Jerusalem, off the coast of Virgin Gorda, is so called because of its seaward appearance as a ruined town.

St. Croix is separated from the rest of the group by the deep Virgin Island trough. The Northside Range in the NW and the East End Range in the E are formed of Cretaceous tuffaceous and volcaniclastic sediments. They are rugged with steep slopes and are separated by a low graben filled with middle Tertiary sediments. The Cretaceous rocks have been affected by two episodes of folding, the earlier with NNE-dipping axial planes and the younger with axial planes dipping ESE. Axial plane cleavage is developed in places. Faulting is mainly normal with N20E and N80E trends. In the W a southward-directed thrust has an estimated displacement of about 5 km. Gravity measurements over the central graben suggest that the Tertiary fill is about 2 km thick, and drilling failed to find Cretaceous rock at over 450 m.

An intrusion of gabbro penetrates the rocks of the Northside Range and one of diorite in the East End Range. Both have produced aureoles of contact metamorphism. There is extensive low-grade metamorphism. Numerous small, deeply weathered dikes can be seen.

Anegada on the northeastern edge of the bank is quite different. It is composed of flat-lying Pleistocene limestone with a maximum elevation of 24 m, and presumably a basement of older volcanics.

Summit levels throughout the Virgin Islands display a marked accordance at a little over 300 m elevation. An old lateritic soil on St. John supports the suggestion that this represents a dissected erosion surface. Widespread intermediate bevels and erosional terraces are also found at about 80 m. Lower-level terraces, raised beaches, or raised reefs are rare. A patch of calcarenite containing shell fragments on the northern coast of Tortola may be a remnant of a Pleistocene beach or may be older. A raised beach and reef has been observed on the northwestern coast of St. Croix and emergent reefs at about 1 m above present sea level are found at several points around the coast.

The Virgin Islands bank itself is thought to be slightly tilted toward the N though it does not seem to have been affected notably by tectonic activity since its formation.

JOHN D. WEAVER

References

Cederstrom, D. J., 1941. "Notes on the physiography of St. Croix, Virgin Islands," *Am. J. Sci.*, **239**(8), 553–576.

———, 1950. "Geology and ground water resources of St. Croix, Virgin islands," *U.S. Geol. Surv. Water-Supply Pap. 1067*, 117p.

Donnelly, T. W., 1966. "Geology of St. Thomas and St. John, U.S. Virgin Islands," *Geol. Soc. Am., Mem. 98*, 85–176.

———, 1972. "Deep-water, shallow-water, and subaerial island-arc volcanism: an example from the Virgin Islands," *Geol. Soc. Am. Mem. 132*, 401–414.

Earle, K. W., 1924. "The geology of the British Virgin Islands," *Geol. Mag.*, **61**, 339–351.

Helsley, C. E., 1960. "Geology of the British Virgin Islands," thesis, Princeton Univ., 219p.

Kemp, J. F., 1926. "Introduction to the geology of the Virgin islands," *N.Y. Acad. Sci., Sci. Surv. P. R. & Virgin Is.*, **4**(1), 1–69.

Macintyre, I. G., 1972. "Submerged reefs of eastern Caribbean," *Bull. Am. Assoc. Petrol. Geologists*, **56**(4), 720–738. Also, Meyerhoff, H. A., 1973, "Discussion," *Ibid.*, **57**(2), 407–411.

Martin-Kaye, P. H. A., 1959. *Reports on the Geology of the Leeward and British Virgin Islands*. St. Lucia: Government Printing, 117p.

Meyerhoff, H. A., 1926–1927. "Physiography of the Virgin islands and of the Porto Rican islands Vieques and Culebra," *N.Y. Acad. Sci., Sci. Surv. P. R. & Virgin Is.*, **4**, 71–141, 145–219.

Vaughan, T. W., 1916. "Some littoral and sublittoral physiographic features of the Virgin and northern Leeward Islands and their bearing on the coral reef problem," *J. Washington Acad. Sci.*, **6**, 53–66.

———, 1923. "Stratigraphy of Virgin islands, Vieques and Culebra islands and eastern Porto Rico," *J. Washington Acad. Sci.*, **13**, 303–317.

Whetten, J. T., 1966. "Geology of St. Croix, U.S. Virgin Islands," *Geol. Soc. Am. Mem. 98*, 177–239.

Cross-references: *Leeward Islands; North America; Puerto Rico; West Indies.*

WAKE ISLAND

Lying in the western Pacific Ocean at 19°17'N and 166°35'E, about 3500 km W of Honolulu and 2500 km NE of Guam, Wake Island is a coral atoll, with three small rim islands, Wake, Wilkes, and Peale, having a total land area of 7.7 km² (2.5 sq mi). There is a broad central lagoon. The land area of emerged coral limestone lies low, and averages 3.5 m above sea level.

The atoll was probably discovered by Mendaña in 1568; being on the direct western approaches to Hawaii, it was annexed by the United States in 1898 and has become a military base with a key airfield. A good air photo of Wake is given by Stearns (1946).

The atoll lies on a very important submarine ridge, the E-W Marcus-Necker seamount chain, colorfully known as the "mid-Pacific Mountains." The sea floor here is probably of Jurassic age, part of the oldest crust in the Pacific (Pitman, et al., 1968). Necker at the eastern end is in the middle of the Hawaiian chain and Marcus (Minami Tori Shima) is situated near the western end, which eventually reaches the Izu (Bonin) Islands, S of Japan. The eastern end of explored by Hamilton (1956).

Marcus, 24°N, 154°E, is an uplifted atoll under Japanese administration. It was described by Bryan (1903-1907).

RHODES W. FAIRBRIDGE

References

Bryan, E. H., 1959. "Notes on the geography and natural history of Wake Island," *Atoll Res. Bull.*, **66,** 22p.

Bryan, W. A., 1903-1907. "A monograph on Marcus Island," *Bernice P. Bishop Mus., Occ. Pap.*, **2,** 77-124.

Fosberg, F. R., 1959. "Vegetation and flora of Wake Island," *Atoll Res. Bull.*, **67,** 20p.

Hamilton, E. L., 1956. "Sunken islands of the Mid-Pacific Mountains," *Geol. Soc. Am. Mem. 64*, 42-52.

Pitman, W. C., III, Herron, E. M., and Heirtzler, J. R., 1968. "Magnetic anomalies in the Pacific and sea-floor spreading," *J. Geophy. Res.*, **73,** 2069-2085.

Stearns, H. T., 1946. "An integration of coral reef hypotheses," *Am. J. Sci.*, **244,** 245-302.

Cross-references: *Marshall Islands; United States-Hawaii.*

WALLIS AND FUTUNA ISLANDS

Both Wallis and Futuna (Horn) Islands are part of *French Polynesia* (q.v.). The Wallis Islands lie between 13°12' and 13°24' and 176°6' and 176°14'W in the southern Pacific Ocean about 350 km W of the Samoan group. They comprise the main island of Uvea and 22 smaller islands and rock islets, all enclosed by a barrier reef. Uvea, 13 × 7 km, has an area of 60 km² (23 sq mi). The island is composed of broad low volcanic domes that merge imperceptibly to form an undulating plateau. The Polynesian population dwell chiefly on the sand flats along the shore. The highest point is Mt. Lulu, altitude about 145 m, near the center of the island. Lakes and marshes occupy six craters. The most spectacular is Lake Lalolalo. It is very deep and bounded by sheer rock walls 40 m high. Rich brown soils cover most of the island to a depth of 2-3 m. No streams exist and gullies are scarce even though rainfall averages 2500 mm (100 in.) per year. Patches of coral are common in the lagoon, the deepest part of which is 50 m. Ocean-going vessels enter it through Honikulu Pass at the southern end of the barrier reef and dock at Gahi.

The high islands are composed of olivine basaltic lavas and pyroclastics, except for one cinder cone and its associated flows of oligoclase andesite on Uvea. The lower islands are composed either of calcareous sand or are erosional remnants of tuff cones and lava domes.

Uvea Island was built by coalescence of lava flows from 19 volcanic vents. The vents comprise 15 flat shield-shaped lava cones, three consolidated ash cones, and one cinder cone. Except for two Holocene lava cones, barely covered with soil, the bulk of Uvea is composed of deeply weathered middle Pleistocene (?) volcanics. Lavas of intermediate age do not exist. Limestone ejecta in the tuff cones indicate that the whole group is built on a submerged reef, probably about 50 m below sea level. The reef, presumably, rests on a truncated Tertiary basaltic volcano. Definite evidence exists of emerged shore lines, at 4.5 and 1.5 m above mean sea level, as well as some benches suggestive of higher stands.

The petrography of Uvea Island is known from MacDonald's study (1945) of specimens collected by the writer in 1943. Alkalic basalts are present. Only four dikes are exposed but their trend and the alignment of the Holocene craters indicate that many of the volcanic fissures trend NW-SE. No evidence of faulting or folding was found by the writer.

The Hoorn Islands (Iles de Horn) comprise Futuna and Olofi. They lie 170 km SW of Uvea at 14°20'S and 178°5'W. Both are pre-Miocene volcanic islands partly surrounded by emerged Miocene sediments. Futuna and Alofi are 8 km apart and are surrounded by fringing reefs.

Futuna is 13 × 8 km and covers 65 km^2. It has a backbone ridge culminating in Mt. Puke of about 700 m high. It is deeply weathered to a laterized soil, and narrow deep valleys cut across the plateaus bordering the central range. The emerged fossiliferous Miocene marine sediments form a circling terrace backed by an ancient marine cliff. A 3.5-m Quaternary marine terrace borders much of the shore and on it are the native villages. The volcanic core is principally andesite in thick flows, generally brecciated, and these overlie dolerites. Augite tuffs occur, notably in Vainifao Valley. Vents are difficult to identify because of weathering and erosion.

Alofi is 10 × 5 km and covers 29 km^2. It is roughly bowl shaped with the emerged Miocene reef forming the rim. It resembles Makatea Island. Mt. Bougainville, the highest point, is about 350 m high. The island is uninhabited. The central core is volcanic, probably andesitic.

Geotectonically the Futuna Islands appear to lie W of the "Andesite (or Sial) line" and so are "continental" islands, whereas the Wallis Islands lie E of the line and are "oceanic" islands.

HAROLD T. STEARNS

References

Aubert de la Rüe, M. E., 1935. "La constitution géologique des îles Wallis et Futuna," *C. R. Acad. Sci.,* **200,** 328–330.

MacDonald, G. A., 1945. "Petrography of the Wallis Islands," *Bull. Geol. Soc. Am.,* **56,** 861–872.

Stearns, H. T., 1945. "Geology of the Wallis Islands," *Bull. Geol. Soc. Am.,* **56,** 849–860.

Cross-references: *Fiji; Gilbert and Ellice Islands; New Hebrides; Western Samoa.*

WEST INDIES

The West Indies, also known as the Antilles, are a chain of islands that lie between North and South American and enclose the Caribbean Sea. The island group reaches from the 12th to the 27th latitude N and from the 59th to the 85th longitude W. Together, they form an arc ("Antillean Arc" or "Caribbean Arc") more than 2500 km in length, which extends from the southern tip of Florida to the northern coast of Venezuela and comprises land areas amounting to 238,000 km^2 (92,000 sq mi).

Geologic research in the West Indies began with Alexander von Humboldt. During the 19th century and the first decades of the 20th century, work was continued by numerous European and North American geologists, most of whom, however, made only short visits to the region. Among them, the names of De La Beche, Gabb, Hill, Lacroix, Tippenhauer, Trechmann, Vaughan, and Woodring are notable. In recent years, various institutions have been active in West Indies geologic research. The more important of these within the West Indies are:

Jamaica: Geological Survey Department, in Kingston; and the Geology Department of the University of the West Indies, also in Kingston.

Puerto Rico: The U.S. Geological Survey, in San Juan; also the University of Puerto Rico, Institute of Caribbean Studies, in Mayaguez.

Cuba: Instituto Cubano de Recursos Minerales, Dpto. Científico de Geología, La Habana; Instituto Nacional de Recursos Hidraulicos, Laboratorio Geológico La Habana; also the Geological Institute at the University of Havana.

Lesser Antilles: Seismic Research Unit, U.C.W.I., St. Augustine, Trinidad; Observatoire du Morne des Cadets, Fonds Saint Denis, Martinique; Observatoire Sismologique, St. Claude, Guadeloupe; Natuurwetenschappelijke Studiekring voor Suriname en de Nederlandse Antillen, The Hague, Netherlands.

The *Transactions of the Caribbean Geological Conference* report regularly on conferences and the progress of research in the Caribbean area.

Geomorphic and Geotectonic Divisions

The West Indies may be divided into the following provinces (Fig. 1):

The Islands
Bahama Islands
Greater Antilles
Lesser Antilles, subdivided into

Leewards Islands		Volcanic Caribbees
Windward Islands	or, alternatively	Limestone Caribbees

South American Offshore Islands

The Caribbean Bathymetric Provinces

Yucatan Basin   Beata Ridge
Cayman Ridge   Venezuela Basin
Cayman Trench   Ridges and Troughs of the Lesser Antilles
Nicaragua Rise   Puerto Rico Trench with the Old Bahama
Colombia Basin   Channel

The Islands. The Bahamas are low, flat islands, composed of reef limestones and dolomites. They comprise 29 inhabited islands, 661 keys (cays), and 2387 rocky pinnacles rising from the Bahamian platform, the greater part of which is covered by water of only 3–4 fm depth (6–7 m). The Bahamian platform is a shelf area, whose structure and history contrast sharply with that of the Antillean orogenic belt. A deep test well in Andros Island revealed Lower Cretaceous reef carbonates at a depth of 4456 m (14,585 ft). From this it is concluded that the major part of this platform was, since the Late Jurassic or Early Cretaceous, a site of subsidence and carbonate sedimentation.

The *Greater Antilles* (Cuba, Jamaica, Hispaniola, Puerto Rico, and the Virgin Islands) are a linear pile of volcanogenic rocks and associated intrusives produced during the late Mesozoic-early Tertiary volcanism and subsequently deformed by folding and faulting. Transgressive sequences were deposited on the flanks after middle Tertiary time.

The *Lesser Antilles,* the emergent crest of a

FIGURE 1. Geomorphic and geotectonic units of the West Indies.

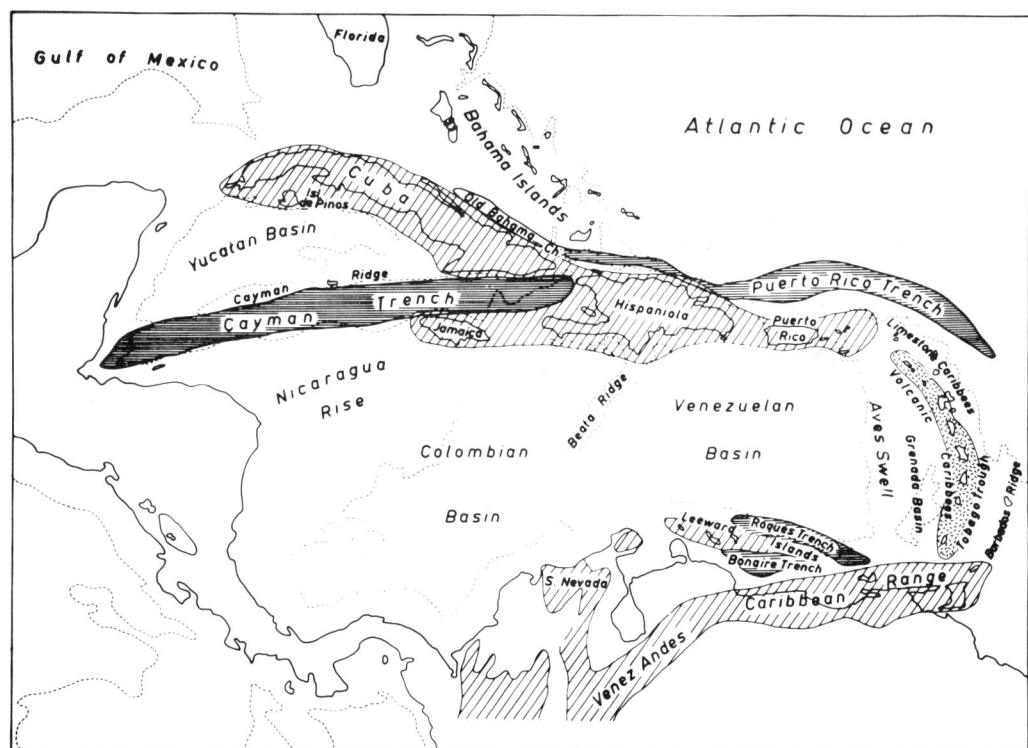

N-S trending volcanic ridge, form the Caribbean Arc proper and are represented by a curvilinear archipelago from Grenada to Sombrero. The islands have classically been divided into the *Limestone Caribbees* (an outer or "non-volcanic" arc) and the *Volcanic Caribbees* (an inner arc) on the basis of the predominant rock type cropping out on the two rows of islands N of Dominica. The northern part of the Lesser Antilles belongs to the Leeward Islands and the southern group to the Windward Islands (i.e. "downwind to a sailing vessel").

At least two distinct long-term phases of volcanic activity produced N-S curved rows of coeval volcanic centers. From Dominica S the two phases are superimposed, but N of Dominica the older volcanic zone of the Limestone Caribbees is offset to the E, thereby creating the double row of islands.

The *Limestone Caribbees* (Marie-Galante to Sombrero, see Fig. 1) consist of andesitic to dacitic flows and volcaniclastics that are primarily of Eocene to Oligocene age, but with at least one radiometrically dated volcanic series of late Jurassic age on La Desirade. The cessation of volcanism around middle Eocene-early Oligocene was followed by erosion, partial truncation, minor hypabyssal intrusive activity, and capping of the volcanic complexes with a thin flat-lying transgressive sequence of late Oligocene-early Miocene age.

The *Volcanic Caribbees* (Saba to Grenada, Fig. 1) N of Dominica consist only of late Miocene or early Pliocene to recent volcanics, but in the single row of islands to the S, the late Tertiary to recent units overlie unconformably the older volcanogenic and sedimentary series.

The island of *Barbados* is geologically distinct from the other Lesser Antillean islands. It is the emergent portion of the submarine Barbados Ridge, which parallels the Lesser Antilles from E of Guadeloupe to the South American continental shelf near Trinidad and Tobago. Field and marine geophysical studies indicate the Barbados Ridge is the result of sedimentary infilling of the Late Cretaceous-early Tertiary Lesser Antillean Trench by longitudinal transport from the S.

The *South American Offshore Islands* belong to two different structural provinces. The eastern islands, Margarita and Tobago, with Trinidad, are part of the Venezuelan Coast Ranges, which are composed of Mesozoic metamorphics and Tertiary sediments. The western islands, principally the Netherlands Antilles or Dutch West Indies, are remnants of a Late Cretaceous geosyncline, whose rocks are similar to those of the Greater Antilles.

The Trenches, Basins, and Ridges. The *Cayman Trough* and the Puerto Rico Trench are both now recognized to be sinistral strike-slip fault zones. However, the exact tectonic style of these features and the nature of any structural connection between them are unresolved problems.

Three smaller trenches lie S of the Venezuelan basin: the *Bonaire Trench, Los Roques Trench,* and the *Cariaco Trench*. Dip-slip faulting has been the dominant tectonic style of these features since the late Cretaceous and any strike-slip motion prior to that time was limited to around a few tens of kilometers.

The crustal layers of the *Nicaragua Rise* and *Beata Ridge* have compressional wave velocities comparable to normal oceanic crust but are more than twice as thick. The continuation of the Central American pre-Mesozoic metamorphic belt through the Nicaraguan Rise into the Greater Antilles has been suggested. Normal faulting of the Caribbean crust in the early Tertiary produced the prominent W-facing escarpment of the Beata Ridge.

The *Aves Ridge* is the third ridge of major proportions in the Caribbean and separates the Lesser Antillean physiographic province from the Venezuelan basin. Its composition, origin, and relationship to the Lesser Antilles Island Arc have yet to be determined. Rock types recovered to date include granodiorite, amygdaloidal basalts, volcanic breccias, pumice, and limestones.

In the *Yucatan basin,* an oceanic crust is found under a cover of sediments, which is intermediate in thickness between the average oceanic layer I and the first sediment layer in the other Caribbean basins.

The crust of the Colombian and Venezuelan basins is intermediate in thickness between that of normal oceanic and continental areas. The main contrast with normal oceanic crust is that instead of a single Layer III there are two layers overlying the mantle—an upper layer 4 km thick (V_p = 6.0–6.3 km/sec) and a lower layer between 5 and 10 km thick (V_p = 7.0–7.3). The composition of the layers comprising the lower crust of the Caribbean basins remains unknown.

Stratigraphic and Orogenic History

The presence of a Paleozoic basement in the Caribbean region has long been assumed, but is now disputed. The metamorphic crystallines of the Isla de Pinos, Cuba, and Hispaniola are considered by several authors to be a continuation of the Paleozoic basement complex of Central America (see *Honduras*, etc.). Others view them as metamorphosed Mesozoic rocks. The "Jamaican Basement Complex" has already been shown to be metamorphosed Cretaceous sediments. It seems probable that metamorphic series of various ages have been involved in the development of the Greater Antilles.

The oldest dated rocks in the West Indies are those of the Middle Jurassic Cayetano and Jagua formations of the Sierra de los Organos, Cuba. They preface the development of the Antillean geosyncline, which following the Lower Cretaceous extended at least from Cuba to the Virgin Islands. The Cretaceous is represented in the Greater Antilles and Leeward Islands by thick deposits of shales, cherts, tuffaceous sandstones, volcanic breccias, agglomerates, and lavas with interbedded limestones. The stratigraphy may be divided with the aid of Rudistacea. The use of these fossils by Chubb in Jamaica was especially successful. The source area of the clastic sediments presents a problem.

At the end of the Cretaceous, the Antillean geosyncline was involved in the Laramide orogeny, during which the focus of orogenic movement slowly migrated E with time. In Cuba the first strong disturbances occurred between the Campanian and Santonian. On Jamaica, Hispaniola, and the Netherlands Antilles, they occurred between the Cretaceous and Eocene. The major orogenic movements did not reach Puerto Rico, however, until after the early Eocene. Late Laramide (Eocene) deformation is also recognized on Jamaica, Cuba, and Hispaniola. The type of structure produced during the Laramide orogeny was largely dependent on the materials involved. Overthrusting occurred in the thick shale and evaporite series of Cuba, with the thrust sheets possibly having dimensions comparable with those of the west alpine nappes. In the tuffs and limestones of Jamaica and Hispaniola, simple folding is found. In Puerto Rico the thick agglomerates and lavas are only weakly folded and most commonly faulted.

The Laramide orogeny was followed in the major part of the West Indies by regional uplift and rapid erosion. In troughs bordering the uplifted areas, thick sequences of conglomerates, sandstones, and shales were deposited; i.e., the Wagwater Formation (Jamaica), the Marigot Formation (Haiti), and the Scotland Beds (Barbadoes). A remnant of the Antillian geosyncline remained in southeastern Cuba and northwestern Hispaniola into the late Eocene.

In the Middle Eocene a new and widespread marine gransgression began which left the Antilles as a few small, scattered islands and deposited a massive, neritic limestone over most of the region. The White Limestone formation of Jamaica, the Plaisance Formation of Hispaniola, and various carbonate unites in Cuba and Puerto Rico were laid down in this transgression. At the same time the Volcanic and Limestone Caribbees were experiencing volcanism, which on the volcanic islands has continued into the present. In the Limestone Caribbees, however, volcanic activity ceased by the Oligocene. Here the eroded base of the volcanics was overlain of Upper Oligocene and Lower Miocene Limestones, hence the name "Limestone Caribbees."

A major deformation began in the Miocene with folding of the Lower Tertiary sediments and uplift of the cores of the present-day islands. Only in detached troughs and along the margin of the islands was there still deposition of shallow-water sediments. These Miocene sediments, which include the Bowden Formation of Jamaica and the Gurabo Formation of Hispaniola, are famous for their abundant and varied marine fauna. On Hispaniola the late Tertiary marine sediments are separated into three graben basins, which strike in an E-W direction through the island.

During the late Miocene and Pliocene, extensive peneplains were formed; i.e., the Guaniguanico peneplain of western Cuba, various erosion surfaces on Hispaniola, and the St. Johns and Caguana peneplains of Puerto Rico and the Virgin Islands. A "cockpit" karst topography was developed in the moist tropical climate on the limestones, excellent examples of which are found in the Sierra de los Organos of western Cuba, the Cockpit country of Jamaica, Los Haitises of the Dominican Republic, the northern part of Puerto Rico, and the Grand Fonds of Guadeloupe.

As a result of the youngest crustal movements in the Antilles, which resulted in faulting and uplift of the islands and sinking in the major troughs, the peneplains now stand at elevations as great as 2000 m. Today they are being rapidly dissected. Figure 2 shows the major faults and grabens: the Cayman, Puerto Rico, and Anegada trenchs in the marine environment, the grabens of Hispaniola, and the strike-slip faults of northern Venezuela on land and offshore. These younger faults control the present-day morphology of the West Indies.

The youngest sediments are Pleistocene coral limestones, which are widespread along the coasts of the islands where they form terraces and raised sea cliffs. Such Pleistocene terraces in the Sierra Maestra of Cuba and northwest Hispaniola today reach elevations of over 300 m and exemplify the strong and relatively young vertical uplifts that have occurred in many parts of the Antilles.

Igneous Activity. The West Indies provide an excellent illustration of the close relation-

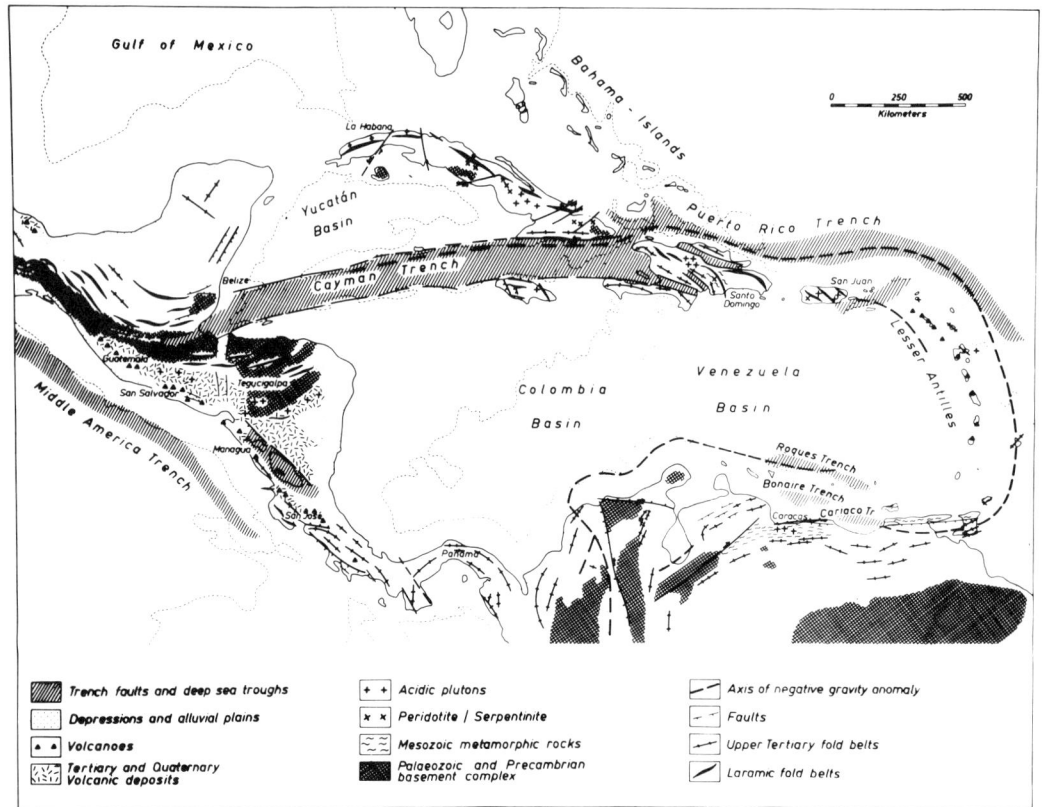

FIGURE 2. Geotectonic map of the West Indies and Central America.

ship between orogeny and the magmatic cycle. Magmatic activity began here in the Cretaceous geosynclinal phase with strong submarine volcanism and outpouring of basalt, andesite, karatophyre, dacite, and spilite. Slightly before and during the Laramide orogeny, great masses of peridotite were intruded, which today form the serpentinite bodies of Cuba, Hispaniola, Puerto Rico, and several of the South American offshore islands. Their exact age and mode of emplacement are still a subject of debate; however, it is most probable that they are, at least in part, of Late Cretaceous age.

Associated with the Laramide orogeny in the Greater Antilles and Netherlands Antilles (Dutch West Indies) are major intrusions of quartz diorite and granodiorite.

Postorogenic volcanism is expressed in massive outpourings of basalt, andesite, bandaite, and dacite, which since the Miocene were concentrated in the Volcanic Caribbees. During the late Tertiary, the Greater Antilles were relatively quiet. In contrast, however, the volcanic Leeward and Windward islands were at this time a center of volcanic activity.

It has long been recognized that the chain of the volcanic Lesser Antilles is closely analogous to the circum-Pacific volcanic belt. The rocks of the Lesser Antilles are characterized by an extremely high content of calcium and sodium and an extremely low content of potassium. Andesites, bandaites, and dacites are dominant.

Recent volcanic activity has been widespread but intermittent. In recorded history, 21 volcanic eruptions have been reported, the most famous being the 1902 eruption of Mt. Pelée on the island of Martinique with its glowing avalanches (nuées ardentes) and gigantic spine. Most of the volcanoes exist today only in the solfataric stage. Characteristic of the volcanic Antilles is the presence of gigantic lava plugs; i.e., the Pitons of St. Lucia (Fig. 3) and Martinique. Sheets of welded tuffs are unknown, calderas, and volcanotectonic depressions like those of Central America are scarce and were not discovered until very recently (Tomblin, 1964). Most of the volcanoes rise immediately from the ocean floor.

There is no evidence of ancient continental crust, which could give the primary source ma-

FIGURE 3. The Pitons of St. Lucia. Typical dacite domes, from which the marginal talus of pyroclastic material has been removed by erosion.

terial for a Pacific-type magma of the Antillean volcanoes. A possible explanation for the occurrence of what are normally sialic rocks in an oceanic environment may be found in the hypothesis that a land mass occupied the area of the Caribbean Sea up to the beginning of the Tertiary. The results of refraction seismographic investigations in the Venezuela basin, however, support the opposite view. They reveal no remnants of a continental crust in the Caribbean basin. This therefore reduces the possibility that the "pacific" magma of the Lesser Antilles was created by melting of an ancient sialic root.

Major Tectonic Elements. The West Indies, with the exception of the Bahamian Platform, are connected to the circum-Pacific orogenic belt, which here extends over 1000 km E from *Central America* (q.v.) into the Atlantic along the Antillean island arc. Although the presence of a Paleozoic basement in Cuba and Hispaniola is uncertain, a Laramide age for most of the basement complex in the Greater Antilles and the Leeward Islands has been firmly established. The structural grain of the Mesozoic basement lies oblique to the strike of recent mountain ranges and the overall trend of the island chain (Fig. 4).

The present tectonic features are the result of late Cenozoic and Holocene events: faulting and uplift in the island chain, sinking in the various Caribbean basins and in the oceanic foredeep, and seismic and volcanic activity. To this extent, the Antillean chain possesses most of the features attributed to a Pacific-type island arc by Gutenberg and Richter and others; the Puerto Rico Trench is a typical oceanic trench or foredeep. It coincides with a belt of shallow earthquakes and strong negative gravity anomalies. The principal structural arc of Late

FIGURE 4. Tectonic stress patterns and sedimentational features of the Caribbean region (by K. O. Emery and E. Uchupi).

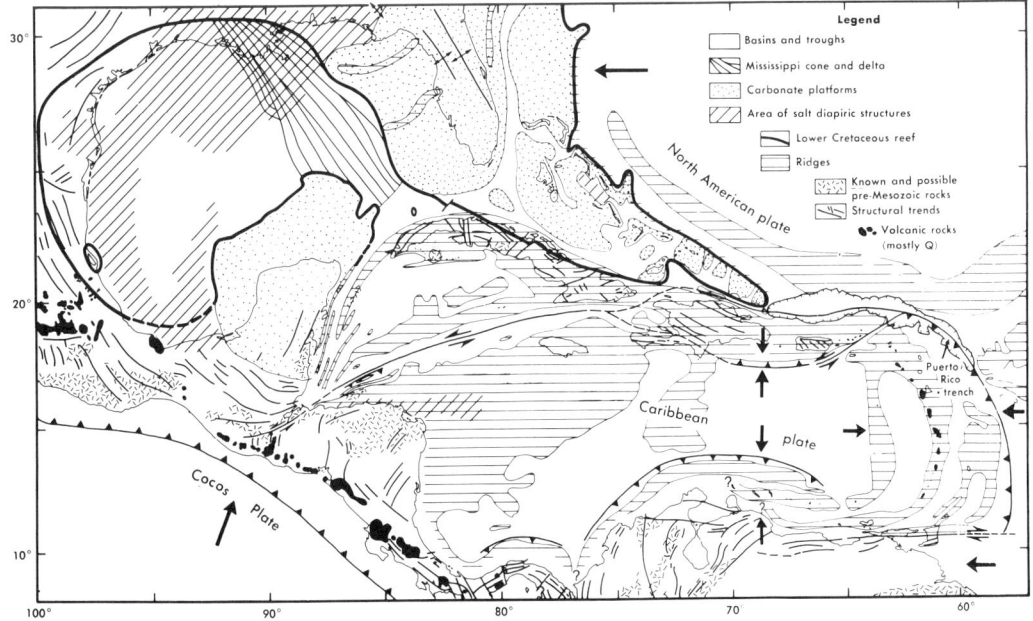

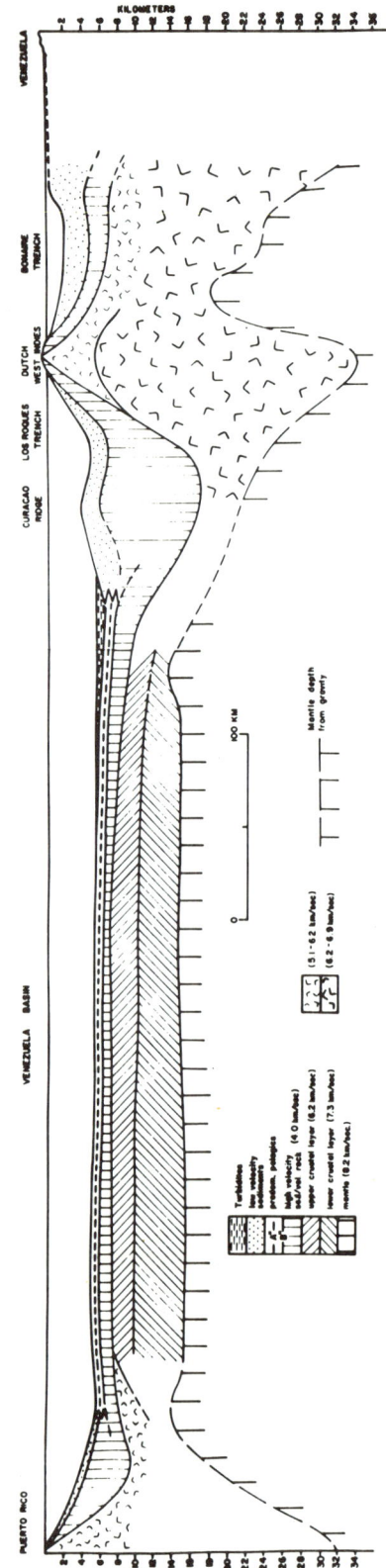

FIGURE 5. Crustal sections of the Caribbean area. (After Officer et al., 1959; Ewing et al., 1960.)

Cretaceous and early Tertiary age is found in the Greater Antilles and the South American offshore islands. The main volcanic arc, accompanied by earthquake shocks at depths of approximately 100 km, is formed by the Volcanic Caribbees. In the basins of the Caribbean Sea, the "backdeep" of the orogenic arc is being built in what appears to be a region of relatively high stability.

The Cayman Trench is a large, ENE-WSW trending graben, which extends into the sinistral strike-slip rift structures of Guatemala and Hispaniola. Numerous earthquakes indicate that the zone is still active.

The overall structural framework of the West Indies has been the subject of numerous hypotheses, the more important of which have been put forth by Van Bemmelen, Bucher, Eardley, Hess, Schuchert, Stille, Vening Meinesz, among others. All these hypotheses, which cannot be discussed here individually, are concerned in part with one of the more important problems: the geological development of the Caribbean basins. The hypotheses are often contradictory. Schuchert considered it to be an old sea basin. Bucher, Butterlin, Chubb, Rutten, and others were of the opinion that until the end of the Cretaceous, a continental land mass occupied this position. The Antillean geosynclines were assumed to have developed on the margins of this land mass. The results of refraction seismograph investigations in the Colombian and Venezuelan basins support a contrary view. They reveal no remnants of ancient continental crust in the Caribbean basin. This therefore reduces the possibility that the Caribbean was created by the foundering of a former continent.

Seismic profiler surveys and sediment coring carried out by the Lamont Geological Observatory have indicated that the interior basins of the Caribbean Sea have been stable, deep-water areas since middle Mesozoic and possibly earlier (Fig. 5). The mapping of a widespread marker of late Mesozoic or earlier Cenozoic age permits certain inferences about the tectonics of the margins of the basins. Vertical, rather than horizontal, forces appear to have been dominant in much of the late deformation. The small variation in the thickness of the Tertiary sediments argues against the concept that sea-floor spreading within the Caribbean Sea itself is responsible for the marginal deformation.

Seismic data, restudied by Molnar and Sykes (1969), support the recent theories of plate tectonics in which large segments of crust have moved coherently with respect to one another as nearly rigid bodies. A Caribbean plate of largely oceanic crust underlies the Caribbean Sea and is bounded by the Middle America arc,

FIGURE 6. Plate tectonic growth of the Caribbean area, at an early Oligocene stage, as modeled by Malfait and Dinkelman (1972).

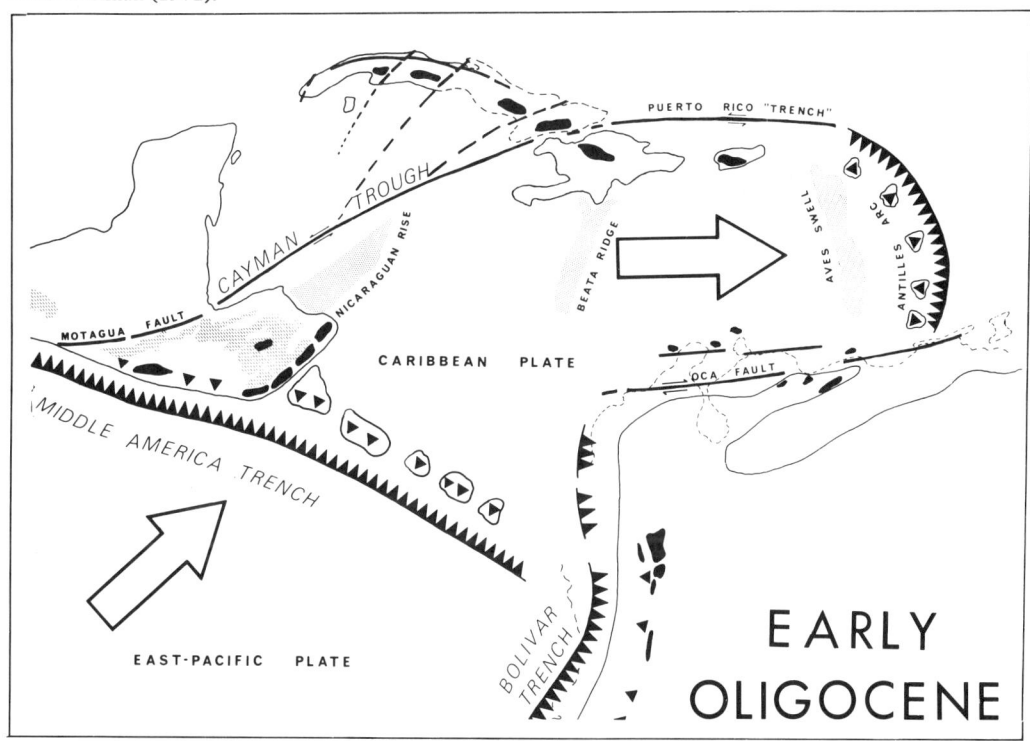

the Cayman Trench, the Antillean arc, and the seismic zone through the northern border of South America. This plate is moving E with respect to the America plate, which today is taken to include both North and South America and the western Atlantic. The America plate is underthrusting the Caribbean in a westerly direction at the Lesser Antilles and near Puerto Rico, producing left-lateral strike-slip lineaments and secondary wedges. Along the northern borders of South America there are corresponding right-lateral transcurrent faults (e.g., Oca Fault). According to a model offered by Malfait and Dinkelman (1972), an intermediate stage, early Oligocene, may be visualized as in Fig. 6. Naturally, this is simply an interpretation and active studies are in progress.

RICHARD WEYL

References

Bader, R. G., et al., 1971. "Leg 4 of the Deep Sea Drilling Project," *Science,* 172(3989), 1197–1205.

Barr, K. W., 1974. "The Caribbean and plate tectonics—some aspects of the problem," *Verhandl. Naturf. Ges. Basel,* 84(1), 45–67.

Bowin, C. O., 1968. "Geophysical study of the Cayman Trough," *J. Geophys. Res.,* 73(16), 5159–5173.

Bunce, E. T., Phillips, J. D., Chase, R. L., et al., 1970. "The Lesser Antilles arc and the eastern margin of the Caribbean sea," *The Sea,* vol. 4. New York: Wiley-Interscience, 359–385.

Butterlin, J., 1956. *La Constitution Géologique et la Structure des Antilles.* Paris: C.N.R.S., 453p.

———, J., 1972. "Regards sur l'origine et l'évolution des unités structurales de la région des Caraïbes," *Bull. Soc. Géol. France.* 14(7).

Cavelier, C., 1968. "Sur la nature superficielle et récente du "substratum volcanique antémiocène" de la Grande-Terre (Guadelope)," *Bull. Soc. Geol. France,* Ser. 7, 9(3), 450–454.

Chase, R. L., and Bunce, E. T., 1969. "Underthrusting of the eastern margin of the Antilles by the floor of the western North Atlantic Ocean and origin of the Barbados Ridge," *J. Geophys. Res.,* 74(6), 1413–1420.

Dengo, G., 1972. "Review of Caribbean serpentinites and their tectonic implications," *Geol. Soc. Am. Mem. 132,* 303–312.

Donnelly, T. W., 1964. "Evolution of eastern Greater Antillean island arc," *Bull. Am. Assoc. Petrol. Geologists,* 48(5), 680–696.

———, ed., 1971. "Caribbean geophysical, tectonic, and petrologic studies," *Geol. Soc. Am. Mem. 130,* 224p.

Eardley, A. J., 1963. *Structural Geology of North America,* 2nd ed. New York: Harper & Row.

Edgar, N. T., Ewing, J. I., and Hennion, J., 1971. "Seismic refraction and reflection in Caribbean Sea," *Bull. Am. Assoc. Petrol. Geologists,* **55**, 833–870.

Ewing, J. I., Edgar, N. T., and Antoine, J. W., 1970. "Structure of the Gulf of Mexico and Caribbean Sea," *The Sea,* vol. 4. New York: Wiley-Interscience, 321–358.

Fink, L. K., Jr., 1972. "Bathymetric and geologic studies of the Guadeloupe region, Lesser Antilles island arc," *Marine Geol.,* **12**, 267–288.

Fox, P. J., Schreiber, E., and Heezen, B. C., 1971. "The geology of the Caribbean crust: tertiary sediments, granitic and basic rocks from the Aves Ridge," *Tectonophysics,* **12**, 89–109.

Gough, D. I., and Heirtzler, J. R., 1969. "Magnetic anomalies and tectonics of the Cayman trough," *Geophys. J. Roy. Astron. Soc.,* **18**, 33–49.

Hatten, C. W., and Meyerhoff, A. A., 1970 (see Škvor, 1969).

Hess, H. H., ed., 1966. "Caribbean geological investigations," *Geol. Soc. Am. Mem. 98.*

Hooker, M., 1969. *Bibliography and Index of the Geology of Puerto Rico and Vicinity, 1966–1968.* San Juan: Geol. Soc. Puerto Rico.

Khudoley, K. M., and Meyerhoff, A. A., 1971. "Paleogeography and geological history of Greater Antilles," *Geol. Soc. Am. Mem. 129,* 199p.

Kumpera, O., and Škvor, V., 1967. "Contribution to the information on the geological development and structure of Cuba and the Caribbean region," *Věstnik Ustředniko Ústavu Geol.* (Prague), **44**, 39–51.

Lagaay, R. A., 1969. "Geophysical investigations of the Netherlands Leeward Antilles," *Verh. Kon. Ned. Akad. Wetensch., Afd. Natuurk.* (Amsterdam), Ser. 1, **25**(2), 86p.

MacDonald, W. D., 1972. "Continental crust, crustal evolution, and the Caribbean," *Geol. Soc. Am. Mem. 132,* 351–362.

MacGillavry, H. J., 1970. "Geological history of the Caribbean," *Proc. Kon. Ned. Akad. Wetensch.,* (Amsterdam), Ser. B, **73**(1), 96p.

Macintyre, I. G., 1972. "Submerged reefs of eastern Caribbean," *Bull. Am. Assoc. Petrol. Geologists,* **56**(4), 720–738.

Malfait, B. T., and Dinkelman, M. G., 1972. "Circum-Caribbean tectonic and igneous activity and the evolution of the Caribbean Plate," *Bull. Geol. Soc. Am.,* **83**, 251–272.

Mattson, P. H., 1965. "Geological characteristics of Puerto Rico. Continental margins and island arcs," *Symp. Geol. Surv. Can. Pap. 66(15),* 124–138.

Molnar, P., and Sykes, L. R., 1969. "Tectonics of the Caribbean and Middle America regions from focal mechanisms and seismicity," *Bull. Geol. Soc. Am.,* **80**(9).

Monroe, W. H., 1968. "The age of the Puerto Rico trench," *Bull. Geol. Soc. Am.,* **79**, 487–494.

Moussa, M. T., and Smith, A. L., compilers, 1970. *Status of Geological Research in the Caribbean,* vol. 6. Puerto Rico: Univ. Inst. Carribbean Sci., 84p.

North, F. K., 1965. "The curvature of the Antilles," *Geol. Mijnb.,* **44**(3), 73–86.

Officer, C. B., et al., 1959. "Geophysical investigations in the Eastern Caribbean: summary of 1955 and 1956 cruises," *Physics and Chemistry of the Earth,* vol. 3. New York: Pergamon Press, 17–109.

Richards, H. G., 1972. "Some aspects of the marine Quaternary of the Caribbean area," *Mem. 6th Conf. Geol. del Caribe,* Margarita, Venezuela, 426–429.

Robson, G. R., and Tomblin, J. R., 1966. *West Indies, Catalogue of Active Volcanoes of the world.* Naples: Intern. Assoc. Volcanol., **20**, 56p.

Schuchert, C., 1935. *Historical Geology of the Antillean-Caribbean Region.* New York: Wiley.

Škvor, V., 1969. "The Caribbean area: a case of destruction and regeneration of a continent," *Bull. Geol. Soc. Am.,* **80**(6). 961–968. (Discussion by Halten, C. W., and Meyerhoff, A. A., 1970, *Ibid.,* **81**(6), 1855–1862).

Tomblin, J. R., 1964. *The Volcanic History and Petrology of the Soufrière region, St. Lucia.* New York, Oxford Univ. Press, 213p.

Weeks, L. A., et al., 1971. "Structural relations among Lesser Antilles, Venezuela, and Trinidad–Tobago." *Bull. Am. Assoc. Petrol. Geologists,* **55**(10), 1741–1752.

Weyl, R., 1966. "Die Geologie der Antillen," *Regionale Geologie der Erde.* Berlin: Gebr. Borntraeger, 410p.

Cross-references: *Antigua; Bahamas; Barbados; Cayman Islands; Cuba; Dominica; Dominican Republic; Grenada; Guadeloupe; Haiti; Jamaica; Martinique; Montserrat; Netherlands Antilles; Puerto Rico; St. Christopher-Nevis-Anguilla; St. Lucia; St. Vincent; Trinidad and Tobago; Virgin Islands.*

WESTERN SAMOA

Western Samoa is an independent state in the South Pacific, covering 2840 km^2 (1097 sq mi). It is comprised of Upolu Island—area 430 sq mi; Savai'i Island—area 730 sq mi; and six small islets. Total population was about 100,000 in 1970. The eastern part of the Samoan chain comes under *American Samoa* (q.v.).

Western Samoa lies between 14°10' and 13°20'S and 170°20' and 172°50'W, about 350 km E of Fiji and 2500 km NE of New Zealand. Upolu Island, upon which Apia, the principal city and capital, is located, is elliptical in outline, measuring 75 × 25 km. It consists of a heavily wooded chain of volcanic cones which reach a maximum height of 1100 m.

Savai'i Island, 75 × 45 km, consists of broad coalescing volcanic domes topped by numerous cones, the highest of which reaches about 1800 m. Volcanic eruptions occurred on Savai'i in 1902 and continuously from 1905 to 1911.

The main mass of Upolu consists of Pliocene(?) volcanics, almost entirely basalts, called the "Fangaloa Volcanics." They were erupted from many cones along a SE-NW rift zone. The oceanic crust here is probably Cretaceous in age. The volcanic accumulations form a shield-shaped dome which became deeply weathered and was cut by deep canyons after eruptions ceased. Eruptions in the Pleistocene and Holocene from cones along the ancient rift poured lavas down the canyons. Most of the flows reaching the coast spread out to form fans on which coral reefs grew.

All islets are tuff cones built by exploding lavas, except Manono Island which is a lava cone.

Savai'i was built over a NW rift, an E rift, and a S rift. It has a core of Fangaloa volcanics of Pliocene(?) age, exposed only in three small areas because it is mostly mantled by middle to late Pleistocene, Holocene, and historic basalts. A major fault cuts the northwestern side of Savai'i.

The Fangaloa lavas range from picrite basalts through hornblende andesites. Trachyte was found at one location on Upolu, indicating that the volcano went through the usual evolution of Pacific volcanoes from tholeiitic basalts to lavas of the alkali suite. Analcite is common in the later lavas.

Living fringing and barrier reefs are common but barrier reefs, notably, are absent from shore lines of steep Pliocene(?) volcanics. Apparently corals were unable to grow fast enough to keep up with the rapid rise in sea level on the steep coasts, when the sea rose eustatically from about minus 130 m at the end of the Pleistocene, but were able to maintain themselves on shelving coasts. Extensive submarine shelves occur at 90 and 180 m, indicating submerged reef levels. Fossiliferous marine tuff deposits of the 7 m eustatic stand of the sea are found on some islets, which suggest a Sangamon age. Marine deposits also indicate 1.5 m and 4.5 m stands of the sea. The same levels were found in American Samoa, indicating that all the Samoan chain has been relatively stable at least since about 120,000 years ago. What appear to be wave-cut benches are found on the oldest rocks at Fangaloa Bay at levels higher than 40 m.

HAROLD T. STEARNS

References

Dana, J. D., 1849. "U.S. Exploring Expedition during the years 1838-42 under command of Charles Wilkes, U.S.N.," *Geology,* 10, 307-336.

Green, R. C., and Davidson, J. M., eds., 1969. "Archaeology in western Samoa," *Auckland Inst. and Mus. Bull.,* 6, vol. 1, 278p.

Hedge, C. E., Peterman, Z. E., and Dickinson, W. R., 1972. "Petrogenesis of lavas from Western Samoa," *Bull. Geol. Soc. Am.,* 83, 2709-2714.

Jensen, H. I., 1907. "The geology of Samoa and the eruptions in Savaii," *Linn. Soc. N.S.W., Proc.,* 31(4), 641-672.

Kear, D., 1967. "Geological notes on Western Samoa," *N.Z. J. Geol. Geophys.,* 10, 1446-1451.

___, and Wood, B. L., 1959. "The geology and hydrology of Western Samoa," *N.Z. Geol. Surv. Bull. N.S.,* 63, 1-92.

Macdonald, G. A., 1944. "Petrography of the Samoan Islands," *Bull. Geol. Soc. Am.,* 55, 1333-1362.

Stearns, H. T., 1944. "Geology of the Samoan Islands," *Bull. Geol. Soc. Am.,* 55, 1279-1332.

Cross-references: *American Samoa; Gilbert and Ellice Islands; Tonga; United States-Hawaii.*

WINDWARD ISLANDS

Together with the *Leeward Islands* (q.v.), the Windwards constitute a part of the Lesser Antilles or Caribbean Arc of the *West Indies* (q.v.). The Windward Islands may be defined geographically as those lying south of latitude 15°N, and include from N to S: Dominica (Br.), Martinique (French), St. Lucia (British, St. Vincent (British), and Grenada (British). With the exception of Martinique they were all formerly part of the British West Indies, but do not include *Trinidad and Tobago* (q.v.) to the SE or *Barbados* (q.v.) to the E.

Geologically the Windward Islands belong entirely to an inner, volcanic arc and are mostly constituted of andesitic lavas and tuffs. There was a main folding in early Miocene times but volcanic activity has continued and both pyroclastics and lavas interfinger with younger formations, mainly limestones, with some deepwater shales. Martinique and St. Vincent still have active centers. (Details are given in the individual island entires.)

The islands lie in the trade wind belt, the winds blowing here mainly from the E and NE. In English usage, "windward" is here taken to mean "downwind", hence the name "Windward for the southerly islands, but in Spanish, French, and Dutch the terms are reversed. The precipitation varies considerably, being low on the low relief islands and on the leeward coasts (to W), but is commonly extremely high on the windward sides and at high elevations. (*Note:* a source of directional confusion exists. On any given island the "windward" side is the NE, but, seen from a sailing vessel in these latitudes, islands lying to the NE are to leeward of an outward-bound ship.)

RHODES W. FAIRBRIDGE

References

Barrabé, L., 1955. "Contribution à l'étude stratigraphique et tectonique des formations sédimentaires de la Martinique," *Rev. Inst. Fr. Petrol. Ann. Comb. Liq.,* **10**(5), 295–308.

Butterlin, J., 1956. *La Constitution Geologique et la Structure des Antilles.* Paris: C.N.R.S., 453p.

Davis, W. M., 1926. *The Lesser Antilles.* New York: Am. Geogr. Soc.

Martin-Kaye, P. H. A., 1969. "A summary of the geology of the Lesser Antilles," *Overseas Geol. Mineral Res.,* **10,** 172–206.

Rea, W. J., 1970. "Andesites of the Lesser Antilles," *Proc. Geol. Soc. Lond.,* **1662,** 39–46.

Robson, G. R., and Tomblin, J. F., 1966. "West Indies," *Catalogue of the Active Volcanoes of the World including Solfatara Fields.* Naples: Intern. Assoc. Volcanol., pt. 20.

Spencer, J. W., 1902. "On the geological and physical development of Dominica, with notes on Martinique, St. Lucia, St. Vincent and the Grenadines," *Quart. J. Geol. Soc. Lond.,* **58,** 341–353.

Tomblin, J. F., 1968. *Geochemistry and Genesis of Lesser Antillean Volcanic Rocks. 5th Caribbean Geol. Conf.,* St. Thomas, V.I.

Trechmann, C. T., 1935. "The geology and fossils of Carriacou, West Indies," *Geol. Mag.,* **72,** 528–555.

Weyl, R., 1966. "Geologie der Antillen," *Regionalen Geologie der Erde,* vol. 4. Berlin: Gebr. Borntraeger, 410p.

Cross-references: *Barbados; Grenada; Leeward Islands; Martinique; St. Lucia; St. Vincent; Trinidad and Tobago; West Indies.*

AUTHOR INDEX

A boldface number or page range indicates an encyclopedia entry by the author; an italic number indicates a work by the author in a reference list; all other numbers indicate text references to the author.

Abbott, D. P., 226, *226*
Abbott, R. T., 120, *121*
Abe, K., *432*
Abou Deeb, J., *21*
Abrard, R., *291*
Ab'Sabér, A. N., 128, 129, 130, *137*
Acevedo-Gonzales, M. J., 255, *255*
Acunha, T. d', 470
Adams, R. D., 337, *337*
Adamson, A. M., *346, 414*
Addicott, W. O., *626*
Adie, R. J., 3, *12, 13,* 273, *274,* 468, 469, 470, 473, 476, 479, *480*
Adkinson, W. L., *403*
Agassiz, A., 291, *291, 292,* 499, *500*
Agassiz, J. L. R., 127
Agassiz, L., 456, 610
Agnew, A. F., *571*
Aguilar-Cortes, E., 512
Aguirre, L., *242, 243*
Ahlfeld, F., *127, 464*
Alberding, H., *654*
Alberta Research Council, 208
Alberta Society of Petroleum Geologists, *192*
Alcock, F. J., 159, 170, *172,* 444, *447*
Alden, W. C., *571*
Alexander, C. S., *364*
Allaart, J. H., *299*
Allan, J. A., 188
Alldredge, L. R., *415*
Allen, C. G., *249*
Allen, C. R., *360*
Allison, E. C., 595, *595*
Almeida, F. F. M. de, 135, *137, 275, 277, 464*
Alpha, A. G., *639*
Alpha, T. R., 514, *521*
Alvardo, B., *425*
Alvarez, G., 470
Alvarez, J., 246, *249*
Amaral, E. J., *607*
Amaral, G., 423, *425*
Ameghino, C., 15
Ameghino, F., 15, 20
American Association of Petroleum Geologists, 628
American Commission on Stratigraphic Nomenclature, 175, *179, 408*

American Geographical Society, 2
American Geological Institute, 137, *513*
Andel, T. H. van, *414, 418, 498*
Anderegg, H., *654*
Anderson, C. A., 555, *556*
Anderson, J. J., 479
Anderson, M. A., **443–447**
Anderson, R. E., *546*
Anderson, R. N., 244
Anderson, R. Y., *556*
Anderson, S. B., *571*
Anderson, T., *348, 448*
Anderson, T. A., *249*
Anderson, T. H., *308*
Andersson, J. G., 273, *274*
Andrada, M. F. R., de, 127
Andrada e Silva, J. B. de, 127
Andrade, G. O. de, *137*
André, E., *448*
Andrews, C. W., 243, *244*
Andrews, J. E., 27, 114, *399*
Andrews, J. T., *156, 408*
Angelelli, V. F., *21*
Angeljan, B. d', *301*
Anisgard, H. W., *654*
Antoine, J., *574, 577, 581, 666*
Aoki, H., *305*
Applin, G. R., *409*
Arctowski, H., 3
Arculus, R. J., 300, *301*
Argand, E., 12
Ariens, P. A., 27
Armstrong, F. C., 634, 637, *639*
Armstrong, R. L., *13, 304, 341, 546*
Arnold, F. K., *447*
Arnould, A., 365, *371*
Arnow, T., 226, *226*
Ashmore, H. T., 570
Asmus, H. E., *137*
Asociación Venezolana de Geología, Minería y Petróleo, 653
Aspinall, W. P., *448*
Atwater, G. I., *577, 581, 607*
Atwater, T., 514, *521,* 546, *546, 626*
Atwood, M. G., *330*
Atwood, W. W., 631, *639*
Atwood, W. W., Jr., 631, *639*
Aubert de la Rüe, E., *283, 284, 291, 346, 390, 414,* 443, 444, *447, 450,* 473, 474, 475, *499, 500, 658*

Aubouin, J., *21, 464*
Audretsch, F. C. d', 372, *382*
Auer, V., *21*
Aumento, F., 171, *172*
Aurich, N., *138*
Austin, P. M., *399*
Australasian Petroleum Company Pty, Lt., 372, *382*
Australian Bureau of Mineral Resources, 55
Australian Mineral Development Laboratories, 278
Avias, J., **365–372,** *371, 414,* 451
Ayres, D. E., *107*
Ayres, W. S., 260, *261*
Azevedo, A. de, *137*
Aziz-ur-Rahman, 100, 402, *402*

Babenroth, D. L., *556*
Bader, R. G., *665*
Bahama Land Resource Survey, 114
Bailey, E. H., *626*
Baillie, R. J., *12*
Bain, J. H. C., *123*
Baird, D. M., *144*
Baker, H. A., 273, *274*
Baker, M. B., *206*
Baker, P. E., 260, *261, 340,* 439, *442,* 466, 468, 469, 470, 471, 472, 478, 480, *480, 496*
Bakker, J. P., 323, *324*
Baksi, A. K., *626*
Baldis, B. A., *21*
Baldwin, E. J., *556*
Baldwin, E. M., *626*
Balk, R., *409*
Ball, G. E., *521*
Ball, M. M., **109–115,** *114*
Ballard, J. A., *359*
Ballard, R. D., *617*
Bally, A. W., 209, 212, 214, *217*
Balme, B. E., *107*
Balsey, J. R., 532, *539*
Bamford, R. W., *382*
Band, R. B., 278
Bandy, M. C., 260, *261*
Bandy, O. L., *418*
Banfield, A. F., 289, *290*
Banks, M. R., **81–91,** *91*
Banner, F. T., *442*
Banwell, C. J., 337, *337*

669

AUTHOR INDEX

Barazangi, M., *409,* 468, 495, 496, *496,* 545, *548*
Barbosa, O., 128, 314, 316, *317,* 322, *324*
Barker, P. F., 273, *274*
Barksdale, H., 532, *539*
Barnes, E., *581*
Barr, K. W., *498, 665*
Barrabé, L., *304,* 348, *348, 667*
Barrero, D., 246, *249*
Barrett, P. J., *12,* **243–244,** *618*
Barrett, W., 258, *259*
Barrie, J., 243, *244*
Barron, C. N., 313, 314, 315, 316, *317,* 319, 321, 322, *324*
Barruol, J., *283,* 315, *317*
Barth, W., *127*
Bartholomew, R. W., 278
Bartlett, G. A., 171, *172*
Bartrum, J. A., 390
Basedow, H., 47
Bass, M. N., *330, 539*
Bassett, W. A., *564, 571*
Bassler, R. S., *539, 539*
Batchelder, G. L., *557*
Bateman, P. C., *626*
Bateson, J. H., *116, 117,* 316, *317,* 321, *324*
Bathurst, R. G. C., *114*
Bauer, E. J., *128*
Bauer, G. R., *496*
Baur, G., 288, *290*
Bayne, C. K., *571*
Beal, M. D., *556*
Beards, R. J., 195, *206*
Bearling, G., 402, *402*
Bebout, D. G., *359*
Becker, M., *450, 493*
Beckman, J. P., *115,* 486, *491*
Beder, R., 15, 419, *425*
Beebe, B. W., *21, 193, 270,* 408, *464, 465,* 521
Beebe, W., 288, *290*
Beets, D. J., *363, 364*
Behre, C. H., Jr., 289, *290*
Bein, J., *45*
Béland, J., 159, *173*
Bell, J. D., *456*
Bell, J. S., *653*
Bell, M. L., *244*
Bell, R., 194
Bell, W. A., 170, *172*
Bellair, P., 480
Bellido, B. E., 427, *432*
Bellizzia, A., 312, 316, 317, *317*
Bellizzia, G. A., *652*
Belsky, C. Y., 486, *491*
Belt, E. S., 170, *172*
Benavides, C. V., 430, *432*
Benavides-Garcia, L., *359*
Bender, M. L., *414*
Bengston, N. A., 328, *330*
Benson, W. E., 451, 486, *491*
Bentley, C. R., 2, *12*
Bentley, R. D., *530*
Berdan, J. M., 198, *206, 617*
Berg, R. R., *639*

Berger, W. H., 486, *491*
Bergstrom, S. M., *208*
Berkland, J. O., *530, 626*
Bermudez, P. J., *255,* 258, *259*
Bernard, H. A., *581*
Bernoulli, D., 113, *114*
Berrangé, J. P., *312,* 313, 314, 315, 316, *317, 318,* 319, 320, 321, 322, 323, *324, 325*
Berry, M. J., 209, *217*
Berry, W., *498*
Berthelsen, A., *299*
Berthois, L., *284, 450*
Bertoni, G. T., 419, 424, *425*
Best, M. G., *556*
Beurlen, K., *137, 425*
Bibikova, E. V., *138*
Bigarella, J. J., **127–138,** *137*
Billings, E., 188, 194
Billings, M. P., *557,* 610, 617
Bird, J. M., 159, 164, *172,* 612, *617*
Bishop, D. G., *399*
Black, L. P., *61*
Black, R. F., 168, *172, 606, 607*
Blackstone, D. L., Jr., *634, 639, 640*
Blackwelder, E., 619, 631
Blake, D. H., *382*
Blake, L. R., *473*
Blanchard, R. L., *347*
Blank, H. R., Jr., *640*
Bleackley, D., 322, 323, *324*
Bligh, W., 447
Blockley, J. G., 103, *107*
Bloom, A. L., 226, *226*
Blow, W. H., *442*
Blundell, D. J., 360, *361*
Bodenbender, G., 15
Boehmer, W. R., 511
Boelrijk, N. A. I. M., 314, 315, 316, *318,* 321, 322, *325,* 363, *364,* 482, 483, 489, *490, 492*
Boettneer, R., 419, 422, *425*
Bogdanoff, A. A., *432*
Bohnenberger, O. H., 228, *236, 251, 252,* 271, *272,* 308, 329, 330, *401*
Bolton, T. E., 195, 197, 198, *206, 207,* 223, *225*
Bonarelli, G., 15
Bond, J. G., 511
Bondam, J., **292–299,** *345*
Bonham, L. D., 345
Bonis, S., 251, *252,* 271, *272,* **305–309,** *308, 401*
Boos, C. M., 633
Booy, T. de, *500*
Borchardt, G. A., 568, *571*
Borchgrevink, C. E., 3
Borello, A. V., *21, 274*
Bornhauser, M., 575, 576, 578, *581*
Bosma, W., 313, 314, 315, 316, *317,* 481, 482, 490, *490, 491*
Bossi, J., 642, *647*
Boucot, A. J., 428, *432, 617*
Bougainville, L. de, 451

Bouguer, P., 261
Boureau, E., 368, *371*
Bourrouilh-Le Jan. F., *244*
Boussingault, J. B., 245
Bouvet de Lozier, J. F. C., 471
Bowes, D. R., *634, 641*
Bowin, C. O., 113, *115,* 227, 258, *259, 340,* 437, *665*
Bowman, R. I., *290*
Boyé, M., *283*
Boyer, R. E., *572–581*
Boyer, W. T., *572–581*
Bracewell, S., 310, 319, *324*
Bracey, D. R., 327, *327, 345*
Brackebusch, L., 15, 456
Bradley, J. S., *114*
Bradley, W. H., *640*
Brady, M. J., *360*
Braithwaite, J. C., *399*
Branisa, L., *127*
Branner, J. C., 127, *137,* 275, 277, 456
Branson, C. C., *618*
Branson, E. B., *606*
Braunstein, J., *533, 539, 581*
Braustein, J., *255*
Bravard, A., 15
Breed, W. J., *556*
Bretz, J. H., 120, *121*
Brew, D. A., *522*
Bricker, O. P., 120, *121*
Bridge, J. S., *226, 449*
Bridgwater, D., *299*
Brien, G. D., *255*
Briggs, R. P., 435, 436, *437*
Brigham, R. J., 194, *206*
Brigham, W. T., *292, 414, 494*
Brineman, J. H., *237, 252,* 329, *330*
Brinkmann, R., *324,* 485, *491*
Bristow, C. R., *270*
British Columbia Department of Mines and Petroleum Resources, 208
Brobst, D. A., *513*
Broch, O. A., *471*
Brodie, J. W., *496*
Brodrick, A. H., *290*
Broecker, W. S., *114,* 115, *115, 347*
Brongniart, A., 245
Bronnimann, P., *115*
Brookhart, J. W., *226*
Brookins, D. G., *617*
Brosge, W. P., *521*
Brothers, R. N., *28,* 337, *337,* 366, *371*
Broughton, J. G., *617*
Brouwer, G. C., **282–284,** 315, **362–365**
Brown, A. S., *187*
Brown, C. B., 309, 319, *324*
Brown, D. A., **28–34,** *34,* 95, *100*
Brown, G. M., 300, *301*
Brown, H. Y. L., 46
Brown, J. S., *327*
Brown, P. M., 536, *539*
Brown, R. D., Jr., *401*

670

Browne, W. R., *34, 45, 61, 382*
Brückner, W. D., 160, 163, *172*
Brüggen, J., 238, *242*
Brune, J. N., *360*
Brunner, C. A., *13*
Brunnschweiler, R. O., *34*
Brunt, M. A., 227, *227*
Bryan, E. H., *657*
Bryan, G. M., 275, *277*
Bryan, W. A., *657, 657*
Bryan, W. B., *383*, 495, *496, 497*
Bryson, R. A., *607*
Bucaram, P., *270*
Bucher, W. H., *364*, 598, *606*, 664
Buckley, D. K., 110, *115*
Buffington, E. C., 514, *521*
Bukry, D., *414*
Bullard, E. C., 112, *114*
Bunce, E. T., *340*, 665
Burbank, W. S., *327*
Burchett, R., *572*
Burchfiel, B. C., *546, 626*
Burckhardt, C., *359*
Burckle, L., *21*, 574, *581*
Bureau of Mineral Resources, Australia, 29, 30, 47, *123, 382*
Burford, R. O., *627*
Burger, D., *318*, 322, *325*
Bürgl, H., 245, 246, 249, *249*, 464
Burk, C. A., *480*, 516, *521*
Burke, D. B., 544, *547, 548*
Burke, H. W., 305, *305, 449*
Burke, K., *335*
Burke, R. O'H., 92
Burmeister, C., 15
Burns, K. L., *91*
Burns, R. E., *27*
Burri, C., *278*
Butera, J. G., *640*
Butler, L. W., 642, 645, 647, *647*
Butterlin, J., 246, 249, 255, **257–259**, *259*, **326–327**, *327, 340,* **348,** *348,* 664, *665, 668*

Cady, W. M., *513, 617*
Caillère, S., *371*
Cailleux, A., *13*
Caldwell, W. G. E., *193*
Caley, J. F', 195, *206*
Calkin, P. E., *13*
Callen, D. J., 24
Calver, J. L., 512
Cameron, A. E., 188
Cameron, C. P., *521*
Campana, B., *80*
Campbell, C. J., **261–270**
Campbell, K. S. W., *13, 34,* 100, *100*
Campos, C. de, 127
Camsell, C., 208
Canfield, R. W., *270*
Cannon, R. T., 313, 314, 317, *318*
Cano, S. del, 472
Cant, R. V., *335*
Canu, F., 539, *539*
Caorsi, J. S., 642, 643, *647*

Capanema, G. S. de, 127
Carey, S. W., *27*, 81, 235, *236*, 377, 378, *382,* 384
Carlile, R. E., *522*
Carlson, C. G., *571*
Carlson, M. P., *571*
Carmichael, I. S. E., 377, *383*
Carnegie, 47
Carnier, K., 419
Carozzi, A. V., *606*
Carpenter, R. H., *236, 330, 530*
Carte Géol. Dét. de la France, 283
Carter, C., *521*
Carter, M. W., 321, *324*
Carter, W. D., *242*
Carvalho, P. F. de, 128
Case, I. E., *249*
Case, J. E., *249, 418, 557*
Casertano, L., *464*
Casey, R. E., *418*
Cassan, J. P., 486, *491*
Castelnau, F. de, 127, 456
Castillo, L. del, 354, *360*
Catala, R. L. A., *414*
Cater, F. W., *556*
Cavelier, C., *304, 665*
Cebull, S. E., *548, 607*
Cebulski, D. E., *107*
Cecioni, G., *242*
Cederstrom, D. J., 532, *539, 656*
Chamberlin, R. T., 1
Chamisso, A. von, 260
Chamot, G. A., 126, *127*
Chapman, F., *390*
Chapman, L. J., *206*
Chappell, J., 381, *382, 383*
Chase, C. G., 279, *282, 345, 415*
Chase, R. L., *340,* 665
Chauveau, J., 499
Chenoweth, P. A., 112, *114*
Chevalier, J. P., **342–343**, *343,* 346, 366, **449–451**, *493,* **498–499**, *499,* **499–500**
Chewings, C., 46
Childs, O. E., *21, 217, 236, 237, 252, 270, 330, 408, 464, 465, 521*
Chilingar, G. V., 401, *401*
Chipping, D. H., *408, 626*
Choubert, B., *283, 284,* **309–318**, *317, 318*
Christensen, M. N., *243*
Christiansen, R. L., *409,* 546, *546, 626, 640*
Christie, R. L., **145–156**
Christman, R. A., 15, *340, 442*
Christoffel, D. A., *399*
Christophersen, E., 342, *342*
Chronic, J., 226, *226, 415,* 429, 433
Chronic, J. H., *640*
Chubb, L. J., *251, 261, 284, 290, 335, 336,* 346, *346,* 413, *414,* 449, *450,* 499, *500,* 664
Church, W. R., *172*

Churkin, M., Jr., *421*
Cingolani, C. A., *21*
Cizancourt, M. de, *115*
Clague, D. A., *414,* 587, 595
Claridge, G. G. C. 402, *402*
Clark, A. H., *242, 243*
Clark, J. R., 203, *207*
Clark, T. H., *144, 193,* **218–225**, *225, 408, 513, 607*
Clark, W. B., 538, *539*
Clark, W. J., *442*
Clark, W. O., 585, 594, *596*
Clarke, D. W., *61*
Clarke, H. A., *274*
Clarke, W. B., 81
Clausen, H. B., 299
Cleary, J. R., *91*
Clemons, R. E., *308*
Cleve, P. T., *340,* 654
Clifford, P. M., 159, *172*
Clifton, B. B., *100*
Clinton, N. J., *556*
Cloos, E., *530, 531*
Closs, D., 644, *647*
Cloud, P. E., Jr., *137,* 305, *305,* 344, *345, 408, 449*
Clough, J. W., *12*
Coates, D. R., 8, *121, 530*
Cobbing, E. J., *428,* 430, *432,* 464
Colbert, E. H., 3, *12, 13, 408*
Cole, W. S., *278, 347, 449, 498*
Coleman, A. P., 202, 203, *206*
Coleman, P. J., *27, 278,* **383–390**, *390,* 402, *402, 414,* **451–456**, *456*
Coleman, R. G., *626*
Colette, B. J., 486, 487, *491*
Colinvaux, P. A., *290*
Colley, H., 384, *390, 456*
Collier, A. J., *571*
Collins, H. R., 512
Collins, R. S., *513*
Collinson, J. W., 3, *13*
Colman, J. A. R., 266, *270*
Colonna-Salard, M., 368, *371*
Colony, W. E., *290*
Colorado Scientific Society, *639*
Colquhoun, D. J., 538, *539*
Columbus, C., 257, 360, 447
Combes, P. J., *371*
Commission de la Carte Géologique de Monde, *464*
Compston, W., *27*
Comte, D., 421, 423, *426*
Condamine, C. M. de la, 261
Condon, M. A., *107*
Condon, T., 619
Condra, G. E., *571*
Coney, P. J., 430, *432, 464*
Coneybeare, C. E. B., *382*
Conn, H., *242*
Conolly, J. R., *4, 21, 27, 28, 34, 45, 91*
Conrad, S. G., 512
Coogan, A. H., *359*
Cook, J., 60, 331, *365,* 402, 467, 468, 494

AUTHOR INDEX

Cook, K. L., *546, 547*
Cook, P. J., 51, 52, *55, 56, 61*
Cooke, C. W., 539
Cooke, D. L., *336*
Cooke, R. U., 242
Cooley, F. E., *556*
Cooper, B. N., *530, 531*
Cooper, J. A., 51, *55*
Cope, E. D., *442*
Corbitt, L. L., *547*
Corcoran, R. E., 512
Cordani, U. G., 276, *277*
Cornwall, I. W., *359*
Corral, J. I., *255*
Cortes, R., *242, 464*
Corvalan Diaz, J., *242,* 243
Corwin, G., *345, 345*
Cossman, M., *348*
Costa Sena, C. da, 127
Cotton, C. A., 390, *399, 414*
Coudray, J., 366, 368, *371*
Coulson, N., *278*
Cousteau, J., 116
Coutinho, P. N., *278*
Couto, J. V., 127
Cox, A., 287, *290*
Cox, D. C., *588, 595*
Cox, E. J., *571*
Cox, S. H., 390
Craddock, C., *13,* 479
Craig, L. C., *556*
Cram, I. H., *408, 513,* 520, *521, 556, 607*
Cramer, F. H., *533, 539*
Cramer, G. H., *117*
Crandell, D. R., *571*
Crawford, A. L., *547*
Crawford, A. R., *13*
Crawford, J. C., 390
Creager, J. S., *521*
Creer, K. M., 12, 21, 464
Crevaux, 309
Crickmay, C. H., 622
Crie, L., 365
Crist, E., *426*
Crohn, P. W., 50, *56*
Croizat, L., 288, *290*
Croneis, C., 607
Crook, K. A. W., *34, 45,* 100, *100*
Crosby, J., *581*
Cross, W., *595*
Crossland, C., *251, 284, 414, 450, 493, 499*
Crowell, J. C., 3, 11, 13, *34, 91,* 273, *274, 626*
Cserna, Z. de, **348–360,** *359*
Cuerda, A. J., *21*
Cull, J. P., *382*
Cullen, D. J., *27, 399*
Cumming, L. M., 170, 172
Curray, J. R., *359*
Curry, W. H., *640*
Curtis, J. W., *382, 456*
Cushman, J. A., *15*

D'Addario, G. W., **46–56**
Daetwyler, C. C., *255*

Dahlstrom, C. D. A., 214, 216, 217, *217*
Daily, B., *80, 91*
Dall, W. H., 539, *539*
Dalmayrac, B., *432, 433,* 428
Dalmus, K. F., *654*
Dalrymple, G. B., 287, *290,* 591, *595*
Daly, R. A., 1, *1,* 179, *497,* 622
Dalziel, I. W. D., *13,* 242, 464, **465–480** *607*
Damberger, H. H., *607*
Damon, P. E., 430, *433*
Damuth, J. E., 275, *277, 318,* 323, *324, 487, 491*
Dana, J. D., *1, 493, 582, 595, 667*
Daneš, J. V., *335*
Daniels, J. L., 104, *107*
Dansgaard, W., *299*
Dardenne, M., *137*
D'Argenio, B., 113, *114*
Darlington, P. J., Jr., *290*
Darnley, A. G., *179*
Darton, N. H., *255,* 562
Darwin, C., 3, 15, 81, 127, 245, *245,* 269, 273, 274, *277,* 288, *290,* 347, 412, 413, 456, 642, *647*
Dasch, P. R., *80*
David, E., 3
David, T. W. E., 3, *34, 45, 61,* 291, 292, *382, 414*
Davidson, J. M., *667*
Davies, G. R., *107*
Davies, H. L., *27,* 378, 380, *382*
Davies, T., *277*
Davies, W. E., *521*
Daviess, S. N., 115, *116*
Davis, A., *61*
Davis, D. A., *588, 595*
Davis, G. A., *626*
Davis, J. F., 512
Davis, M. B., *531*
Davis, S. N., *571*
Davis, W. M., *226,* 245, *245, 257,* 284, 291, 301, 332, 340, *341, 342, 343,* 365, *414, 371, 433, 434,* 443, *443, 494, 500,* 524, 543, *668*
Davoren, P. J., *34*
Dawes, P. R., *300*
Dawson, G. M., 144, 179, 188, 208
Day, W. H., *432*
Dean, W. E., Jr., *556*
DeBuissonjé, P. H., *364*
Deffeyes, K. S., 363, *364*
Defoe, D., 336
Deiss, C. F., 629, *640*
DeJong, K. A., *173*
De La Beche, H. T., *658*
Delaney, J. R., *290*
Delaney, P. J. V., 642, 643, 646, *647*
Den, D., *333*
Denburg, J. van, 288, *290*
Deneufbourg, G., *499, 499*
Dengo, G., **116–117,** *236,* 251,
252, 271, *272, 308,* 329, 330, *330,* 401, *665*
Denham, D., *27,* 122, *123,* 377, *382,* 452, *456*
Denizot, M., *499, 499*
Denmead, A. K., 60, *61*
Dennis, J. G., *598, 607*
Denton, G. H., *13*
D'Entrecasteaux, J. A. B., 81
Derby, O., 127, 456
Derbyshire, E., *91*
Deslongchamps, E., 365, *371*
Detrick, R., *415*
Deutsch, S., *21*
Dewey, J. F., 159, 164, *172,* 612, *617*
Dickerson, R. E., *116*
Dickins, J. M., *61*
Dickinson, W. R., *304, 341, 513, 667*
Dieffenbach, 390
Dietz, R. S., 110, 113, 114, *114, 332, 414,* 478, 584, *581,* 587, *595*
Dill, R. F., *116*
Dillon, W. P., *116*
Dimroth, E., *179*
Dineley, D. L., *156*
Dinkelmann, M. G., 235, 236, *236,* 329, *330,* 401, *401, 419,* 665, *666*
Dirección de Geología (Venezuela), 653
Diver, B. B. van, *617*
Dixon, C. G., 116, 228, *236,* 319, *324, 325*
Doan, D. B., *305*
Dobrin, M. B., *347,* 414
Dobrovolny, E., *522*
Dodd, J. R., *539*
Doell, R. R., *595*
Doering, A., 15
Dolfuss, A., 305
Doll, C. G., 512
Domeyko, I., 456
Donath, F. A., *408, 415*
Dondoli, C., *252*
Donnell, J. R., *640*
Donnelly, T. W., 259, *304, 336, 341, 442,* 655, *656*
Doornbos, D. J., 486, 487, *491*
Doran, E., Jr., 109, *227, 500*
Dorman, H. J., 468
Dorst, J., *290*
Dort, W., Jr., *571, 640*
Dott, R. H., Jr., *607,* **618–627,** *626*
Douglas, E. M., *513*
Douglas, J. G., *81, 100*
Douglas, R. J. W., 144, 145, 156, *172, 173,* 174, *179,* 187, 193, 195, 198, *206, 207, 217,* 225, *408*
Doumenge, F., *284,* 366, *450*
Doutch, H. F., *61*
Dow, D. B., 104, *107,* 374, *382*
Dowling, D. B., 144, 188
Doyle, H. A., *27*

Doyle, R. G., 511
Dozy, J. J., 374, 381, *382*
Drake, C. L., 112, *115,* 168, *172,* 456, 480
Drapeau, G., *173*
Dreeszen, V. H., 511
Dreimanis, A., 201, 202, 204, 205, *206, 207, 208, 225*
Dresher, W. H., 510
Drooger, C. W., *341, 364, 442*
DuBar, J., 537, *539,* 540
Duberal, R. F., 278, 280, *282*
Du Bois, G. C., 309
Ducloz, C., *255*
Duffield, W. A., 594, *596*
Dunbar, C. O., 223, *408*
Dunn, D. E., *531*
Dunn, P. R., *34,* 50, *55, 56, 80*
Dunne, J. C., 470
Durham, J. W., **284–290,** *290,* 595, *595*
Durif, P., 486, *491*
Dürr, F., 271, *272*
Dury, D. H., *607*
Du Toit, A. I., 12, 274, *274*
Dutro, J. T., Jr., 417, 514, *521*
Dutton, C. E., 549

Eade, J. V., *399,* 494, *497*
Eames, F. E. *442*
Eardley, A. J., **403–409,** *408,* 542, *547,* 556, 622, 628, 632, 633, 637, *640,* 664, 665
Earle, K. W., *257,* 15, *301,* 442, *448,* 654, *655*
Easterbrook, D. J., *626*
Eastler, T. W., *172*
Easton, W. H., 586, *595*
Eastwood, W. P., *571*
Eaton, G. P., *547*
Eaton, J. P., *595*
Eberlein, G. D., *521*
Ebert, H., 131, *137*
Eckel, E. B., 419, 422, *426*
Eckelman, F. D., *531*
Edgar, N. T., 326, *666*
Edwards, K. L., *654*
Eggler, D. H., *227*
Eibl-Eibesfeldt, I., *290*
Eifler, G. K., Jr., *581*
Elders, W. A., *626*
Elias, M. K., *571*
Elliot, D. H., 3, *13,* 466, 467, 468, 469, 470
Ellis, A. F., *292, 362*
Elsasser, W. M., 43
Elston, D. P., *557*
Elston, W. E., *547*
Embleton, B. J. J., 27, *27, 34*
Emerson, B. K., 610
Emery, D. P., *499*
Emery, K. O., 113, *115,* 342, *347, 414,* 438, *540,* 663
Emiliani, C., 236, *236*
Engel, A. E. J., *252, 640*
Englert, S., 260, *261*
Epp, D., *333*

Epstein, A. G., *530*
Epstein, J. B., *530*
Erikson, A. J., 227, *227*
Ernst, W. G., *626*
Erwin, R. B., 513
Escalante, G., *252*
Escher, A. E., *300*
Eschwege, W. L., 127, 456
Espirat, 366
Eugster, H. P., *640, 641*
Evans, J., 456
Evans, J. H., *61*
Evenson, E. B., *607*
Everett, J., 112, *114*
Ewart, A., 45, *45, 497*
Ewing, J. I., *114,* 664, *666*
Ewing, M., *21,* 118, *121,* 574, 577, *581,* 642, 647, *647*
Exon, N. F., *61*
Eyk, J. J. van der, 486, *491*

Fagan, J. M., 530
Fahrig, W. H., 177, *179*
Faill, R. T., *530*
Fairbairn, H. W., 160, *172,* 311, *318, 618*
Fairbridge, R. W., **2–12,** *13, 107,* 109–115, *114,* **115–116,** 121–123, *138,* **225–226,** *244,* **244–245,** *245,* **250–251,** *251–252,* 257, *257–259,* **270–272,** *273–274,* **274–278,** 280, *282,* **284,** **290–291,** *292,* **300–301,** *318,* 323, *324,* **327–330,** 336, **339–341,** *341–342,* **345–346,** **346–347,** *347,* **362,** 366, *371,* 380, *382,* **400–401,** **410–415,** *414,* 416, 433, *433–434,* 443, *447–448,* 449, 451, 487, *491,* **493–494,** *494–497, 497,* **497–498,** **498–499, 499–500,** 585, *595,* **627–641,** *657,* **667–668**
Falconer, D. J., 642, *647*
Falke, H., *261*
Falls, D. F., *243*
Faniran, A., *45*
Farrar, E., *242*
Favera, J. C. della, *138*
Fearn, 362
Feden, R. H., *359*
Fenneman, N. M., *513,* 541, *547, 556,* 559, 560, 561, *571,* 627, 628, *640*
Ferdon, E., 260, *261*
Ferm, J. C., *45*
Fernández, J., 449
Ferrar, H. T., 473
Ferreira, E. O., *138*
Ferrigno, K. F., 200, *207*
Feruglio, E., *21*
Fiedler, V. G., *653*
Fieldes, M., 402, *402*
Fiji Geological Survey, 278
Fillon, R. H., *408*
Fink, L. K., Jr., **14–15,** **301–304,** *304,* 340, *341,* 348, *382,* **339–341, 439–443,** *442, 666*

Firman, J. B., **61–80,** *80*
Fischer, A. G., 344, *345,* 427, 429, *432*
Fisher, D., 55
Fisher, D. W., 196, *617*
Fisher, G. W., *408,* 530, 531
Fisher, N. H., 121, 122, *123,* 383
Fisher, R. L., *242,* 260, *261*
Fisher, S. P., *418*
Fisher, W. L., 512
Fisk, H. N., *581*
Fitch, T. J., *332*
Fittkau, E. J., *426*
Fitzsimmons, J. P., **548–557**
Flavio, F., 277
Flawn, P. T., *408*
Fleck, R. J., *21, 547*
Fleming, C. A., *243,* 397, 398, *399*
Flett, J. S., 448
Flinn, D., *300*
Flint, R. F., *408, 571, 607*
Flinter, B. H., 45
Flores, G., *116, 236*
Folk, R. L., *55,* 227
Foord, E. E., *540*
Forbes, B. G., *80*
Forbes, D., 456
Forbes, R. B., *521*
Ford, A. B., 12, *13*
Ford, T. D., *556*
Forjonel, J., *653*
Forman, D. J., 51, 52, *55, 56*
Forman, M. J., 577, *581*
Forrest, J., 46
Fortier, Y. O., *156*
Fosberg, F. R., *347,* 512, *657*
Fountain, R. J., *456*
Four Corners Geological Society, *557*
Fox, F. G., *408*
Fox, J. P., *251*
Fox, P. J., 329, *330,* 486, *491,* 666
Frakes, L. A., 3, 11, *13, 34,* 91, 273, *274*
Francheteau, J., *464*
Franklin, E. H., *100*
Franklin, J., 188
Freeland, G. L., *521*
Fremont, J. C., 541
Frerichs, W. E., 487, *491*
Frey, D. G., *513,* 522, *540,* 556, *581, 618, 641*
Frey, M. G., *530*
Fricke, W., *127*
Fritts, C. E., *521*
Frolor, A. L., *13*
Frye, J. C., *571, 607*
Fuenzalida, H., 238, *242*
Fuller, J. G. C. M., *571*
Furon, R., *414*
Furrazola-Bermudez, G., 254, *255*

Gabb, W. M., 228, 258, *259,* 363, 658
Gabrielse, H., **179–187,** *187,* 210, *217*
Gadd, N. R., *225*

AUTHOR INDEX

Gallowny, J. J., *437*
Gansser, A., 316, *318*, *464*
Garaya, J., 427, *433*
Gardiner, J. S., 278, *282*, *292*
Gardner, J., 539, *540*
Gardner, J. V., *61*
Garfunkel, Z., *626*
Garner, H. F., 490, *491*
Garnier, J., 365, *371*
Garrison, R. E., *626*
Gartner, S., 236, *236*
Gaskell, T. F., 292, *292*
Gastil, G., *626*
Gatehouse, C. G., *13*
Gates, G. O., 414, 514, *521*
Gates, R. M., *618*
Gealey, W. K., *272*
Gebelein, C. D., *121*
Gee, R. D., *91*
Gees, R. A., 118, *121*
Gemuts, I., 104, *107*
Geological Society of America, *639*
Geological Society of Australia, 23, 27, 34, 52, 55, 100
Geological Survey of Canada, 180, 187, *193*, 204, 205, 208
Geological Survey of Greenland, *300*
Geological Survey of South Australia, 61
Geological Survey of Western Australia, 103, 104, *107*
George, 47
George, H. K., *324*
Geraghty, J. J., *513*
Gerhard, J. E., *581*
Gerlach, T. M., *290*
Gerth, H., *21*, *464*
Gibbs, B., *448*
Gibbs, P. E., *251*
Gibson, T. G., *540*
Giddings, J. L., 514, *522*
Gierloff-Emden, H. G., *272*, *359*
Giglioli, M. E. C., 227, *227*
Gilbert, C. M., *547*
Gilbert, G. K., 541, 543, 545, *547*, 549, 619
Giles, E., 46
Giletti, B. J., *432*
Gill, E. D., *34*, **91–100**, *91*, *100*
Gill, J. B., *27*
Gillmann, M., 486, *491*
Gilluly, J., *408*, *513*, *540*, 543, *547*, 556, 622, *626*, *640*
Ginsburg, R. N., 112, *114*, 115, 119, *121*
Girod, M., 466, 470, 471, 472, 473, 474, 475, 476, 477, 478, *480*
Glaessner, M. F., 61, *80*, *382*, 451
Glaister, R. P., *144*, *193*, *217*
Glasser, E., 365, *371*
Glenister, B. F., *107*
Glikson, A. Y., *61*
Glockhoff, C., 114, *114*
Glover, J. E., 103, *107*
Glover, L., III, *437*, *531*
Goddard, E. N., *640*

Goldberry, R., *45*
Goldschlag, M., 419, *426*
Goldthwait, R. P., *521*, *606*, *607*
Golson, J., *34*
Goñi, J. C., 642, 643, *647*
Gonord, H., **365–372**, *371*
Gonzaga de Campos, L. F., 127, 128
Gonzales, E., *249*
González Bonorino, F., **15–21**, *21*, *242*
Gonzalez-Ferran, D., *243*, 469
Gonzalez-Reyna, J., *359*
Goodwin, R. H., *28*
Gorceix, H., 127, 456
Gordon, M., Jr., 258, *259*
Gordy, P. L., *209*, 212, 214, *217*
Goreau, T., *335*
Gorini, M., **274–278**, *277*
Gostin, V. A., *80*
Gough, D. I., *227*, *666*
Gould, C., 81
Gould, S. J., 111, 120, *121*
Graf, C. H., *653*
Graham, A., *236*
Gramlick, J. W., *595*
Grange, J., 3
Grange, L. E., *251*
Grantham, D. H., 322, *324*
Grasty, R. L., *179*, *243*
Graty, A. M. du, 419
Grauch, R. I., *653*
Green, D. C., *61*, *91*
Green, G. R., *91*
Green, R. C., *667*
Green, T. H., *402*
Greenwood, R. H., *61*
Gregory, A. C., 3, 46
Gregory, H. E., *251*, *342*, 411, *433*, 549, *556*
Gregory, J. W., 412, 414, *414*
Gregory, M. R., *399*
Gresens, R. L., *626*
Gressitt, J. L., *13*, *415*
Griffiths, J. C., 323, *325*
Griffiths, J. R., *28*, *34*, *91*, *100*, 398, *399*
Grigg, R. W., 594, *595*
Grikurov, G. E., *4*, *13*
Grimes, K. C., *61*
Grimsdale, T. F., *292*
Grindley, G. W., *13*, *399*
Grodidier, E., 486, *491*
Groeber, P., 15
Groeneweg, W., 313, 314, 315, 316, *317*, 481, *490*
Groff, S. L., 511
Groot, C. R., *21*
Groot, J. J., *21*, 538, *540*
Gross, M. G., *415*, 591, *595*
Gross, M. G., Jr., 120, *121*
Grosse, E., 245
Grover, J. C., 451, *456*
Grow, J. A., 514, *521*
Grubb, P. L. C., *61*
Grüger, E., 606, *607*
Grunevald, H., 348, *348*

Grushinsky, N. P., *13*
Gubler, Y., *270*, *371*
Guerin, S., 367, *371*
Guest, J. E., *242*, *243*
Guest, N. J., *497*
Guilcher, A., *284*, 366, *414*, *450*
Guillon J-H., 366, *371*
Guimarães, D., *128*, *138*, *278*
Gunn, B. M., *13*, *348*
Gunning, H. C., *187*
Guppy, H. B., 245, *245*, *451*
Gussow, W. C., 170, *172*, *640*
Gutenberg, B. 663
Guyana Geological Survey, *324*
Guzman, E. J., *359*
Gwinn, V. E., *530*
Gyrc, G., 417, 514, *521*

Haantjens, H. A., *382*
Haast, J. von 390
Hack, J. T., 537, *540*, *556*
Hackman, B. D., *451*
Hadley, J. B., *13*, *530*, *618*
Haeberle, F. R., *343*
Hagan, W. W., *511*
Hage, C. O., 188
Hagner, A. F., *641*
Hague, J. D., *342*, *433*
Hainer, J. L., *571*
Halbouty, M. T., *408*, *581*
Hall, I. H. S., *116*
Hall, J., 512, 610
Hallam, A., 12, *13*
Halle, T. G., 273, *274*
Halley, E., 275
Halpern, M., *242*, 642, 643, 644, *648*
Halunen, A. J., Jr., *456*
Ham, C. K., *270*, 428, *432*, *464*
Ham, W. E., 598, *607*
Hambleton, W. W., 511
Hamblin, W. K., *556*
Hambruch, P., *362*
Hamilton, E. L., 414, *414*, 415, 585, 587, 591, 595, 657, *657*
Hamilton, W., *13*, *43*, *432*, *464*, *547*, *626*
Hammen, T. van der, *249*, *318*, 322, 323, *325*, 483, *491*
Hanzawa, S., *333*
Harbison, A., 539, *540*
Harder, E. C., *324*
Hare, M. G., *607*
Harper, C. T., *304*, 340, *341*
Harrington, C. R., *521*
Harrington, H. J., 2, 3, 5, 9, 11, *13*, *17*, *21*, *138*, *242*, *399*, 419, 422, *426*, **456–465**, *465*, 469, 642, *647*
Harris, D. R., *15*
Harris, G. D., *498*
Harris, L. D., *530*
Harris, W. J., 96, *100*
Harris, W. K., 74, *80*
Harrison, C. G. A., **109–115**

674

Harrison, J. B., 309, 319, *324*
Harrison, J. E., *640*
Harrison, R. S., *116*
Hart, P. J., *456*
Hart, S. R., *278*
Hartman, C. W., *521*
Hartshorn, J. H., 610, *618*
Hartt, C. F., 127, 456
Hasselborough, F., 473
Hasui, Y., 421, 423, *426*
Hatcher, R. D., Jr., *530*
Hatherton, T., *12, 13, 304, 341*
Hatten, C. W., 114, *114,* 254, *255, 256, 666*
Hattin, D. E., *571*
Haun, J. D., *572,* 628, 634, 635, 637, 638, *640, 641*
Hawkes, D. D., 316, *318,* 322, *324, 325*
Hay, R. F., 250, *251, 399*
Hay, R. L., 226, *226,* 448
Hayden, F. V., 619
Hayes, D. E., 4, *13, 27, 242,* 470, 473, 486, *491*
Hayes, P. T., *547*
Haynes, S. J., *242*
Hays, J., 48, *55*
Healy, J., 252, *252,* 337, *337, 399*
Heaphy, 390
Hebeda, E. H., 314, 315, 316, *318,* 321, 322, *325,* 363, *364,* 483, *492*
Hector, J., 188, 390
Hedberg, F., *653*
Hedberg, H. D., *364, 653*
Hedge, C. E., *667*
Heezen, B. C., 2, 329, *330,* 486, *491, 540, 595, 666*
Heidecker, 44
Heilprin, A., *348*
Heim, 365
Heine, K., *360*
Heinz, 365
Heirtzler, J. R., *27, 28, 227,* 287, *290,* 574, *581,* 657, *657, 666*
Heiskanen, W. H., *333*
Helbig, K. M., *330*
Helby, R. J., *45*
Helmreichen, V. von, 456
Helsley, C. E., *641,* 655, *656*
Helwig, J., *127, 172*
Heming, R. F., *382*
Heminway, C. E., *437*
Henderson, E. P., 202, *207*
Hendry, C. W., 510
Henningsen, D. W., *252*
Hennion, J., *666*
Henriken, N., *299*
Henry, V. J., Jr., *540*
Henyey, T. L., *626*
Herm, D., *242*
Hermes, J. J., 372, *382, 383*
Hernandez-Sanchez, M. S., *360*
Heron, S. D., 534, *540*
Herrera, A., *21*
Herrera, L. J., Jr., *270,* 428, *432, 464*

Herrero Ducloux, A., 20, *21,* 465
Herrick, S. M., 538, *539*
Herron, E. M., *261,* 287, *290,* 449, *449,* 657, *657*
Hershey, O. H., 416, *419*
Hershey, R. E., 512
Herzen, R. P. von, *456*
Hesp, W. R., *45*
Hess, H. H., 227, *227,* 345, *345, 414,* 434, 653, 664, *666*
Hettner, A., 245, 456
Heusser, J., 456
Hewitt, D. F., 195, *207*
Heyerdahl, T., 260, *261, 414,* 498
Heyl, A. V., *408,* 598, *607*
Hibsch, J. E., 419
Higgs, D. V., *408*
Hill, D., **56–61,** *61*
Hill, J. V., *207*
Hill, O. M., *448*
Hill, R. T., *335,* 658
Hill, W. T., *530*
Hillebrandt, V. A. von, *433*
Hills, E. S., 92, 98, *100*
Hills, L., 81
Hind, H. Y., 188
Hind, M. C., *45*
Hinde, G. J., 202, *207*
Hintze, L. F., *640*
Hirst, J. A., 278
Hitchcock, E., 610
Hobson, G. D., 204, *207*
Hochstein, M. P., *251*
Hochstetter, F. von, 390
Hockley, J. J., *45*
Hodden, J. C., 110, 113, *114*
Hodge, W., 257
Hodgson, E. A., *55*
Hoffmeister, J. E., 278, 280, 281, *282, 415,* 496, *497*
Hoffstetter, R., *236, 252, 330, 401, 647*
Holden, E. S., *342*
Holden, J. C., 114
Ho Len Fat, A. G., 490, *491*
Holland, C. H., *408, 513, 522, 557*
Holland, J. G., 300, *301*
Holland, W. N., *45*
Holliday, B. W., *541*
Hollingworth, S. E., *242*
Hollister, C. D., *114*
Holmes, W. H., 549
Holtedahl, O., 471, 473, 479, 480
Hood, D. W., *522*
Hooker, J. D., 275
Hooker, M., *437, 666*
Hope, G. S., 374, *383*
Hopkins, B. M., *100*
Hopkins, D. M., *408,* 514, 515, 517, *521, 522*
Hopkins, E. A., *426*
Hopley, D., *34, 61, 91*
Hopson, C. A., *531*
Hormann, P. K., *21*
Horne, G. S., *172, 330*
Hornibrook, N. de B., *400*
Hose, H. R., *335*

Hoshino, M., *305*
Hough, J. L., 205, *207, 408, 607*
Hough, L. W., 511
Houtz, R. E., 278, *282*
Hovey, E. O., *257, 448*
Howard, A. D., Jr., *571*
Howard, P. F., *34*
Howchin, W., 61, *80*
Howe, E., 416, *419*
Howe, W. B., 511, *571*
Howie, R. D., 170, *172*
Howland, A. L., *415, 451*
Hoy, N. D., *500*
Hoylman, H. W., 401, *401*
Hoyt, J. H., *540*
Hsü, K. J., *414, 626*
Hubach, E., 245
Hubert, J. F., *640*
Huene, F. von, 645, *647*
Huene, R. von, *243, 522*
Huffman, O. F., *626*
Hugh, K. E., *401*
Hughes, C. J., 16, *172*
Hughes, I. G., *335*
Hughes, R. J., *45*
Humboldt, A. von, 245, 261, 268, 456, 618, 627, 658
Hume, G. S., 188
Humphrey, A. W., 81
Humphrey, W. E., *581*
Hunt, C. B., 628, *640, 513, 556*
Hunt, T. S., 194
Hurley, P. M., 311, 312, 314, 315, *318,* 320, 321, *325,* 642, *647*
Hurley, R. J., *114*
Hunsted, J. E., 533, *540*
Hutchinson, G. E., *342, 362, 414, 433*
Hutchinson, R. D., 160, 162, *173,* 444, *447*
Hutchinson, W. W., 186, *187*
Hutton, C. O., 441, 442, *442*
Hutton, J. T., 402, *402*
Hyndman, R. D., *300, 409*
Hyne, N. J., *626*

Ibbotson, P., 278
Ichikawa, M., *333*
Iddings, J. P., *450*
Ijzerman, R., 480, *491*
Illing, L. V., 112, *114*
Illing, V. C., *498*
Iltchenko, L. M., 469
Imlay, R. W., *360*
Ingerson, E., *347*
Instituto de Investigaciones Geológicas, 238, *242*
Instituto Geográfico Nacional, 307, *308*
Instituto Nacional de Geologiá y Minería, 15
Instituto Nacional de Investigaciones Geologico Mineras, 245
Intermountain Association of Petroleum Geologists, *557*

AUTHOR INDEX

Irving, E. M., 228, 234, *237*, 252, 308, 330, *401*, *419*
Irving, W. N., *521*
Irwin, W. P., *626*
Isachsen, Y. W., *617*
Isacks, B., *27*, 468
Isbister, A. K., 188
Isshiki, N., *333*
Itturalde-Vincent, M. A., 255, *255*
Ives, J. C., 549
Izett, G. A., 568, *571, 572*

Jack, R. L., 70, *80*
Jackson, D. B., 594, *596*
Jackson, E. D., 591, *595*
Jacob, K. H., 272, *272*
Jacoby, W. R., 209, *217*
Jaggar, T. A., *497, 582, 595*
Jago, J. B., *91*
Jakes, P., *27*, 304, 341, 345, *345*
James, D. E., 430, 431, *433, 465*
James, H. L., *408*
James, N. P., *116*
Janssen, J. J., 313, *318*
Jardine, S., 486, *491*
Jarrard, R. D., *414*, 587, *595*
Jeannel, R., 472
Jeletzky, J. A., *156*
Jell, J. S., *34*
Jenkins, D. A. L., 381, *382*
Jenks, W. F., *243, 324,* **426-433**, *433, 465, 647*
Jennings, J. N., *45, 55, 100*
Jensen, H. I., *667*
Jérémine, E., 499, *500*
Jessup, R. G. C., *382*
Jessup, R. W., *80*
Johns, R. H., *626*
Johnsen, S. J., *299*
Johnson, A. J., *547*
Johnson, D. A., *414*
Johnson, F. W., *654*
Johnson, G. H., *541*
Johnson, H., *640*
Johnson, J. H., *347*
Johnson, M. E., *540*
Johnson, P. R., *521*
Johnson, R. W., *27*
Johnson, T., 122, *123*
Johnson, W. D., 559, *571*
Johnson, W. R., *540*
Johnston, R. M., 81
Johnston, W. A., 195, *207*
Johnstone, M. H., *80*
Joleaud, L., *415*
Jolivet, J., *304*
Jones, B. G., *55*
Jones, D. L., *626*
Jones, F. O., *138*
Jones, J. G., *27*, 402, *402, 521*
Jones, J. K., Jr., *571*
Jones, P. J., *34*
Jordan, R. R., 510, 534, *540*
Josey, W. L., *117*
Joyce, E. B., *100*
Judd, J. B., *114*

Judson, S., *540*
Jukes, J. B., 81
Julian, B., *627*
Julien, A. A., *442*
Julivert, M., *249*
Jung, P., 301
Junk, W., *13*
Jutson, J. T., *102, 107*

Kalliokoski, J., 311, 312, 314, 316, 317, *318*
Kamerlings, P., *80*
Kamp, R. C. van de, *547*
Kaneoka, I., *333*
Kanter, J. J., *419*
Kapitsa, A. P., *13*
Kaplan, P. A., *292*
Karig, D. E., *27*, 43, **344-345** *345, 415*, 494, 495, *497*
Karim, M., 401, *401*
Karlstrom, T. N. V., *521*
Karpoff, M. R., 421, *426*
Karrow, P. F., 201, 202, 203, 204, 206, 207, 208, 225
Karsten, H., 245, 456
Kassem, T., 246, *249*
Katili, J. A., *382*
Katsumata, M., *333*
Katz, H. R., *242*
Katz, M. B., *13*
Kausel, E., *242*
Kay, M., 163, 167, *173*, 299, 300, 368, *408, 622*, 623, *626*
Kaye, C. A., *437*
Kaye, P., *107*
Kays, M. A., *626*
Kear, D., *667*
Kegel, W., 135
Kehle, R. O., 578, *581*
Keidel, J., 15
Keller, F., Jr., *415*
Keller, G. R., *607*
Kelley, D. G., 169, 170, *173*
Kelley, V. C., *556*
Kellum, L. B., 537, *540*
Kemp, J. F., *656*
Kemp, S., 468
Kennedy, M. J., 168, *173, 174*
Kennedy, T. F., *402*
Kennett, J. P., *13*
Kent, H. C., 628, 634, 635, 637, 638, *640*
Kent, P. E., *250*
Kerguelen, Y. de, 474
Kersen, J. F. van, *325*
Kesel, R. H., *252*
Kesler, S. E., *117, 308,* 326, *327, 360*
Keyes, C. R., *556*
Khoury, S. G., 634, *641*
Khudoley, K. M., 253, 254, 255, *255, 666*
Kidwell, A. L., *255*
Kiel, H., *341, 365, 442*
Kienle, C. F., *117*
Kilian, C., 365

Kilkenny, J. E., *626*
Kimmel, M. A., 586, *595*
Kindle, C. H., 161, 162, *173*
Kindle, E. M., 188
King, A. P., *330*
King, C., 541, 619
King, D., 75, *80*
King, E. R., *408, 607*
King, L. C., 310, 487, *491*
King, L. H., 172, *173*
King, P. B., 209, *217*, 360, *408*, *513,* **522-531***, 531,* **608-618***,* 628, 631, *640*
King, R. E., *408*
Kirk, B., 188
Kirk, H. K., *282, 403*
Kirkland, D. W., *556, 581*
Kistler, R. W., *408, 626*
Kitching, J. W., 3, *13*
Klein, G. de V., *138,* 171, *173*
Klein, H., *500*
Kling, S. A., *117*
Klinge, H., 271, *272*
Kluiving, R. B., 372, *382*
Knebel, H. J., *521*
Knopoff, L., *456*
Knox, G. A., *13*
Knox, J. C., *607*
Kobayashi, K., *305*
Koch, E., *433*
Koch, P., *343,* 368, *371*
Kojima, G., 593, *595*
Kord, B., *653*
Koschmann, D. G., 258, *259*
Köster, R., **237-242**
Kottlowski, F. E., 512, *556*
Kozary, M., *255*
Krause, D. C., 378, *382*
Kraut, F., *371*
Krinsley, D. H., *531*
Krommelbein, K., 255, *255*
Krook, L., **480-492,** *491*
Krummerbacher, R., *249*
Krushensky, R. D., *252*
Ku, T. L., 115, *115*, 586, *595*
Küchler, A. W., *513*
Kuenen, P. H., 347, *347*
Kugler, G., 323, *325*
Kugler, H. G., *498*
Kuhlmann, D. H. H., *255*
Kulp, J. L., *531*
Kumarapeli, P. S., 171, *173*
Kummel, B., *409*, 628, *640*
Kumpera, O., *666*
Kuno, H., *333*, 345, *345*
Kurie, A. E., *556*
Kurtz, F., 15

Laborde, D., *348*
Lacroix, A., *261, 284,* 346, *346, 348,* 365, *371,* 383, *415,* 450, *493, 500,* 658
Ladd, H. S., 278, 280, 281, *282, 292*, 392, 347, *347*, 414, *414, 415,* 449, *497,* 591, *595*
LaFehr, T. R., *626*

Lagaay, R. A., 363, *364, 666*
Lahr, J., *522*
Laing, R. M., 402, *402*
Lambert, R., 642, *647*
Lamego, A. B., 128
Lamego, A. R., 131, *138*
LaMoreaux, P. E., 510
Land, L. S., 111, 120, *121*
Landes, K. K., *409, 513, 572*
Landis, C. A., *400*
Landry, S. O., *530*
Landsberg, H. E., *546*
Langenegger, O., 474
Langseth, M., *333*
Langway, C. C., *299*
Lankform, R. R., *581*
Lanphere, M. A., *409, 521, 522, 591, 595, 626*
Larrea, C. M., *290*
Larsen, C. A., 3, 468
Larson, R. L., *345, 409, 415*
LaSalle, P., *207*
Laseron, C. F., *34*
Lattimore, R. K., *653*
Laubacher, G., 428, *433*
Laughlin, A. W., 430, *433*
Launey, J., *343,* 370, *371*
Laurence, R. A., *531*
Lawrence, D., 537, *540*
Lawson, A. C., 619
Leanza, A. F., *21*
LeBlanc, R. J., *581*
Le Conte, J., 545, *547,* 619
Leech, G. B., *409*
Leeper, G. W., *100*
Leeson, I., *415*
Legget, R. F., *206*
LeGrand, H. E., *540*
Lehmann, H., 255, *255*
Lehner, P., 575, 576, 577, *581*
Leitch, E. C., 36, *45,* 46, *55,* 128, *607*
Lelong, F., *284*
Le Maitre, R. W., 470, 473
Leonard, A. B., *571*
Leonardos, O. H., 128
Leontyev, O. K., *292*
Leopold, E. B., 226, *226, 347, 415*
Le Pichon, X., *300,* 377, *382, 409, 464*
Leverson, A. I., 635, *640*
Levi, B., *242*
Lewis, A. N., 81
Lewis, C. F. M., 202, *207*
Lewis, G. E., *270*
Lewis, J. F., *304,* 334, 335, 336, *341, 442, 448*
Lewis, J. W., *255*
Lewis, K. B., *338*
Lewis, V. A., *595*
Leyden, R., 642, *647, 647*
Liberty, B. A., 195, *207*
Liddle, R. A., 309, *498*
Lidiak, E. G., *304, 341*
Lidz, B., 112, *114,* 236, *236*
Like, H., *261*
Lillie, A. R., *28,* 366, *371*

Lilly, H. D., 159, *173*
Lima, J. C., *21*
Lind, A. P., 109, 111, *114*
Lindgren, W., 619
Lindner, A. W., *107*
Lindsay, J. F., *618*
Lipe, W. D., *530*
Lipman, P. W., *409,* 546, *546, 547, 626, 640*
Lisboa, M. A., 127
Lister, J. J., 496, *497*
Litchfield, 47
Livingston, V. E., Jr., 513
Ljunggren, P., 125
Lloyd, E. F., 337, *337*
Lloyd, J. J., 236, *401*
Lloyd, R. M., 112, *114*
Lobeck, A. K., *513*
Lobo, B., *278*
Lockman-Balk, C., *556* 634, *640*
Locke, D. H., *144, 409, 513*
Lodder, W., 363, *364*
Loemban-Tobing, 315
Löffler, E., *121*–*123, 372*–*383, 382*
Logan, B. W., *107*
Logan, R. W., *327*
Logan, W., 144, 194, 195, *207,* 610
Lohmann, H. H., *123*–*127, 127,* 228, *465*
Lomnitz, C., *242,* 360, *465*
Long, H. K., *513*
Long, J. V. P., *278*
Long, W. E., 3
Longwell, C. R., 546, *547*
Lormand, 368
Lotschert, W., 255, *255*
Love, J. D., 630, *640*
Lovering, J. F., *100*
Lovering, T. S., *640*
Lovlie, R., *531*
Low, A. P., 144
Low, D., *596*
Lowden, J. A., *409*
Lowder, G. G., 377, 383
Lowenstam, H. A., *607*
Lowry, D. C., *80, 107*
Lowry, W. D., *530, 531*
Lozo, F. E., *581*
Lucas, G., 366
Lucchita, I., *547*
Lucia, F. J., *364*
Ludbrook, N. H., 61
Ludwig, W. J., *333,* 642, *647, 647*
Lueninghoener, G. C., *572*
Lugn, A. L., *557*–*572, 572*
Lum, D., *595*
Lund, P. W., 127, 456
Lund, R. J., *607*
Luyendyk, B. P., 113, *115, 383, 390*
Lynts, G. W., *114*

Maack, R., 128, *138*
Mabbutt, J. A., *55, 100*
Mabesoone, J. M., 275, *278*

McAndrew, J., *34, 45, 91, 100*
MacArthur, R. H., *336*
McBirney, A. C., **284**–**290**, *290, 329, 330*
McCallum, J. S., 112, *114*
McCandless, G. C., 314, 315, *318*
McCarthy, G. R., 532, *540*
McCartney, W. D., 159, *173*
McColl, D. H., *80, 100*
McConnell, R. B., 179, 188, 208, 309–318, *318,* **318**–**325**, *324,* 482, 487, *491, 492*
McCormick, J. E., *530*
McCourt, G. B., *193*
McCoy, F. W., Jr., 2
McCrossan, R. G., *144, 193,* 217
McDonald, B. C., 158, *173,* 207, 225
MacDonald, C. A., *415*
MacDonald, D. F., 416, *419*
McDonald, D. R., *107*
MacDonald, G. A., *1,* 588, 593, *595, 596, 658, 658, 667*
MacDonald, G. E., *596*
MacDonald, K. C., *383*
MacDonald, W. D., *249,* 259, *259, 327, 327, 666*
MacDougall, I., 13, 280, 382, *383, 402, 402, 456,* 591, *595*
McEvoy, J., 208
McGee, B. A., 144
McGerrigle, H. W., 167, *173*
McGhee, R., *156*
MacGillavry, H. J., *255, 666*
MacGregor, A. G., *341, 348, 361*
MacGregor, D. J., *512*
MacGregor, I. D., *595*
McGrew, P. O., *330*
MacIlvaine, J. C., *419*
McIntire, W. G., *15*
Macintyre, I. G., 115, *116,* 656, *666*
McIntyre, L. B., *547,* 550, 551, *556*
McIver, N. L., 171, 172, *173*
McKay, A., 390
McKee, B., *626*
McKee, E. D., 226, *226, 409, 415, 556, 640*
McKee, E. H., 546, *547,* 555, *556*
McKenzie, D., *627*
MacKenzie, D. E., *27*
MacKenzie, F. T., 111, **117**–**121**, *121*
McKenzie, G. D., 204, *207, 521*
MacKenzie, S. C., 323, *325*
McKenzie-Pollock, L., *156*
Mackin, J. H., 541, 545, *547,* 631, *640*
McKnight, E. T., *556*
MacLachlan, J. C., *640*
McLaughlin, D., Jr., *249*
McLaughlin, D. H., 426
MacLean, B., *173*
McLearn, F. H., 188
McLeod, I. R., *34*
MacLeod, W. N., *107, 107*
McLintock, A. H., *400*

McMillan, D. T., 512
MacMillan, J. G., 642, *647*
McNair, A. H., *156*
Macomb, J. N., 549
Macpherson, E. O., 412, *415*
MacPherson, J. B., *225*
McReath, I., *469*
McWhae, J. R. H., *107*
Madigan, C. T., 3
Mädler, K., *423*
Magellan, F., 344
Maggio, C., *359*
Maher, J. C., *409, 572*
Maitland, A. G., 100
Malahoff, A., *342*, 384, 387, 388, *390*, 456, 587, *595*
Maldonado-Koerdell, M., *330*, 438, *438*
Malfait, B. T., 235, 236, *236*, 329, *330*, 401, *401*, 419, 665, *666*
Maloney, N. J., *653, 654*
Mankin, C. J., 512
Manser, W., *91*
Mansfield, G. R., 630, *640*
Marchant, S., *270*
Marcou, J., 549
Maresch, W. V., *654*
Marine, I. W., *540*
Mariner, R. H., *641*
Marion-Dufresne, N., 473
Marks, J. G., *270*
Marlow, M. S., *243*, 514, *521*
Marocco, R., 428, *433*
Marsden, M. A. H., *34, 61, 100*
Marshall, B., *91*
Marshall, P., 250, *251, 342*, 412, *415, 434, 451*, 493, *493, 494, 500*
Martin, A. J., *521*
Martin, K. R., 337, *337*
Martin, P. S., *556*
Martin, R., *654*
Martin-Bellizzia, C., 312, 314, *317*, 318
Martinez, A. R., *654*
Martinez, C., 428, *433*
Martini, I. P., 198, *207*
Martin-Kaye, P. H. A., 15, *15, 257, 304, 321, 325*, 339, *341, 364*, 439, 440, *442, 443*, 654, *655, 668*
Martison, N. W., *107*
Marvin, R. F., *618*
Mason, B., *100*
Mason, D. R., *55*
Masson Smith, D., 322, *325*
MasVall, J., *654*
Mather, J. D., **14-15**, 110, **115**, 227, *227*, **500-501**
Matley, C. A., 227, *227, 336*
Matson, N. A., *522*
Matsuda, T., *333*
Matthews, J. L., 3, *13*
Matthews, R. K., 115, *115, 116*
Matthews, W. H., III, *144*, 513
Mattick, R. E., *540*
Mattson, P. H., *437, 666*

Maury, C. J., *498*
Mawson, D., 3, 67, *80*, 383, *390*, 473, 479
Maxey, L. R., *531*
Maxwell, A. E., *340*
Maxwell, J. A., *531*
Maxwell, J. C., 227, *227, 498, 627*
Maxwell, W. G. H., *34, 61*
May, H. G., *305*
May, J. P., *581*
May, P. R., 483, *492*
Mayer, A. E. S., *242*
Mayor, A. G., 1, *2*
Mayr, E., *274*
Meade, M., 123
Medvedev, V. S., *292*
Megard, F., 428, *432, 433*
Mehringer, P. J., Jr., *556*
Meinesz, F. A. V., *664*
Meister, E. M., *138*
Melcher, C. C., *318*
Melhado, A. R., *330*
Melo, U. de, *138*
Melson, W. G., *252*, 259, *259, 278*, 327, *327*
Melville, R. V., 12
Menard, H. W., *282*, 347, *347, 403*, 412, 413, 414, *415*, 546, *547, 574, 581*, 584, 587, *595, 595*
Mencher, E., *654*
Mendaña, A., 451, 657
Mendez Alzola, R., 642, 643, *647*
Mendoza, V., 315, *318*
Menendez, C. A., 419, 422, *426*
Menzer, F. J., Jr., *627*
Merriam, D. F., *572, 607*
Merriam, J. C., 619
Merrill, G. P., *409*
Merrill, R. K., *556*
Mersay, F. de, 419
Mesolella, K. J., 115, *115, 116*
Metraux, A., 260, *261*
Metson, A. J., 337, 338
Meusakka, J. L., *415*
Meyer, H. C., 125
Meyer, J., *272*
Meyer, R. F., *409*
Meyer, R. P., *607*
Meyer-Abich, H., 271, *272*
Meyerhoff, A. A., 114, *114, 237*, 254, 255, *255, 256, 666*
Meyerhoff, H. A., *237, 255*, 327, *327*, 434, 531, *656*
Meyers, N. A., *61*
Migdisof, A. A., *409*
Malankovitch, M., 115
Milch, L., *426*
Miller, D. J., 514, *521*
Miller, D. N., Jr., 513
Miller, D. W., *513*
Miller, H., *242*, 465
Miller, J. B., *654*
Milligan, J., 81
Milliman, J. D., 113, *115, 540, 595*
Mills, R. A., *237, 330, 401*
Milnes, A. R., *80*
Milsom, J., *383*

Milton, B. E., *56*
Minard, J. P., *618*
Minato, M., 331, *333*
Mingramm, A., 419, 423, *426*
Ministerio de Minas e Hidrocarburos, Dirección de Geología (Venezuela), *652, 654*
Mitchell, A. H., *304*
Mitchell, A. H. G., 384, 385, 387, *390*
Mitchell, J., *21*
Mitchell, J. G., *21*
Mitchell, J. S., *556*
Mitchell, R. C., *255*, 326, 327, *327, 341, 437*
Mitchell-Thome, R. C., 274, *274*, 275, 276, 277, *278*
Mogi, A., *333*
Molenaar, C. M., *556*
Molengraaff, G. J. H., *341*, 362, 363, 364, *364*
Mollan, R. G., *61*
Molnar, P., 122, *123*, 251, *252, 341, 437*, 468, 664, *666*
Monger, J. W. H., 186, *187*, 210, *217*
Monroe, W. H., *666*
Montagne, D. G., 483, *492*
Montserrat, E. de, 305
Moody, J. D., **228-237**
Moore, B. R., *100*
Moore, C. H., 227, *227*
Moore, G. W., 354, *360*
Moore, H. G., 593, *595*
Moore, J. C., 514, *521, 522*
Moore, J. G., 544, 546, *547*, 594, *595*
Moore, R. C., *572*
Moore, W. H., 511
Moore, W. S., *114*
Mooser, F., *360*, 438, *438*
Moraes, L. J. de, 128
Moraes Rego, L. F. de, 128
Moran, M. C., *330*
Mordock, L. W., *653*
Morgan, W. J., *243*
Morgan, W. R., *61*
Morgridge, D. L., 520, *522*
Morinigo, G. V., 419, 423, *426*
Morley, E. W., *450*
Mörner, N. A., 203, *207*
Morrison, R. P., *465*
Morrow, E. H., *607*
Mortimer, C., *242, 243*
Mott, R. J., *207*
Mougin, J. L., 499, *499*
Mount, T. J., 63, *80*
Mountjoy, E. W., *116*, **208-217**, *217, 301*
Moussa, M. T., *437, 666*
Mudge, M. R., *641*
Muehlberger, W. R., *641*
Mueller, G., *243*
Muir, J., 620
Mukherji, K. K., 197, *207*
Mulcahy, M. J., *80*
Mulder, F. G., 483, *492*

Mulders, M. A., 483, *491*
Muller, G., *114*
Mulloy, W., 260, *261*
Mulvaney, D. J., 34
Muñoz Cristi, J., 238, *243*
Murata, K. J., 252, *593*, *596*
Marauchi, S., *333*
Murray, A., 194
Murray, C. G., *522*
Murray, G. E., *409*, *540*, 574, 575, 576, 578, *581*
Murray, J., 3
Musich, L. F., 486, *491*
Myers, W. B., *547*, *626*

Naeser, C. W., *572*
Nagata, T., *13*
Nagera, J. J., 15
Nagle, F., *259*, *304*, 340, *341*
Nairn, A. E. M., *13*, *480*, *547*
National Research Council (Canada), 144
Naughton, J. J., *595*
Neale, E. R. W., 159, 162, 168, *173*, *174*
Neathery, T. L., *530*
Nederloff, M. H., *127*
Nelson, C. A., *80*
Nelson, C. H., 514, *522*
Nelson, H. W., 248, *249*
Nelson, P. H. H., 468
Nelson, S. J., *145*, *193*
Nesbitt, R. W., *80*
Neuman, A. C., 112, *114*, 120, *121*
Neumann, W. S., 119
Newberry, J. S., 549
Newell, N. D., *415*, 114, 428, 429, *433*, 499, 644, *648*
Newell, W. L., *618*
Newhouse, W. H., 309
Newman, W. S., 118, *121*
New Mexico Geological Society, *557*
Niblett, E. R., 209, *217*
Nicholas, T., 51, *55*
Nichols, R. L., *13*
Nicolaysen, L. O., *607*
Nicoll, R. S., 198, *207*
Nikiforov, L. G., *292*
Noakes, L. C., *34*, 47
Noble, D. C., 546, *547*
Noble, E. A., 512
Noble, J. A., *641*
Noe-Nygaard, A., *299*
Noesmoen, A., 366. 368, *371*
Nolan, T. B., 541, *547*
Noorthoorn van de Kruijff, J. F., 483, *492*
Nordenskjöld, O., 3
Nordlie, B. E., *290*
Norford, B. S., 198
Norris, D. K., *156*
Norris, R. M., 80, 260, *261*
North, F. K., *522*, *666*
Nota, D. J. G., 486, *492*

Nougier, J., 466, 470, 471, 473, 474, 475, 476, 477, 478, 480
Nugent, L. E., Jr., 225, *226*, 291, *292*, *347*, 362
Nugent, N., *498*
Nuñez-Jimenez, A., *256*
Nuttal, W. L. F., 243, *244*
Nygren, W. E., *270*

Oakeshott, G. B., *627*
Oaks, R. Q., Jr., *540*
Obelliane, J. M., *284*, *390*, *451*, *493*
Obermuller, A. G., 244, *244*
Ocola, L., *243*, *607*
Odiorne, H. H., 633
O'Driscoll, E. P. D., 70, *80*
Officer, C. B., 118, 121, *664*, *666*
Ogasawara, S., 331
Ogden, T. A., *345*
O'Herne, L., 310, 316, *318*, 481, *492*
Ohlen, H. R., *547*, 550, 551, *556*
Okulitch, V. J., 195, *207*
Olgers, R., *61*
Oliveira, A. I., 128, 133, 135, *138*
Oliveira, E. P. de, 127, 128
Oliver, J., 27
Oliver, W. A., Jr., *618*
Ollier, C. D., *100*, *123*
Olsen, W. S., 114
Olson, E. C., *330*
Olson, N. K., 512
Olsson, A., 539, *540*
Olsson, A. A., 228, *419*
Omura, A., *116*
O'Neil, T. J., 586, *595*
Oosterbaan, W. E., 481, 482, 490, *490*
Opdyke, N. D., *249*, *531*
Orbell, G. E., *497*
Orbigny, A. d', 15, 123, 127, 245, 456, 642, *647*
Organist, D., 538, *540*
Orheim, O., 469
Oriel, S. S., 634, 637, 639
Orloff, O., *371*
Orme, A. R., *360*
Ortynski, I., *270*
Osborne, F. F., *618*
Ospina, T., 245, 247, *249*
Ostergaard, J. M., *497*
Ostrom, M. E., 513
Oswald, D. H., *193*
Otuka, Y., *333*
Otulana, H. I., *300*
Oudemans, W., 372, *382*
Ovenshine, A. T., *522*
Oversby, B., 95, 97, *100*
Owen, L., 292, *292*
Owens, J., 534, *540*
Ower, L. H., *117*

Pacheco, J., 128

Packham, G., **34–43**, *45*, **45–46**, 97, *100*
Padula, E., 419, 423, *426*
Padula, V. T., *138*
Page, B. M., *627*
Page, R. A., Jr., *522*
Page, R. W., 382, *383*, *456*
Paine, A. G. L., *61*
Pakiser, L. C., 545, *547*, *556*
Palmer, A. R., *13*, *557*
Palmer, H. C., *259*
Palmer, K. V. W., 536, *540*
Palmer, N., 468
Palmer, R. H., *256*
Pantot, G., *300*
Paredes, J., 428, *433*
Park, C. F., *243*
Park, J., 390
Parker, F. L., *414*
Parkin, C. W., 61, 67, 68, *80*
Parkin, L. W., *34*, *80*
Parks, W. A., 195
Parodiz, J. J., *648*
Parrott, W. R., *540*
Parsons, W. H., 630
Paskoff, R., *242*, *243*
Pastore, F., 15
Paterson, H. L., 45, *45*
Paterson, W. S. B., *145*
Paton, R. F., 322, *324*
Patterson, J. M., *654*
Patton, J. B., 511
Patton, W. W., Jr., *521*
Paula Oliveira, F., de, 127
Pautot, G., *409*
Payne, T. G., 417, 514, *521*
Peale, A. C., 549
Peck, D. L., *593*, *595*
Peckham, S. F., *498*
Peirce, H. W., *557*
Pels, A., 45
Pelton, P. J., *546*
Perales-Calderon, F., *433*
Perkins, B., Jr., *414*
Perret, F. A., *341*, 348
Perry, E. S., *641*
Perry, R. A., *34*
Perry, W. J., *107*
Pessagno, E. A., *419*, *437*
Peter, G., 574, *581*
Peterman, Z. E., *383*, *627*, *667*
Petersen, U., 332, *426*, *433*, *465*
Peterson, D. N., *547*
Peterson, D. W., *594*, *595*, *596*
Peterson, J. A., *34*, 91, 374, *383*, *641*
Pettijohn, F. J., 176, *179*, *530*
Pettijohn, W. A., *572*
Péwé, T. L., 514, 515, *522*, *556*, *557*
Pflaker, G., *522*
Phillips, J. D., *340*, *665*
Phillips, K. A., 278, *282*
Phillips, R. L., *594*, *595*
Phillips, R. P., *626*
Phillips, W. E. A., *173*
Pichler, H., *249*

AUTHOR INDEX

Pickering, S. M., Jr., 510
Pickett, J. W., *34*
Pierce, W. G., 629, *641*
Pijpers, P. J., 363, *364*
Pimm, A. C., 486, *491*
Pinchow, R. P., *448*
Pinet, P. R., 227, *227*, *237*
Pinson, W. H., 311, *318*
Piper, D. J. W., 227, *227*
Pirie, J. H., 468
Piroutet, M., 365, 368, *371*
Pirsson, L. V., 118, *121*, 470
Pitcher, M. G., *156, 409, 522*
Pitcher, W. S., 428, 430, *432*
Pitman, W. C., III, 657, *657*
Plafker, G., *243, 401*
Playford, P. E., **100–108**, *107*
Plessis, Y., *499, 499*
Plumb, K. A., 50, *55*
Plummer, R. G., 135
Plumstead, E., *13*
Pollock, C. A., 198, *207*
Pomeyrol, R., *371*
Pommerene, J. B., *138*
Pond, P., 188
Pons, L. J., *324*, 485, *491*
Poole, W. H., **156–174**, *172, 173*
Porter, D. M., 287
Porter, J. W., *571*
Porter, S. C., *596*
Porto, R., *137*
Portugal, J. A., *433*
Posadas, V. G., 314, *318*
Postuma, J. A., *456*
Pöthe de Baldie, E. D., 419, 422, *426*
Potratz, H. A., *347*
Potter, H. C., *498*
Potter, R. R., 159, *173*
Powell, G., 468
Powell, J. W., 541, *549*
Power, F. D., *292, 342, 362, 415, 433*
Power, P. E., *61*
Prather, R. W., *193*
Pratt, R. M., *618*
Pratt, W. P., *513*
Prest, V. K., *156, 158, 173,* 204, 205, *207*
Prestat, B., 486, *491*
Price, R. A., *145,* 209, 212, 213, 214, *217, 641*
Pride, D. E., *641*
Priem, H. N. A., *284,* 314, 315, 316, *318,* 322, *325,* 363, *364,* 482, 483, *492*
Priestley, R., 3, *13, 480*
Primer Congreso Venezolano de Petróleo, *652*
Proctor, P. D., 513, *522*
Prodehl, C., *513,* 545, *547*
Prostka, H. J., *409,* 626, 640, *641,*
Prowazak, S. von, 225, 226
Pryor, W. A., *607*
Pudsey-Dawson, P., 451
Pulvertaft, T. C. R., *300*
Purdy, E. G., *115*

Puri, H. S., *581*
Pushkar, P., *304, 341*
Putman, D. F., *206*
Putzer, H., *138, 270,* **419–426**, *426*

Quensel, P. D., 336, *336*
Quillian, R. G., *207*
Quilty, P. G., *80, 91*
Quinn, A. W., *618*
Quirt, G. S., *242*

Raasch, G. O., *156, 300*
Rad, U. von, 486, *491*
Radelli, L., *127,* 246, 249, *465*
Rainwater, E. H., *581*
Raisz, E., 349, *360,* 541, 547, 242, *347, 497*
Ramos, J. R. de A., 314, 316, *317*, 322, *324*
Rand, J. R., 12, 311, 312, *318*
Randal, M. A., 50, *56*
Ranford, L. C., 51, 52, *56*
Rankama, K., *34, 80,* 299, *409, 531, 641*
Ranson, G., *499*
Ratcliff, N. M., *531*
Rathbun, M. J., *541*
Ratte, 365
Raumondi, A., *426*
Ravich, M. G., 4, *13*
Ray, C., *501*
Raymond, L. A., *530*
Rea, W. J., **300–301**, *301,* **360–361**, *361, 443, 668*
Read, J. F., *107*
Reading, H. G., *304, 341*
Recy, J., *343,* 370, *371*
Redfield, A. H., *330*
Reed, B. L., *409, 522*
Reed, E. C., *571, 606*
Reed, J. C., Jr., *513, 530, 641*
Reed, P. R. C., *292*
Reesor, J. E., *217*
Reeves, F., 629, *641*
Reger, R. D., 515, *522, 557*
Rehder, H., 227, *227*
Reid, K. O., *91*
Reid, W. M., 578, *581*
Reiter, H., 3
Renard, A., *278*
Renger, F., *138*
Renz, H. H., *498, 654*
Repelin, J., *499, 499*
Revelle, R., *347*
Revert, E., *348*
Rex, R. W., *414*
Rexroad, C. B., 198, *207*
Reyes, L., 427, *433*
Reynal, A. de, *304*
Reyner, M. L., 70, *80*
Reynolds, M. W., *547*
Reynolds, P. H., *409*
Rice, R. F., *640*
Rich, J. L., *531*
Richard, L. V., *617*

Richards, A., *290*
Richards, A. F., *438*
Richards, H. G., 120, *121, 138,* **226–227**, *227,* **260–261**, *501, 522,* **531–541**, *539, 540, 666*
Richards, J. R., *61,* 378, *383*
Richards, M. G., *115*
Richardson, J., 188
Richmond, G. M., 631, *641*
Richter, C. F., 663
Richter, D. H., *522,* 593, *596*
Richter, R., 273
Rickard, L. V., 198, *207*
Rickard, M. J., 278
Rickwood, F. K., 451, *522*
Ridge, J. D., *409, 513, 531, 607, 641*
Ridlon, J. B., *243*
Rigassi-Studer, D., *256*
Rigby, D., 45
Rigby, J. K., *114, 641*
Ripper, I. D., *383*
Ritchie, G. S., *342*
Ritter, D. F., *540*
Rittman, A., *442*
Riva, J., *547*
Roberts, F. H. H., Jr., 638, *641*
Roberts, H. G., 50, *55, 107*
Roberts, H. H., 227
Roberts, J., *34*
Roberts, R. J., 228, 234, *237,* 252, 308, 330, 401, 419, 545, 547, *627,* 634, 636, 637, *641*
Roberts, T. G., *429, 433*
Robie, R. H., *654*
Robin, G. de Q., *13,* 480
Robinson, E., **334–336**, *336*
Robinson, G. P., 384, *390*
Robson, G. R., *257, 257,* 300, *301, 304, 341,* 361, *361,* 448, *666, 668*
Rocky Mountain Association of Geologists, 628, *639, 640, 641*
Rod, E., *382*
Rodda, P., **278–282**
Rodgers, J., 159, 162, 166, *173, 531, 618*
Rodriquez-Torres, R., *626*
Roeder, D. H., *409, 641*
Roever, E. W. F. de, *492*
Rogers, H. D., 524
Rogers, J. J. W., *304, 341*
Rogers, J. W., *581*
Rogers, W. B., 524
Rohmeder, W., 419, *426*
Rold, J. W., 510
Rolff, P. A. M. A., *138*
Roliff, W. A., 223, *225*
Ronov, A. B., *409*
Roobol, M. J., *348*
Rose, E. R., 159, *173*
Rose, H. J., Jr., *243*
Rose, J. C., *456*
Rose, P. R., *581*
Rose, W. I., 272, *272,* 308
Rosenfeld, J. L., *618*
Rosenkrantz, A., *300*

Rosier, G. F., 131
Ross, C. A., *117*
Ross, C. P., *641*
Ross, D. A., *419*
Ross, D. C., *627*
Ross, J., 3
Ross, P. J., *539*
Ross, R. J., Jr., *548*
Ross, S. H., *641*
Roth, P. H., 486, *491*
Rothrock, D. P., *641*
Routhier, P., 365, 366, 368, 369, 370, *371*
Rowland, R. W., *521*
Rubenach, M. J., 473, 475
Rubin, M., *522*
Ruedemann, R., *409*
Rüegg, W., 429, *433*
Ruellan, F., 132, *138*
Ruhe, R. V., *556*
Ruiz, C., *243*
Runnegar, B., *648*
Russell, L. S., 188
Russell, M., *347*
Russell, R. J., *15, 581*
Rutherford, R. L., 188
Rutland, R. W. R., *242, 243*
Rutsch, R., *498*
Rutten, K. W., 486, 487, *491*
Rutten, L., *256*
Rutten, L. M. R., 362, 363, *364*
Rutten, P. B., 664
Ruxton, B. P., 380, *383*
Ruy Derze, G., *137*

Sachet, M. H., 244, *244*, 412
Sachs, P. L., *498*
Saenz, R., *252*
Sagatzky, J., *390*
St. Clair, D., 289, *290*
Saint Clair, S., *607*
Salas, G. P., *360*
Salay Gomez, 260
Salvador, A., **649–654**
Salvat, B., 499, *499*
Sanderson, J. O. G., 188
Sanford, B. V., **192–208**, *207, 208*
Santa Cruz, J. A. de, 123
Santamaria, F., *654*
Sapper, K., 228, *237, 257,* 271, *272, 301, 361, 442, 443*
Sarasin, F., *343*
Sarcia, J., *499*
Sarpi, E., *172*
Sass, J. H., 545, *548*
Satterthwaite, G. E., *250*
Sauer, W., *270, 465*
Saull, V. A., 171, *173*
Saunders, J. B., *498*
Savage, C. N., *641*
Savage, J. C., *243, 627*
Sawkins, J. G., 309, 319, *324,* 456
Sayles, R. W., *121*
Sbar, M. L., *409,* 545, *548,* 549, *557, 641*
Scarpelli, W., 314, *318*

Schafer, J. P., 610, *618*
Schaub, H. P., 363, *364*
Scheibe, E. A., *138*
Scheibe, R., 245
Scheibner, E., *28,* 37, **43–46**, *45*
Scheibnerova, V., *34,* 61
Schenk, P. E., 159, 161, 164, 170, *173*
Schilling, J. H., 511
Schlanger, S. O., *226, 305,* 344, *345, 347, 414*
Schlatter, L. E., *127*
Schlee, J., *618*
Schmalz, R. E., 291, 292
Schmidl, V., 419
Schmidt, R. G., 305, *305,* 412, *415, 449*
Schmincke, H. U., *627*
Schmoll, H. R., *522*
Schneider, A. F., 606, *607*
Schneider, R. C., *557*
Schofield, J. C., **250–251**, *251,* 281, **401–402,** *402,* 496, *497*
Scholl, D. W., *243,* 514, *521, 522*
Scholle, P. A., *117*
Schols, H., 310
Scholten, R., *173*
Scholz, C. H., *400, 409,* 545, *548*
Schönberger, J. M. H., *492*
Schöpf, J. D., *409*
Schopf, J. M., 12
Schouten, J. A., 486, 487, *491*
Schreiber, E., 329, *330,* 666
Schroeder, J. H., 119, *121*
Schubert, C., *654*
Schuchert, C., 223, 228, *237, 301, 341, 419,* 618, 622, 664
Schultz, L. G., *572*
Schuster, G., 419
Schwab, F. L., *409*
Schwab, K., *21*
Sclater, J. G., *244, 415,* 495, *497*
Scorza, E. P., 276, *278*
Scott, K. M., *243*
Scott, P. A., *45*
Scripps (Institution of Oceanography) Expedition *Camarsel,* 226
Scurlock, J. R., *557*
Seager, W. R., *548*
Searle, E. J., **337–338**, *337*
Seddon, G., *34*
Segerstrom, K., *243, 243*
Seiglie, G. A., 258, *259, 437*
Selfridge, T. O., 416, *419*
Selkirk, A., 336
Sellow, F. von, 642, *647*
Selwyn, A. R. C., 81, 188
Senn, A., *116*
Sepp, A., 419
Seward, A. C., 273, *274*
Shackleton, E., 3, 469
Shagam, R., *654*
Shapiro, E., 534, *540*
Sharp, R. P., *557, 627*
Sharples, S. P., *501*
Shattuck, G. B., 539, *541*
Shaw, D. R., *548*

Shaw, E. W., *217*
Shaw, H. R., *596, 627*
Sheldon, R. P., *641*
Shepard, F. P., *114,* 226, *226,* 347, *347, 415, 596, 607*
Shepherd, J. B., 361, *361,* 448
Sheridan, R. E., 112, *115, 540*
Sherwood, C. B., *500*
Shideler, G. L., *541*
Shilts, W. W., *225*
Shinn, E. A., 112, *115*
Shoemaker, E. M., *557*
Shor, G. G., *282, 403,* 592, *596*
Short, K. C., 312, 317, *318*
Shotton, F. W., 360, *361*
Shultz, C. H., 469
Shurbet, D. H., *548*
Shutler, R., Jr., *390*
Siemers, C. T., *539*
Sigurdsson, H., 257, *257,* 300, *301, 448*
Silberman, M. L., *547*
Sillitoe, R. H., *242, 243*
Silva Coutinho, J. M. da, 127
Silver, E. A., *595,* 622, *627*
Simmons, F. S., *419*
Simmons, G. C., *138*
Simmons, W. A., *117*
Simon, J. A., 511
Simpson, D. W., *91*
Simpson, F. A., *13*
Simpson, H. E., *572*
Sinden, J. A., *34*
Singewald, J. T., *465*
Singh, S., 312, 313, 315, *318,* 320, 321, *325*
Sinoto, Y. H., *261*
Siple, G. E., *540*
Skeats, E. W., *292*
Skinner, W. R., 634, *641*
Skottsberg, C., 336, *336*
Škvor, V., *256,* 666
Skymer, T., 120, *121*
Slater, R. A., *28*
Slatyer, R. O., *34*
Slaughter, A., 511
Slevin, J. R., *290*
Sloss, L. L., 563, *572,* 599, 600, 601, 602, 603, 604, *607,* 634, 635, *641*
Slossen, J. E., 510
Sly, P. G., 202, *207*
Smart, J., *61*
Smart, P. G., 45, *45*
Smedes, H. W., *641*
Smit, D. E., 162, *173*
Smith, A., 12, *13*
Smith, A. C., *278*
Smith, A. G., 112, *114*
Smith, A. L., *348, 666*
Smith, B. L., *640*
Smith, C. I., *581*
Smith, C. Y., *145*
Smith, I. E., *27,* 378, 380, *382, 383*
Smith, J. P., 619
Smith, K. G., 51, *56*
Smith, L., 171, *172*

AUTHOR INDEX

Smith, P. B., *548*
Smith, P. S., *522*
Smith, R., *80, 409, 572*
Smith, R. B., *641*
Smith, S. M., *415*
Smith, T. J., *243*
Smith, W., 469
Smith, W. B., Jr., 520, *522*
Smith, W. C., *500*
Snavely, B. L., *414*
Snavely, P. D., Jr., *627*
Snelling, N. J., *242,* 278, 313, 314, 315, 316, *318,* 319, 320, 321, 322, *325,* 428
Snow, P. A., *497*
Snow, W. P., *426*
Snyder, F. G., *607*
Sociedad Venezolana de Ingenieros de Petróleo, *654*
Socolow, A. A., 512
Sollas, W. J., 291, *292, 414*
Solomon, M., *28, 91*
Somerville, B. T., 494, *497*
Sommer, F. von, 100
Soper, N. J., *300*
South Australian Dept. of Mines, 62, 68, 73
Souther, J. G., 210, *217*
Spath, L. F., *497*
Speight, R., 390, 402, *403*
Spencer, A. M., *250, 270, 282*
Spencer, J. W., *257, 442, 443, 668*
Spieker, E. M., *557*
Splettstoesser, J. F., 3, *13*
Spooner, C. M., *284,* 313, *318,* 320, *325, 618*
Sprigg, R. C., 67, 69, *80, 399*
Sproll, W. P., 110, 113, *114*
Sproule, J. C., and Associates Ltd., *156*
Sprunt, A. N., *501*
Spry, A. H., *91*
Spurr, J. E., 514, *522*
Stacey, R. A., 209, *217*
Stainforth, R. M., 323, *325*
Standard, J. C., *45*
Stanley, D. J., 117, *121*
Stanton, R. L., *46, 371,* 451, *456*
Stapp, W. L., *581*
Stappenbeck, R., 15
Stark, J. T., 226, *226,* 305, *415, 451*
Stauffer, C. R., 195, *207*
Staverman, W. H., 486, 487, *491*
Stearn, C. W., **139–145**, *144, 193, 408, 513, 607*
Stearns, H. T., **1–2**, *2,* **304–305**, *305, 371, 415,* **448–449**, **494–497**, *497,* **581–596**, *595, 596,* **657–658**, *657, 658,* **666–667**, *667*
Stearns, R. G., *531*
Steenken, W. F., 312, 317, *318*
Steers, J. A., *336*
Stehli, F. G., *13, 480*
Stehman, C. F., *114*
Steinen, R. P., *116*

Steiner, M. B., *641*
Steinhart, J. S., *243*
Steinmann, G., *433*
Stelck, C. R., **187–192**
Stelzner, A., 15, 456
Stephens, C. G., 402, *402*
Stephenson, H. K., *347*
Stephenson, L. B., *541*
Stephenson, P. J., 475, 476
Stern, T. W., *243*
Stevens, G. R., *13,* **390–400**, *400*
Stevens, N. C., 45, *45*
Stevens, R. K., 161, *172, 173*
Stevenson, J. S., *409*
Stewart, D. P., *617, 618*
Stewart, G. A., 209, 212, 214, *217*
Stewart, G. W., 511
Stewart, H. B., Jr., 280, *282, 414,* 494, *497*
Stewart, I. C. F., 63, *80*
Stewart, J. H., *409, 548, 641*
Stewart, J. W., 428, *433*
Stewart, R. H., **416–419**
Stibane, F. R., **245–250**, *249, 250*
Stice, G. D., *2*
Stiefel, J., *243*
Stille, H., *409,* 451, *465,* 622, *641, 664*
Stipp, J. J., *304,* 340, *341*
Stirewalt, G. I., *531*
Stirton, K. A., *272*
Stockley, G. M., 324, *325*
Stockman, K. W., 112, *114*
Stockwell, C. H., *156, 173,* **174–179**, *179, 193, 409*
Stoddart, D. R., *117, 251*
Stoecklin, J., 419
Stoiber, R. E., 272, *272*
Stokes, W. L., *557, 641*
Stone, D. P., *521*
Stone, E. L., Jr., *415*
Stoneley, R., 247, *250,* 514, *522*
Stormer, J. C., Jr., *548,* 555, *557*
Stout, T. M., *572*
Stover, C. W., *290*
Strahler, A. N., *556, 557*
Straley, H. W., 532, *540*
Strong, D. F., 162, *173*
Strusz, D. L., *46*
Strzelecki, 81
Stuart, J. McD., *46*
Stuart, W. D., 118, *121*
Stuart, W. J., Jr., *80*
Stuiver, M., *13*
Stumm, E. C., 200, *207*
Stump, E., **2–13**
Stutzer, O., 245
Suess, E., 3
Suggate, R. P., 398, *400*
Sugimura, A., **331–333**
Summerhayes, C. P., *251*
Summerson, C. H., *409, 618*
Sumner, J. R., *627*
Supko, P. R., 486, *491*
Suppe, J., *627*
Surdam, R. C., *640, 641*
Suszcyznski, E. F., *138*

Suter, H. H., *498*
Sutherland, F. L., *91*
Sutter, J., *546*
Sutton, A. H., *607*
Sutton, J., *480,* 640
Swain, F., 533, *541*
Swallow, J. C., 292, *292*
Swann, D. H., *409*
Swanson, D. A., 594, *595, 596*
Swanson, F. J., *290*
Swarth, H. S., *290*
Swartz, J. H., *415*
Sweet, G., 291, *292, 414*
Sweet, W. C., *208*
Sweeting, M. M., *336*
Swett, K., 162, *173*
Swift, D. J. P., 117, *121, 541*
Sykes, L. R., *27,* 251, *252,* 333, *341, 409, 437,* 664, *666*
Symons, J. E., 81
Szabo, B. J., *522*

Taber, S., 255, *256*
Tailleur, I. L., *521*
Takai, F., *333*
Talent, J. A., *34*
Taliaferro, N. L., *624*
Talwani, M., 4, 112, 113, *115, 497, 548*
Tamayo, J. L., *360*
Tardy, M., *360*
Tarling, D. H., 250, *251,* 300, *497*
Tasman, A., 46
Tate, G. H. H., 47, *325*
Tatel, H. E., 545, *548*
Taubeneck, W. H., *641*
Tayama, R., *226,* 347, *415,* 449
Taylor, A., *156*
Taylor, N. H., 402, *402*
Teal, J. J. H., 244, *244*
Tenison Woods, J. E., 47
TePunga, M. T., *400*
Terasmae, J., **192–208**, *207, 208*
Tercinier, G., *284*
Terman, T. H., 345, *345*
Terry, R. A., 228, *237,* 419
Tester, A. C., 347, *347, 415*
Textoris, D. A., *606*
Tharp, M., 2
Thatcher, W., *360*
Thom, W. T., Jr., *571*
Thomas, D. E., *100*
Thomas, H. D., *15, 341, 641*
Thomas, H. F., 510
Thomas, N. A., *243*
Thomas, W. A., 603, *607*
Thompson, G., *278*
Thompson, G. A., 544, 545, *548*
Thompson, J. B., Jr., *618*
Thompson, J. E., 121, *123, 383*
Thompson, R., *21*
Thompson, R. B., 451
Thomson, J. A., 337, *497*
Thornbury, W. D., *409,* 513, 557, 628, *641*
Thornton, 46

Thorsteinsson, R., *156*
Thurber, D. L., *114,* 115, *115, 347*
Tilley, C. E., 468
Tippenhauer, G., 658
Tissot, B., 366, *371*
Tjia, H. D., *383*
Tobisch, O. T., *531*
Tocher, D., *522*
Todd, R., 278, 591, *596*
Tomasi, P., 428, *433*
Tomblin, J. F., 257, *257,* 300, *301, 304, 341,* 361, *361,* 442, *442, 472, 498, 668*
Tomblin, J. R., 662, *666*
Tonkin, P. C., *56*
Townsend, R. C., *347*
Tozer, E. T., *156*
Tracey, J. I., Jr., 226, *226,* 305, 344, 345, *345, 347, 414, 415,* 591, *595*
Travers, W. B., *627*
Travis, R. B., *433*
Trechmann, C. T., *15, 116,* 301, *336, 442, 498,* 658, *668*
Trendall, A. F., 103, *107, 138,* 467, 468
Tricart, J., 271, *272*
Trichet, J., *415, 499*
Trowbridge, A. C., *607*
Trueman, N. A., 243, *244*
Trümpy, D., 245, 246, *250*
Tschopp, H. J., 264, 265, 268, *270*
Tsuya, H., *333*
Tucker, D. R., *581*
Tuell, E. J., *521*
Tugarinov, A. I., *138*
Turekian, K. K., *13, 207, 299*
Turneaure, F. S., *127*
Turner, J. C. M., *21,* **642–647**
Turner, S., *531*
Tuthill, S. J., 511
Tuve, M. A., 545, *548*
Twenhofel, W. H., 223, *225*
Tychsen, P. C., *571*
Tyler, S. A., *138*
Tyrrell, J. B., 144, 188, 469

Uchupi, E., 113, *115,* 237, *617, 618,* 663
Ulrych, T. J., *409*
Umpierre, M., 642, 643, 644, *648*
University of Miami, Rosenstiel School of Marine and Atmospheric Science, 114
Unterman, B. R., *641*
Unterman, G. E., *641*
Upchurch, S. B., 119, *121*
Updike, R. G., *557*
Urien, C. M., *21*
Urville, D. d', 3
U.S. Geological Survey, 328, 514, *522,* 584
Ushakov, S. A., *13*
U.S. Hydrographic Office Charts, 280
U.S. Navy Department, *416*

U.S. Office of Geography, *415*
Utah Geological Society, *557*
Utsu, T., *333*
Uyeda, S., *333*

Vacher, H. L., **117–121,** *121*
Vaksvik, K. N., *596*
Valencio, D. A., *21*
Valeton, I., 323, 324, 325
Van Bemmelen, R. W., 664
Van den Bold, W. A., *327,* 340
Van den Boom, G., 419, 424, *426*
Van der Linden, W. J. M., 280, *282,* 370, 371, *402*
van Deventer, J., *456*
Van Mieghem, J., *13, 546*
Van Oye, P., *13*
Van Zinderen Bakker, E. M., *480*
Varne, R., 473, *475*
Vaughan, T. W., 121, *227,* 258, *259, 498,* 656, 658
Vedder, J. G., *116*
Veeh, H. H., 250, *251,* 381, *383, 415,* 586, *596*
Veen, A. W. L., 485, *492*
Veevers, J. J., 12, *28,* 54, *56, 108,* **402–403,** *402*
Vegara, M., *243*
Vélain, C., 456, 473, 475
Veldkamp, J., 483, *492*
Vellard, J., 424, *426*
Veltheim, R. V., 278
Vening Meinesz, F. A., 664
Verdurmen, E. A. T., 314, 315, 316, *318,* 322, *325,* 326, 482, 483, *492*
Verhofstad, J., 315, *318,* 488, *492*
Vernon, R. O., *581*
Verrill, A. E., *121*
Verschure, R. H., 314, 315, 316, *318,* 321, *322, 325,* 363, *364,* 482, 483, *492*
Versey, H. R., *335, 336*
Verstappen, H. T., 374, *383*
Verwoerd, W. J., 472, 474
Vevers, H. G., *251*
Vicente, J.-C., *465*
Viele, G. W., 345, *345*
Vinha, C. A. G. da, *137*
Vinson, G. L., *237, 252, 309,* 329, *330*
Visher, G. S., *557*
Visher, S. S., *513*
Visser, W. A., **372–383,** *383*
Vivar, D. V. D., *426*
Vivar, V. D., 419, 423, *426*
Voges, A., 125, *127*
Vogt, P. R., *28, 243,* 298, *300, 327, 327*
Voisey, A. H., *46*
Volk, R. W., *570*
Volkheimer, W., *21*
Voronov, P. S., *13*
Vuagnat, M., *255*
Vuilleumier, B. S., *465*

Waage, K. M., *408*
Wade, M., *80*
Wager, L. R., *448*
Wagner, F. J. E., *225*
Wahrhaftig, C., 514, 516, 522
Walcott, C. D., *557,* 618
Walcott, R. I., *145,* 538
Walker, R. G., *531*
Wallace, A. R., 414, *415*
Wallace, R. E., **541–548**
Wallace, R. M., *138*
Walper, J. L., 113, *115, 581*
Walpole, B. P., 50, *56, 61*
Walthall, B. H., *581*
Walther, K., 642, 644, *648*
Walton, J., 273, *274*
Walton, M., 511
Wanless, R. K., *145,* 177, *179,* 409, *607*
Wanoa, R. J., *338*
Ward, C. R., *46*
Ward, P. L., 272, *272, 401*
Ward, W. C., *360*
Ward, W. T., *100,* 539, *596*
Warden, A. J., **383–390,** *390*
Waring, G. A., *498*
Warner, R. F., *46*
Warren, G., *13*
Warren, P. S., 188
Warren, R. G., *34*
Washburne, C., 128
Washington, H. S., *278,* 330, 449, *449*
Wasserburg, G. J., *626*
Watanabe, T., *333*
Waterhouse, J. B., *372*
Waters, A. C., *627*
Watson, B. N., 430, *433*
Watson, W. O., Jr., 511
Watters, W. A., *243*
Wayne, W. J., *607*
Weaver, J. D., **434–437,** *437,* **654–656**
Weaver, K. N., *530*
Webb, E. J., *531*
Webb, G. W., 168, *173*
Webb, J. B., *193*
Webb, T., III, *607*
Webby, B. D., *46*
Weber, J. N., *292*
Wedow, H., Jr., *530*
Weeks, C. G., *100*
Weeks, L. A., *653, 666*
Weeks, L. G., *465,* 654
Weeks, L. J., 159, 160, 161, 166, *173,* 444, *447*
Wegen, G. van der, 374, 381, *383*
Wegener, A., 3, 299
Weidick, A., *300*
Weimer, R. J., *572, 641*
Weisbord, N. E., *116,* 654
Weischet, W., *242*
Weller, J. M., **596–607,** *607*
Weller, S., 538, 539, *541, 607*
Wells, A. T., 51, 52, *55, 56, 108*
Wells, N., 402, *402*
Welter, C. C., 490, *491*

AUTHOR INDEX

Wentworth, C. K., *415*
Wertz, J. B., 545, *548, 641*
Westermann, J. H., *341,* 363, 364, *365, 442*
Westerndorp, F. J. van, 401, *402*
Wetzel, W., 241, *243*
Weyl, P. K., *364*
Weyl, R., 228, *237, 252, 257, 272, 304, 309, 341,* 364, *365, 401, 437, 443,* **658–666,** *666, 668*
Wharton, W. J., 244, *244*
Wharton, W. J. L., *415*
Wheeler, H. E., *409,* 602, *607,* 619, 622
Wheeler, J. O., *187,* 211, 213, *217*
Wheeler, W. H., *534, 540*
Whetten, J. T., *627, 655, 656*
Whipple, A. W., 549
White, A. J. R., *27, 304, 341,* 345, *345*
White, D. E., *627*
White, I. C., 127, 524
White, M. E., *61*
White, W. S., *607, 618*
Whiteman, A. J., *114*
Whitney, J. D., 619, 622
Whittaker, E. S., 188
Whittington, H. B., 161, 162, *173*
Wickenden, R. T. D., 188
Widmer, K., 444, 447, 512
Wiens, H. J., *284, 291, 415,* 499
Wiggins, I. L., 287, *290*
Wigley, P., *15*
Wijmstra, T. A., 322, 323, *325,* 483, 485, 486, *491, 492*
Wilcox, R. E., 568, *571, 572*
Wilhelm, O., *581*
Wilhelmy, H., 419, *426*
Wilkes, C., 3
Will, H., 467
Williams, E., 313, 314, 317, *318,* 319, 320, 321, *325,* 482, *492*
Williams, G. E., *80,* 100

Williams, H., 159, 162, 168, 169, *173, 174,* 252, 271, *272,* 284, 286, 288, *290, 309,* 329, *330, 415, 451,* 493, *493, 557*
Williams, J. A., 510
Williams, J. B., *336*
Williams, M. Y., 188, 195, 198, *208*
Williams, N. F., 510
Williams, P. W., 380, *383*
Williams, R. E. G., 360, *361*
Williamson, J. D. M., *581*
Willis, B., 449, *449,* 524, 619
Willman, H. B., *606, 607*
Willmont, W. F., 378, *383*
Wills, W. J., 92
Wilson, A. E., 196, *208*
Wilson, C. C., *498*
Wilson, E. D., *548, 557*
Wilson, E. O., *336*
Wilson, H. H., *309,* 330
Wilson, J. J., 427, *433*
Wilson, J. L., *360,* 598, *607*
Wilson, J. T., 168, *174,* 277, 546, *548,* 596
Wilson, L., **502–513**
Wilson, R. B., *80*
Winder, C. G., **192–208,** *207, 208*
Windhausen, A., *425*
Winterer, E. L., 376, *383, 412, 415*
Wiseman, J. D. H., 277, *278*
Wolcott, P. P., *654*
Wolf, T., *290,* 456
Wolfart, R., 125, 419, 421, 422, *426*
Woo, D. F., 361, *361*
Wood, B. L., 250, *251,* 399, *415,* 667
Woodhead, P. M. J., *292*
Wood-Jones, F., *245*
Woodring, W. P., *237, 327,* 416, *419,* 658
Woodward, H. P., 168, 172
Woodward, L. A., *547*
Woodward, P. W., *596*

Woollard, G. P., *13,* 384, *456, 531, 541,* 587, *595*
Woolnough, W. G., 47
Wopfner, H., *81*
Wordie, J. M., 469
Wright, A. C. S., 337, *338,* 401, *402*
Wright, F. F., **513–522**
Wright, H. E., Jr., *513, 522, 540, 556, 581, 618, 641*
Wright, J. D., 200, *207*
Wright, R., *278*
Wuenschel, P. C., 118, *121*
Wulf, G. R., *572*
Wyatt, B. W., *91*

Yasui, M., *333*
Yawata, I., *261*
Yoder, H. S., Jr., *592, 596*
Yoshiwara, S., 331, *333*
Young, D. A., *531*
Young, G. A., *654*
Young, K., 579, *581*
Young, R. N., **109–115,** *115*
Young, R. W., *46*

Zambrano, J. J., *21*
Zans, V. A., *336*
Zartman, R. E., *618, 641*
Zashu, S., *333*
Zeil, W., *127,* 238, *243, 465*
Zen, E., *409, 618*
Zentilli, M., *242*
Zhivago, A. V., *13*
Zietz, I., *408, 607*
Zijderveld, J. D. A., *483, 492*
Zonneveld, J. I. S., 363, *365,* 485, 486, 488, 489, *492*
Zullo, V. A., 286
Zuloaga, G., 309
Zumberge, J. H., *607*
Zurbrigg, H. F., 178, *179*

Thorsteinsson, R., *156*
Thurber, D. L., *114*, 115, *115, 347*
Tilley, C. E., *468*
Tippenhauer, G., *658*
Tissot, B., 366, *371*
Tjia, H. D., *383*
Tobisch, O. T., *531*
Tocher, D., *522*
Todd, R., *278*, 591, *596*
Tomasi, P., 428, *433*
Tomblin, J. F., 257, *257*, 300, *301, 304, 341*, 361, *361*, 442, *442, 472, 498, 668*
Tomblin, J. R., *662, 666*
Tonkin, P. C., *56*
Townsend, R. C., *347*
Tozer, E. T., *156*
Tracey, J. I., Jr., 226, *226*, *305*, 344, 345, *345, 347, 414, 415*, 591, *595*
Travers, W. B., *627*
Travis, R. B., *433*
Trechmann, C. T., *15, 116, 301, 336, 442, 498, 658, 668*
Trendall, A. F., 103, *107, 138*, 467, 468
Tricart, J., 271, *272*
Trichet, J., *415, 499*
Trowbridge, A. C., *607*
Trueman, N. A., 243, *244*
Trümpy, D., 245, 246, *250*
Tschopp, H. J., 264, 265, 268, *270*
Tsuya, H., *333*
Tucker, D. R., *581*
Tuell, E. J., *521*
Tugarinov, A. I., *138*
Turekian, K. K., *13, 207, 299*
Turneaure, F. S., *127*
Turner, J. C. M., *21*, **642–647**
Turner, S., *531*
Tuthill, S. J., *511*
Tuve, M. A., 545, *548*
Twenhofel, W. H., 223, *225*
Tychsen, P. C., *571*
Tyler, S. A., *138*
Tyrrell, J. B., 144, 188, 469

Uchupi, E., 113, *115*, 237, *617, 618*, 663
Ulrych, T. J., *409*
Umpierre, M., 642, 643, 644, *648*
University of Miami, Rosenstiel School of Marine and Atmospheric Science, 114
Unterman, B. R., *641*
Unterman, G. E., *641*
Upchurch, S. B., 119, *121*
Updike, R. G., *557*
Urien, C. M., *21*
Urville, D. d', 3
U.S. Geological Survey, 328, 514, *522*, 584
Ushakov, S. A., *13*
U.S. Hydrographic Office Charts, 280
U.S. Navy Department, *416*

U.S. Office of Geography, *415*
Utah Geological Society, *557*
Utsu, T., *333*
Uyeda, S., *333*

Vacher, H. L., **117–121**, *121*
Vaksvik, K. N., *596*
Valencio, D. A., *21*
Valeton, I., 323, 324, 325
Van Bemmelen, R. W., 664
Van den Bold, W. A., *327*, 340
Van den Boom, G., 419, 424, *426*
Van der Linden, W. J. M., 280, *282*, 370, 371, *402*
van Deventer, J., *456*
Van Mieghem, J., *13, 546*
Van Oye, P., *13*
Van Zinderen Bakker, E. M., *480*
Varne, R., 473, 475
Vaughan, T. W., 121, *227*, 258, 259, *498, 656, 658*
Vedder, J. G., *116*
Veeh, H. H., 250, *251*, 381, *383, 415*, 586, *596*
Veen, A. W. L., 485, *492*
Veevers, J. J., 12, *28*, 54, *56*, *108*, **402–403**, *402*
Vegara, M., *243*
Vélain, C., 456, 473, 475
Veldkamp, J., 483, *492*
Vellard, J., 424, *426*
Veltheim, R. V., *278*
Vening Meinesz, F. A., 664
Verdurmen, E. A. T., 314, 315, 316, *318*, 322, *325*, 326, 482, 483, *492*
Verhofstad, J., 315, *318*, 488, *492*
Vernon, R. O., *581*
Verrill, A. E., *121*
Verschure, R. H., 314, 315, 316, *318*, 321, *322*, *325*, 363, *364*, 482, 483, *492*
Versey, H. R., *335, 336*
Verstappen, H. T., 374, *383*
Verwoerd, W. J., 472, 474
Vevers, H. G., *251*
Vicente, J.-C., *465*
Viele, G. W., 345, *345*
Vinha, C. A. G. da, *137*
Vinson, G. L., *237, 252, 309*, 329, *330*
Visher, G. S., *557*
Visher, S. S., *513*
Visser, W. A., **372–383**, *383*
Vivar, D. V. D., *426*
Vivar, V. D., 419, 423, *426*
Voges, A., 125, *127*
Vogt, P. R., *28, 243*, 298, *300*, *327, 327*
Voisey, A. H., *46*
Volk, R. W., *570*
Volkheimer, W., *21*
Voronov, P. S., *13*
Vuagnat, M., *255*
Vuilleumier, B. S., *465*

Waage, K. M., *408*
Wade, M., *80*
Wager, L. R., *448*
Wagner, F. J. E., *225*
Wahrhaftig, C., 514, 516, 522
Walcott, C. D., *557*, 618
Walcott, R. I., *145*, 538
Walker, R. G., *531*
Wallace, A. R., 414, *415*
Wallace, R. E., **541–548**
Wallace, R. M., *138*
Walper, J. L., 113, *115, 581*
Walpole, B. P., 50, *56, 61*
Walthall, B. H., *581*
Walther, K., 642, 644, *648*
Walton, J., 273, *274*
Walton, M., 511
Wanless, R. K., *145*, 177, *179*, 409, *607*
Wanoa, R. J., *338*
Ward, C. R., *46*
Ward, P. L., 272, *272*, *401*
Ward, W. C., *360*
Ward, W. T., *100*, 539, *596*
Warden, A. J., **383–390**, *390*
Waring, G. A., *498*
Warner, R. F., *46*
Warren, G., *13*
Warren, P. S., 188
Warren, R. G., *34*
Washburne, C., 128
Washington, H. S., *278*, 330, 449, *449*
Wasserburg, G. J., *626*
Watanabe, T., *333*
Waterhouse, J. B., *372*
Waters, A. C., *627*
Watson, B. N., 430, *433*
Watson, W. O., Jr., 511
Watters, W. A., *243*
Wayne, W. J., *607*
Weaver, J. D., **434–437**, *437*, **654–656**
Weaver, K. N., *530*
Webb, E. J., *531*
Webb, G. W., 168, *173*
Webb, J. B., *193*
Webb, T., III, *607*
Webby, B. D., *46*
Weber, J. N., *292*
Wedow, H., Jr., *530*
Weeks, C. G., *100*
Weeks, L. A., *653, 666*
Weeks, L. G., *465, 654*
Weeks, L. J., 159, 160, 161, 166, *173*, 444, *447*
Wegen, G. van der, 374, 381, *383*
Wegener, A., 3, 299
Weidick, A., *300*
Weimer, R. J., *572, 641*
Weischet, W., *242*
Weller, J. M., **596–607**, *607*
Weller, S., 538, 539, *541*, 607
Wells, A. T., 51, 52, *55*, 56, *108*
Wells, N., 402, *402*
Welter, C. C., 490, *491*

Wentworth, C. K., *415*
Wertz, J. B., 545, *548*, *641*
Westermann, J. H., *341*, 363, 364, *365*, *442*
Westerndorp, F. J. van, 401, *402*
Wetzel, W., 241, *243*
Weyl, P. K., *364*
Weyl, R., 228, *237*, *252*, *257*, *272*, *304*, *309*, *341*, 364, *365*, *401*, *437*, *443*, **658–666**, *666*, *668*
Wharton, W. J., 244, *244*
Wharton, W. J. L., *415*
Wheeler, H. E., *409*, 602, *607*, 619, 622
Wheeler, J. O., *187*, 211, 213, *217*
Wheeler, W. H., *534*, *540*
Whetten, J. T., *627*, *655*, *656*
Whipple, A. W., 549
White, A. J. R., *27*, *304*, *341*, 345, *345*
White, D. E., *627*
White, I. C., 127, 524
White, M. E., *61*
White, W. S., *607*, *618*
Whiteman, A. J., *114*
Whitney, J. D., 619, 622
Whittaker, E. S., 188
Whittington, H. B., 161, 162, *173*
Wickenden, R. T. D., 188
Widmer, K., 444, 447, 512
Wiens, H. J., *284*, *291*, *415*, 499
Wiggins, I. L., 287, *290*
Wigley, P., *15*
Wijmstra, T. A., 322, 323, *325*, 483, 485, 486, *491*, *492*
Wilcox, R. E., 568, *571*, *572*
Wilhelm, O., *581*
Wilhelmy, H., 419, *426*
Wilkes, C., 3
Will, H., 467
Williams, E., 313, 314, 317, *318*, 319, 320, 321, *325*, 482, *492*
Williams, G. E., *80*, *100*

Williams, H., 159, 162, 168, 169, *173*, *174*, 252, 271, 272, *284*, 286, 288, *290*, *309*, 329, *330*, *415*, *451*, 493, *493*, *557*
Williams, J. A., 510
Williams, J. B., *336*
Williams, M. Y., 188, 195, 198, *208*
Williams, N. F., 510
Williams, P. W., 380, *383*
Williams, R. E. G., 360, *361*
Williamson, J. D. M., *581*
Willis, B., 449, *449*, 524, 619
Willman, H. B., *606*, *607*
Willmont, W. F., 378, *383*
Wills, W. J., 92
Wilson, A. E., 196, *208*
Wilson, C. C., *498*
Wilson, E. D., *548*, *557*
Wilson, E. O., *336*
Wilson, H. H., *309*, *330*
Wilson, J. J., 427, *433*
Wilson, J. L., *360*, *598*, *607*
Wilson, J. T., 168, *174*, 277, 546, *548*, *596*
Wilson, L., **502–513**
Wilson, R. B., *80*
Winder, C. G., **192–208**, *207*, *208*
Windhausen, A., *425*
Winterer, E. L., 376, *383*, *412*, *415*
Wiseman, J. D. H., 277, *278*
Wolcott, P. P., *654*
Wolf, T., *290*, *456*
Wolfart, R., 125, 419, 421, 422, *426*
Woo, D. F., 361, *361*
Wood, B. L., 250, *251*, *399*, *415*, *667*
Woodhead, P. M. J., *292*
Wood-Jones, F., *245*
Woodring, W. P., *237*, *327*, 416, *419*, 658
Woodward, H. P., 168, 172
Woodward, L. A., *547*
Woodward, P. W., *596*

Woollard, G. P., *13*, 384, *456*, *531*, *541*, 587, *595*
Woolnough, W. G., 47
Wopfner, H., *81*
Wordie, J. M., 469
Wright, A. C. S., 337, *338*, 401, *402*
Wright, F. F., **513–522**
Wright, H. E., Jr., *513*, *522*, *540*, *556*, *581*, *618*, *641*
Wright, J. D., 200, *207*
Wright, R., *278*
Wuenschel, P. C., 118, *121*
Wulf, G. R., *572*
Wyatt, B. W., *91*

Yasui, M., *333*
Yawata, I., *261*
Yoder, H. S., Jr., *592*, *596*
Yoshiwara, S., 331, *333*
Young, D. A., *531*
Young, G. A., *654*
Young, K., 579, *581*
Young, R. N., **109–115**, *115*
Young, R. W., *46*

Zambrano, J. J., *21*
Zans, V. A., *336*
Zartman, R. E., *618*, *641*
Zashu, S., *333*
Zeil, W., *127*, 238, *243*, *465*
Zen, E., *409*, *618*
Zentilli, M., *242*
Zhivago, A. V., *13*
Zietz, I., *408*, *607*
Zijderveld, J. D. A., *483*, *492*
Zonneveld, J. I. S., 363, *365*, 485, 486, 488, 489, *492*
Zullo, V. A., 286
Zuloaga, G., 309
Zumberge, J. H., *607*
Zurbrigg, H. F., 178, *179*

SUBJECT INDEX

Boldface entires represent article titles; boldface page numbers indicate first page of main articles.

Aa, 582
Abajo Mts., 551
A.B.C. Is., 362
Abingdon Is., 289
Abrolhos Terrace, 370
Absaroka Fault, 629
Absaroka sequence, 603, 635
Absaroka Volcanic Field, 508, 630
Abuillot Fm., 326
Acacia, 87
Acadian Geosyncline, 159, 161
Acadian Orogenic Belt, 405
Acadian Orogeny, 159, 165, 167, 528, 602, 612
Acaena, 465
Acanthocardia, 368
Acarai Mts., 489
Acrolepis, 644
Acrophyllum, 199
Acrospirifer, 199
Adampada Falawatra Group, 312, 313
Adelaidean System, 28, 50
Adelaide Geosyncline, 67
Adirondack Mts., 196, 608
Admiralty Is., 121, 372
Aemulophyllum, 199
Africa, 164
African shelf, 161
Agate Springs fossil quarries, 566
Aguijan, 344
Ahuachapan, 272
Aiguá Series, 643
Aitutaki, 250
Akawaian period, 314
Alabama, 510
Alae Gater, 593
Alaska, 502, 510, **513**
Albemarle, 288, 289
Albemarle Basin, 532
Alberta Basin, 189
Albertan Series, 219
Alberta Syncline, 192
Aleamaru, 290
Alejandro Selkirk, 336
Aleutian Is. Chain, 515
Alexa Bank, 280
Alexander Is., 9
Alexandrian Series, 219
Algoman Granite, 599
Algonquin Arch, 193, 197, 198
Alice Springs, 48
Alifan Limestone, 305
Allegheny Orogeny, 168, 528, 613

Allegheny Plateau, 522
Allegheny Trough, 193, 195, 197
Allochthon, 159, 163, 354, 616
Alpine Fault, 392
Alpine glaciers, 10, 464
Alpine Schist, 393
Alpinotype Orogeny, structures, 633, 637
Alta Verapaz, 308
Altiplano, 629
Altiplano Massif, 124
Alutom Fm., 305
Alveolites, 198
Amabel Group, 198
Amaila Falls, 319
Amapá Series, 312
Amargosa desert, 542
Amarillo Mt. Range, 562, 570
Amazonas Basin, 131
Amazon Basin, 269
Amazon mud, 486
Amazon River, 130, 457
Amblyrhyncus, 287
Ambrym, 384
American Samoa, **1**
American Virgin Is., 653
Amherstburg Fm., 199, 200
Ammonoosuc Fault, 615
Amotape Chanchan Belt, 263, 268
Amotape Fm., 264
Amphigenia, 199
Amphistegina, 343
Amphitheater-headed valleys, 585
Amsterdam Is., 472
Amuku Mts., 319
Anabala, 422
Anaco, 651
Anadarko Basin, 598
Ana Maria Depression, 253, 254
Anatahan, 345
Ancestral Front Range, 635
Ancestral Man, 638
Ancestral Rocky Mts., 404, 563, 631
Ancon Field, 270
Andean batholith, 430
Andean Coastal Slope, 427
Andean Cordillera, 15, 463
Andean folding, movements, Orogeny, 9, 124, 239, 265
Andean Foothills, 245
Anderson Basin, 189
Andes, Andean Belt, 245, 427, 460, 648

Andesite (Sial) line, 365, 410, 494, 658
Andine Geosyncline, 422
Andros Is. 109, 659
Anegada Passage, 339
Anerdon Member, 200
Anguilla, **439**
Anhydrite, 191, 198, 234, 423, 575
Anodontophora, 367
Antarctica, **2**
Antarctic Peninsula, 10, 465
Antarctosaurus, 645
Anticlinal Theory, 194
Anticosti Basin, 218
Anticosti Is., 162, 166, 224
Antigua, **14**
Antillean Arc, 658
Antillean Geosyncline, 661
Antillite, 258
Antimony, 126, 464
Antipodes, 390
Antler Orogeny, 508, 543, 623, 635
Antrim Plateau, 23
Anzoátegui, 652
Aoba, 384
Aoga-shima, 332
Apa Mts., 421
Apatue Dolerite, 481
Aphebian period, 140, 176, 210, 599
Apoteri Volcanic Fm., 320, 322
Appalachian front, 223
Appalachian Geosyncline, 159, 168
Appalachian glaciation, 158
Appalachian Mt. System, 140
Appalachian Orogenic Belt, 156
Appalachian Orogeny, 354
Appalachian Plateaus, 522
Appalachian Region (U.S.), **504**, **522**, **608**
Appalachian Revolution, 168
Aquari Fm., 642
Aquia Fm. 536
Aquidauana Sandstone, 422
Arafura Sea, 46, 372
Arapey lavas, 642
Araucaria, 402
Arbuckle Mts., 597
Archaeocyathus, 184
Archean period, 140, 149, 293, 311, 634
Archeocyatha, 69
Archeofijia, 280

685

SUBJECT INDEX

Arches National Monument, 555
Archidiskodon, 568
Archipelago de los Carrarreos, 252
Archipelago de Sabana, 252
Arctic Archipelago, 145
Arctic beds, 191
Arctic Coastal Plain, 143, 145, 154, 189, 517
Arcto-Tertiary floras, 624
Ardennic, Ardennes phase, 602, 635
Areniscas con Dinosaurios, 645
Argentina, 15
Argyrosaurus, 645
Arikaree Fm., 566
Arisaig Group, 166
Arizona, 510
Arkansas, 510
Arkansas Basin, 598
Armina Fm., 316, 481
Armina Series, 312
Arnioceras, 367
Arroyoa, 543
Arthrophycus, 198, 422
Aru (Aroe) Is., 380
Aruba, 362
Arunta Complex, 15
Asbestos, 143, 179, 617
Ascension Is., 226
Aseismic Ridge, 476
Asencio Fm., 645
Ash Hollow Fm., 569
Asphalt, 253, 497
Assyntic Orogeny, 420
Astrolabe Reefs, 343
Asuncion, 423
Asuncion Agrihan, 345
Atabupo, 457
Atacama Trench, 241
Atacocha, 432
Atafu, 494
Atane Flora, 298
Athabasca Lake, 140
Athabasca River, 192
Athabasca Tar Sands, 143
Atiu, 250
Atlantic Coastal Plain, 608
Atlantic Coastal Province (U.S.), 504, **531**
Atlantic faunal realms, 168
Atlantic Ocean Basin, 405
Atlantic Pacific Muck Fm., 418
Atlantic Pacific seaway, 236
Atlantic Provinces (Canada), 156
Atlantic-type coast, 181
Atlantic type rocks, 412
Atlantic Uplands, 157
Atokan Strata, 603
Atomodesma, 366
Atrypa, 200
Auckland, 390
August Town Fm., 335
Aukena, 290
Aure Trough, 25, 375
Austin, 579

Australasia–Regional Review, 21
Austral fauna, 643
Australia, 28
Australia–New South Wales, 34
Australia–Northern Territory, 46
Australian Plate, 279, 377
Australian Shield, 374
Australia–Queensland, 56
Australia–South Australia, 61
Australia–Tasmania, 81
Australia–Victoria, 91
Australia–Western Australia, 100
Austral Islands, 499
Australocoelia, 422, 643
Australo-Melanesian Trough, 368
Australospirifer, 422
Australostrophia, 422
Avalon Geosyncline, 159
Avalonian Orogeny, 160, 612
Avalon Peninsula, 159
Avalon Platform, 161, 162, 166
Avanvero Dolerite, 481
Avenue of Volcanoes, 268
Aves Ridge, 660
Awatere, 394
Axel Heiberg Is., 149, 152
Ayavacas Limestone, 430
Ayers Rock, 48
Azarati Fm., 646
Azorella, 465
Azuero Peninsula, 417

Babelthuap, 225
Bacchus March, 97
Bachaquero, 651
Backbone of the Americas, 625
Baculites, 368
Badlands, 192, 506, 566
Baffin Is., 149, 151
Bahamas, 109, 500, 658
Bahamian Platform, 659
Bahia Honda tectonic unit, 253
Baikalian Orogeny, 4
Baja California, 618
Bajada, 542
Bajio, 351
Baker Is., 433
Bakhuis Mts., 489
Balanus, 498
Balcones Escarpment, 561, 578
Baliceaux, 447
Balleny Is., 478
Balsam Block, 271
Bambui Episode, 459
Banco Gorda, 328
Banff, 208
Bangamall Basin, 101
Banks Is., 149, 384
Banos Cotopani Fault, 268
Bao River, 258
Baragwanathia, 97
Barama Group, 313, 320, 321

Barama Mazaruni Supergroup, 314
Barbados, 115, 660
Barbuda, 14
Barinas Basin, 648, 651
Barite, 253, 521
Barking Sands, 585
Barkly Tableland, 46
Barreiras Series, 136
Barrier Ranges, 64
Barrier reefs, 365
Bartica Gneiss, 313
Bartlett Trough, 227, 328
Basaltic Shield Volcano, 1
Basement, 310, 348, 460, 528
Basement Complex, 240, 453
Basement faults, 599
Basin-and-Range block faulting, 624
Basin and Range Province (U.S.), 171, 350, 504, **541**
Basin and Range structure, 541
Basin fill, 633
Basse-Terre, 301
Bass Is., 24, 198
Bass Is. Fm., 198
Bass Islets, 499
Basslerata, 197
Bass Strait, 81
Bathurst, 164
Bathyurus, 197, 224
Battery Point, 167
Bauru Sandstone, 424
Bauxite, 55, 56, 258, 282, 317, 322, 324, 326, 334, 456, 464, 485, 490, 580, 594, 652
"Bauxite hiatus," 483
Bauxitization, 489
Bayano-Chucunaqui-Tuira Geosyncline, 417
Bay du Nord Group, 167
Bay Is., 329
Bay of Fundy, 158
Bay of Harbours beds, 273
Bayonnaise Rocks, 332
Beach deposits, 202
Beacon Supergroup, 7
Beardmore Orogeny, 7
Bearpaw Mts., 629
Bear Province, 140
Beartooth Orogeny, 634
Beata Ridge, 659
Beauceville, 164
Beaufort Basin, 189, 532
Beaufort Fm., 536
Beauharnois Dolomite, 220
Beautemps Beaupre Is., 343
Beaverlodge, 178
Becscie Fm., 224
Bedford, 200, 201
Beekmantown Dolomite, 219, 220
Belep Is., 364
Belize (British Honduras), 116, 357
Belleoram granite, 170
Bellingshausen Is., 450, 468
Bellona Is., 455

686

SUBJECT INDEX

Belosepia, 348
Beltian Trough, 403
Belt Series, Supergroup, 628, 634
Benambrian Orogeny, 29, 95
Benettiteae, 423
Beni, 457
Benioff Zone, 43, 122, 317, 384, 520
Bentonite, 563, 571, 580
Bequia, 447
Berea, 200, 201
Beringia, 516, 517
Bering Sea, 513
Bering Strait, 513
Berkshire Hills, 608
Bermuda, 117
Beryllium, 464
Big Bend, 561
Bighorn Mts., 629
Big Is., 582
Bikini, 346
Billingsarea, 221
Bindloe Is., 289
Bioherm, 200, 224, 603
Biostromes, 197
Birch Creek Schists, 518
Birnie, 433
Bishop Ash, 625
Bismarck Archipelago, 121, 372, 451
Bismuth, 464
Bison, 638
Black Creek Fm., 534
Black Hills, 406
Blackriveran stage, 219
Black River Escarpment, 201
Black River Group, 196, 219, 221, 224
Black Rock, 542
Black (iron) sand, 391
Black shales, facies, 191, 200, 222, 224, 490, 603
Blanco Fm., 568
Blind River, 178
Block-faulted plateau, 5
Block-faulted structures, 529, 633
Block movements, 637
Blomidon, 171
Blue Holes, 116
Blue Mountain Fm., 196, 334, 620
Blue Pinion Fm., 444
Blue Ridge, 504
Blue Ridge Province, 523
"Blue–schists," 621
Bluff Butte, 566
Bluff Fm., 227
Bois Blanc Fm., 199, 200
Bokkeveld beds, 273
Bolboporites, 221
Bolívar dislocation zone, 313, 317
Bolívar Geosyncline, Trench, 236, 418
Bolivia, 123
Bolsas, 252
Bolsones, 328

Bolts Blanc Escarpment, 202
Bonaire, 362, 660
Bonao, 258
Bonaparte Gulf Basin, 29, 51, 104
Bonidoro Series, 312, 315
Bonin Is., Ridge, 331, 344
Bonita Peak, 488
Bonnet Plume Basin, 213
Bonya Limestone, 305
Book Cliffs, 551
Boothia, 145, 151
Bora-Bora, 449
Borderland, 622
Borokua, 454
Borus, 645, 646
Boston Basin, 611
Botryococcus, 423
Botucatú Fm., 134, 423
Bougainville, 123, 372, 382, 451
Boulder batholith, 628
Bounty Is., 390
Bouvet Is., 471
Bovais Lake anticline, 189
Bowden, Bowden Fm., 335, 661
Bowen, 24
Bowen Basin, 58
Bowen Orogeny, 45
Bowser Basin, 185, 186
Brachyprion, 224
Bradore Group, 161, 223
Bradtville Drift, 204
Brazil, 127
Brazilian Cycle, 459
Brazilian Shield, 18, 428, 457
Bretonic Orogeny, phase, 602
Brevard Zone, 527
Brevispirifer, 200
Bridgeman Is., 469
Brightseat Fm., 536
Brimstone Hill Fm., 339, 440
Bristol Is., 468
Britannia, 187
British Columbia, 208
British Honduras, 116
British Solomons, 451
British Virgin Is., 653
British Windward Is., 300
Brito-Arctic Province, 298
Brockocystis, 198
Brockville, 202
Bronson Hill anticlinorium, 615
Brontothere, 192
Brooks Range, 515
Browning Orogeny, 24
Browns Mt. Group, 162
Bruce Peninsula, 197, 198, 199
Brule, 564
Bryce Canyon, 550
Buchia, 368
Buckle Is., 478
Buena Vista Fm., 423, 642, 644
Buff Bay, 375
Building stone, 617
Buka, 123, 372, 451
Bulimus, 645, 646

Buried Escarpment, 200
Buried Ranges, 350
Burro-burro Group, 321
Butler's Mt. Dome, 440
Butte, 552, 555, 639
Byssonychia, 197

Caaguzú, 423
Cabezas Fm., 258
Cabot Fault, 168
Cabot Head Shale, 197
Cabot Strait, 157
Cacaguatique, 272
Cachabi Fm., 267
Caguana peneplain, 661
Caicos Islands, 500
Caiuá Fm., 424
Cajamarca Series, 246
Calcarina, 343
Calcrete, 76
Caldera, 337
Calderas, 345, 346, 377, 384, 475, 584
Caldonian Orogeny, 264, 294, 602, 635
Calentura Limestone, 265
Calgary Arch, 189, 190
Caliche, 561
California, 510
California Coast Range, 621
Callo Group, 264
Calmonia, 643
Caloosahatchee, 538
Calvert Fm., 537
Calymene, 422
Camachian Stage, 219
Camacho Fm., 645
Camalotal, 270
Camaqua Series, 132
Camarotoechia, 198, 422
Campbell Is., 390
Campbell Plateau, 26, 398
Campos, 419
Canada, 139
Canada–Arctic Archipelago, 145
Canada–Atlantic Provinces, 156
Canada–Canadian Shield, 174
Canada–Cordilleran Region, Interior and Western Belts, 179
Canada Hills Volcanics, 440
Canada–Northern Great Plains Province, 187
Canada–Ontario Basin, 193
Canada–Rocky Mountains and Eastern Cordilleran Region, 208
Canada–St. Lawrence Lowlands of Quebec, 218
Canada Tungsten, 187
Canadian Cordillera, 179, 208
Canadian Epoch, 196
Canadian River, 558
Canadian Series, 602

SUBJECT INDEX

Canadian Shield, 139, 145, 174, 187, 209, 404, 502, 612
Canal Zone, 417
Candlemas Is., 468
Canelones graben, 647
Canga, 424
Canning Basin, 23, 102
Cannonball Member, 564
Canon del Sumidero, 356
Canso Group, 170
Canterbury Plains, 393
Canton Atoll, 433
Canyon de Chelly, 555
Caovan, 447
Cape Breton Is., 158, 170, 177
Cape Cod, 507, 608
Cape Fear Arch, 532
Cape May Fm., 538
Cape May Slope, 532
Capides, 274
Capitol Reefs Monument, 555
Cap Rock, 576
Caracas Sandstone, 275
Caraibe, 312
Cardiomys, 645
Cardium, 192, 645
Cariaco Trench, 660
Caribbean Arc, 658
Caribbean Belt, 300
Caribbean Mts., 648
Caribbean Plate, 664
Caribbean Sea, 456
Caribbean Series, 497
Cariboo district, 143
Cariboo Mts., 213
Cariboo Orogeny, 212
Carichapo Group, 313, 315
Cariri Orogenic Belt, 459
Carmacks, 186
Carnarvon, 23
Carnarvon Basin, 102
Carnegie Ridge, 286
Carolina Bays, 507
Carolina Slate Belt, 525
Caroline Atoll, 341
Caroline Islands, 225, 346
Carolinidan Fold Belt, 294
Caroni River, 651
Carpentarian System, 23, 28, 50
Carriacou, 300
Carrizal, 272
Carrizo Mts., 551
Carson, 542
Carson Sink, 542
Caryocrinites, 198
Casapalca, 432
Cascadan Orogeny, 624
Cascade Mts., 182, 509, 619
Cascadia, 622
Cassiar, 186
Cassiar Orogeny, 623
Cassiquiere, 457
Castle Hayne Fm., 536
Catahoula, 579, 580
Cataract Group, 197, 198
Catastrophic flooding, 629
Catazyga, 197, 222

Cat Is., 111
Catskill Mts., 608
Caute Depression, 253
Cauto Tectonic Unit, 253
Cayenne, 312
Cayetano, 661
Cayman Brac, 226
Cayman Islands, 226
Cayman Ridge, 659
Cayman Trench, Trough, 227, 235, 253, 305, 328, 334, 407, 659
Cayo Coco Tectonic Unit, 253
Cayos, 328
Cays, 348
Cay Sal, 252
Cayugan Series, 219, 602
Cedar Breaks, 555
Cedar Hills Phase, 637
Celtis, 565
Centerfield, 200
Central America—Regional Review, 228
Central City, 639
Central Georgia Uplift, 532
Central Lowlands, 505
Central Mesa, 350
Central St. Lawrence Lowland, 193
Centre Hills, 360
Centro, 651
Ceraurus, 221, 224
Cercado Fm., 258
Cerrezuelo Fm., 643
Cerro Aconcagua, 457
Cerro Bolivar, 651
Cerro Bonete, 17
Cerro Chachagón, 227, 400
Cerro de Pasco, 430
Cerro de Punta, 434
Cerro Guezapa, 272
Cerro Nejapa, 272
Cerro Rerro Sandstone, 422
Cerro Rico, 123
Cerros de Amotape, 427
Cerros de Clumi, 228
Cerros de Culmi, 327
Cerros de Illescas, 427
Ceuta, 651
Cevicos, 258
Chaco, 419, 457
Chaco-Pampas, 15
Chaco Trough, 420, 425
Chadron, 564
Chaleur Bay, 158
Chaleur Uplands, 160
Challenger Deep, 345
Chamberlains Brook Fm., 445
Champlainian, 219, 602
Champlain Sea, 201, 204
Channeled scablands, 625
Channel of Chacao, 239
Chapadas, 129
Chapiza Fm., 264
Chapleton Fm., 335
Charles Is., 289
Charleston, 546
Charlestonian Phase, 67
Charnockite, 7

Chateauguay Fm., 220
Chateauguay Sandstone, 218
Chatham, 285
Chatham Is., 390
Chatham Rise, 26
Chatham Sag, 195
Chattonooga, 603
Chazy, 218, 219
Chazyan Stage, 219
Chedabucto Bay, 171
Cheirurus, 198
Chelonia, 134
Cheniers, 485
Cherry Is., 455
Cherrytree Stadial, 204
Cherts, 623
Cheshire Quartzite, 612
Chesterian beds, 603
Chiapas Depression, 233
Chiapas Massif, 230, 232
Chibougamau, 178
Chicotte Fm., 224
Chile, 237
Chilean-Argentine Trough, 462
Chimborazo, 266
Chiquitos Belt, 120
Chiripo Grande, 228
Chirqui lagoon, 231
Chochal Fm., 233
Choctawhatchee Fm., 537
Choiseul, 452
Chonetes, 273
Chonophyllum, 224
Choptank Fm., 537
Christmas Island, 243, 341
Chrome, 253
Chromite, 391
Chromium, 370
Chugach Range, 515
Chugwater Fm., 636
Chuquicamata, 242
Churchill, 175, 209
Churchill Province, 140, 148, 210
Chuy Fm., 646
Cibao River, 257
Cimmerian (Kimmerian) beds, 274
Cincinnatian Series, 219, 602
Cincinnati Arch, 521, 597
Cinder cones, 588
Cinders, 595
Ciraruco Fm., 315
Cirque glaciers, 608
Citronelle, 586
Cladophlebis, 87
Claiborne, 536, 579
Clarence-Moreton Basin, 36, 41
Clarion Fracture Zone, 438
Clastic wedge, 210, 212, 356
Clay, 531, 580
Clay plains, 202
Clayton, 534
Climacograptus, 422
Climactichnites, 220
Climax, 639
Clinton, 219
Clipperton Fracture Zone, 230, 244, 251

Clipperton Island, 244
Clipperton Zone, 401
Cloridorme Fm., 164
Coahuilian Orogeny, 354
Coahula Series, 579
Coal, 56, 81, 91, 107, 143, 152, 170, 186, 187, 192, 213, 217, 241, 359, 370, 391, 464, 529, 580, 625, 639, 652
Coal swamps, 637
Coastal Limestone, 105
Coastal Plain, 504, 574
Coastal Ranges, 141
Coastal Range (Chile), 239
Coast Fm., 258
Coast Mts., 181
Coast Ranges, 619
Coatepeque Caldera, 272
Cobalt, 178, 370
Coban, 308
Cobbler's Creek Orogeny, 38
Coboconk Fm., 196, 197
Cobourg Fm., 197
Cobriza, 432
Cochinus Depression, 253
Cockpit, 334, 661
Cockscomb, 116
Cocos (Keeling) Islands, 231, 245
Cocos Ridge, 236, 286
Coelospira, 198
Coeur d'Alene, 625
Coeroeni Group, 313, 481
Coesewijne Fm., 485
Cohansey Sand, 537
Collenia, 353
Collingwood Fm., 196
Collision, 45
Colombia, 245
Colombia Basin, 659
Colonche-Chongon Hills, 269
Colorado, 510, 558
Colorado Mineral Belt, 639
Colorado Plateau Province (U.S.), 504, **548**
Colpoy Bay, 198
Columbia Lava Field, 407, 619
Columbia Mts., 179, 208, 213
Columbian Orogeny, 142
Columbia Plateau, 509
Columbia Seamounts, 276
Columbus, 602
Colville uplift, 189
Comanchean Series, 579
Comayagua Graben, 271
Como Bluff, 636
Comstock silver, 625
Concepción, 300
Concepción Las Minas, 308
Conchidium, 198
Concretions, 200
Cone karst, 334
Connecticut, 510
Connecticut Valley, 611
Conodont Hash, 200
Conolphus, 287
Consolación del Norte Fault, 253
Continental Divide, 208, 508

Continental Drift, 3
Continental Shelf, 574
Cook Inlet, 515
Cook Islands, 250, 468
Cookson Fm., 161, 164
Cook Strait, 390
Coolgardie, 100
Cooper Basin, 70
Copi, 93
Copper, 55, 126, 143, 178, 187, 253, 270, 281, 359, 370, 382, 391, 400, 418, 432, 455, 521, 530, 555, 617, 625
Coppermine arch, 189
Coppermine Series, 149
Coral Sea, 24, 56, 376
Corantijn Group, 481
Corantijn River, 489
Cordillera, 124, 457, 618
Cordillera Blanca, 427, 464
Cordillera Central, 245, 434
Cordillera de Amambay, 424
Cordillera de Monte Cristi, 258
Cordillera de Nieba, 258
Cordillera de Talamanca, 230, 251
Cordillera de Ybytyruzú, 423
Cordillera Frontal, 16
Cordillera Isabelia, 400
Cordilleran Fold Belt, 306
Cordilleran Geosynclinal Belt, 543, 622, 630
Cordilleran Glaciation, 188
Cordilleran Region (Canada), 143, 179, **208**
Cordilleran System, 627
Cordillera Occidental, 245, 427
Cordillera Real, 263, 268
Cordillera Septentrional, 257
Cordillera Volcanica, 230, 251
Cordobés Fm., 643
Corn Belt, 505
Coropina, 283
Coropina Fm., 485
Corridor, 191
Cortez-Vinta Axis, 630
Corumba Series, 421
Costa Brava, 352
Costa Rica, 251
Costricklandia, 198
Coswine, 283
Cotapaxl, 266
Cotton Valley Group, 579
Coulees, 188, 509, 629
Covey Hill Sandstone, 220
Cowra Trough, 38
Craton, 140, 251, 502, 506, 623, 630, 636, 649
Crazy Mt. Basin, 629
Creek sequence, 602
Crinoidea, 422
Cripple Creek, 639
Croatan Fm., 537, 538
Croixan Series, 219
Crozet Is., 473
Cruziana, 123
Cryptolithus, 197, 221, 222
Crpytophragmus, 197

Cryptovolcanic structures, 177, 598
Cryptozoon, 221, 224
Cuba, 252, 658
Cuban geosyncline, 258
Cuban phase, 254
Cuche Shale, 497
Cuchivero suite, 313
Cuchumatanes, 232
Cuestas, 508, 548
Cuiabú Series, 421
Culebra, 434
Culpepper Is., 289
Cumberland, 171
Cumberland Bay sequence, 467
Cumberland Group, 170
Cumberland Lowland, 157
Cumberland Plateau, 522
Curaçao, 362
Curitiba surface, 137
Cutler Fm., 552
Cuvieronius, 424
Cuyuni Fm., 313, 314, 320, 321
Cycladeae, 423
Cyclic oscillations, 636
Cyclic sequences, 578
Cyclolobus, 367
Cyclothem, 603
Cypress Hills Fm., 564
Cypress Hills Plateau, 564
Cyrtia, 224
Cyrtodonta, 224

Dacli Limestone, 234
Dadoxylon, 423, 644
Daemonelix, 566
Dagua Group, 247, 264
Dakota Group, 636
Dalbana Fm., 315
Dalbana Rhyolite, 481
Dalmanella, 221, 224, 422
Danger Is., 250
Darling Fault, 102
Darling-Warrego Basins, 33
Darwin Is., 289
Darwinian theory of atolls, 291
Darwin Rise, 347, 413
Darwin-type volcanic island, 450
Datil section, 551
Daule Quininde-Jipijapa Basin, 268
Davenport Geosyncline, 50
Death Valley, 542, 619
Deception Is., 469
Décollement, 214, 216, 506
Deglaciation, 156, 202
De Goeje Gabbro, 481
Delamerian Orogeny, 29, 63, 69
Delaware, 510
Deltaic Systems, 572
Demerara Fm., 283, 485
Demerara Rise, 486
Demon Block, 40
Denali Fault, 517
Densin Yama Fm., 448

SUBJECT INDEX

D'Entrecasteaux Fracture Zone, 384
Denver Basin, 560
Deschambault Limestone, 221
Desejado sequence, 276
Desirade Fault Zone, 304
Desmoinesian beds, 603
Detachment thrusts, 629
Devil's Egg, 488
Devil's Is., 282
Devil's Tower, 559, 565
Devon Is., 142
Diabase, 9, 95, 248, 274, 363 (*See also* Dolerite)
Diamictite, 210
Diamond, 317, 324, 464, 490, 652
Diapiric salt, 102, 574
Diapirism, 63, 355
Diatreme breccia, 222
Dicoelosia, 198
Dicroidium, 8, 31, 59, 87, 462
Digonus, 643
Dinosaur, 192, 555
Dinosaur National Monument, 628
Dinosuchus, 136
Diplograptus, 422
Diprotodon, 60, 87
Discocyclina, 368
Disturbed Belt, 628, 629
Dolerite, 32, 274 (*see also* Diabase)
Doleroides, 197
Dolomite, 363
Dolomitization, 291
Dolores Fault, 261
Dolores-Guayaquil megashear, 269
Dolorthoceras, 644
Dominica, 257, 660, 667
Dominica Fault, 304
Dominican Republic, 257
Don Valley Brickyard, 203
Dresbachian Stage, 219, 600
Drift-free areas, 606
Driftless Area, 505
Dromicus, 287
Drumheller Badlands, 188, 192
Drumlins, 202
Duarte Fm., 258
Ducie Is., 434
Duff Is., 455
Dummer Moraine, 205
Duncan Is., 289
Durazno Group, 643
Dundee Fm., 200
Dunes, 78, 120, 202, 464, 561, 584
Dunite, 1
Duplin Fm., 537
Durand Reef, 342
Durango Group, 579
Durazno Group, 643
Duricrust profile, 74, 326
Dutch West Indies, 362, 660
Dyer Dolomite, 635
Dwyka-type glacial beds, 273

Eagleford, 579
Eagle Mills, 579

Earthquake epicenters, 528
Earthquakes, 544, 592, 625, 639
Eastend (pottery clays), 192
Easter Island and Sala y Gomez, 260, 346
Easter Is. Fracture Zone, 449
Eastern Townships, 162
Eastern Venezuela Basin, 648
East Greenland Fold Belt, 294
East Pacific Rise, 244, 289, 358
East Texas embayment, 574
Eau Claire Fm., 196
Eaurupik Is., 226
Ecuador, 261
Edaphophyllum, 199
Edenian Stage, 219
Edgecliff Member, 200
Ediacara fauna, 69
Edmonton (Alberta), 188
Edmonton Fm., 192
Edwards Plateau, 557
Efate Is., 383
Eismitte, 299
El Baru volcano, 416, 417
El Baúl Swell, 648
El Callao, 652
El Dorado (Guiana Shield), 309
Eleanore Bay Group, 297
Elephant (fossil), 568
El Espirito (gold mine), 418
El Gallo (mine), 418
Elk Point Basin, 189, 190
Ellesmere Is., 142, 149
Ellesmerian Orogeny, 142, 152, 153, 184, 272
Ellice Islands, 291
Ellis Bay Fm., 224
Ellsworth Mts., 7
El Misti (volcano), 430
El Niño Current, 285
Elomichthys, 644
El Pao, 651
El Rosário, 234
El Salvador, 242, 270
El Teinente (mine), 242
El Valle (volcano), 417
Embar Fm., 636
Emerald, 464
Emerged reef, 591
Emerged shorelines, 586
Emperor Chain, 587
Enchanted Is., 285
Encrucijada Granite, 313
Enderby Land, 433
Endorheic (drainage), 509
English Head Fm., 224
English's Crater, 360
Eniwetok (Enewetak), 292, 346
Ennery Fm., 326
Entrada Sandstone, 636
Entre Ríus Fm., 645
Eocambrian beds, 161
Eocodia, 198
Eodinobolus, 197
Eofletcheria, 221
Eolianite, 105, 363, 368, 402, 585, 643

Eophacops, 422
Equus, 646
Eramosa Fm., 198
Erian Series, 219
Eridophyllum, 200
Ernabellan Phase, 66
Erodona, 646
Eromanga Basin, 58
Erosion, 134
Erratic boulders, 3
Erromango Is., 384
Escobar Series, 421
Escuminac Shale, 170
Esmeralda Bank, 345
Esmeraldas-Muisne Basin, 269
Espinhaçian Cycle, 458
Espinhaço Geosyncline, 131
Espiritu Santo, 234, 455
Esquistos de Santa Marta group, 247
Esterhazy, 192
Estevan, 188
Estrada Nova Fm., 423, 644
Eucalyptus, 87
Eucla Basin, 72
Eugeosynclinal rocks, 613, 622
Eurafrica, 507
Eurekan Orogeny, 143, 148, 153
Eurydesma, 59, 83, 462
Eusebia Ayala Sandstone, 422
Eustatic shorelines, 1
Euxinic (facies), 222
Evaporites, 151, 165, 306, 356, 525, 579
Everardian Phase, 66
Evington Group, 525
Evolution, 288
Exogeosynclinal basin, 636
Exorheic (drainage), 509, 541
Exuma (bank), 109

Facies, 380
Fairweather Range, 515
Fais (Feis) (coral platform), 226
Fakaofo (atoll), 494
Falawatra Group, 458, 481
Falcon Is., 494
Falcón Basin, 651
Falkland Islands (Islas Malvinas), 18, 273
Falkland Plateau, 273
Fall Zone, 523, 531
Fall Zone Peneplain, 533
Falsino River Granulites, 312
Fangaloa Volcanics, 666
Fanglomerates, 613
Fanning Atoll, 341, 342
Farallon de Pajaros, 345
Faraulep (atoll), 226
Faros, 342
Fatu Hiva Is., 346
Fault block, 543
Favosites, 198
Feldspar, 530
Fenua Ura (atoll), 450
Fernandina Is., 285, 288

690

SUBJECT INDEX

Fernando de Noronha Rocas, Trindade, Martin Vaz, and Saint Paul Rocks, 274
Fernando de Noronha Fracture Zone, 275
Ferrazia cardinalis, 644
Ferricrete, 368
Ferruginous laterites, 370
Fiji, 25, 277, 283
Fiji orocline, 384
Fina-sisu Fm., 448
Findlay Arch, 193, 598
Fiordland, 392
Fiords (fjords), 139, 181, 241, 392
Fissure eruptions, 588
Fitzroy Valley, 106
Fitzroya, 274
Fjords (*see* Fiords)
Flathead Valley, 213
Flatirons, 636
Flattop (peneplain), 631
Flaxville Fm., 565
Fleming, 580
Flemingites, 367
Fletcheria, 198
Fleur de Lys Group, 161
Flexicalymene, 221
Flinders Ranges, 67
Flin Flon, 178
Flint Is., 341, 342
Flood basalts, 509
Floreana Is., 289
Florentino, 258
Florida, 510
Florida group, 456
Florissant Fossil Beds, 637
Fluorite, 359, 606
Fluvoglacial deposits, 59
Flysch, 115, 258, 353, 363, 429, 517, 623
Flysch facies, 151
Flysch wedge, 43
Foerstephyllum, 221
Folsom, 638
Fontaine, 221
Foothills, 141
Forari, 389
Forest City Basin, 597
Forteau Fm., 223
Fort Monroe High, 532
Fort Nelson Basin, Lowlands, 188, 189
Fort Norman, 213
Fort Union Fm., 564
Fossil cliffs, 493
Fossil Hill Fm., 198
Fountain Fm., 636
Four Corners area, 556
Four Mile Dam, 200
Fox Hills Creek Bed, 563, 564
Franciscan Complex, 621
Franconian Stage, 196, 219, 600
Franklin Geosyncline, 142, 145, 149, 151, 152, 297
Franklin Mts., 189, 208, 216
Fray Bentos Fm., 645
Frayle Muerto Fm., 642, 644

Fredericksburg, 579
Fredricksen Is., 468
Fredericton Trough, 166
Fremont Limestone, 634
French Guiana, 282, 309, 480
Frenchman Orogeny, 81
French Polynesia, 284, 345, 493, 498, 499
Friendly Is., 494
Frontenac Arch, 193, 196
Frontenac Quartzite, 167
Front Range, 631
Fulanga Is., 281
Fullerton Fm., 568
Fulton River (Alaska), 520
Funafuti I., 291
Fundy Geosyncline, Epieugeosyncline, 159, 169
Furnas Fm., 643
Fusispira, 197
Futuna Island, 657

Gaferut, 226
Galapagos Fracture Zone, 269
Galápagos Islands, 269, 285, 346
Galibi (granite), 312
Galvez Bank, 344
Gambier (Mangareva) Islands, 290
Gananoque, 202
Gander Lake Group, 161
Gangamopteris, 70, 83, 97
Gangplank, 557
Gardar sediments, 294
Gardner (atoll), 433
Garimpeiros, 128
Garlock Fault, 545, 621
Garnierite, 365
Gas, 192, 218, 270, 359, 432, 529, 556, 572, 580, 625, 639, 997
Gaspé Limestone, 166
Gatun Fm., 251
Gausava (ultrabasic), 455
Gawler Block, 62
Geelvink Bay, 379
Geisonoceras, 224
Gems, 400
Genovesa Is., 289
Geochelone, 287, 363
Georges Bank, 608
Georgia, 510
Georgian Bay, 193, 197
Georgina Basin, 29, 51
Geospizidae, 287
Geothermal energy, 272, 639
Germanotype movements, 462, 637
Geysers, 630
Gilbert and Ellice Islands, 291
Gilsonite, 556
Gippsland Basin, 33
Girdle of Fire, 412
Glacial deposits, 202
Glacial Lake Agassiz, 188
Glacial Lake Arkona, 205
Glacial Lake Chicago, 205
Glacial lakes, 202

Glaciers, Glaciation, 143, 155, 174, 241, 374, 380, 392, 461, 462, 479, 506, 567, 608, 625, 630, 638
Glacier National Park, 627
Glacio-isostatic rebound, 202, 625
Glaciomarine Deposits, 7
Glenarm Series, 525
Glen Falls Fm., 444
Globigerina, 343, 368, 380
Glossopteris, Glossopteris flora, 8, 19, 31, 59, 70, 83, 105, 134, 273, 462
Glyptodon, 424, 646
Gnamma holes, 103
Goat Island Fm., 198
Godthaab, 292
Godwin Gut, 339
Golconda thrust, 543
Gold, 56, 91, 107, 126, 143, 166, 177, 186, 217, 253, 281, 324, 370, 372, 391, 400, 418, 455, 490, 521, 530, 619, 625, 639, 652
Golden Lane, 359
Golden Valley Fm., 564
Goldenville Fm., 161
Gold Rush, 620
Golf ball sand, 196
Golfo de Barabano, 252
Gonçalo Alvarez Is., 470
Gondwana facies, rocks, 134, 273, 422, 647
Gondwana glaciation, 126
Gondwanaland, 11, 24, 134, 398
Gondwanian Orogeny, 466
Gondwanide Orogeny, 8
Gonioceras, 197
Goose Egg Fm. 636
Goosenecks, 552
Gough Is., 470
Goyllarisquisga group, 429
Graben, 578, 615
Graben, Honduras, 230
Grampian Mts., 97
Grand Banks, 172
Grand Canyon, 509, 549, 628
Grand Canyon disturbance, 543
Grand Cayman, 226
Grand Coulee, 625
Grand Baie Beds, 301
Grand Etang, 300
Grande-Terre, 301
Grand Fonds, 661
Grand Hogback, 551
Grand Lake Brook Group, 161
Grand Soufrière, 257
Grand Teton National Park, 627
Grand Turk, 500
Granite Mts. Uplift, 629
Granites caraibes, 314
Granites galibi, 283
Granites guyanais, 283, 314
Granites-Tanami Complex, 50
Granite wash, 191
Granulite facies, 7
Graptolitic rocks, 152

691

SUBJECT INDEX

Gravel, 571, 580, 606
Gravity anomalies, 528
Gravity sliding, 544, 546, 578
Graywacke, 623
Graywacke-Shale Fm., 468
Great American Desert, 558
Great Artesian Basin, 24, 36, 41, 56, 63
Great Barrier Is., 390
Great Barrier Reef, 24, 56
Great Basin, 407, 541
Great Bear Lake, 140
Great Carolina Ridge, 532
Great Corn, 231
Great Dividing Range, 93
Greater Antilles, 407, 658
Great Glen Fault, 168
Great Lakes, 193
Great Plains, **Great Plains Province (U.S.)** 504, 506, **557**, 627
Great Salt Lake, 542
Great Serpentine Belt, 31
Great Slave Lake, 140
Great Smokies, 507
Great Valley, 407, 507, 619
Green Is., 583
Greenland, 145, 149, 292
Green Mts., 608
Green River Basin, 629
Green Schist facies, 7
Greenstone belts, 102
Grenada, 300, 660, 667
Grenadine Group, 300, 447
Grenville Province, rocks, Series, 140, 159, 162, 169, 175, 193, 195, 353, 612
Grenvillian, 403, 599
Grimsby (sandstones), 197
Growth faults, 578
Grupo Diabasico, 248
Guadalcanal, 451
Guadeloupe Region, Guadeloupe Is., **301**, 660
Guadalupe Fm., 265
Guajira, 648
Guam, 304, 344
Guanacaste, 251
Guaniquanicco Peneplain, 661
Guano, 242, 244, 342, 346, 433, 500
Guanoco (asphalt lake), 650
Guantanamo depression, 253, 254
Guárico, 651
Guatemala, 305
Guatemala Trench, 230
Guayana Shield, 261
Guayaquil, 261
Guayaquil Fm., 265
Guelph Fm., 265
Guelph Fm., 198
Guguan (volcano), 345
Guiana Basin, 481
Guiana Granites, 283
Guiana Marginal Plateau, 486
Guiana Shield—Regional Review, 246, 282, **309,** 318, 457, 648
Guichon Fm., 644

Guija (lake), 270
Gulf Coastal Plain, 351, 356
Gulf Coastal Province (U.S.), 504, 572
Gulfian beds, 579
Gulf of Alaska, 515
Gulf of Carpentaria, 46, 56
Gulf of Fonseca, 230, 271, 328
Gulf of Guayaquil Province, 269
Gulf of Mexico, 573
Gulf of Mexico Diothesis, 235
Gulf of Paria, 650
Gulf of St. Lawrence, 158, 162
Gulf Series, 579
Gull River, 196
Gunning-Wyangala Batholith, 39
Gun River, 224
Gurabo Fm., 258, 661
Gurian Orogenic Cycle, 458
Guyana, 309, 312, **318,** 480
Guyana Shield, 648
Guyana-Venezuela Basin, 315
Guyots, 412
Gypsina, 343
Gypsum, gypsum-anhydrite, 151, 153, 206, 210, 239, 258, 334, 362, 364, 370, 400, 423, 563, 580

Ha'apai Is., 494
Habirites, 313
Hachijo (volcano), 332
Hadrosaurus, 538
Hadrynian period, 140, 149, 159, 209
Hagman Fm., 448
Haimaraka Fm., 313, 314, 320, 321
Haiti, 326
Halemaumau Caldera, 584
Halifax Fm., 161
Halmahera Is., 377
Halobia, 367
Halokinesis, halokinetic, halokinogenesis, 551, 575, 576
Halysites, 198, 224
Hamersley Ranges, 23, 28, 101
Hamilton Escarpment, 202
Hamilton Group, 200
Hanesavo Is., 456
Hansonburg, 555
Harding Sandstone, 634
Harney lava plateau, 620
Harris-Bugby Centre, 360
Hatteras Low, 532
Hawaii, 502, 510, 511, **581**
Hawaiian Ridge, 587
Hawaiite, 592
Hawke's Bay, 396
Hawkesbury Sandstone, 32, 41
Hawthorne Fm., 537
Haystack Hills, 435
Heard Is., 474
Heat-flow, 528, 545
Heavy mineral sands, 107
Hedley Camp, 187
Helderbergian Series, 219, 602

Helen mine, 179
Helikian period, 140, 159
Helium, 556, 570
Hell, 227
Hell Creek Bed, 563, 564
Hells Canyon, 628
Helopora, 198
Hemingford Group, 566
Henderson Is., 434
Henry Mts., 551
Hercynian Orogeny, 264, 428
Hermitage Bay Fault, 168
Hermosa Fm., 552, 636
Hervey Is., 250
Hexagonaria, 200
Hiatella, 202
Hidalgoan Orogeny, 356
Hidalgo Fm., 326
Hidalgo Limestone, 258
High Cordillera (Chile), 239
Highlandcroft, 614
High Plains, 506, 559
High Plateaus of Utah, 550
Hilo Rock, 595
Hispaniola, 257, 326, 659
Hiva Oa, 346
Hodgkinson Basin, 56
Hogbacks, 508
Holdrege Fm., 568
Hollin Fm., 265
Holocene sedimentation, 572
Holyrood granite, 159
Homolonotus, 273
Homotelus, 224
Honduras, 327
Honduras graben, 230
Honduras Massif, 230, 271, 328
Honduras Sphenochasm, 235
Hoorn Is., 658
Hope-Kaikoura faults, 394
Hopi, 555
Horizon Deep, 495
Hornerstown, 534
Horn Is., 657
Horsts, 544
Horton Group, 170
Hot spot, 587
Houston, 580
Howland Is., 433
Huahine Is., 450
Huancabamba deflection, 261
Hudson Bay, 174, 188
Hudson Highlands, 608
Hudsonian beds, 209
Hudsonian Orogeny, 149, 403
Hudson River, 608
Huehuetenango, 308
Hull Is., 433
Humboldt Current, 346
Humboldt (sink), 542
Hungry Hollow Fm., 200
Hunter-Bowen Orogeny, 24, 31, 40
Hunter Fracture Zone, 279
Hunter Is., 383
Hunter Orogeny, 45
Hunter Ridge, 384

Hunter Valley, 41
Huon Is., 365

Ice sheets, 605
Ice-wedge, 517
Ichthyostega, 297
Idaho, 511
Idaho Batholith, 628
Idaho-Northern Washington Mt. Complex, 620
Idaho-Wyoming Thrust Belt, 637
Ifaluk Atoll, 226
Iguacu Falls, 130
Ile aux Cochons, 473
Ile aux Pinguoins, 473
Ile de Cayenne Group, 313
Ile de la Possession, 473
Ile de L'Est, 473
Ile de l Ouest, 474
Ile de Quatre, 447
Ile Roche, 474
Iles-au-vent, 348
Iles de Horn, 658
Iles du Vent, 449
Iles Loyauté, 342
Iles Nuageuses, 474
Iles Sous-le-Vent, 449
Illinoian time, 569, 605
Illinois Basin, 597
Illinois (state survey), 511
Ilopango Depression, 270, 272
Imataca Complex, 311, 313
Imatacan Orogeny, 311, 320
Inaccessible Is., 479
Indefatigable Is., 285, 289
Independencia facies, 423
Indiana (state survey), 511
Indonesia, 372
Inklinian orogeny, 141
Inklin thrust faults, 183
Inland Ice, 299
Inman Valley, 70
Innamincka Red Beds, 70
Inner Volcanic Arc, 339
Innuitian Orogenic Belt, 142, 145
Inoceramus, 298, 368
Inselberg landscape, 350
Interglacial sediments, 202
Interglacial strand lines, 241
Interior Basin Province, 218
Interior Lowlands, 559
Intermountain Seismic Belt, 549
Interstadials (Ontario), 203
Intraoceanic island arcs, 345
Inuvik Minto Arch, 189
Iodine, 242, 464
Iowa (state survey), 511
Irati Fm., 644
Irati oil shales, 134
Irian Jaya (West New Guinea), 372
Iron, 55, 126, 143, 187, 192, 217, 253, 258, 317, 370, 425, 464, 490, 529, 580, 651
Irondequoit Fm., 198
Iron Knob, 79
Ironshore Fm., 227

Isabela Is., 288, 289
Isla Carion, 438
Isla de Pascua, 260
Isla de Pinos, 252
Isla de Pinos tectonic unit, 253
Island arcs, 494
Island arc system, 345
Isla San Benedicto, 438
Islas de los Ladrones, 344
Islas de los Desventurados, 449
Islas Malvinas (Falkland Islands), 273
Isla Socorro, 438
Isles des Saintes, 301
Isles Marquises, 345
Isotelus, 221
Isthmian embayment, 574
Isthmian link, 231, 401, 414, 416
Isthmus of Panama, 417
Isthmus of Tehuantepec, 349
Itabirites, 311
Itacolumi Series, 132
Itacurubi, 424
Itaituba Series, 134
Itapucumi Limestone, 420, 424
Itarare Fm., 674
Itelau Reef, 226
Ituzaingo Fm., 424
Iwo Islands, 331
Iwojima, 331
Iwokrama Igneous Complex, 321
Izabal Depression, Graben, 232, 234
Iztaccihuatl Mt., 359
Izu-Bonin Is., 331
Izu-Ogasawara-Iwo Islands (Nampo-Shoto), 331
Izu Mariana Arc, 331

Jackson Hole, 630
Jackson uplift (dome), 575
Jackson (Upper Eocene) Series, 534, 536, 579
Jade, 521
Jalisco-Nicoya Fault, 358
Jalpatuqua Fault, 307
Jama Hills, 269
Jamaica, 334, 658
Jamaican Basement Complex, 660
James Is., 285, 289, 524
Jaqua Fm., 661
Jardines de la Reina, 252
Jarum Limestone, 305
Jarvis Is., 341
Jasper National Parks, 208
Jemez Mts., 555
Jervis Is., 289
Johnson Deep, 383
Johnston Is., 341
Joseph Bonaparte Gulf, 47
Jost Van Dyke Is., 653
Juan Fernández, 336
Juan Fernandez Ridge, 449
Jucuaran Block, 271
Jukesian Movement, 83
Juliana Peak, 488

Jungfern Passage, 339
Jupiter Fm., 224
Juratype, 504
Juvenile water, 588

Kagawong Fm., 19
Kaieteur Falls, 319
Kaieteur Fm., 316
Kaieteur (planation surface), 310
Kaikoura Orogeny, 25, 393
Kaiser-Wilhelm-land, 372
Kalgoorlie District, 100
Kanawha River, 524
Kangaroo Is., 63
Kanimblam Orogeny, 24, 40
Kansan glaciation, 568, 605
Kansas, 558
Kansas (state survey), 511
Kanuku-Bakhuys Mts. Dislocation Zone, 317
Kanuku Complex, 313, 316, 319, 320, 458
Kao Is., 494
Kapingamaringi, 226
Kap Washington Group, 299
Karroo, 273
Karst, 111, 122, 226, 244, 250, 334, 363, 401, 507, 661
Kaskaskia sequence, 601, 635
Kawhia Syncline, 395
Kazan Is., 331
Kechika Trough, 181
Keele Range, 184
Keeling Islands, 245
Keewatin Glaciation, 188
Keewatin Series, 599
Keilor Cranium, 100
Kekerengu fault, 394
Kempsey Block, 40
Kenai Range, 515
Kenoran orogenic belt, 403
Kentucky (state survey), 511
Kentville Fm., 166
Kerguelen Is., 474
Kermadec Islands, 337, 390
Ketilidian Mobile Belt, 294
Kettle Point Fm., 200
Keuper Series, 298
Keweenawan Series, 599
Keyuwini Group, 313
Kilauea (volcano), 584
Kilinailau Atoll, 455
Killarney Granite, 599
Kimball Fm., 565
Kimban Phase, 66
Kimberley Block, District, 100, 101
Kinderhookian beds, 603
King Is., 81
King George Is., 469
Kingman Reef, 341
King Salmon thrust fault, 183
Kingston, 202
Kirkfieldian Stage, 219
Kirkland Lake, 178
Kirkwood Fm., 536
Kishenehn Fm., 213

Klamath Mts., 509, 619
Klippen hypothesis, 163, 168
Klippe (Taconic), 612
Klondike, 144, 186
K'mudku mylonite episode, 313, 316, 322
Knife Lake Series, 599
Knip Group, 363
Knoydart beds, 166, 167
Kodiak Is., 515
Kolombangara (volcano), 452
Koloula Diorite, 455
Kome flora, 298
Kona, 582
Kootenay Arc, 208, 213
Kopinang (planation surface), 310
Koro Is., 281
Koror Is., 225
Kosciusko Epoch, 33, 64, 92
Kosciusko Granite, 39
Kosmaticeras, 368
Kozu (volcano), 332
Kronprins Christian Land, 298
Kulgeran Phase, 66
Kure Is., 583
Kusaie Is., 225
Kuyuwini Group, 321
Kuyuwini igneous suite, 320
Kwajalein Atoll, 346
Kwitaro Group, 313, 315, 320, 321
Kyanite, 490, 530

Labrador trough, 143, 219
La Brea oil sands, 498
Laccolite, 549
Laccoliths, 509, 564, 593
Lachine Shale, 221
Lachlan Fold Belt, 36
Lachlan Geosyncline, 30
La Crête Fm., 326
La Désirade, 301, 660
Lafayette Gravel, 605
Laforian Series, 273
Lag deposits, 605
Lagerenza (wildcat drillhole), 422
Lagoons, 561
Laguna Caratasca, 328
Lagunillas (oil field), 650
La Guyane Française, 282
Lahillia, 368
Lake Algoma, 206
Lake Algonquin, 204, 205
Lake Amadeus, 46
Lake Amatitlán, 306
Lake Arenal, 228
Lake Atitlán, 228
Lake Ayarza, 306
Lake Barlow-Ojibway, 206
Lake Bonneville, 542
Lake Camalotal, 228
Lake Caratasca, 228
Lake Champlain, 608
Lake Enriquillo, 258
Lake Erie, 193, 198
Lake Eyre, 33, 56
Lake Gatún, 228

Lake Guija, 228
Lake Huron, 193
Lake Ilopango, 228
Lake Iroquois, 203, 204, 205
Lake Izabal, 228
Lake Laguna, 228
Lake Lahontan, 542
Lake Lalolalo, 657
Lake Madden, 228
Lake Managua, 228
Lake Maumee, 205
Lake Missoula, 629
Lake Nicaragua, 228, 400
Lake Nipissing, 205, 206
Lake Ontario, 193
Lake Petén, 228
Lake Phillipson Beds, 70
Lake St. Claire, 206
Lake Simcoe, 205
Lake Superior, 596
Lake Titicaca, 124
Lake Torrens Lineament, 63
Lake Whittlesey, 205
Lake Winnipeg, 140, 188
Lake Winnipegosis, 188
Lake Yojoa, 228
Lake Ypacaraí, 421
La Libertad Arch, 233
Lama (oilfield), 651
Lamar (oilfield), 651
Lambian Basins, 30
Lamotrek, 226
Lancaster Reef, 499
Lance Creek Beds, 654
Land-bridge, 33, 414
Landenian Fm., 536
Landslide, Landslide-block, 202, 546, 585
Langlade Is., 444
La Palma, 271
La Paloma Fm., 643
La Paz, 651
La Pedreria Fm., 643
Laplatasaurus, 645
La Plata tectonic trend, 643
Laramian orogenic belt, 406
Laramide Orogeny, 206, 235, 265, 307, 326, 401, 504, 508, 519, 544, 624, 637, 661
Laramide overthrusts, 630
Laramie Fm., 637
Larder Lake-Malaritic Belt, 178
La Salle Uplift, 575
La Sal Mts., 551
Las Cahobas Fm., 327
Lascano Series, 643
La Soufrière (volcano), 447
Las Vegas Shear Zone, 545
Las Villas Tectonic Zone, 253
Late Is. (volcano), 494
Later Basin-Huntley Fault Zone, 629
Laterite, lateritic duricrust, 46, 56, 346, 424, 493
Lateritization, 489
Latrobe Valley, 94
La Trocha deep fault, 253

Lau Basin, 495
Lau Group, 278
Laura Basin, 58
Laurentian Upland, 502, 505
Laurentide ice sheet, 143, 155
Lau Ridge, 26, 495
Laurie Is., 468
Lava fountain, 588
Laval Calcarenite, 221
Laval Fm., 221, 224
Lavalleja Series, 642
La Vega Real Valley, 257
Lavongai Is., 121
Laysan Is., 583
Leaching, 198
Lead, 55, 56, 91, 126, 143, 178, 192, 197, 217, 234, 359, 370, 391, 432, 464, 530, 606, 625, 639
Leadville limestone, 635
Leaside Till, 203
Lebo Shale, 564
Lecanospira, 221
Le Cap Is., 443
Leda Clay, 202, 222
Leeward group, 449
Leeward Islands, 301, 339, 439, 591, 659, 660
Leigh Creek Coalfield, 32, 71
Leinzia, 423
Leiorhynchus, 200
Lelydorp Sands, 485
Lempa Valley, 230, 271
Leperditia, 198
Lepidotus, 644
Leptaena, 224
Leptobolus, 224
Leptocoelia, 273
Leray Limestone, 221
Les Apotres, 473
Leskou Is., 468
Lesser Antillean Arc, 14
Lesser Antilles, 257, 300, 340, 363, 658
Leucitite, 106
Lewis and Clark faults, 629, 630
Lewis overthrust, 406, 629
Liard Plateau, 208
Liard River, 188
Libocedrus, 274
Lifou Is., 343
Lightening Ridge, 42
Lignite, 580, 639
Lihir Group, 451
Limestone Antilles, 363
Limestone Caribbess, 14, 339, 659
Limestone-volcanic (outer) islands, 345
Limonite, 416
Lindenmeier (fossil site), 638
Lineaments, 183, 193, 201, 230, 275, 277, 665
Line Islands, 341
Lingula, 200, 643
Lingulepis, 220
Liocalymene, 198
Lions Head Fm., 198

Lipalian, 634
Lisianski Is., 583
Litchfield Complex, 50
Lithothamnium, 275, 363, 364
Little Barrier Is., 390
Little Cayman, 226
Little Falls Fm., 195, 196
Livingston Is., 469
Llanoria, 235
Llanos, 245, 309, 457
Llanos Basin, 648
Lockportian Stage, 219
Loess, 20, 424, 606
Loess Hill, 569
Loess Plain, 569
Logan's Line, 159, 168, 222, 610
Long Is., 507, 531, 608
Longitudinal Valley (Chile), 239
Long Point Fm., 164
Lopevi (volcano), 384
Lopez (wildcat drillhole), 422
Lophospira, 222
Lord Howe Atoll, 455
Lord Howe Rise, 26, 42, 398
Lorraine Group, 219, Group, 222, 224
Los Haitises, 661
Los Roques Trench, 660
Lotbinière Shale, 221
Louann salt (Jurassic), 578
Louark Group, 579
Louisiade Archipelago, 381
Louisiana (state survey), 511
Love Cove Group, 161
Loveland Fm., 569
Low Is., 469
Lower California, 349
Lowlands Fm., 364
Lowville Limestone, 221
Loyalty Islands, 342, 365
Lucas Fm., 200
Lucina, 644
Luling graben system, 578
Lynchburg Fm., 525
Lynn Lake, 178
Lyons Sandstone, 636
Lystrosaurus, 8

McDonald Is., 474
McAdam Lake Fm., 167
Macal Series, 232, 233
Macasty Fm., 224
Macaya Fm., 258, 326
McDonald Is., 474
MacDonnell Ranges, 46
Machu Picchu, 431
Macizos Antiguos, 246
McKean Is., 433
MacKenzie delta, 189
Mackenzie district, 179
Mackenzie-Great Bear Lowlands, 188
MacKenzie Mts., 179, 181, 189, 208
MacKenzie River, 181, 188
Mackinac Is., 199

Mackinaw interstadial, 205
Maclurites, 197
McMurdo Volcanic Group, 479
McMurray tar sands, 190
Macoma, 202
Macquarie Is., 473
Macrauchenia, 424, 646
Mactra, 646
Macuma Fm., 264
Madame Joie Fm., 326
Madison Limestone, 562
Madre de Dios, 430
Madrejon (wildcat drillhole), 422
Madro-Tertiary, 625
Maéwo Is., 383
Magdalena (district), 555
Magdalen Is., 158
Magog Belt, 164
Magothy Fm., 534
Magur Atoll, 226
Maiao Is., 449, 450
Maimon Fm., 258
Maine (state survey), 511
Maitaia, 366
Makarapan Granite, 313
Makatea Is., 250, 499, 658
Malaspina glacier, 515
Malbaie Fm., 167
Malden Is., 347
Maligne Lake, 215
Malinche Mt., 359
Malocystites, 221
Malvinas (Falkland Islands), 18, 273
Malvinokaffric fauna, province, 273, 425, 643
Managuq Depression, 400
Manasquan-Shark River Fm, 536
Manchioneal Fm., 335
Mangaia Is., 250
Manganese, 55, 253, 324, 344, 359, 370, 383, 389, 391, 418, 425, 448, 464, 529, 652
Mangareva Islands, 290
Mango Is., 281
Mangrullo Fm., 642, 644
Manihiki Atoll, 250
Manitoba Escarpment, 188, 191
Manitoba Lake, 188
Manitoulin Is., 193, 197
Manono Is., 667
Manta Basin, 269
Mantled gneiss dome, 615
Manto-type deposits, 359
Manus Is., 121
Mao Fm., 258
Maorian Geosyncline, 25
Maquoketa Fm., 602
Maracaibo Basin, 648
Maracaibo Platform, 649
Maranhao-Piaui Plateau, 129
Marañon Portal, 261, 265
Mara (oilfield), 651
Marathon Mts., 405
Marau Is., 455
Marble, 530
Marcellus Shale, 200

Marchena Is., 289
Marcus Atoll, 657
Marcus-Necker Seamount Chain, 657
Maré Is., 342
Margarita Is., 660
Marginal bulge, 158
Maria Is., 499
Mariana Arc, 225
Mariana Islands, 304, 344, 448
Mariana Limestone, 448
Maridale Marl, 497
Marie Galante, 301, 650
Marigot Fm., 326, 661
Marinoan (Wilpena Group), 68
Marion Is., 472
Maritime Disturbance, 159, 168, 171
Maritime Plain, 158
Maritime Provinces, 156
Marlborough Basin, 58
Marlborough Schist, 393
Marowijne Group, 481
Marquesas Islands (Isles Marquises), 345
Marshall Islands, 346
Marshall Line, 412
Martinique, 348
Martinique Is., 662, 667
Martin Vaz Island, 274
Marua Is., 450
Marudi Mts., 319
Marvels River Fm., 445
Maryburian Orogeny, 33
Maryland (state survey), 511
Masahuat (volcano), 272
Massachusetts (state survey), 511
Matakaoa Volcanics, 394
Matansa Fm., 448
Matapedia Belt, 164, 165
Matapi Spilite, 481
Matawan Group, 534
Mato Grosso, 132, 422
Mato Tepee (pluton), 564
Matthew Is., 383
Mauke Atoll, 250
Mauna Kea Mt., 584
Mauna Loa Mt., 584
Maupiti atoll, 450
Mayaguez, 435
Maya Massif Mts., 116, 232, 306
Mayan civilization, 327
Maya Series, 233
Mayero Is., 447
Mayor Is., 390
Maysvillian Stage, 219
Mazama Ash, 625
Mazaruni Group, 310, 313, 314, 320, 321
Mazatzal Revolution, 543
Meades Ranch, 502
Meadowcliffe Till, 203
Meaford-Dundas Fm., 196, 197
Meaford Fm., 197
Mecejana phonolite, 275
Medicine Bow Mts., 631
Medicine Hat (pottery clay), 192

695

SUBJECT INDEX

Medina Group, 198, 219
Medinan period, 602
Megacycles, 635
Megallanes Geosyncline, 240
Megalomus, 198
Megalonychops, 645
Megatherium, 424
Meguma Group, 159, 161, 166
Mehendjo-Daro, 410
Mehetia Is., 449
Melanesia, 25, 279, 365, 410
Melanesian border plateau, 280, 455
Melanesian megashear, 378
Melange, 621
Melville Is., 156
Mendana Is., 346
Mendoza (wildcat drillhole), 422
Mene Grande, 650
Mercedes Fm., 645, 651
Mercury, 187, 359, 391, 464, 521
Merir Is., 226
Merrimack Synclinorium, 611, 615
Meso-cordilleran Geosyncline, 634
Mesopotamian Plateau, 16
Mesosaurus, 134, 423, 644
Mesaverde Fm., 637
Messinian period, 487
Metabentonite, 196
Metamynodon Channels, 565
Metapán, 231, 271
Metapi Fm., 315
Meteorite crater, 598
Mexia-Talco Systems, 578
Mexican Basin, 405
Mexican geosyncline, 355
Mexican geotectonic cycle, 354
Mexican Highland, 541
Mexico, 348
Mexico-Mesoamerica Trench, 358
Mica, 464, 530
Michigan, 193, 198, 597
Michigan (state survey), 511
Micro-continent, 412
Micronesia, 344, 346, 410, 416
Microplates, 365
Mid-American Trench, 305, 401
Mid-Atlantic Ridge, 405, 465
Midden Curaçao Group, 363
Middendorf Fm., 534
Middleback Ranges, 66
Middle River, 166
Middle Rocky Mts., 627
Mid-Pacific atolls, 347
Mid-Pacific Mts., 657
Midway Is., 591
Midway-Sunset-Buena Vista Field, 625
Midwestern Region (U.S.), 504, 596
Migmatite, 7
Migration Route, 418
Mikura (volcano), 332
Milford Group, 214
Milford Sound, 392
Millepora, 346

Mimella, 221
Minami Tori Shima, 657
Minas Basin, 171
Minas Gerais, 457
Minas Series, 132, 642
Mineral Belt, 639
Minetrigonia, 367
Minette dikes, 192
Mingan Is., 218, 224
Mingan Limestone, 224
Minnekahta Fm., 562
Minnesota (state survey), 511
Minto Arch, 149, 171
Minturn Fm., 636
Miogeosynclinal rocks, 613, 622
Miquelon, 443
Miramichi Geanticline, 161
Misima Is., 381
Misiones Fm., 423
Misool Is., 377
Mississippi Arch, 597
Mississippi Empayment, 597
Mississippi River, 188, 572
Mississippi (state survey), 511
Missoula Group, 634
Missouri coteau, 188
Missouri River, 558
Missouri (state survey), 511
Mitiaro Atoll, 250
Miyake (volcano), 332
Moala Group, 281
Moat, 587
Moenkopi (Shale), 552
Mogollon Rim, 550
Mogote, 252, 435
Mohawk Valley, 542
Mohelia atoll, 450
Molasse facies, 210, 258, 354, 375, 636
Molokai Fracture Zone, 587
Moluccan Geosyncline, 374
Molybdenum, 187, 270, 464, 625, 639
Mona Is., 434
Mona Passage, 257
Monkey Hill, 440
Monmouth Group, 534
Monophora, 645
Monos Shale, 422
Monotis Zone, 367
Monroe Uplift, 575
Montagu Is., 468
Montana Gap, 629
Montana Lineament, 629
Montana (state survey), 511
Montastrea, 440
Monteregian intrusives, plugs, plutons, 171, 223, 611
Montevideo, 642
Montreal Basin, 218
Montreal Fm., 211
Montserrat, 360
Monument Upwarp, 552
Monument Valley, 552
Moorea Is., 449
Moorehouse Member (of Onondaga Fm.), 200

Moose Channel Fm., 213
Mopeha atoll, 450
Moraine (on Mauna Kea), 585
Morehouse Fm., 579
Morne Delmas Fm., 327
Morne Diablotin, 257
Morne Ginne (Peak), 443
Morne Trois Pitons, 257
Morocco, 167 (*see also* Vol. VIII, Pt. 2)
Morococha, 432
Morotiri (Bass Islets), 499
Morrison Fm., 562, 636
Morrison River Fm., 444
Morro Vermelho Fm., 276
Mortar Beds, 565
Moruiti Is., 450
Moskitia Shelf, 232
Mosquitia Coast, 328
Mosquito Coastal Plain, 210
Moss rock, 595
Motagua Fault, Shear Zone, 232, 230, 235, 306
Mother Lode, 620
Motu, 250
Motu-Iki (atoll), 346, 450
Motu One, 450
Mountain Meadow Surface, 565
Mt. Aconcagua, 16
Mt. Bagana, 452
Mt. Baker, 619
Mt. Belinda, 468
Mt. Bougainville, 658
Mt. Carbet, 348
Mt. Carstensz, 374
Mt. Cook, 393
Mt. Duff, 290
Mt. Egmont, 395
Mt. Flora Beds, 8
Mt. Gambier Sunkland, 94
Mt. Hood, 619
Mt. Isa Geosyncline, 28
Mt. Izalco, 270
Mt. Kasi, 282
Mt. Katahdin, 608
Mt. Kitchener, 209
Mt. Kusale, 226
Mt. Lamlam, 305
Mt. Lassen, 619
Mt. Logan, 180
Mt. Lulu, 657
Mt. McKinley, 502
Mt. Misery Peak, 440
Mt. Momotombo, 400
Mt. Morgan, 31
Mt. Olga, 48
Mt. Orohena, 493
Mt. Painter, 67
Mt. Pelée, 348, 447, 662
Mt. Popamanisiu, 451
Mt. Princeton pluton, 633
Mt. Puke, 658
Mt. Rainier, 619
Mt. Robson, 208, 215
Mt. Rotrima, 310
Mt. St. Catherine, 300
Mt. San Miguel, 270

Mt. Shasta, 619
Mt. Simon, 196
Mt. Soufrière, 339
Mt. Tasman, 393
Mt. Totolau, 226
Mt. Warning, 42
Mt. Washington, 608
Mt. Whitney, 502, 619
Mt. Zeil, 46
Muck, 418
Multicolored Tuff Fm., 366
Muruwa Fm., 321
Murchison, 100
Murravian Basin, 98
Murray Basin Plains, 24, 42, 72, 98, 93
Murray Fracture Zone, 621
Murrumbidgee Batholith, 39
Mururoa Is., 497
Muschelkalk Fm., 298
Musgrave Block, 50, 62
Muskeg, 515
Mustique Is., 447
Mya, 202
Myochlamys, 645
Myojinsho (volcano), 332
Mytilus, 367

Nagssugtoquidian Mobile Belt, 294
Nahone Range, 182
Nain (region), 175
Najavo Sandstone, 636
Nambu (mine), 282
Nampo-Shoto, 331
Nanaimo Basin, 186
Napo Fm., 265
Nappes, 629
Narborough Is., 285, 288
Naricual (Venezuela), 652
Narrangansett Basin, 611
Nashville Dome, 597
Nassau Is., 250
Nassau Range, 374
Nassian Orogeny, 142
Natural Bridges Monument, 555
Nauru, 362
Navajo Country, 551
Navarro Fm., 579
Nazca (plate), 429, 430, 449
Ndeni Is., 455
Ndreketi Basin, 281
Neahga Fm., 198
Nebo (thrust), 546
Nebraskan glaciation, 568, 605
Nebraskan glaciation, 568, 605
Necker Is., 583
Nedrow Member, 200
Needsmust Quarry Andesite, 440
Neiba Fm., 326
Neilson Reef, 499
Nelson Syncline, 393
Neocaledonian Archipelago, 365
Neogiron Fm., 246
Neoglacial Activity, 359
Neohelikian beds, 149, 599
Neotectonics, 639

Neovolcanic Plateau, 350
Nepean beds, 196
Nepan Is., 402
Netherlands Antilles, 362, 660
Netherlands Leeward Is., 362
Netherlands Windward Is., 362
Neumecklenburg, 372
Neupommern Is., 372
Neuville Fm., 221
Neuvo Leon Group. 579
Nevadan Orogeny, 463, 508, 519, 624, 636
Nevada (state survey), 511
Nevado Huascarán (peak), 427
Nevis, 439
Nevis Peak Volcano, 440
New Albany Fm., 603
New Britain, 121, 372
New Brunswick, 156
New Caledonia, 25, 342, 365
New England Batholith, 31
New England Fold Belt, 36, 40
New England Region (U.S.), 504, 608
Newfoundland, 156
Newfoundland Highlands, 157
New Georgia, 452
New Guinea, 25, **372,** 451
New Hampshire (state survey), 511
New Hanover (Lavongai), 121
New Hebrides, 25, **383,** 455
New Hebrides Trench, 279
New Ireland, 121, 372
New Jersey (state survey), 512
New Madrid Fault, 599
New Mexico (state survey), 512
New Providence Is., 109, 231
New (Kanawha) River, 524
New South Wales, 34
New York (state survey), 512
New Zealand, 390
New Zealand Alpine Fault, 494
New Zealand Geosyncline, 397
Ngalia Basin, 51
Ngatik (atoll), 226
Niagara Escarpment, 193, 196, 197, 202
Niagara Gorge, 204
Niagaran Series, 602, 219
Niagara River, 199
Nicaragua, 400
Nicaraguan Volcanic Upland, 230, 251, 329, 400
Nicaragua Rise, 659
Nickel, 143, 178, 253, 258, 370, 383, 455, 625
Nickerie Fm., 458, 481
Nicolet River Fm., 222
Nicolet Stadial, 203
Nicoya Complex, 231, 234, 251
Nipe Depression, 253, 254
Nightingale Is., 470
Nii-jima, 332
Niobium, 464
Niobrara Limestone, 637
Niobrara Shale, 566
Nishino-shima Is., 332

Niue Island, 401
Niugini, Niu Guini (New Guinea), 372, 451
Nizao River, 258
Noeggerathiopsis, 83
Noranda (district), 178
Norfolk Island, 402
Norfolk Ridge, 26
Normal accidents, 414
Norte Chico (Chile), 239
Norte Grande (Chile), 239
North America, 403
North American Cordillera, 141
North American Platform, 502
North Bartlett Fault, 253
North Bay spillway, 204
North Carolina (state survey), 512
North Creek Sand, 431
North Dakota (state survey), 512
Northern Great Plains Province (Canada), 187
Northern Rocky Mts., 627
Northern Rocky Mt. Trench, 180
Northern Territory (Australia), 46
North Greenland Fold Belt, 297
North Mt. Basalt, 171
North Park Basin, 631
North Platte River, 558
North Slope, 517
Northwest Territories, 208
Notiochonetes, 422
Notiotherium, 87
Notre Dame Trough, 159, 161, 166, 180
Nova Olinda Basalt, 312
Nova Scotia, 156
Nova Scotia Highlands, 157
Nuclear Central America, 233, 329, 401
Nuées Ardentes, 348, 662
Nuku Hiva Is., 346
Nukumanu Is., 455
Nukunonu (atoll), 494
Nukuoro (atoll), 226
Nullagine Basin, 101
Nullaginian System, 28
Nullarbor Limestone, 74
Nummulites, 115
Nymboida Coal Measures, 41

Oaxacan Structural Belt, 352
Oca Fault, 665
Ocala Limestone, 536
Ocala Peninsular Uplift, 533, 573
Oca-Pilar Faults, 261
Ocean Is., 292, 583
Oceania, 410, 413
Oceanic type volcanics, 412
Oceanization, 574
Ocoee Series, 525
Odontostamus, 645
Oeno Is., 434
Officer Basin, 23, 102
Oficina (oilfield), 651
Ofu Is., 1
Ogallala Beds, Group, 559, 566

SUBJECT INDEX

Ogasawara Islands, 331
Ogilvie Arch, 181
Ogilvie Mt., 208
Ohio River, 525
Ohio (state survey), 512
Oil, 218, 270, 359, 372, 391, 418, 425, 432, 464, 497, 529, 556, 572, 580, 625, 639
Oil-bleeders, 571
Oil sands, 192
Oil shales, 81, 392, 556, 636
Okanagan Highlands, 620, 628
Okefenokee Embayment, 532
Oklahoma panhandle, 562
Oklahoma (state survey), 512
Old Faithful, 630
Old Sea Clays, 485
Olenellus, 184
Olistoliths, 234
Olistostrome, 115, 266
Oliverian Plutonic Series, 614
Olofi Is., 658
Olosega Is., 1
Olympic Mts., 625
Omineca geanticline, 181, 185
Onondaga Fm., 200
Ontarian Stage, 219
Ontario Basin, 193
Ontong Java Plateau, 452
Onverdacht Fm., 485
Oodnadatta Fm., 72
Operculina, 343
Oranje Mts., 489
Oranje Stad Fm., 363
Orapu (Series), 312
Orbiculoides, 200
Oregon (state survey), 512
Orella Member, 566
Oriçangas, 323
Oriente Basin, 261
Oriente Tectonic Unit, 253
Orinoco River, 457, 648
Oriomo Trend, 374
Oriskany Fm., 199
Oriskany Sandstone, 200, 602
Orizabu (volcano), 351
Oroclinal Fold, 517
Orthoceras, 422
Osburn Lineament, 630
O-shima (volcano), 332
Orthophragmina, 368
Ostrea, 645
Otago Schists, 393
Otapiria, 367
Otiran Stage, 399
Otozamites, 59
Ottawa Valley, 196, 205
Ottley's Mt., 440
Otway Basin, 33
Otway Group, 72
Ouachita Fold Belt, 575
Ouachita Mts., 405, 506, 597
Ouareau (limestone), 221
Ouray (mineral belt), 639
Outer Limestone Arc, 339
Ouvea Is., 343
Owen Stanley Range, 25, 376

Owl Creek (regression), 602
Oxoplecia, 197
Ozark Dome, Mts., Plateau, 506, 596, 597

Pacaraima Mts., 310
Pachuca (district), 359
Pacific-Antarctic Ridge, 473
Pacific Coastal Plain, 230
Pacific Coast Range, 269
Pacific Continent, 336, 414
Pacific Cordillera, 228
Pacific Cordilleran Region (U.S.), 504, 618
Pacific faunal realms, 168
Pacific Islands Trust Territory (U.S.), 416
Pacific Ocean, 412
Pacific Plate, 279, 292, 377, 406, 624
Pacific province, 454
Pacific seas, 190
Pacific suite, 412
Pacitrigonia, 368
Pagan Is., 345
Pahasappa (limestone), 562
Pahoehoe (lava), 582
Painted Desert, 555
Pakaraima Mts., 319
Palacio Member, 645
Palacios Depression, 253
Palau Is., 225
Paleocurrent structures, 198
Paleofavosites, 198
Paleogeographic maps, 414
Paleohelikian beds, 599
Paleo-Indian culture, 638
Paleo-Pacific Ocean, 44
Paleoplain, 505
Paleosols, 606
Palermo Beds, 644
Pali (cliffs), 585
Palinspastic reconstruction, 213
Palisades, Palisadian Disturbance, Taphrogeny, 171, 274, 529, 617
Palmerston Atoll, 250
Palmiramys, 645
Palmyra Is., 341
Pamelia Group, 221
Pamlico Fm., 538
Pampa de Junin plain, 427
Pampas, 457
Pampas Ranges Massif, 463
Pampeano Fm., 424
Pampean Ranges, 460
Panama, 416
Pan-American orogenic cycle, 459, 643, 647
Panamint (desert), 542
Panguna (porphyry copper deposit), 456
Panhandle (Oklahoma), 566
Panochthus, 646
Pão de Açucar, 421
Papua New Guinea, 372
Papuan Geosyncline, 375, 398

Para-Andean depression, 427
Para clays, 485
Paradox Basin, 551
Paradox fold, 549
Paradoxides, 444
Paraechmina, 198
Paraguana, 648
Paraguay, 419
Paraguay-Araguala orogenic belt, 459
Paraguay River, 457
Paraiba Geosyncline, 131
Paraiba Orogenic Belt, 459
Paraliageosyncline, 633
Paramaka Fm., 312, 481
Paramaca Series, 315
Paraná Basin, 134, 273, 647
Paraná River, 419, 457
Paraspirifer, 200
Paredão volcano, 276
Park City Group, 636
Parkes Platform, 38
Park of Pillars, 196
Park Range Uplift, 631
Parks, 508, 637
Parnaiba Basin, 131, 134
Paso Agular Fm., 642, 644
Passa Dois Series, 423
Pastora Group, 313, 315
Patagonian-Fuegian Cordillera, 464
Patagonian Massif, Tableland, 239, 460
Patagonian Trough, 463
Patagonia (plateau region), 15
Pauloenus, 363
Paxixi Member, 423
Payande Fm., 249
Peace River, 208
Peace River Arch, high, 140, 192
Peace River Lowlands, plains, 188, 557
Pearlette Ash, 568
Pecos River, 558
Pecten Conglomerate, 8
Pedimentation, 488
Pediments, 638
Pediplains, 309, 319
Peedee Fm., 534
Peggy Ridge, 495
Pegmatite, 490, 617
Pelean-type, Peleé type, eruption, 348, 454
Peleliu Is., 225
Pelly Cassiar Platform, 181
Pelotas Basin, 647
Pembina field, 192
Peneplain, 134, 145, 365, 416, 488, 534, 585, 605, 617, 631
Peneplanation, 36, 195, 283, 285, 370, 617
Penguin movement, 81
Peninsular Arch, 532
Pennsylvania (state survey), 512
Penrhyn Atoll (Tongareva), 250
Pentameroides, 198
Pentamerus, 198
Pentecost Is., 383

Pepinos, 435
Peravillo Fm., 258
Periglacial features, 634
Permafrost, 515
Permian glaciation, 31, 102
Permo-Carboniferous glaciation, 3, 11
Perodin Fm., 326
Peron Terrace, 370
Perth Basin, 102
Peru, 426
Peru-Chile Trench, 427
Peruvian Current, 346
Peruvian Trench, 269, 286
Petén Lowland, 232, 306
Peter I Is., 479
Petitot (anticline), 189
Petrie Reef, 343
Petrified Forest, 555
Petrified wood, 555
Petroleum, 107, 143, 217, 242, 253, 270, 432, 497, 520, 539, 650
Phacops, 224
Phaenopora, 198, 224
Philip Is., 402
Philippine Sea, 344
Phillippine Is. (*see* Vol. VIII, Pt. 2)
Phoenix Islands, 433
Pholadomorpha, 222
Phosphate, 55, 226, 243, 275, 344, 362, 448, 456, 539, 652
Phosphoria Fm., 636
Phosphorite, 359
Phyllodactylus, 287
Pico da Neblina, 129
Pico Duarte, 257
Pictou Group, 158, 170, 171
Picuiba (wildcat), 422
Piedmont Province, 504, 523
Piedras de Afilar series, 642, 643
Piercement domes, structures, 142, 153
Pierre Shale, 563, 566, 637
Pigailoe (atoll), 226
Pilbara Block, (province), 22, 101
Pilbara System, 28
Pillow lavas, 519
Pimichin (River), 457
Pinar del Río unit, 253
Pinchi Fault Zone, 182
Pinchi Lake, 187
Pine Ridge, 566
Pines Is., 365
Piñon Group, 264
Pinta Is., 289
Pinzonella, 423
Pinzon Is., 289
Pirate Cave Fm., 167
Pirizal (well), 423
Pitcairn Group, 290
Pitcairn Islands, 346, 433
Pitchblende, 639
Pitch Lake, 497
Pitons, 443, 662
Placers, 625
Placetas zone, 254

Plaisance Fm., 326, 661
Planet Deep, 453
Planorbis, 645, 646
Planorbulina, 343
Plate collisions, 504
Plate tectonic theory, 506, 612
Platforms, 140
Platina, 457
Platinum, 464, 521, 529
Playa lake, 542
Plectatrypa, 197
Pleistocene glaciation, 36
Plethorhyncha, 199
Plum Point, 203
Plum Point-Port Talbot, 209
Poa, 465
Poas Volcano, 251
Pocillopora, 346
Podocarpus, 274
Point à Pierre Fm., 497
Point Blanche Fm., 364
Polar pack ice, 145
Polar shift, 347
Poljes, 252
Polochic Fault Zone, 306
Polochic Motagua-Cayman Magashear, 329
Polynesia, 284, 410
Polynesians, 582
Ponape Is., 225
Ponta Grossa beds, 643
Poopo Basin, 427
Popanoceras, 367
Popocatépetl (volcano), 351
Porcupine Lake, 178
Porites, 346
Porphyry Fm., 240
Port Huron Stadial, 203, 205
Port Lambton Group, 200
Port Philomel Beds, 273
Port Talbot (time-stratigraphic unit), 203
Port Willunga Beds, 74
Possession Is., 3
Postglacial Isostatic rebound, 205, 616
Postglacial lakes, 191
Postorogenic facies, 613
Potash, 197
Potomac River, 524
Potomac Fm., 534
Potrerillos (copper mine), 242
Potsdam Fm., Sandstone, 195, 196, 219, 220, 612
Potsdamian beds, 599
Pound Quartzite, 69
Powder River Basin, 563
Powell Is., 468
Prasopora, 197
Preandean geosyncline, 239
Precambrian Morainic boulders, 3
Precordillera, 16
Precordilleran Trough, 461
Preglacial drainage, 174
Premonitory movements, 369
Pre-Newark surface, 617
President Thiers Bank, 499

Pridoli Is., 198
Prince Edward Is., 156, 158, 472
Principal Cordillera, 463
Probillingsites, 197
Proetus, 222, 422
Propachyrucos, 645
Protichnites, 220
Prosserella, 200
Proto-Atlantic Ocean, 168, 461, 507
Proto-Kanuku Complex, 320
Protoslavinia, 134
Providenciales Is., 500
Pseudaucella, 367
Pseudogygites, 197
Pseudo-karst, 488
Puerto Rico, 434, 512, 658
Puerto Rico Trench, 437, 659
Puget Sound, 618
Pulap (atoll), 226
Puluwat (atoll), 226
Pumbuiza Fm., 263
Pumicestone, 362
Puna de Atacama, 239
Puna (region), 16
Purcell Anticlinorium, 213
Purcell Argillite, 217
Purcell Fault, 214
Pustulacia, 422
Pulo-Anna (Is.), 226
Pycnostylus, 198
Pyramid Lake, 542
Pyramus, 423, 644

Quarry Creek Orogeny, 38
Quasicraton, 365, 451, 495
Quaternary glaciation, 10
Quebec, 156
Quebec Basin, 218
Quebec Group, 161, 164
Queen Charlotte Is., 182, 187
Queen Maud Land, 7
Queensland, 56
Queensland Plateau, 24
Queenston, 219
Queenston Fm., 196, 197
Quequay Limestone, 645
Querandi Fm., 646
Quetane Series, 246
Quick Clay Phenomena, 202
Quill volcano, 364
Quiriquina Beds, 240
Quiriquire (oilfield), 651
Quito Cuenca Depression, 268
Quixaba Plain, 275

Rabaul, 122
Rabida Is., 289
Racklan Orogeny, 183, 210
Radak chain, 346
Radiolarite, 115
Rafinesquina, 197, 221
Raiatea Is., 449
Rairoa (atoll), 498
Raivavae Is., 499

SUBJECT INDEX

Ralek chain, 346
Ramu-Markham Valley fault, 377
Random Fm., 444
Rangiroa (atoll), 498
Rangitata Orogeny, 25, 392
Rangitikei Basin, 395
Rano Raraku (volcanic crater lake), 260
Raokumara Peninsula, 394
Raoul Is., 337
Rapa Is., 499
Rapa Nui Is., 260
Rapitan Fm., Group, 211, 217
Raritan Fm., 534
Rarotonga Is., 250
Ratak chain, 346
Raton Basin, 570
Receptaculites, 197
Red beds, facies, 191, 211, 419, 423, 613, 636
Red Race Track, 563
Redonda Is., 14
Red Rose (mine), 187
Reef carbonates, 191
Reefs, 192, 198, 217, 221, 234, 250, 298, 328, 341, 346, 355, 362, 365, 381, 401, 412, 413, 418, 434, 499, 585
Reg, 542
Relief thrusts, 629
Remedios tectonic unit, 253
Renconcavo Basin, 135
Rensellaeria, 199
Resserella, 197, 198
Revillagigedo Islands, 438
Reynales-Fossil Hill Fm., 198
Rewa River, 278
Rhacopteris, 59
Rhizo-concretions, 567
Rhynchocephalia, 134
Rhytimya, 224
Richardson Mts., 179, 181, 208, 216
Richmond Fm., Group, 197, 224, 335
Richmondian Stages, 219
Riddle (nickel deposit), 625
Ridge Province, 522
Rift, Rift zone, 140, 588
Rimatara, 499
Rimroad Gravel, 565
Rim syncline, 576
Rincon Fm., 363
Rio Aconcagua, 239
Rio Bio Bio, 239
Rio Bonito Fm., 644
Rio Branco territory, 316
Rio Cauca, 245
Rio Chixoy, 306
Rio das Velhas Series, 132
Rio de la Pasíon, 306
Rio de la Plata, 419, 457, 642
Rio Dulce, 234
Rio Grande, 351
Rio Grande Embayment, 574
Rio Grande Trough, 552
Rio Hondo Fault Zone, 233

Rio Magdalena, 245
Rio Negro, 457, 642
Rio Verde Arch, 269
Rivas Arch, 231, 234
River John Group, 167
River Plate, 457
Riversdale Group, 170
Rivière Gauche Fm., 327
Rivière Grise Fm., 327
Roan Cliff, 551
Robinson Crusoe Is., 336
Roca Partida Is., 438
Rocas Islands, 274
Roche Bluff Fm., 339
Roche Salamanca, 474
Rock glaciers, 634
Rocklandian Stage, 219
Rocky Mt. front, 508
Rocky Mt. Trench, 141, 185, 208, 213, 407, 549, 638
Rocky Mts., 141, 208, 406, 631
Rocky Mountain Province (U.S.), 504, 627
Rolling Downs Group, 42
Romanche Fracture Zone, 269, 277
Romanche Trench, 277
Romane Fm., 223
Rondonia Craton, 458
Roraima Fm., 310, 321, 458, 481
Roraima intrusives, 312, 320
Rosário mining district, 231
Rosebel Fm., Series, 312, 316, 481
Rose Is., 1
Ross Orogeny, 7
Ross River, 187
Rostricellula, 221
Rota Is., 344
Rotation, 377, 520
Rotation (of the Pacific), 621
Rotuma Is., 278
Rough Creek Fault Zone, 603
Rough Range (Carnavon Basin), 33
Ruby Valley (Nevada), 542
Rumble III (volcano), 337
Rum Jungle, 28, 49
Rupununi (planation surface), 310
Rupununi savannas, 323
Rurutu Is., 499
Russell Is., 452

Saba Is., 362, 660
Sabana Grande, 258
Sabana of Bogotá, 246
Sabine uplift, 575
Sable Is., 142, 172
Sacramento Section (of Basin-Range), 541
Sahna Basin, 598
Sahnioxylon, 368
St. Alban Limestone, 221
St. Bartholomew Is., 301
St. Barthelemy Is., 301
St. Christopher-Nevis-Anguilla, 439
St. Clair River, 200
St. Croixan Series, 599
St. Croix Fm., 498

St. Elias Mts., 180, 184, 515
St. Eustatius Is., 362
St. George Dolomite, 162
St. Germain Complex, 222
St. Johns Peneplain, 661
St. Kitts, 439
St. Lawrence Lowlands, 140, 218
St. Lawrence Platform, 159, 162, 166, 169
St. Lawrence Seaway, 202
St. Lawrence Valley, 202
St. Lucia, 443, 662, 667
St. Maarten Is., 362
St. Martin Is., 301
St. Marys Fm. (Canada), 200
St. Mary's Fm. (U.S.), 537
St. Matthias Is. Group, 121
St. Paul Is., 472
Saint Paul Rocks, 274
St. Paul's Fracture Zone, 277
St. Peter (Sandstone), 602
Saint Pierre and Miquelon, 443
St. Pierre Interstadial, 203
St. Vincent, 300, 447, 667
Saipan, 344, 416, 448
Sala y Gomez, 260
Salina, 198, 542
Saline Valley, 542
Salisbury Embayment, 532
Salitrales, 424
Salmon River Mts., 628
Salopina, 198
Salt, 107, 192, 198, 206, 234, 241, 258, 356, 500, 578, 602, 606
Salt Cay, 500
Salt collapse structures, 189
Salt diapir, dome, 126, 351, 575
Salton trough, 541
Salto Sandstone, 645
Saltpeter, 242, 464
Salt Pond Peninsula, 439
Salvador, 270
San Ambrosio and San Félix Islands, 449
Samoa Western, 666
Samoa, American, 1
San Andres Is., 231
San Andreas fault, 504, 545, 621
San Cristóbal Fm., 423
San Cristóbal Is., 285, 452
San Cristóbal trench, 452
Sand, 539, 580, 606
Sand barrens, 507
San Diego de los Banos tectonic unit, 253
Sand Hills region, 569
Sand Is., 499
Sand plains, 202
Sandy Point Hills, 440
San Eduardo Fm., 266
San Félix Islands, 449
San Francisco peaks, 549
Sangamon, 538, 586
Sangay Volcano, 268
Sangre de Cristo Mts., Ranges, 631
San Gregorio Fm., 643
San Juan Basin, 551

SUBJECT INDEX

San Juan lowland, 234, 251
San Juan River, 552
San Juan Uplift, 631
Sankakuyama Fm., 448
San Marcos arch, 575
San Miguel (volcano), 272
San Pedro rocks, 275
San Salvador Is., 285
San Salvador (volcano), 272
Santa Ana (volcano), 272
Santa Catharina supergroup, 134
Santa Clara zone, 254
Santa Cruz Is., 285, 289, 383, 455
Santa Elena Peninsula, 269
Santa Lucía graben, 647
Santa Lucian Orogeny, 624
Santa Rosa Fm., 233
Santa Rosa reef, 344
Santa Ysabel Is., 452
Santiago, 237
Santiago Fm., 264
Santiago Is., 289
Santiaguito (volcano), 307
Santo Is., 383
San Vicente (volcano), 272
Saõ Bento Series, 423
Saõ Francisco basin, 131
Saõ Francisco River, 130
Saõ Francisco shield, 458
Saõ Luiz cratonic block, 457
Saõ Paulo Rocks, 276
Sappa Fm., 568
Saprolite, 487
Sardian Orogeny, 461
Sarstún (stream), 306
Sarstun trough, 232, 234
Saskatchewan Plain, 188
Sauce batholith, 643
Sauk Sequence, 600, 634
Saunders Is., 468
Savage Is., 401
Savai'i Is., 666
Savanna Belt, 485
Saxiana sand, 222
Sawatch Sandstone, 634
Scablands, 509, 629
Scalesia, 287, 288
Scarborough Fm., 203
Scarp, 544
Schefferville (Labrador), 171
Schoharie Fm., 200
Schwartzwalder Mine, 639
Scilly Is., 450
Scotia Arc, 17, 273, 465
Scotian Shelf, 161, 171
Scotia Ridge, 465
Scotland beds, 661
Scott Is., 479
Scotts Bluff Butte, 566
Scutellum, 198
Sea cliffs, 591
Searles Valley, 542
Seatura shield volcano, 281
Seborucco, 255
Sechura Desert, 427
Second prairie, 188
Sediment trap, 574

Selenium, 464
Selva, 122
Selwyn Basin, 181, 184
Selwyn's rock, 70
Seneca Member, 200
Senyavin Is., 226
Sequoia, 624
Série de l'Orapu, 283
Série de Paramaca, 283
Série du Bonidoro, 283
Serra Acarac, 310
Serra do Mar, 131
Serra do Navio Group, 312
Serra Geral Fm., 423, 644
Serra Tumucumaque, 310
Serou di Cueba Fm., 363
Serou Teintje, 363
Serpentinite, 43
Serros de Sal, 258
Sevier Desert (Utah), 542
Seymour Is., 3
Seymour Is., series, 8
Shadow Lake, 196
Shag Is., 474
Shakwak trench, 180, 182
Shermanian Stage, 219
Sherman ridge, 629, 631
Shichito-Iwojima Ridge, 331
Shickshock Zone, 161
Shiki Is., 226
Shikine (volcano), 332
Shinarump Conglomerate, 552
Shinchiyo Is., 226
Ship Rock (volcanic feature), 555
Shortland Is., 452
Shuswap complex, 209
Sierra de Jacuaran, 272
Sial Line, 658
Sierra de Jatibonicao fault, 253, 254
Sierra de las Animas, 642
Sierra de los Organos, 252, 661
Sierra de Minas Viejas, 355
Sierra Màdre del Sur, 351
Sierra Madre del Sur province, 351
Sierra Madre Occidental, 350
Sierra Madre Oriental fold belt, 349
Sierra Nevada, 406, 509, 618
Sierras Bonaerenses, 16
Sierras of Buenos Aires, 18
Sierras Pampeanas, 16
Sierras Subandinas, 16
Signy Is., 468
Sigsbee escarpment, 574
Sigsbee Knolls, 574
Silcrete, 63, 74
Silica sands, 606
Silicified fossils, 637
Silver, 55, 56, 91, 123, 178, 187, 217, 234, 253, 281, 359, 391, 400, 432, 625, 639
Silver Hill, 360
Silverton, 639
Simpson Desert, 46
Sinclinorio del Cauca, 248
Sinclinorio del Magdalena, 248
Sinclinorio del Pacífico, 248

Sinclinorio pre-Andino, 248
Sir Timothy's Hill andesite, 440
Skarn, 165, 187
Skeena arch, 187
Skolithos, 226, 422
Skull Creek Shale, 636
Slave Province, 140, 175, 209, 210
Slot, The, 452
Smilodon, 646
Smoke Creek desert, 542
Snake Range thrust, 546
Snake River plains, 620
Snares Is., 390
Snoqualmie batholith, 624
Snowhill Is. Series, 8
Snow Is., 469
Snowy Mts., 33
Soapstone, 521
Society Islands, 449, 493
Soebi Blanco Fm., 363
Solander Is., 390
Solenoprora, 221
Solfatara-stage volcanism, 630
Solomon Chain, 123
Solomon Islands, 25, 372, 383, **451**
Solomon Sea, 377
Solomon trench, 452
Sombrerito Fm., 258
Sombrero Is., 442, 660
Somerset Is., 149
Sonoma Orogeny, 543
Sonoran Desert, 541
Sonoran Orogeny, 354
Sonsorol Is., 226
Sorol (atoll), 226
Sorong Fault Zone, 377
Sourfrière Hill, 360
Soufrière volcano, 301
Souixia, 635
South America, 456
South Australia, 61
South Carolina (state survey), 512
South East Range, 439
Southern Alps, 393
Southern Hills aulacogene, 461
Southern Rocky Mts., 627
South Georgia, 273, 466
South Honshu Ridge, 345
Southland Syncline, 393
South Orkney Is., 468
South Platte River, 558
South Sandwich Is., 273, 468
South Savanna Granite, 314
South Sea Is., 410
Southwold Drift, 204
Sowerbyella, 221, 224
Spanish Peaks, 633
Spencer Gulf, 62
Spenochasm (Honduras), 329
Spine, 662
Spirifer, 273
Spiriferina, 367
Spriggina, 69
Springlea, 465
Squantum Tillite, 613
Squaw Bay Fm., 200
Stable shelf, 374

Stakes, 558
Star Basin, 58
Starbuck Is., 342
State geologists (U.S.), 510
Statia Bank, 439
Stenarcestes, 367
Stephanocrinus, 198
Stepping stones, 414
Stewart Is., 390, 455
Stikine Plateau, 184
Stone rings, 524
Stone rivers, 274
Ston Fm., 315, 481
Stony Creek (oil and gas field), 170
Strait of Belle Isle, 218, 223
Stratiform shield volcanoes, 9
Stratigraphic nomenclature, 175
Stratocones, 345
Stratovolcanoes, 351, 364, 375, 377, 471, 472
Streptelasma, 222
Strike-slip faulting, movements, 377, 545, 631
Strip mining, 580
Stromaphoropsis, 645
Stromatocerium, 197, 221
Strophocheilus, 645
Strophomena, 222, 224
Stuart City barrier reef, 579
Sturge Is., 428
Sturtian (Glaciation), 68
Styx Basin, 58
Subandean Foreland, 428
Sub-Andean trough, 246
Sub-Antarctic Islands, 465
Subduction, Subduction zone, 43, 167, 236, 251, 329, 345, 406, 494, 546, 621
Subhercynian Movements, Orogeny, Phase, 254, 265
Submarine canyons, 376, 591
Submarine eruption, 332
Submerged shelves, 586
Subsummit peneplain, 631
Sudbury Basin, 143
Sugar Loaf Butte, 565
Sulas, 489
Sulfide mineralization, 617
Sulfur, 242, 331, 364, 464, 580
Sullivan Mine, 143, 217
Summerville Basin, 531
Summit peneplain, 544, 631
Sunbury Fm., 200, 201
Sundance Sea, 562, 636
Sunday Is., 337
Sungei (soengei), 380
Sunnybrook Till, 203
Sunset Crater National Monument, 555
Supamo Complex, 313, 314
Superior Is., 505
Superior Province, 140, 175
Sur Chico Mt., 239
Sur Grande Mt., 239
Surinam, 309, 480
Surumu suite, 312
Susquehanna River, 524

Sustut Basin, 185
Suswap Complex, 213
Suta Fault, 455
Sutton Mts., 608
Suva Is., 278
Suwannee River Basin, 533
Suwanne Strait, 533
Suwarrow Atoll, 250
Sverdrup Basin, 143, 145, 148, 151
Swains Is., 494
Swan Is., 328
Swatch Uplift, 631
Sweetgrass Arch, 189
Sweetgrass intrusions, 192
Sweetwater Arch, 629
Swift Current Plateau, 192
Sydney, 24, 171, 433
Sydney Basin, 36, 41
Sylvania Member, 200
Synectonic facies, 151
Syntrophia, 224

Tabberabberan Orogeny, 24, 30, 39, 44, 96
Tablazo Fm., 267
Table Head Limestone, 163
Tabular Proterozoic beds, 311, 316
Táchira (state), 652
Taconian (Taconic) Orogeny, 159, 162, 222, 353, 405, 461, 528, 610, 612
Taconic Mts., 608
Taconite, 179
Tacuarembo Sandstone, 642
Tacurú, 424
Tafelberg (Surinam), 488
Tafoni, 323
Tagpochau Limestone, 448
Tagula Is., 381
Tahiti, 284, 449, 493
Tahiti-Iki, 493
Tahiti-Nui, 493
Tahltanian Orogeny, 141, 212
Tahoa (Cone), 450
Taiarapu (cone), 493
Taiga, 515
Taiga Ranges, 181
Takutea (atoll), 250
Takutu Fm., 313, 320, 322
Takutu rift valley, 310, 319
Talamana Mts., 251
Talara (oil field), 427
Talgai skull, 60
Talus stream, 634
Tampa Limestone, 537
Tampico Embayment, 574
Tamworth Group, 40
Tanapag Fm., 448
Tanna Is., 384
Tantalum, 464
Tantalus Basin, 185
Tanu Is., 455
Tapaje Group, 312
Taphrogeny, Taprogenic stage, troughs, 463, 529, 533, 544
Taphrogeosyncline, 525

Taranaki (region), 395
Taravai Is., 290
Tarawa Atoll, 291
Tasman Geosyncline, 23, 56, 92, 378
Tasmania, 81
Tasmanites, 200
Tasman orogenic belt, 374
Tasmantia, Tasmantis, 367, 451
Ta'u Is., 1
Tavera dam, 258
Taveuni Is., 281
Tavua (goldfield), 282
Taylor Fm., 579
Tehuantepec (Isthmus of), 349, 574
Tejas Sequence, 637
Telkawa (coal deposits), 186
Telluride, 639
Te Manga peak, 250
Tena Fm., 265
Ten Mile Creek, 200
Tennessee River, 524
Tennessee (state survey), 512
Tentaculites, 422
Te Pita te Henua, 260
Tepui, 310, 319
Terevaka (volcano), 260
Termi (branch), 457
Terrace Hills, 568
Terraces, 606
Terraia, 644
Terrestrial snails, 347
Tetagouche Belt, 161, 163
Tethyan fauna, 536
Tethys (Alpine), 113
Tetiaroa Is., 449, 450
Teton Mts., 630
Tetradium, 197, 221
Tetrapod reptile, 3
Tetreauville Limestone, 221
Texas lineament, 354, 545
Texas panhandle, 558
Texas (state survey), 512
Theresa Fm., 195, 196
Thinnfeldia, 134
Thirty-eighth (38th) Parallel Lineament, 598
Tholeiite basalts, 9, 588
Thomasatia, 197
Thomonde Fm., 327
Thompson Mine, 178
Thorium, 464
Thorncliffe Fm., 203
Thorold Fm., 198
Three King's Is., 390
Thule (Greenland), 149, 298
Thule Is., 468
Tía Juana (oilfield), 650
Tiebaghi (chromium mine), 370
Tierra colorada, 419
Tierra del Fuego, 17
Tiga Is., 343
Tigre Is., 272
Tikopia Is., 455
Till, Till sheet, 156, 606
Tillite, 134, 273, 297, 422, 643

SUBJECT INDEX

Tilt meters, 592
Time classification, 174
Timoe (atoll), 291
Timor Sea, 46
Tin, 55, 91, 126, 464
Tinian Is., 344
Tintinna Trench, 141, 180, 185
Tioga bentonite, 200
Tippecanoe Sequence, 601, 634
Titanium, 530
Titanosaurus, 645
Titanothere, 192
Titicaca Depression, 427
Tiyuyacu Fm., 266
Toadstool Park, 565, 566
Tobago, 495, 660
Tobi Is., 226
Tocantins-Tapajos craton, 458
Todos Santos Red Beds, 233
Tofua (volcano), 494
Tokelau Islands (Union Islands), 493
Toledo Fm., 234
Tol Is., 226
Tonawandian Stage, 219
Tonga, 494
Tonga chain, 26
Tonga-Kermadec Trench, 279, 383
Tongatapu (Tongatabu) Is., 494
Tonga Trench, 495
Toquepala (copper mine), 432
Tori (volcano), 332
Torngat Mts., 139
Toronto, 202
Torreón-Monterey Fracture Zone, 354
Torrensian (Series), 68
Torres Strait, 372
Torres Trench, 384
Tortola Is., 653
Tosagua Arch, 269
To-shima (volcano), 332
Touche Is., 401
Toxodon, 424, 646
Tower karst, 435
Tower Is., 289
Trans-Amazonian Orogenic Cycle, 283, 311, 314, 320, 458, 481, 642, 646
Transantarctic Mts., 7
Transcontinental Arch, 140, 404, 635
Transcurrent faults, 168, 216, 227, 235, 268, 354, 394, 621, 665
Transform faults, 236, 495, 621
Trans-Mexico Volcanic Belt, 358
Transverse Ranges, 621
Trempealeauan Stage, 196, 219, 600
Trent Fm., 537
Trenton Group, 164, 218, 219, 221
Trentonian beds, 602
Tres Islas Fm., 642, 644
Triarthrus, 197, 224
Triassic basins, 531
Triassic rift zone, 507
Trimerella, 198

Trimerus, 422
Trindade, 274
Trinidad and Tobago, 497, 660
Trinidad tectonic unit, 253
Trinity Fm., 579
Trinity Trough, 159
Triple point, 279
Triplesia, 197, 224
Tristan da Cunha, 473
Tristate district, 606
Trois Rivières Fm., 326
Tropidurus, 287
Truckee River, 542
Truk Is., 225
Truro (Nova Scotia), 171
Trust Territory of the Pacific Islands, 410
Tsunami deposits, 585
Tuamotu Islands, 498
Tubai Is., 450
Tubarão Series, 134, 422
Tubuai (Austral) Islands, 499
Tubuai-Manu, 450
Tuhua Orogeny, 345
Tumac Humac Mts., 310, 489
Tundra, 515, 606
Tungsten, 55, 187, 400, 464, 530, 625
Turbidites, 38
Turks and Caicos Islands, 500
Turner Valley, 216, 217
Turritella, 536
Tuscaloosa Fm., 534
Tutelo sequence, 602
Tutuila Is., 1
Tutvutha Is., 281
Tuy Cariaco Basin, 648
Tweed Complex, 42
Tyaughton Trough, 182, 185
Tyennan Geanticline, 81
Tyndall Stone, 192
Typotherium, 646
Tyrconnell Fm., 204

Ua Pu Is., 346
Uinta Mts., 550, 629
Ulawa Is., 454
Ulithi Is., 226
Ulsterian Series, 219
Ultrabasic complexes, 369
Ulua Basin, 329
Ulua Fonseca Trend, 330
Ulol, 226
Umberatana Group, 68
Umutac Fm., 305
Uncompahgre uplift, 551
Underthrusting, 430
Unea Is., 122
Union Is., 447
Union Islands, 493, 494
United States–Alaska, 513
United States–Appalachian Region, 522
United States–Atlantic Coastal Province, 531

United States–Basin and Range Province, 541
United States–Colorado Plateau Province, 548
United States–Great Plains Province, 557
United States–Gulf Coastal Province, 572
United States–Hawaii, 581
United States–Midwestern Region, 596
United States–New England Region, 608
United States of America - General, 502
United States–Pacific Cordilleran Region, 618
United States–Rocky Mountain Province, 627
United States Trust Territory of the Pacific Islands, 502
Uplifted Atoll, 433, 434, 442, 455, 499
Upolu Is., 666
Uracas Is., 345
Uraçuian Minas Cycle, 458
Uranium, 55, 143, 178, 391, 464, 555, 580, 625, 639
Urozean, 412
Uruguay, 642
Uruguay River, 457
Uruguaysuchus, 645
Uruguay-Tandilia Craton, 458
Usulutan (volcano), 272
Utah Lakes, 542
Utah (state survey), 512
Utah-Wyoming Shelf, 634
Ute Mts., 551
Utica Group, 164, 219, 221, 224
Uvea Is., 657

Valders Stadial, 205
Valdesia dam, 258
Valdivia (Chile), 239
Valentine Fm., 569
Valles Caldera, 633
Valley and Ridge Province (US–Appalachia), 522
Vanadium, 464, 555
Vancouver Is., 183, 187
Vanua Levu Is., 278
Vanua Mbalavu Is., 281
Vargas-Peña Shale, 422
Varvites, 422
Vatukoula (goldfield), 282
Vauvréal Fm., 224
Vavau Is., 494
Vavitu Is., 499
Venezuela, 648
Venezuela Basin, 659
Venezuelan Caribbean Is., 648
Venezuelan Peruvian Trough, 462
Ventana Ranges, 19
Vermont (state survey), 512
Verulan Fm., 197
Vicksburg, 579

SUBJECT INDEX

Victoria, 91
Victoria Is., 149
Victoria Peak, 228
Vieques Is., 434
Vincentown Fm., 534, 536
Vindication Is., 465
Virgiana, 198, 224
Virgin Islands, 653, 659
Virginia City, 639
Virginia (state survey), 512
Visokoi Is., 468
Viti Levu Is., 278
Vitric tuff, 590
Vityaz Trench, 455
Vivian Fm., 265
Vizcaino Fm., 646
Volcán de Boqueron, 228
Volcán de Chiriqúi, 228
Volcán de Santa Ana, 228
Volcán de Tajumulco, 228, 306
Volcán de Viejo, 401
Volcán Fuego, 307
Volcanic Caribees, 339, 659
Volcanic domes, 588
Volcanic (inner) chain, 345
Volcanosedimentary formations, 368
Volcán Santa Maria, 307
Volcán Wolf, 285
Vostok Is., 342
Vulcan (volcano), 122

Waccamaw Fm., 537, 538
Waehneroceras, 367
Wagenoceras, 367
Wagwater Fm., 335, 661
Waikiki, 582
Waimanalo, 586
Wainimbuka River, 278
Wairau Fault, 393
Waitemata Group, 399
Wake Island, 346, 657
Walker Line, 545
Walker River, 542
Wallis and Futuna Islands, 657
Walloon Coal Measures, 41
Walpole Is., 342
Walter Lake, 542
Wapsipinicon (limestone), 602
Warm loess, 20
Warramanga Geosyncline, 50
Warrumbungle Mts., 42
Wartatakan Phase, 67
Wasatch Fm., 633
Wasatch Line, 634, 637
Wasatch Mts., Range, 546, 629
Washikemba Fm., 363
Washington Is., 342
Washington (state survey), 513
Water gaps, 524

Waucoban Series, 219
Wave-hill (feature), 78
Wenman Is., 289
Wernecke Mt., 181, 184, 208
Werner Fm., 579
West Antarctica, 7
West Drift, 494
Western Samoa, 666
West Indies, 658
West Irian, 372
West Mariana Ridge, 345
West Texas Permian Basin, 563
West Virginia (state survey), 513
Whirlpool sandstone, 197
Whitehorse Trough, 185
White Is., 390
White limestone, 335
White Limestone Fm. (Jamaica), 661
Whitella, 222
White Mt. Peak, 542
White Mt. Plutonic-Volcanic Series, 614
White Mts., 608
Whitemud beds, 192
White River Beds, 564
White River Group, 565
White River time, 560
White River uplift, 631
White Rock Fm., 162, 164
White Wall Fm., 364
Wianamatta group, 41
Wiarton Fms., 198
Wichita Mts., 596, 597
Wilcox Fm., 534, 579
Wild Cat Ridge, 566
Wildflysch, 497
Willard Thrusts, 546
Williston Basin, 191, 561, 635
Willyama Block, 67
Willyama Complex, 43
Wilouran Series, 68
Wilpena Group, 68
Wilson Bluff Limestone, 74
Winanyan Phase, 66
Windermere (sediments), 209
Wind River Mts., 629
Windsor Group, 170
Windward Group, 449
Windward Islands, 659, 660, 667
Winton Fm., 72
Wisconsinan period, 203, 606
Wisconsin Arch, 597
Wisconsin (state survey), 513
Wiso Basin, 51
Witteberg beds, 273
Woleai Atoll, 226
Wolfcampian (Series), 604
Wolf Is., 289
Wolfville Fm., 171
Wolverine Shoal, 337
Woodbine Fm., 579

Wrangel Is., 515
Wyomide Orogenic Belt, 630
Wyoming Basin, 508, 627
Wyoming (state survey), 513

Xaraiés Member, 424
Xenotes, 352

Yaguari Fm., 644
Yague Group, 258
Yalakom Fault Zone, 182
Yampi Sound, 28
Yap Is., 225
Yardangs, 542
Yarrol Basin, 58
Yasawa Group, 280
Yasour (volcano), 384
Yauricocha, 432
Yayantique (volcano), 272
Yellow Limestone, 335
Yellowstone National Park, 627
Yellowstone Plateau, 508
Yellowstone River, 558
Yilgarn Block, province, 22, 101
Yilgarn System, 28
Yoho National Park, 208
York River, 167
York Till, 203
Yorktown Fm., 537
Young Is., 478
Yucatan Basin, 659
Yucatan Block, 39
Yucatan Peninsula, 349
Yucatan Shelf, 234
Yukon River, 141, 186, 516
Yukon-Tanana Upland, 518
Yukon Territory, 179, 208

Zacatecas Event, 354
Zacatecas Fracture Zone, 354
Zacate Grande, 272
Zanderij Fm., 283, 323
Zapla tillite, 126
Zavodovski Is., 468
Zaza tectonic unit, 253
Zeehan, 91
Zeugogeosyncline, 633
Zinc, 55, 56, 91, 126, 143, 178, 187, 192, 217, 234, 359, 370, 391, 432, 464, 530, 555, 606, 625, 639
Zion Canyon, 550
Zion Park, 555
Zona Central, 239
Zuni Sequence, 636
Zwischengebirge, 629
Zygobolba, 198
Zygospira, 221, 222

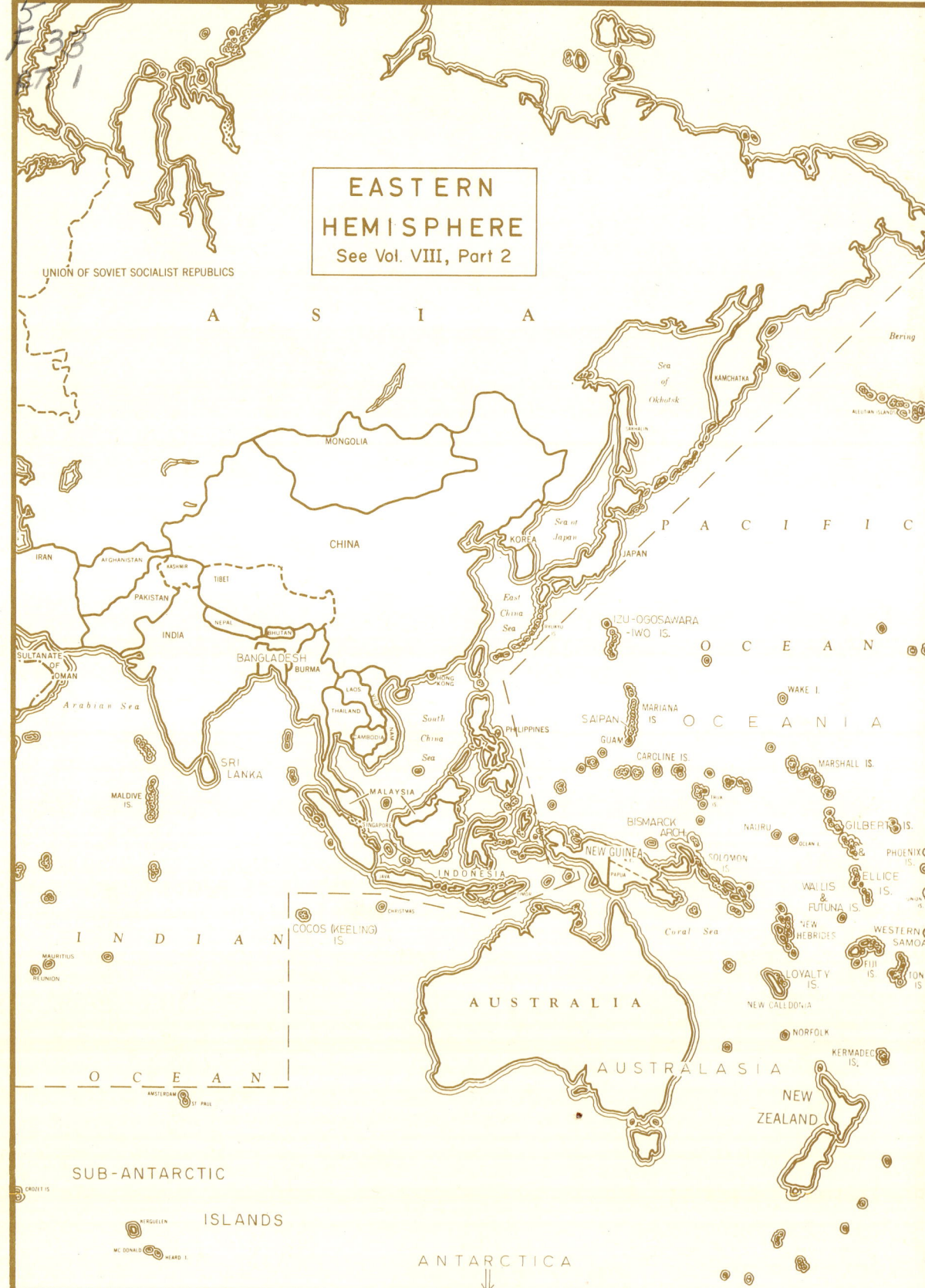